MICROENVIRONMENTS IN THE LYMPHOID SYSTEM

ADVANCES IN EXPERIMENTAL MEDICINE AND BIOLOGY

Recent Volumes in this Series

Volume 180
OXYGEN TRANSPORT TO TISSUE–VI
Edited by Duane Bruley, Haim I. Bicher, and Daniel Reneau

Volume 181
GENE EXPRESSION AND CELL-CELL INTERACTIONS IN THE DEVELOPING NERVOUS SYSTEM
Edited by Jean M. Lauder and Phillip G. Nelson

Volume 182
GENE EXPRESSION IN MUSCLE
Edited by Richard C. Strohman and Stewart Wolf

Volume 183
DRUGS AFFECTING LIPID METABOLISM VIII
Edited by David Kritchevsky, William L. Holmes, and Rodolfo Paoletti

Volume 184
MECHANISMS OF CELL-MEDIATED CYTOTOXICITY II
Edited by Pierre Henkart and Eric Martz

Volume 185
IMMUNOBIOLOGY OF PROTEINS AND PEPTIDES III: VIRAL AND BACTERIAL ANTIGENS
Edited by M. Z. Atassi and Howard L. Bachrach

Volume 186
MICROENVIRONMENTS IN THE LYMPHOID SYSTEM
Edited by G. G. B. Klaus

MICROENVIRONMENTS IN THE LYMPHOID SYSTEM

Edited by
G. G. B. Klaus
National Institute for Medical Research
London, England

PLENUM PRESS • NEW YORK AND LONDON

Library of Congress Cataloging in Publication Data

International Conference on Lymphatic Tissues and Germinal Centers in Immune Reactions (8th: 1984: Babraham, Cambridgeshire)
Microenvironments in the lymphoid system.

(Advances in experimental medicine and biology; v. 186)
"Proceedings of the Eighth International Conference on Lymphatic Tissues and Germinal Centers in Immune Reactions, held August 14–17, 1984, in Babraham, Cambridge, United Kingdom"—Verso t.p.
Includes bibliographical references and index.
1. Lymphocytes—Congresses. 2. Immune response—Regulation—Congresses. 3. Lymphatics—Congresses. I. Klaus, G. G. B. II. Title. III. Series.
QR185.8.L9155 1984 616.07′9 85-559

ISBN-13: 978-1-4612-9495-5 e-ISBN-13: 978-1-4613-2463-8
DOI: 10.1007/978-1-4613-2463-8

Proceedings of the Eighth International Conference on Lymphatic Tissues and Germinal Centers in Immune Reactions, held August 14–17, 1984, in Babraham, Cambridge, United Kingdom

Softcover reprint of the hardcover 1st edition 1985

A Division of Plenum Publishing Corporation
233 Spring Street, New York, N.Y. 10013

W. L. FORD

This volume is dedicated to the memory of W. L. Ford, Professor of Immunology, University of Manchester Medical School, Manchester, United Kingdom, who was killed in a car accident in Australia on 18th November, 1984. Bill Ford was one of the most outstanding immunologists of his era, one who made numerous major contributions to our understanding of lymphocyte migration and recirculation – a man truly dedicated to *in vivo* immunology. He was a regular, enthusiastic, and invaluable participant in the Germinal Center Conferences, including the meeting in Cambridge. His many friends and colleagues will miss him greatly.

Gerry Klaus

PREFACE

Since 1966, at roughly three-yearly intervals, an international group of immunologists has met somewhere in Europe to discuss the latest developments in our understanding of the mechanisms governing the functioning of the immune system in vivo. These meetings have become known as the International Conferences on Lymphatic Tissues and Germinal Centers in Immune Reactions, or for the regular devotees, simply as the Germinal Center Conferences (GCC). This volume represents the proceedings of the 8th GCC, which was held in Babraham, near Cambridge, UK, between the 14th and 17th August, 1984.

When one considers how cellular immunology has become increasingly dominated by in vitro methodology over the past twenty years, it may seem remarkable that these conferences have survived at all, let alone prospered. However, I for one do not find this surprising, since I suspect that the exquisitely complex architecture and microenvironments of the lymphoid system will never be fully understood through in vitro studies. If "the proper study of mankind is man", then surely ultimately the proper study of all the interacting elements which comprise the immune system has to be in vivo.

This belief is shared by a substantial number of immunologists, as the contents of this volume will attest. Although the GCC were originally devoted to unravelling the mysteries of the germinal center response (and these are still not fully resolved), over the years the scope of the meetings has inevitably broadened. Thus, the present conference covered not only the development of B lymphocytes, but also T cell differentiation, the structural basis of immune induction and immunoregulation by products of lymphoid cells. Finally, to add a slightly applied finishing touch to the proceedings, there were sessions devoted to human lymphocyte subpopulations, and to the potentials of monoclonal antibodies as therapeutic agents. I hope that a perusal of this volume will convince you that in vivo cellular immunology is still alive and well, and that the Germinal Center Conferences continue to provide a unique forum for work in this important field.

The success of the present meeting resulted from the combined efforts of many people, notably John Humphrey, Su Metcalfe, Herman Waldmann, and Peter Irving and his staff at Conference Contact. Bruce Roser deserves a special mention for all that he did as secretary of the Organising Committee, and for ensuring that everything ran smoothly at Babraham. I would like to thank all these people for their help and friendship.

G.G.B. Klaus

CONTENTS

B-CELL DIFFERENTIATION, SUBPOPULATIONS, THE ROLE OF GERMINAL CENTERS

Prospects for Understanding the Role of Lymphoid Microenvironments in B Cell Differentiation in Vivo 3
E.C. Butcher

Pre-B Cells in Rat Bone Marrow: Identification, Surface Markers and Isolation 9
D. Opstelten, G.J. Deenen, T. Jaarsma, J. Rozing and S.V. Hunt

T Cell Markers on Pre-B cells in the Rat 17
L.M.B. Vaessen, P. Joling, P. Nieuwenhuis, M.J. van Haperen and J. Rozing

Ig Light Chain Gene Rearrangement and Chromosomal Abnormality in the LAMA Early B-Lineage Tumour of the Rat 27
D. Opstelten, G.J. Deenen, B. de Jong V.J.S. Idenburg and S.V. Hunt

In Vivo Regulation of B Lymphocyte Production in the Bone Marrow: Effects and Mechanism of Action of Exogenous Stimuli on Pre-B Cell Proliferation and Lymphocyte Turnover 35
D.G. Osmond, G.M. Fulop, D. Opstelten and C. Pietrangeli

Influence of the Xid Mutation on B Lymphocyte Development in Adult Mice 47
J. Witwowski, L.M. Forrester, J.D. Ansell and H.S. Micklem

In Vitro Growth and Differentiation of Rat B Cell Subpopulations 57
C. de Groot, R. van Schie, J. Wormmeester and M.L. Kapsenberg

B Cell Production in Adult Rats 65
H. Bazin, B. Platteau, I.C.M. MacLennan N.S.A. Stuart, M. Khan and G.D. Johnson

The Paradox of High Rates of B Cell Production in Bone Marrow and the Longevity of Most Mature B cells 73
D.S. Kumararatne, D. Gray, I.C.M. MacLennan J. Lortan, B. Platteau and H. Bazin

Monoclonal Antibodies to Rat B Lymphocyte (Sub-)Populations 81
F.G.M. Kroese, D. Opstelten, A.S. Wubbena G.J. Deenen, J. Aten, E.H. Schwander L. de Leij and P. Nieuwenhuis

Effect of Early Embryonic Bursectomy on Lymphocyte Differentiation and Maturation 91
D.L. Ewert, C. Oates and C.H. Chen

Evidence for the Existence of Two Distinct Types of Peyer's Patches in Sheep 101
J.D. Reynolds, R. Pabst and G. Bordmann

Evidence of Differences Between Peyer's Patches and Germinal Centers 111
J.D. Reynolds

Differentiation of B Lymphocytes in Sheep. II. Surface Phenotype of B Cells leaving the 'Bursa-Equivalent' Lymphoid Tissue of Sheep, Ileal Peyer's Patches 119
M. Miyasaka, J. Reynolds, L. Dudler M-F. Beya, W. Leiserson and Z. Trnka

Origin and Fate of IgE-Bearing Lymphocytes: Modulation of IgE Isotype Expression on Peyer's Patch Cells 127
H.G. Durkin, S.M. Chice and S.I. Jiang

Immunoglobulin Isotypes and Antibody Specificity Repertoire of "Spontaneously" Occurring Ig-Secreting Cells in Germfree Mice Fed Chemically Defined UltraFiltered, Antigen-Free Diet 131
H. Hooijkaas, N. Bos, R. Benner, J.R. Pleasants and B.S. Wostmann

Comparative Analysis of the Development of B Cells in Marginal Zones and Follicles 139
I.C.M. MacLennan, H. Bazin, D. Chassoux D. Gray and J. Lortan

Germinal Center Cells: Antigen Specificity, Heavy Chain Class Expression and Evidence of Memory 145
G. Kraal, I.L. Weissman and E.C. Butcher

The Development of Specific Antibody Producing Cells in the Spleen of Rabbits During the Primary and Secondary Immune Response 153
N. van Rooijen, N. Kors and R. van Nieuwmegen

The Ontogenetic Development of the Follicular Dendritic Cell in Rat Spleen 161
C.D. Dijkstra, E.A. Döpp and E.W.A. Kamperdijk

Origin and Turnover of Follicular Dendritic Cells and Marginal Zone Macrophages in the Mouse Spleen 167
J.H. Humphrey and V. Sundaram

Mouse Lymph Node Follicular Dendritic Cells: Quantitative Analysis and Isolation 171
E. Heinen, C. Kinet-Denoël, D. Radoux and L.J. Simar

Antigen/Antibody Retention by Follicular Dendritic Cells 185
D. Radoux, E. Heinen, C. Kinet-Denoël and L.J. Simar

Antigenic Phenotype of Rodent Follicular Dendritic Cells in Vitro and *in Situ* 193
M.H. Kosco, A.K. Szakal and J.G. Tew

B Cell Differentiation: B Cell Subsets and Factors, Organs or Structures involved in B Lymphocyte Genesis 203
P. Nieuwenhuis

THYMUS AND T CELL DIFFERENTIATION

Precursors of T-Cell Mitogen Reactive Cells and Their Suppressors in the Mouse Fetal Liver 215
M. Rosenberg, L. Abel, T. Kukulansky, N. Sharon and A. Globerson

Thymus Homing Clonogenic Bone Marrow Cells 223
S. Ezine, I.L. Weissman, R.V. Rouse

The Role of the Thymic Cortex and Medulla in T Cell Differentiation 229
R. Scollay, P. Andrews, R. Boyd and K. Shortman

Intrathymic Differentiation in the Rat 235
P. Joling, F.J. Tielen, L.M.B. Vaessen C.J. Hesse and J. Rozing

Stromal Cell Populations in the Developing Thymus of Normal and Nude Mice 245
E.J. Jenkinson, R. Kingston and J.J.T. Owen

New Surface Antigens on Cells in the Early Phase of T-Cell Differentiation, Detected by Monoclonal Antibodies 251
L.M.B. Vaessen, P. Joling, F. Tielen and J. Rozing

Studies on the Role of the Thymus in the Selection and Clonal Expansion of T-Cell Precursors 261
R.K. Jordan, J.H. Robinson, A.L. Bentley K.C. House and N.A. Hopkinson

Culture of Mouse Thymic Epithelial Cells and Studies of Age-Related Changes 273
M. Small, L. Barr-Nea and M. Aronson

Heterogeneity of Human Thymus Epithelial Cells Revealed by Monoclonal Anti-Epithelial Cell Antibodies 283
M.A. Ritter, H.J. Schuurman, W.A. MacKenzie R.A. de Maagd, K.M. Price, R. Broekhuizen and L. Kater

Heterogeneity of the Human Thymus Epithelial Microenvironment at the Utrastructural Level 289
M.D. Kendall, F.P. van de Wijngaert, H.J. Schuurman, L.H.P.M. Rademakers and L. Kater

T Cell Differentiation Within TNC Related Lympho-Epithelial Complexes in the Thymus Cortex 299
W. Leene, R. de Waal Malefijt, P.J.M. Roholl and K.A. Hoeben

The Relation Between Thymic Interdigitating Cells and Dendritic Cells 307
E.W.A. Kamperdijk, M. van den Berg, M. Kapsenberg, A.M. Duijvestijn and E.C.M. Hoefsmit

"T-Cells" in Nude Rats 313
L.M.B. Vaessen, R. Broekhuizen, J.G. Vos, H.J. Schuurman and J. Rozing

T-Cells and T-Cell Subsets in Rat Bronchus-Associated Lymphoid Tissue (BALT) in Situ and in Suspension 323
G. v.d. Brugge-Gamelkoorn and T. Sminia

K Cell Activity is Independent of Both Thymus and Spleen 331
D.M. Chassoux, J.P. Kolb and I.C.M. MacLennan

Use of Transgenic Mice with Human MHC Class 1 Gene (HLA-A2) 341
T. Matsunaga, O. Cabilly-Horesh, D. Terz, A. Lee, L. Christian S. Ohno and A.P. Biro

Effect of Sympathectomy on the Maturation of Fetal Thymocytes Grown Within the Anterior Eye Chambers in Mice 349
U. Singh

The Effect of Dioctyltin Dichloride on the Thymus and T Cell Differentiation in the Rat 357
S. Nicklin, M.P. Scott, J. Evans and K. Miller

STRUCTURAL ASPECTS OF IMMUNE INDUCTION

Antibody Responses to T-Dependent Antigens: Contributions of Dendritic Cells and Helper T Lymphocytes 369
K. Inaba and R.M. Steinman

Development of Dendritic Cells from Rat Bone Marrow 377
W.E. Bowers and M.R. Berkowitz

The Relation Between Rat Lymph Node Interdigitating Cells in Situ and Dendritic Cells in Suspension 383
E.W.A. Kamperdijk, M. van den Berg, M. Kapsenberg and E.C.M. Hoefsmit

Dendritic Cells and Macrophages as Antigen Presenting Cells in Ovalbumin-Induced T Lymphocyte Proliferation In Vitro 389
M.L. Kapsenberg, F.E.M. Stiekema and M.B.M. Teunissen

Veiled Cell Lymphocyte Interaction 395
B.M. Balfour, L. Buttifant, J. O'Brien, J. Clarke and S.C. Knight

A Comparison Between Antigen-Laden Cells and Dendritic Cells in Afferent and Efferent Lymph 401
M. Drayson, D.R. Canning and E.B. Bell

The Heterogeneity of Mononuclear Phagocytes in Lymphoid Organs: Distinct Macrophage Subpopulations in Rat Recognized by Monoclonal Antibodies ED1, ED2 and ED3 409
C.D. Dijkstra, E.A. Döpp, P. Joling and G. Kraal

Macrophages Migrate from the Marginal Zone into the Germinal Centre of the Rodent Spleen 421
K. Matsuno, K. Miyakawa and M. Kotani

Splenic Dependence of the Antibody Response to the TI-2 Antigen DNP-Ficoll: Effect of the Ficoll Carrier on the Tissue Distribution of 125-I-Ficoll Conjugates 427
P.L. Amlot, D. Grennan and J.H. Humphrey

Evidence That Static but not Recirculating B Cells are Responsible for Antibody Production Against DNP on Neutral Polysaccharide, a TI-2 Antigen 437
D. Gray, I.C.M. MacLennan, B. Platteau, H. Bazin, J. Lortan and G.D. Johnson

T Lymphocytes in Rat Lymphoid Follicles are a Subset of T Helper Cells 443
F.G.M. Kroese, A.S. Wubbena, P. Joling, and P. Nieuwenhuis

Functional Differences Associated with Quantitative Distribution of Ia Antigens on Mouse B cells 451
E.M. Andrew, N.M. Mackenzie and R.M.E. Parkhouse

Evidence for a Clonal Development of Specific Antibody Forming Cells in the Spleen 457
N. van Rooijen, N. Kors and R. van Nieuwmegen

Modification of Immune Induction by Adjuvant 463
A.T.J. Bianchi, Ph.J. van der Heijden and B.A. Bokhout

Footpad Responses in Nude Mice Immunized with Chicken Erythrocytes 471
J.D. Jackson, G.A. Perry and J.G. Sharp

Cellular Characteristics of the Rosette Forming Cells in the Mouse Spleen 477
H.T. Chung and T.Y. Ha

The Bursa of Fabricius as a Trapping Mechanism for Environmental Antigens 487
S. Ekino, T. Urano, H. Fujii and M. Kotani

Maternal Autoimmune Disease Influences Self-Tolerance in Offspring 495
J. Gibson, K.L. Wong and A. Basten

The Role of Suppressor T Lymphocytes in the Maintenance of Tolerance to Alloantigens in Bone Marrow Chimeras Prepared with Total Lymphoid Irradiation 503
S. Slavin, M. Weigensberg, S. Bar, and S. Morecki

The Role of Memory Suppressor T Cells in Self Tolerance: Induction *in Utero* and in Athymic Mice 511
A. Basten, J. Gibson, R.H. Loblay, K.L. Wong and B. Fazekas de St. Groth

The Long-Lasting State of Specific Nonresponsiveness Induced by I.V. Immunization with Alloantigens is Due to the Generation of Recirculating Suppressor T Cells 521
A.T.J. Bianchi, L.M. Hussaarts-Odijk, H. Bril, H. de Ruiter and R. Benner

Pathogenesis of the Intestinal Phase of the Graft-Versus-Host Reaction in F_1 Hybrid Mice 531
A.McI. Mowat, A. Borland and D.M.V. Parrott

Graft Versus Host Reaction in the Rat: Effect of Cyclosporin A and Cyclophosphamide on Lymphocyte Subsets Involved in GVHR 539
A.A. van den Broek, R.J.M. Stet, G.G. van Eden, J. Kuiper, B. Meedendorp and C. Thomas

Graft Versus Host Disease in the Rat:
Ia Expression on Target Tissues 545
R.J.M. Stet, C. Thomas and
P. Nieuwenhuis

Induction of Class I and Class II
Antigens During Rejection of
Allografted Human Hearts 555
M.L. Rose, M.I. Coles, R.J. Griffin
and M.Y. Yacoub

Enhanced Regeneration of Transplanted
Splenic Tissue by Increased
Work load 565
R. Pabst, R. Hafke and J. Hillebrand

Retinoid Effects on Murine Immune
Responses 571
M. Drzymala, D.R. Katz, J.A. Turton
and R.M. Hicks

Effect of Avridine on Enteric Antigen
Uptake and Mucosal Immunity to
Reovirus 579
A.O. Anderson and D.H. Rubin

LYMPHOID CELL TRAFFIC PATTERNS

Regulation of the Size of the Recirculating
B Cell Pool of Adult Rats 593
J. Lortan, D. Gray, D.S. Kumararatne,
B. Platteau, H. Bazin and
I.C.M. MacLennan

The Influence of Complement Receptors
on the Localization of Lymphocytes
In Vivo 603
G. Kraal and I. van Hoogstraten

Interaction of High Endothelial Venules
With T and B Cells After Antigenic
Stimulation 609
G. Kraal and A. Twisk

Human Lymphocyte-High Endothelial Venule
Interaction: Functional and
Molecular Characterization 615
S. Jalkanen, R. Bargatze, L. Herron
and E.C. Butcher

T Cell Clones: A Model of a Non-recirculating Phase of T Cell Differentiation 621
M.O. Dailey, W.M. Gallatin, G. Kraal E.C. Butcher and I.L. Weissman

Radiosensitivity of the Circulation of DTH Effector T Lymphocytes: An Accurate Estimate Using a Quantitative Assay 629
G. Milon and G. Marchal

Evidence for Local Neuromodulation of T Cell Migration In Vivo 637
C.A. Ottaway

Lymphocyte Migration From Peyer's Patches by Diapedesis Through M Cells into the Intestinal Lumen 647
R.L. Owen and M.F. Heyworth

Emigration of Newly Formed Lymphocytes from Ileal Peyer's Patches in Lambs 655
R. Pabst and J.D. Reynolds

Patterns of Migration of Labelled Blood Lymphocyte Subpopulations: Evidence for Two Types of Peyer's Patch in the Young Pig 661
R.M. Binns and S.T. Licence

Interaction Mechanism of Clustering Lymphocytes: Specificity Molecular Properties and its Relation to Migration 669
A. Hamann, D. Jablonski-Westrich and H.-G. Thiele

Exploiting Lymphocyte Traffic to Deliver Radioactivity or Ricin to Lymphatic Tissues 675
W.L. Ford, A.J.S. Davies, M. Birch, J.A. Forrester, D. McIntosh, H. Sharma, S. Sparshott and C. Wood

IMMUNOREGULATION BY LYMPHOID CELL PRODUCTS

TMF, a Thymic Epithelial Glycoprotein: Chemotaxis of Hematopoietic Precursor Cells 683
E.F. Potworowski, K.W. Pyke, T. Brodniewicz-Proba and M. Zelechowska

Characterization of T Cell Growth Factor Responsive Thymocytes in Guinea Pigs 693
O. Söder, G. Sandberg, U. Ernström And J. Löthwall

Lymphokine Production by T Lymphocytes Isolated From Murine Small Intestinal Epithelium 701
S.B. Dillon and T.T. MacDonald

Resistance to the Induction of T Cell-Mediated Autoimmunity Correlates with Lower IL 2 Production 713
M. Mostarica-Stojković, L. Ejdus-Konstantinović, M. Kostić and M.L. Lukić

Detection of Antibody-Induced Feedback Suppression in Spleen Cell Populations of Primed and Unprimed Mice 721
A.L. Rothermel and C.E. Calkins

IgD-Induced Enhancement of the Immune Response 729
R.F. Coico, D. Wallace, B. Xue, J. Rosen B. Pernis, G.J. Thorbecke and G.W. Siskind

Regulation of Isotype Expression in the Immune Response to Type-2 Thymus Independent Dextran Antigens 739
K. Struckmann, W. Schuler and E. Kölsch

Organ-Selective Impairment of Phagocytosis by Cadmium 749
D.L. Vredevoe and L. Levy

Quantitation of Murine IgE in an Automatic ELISA System 757
H.F.J. Savelkoul, P.W.C. Soeting, R. Benner and J. Radl

Double Immunoenzymatic Staining Employing Rat and Mouse Monoclonal Antibodies 767
S. Poppema, A. Boes, L. de Ley, T. Rozeboom, W. Timens, E. Schwander and L. Visser

ANTIBODIES AS THERAPEUTIC TOOLS

Monoclonal Antibodies as Therapeutic Tools in Vivo 779
G.F. Rowland

Immunosuppression With Monoclonal Antibodies - Rules for Effective Serotherapy 789
S.P. Cobbold, G. Martin, P.E. Lovat and H. Waldmann

Interaction of Rat Monoclonal Antibodies With Human Killer Cells 797
M. Clark, G. Hale and H. Waldmann

Therapeutic Potential of Monoclonal Antibodies to the Leucocyte Common Antigen 805
C. Bindon, G. Hale and H. Waldmann

Elimination of Graft Versus Host Disease in Matched Allogeneic Leukemic Transplant Recipients Using CAMPATH-1 813
S. Slavin, R. Or, L. Weiss, D. Steiner-Zalz, G. Cividalli, C. Brautbar, Z. Weshler, E. Galun, S. Samuels, D. Galili, M. Ephros, M. Shapiro, G. Hale and H. Waldmann

The Immunotherapeutic Effect of Anti Lyt-1 Antibodies on Local and Metastatic Tumor Growth 819
N. Hollander

Chemotactic Anti-Tumour Antibodies: In Vitro Results With Four Different Antibody Preparations 827
R. Obrist, J.P. Obrecht, R.C. Bast, Jr. and A.L. Sandberg

A Monoclonal Antibody With Selectivity For Human *Kappa* Myeloma and Lymphoma Cells Which has Potential as a Therapeutic Agent 833
K.Z. Walker, H.A. Boux, G.E. Hayden, C.C. Goodnow and R.L. Raison

Derivatives of Anti-Idiotype Antibodies in the Treatment of B-Cell Lymphoma 843
M.J. Glennie and G.T. Stevenson

Protective Effects of Monoclonal Antibodies and Immunopurified Glycoproteins in Primary HSV-1 Infection 855
M.L. Lukic, F.Y. Liew and W.L. Chan

A New Indicator of Human Malignant Tumour Found on Erythrocytes 863
S. Metcalfe

Rat Monoclonal Antibodies for Bone Marrow Transplantation - the CAMPATH Series 869
H. Waldmann, S. Cobbold, A. Wilson, M. Clark, S. Watt, G. Hale and H. Tighe

HUMAN LYMPHOCYTE SUBPOPULATIONS IN HEALTH AND DISEASE

Changes of Lymphoid Microenvironments in Human Diseases: A Short Review 879
G. Janossy and M. Bofill

The Small Subpopulation of T Cell Marker$^+$/TdT$^+$ Cells in the Human Bone Marrow may Represent Prothymocytes 889
J.J.M. van Dongen, H. Hooijkaas, W.M. Comans-Bitter, K. Hahlen and G.E. van Zanen

M$^+$ Lymphocytes of the Human Foetal Liver are Precursors of T Lymphocytes 901
B.D. Jankovic, K. Isakovic and T. Matejic

M Receptor and T-Cell Differentiation in the Human Foetal Liver: An *In Vitro* Study 909
K. Isakovic, T. Matejic and B.D. Jankovic

Is Ia Antigen Expression by Thymic Epithelial Cells Constitutive? 915
R.V. Rouse

The Thymus in "Bare Lymphocyte" Syndrome 921
H.-J. Schuurman, F.P. van de Wijngaert, J. Huber, B.J.M. Zegers, R.K.B. Schuurman, J.J. Roord and L. Kater

The Thymus in Myasthenia Gravis: An Immunohistologic Study 929
M.J. Kornstein, J.J. Brooks, A.O. Anderson, A.I. Levinson, R.P. Lisak and B. Zweiman

B Cell Function in Acute Malaria 937
D.S. Kumararatne, P. Chiodini, C. Ellis, P. Richardson, T.G. Gentle, L. Walker and R.P. Stokes

The Origin of B-Chronic Lymphocytic Leukaemia (B-CLL) and its Relationship to Hairy Cell Leukaemia (HCL) 949
F. Caligaris-Cappio, L. Bergui, G. Pizzolo, M. Chilosi, L. Tesio, D. Campana, G. Semenzato, F. Malavasi and M. Gobbi

Increased Representation of a Subpopulation of Early B Cells in the Peripheral Blood of Patients with Rheumatoid Arthritis 957
C. Plater-Zyberk, K. Lam, T.D. Kennedy, R.N. Maini and G. Janossy

Nodular Lymphocyte Predominance Type of Hodgkin's Disease is a B Cell Lymphoma 963
S. Poppema, W. Timens and L. Visser

Two Monoclonal Antibodies Raised Against a Burkitt Lymphoma Cell Line Recognise Separate Components of Lymphoid Follicles 971
L.J. Murray and J.A. Habeshaw

Isolation of Germinal Centre Cells is Greatly Improved by Using the Protease Dispase to Prepare Cell Suspensions 979
N. Willcox, M. Schluep, M. Bofill and J. Newsom-Davis

Follicular Dendritic Cells Isolated From Human Tonsils 985
C. Kinet-Denoël, E. Heinen, D. Radoux and L. Simar

Follicular Dendritic Cells in Human Germinal Centres and Lymphomatous Follicles: A Morphological Comparison 993
L.H.P.M. Rademakers, J.P.J. Peters, J.M.M. Roelofs and J.A.M. van Unnik

The Nature of the Folliculoid Structures in Eosinophilic Lymphfolliculoid Granuloma (Kimura's Disease) 999
S. Takasaki, H. Tanaka, K. Takaki, S. Fukunaga and E. Ishikawa

Normal and Neoplastic Langerhans Cells: Phenotypic Comparison With Other Types of Macrophages 1009
E. Ralfkiaer, H. Stein, N. Ralfkiaer, K. Hou-Jensen and D.Y. Mason

$Ia^{+}T10^{+}$ Blast Cells in Acute Lymphoblastic Leukaemia (ALL) are Precursors of T^{+} and B^{+} Cells 1017
A.K. Miljković, K. Isaković, T. Matejić and B. Stojanović

A Study of Lymph Node Lymphocytes in Rheumatoid Arthritis 1027
D. Symmons, P.A. Bacon, M. Salmon and A.D. Blann

Benign and Malignant Lymphocytic Skin Infiltrates: Immunological Patterns Recognized by Immunohistochemical Staining with Monoclonal Antibodies 1033
E. Ralfkiaer, K. Hou-Jensen, D.Y. Mason, H. Stein, G.L. Wantzin and K. Thomsen

Differential Expression of Class II MHC Antigens on Cells Derived from Human Tissue 1043
B. McDowell, A.T. Cifelli, J.D. Nunn and D.R. Katz

Beta-Adrenergic Regulation of the Blood Lymphocyte Phenotype Distribution in Normal Subjects and Splenectomized Patients 1051
R. Landmann, M. Durig, F. Gudat, M. Wesp and F. Harder

Characterisation of Mononuclear Cell Infiltrates by Monoclonal Antibodies in Normal and Neoplastic Colorectal Tissue 1063
D. Heinemann, H.C. Umpleby, M.O. Symes and R.C.N. Williamson

Human Lymphocyte Subpopulations in Lymphadenopathies in Homosexual Men 1069
P. Rácz, A.C. Feller, K. Tenner-Rácz, B. Luther, K. Lennert, M. Dietrich and P. Kern

Index 1077

SESSION 1

B-CELL DIFFERENTIATION, SUBPOPULATIONS, THE ROLE OF GERMINAL CENTERS

CHAIRPERSONS: E.C. Butcher
P. Nieuwenhuis

PROSPECTS FOR UNDERSTANDING THE ROLE OF LYMPHOID MICROENVIRONMENTS IN B CELL DIFFERENTIATION *IN VIVO*

Eugene C. Butcher

Department of Pathology
Stanford University Medical Center
Stanford, CA, and the
Veterans Administration Medical Center
Palo Alto, CA USA

In this introduction to the section on B cell differentiation, I will focus on one of the traditional subjects of the Germinal Center Conferences --- the role of lymphoid tissue microenvironments, and in particular of germinal centers, in controlling the development and differentiation of functional B cell subsets. Gerry Klaus, in his introductory remarks in our abstract books, emphasized the fact that germinal centers seem rather an esoteric topic to many immunologists today --- including, undoubtedly, some attending this conference for the first time. This was not always the case, however. Twenty years ago the role of microenvironments in immune responses was a central concern of the immunology community, perhaps equivalent in general interest to the fascination today with lymphokines or the major histocompatibility complex. In this introduction, I will present my own (probably distorted) perspective on why interest in this area has so declined, and why investigations of *in vivo* microenvironments should hold renewed fascination to immunologists in the future.

The appeal of studies of the role of lymphoid architecture in *in vivo* immune responses 20 years ago, which led to the first of these Germinal Center Conferences in 1966, is obvious. At the time, histologic and morphologic techniques were among the most powerful tools available, and with these methods the lymphoid microenvironment was highly visible, as were the changes that occur in lymphoid architecture following antigenic stimulation. Germinal centers, well delineated clusters of immunoblasts seen in cortical follicles of stimulated lymph nodes, were particularly fascinating not only because they are obviously a highly organized microenvironment --- electron microscopic studies revealed characteristic follicular dendritic cells (FDC's) with complex processes interacting with the dividing germinal center lymphoblasts, as well as tingible body macrophages apparently engulfing germinal center cells --- but also because they are a visible consequence of immunization with most complex antigens. The appearance of germinal centers as discrete microenvironments following immunization, along with the subsequent demonstration that FDC demonstrate a special capacity to trap and retain injected antigen, together made it obvious that germinal centers must involve the accumulation and stimulation of antigen-specific lymphocytes, and must constitute a microenvironment of central importance in immune responses.

In light of their obvious fascination, why then did the study of germinal centers (and lymphoid architecture in general) fall out of fashion? The decline in interest is chronicled even in the very supportive context of the proceedings of these Germinal Center Conferences. From an initial high of roughly 35 papers dealing directly with germinal centers in the first conference (1966), the number of articles discussing germinal centers plummetted to an abysmal low of three in the fifth conference (1978). The decline in investigation on *in vivo* microenvironments did not, however, reflect a loss of interest so much as a shift in direction by immunologists. During this same period powerful new *in vitro* methods for dissecting the immune response were being developed. In addition, the functional heterogeneity of lymphocytes was being unfolded --- between the first and second germinal center conferences, the existence of thymus derived (T) and bursa derived (B) lymphocytes was described, and the definition of functional T cell subsets followed soon thereafter. Markers were identified allowing discrimination of these subpopulations, and the study of these newly defined lymphocyte subsets in functional *in vitro* systems became irresistible.

The development and application of *in vitro* methods to the study of B cell differentiation has by no means ceased --- within the past few years, we have seen the description of growth factors acting to control the proliferation and differentiation of B cells, and the development of methods allowing clonal growth of normal pre-B and other B cell populations *in vitro*. Furthermore, with the development and increasing application of monoclonal antibody techniques, we can only expect to see the definition of more and more B cell subsets, or stages of B cell differentiation, distinguished by lymphocyte surface antigen expression. One has the fear, then, that we may simply be developing more and more knowledge about the *in vitro* responses of more and more B cell populations whose existence as distinct functional populations *in vivo* is less and less clear, and whose physiologic relevance is totally obscure. It is my hope, however, that this point of view is falsely pessimistic. In fact, I think it is likely that as markers of lymphocyte and accessory cell populations become more and more sophisticated, our interest in and ability to understand the role of lymphoid architecture *in vivo* should increase rather than declining yet further. Several examples will serve to support this conviction:

A) By using dual fluorescence flow cytometry, Hardy *et al* (1) and others have been able to define at least three discrete surface μ^+ B cell populations in the mouse spleen -- $\mu^+ \delta^+$, $\mu^{+++} \delta^{lo}$, and $\mu^+ \delta^-$. These antigenically defined populations were shown to be distinct by several criteria --- both functional characteristics *in vitro*, and presence in different lymphoid organs. If these antigenically distinct populations actually represented physiologically different B cell subsets *in vivo*, then it seemed reasonable to me to ask whether they might exist in distinct splenic microenviroments. A synthesis of the FACS data with histologic and immunohistologic studies of the spleen [the work of Gray *et al* on μ^{+++} δ^- marginal zone B cells in the rat (2); and our own analyses of the mouse spleen (RA Reichert, unpublished observations)] led to the conclusion that the three populations described above represent (with reasonable certainty, and probably some overlap) follicle B cells (μ^+, δ^+), marginal zone B cells (μ^{+++}, δ^{lo}), and germinal center B cells plus plasma blasts (μ^+, δ^-). In my mind, the existence of these cells in distinct microenvironments lends needed support to the physiologic significance of the differences in the FACS-defined populations. It should now be possible to synthesize our understanding of the cellular elements comprising the marginal zone and follicular B cell microenvironments with *in vitro* functional analyses of their isolated B cell populations, thus leading to a better comprehension of the role of these distinct splenic B cell environments in *in vivo* immunity.

B) Another example of the application of novel markers to the understanding of lymphoid architecture stems from our own studies of the molecular basis of lymphocyte traffic *in vivo*. Monoclonal antibodies [MEL-14 in the mouse (3), LSRE-1 in the human (4), or the antibodies against rat receptors described by Woodruff and Chin in this conference] have been produced against lymphocyte surface "homing receptors" for high endothelial venules -- receptors essential for lymphocyte exit from the blood, and hence for the control of lymphocyte migration and recirculation. These antibodies can be used to define those specific microenvironments frequented by circulating B and T lymphocyte populations -- lymph node paracortex and primary follicles, and the white pulp of the spleen; and perhaps more strikingly to define lymphoid micro-enviroments from which circulating lymphocytes are excluded or do not long reside -- germinal centers, portions of the lymph node medulla, and the marginal zone of the spleen. The expression of only low levels of lymph node homing receptor by marginal zone B cells (RA Reichert and EC Butcher, personal observations) may help explain their absence as a significant population in lymph nodes and thus may relate, through coregulation with other surface elements, to the sessile nature of these unique splenic B cells.

C) Perhaps the most powerful example of the ability of a novel marker to advance our understanding of a specific lymphoid microenvironment is the recent application of peanut agglutinin (PNA), a plant lectin with specificity for terminal beta galactosyl residues, to the analysis of germinal center B cells. Just before the 7th Germinal Center Conference in Groningen, Marlene Rose and her coworkers described the selective binding of PNA to germinal center lymphoid cells (GCLC) in mouse Peyer's patches (5). This observation was subsequently confirmed and extended by two groups who independently made the same observation --- Raedler *et al* (6) showed that PNA^{hi} lymphocytes from antigen-stimulated lymph nodes display the electron microscopic morphology of GCLC; and our group confirmed in morphometric analyses that nearly all GCLC are PNA^{hi} (7) and in cell sorting experiments showed that plasma cells and most plasmablasts were PNA^{lo} (8). Thus, although some PNA^{hi} lymphocytes are T cells, within the B cell lineage PNA is a remarkably specific marker for germinal center cells. As such it has allowed the analysis of several important questions about the nature of germinal centers that could not be addressed previously. For example, we have studied the surface antigenic phenotype of GCLC, obtaining evidence that GCLC are a <u>phenotypically unique population</u> of proliferating B cells, differing from surrounding small B lymphocytes not only by virtue of their abundant surface levels of receptors for PNA, but also in their expression of high levels of surface Ia antigens, low levels of H-2K antigen, of significant levels of Bla-1, absence of ThB and (as mentioned above) their lack of surface homing receptor expression. It is interesting that, although the PNA^{hi}, δ^-, homing receptor$^-$ phenotype is established by the earliest time germinal centers are identifiable, the changes in H-2 antigens, loss of ThB and expression of Bla-1 are gradual changes which become fully established only after several days (G Kraal, R Hardy, I Weissman and EC Butcher, in preparation), suggesting that GCLC may derive from a ThB^+, Bla^- precursor -- the follicular B cell? The availability of PNA as a marker has also permitted us to demonstrate that many (and perhaps all) GCLC express surface antibody specific for the immunizing antigen (8); and has allowed Coico *et al* (9), and ourselves (Kraal, Weissman and Butcher, this conference) to confirm that GCLC are capable of functioning as memory cells, transferring specific IgG immune responsiveness to irradiated adoptive recipients. Thus studies using PNA have provided direct evidence that germinal centers are sites of proliferation and differentiation of antigen-specific B cells and, as proposed many years ago by Thorbecke and others, that these unique lymphocytes are indeed able to function as B memory cells.

Many questions still remain, of course. Some will be approachable using PNA as a marker, others will require the development of novel markers defining the various cellular elements of germinal centers. For example, what is the immediate precursor of the GCLC? Perhaps the most common precursor must be a δ^+ μ^+ (follicular, circulating) B cell, although it is clear that blast cells and probably memory cells can also enter the germinal center compartment (10). What controls the proliferation, and regulates the extensive cell selection occuring in germinal centers? Is the germinal center microenvironment in mucosal versus non-mucosal sites responsible for selecting or directing the expression of particular immunoglobulin isotypes? Or of particular organ-specific homing receptors? Are GCLC's undergoing somatic mutation as part of an affinity maturation program controlled by the germinal center microenvironment? What is the role of the T cells found in germinal centers? What is the role of the FDC? We know that FDC bind antigen-antibody complexes, and appear to initiate the nidus about which B cells initially cluster to form the earliest recognizable germinal center. Markers and isolation methods for FDC are now available, as described by Humphreys and Sunderam in this session, and Kosco *et al* later in the conference. These novel methods will hopefully lead to a clearer understanding of the role of this fascinating cell. Finally, what controls the entry of B cells into germinal centers, holds germinal center B cells together in a tight coherent cluster, and what molecular mechanisms control the subsequent return of presumably mature B effector or memory cells to the circulating lymphocyte pool? Although it is possible that altered carbohydrates (such as the receptor for PNA) may interact with specific lectins on FDC to mediate the cohesion of GCLC, such possibilities remain mere speculation. Only more and more precise markers --- not only of cellular elements but also of interstitial and secreted factors --- will allow us to isolate the various components in the complex GC microenvironment, to analyze their functional properties and then to synthesize our understanding of these separate elements to yield a clearer picture of their unified operation and thus of the role of the germinal center in *in vivo* immunity.

These examples, I believe, justify a cautious optimism concerning our ability to answer some of the basic functional questions about lymphoid architecture and its role in immune responses. The continuing development of markers of B cells, and of accessory cells, will facilitate rather than hinder efforts to bridge the gap between our understanding of lymphocyte function *in vitro*, and our rudimentary comprehension of the operation of the immune system within the structured microenvironments of lymphoid tissues *in vivo*. This transfer of knowledge between *in vitro* and *in vivo* systems will not occur, however, unless we as immunologists are reminded periodically of its importance. We must therefore be thankful for these Germinal Center Conferences, which provide a social and intellectual microenviroment that is uniquely suited for the synthesis of *in vivo* and *in vitro* immunology.

REFERENCES

1. Hardy, R.R., K. Hayakawa, D.R. Parks, L.A. Herzenberg, and L.A. Herzenberg, Murine B cell differentiation lineages, J. Exp. Med. 159:1169 (1984).
2. Gray, D., I.C.M. MacLennan, H. Bazin, and M. Khan, Migrant $\mu^+\delta^+$ and static $\mu^+\delta^-$ B lymphocyte subsets, Eur. J. Immunol. 12:564 (1982).
3. Gallatin, W.M., I.L. Weissman, and E.C. Butcher, A cell surface molecule involved in organ-specific homing of lymphocytes, Nature 303:30 (1983).
4. Jalkanen, S.T., R. Bargatze, L. Herron, and E.C. Butcher, manuscript in preparation.
5. Rose, M.L., M.S.C. Birbeck, V.J. Wallis, J.A. Forrester, and A.J.S. Davies, Peanut lectin binding properties of germinal centers in mouse lymphoid tissues, Nature 284:364 (1980).
6. Raedler, A., E. Raedler, R. Arndt, and H.G. Thiele, Centroblasts and centrocytes display receptors for peanut lectin, Immun. Lett. 2:335 (1981).
7. Butcher, E.C., R.V. Rouse, R.L. Coffman, C. Nottenburg, R.R. Hardy, and I.L. Weissman, Surface phenotype of Peyer's patch germinal center cells, J. Immunol. 129:2698 (1982).
8. Kraal, G., I.L. Weissman, and E.C. Butcher, Germinal Center B cells: antigen specificity and changes in heavy chain class expression, Nature 298:377 (1982).
9. Coico, R.F., B.S. Bhogal, G.J. Thorbecke, Relationship of germinal centers in lymphoid tissue to immunologic memory, J. Immunol. 131:2254 (1983).
10. Nieuwenhuis, P., N.A. Gastkemper, and D. Opstelten,, Histophysiology of follicular structures and germinal centres in relation to B cell differentiation, In: "Microenvironments in haemopoietic and lymphoid differentiation," Pitman Medical, London. (Ciba Foundation Symposium 84) p. 246-264.

PRE-B CELLS IN RAT BONE MARROW: IDENTIFICATION, SURFACE MARKERS AND ISOLATION

D. Opstelten, G.J. Deenen, T. Jaarsma, J. Rozing* and S.V. Hunt**

Dept. of Histology, State University, Oostersingel 69/1 Groningen, *Dept. of Internal Medicine I, Erasmus University Rotterdam and **Sir William Dunn School of Pathology Oxford University, Oxford

INTRODUCTION

In adult rats, as in other species, bone marrow has the highest potential for B-lymphocyte genesis as determined by a long-term repopulation assay for B-lymphocyte stem cells (1). These cells carry the Thy-1 and W3/13 antigens (2, 3). Pre-B cells characterized by the presence of μ heavy chains in the cytoplasm (cμ) but not on the cell surface (sμ) have been described in mice (4), men (5) and rabbits (6) but not in rats, and appear to be the immediate precursors of B-lymphocytes (7-10). The relation between pre-B cells and B-lymphocyte stem cells is not clear, however. Therefore, the aims of the present study were to identify $c\mu^{+}s\mu^{-}$ pre-B cells in rat bone marrow and to find surface markers that would permit isolation of viable pre-B cells to subsequently assay their functional potential *in vivo*.

MATERIALS AND METHODS

Animals: $(AO/G \times BN/G)F_1$ ♂ hybrid rats ($RT1^{u/n}$) were obtained from stocks held at the Dept. of Histology, Groningen, and maintained under conventional conditions. PVG ($RT1^{c}$) ♂ rats were from the Sir William Dunn School of Pathology, Oxford, and maintained under specific pathogen free conditions.
Antibodies: Rabbit antibody to rat μ chains (11) and monoclonal antibody (MAb) OX-22 (12) were a kind gift of Dr. A.F. Williams, MRC CIU, Dunn School of Pathology, Oxford. Horse anti rabbit Ig antibodies conjugated to FITC or TRITC and rabbit anti mouse Ig antibody conjugated to FITC were prepared as described (11). HIS MAb's were prepared as described (13). MAb ER 14 was produced as

described (14). Affinity purified goat anti mouse Ig antibody conjugated to FITC was from Kirkegaard and Perry Laboratories, Gaithersburg, MD, USA. For detection of terminal deoxynucleotidyl transferase (TdT) the immunofluorescent assay of Bethesda Research Laboratories (Gaithersburg, MD, USA) was used. All first stage antibodies were diluted in phosphate buffered saline (PBS), second stage antibodies in PBS containing 1% (v/v) normal rat serum.

Cell samples: Bone marrow cells were obtained as described for mice (9). Nucleated cells recovered per femur were $8.1 \pm 0.7 \times 10^7$ for 5 wk old, and $9.8 \pm 0.2 \times 10^7$ for 10-12 wk old AOxBN; 5 wk old PVG contained $6.2 \pm 0.3 \times 10^7$ cells/femur.

Labeling of B-lymphocytes and pre-B cells: The labeling procedure was as described for mice (9), with the exception of an indirect instead of a direct labeling method for μ chains.

Labeling of cell surface antigens with MAb: Viable bone marrow cells were incubated with mouse anti rat MAb's, washed twice through PBS/3% bovine serum albumine and then stained by incubation with RαM-FITC or GαM-FITC (30', 0°C). Cells were washed, cytocentrifuged, fixed and stained for μ chains using HαR-TRITC as second stage antibody.

Fluorescence microscopy: Preparations were mounted in glycerol/PBS containing an anti-fading agent. An epifluorescence microscope (Leitz Wetzlar, FRG) equipped with a HBO-100 mercury lamp, phase-contrast optics and a x100 oil immersion phase-contrast objective was used for analysis. B-lymphocytes (fluorescing green and red) and pre-B cells (fluorescing red only) were scored as described for mice (9). In preparations double-stained with MAb and anti-μ each cell was scored for presence of fluorescein or rhodamine alone, or both, or neither, thus classifying them into four subsets. In this analysis (Table 2), the distinction between pre-B cells and B-lymphocytes (both fluorescing red) was made on the basis of fluorescence intensity, being considerably higher in the latter, and cell size, the former being partly larger. In all cases the number of cells analysed was such that at least 100 μ-chain positive cells were scored.

Preparation of anti rat Ig-coated erythrocytes: For Exp. 1, rabbit anti rat/mouse Ig-coated red blood cells (RBC) were prepared as described (15). For Exp. 2, ox RBC were coated with sheep anti rat Ig as described (16).

Removal of sIg^+ cells from marrow samples: Marrow cells were incubated with anti Ig-coated RBC followed (in Exp. 1) by incubation with mouse Ig-coated RBC and sedimentation as described (15) or (in Exp. 2) by centrifugation through a Percoll solution (1.090 g/ml).

Fluorescence activated cell sorting: The rosette-depleted marrow cells were incubated with HIS 24 followed by GαM-FITC and sorted on a FACS-II (Becton Dickinson, Sunnyvale CA, USA). For phenotypic evaluation sorted cells (HIS 24 bright vs. HIS 24 dull) were cytocentrifuged and fixed.

RESULTS

μ^+ B-lineage Cells in Rat Bone Marrow

Using double immunofluorescence to separately detect surface and intracellular μ chains, both $s\mu^+$ and $c\mu^+s\mu^-$ cells were detected in rat bone marrow (Table 1). Judged by phase contrast microscopy, all μ^+ cells appeared lymphoid in morphology; $s\mu^+$ cells were mainly small and $c\mu^+s\mu^-$ cells were heterogeneous in size. Plasmablasts and plasmacells were rare (< 1%) and easily discriminated from μ^+ early B-lineage cells by their brilliant fluorescence. In marrow of 5 wk old adolescent rats, μ^+ cells constituted 25.0±6.8% of nucleated cells. $C\mu^+s\mu^-$ pre-B cells were far more numerous than $s\mu^+$ B-lymphocytes, in a ratio of 4:1. At the age of 10-12 wk, both the incidence and absolute number of $c\mu^+s\mu^-$ pre-B cells but not $s\mu^+$ B-lymphocytes had declined (Table 1).

Table 1

μ^+ B-lineage Cells in Bone Marrow of 5 wk old AOxBN ♂ Rats

	Incidence (% of total nucl. cells)		Absolute number per femur ($x10^{-6}$)	
	5 wk	10-12 wk	5 wk	10-12 wk
total μ^+ cells	25.0±6.8	15.0±2.8	20.7±6.4	14.5±2.5
$c\mu^+s\mu^-$ pre-B cells	20.2±5.3	11.3±3.2	16.7±5.0	10.9±3.0
$s\mu^+$ B-lymphocytes	4.9±2.2	3.8±0.4	4.1±1.7	3.7±0.4

Surface Antigens of Rat Pre-B Cells

To search for a tool to isolate pre-B cells, μ^+ B-lineage cells in bone marrow of 5 wk old rats were analysed by double immunofluorescence microscopy for 10 monoclonal antibodies differing in their specificity for rat lymphocyte subsets (Table 2). The MAb's ER 14, HIS 14, HIS 24 and HIS 25 stained virtually all μ^+ marrow cells; the intensity of binding per cell was ER 14 > HIS 24 > HIS 25 > HIS 14. To select the most suitable for pre-B cell isolation, reactivity of these MAb's with marrow cells other than μ^+ was also taken into account. HIS 14 bound to μ^+ and TdT^+ cells, a minor (2-5 %) μ^- lymphoid cell subset in rat bone marrow (D. Opstelten et al., preliminary observations). Of the ER 14^+, HIS 24^+ and HIS 25^+ marrow cell populations, 35, 15 and 15%, respectively, were μ^- cells. For all three, these included TdT^+ cells and few large cells with unclassifiable phase contrast morphology. ER 14, moreover, appeared to bind also to some myeloid progenitors identified by their high cytoplasm to nucleus ratio.

Table 2

Surface Antigens of B-lineage Cells in Marrow of 5 wk old AOxBN ♂ Rats

MAb	Description of specificity	Weak μ⁺ pre-B cells	strong μ⁺ B lymphocytes
ER 14	Thyl.1	■	■
HIS 14	peripheral B-lymphocytes	■	■
HIS 24	B-ly but not marginal zone	■	■
HIS 25	peripheral B-ly, subset T-ly	■	■
HIS 17	W3/13-like	◪	□
HIS 22	splenic corona cells only	◪	■
OX-22	peripheral B, T, NK cells	◪	■
HIS 11	peripheral B-lymphocytes	◪	■
HIS 19	MHC class II (Ia)	□	◪
HIS 8	Ig light chain	□	■

■ > 90% ◪ 10-90%, □ < 10% of the cells reactive.

On the basis of these data on B-lineage specificity and binding intensity, HIS 24 was chosen to isolate pre-B cells from rat bone marrow.

Isolation of Pre-B Cells From Rat Bone Marrow

As a first step, sIg^+ B-lymphocytes and plasmacells were removed from bone marrow cell suspensions by rosetting with α-Ig coated RBC followed by sedimentation. As shown in Table 3, the

Table 3

Removal of Marrow B Cells and Plasmacells by Rosetting

Phenotype	Incidence (%) before Exp.no. 1	before 2	after 1	after 2
surface μ chain⁺	7	8	<0.5	<0.5
k light chain⁺	10	8	<0.5	<0.5
ck⁺⁺⁺ plasmacells	<1	<0.5	<0.001	<0.001
HIS 24 antigen⁺	46	39	41	48

resulting marrow cells were virtually devoid of B cells and plasmacells, while HIS 24^+ cells were still present. The depleted marrow cells were then sorted into HIS 24^+ and HIS 24^- subsets by FACS (Table 4). The quality of sorting was slightly better in

Table 4

Phenotype of HIS 24^+ Marrow Cells Isolated by FACS

Phenotype	Exp.no.	Total marrow cells 1	Total marrow cells 2	FACS-separated HIS 24^+ cells 1	FACS-separated HIS 24^+ cells 2
HIS 24 antigen$^+$		46	39	98	91
k^+ B-lymphocytes		10	8	<0.1	<0.1
μ^+k^- pre-B cells		31	27	91	74
TdT$^+$ cells		n.d.	4	n.d.	16

Exp. 1 due to more clear-cut staining by HIS 24 (Fig. 1). Assay for presence of μ and k chains revealed strong enrichment of pre-B cells.

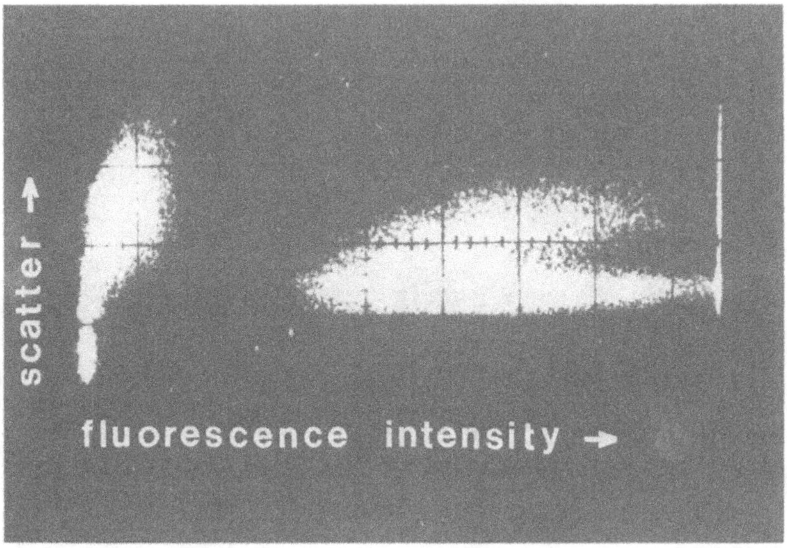

Fig. 1. Fluorescence-scatter diagram of sIg^+ cell-depleted marrow stained with HIS 24 (Exp. 1). Note a subset of strongly fluorescent cells heterogeneous in size, but absence of fluorescence on the largest and smallest marrow cells.

TdT$^+$ cells were enriched as well and accounted for most of the μ^- HIS 24$^+$ cells in the HIS 24$^+$ fraction. In Exp. 2 using bone marrow of AOxBN rats, morphologic analysis of the HIS 24$^+$ fraction revealed contamination with only 2% recognizable myeloid and 1% recognizable erythroid cells. Four % were not classifiable large cells, and all others (92%) were lymphoid cells heterogeneous in size with mitotic nuclei among the large (Fig. 2). In PVG rats (Exp. 1) the distinction between erythroid and lymphoid cells was unclear. As a consequence, in Exp. 1 morphologic analysis of the HIS 24$^+$ fraction was limited to contaminating myeloid and late erythroid cells, which were again only 3%.

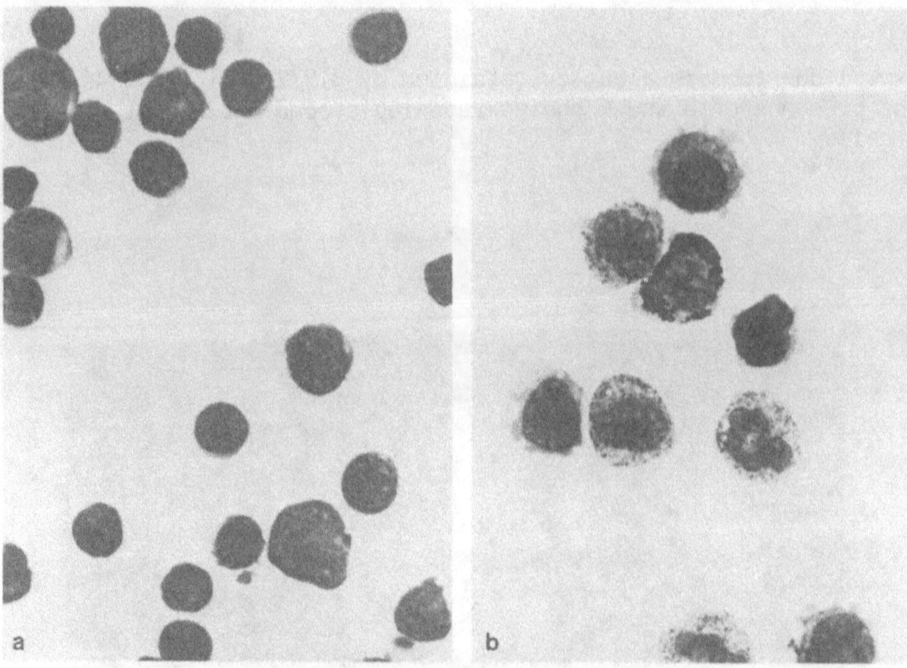

Fig. 2. Giemsa-stained cytocentrifuge preparations of FACS-separated marrow cells (Exp. 2; x 750). a) HIS 24$^+$ cells: note abundance of cells, heterogeneous in size, with lymphoid morphology. b) HIS 24$^-$ cells: note myeloid morphology.

DISCUSSION

This study is the first to demonstrate cμ^+sμ^- cells in bone marrow of the rat. In morphology, size distribution and cytoplasmic μ chain concentration rat cμ^+sμ^- cells were comparable to cμ^+ sμ^- pre-B cells described in mice (see e.g. 4, 7, 9).

However, in contrast to mice, they outnumbered sμ^+ marrow B lymphocytes 4 to 3 times (depending on age). This is due to a low incidence of sμ^+ rather than high incidence of cμ^+sμ^- cells in rat as compared to mouse (9). The hint of the age dependency of marrow pre-B cell numbers reported here is reminiscent of the burst of Thy-1 positive cells in weanling rats (19) and suggests that it would be worthwhile performing a full study of pre-B cell ontogeny in rat.

Surface marker analysis of μ^+ marrow B-lineage cells with a panel of MAb's revealed several to be potentially useful for studies of early B cell differentiation in the rat. μ^+ B-lineage specificity was highest for HIS 14, but this MAb was not used for pre-B cell isolation by FACS because of its low binding intensity. Remarkably, among the panel tested all MAb's binding to pre-B cells (-subsets) also stained TdT$^+$ marrow cells (-subsets) (D. Opstelten et al., preliminary observations). This may suggest that TdT$^+$ and cμ^+sμ^- cells, separate populations in rat bone marrow (D. Opstelten et al., preliminary observations) are closely related. As to the functional capacity of marrow TdT$^+$ cells the postulate for mice, that they are the most likely candidates for prothymocytes could not be substantiated for rats (18). However, further studies on possible T and/or B stem cell activities of TdT$^+$ marrow cells are required (19).

Of the four pre-B cell binding MAb's (ER 14, HIS 14, HIS 24, HIS 25) all but HIS 14 also stained some large unclassifiable cells; these cells accounted for $\sim$1% in normal marrow and could be enriched to 16% in a FACS isolated large HIS 24$^+$ marrow cell subset (D. Opstelten et al., preliminary observations). It would be of interest to study the functional potential of these cells, in particular their possible (more or less restricted) stem cell nature.

The FACS isolation of HIS 24$^+$ marrow cells was of high quality, the contamination with recognizable myeloid and erythroid lineage cells being very low and enrichment for pre-B cells and TdT$^+$ cells being very high. This opens up the possibility for functional studies of rat marrow lymphoid cells and subsets by use of HIS 24 alone or in combination with some of the other MAb's described here. Moreover, a study of early B cell differentiation events at the molecular level, in particular Ig gene rearrangements, may now be feasible as well.

ACKNOWLEDGEMENTS

The authors wish to thank F.J. Tielen for technical assistance (Exp.no. 2), L. Baldwin for operating the FACS (Exp. no. 1), P. v.d. Sijde and D. Huizinga for photography and A. Zandstra for preparing the manuscript. This study was supported in part by the Netherlands Organization for the Advancement of Pure Research (Z.W.O.) (Grant no. R 93-113), and by the British Council.

REFERENCES

1. S.V. Hunt. In: "Immunobiology of Bone Marrow Transplantation", S. Thierfelder, H. Rodt and H.J. Kolb, eds., Springer Verlag, Berlin (1980) pp 193-203.
2. S.V. Hunt. Eur. J. Immunol. 9: 853 (1979).
3. M.J.S. Dyer and S.V. Hunt. J. Exp. Med. 154: 1164 (1981).
4. M.C. Raff, M. Megson, J.J.T. Owen and M.D. Cooper. Nature 259: 224 (1976).
5. W.E. Gathings, A.R. Lawton and M.D. Cooper. Eur. J. Immunol. 7: 804 (1977).
6. A.R. Hayward, M.A. Simons, A.R. Lawton, R.G. Mage and M.D. Cooper. J. Exp. Med. 148: 1367 (1978).
7. K.S. Landreth, C. Rosse and J. Clagett. J. Immunol. 127: 2027 (1981).
8. R.L. Coffman. Immunol. Rev. 69: 5 (1983).
9. D. Opstelten and D.G. Osmond. J. Immunol. 131: 2635 (1983).
10. D.G. Osmond, F. Melchers and C.J. Paige. J. Immunol. 133: 86 (1984).
11. A.F. Williams. Eur. J. Immunol. 5: 883 (1975).
12. G.P. Spickett, M.R. Brandon, D.W. Mason, A.F. Williams and G.R. Woollett. J. Exp. Med. 158: 795 (1983).
13. F.G.M. Kroese, D. Opstelten, A.S. Wubbena, G.J. Deenen, E.H. Schwander, L. de Ley and P. Nieuwenhuis. This volume.
14. J. Rozing, P. Joling, P. Nieuwenhuis and L.M.B. Vaessen. This volume.
15. S.V. Hunt. In: "Handbook of Experimental Immunology". D.M. Weir, ed., Blackwell, Oxford. Edition 1984, in press.
16. M. v.d. Giessen, S. Postma and T.H. The. Submitted to Scand. J. Immunol.
17. S.V. Hunt, D.W. Mason and A.F. Williams. Eur. J. Immunol. 7: 817 (1977).
18. D.L. Greiner, I. Goldschneider and R.W. Barton. J. Exp. Med. 156: 1448 (1982).
19. I.D. Goldschneider, T. Metcalf, T. Mandel and F.J. Bollum. J. Exp. Med. 152: 438 (1980).

T CELL MARKERS ON PRE-B CELLS IN THE RAT

L.M.B. Vaessen, P. Joling, P. Nieuwenhuis,
M.J. van Haperen and J. Rozing

Dept. of Internal Medicine I, Renal Division
Erasmus University Rotterdam.Dept. of Histology
University of Groningen, The Netherlands

INTRODUCTION

In order to prepare monoclonal antibodies against cells in the early stages of B cell differentiation in the rat, mice were immunised with leukaemic pre B cells [1]. The spleens of such mice were subsequently used for hybridisation with the SP2/0 fusion line. These fusions resulted in a panel of monoclonal antibodies reacting with determinants on rat lympho-haemopoietic cells. Based on data obtained in flow cytometry and immune peroxidase histology two clones were selected. This paper gives a further characterisation of the reactivity pattern of those two antibodies.

MATERIALS AND METHODS

Animals

Balb/c mice (10-14 weeks old) were used for immunisation and ascites production. AKR mice, 12 weeks of age, were used in THY-1 binding studies.All mice were bred in our own colony at the Erasmus University Rotterdam.(AOxBN)F1 hybrid rats were obtained from the Dept. of Histology, University of Groningen. Lewis rats were purchased from Olac, Bicester, United Kingdom. All rats were used at 10-20 weeks of age.

Antibodies

Clones W3/13 (anti-T cell), OX7 (anti-THY-1) and OX19

(anti-T cell) were kindly provided by Dr.A.F.Williams, Oxford, United Kingdom. Clone MARK-1 (anti-rat kappa chain) was a kind gift of Dr. H.Bazin, Brussels, Belgium.

Fusion

Balb/c mice were immunised with leukaemic pre B cells twice. A detailed characterisation of this tumour has been published elsewhere [1]. Six weeks after the initial immunisation with 10^7 tumour cells intraveneously (iv), the mice were boosted with 5x10^7 tumour cells iv. Four days later the spleens of these mice were removed aseptically and used for fusion. 2x10^7 Spleen cells were fused with 5x10^7 SP2/0 cells according to the method as described by Fazekas de St. Groth and Scheidegger [2]. The fusion mixture was plated in selective HAT-medium and followed for growing hybrids. Growing clones were screened for antibody production, selected and recloned twice, using a limiting dilution procedure. Finally hybrid cells were injected intraperitoneally into Balb/c mice for ascites production.

Radio immunoassay (RIA)

The initial screening for antibody production was done in a RIA on target cells coated to the wells of a flexible titration plate as described in detail previously [3].

Fluorescence Activated Cell Sorter (FACS) analysis

Selected antibodies were analysed on the FACS in a direct immune fluorescence assay.Antibodies were, after DEAE purification, directly conjugated to FITC (Sigma, USA) using standard methods as described by Goding [4]. Fluorescent cells

Fig. 1. Screening of immunoglobulin containing hybrid culture supernatants on tumour target cells.

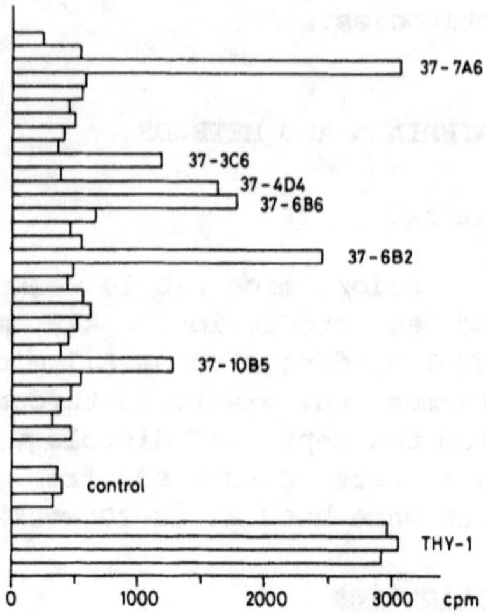

were analysed on a FACS II (Becton and Dickinson, USA), equipped with a ND 100 pulse height analyser (Nuclear Data, USA) interphased with a PDP 11/70 computer (Digital, USA) for further analysis.

THY-1 binding

Binding of monoclonal antibodies to the THY-1 glycoprotein was studied in three ways. First, crossreactivity of the antibodies was determined in a fluorescence staining test on thymocytes of AKR mice. These mice carry the same THY-1 allele as rats [5]. Second binding of the antibodies was measured in a micro-ELISA system, in which purified rat THY-1 was coated to the bottom of microtiter wells and the presence of bound antibodies was established with a β-galactosidase rabbit-anti-mouse serum in combination with the substrate ortho-phenyl-diamine (OPD). Purified rat THY-1 was kindly provided by Dr.A.F.Williams, Oxford, England. Finally, this purified THY-1 was also used in a blocking assay. Therefore antibodies were preincubated with THY-1 and remaining activity was measured in a fluorescence assay on rat thymocytes.

Immuneperoxidase histology

Specific immune histology was done on 5 micrometer cryostate sections of the spleen. After incubation with the monoclonal antibody, the sections were washed and then incubated with horse-radish peroxidase conjugated rabbit-anti-mouse immuneglobulin (Miles, England). After washing the substrate

Table 1. Reactivity of monoclonal antibodies in various lymphoid organs

Antibody	Reactivity	Percentage[a]			
		Bone marrow	Thymus	Spleen	Lymph node
37-4D4	?	31	96	3	4
37-7A6	?	56	96	41	58
OX19	anti-T	3	93	36	53
MARK-1	anti-B	13	nd	32	33

a. Percentage positive cells determined in lymphoid organs of adult LEWIS rats by FACS analysis.
nd= not done

diamino benzidine tetrahydrochloride (DAB) was added. The preparations were mounted finally in depex and examined under a standard Zeiss microscope.

Immunoprecipitation

The molecular weight of the cell surface antigens, recognised by the various monoclonal antibodies, was determined on thymocytes. Thymocytes were radiolabelled with 125-Iodine with Iodogen and lysed with NP40. Immuneprecipitation was done according to Agthoven et al [6]. Molecular size of the antigens was determined by running the precipitate on a 7-20% gradient SDS-PAGE gel, followed by autoradiography.

Intracellular (cIg) and surface immunoglobulin (sIg) staining

FACS sorted cells were analysed for cIg and sIg expression using a direct immune fluorescence assay as described previously [7]. TRITC conjugated MARK-1 antibodies were used for the detection of Ig. Preparations were analysed on a Zeiss fluorescence microscope equipped with a IV F vertical illuminator and the appropriate filter combination for red/green distinction.

In vitro culture assay

Sterile, FACS sorted cells, depleted for sIg positive cells, were cultured for 96 hrs in RPMI-1640 medium supplemented with 10% autologous rat serum at 37°C and 5% CO_2. At various times small aliqouts were harvested and analysed for the expression of

Table 2. THY-1 binding properties of monoclonal antibodies

Antibody	Binding		Blocking[c]
	AKR-thymus[a]	THY-1[b]	THY-1
OX7	+	+	+
37-4D4	+	+	+
37-7A6	–	–	–

a All antibodies were directly conjugated with FITC and analysis was done on a FACS II.
b Binding of antibody to purified THY-1 was measured in a micro-ELISA.
c Blocking with purified THY-1 of the antibody interaction with rat thymocytes.

sIg. Analysis was done using FITC labelled MARK-1 antibodies for the detection of sIg and the FACS for quantitation.

RESULTS

Immunisation of mice with rat leukaemic pre B cells and subsequent fusion of the spleens of such mice with SP2/0 cells, resulted in a large number of growing hybrids. Approximately 30 of these produced immunoglobulins and could be established as stabile cell lines. When screened in a RIA on the tumour cells, which had been used as immunogen, 6 of them reacted positively on these target cells (Fig.1). In further screenings using normal tissues of (AOxBN)F1 hybrid rats, only clones 37-4D4 and 37-7A6 proved to recognise lymphoid and haemopoietic tissues, whereas the remaining four did not (Data not shown). Therefore clones 37-4D4 and 37-7A6 were selected, grown and used for further analysis.

When compared with a T cell reagent (OX19) and a B cell reagent (MARK-1) on a variety of organs in FACS analysis neither 37-7A6 nor 37-4D4 gave a pattern similar to these markers (Table 1). Clone 37-4D4 reacted predominantly with cells in the thymus and the bone marrow, with no reactivity in peripheral lymphoid organs. Clone 37-7A6 on the other hand did also recognise cells

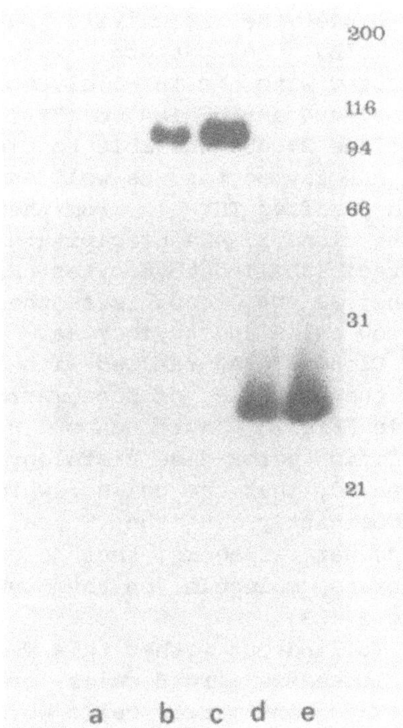

Fig. 2. Immunoprecipitation of ^{125}I-labelled thymocytes on SDS-PAGE. Lane a=SP2/0 ascites; b= W3/13; c= 37-7A6; d= OX7; e= 37-4D4

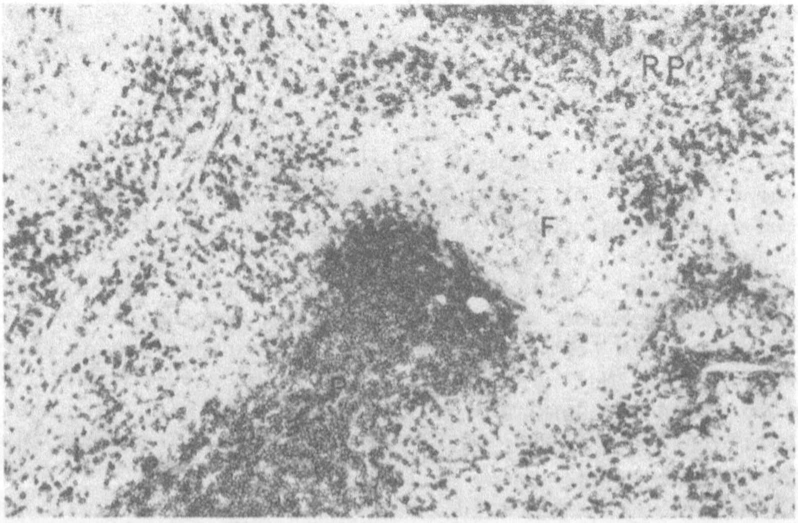

Fig. 3. Peroxidase histology of clone 37-7A6 on spleen.
F= Follicle; P= PALS; RP= Red pulp

in the bone marrow and the thymus, but also a distinct percentage of the cells in spleen and lymph nodes.

Since the reactivity pattern of clone 37-4D4 closely resembled the pattern, which has been described for clone OX7, reacting with the THY-1 glycoprotein of the rat [8], the binding of 37-4D4 antibodies to THY-1 was determined. As shown in Table 2, clone 37-4D4 was able to bind both to purified rat THY-1 and to AKR thymocytes, as well as to block the reaction of clone OX7 with purified THY-1. From these data as well as from the fact, that clone 37-4D4 precipitates a 25 kD molecule from the surface of radiolabelled thymocytes (Fig.2), it can be concluded, that clone 37-4D4 recognises the 25 kD THY-1 glycoprotein on bone marrow cells and thymocytes.

Clone 37-7A6 reacted with a marker of 95 kD molecular weight on the surface of bone marrow cells, thymocytes and peripheral cells (Fig.2). Although pre B cells had been used as immunogen, specific peroxidase histology with antibody 37-7A6 on the spleen revealed, that the cells, which are recognised, were located predominantly in the T cell area of this organ (Fig.3). From these data it seems, that clone 37-7A6 reacts with the same membrane molecule on thymocytes and T cells as clone W3/13 does (Fig.2) [9].

To find out wether this T cell determinant is only expressed by leukaemic pre B cells, or also by normal pre B cells, 37-7A6 positive bone marrow cells were sorted with the FACS and analysed

Table 3. Association of clone 37-7A6 with sIg and cIg in the bone marrow

Cells	Percentage of positive cells			
	Unsorted[a]	Purity[a] after sort	sIg[b]	cIg[c]
37-7A6 positive	58	93	0.5-1	65
37-7A6 negative	42	95	10	nd

a. Antibody 37-7A6 was directly conjugated to FITC and analysis was done on a FACS II.
b. Determined by IF-microscopy using TRITC-MARK-1 for sIg staining.
c. Determined by IF-microscopy after cytospin and staining for cIg using TRITC-MARK-1.
nd= not done

for the presence of cIg, a marker for pre B lymphocytes. As shown in Table 3. more than half of the 37-7A6 positive bone marrow cells were positive for cIg. Although the vast majority of the cIg staining was of the pre B cell type, 37-7A6 also reacted strongly with cells with the characteristic cIg staining type of plasma cells.

The existence of normal pre B cells in the bone marrow, displaying the "T cell marker" 37-7A6 could also be confirmed in an experiment, in which 37-7A6 positive bone marrow cells were sorted under sterile conditions and cultured in vitro for 96 hours. Repeated analysis for the appearance of sIg positive B cells at various time points clearly showed, that within the culture 37-7A6 positive, sIg negative bone marrow cells could give rise to sIg positive B lymphocytes (Fig.4).

DISCUSSION

Using leukaemic pre B cells as immunogen two monoclonal antibodies were prepared, which reacted not only with the tumour cells, but also with normal rat lympho-haemopoietic tissues. Of these, clone 37-4D4 could be shown to react with the 25 kD THY-1 glycoprotein [8]. From previous work by several authors [10,11] it is known, that this molecule is present on thymocytes and early haemopoietic precursor cells, including pre B cells, in the rat. The 95 kD cell surface molecule recognised by clone 37-7A6 on the other hand, which is likely to be the same molecule as detected by clone W3/13, is considered to be specific for peripheral T cells, haemopoietic stem cells and precursor thymocytes, but not for pre B lymphocytes [9,12]. However, the data presented in this paper clearly indicated, that the determinant, which is recognised by clone 37-7A6, is present not only on leukaemic pre B cells, but also on normal pre B lymphocytes. These findings seem to be in contrast with those of Dyer and Hunt [12], who concluded, after transfer experiments of W3/13 positive sorted bone marrow cells into congenic B and T allogeneic hosts, that W3/13 is present on the direct precursors of the thymocyte, but probably not on precursor B lymphocytes. Possibly clones 37-7A6 and W3/13 recognise different epitopes on the same 95 kD molecule. A differential expression of these epitopes during B cell differentiation may then have contributed to these contradictory observations. Also pure technical aspects could be involved. Whereas Dyer and Hunt [12] sorted out only the brightest positive cells, FACS experiments in this paper also included the dull fraction. In that case, the pre B cell is likely to be found in the 37-7A6 dull fraction. So a marker, which in peripheral lymphoid organs is specific for T cells is also found on precursor B lymphocytes.

The occurence of "T cell markers" on early cells of the B cell lineage in rats is not unique. Similar findings have been reported both in mice [13,14] and in man [15,16]. Originally these markers, such as Lyt1 in mice and Leu1 in man, could only be demonstrated on leukaemic B cells [13,15]. Recently, however, it was shown that also a proportion of normal B cells express these determinants [14,16]. From the observations, that such Lyt1 B cells in mice are more abundant in neonatal animals, as compared to adult animals, and that they express more IgM and less IgD [17], it can be concluded that this subset of B cells

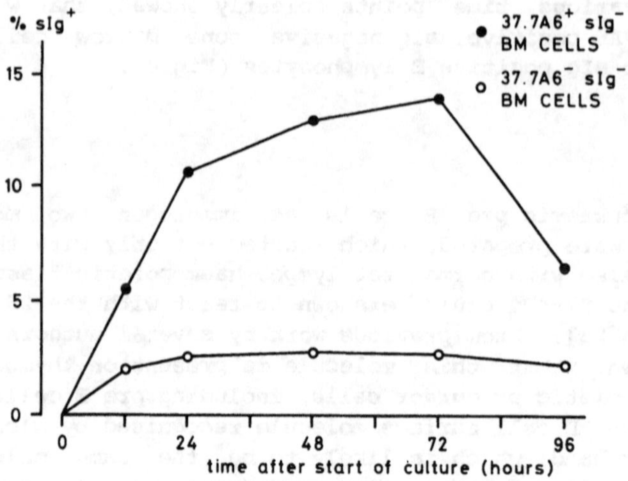

Fig. 4. Appearance of sIg positive cells during in vitro culture

represents an immature stage of B cell differentiation. Clone 37-7A6 reacts also with a low, but definite proportion of normal B cells in the rat (F.Kroese, unpublished results). Such a comparable expression of surface markers, which are present on cells throughout the T cell differentiation lineage, on part of the B cells and their precursors in these three species, indicate a similar genetic regulation. Antibodies 37-7A6 and 37-4D4 and the rat as experimental model may well help to answer such questions.

Antibody 37-7A6, finally, has been renamed ER-1, and antibody 37-4D4 has been renamed ER-14.

ACKNOWLEDGEMENTS

Part of this investigation has been subsidised by the Dutch Renal Foundation (Grant C245), Amsterdam, The Netherlands. The authors wishes to express their gratitude to Prof.Dr.J.Jongkind for the use of his FACS equipment.

REFERENCES

1. J.Rozing, L.M.B. Vaessen, L.Faber, M.van Oven, L.de Vries-Bos, B.de Jong and P.Nieuwenhuis.,Pre B cell leukaemia in the rat, Adv.Exp.Biol.Med.,149: 111 (1982)
2. S.Fazekas de St.Groth and D.Scheidegger,Production of monoclonal antibody:Strategy and tactics, J.Immunol.Meth., 35:1 (1980)
3. J.Rozing,P.Joling, W. van Ewijk and C.J.Hesse,Comparison of screening techniques related to the production of monoclonal antibodies,in:"Protides of the biological fluids", Ed.H.Peeters,Pergamon Press,Oxford,New York,Vol.29:637 (1982)
4. J.W.Goding.,Conjugation of antibodies with fluorochromes: modifications to the standard methods, J.Immunol.Meth., 13:215 (1976)
5. T.C.Douglas.,A rat analog of the mouse theta antigen, Transpl.Proc. 5:79 (1973)
6. A.Agthoven, C.Terhorst, E.Reinherz and S.F.Schlossman, Characterisation of T cell surface glycoproteins T1 and T3 present on all human peripheral T lymphocytes and functionally mature thymocytes, Eur.J.Immunol., 11,18 (1981)
7. J.Rozing, N.H.C.Brons and R.Brenner, Effects of splenectomy on the humoral immune system. A study in neonatally and adult splenectomized mice, Immunology, 34: 909 (1978)
8. D.W.Mason and A.F.Williams, The kinetics of antibody binding to membrane antigens in solution and at the cell surface, Biochemical J. 187:1 (1980)

9. A.F.Williams, G.Galfre and C.Milstein,Analysis of cell surfaces by xenogeneic myeloma-hybrid antibodies:differentiation antigens of rat lymphocytes, Cell 12:663 (1977)
10. I. Goldschneider, L.K. Gordon, and R.J. Morris,Demonstration of THY-1 antigen on pluripotent hemopoietic stem cells in the rat, J. Exp. Med. 148: 1351 (1978)
11. D.L. Greiner, I. Goldschneider, and R.W. Barton, Identification of thymocyte progenitors in hemopoietic tissues of the rat. II.Enrichment of functional prothymocytes on the Fluorescence Activated Cell Sorter, J. Exp.Med. 156: 1448 (1982).
12. M.J.S.Dyer and S.V. Hunt, Committed T-lymphocyte stem cells of rats. Characterization by surface W3/13 antigen and radiosensitivity, J. Exp. Med. 154: 1164 (1981)
13. L.L. Lanier, N.L. Warner, J.A. Ledbetter, L.A. Herzenberg, Expression of Lyt-1 antigen on certain murine B cell lymphomas,J. Exp. Med. 153: 998 (1981)
14. V. Manohar, E.Brown, W.M. Leiserson, and T.M. Chused, Expression of Lyt-1 by a subset of B lymphocytes, J. Immunol. 129: 532 (1982)
15. C.Y. Wang, R.A. Good,P. Amirato, G. Dymbort, and R.L.Evans, Identification of a p69/70 complex expressed on human T cells sharing determinants with B-type chronic lymphatic leukemia cells, J. Exp. Med. 151: 1539 (1980)
16. F.Caligaris-Cappio, M. Gobbi, M. Bofill, and G. Janossi, Infrequent normal Blymphocytes express features of B-chronic lymphocytic leukemia, J. Exp. Med. 156: 623 (1982)
17. K. Hayakawa, R.R. Hardy, D.R. Parks, and L.A. Herzenberg, The "Ly-1 B" cell subpopulation in normal, immunodefective, and autoimmune mice, J. Exp. Med. 157: 202 (1983)

Ig LIGHT CHAIN GENE REARRANGEMENT AND CHROMOSOMAL ABNORMALITY IN THE LAMA EARLY B-LINEAGE TUMOUR OF THE RAT

D.Opstelten, G.J.Deenen, B.de Jong*, V.J.S.Idenburg* and S.V.Hunt**

Dept. of Histology and *Human Genetics, State University Groningen and **Sir William Dunn School of Pathology Oxford University

In the rat, few lymphoid tumours have been described so far. At the previous Germinal Centre Conference, J. Rozing and coworkers (1) reported on LAMA, a transplantable tumour that arose spontaneously in a ♂ $(AOxBN)F_1$ rat. From this study of surface antigens and mitogen reactivity it was tentatively proposed that LAMA was a pre-B leukemia, opening up possibilities in the rat for studies of B cell differentiation events at the molecular and cellular level. On the other hand, LAMA might serve as an *in vivo* tumour model in the rat. Here we present data on 1) the antigenic phenotype, 2) Ig light chain gene rearrangement and 3) the karyotype of LAMA.

MATERIALS AND METHODS

Cells

LAMA was propagated by serial subcutaneous transfer of $1x10^4$ tumour cells in 3-4 months old ♂ $(AO/GxBN/G)F_1$ hybrid $(RT1^{u/n})$ rats. Cells were harvested from solid tumours 5-6 wk after inoculation. For antigenic phenotyping by use of fluorescence activated cell sorter (FACS), cells were harvested from spleens of rats injected i.v. with 10^6 LAMA cells per 200 gr body weight 17 days previously (terminal phase).

Thoracic duct lymphocytes (TDL) were obtained as described (2). T cells were killed by incubation with W3/13 followed by rabbit anti-mouse Ig (Nordic, Tilburg, The Netherlands) and guinea pig complement, and separated from live cells by Percoll density gradient centrifugation. The resulting viable cell population contained > 95% surface Ig positive (sIg^+) cells.

Antibodies

Rabbit anti-μ chain antibody (3) and OX monoclonal antibodies (MAb) were a kind gift of Dr. A.F. Williams, Dunn School of Pathology, Oxford. HIS MAb's were produced as described (4). Rabbit anti rat δ chain antibody was kindly provided by Dr. H. Bazin/J. Rozing. Dr. J. Rozing also donated ER MAb's. Mouse anti rat k chain antibody MARK-1 was a gift from Dr. H. Bazin. PNA-FITC was from E.Y. Laboratories (San Mateo, U.S.A.), rabbit anti TdT antibody and goat anti rabbit Ig-FITC from Bethesda Research Laboratories (Gaithersburg, MD, USA), W3/13 and W3/25 from Serotec (Blackthorn, Bicester, Oxon, England). Horse anti rabbit Ig (FITC- or TRITC-conjugated) were produced as described (3). Goat anti mouse Ig (FITC-conjugated) was from Kirkegaard and Perry Laboratories (Gaithersburg, MD, USA), rabbit anti mouse Ig (peroxidase-conjugated) from Dakopatts (Glostrup, Denmark). Second stage antibodies were diluted in phosphate-buffered saline containing 1% (v/v) normal rat serum.

Experimental techniques

Antigenic phenotyping: LAMA cells were separately labeled for surface and cytoplasmic Ig chains by the double immunofluorescence staining technique described for μ chains in mice (5). Non-Ig antigens were detected by standard immunofluorescence and/or -peroxidase staining procedures on either fixed or viable tumour cells (4, 6). Cytocentrifuged cell preparations were judged for their staining characteristics (proportion and intensity of positive cells) by fluorescence or conventional microscopy, or cytofluorometry using a FACS.

Ig k light chain gene rearrangement: essentially, the protocol described by Tsukamoto et al. (7) was used. Briefly, high molecular weight genomic DNA was isolated and EcoR1-digested, followed by gel electrophoresis, Southern transfer, hybridization to ^{32}P-J_k and C_k probes and radioautography. J_k and C_k probes (prJ1/2 and prC_k, ref. 7) were a kind gift of Dr. G. Gutman, Dept. of Microbiology, University of California at Irvine, USA.

Karyotyping: standard procedures were used; chromosome preparations were G-banded by acetic saline or trypsin using Giemsa, or Q-banded by fluorescence using Hoechst 33258. Rat chromosomes were identified according to the Committee for a standardized karyotype of Rattus Norvegicus (8). The break points on the chromosome were located according to the nomenclature of Levan (9).

RESULTS AND DISCUSSION

Phenotype

Lama displayed μ heavy and k light chains in the cytoplasm of all tumour cells with a staining intensity comparable to that in normal bone marrow early B-lineage cells, but variable (1-90%) and very weak staining of μ and k on the cell surface which might either be exogenously acquired or endogenously synthesised (Table 1). Delta heavy chains were absent.

Table 1
Phenotype of LAMA tumour cells: Ig light and heavy chains.

Ig chain	Antibody	LAMA cell cytoplasm	LAMA cell surface
μ	RαRa- μ	+	variable; very weak
k	MaRK-1; OX-12; HIS 8	+	variable; very weak
δ	RαRa- δ	-	-

Reactivity of LAMA with a panel of monoclonal antibodies (MAb) to rat non-Ig determinants (Table 2) was consistent with a B-lineage nature for this tumour. LAMA carried Thy-1.1, and the B cell form of the leukocyte-common antigen, recognised by MRC OX-22 (10). LAMA reactivity with the B-lineage specific HIS MAb's (4, 6) (11^-, 22^-, 14^+, 24^+, 25^+) resembled that of normal marrow pre-B cells more than of B-lymphocytes. The lack of reaction with MRC OX-8, MRC OX-19 and W3/25 which label thymocytes and mature T cells and their subsets (11) argues against a mature T cell phenotype for LAMA. MRC OX-2 and W3/13 raised originally against thymocyte antigens (5), have a fairly widespread distribution on several leukocyte lineages and do not help to place LAMA accurately. The variable MHC class II (Ia) and transferrin receptor expression may reflect physiological conditions of cell activation and proliferation, respectively, at the time the tumour cells were examined.

These phenotypic data suggest that LAMA represents a malignant arrest in an early stage of B-lymphocyte differentiation, perhaps nearing a B-lymphocyte since kappa chains are expressed cytoplasmically.

Ig k light chain gene rearrangement

Figure 1 shows a radioautograph of EcoR1 digested LAMA DNA probed for the joining and constant region of the k light chain gene. Compared to liver and B-TDL DNA, which had the J_k fragment respectively all or partly in germline configuration (1.6 kb), germline J_k was completely absent in LAMA DNA. Instead, LAMA contained J_k sequences in a larger EcoR1 restriction fragment ($\sim$ 4.8 kb).

The absence of germline J_k in LAMA indicates k gene rearrangement in both alleles. However, since two rearrangement events are highly unlikely to yield identical sized new fragments containing the sequences probed for, the detection of only one new J_k fragment was surprising. This may indicate that one k gene was rearranged produc-

Table 2 Phenotype of LAMA tumour cells: non-Ig antigens.

Specificity	MAb/marker	Reference	LAMA a)	Rat marrow b) pre-B cells	Rat marrow b) B-lymphocytes
MHC class I	OX-18	20	■	■	■ weak
MHC class II	OX-3, OX-4, OX-17 HIS 19, ER 13	14 15	◪	□	◪
transferrin receptor	OX-26	16	◪ weak	◪ weak	□
large sialoglycoprotein	W3/13	13	■	◪ weak	□
widespread	OX-2	12	■	■ weak	■ weak
Marrow myeloid cells	PNA	c)	◪ weak	□	□
Thy-1.1	OX-7 ER 14	17 18	■ strong	■	■
Marrow lymphoid (progenitor?) cell	TdT	6; 19	□ (<5% weakly +)	□	□
Pan T	OX-19	13	□	◪ weak	◪ weak
T helper, macrophage	W3/25	13	□	□	□
T suppressor, NK cells	OX-8	13	□	□	□
Leuc. common, B cell form	OX-22	12	■	◪ weak	■
all peripheral B	HIS 11	4	□	◪ weak	■
all peripheral B	HIS 14	4	■ weak	■ weak	■
all peripheral B	HIS 25	4	■	■	■
B minus marginal zone	HIS 24	4	■	■	■
corona cells only	HIS 22	4	□	◪ weak	■

■ > 90%; ◪ 10-90%; □ < 10% of cells reactive.

a) OX-4, -7, -17, -18, -19, -22 and -26 binding was studied on LAMA cells from spleen, all others on LAMA cells from subcutaneous tumours (see M & M).

b) see ref.6 for experimental details, and part of the data.

c) D. Opstelten et al., manuscript in preparation.

tively, leading to internal expression of the k gene, while the other allele has been lost.

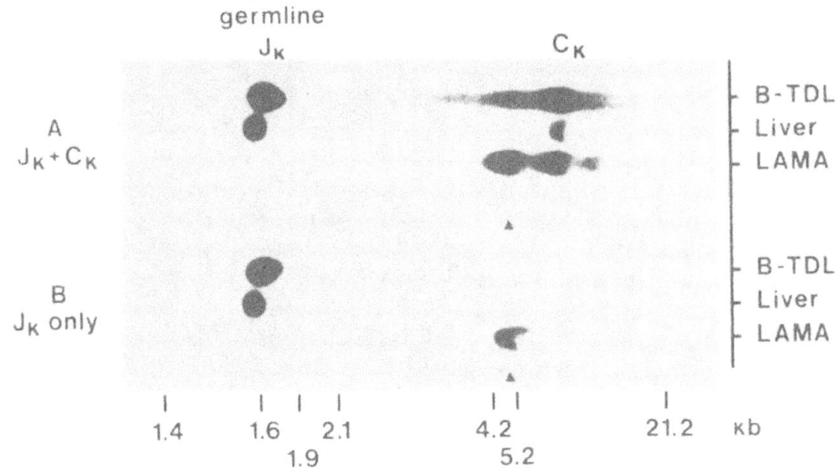

Fig. 1: Ig k light chain gene rearrangement in LAMA DNA. For experimental details see M & M. A: DNA probed for J_k and C_k; B: probed for J_k only. Note absence of germline J_k in LAMA DNA.

Karyotype

As shown in Fig. 2, karyotype analysis revealed an inverted duplication in chromosome ≠ 1. Breakage and reunion had occurred at band q22 of chromosome ≠ 1 with segment q42 -> q32 being inverted and displaced duplicated. No other chromosomal aberrations were found.

Interestingly, a breakpoint at band q22 of chromosome ≠ 1 has been described for four other, chemically induced, rat tumours as well (9, 21-23). This may indicate the location of an oncogene near 1q22, considering its likely involvement in malignant cell transformation (for review see 24).

CONCLUDING REMARKS

The possible loss of one of the k gene alleles in LAMA may be related to the aberration in chromosome ≠ 1. The chromosomal localisation of Ig genes being unknown in rat (25), not only an oncogene but also Ig k gene location near 1q22 may be hypothesized. Isolation of the abnormal chromosome ≠ 1 from LAMA will allow molecular hybridization studies with k and oncogene probes to test these hypotheses.

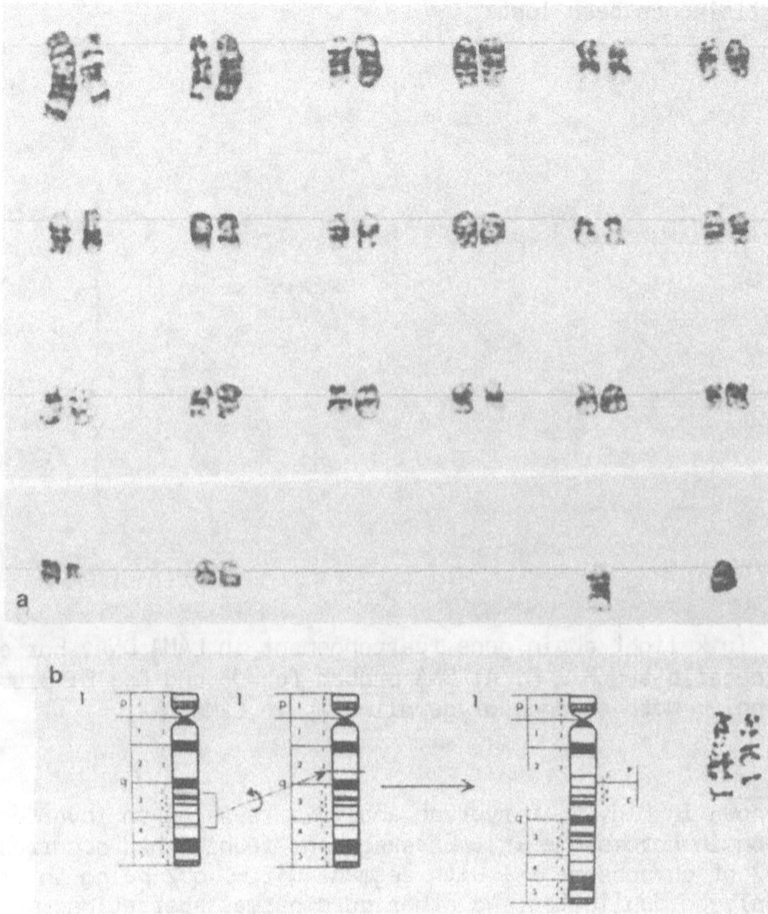

Fig. 2: Karyotype of LAMA. a: chromosome map. Note one abnormally long chromosome ≠ 1. b: schematic representation of the aberration in chromosome ≠ 1.

dup.(1)(p ter → q22 :: q42?→ q32 :: q22 → q ter)

ACKNOWLEDGEMENTS

We thank Dr. K. Umezawa and G. Harrod for assistance in the DNA experiments, P. van der Sijde and D. Huizinga for photography and A. Zandstra for preparing the manuscript. This study was supported in part by the Netherlands Organization for the Advancement of Pure Research (Z.W.O.) (Grant no. R 93-113), and by the British Council.

REFERENCES

1. J. Rozing, L.M.B. Vaessen, L. Faber, M. van Oven, L. de Vries-Bos, B. de Jong and P. Nieuwenhuis. Adv. Exp. Med. Biol. 149: 111 (1982).
2. W.L. Ford and S.V. Hunt. In:" Handbook of Experimental Immunology", D.M. Weir, ed., Blackwell, Oxford. p. 23.2 (1973).
3. A.F. Williams. Eur. J. Immunol. 5: 883 (1975).
4. F.G.M. Kroese, D. Opstelten, A.S. Wubbena, G.J. Deenen, E.H. Schwander, L. de Leij and P. Nieuwenhuis. This volume.
5. D. Opstelten and D.G. Osmond. J. Immunol. 131: 2635 (1983).
6. D. Opstelten, G.J. Deenen, T. Jaarsma, J. Rozing and S.V. Hunt. This volume.
7. A. Tsukamoto, I.L.Weissman and S.V. Hunt. EMBO J. 3: 975 (1984).
8. Committee for a standardized karyotype of Rattus Norvegicus. Cytogenet. Cell Genet. 12: 199 (1973).
9. G. Levan. Hereditas 77: 37 (1974).
10. G.P. Spickett, M.R. Brandon, D.W. Mason, A.F. Williams and G.R. Woollett. J. Exp. Med. 158: 795 (1983).
11. D.W. Mason, R.J. Brideau, W.R. McMaster, M. Webb, R.A.H. White and A.F. Williams. In: "Monoclonal Antibodies", R.H. Kennett, T.J. McKearn, K.B. Bechtol, eds., Plenum, New York, p. 251 (1980).
12. A.N. Barclay. Immunology 44: 727 (1981).
13. A.F. Williams, G. Galfrè and C. Milstein. Cell 12: 663 (1977).
14. W.R. McMaster and A.F. Williams. Immunol. Rev. 47: 117 (1979).
15. R.J.M. Stet, J. Rozing, G.D. Majoor, F.G.M. Kroese, D. Opstelten and P. Nieuwenhuis. Transplant. Proc. 1985, in press.
16. W.A. Jefferies, M.R. Brandon, A.F. Williams and S.V. Hunt. Immunology, submitted.
17. D.W. Mason and A.F. Williams. Biochem. J. 187: 1 (1980).
18. J. Rozing, P. Joling, P. Nieuwenhuis and L.M.B. Vaessen. This volume.
19. D.L. Greiner, I. Goldschneider, and R.W. Barton. J. Exp. Med. 156: 1448 (1982).
20. T. Fukummoto, W.R. McMaster and A.F. Williams. Eur. J. Immunol. 12: 237 (1982).
21. G. Levan, U. Ahlström and F. Mitelman. Hereditas 77: 263 (1974).
22. U. Ahlström. Hereditas 78: 235 (1974).
23. J. Kovi, H.P. Morris and E. Kovi. J. Natl. Cancer Inst. 58: 377 (1977).
24. G. Klein. Cell 32: 311 (1983).
25. J. Sümegi, J. Spira, H. Bazin, J. Szpirer, G. Levan and G. Klein. Nature 306: 497 (1983).

IN VIVO REGULATION OF B LYMPHOCYTE PRODUCTION IN THE BONE MARROW: EFFECTS AND MECHANISM OF ACTION OF EXOGENOUS STIMULI ON PRE-B CELL PROLIFERATION AND LYMPHOCYTE TURNOVER

D.G. Osmond, G.M. Fulop,[1] D. Opstelten[2] and C. Pietrangeli

Department of Anatomy, McGill University
Montreal, Quebec, Canada H3A 2B2

INTRODUCTION

Under normal *in vivo* conditions, mouse bone marrow continuously produces many primary B lymphocytes (1, 2). The mechanisms which determine the magnitude of this process, however, are poorly understood.

Within the bone marrow, rapidly proliferating pre-B cells containing free cytoplasmic μ chains (cμ) give rise to rapidly renewed small lymphocytes, non-dividing pre-B cells which subsequently develop into B lymphocytes expressing surface μ chains (sμ) in IgM molecules (Fig. 1) (3-10). Various non-lymphoid cells and their products have been shown to influence bone marrow B lymphocyte genesis *in vitro* (11-14). *In vivo*, we have shown that exogenous agents, notably an injection of sheep red blood cells (SRBC), can stimulate increased bone marrow lymphocyte production, suggesting that external environmental agents may have an important modifying influence on B lymphocyte genesis (15-17). These latter studies, however, have been based solely on the turnover of bone marrow small lymphocytes without specific lineage identification, while kinetic effects on precursor cells have not been examined.

Present addresses:
[1]The Ontario Cancer Institute, Toronto, Ontario, Canada
[2]Department of Histology, University of Groningen, Groningen, The Netherlands

The present studies have aimed firstly to examine the mechanisms mediating the effects of a single injection of SRBC on marrow B lymphocyte production, particularly concerning the nature and site of cells mediating stimulation of marrow lymphocytopoiesis. In view of earlier indications that macrophages may be implicated (16), the sensitivity of the effect to systemically administered silica, a macrophage poison (18-20), has been examined. To determine whether the effect acts directly on the bone marrow or via peripherally located cells, experiments have tested possible involvement of the spleen. Secondly, to simulate the chronic effects of environmental stimulation, multiple SRBC injections have been given for 4 wk. Throughout the present work, two *in vivo* assays have provided quantitative indices of B lymphocyte production in the bone marrow, 1) a radioautographic assay of small lymphocyte renewal, using DNA labeling with ^{3}H-thymidine (4, 15, 16), 2) an immunofluorescence assay of pre-B cell proliferation, based on specific identification of pre-B cells by cμ labeling (5) and on recently characterized methods for analysing the mitotic activity of pre-B cells by a metaphase arrest technique (7, 9).

MATERIALS AND METHODS

Mice

C3H/HeJ male mice (Jackson Laboratories, Bar Harbor, ME) were used at 8-10 wk of age.

Antisera

Fluorescein isothiocyanate (FITC)- and tetramethyl rhodamine isothiocyanate (TRITC)-conjugated affinity purified goat antibodies to mouse μ chains (α-μ FITC and α-μ TRITC; Kirkegaard and Perry Laboratories, Gaithersburg, MD) were ultracentrifuged before use.

Isotopes

Methyl-^{3}H-thymidine (New England Nuclear Corp., Boston, MA) was used at a specific activity of 6.7 Ci/mmol.

Stimulants

Sheep red blood cells (Institut Armand Frappier, Laval-des-Rapides, P.Q.) were washed and suspended in sterile saline ($8x10^{8}$ cells/ml).

Silica

Silica particles (0.012 μm; Sigma Chemical Co., St. Louis, MO) were suspended in sterile saline (10 mg/ml).

Splenectomy

Splenectomy (Sx) and sham-splenectomy (Sham Sx) were done under ether anaesthesia, as described (21).

Bone Marrow Cell Suspensions

Cell suspensions of femoral bone marrow pooled from groups of 3 mice were prepared in minimum essential medium with 10% newborn calf serum and were counted as described previously (3, 4).

Radioautographic Assay of Small Lymphocyte Renewal

Bone marrow cells were sampled at a constant interval of 24 h after administering a single intraperitoneal dose of ^{3}H-thymidine (1 μCi/g body wt) at 9.00 a.m. In radioautographic smears, prepared and processed as described previously (4), the incidence of small lymphocytes and the proportion and absolute number of these cells labeled by ^{3}H-thymidine (3 grains, 28d exposure) were determined. The number of labeled small lymphocytes per femur gave a quantitative index of the renewal of all small lymphocytes which include late pre-B cells, B lymphocytes and some non B cells (3, 6, 7, 9, 22).

Immunofluorescent Stathmokinetic Assay of Pre-B Cell Proliferation

Bone marrow cells were sampled at a constant interval of 2 h 40 m after giving vincristine sulphate (Oncovin, Eli Lilly, Toronto, Ont.) intraperitoneally at 8.a.m. (1 mg/kg body wt). The cells were exposed to α-μ FITC in suspension to label sμ followed by α-μ TRITC in fixed cytocentrifuge preparations to label cμ (±sμ), as described (7, 9). By epifluorescence and phase contrast microscopy the incidence of $c\mu^+s\mu^-$ pre-B cells and the proportion and absolute number of these cells arrested in metaphase were determined (7, 9). The number of $c\mu^+s\mu^-$ cells in metaphase gave a direct measure of the proliferation of specifically identified pre-B cells.

RESULTS

Stimulation of Bone Marrow B Lymphocyte Production by a Single Injection of SRBC

In control mice given an injection of saline the normal active production of bone marrow B lymphocytes was revealed by the substantial numbers of radioautographically labeled small lymphocytes 1d after ^{3}H-thymidine injection and of $c\mu^+s\mu^-$ pre-B cells arrested in metaphase 2 h 40 m after vincristine injection (Table 1).

Both of these parameters, reflecting small lymphocyte renewal and pre-B cell proliferation, respectively, were substantially increased after an intraperitoneal injection of SRBC (4×10^8 cells) (Table 1). The post-SRBC assay intervals used in the present work were selected for routine use from the results of temporal studies showing elevated values from 3d to 7d in small lymphocyte turnover (15, 16) and in pre-B cell mitoses (D. Opstelten, C. Pietrangeli, and D.G. Osmond, unpublished observations).

Table 1. Effect of an Injection of SRBC on Marrow Lymphocyte Production and Pre-B cell Proliferation[1]

Treatment	Lymphocyte Renewal[2]		Pre-B Cell Proliferation[3]	
	Labeling Index of Small Lymphocytes (%)	Labeled Small Lymphocytes per Femur ($\times 10^{-5}$)	Metaphase Index of $c\mu^+s\mu^-$ Cells (%)	$c\mu^+s\mu^-$ Cells in Metaphase per Femur ($\times 10^{-5}$)
Saline	21.8 ±1.1	10.2 ±1.6	12.5 ±0.5	2.4 ±0.1
SRBC	36.7 ±4.9	17.4 ±0.4	28.9 ±0.1	6.8 ±0.9

[1] Mean ±SE from 3 sets of experiments, each using pooled cells from 3 mice
[2] Radioautographic labeling of small lymphocytes 1d after ^{3}H-thymidine, 4d after SRBC or saline
[3] Immunofluorescent labeling of $c\mu^+s\mu^-$ cells 2 h 40 m after vincristine, 6d after SRBC or saline

Silica Sensitivity of SRBC Stimulation of Marrow B Lymphocytopoiesis

In further experiments, detailed in Table 2, mice pretreated with a saline injection 2h before SRBC showed the usual stimulation of marrow small lymphocyte ^{3}H-thymidine labeling and of pre-B cell mitoses. Both of these increases were completely abrogated, however, in mice which had been pretreated with silica (3 mg, ip) 2h before administering the SRBC. That this did not reflect simply a direct toxic effect of silica has been indicated by the observation that silica pretreated mice do show some responsiveness to another potential stimulant of marrow lymphocyte production, bovine serum albumin (C. Pietrangeli and D.G. Osmond, unpublished observations).

Table 2. Effect of Silica Pretreatment on SRBC Stimulation of Marrow Lymphocyte Production and Pre-B Cell Proliferation[1]

Treatments[2]		Lymphocyte Renewal[3]		Pre-B Cell Proliferation[4]	
1	2	Labeling Index of Small Lymphocytes (%)	Labeled Small Lymphocytes per Femur ($\times 10^{-5}$)	Metaphase Index of $c\mu^{+}s\mu^{-}$ Cells (%)	$c\mu^{+}s\mu^{-}$ Cells in Metaphase per Femur ($\times 10^{-5}$)
Saline	Saline	23.2±1.3	10.9±1.3	12.4±0.7	2.4±0.1
Saline	SRBC	39.3±2.5	17.6±0.4	28.9±0.7	6.9±1.2
Silica	Saline	20.3[5]	3.1[5]	13.0±1.2	1.3±0.2
Silica	SRBC	10.7±5.9	2.6±2.5	15.1±2.2	1.1±0.3

[1] Mean±SE from duplicate sets of experiments, each experiment using pooled cells from 3 mice
[2] 2 ip injections, 2h apart
[3] 1d after ^{3}H-thymidine, 4d post-treatment
[4] 2 h 40 m after vincristine, 6d post-treatment
[5] Single experiment using pooled cells from 3 mice

Spleen Dependence of SRBC Stimulation of Marrow B Lymphocytopoiesis

Groups of mice were subjected to either Sx or Sham Sx at 6 wk of age and assayed 4 wk later, as shown in Table 3. Sham Sx mice showed normal values for bone marrow small lymphocyte renewal and pre-B cell proliferation as well as a well-marked increase in both of these parameters after SRBC injection. Sx mice, in contrast, failed to respond to SRBC by either parameter.

Table 3. Effect of Prior Splenectomy on SRBC Stimulation of Marrow Lymphocyte Production and Pre-B Cell Proliferation[1]

Surgery[2]	Treatment	Lymphocyte Renewal[3]		Pre-B Cell Proliferation[4]	
		Labeling Index of Small Lymphocytes (%)	Labeled Small Lymphocytes per Femur ($x10^{-5}$)	Metaphase Index of $c\mu^+s\mu^-$ Cells (%)	$c\mu^+s\mu^-$ Cells in Metaphase per Femur ($x10^{-5}$)
Sham-splenectomy	Saline	21.7±3.2	10.7±2.7	13.1	1.9
	SRBC	31.7±5.0	18.8±3.1	28.4	3.9
Splenectomy	Saline	20.1±2.7	9.7±1.0	12.8	1.8
	SRBC	20.2±0.9	6.1±0.2	7.6	1.2

[1] Mean±SE from 2 sets of experiments for lymphocyte renewal, 1 set of experiments for pre-B cell proliferation. All experiments used pooled cells from 3 mice
[2] Surgery performed 4 wk previously
[3] 1d after ^{3}H-thymidine, 4d post-treatment
[4] 2 h 40 m after vincristine, 6d post-treatment

Silica-Sensitive and Spleen-Dependent Phases of SRBC Stimulation

Mice were either treated with silica or subjected to Sx at intervals *after* the injection of SRBC to examine the length of time for which silica-sensitive and spleen-dependent mechanisms were necessary to mediate the SRBC stimulation of marrow lymphocyte production. As shown in Table 4, either silica treatment or Sx abrogated the SRBC stimulation of bone marrow small lymphocyte turnover even when delayed until 1d after SRBC administration. When performed 2d after SRBC injection, however, silica treatment no longer prevented the marrow lymphocyte response, whereas Sx was still as effective in so doing as at earlier intervals. Thus, the stimulation of marrow lymphocyte production requires the operation of a silica-sensitive mechanism for at least 1d but less than 2d after injecting SRBC, while the presence of the spleen is needed for 2d or more.

Table 4. Silica-Sensitive and Spleen-Dependent Phases of SRBC Stimulation of Marrow Lymphocyte Production[1]

Treatments	Lymphocyte Production[2] Labeled Small Lymphocytes per femur ($x10^{-5}$) in mice given: Silica	Splenectomy
SRBC alone	17.3	18.7
SRBC followed by silica or splenectomy after: 1d	6.3	6.2
2d	19.4	5.4

[1] Data derived from pooled cells from 3 mice in each experiment
[2] 1d after ^{3}H-thymidine, 4d after SRBC administration

Effects of Multiple Injections of SRBC on Bone Marrow B Lineage Cells

To produce a sustained elevation in exposure to exogenous stimuli, groups of mice were injected ip with either SRBC ($1.6x10^7$ cells/g body wt) or saline 3 times per week from 4 wk to 8 wk of age. Radioautographic analysis during continuous perfusion of ^{3}H-thymidine for 0.5d to 3d showed that the turnover rate of the marrow small lymphocyte compartment in the SRBC injected mice was the same as in saline injected controls (2.5%/hr) but the population size of the compartment was expanded and the total small lymphocyte production per femur per unit time was increased by approximately 30% (G. Fulop and D.G. Osmond, data not shown).

By immunofluorescence assay, the multiple SRBC injections were found to produce a substantial increase in the absolute number of pre-B cells entering mitosis, compared with control mice given repeated injections of saline (Table 5). The SRBC injected mice also showed a marked increase in the total number per femur of both pre-B cells and B lymphocytes (Table 5) in contrast with the minimal changes in total cell population sizes seen after a single SRBC stimulus.

Table 5. Pre-B Cells and B Lymphocytes in the Bone Marrow of Mice given Multiple Injections of SRBC[1]

Treatment[2]	$c\mu^+s\mu^-$ Cells				$s\mu^+$ Cells	
	Metaphase Index[3] (%)	Metaphase Cells per Femur[3] ($x10^{-5}$)	Incidence (%)	Cells per Femur ($x10^{-5}$)	Incidence (%)	Cells per Femur ($x10^{-5}$)
Saline	8.7	1.4	11.0	16.2	7.3	10.7
SRBC	13.1	4.7	25.3	36.2	19.6	28.1

[1] Data derived from pooled cells from 3 mice in each experiment
[2] 4 wk old mice injected 3 times/wk with saline or SRBC for 4 wk
[3] 2 h 40 m after vincristine

DISCUSSION

A single administration of a foreign particulate agent, SRBC, has been shown in the studies summarized here to be followed by an increase in bone marrow B lymphocyte production, as demonstrated by 2 different but complementary *in vivo* assays. The radioautographic assay, as used previously (15, 16), provides an index of the turnover of the small lymphocyte cell category as a whole, which includes a mixture of late non-dividing pre-B cells, B lymphocytes and some non-B cells (3, 6, 7, 22). The immunofluorescence assay, on the other hand, provides a direct quantitation of the rate of entry into mitosis of pre-B cells, specifically identified by the presence of free $c\mu$ chains. The close accord in all experiments between the results of the 2 assays, which involve 2 different methodologies and principles and which were performed on separate groups of mice, testify to the consistency of the SRBC stimulatory effect, while both assays reflect events occurring under essentially physiological conditions *in vivo*. The results demonstrate clearly that an exogenous agent can stimulate B lineage cell production in the bone marrow of the intact animal. The mechanism of this effect, however, has been less clear.

The stimulatory effect of SRBC on bone marrow small lymphocyte renewal has been found previously to occur in mice lacking mature B lymphocytes, T lymphocytes or immune responsiveness, but to be abolished by treatment with silica (16). In addition, the present work demonstrates the silica sensitivity of the SRBC stimulatory effect, as assayed by pre-B cell proliferation. In view of the ability of silica particles in the size and dose used to delete macrophage function *in vivo* (18-20) these results strongly implicate macrophages as an obligatory cell type needed to mediate the SRBC effect, possibly by processing the SRBC and/or by initiating a stimulatory sequence of events. Such a cellular mediation would accord with the concept of an antigenically non-specific stimulation of B lymphocyte production (16).

The lack of a stimulatory effect by SRBC when given to splenectomized mice indicates that some necessary components of the response are mediated peripherally in the spleen under the conditions of the present experiments, rather than being a central effect mediated by cells in the bone marrow itself. Together, the foregoing results strongly implicate macrophages within the spleen as cells responsible for mediating the effect of ip injection of SRBC on bone marrow B lymphocyte production. If so, a distinct period of time appears to be needed by these cells to initiate successfully the bone marrow lymphocyte reaction *in vivo*. Radiolabeled SRBC after ip injection in mice take approximately 4h to enter the blood stream and then to leave the circulation (C. Pietrangeli and D.G. Osmond, unpublished observations), presumably mainly by clearance in the spleen. Subsequently, the existence of an early phase in the bone marrow response which is both silica-sensitive and spleen-dependent suggests that splenic macrophages play an essential role for at least the first day or so. A subsequent phase at 2d which is spleen-dependent but no longer silica-sensitive suggests that further events initiated in the spleen no longer need the active participation of macrophages. Speculatively, such events might involve either the formation of factors or the activation of cells which would in turn trigger the wave of B lymphocyte genesis in the bone marrow. Various non-lymphoid cells within the bone marrow itself (macrophages, adherent cells, stromal cells) can help to support the production and/or maturation of B lymphocytes in short and long term bone marrow cultures (11-14). This property may represent the endogenous local regulation by accessory cells of the induction of B lymphocyte genesis and of a basal level of production, however, rather than the mediation of exogenous stimuli. Possibly, under *in vivo* conditions the level of B lymphocytopoiesis in the bone marrow may be influenced by diverse activities of non-lymphoid cells both in the bone marrow itself and, as suggested by the present work, in the spleen.

The acute exposure to foreign material represented by a single ip injection of SRBC induces a transient wave of increased B lymphocyte genesis in the bone marrow, lasting approximately 1 wk (15). In conventionally treated animals *in vivo*, on the other hand, bone marrow lymphocyte production is maintained in a kinetic steady state set at a level of cell population size and production rate generally characteristic of post-natal age (4). The present results of chronic exposure to foreign agents by multiple injections of SRBC for 4 wk in young mice are important in showing that B lymphocyte genesis had apparently become set at a new kineticsteady state with increased B lineage population sizes and absolute production rates. Such conditions may more closely mimic the effects of exposure to natural exogenous agents, which include products of intestinal flora, other microorganisms and dietary constituents. The findings strengthen the view that, although theprimary differentiation of B lymphocytes in the bone marrow can proceed independent of antigenic influences in accord with the concept of clonal selection, the level of B lymphocyte production may be boosted non-specifically by environmental stimuli.

SUMMARY AND CONCLUSIONS

The present studies demonstrate that a single administration of an extrinsic agent (SRBC) can stimulate increased production of B lymphocytes in mouse bone marrow as revealed by 2 *in vivo* assays which quantitate pre-B cell proliferation and small lymphocyte renewal, respectively. The mechanisms mediating this stimulatory effect are sensitive to silica *in vivo* and require the presence of the spleen. Early events are both silica-sensitive and spleen-dependent, while a subsequent stage appears still to be spleen-dependent but not silica-sensitive. Sustained exogenous stimulation by multiple SRBC injections for 4 wk in young mice produces an expanded population size and increased production of pre-B cells and B lymphocytes in the bone marrow, apparently an elevated kinetic steady state of B lymphocyte production.

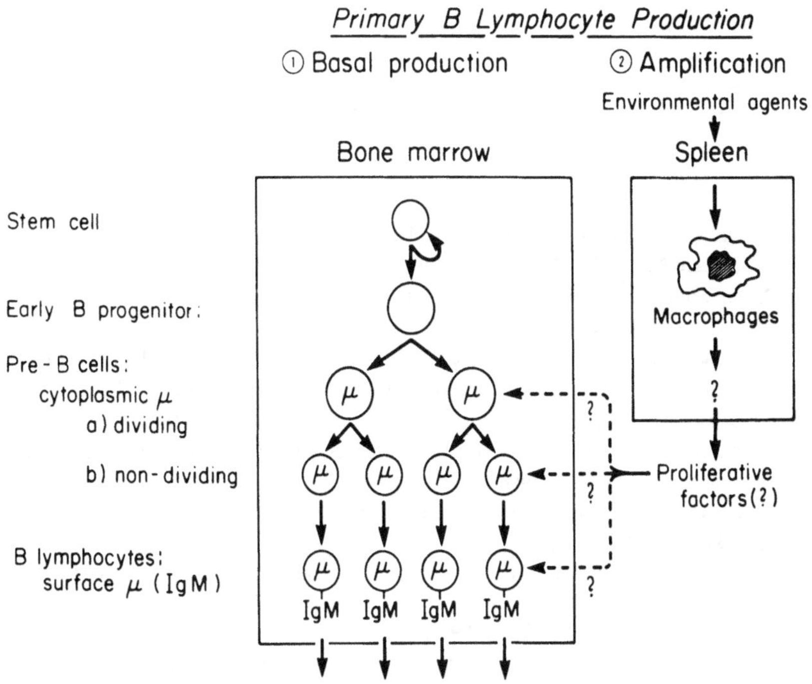

Fig. 1. Scheme of B lymphocyte production in the bone marrow. See text for details.

As depicted schematically in Figure 1, the results suggest that the magnitude of bone marrow B lymphocyte production _in vivo_ may reflect a basal level, putatively regulated by microenvironmental and other endogenous factors, which is amplified by exogenous environmental stimuli mediated by the action of macrophages located in the spleen. Further questions about such an environmental amplification (Fig. 1) concern the nature of later events in the spleen, the identity of putative stimulatory factors or cells circulating from the spleen to the bone marrow, the receptive target cell stages in the bone marrow and the consequences of this process with respect to the size and diversity of B lymphocyte clones and of primary humoral immune responses _in vivo_.

ACKNOWLEDGEMENTS

This work was supported by the National Cancer Institute of Canada and the Medical Research Council of Canada.

REFERENCES

1. D.G. Osmond, Mongr. Allergy 16:157 (1980).
2. D.G. Osmond, in "In Vivo Immunology: Histophysiology of the Lymphoid System", pp. 3-10, A.A. van den Brock, and M.G. Hanna, eds., Plenum Press, New York (1982).
3. D.G. Osmond, and G.J.V. Nossal, Cell, Immunol. 13:132 (1974).
4. S.C. Miller, and D.G. Osmond, Cell Tissue Kinet. 8:97 (1975).
5. M.D. Cooper, and A.R. Lawton, in: "Cells of Immunoglobulin Synthesis", pp. 411-425, B. Pernis and H.J. Vogel, eds., Academic Press, New York (1979).
6. K.S. Landreth, C. Rosse, and J. Clagett, J. Immunol. 127:2027 (1981).
7. D. Opstelten, and D.G. Osmond, J. Immunol. 131:2635 (1983).
8. K.S. Landreth, P.N. Kincade, G. Lee, and E.S. Medlock, J. Immunol. 131:572 (1983).
9. D.G. Osmond, and J.J.T. Owen, Immunology 51:333 (1984).
10. D.G. Osmond, F. Melchers, and C.J. Paige, J. Immunol. 133:86 (1984).
11. P.W. Kincade, G. Lee, C.J. Paige, and M.P. Sneid. J. Immunol. 127:255 (1981).
12. C.A. Whitlock, and O.N. Witte, Proc. Natl. Acad. Sci. USA 79:3608 (1982).
13. K. Dorshkind, and R.A. Phillips, J. Immunol. 129:2444 (1982).
14. P.P. LeBouteiller, and G.L. Asherson, in: "In Vivo Immunology: Histophysiology of the Lymphoid System", pp. 31-38, P. Nieuwenhuis, A.A. van den Brock, and M.G. Hanna, eds., Plenum Press, New York (1982).
15. G. Fulop, and D.G. Osmond, Cell. Immunol. 75:80 (1983).
16. G. Fulop, and D.G. Osmond, Cell. Immunol. 75:91 (1983).
17. G. Fulop, J. Gordon, and D.G. Osmond, J. Immunol. 130:644 (1983).
18. B.M. Allison, J.S. Harington, and M. Birbeck, J. Exp. Med. 124:141 (1966).
19. M.H. Levy, and E.F. Wheelock, J. Immunol. 115:41 (1975).
20. H. Van Loveran, M. Snock, and W.D. Otter, J. Reticuloendothel. Soc. 22:523 (1977).
21. R. Benner, F. Meima, G.M. van der Meulen, W.B. van Muiswinkel, Immunology 26:247 (1974).
22. M.D. Rahal, and D.G. Osmond, Cell. Immunol. 87:in press (1984).

INFLUENCE OF THE *XID* MUTATION ON B LYMPHOCYTE DEVELOPMENT IN ADULT MICE

J. Witkowski, L.M. Forrester, J.D. Ansell and H.S. Micklem

Department of Zoology, University of Edinburgh
Edinburgh EH9 3JT, United Kingdom

INTRODUCTION

The genesis and maturation of murine B lymphocytes in the bone marrow is marked by the sequential appearance of the B-220 molecule, cytoplasmic u chains, surface IgM and IgD [1,2]. In the spleen at least three distinct populations of B cells can be identified on the basis of the relative concentrations of surface IgM and IgD [3]. One of these, carrying high IgD and low IgM concentrations, is absent in CBA/N mice, which are homozygous for the X-linked immunodeficiency (*xid*) trait [3]. Transplantation of normal bone marrow or foetal liver cells into CBA/N mice results in normal B cell development, implying that the *xid* defect is intrinsic to the cells and not microenvironmental [4]. Splenic B lymphocytes in female *xid/+* heterozygotes were recently found to express mainly the non-*xid* carrying chromosome in a population associated with IgG3 production [4]: presumably those stem cells in which the *xid*-carrying X-chromosome was active showed at least a partial failure to differentiate into B lymphocytes. Thus, the defect appeared to act directly on B cells rather than on some other cell type influencing B cell development.

In *xid/+* female heterozygotes, the two electrophoretically distinct forms of the X-linked enzyme phosphoglycerate kinase (PGK-1A and PGK-1B) can be used as markers of the normal or *xid*-carrying X-chromosome respectively [5]. Using these markers, we are now approaching the following questions: (1) how early in B cell development does the *xid* trait express itself? (2) Does the trait affect all B cells? (3) Is the development of other haematopoietic lineages affected? Partial answers, based on the expression of *xid*-carrying and normal X-chromosomes in haematopoietic tissues and in

FACS-sorted bone marrow and peripheral lymphoid cells, are advanced in this paper.

MATERIALS AND METHODS

Mice. Mice of the inbred CBA/Ca strain and certain lines derived from it were used: CBA/N, CBA/Ca-$Pgk\text{-}1^a$, PGK-1-heterozygous F_1 hybrids of the above, and CBA/Ca-$Gpi\text{-}1^a$. With the exception of the CBA/Ca-$Pgk\text{-}1^a$ congenic line and the hybrids, all other mice have the PGK-1B phenotype. GPI-1 (glucose phosphate isomerase) is an autosomally encoded enzyme with electrophoretic variants; normal CBA/Ca and CBA/N mice, and the $Pgk\text{-}1^a$ derivatives are $Gpi\text{-}1^b$ [6]. All mice were bred and maintained in our own animal facilities, and used when 3-4 months old. We are indebted to Dr M. J. Marshall for the gift of a breeding nucleus of the CBA/Ca-$Gpi\text{-}1^a$ mice.

Antibodies. The following were used. 1. Rat anti-mouse B-lineage cell antigen monoclonal antibody 14.8 (kindly provided by Dr P.W. Kincade). 2. Fluorescein isothiocyanate (FITC)-labelled goat anti-rat immunoglobulin antibody (Miles-Yeda), solid phase-absorbed with mouse immunoglobulin (used as a second step in 14.8 antibody staining). 3. FITC-labelled goat anti-mouse immunoglobulin (Miles-Yeda).

Cells. Animals were anaesthetized with chloroform vapour and blood was collected by heart puncture into a heparinized syringe. Cell suspensions were prepared from spleen,thymus, subcutaneous and mesenteric lymph nodes and bone marrow in RPMI 1640 medium supplemented with 0.3% bovine serum albumin (RPMI-BSA). Erythrocytes were eliminated from suspensions of spleen and bone marrow cells and a portion of the blood by hypotonic shock. Suspensions were made up to a concentration of 10^7 viable cells per ml in RPMI-BSA.

Fluorescent staining. 4 x 10^6 peripheral blood leucocytes, bone marrow, spleen or lymph node cells were stained with 10ul of FITC-anti-mouse Ig antibody to reveal cells carrying surface immunoglobulin (sIg). The same number of bone marrow cells were incubated first with 20ul of purified 14.8 monoclonal antibody and then stained with FITC-anti-rat-Ig antibody. All staining steps were performed for 40 mins on melting ice in a volume of 0.4ml RPMI-BSA containing 0.1% sodium azide and were followed by three washes in cold azide- containing medium.

Bone marrow repopulation. CBA/Ca-$Gpi\text{-}1^a$ male mice were lethally gamma-irradiated (10.5Gy, whole body, from a ^{137}Cs source). Within 30mins 10^7 bone marrow cells were injected intravenously into the lateral tail vein . The bone marrow cell suspension was a mixture consisting either of $4x10^6$ CBA/Ca-$Pgk\text{-}1^a$ plus $6x10^6$ CBA/N ($Pgk\text{-}1^b$) cells or of $4x10^6$

CBA/Ca-Pgk-1[a] plus 6x10^6 CBA/Ca (Pgk-1[b]) cells. The first group is referred to as "experimental" and the second as "control". Thus, in the experimental group, the xid-carrying cells were marked with PGK-1B and were in an initial 3:2 majority. The use of the congenic strain, CBA/Ca-Gpi-1[a], as the recipient in these experiments allowed us to estimate the host contribution (if any) to the cell populations analysed, since all donor cell populations have a GPI-1B phenotype [6]. At two-week intervals after irradiation animals from both groups were killed and cell suspensions from bone marrow, blood, spleen and lymph nodes were prepared and stained as described above. Host activity was significant only at 2-4 weeks post-irradiation and was allowed for in calculating the ratio of the two donors.

FACS analysis and sorting. Stained cell suspensions were analysed with a fluorescence activated cell sorter (FACS- IV, Becton Dickinson, Sunnyvale, California), on the basis of forward and right angle light scattering properties [7] and fluorescence intensity. Between 50 and 100 thousand cells were then sorted for enzyme analysis.

Measurement of PGK-1 and GPI-1 alloenzymes. These techniques were performed as described in full elsewhere [8]. Briefly, cell lysates were electrophoresed on cellulose acetate membranes for 45 minutes at +4°C (12.5mA/membrane, constant current) in the case of PGK-1 or for 1h 20mins (350V/membrane, constant voltage) in the case of GPI-1. The activity of both enzymes was visualized in the membranes by means of linked enzyme reactions leading to the reduction of the dye thiazolyl blue to its formazan derivative; alternatively, with PGK-1, ^{14}C-labelled glucose was included in the substrate and ^{14}C-labelled products were detected by autoradiography. Both colour and autoradiographic methods yield similar results for the measurement of PGK-1 activity; autoradiography, being more sensitive, was used for the analysis of FACS-sorted cell samples. The relative proportions of PGK-1A and B were then measured on an automated scanning and integrating densitometer (Chromoscan 3, Joyce-Loebl, Gateshead, England). Measurements of replicate samples showed an accuracy within 5% above or below the mean. All experimental samples were run twice or more.

RESULTS

The ratios of PGK-1A to B activity in spleen cells and in sorted B lymphocyte populations of xid/+ heterozygotes, were consistently higher than in sorted non-B cells and unsorted cells from several non-splenic haematopoietic tissues (representative data are shown in Table 1). This shows that disproportionately few B cells had the xid-carrying chromosome active. Some splenic and

Table 1 Percent PGK-1A present in unsorted cell suspensions and purified cell populations of *+/+* heterozygotes (mice 1 and 2) and *+/xid* heterozygotes (mice 3 and 4).*

	+/+ Pgk-1^{ab}		*+/xid Pgk-1*^{ab}	
	1	2	3	4
erythrocytes	64	70	72	46
thymus	58	74	72	46
spleen (unsorted)	57	63	83	61
spleen sIg$^-$	61	70	76	ND
sIg$^+$	61	66	100	ND
lymph node (unsorted)	60	72	70	44
lymph node sIg$^-$	60	60	68	43
sIg$^+$	60	61	100	100
blood lymphocytes				
sIg$^-$	58	ND	68	41
sIg$^+$	56	ND	90	75
bone marrow	52	70	71	44
bone marrow sIg$^-$	53	64	71	34
sIg$^+$	54	62	86	59
14.8 dull	52	66	70	41
14.8 bright	53	65	86	49

* In the *+/xid* mice, PGK-1A was associated with the activity of the non- *xid*-carrying X-chromosome. Values showing a clear bias against the expression of the *xid* (and *Pgk-1*^{b}) X chromosome are underlined. Note that these all relate to samples containing a large proportion of B lymphocytes. Broadly similar results were obtained in a further 10 *+/xid* and 3 *+/+* mice.
ND = not done sIg = surface immunoglobulin

lymph node B cell samples showed exclusively PGK-1A activity, indicating that cells with the *xid*-bearing chromosome active were virtually absent. In contrast, the sIg-positive cells in the bone marrow showed only a slight excess of PGK-1A activity compared to the sIg-negative cells from the same bones. After staining bone marrow cells with the 14.8 monoclonal antibody, two positive fluorescent populations were observed: about 30-40% of the nucleated cells stained above background and a minority (5-10% of the nucleated cells) stained brightly. We have found (unpublished data) that the 14.8- 'bright' subset includes most of the sIg- positive cells, while 14.8- 'dull' cells are nearly all sIg- negative, presumably pre-B [9] cells. (Less than 3% of bone marrow 14.8-positive cells were T lymphocytes as judged by staining with monoclonal antibodies to Thy- 1, Ly-1 and Lyt-2.) Like marrow sIg-positive cells, the 14.8- 'bright' cells showed only a slight excess of PGK-1A activity. The 14.8-'dull' cells had the same relative PGK-1A activity as the non- stained (i.e. non B-lineage) cells from the same bones.

In contrast to the xid/+ animals, +/+ Pgk-lab females showed little variation in PGK-1A/B ratios between the tissues studied and between sorted B and non-B cells (Table 1).

Results of PGK-1 analyses of samples from irradiated and repopulated mice are shown in Fig.1. Differences in the patterns of mosaicism between the two groups of animals reflect the repopulating abilities of the donor cells. In the control group only small differences were seen between sorted B and non-B cells and unsorted populations. In the experimental group, which received mixed CBA/N and CBA/Ca-Pgk-1^a bone marrow, the sIg-positive cells in the blood showed a progressive increase in the proportion of PGK-1A, that is, by inference, a disappearance of CBA/N-derived cells. The lymph node B cells of experimental animals showed a marked deficit of CBA/N-derived cells throughout the period of observation; this deficit was total in the animals assayed at 6 and 8 weeks after irradiation. Observations on unfractionated spleen cells suggested that the splenic B population was also deficient in CBA/N-derived cells, but sIg-positive and negative cells were not studied separately to test this. Deficits of CBA/N-derived cells were not clearly evident in sIg-negative lymphocytes of blood or lymph nodes, although the data do not rule out a marginal effect. The 14.8-'dull' (pre-B) bone marrow population in the experimental mice showed little or no difference from the 14.8-negative (non-B) cells from the same bones or from either population in controls. The numbers of the 14.8 "bright" cells in the repopulated bone marrow were too low to collect for enzyme analysis.

DISCUSSION

Due to the phenomenon of X-chromosome inactivation [10] only one X-chromosome is active in female cells. Thus, in normal female mice heterozygous for neutral polymorphisms on the X-chromosome, a balanced mosaicism exists for the products of the polymorphic genes. If a mutation on the X-chromosome directly affects a particular cell population, being, for example, lethal at a particular developmental stage or affecting the ability of the cells to proliferate, this will result in the mosaicism becoming unbalanced in the affected cell population. If, on the other hand, the mutation acts indirectly via other cells or microenvironmental factors, one would not expect mosaicism in the final target cells to be affected.

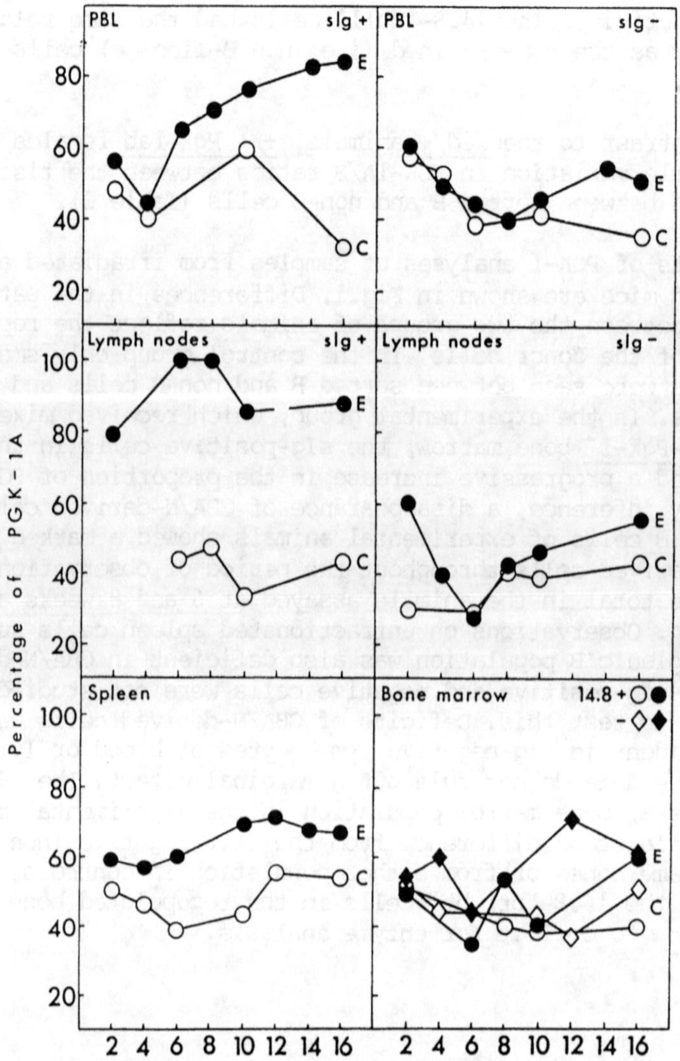

FIGURE 1.
Reconstitution of various cell populations of lethally irradiated mice injected with either a mixture of +/+ Pgk-1^a and +/+ Pgk-1^b (control-C) or +/+ Pgk-1^a and xid/xid Pgk-1^b (experimental-E) bone marrow. In the E group PGK-1A identifies non-xid-bearing cells.

We have used variants of the X-linked enzyme PGK-1, which differ in their electrophoretic mobilities, to monitor the activity of the xid-carrying X-chromosome in lymphoid cells of xid/+ female heterozygotes. A similar strategy was adopted by Nahm et al.[5], but the present methods for visualizing and measuring the alloenzymes are more quantitative [8].The data demonstrate that in the peripheral lymphoid organs and blood, surface immunoglobulin (sIg)-bearing lymphocytes carrying an active xid gene are severely depleted compared to those with the normal gene active. In CBA/N mice, which are homozygous for xid, it has been demonstrated that the subpopulation most severely affected is one carrying high levels of IgD and low levels of IgM [3,4]. Since this is the predominant B cell type in the periphery, it is possible that the enzyme data reflect the depletion of cells carrying an active xid gene specifically from this sub-population . The presence of some xid-expressing B cells was clearly demonstrated in blood and (in some individuals) spleen and lymph nodes. However, the present data do not exclude some (presumably smaller) effects of the xid mutation on the development or function of other B cell subpopulations.

Bone marrow B lymphocytes showed relatively slight imbalances of PGK-1A/B ratio in xid/+ heterozygotes. In the 14.8-'dull'(pre-B) cells PGK-1A values were not elevated. In the 14.8-'bright' and the sIg-positive cells, some excess of cells with the normal chromosome active was apparent, but it was far from complete. These data suggest that the xid gene acts at a late stage of differentiation of B cells in the marrow, but some effect on a minor population of pre-B cells is not excluded. It is probable that some of the sIg-positive, 14.8-bright cells found in bone marrow were immigrants from the periphery, and possible that immigrants were sufficient in number to account for the imbalance of X-chromosome expression in this population.

In irradiated mice that had been repopulated with a mixture of CBA/N and normal (PGK-1A marked) marrow the patterns of mosaicism came to resemble those seen in the xid/+ heterozygotes: there was an excess in the proportion of non-xid B cells in the periphery, but not in the marrow. The observation that xid B cells decreased proportionally in blood over a period of several weeks may indicate either that relatively immature B cells were migrating out of the bone marrow during this period or that the normal numerical balance between mature B cell subpopulations was upset. The slight decrease in splenic xid cells over the same period is compatible with the idea that immature B cells migrate from marrow to spleen during the early weeks of repopulation and then gradually cease to do so. However, interpretation of the spleen data is clouded by the probable presence of splenic erythropoiesis during the post-radiation recovery period and by the absence of data on sorted splenic B cells. In the lymph nodes xid B cells were already absent by 6 weeks, suggesting that most recirculating B cells belonged to the xid-affected subpopulation(s).

In summary, the data presented show that xid acts directly on B cells and that its expression is unaffected by the concomitant presence of non-xid cells . It apparently acts on the differentiation and development of certain B cells, probably at a stage after the acquisition of surface IgM. No effect was demonstrable on bone marrow pre-B cells or on non-B lymphocytes. Further analysis of the patterns of mosaicism of PGK-1 alloenzymes in subpopulations of B cells separated on the basis of their quantitative expression of IgM, IgD, B-220 and other markers of B cell development [11] may indicate more precisely the subpopulation(s) in which the X-linked immunodeficiency trait is active, and provide some information about the lineage relationships between these subpopulations.

ACKNOWLEDGEMENTS

This work was supported by the Cancer Research Campaign and the Medical Research Council. J.W. was the recipient of a Fellowship from the Wellcome Trust. L.M.F. holds an M.R.C. Postgraduate Studentship. We are indebted to Robert Gray and Andrew Sanderson for advice and skilled assistance with the cell sorting.

REFERENCES

1. P. W. Kincade. Formation of B lymphocytes in fetal and adult life. Adv. Immunol. 31:177 (1981).
2. R. L. Coffman and I. L. Weissman. Immunoglobulin gene rearrangement during pre-B cell differentiation. J. mol. cell. Immunol. 1:31 (1983).
3. R. R. Hardy, K. Hayakawa, J. J. Haaijman, L. A. Herzenberg and L. A. Herzenberg. B cell subpopulations identified by two-color fluorescence analysis. Nature 297:589 (1982).
4. I. Scher. The CBA/N mouse strain: an experimental model illustrating the influence of the X chromosome on immunity. Adv. Immmunol. 33:1 (1982).
5. M. H. Nahm, J. W. Paslay and J. M. Davie. Unbalanced X chromosome mosaicism in B cells of mice with X-linked immunodeficiency. J. exp. Med. 158:920 (1983).
6. M. C. Green, ed. "Genetic Variants and Strains of the Laboratory Mouse". Gustav Fischer Verlag, Stuttgart (1981).
7. A. W. S. Ritchie, R. A. Gray and H. S. Micklem. Right angle light scatter: a necessary parameter in flow cytofluorimetric analysis of human peripheral blood mononuclear cells. J. imm. Meths. 64:109 (1983).
8. J. D. Ansell and H. S. Micklem. Genetical markers for

following cell populations, *in*: "Handbook of Experimental Immunology", 4th ed. (D. M. Weir, L. A. Herzenberg, C. C. Blackwell and L. A. Herzenberg, eds.). Blackwells, Oxford (in press).

9. M. P. Scheid, K. S. Landreth, J. W. Tung and P. W. Kincade. Preferential but nonexclusive expression of macromolecular antigens on B-lineage cells. *Immunol. Revs.* 69:141 (1982).
10. M. F. Lyon. Mechanisms and evolutionary origins of variable X chromosome activity in mammals. *Proc. Roy. Soc. B* 187:243 (1974).
11. R. R. Hardy, K. Hayakawa, D. R. Parks and L. A. Herzenberg. Demonstration of B cell maturation in X-linked immunodeficient mice by simultaneous three-color immunofluorescence. *Nature* 306:270 (1983).

IN VITRO GROWTH AND DIFFERENTIATION OF RAT B CELL SUBPOPULATIONS

Cornelis de Groot, Rob van Schie, Jan Wormmeester and Martien L. Kapsenberg

Department of Histology and Cell Biology, Faculty of Medicine, Academic Medical Center, Meibergdreef 15 1105 AZ Amsterdam, The Netherlands

INTRODUCTION

In previous studies we have shown the existence of at least two subpopulations of B lymphocytes in rabbit spleen, different in size, morphology and redistribution behaviour of their surface immunoglobulin (sIg). One of these subsets consists of medium-sized lymphocytes that form caps with anti-immunoglobulin antibodies. The other subset consists of small lymphocytes that do not form sIg-caps, but show rapid endocytosis in the cap configuration [1,2]. Bazin et al.[3] and Gray et al.[4] showed in rat spleen that marginal zone lymphocytes carry sIgM but no sIgD, while the small B lymphocytes in the corona of the follicles carry both sIgM and sIgD. In preliminary studies with rats, we have found analogies with our previous work with rabbits, i.e. sIgM could be brought to cap formation in about one half of the B cell population, while sIgD could not be found in caps to a considerable amount and disappeared rapidly from the cell surface, presumably by endocytosis. In the present study we have tried to separate both subsets to study their responsiveness to B cell mitogens in vitro.

MATERIALS AND METHODS

Animals and cells

Male Wistar rats, 7-9 weeks old (175-225 grams) were used. Spleen lymphocytes were prepared by pressing tissue fragments through 6 layers of 20 denier nylon gauze and centrifugation at 4°C on a layer of ficoll-metrizoate (1.095 g/ml) for 30 min at 1200 g. Cells from the interface were collected and contained about 95% lymphocytes. Viability was 85-95%.

T cell depletion was done by labelling with biotinylated W3/13 monoclonal antibody, rosetting with avidin-coated sheep erythrocytes and centrifugation on ficoll-metrizoate (1.095 g/ml) as described previously[5]. The resultant population contained 85 ± 7% sIgM+ve B cells and 1.3 ± 1.5 T cells.

Separation of small (B_s) and large (B_l) B lymphocytes was done essentially according to Salisbury et al.[6]. $2.10^7 - 2.10^8$ B lymphocytes were suspended in 30 ml Percoll (1.072 g/ml), pipetted carefully onto a layer of 5 ml Percoll (1.095 g/ml), overlayered by 5 ml of Percoll (1.030 g/ml) and centrifuged 45 min at $1250g_{max}$ and room temperature. Cells were collected from both interfaces and washed.

Flow cytofluorometry was done with an Ortho Cytofluorograph with a 4W Argon laser, operating at 488nm (500mW at this line). Integrated 90° scatter was taken to estimate cell size.

In some experiments, macrophages were depleted by adherence of 10^6 cells per ml to Nunc plastic Petri dishes for 90 min at 37°C. Non-adherent cells were collected by rinsing the dishes with Iscove's modified Dulbecco's medium (IMDM) + 5% foetal calf serum. If macrophages werd added, peritoneal cells were taken from the spleen donor by washing the peritoneal cavity with 20 ml Earle's balanced salt solution (EBSS).

Cell culture conditions

Lymphocytes were cultured in triplo in flatbottom culture plates (Falcon 3072, 96 wells Microtest III) at 10^5 cells per well in 200 µl IMDM supplemented with 7.5% foetal calf serum and 5.10^{-5}M 2-mercaptoethanol. In some cultures T cell supernatant and/or macrophages were added. To study proliferation, cultures were kept for 48 h at 37°C in 5% CO_2. 0.25 µCi ^{3}H-thymidin was added to each well and cells were harvested after 16 h with a Titertek automatic cell harvester and radioactivity was counted in a LKB Wallac liquid scintillation counter. To study plasma cell differentiation, cells were cultured for 5 days and tested in a Protein A-plaque assay as described by Gronowicz et al.[7] with some modifications according to Van Oudenaren et al.[8]. Protein A-coated sheep erythrocytes and a rabbit anti rat IgM antiserum at optimal dilution were used to estimate the number of IgM-producing plasma cells. Guinea pig serum, adsorbed with sheep erythrocytes and a Protein A-coated Sepharose column, was a kind gift from Dr. H. Savelkoul (Erasmus University, Rotterdam).

Lipopolysaccharide (LPS), batch 707997 from E.coli (0111:B4) was purchased from Difco (Detroit, Michigan). Dextran sulphate (DXS, M.W. 500.000) was obtained from Sigma (St. Louis, Missouri). T cell supernatant was obtained from rat spleen cells, cultured for 4 h in the presence of 10^{-4} mg/ml PMA, washed twice, recultured for 8 h in the presence of 2 µg/ml ConA, washed 5 times and finally cultured 20 h in serum free medium.

RESULTS

Cell separations

B lymphocyte subsets were separated in two stages. First, T cells were depleted by labelling with biotinylated W3/13 monoclonal antibody and rosetting with avidin-coated sheep red blood cells, followed by centrifugation on Ficoll-metrizoate (1.095 g/ml) as described previously[5]. The resultant population consists of at least 85% sIgM+ve cells, contaminated with less than 1.5% T cells. On the basis of May Grünwald-Giemsa- and acid phosphatase-stained cytospin preparations, 5-7% macrophages could be recognized. In the second step, the T cell depleted population was separated on a discontinuous Percoll gradient[6]. Cells at the interface between 1.072 and 1.096 g/ml and at the interface between 1.030 and 1.072 g/ml were collected and washed twice. The resultant populations of high and low density cells were examined by flow cytofluorometry and May Grünwald-Giemsa- and acid phosphatase staining. High density cells were predominantly small lymphocytes (Fig. 1a) with minor contamination of other cell types (< 1% macrophages). Low density cells were mainly medium-sized lymphocytes. However, flow cytofluorometry showed that some small lymphocytes were present as well, and acid phosphatase staining revealed that macrophages had been selectively collected into this preparation (up to 14%). In some experiments we have attempted to further deplete macrophages from these subpopulations by adherence to plastic Petri dishes.

Mitogenic stimulation of small and medium-sized B cells

When rat spleen lymphocytes or T cell depleted spleen lymphocytes were cultured *in vitro* in the presence of LPS and DXS, optimal proliferation was obtained after 64 h culture with a combination of 50 μg/ml LPS and 18 μg/ml DXS. The mitogenic effect of LPS and DXS appeared synergistic, *i.e.* ^{3}H-thymidin incorporation was 2-3 times higher with the combination of the mitogens than with the mitogens separately.

Further enhancement of the proliferative response was observed when T cell factors from a supernatant of ConA-stimulated spleen cells were present. This was especially evident when B cells were cultured (Fig. 2).

When subpopulations of small and medium-sized B cells were examined, variable results were obtained. It was often observed that the B_l subset proliferated poorly and that after 2 or 5 days of culture high numbers of macrophages were present. We attempted to deplete macrophages (mØ) from the B_l and B_s subset by adherence to plastic prior to stimulation and, alternatively, we added 4% peritoneal exudate cells (Fig. 3). The addition of macrophages led to a decrease in the proliferation of small B cells, but not of the medium-sized B cells. The amount of decrease in the response of B_s varied from experiment to experiment. When adherent cells

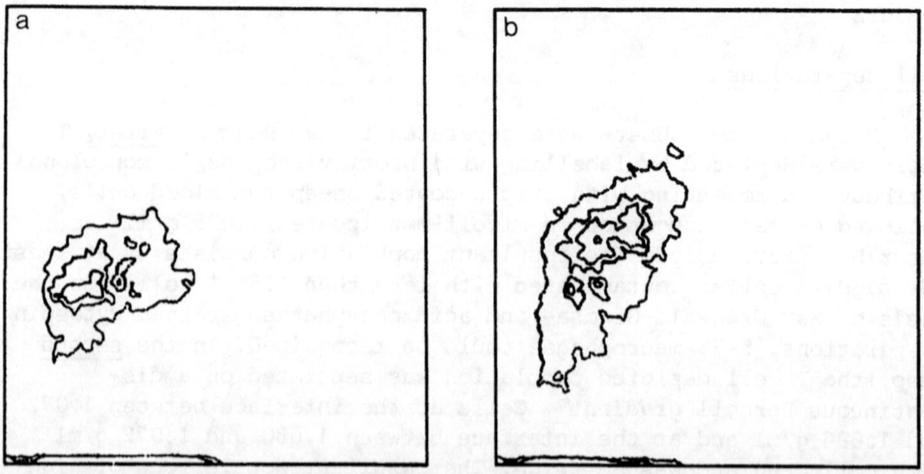

Fig. 1. Flow cytofluorometric profiles of high density (a) and low density (b) B lymphocytes. Abcissa shows green fluorescence and ordinate shows integrated 90°light scatter representing cell size. Amplification on both scales is logarithmic.

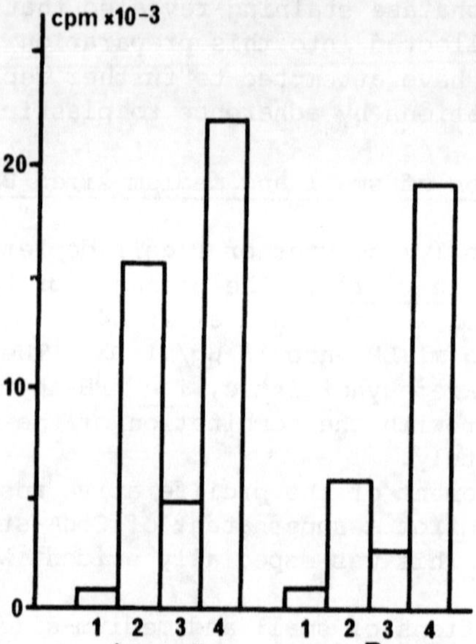

Fig. 2. Mitogenic stimulation by LPS (50 µg/ml) and DXS (18 µg/ml) of rat spleen lymphocytes (left) and B lymphocytes (right) cultured for 48 h + 16 h with 0.25 µCi ^{3}H-thymidin per well. Data are expressed as the mean cpm of triplicate cultures from a representative experiment. Numbers refer to: (1) control; (2) LPS/DXS; (3) T cell supernatant; (4) supernatant + LPS/DXS.

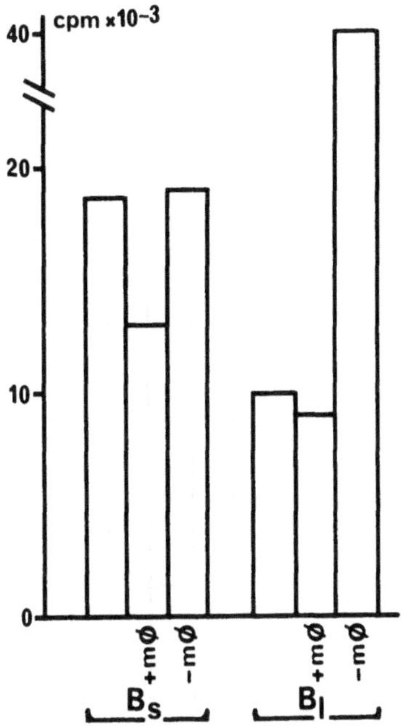

Fig. 3. Mitogenic stimulation by LPS (50 µg/ml) and DXS (18 µg/ml) of rat spleen B_s and B_l lymphocytes cultured for 48 h + 16 h with 0.25 µCi ^{3}H-thymidin per well. Data are expressed as the mean cpm of triplicate cultures from a representative experiment.

were removed, the proliferation of B_s remained unaltered. This was consistent with our finding that very few macrophages were present in this subpopulation. However, when adherent cells were removed from medium-sized B cells, very high proliferative responses could be obtained (Fig. 3).

The production of IgM-secreting cells from B_s and B_l cells

Using a Protein A-plaque assay[7,8] the number of IgM-secreting plasma cells in LPS/DXS-stimulated cultures was estimated. Again, responses of small and medium-sized B cells were found variable. It was found that both subsets had the potency to generate IgM-plaque forming cells after 5 days culture with LPS/DXS or with LPS/DXS + T cell supernatant. In general, the number of plaque forming cells was higher when the supernatant was present (Fig. 4). Addition of macrophages severely inhibited the response of B_s, but the response of B_l was hardly affected. However, when adherent cells were depleted from the medium-sized B cells, very high numbers of cells were found with a high frequency of IgM-secreting cells.

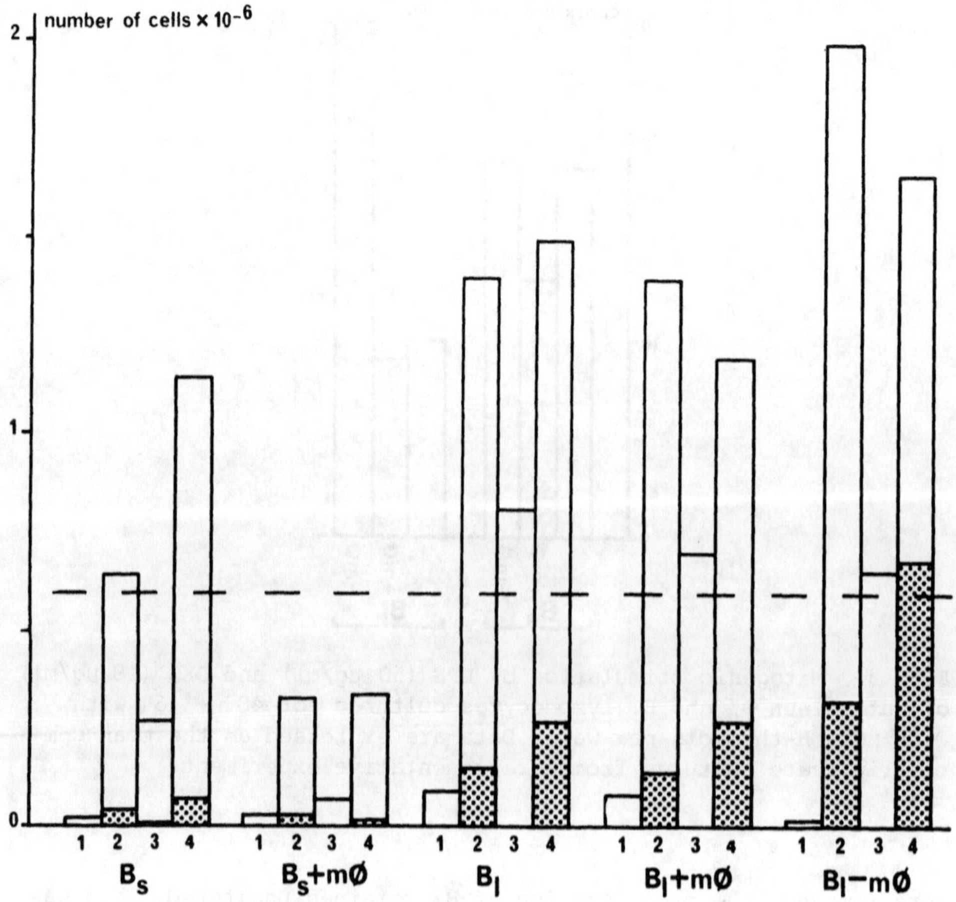

Fig. 4. Mitogenic stimulation of rat spleen B_s and B_l cells, cultured for 5 days with: (1) control; (2) LPS/DXS; (3) T cell supernatant; (4) supernatant + LPS/DXS. Dashed line indicates starting number of cells. Open bars represent total number of cells; filled bars represent number of IgM-secreting cells.

DISCUSSION

We have tried to elicit a proliferation and a differentiation with a combination of the T cell-independent mitogens LPS and dextran sulphate. At optimal concentrations of 18 µg/ml DXS and 50 µg/ml LPS we have obtained high mitogenic responses. When considering these results with the results of Rosenberg et al.[12], using anti-κ stimulation, and those of Naot et al.[13], using LPS or Mycoplasma neurolyticum mitogen, the proliferative response to LPS/DXS can be found some orders of magnitude higher. The response of rat lymphocytes to LPS/DXS occurs in the absence of T cells, but it should be noted that the presence of T cells or of a supernatant from cultured T cells results in some enhancement of the stimulation.

The present study was undertaken to find out possible functional differences between small and medium-sized B lymphocytes in rat spleen, which can be considered to represent the recirculating sIgDsIgM+ve corona B lymphocytes and the sessile sIgM+ve marginal zone B cells respectively[3,4,9]. Controversial data are available on the responsiveness of sIgD+ve and sIgD-ve cells from the mouse spleen. Marshall-Clarke et al.[10] have provided that both subsets are able to respond to thymus-independent antigens. However, McFadden and Vitetta[11] have reported that virtually all cells responding to the thymus-independent antigen TNP-Brucella abortus were sIgD+ve.

When B cells were separated into high and low density cells, representing small and medium-sized B cells respectively, it appeared that both subsets could become stimulated both to proliferation and to plasma cell differentiation. However, the height of the responses varied from experiment to experiment. One of the factors that influenced the response was the number of macrophages present. B_s cells were very poor in macrophages, but B_l cells were selectively enriched with macrophages. It appeared that adherent cells could exert a suppressive effect in our cultures. Yet, the number of macrophages cannot fully explain the variable results. The handling of cells prior to culture may introduce another factor. We have the impression that medium-sized B cells are relatively fragile and may become low- or non-responsive in culture. This implicates that the resultant response of B_l cells could be ruled by both the number of responsive medium-sized cells and the number of small B cells present as a contamination. The influence of cell separation media and procedures is now in further study.

ACKNOWLEDGEMENTS

The authors are indebted to Dr. Huub Savelkoul (Erasmus University Rotterdam) for his advice in the Protein A-plaque assay and to Dr. Jacob Aten (Dept. of Radiobiology) for the Cytofluorograph analyses.

REFERENCES

1. C. de Groot and J. Wormmeester, Capping of surface immunoglobulin on rabbit and mouse lymphocytes. I. Kinetics, Eur. J. Cell Biol. 24: 9 (1981).
2. C. de Groot, J. Wormmeester, and C. Mangnus-Smet, Capping of surface immunoglobulin on rabbit and mouse lymphocytes. II. Cytoskeletal involvement in different subpopulations, Eur. J. Cell Biol. 25: 202 (1981).
3. H. Bazin, D. Gray, B. Platteau, and I.C.M. McLennan, Distinct δ^+ and δ^- B-lymphocytes in the rat, Ann. N.Y. Acad. Sci. 399: 157 (1982).

4. D. Gray, I.C.M. McLennan, H. Bazin, and M. Khan, Migrant $\mu^+\delta^+$ and static $\mu^+\delta^-$ B lymphocyte subsets, Eur. J. Immunol. 12: 564 (1982).
5. J. Wormmeester, F.E.M. Stiekema, and C. de Groot, A simple method for immunoselective cell separation with the avidin-biotin system, J. Immunol. Meth. 67: 389 (1984).
6. J.G. Salisbury, J.M. Graham, and C.A. Pasternak, A rapid method for the separation of large and small thymocytes from rats and mice, J. Biochem. Biophys. Meth. 1: 341 (1979).
7. E. Gronowicz, A. Coutinho, and F. Melchers, A Plaque-assay for all cells secreting Ig of a given type or class, Eur. J. Immunol. 6: 588 (1976).
8. A. van Oudenaren, H. Hooijkaas, and R. Benner, Improvement of the Protein A plaque assay for immunoglobulin secreting cells by using immunoglobulin-depleted guinea pig serum as a source of complement, J. Immunol. Meth. 43: 219 (1981).
9. D.S. Kumararatne, H. Bazin, and I.C.M. McLennan, Marginal zones: the major B compartment of rat spleens, Eur. J. Immunol. 11: 858 (1981).
10. S. Marshall-Clarke, K.D. Keeler, and R.M.E. Parkhouse, The expression of surface IgD on B cells responsive to thymus-independent and thymus-dependent antigens and its requirement for B cell triggering, Immunology 48: 393 (1983).
11. S.F. McFadden and E.S. Vitetta, sIgD-negative B cells from neonatal mice do not respond to the thymus-independent antigen TNP-BA in limiting dilution cultures, J. Immunol. 132: 1717 (1984).
12. J.S. Rosenberg, S.C. Gilman, and J.D. Feldman, Regulation of rat B cell responses by T cell subsets and IL2. J. Immunol. 129: 996 (1982).
13. Y. Naot, S. Merchav, and H. Ginsburg, Mycoplasma neurolyticum: a potent mitogen for rat B lymphocytes, Eur. J. Immunol. 9: 185 (1979).

B CELL PRODUCTION IN ADULT RATS

H. Bazin[1], B. Platteau[1], I.C.M. MacLennan[2], N.S.A. Stuart[2], M. Khan[2], and G.D. Johnson[2]

[1]Experimental Immunology Unit, University of Louvain, 1200 Brussels, Belgium. [2]Dept. of Immunology, University of Birmingham Medical School, Birmingham B15 2TJ

INTRODUCTION

The sequence of B cell development during ontogeny has been investigated by a number of groups (1,2). However, these studies mainly concentrate on events before the full development of secondary lymphoid organs. B cells continue to be produced throughout life in the bone marrow of rodents (3,4). Recent experience with bone marrow transplantation confirms this ability for adult B cell production in humans. We have set out to study the sequence and rate of B cell entry from primary to secondary lymphoid organs in adult rats. This has been achieved by suppressing the development of B cells by administration from birth for 6 weeks rabbit-anti-rat-IgM and IgD. The technique used has been described previously (5). The rats received their last injection of anti-rat-IgM and IgD at 43 days of age and suppression was stopped two days later by injection of 4.5 mg each of rat IgM and IgD. At the time of stopping suppression and at intervals thereafter rats were killed and the extent of B cell and plasma cell population of spleen, lymph nodes and gut-associated lymphoid tissue was assessed by immunohistology.

RESULTS

Degree of Suppression of B Cell Development Achieved

The combined administration of anti IgM and IgD produced

profound suppression of B cell numbers in spleen and lymph nodes. This was assessed in 13 rats killed at the time of stopping suppression or within 12 hours of injection of IgM and IgD. In 5 of these rats no B cells were found. The remainder had occasional isolated B cells in the outer aspect of the periarteriolar lymphocytic sheath of the spleen with surface (S) IgM or IgM plus IgD. No B cells were detected outside the splenic marginal sinus or in lymph nodes. Conventional histology confirmed that there were no accumulations of lymphocytes which had not been detected by immunohistology in the B cell areas of spleens or lymph nodes.

The Rate and Sequence of B Cell Appearance in Spleen and Lymph Nodes After Stopping Suppression

The rate of B cell development was sufficient to result in the near-normal appearance of secondary lymphoid organs within 8 days of stopping suppression (Fig. 1a and b). At this stage there were both primary and secondary follicles and the marginal zones were full of SIgM+ve SIgD-ve cells. In more detail the sequence of appearance was:

a) Appearance of B cells in the red pulp of the spleen. Of the tissues examined this was the first site where B cells were seen. They were already present in 2 of 3 rats examined 24 hours after stopping suppression (Figs. 2a, b, and c). The surface immunoglobulin isotypes expressed were either IgM alone or IgM plus IgA. These cells presumably represent B cells produced *in situ* in the primary lymphoid tissue of the red pulp. Very few SmIgD+ve cells were seen in the red pulp until the 4th day after stopping suppression.

b) Development of splenic follicles. SmIgD+ve B cells were present on the inside of the marginal sinus by 48 hours after stopping suppression. Most of these cells also expressed IgM and some IgA. By the third day after stopping suppression there were well developed splenic primary follicles. Secondary follicles i.e. follicles with germinal centres were first identified at 8 days after stopping suppression. The presence of follicular dendritic cells was assessed by using OX-2 which identifies antigens on these cells (6). OX-2 +ve follicular dendritic cells were absent from animals killed immediately after termination of suppression and did not appear until the fifth day after suppression, three days before germinal centres were present and two days after the development of primary follicles.

c) Lymph node follicles. The relative tempo of follicle development in lymph nodes and spleen was assessed by examining both these tissues from the same animals. The appearance of B cells in the cortex of lymph nodes was consistently delayed until

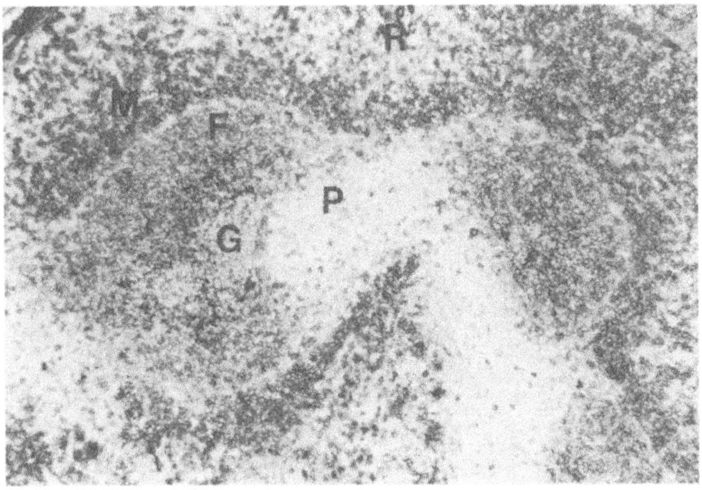

a

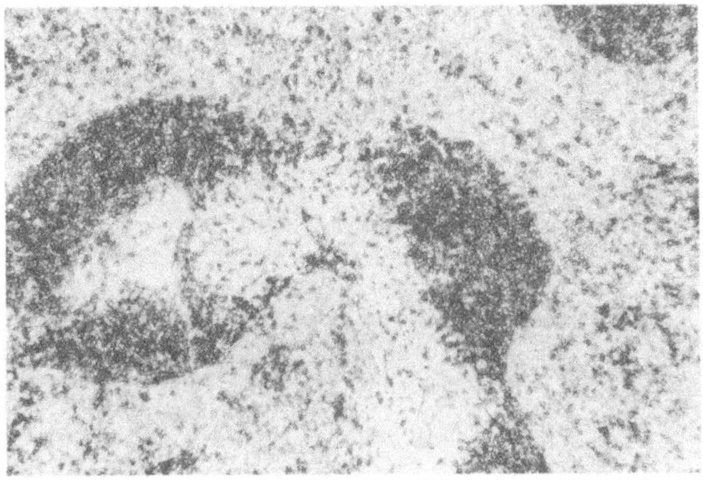

b

Fig. 1. Serial frozen sections of spleen from a rat 8 days after stopping suppression with rabbit-anti-rat-IgM and IgD. (a) stained for rat IgM (b) stained for rat IgD. R = red pulp M = marginal zone F = follicular mantle G = germinal centre P = periarteriolar lymphocytic sheath.

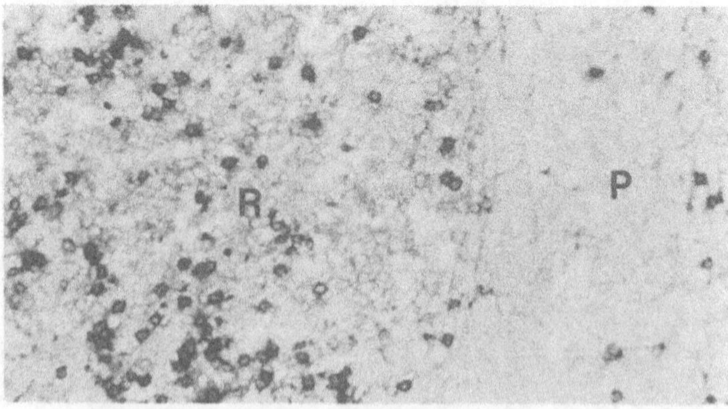

a

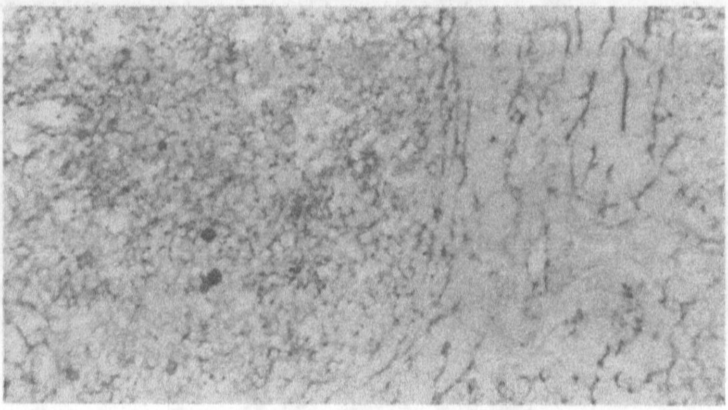

b

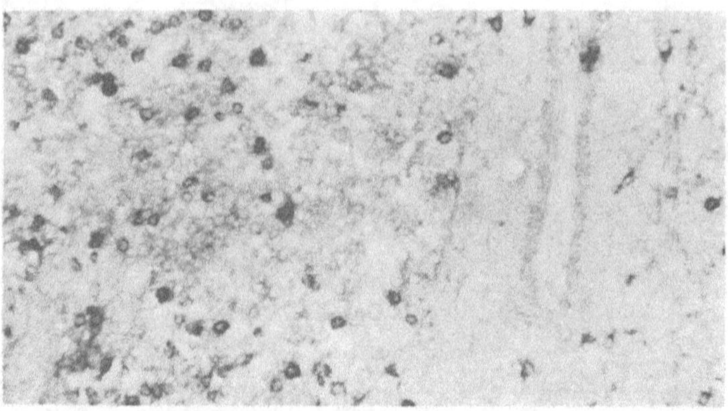

c

Fig. 2. Serial frozen sections of spleen from a rat 24 hours after stopping suppression. (a) stained for rat IgM (b) stained for rat IgD (c) stained for rat IgA. Symbols as in Fig. 1.

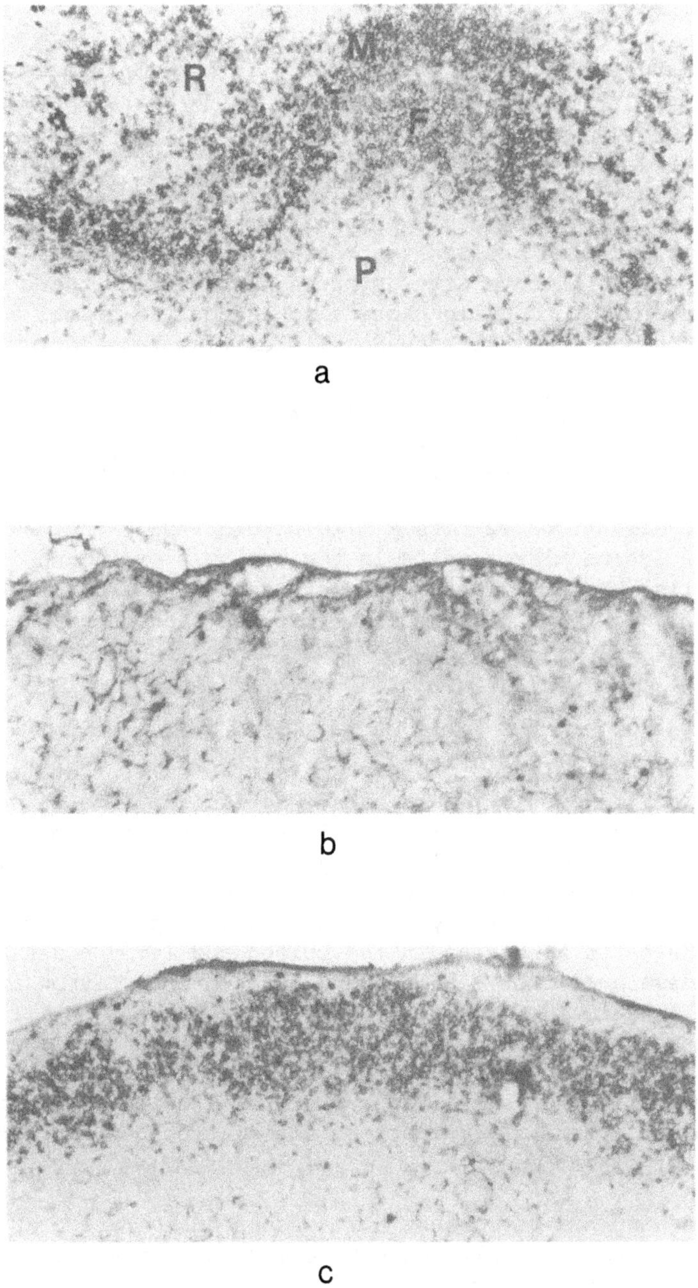

Fig. 3. (a) Section of spleen from a rat 3 days post suppression stained for rat IgM. (b) section of lymph node cortex of the same rat as in 3a also stained for rat IgM (c) section of lymph node cortex of a rat 4 days post suppression stained for IgM.

24 hours after that in the spleen i.e. at 4 days as opposed to 3 days (Fig. 3a, b, c).

d) Marginal zones of the spleen. In 1 of 3 animals at 24 hours after suppression and all animals at 48 hours there were SIgM+ve and SIgM with IgA+ve cells in the marginal zones. Few cells outside the marginal sinus were SmIgD+ve. On the 3rd and 4th days the picture was essentially the same but on days 5 and 6 numbers increased so that after day 6 marginal zone numbers appeared normal. These cells had SIgM and SIgA expression as in normal adult rats. However, even at 14 days after stopping suppression there were minimal marginal zone cells expressing IgG2c. In normal adults up to 20 percent of marginal zone B cells express this isotype.

e) Plasma cells. IgM-containing plasma cells were seen in the spleen and lymph nodes by 4 days after stopping suppression and reached normal levels by the tenth day. IgA plasma cells were detected in the lamina propria of the gut by day 6 and the spleen by day 10. (In 11 of 22 rats killed before 3 days there were small numbers of IgA+ve plasma cells in the lamina propria and in 13 IgD+ve cells. Paradoxically none were seen in the 6 rats killed at 4 and 5 days. IgM plasma cells were not seen in suppressed rats).

DISCUSSION

This study indicates the high rate of B cell production which can be achieved by adult rats. Sufficient B cells can be produced within a week to fill the B cell areas of secondary lymphoid organs. On morphological grounds it seems unlikely that more than small numbers of B cells were present with capped immunoglobulin in the secondary lymphoid organs of suppressed rats. Most of the cells appearing in the spleen and lymph nodes have probably developed from B cell precursors in bone marrow and splenic red pulp. The rapid tempo of appearance of B cells in all compartments makes it improbable that memory B cell formation was an important factor in the population of either the follicles or the marginal zones. The delayed appearance of B cells expressing IgG2c in the marginal zone may reflect an association between expression of this isotype and memory.

Osmond and his co-workers have assessed the rate of B cell generation in normal adult mouse bone marrow. They have estimated the production of as many as 5×10^7 B cells per day (7). As the total mouse peripheral B cell pool is probably only in the order of 2×10^8 B cells their data suggest that the rates of B cell production we have observed in this study are not unusual.

REFERENCES

1. E. R. Abney, M. D. Cooper, J. F. Kearney, A. R. Lawton, and R. M. E. Parkhouse. J. Immunol. 120:2041 (1978).
2. E. S. Vitetta, A. Melchers, M. McWilliams, J. Phillips-Quaglietta, M. E. Lamm, and J. H. Uhr. J. Exp. Med. 141:206 (1975).
3. D. G. Osmond, and G. J. V. Nossal. Cell Immunol. 13:132 (1974).
4. R. Aspinal, and J. J. T. Owen. Immunol. 48:309 (1983).
5. H. Bazin, B. Platteau, A. Beckers, and R. Pauwels. J. Immunol. 121:2083 (1978).
6. A. N. Barclay. Immunol. 44:727 (1981).
7. D. G. Osmond, M. T. E. Fahlman, G. M. Fulop, and D. M. Rahal. CIBA Foundation Symposium 84:68 (1981).

THE PARADOX OF HIGH RATES OF B CELL PRODUCTION IN BONE MARROW AND THE LONGEVITY OF MOST MATURE B CELLS

D.S. Kumararatne, D. Gray, I.C.M. MacLennan,
J. Lortan, B. Platteau* and H. Bazin*

Dept. Immunology, University of Birmingham
Medical School, Birmingham B15 2TJ. *Experimental
Immunology Unit, University of Louvain, 1200 Brussels
Belgium

INTRODUCTION

In mammals, the bone marrow is the principal site of primary B cell generation in post-natal life (1). Studies in mice indicate the the physiological rate of B cell production in bone marrow is sufficient to replenish the peripheral B lymphocyte pool within 4 days (2). The generative potential of B cell precursors is further emphasised by the ability of rats treated from birth with anti-IgM and anti-IgD antibodies, to fully repopulate secondary lymphoid organs with B cells, within a span of eight days following the cessation of antibody-induced suppression (3). These observations alone are consistent with the view that mature B cells have a relatively short half-life of a few days (4,5). However, only 25% of recirculating and static B cells of the splenic white pulp of rats are labelled by five days of ^{3}H-thymidine infusion, indicating that most B cells within secondary lymphoid organs are not newly formed cells (6). The studies of Sprent and Basten (7) also support such a viewpoint. These data (6,7), taken together with the observations on the rate of B cell production in bone marrow (2), imply that only a small proportion of B cells produced in adult bone marrow become part of the mature lymphocyte pool.

The rate of entry of immature B cells derived from bone marrow into the mature B lymphocyte pool and the survival of adoptively transferred B cells has been investigated using congenic rats differing in their kappa (k)-allotype. This constitutes a genetic marker for tracing the cellular lineage of adoptively transferred B cells and their precursors. Since more than 95% of immunoglobulin in rats has the k-isotype, the k-allotype is a marker for almost all transferred B cells in this system (8). This study was carried out in two ways: First, by studying the repopulation of mature B cell compartments in the spleens of lethally irradiated rats reconstituted with k-allotype-marked bone marrow and/or thoracic duct lymphocytes. Secondly, recirculating follicular B cell (RF B cell) chimaeras were obtained by parabiosis of intact adult rats differing in their k-allotype (9). The decay of the degree of B cell chimaerism was studied after separation of the parabionts from each other.

MATERIALS AND METHODS

Animals

Adult, highly inbred PVG/Ciu rats (8), and the PVG-RI-a^a (DA) strain (8) of animals bred under specific pathogen-free conditions from parents were kindly donated by Dr. Simon Hunt, MRC Cellular Immunology Unit, Oxford. These rat strains are congenic, but differ in their immunoglobulin k-light chain allotype, which are K-1a and K-1b, respectively. The two strains will be referred to as PVG-1a and PVG-1b, in turn. They were used when 10-16 weeks old, unless otherwise stated, and were age and sex matched in each experiment.

Irradiation

Rats were exposed to 8Gy gamma irradiation over a period of 5 minutes, using an ACEL Beam ^{60}Co source at the Dept. of Radiation Physics, University of Birmingham.

Cell Collection

a) Thoracic duct lymphocytes (TDL) were collected at 4°C, in the first 48 hr after cannulation of 8-10 week old rats (9). TDL were injected into reconstituted animals at 12 hr intervals. Viability, as assessed by trypan blue exclusion, was over 95%.
b) Bone marrow cells were collected from long bones of hind limbs of adult rats, into ice-cold RPMI-1640 and injected intravenously into recipients with minimum delay.

Parabiosis

This was carried out in Brussels between age and sex-matched adult Lou/cK1a and Lou/cK1b rats (10), which bear the IgK-1a and K-1b allotypes, respectively. The animals were separated after 10 days of parabiosis, which period is sufficient to achieve circulating B cell chimaerism (10).

Cell Reconstitution

Details are indicated in Results. Cells were injected iv, and donors and recipients were age and sex matched.

Antibodies

Purified polyclonal rabbit anti-rat u and δ, and monoclonal mouse anti-rat kappa la (MARK-3) antibodies (10) were made in the Experimental Immunology Unit, University of Louvain, and conjugated to fluorochromes by Dr. D. Johnson, Dept. Immunology, University of Birmingham (9). A rat anti-rat kappa lb was raised in the Dept. of Immunolgy, University of Birmingham (10). Peroxidase-conjugated swine anti-rabbit Ig and swine anti-mouse Ig respectively, used for immunoperoxidase staining, were purchased from Dakopatts, Copenhagen, Denmark.

Immunohistology

The double immunofluorescence techniques and the immunoperoxidase methods used to detect surface membrane immunoglobulin k-allotype and/or isotype are fully described elsewhere (10).

Analysis of Chimaerism

Serial acetone-fixed cryostat sections (5u) from lymphoid tissues were simultaneously stained (10) with rhodamine conjugated anti-u (or anti-δ), and fluorescein conjugated anti-allotype antibodies (anti-K-la or lb). These sections were examined using a Zeiss microscope equipped for incident light illumunation with a HBO50 mercury lamp. The proportion of K-la-positive cells among surface positive cells of the follicles was determined by counting at least 500 cells. This is an estimate of the degree of chimaerism of the recirculating lymphocyte pool (10). Similarly, the proportion of K-la positive cells among surface IgM-positive cells of the marginal zone was used to estimate the degree of chim-aerism in the non-recirculating (10) lymphocyte pool.

RESULTS

The Longevity of Recirculating B Cells

The large majority of B cells found in the thoracic duct lymph are mature, recirculating cells (6,9). To study their survival after adoptive transfer $2X10^8$ PVG-la TDL were injected iv into lethally irradiated (800 cGy) adult PVG-lb rats. Fig. 1 shows the

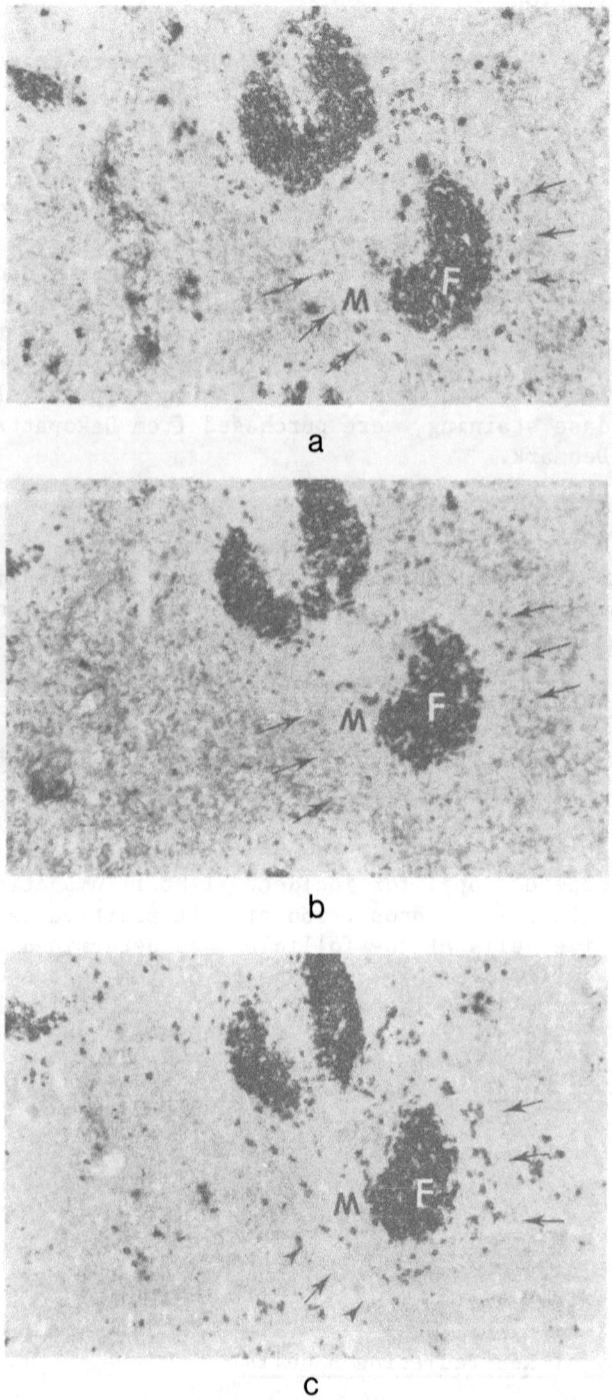

Fig. 1. Serial sections of spleen from irradiated PVG-1b animals given PVG-1a TDL, killed 3 days post cell transfer: 1a, 1b and 1c: stained for surface IgM, IgD and K-1a allotype, respectively. F = follicle, M = marginal zone, P = PALS. Arrows indicate the limit of the marginal zone.

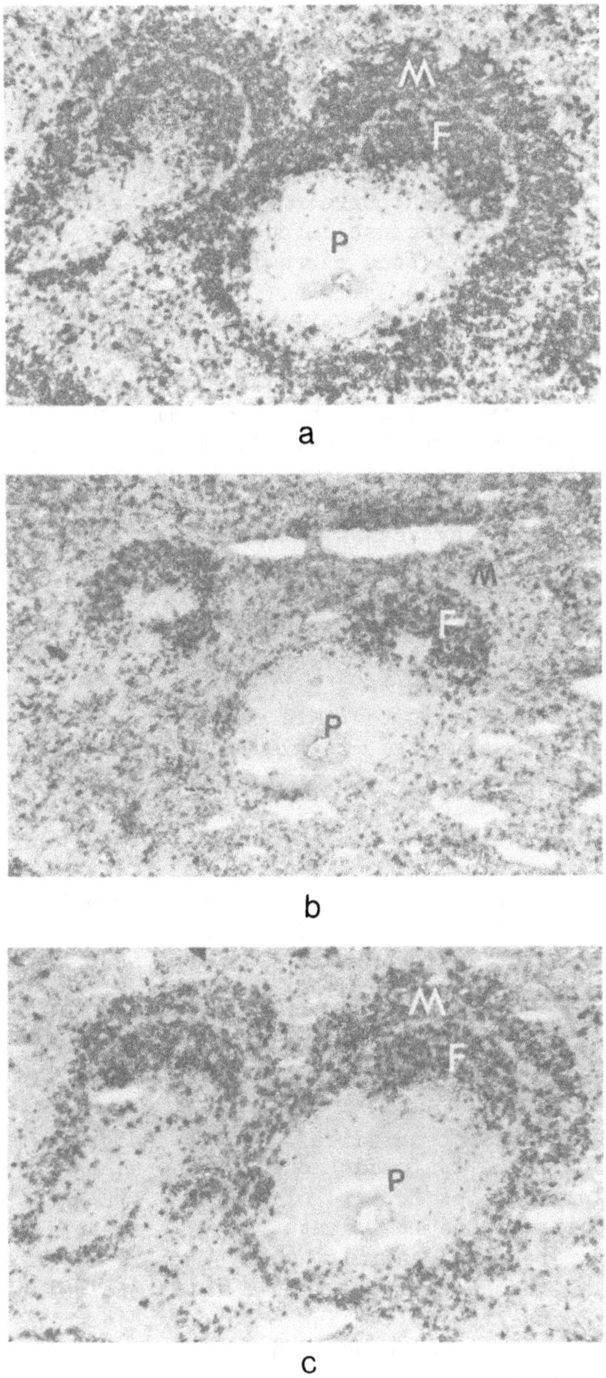

Fig. 2. Serial sections of spleen from irradiated PVG-1b animals reconstituted with PVG-1a TDL and syngeneic bone marrow, killed 6 weeks post-transfer. 2a, 2b, 2c: stained for surface IgM, IgD and K-1a allotype, respectively. See Fig. 1 for explanation of symbols.

appearance of splenic tissue of rats killed 3 days after TDL transfer. It can be seen that B cell areas of the recipient spleen predominantly contain surface membrane u+ δ + lymphocytes, almost all of which the donor (K-la) allotype. The marginal zones, however, contained very few B cells: these too expressed the K-la allotype.

In the next experiment, recirculating B cell chimaeras constructed as above were reconstituted (48 hr after the PVG-Kla TDL transfer) with syngeneic bone marrow cells (obtained from the hind limbs of two adult PVG-Klb animals). Fig. 2 shows spleen sections from rats killed 6 weeks after allotype-marked TDL transfer. These show well-developed follicles and marginal zones replete with B cells. The follicular cells bore the u+ δ + surface phenotype characteristic of recirculating B cells, while the marginal zone cells were surface membrane u+ δ -. Substantial numbers (15-30%) of B cells in both follicles and marginal zones bore the allotype of the TDL donors. Hence, a substantial proportion of B cells found in TDL (or their progeny) can survive as long as 6 weeks after adoptive transfer into irradiated animals.

The Survival of Mature Recirculating B Cells in Intact Animals

Recirculating follicular B cell chimaeras were obtained by parabiosis of intact adult Lou/CKla rats with Lou/cKlb rats (10). These animals were separated after 10 days of parabiosis and were sacrificed at that stage, or 15 days later. Table 1 shows that after 10 days of parabiosis the degree of chimaerism among RF B cells is 30-30%, while that in the marginal zone is only 1-2%. Fifteen days after separation the degree of follicular chimaerism had not decreased, while the marginal zone chimaerism was about 3-5%. These results reinforce the data from ^{3}H-thymidine labelling studies cited in the Introduction (6,7), and indicate that most RF B lymphocytes are not newly generated cells.

DISCUSSION

The results presented indicate that a substantial proportion of recirculating B cells found among TDL (or their progeny) can survive for long periods (upward of 6 weeks) after adoptive transfer into irradiated recipients. The persistence of RF B cell chimaerism in parabiosed animals, when examined 15 days after separation, shows that recirculating B cells are relatively long-lived even in intact animals. These data are in agreement with the estimates of B cell lifespan derived from ^{3}H-thymidine labelling studies (7). Our results do not agree with the conclusions of de Freitas and Coutinho (5), who estimated the rate of turnover of LPS-responsive splenic B cells to be 50% per day. It is quite likely that their results were due to rapid elimination of congenic

Table 1. Summary of Cell Distribution in Rat Parabiosis Experiment

Time after separation of parabionts (days):	% Donor allotype-positive cells in spleen:*	
	Marginal zone:	Follicles:
0	1.3 ± 0.2 (n = 8)	25.8 ± 1.3 (n = 8)
15	3.7 ± 0.3 (n = 12)	24.7 ± 0.6 (n = 12)

Rats were parabiosed for 10 days, and then separated.
*Mean % $\pm$ standard error, for n animals is shown.

B lymphocytes when adoptively transferred to intact animals (see Lortan et al. (12), in this volume for further discussion of this point).

Data establishing high rates of B cell production in the bone marrow are extensively documented (2,3,). This, taken together with the findings supporting the longevity of mature B cells (summarised above), imply that only a small proportion of B cells newly generated in adult bone marrow enters the mature lymphocyte pool found in secondary lymphoid organs. This conclusion is further strengthened by data suggesting that a proportion of cells entering the mature B cell pool are derived from germinal centres (13). A corollary of this line of argument is that a large proportion of immature B cells produced in adult bone marrow are rapidly eliminated. The site of destruction of these cells is not known. It is possible that entry of immature B cells (produced in the bone marrow) into the mature B cell pool of secondary lymphoid organs is influenced by antigen (14); the chance of an immature B cell entering the mature compartment being improved by exposure to the appropriate antigen (13). Such a scheme would have the obvious advantage of generation of antibody diversity by the high rate of B-lymphopoiesis in bone marrow, coupled with appropriate selection of these cells into the mature lymphocyte pool by expos-ure to antigen.

Finally, the relevance of our findings to the ontogeny of marginal zone and RF B cells is further discussed by MacLennan et al. in this volume (11).

REFERENCES

1. D.G. Osmond and G.J.V. Nossal, Cell. Immunol. 13: 132 (1984).
2. D. Opstelten and D.G. Osmond, J.Immunol. 131: 635 (1983).
3. H. Bazin, B. Platteau, I.C.M. MacLennan, N.S.A. Stuart, M. Khan and G.D. Johnson, This volume p. 65.
4. S. Strober, J. Exp. Med. 136: 851 (1972).
5. A.A. de Freitas and A. Coutinho, J. Exp. Med. 154: 994 (1981).
6. D.S. Kumararatne and I.C.M. MacLennan, Eur. J. Immunol. 11: 865 (1981).
7. J. Sprent and A. Basten, Cell. Immunol. 3: 361 (1975).
8. S.V. Hunt and M.H. Fowler, Cell Tissue Kinet. 14: 445 (1981).
9. W.L. Ford, In: Handbook of Experimental Immunology, D.M. Weir, (ed), Blackwell Scientific Publications, Oxford (1979).
10. H. Bazin, F. Cormont and L. DeClerq, J. Immunol. Methods 71: 9 (1984).
11. I.C.M. MacLennan, H. Bazin, D. Chassoux, D. Gray and J. Lortan, This volume, p. 139.
12. J. Lortan, D. Gray, D.S. Kumararatne, B. Platteau, H. Bazin and I.C.M. MacLennan, This volume, p. 73.
13. P. Nieuwenhuis In: B Lymphocytes Today, Elsevier Medical Press, Amsterdam, (1982), p. 95.
14. K.D. Shortman, M.C. Howard and J.A. Baker, J. Immunol. 121: 2060 (1979).

MONOCLONAL ANTIBODIES TO RAT B LYMPHOCYTE (SUB-)POPULATIONS

F.G.M.Kroese, D.Opstelten, A.S.Wubbena, G.J.Deenen
J.Aten, E.H.Schwander*, L.de Leij* and P.Nieuwenhuis

Dept. of Histology and *Dept. of Clinical Immunology
State University Groningen, Oostersingel 69/1, Groningen
The Netherlands

INTRODUCTION

The study of B lymphocyte differentiation is greatly facilitated by the use of suitable monoclonal antibodies (MAb's). For mice and men a variety of MAb's binding to cell surface determinants of B lineage cells has been described (1). However, as yet almost no B lineage specific MAb's have been described for the rat. Since this animal appears very appropiate for studies of early phases of B lymphocyte differentiation (2) as well as late phases, in particular with respect to marginal zone cells (3), we produced MAb's against surface antigens of rat B lineage cells.

MATERIALS AND METHODS

Animals, tissues and cell suspensions.

Balb/c mice were obtained from TNO, Rijswijk, The Netherlands. Young adult AO/G rats were bred in our own colony. Small fragments of spleen, cervical lymph node, Peijer's patch, thymus, liver, kidney and brain of rats were deep frozen in liquid freon for immunoperoxidase staining of frozen sections. Single cell suspensions of spleen and cervical lymph nodes were prepared by gently mashing of small tissue fragments through a nylon gauze; bone marrow cells were isolated as described (4); thoracic duct lymphocytes were obtained as described (5). Cells were cytocentrifuged and fixed in acetone (10', room temperature) for immunoperoxidase staining, or in 5% (v/v) acetic acid in ethanol (12',0°C) for double immunofluorescence microscopy.

Immunization

Single cell suspensions of rat Peijer's patches (PP) were prepared as described (6). To enrich for germinal centre cells, PP cells were subjected to Percoll density gradient centrifugation. The low density cell fraction (1.062 - 1.068 g/ml) was used to immunize a Balb/c mouse, by injecting i.p. 10^7 cells in Freund's complete adjuvant followed one month later by 10^7 cells in incomplete adjuvant. After another month a second booster of 10^7 PP cells was given i.v. Three days later the spleen was removed for fusion.

Fusion protocol

The fusion protocol of Kennett (7) was used as described (8). Briefly, mouse spleen cells were fused with the murine myeloma cell-ine X63 Ag8. The fusion mixture was divided over 600 wells. All wells were tested for antibody production by indirect immunoperoxidase staining of frozen sections of rat spleen. Positive wells were selected and recloned by limiting dilution.

Immunoperoxidase staining

Frozen sections of tissue were overlayered with undiluted hybridoma culture supernatant (60', room temperature). After washing, sections were incubated for 30' with peroxidase conjugated rabbit anti mouse Ig (Dakopatts, Denmark) diluted 1:40 with PBS and supplemented with 1% (v/v) normal rat serum (NRS). After washing, the sections were stained for peroxidase activity with 3.3'-diaminobenzidine tetra-HCl (Sigma, St. Louis, U.S.A.) (0.5 mg/ml) in 0.05 M Tris/HCl buffer (pH 7.0) containing 0.01% H_2O_2. The color of the reaction product was intensified by incubation for 5' in 0.5% (w/v) $CuSO_4$ in 0.9% (w/v) NaCl. Sections were then counterstained with hematoxylin according to Mayer.

Determination of the isotype of mouse anti rat MAb's

MAb isotypes were determined by immunoperoxidase staining as described above. In this assay frozen sections of spleen, incubated with hybridoma culture supernatant, were incubated with one of the peroxidase conjugated sheep antibodies to mouse IgM, IgG, IgG1, IgG2a, IgG2b, IgG3 or IgA (Serotec, Bicester, U.K.), diluted 1:25 in PBS.

Double immunofluorescence staining

Fixed cytocentrifuge preparations were overlayered with hybridoma cell culture supernatant (60', room temperature). After washing in PBS (3 x 10'), slides were incubated (30', room temperature) with affinity purified goat anti mouse Ig conjugated to dichlorotriazinylaminofluorescein (DTAF), diluted 1:20 in PBS/1% (v/v) NRS. After washing, rat Ig was stained with affinity purified goat anti rat Ig (H+L chains) conjugated to tetramethylrhodamineisothiocyanate (TRITC), diluted 1:20 in PBS (30', room temperature). In one experiment slides were incubated with rabbit anti rat IgD (a kind gift of Dr. H. Bazin, and purified by Dr. J. Rozing) diluted 1:10

in PBS/1% (v/v) normal mouse serum (NMS) followed by affinity purified goat anti rabbit Ig TRITC, diluted 1:20 in PBS/1% (v/v) NRS/1% (v/v) NMS. Fluorochrome conjugated antibodies were obtained from Kirkegaard and Perry Lab., Gaithersburg, MD, U.S.A.. Double immunofluorescence staining of bone marrow cells was as described (4). Slides were embedded in glycerol/PBS containing an anti-fading agent (9) and stored in the dark at 4°C. Analysis was done with an epifluorescence microscope (Leitz Wetzlar, FRG) equipped with filter systems to separately detect DTAF and TRITC, using a x100 objective. At least 100 fluorochrome positive lymphoid cells were scored.

RESULTS

Hybridoma cells were selected and cloned when producing antibodies reacting with B cell areas of the spleen (viz. marginal zone, and follicles consisting of corona and germinal centre). These so-called HIS MAb's (HIS stands for HIStology Department, Groningen) were grouped according to their reactivity with one ore more of the B cell compartments in the spleen: group I MAb's reacted with all B cell areas, group II with corona and germinal centre, not with marginal zone, and group III with corona only (Table 1). For each group reactivity with tissue sections and cell suspensions are described. A separate section at the end describes reactivity with bone marrow pre-B and B cells.

Table 1
Selected Monoclonal Antibodies

	Reactivity with frozen sections of spleen	MAb name	Isotype
Group I	All B cell areas	HIS 8	IgG1
		HIS 11	IgG1
		HIS 14	IgG1
		HIS 25	IgG1
Group II	Corona and germinal centre, <u>not</u> marginal zone	HIS 24	IgG2b
Group III	Corona only	HIS 22	IgM

Group I: HIS 8, HIS 11, HIS 14 and HIS 25.

Tissue Sections

Spleen: Virtually all lymphoid cells of the marginal zone, corona and germinal centre were stained (Figs. 1a and b). Reactivity

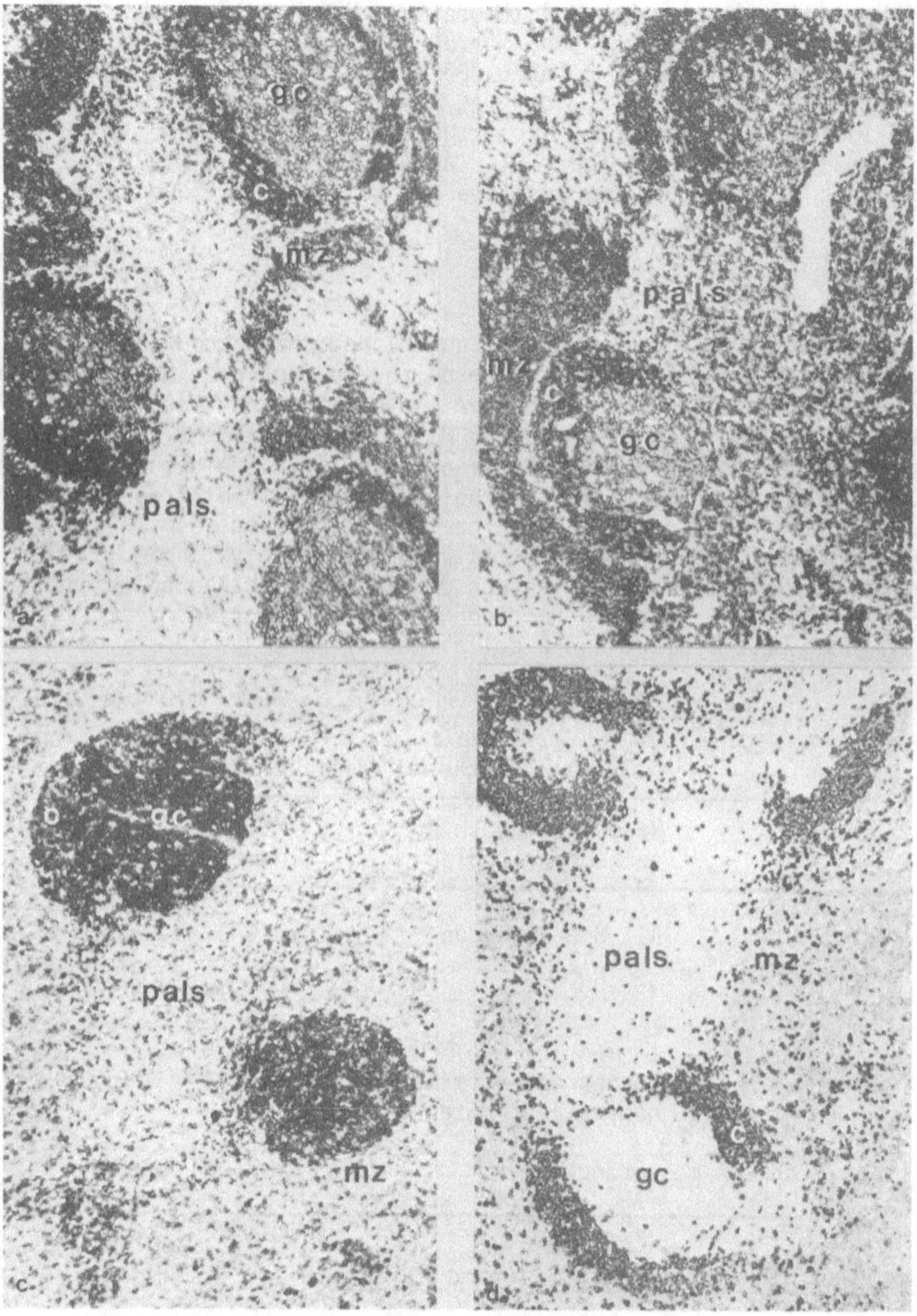

Fig. 1: Immunoperoxidase staining patterns of HIS MAb's with frozen sections of rat spleen: a) HIS 14, b) HIS 25, c) HIS 24 and d) HIS 22. c= corona, gc = germinal centre, mz = marginal zone, pals = periarteriolar lymphocyte sheath.

with lymphoid cells of the periarteriolar lymphocyte sheath (PALS; T cell area) was variable. HIS 8 and HIS 14 (Fig. 1a) stained few PALS lymphocytes. Slightly more positive cells were observed in PALS with HIS 11, whereas HIS 25 (Fig. 1b) stained approximately half of the PALS. With each Mab, positive cells were also found scattered in the red pulp. In addition, HIS 8 gave high background staining of the red pulp, as well as scattered heavily (presumably cytoplasmically) stained cells.

Lymph node and Peijer's patch: Follicles (corona and germinal centre) were stained. In the T-cell areas, few cells, concentrated around high endothelial venules, were stained with HIS 8, HIS 11 and HIS 14, while many HIS 25^+ cells were observed.

Thymus: Scattered HIS 8^+ and HIS 11^+ cells were found predominantly in the medulla. HIS 14 and HIS 25 stained slightly more cells in the medulla, but also some in the cortex.

Kidney, liver, brain: HIS 11, HIS 14 and HIS 25 did not stain these tissues, while HIS 8 gave high background staining of the connective tissue as well as specific staining of the cytoplasm of neurons in brain.

Intestine: Many HIS 8^+ (strongly), HIS 14^+ and HIS 25^+ cells were found in the lamina propria of the villi, with only a few HIS 11^+ cells.

Cell Suspensions

Table 2 shows the incidence of HIS MAb binding cells in spleen and lymph nodes. In agreement with histological observations on spleen, variable proportions of positive cells were found: HIS 25 > HIS 11 > HIS 14 and HIS 8.

Table 2
Incidence of HIS positive cells in suspensions of rat spleen and lymph nodes determined by immunoperoxidase staining of cytocentrifuged cells.

MAb	Incidence (%) Spleen	Incidence (%) Lymph node
HIS 8	37 ± 3*	32 ± 4
HIS 11	48 ± 5	42 ± 1
HIS 14	43 ± 1	27 ± 1
HIS 25	59 ± 2	56 ± 5
HIS 24	40 ± 2	25 ± 1
HIS 22	n.d.	n.d.

* mean ± s.d.; n = 4
n.d. = not done

The predominant anti B lymphocyte character was further established by double immunofluorescence staining of spleen cell cytocentrifuge preparations (Table 3), showing that virtually all Ig^+ (B) lymphocytes carried these HIS determinants. However, variable proportions of Ig^- cells were HIS positive as well: while HIS 8 and HIS 14 binding was restricted to Ig^+ lymphocytes, HIS 11 and HIS 25 reacted with 16% and 39% of Ig^-(non-B) cells respectively.

Table 3
Double immunofluorescence staining of spleen cells by group I HIS-MAb's and anti-rat Ig.

spleen cell subset	Incidence (%) HIS 8 (n=3)	HIS 11 (n=4)	HIS 14 (n=2)	HIS 25 (n=1)
Ig^+ HIS^+	31 ± 2*	37 ± 5	38 ± 2	43
Ig^+ HIS^-	3 ± 5	0	2 ± 1	0
Ig^- HIS^+	1 ± 1	11 ± 1	2 ± 1	22
Ig^- HIS^-	65 ± 3	53 ± 4	60 ± 1	35

* mean ± s.d.

Group II: HIS 24

Tissue Sections

Spleen: Corona and germinal centre but not marginal zone were stained (Fig. 1c). Scattered positive cells were found in PALS and red pulp.

Lymph node and Peijer's patch: All follicular cells were stained, as well as few positive cells in T cell areas.

Thymus: Few positive cells were observed in the medulla.

Liver, kidney, brain: No reactivity was found.

Intestine: Few positive cells were observed in the lamina propria of the villi.

Cell Suspensions

Table 2 shows the incidence of HIS 24^+ cells in the spleen and lymph node (40% and 25% respectively). Double immunofluorescence studies showed HIS 24 antigen to be predominantly on Ig^+ cells (Table 4). Remarkably, in spleen about 25% of Ig^+ lymphocytes were not labeled by HIS 24, while in lymph nodes and thoracic duct lymph the proportion of Ig^+ B lymphocytes that did not carry HIS 24 antigen was much lower: 12% and 5% respectively (Table 4).

Table 4
Double immunofluorescence staining by HIS 24 and anti-rat Ig.

cell subset	Incidence (%) spleen (n=14)	lymph node (n=4)	TDL (n=2)
Ig^+ HIS 24^+	26 $\pm$ 3*	21 $\pm$ 3	39 $\pm$ 4
Ig^+ HIS 24^-	8 $\pm$ 2	3 $\pm$ 1	2 $\pm$ 0
Ig^- HIS 24^+	8 $\pm$ 2	8 $\pm$ 1	4 $\pm$ 1
Ig^- HIS 24^-	58 $\pm$ 3	68 $\pm$ 4	55 $\pm$ 6

* mean $\pm$ s.d.

Group III: HIS 22

Tissue Sections

Spleen: Corona, but not germinal centre and marginal zone were stained (Fig. 1d). Scattered positive cells were found in red pulp and marginal zone and only very rarely in the germinal centre and PALS.

Lymph node and Peijer's patch: Corona was stained, with scattered positive cells in T cell areas and germinal centres.

Thymus: Scattered positive cells were found in the medulla.

Liver, kidney, brain: No reactivity was found.

Intestine: Very few positive cells were found in the lamina propria of the villi.

Cell suspensions

Double immunofluorescence studies showed that the HIS 22 antigen was present on 70% of Ig^+ cells and was virtually absent on Ig^- cells (Table 5). Because the corona-only staining pattern of lymphoid follicles (see above) resembled anti-IgD staining patterns (10), double immunofluorescence studies using HIS 22 and anti-IgD were performed (Table 5) and showed an almost complete overlap of HIS 22^+ and IgD^+ cell populations in spleen.

Table 5
Double immunofluorescence staining of spleen cells by HIS 22 and anti-rat Ig (H + L) or anti-rat IgD.

spleen cell subset	Incidence (%) anti-Ig (H+L) (n=5)	anti-IgD (n=2)
Ig^+ HIS 22^+	23 ± 4*	28 ± 8
Ig^+ HIS 22^-	9 ± 2	1 ± 0
Ig^- HIS 22^+	1 ± 1	1 ± 0
Ig^- HIS 22^-	67 ± 4	71 ± 8

* mean ± s.d.

Reactivity of HIS MAb's with Early B lineage cells.

To study HIS MAb reactivity with immature B lymphocytes and pre-B cells, bone marrow cells were double labeled with HIS MAb's and anti-u (see M & M). In this assay u^+B lymphocytes and pre-B cells were discriminated on the basis of u chain density, being considerably higher in the former (4). All HIS MAb's were found to react with strongly u^+ B lymphocytes but they were heterogeneous with respect to weakly u^+ pre-B cells. Virtually all pre-B cells carried HIS 14, HIS 24 and HIS 25 antigens on their cell surface, while approximately half and one-third of the pre-B cells were HIS 11^+ and HIS 22^+, respectively. HIS 8 did not react with pre-B cells.

DISCUSSION

The HIS MAb's described in this study are predominantly reactive with B lineage cells of the rat. The differential staining of mature and immature B lineage cells by these MAb's illuminates their usefulness for B lymphocyte differentiation studies.

Most of the HIS MAb's were also reacting with a (small) proportion of Ig^- (non-B) cells (HIS 11, HIS 24, HIS 25) and/or with the cell surface of pre-B cells lacking Ig on their cell membrane (HIS 11, HIS 14, HIS 22, HIS 24, HIS 25). Thus only HIS 8 may recognize a rat Ig determinant: the staining pattern of lymphoid organs as well as the results of double immunofluorescence studies were suggestive of a rat light chain specificity. Indeed, HIS 8 was found to bind to both IgG or IgM coated sheep red blood cells (not shown).

HIS 24 antigen was found on a subset of Ig^+ B lymphocytes: it was absent on marginal zone cells of the spleen. Thus in spleen cell suspensions marginal zone cells may be specifically identified as Ig^+ HIS 24^- cells, allowing isolation of these cells. The low incidence of Ig^+ HIS 24^- cells in thoracic duct lymph reported here is in agreement with the apparent non-migratory nature of marginal zone cells (10).

In peripheral lymphoid organs, HIS 22 staining is similar to anti-δ chain staining. In bone marrow, however, one-third of pre-B cells carry HIS 22 on their cell surface, apparently not consistent with a δ chain specificity. Further biochemical characterization of the HIS 22 antigen will elucidate this point.

Since HIS 11, HIS 14, HIS 22, HIS 24 and HIS 25 recognize determinants that have not yet been described, they represent new and powerful tools to study both early and late B lymphocyte differentiation events in the rat.

ACKNOWLEDGEMENTS

The authors wish to thank P. van der Sijde for photography and A. Zandstra for typing the manuscript.

REFERENCES

1. I.F.C. McKenzie and H. Zola. Immunol. Today 4: 10 (1983).
2. S.V. Hunt. In: "Immunobiology of bone marrow transplantation", S. Thierfelder, H. Rodt and H.J. Kolb, eds), Springer Verlag, Berlin p. 193 (1980).
3. I.C.M. MacLennan, D. Gray, D.S. Kumararatne and H. Bazin. Immunol. Today 3: 305 (1982).
4. D. Opstelten, G.J. Deenen, T. Jaarsma, J. Rozing and S.V. Hunt. This volume.
5. W.L. Ford and S.V. Hunt. In: "Handbook of Experimental Immunology", D.M. Weir, ed. Blackwell, Oxford, p. 232 (1973).
6. G.J. Deenen, D. Opstelten and P. Nieuwenhuis. Cell Tiss. Res., in press (1984).
7. R.H. Kennett. In: "Monoclonal Antibodies", R.H. Kennett, T.J. Mc Kearn, eds., Plenum Press, New York, p. 365 (1980).
8. L. de Leij, S. Poppema, J. Klein-Nulend, J.G. ter Haar, E. Schwander and T.H. The. Eur. J. Cancer Clin. Oncol. 20: 123 (1984).
9. G.D. Johnson, R.S. Davidson, K.C. McNamee, G. Russell, D. Goodwin and E.J. Holborow. J. Immunol. Meth. 55: 231 (1982).
10. D. Gray, I.C.M. MacLennan, H. Bazin and M. Khan. Eur. J. Immunol. 12: 564 (1982).

EFFECT OF EARLY EMBRYONIC BURSECTOMY ON LYMPHOCYTE DIFFERENTIATION AND MATURATION

Donald L. Ewert, Carolyn Oates, and Chen-lo H. Chen

The Wistar Institute, Philadelphia, Pennsylvania 19104
Moredun Institute, Edinburgh, Scotland; and Tumor Institute, University of Alabama, Birmingham, Alabama

INTRODUCTION

The hypothesis that the avian bursa of Fabricius is essential for the development of B lymphocytes has been questioned by studies in which the primordium of the bursa was removed from the embryo between 54 and 60 hr of development. Such early embryonically bursectomized chickens have been reported to develop B cells as well as serum immunoglobulins (Ig) (1-7). However, some controversy exists over the ability of these early embryonically bursectomized chickens to produce a humoral immune response (1, 3, 4, 7). Also, because the frequencies of Ig^+ and Ia-like (B-L) antigen positive cells in lymphoid tissues of normal chickens given in these reports (8, 9) are at variance with past reported frequencies questions inevitably arose concerning the validity of the frequencies obtained in bursectomized chickens. To clarify previous findings of us (10) and others, we report here the results of an analysis of the ontogeny of B cell development, Ig expression, and antibody synthesis following removal of the bursal primordium using a modification of the embryonic bursectomy techniques.

MATERIALS AND METHODS

Bursectomy

Fertile outbred eggs were obtained from a Gold Kist Co., Leeds, AL. At approximately 60 hr of incubation [stage 17(11)], just prior to the closure of the amniotic membrane, the tail bud was burned off caudally to the leg buds using an opthalmic cautery needle. This procedure minimized bleeding. Other details of the procedure were as previously described (1, 2).

Immunofluorescence Analysis

The preparation of cell suspension from lymphoid tissues, the reaction of viable cells with fluorochrome conjugated antibody, the characterization of immunofluorescent reagents reactive with chicken IgG, IgM, and Ia-like (B-L) antigens, and the microscopic analysis of antibody reactive cells have been described elsewhere (8, 9).

Ig and Antibody Analysis

Relative levels of circulating Ig were determined by radial diffusion in gel (12) using affinity-purified monospecific antibodies to chicken IgG or IgM (9). The precipitin ring diameter was measured for each sample and the data represented as the percent of the ring diameter obtained with a standard adult serum. Antibody levels against sheep red blood cells (SRBC) and natural agglutinins to rabbit red blood cells (RRBC) were detected by a standard double dilution microtiter hemagglutination assay using 96-well V-bottom microtiter plates. 0.025 ml phosphate buffered saline, 0.025 ml of diluted serum and 0.025 of a 1% suspension of erythrocytes was added to each well and incubated for 1 hr at room temperature.

Two-dimension Gel Analysis

Serum gammaglobulins, precipitated from serum of 12-week-old chickens with 10% Na_2SO_4, were radiolabeled with ^{125}I (13), reacted with rabbit anti-chicken light chain antibody and adsorbed to *S. aureus*. After solubilization of the immune complex, the Ig light chains were isolated by SDS-PAGE under reducing conditions and analyzed by isoelectric focusing as previously described (14).

Stimulation of Lymphocytes

10^6 peripheral blood leukocytes (buffy coat cells) in 180 µl RPMI-1640 medium containing 5% fetal bovine serum and 20 µl of mitogen were added to each well of a round bottom 96-well microtiter plate. Mitogens were used at 1:4, 1:16, 1:64 of stock [concanavalin A (Con A) = 6 mg/ml, phytohemagglutinin (PHA) = as supplied by Difco, Detroit, MI, USA]. Cells were incubated in a humidified atmosphere containing 5% CO_2 for 2 days and pulsed for 18 hr with [^{3}H]thymidine (6.7 Ci/mmol) and the incorporated activity counted.

RESULTS

Completeness of Bursectomy

Each bursectomized chicken or embryo was examined, upon sacrifice, for gross or histological evidence of a bursa or lymphocyte accumulation in the caudal region of the intestinal tract. Successful bursectomy was found to correlate with the absence of a tail and feathers in the tail region. Also no extra bursal accumulations of lymphoid cells were evident in the distal region of the intestinal tract of bursectomized chickens. Data obtained from chickens showing evidence of partial bursal development were not included in this study.

B Cell Expression

The frequency of IgM^+ B cells in peripheral lymphoid tissues was determined by immunofluorescence analysis of cell suspensions using FITC conjugated goat anti-IgM (μ chain) antibody. Cells from spleen and bone marrow of bursectomized embryos obtained a few days before hatching were found to contain approximately the same frequency of B cells as those from the same tissues of age-matched unoperated embryos (Table 1). At three days after hatching the frequency of B cells in the lymphoid tissues of non-bursectomized chicks had increased markedly but no increase was detected in the bursectomized chickens (Table 1). To follow the ontogeny of B cells and circulating Ig and antibody, six embryonic bursectomized chickens and seven control chickens were analyzed

Table 1. Percent B Cells (Ig M^+) in Tissues of Bursectomized (Bx) and Control (Co) Chickens Before and After Hatching

	Before hatch (18 d)		After Hatch (3 d)		After Hatch 12 wk	
	Bx	Co	Bx	Co	Bx	Co
Spleen	.15/.08	.25/.15	<0.1/.03	5.3/4.1	0.5	24
Bone Marrow	.04/.03	.05/.04	<0.1/.01	0.2/0.3	1.2	2.3

Table 2. Percent Ia^+ and IgM^+ peripheral blood leukocytes

Age (wk)		3		6		9		12	
		μ^+	Ia^+	μ^+	Ia^+	μ^+	Ia^+	μ^+/γ^+	Ia^+
	481	<.5	1.0	0.4	2.0	0.6	.9	0.2/.01	1.0
	482	<0.1	0.7	0.02	0.7	-	-	-	-
Bx	483	<0.3	1.0	<.04	1.0	0.13	1.0	0.5/.08	2.5
	484	<0.5	0.5	.02	0.5	<.01	.7	.08/<.03	1.2
	485	<0.1	3.0	.04	1.0	.76	1.5	.68/.04	.88
	486	<0.1	3.0	.01	1.0	.36	2.0	0.3/0.2	0.9
Controls	--[a]	25.5	26.0	17.5	19	18	20	13/1.5	17

[a] Mean of three unoperated hatchmates.

during a 12-week period after hatching. Peripheral blood leukocytes (PBL) were analyzed at 3-week intervals for the frequency of cells expressing IgM and Ia-like antigens (Table 2). Although undetectable at 3 weeks post hatch, IgM^+ B cells were detected by 6 weeks of age, and continued to increase in frequency to 12 weeks, but the maximum level attained was well below (> 1/20) that of the normal chickens.

<u>Ig Levels and Light Chain Heterogeneity</u>

At 3 weeks of age, the level of IgM and IgG in serum of bursectomized chickens was either not detectable or below that of nonoperated age-matched chickens (Table 3). However, by 12 weeks, most operated chickens had levels of IgM and IgG comparable to that of controls. One exception was a chicken (No. 484) that evidenced a delayed development of IgM and IgG. Two-dimensional gel electrophoresis of serum Ig light chains was performed to indirectly assess the diversity of B cells present in bursectomized chickens as described above. The number and location of bands resolved by this method were similar in both operated and control birds except for bird 484 (Figure 1).

Table 3. Ig Levels

	IgM			IgG		
Age wks.	3	6	12	3	6	12
481	48[a]	71	85	39	63	99
482	34	Died	--	36	--	--
483	65	75	92	40	61	124
Bx 484	0	72	77	0	0	64
485	54	95	110	28	60	98
486	54	58	66	33	80	114
Controls	74 8.1[b]	96.4 11	79 8.4	53 6.2	66.9 14	110.3 6.8

[a]Percent of preciptin ring (diameter) obtained by cont. serum.
[b]Mean and standard deviation.

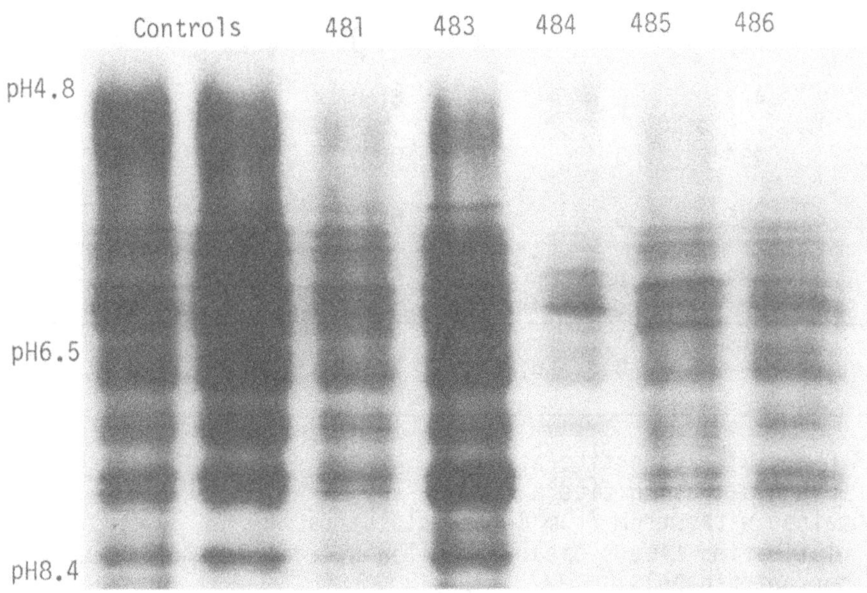

Fig. 1. Isoelectric-focusing analysis of Ig light chains.

Antibody Synthesis

To determine if B cells present in the embryonically bursectomized birds were capable of responding to antigenic stimulation and antibody synthesis the chickens were given three consecutive i.v. injections of SRBC at 30, 51 and 74 days of age. Serum, drawn 10 days after each injection was assayed for antibody by hemagglutination. As shown in Table 4, although antibody levels increased more slowly in operated birds during the course of the experiment and, except for one bird which did not produce detectable anti-SRBC antibody, by 12 weeks most birds produced levels of antibody comparable to that of the controls. Spontaneously produced antibodies to rabbit erythrocytes were much lower in the bursectomized chickens. Also some, but not all, of the antibody produced by the bursectomized chickens was sensitive to 2 mercaptoethanol treatment suggesting both IgM and IgG antibodies were present.

Table 4. Antibody Production

	No. Bird	Pre-immune	Post-immune (SRBC)[a] 1°	2°	3°	RRBC[b]
	481	0	0	4(1)[c]	11	2
	483	0	4	6(5)	10	3
Bx	484	0	0	0(0)	0	0
	485	0	2	5(4)	8	2
	486	0	0	6(2)	5	2
Control	---	0	6.8±0.7[d]	7.5±2	8.8±1.5	8.4±1.6

[a] SRBC agglutination titers ($\log_2$)
[b] RRBC agglutination titers ($\log_2$) obtained with serum from 8-week old birds.
[c] Agglutination titers obtained following treatment of serum with 2-mercaptoethanol.
[d] Mean and standard deviation.

In Vitro Stimulation of Lymphocytes

PBL from 12-week-old chickens were cultured with mitogens, PHA, Con A, or SRBC to determine the effect, if any, of bursectomy on the response of lymphocytes to mitogens and antigens. As shown in Figure 2, the response of lymphocytes from the embryonically bursectomized group to mitogens and antigens after antigenic priming was as vigorous as that of the non-bursectomized chicken lymphocytes.

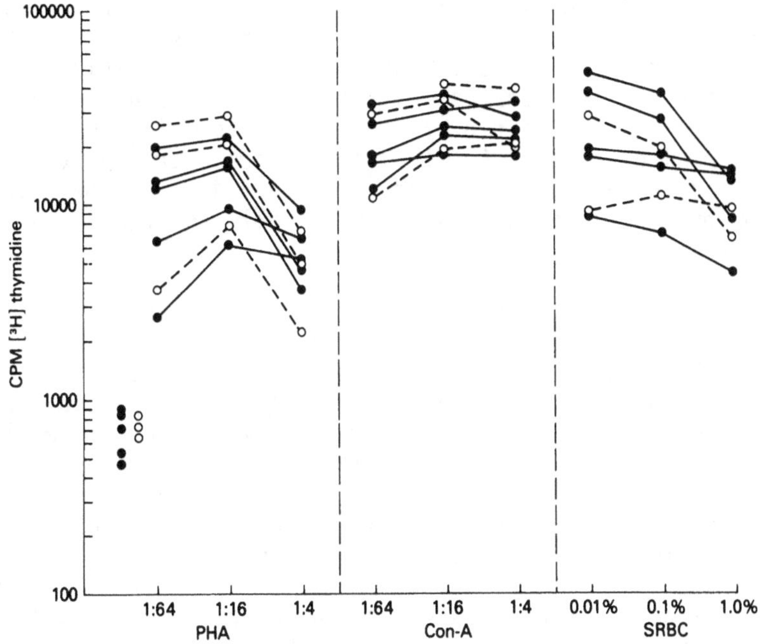

Fig. 2. In vitro responses of PBL from 12-week-old embryonically bursectomized (•) and control (o) chickens stimulated with mitogens PHA or Con A or SRBC.

DISCUSSION

The data presented here indicates, as previously reported, that chickens operated as 60-hr embryos to remove the tailbud, and lacking any bursa, develop B cells that differentiate, as evidenced by the expression of IgG, and mature to secrete IgM and IgG. Furthermore, destruction of the tailbud region does not appear to alter the development of lymphocytes capable of responding to PHA or Con A mitogens, which primarily affect T-lymphocytes or to particulate antigens. In contrast to previous studies (2, 6), we found that the frequency of IgM^+ B cells in peripheral blood never exceeded 1/20th of the normal level of 20-25%. Since previous work has shown that >90% of the peripheral B cells express IgM (9), this frequency is most likely a close estimate of the total number of B cells present. Also, the frequency of IgG^+ PBL was reduced from the 1-2% found in normal chickens to, at most, 0.2% by the bursectomy. The number of Ia^+ PBL is consistent with the frequency of Ig^-Ia^+ cells in normal chickens (8) suggesting that the development of this population is not affected by bursectomy. These findings are in contrast to the previous report of significantly higher frequencies of both IgM^+, IgG^+ and Ia^+ cells in tissues of both normal and bursectomized chickens (6). Our findings therefore underscore the ability of a limited number of B cells to support synthesis of near normal levels of circulating Ig. Furthermore, we have shown that the serum Ig of bursectomized chickens exhibits a comparable degree of heterogeneity of light chains, as evidenced by the number and similarity of bands obtained by isoelectric focusing analysis, as that of normal chickens. A shift in the isoelectric pattern of light chains in one bursectomized bird, no. 484, may be related to the observed delayed expression of B cells and Igs in this chicken relative to other bursectomized chickens.

Antibody reactive with SRBC at levels comparable to that of control chickens was detected in all but one bursectomized chicken. Although detection of an anti-SRBC response in embryonically bursectomized chickens has previously been reported, confirmation of the effectiveness of the bursectomy was not made (5), leaving open the possibility that the antibody response was a consequence of an incomplete bursectomy. However, in other studies in which complete bursectomy was confirmed, antibody responses to several soluble and particulate antigens were not detected (7). In this study, the failure of one bursectomized chicken, no. 484, to exhibit a detectable anti-SRBC response may be related to the observed lower frequency of B cells and the delayed development of B cells detected in this chicken relative to other bursectomized chickens. Taken together, these data suggest that the B lymphocytes in early embryonically bursectomized chickens are capable of responding to appropriate antigenic stimulation but that the B cell numbers and diversity may be severely restricted relative to the repertoire of normal chickens.

The central question of whether the B cells and B cell functions detected in these bursectomized chickens developed in the complete absence of a bursa of Fabricius is arguable. Failure to detect a bursa at the end of the 12-week experiment, when normal bursal involution is beginning, suggests but does not prove that none was present, since complete involution of a small remnant may have reached completion by 12 weeks of age. However, the same argument could not be made for the 18d embryos or 3d chicks that were examined and found to lack a bursa, although humoral immune function could not be assessed. An alternative possibility is that normally the induction of an environment within an embryo capable of supporting stem cell differentiation to B cells is restricted to the bursa and occur for a limited period of development prior to the receptive period of stem cell homing to the bursa (8-14d). If the bursal primordium is removed prior to this period of induction another microenvironment may be potentiated that can support B cell differentiation. Such a model would explain both the presence of B cells and Ig in early embryonically bursectomized chickens and the complete agammaglobulinemia obtained by late embryonic (17d) bursectomy (15).

It should be noted that this work was in progress prior to recent publication of similar work involving early embryonic bursectomy (5, 6, 7).

This work was supported in part by grant IM-320 from The American Cancer Society, grant PCM-8215968 from The National Science Foundation and by a Yaniagiwa-Yoshida Memorial Institute Cancer Study grant awarded by The International Union Against Cancer.

REFERENCES

1. R.C. Fitzsimmons, E.M.F. Garrod, and I. Garnett. Cell Immunol. 9:377 (1973).
2. B.D. Jankovic, Z. Knezevic, K. Isakovic, K. Mitrovic, B.M. Markovic, and M. Rajcevic. Eur. J. Immunol. 5:656 (1975).
3. B.D. Jankovic, K. Iskovic, B.M. Markovic, M. Rajcevic, and Z. Knecexic. Exp. Hematol. 4:246 (1976).
4. B.D. Jankovic, K. Iskovic, B.M. Markovic, and M. Rajcevic. Immunology 32:689 (1977).
5. K. Granfors, C. Martin, O. Lassila, R. Suvitaival, A. Toivanen, and P. Toivanen. Clin. Immunol. Immunopath. 23:459 (1982).
6. S. Jalkanen, K. Granfors, M. Jalkanen, and P. Toivanen. J. Immunol. 130:2038 (1983).
7. S. Jalkanen, K. Granfors, M. Jalkanen, and P. Toivanen. Cell. Immunol. 80:363 (1983).
8. D.L. Ewert, and M.D. Cooper. Immunogenetics 7:521 (1978).
9. C.-L. Chen, J.E. Lehmeyer, and M.D. Cooper. J. Immunol. 129:2580 (1982).
10. D.L. Ewert, and M.D. Cooper. Fed. Proc. 38:1367 (1979).
11. V. Hamburger, and H.L. Hamilton. J. Morphol. 88:49 (1951).
12. J.L. Fahey, and E.M. McKelvey. J. Immunol. 94:84 (1965).
13. P.J. Fraker, and J.C. Speck. Biochem. Biophys. Res. Comm. 80:849 (1978).
14. J.M. LeJeune, D.E. Briles, and J.F. Kearney. J. Immunol. 129:673 (1982).
15. M.D. Cooper, W.A. Cain, P.J. VanAlten, and R.A. Good. Int. Arch. Allergy 35:242 (1969).

EVIDENCE FOR THE EXISTENCE OF TWO DISTINCT TYPES OF PEYER'S PATCHES IN SHEEP

John Reynolds[1,3], Reinhard Pabst[2] and Gerard Bordmann[3]

[1]Department of Medical Physiology, University of Calgary Alberta, Canada. [2]Center of Anatomy, Medical School of Hannover, West Germany. [3]Basel Institute for Immunology Switzerland

INTRODUCTION

Johann Peyer included a description of the Peyer's patches (PP) in sheep as part of his monograph published in 1677 which provided the first detailed account of the "intestinal glands" that are found in all higher vertebrates. The inclusion of sheep in the list of species in which he noted PP was probably as a consequence of one of his other studies on the mechanisms of rumination. The PP in sheep have received attention periodically since that time. In 1928 Carlens published an account of the PP in domestic animals which included their distribution, ontogeny and age-related changes. We have recently reexamined these topics in the sheep and confirmed Carlens' observations which showed that sheep PP have a number of important features (Reynolds and Morris, 1983). For example, the bulk of the PP tissue in lambs, as in other species, is concentrated in the terminal ileum. In this part of the sheep intestine, however, the PP do not form small discrete areas of lymphoid tissue but instead there is a single PP that almost completely surrounds the circumference of the gut and forms a continuous band that extends throughout the last 1.5-2.0 meters of small intestine (Carlens, 1928; Reynolds and Morris, 1983); about 90% of all the PP tissue in lambs is in the terminal ileum (Reynolds and Morris, 1983). The fact that PP are so heavily localized in this well defined region of the gut in sheep has facilitated a number of studies in which various surgical manipulations were used to examine questions of PP development and function (Reynolds and Morris, 1983, 1984; Reynolds and Pabst, 1984; Pabst and Reynolds, this volume). Most of our studies have concentrated on the ileocecal PP (IPP) but where possible we also studied some of the 30-40 smaller PP (each 2-12cm long) that are scattered throughout the proximal ileum and the jejunum (JPP). These experiments drew our attention to the fact that the properties of the JPP differed in a number of significant ways from those of the IPP. This raises the possibility that sheep might have two distinct types of PP. The differences between the PP in the ileum and those in the rest of the small intestine are outlined below.

MICROANATOMY

The two types of PP can be easily distinguished by their microanatomy (Fig. 1). The histological appearance of the IPP is dominated by the lymphoid follicles that are tightly packed together and separated only at their sides by the connective tissue sheath that surrounds the greater part of each follicle. The follicles are elongated structures that extend for up to 3mm in the submucosa and then narrow at the apex before merging with the dome region via a gap in the muscularis mucosa. The area bounded by the sides of the adjacent follicles where they narrow and the muscularis mucosa forms a small triangular region of interfollicular tissue. The dome regions in the mucosa are relatively small and often appear to be submerged under the surrounding intestinal villi.

The follicles in the JPP are more pear-shaped and in general smaller than those in the IPP. They are also surrounded at the base and sides by a connective tissue capsule but rarely does one follicle abut the adjacent follicle because of the large tracts of interfollicular tissue that separate them. The interfollicular tissue is a major feature that usually accounts for half or more of the volume of the JPP. The dome region is also substantial and can, in some cases, be as large as the follicle itself. The encapsulation of the follicle is also less complete than in the IPP due to a region below the muscularis mucosa where the follicle and interfollicular area are separated only by a band of smaller lymphocytes that form a corona similar to that described in rabbit PP (Waksman, Ozer and Blythman, 1973).

The rabbit also has 2 distinct types of gut-associated lymphoid tissues based on histological appearance (Waksman, Ozer and Blythman, 1973) which can be compared directly with the PP in sheep. The appendix is similar to the IPP while the PP in the rabbit resembles the JPP in sheep.

If the relative proportions of the compartments are overlooked then an examination of the cellular composition of the follicle, the interfollicular tissue and the dome region shows that, at least at the microscopic level, there are no apparent differences between the IPP and the JPP. In addition cells enter mitosis at the same rate in the two types of PP (Reynolds, unpublished result).

Nevertheless, each of the 3 main components of a PP have quite distinct characteristics; the follicle is a site of lymphopoiesis, the interfollicular area is a site of lymphocyte traffic and the dome is a site of antigen entry. It would therefore be surprising if the IPP and JPP, with such substantial differences in the proportions of these compartments, did not also have different properties.

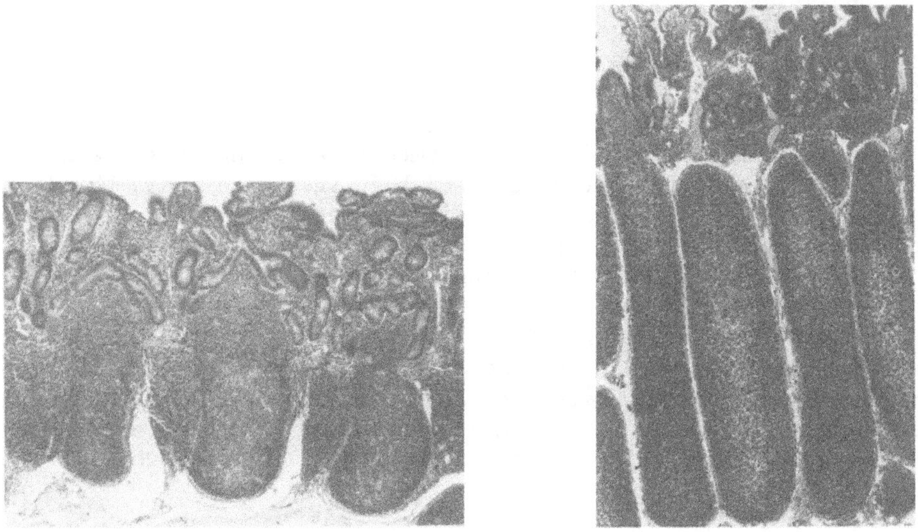

Fig. 1. The microanatomy of JPP (left) and IPP (right) in lambs at 8 weeks of age. Note the differences in the relative proportions of the follicle, dome and interfollicular tissue. Hematoxylin-eosin (x 40).

THE NATURE OF THE PP LYMPHOCYTES

Cell suspensions of the 2 types of PP were prepared from 30 lambs at 8-12 weeks of age and the cells were analyzed by the technique of indirect immunofluorescence. The primary antibodies were a monoclonal mouse anti-sheep T cell (Miyasaka et al., 1983) and a rabbit $F(ab')_2$ anti-sheep immunoglobulin (kindly supplied by Dr. J.V. Peppard); the developing antibodies were FITC-conjugated $F(ab')_2$ fragments (Cappel, West Chester, PA, USA).

There are substantial differences between the IPP and the JPP with respect to the relative proportions of T and B lymphocytes. B cells account for a greater proportion of the cells in the IPP (73±11.8%) than in the JPP (59±12%), whereas T cells amount to only a tiny fraction of the cells in the IPP (1.6±1.0%) but a substantial fraction of the JPP cells (32±7.3%). These results can be attributed to the microanatomical differences outlined above. The application of the immunoperoxidase technique to examine cryostat sections of sheep PP has shown that, as in the other species that have been examined (Butcher et al., 1982), the PP follicles are filled with B lymphocytes whereas the interfollicular tissue contains mostly T lymphocytes (results not shown). The paucity of the interfollicular tissue in the IPP and its relative abundance in the JPP correlates well with the observed differences in the composition of the lymphocyte populations.

LYMPHOCYTE MIGRATION TO PP

The magnitude of the influx of lymphocytes is another feature which distinguishes the IPP from the JPP in sheep and again this is almost certainly related to the differences in PP microanatomy. Lymphoid cells from a variety of sources and a number of species have been radiolabeled and following i.v. injection their fate was determined by autoradiography. It became apparent from these studies that lymphocyte extravasation in PP is largely restricted to the interfollicular tissue and the dome region (Perey and Milne, 1975; Reynolds, 1976; Fossum, Smith, and Ford, 1983). Cell migration to the follicle is negligible by comparison and we were unable to find any autoradiographic evidence that it occurs at all in sheep (Reynolds, 1976).

Two different techniques were used to quantitate the magnitude of the cell migration to IPP and JPP in 6-10 week old lambs. In both experiments the cell populations chosen for study were composed entirely of migratory cells i.e., the cells had decided to migrate of their own accord. The lymphocytes being carried from the intestine (including the PP) in the lymph were the focus of the first study (Reynolds, 1976). They were collected by cannulating the intestinal lymphatic using the technique described by Lascelles and Morris (1961). The lymphocytes were labeled _in vitro_ with ^{3}H-cytidine, reinjected i.v. and their localization determined 36h later. The results are given in Table 1. The radioactivity associated with the JPP was about 4-fold greater than in the IPP. A comparison of the radioactivity/g in different tissues showed that the IPP was comparable to the non-lymphoid tissues such as the kidney and liver, whereas the JPP were of the same order of magnitude as the lymph nodes. These differences were confirmed by autoradiography. In the second study the migration of cells from one PP to another was followed. An extracorporeal perfusion system was used to restrict labeling to a region of the ileum that included about 90% of the length of the IPP. In this way all the cells in the perfused region were labeled with fluorescein isothiocyanate (FITC) and their migration to the non-perfused tissues, including the JPP and part of the IPP, could then be followed. Details of the technique used are given elsewhere (Reynolds and Pabst, 1984; Pabst and Reynolds, this volume). Cell suspensions of the tissues were prepared 24h after the perfusion and the concentration of FITC-positive cells was determined. The results are given in Table 1. As in the first study, a similar number of labeled cells had localized in the JPP as in the lymph nodes, while the non-perfused region of the IPP contained about 4-fold fewer cells than these other sites. The thymus was the only other major lymphoid tissue that contained a low concentration of emigrant PP cells.

Thus, the IPP, unlike the JPP, is to a large degree outside the circuit taken by the migratory lymphocytes during their wanderings around the lymphoid tissues. The failure of cells to enter the follicle and the paucity of interfollicular tissue in the IPP are certainly the major reason for this.

Table 1. Cell Migration to PP in Lambs at 6-10 Weeks of Age.

Tissue	dpm/g tissue[a]	FITC-pos cells/10^6 cells[b]
IPP	6,000	580±430[c]
JPP	25,000	2,160±643
Lymph nodes	25,000-48,000	1,850±750
Kidney, Lung, Liver	4,000-10,000	not tested
Thymus	4,000	70±10

a. ^{3}H-cytidine labeled intestinal lymph cells reinjected i.v. and their localization determined 36h later with the aid of a Packard TriCarb Oxidiser to process the tissue samples.
b. Most of the IPP selectively labeled with FITC by extracorporeal perfusion and the localization of the cells that emigrated from this site was determined 24h later by quantitating fluorescent cells in cell suspensions.
c. Counts are from a non-perfused length of IPP.

THE ONTOGENY OF PP

The most compelling evidence to suggest that the IPP and the JPP may be different comes from an examination of PP development throughout the life of the animal. Sheep have a gestation period of about 150 days and the rudimentary PP are first detected about 70 days of gestation. The sites are easily recognizable as accumulations of small lymphoid cells which, in some cases, are organized into follicles (Fig. 2). At this stage of development the PP cannot be seen with the naked eye and because of the random location of the JPP they can only be detected by a systematic microscopic examination. The position of the IPP, however, can always be located with certainty because the terminal ileum is destined to become packed with PP tissue. An examination of this region of the gut in fetal lambs of different ages showed that the development of the primitive IPP does not occur until around the 110th day of gestation. By this time the JPP have already reached an advanced stage of development which includes the establishment of follicular lymphopoiesis (Reynolds and Morris, 1983). This difference in ontogeny, and the differences in postnatal development mentioned in the next section, raises the possibility that the factors that influence the IPP differ in some way from those that influence the JPP.

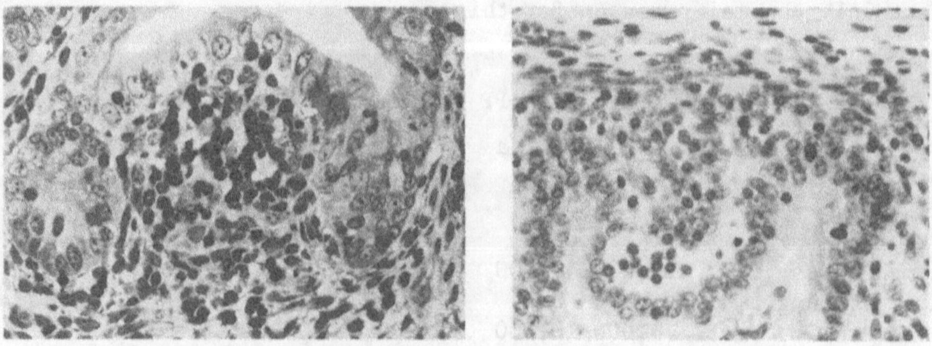

Fig. 2. The development of PP in the jejunum at 75 days of gestation (left) and in the ileum at 110 days of gestation (right). The IPP showed no signs of developing at 75 days and it did not start to appear until around 110 days. The JPP at this stage contained well developed follicles that were lymphopoietic. Methacrylate, Nochts azure-eosin stain (x300).

THE POSTNATAL DEVELOPMENT OF PP

PP reach their maximum size in lambs about 8 weeks after birth. After about 12 weeks of age the follicles in the IPP become noticeably smaller. The follicles give the appearance of contracting towards the dome regions, the cellularity of the follicles decreases and by about 18 months of age all that remains of the IPP are a few wizened follicles and small amounts of residual interfollicular tissue and dome regions (Fig. 3; Reynolds and Morris, 1983). While the terminal ileum loses all macroscopic evidence of the 2 meter long PP the JPP remain visible to the naked eye even in old sheep. The microscopic appearance of the JPP also remains little altered throughout the life of the animal and the follicles continue to be active sites of lymphocyte production long after the follicles have disappeared from the IPP (Fig. 3; Reynolds and Morris, 1983). The older studies of Carlens (1928) showed that basically similar trends occur in cattle and sheep.

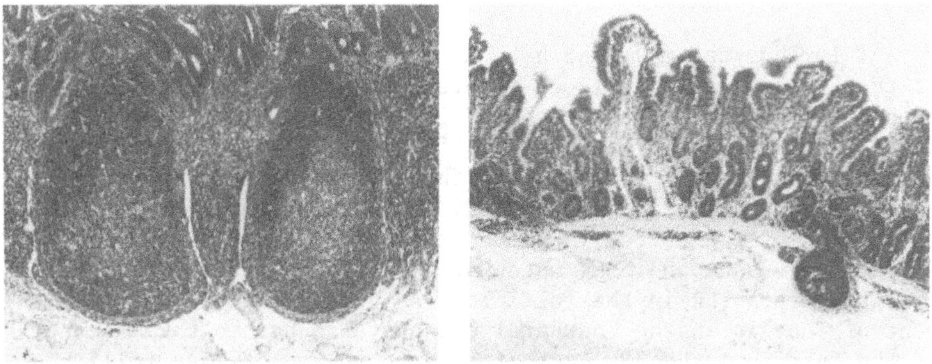

Fig. 3. The microanatomy of the JPP (left) and IPP (right) in sheep at 18 months of age. The structure of the JPP remains largely unaltered but the follicles have almost completely disappeared from the IPP; some domes and interfollicular tissue remain. Hematoxylin-eosin (x 40).

CONCLUSION

The smaller discrete PP in the jejunum and proximal ileum of sheep contain a total of about 10,000 follicles, start to develop about mid-gestation and remain lymphopoietic throughout life. The continuous 1.5-2 meter long PP in the terminal ileum contains about 90% of the total PP tissue in young lambs and about 100,000 follicles but its presence, and hence its function, is only restricted to a short period in the life of the animal (Reynolds and Morris, 1983). The IPP, because of its size, has been easier to study than the JPP in sheep. The IPP is known to be the main source of B lymphocytes in lambs (Gerber, 1979; Reynolds, unpublished) since removal of the terminal ileum from perinatal lambs causes the animals to develop severe B cell deficiencies; removal of this region from older animals has a much less dramatic influence on B cell numbers (Gerber, 1979). Of the cells that leave the IPP many are small B cells that migrate to all lymphoid tissues (Reynolds and Pabst, unpublished results), others are destined to become plasma cells that can be found in the lamina propria of the gut within 24h of their production in the PP (Pabst and Reynolds, this volume). As yet not enough is known about the JPP to determine whether they might have a different role. It is also possible that both types of PP might have similar functions but that some factor(s) that is (are) crucial to PP function might be limiting for the IPP but not for the JPP. The IPP and the JPP would certainly experience different influences from the local environment such as differences in gut contents and the absorption of metabolites e.g., bile salts and vitamin B_{12} are absorbed only across the ileal mucosa. There is currently no evidence either for or against the involvement of factors such as these.

T lymphocytes may also have an influence on the life-history of the PP in sheep. It is of interest to note that the IPP involutes around the same time as the thymus is involuting (Reynolds and Morris, 1983). Either both organs involute in response to some external (hormonal?) influence or thymic involution may itself initiate the involution of the IPP. T cells account for a tiny proportion of the IPP cells and most are restricted to the interfollicular tissue and dome which are separated from the site of B cell proliferation in the follicle. Any discussion of the involvement of T cells in the postnatal changes also has to take account of the fact that the JPP, with its large volume of interfollicular tissue and dome, large numbers of T cells, and extensive cell migration, is able to maintain follicular lymphopoiesis in spite of thymic involution. It could be argued that if the follicular lymphopoiesis is somehow dependent on the presence of T cells, then the changes in the quality and/or quantity of T cells following thymic involution might not influence the JPP, with its typically large content of T cells, whereas the IPP in young lambs with its relative paucity of T cells may not have the same capacity to survive. There is currently no satisfactory interpretation of the differences between the two types of PP in sheep and the reasons will remain poorly understood until more is known about the role of the T cell in the PP and about the nature of the interactions that occur between the different compartments of the PP.

ACKNOWLEDGEMENTS

The Basel Institute for Immunology was founded and is supported by F. Hoffman-La Roche AG, Basel. JR is currently supported by the Medical Research Council of Canada and the Alberta Heritage Foundation for Medical Research. RP is supported by grants from the Deutsche Forschungsgemeinshaft.

REFERENCES

Butcher, E.C., Rouse, R.V., Coffman, R.L., Nottenburg, C.N., Hardy, R.Y., and Weissman, I.L., 1982, Surface phenotype of Peyer's patch germinal center cells: implication for the role of germinal centers in B cell differentiation. J. Immunol 129:2698.

Carlens, O., 1928, Studien ueber das lymphatische Gewebe des Darmkanals bei einigen Haustieren, mit besonderer Beruecksichtigung der embryonalen Entwicklung, der Mengenverhaeltnisse und der Altersinvolution dieses Gewebes im Duenndarm des Rindes., Z. Anat. Entwicklungsgesch., 86:393.

Fossum, S., Smith, M.E., and Ford, W.L., 1983, The migration of lymphocytes across specialized vascular endothelium VII. The migration of T and B lymphocytes from the blood of the athymic, nude rat. Scand. J. Immunol., 17:539.

Gerber, H.A., 1979, "Functional Studies on Gut-Associated Lymphoid Tissues". Ph.D. thesis, Australian National University.

Lascelles, A.K., and Morris, B., 1961, Surgical techniques for the collection of lymph from unanaesthetized sheep. Q. J. Exp. Physiol., 46:199.

Miyasaka, M., Heron, I., Dudler, L., Cahill, R.N.P., Forni, L., Knaak, T., and Trnka, Z., 1983, Studies of the differentiation of T lymphocytes in sheep. I. Recognition of a sheep T-lymphocyte differentiation antigen by a monoclonal antibody T-80, Immunology, 49:545.

Perey, D.Y.E., and Milne, R.W., 1975, Rabbit gut-associated lymphoid tissue. Major pathway for thoracic duct lymphocyte circulation. Lab Invest., 18:15.

Peyer, J.C., 1677, "Exercitatio anat. de glandulis intestinorum earumque usu et affectionibus", Schaffhausen.

Reynolds, J.D., 1976, "The Development and Physiology of the Gut-Associated Lymphoid Tissues in Lambs" Ph.D. thesis, Australian National University.

Reynolds, J.D. and Morris, B., 1983, The evolution and involution of Peyer's patches in fetal and postnatal sheep. Eur, J. Immunol., 13:627.

Reynolds, J.D., and Morris, B., 1984, The effect of antigen on the development of Peyer's patches in sheep. Eur. J. Immunol., 14:1.

Reynolds, J.D., and Pabst, R., 1984, The emigration of lymphocytes from the Peyer's patches in sheep. Eur. J. Immunol., 14:7.

Waksman, B.H., Ozer, H., and Blythman, H.E., 1973, Appendix and γM antibody formation. VI. The functional anatomy of the rabbit appendix. Lab. Invest., 28:614.

EVIDENCE OF DIFFERENCES BETWEEN PEYER'S PATCHES AND GERMINAL CENTERS

John D. Reynolds

Department of Medical Physiology, The University of Calgary, 3330 Hospital Drive NW, Calgary, Alberta Canada

INTRODUCTION

Ever since the first description of Peyer's patches (PP), which was probably by Marco Severino in his treatise on comparative anatomy published in 1645 (cited by Schmidt, 1959), there have been changing views on the function of these lymphoid tissues found along the intestinal tract. The first detailed study by Peyer (1677) led to the concept, which survived for almost 2 centuries, that PP were involved in mucous secretion. The attention of the microscopists in the second half of the 19th century, however, correctly identified PP as sites where lymphocytes are being produced. In 1851 Bruecke described the "milky white area" which was the follicle filled with lymphocytes. In 1862 His described what he called "vacuoles" in the lymph nodes which appeared to be similar to the follicles in the PP. Then in 1885 Flemming noted that these areas in the lymph nodes and the PP were both sites of high mitotic activity. He coined the term "germinal center" (GC) and correctly suggested that they are breeding sites of lymphocytes. Flemming also noted that GC are only temporary structures in lymph nodes, they arise, operate for varying periods of time and then disappear. Hellman (1943) subsequently observed that the GC could be made to appear in lymph nodes by injecting various toxins and so the idea arose that GC develop in response to antigenic stimulation.

The PP follicles are still generally considered to be GC in the wall of the gut and are often referred to as GC. They do indeed share features in common with GC; both are histologically very similar and are composed predominantly of B lymphocytes that have a high rate of proliferation. However, the result of studies in sheep have drawn attention to some fundamental differences between the follicles in PP and the GC found in lymph nodes and the spleen. These differences relate to the role played by antigen in the initiation of the lymphocyte proliferation. The association between antigenic stimulation and the initiation of GC formation has received ample verification since the original observations by Hellman, but the role played by antigen in the genesis of the PP follicles is less certain and this was the topic of the studies outlined here.

PP BECOME MORPHOLOGICALLY MATURE IN FETAL SHEEP IN THE ABSENCE OF ANTIGEN

The ontogeny of the lymphoid apparatus has now been examined in over 150 fetal lambs of different gestational ages. Under normal circumstances the fetal lamb is totally shielded from the antigens in the external environment (Brambell, 1970) and so the development of its lymphoid tissues during this phase is under endogenous control. Under these conditions all the lymphoid organs develop but there are big variations in the degree of maturation reached by the individual organs. The lymph nodes contain a narrow band of cortex and a sparsely organized medulla while in the spleen the periarterial lymphocytic sheath develops (Silverstein and Prendergast, 1971; Cole and Morris, 1973; Reynolds, 1976). These lymphoid organs appear unreactive, there are no plasma cells, few lymphoid blast cells and, most importantly, GC are never found. The conditions that fail to support the establishment of GC are nevertheless quite appropriate for the development of the follicles in PP (Reynolds and Morris, 1983). Rudimentary PP first appear in fetal lambs about half-way through the 150-day period of gestation and by 100 days of gestation the follicles have formed and are filled with proliferating lymphocytes. PP appear histologically mature during the last 20 days before birth (Fig. 1). It is also of interest to note that the thymus is the only other lymphoid organ that reaches a comparable level of maturity during intra-uterine life.

LYMPHOCYTES PROLIFERATE IN PP FOLLICLES AT THE SAME RATE BEFORE AND AFTER BIRTH

The rate at which lymphocytes are proliferating in the PP follicles has been examined to determine whether there is any change as a result of the transition from the prenatal environment which lacks immunogen to the postnatal one which is crawling with immunogen. For this purpose a stathmokinetic analysis of the PP was done by injecting vincristine sulphate ("Oncovin", Lily, Giessen, FRG) intravenously into fetuses and lambs and determining what proportion of the lymphocyte population had been arrested in metaphase at hourly intervals up to 4 hours after the injection.

An analysis of the data in Table 1 has shown that no distinction can be made between the rate at which the follicular lymphocytes are proliferating in fetuses and in lambs after birth. For each of the experiments (4 fetuses at about 1 week before birth and 2 lambs about 7 weeks after birth) the proportion of the follicular cells in metaphase increased in a linear fashion for the first 3 hours after the injection of vincristine. It was therefore possible to determine from the slope of the line the rate at which the cells entered mitosis; for the PP in fetuses it was found to be 4.8% per h and for the PP in postnatal lambs it was 5.5% per h. These values were not significantly different. Thus the PP follicles, that are often referred to as GC, contain lymphocytes that are entering mitosis at the same rate in the fetus, when no GC could be found in the lymph nodes or the spleen, as in the PP of lambs after birth when GC abound in these other lymphoid tissues. An example of a follicle from a PP taken from a fetus 3 hours after the vincristine injection is shown in Figure 2 to help emphasize the extent of the proliferation.

PP CONTAIN A SIMILAR NUMBER OF FOLLICLES BEFORE AND AFTER BIRTH

The number of GC in a lymphoid tissue provides a good indication of the degree of antigenic stimulation being experienced by that tissue (Cottier et al., 1980). The mesenteric lymph nodes (MLN) and the prescapular lymph nodes (PLN) were also examined in lambs in the perinatal period so that their development could be compared with that of the PP. In the case of the MLN the transformation occurred particularly rapidly after birth. Small foci of proliferating lymphocytes appeared in the cortex by 24 hours after birth and typical GC could be observed by the 10th day of postnatal life; large numbers of GC subsequently developed. At the time of birth the MLN weighed about 1.7g, which was a little less than the PLN at 2.3g. The relative sizes of these lymphoid organs changed dramatically during the first month after birth. The PLN grew at about half the rate of total body growth (to 3.7g) while the MLN over the same period grew at about 5 times the rate of the total body growth (to 20g). This rapid rate of growth of the MLN after birth clearly reflects the extent of the antigenic challenge originating from the gut. We therefore wanted to know whether the number of follicles in the PP of sheep alters in response to the establishment of this massive antigenic load in the gut. The small intestines from lambs ranging in age from 2 weeks before birth to 9 weeks after birth was examined and all the PP tissue was excised and weighed (Reynolds and Morris, 1983). A number of representative examples of PP from the jejunum and a number of weighed strips of the 1-2 meter long PP in the ileum were placed in a solution of 1% acetic acid for about 1 hour to help make the follicles more prominent. These pieces were then placed between 2 glass slides and the number of follicles was counted

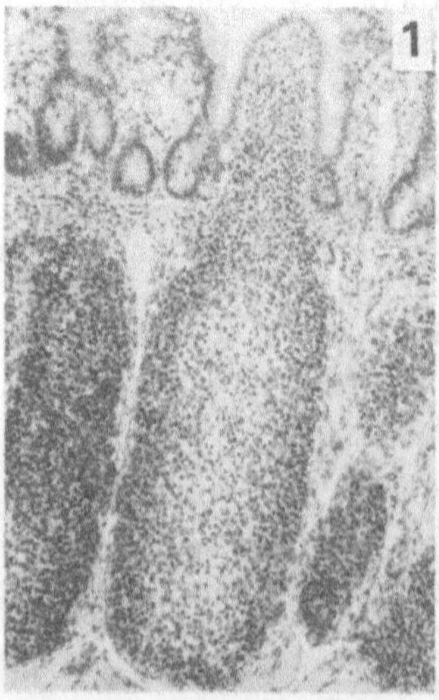

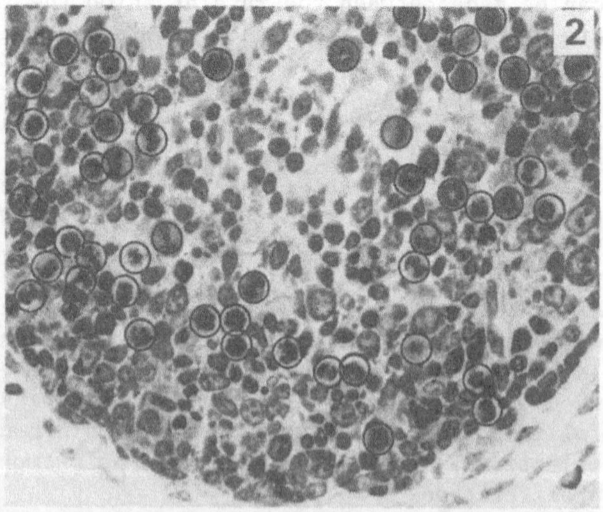

Fig. 1. A PP from the terminal ileum of a fetal lamb at 135 days of gestation. All the features of a mature PP are present. Paraffin section stained with hematoxylin-eosin (x 40).

Fig. 2. A PP follicle from the jejunum of a fetal lamb 3h after the i.v. injection of vincristine. The cells which are circled had entered metaphase during the 3h interval; they comprise about 15% of the total population of cells in the follicle (x480). Additional details are given in Table 1.

Table 1. Stathmokinetic Analysis of Lymphocyte Proliferation in the PP of Fetuses (1 week before birth) and Lambs (8 weeks after birth).

	Percentage of Follicle Cells in Metaphase[b]				
Hour (before or after vincristine)[a]	0	1	2	3	4
Fetuses[c]	1.5 ±0.24	6.6 ±0.21	12.0 ±0.48	17.2 ±1.58	16.1 ±0.58
Lambs[d]	1.7 2.4	5.2 7.8	9.6 14.6	16.2 21.9	16.2 19.7

a. vincristine injected i.v. into anesthetized animals at a dose of 0.4mg/kg body weight.
b. metaphase index determined in the follicle cortex in 1μ glycol-methacrylate sections stained with Nocht's azure-eosin.
c. the values are given as mean±standard deviation.
d. the values are given for each animal.

with the aid of a dissecting microscope. The total number of follicles in all the PP of the animal could then be estimated. Two weeks before birth or at the time of birth the PP contained a total number of 116,000 and 134,000 follicles respectively, whereas in 4 lambs from 3-9 weeks of age there were between 83,000 and 145,000 follicles (Reynolds and Morris, 1983). Thus there was no evidence to suggest that the influx of antigen at birth caused a change in either the histological appearance of the follicles or in the number of follicles.

The follicles in the PP of sheep are also unlike GC in that the follicles are almost completely surrounded by a connective tissue capsule which forms a clear boundary with the surrounding tissue (Fig. 2). PP follicles lack the transient nature of the GC. A follicle develops during fetal life, enlarges to its greatest size about 8 weeks after birth and then maintains its structure for at least another 13 months in the case of the PP in the ileum and for much longer for the PP in the more proximal parts of the intestine (see Reynolds, Pabst and Bordmann, this volume). When the PP are examined in sheep of any particular age all the follicles give the appearance of being at the same stage of development. In lambs from birth to 2 months of age for example there is no evidence of follicles that are just being formed or of others that are in the process of involution. The productive life-span of a GC as a source of lymphocytes may be of the order of 1 month (see Nieuwenhuis et al., 1981).

ANTIGEN INJECTION INTO FETAL LAMBS INITIATES GC FORMATION BUT FAILS TO ALTER THE APPEARANCE OF THE DEVELOPING PP FOLLICLES

The fact that highly lymphopoietic follicles develop in the PP of fetal lambs (and some other species) has been interpreted by some to indicate that the lumen of the fetal gut may contain fetal material that has been modified by some process to make it become antigenic (i.e., altered self). This material, it is argued, could provide the stimulus for proliferating lymphocytes. Such a notion, however, seems unlikely since there is no other evidence to suggest that there is any antigenic material in the gut lumen. There are, for example, no plasma cells in the gut or elsewhere. In addition, the MLN as was pointed out previously are sensitive indicators of the presence of antigen in the gut. Material can be absorbed intact from the fetal gut (Orlic and Lev, 1973) and if antigen was present in the gut it would be expected to be conveyed directly to the MLN via the lymphatic vessels. The MLN in the fetal lamb, however, appear completely unreactive although they do have the potential to respond to an antigenic challenge. The impact of antigen on the PP and the MLN of late-term lambs has been examined by injecting various antigens (killed *B. abortus*, ferritin or sterile maternal colostrum) into the gut of the fetus by surgically exposing the fetus under aseptic conditions and subsequently allowing the lamb to continue its intrauterine development (Reynolds and Morris, 1984). The injected fetuses were examined 5-26 days later and the development of the PP (assessed by their histological appearance and composition of proliferating lymphocytes) was found to be unaffected by the presence of antigen in the adjacent lumen (Fig. 3; Reynolds and Morris, 1984). The MLN by contrast underwent dramatic reactive changes similar to those that occur normally after birth; the response included the development of GC with all the usual characteristics (Fig. 4). Thus the MLN in fetal lambs have the capacity to respond in an unequivocal way to the presence of antigen in the gut lumen, and the fact that they do not normally do so is taken as evidence of the lack of antigen in the gut of fetal lambs.

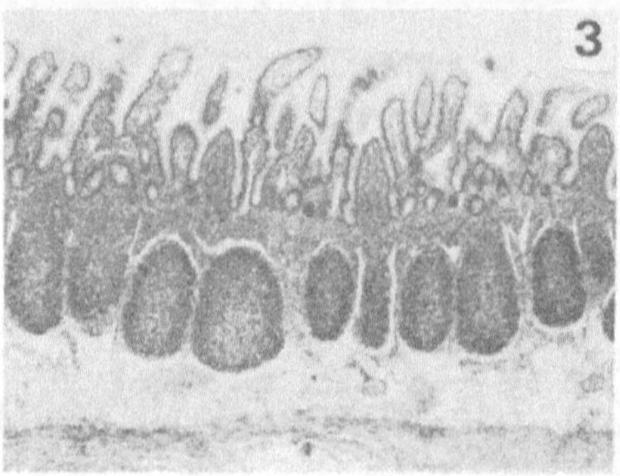

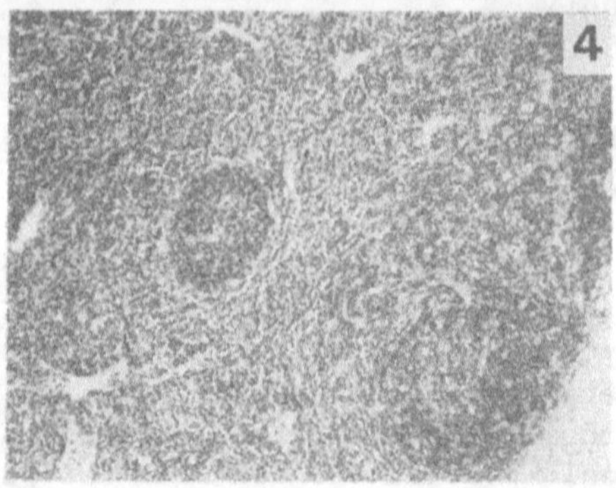

Fig. 3. A PP from the ileum of a fetal lamb 7 days after the injection of $5x10^8$ killed B. abortus organisms into the lumen of the gut adjacent to the PP. The antigen had no apparent impact on the PP development including the extent of the follicular lymphopoiesis. Paraffin section stained with hematoxylin-eosin (x100).

Fig. 4. The MLN from the same experiment described in Fig. 3. The lymph node received lymph from the ileal PP at the site where antigen had been injected into the gut lumen. Germinal center formation has been initiated. Paraffin section stained with hematoxylin-eosin (x 80).

CONCLUSIONS

These studies provide evidence that lymphopoiesis in PP follicles on the one hand, and GC in lymph nodes and the spleen on the other, is initiated by quite separate mechanisms. The extent of the PP follicular lymphopoiesis shows that at least in the broadest sense these structures are truly "germinal centers". Nevertheless the other characteristics of the follicles in PP seem to make it inappropriate to place them in the same category as the GC in the lymph node and spleen. This is not to infer that antigen has no role in the events occurring in PP. There is a lot of evidence which shows that antigen does interact with lymphoid cells in PP follicles (see reviews by Cebra et al., 1982; Woloschak and Tomasi, 1983) including some of our own studies in 8-10 week old lambs in which large numbers of labeled plasma cells were detected in the lamina propria of the proximal gut as soon as 24 hours after dividing cells in the ileal PP were selectively labeled by extracorporeal perfusion (Pabst and Reynolds, this volume). It may be that the B cells in the follicles are induced to divide by something other than antigen but that if antigen should meet the appropriate B cell in the PP then it might initiate the terminal differentiation of the cell into an antibody-secreting plasma cell. The follicular lymphopoiesis that occurs in the fetus in the absence of antigen does not result in plasma cell formation.

The follicles in the bursa of Fabricius also resemble GC though they are rarely referred to as such and the lymphopoiesis is not generally considered to be driven by external antigen (Lydyard, Grossi and Cooper, 1976). The nature of the proliferative stimulus in the bursa and in the PP of sheep remains unknown although similar processes may be occurring in both these sites.

ACKNOWLEDGEMENTS

Research currently supported by the Medical Research Council of Canada and the Alberta Heritage Foundation for Medical Research. I am also indebted to Dr. Bede Morris and to the members of the Basel Institute for Immunology for their support.

REFERENCES

Brambell, F.W.R., 1970, "The Transmission of Passive Immunity from Mother to Young," North-Holland Publ. Co., Amsterdam.

Bruecke, E., 1851, Ueber den Bau und die Physiologische Bedeutung der Peyerischen Druesen. Denkschriften der kaiserlichen Akademie der Wissenschaften. Wein, 2:21.

Cebra, J.J., Fuhrman, J.A., Gearhart, P.J., Hurwitz, J.L., and Shahin, R.D., 1982, B lymphocyte differentiation leading to a commitment to IgA expression may depend on cell division and may occur during antigen-stimulated clonal expansion, in: "Recent Advances in Mucosal Immunity," W. Strober, L.A. Hanson, and K.W. Sell, eds, Raven Press, New York.

Cole, G.J., and Morris, B., 1973, The lymphoid apparatus of the sheep: its growth, development and significance in immunologic reactions. Adv. Vet. Sci. Comp. Med., 17:225.

Cottier, H., Hess, M.W., and Keller, H.-U., 1980, Structural basis for lymphoid tissue functions: established and disputable sites of antigen-cell and cell-to-cell interactions in vivo. Monogr. Allergy, 16:50.

Flemming, W., 1885, Studien uber Regeneration der Gewebe. Arch. F. Micr. Anat., 24:50.

Hellman, T., 1943, Lymphgefaesze, Lymphknoetchen und Lymphknoten. Handbuch f. d. Mik. Anat. d. Menschen, 6/1:233.

His, W., 1860-1862, Beitraege zur Kenntnis der zum Lymphsystem gehoerigen Druesen., Zeitschr. F. Wiss. Zool., 10:333, 11:65.

Lydyard, P.M., Grossi, C.E., and Cooper, M.D., 1976, Ontogeny of B cells in the chicken. 1. Sequential development of clonal diversity in the bursa., J. Exp. Med., 144:79.

Nieuwenhuis, P., Gastkemper, N.A., and Opstelten, D., 1981, Histophysiology of follicular structures and germinal centres in relation to B cell differentiation, in: "Ciba Foundation Symposium" 84:246, Pitman, London.

Orlic, D., and Lev, R., 1973, Fetal rat intestinal absorption of horseradish peroxidase from swallowed amniotic fluid. J. Cell Biol., 56:106.

Peyer, J.C., 1677, "Exercitatio anat. de glandulis intestinorum earumque usu et affectionibus," Schaffhausen.

Reynolds, J.D., 1976, "The Development and Physiology of the Gut-Associated Lymphoid Tissues in Lambs" Ph.D. thesis, Australian National University.

Reynolds, J.D., and Morris, B., 1983, The evolution and involution of Peyer's patches in fetal and postnatal sheep. Eur. J. Immunol., 13:627.

Reynolds, J.D., and Morris, B., 1984, The effect of antigen on the development of Peyer's patches in sheep. Eur. J. Immunol., 14:1.

Schmidt, J.E., 1959, "Medical Discoveries," Charles C. Thomas, Springfield.

Silverstein, A.M., and Prendergast, R.A., 1971, The maturation of lymphoid tissue structure and function in ontogeny. Adv. Exp. Med. Biol., 29:383.

Woloschak, G.E., and Tomasi, T.B., 1983, The immunology and molecular biology of the gut-associated immune system. CRC Critical Reviews in Immunology, 4:1.

DIFFERENTIATION OF B LYMPHOCYTES IN SHEEP. II.
SURFACE PHENOTYPE OF B CELLS LEAVING THE 'BURSA-EQUIVALENT' LYMPHOID TISSUE OF SHEEP, ILEAL PEYER'S PATCHES

M. Miyasaka, J. Reynolds, L. Dudler, M-F. Beya,
W. Leiserson and Z. Trnka

Basel Institute for Immunology, Basel, Switzerland

INTRODUCTION

It has been shown that there are two different types of Peyer's patches, jejunal Peyer's patches (JPP) and ileocecal or ileal Peyer's patches (IPP) in sheep[1]. IPP have been proposed to be 'bursa-equivalent' lymphoid tissue of sheep and bear some characteristics of the primary lymphoid organ[2]. Like the bursa of Fabricius of chickens, IPP consist of numerous follicles of B cells whereas T cells account for only less than 1% of total cells. The majority of IPP cells express small amounts of surface IgM (low $sIgM^{+}$ or sIg^{lo})[3,4,5]. These $sIgM^{lo}$ cells are immature; they respond poorly to B cell mitogens such as LPS and protein A, they proliferate rapidly _in vivo_ expressing receptors to peanut agglutinin (PNA); they can also be induced to express large amounts of sIgM upon stimulation with phorbol myristate acetate (PMA)[4,5]. The surgical resection of a large proportion of the IPP before birth or during the first week of postnatal life results in a severe reduction in the number of circulating B cells in lambs[2]. Foetal IPP show extensive lymphopoiesis in the obvious absence of maternal or exogenous antigen[1,6]. IPP undergo a significant degree of postnatal involution[1] as seen in the bursa of Fabricius. The extent of emigration of newly-formed B cells from the IPP of young lambs is sufficient to replace the B cell pool in the peripheral blood within 2 to 5 days[7]. By contrast, bone marrow, a primary site for B cell production in other mammalian species is not very lymphocytic in sheep and less than 4% of the nucleated cells are lymphoid[8].

One peculiar feature of the IPP is the fact that not only extensive lymphopoiesis but also considerable cell death takes place. Abundant pyknotic figures can be observed in IPP follicles and stathmokinetic studies confirmed the presence of large-scale cell death in IPP[9]. In order to gain insight into the nature of this obvious 'ineffective' lymphopoiesis, we have compared the cells leaving the IPP with those remaining and also with those already disseminated into the periphery.

We demonstrate in this paper that a considerable proportion of cells leaving IPP do not possess the phenotype of mature B cells but that they are phenotypically similar to IPP cells. These are presumably immature B cells leaving IPP and may require further maturation in the periphery before they become mature B cells.

MATERIALS AND METHODS

Animals

One to three month-old White Alpine and Black Jura lambs of both sexes were obtained from Versuchsbetrieb Sennweid, Olsberg, Switzerland.

Preparation of IPP single cell suspension

This was done as described previously[5]. In brief, a segment of terminal ileum was removed from young lambs and rinsed exhaustively in RPMI 1640. After trimming off non-lymphoid tissues, IPP follicles were scraped from the segment, chopped into fine pieces and resuspended by repeated jet-pipetting in RPMI 1640 containing DNAase. Dead cells were removed by use of a Percoll discontinuous gradient. Cells were washed twice prior to use.

Collection of cells leaving IPP

Most of the cells that leave the IPP do so via the lymph[7]. Although the efferent lymphatics from IPP are numerous, they are too small for cannulation. However, if the mesenteric lymph nodes to which the lymphatics enter are removed the lymphatics often fuse together to form a single lymphatic vessel that is large enough to be cannulated. The nodes were removed from newborn lambs as described[7] and approximately 6 weeks later a fused lymphatic coming from the IPP was cannulated using a vinyl cannula.

Separation of B cells from ileal efferent lymph

Ileal efferent lymph cells were washed twice and incubated with anti-sheep pan T cell monoclonal antibody, ST-1[10] for 30 min on ice, washed once and subsequently treated with rabbit complement for 30 min at 37°C. Dead cells were removed by a Percoll discontinuous gradient[11]. More than 98% of the cells recovered were viable as assessed by a trypan blue exclusion assay and less than 3% of them were ST-1^+ (T cells) when they were stained by FITC-goat $F(ab')_2$ anti mouse Ig (Cappel).

Immunofluorescence staining and FACS analysis

This has been described previously in detail[11]. All antibodies used were titrated for optimal staining with immunofluorescence method and stained cells were analyzed by a Becton Dickinson FACS II.

Reagents

Fluorescein isothiocyanate (FITC) conjugates of $F(ab')_2$ rabbit anti sheep Ig and $F(ab')_2$ goat anti mouse Ig were obtained from Cappel Laboratories (Cochranville, PA). Monoclonal antibody DA6.78 which recognizes β chain of human Ia antigens was a gift of Dr. R. Palacios, Basel Institute for Immunology. Monoclonal antibodies ST-1 and SB-1 were prepared in our laboratory as described previously[11]. Fluoresceinated peanut agglutinin (FITC-PNA) was purchased from Vector laboratories (Burlingame, CA).

RESULTS AND DISCUSSION

Cells observed in lymph coming from IPP

Efferent lymph was obtained from the part of the ileum where more than 80% of the tissue is IPP. As shown in Fig. 1, the most distinct feature of the ileal efferent cells is their morphological heterogeneity. While small lymphocytes, which are the major population of cells in efferent lymph from non-stimulated lymph nodes, are also abundant in the ileal efferent lymph, a considerable proportion of the efferent cells were lymphocytes of medium to large size some of which were dividing at the time of examination. The existence of non-lymphoid cells such as macrophages and veiled cells or dendritic cells was also a salient feature. Approximately 60% of the ileal efferent cells were T cells as identified by ST-1 monoclonal antibody[10]. The majority of the T cells were small lymphocytes with dense chroma-

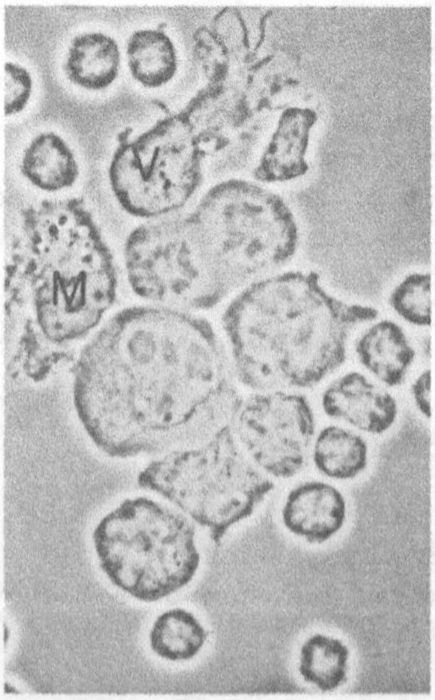

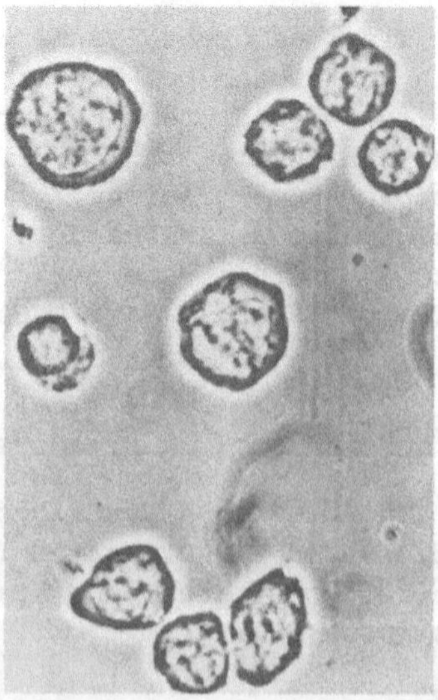

Fig.1. Ileal efferent lymph cells. M-Macrophage, V-Veiled cells.

Fig.2. Ileal efferent B cell fraction.

tins. It is likely that the majority of these T cells are in fact recirculating T cells derived not from the IPP but the lamina propria. T cells constitute less than 1% of IPP[11]. Among cells which bear surface immunoglobulin (sIg^+), considerable size heterogeneity was observed. Since this population is of special interest to us, we enriched this B cell population by removing T cells from ileal lymph preparation by treating cells with ST-1 monoclonal antibody and complement *in vitro* (Fig. 2).

B cells present in the lymph coming from IPP

After the T cell killing, only 3% of the cells in the ileal efferent B cell fraction were found to be T cells, 1% accounting for non-lymphoid cells such as macrophages and veiled cells. More than 95% of the cells bore detectable levels of sIg, IgM being the major isotype of immunoglobulin expressed on the cell surface. As shown in Fig. 3, the fluorescence intensity of the ileal efferent B cells stained with FITC-conjugated rabbit $F(ab')_2$ anti-sheep Ig was noticeably less than that of small B cells obtained from the peripheral blood or efferent lymph of non-stimulated lymph nodes. Examination of stained B cells by a fluorescence microscope revealed that almost all cells of medium to large size were stained dull (sIg^{lo}), whereas small cells were much brighter and the intensity was indistinguishable from those obtained with cells from the blood or normal lymph (sIg^{hi}). The dull staining of larger cells is reminiscent of IPP cells[5] which are $sIgM^{lo}$.

We have previously reported that a mouse monoclonal antibody against β chain of human Ia antigens DA6.78 crossreacted with ovine B cells detecting Ia-like antigens on the cell surface[5]. Flow cytometry analysis of various populations of B cells stained with DA6.78 showed that ileal efferent B cells possess higher levels of Ia-like antigens than IPP cells and the levels were comparable to those of circulating B cells.

Ileal efferent B cells were also different from other populations of B cells in expression of other differentiation markers such as a monoclonal antibody, SB-1 and peanut agglutinin (PNA). Our preliminary studies showed that SB-1 binds to ovine B cells but not to T cells and an indication was obtained that SB-1 binds to some surface antigens which increase in amounts during the course of differentiation of B cells[13]. As shown in Fig. 3, the SB-1 binding was most intense in activated B cells (LPS blasts), least intense in IPP cells, and intermediate in circulating B cells. Interestingly the staining intensity of ileal efferent B cells was slightly stronger than cells present in IPP but considerably weaker than circulating B cells.

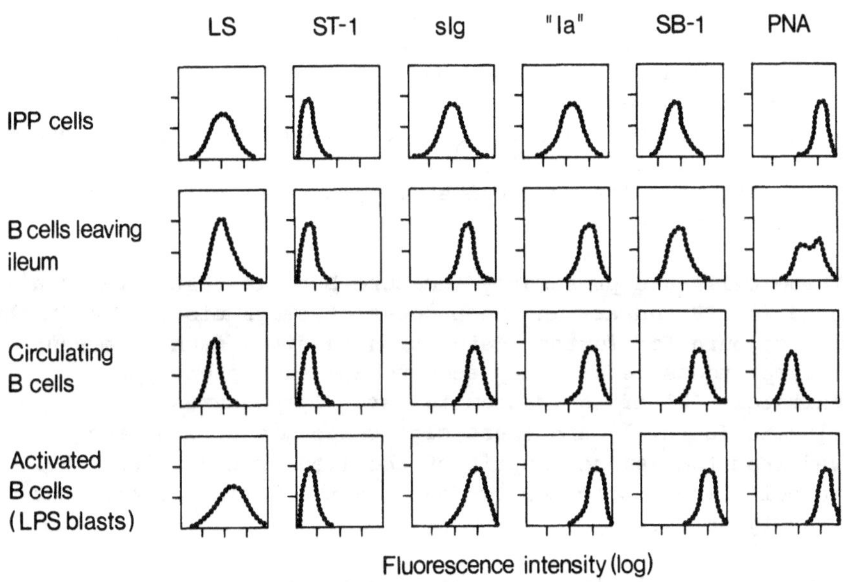

Fig. 3. FACS analysis of various populations of B cells stained with various antibodies and peanut agglutinin. Logarithmic fluorescence intensity is plotted on the abscissa in arbitrary units, and linear cell frequency on the ordinate. LS; forward light scatter.

Peanut agglutinin (PNA) has been reported to show differential binding to various populations of ovine B cells, the binding being most intense in immature B cells and least prominent in small resting B cells[5]. Activated B cells are also strongly positive for PNA[14]. Compatible with previous findings, IPP cells and efferent lymph lymphocytes stimulated with LPS showed very intense staining by fluoresceinated PNA whereas circulating B cells showed only dull staining (Fig. 3). In contrast, B cells obtained from ileal efferent lymph, showed a bimodal staining pattern. Separation of cells, however, by using a FACS on the basis of forward light scatter which is indicative of cell size, revealed that the more intense staining was that of large to medium-sized lymphocytes whereas the dull staining was that of small cells.

Thus, several points have emerged from the phenotypic analysis of B cells in ileal efferent lymph. In lymph coming from the part of the ileum mainly covered with IPP, there are at least two different subpopulations of B cells, as shown in Fig. 4. One population of cells has the following characteristics. They are;

1) medium to large lymphocytes and,
2) the sIg expression is low (sIg^{lo}), whereas circulating B cells are sIg^{hi},
3) levels of sIa-like antigen on the cell surface is about the same as that of circulating B cells (sIa^{hi}),
4) majority of them express small amounts of antigen recognized by SB-1 monoclonal antibody ($SB\text{-}1^{lo}$) which appears to recognize B cell differentiation antigen. The antigen is most abundant in activated B blasts and less so in circulating B cells,
5) unlike circulating B cells, they bind high levels of peanut agglutinin (PNA^{hi}).

These cells are presumably immature B cells that have been produced in IPP and are in the process of emigration. They might require sojourn for further maturation in sites such as mesenteric lymph nodes before they acquire a mature phenotype. Approximately 10% of the medium to large lymphocytes bore cytoplasmic Ig (cIg^+) and these cIg^+ cells might have been derived from the lamina propria of the ileum but not from IPP. Plasma cells are most commonly found in the lamina propria but very rarely in IPP.

The other population consists of cells that are phenotypically indistinguishable from circulating B cells. Their characteristics are;

1) small lymphocytes,
2) the sIg expression is high (sIg^{hi}),

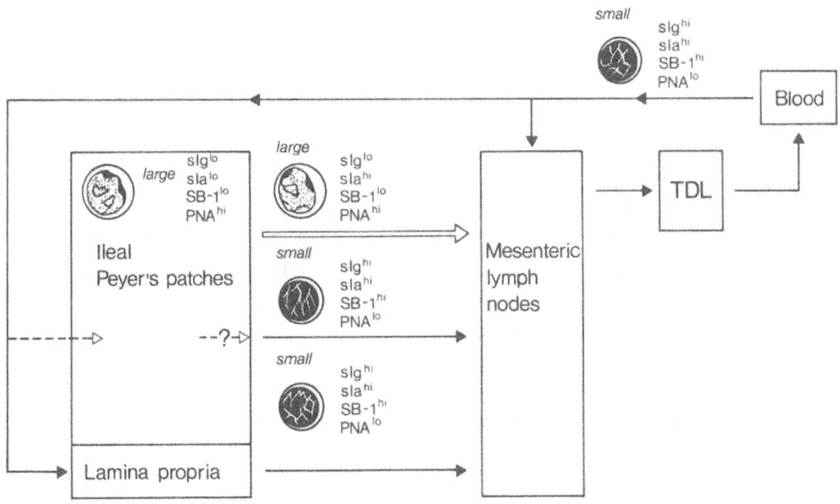

Fig.4. Possible differentiation and migration pattern of sheep B lymphocytes. See text for details.

3) levels of sIa-like antigen is high (sIa^{hi}),
4) binding of SB-1 monoclonal antibody is high (SB-1^{hi}), and
5) PNA binding is low (PNA^{lo}).

We are currently investigating whether or not this small cell population consists entirely of recirculating B cells or whether some of them were in fact produced in IPP and are in the process of emigrating to the periphery bearing a mature phenotype.

Availability of more monoclonal antibodies against various B cell differentiation antigens together with the use of surgical techniques as described above would enable us to carry out a more detailed examination of migration and differentiation of B lymphocytes *in vivo*.

ACKNOWLEDGEMENT

We thank Drs. T. Ezaki, W. Weber and Ms. L. Gut for critical reading and advice in the preparation of the manuscript and Ms. L. Gisler for excellent typing. The Basel Institute for Immunology was founded and is supported by F. Hoffmann-La Roche and Co., Ltd, Basel, Switzerland.

REFERENCES

1. J.D. Reynolds and B. Morris. The evolution and involution of Peyer's patches in fetal and postnatal sheep. Eur. J. Immunol. 13:627 (1983).
2. H.A. Gerber. Functional studies on gut associated lymphoid tissues. Ph.D. Thesis, Australian National University, Canberra (1979).
3. J. Reynolds. Peyer's patches and the ontogeny of B lymphocytes in sheep. In: "The immune system, vol. 1", C.M. Steinberg and I. Lefkovits, eds., Karger, Basel (1981).
4. M. Miyasaka and L. Dudler. Differentiation of ovine immature B cells upon exposure to phorbol myristate acetate. Int. Arch. Allergy Appl. Immunol. 74:281 (1984).
5. M. Miyasaka, L. Dudler, G. Bordmann, W.M. Leiserson, H.A. Gerber, J. Reynolds and Z. Trnka. Differentiation of B lymphocytes in sheep. I. Phenotypic analysis of ileal Peyer's patch cells and the demonstration of a precursor population for sIg^+ cells in the ileal Peyer's patches. Immunology, in press.
6. J.D. Reynolds and B. Morris. The effect of antigen on the development of Peyer's patches. Eur. J. Immunol. 14:1 (1984).
7. J.D. Reynolds and R. Pabst. The emigration of lymphocytes from Peyer's patches in sheep. Eur. J. Immunol. 14:7 (1984).
8. M. Miyasaka, L. Dudler, R. Pabst and Z. Trnka. Migration of bone marrow lymphocytes in sheep: In situ labelling of bone marrow lymphocytes with fluorescein isothiocyanate. Manuscript in preparation.
9. J. Reynolds, unpublished observations.
10. M.-F. Beya, L. Dudler, M. Miyasaka and Z. Trnka. Studies on the differentiation of T lymphocytes in sheep. II. Characterization of anti-pan T monoclonal antibodies. Manuscript in preparation.
11. M. Miyasaka and Z. Trnka. Sheep as an experimental model for immunology - immunological techniques in vitro and in vivo. In: "Immunological Methods III", I. Lefkovits and B. Pernis, eds., Academic Press, New York, in press.
12. M. Miyasaka, I. Heron, L. Dudler, R.N.P. Cahill, L. Forni, T. Knaak and Z. Trnka. Studies on the differentiation of T lymphocytes in sheep. I. Recognition of a sheep T-lymphocyte differentiation antigen by a monoclonal antibody T-80. Immunol. 49:545 (1983).
13. M.-F. Beya and M. Miyasaka, unpublished observations.
14. M.L. Rose and F. Malchiodis. Binding of peanut lectin to thymic cortex and germinal centres of lymphoid tissue. Immunol. 42:583 (1981).

ORIGIN AND FATE OF IgE-BEARING LYMPHOCYTES: MODULATION OF IgE ISOTYPE EXPRESSION ON PEYER'S PATCH CELLS

Helen G. Durkin, Seto M. Chice and Su Ian Jiang

Department of Pathology, State University of New York, Downstate Medical Center, Brooklyn, New York USA

The absence of precursor B lymphocytes bearing membrane bound IgE (sIgE) from bone marrow (1,2) and Peyer's patches of conventionally raised (C) adult rats, and the presence of high numbers of these cells in PP (2), but not bone marrow (2) of germfree (GF) rats suggests that environmental antigens, such as bacteria, may play a role in "natural" regulation of IgE isotype expression. We have now succeeded in modulating IgE isotype expression on GF rat PP cells so that they resemble those of C rats. The data suggest that exposure to gut flora influenced these changes, either directly or through the mediation of sThy 1+ regulatory T cells (see 3,4).

MATERIALS AND METHODS

Germfree Sprague-Dawley rats (males, 8-10 weeks old) were anaesthetized, 2-3 PP removed, and their cell surface markers analyzed using indirect immunofluorescence (2). The rats were then either thymectomized or fed standard chow for one day, at which time they were killed and the remaining PP removed for analysis. In other studies, purified (>98%) subpopulations of sIgE+A$^-$ or sIgA+E$^-$ cells were isolated from PP using the "panning" technique (5), labelled (>98%) with ^{3}H-adenosine (6), and injected i.v. into GF or C recipients. The quantitative distributions of donor origin cells accumulating in recipient spleen at 24 hr was determined by combined autoradiographic and double immunofluorescence labelling for IgE and IgA surface markers. Immunofluorescence labelling was carried out using living recipient spleen cells in suspension, after which the cells were applied to slides. The relative numbers and locations of cells bearing sIgE and/or sIgA were determined, and the slides processed for autoradiography (6).

The per cent of radiolabelled cells bearing sIgE and/or sIgA was determined by comparing the numbers and locations of radiolabelled cells with immunofluorescence data previously obtained.

Indirect immunofluorescence assay of cells for membrane bound immunoglobulin was carried out with anti-IgM, A and E antisera prepared by Dr. Hervé Bazin, University of Louvain, Brussels, Belgium (7,8), as previously described (2). T cell surface markers were studied using anti-RT 7.1 (formerly ART 1a) and anti-Thy 1 antisera, prepared as described (9-11 see also 4). Fluorescein and rhodamine conjugates (goat anti-rabbit or rabbit anti-goat Ig) were purchased from N.L. Cappel Laboratories, Inc. Data are expressed either as per cent total cells in suspension or per cent total radiolabelled cells; >5000 cells were counted.

RESULTS

1. Modulation of IgE isotype expression on PP cells by thymectomy or feeding. In agreement with our earlier studies (2), GF rat PP contained high numbers of cells with sIgE (~20%), whereas sIgE+ cells were virtually absent from PP of C rats (Table I). The numbers of sIgA+ cells were also much higher in PP of GF (26-45%) when compared with C (17%) rats. About one-half of sIgE+ cells were also sIgA+. PP of GF rats also contained high numbers of T cells bearing sThy 1 (suppressor phenotype) (3,4) (20%),which which were identified as either sThy 1+ (RT 7.1+) or sThy 1+ (RT 7.1^-). Such cells were virtually absent (1%) from PP of C rats. Further, the numbers of sRT 7.1+ (Thy 1^-) (helper phenotype) (10, see also 3,4) T cells were reduced in GF (~15%) when compared with C rat PP (35%). The cellular composition of GF rat PP could be converted to that resembling a C rat within 1 day after either thymectomy or feeding standard chow (Table I), with each procedure resulting in (a) loss of sIgE+ B cells and sThy 1+ T cells, (b) increase in sIgM+ B cells and RT 7.1+ (Thy 1^-) T cells, and (c) decrease in sIgA+ B cells. Neither treatment increased the numbers of either sIgE+ or sThy 1+ cells detected in bone marrow, other lymphoid organs or blood (data not shown), excluding the possibility that the absence of such cells from PP was the result of their redistribution to other sites.

2. Migration studies: Modulation of sIgE isotype expression on radiolabelled GF rat PP cells after transfer to C rats. Radioactivity was found principally in spleen and mesenteric lymph node of C and GF rats 24 hr after transfer of purified subpopulations of either sIgE+A^- or sIgA+E^- PP cells (data not shown); 2-4% and 3-5% of total spleen cells were radiolabelled after transfer of the respective subpopulations. About 60% of radiolabelled donor sIgE+A^- cells transferred to C recipients demonstrated phenotypic changes in that ~30% of the radiolabelled cells localizing in spleen now simultaneously expressed IgE and IgA surface markers,and

TABLE I

CHANGES IN CELLULAR COMPOSITION OF GERMFREE RAT PEYER'S PATCHES AFTER THYMECTOMY (Tx) OR FEEDING STANDARD CHOW[1)]

		Surface Markers (% Total Cells)							
		RT+*	RT+	RT⁻	Ig-Bearing Cells				
Day	Treatment	Thy⁻	Thy+	Thy+	M	G	A	E	EA
Rat No. 1									
0	3 PP removed + Tx	12	18	4	6	5	45	20	11
1	Remaining PP removed	26	2	0	20	8	20	1	1
Rat No. 2									
0	3 PP removed + sham Tx	18	18	4	5	4	29	14	7
1	Remaining PP removed	16	17	4	4	5	26	14	8
Rat No. 3									
0	3 PP removed + standard chow	23	20	5	10	4	38	29	13
1	Remaining PP removed	30	1	1	21	7	20	1	1
Rat No. 4									
0	3 PP removed + autoclaved chow	19	11	4	7	5	26	21	8
1	Remaining PP removed	20	15	3	8	5	31	25	11
Conventional Rat									
	None	35	1	1	14	5	17	1	1

1) Representative experiments, 1 of 3, each group, with comparable data. Indirect immunofluorescence, cell suspensions.
* RT = RT 7.1

~35% acquired the sIgA+E⁻ phenotype (Fig. 1). In direct contrast, <10% of the radiolabelled donor sIgE+A⁻ cells localizing in spleen of GF recipients exhibited these phenotypic alterations (data not shown). Further, none of the radiolabelled donor sIgA+E⁻ cells localizing in spleen of either C (Fig. 1) or GF rats acquired the IgE phenotype.

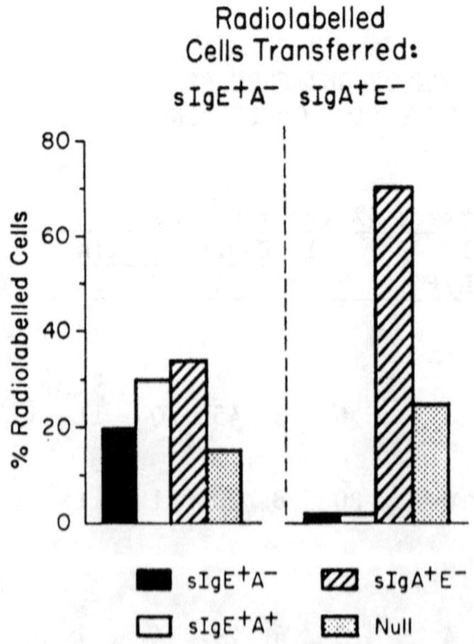

Figure 1. Changes in surface phenotypes of GF rat PP cells localizing in C rat spleen. Recipients killed at 24 hr after i.v. transfer of GF rat PP ^{3}H-adenosine labelled (>98%) sIgE+A$^-$ cells (left panel) or sIgA+E$^-$ cells (right panel); such subpopulations were purified (>98%) by the "panning" technique. Surface markers analyzed using double indirect immunofluorescence techniques. Data expressed as % total radiolabelled cells.

Thus, these findings are consistent with the idea that exposure to gut flora influenced IgE isotype switching in PP, either directly or through the mediation of sThy 1+ regulatory cells. It is, therefore, possible that elucidation of a mechanism for specific suppression of IgE precursor cells will provide a new means by which IgE antibody production can be controlled.

REFERENCES

1. J.F.Urban,K.Ishizaka and H.Bazin.1980. J.Imm. 124: 527.
2. H.G. Durkin, H. Bazin and B.H. Waksman,1981. J.Exp.Med.154:640.
3. H.G. Durkin, and I. Goldschneider, 1984. Annals New York Acad. Sci., in press.
4. H.G. Durkin and I.Goldschneider,1984. J. Immunol., submitted.
5. L.J.Wysocki and V.L.Sato,1978. Proc.Nat.Acad.Sci.USA 75:2844.
6. H.G. Durkin, J.M.Carboni and B.H.Waksman. 1978. J.Imm.121:1075.
7. H.Bazin, A.Beckers, and P.Querinjean,1974. Eur.J. Imm. 4:44.
8. H.Bazin and A.Beckers, 1976. In Molec. and Biol. Aspects of the Acute Allergic Reaction. S.G.O. Johannson, et al., eds. Plenum Publishing Corp., New York, p. 125.
9. D.M. Lubaroff, 1973. Transpl. Proc. 5: 115.
10. D.M. Lubaroff, 1983. Transpl. Proc. 15: 1625.
11. J.M. Crawford and I.Goldschneider,1980. J. Imm. 124: 969.

IMMUNOGLOBULIN ISOTYPES AND ANTIBODY SPECIFICITY REPERTOIRE OF "SPONTANEOUSLY" OCCURRING ("BACKGROUND") IMMUNOGLOBULIN-SECRETING CELLS IN GERMFREE MICE FED CHEMICALLY DEFINED ULTRAFILTERED "ANTIGEN-FREE" DIET

Herbert Hooijkaas, Nico Bos, Robbert Benner, Julian R. Pleasants[1] and Bernard S. Wostmann[1]

Department of Cell Biology and Genetics, Erasmus University, Rotterdam, The Netherlands, and [1]Lobund Laboratory University of Notre Dame, Notre Dame, Indiana, USA

INTRODUCTION

"Spontaneously" occurring ("background") immunoglobulin (Ig)-secreting cells have been observed in all lymphoid organs of non-immunized mice. Their frequency was found to depend on the genetic background, age and antigenic load of the mice studied (Benner et al., 1982). Even in germfree (GF) mice "background" antibody-secreting cells occur, as has been assessed in antigen-specific plaque assays with sheep red blood cells (SRBC) as targets (Nordin, 1968; Hof et al., 1975; Benner et al., 1981b). The total number of IgM-secreting cells as detected with the protein A plaque assay is virtually normal in GF mice, but the numbers of IgG- and IgA-secreting cells are severely decreased as compared to conventionally raised mice (Benner et al., 1981b). The latter indicates that exogenous antigenic and mitogenic stimulation is a major driving force for the "background" IgG- and IgA-synthesis. For IgM this is unclear, because of the crucial question whether the GF animals used in those experiments can really be regarded as devoid of exogenous stimuli. It has been shown that a substantial mitogenic and/or antigenic activity can be exerted by food substances or dead bacteria in the diet (Wostmann et al., 1971). With the successful breeding of GF mice fed an ultrafiltered solution of chemically defined (CD) low molecular weight nutrients, these exogenous stimuli are reduced to the presently achievable minimum (Pleasants et al., 1981; Wostmann et al., 1982; Pleasants, 1984). These mice seem to be the experimental animals of choice to settle the question whether or not exogenous antigens and/or mitogens play an essential role in the development of "background" Ig-secreting cells and their antibody specificity repertoire. Previous studies using these GF-CD mice revealed both the occurrence of IgM in the serum (Wostmann et al.,

1971) and cytoplasmic IgM containing cells (c-IgM cells) in the spleen (Benner et al., 1981c). IgG was absent (Wostmann et al., 1971) or hardly detectable (Hashimoto et al., 1978) in the serum and the number of c-IgG cells in the spleen was 10-fold reduced (Benner et al., 1981c). In addition, IgA was not detected in the serum (Hashimoto et al., 1978) and c-IgA cells could not be detected in spleen, bone marrow (BM) and mesenteric lymph nodes (MLN).

Insight into the occurrence of "spontaneous" Ig production might be relevant for contemporary concepts about the internal regulation in the immune system (Jerne, 1974; Coutinho, 1980; Coutinho et al., 1983). We therefore decided to search for data on the immunological activity of mice deprived from exogenous antigenic and/or mitogenic stimuli. To this end IgM-, IgG- and IgA-secretion was studied in spleen, BM and MLN of GF-CD and conventional (CV) mice, fed natural ingredient (NI) (Kellogg and Wostmann, 1969), employing the protein A plaque assay. Furthermore, the antibody specificity repertoire of the IgM-secreting cells in the spleen was analyzed using plaque assays, specific for differently haptenated SRBC.

MATERIALS AND METHODS

Mice

Male and female C3H/HeCr mice were reared and maintained in the Lobund Laboratory, University of Notre Dame, Notre Dame, Indiana, USA under CV or GF conditions (Wostmann et al., 1982). The CV mice were fed NI L-485 (Kellogg and Wostmann, 1969) and the GF mice CD ultrafiltered "antigen-free" diet which consisted of the water-soluble L-489 E14Se diet and the lipid supplement LADEK 69E6 (Pleasants, 1984). The ashless filter paper bedding was consumed by the mice and proved important for normal intestinal function (Pleasants et al., 1981), so it must be considered as part of the diet. In the experiments 8- to 12-wk-old mice were used and the data obtained from male and female mice were pooled. All mice tested were in healthy condition and showed no abnormalities on dissection. It must be noted, however, that all so-called GF mice carry a leukemogenic virus and that at present no strictly GF mice are known to exist (Kajima and Pollard, 1968; Wostmann et al., 1982).

Cells

Spleen, BM and MLN cell suspensions were prepared as described (Benner et al., 1981a).

Plaque assays

The target cells for the protein A plaque assay (Gronowicz et al., 1976; Van Oudenaren et al., 1981) and the antigen-specific plaque assays were prepared and the plaque assays were performed as has been described (Hooijkaas et al., 1983). For the antigen-specific plaque assays, nitroiodophenyl (NIP), 4-hydroxy-3,5-dinitrophenyl (NNP) and 2,4,6-trinitrophenyl (TNP) were coupled to SRBC with 0.4, 0.2 and 30 mg of the hapten per ml of washed packed SRBC, respectively. They are referred to as $NIP_{0.4}$-, $NNP_{0.2}$- and TNP_{30}-SRBC, respectively. Throughout all experiments SRBC from a single donor were used.

Calculations and statistics

The total number of Ig-secreting cells of a given class per organ was calculated using the number of Ig-secreting cells in the protein A plaque assay and the total cell yield per organ. Ig-secreting cells in the BM were determined in the femoral BM. Following Benner et al. (1981b) we have adopted a conversion factor 7.9 to calculate the number of Ig-secreting cells in the total BM from the number of Ig-secreting cells found in two femurs. The frequencies of "background" antigen-specific IgM-secreting cells in spleen and BM were determined in the appropriate plaque assays and calculated as the ratio of specific IgM antibody-secreting cells to the total number of IgM-secreting cells. Statistical evaluation was performed with the Student t test.

RESULTS

The numbers of background Ig-secreting cells in spleen, BM and MLN of CV-NI and GF-CD mice are given in Table 1. In the spleen of GF-CD mice the numbers of IgM-, IgG- and IgA-secreting cells were significantly reduced as compared with CV-NI mice. In the BM the numbers of IgG- and IgA-secreting cells were 3.5 and 9 times lower than in CV-NI controls, whereas the number of IgM-secreting cells was not significantly different. In only 4 out of 10 GF-CD mice MLN could be detected and from the mice that did possess MLN the data are given. In these animals the numbers of IgM-, IgG- and IgA-secreting cells were about 3, 60 and 10 fold reduced, respectively, as compared with the CV-NI controls.

The results given in Table 2 show that in GF-CD mice the frequencies of IgM-secreting cells specific for $NIP_{0.4}$-SRBC, $NNP_{0.2}$-SRBC and TNP_{30}-SRBC were not significantly different from those found in CV-NI controls.

DISCUSSION

The data presented in this paper indicate that complete deprivation of GF mice from exogenous antigens cannot prevent the generation of Ig-secreting cells, although their number was drastically reduced in comparison with CV mice. Especially the number of Ig-secreting cells in the gut-associated lymphoid tissue of GF-CD mice is severely diminished. In less than half of the GF-CD mice MLN could be detected, whereas in mice who did possess MLN highly reduced

Table 1. Numbers of background Ig-secreting cells in the spleen and BM of conventional C3H/HeCr mice fed natural ingredient (CV-NI) and germfree C3H/HeCr mice fed chemically defined diet (GF-CD)

Organ	Ig-isotype	Ig-secreting cells/organ/10^{-3} CV-NI	GF-CD
Spleen	IgM	780(743-818)[a]	337(328-347)[b]
	IgG	48(45-51)	29(28-30)[b]
	IgA	53(49-57)	17(16-18)[b]
BM	IgM	60(56-64)	62(60-63)
	IgG	28(26-30)	8.1(7.6-8.6)[b]
	IgA	26(24-28)	3.0(2.8-3.2)[b]
MLN	IgM	0.6(0.5-0.7)	0.19(0.11-0.33)[b,c]
	IgG	0.6(0.5-0.6)	0.01(0.003-0.03)[b]
	IgA	1.2(0.9-1.4)	0.11(0.03-0.35)[b]

a. Figures represent the geometric mean $\pm$ 1 SE of 10 individually tested mice.
b. significantly different ($p < 0.001$).
c. In the GF-CD group, MLN were found in only 4 out of 10 mice.

Table 2. Relative frequencies of background IgM-secreting cells specific for $NIP_{0.4}$-SRBC, $NNP_{0.2}$-SRBC and TNP_{30}-SRBC in the spleen of conventional C3H/HeCr mice fed natural ingredient (CV-NI) and germfree C3H/HeCr mice fed chemically defined diet (GF-CD).

Antigen	CV-NI	GF-CD
$NIP_{0.4}$-SRBC	1 in 159 (148-168)[a]	1 in 139 (133-146)
$NNP_{0.2}$-SRBC	1 in 636 (603-671)	1 in 577 (545-611)
TNP_{30}-SRBC	1 in 84 (73-94)	1 in 106 (95-116)

a. Figures represent the mean ratio of specific IgM-antibody secreting cells to the total number of IgM-secreting cells as determined in the protein A plaque assay. The geometric mean $\pm$ 1 SE has been calculated. In the spleen of the CV-NI and GF-CD mice (n=10) 3119 (3013-3237) and 3290 (2869-3709) IgM-secreting cells per 10^6 nucleated cells were found, respectively.

numbers of Ig-secreting cells were found. In the spleen of GF-CD mice, IgM-, IgG- and IgA-secreting cells were decreased to about 50% of the value in CV-NI mice. In the BM of GF-CD mice, on the other hand, only the IgG- and IgA-secreting cells were decreased in number. The low numbers of IgG-secreting cells are in accordance with the finding that IgG is hardly detectable in the serum of GF-CD mice (Hashimoto et al., 1978). On the other hand, serum IgA or IgA-producing cells have not been detected in GF-CD mice before (Hashimoto et al., 1978; Benner et al., 1981c). The underlying cause of this discrepancy is probably that in the immunofluorescence assay only plasmablasts and plasma cells are identifiable with certainty. However, earlier stages of the plasmacytic series are also able to produce and secrete Ig (Melchers and Andersson, 1974), which makes them detectable in the protein A plaque assay.

Although we have only tested a minor fraction of the total antibody specificity repertoire our data also point to the remarkable conclusion that exogenous stimuli do not affect the antibody specificity repertoire as expressed in "background" IgM-secreting cells. This suggests that available antibody repertoires are established independently of exogenous antigenic and/or mitogenic stimulation. It has been reported, however, that the production rate of B lymphocytes in the BM of GF mice is markedly reduced as compared with CV controls (Osmond et al., 1981). Moreover, it was shown that administration of extrinsic agents, such as antigens or irritants, can stimulate the production rate of virgin B cells in the BM of CV reared mice (Fulop and Osmond, 1982). Antigenic challenges of GF and CV mice, on the other hand, revealed that GF mice are capable of generating adequate immune responses as measured either by antibody levels in the serum (Olson and Wostmann, 1964) or by the number of antibody-secreting cells in the spleen (Nordin, 1968; Hof et al., 1975). This indicates that GF mice are fully B cell immunocompetent, which might be reflected in the specificity repertoire of the "background" Ig-secreting cells. Thus, exogenous antigens and/or mitogens are not necessary for the generation of the B cell repertoire. It is therefore tempting to suggest that endogenous stimulation must account for the activation of "background" Ig-secreting cells and in general for the selection of particular repertoires of B lymphocytes. This places the driving forces for the intrinsic activity of the immune system within the individual itself, which is in direct support of Jerne's network ideas (Jerne, 1974) postulating autonomy and completeness in a system embodied by the interaction between complementary molecules including antibodies (Coutinho, 1980), T cell receptors and receptors controlling the growth and differentiation of B cell precursors and the activation of mature B lymphocytes (Coutinho et al., 1983). It can be speculated that these basic stimuli could be exerted by idiotypes derived from maternal Ig's transferred to the pre- or neonates via placenta or milk (Weiler et al., 1977; Wikler et al., 1980; Bernabé et al., 1981) or by any other self antigens. Especially the maternal influence is an intriguing possibility, since it has been shown that in GF colostrum-deprived piglets no serum Ig's could be detected, probably due to the six-layered epitheliochorial placenta, impermeable to antigens and Ig's (Sterzl, 1967; Kim, 1975). In addition, no "background" antibody-forming cells against SRBC could be detected in their lymphoid organs, although the piglets proved to be normally immunocompetent upon antigenic challenge (Sterzl, 1967; Kim, 1975). If these findings could be confirmed in protein A plaque assays it would imply that in the normal physiological situation, maternal Ig's could play an important role in the generation of background Ig-secreting cells (Tlaskalova-Hogenova, 1983) and their specificity repertoire.

ACKNOWLEDGMENTS

We gratefully acknowledge Mrs. C.J.M. Meijerink-Clerkx for typing the manuscript and all members of the Lobund Laboratory who in some way contributed to this work.

REFERENCES

Benner, R., Van Oudenaren, A., and Koch, G., 1981a, Induction of antibody formation in mouse bone marrow, in: "Immunological Methods, Vol. 2", I. Lefkovits and B. Pernis, eds., Academic Press, New York, p. 247.

Benner, R., Rijnbeek, A.-M., Bernabé, R.R., Martinez-Alonso, C., and Coutinho, A., 1981b, Frequencies of background immunoglobulin-secreting cells in mice as a function of organ, age and immune status, Immunobiol., 158:225.

Benner, R., Van Oudenaren, A., Haaijman, J.J., Slingerland-Teunissen, J., Wostmann, B.S., and Hijmans, W., 1981c, Regulation of the "spontaneous" ("background") immunoglobulin synthesis, Int. Archs. Allergy appl. Immunol., 66:404.

Benner, R., Van Oudenaren, A., Björklund, M., Ivars, F., and Holmberg, D., 1982, "Background" immunoglobulin production: measurement, biological significance and regulation. Immunol. Today, 3:243.

Bernabé, R.R., Coutinho, A., Cazenave, P.-A., and Forni, L., 1981, Suppression of a "recurrent" idiotype results in profound alterations of the whole B-cell compartment, Proc. Natl. Acad. Sci. USA, 78:6416.

Coutinho, A., 1980, The self-nonself discrimination and the nature and acquisition of the antibody repertoire, Ann. Immunol. (Inst. Pasteur), 131 D:235.

Coutinho, A., Forni, L., Holmberg, D., and Ivars, F., 1983, Is the network theory tautologic?, in: "Genetics of the Immune Response", E. Möller and G. Möller, eds., Plenum Press, New York, p. 273.

Fulop, G.M., and Osmond, D.G., 1983, Regulation of bone marrow lymphocyte production. III. Increased production of B and non-B lymphocytes after administering systemic antigens, Cell Immunol., 75:80.

Gronowicz, E., Coutinho, A., and Melchers, F., 1976, A plaque assay for all cells secreting Ig of a given type or class, Eur. J. Immunol., 6:588.

Hashimoto, K., Handa, H., Umehara, K., and Sasaki, S., 1978, Germfree mice reared on an "antigen-free" diet, Lab. Anim. Sci., 28:38.

Hof, H., Finger, H., and Karle, E., 1975, Influence of Bordetella pertussis and bacterial endotoxins on the immunological reactivity of germfree mice, Zbl. Bakt. Hyg. I. Abt. Orig. A, 232:73.

Hooijkaas, H., Preesman, A.A., Van Oudenaren, A., Benner, R., and Haaijman, J.J., 1983, Frequency analysis of functional immunoglobulin C and V gene expression in murine B cells at various ages, J. Immunol., 131:1629.

Jerne, N.K., 1974. Towards a network theory of the immune system, Ann. Immunol. (Inst. Pasteur), 125 C:373.

Kajima, M., and Pollard, M., 1968, Wide distribution of leukaemia virus in strains of laboratory mice, Nature, 218:188.

Kellogg, T.F., and Wostmann, B.S., 1969, Stock diet for colony production of germfree rats and mice, Lab. Anim. Care, 19:812.

Kim, Y.-B., 1975, Developmental immunity in the piglet, in: "Immunodeficiency in Man and Mammals, Birth Defects: Original Article Series, Vol. 11, no. 1", D. Bergsma, R.A. Good and J. Finstadt, eds., Sinauer Associates Inc., Sunderland, p. 549.

Melchers, F., and Andersson, J., 1974, IgM in bone marrow derived lymphocytes. Changes in synthesis, turnover and secretion, and in numbers of molecules on the surface of B cells after mitogenic stimulation, Eur. J. Immunol., 4:181.

Nordin, A.A., 1968, The occurrence of plaque forming cells in normal and immunized conventional and germfree mice, Proc. Soc. Exp. Biol. Med., 129:57.

Olson, G.B., and Wostmann, B.S., 1966, Cellular and humoral immune response of germfree mice stimulated with 7S HGG or Salmonella typhimurium, J. Immunol., 97:275.

Osmond, D.G., Fahlmann, M.T.E., Fulop, G.M., and Rahal, D.M., 1981, Regulation and localization of lymphocyte production in the bone marrow, in: "Microenvironments in Hemopoietic and Lymphoid Differentiation, CIBA Foundation Symposium 84", Pitman Medical London, p. 68.

Pleasants, J.R., 1984, Diets for germfree animals. Part 2. The germfree animal fed chemically defined ultrafiltered diet, in: "The Germ-free Animal in Biomedical Research. Laboratory Animal Handbooks Vol. 9", M.E. Coates and B.E. Gustafsson, eds., Laboratory Animals Ltd., London, p. 91.

Pleasants, J.R., Bruckner-Kardoss, E., Bartizal, K.F., Beaver, M.H., and Wostmann, B.S., 1981, Reproductive and physiological parameters of germfree C3H mice fed chemically defined diet, in: "Recent Advances in Germfree Research", S. Sasaki, A. Ozawa, and K. Hashimoto, eds., Tokai University Press, Tokyo, p. 333.

Sterzl, J., 1967, Factors determining the differentiation pathways of immunocompetent cells, Cold Spring Harbor Symp. Quant Biol., 32:493.

Tlaskalova-Hogenova, H., Sterzl, J., Stepankova, R., Dlabac, C.V., Vetvicka, V., Rossmann, P., Mandel, L., and Rejnek, J., 1983, Development of immunological capacity under germfree and conventional conditions, Ann. N.Y. Acad. Sci., 409:96.

Van Oudenaren, A., Hooijkaas, H., and Benner, R., 1981, Improvement of the protein A plaque assay for immunoglobulin secreting cells by using immunoglobulin-depleted guinea pig serum as a source of complement. J. Immunol. Methods, 43:219.

Weiler, I.J., Weiler, E., Sprenger, R., and Cosenza, H., 1977, Idiotype suppression by maternal influence, Eur. J. Immunol., 7:591.

Wikler, M., Demeur, C., Dewasme, G., and Urbain, J., 1980, Immunoregulatory role of maternal idiotypes. Ontogeny of immune networks, J. Exp. Med., 152:1024.

Wostmann, B.S., Bruckner-Kardoss, E., and Pleasants, J.R., 1982, Oxygen consumption and thyroid hormones in germfree mice fed glucose-amino acid liquid diet, J. Nutr., 112:552.

Wostmann, B.S., Pleasants, J.R., and Bealmear, P., 1971, Dietary stimulation of immune mechanisms, Fed. Proc., 30:1779.

COMPARATIVE ANALYSIS OF THE DEVELOPMENT OF B CELLS IN MARGINAL ZONES AND FOLLICLES

I.C.M. MacLennan[1], H. Bazin[2], D. Chassoux[3], D. Gray[1], and J. Lortan[1]

[1]Dept. of Immunology, University of Birmingham Medical School, Birmingham B15 2TJ. [2]Experimental Immunology Unit, University of Louvain, 1200 Brussels, Belgium [3]Institut de Recherches Scientifiques sur le Cancer 94800 Villejuif, France

INTRODUCTION

Two distinct B cell compartments can be identified in rat spleen. The follicles which contain small recirculating B cells which express surface IgM and IgD and the marginal zones, which are populated with static IgM+ve IgD-ve B cells (1,2). Increasing evidence implicates marginal zone B cells in responses to T cell-independent type 2 (TI-2) antigens and recirculating cells in responses to T cell-dependent antigens (3). This paper describes observations which give insight into the development and maturation of the recirculating follicular (RF) B cells and marginal zone (MZ) B cells.

RESULTS AND DISCUSSION

Evidence That RF B Cells Can Develop in Animals Lacking MZ B Cells

Administration of anti IgM antibody from birth has been used by many workers to suppress the development *in vivo* of cells expressing surface membrane immunoglobulin since its introduction by Kincade et al (4). A minority of euthymic rats treated in this way for the first 6 weeks of life have B cells in the follicles of secondary lymphoid organs (5). This has been observed in 12 rats. None of these rats' spleens showed morphological or immunohistological evidence of marginal zone B cells. The follicular B cells in these animals have strong surface IgD expression and weak expression of IgM (Fig. 1a & b). The evidence provided by these 12 rats indicates that RF B cells can develop independently from MZ B cells.

MZ B Cell Development in Rats Lacking RF B Cells

The data here are less clear cut than those in the previous section. If rats are treated from birth with rabbit-anti-rat IgD, they have marginal zones fully populated with B cells which are surface SIgM+ve IgD-ve (Fig. 2a & b) (6). The spleens do have some B cells in their follicles but these are also SIgM+ve IgD-ve. These spleens resemble those of rats which have been depleted of recirculating B cells (1). Lymph node follicles of anti IgD-treated rats, on the other hand, contain very few B cells (1). More than 90% of the lymphocytes in normal lymph node follicles have the RF B cell phenotype i.e. are both SIgM and SIgD+ve.

Evidence That RF B Cells Can Develop into MZ B Cells

Although the above data provide suggestive evidence that MZ B cells can develop independently from RF B cells recent experiments strongly indicate that some MZ B cells in adult rats are derived from recirculating precursors. Two sets of experiments point to this conclusion. These experiments involve B cell chimaerism between congenic strains of rats differing in kappa allotype. The first experiments used PVG/c^{K1b} hosts, which received 8 Gy whole body γ irradiation. The rats were reconstituted the following day with 3 x 10^8 PVG/c^{K1a} thoracic duct lymphocytes (TDL) and PVG/c^{K1b} bone marrow cells from two femora and two tibias 24 hours later. Three days after reconstitution with allotype-marked TDL the spleens of the recipients contained follicles well populated with SIgM+ve SIgD+ve B cells of donor allotype. Only occasional SIgD-ve cells were seen in the marginal zone (Fig. 1 in ref. 7). In control irradiated rats, given no TDL but kappa allotype-marked congenic bone marrow cells, full repopulation of follicles occurred between 3 and 6 weeks after reconstitution. The test rats receiving marked TDL showed between 20 and 40 percent MZ B cells of donor origin at 6 weeks. A similar level of RF B cell chimaerism remained at that stage (Fig. 2 in ref. 7). Either these MZ B cells of donor origin were derived from RF B cells or the donor TDL contained about one-third as many stem cells as the syngeneic bone marrow. The latter explanation seems unlikely.

The second set of experiments involved parabiosis between congenic rats differing at kappa allotype (3). Here the rats used

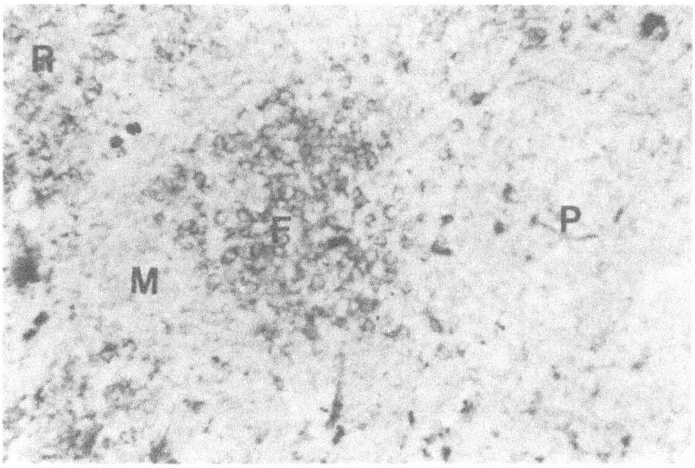

a

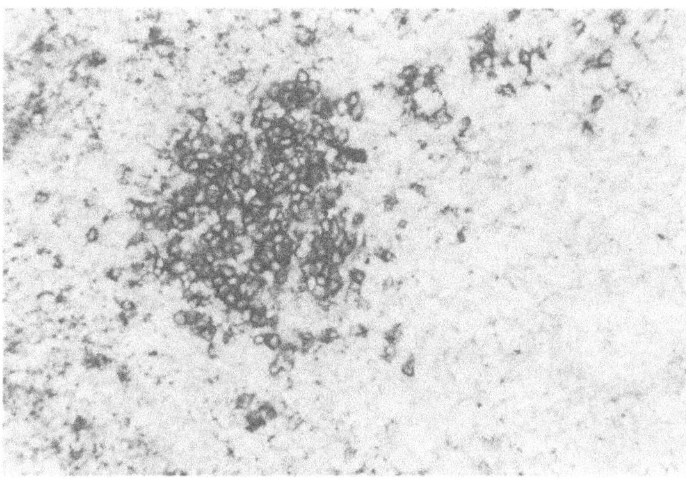

b

Fig. 1 Serial sections of the spleen of a euthymic rat treated from birth with rabbit anti rat IgM. The sections show the presence of follicular B cells without marginal zone B cells. (a) stained for rat IgM (b) stained for rat IgD. M = Marginal zone F = follicle P = periarteriolar lymphocytic sheath R = red pulp

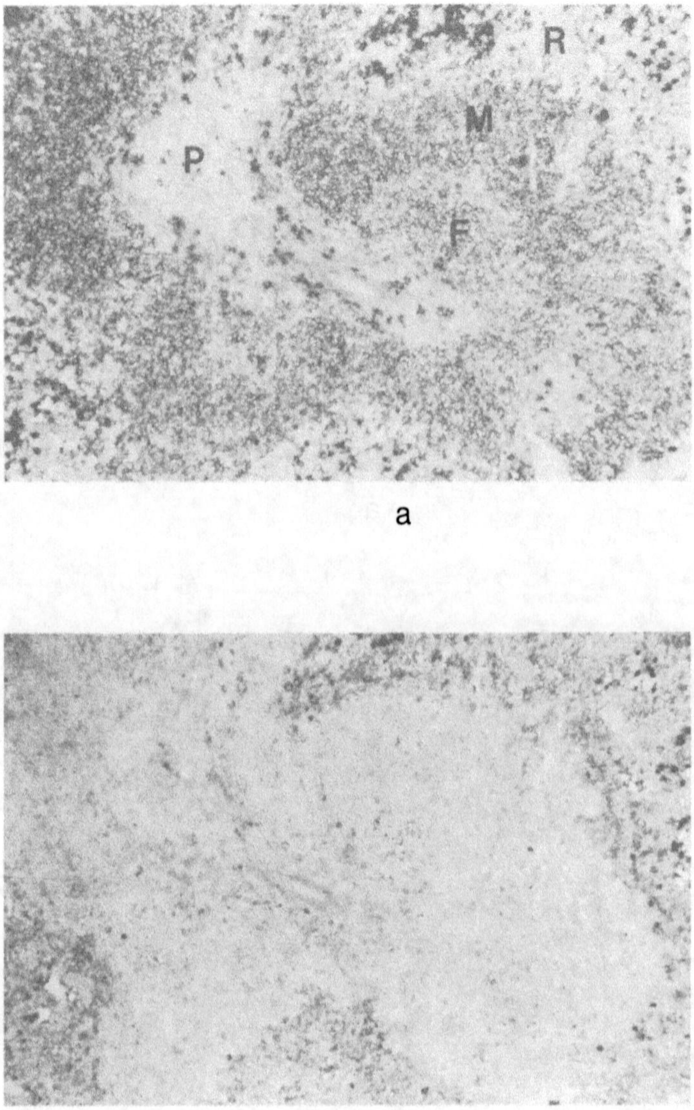

Fig. 2 Serial sections of the spleen of a congenitally athymic rat treated with rabbit anti rat IgD. (a) stained for rat IgM (b) stained for rat IgD. M = Marginal zone F = follicle P = periarteriolar lymphocytic sheath R = red pulp

were on a Lou/C background. Cross circulation achieved by parabiosis allows exchange of RF but not MZ B cells. One would, however, also expect some MZ B cell chimaerism by natural turnover and repopulation from primary lymphoid organs. Parabiosis was established for 10 days. At the end of this period the rats were separated. Some were sacrificed at this stage and some were killed 15 days after separation. The level of chimaerism in both groups was assessed by double fluorescence. The level of RF B cell chimaerism did not fall significantly in the 15 days after separation. However, MZ B cell chimaerism increased from a median of 1.5% to 4%. These data give some indication of the rate of RF B cell conversion to MZ B cells. If 25% RF B cells give rise to approximately 2% of MZ B cells in 15 days, about 0.5% MZ B cells in intact animals per day are derived from RF B cells. This rate of conversion seems to be largely independent of the size of the MZ B cell pool for similar rates of conversion were observed in the TDL transfer experiments quoted above where the MZ B cell pool was grossly depleted for the first 3 weeks after TDL transfer. This slow tempo of RF to MZ B cell conversion is not sufficient to explain the rates of MZ B cell population during ontogeny (8). Also where rats' B cell production is suppressed from birth with rabbit anti-rat IgM plus anti-IgD, MZ B cells have been shown to reach normal levels within a week of stopping suppression (9).

CONCLUSIONS

The results of the experiments quoted in this paper can be used to draw a tentative scheme for MZ B cell and RF B cell development. Primary development of RF B cells can occur in the absence of MZ cells. It also is probable that MZ B cells can be derived directly from stem cells. However, RF B cells can give rise to MZ B cells. This conversion may well be secondary to antigenic stimulation of RF B cells. No data are available, which indicate that MZ B cells can convert to RF B cells.

REFERENCES

1. D. Gray, I. C. M. MacLennan, H. Bazin, and M. Khan. Eur. J. Immunol. 12:562 (1982).
2. I. C. M. MacLennan, D. Gray, D. S. Kumararatne, and H. Bazin. Immunology Today 3:305 (1982).
3. D. Gray, I. C. M. MacLennan, B. Platteau, H. Bazin, J. Lortan, and G. D. Johnson. Evidence that static but not recirculating B cells are responsible for antibody production against dinitrophenol on neutral polysaccharide. This volume.
4. P. W. Kincade, A. R. Lawton, D. E. Brockman, and M. D. Cooper. Proc. Natl. Acad. Sci. USA. 67:1918 (1970).

5. D. Chassoux, J. P. Kolb, H. Bazin, and I. C. M. MacLennan. Immunology 50:327 (1983).
6. H. Bazin, D. Gray, B. Platteau, I. C. M. MacLennan. Ann. N. Y. Acad. Sci. 339:157 (1982).
7. D. S. Kumararatne, D. Gray, I. C. M. MacLennan, J. Lortan, B. Platteau, and H. Bazin. The paradox of high rates of B cell production in bone marrow and the longevity of most mature B cells. This volume.
8. C. Dijkstra, and E. A. Dopp. Cell Tissue Res. 229:351 (1983).
9. H. Bazin, B. Platteau, I. C. M. MacLennan, M. Khan, and G. D. Johnson. This volume.

GERMINAL CENTER CELLS: ANTIGEN SPECIFICITY, HEAVY CHAIN CLASS EXPRESSION AND EVIDENCE OF MEMORY

G. Kraal, I.L. Weissman[1] and E.C. Butcher[1]

Dept. of Histology, Free University, Amsterdam, The Netherlands; [1] Dept. of Pathology, Stanford Medical Center, Stanford University, Stanford Ca 95305, U.S.A.

INTRODUCTION

Germinal centers arise as discrete sites of B-cell differentiation in B cell areas of lymphoid tissue after antigenic stimulation. In the mouse germinal center cells can be confidently identified in cell suspensions as PNA^{hi} B cells, binding 10-30 times as much PNA as other B and T cells or plasma cells (PNA^{lo}) and are furthermore characterized by a lack of surface IgD[1-3]. In Peyer's patches the majority of germinal center cells bear the IgA isotype whereas in lymph nodes the predominant isotype after primary stimulation is IgM[2,3]. Surface IgG expression was observed after secondary stimulation of the lymph nodes, suggesting an important role for the germinal center in heavy chain class switching[2]. The germinal center population contains the majority of antigen-binding cells[2] and in a study on the localization of germinal center cells it has been demonstrated that they lack normal migratory and homing mechanisms most small lymphocytes possess[4]. Using two color FACS analysis we studied the stages in germinal center development and looked at the memory potential of these cells using an adoptive transfer system.

MATERIALS AND METHODS

Mice

BALB/c mice were used throughout the experiments.

Antigens and immunization

For the kinetic studies mice were immunized in the front footpad with either 10^7 SRBC or Keyhole limpet hemocyanin (KLH) in saline. For the transfer experiments B cell donor mice were primed with 100 μg alumin precipitated DNP-7-KLH in the front footpads. T cell donors were carrier-primed i.p. with 100 μg KLH.

Cell labeling reagents

FITC-conjugated as well as unconjugated PNA were purchased from E.Y. Labs. Unconjugated PNA was coupled to Texas Red (TR) as described elsewhere[3]. Affinity purified rabbit antiserum against KLH was also directly coupled to Texas Red. Monoclonal antibodies 10.4.22 (anti-Igh-5a)[5] and MEL-14 (recognizing a cell surface determinant assoicated with the receptor for peripheral lymph node HEV)[6], were used as biotin conjugates in combination with FITC-conjugated avidin (Vector Labs, Ca).

Suspension staining

Cell suspensions from axillary and brachial lymph nodes were stained with 10 μl/10^6 of first stage antibodies for 15 min. The cells were washed three times and resuspended in second stage reagents. For PNA staining TR-PNA or FITC-PNA was added at 25 μg/ml halfway through the period of incubation with second stage reagents.

Two-color immunofluorescence analysis

A dual-laser FACS equipped with logarithmic amplifiers for the fluorescence channels was used[7,8]. Dead cells were excluded using propidium iodine. Texas Red, Fluorescein and scatter measurements for each cell were collected and stored using a VAX computer. In most cases, measurements were collected on a total of 30,000 (live) cells[7,8].

Cell transfer and memory assay

Seven days after primary or secondary immunization with DNP-KLH axillary and brachial lymph node suspensions were treated with anti-Thy-1 plus complement. Thereafter the cells were stained with either PNA or MEL-14 and sorted in negative and positive populations using the FACS. Positive, negative and unselected cells were transferred in lethally irradiated recipients with 10 μg DNP-KLH and 2.5×10^6 nylon-wool passed KLH-primed T cells. Seven days after transfer the anti-DNP activity in the sera of the recipients was determined by RIA[9] as index of the amount of B cell memory present in the donor B cell population.

RESULTS

During the time course after stimulation it was shown that the loss of IgD correlates very well with the appearance of PNA recep-

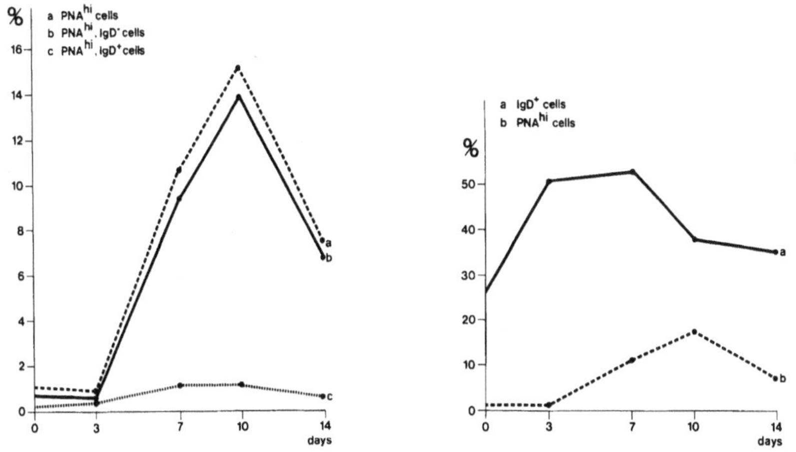

Figs. 1 and 2. <u>PNA^{hi} cells are IgD$^-$.</u>
Animals were immunized with KLH in the front footpads 3, 7, 10 or 14 days prior to sacrifice, and cell suspensions of the draining brachial lymph nodes were stained with Texas Red-PNA and fluorescein-indicated anti-IgD. The percentage of cells staining with one or both of the reagents was calculated by computer-integrations of the stored data from two-color FACS analysis. Calculations were done on 30,000 viable cells per sample and no size selection was made. Staining conditions and data integration-boundaries were kept constant for a given experiment. The data are from one of many representative experiments. Similar data were obtained after immunization with SRBC.

tors (Fig. 1) and that previous to the appearance of germinal center cells there is a strong increase of IgD bearing cells, suggesting that after stimulation specific retention of IgD B cells takes place in the lymph nodes (Fig. 2).

Antigen-binding cells in the germinal center population

It was found that in KLH-stimulated lymph nodes almost all cells that bound KLH specifically were PNA^{hi} (Fig. 3). At seven days after immunization they constitute up to 25% of the PNA^{hi} germinal center cells. That these cells are indeed IgD$^-$, B cells were demonstrated by double immunofluorescence using anti-B220 and anti-IgD in combination with KLH binding (data not shown).

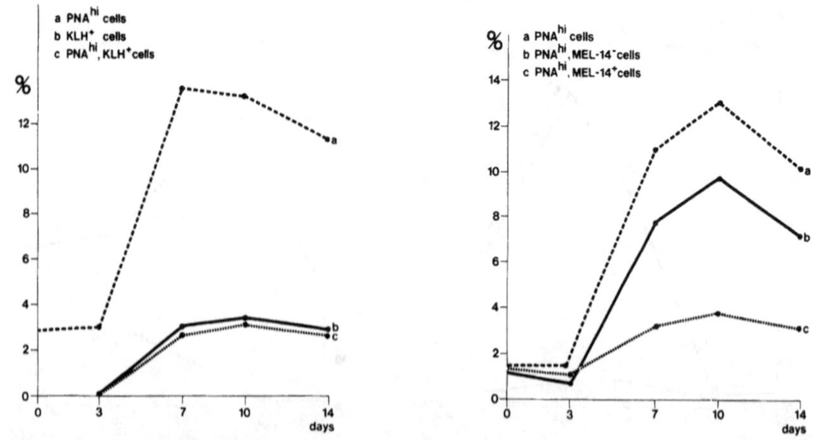

Fig. 3. KLH-binding cells are PNA^{hi}
On various days after immunization with KLH lymph node cells were stained for KLH and PNA and the percentage of cells staining with KLH was determined by integration of the contour plots, subtracting the background binding of KLH exhibited by cells from control animals immunized with SRBC at the same time points.

Fig. 4. The percentage of PNA^{hi} cells bearing MEL-14 on various days after primary immunization with KLH. See legend Fig. 1 for data analysis.

Germinal center cells do not stain with MEL-14

The monoclonal antibody MEL-14 recognizes determinants associated with the putative homing receptor and reacts with almost all IgD$^+$ B cells in the lymph node. After stimulation the majority of germinal center cells do not express the HEV receptor although a small portion of the PNA^{hi} cells does stain with MEL-14 (Figs. 4,5).

Memory transfer by germinal center cells

PNA^{hi} or B cells were sorted from lymph nodes that had been primed with DNP-KLH one week previously and transferred into irradiated recipients together with carrier-primed T helper cells. The serum titers of the recipients showed that most of the cells capable of transferring a secondary response are in the PNA^{hi} germinal center population (Fig. 6). The bars in Fig. 5 reflect a direct comparison of the various populations in the lymph node: per recipient 3-4 times as many PNA^{lo} as PNA^{hi} cells was transferred. The predominant isotype found in these transfers was IgG1, no IgM could be demonstrated and only small amounts of IgG2a and IgG3. When germinal centers are the site of memory-cell differen-

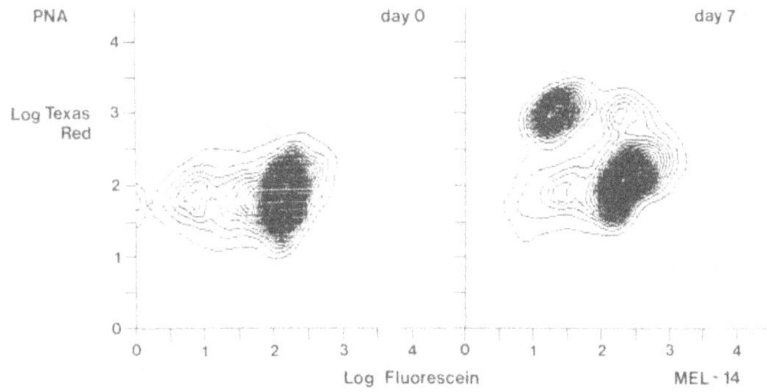

Fig. 5. The majority of PNA^{hi} cells are $MEL\text{-}14^{-}$
Two-color FACS contour plot of peripheral lymph node cells 0 and 7 days after primary immunization with KLH.

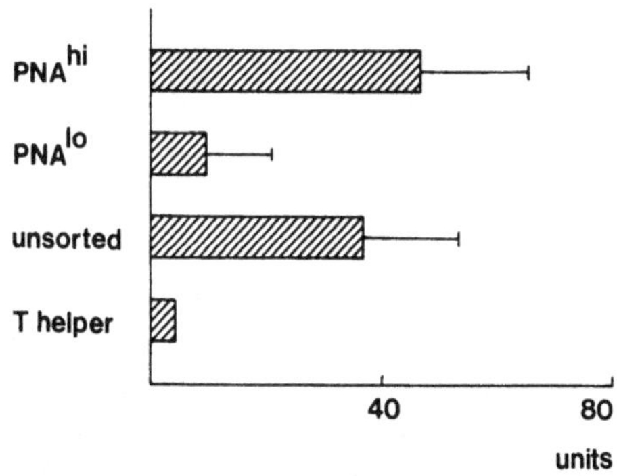

Fig. 6. Transfer of memory after primary immunization
Lymph node cells from animals immunized (DNP-KLH) in the footpad 1 week previously were depleted of T cells and sorted using a FACS. The sorted populations were transferred together with carrier-primed helper T cells and antigen into irradiated recipients. One week later the sera of the recipients were tested with a solid-phase radio-immuno-assay for the presence of anti-DNP antibodies. The anti-DNP response is expressed in arbitrary units relative to an adoptive secondary standard serum. All responses are IgG1 (Igh-4a) responses. The results are from a representative experiment. Each bar represents the mean (± SE) of 3-5 recipients.

tiation, the memory cells will eventually leave the germinal center, re-expressing MEL-14 in order to recirculate. To test this transfer experiments were performed in which cells were sorted for MEL-14 6 weeks after primary immunization (Fig. 7). At this time memory cells

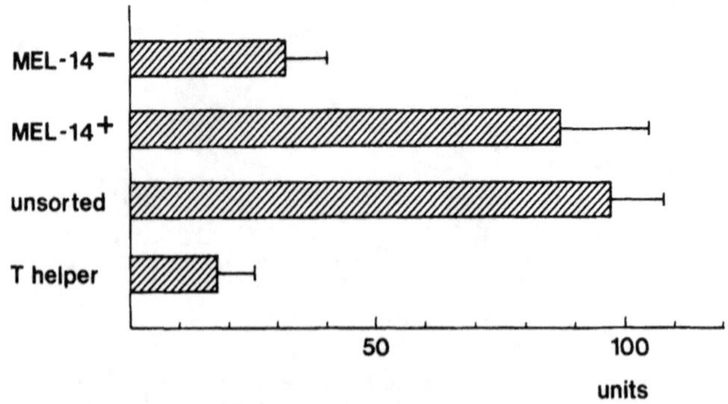

Fig. 7. Transfer of memory after primary immunization. Transfer of $MEL\text{-}14^+$ and $MEL\text{-}14^-$ cells
See legend Fig. 1 for experimental protocol and data analysis.

will have left the germinal center and memory cells are now found in the MEL-14 positive population. This illustrates that memory cells can re-express the homing receptor after differentiation in the germinal center.

DISCUSSION

The hyperplasia of the B cell compartment after antigenic stimulation of the lymph node as demonstrated by the increase of IgD^+ cells precedes the actual germinal center formation. This increase is not caused by changes in the HEV specificity[10], but results from proliferation and specific retention inside the lymph nodes.

Germinal center formation starts near the paracortical T cell area in B cell follicles densely populated with surface IgM^+, IgD^+ cells. It is therefore likely that the germinal center starts as a differentiation of IgM^+, IgD^+ cells at certain sites in the B cell area, but the exact identity of the germinal center precursor cell has not been established. If the precursor of the germinal center cell is a small, recirculating IgD^+ B cell the set of events leading to the formation of a germinal center involve major phenotypic changes in these cells such as the appearance of PNA receptors and the loss of surface IgD and MEL-14 receptors that occur almost simultaneously. In this report we have established that the germinal center formation will lead to the production of memory cells. One week after a primary immunization the majority of memory cells can be found within the germinal centers. They are PNA^{hi} and lack IgD and MEL-14. At later time points after primary immunization memory cells now carry the receptor for MEL-14, strongly suggesting that memory cells undergo a transient non-migratory phase of differentiation inside the germinal center but will re-express the homing receptor on their surface just before or after leaving this site.

The isotype of the adoptive memory response is predominantly IgG, demonstrating that second antigen contact, typical of a memory related immune response, is necessary. The observed immunoglobulin switch and the expression of the various markers, in particular the switch-off switch-on of the MEL-14 clearly illustrate the complex set of events that take place inside the germinal centers and emphasize the role they play in B cell differentiation.

ACKNOWLEDGEMENTS

During this work G.K. was a fellow of the Fogarty International Center. E.C.B. is a Scholar of the Leukemia Society of America. This work was supported by grant AI 09072 from the National Institutes of Health and by a grant from the Veterans Administration.

REFERENCES

1. M.L. Rose, M.S.C. Birbeck, V.J. Wallis, J.A. Forrester, and A.J.S. Davies. Nature 284:364 (1980).
2. G. Kraal, I.L. Weissman, and E.C. Butcher. Nature 298:377 (1982).
3. E.C. Butcher, R.V. Rouse, R.L. Coffman, C.N. Nottenburg, R.R. Hardy, and I.L. Weissman. J. Immunol. 129:2698 (1982).
4. R.A. Reichert, W.M. Gallatin, I.L. Weissman, and E.C. Butcher. J. Exp. Med. 157:813 (1983).
5. V.T. Oi, P.P. Jones, J.W. Goding, L.A. Herzenberg, and L.A. Herzenberg. Curr. Top. Microbiol. Immunol. 81:115 (1978).
6. W.M. Gallatin, I.L. Weissman, and E.C. Butcher. Nature 304:30 (1983).
7. R.R. Hardy, K. Hayakawa, J. Haaijman, and L.A. Herzenberg. Nature 297:589 (1982).
8. D.R. Parks, R.R. Hardy, and L.A. Herzenberg. Immunol. Today 4:145 (1983).
9. T.T. Tsu, L.A. Herzenberg. In: Selected methods in cellular immunology (Mishell and Shiigi, eds.), Freeman & Co Publ. San Francisco, p. 373 (1980).
10. G. Kraal, and A. Twisk. Eur. J. Immunol. 14:586 (1984).

THE DEVELOPMENT OF SPECIFIC ANTIBODY PRODUCING CELLS IN THE SPLEEN OF RABBITS DURING THE PRIMARY AND SECONDARY IMMUNE RESPONSE

N. van Rooijen, N. Kors and R. van Nieuwmegen

Immunocytochemistry Unit, Medical Faculty, Department of Histology, Free University, P.O. Box 7161
1007 MC Amsterdam, The Netherlands

INTRODUCTION

Until recently, no suitable method was available to detect specific antibody producing cells in the lymphoid tissues, except for cells producing antibodies against the antigen horseradish peroxidase (HRP). The major disadvantage of HRP is that it is a weak antigen and has to be administered together with a strong adjuvant such as aluminum phosphate[1] or Freund's complete adjuvant[2]. An intravenous injection e.g., in order to study the immune response in the spleen could not be given. Moreover, the antigen HRP when adsorbed to aluminum phosphate or emulsified in Freund's complete adjuvant may be released from the injection site only slowly. In that case it is difficult to determine, at what time interval after injection the primary immune response finishes and the secondary immune response starts. Recently we introduced a new approach for the detection of specific antibody producing cells in lymphoid tissues. An antigen-HRP conjugate or an antigen-alkaline phosphatase (AP) conjugate was used for detection of cells containing specific antibodies in their cytoplasm[3-11]. The aim of the present contribution was to review our recent findings on the development of specific antibody producing cells during the primary and secondary immune response in the spleen. Specific antibody producing cells were found in various compartments of the white pulp of the spleen.

MATERIALS AND METHODS

Experimental design

Young male chinchilla rabbits were intravenously injected with human serum albumin (HSA), bovine gamma globulin (BGG), penicilloyl-

BGG (Pen-BGG) or TNP-Keyhole limpet haemocyanin (TNP-KLH). At different time intervals after injection, a spleen biopsy was taken from the rabbits. One till two months after the primary injection a booster injection was given with the same antigen as used for priming. At different time intervals hereafter, another spleen biopsy was taken from the rabbits. Specific antibody forming cells against the antigens HSA and BGG or against the haptens penicilloyl and TNP were detected using an antigen (hapten)-enzyme conjugate followed by enzymehistochemistry. HSA-HRP conjugate was used for detection of anti-HSA antibody producing cells[3-5]. Pen-HSA-HRP conjugate was used for detection of anti-penicilloyl antibody producing cells[6,7]. BGG-AP conjugate was used for detection of anti-BGG antibody producing cells[7-9]. TNP-AP conjugate was used for detection of anti-TNP antibody producing cells[10,11].

<u>Detection of specific antibody producing cells using an antigen (hapten)-enzyme conjugate</u>

Pieces of spleen were frozen in liquid nitrogen and afterwards cryostat sections of 8 µm thickness were cut. Sections were air-dried overnight. They were fixed for 10 min at 20°C with acetone (pro analyse) to which H_2O_2 was added (0.02% final concentration). They were rinsed 3x in PBS (pH 7.4). The sections were incubated overnight horizontally at 4°C with the antigen (hapten)-HRP conjugate in PBS + 0.05% BSA. Afterwards they were washed with PBS and treated for 8 min with 3-amino-9-ethylcarbazole (AEC) + H_2O_2[12] or with 3,3'-diaminobenzidine-tetra-HCl (DAB) + H_2O_2[13]. Sections were briefly counterstained with haematoxylin and mounted in glycerine-gelatine. Sections were incubated with antigen (hapten)-AP conjugate as described above. They were then washed 3 times with PBS and treated for alkaline phosphatase activity according to Burnstone (see Pearse, ref. 14) using Fast-Blue-BB salt for 60 min at 37°C. They were washed, counterstained and mounted as described above.

RESULTS

<u>Localization of specific anti-HSA antibody producing cells in the spleen during the primary immune response</u>

In control animals, not injected with HSA, no specific anti-HSA antibody producing cells were found. In the spleen of rabbits, considerable numbers of specific antibody producing cells could be detected already at 4 days after injection of 10 mg of HSA. The bulk of these cells localized in the coaxial sheaths of lymphoid tissue around the terminal arteriolar branches (Fig. 1) and in the outer part of the periarteriolar lymphocyte sheaths (PALS; Fig. 2). Their numbers increased in the following days but decreased rapidly after one week so that only very few, if any, specific antibody producing cells could be detected on these sites at two weeks after injection.

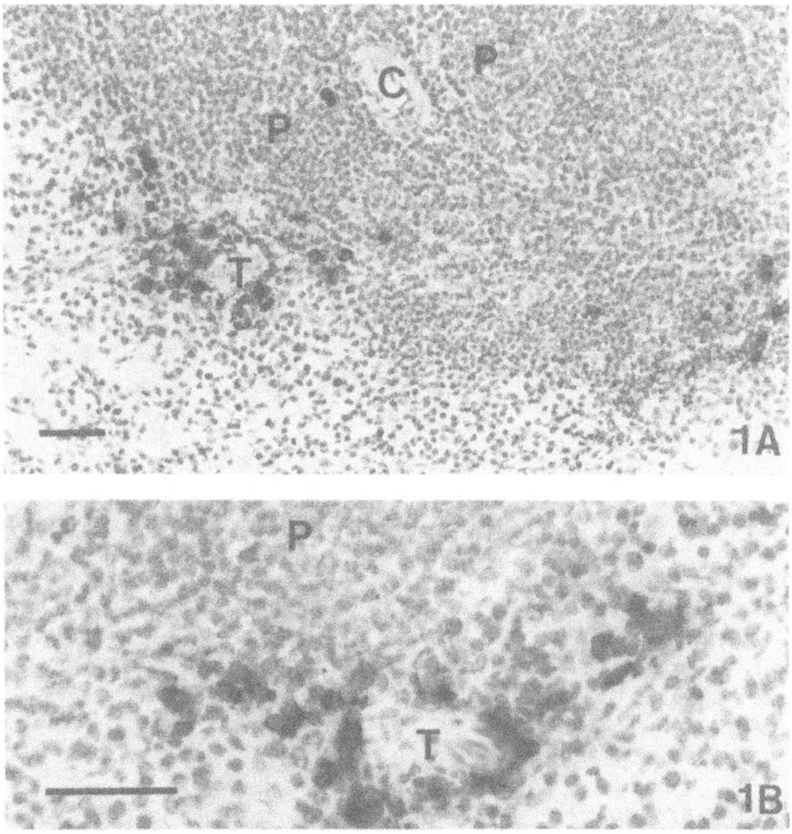

Fig. 1. Anti-HSA antibody producing cells in coaxial sheaths of lymphoid tissue around a terminal arteriolar branch in the spleen of a rabbit at 6 days after a single intravenous injection of 10 mg HSA. A: survey; B: magnification. C = central arteriole; P = periarteriolar lymphocyte sheath; T = terminal arteriolar branch. Bar = 100 um.

Relatively few specific antibody producing cells were detected in the red pulp and in the inner PALS. Some specific antibody producing cells were also found in the follicles of the white pulp, from day 5 after antigen injection. The numbers of antibody producing cells per follicle varied in a wide range. Most follicles showed no or only a few cells per section, whereas in other follicles of the same animal, several specific antibody producing cells could be detected.

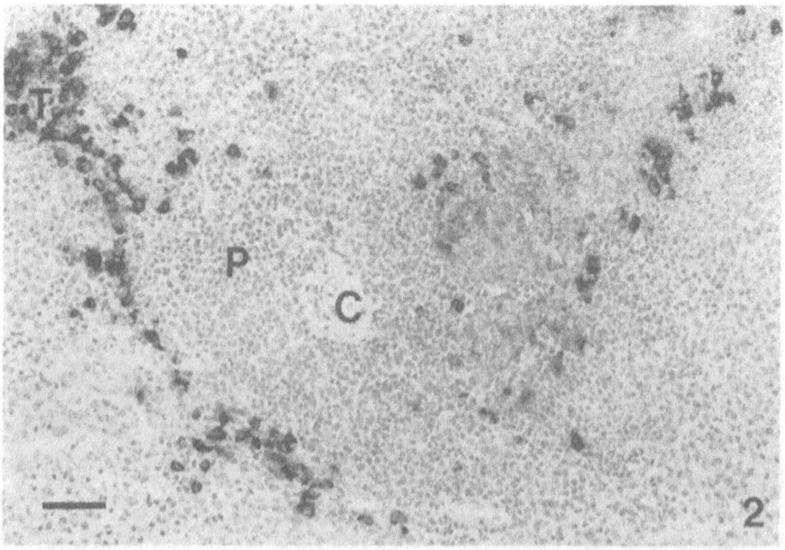

Fig. 2. Anti-HSA antibody producing cells in the outer periarteriolar lymphocyte sheath (P) and in a coaxial sheath of lymphoid tissue around a terminal arteriolar branch in the spleen of a rabbit at 6 days after a single intravenous injection of 10 mg HSA. C = central arteriole; T = terminal arteriolar branch. Bar = 100 μm.

Localization of specific anti-HSA antibody producing cells in the spleen during the secondary immune response

We found only sporadically anti-HSA antibody producing cells at one till two months after a primary injection of HSA (before the booster injection). Two days after the booster injection of HSA anti-HSA antibody producing cells could be detected in the spleens of most of the rabbits. Large numbers of specific antibody producing cells were found in the spleens of all rabbits at 4 days after the booster injection, whereas their number had decreased again after 8 days. The bulk of the antibody producing cells were found in the coaxial sheaths of lymphoid tissue surrounding the terminal arterioles in the spleen as we found during the primary response. Only relatively few cells were found in the PALS. In the spleens of part of the rabbits, a concentration of antibody producing cells was found in the follicles (Fig. 3). Some of these follicles contained many specific antibody producing cells, while other follicles contained no cells or only few of them. In follicles with germinal centers specific antibody producing cells were found in the periphery of these centers (Fig. 3A), while other follicles without germinal centers contained anti-HSA producing cells concentrated in the follicle centers (Fig. 3B).

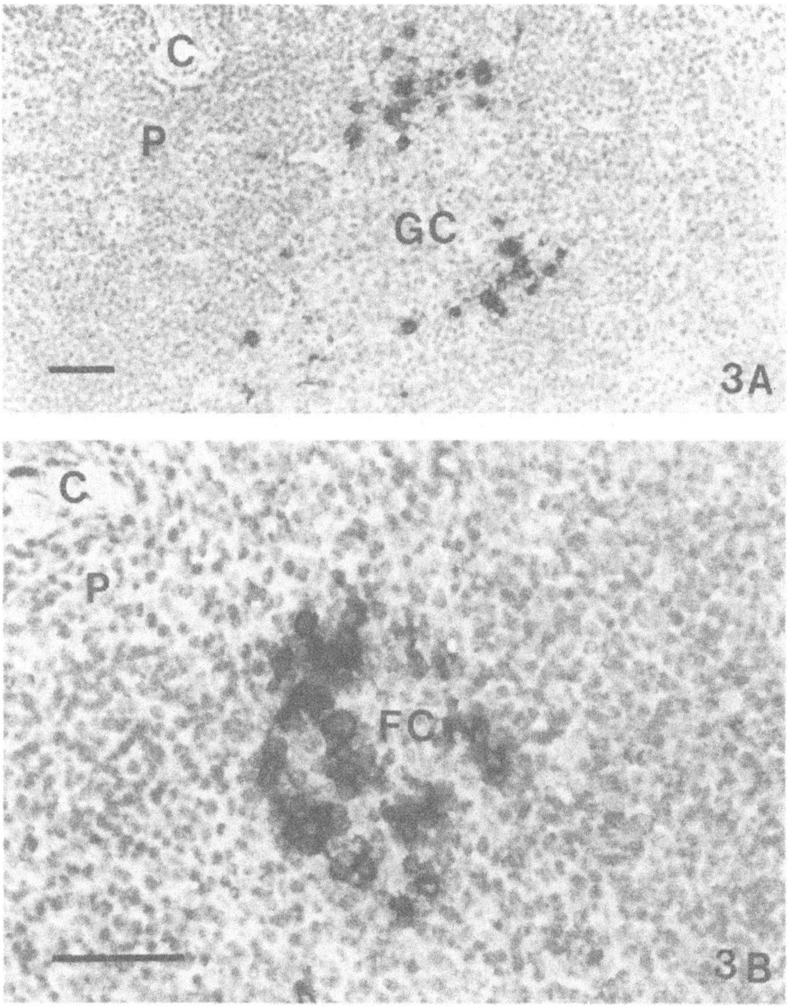

Fig. 3. Anti-HSA antibody producing cells in follicles in the spleen of a rabbit at 4 days after a booster injection of 120 μg HSA. A: specific antibody producing cells in the periphery of a germinal center; B: specific antibody producing cells in a follicle center. C = central arteriole; P = periarteriolar lymphocyte sheath; FC = follicle center; GC = germinal center. Bar = 100 μm.

Localization of anti-BGG antibody producing cells, anti-penicilloyl antibody producing cells and anti-TNP antibody producing cells in the spleen

The localization patterns of anti-BGG antibody producing cells, anti-penicilloyl antibody producing cells and anti-TNP antibody producing cells in the spleen of rabbits during the primary and secondary immune responses against these antigens or haptens did not differ from the patterns of anti-HSA antibody producing cells described above.

DISCUSSION

The distribution pattern of specific antibody producing cells during the secondary immune response did not differ markedly from that observed during the primary response against the same antigen. We found relatively more cells in the outer parts of the PALS during the primary immune response and relatively more cells in the follicles during the secondary immune response. During both the primary and the secondary immune response, most cells were found in the small coaxial sheaths of lymphoid tissue surrounding the terminal arteriolar branches. Specific antibody forming cells have been frequently studied in vitro; by contrast; only few studies deal with antibody-forming cells in vivo. Nearly all in vivo studies concerned cells producing antibodies against the antigen HRP, since these cells can be demonstrated very easily after incubation with the enzyme itself. The most important disadvantage of HRP as an antigen is that it is a very weak antigen and therefore has to be administered together with a strong adjuvant. For that reason, a primary immune response against HRP cannot be studied in the spleen. The animals cannot be primed intravenously since the antigen has to be administered together with the adjuvant. In a recent study Geldof et al.[2] studied the secondary immune response against HRP in the spleen of mice. The mice were primed subcutaneously with HRP in Freund's complete adjuvant and boosted intravenously with the same antigen without the adjuvant. The distribution pattern of anti-HSA producing cells during the secondary immune response against HSA in rabbits differs markedly from the distribution pattern of anti-HRP producing cells observed during the secondary response against HRP in mice[2]. By contrast with the results of Geldof et al.[2], we found specific antibody producing cells in the follicles of the spleen. Since it is not probable that the different localization patterns are caused by the different antigens or animals used in both studies, we believe that the method of administration of the antigens may be responsible for the differences in both studies. Contrary to HSA injected intravenously during our experiments, HRP had to be administered subcutaneously together with FCA for priming. Primed cells may then be expected mainly in the lymph nodes, and may later travel to the spleen, where they can be found in the white pulp, except for the follicles. After intravenous priming, primed cells may well develop in the follicles. Sordat et al.[1] demonstrated anti-HRP antibody forming cells in follicles in the lymph nodes of mice after footpad injection of HRP together with a strong adjuvant.

REFERENCES

1. B. Sordat, M. Sordat, M.W. Hess, R.D. Stoner, and H. Cottier. J. Exp. Med. 131:77 (1970).
2. A.A. Geldof, M. v.d. Ende, E.M. Janse, and T. Sminia. Cell Tissue Res. 231:135 (1983).
3. N. van Rooijen, N. Kors, R. van Nieuwmegen, and P. Eikelenboom. Anat. Rec. 206:189 (1983).
4. N. van Rooijen, R. van Nieuwmegen, N. Kors, and P. Eikelenboom. Anat. Rec. 208:579 (1984).
5. N. van Rooijen, N. Kors, and R. van Nieuwmegen. J. Histochem. Cytochem. 32:677 (1984).
6. N. van Rooijen, N. Kors, D.M. Boorsma, P. de Haan, and R. van Nieuwmegen. Immunol. Commun. 13:29 (1984).
7. N. van Rooijen, N. Kors, D.M. Boorsma, P. de Haan, and R. van Nieuwmegen. Immunology 51:417 (1984).
8. N. van Rooijen, N. Kors, and R. van Nieuwmegen. Anat. Rec. 209:385 (1984).
9. N. van Rooijen, N. Kors, and R. van Nieuwmegen. J. Histochem. Cytochem. 32 (1984), in press.
10. E. Claassen, and N. van Rooijen. Submitted for publication.
11. N. van Rooijen, N. Kors, and E. Claassen. J. Histochem. Cytochem. in press.
12. R.C. Graham, U. Lundholm, and N.J. Karnovsky. J. Histochem. Cytochem. 13:150 (1965).
13. R.C. Graham, and N.J. Karnovsky. J. Histochem. Cytochem. 14:291 (1966).
14. A.G.E. Pearse. Histochemistry. Theoretical and Applied. Vols. I and II, Churchill Livingstone, Edinburgh, U.K.

THE ONTOGENETIC DEVELOPMENT OF THE FOLLICULAR DENDRITIC CELL IN RAT SPLEEN

C.D. Dijkstra, E.A. Döpp and E.W.A. Kamperdijk[1]

Department of Histology, [1] Department of Electron Microscopy, Medical Faculty, Free University Amsterdam, The Netherlands

INTRODUCTION

Secondary follicles of lymphoid organs contain a highly specialized non-lymphoid cell type, the so-called follicular dendritic cells (FDC). FDC are cells with extraordinary long slender cell processes, which form an intricate labyrinth of invaginations[11,22] in which antigen-antibody complexes can be trapped and retained for several weeks[2]. This trapping is believed to play a role in the induction of immunological memory[17] or as feedback mechanisms regulating the immune response[24].

The origin of the FDC is still a subject of discussion. Several authors[10,13,26,28] believe that FDC are differentiated forms of fibroblastic reticulum cells (FRC), which build the basic stroma of lymphoid organs[16]. Arguments for this vision are the morphological characteristics of FDC (e.g. close relationship with reticulin fibers, desmosomes, formation of a network) and the existence of transitional forms between FRC and FDC e.g. during the ontogenetic development[10,13,26,28]. In these studies, however, no specific marker was used. Other authors[9,12] suggest that FDC belong to the mononuclear phagocyte system[8] since they share phenotypical characteristics with cells of the monocyte/macrophage line like C3 receptors, HLA-DR antigens, class II MHC and some macrophage-related antigens[9]. Recently, monoclonal antibodies have been described that recognize FDC: Ox2[1] (rat) and R4/23[21] (man) react with only FDC and a subset of B cells. KiM4[23] (man) reacts with FDC and with a subset of the endogenous peroxidase positive mononuclear cells in the peripheral blood, also suggestive for the fact that FDC belong to the monocytic line being characterized by endogenous peroxidase activity.

It has been demonstrated that intravenously injected HRP-anti-

HRP complexes are retained exclusively by FDC and no other cell type[2]. Therefore, to obtain evidence for the origin of the FDC we used the highly specific capacity of these cells to trap and retain immune complexes, to study the ontogenetic development of the FDC's. During ontogeny it appeared that such trapping of HRP-anti-HRP occurred in primary follicles (day 21 after birth), i.e. before the development of germinal centers (day 28 after birth[3]). Mature FDC's, however, based on morphological criteria, could not be found in primary follicles[26]. In the present study the morphological characteristics of the immune complex trapping cell in primary follicles are studied at ultrastructural level.

MATERIALS AND METHODS

Animals

Wistar rats originally obtained from the Centraal Proefdierenbedrijf T.N.O., (Zeist, The Netherlands) were bred randomly in our laboratory.

Experimental design

The spleens of 21-day old and of 8-week old rats were investigated 24 h after intravenous injection of 0.2 ml of a solution of HRP complexed with rabbit anti-HRP (PAP, Dako, Denmark). For each experiment at least two rats were used.

Preparation of tissue specimens

Following anesthesia with Nembutal (SA, Abbott), the rats were perfused briefly with Meyler's saline at 37°C, the ionic strength of the saline having been adjusted to that of rat serum[20]. After bleaching of the spleen, perfusion fixation followed with 1.5% glutaraldehyde in 0.09 M buffer (pH 7.1) for 8 min. For technical details of the perfusion method, see Veerman et al.[27]. After the perfusion fixation spleens were removed and cut into pieces of about 0.5 mm thickness. The pieces were fixed for 2-3 h at 4°C in a solution of 2% glutaraldehyde in 0.2 M sodium cacodylate buffer, pH 7.4. These prefixed tissue specimens were cut on a Vibratome into sections of 100 µm thickness, which were incubated with 3,3'-diaminobenzidine-tetra-HCl (DAB, Sigma, USA), 0.5 mg per ml 0.05 M Tris-HCl buffer, pH 7.6, containing 0.01% H_2O_2. Sections were then thoroughly washed, postfixed in 1% osmium tetroxide in 0.2 M sodium cacodylate buffer for 1 h at 0-4°C, dehydrated with ethanol and embedded in Araldite. From the 100-µm thick sections the easily recognizable follicular areas, where the trapped immune complexes are visible, were selectively trimmed out. Then semithin sections of 1 µm were cut and stained with toluidine blue for light microscopy. Ultrathin sections were stained with uranyl acetate and lead citrate and examined with a Philips EM 301 electron microscope at 40 or 60 kV.

RESULTS

The rat spleen consists of two major areas, the hemopoietic red pulp and the lymphoid white pulp. The white pulp consists of three compartments: 1) the periarteriolar lymphocyte sheath (PALS). 2) the follicles and 3) the marginal zone surrounding both PALS and follicles. Germinal centers do not occur before day 28 after birth[4,26] so all follicles present at day 21 may be considered as being primary follicles.

HRP-anti-HRP complexes, 24 h after i.v.-injection, can be recognized both in primary and in secondary follicles at the light-microscopical level as irregular strands of dark brown granules[3,6] at the ultrastructural level as extracellularly located electron-dense material[2].

Primary follicles

In the primary follicles of spleens from 21-day-old rats reticulum cells (RCs) could be observed (Fig. 1). The nuclei of these cells were oviform, sometimes triangular or polygonal in profile; they usually had an euchromatic structure and occasionally contained a nucleolus. The nucleus was surrounded by a small rim of cytoplasm. There was no distinct cytocenter. Some ribosomes and rough endoplasmic reticulum were present, surrounding the mitochondria. The long slender cell processes were 5-10 μm in length and sometimes contained within their cytoplasm a stack of Golgi vesicles. A well-developed smooth endoplasmic reticulum was also present, consisting of vesicles and tubules. The cell processes occasionally contained microtubules and microfilaments, usually arranged parallel to the cell membrane. These RCs enwrapped collagen fibers or were in close apposition to them; thereby being very similar to the reticulum cells present in the marginal zone. However, the RCs in the primary follicles showed a number of small membrane folds, which had a diameter of 80-150 nm. These were located in small patches on the cell body or on the cytoplasmic processes. Usually electron-dense material was seen in these foldings representing the extracellularly intravenously injected HRP-anti-HRP complexes. The primary follicles never showed characteristic follicular dendritic cells (FDC).

Secondary follicles

Well-developed follicular dendritic cells (FDC) with slender cytoplasmic processes forming a web-like structure were only observed in the germinal centers of spleens from young adult rats. The nuclei of these cells had an euchromatic structure and were triangular or quadrangular in profile. The cytoplasm consisted of a thin layer surrounding the nucleus with extensions forming the long cell processes. It contained some mitochondria, small strands of rough endoplasmic reticulum, a few stacks of Golgi cisternae and

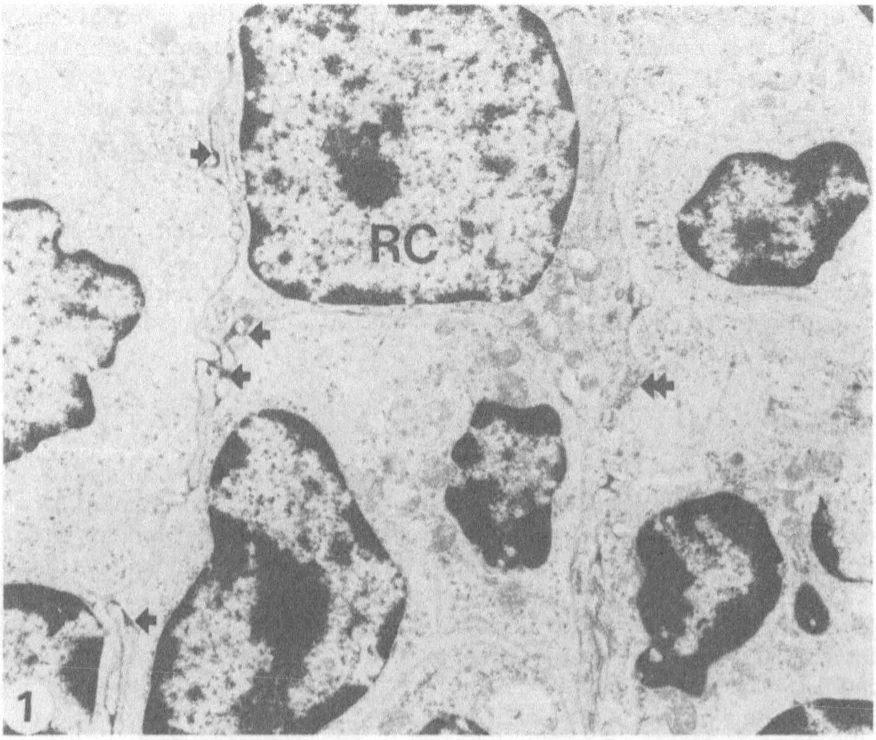

Fig. 1. Immune complex trapping cell in a primary follicle; spleen of 21-day old rat. The cell processes show membrane infoldings marked by extracellular HRP-anti-HRP complexes (arrows). The cell is in close apposition to collagen fibers (double arrow).

many other smooth vesicles and tubules. The entire surface of these FDC's consisted of an intricate system of plasma-membrane invaginations embedded in a homogeneous, electron-dense substance of the injected HRP-anti-HRP complex. Desmosome-like structures, normally present on FDC[13] could not be recognized due to the presence of these immune complexes.

DISCUSSION

The present study shows that HRP-anti-HRP complexes in primary follicles are trapped and retained on cells with characteristics of reticulum cells of mesenchymal origin. These reticulum cells show no signs of phagocytic ability, form networks and are in close relationship with reticulin fibers, characteristics which distinguish them from macrophages and lymphoid cells[16]. These reticulum cells are the immature forms of FDC, because, like the FDC, they are capable to retain HRP-anti-HRP immune complexes. So the results present evidence that FDC originate from reticulum cells of mesenchymal origin. Several authors draw the same conclusion from their ultrastructural studies of the development of the FDC[10,13,26,28], based on the observation of transitional forms between reticulum cells and FDC. Our results, however, are in contrast with those of authors who believe that FDC belong to the monocyte macrophage line[9,12,23]. Against their arguments the following remarks can be made.

1) In mice none of the monoclonal antibodies against macrophages (Mac-1, Mac-2, Mac-3; F4/80) react with FDC[7,14]. In rat the monoclonal antibodies which exclusively recognize macrophages (ED2, ED3) do not react with FDC either[5].

2) Of the monocyte/macrophage lineage the stem cell is known to be located in the bone marrow. Recently Humphrey et al.[15] used F1→P bone marrow chimeras between CBA and B10 mice to show that even after 1 year FDC are not derived from bone marrow stem cells.

3) Not only in ontogeny but also in phylogeny indications can be found that immune complex trapping cells are derived from reticulum cells rather than from monocytes or macrophages. Antigen trapping in carp occurs in association with reticular fibers[25] and in the splenic white pulp of the clawed toad and of the red-eared turtle on dendritic cells which are clearly distinguishable from macrophages[18].

4) Last but not least FDC do not show any phagocytic capacity. They contain no phagolysosomes[19], do not ingest or degrade the entrapped complexes, do not show carbon uptake, nor does the enzyme pattern of FDC point to a phagocytic function[6].

We therefore conclude that FDC are differentiated forms of reticulum cells from mesenchymal origin.

REFERENCES

1. A.N. Barclay. Immunology 44:727 (1981).
2. L.L. Chen, J.C. Adam, and R.M. Steinman. J. Cell Biol. 77:148 (1978).
3. C.D. Dijkstra, N.J. van Tilburg, and E.A. Döpp. Cell Tissue Res. 223:545 (1982).
4. C.D. Dijkstra, and E.A. Döpp. Cell Tissue Res. 229:351 (1983).
5. C.D. Dijkstra, E.A. Döpp, P. Joling, and G. Kraal. Proc. 8th GCC, Cambridge (1984).
6. P. Eikelenboom. Cell Tissue Res. 195:445 (1978).
7. T.J. Flotte, T.A. Springer, and G.J. Thorbecke. Am. J. Pathol. 111:112 (1983).
8. R. van Furth, Z.A. Cohn, J.G. Hirsch, J.H. Humphrey, W.G. Spector, and H.L. Langevoort. Bull WHO 46:845 (1972).
9. J. Gerdes, H. Stein, D.Y. Mason, and A. Ziegler. Virch. Arch. B 42:161 (1983).

10. P. Groscurth. Pathol. Rec. Pract. 169:235 (1980).
11. M.G. Hanna, and A.K. Szakal. J. Immunol. 101:949 (1968).
12. E. Heinen, C. Lilet-Leclerq, D.Y. Mason, H. Stein, J. Boniver, D. Radoux, C. Kinet-Denoël, and L.J. Simar. Eur. J. Immunol. 14:267 (1984).
13. U. Heuserman, K-H. Zurborn, L. Schroeder, and H.J. Stutte. Cell Tissue Res. 209:279 (1980).
14. D.A. Hume, A.P. Robinson, G.G. MacPherson, and S. Gordon. J. Exp. Med. 158:1522 (1983).
15. J.H. Humphrey, D. Grennan, and V. Sundaram. Eur. J. Immunol., in press (1984).
16. Y. Imai, K. Terashima, M. Matsuda, M. Dobashi, K. Maeda, and T. Kasajima. Rec. Adv. RES. Res. 21:51 (1983).
17. G.G.B. Klaus, J.H. Humphrey, A. Kunkl, and D.W. Dongworth. Immunol. Rev. 53:3 (1980).
18. F.G.M. Kroese, P. Eikelenboom, and N. van Rooijen. Dev. Comp. Immunol. Supp. 2:69 (1982).
19. C. Lilet-Leclerq, D. Radoux, E. Heinen, C. Kinet-Denoël, J-O. Degraigne, M-P. Houben-Defresne, and L.J. Simar. J. Immunol. Meth. 66:235 (1984).
20. F.J. Meyler. Acad. Thesis, Amsterdam, The Netherlands (1960).
21. M. Naiem, J. Gerdes, Z. Abdulaziz, H. Stein, and D.Y. Mason. J. Clin. Pathol. 36:167 (1983).
22. G.J.V. Nossal, A. Abbot, J. Mitchell, and Z. Lummus. J. Exp. Med. 127:277 (1968).
23. M.R. Parwaresh, H.J. Radzun, M-L. Hansmann, and K-P. Peters. Blood 62:585 (1983).
24. N. van Rooijen. In: Phylogeny of immunological memory, M.J. Manning, ed., Elsevier/North-Holland, Amsterdam, New York, Oxford (1980).
25. C.J. Secombes, M.J. Manning, and E.A. Ellis. Immunol. 47:101 (1982).
26. A.J.P. Veerman. Z. Immun. Forsch. Immunobiol. 150:45 (1975).
27. A.J.P. Veerman, E.C. Hoefsmit, and H. Boere. Stain Technol. 49:111 (1974).
28. A. Villena, A. Zapata, J.M. Rivèra-Pomar, M.G. Barrutia, and J. Fongua. Cell Tissue Res. 229:219 (1983).

ORIGIN AND TURNOVER OF FOLLICULAR DENDRITIC CELLS AND MARGINAL ZONE MACROPHAGES IN THE MOUSE SPLEEN

John H. Humphrey and Vijaya Sundaram

Department of Medicine, Royal Postgraduate
Medical School, London W12 OHS, U.K.

INTRODUCTION

By constructing bone marrow chimaeras between F1 mice and either parent it becomes possible to determine whether any particular accessory cell type subject to renewal is derived from stem cells present in the donor bone marrow (b.m.). Necessary conditions are that the cell type can be identified and clearly distinguished from other cells, and that appropriate reagents are available to distinguish between cells of the F1 donor and parent recipient origin.

Although macrophages and the dendritic cells which appear under various guises (Langerhans cells, veiled cells, interdigitating cells) are undoubtedly b.m. derived, the origin of follicular dendritic cells (f.d.c.) remains uncertain. They have not been included in the Mononuclear Phagocyte classification, and it has been inferred from electron microscopic studies that they may be derived from reticulum cells (e.g. 1). A major problem in characterising them by means of antibody markers is their very intimate contact with B lymphocytes, which make distinction from the latter in respect of surface markers almost impossible in normal histological preparations. Isolation of f.d.c. for study in cell suspension has also been difficult, because their dendritic processes are so delicate that they are destroyed by mechanical separation techniques. However, as was reported at the 7th Germinal Centre Conference, f.d.c. identifiable by surface-bound aggregated FITC-HGG can be separated in suspension from mouse spleens by digestion with collagenase and Dispase (2). Although a few f.d.c. occur singly in such preparations the majority are in the form of cell clusters enclosed in coalesced dendritic processes. Since no other cells in the suspension are labelled with FITC, the f.d.c. are easily identified.

ORIGIN OF FDC

Chimaeras were prepared by injecting 15 X 10^6 b.m. cells, pretreated with anti-Thy 1 and complement, from (CBAXB10)F1 mice i.v. into 8 week old recipients of either parental strain one day after 8.5 Gy whole body irradiation. After 3 to 14 months, chimaeric or parental mice received 2 mg. agg. FITC-HGG i.v. Spleen cell suspensions were prepared 2-3 days later. Donor and recipient cells were identified by means of biotinylated monoclonal anti-$H2^k$ or anti-$H2^b$ and TRITC-avidin. Recipient cells were stained only by the appropriate antibody, whereas F1 donor cells were stained by both.

The results, reported more fully in (3), were quite clear-cut. As expected, f.d.c. and all the cells of either parental strain were stained only by the approp-riate anti-H2, whereas those from the F1 hybrids were stained by both. In the chimaeric mice, at all the time intervals examined, free cells and most of the cells in the clusters were stained by anti-$H2^k$ and anti-$H2^b$ (i.e. were of donor origin) but the f.d.c. were stained only by the anti-H2 specific for the recipient. Apart fron the unlikelt possibility that the f.d.c. are derived from b.m. cells prior to 8 weeks, this finding indicates that f.d.c. are not b.m. derived.

Most of the cells within the clusters were of donor origin. They were presumably B lymphocytes associated with the f.d.c. in vivo and trapped in the coalesced dendritic processes during isolation. In the hope of being able to culture the clusters in vitro we attempted to separate them on a fluorescence activated cell sorter. The large clusters as well as other cell aggregates had first to be removed by filtration through a 50 um mesh to avoid blocking the inlet. However, by gating for size and fluorescence intensity clusters containing up to 10 cells have been separated reasonably cleanly. So far, too few have been collected to attempt to culture them, but clusters collected and fixed on glass slides have been stained with anti-mouse u and anti-k. Both the f.d.c. membranes and the cells within the clusters were strongly stained, confirming that the latter were B lymphocytes.

SPLEEN MARGINAL ZONE MACROPHAGES

Macrophages in the marginal zone (MZ) of the spleen white pulp and those lining the marginal sinus and septa and some in the medulla of peripheral lymph nodes selectively take up various TI-2 immunogens (hapten conjugates of Ficoll (FIC), hydroxyethyl starch (HES), or dextran 2000) after i.p. or i.v. injection. Such conjugates are degraded very slowly in vivo and persist in the body for very long periods of time (4). A rapid and long-lasting antibody response (mainly, but not exclusively IgM) ensues, which in the case of FIC and HES is exclusively against the attached hapten. MZ macrophages in our experience do not express Ia determinants, but they almost certainly present TI-2 antigens to the B lymphocytes which respond to them (5,6). They are morphologically distinct from other macrophages (4) and are distinguishable by a monoclonal antibody which reacts exclusively with them (7).

When fluorescent conjugates of FIC or HES are used MZ macrophages are readily identified in enzyme digested spleen cell suspensions because they are brightly fluorescent, apart from ordinary macrophages which may have temporarily ingested antigen-antibody complexes. They are also larger than ordinary macrophages, which can be identified in turn by a dull red fluorescence, probably due to lipofuscin. Consequently, it was possible to use chimaeras similar to those already described to determine the turnover rate of MZ macrophages in vivo.

Spleen cell suspensions were prepared, at various times after establishing chimaerism, from mice which had received 100 ug FITC-FIC, or FITC-HES i.p. 1-3 days before, and were stained with biotinylated anti-H2. In chimaeras established for 1 mo. or more, red pulp macrophages were all donor derived. The proportions of donor derived MZ macrophages at various times are set out below. Each value represents an individual mouse.

Chimaerism (months):	% MZ macroph. donor origin:
One	8, 16, 18
Three	51, 51, 56
Six	88, 94, 94, 94

This is consistent with a replacement rate of approximately 25% per month.

DISCUSSION

Our findings indicate that mouse spleen f.d.c. are not derived from b.m. stem cells, and that f.d.c. clusters isolated from mouse spleens (and presumably also from human tonsils, ref. 8) envelope B cells. Although thee results do not indicate the origin of f.d.c. they add weight to the suggestion that they are derived by differentiation from reticulum cells (1). They also imply that caution is needed in interpreting observations made with monoclonal antibodies in fixed preparations of cell clusters, because only by electron microscopy is it possible to be certain whether the antibodies react with the f.d.c., or the trapped B cells. Although we had earlier concluded (2) that freshly isolated f.d.c. in suspension did not express detectable Ia determinants, this is now disputed (8,9). We hope that the issue will be clarified at this meeting.

The observations that MZ macrophages are replaced much more slowly than others in the spleen confirms that they are in some way specialised, and is compatible with their retention of poorly degradable TI-2 antigens. We did not examine turnover in lymph nodes, and therefore do not know whether the corresponding cells there behave similarly. A slow replacement rate does not in itself account for prolonged retention of TI-2 antigens, since those which we tested are not retained by liver Kupffer cells, which also normally turn over very slolwy. Both f.d.c. and MZ macrophages are examples of specialised antigen-presenting cells, functioning in specialised microenvironments, which determine the biological outcome of antigenic stimulation.

REFERENCES

1. C.D. Dijkstra, E.W.A. Kamperdijk, and E.A. Dopp. The ontogenetic development of the follicular dendritic cell, Cell Tissue Res. 236: 203 (1984)
2. J.H. Humphrey, and D. Grennan, Isolation and properties of follicular dendritic cells, Adv. Exp. Med. Biol. 149: 823 (1981)
3. J.H. Humphrey and D. Grennan, The origin of follicular dendritic cells in the mouse and the mechanism of trapping of immune complexes by them, Eur. J. Immunol. (In Press)
4. J.H. Humphrey and D. Grennan, Different macrophage populations distinguished by means of fluorescent polysaccharides. Recognition and properties of marginal zone macrophages. Eur. J. Immunol. 11: 221 (1981)
5. D. Gray, J. Lortan, I.C.M. MacLennan, G.D. Johnson, B. Platteau and H. Bazin, Evidence that static but not recirculating B cells are responsible for antibody against DNP on neutral polysaccharide. (This Volume)
6. P.L. Amlot, D. Grennan and J.H. Humphrey, Splenic dependence of the antibody response to TI-2 antigens (Submitted for publication)
7. E. van Vliet, M. Melis and W. van Ewijk, Marginal zone macrophages in the mouse spleen identified by a monoclonal antibody. 1. Anatomical correlation with a B cell subpopulation, J. Histochem. Cytochem. (In Press)
8. E. Heinen, D. Radoux, D.Y. Mason, H. Stein, C. Kinet-Denoel, C. Lilet-Leclerq and J. Simar, Morphological and immunocytochemical characterisation of human follicular dendritic cells isolated from tonsils, (This Volume)
9. M.H. Kosco, A.K. Szakal, R. Gieringer and J.G. Tew, Cytological features of mouse follicular dendritic cells, (This Volume)

MOUSE LYMPH NODE FOLLICULAR DENDRITIC CELLS: QUANTITATIVE ANALYSIS AND ISOLATION

Ernst Heinen, Cécile Kinet-Denoël, Daniel Radoux and Léon Simar

Institute of Human Histology, University of Liège
20, rue de Pitteurs, B-4020 Liège, Belgium

INTRODUCTION

Follicular dendritic cells (FDC) are peculiar non-lymphoid cells only present in lymph follicles especially in germinal centers. They retain antigens in form of immune complexes on the surface of their membranes (1, 2, 3). Few data are available concerning their nature, origin and function. Their study is difficult as they represent only a low percentage of the cell population inside the germinal centers. Recently we achieved the isolation of FDC from human tonsils and adenoids (4). After isolation human FDC form spherical structures where they envelope lymphoid cells (mainly B cells, some T helper cells) with their cytoplasmic extensions. Isolated tonsillar FDC bear IgM, IgA, IgG and IGE but no IgD on their surface. They share also some antigens with macrophages (HLA-DR and Leu-M3) and express, at high density, C_3b and Fc receptors (5).

In the present work we applied the same isolation procedure on mouse lymph nodes. This is a technical difficult performance as mouse lymph nodes are small. But, if achieved, it would provide a material permitting a better analysis of this cell type especially after specific antigen stimulation which cannot be performed on

---

Abbreviations used: BSA: bovine serum albumin
FDC: follicular dendritic cell
GAM-G20: goat anti-mouse immunoglobulins bound to 20 nm gold particles
PBS: phosphate buffer solution
SRBC: sheep red blood cells

human cells. Further we would be able to compare, in identical conditions, mouse lymph node FDC with human tonsillar FDC.

We applied the same isolation procedure as for tonsils on sheep red blood cells (SRBC) stimulated mice. Before isolation we determined best experimental conditions i.e. appearance of large germinal centers. We previously observed retention of immunoglobulins by isolated tonsillar FDC (Heinen et al. submitted), so we tested anti-mouse immunoglobulins antibodies to label FDC isolated from mice.

MATERIAL AND METHODS

Mice: Six weeks old female C57Bl/6 mice were used for the present work.

Antigen stimulation: C57Bl/6 mice were stimulated by an intraperitoneal injection of 150.10^6 sheep red blood cells (Bioferme; Belgium). After 3 or 4 weeks a secondary stimulation was performed. Mice were then used 4 to 20 days after this stimulation.

Cell separation. We used 2 or 3 mice for each experimental condition to isolate FDC from lymph nodes. Inguinal, axillary, mesenteric and aortic lymph nodes were dissected in a refrigerated physiologic solution (Tyrode-glycosol) containing 0.4% bovine serum albumin (BSA). Adipose tissue was removed and the capsule opened by incisions; the hilus and the medullary zone were removed (under a biomicroscope at 20x magnification). Digestions were performed, according to a technique previously described (4), in four steps: the two first in Dulbecco's minimum medium containing 0.4% BSA and 0.05% collagenase (Boehringer). The remaining fragments were then incubated twice in a phosphate buffer (0.1 M; PBS; pH 7.2) solution containing 0.4% BSA, 0.05% collagenase, 0.05% dispase (Boehringer) and 0.004% deoxyribonuclease (Worthington). Freed cells were pooled, harvested at 1,300 rpm for 3 min in a Janetzky T 32C centrifuge and layered on a 1.25% BSA-PBS solution. Sedimentation at unit gravity was done during 30 min at 4°C. The supernatant solution was sucked off and both fractions were centrifuged.

Electron microscopy preparation

Lymph node fragments or pellets obtained after sedimentation were fixed in 2.5% glutaraldehyde in 0.1 M cacodylate buffer (pH 7.2), 30 min at 4°C. After rincing the samples were postfixed in OsO_4 (2%), dehydrated and embedded in Epon 812. Ultrathin sections were cut with a LKB ultratome or a Reichert OMU3 microtome. The sections were treated with lead citrate and uranyl acetate for contrast and observed in a Philipps 301 microscope at 80 KV.

In order to stain material retained by FDC, some preparations were fixed 2 h in a glutaraldehyde (2.5%) and ruthenium red (1%) solution, rinced in a cacodylate buffer solution containing 1% ruthenium red and postfixed 2 h in OsO4 (1%) mixed to ruthenium red (1%) just before use.

Morphometric measurements

Ultrathin sections were analysed according to methods described by Weibel et al., (6) and Simar and Weibel, (7). In short, 15 mice lymph nodes were sectioned; 20 photographs were taken randomly in each case inside the germinal centers and analysed in a test system (Leitz ASM1) to estimate the surface density expressed in u^2/u^3.

Immunogold labelling

Cells obtained after sedimentation were incubated in Dulbecco's minimum medium solution containing 0.4 BSA and goat anti-mouse immunoglobulins bound to colloidal gold particles of 20 nm diameter (GAM-G20; Janssen Pharmaceutica; 1/20) at 4°C during 30 min. Thereafter cells were rinsed twice and fixed for electron microscopy. Controls were made in abscence of specific antibodies.

RESULTS

Morphological aspects of FDC in situ

Germinal centers, after secondary stimulation with SRBC, show a highly developed FDC network surrounding centrocytes and centroblasts (fig. 1). Their indented nucleus contained dispersed chromatin delineated at the periphery by a thin rim of heterochromatin in contact with a well-developed lamina densa. The cytoplasmic extensions formed dendritic or lamellar evaginations or deep channels retaining electron dense material composed of amorphous or particulate virus like structures. This material was strongly contrasted by ruthenium red fixation. Some desmosome like junctions could be found (fig. 2).

This morphological aspect of FDC was variable: FDC appeared as in fig. 1, well-developed or, in contrary, covered with poorly developed cytoplasmic evaginations and bearing a rather round-shaped nucleus; others presented a highly indented nucleus containing dense chromatin blocks. Finally someone seemed degenerating or shared similar aspects with dark cells.

Determination of the percentage of FDC inside the germinal centers

Ruthenium red fixed lymph node sections of SRBC stimulated

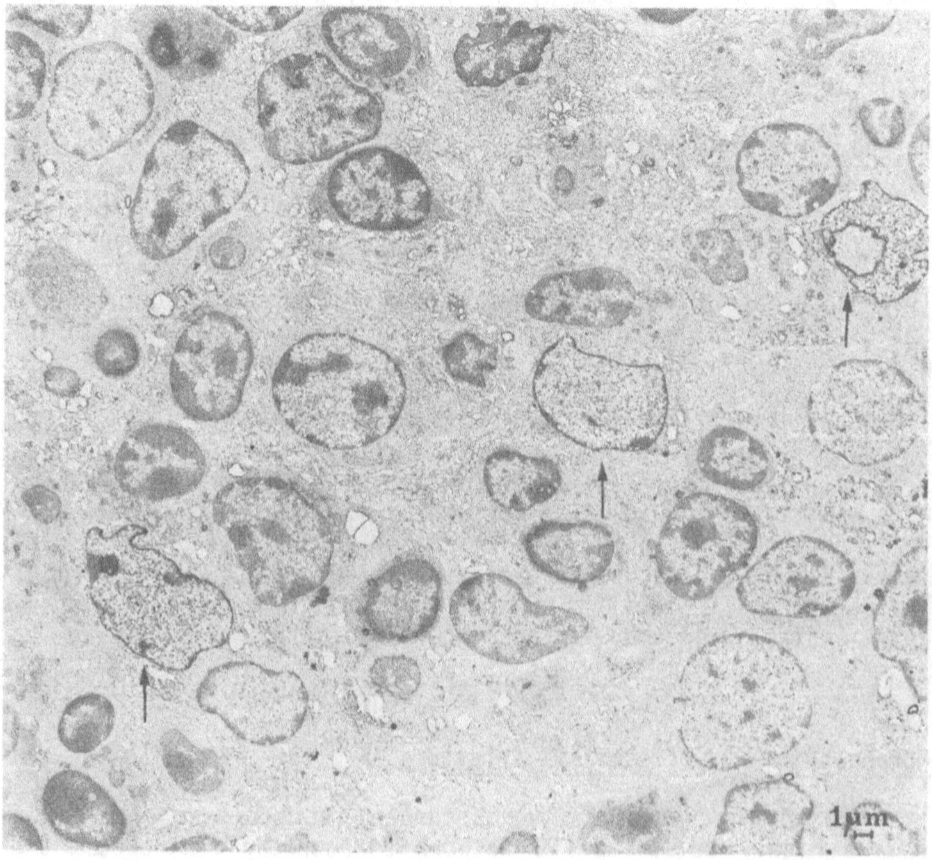

Fig. 1. Ultrastructural aspect of a germinal center: lymphoid cells are surrounded by cytoplasmic extentions of FDC (arrows) (x 2,700).

Table 1. Determination of minimal and maximal percentages of FDC inside the germinal centers or in the intermediary zone between the germinal centers and the corona in lymph nodes of mice stimulated by SRBC. T_0: 3 weeks after primary stimulation, T_{24h}: 24 h after secondary stimulation.

	T_0			T_{24h}		
	Number of counted cells	% de FDC min	% de FDC max	Number of counted cells	% de FDC min	% de FDC max
Germinal Centers	3.219	1,8 ± 0,9	2,6 ± 1,2	5.037	1,8 ± 0,7	2,2 ± 0,6
Intermediary zone between germinal center and corona	1.774	0,8 ± 0,7	1,2 ± 0,7	2.449	0,9 ± 0,7	1,4 ± 0,8

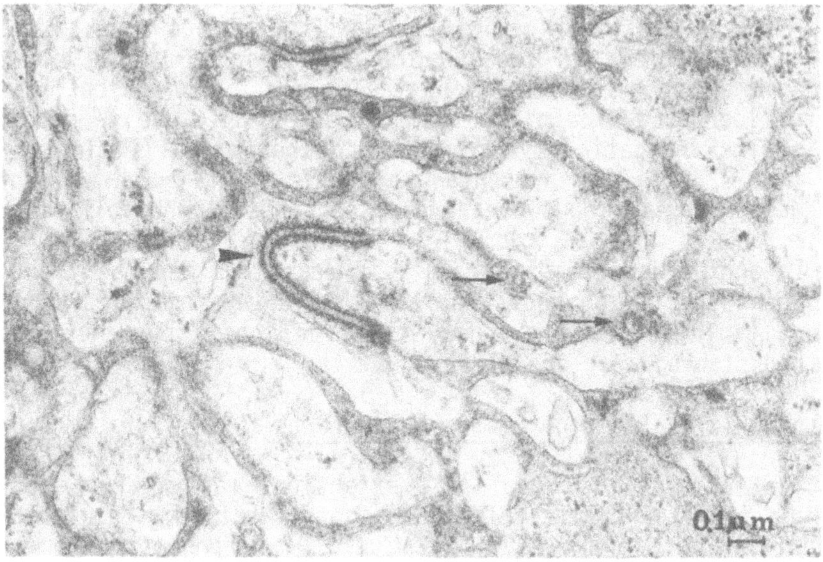

Fig. 2. High magnification of FDC cytoplasmic extensions. Dense extracellular material containing virus like particles (arrows) and a desmosome like junction (arrowhead) are visible (x 54,000).

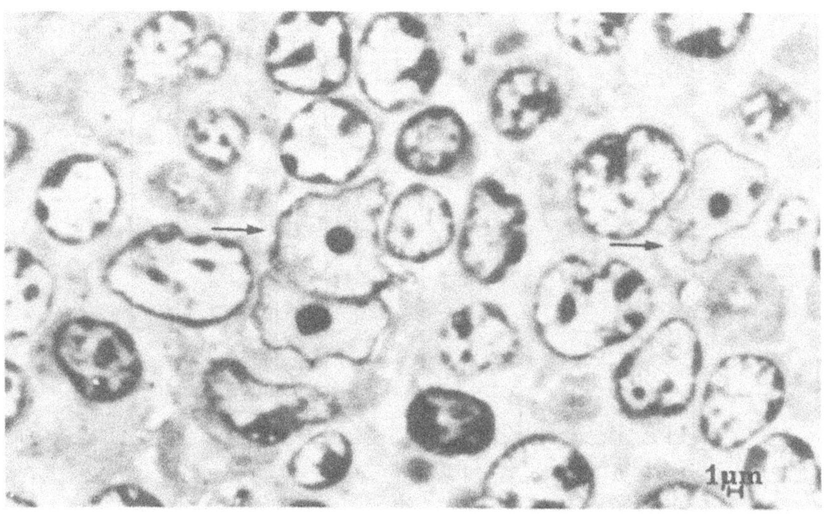

Fig. 3. Semi-thin section of a germinal center showing the presence of two dendritic cells (arrows). This observation was made with a phase contrast microscope after ruthenium red staining (x 2,200).

mice were examined in semithin epon sections. At high magnifications it was possible to discern clear indented nuclei lightly stained with ruthenium red (fig. 3). Their relative numbers in comparison to the other cells were determined in two parts of the lymphoid follicles: in the germinal centers or at the junction between the germinal centers and the corona. Since at optical level the identification of FDC is not always absolute we presented the results indicating the minimal and the maximal estimated values (table I). FDC represented about 2% of the germinal center, were unfrequent in the corona and absent outside the follicles.

In another experimental analysis we determined the frequency of FDC in other lymph nodes and at other delays after SRBC stimulation (table 2). As previously observed the percentages of FDC remained stable, near the 2% value.

Morphometric analysis of FDC cell membranes

The increase of cytoplasmic extensions was clearly apparent during the first 48 hours after a secondary stimulation with SRBC. We analysed this increase with the aid of morphometrical methods in comparing the cell surface of lymphoid cells with those of FDC

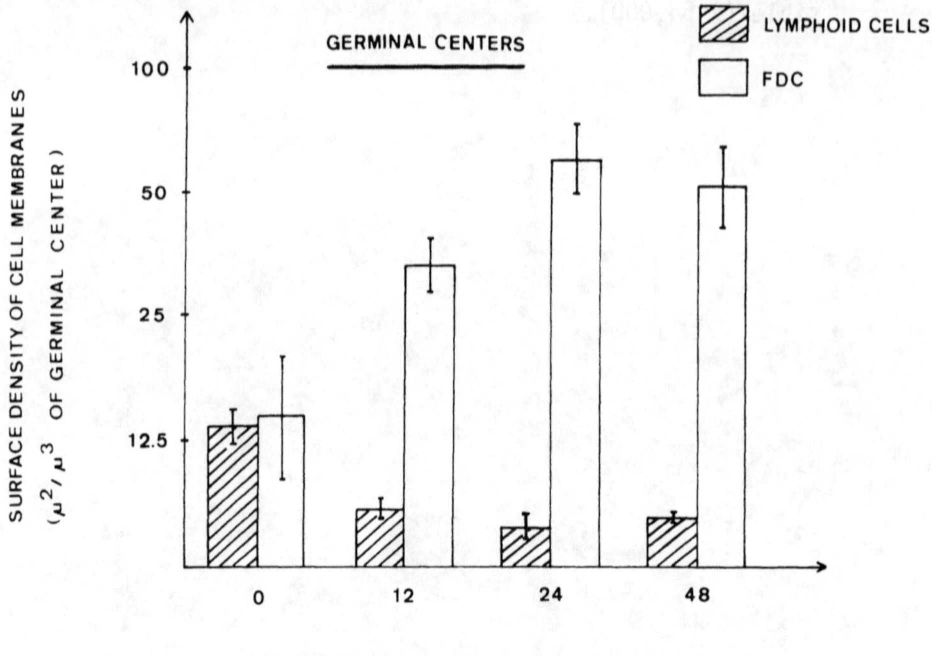

Fig. 4. Evolution of the surface density of lymphoid cells and FDC during the first 48h after secondary stimulaion with SRBC.

Table 2. Determinations of minimal and maximal percentages of FDC inside germinal centers of inguinal and mesenteric lymph nodes in controls or during secondary stimulation with SRBC. T_0 are mice sacrificed 3 weeks after primary stimulation.

Lymph nodes	Experimental conditions	Counted Cells	% of FDC min	max
inguinal	untreated mice	981	2,4 $\pm$ 0,9	3,2 $\pm$ 1,4
inguinal	T_0	1.704	1,9 $\pm$ 1,1	2,5 $\pm$ 1,5
mesenteric	T_0	1.515	1,6 $\pm$ 0,4	2,8 $\pm$ 0,8
inguinal	24h	2.637	1,8 $\pm$ 0,8	2,3 $\pm$ 0,7
mesenteric	24h	2.400	1,7 $\pm$ 0,4	2,3 $\pm$ 0,3
inguinal	48h	414	1,6 $\pm$ 0,3	2,5 $\pm$ 0,3
inguinal	8d	312	1,2 $\pm$ 0,6	3,0 $\pm$ 1,0
inguinal	32d	506	1,4 $\pm$ 0,8	2,0 $\pm$ 1,1

(fig. 4). Three weeks after primary stimulation the density of cell surface membranes was similar in the two cell populations (about 12.5 u^2/u^3). During secondary stimulation, the surface density of FDC increased at 50 u^2/u^3 (48 h) whereas that of lymphocytes diminished to 4.5 u^2/u^3. So, FDC representing only 2% of the germinal center cells thus bore 12 times more cell surfaces than the lymphoid population.

Isolation of FDC from lymph nodes

Inguinal, axillary, mesenteric and aortic lymph nodes dissected from SRBC stimulated mice were dissociated by enzymatic digestions. Freed cells were separated at unit gravity on a BSA gradient. When examined at phase contrast microscopy sedimented cells were composed of free lymphocytes and macrophages but also of cell clusters where distinctly lymphocytes were imprisoned (fig. 5).

At ultrastructural level these clusters showed the presence of peculiar cells enveloping lymphoid cells (fig. 6). Their nucleus exhibit variable aspects either indented and containing blocks of dense chromatin or spherical and with essentially dispersed chromatin. In all cases the nuclei were surrounded by a well-developed lamina densa (fig. 7). Fine cytoplasmic extensions enveloped lymphoid cells; these extensions were joined by desmoso-

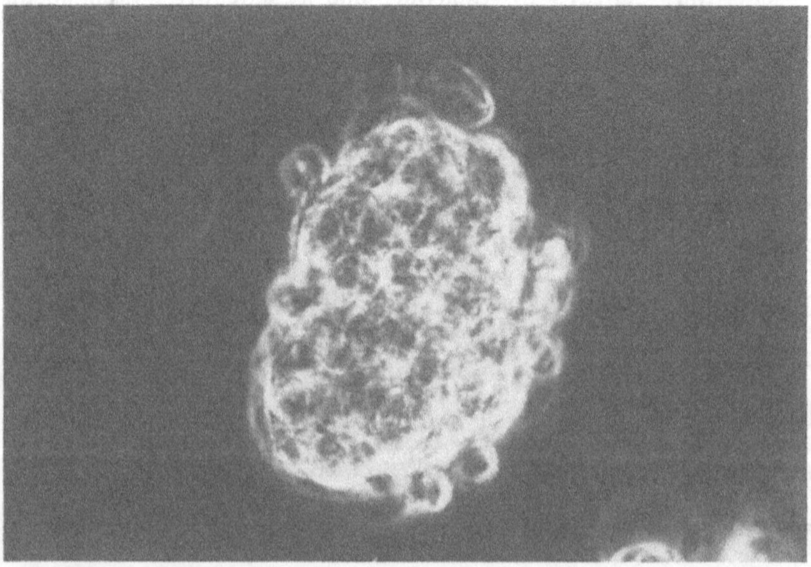

Fig. 5. Phase contrast observation of an isolated FDC cluster. Lymphoid cells are grouped in the central part of the central part of the cluster (x 1200).

me like junctions (fig. 7). Adherent lymphocytes were frequently found at the periphery of these cell clusters. The perinuclear cytoplasm contained mitochondria, a Golgi complex, some rough endoplasmic reticulum and free ribosomes; phagosomes or lysosomes were rare.

Isolated FDC clusters were mostly formed of one FDC surrounding 1 to 11 (mean 5) lymphocytes. Some clusters were build up by 2 or more FDC. Most of the FDC were mononucleate but bi-or trinucleate were also found. Generally lymphoid cells adhered at the periphery of the FDC clusters: 0 to 9 (mean 4) were counted at the periphery of clusters build up by one FDC. About 13% of these clusters were trypan blue dye positive. Per mouse were obtained about 5,000 FDC clusters.

Other multicellular structures were fragments of blood vessels; in some cases fragments of high endothelial venules were found. Among free cells we identified lymphocytes, plasmocytes, macrophages and interdigitating cells. These lasts were isolated or grouped (fig. 8). Some macrophages enveloped without apparent signs of phagocytosis single lymphoid cells.

Characterization of isolated FDC

Isolated FDC were incubated in presence of goat anti-mouse

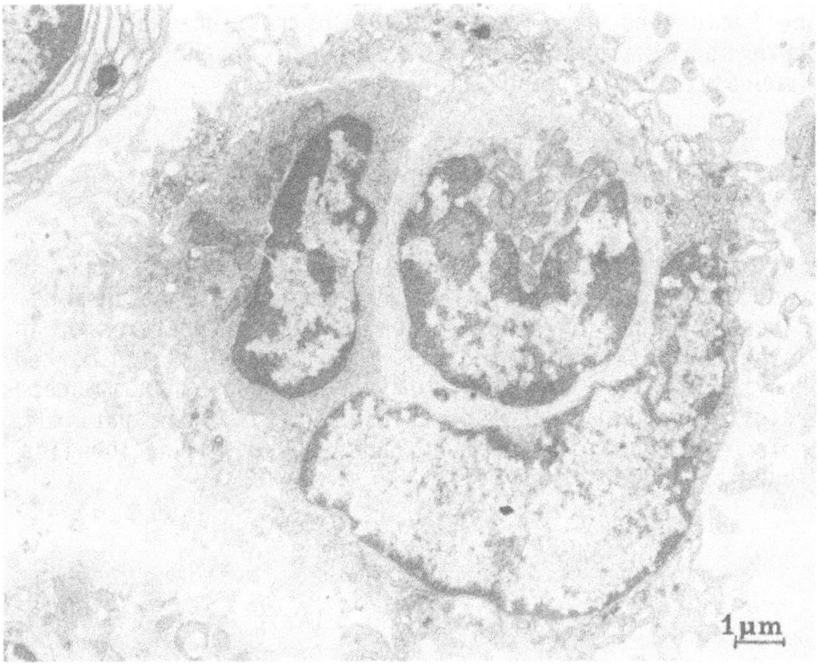

Fig. 6. Ultrastructural aspect of a FDC cluster. A FDC surrounds with its cytoplasmic extensions lymphoid cells (x 7,500).

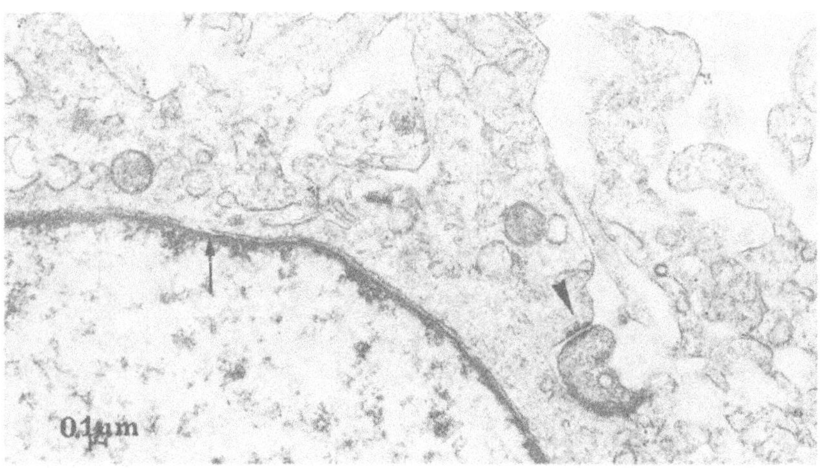

Fig. 7. High magnification of an isolated FDC labelled with GAM-G20. The nucleus contains dispersed chromatin and an apparent lamina densa (arrow). The cytoplasmic extensions, joined by a desmosome like junction (arrowhead), bear gold particles. (x 29,000).

immunoglobulins labelled with fluorescein (fig. 9 a and b) or colloidal gold (fig. 7 and 10). Among multicellular structures only FDC cell clusters were labelled. At ultrastructural level clear positive correlations were found between morphological criteria and immunocytochemical labelling.

We examined after fluorescence labelling the cells of the upper level of the sedimentation gradient and found between numerous free cells some FDC clusters but also some free FDC.

At ultrastructural level gold particles were found mainly adherent at the periphery of the FDC cytoplasmic extensions. They were located either on dendritic or lamellar structures or in narrow invaginations (fig. 7 and 11). Gold particles also marked some lymphocytes or were found inside phagolysosomes of macrophages. Interdigitating cells showed no presence of gold particles. In controls gold particles without anti-mouse immunoglobulins were not found on FDC or other cells.

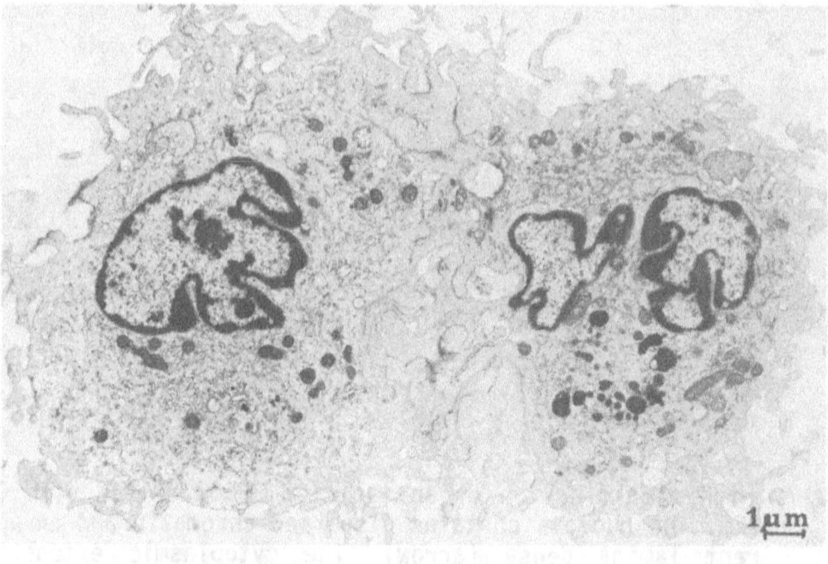

Fig. 8. Ultrastructural aspect of isolated interdigitating cells found after sedimentation. Two cells are intimately joined together. (x 6000).

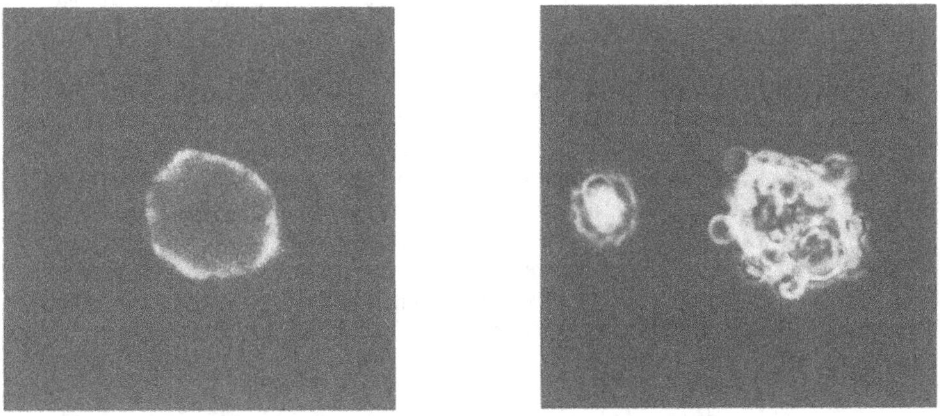

a b

Fig. 9 (a and b). Fluorescence and phase contrast microscope observations of FITC labelled goat anti-mouse immunoglobulins treated isolated FDC. Only FDC clusters are fluorescent. (x 1000).

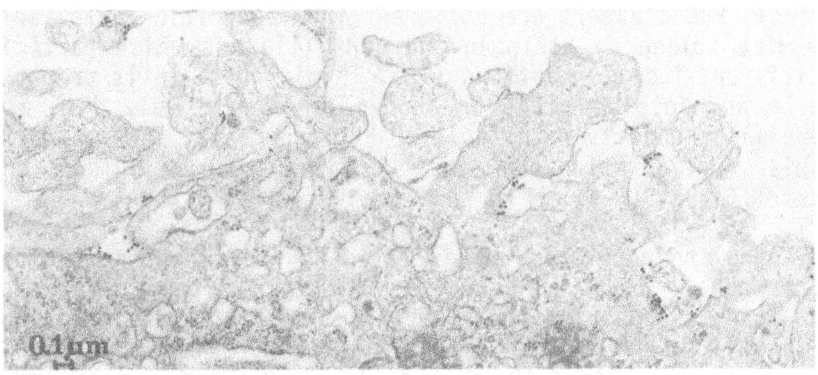

Fig. 10. Ultrastructural aspect of FDC cytoplasmic extensions retaining GAM-G20 particles. (x 36,500).

DISCUSSION

According to our estimations on SRBC stimulated mouse lymph nodes the percentage of FDC inside germinal centers is low (about 2% of total cells); this percentage remains relatively stable in different lymph nodes or experimental conditions. A regulatory mechanism thus seems to exist determining the proportion of lymphoid cells to FDC inside de germinal centers. According to Rademakers et al., (8) FDC represent about 12% of the volume of the germinal centers. As the germinal centers increase in size after secondary stimulation (9) and the percentage of FDC remains stable we can conclude that the number of FDC increases after secondary stimulation.

Secondary stimulated mice are thus convenient for the isolation of FDC inasmuch these FDC clearly increase their cytoplasmic extensions. Our morphometric measurements indicate they bear, after stimulation, 12 times more cell surface than the lymphoid cell population. After stimulation lymphoid cells bore a lesser developed cell surface than previously, this was due to the increase of their volume after blastic transformation. The strong increase of FDC membranes indicates their importance in the humoral response inasmuch as FDC are able to retain a large amount of immune complexes (1, 2, 3).

Enzymatic dissociation gave rise to multicellular clusters similar to those observed after human tonsillar dissociation (4). The isolated FDC bear the same morphological characteristics as in situ and retain at their surface immunoglobulins at high density. These last are an useful tool to label specifically FDC clusters and to distinguish them from fragments of vessels. The FDC clusters show a peculiar arrangement where FDC envelope lymphoid cells. Humphrey and Grennan (10) prepared from mice spleens FDC which they described as being multinucleate syncytia; this could be due to misinterpretation of the imprisoned lymphocytes or to an artefact. FDC clusters are different from dendritic cells isolated from mice spleen by Steinman and Cohn (11) and which participate in different T cell reactions (12). Thymic nurse cells prepared by Wekerle and Ketelson (13) or Houben-Defresne et al. (14) shared some similar aspects with isolated FDC especially the round multicellular configuration. But their origin, nature and function are quite different.

Among the population of isolated FDC different morphological aspects can be found. FDC represent either a round clear nucleus or an indented one rich in heterochromatin blocks. This may be due to alterations during the isolation procedure or, more probably, reflect the heterogeneity in the FDC population inside germinal centers. Previously we reported in mice the presence of FDC with disorganized cytoplasmic extensions several days after secondary stimulation (3) and in the present paper we observed different nuclear and cytoplasmic aspects of FDC in situ. FDC found inside germinal centers apparently represent a population of cells at various differentiation stades. Questions remain unanswered concerning the undifferentiated FDC. Even if Szakal et al. (15) described an early retention of peroxydase-anti-peroxydase complexes beneath the marginal sinus and Parwaresch et al. (16) found a precursor labelled with an anti-FDC monoclonal antibody in the blood, clear conclusions concerning the origin of FDC cannot be drawn. Maybe that isolation will contribute to clarify this.

Acknowledgements

We are grateful to the Fonds de la Recherche Scientifique Médicale and to the Leukemia Fund of their Financial support. The authors wish to thank Ms Skivee and Ms Germai for their technical assistance and Ms Dubois for preparation of the manuscript.

REFERENCES

1. M.G. Hanna, and A.K. Szakal. J. Immunol. 101: 949-962 (1968).
2. G.J.V. Nossal, A. Abbot, J. Mitchell, and Z. Lummus. J. Exp. Med. 127: 277-296 (1968).
3. D. Radoux, E. Heinen, C. Kinet-Denoël, E. Tihange, and L.J. Simar. Cell. Tissue Res. 235: 267-274 (1984).
4. Ch. Lilet-Leclercq, D. Radoux, E. Heinen, C. Kinet-Denoël, J.O. Defraigne, M.-P. Houben-Defresne, and Simar. J. Immunol. Meth. 66: 235-244 (1984).
5. E. Heinen, Ch. Lilet-Leclercq, D.Y. Mason, H. Stein, J. Boniver, D. Radoux, C. Kinet-Denoël, and L.J. Simar. Europ. J. Immunol. 14: 267-273 (1984).
6. E.R. Weibel, G.S., Kistler, and W.F. Scherle. J. Cell Biol. 30: 23-38 (1966).
7. L.J. Simar, and E.R. Weibel. In "Stereological Methods" (E.R. Weibel, ed.). pp. 337-342. Academic Press, London, 1979.
8. L.M. Radermakers, J.P., Peters, J.A., and Unnik van. Virchows Arch. (Cell Pathol.) 44: 85-94 (1983).
9. H. Buerki, H. Cottier, M.W. Hess, J. Laissue, R.D. and Stoner. J. Immunol. 112: 1961-1970 (1974).
10. J.H. Humphrey, and D. Grennan. Adv. Exp. Med. Biol. 149: 823-827 (1982).
11. R.M. Steinman, and Z.A. Cohn. J. Exp. Med. 137: 1142 (1973).
12. R.M. Steinman, B. Gutchinou, M.D. Witmer, and M.C. Nussenzweig. J. Exp. Med., 157: 613-628 (1983).
13. H. Wekerle, and U.-P. Ketelsen. Nature, 283: 402-404 (1980).
14. M.P. Houben-Defresne, A. Varlet, G. Goffinet, and J. Boniver. Leukemia Res. 6: 231-241 (1982).
15. A.K. Szakal, K.L. Holmes, and J.G. Tew. J. Immunol. 131: 1714-1727 (1983).
16. M.R. Parwaresch, H.J. Radzun, A.C. Feller, K.P. Peters, and M.L. Hansmann. J. Immunol. 131: 2719-2725 (1983).
17. T. Sminia, E.M. Janse, and M.M. Wilders. Scand. J. Immunol. 16: 481-485 (1982).
18. A.N. Barclay. Immunology 42: 593-601 (1981).
19. J. Gerdes, H.S. Stein, D.Y. Mason, D.Y. and A. Ziegler. Virchows Arch. (Cell Path.) 42: 161-172 (1983).

ANTIGEN/ANTIBODY RETENTION BY FOLLICULAR DENDRITIC CELLS (FDC)

Daniel Radoux, Ernst Heinen, C. Kinet-Denoël and L.J. Simar

Institute of Human Histology - University of Liège
20, rue de Pitteurs,
B-4020 Liège Belgium

INTRODUCTION

It is now well established the cells responsible for the long term retention of antigens inside the lymphoid follicles are the follicular dendritic cells (FDC). Rooijen (1) demonstrated the same lymphoid follicle could trap two non-cross-reacting antigens. But it remained to be determined whether an individual FDC is specific for one antigen or could also retain other non-cross-reacting ones. In that aim, two antigens visible at electron microscopic level (bovine serum albumin, BSA) linked to gold particles and ferritin) have been injected simultaneously into preimmunized mice in order to obtain the retention of these two markers.

Some indirect arguments suggested that antigens are trapped in form of immune complexes (I.C.): the presence of specific antibodies is necessary to observe a trapping (2, 3) I.C. formed of $F(ab')_2$ fragments are not retained in germinal centres (3). Moreover when different isotypes are used to form I.C., their retention is variable (4). To allow a clear visualization of both components of the retained complexes, gold particles of different sizes were used. The antigen (BSA) was linked to gold particles of 5 nm diameter and the specific antibody (anti-BSA) to 20 nm gold particles. The I.C. were formed in vitro and then injected into mice.

Two receptors seem to play a part in the retention of I.C.: C_3b and Fc receptors. To determine if the I.C.-gold complexes activate the complement and so fix C_3b fragments susceptible to attach on C_3b receptors we measured the haemolytic activity of complement after preincubation with I.C.-gold.

MATERIALS AND METHODS

1. Injection of BSA or/and ferritin

a) In non-immunized mice: BSA conjugated to colloidal gold (20 nm: Janssen Pharmaceutica; 50 ug/mouse) or horse spleen ferritin (100 ug/mouse) were injected into C57bl/6J mice at the base of the tail or in the hind footpath. They were sacrificed on day 1, 2 and 5.

b) In immunized mice: after a secondary immunization against BSA or ferritin mice received 6 days later either BSA-gold (50 ug/mouse) or ferritin (100 ug/mouse) and were sacrificed on day 1, 2, 5, 9, 33 and 70 for BSA and on day 2 for ferritin. A double immunization schedule was obtained by injecting simultaneously BSA and ferritin at the dose indicated above to the same animals (four injections) followed by two injections of BSA-gold and ferritin in saline solution at 5 day intervals. Animals were sacrificed on day 2, 4, 8, 16 and 31.

2. Injection of I.C.-gold

Anti-BSA were produced by hyperimmunizing mice. Repeated blood collections were performed on each mouse. The isolation of anti-BSA was carried out by affinity chromatography using CNBr-activated Sepharose 4B Column. Purity controls were checked by polyacrylamide gel electrophoresis in SDS and immunoelectrophoresis. Anti-BSA antibodies were linked to gold particles of 20 nm diameter and BSA to 5 nm by De Mey (Janssen Pharmaceutica; Beerse, Belgium). I.C.-gold were prepared in vitro and injected into mice. 1, 2 and 5 days later lymph nodes were fixed for electron microscopy.

3. Haemolytic activity of complement

Rosenberg and Tachibana's method (5) was used in this study. Blood was collected and coagulated for 3h at room temperature. SRBC were washed in GVB (veronal buffer with 1% gelatin), labelled with ^{51}Cr and sensitized with rabbit haemolytic serum. BSA-gold, anti-BSA-gold and I.C.-gold were preincubated with serum before measuring haemolytic activity of complement. All relevant controls were included with each set. The results are expressed as the percentage of lysis.

RESULTS

1. Injection of BSA and/or ferritin

a) In non-immunized mice: the markers (BSA-gold or ferritin) were observed in large quantities in macrophages of lymph nodes but were not present on the surface of FDC after 1, 2 or 5 days.

b) In immunized mice: 24h after the injection of BSA-gold, numerous particles were present in the electron dense material in deep plicae or narrow channels of the cytoplasmic FDC processes (fig. 1). Some rare particles were found located between FDC and lymphocytes. No particles were found extracellularely elsewhere as in the germinal centers. Gold particles were also observed in macrophages in the different lymph node areas. However, the marker was not detected inside the tingible body macrophages.

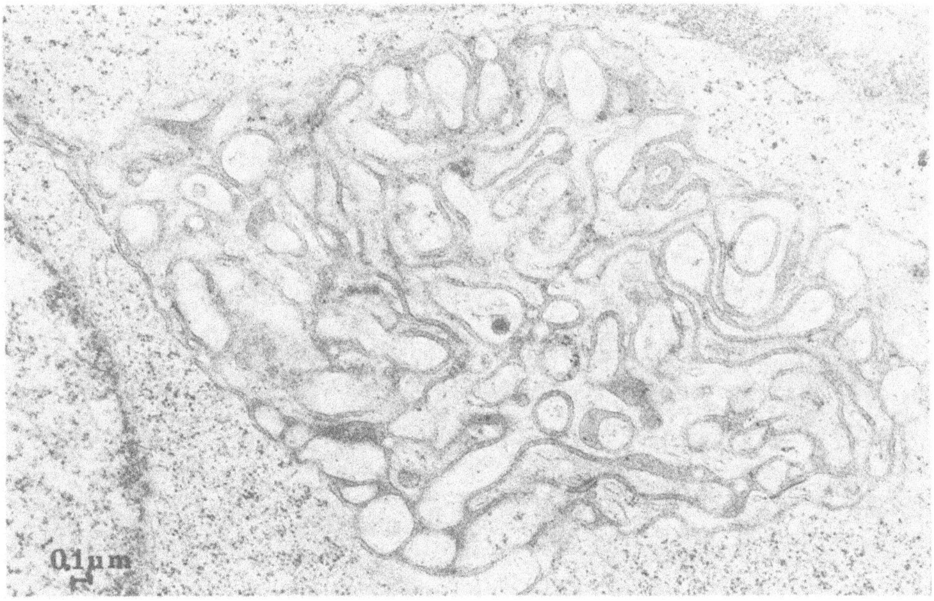

Fig. 1. Ultrastructural detection of BSA labelled with colloidal gold particles in the FDC cytoplasmic extensions network.

In some germinal centers gold particles were occasionally seen adjacent to cell extensions of dark cells or in contact with collagen fibres. More specially gold particles were sometimes in vicinity of fibroblastic like cells lining collagen fibres (fig. 2).

After 48h and 9 days the localizations of the labelled antigen remained unchanged but the frequency of BSA-gold decreased with time. On day 33 and 70, no particles were detected. The localization of ferritin was similar to that of BSA-gold. However, peculiar features were observed with that antigen. Ferritin molecules were regularly aligned in one row in the dense material

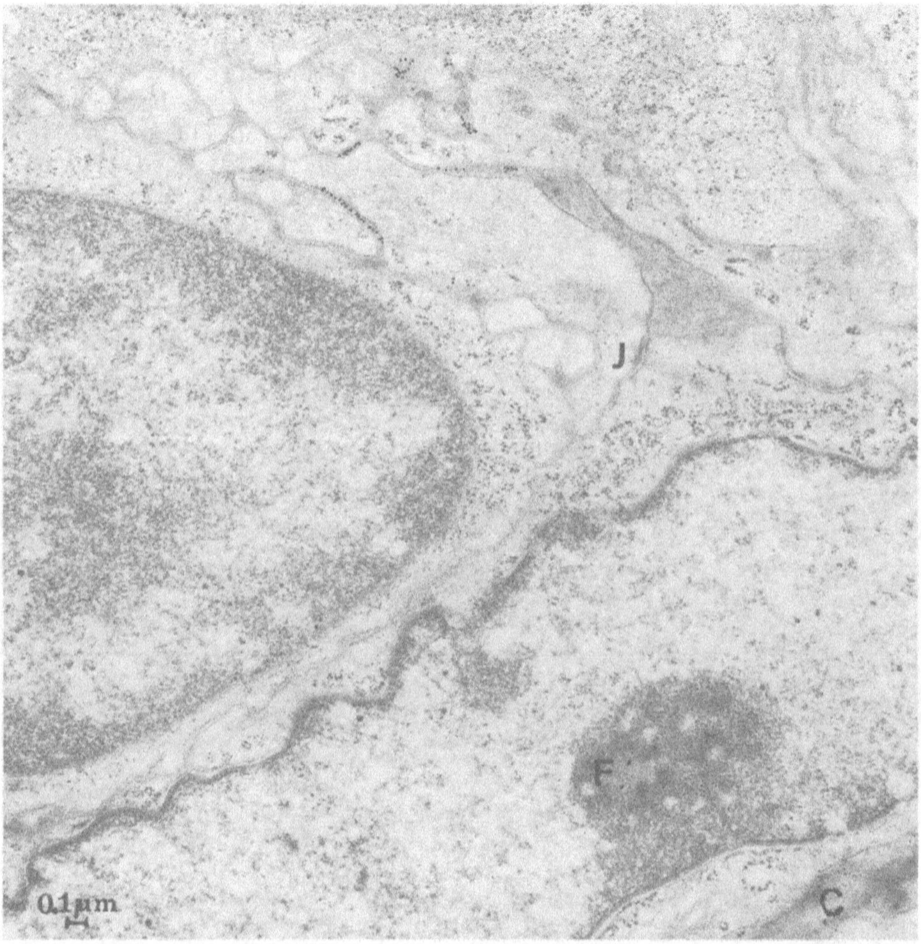

Fig. 2. BSA-gold particles in vicinity of a fibroblastic-like cell (F). Note the presence of collagen bundles (C) and their junctions with adjacent cells (J)

between the cytoplasmic folds of FDC. The electron-dense core of each ferritin was located at an equal distance between two adjacent membranes, so the width of that space was almost uniform (fig. 3a).

After injections of both antigens, adjacent ferritin and BSA-gold were observed together inside the folds of the same FDC 2, 4, 8 and 16 days after the last challenge (fig. 3 b). Ferritin was always more numerous than BSA-gold. After 31 days the two markers were only rarely detected.

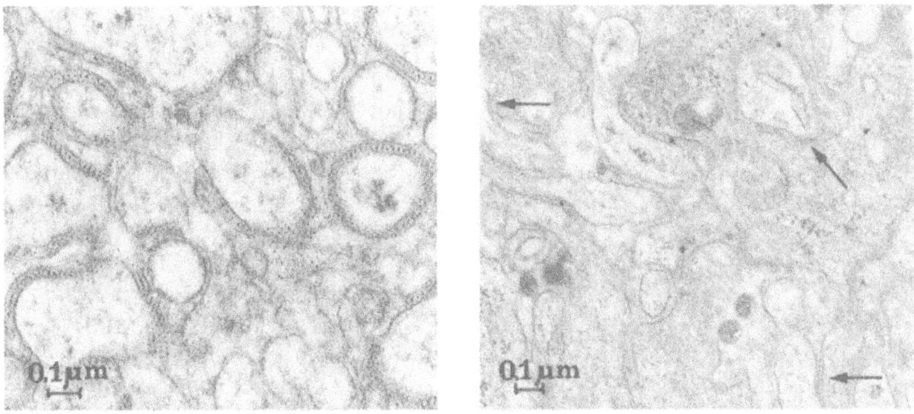

a b

Fig. 3 a and b. a: FDC cytoplasmic extensions after ferritin injection: it is located at equal distance between the membranes.
b: 2 days after simultaneous injections of BSA-gold and ferritin (arrows): both markers are found in contact with the same FDC.

2. Injection of I.C.-gold

Anti-BSA antibodies linked to 20 nm gold particles injected in the form of I.C. (with BSA-gold: 5 nm) were found in popliteal lymph nodes 1, 2 and 5 days after injection (fig. 4). I.C.-gold were located in close contact with the dendritic cytoplasmic extensions of FDC as well as in macrophages of different areas. Some 20 nm particles were seen alone but most of them were in contact with 5 nm particles. Free injected anti-BSA antibodies were observed in the same locations as I.C.-gold.

3. Haemolytic activity of complement

In each set of the complement activation tests the standart curves showed sigmoid lines. After preincubation of serum with colloidal gold, anti-BSA-gold or I.C.-gold, we observed only a light decrease (5 to 15%) of the complement haemolytic activity.

DISCUSSION

Thanks to the two antigens visible at the ultrastructural level (BSA-gold and ferritin) we have been able to localize with high precision the antigens (6, 7). We showed that retention required the presence of specific antibodies. Indeed, without preimmunization both antigens were not retained on the surface of FDC.

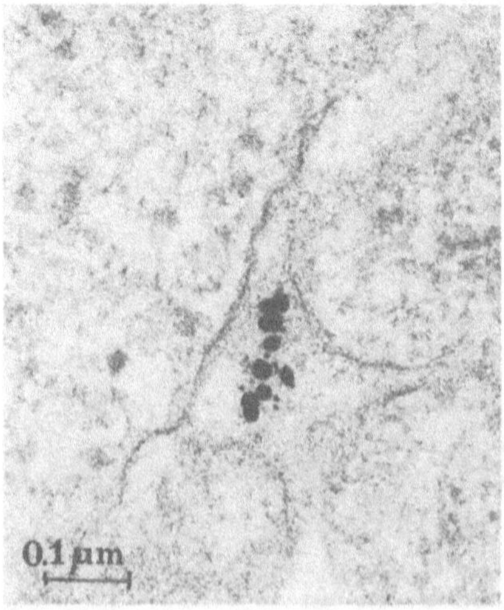

Fig. 4. Gold labelled immune complexes retained by FDC in germinal centers. Large particles bear anti-BSA antibodies; small particles bear BSA. Both kinds of particles are in close contact together.

The contact of BSA-gold with some fibroblastic-like cells could favor the hypothesis of authors considering FDC as differentiated from reticulum cells (8). But immunolabelling of isolated FDC showed the presence of macrophage specific antigens on their surface (H). Further investigations are thus necessary to determine their actual origin.

The peculiar arrangement of ferritin in one line between the two adjacent FDC membranes could be due to the fixation of the antigen on both cell membranes by the aid of specific antibodies. The regularity of the FDC membranous network could result from the bridging of membranes by retained antigens.

At electron microscopic level we demonstrated that the same FDC can trap simultaneously two different antigens. It is likely that all antigens in presence of specific antibodies could be retained on FDC. So, our data support the hypothesis of Mandel and coworkers (9) who proposed that FDC can fix different I.C.

In the present study we investigated the fate of homologous antibodies labelled with colloidal gold. We showed that anti-BSA antibodies linked to gold and injected in association with BSA-gold (I.C.-gold) were located in form of I.C. in close contact with the FDC external membranes. It is the first time that both components of I.C. could be localized simultaneously and ultrastructurally on FDC (10). The results of haemolytic activity of complement preincubated with I.C.-gold indicated that I.C.-gold dit not activate complement. So in our experimental conditions the retention of I.C.-gold seems not be due to the C_3b receptor. The Fc receptors thus appears as being mainly responsible for I.C. retention by FDC; this was supported by results obtained on isolated tonsillar FDC which could fix I.C. in absence of serum (11).

ACKNOWLEDGMENTS

The authors are grateful to the "Fonds de la Recherche Scientifique Médicale" for their generous financial support (Grant nr 3.4518.81).

REFERENCES

1. N. Rooijen van. Acta Morpho. Neerl. Scand. 16: 121-127 (1978).
2. Z.L. Herd and G.L. Ada. Aust. J. Exp. Med. Sci. 47: 63 (1969).
3. G.G.B. Klaus, J.H. Humphrey, A. Kunkl and D.W. Dongworth. Immunol. Rev. 53: 3-28 (1980).
4. R.P. Phipps, G.F. Mitchell, T.E. Mandel and J.G. Tew. Immunology 40: 459 (1980).
5. L.T. Rosenberg and D.K. Tachibana. J. Immunol. 89: 861 (1962).
6. E. Heinen, D. Radoux, C. Kinet-Denoël and L.J. Simar. J. Immunol. Meth. 59: 61 (1983).
7. D. Radoux, E. Heinen, C. Kinet-Denoël, E. Tihange and L.J. Simar. Cell Tissue Res., 235: 267 (1984).
8. U. Heusermann, K.H. Zurborn, L. Schroeder, and H.J. Stutte. Cell and Tissue Res. 209: 279-294 (1980).
8. T.E. Mandel, R.P. Phipps, A.P. Abbot and J.G. Tew. Immunology 43: 353 (1981).
10. D. Radoux, C. Kinet-Denoël, E. Heinen, M. Moeremans, J. De Mey and L.J. Simar. (submitted for publication).
11. E. Heinen, D. Radoux, C. Kinet-Denoël, M. Moeremans, J. De Mey and L.J. Simar. (submitted for publication).

ANTIGENIC PHENOTYPE OF RODENT FOLLICULAR DENDRITIC CELLS *IN VITRO* AND *IN SITU*

Marie H. Kosco, Andras K. Szakal, and John G. Tew

Departments of Anatomy; Microbiology and Immunology
Medical College of Virginia, VCU, Richmond, VA 23298

INTRODUCTION

Follicular dendritic cells (FDCs) were first described at the transmission electron microscopic level in the lymph node by Nossal et al. (1968) and in the spleen by Szakal and Hanna (1968). Until recently, the study of FDCs was limited to *in situ* investigations due to the lack of adequate isolation and enrichment techniques. In 1982, Schnizlein et al. developed a procedure in our laboratory which produced a single cell suspension specifically enriched for FDCs from lymph nodes of mice. This procedure provided an opportunity to establish diagnostic morphological features of intact FDCs *in vitro*, using Nomarski optics, scanning EM (SEM) and transmission EM (TEM). The SEM of isolated FDCs revealed cells with a "sunburst-like" appearance due to numerous long thin dendrites emanating from a spherical or ovoid cell body. Nomarski optics and TEM confirmed the presence of immune complexes on the processes of FDCs by using the histochemically detectable antigen (Ag), horseradish peroxidase (HRP), and revealed features consistent with those observed in lymphoid sections. The cytological evaluation of FDCs *in vitro* provided functionally supported morphological criteria by which these cells could be distinguished from leukocytes and other dendritic cells (Szakal, Gieringer, Kosco and Tew, 1984).

The present study was undertaken to evaluate the antigenic phenotype of FDCs. Immunocytochemical techniques were utilized in conjunction with monoclonal antibodies (Ab) in order to demonstrate cell surface markers. In preliminary work it was difficult to establish whether the labeling of Ia was attributable to nonspecific binding of reagents or not. However, we are now confident that FDCs can express Ia. For detection at the light microscopic level, isolated cells and cryostat sections from mice or rats were labeled indirectly using anti-Ia monoclonal Abs. In one case involving Ox4, isolated cells and tissue were also directly stained with fluoresceinated anti-Ia Ab. Tissue was also prepared for TEM in order to demonstrate the distinction between Ia on lymphocyte versus FDCs _in situ_. Results from these experiments indicated that FDCs expressed Ia _in situ_ and _in vitro_.

In addition, FDCs were shown to express surface markers such as H-2K and common leukocyte antigen (CLA), but lacked Thy-1, Mac 1, Mac 2, Mac 3 and other macrophage antigens. Using a monoclonal Ab or rosetting with IgG coated sRBCs, FDCs were also shown to have Fc receptors. Thus, the antigenic phenotype of FDCs together with their characteristic morphology, their ability to trap and retain Ag on their cell surface, their nonphagocytic and nonadherent properties identifies them as a unique cell population whose role in the immune system is just beginning to be understood.

MATERIALS AND METHODS

Isolated cells and tissue sections were obtained from lymph nodes of young adult C3H, Balb/c, or C57BL/6 mice as well as Wistar rats. The animals were passively immunized with sheep or rabbit anti-HRP or rabbit anti-HSA. Twenty-four hours later, the animals were injected in each footpad with the corresponding Ag, HRP or HSA, and the lymph nodes were collected the following day. Sheep or rabbit Abs were used in the passive immunizations in order to form immune complexes _in vivo_ which would not react with anti-rodent immunoglobulin reagents. For studies designed for evaluation by light microscopy, HRP was used as the histochemically detectable Ag. Development of HRP yielded visible immune complexes (HRP-anti-HRP) retained on the plasma membrane of FDCs. This in turn provided a marker for identification of FDCs in isolated cell preparations or sections. The Ag, HSA, was utilized for TEM studies in order not to interfere with the visualization of the HRP-conjugated secondary Ab used in the demonstration of Ia surface determinants.

Monoclonal Abs which recognize Ia determinants of the rodent system were used with indirect or direct immunofluorescence (IF) or immunoperoxidase techniques. Four of these monoclonal Abs were used in the mouse system with FITC or HRP conjugated secondary Abs; MRC Ox4, MK-D6, and 10.2-16 for IA, and 14.4-4 for IE. The monoclonal Ab, Ox4, was also obtained directly coupled to fluorescein to eliminate the need for a secondary Ab and to obtain more direct evidence of Ia expression. The mouse monoclonal Ab, Ox4, which recognizes rat Ia determinants, also cross-reacts with mouse Ia gene products of the k or s haplotypes. Therefore, this

Ab was utilized for experiments involving both rodent systems.

Additional monoclonal Abs were used to identify other antigenic characteristics of FDCs. Antibodies from ascites fluid or from cell culture supernatants of the clones 11-4.1, 2.4G2, 5A-8, and M1/93 were used to demonstrate the expression of H-2K, Fc receptors, Thy 1, and common leukocyte Ags, respectively. A series of monoclonal Abs which identify surface Ags of macrophage populations, i.e. Mac 1, Mac 2, Mac 3, and F4/80, were also tested. The secondary Abs used were peroxidase- or fluorescein-conjugated affinity purified $F(ab')_2$ fragments of IgG. These included goat or rat anti-mouse and mouse anti-rat IgG, specific for either the Fc region or heavy and light chains. Opsonization of sheep RBCs with IgG provided an additional tool for demonstrating Fc receptors.

Enriched preparations of FDCs from single cell suspensions of lymph nodes were obtained according to the procedure developed by Schnizlein, Szakal, and Tew (1984). Briefly, this involved the enzymatic (i.e. dispase, collagenase, and DNAase) dissociation of lymph nodes from passively immunized mice, isolation of the non-adherent population using a discontinuous BSA density gradient and a one hour adherence step for elimination of macrophages. The nonadherent suspension was kept on ice for cytochemistry or plated out for culture. Following appropriate processing, the cell preparations were cytobucketed onto poly-L-lysine coated slides, fixed in paraformaldehyde/glutataldehyde fixative, and evaluated using Nomarski differential interference contrast microscopy and epifluorescence.

Lymph nodes for cryostat sectioning were also obtained from mice and rats passively immunized for HRP. Serial sections were cut and collected individually on poly-L-lysine coated slides. By alternating the processing of adjacent sections, the tissues were reacted for either Ia expression or Ag (HRP) localization. The localization of HRP-immune complexes in lymphoid follicles aided in the visualization of FDCs. These areas of distinct peroxidase (PO) activity patterns in the nodules were used as land marks for evaluation of slides processed for Ia. A "dendritic" pattern of fluorescent staining in the follicles was sought for the demonstration of Ia on FDCs.

Specimens for TEM were obtained from rat lymph nodes. The monoclonal Ab, 0x4 (mouse anti-rat IgG), and Fc region specific, HRP-conjugated $F(ab')_2$ fragments of rat anti-mouse IgG, were used as the reagents to demonstrate the expression of Ia on FDCs. Controls were incubated with secondary Ab only. For light microscopy and TEM, cytochemical localization of HRP was accomplished using diaminobenzidine hydrochloride (DAB) and H_2O_2 according to the procedure specified by Szakal, Holmes and Tew, (1983).

RESULTS

Demonstration of the expression of Ia was first pursued with FDC enriched cell preparations from lymph nodes of passively immunized mice. Three strains of mice and four monoclonal Abs (see Table 1) were utilized in conjunction with direct and indirect IF techniques. The results indicated that freshly isolated FDCs could express Ia and would continued to do so after 4 days of culturing. Cryostat sections obtained from lymph nodes of passively immunized mice and rats also presented evidence for Ia localization on FDCs.

Table 1 is a summary of the experiments conducted to evaluate the expression of Ia by FDCs. The monoclonal Abs recognized various Ia gene products while each mouse strain produced different Ia determinants. In turn, each combination was tested using indirect IF. The FITC-conjugated monoclonal Ab, Ox4, also provided direct evidence of Ia molecules on isolated FDCs and in cryostat sections. In each case involving reaction of isolated FDCs from mouse strains whose haplotype matched the haplotype specificity of a monoclonal Ab, Ia molecules on the cell body and dendrites were positive for IF. The controls for nonspecific reactivity of inappropriate monoclonal Abs and surface determinants were always negative.

Table 1. Distribution of Ia on FDCs from Mouse and Rat Lymph Nodes[1]

			Ag Recognized on FDCs			
Clone[2]	Antibody Specificity	Antigen Labeling	Balb/c (Ia^d)	C3H (Ia^k)	C57BL/6 (Ia^b)	Rat (Ia)
MRC Ox4	$I\text{-}A^{k/s}$	Direct	-	+	-	+
		Indirect	-	+	-	+
14-4.4	$I\text{-}E^{k/d}$	Indirect	+	+	-	NT[3]
10-2.16	$I\text{-}A^{k}$	Indirect	-	+	-	NT
MK-D6	$I\text{-}A^{d}$	Indirect	+	-	-	NT

[1]Mouse - Axillary, Brachial & Popliteal lymph nodes;
Rat - Popliteal lymph nodes;
[2]Positive Control: Con A elicited Peritoneal Macrophages;
[3]NT, Not Tested;

Cryostat sections were also cut from mice and rats passively immunized for HRP. The tissue from rat was reacted with 0x4, a mouse anti-rat Ia monoclonal Ab. Immunofluorescence labeling was accomplished both indirectly with a rat anti-mouse secondary Ab and directly with fluoresceinated 0x4. For both mouse and rat tissues, serial cryostat sections were developed for either Ia expression or Ag (HRP) localization (see methods). The specimens demonstrating reaction between rodent tissue and the appropriate monoclonal Ab often revealed a reticular or web-like pattern of staining in the follicles. This staining corresponded to areas of Ag (HRP) retention on adjacent slides. The staining was negative for controls involving incubation with secondary Ab only and between inappropriate pairs of tissue strain and monoclonal Ab.

Finally, in order to conclusively demonstrate Ia molecules on FDCs _in situ_, specimens were prepared for examination at the ultrastructural level. Paraformaldehyde/glutaraldehyde fixed sections from popliteal lymph nodes of HSA-immune rats were incubated with 0x4 and an HRP-conjugated rat anti-mouse secondary Ab. The PO activity was developed in a DAB substrate to provide an electron dense (black) marker for the visualization of Ia. Figure 1 demonstrates the immunocytochemical localization of Ia antigenic determinants on the plasma membrane of a typical FDC (Fig. 1a), dendritic processes (Figs. 1a, b and c) and B lymphocytes (Figs. 1a, b, c, and d). The specificity of the immunocytochemical reaction could be ascertained clearly around capillaries in the lymphoid follicles (Fig. 1d). In these regions the plasma membrane of B lymphocytes was labelled but the surface of endothelial cells and pericytes were negative. These regions provided excellent internal controls for the specificity of labelling. In addition, control specimens incubated with the PO-conjugated secondary Ab only were also negative (not shown).

Additional antigenic determinants of FDCs were evaluated using single cell suspensions and cryostat sections from mouse lymph nodes. These surface markers were demonstrated using indirect IF in conjunction with various monoclonal Abs. Table 2 lists the results of these experiments, summarizing the antigenic phenotype of lymph node FDCs. FDCs were determined to be positive for H-2K and common leukocyte Ags. They were shown to express Fc receptors using the monoclonal Ab, 2.4G2, and the results confirmed using rosetting assays with IgG coated sRBCs. Cell surface antigens determined to be absent included the T-cell Ag, Thy-1, as well as four Ags reported to be expressed by macrophage populations, Mac-1, -2, -3, and the determinant recognized by F4/80. Appropriate controls, thymocytes for Thy-1, and thioglycollate elicited or resident peritoneal macrophages for the anti-macrophage Abs, were positive. Controls for IF of tissue sections corresponded to what has been reported in the literature as positive staining patterns for the lymph node and thymus.

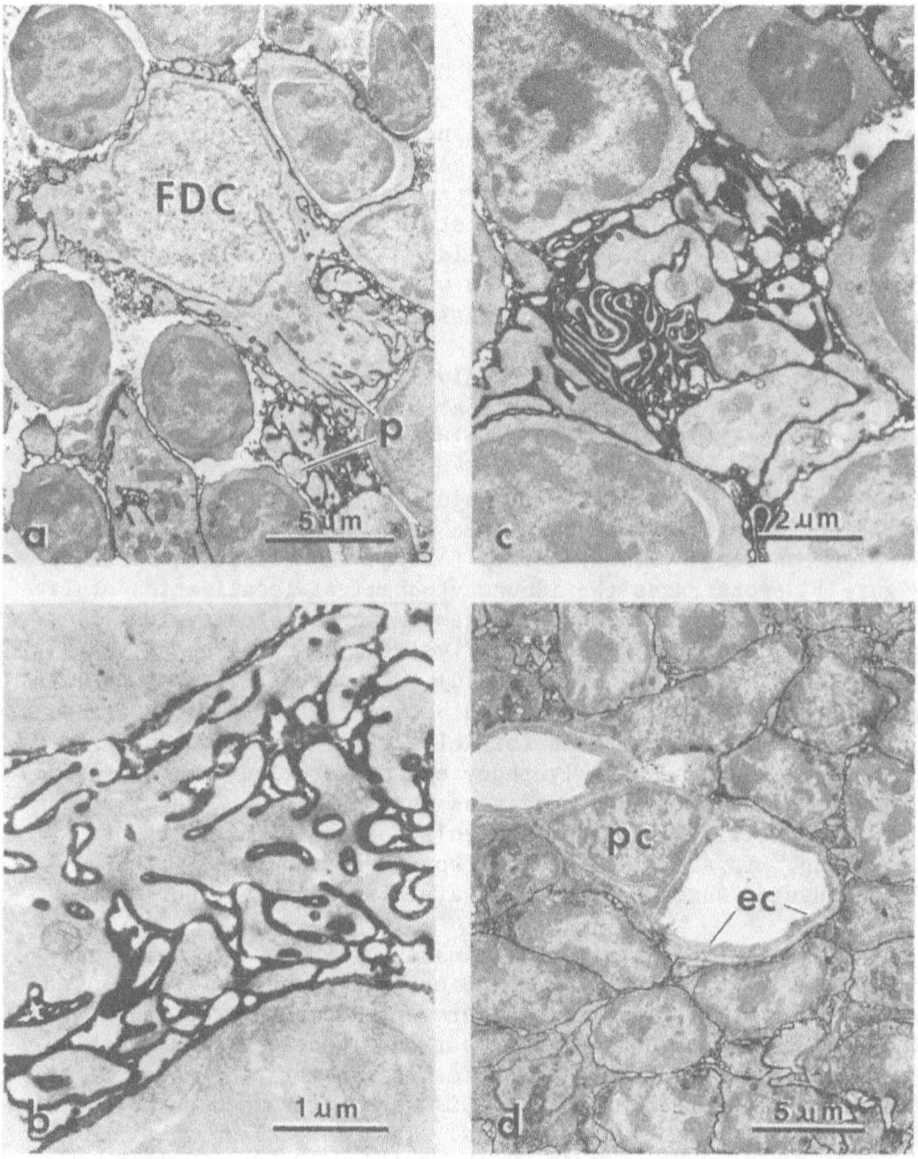

FIG. 1: Immunocytochemical localization of Ia antigenic determinants (black) on the plasma membrane of FDCs and B lymphocytes in the follicles of rat popliteal lymph nodes with Ox4 and PO-conjugated rat anti-mouse IgG. (a) Typical FDC cell body with euchromatic nucleus and origin of dendritic processes, p. (b) and (c) Cross-sections through anastomosing convolutions of dendrites surrounded by B-lymphocytes. (d) Internal control; Note specificity (label on B-cells and absence of label on endothelial cells, ec, and pericyte, pc.)

Table 2. Summary of FDC Antigenic Phenotype

Clone	Antigen Recognized	Expression[1]
11-4.1	H2-K	+
M1/9.3	Leukocyte Common	+
MRC 0x4 14-4.4 10-2.16 MK-D6	Ia	+[2]
2.4G2	Fc Receptor	+[3]
5a-6	Thy-1	-
M1/70	Mac-1	-
M3/38	Mac-2	-
M3/84	Mac-3	-
F4/80	MØ Determinants	-

[1]Determined using indirect IF *in situ* on cryostat sections and isolated FDC preparations of mouse lymph nodes;
[2]Also determined by EM immunocytochemistry in mice and rats *in situ*;
[3]Also confirmed using IgG coated sRBCs;

DISCUSSION

The study of FDCs *in vitro* was hampered for several years by not having adequate techniques to preserve their delicate, fine processes and to isolate them from other cell types. Enzymatic dissociation of the stroma of lymphoid organs has dramatically increased the ability to obtain intact FDCs. As early as 1977, our laboratory has made use of the enzyme, collagenase. Dr. Humphrey's laboratory in 1982 reported that dispase was also helpful. Since this time, we have found dispase, collagenase and DNAase to be important in the isolation procedure as did Lilet-Leclercg and coworkers (1984) in their isolation of human tonsillar complexes of FDCs and lymphocytes. Recent technology such as the development of a discontinuous BSA gradient separating FDCs at a particular density, has provided another critical tool (Schnizlein, Szakal and Tew, 1984). In our laboratory, these advances brought forth a technique which preserved morphological integrity and provided an FDC enriched single-cell suspension. This in turn has provided a means to study the morphological and antigenic characteristics of these cells *in vitro*. The diagnostic cytological features of FDCs established using Nomarski optics, SEM, and TEM include numerous intertwining dendrites made up of filiform or beaded processes radiating for varying distances from the cell body (Szakal et al., 1984). These dendritic processes have been shown to trap and retain Ag on their plasma membranes as demonstrated by the *in vivo* localization of the cytochemically detectable Ag, HRP (Szakal, Holmes, and Tew, 1983). Follicular dendritic cells were also shown to retain HRP-immune complexes during and after isolation (Szakal et al., 1984). Utilizing these characteristics, specifically morphology and the retained antigen as a marker, FDCs can be identified *in vitro* and their antigenic phenotype determined.

Studies conducted to demonstrate Ia expression on the plasma membranes of FDCs included preparations of isolated cells, cryostat sections for light microscopy and fixed vibratome sections prepared for transmission EM. Mouse and rat lymph nodes were utilized in conjunction with monoclonal Abs and IF techniques. Isolated FDCs were shown to label specifically for Ia using both an indirect IF method involving a secondary Ab as well as a direct method using monoclonal Ab conjugated to fluorescein. Work done with sections of lymph nodes also provided evidence for *in situ* expression of Ia on FDCs. Evaluation was continued at the ultrastructural level in order to distinguish the labeling found on FDC membranes versus other Ia positive cells present in lymphoid follicles. Indeed figure 1 demonstrates the distinct labeling of Ia molecules on the cell body of FDCs as well as on their convoluted processes which are not associated with B-cell membranes. Controversy has arisen regarding the issue of Ia expression by FDCs. Barclay (1981) reported that Ia expression by splenic and lymph node FDCs was questionable in the rat system. Furthermore, Humphrey (1982) was unable to detect Ia on mouse splenic FDCs. However, a recent report by Gerdes et al. (1983) describing dendritic reticular cells in human lymphoid tissue presents evidence for the expression of an Ia-like Ag. This study is in agreement with our findings that lymph node FDCs can express Ia molecules. Whether this is a constitutive trait or inducible has not been established.

Results of the morphological and antigenic phenotyping studies facilitates the comparison of FDCs to other cells involved in the immune response. The FDCs have several characteristics which identifies them as a separate population from macrophages and other dendritic cells. FDCs differ from macrophages in that they do not express the antigens identified by the Mac series of monoclonal Abs nor do they react with F4/80. Furthermore, FDCs are nonadherent, nonphagocytic, and retain Ag on their cell surface for long periods of time. In comparison to lymphoid dendritic cells (LDCs), FDCs are larger and demonstrate a much more complex dendritic morphology. They have been shown to express receptors for the Fc region of IgG and for complement (Humphrey and Grennan, 1982) while LDCs do not (Nussenzweig et al., 1981). In their surface determinants, FDCs are similar to Langerhans cells (LCs) in that LCs possess Ia antigens, Fc and complement receptors, but differ in that FDCs do not react with the monoclonal Abs, F4/80, anti-Mac 2 or 3 (Haines et al., 1983). In addition, FDCs do express antigens common to cells of the lymphoid system (i.e. H2-K and CLA). The relationships between cells of the dendritic family are still not clearly defined but an argument suggesting that each population of LCs, LDCs, and FDCs, are distinct cell types with their own characteristic surface markers and morphology can be supported. Identifying the stem cell origin of FDCs (currently in progress) will be helpful for understanding their ontogeny and defining their relationships to other lymphoid cells.

REFERENCES

Barclay, A.N., 1981, Different reticular elements in rat lymphoid tissue identified by localization of Ia, Thy-1 and MRC Ox2 antigens, Immunology, 44:727.

Gerdes, T., Stein, H., Mason, D.Y., and Ziegler, A., 1983, Human dendritic reticulum cells of lymphoid follicles: Their antigenic profile and their identification as multinucleated giant cells, Virclows. Arch., 42:161.

Haines, K.A., Flotte, T.J., Springer, T.A., Gigli, I., and Thorbecke, G.J., 1983, Staining of Langerhans cells with monoclonal antibodies to macrophages and lymphoid cells, Proc. Natl. Acad. Sci, 80:3448.

Humphrey, J.H., and Grennan, D., 1982, Isolation and properties of spleen follicular dendritic cells, Adv Exp Med Biol, 149:823.

Lilet-Leclercq, C., Raddoux, D., Heinen, E., Kinet-Denoel, C., Defraigne, T., Houben-Defresne, M., and Simar, L.T., 1984, Isolation of Follicular Dendritic Cells from Human Tonsils and Adenoids. I. Procedure and Morphological Characterization, J. of Immunological Methods, 66:235.

Nossal, G.J.V, Abbot, A., Mitchell, J., and Lummus, Z., 1968, XV. Ultrastructural features of antigen capture in primary and secondary lymphoid follicles, J. Exp. Med., 127:277.

Nussenzweig, M.C., Steinman, R.M., Unkeless, J.C., Witmer, M.D., Gutchinov, B., and Cohn, Z.A., 1981, Studies of the cell surface of mouse dendritic cells and other leukocytes, J. Exp. Med., 154:168.

Schnizlein, C.T., Holmes, K.L., Keesling, N.L., and Tew, J.G., 1982, Identification, enrichment and partial characterization of follicular dendritic cells from mouse lymph nodes in vitro, RES Abstract #70, p. 75.

Schnizlein, C.T., Szakal, A.K., and Tew, J.G., 1984, Identification of FDCs in cell suspension, Submitted J. Immunol.

Szakal, A.K. and Hanna, M.G., 1968, The ultrastructure of antigen localization and virus-like particles in mouse spleen germinal centers, Exptl. Mol. Pathol., 8:75.

Szakal, A.K., Holmes, K.L., and Tew, J.G., 1983, Transport of immune complexes from the subcapsular sinus to lymph node follicles on the surface of nonphagocytic cells including cells with dendritic morphology, J. Immunol., 131:1714.

Szakal, A.K., Gieringer, R.L., Kosco, M.H., and Tew, J.G., 1984, Isolated Follicular Dendritic Cells: Cytochemical antigen localization, Nomarski, SEM and TEM morphology, Submitted J. Immunol.

ACKNOWLEDGEMENTS

The authors would like to express their gratitude to Dr. R. Steinman for providing F4/80, Dr. J. Unkeless for 2.462 and Dr. L. Schook for various anti-Ia monoclonal antibodies used in this study. In addition, the authors gratefully acknowledge the expert technical assistance of Ms. Judith Williamson and Mr. Alan Monfalcone.

B CELL DIFFERENTIATION: B CELL SUBSETS and FACTORS, ORGANS OR STRUCTURES INVOLVED IN B LYMPHOCYTE GENESIS

Paul Nieuwenhuis

Department of Histology, University of Groningen
Groningen, The Netherlands

B CELL SUBSETS in the SPLEEN

Cell constituents of rat spleen follicular structures

Follicular structures are made up of both lymphoid and non-lymphoid cells. In normal adult rats these structures usually consist of (i) a germinal centre surrounded by (ii) a lymphocyte corona (or mantle zone) and, beyond the marginal sinus (iii) a so called marginal zone bordering the red pulp. Follicular structures are predominantly made up of B cells but special subsets of T cells have been identified in the lymphocyte corona and light zone of the germinal centre (see e.g. Kroese et al. (1)). For non-lymphoid cells, non- phagocytic, presumably antigen presenting, follicular dendritic cells (FDC) have been identified in the germinal centre light zone, whereas phagocytic cells are found in the marginal zone (MZ macrophages) and in highly active germinal centres (tingible bodies macrophages). We will now concentrate on follicular B cells only.

By Kroese et al. (2) a series of monoclonal antibodies against rat B cells was presented, which could be subdivided into three groups reacting (i) with all B cells plus variable proportions of T cells (HIS 8, HIS 11 and HIS 25), (ii) with germinal centre and lymphocyte corona but not with marginal zone (HIS 24), and (iii) with lymphocyte corona only (HIS 22). Among these, HIS 8 seems to be specific for light chains, whereas HIS 22 might turn out to be an anti IgD Mab. Comparing these markers with the presence of sμ and sδ we get the following picture:

	sμ	HIS 8	sδ	HIS 22	HIS 24
Marginal zone	+	+	-	-	-
Lymphocyte corona	+	+	+	+	+
Germinal centre	low +	+	-	-	+

Do B cells from lymphocyte corona and marginal zone constitute two separate cell lineages?

By MacLennan et al. (3) evidence was presented that marginal zone B cells develop independently from recirculating corona type follicular B cells. Upon transfer of K allotype marked B cells into cyclophosphamide treated (= B cell depleted) recipients, high levels of donor cell chimerism in follicular B cells but not in marginal zone was observed. In time, marginal zone repopulated with B cells of host allotype only, indicating that marginal zone B cells do not derive from recirculating follicular B cells. This, however, does not exclude transition the other way around, i.e. $\mu^+\delta^-$ marginal zone cells turning into $\mu^+\delta^+$ follicular B cells. The latter would be compatible with data from blocking experiments using either anti μ or anti δ : treatment from birth with anti δ does not prevent the development of marginal zone B cells, whereas treatment with anti μ prevents both marginal zone and lymphocyte corona to develop, albeit with some exceptions (see MacLennan et al. (3)).

When B cell development was suppressed by treatment with anti $\mu + \delta$ antibodies from birth for six weeks, spleen from animals thus treated did not show any B cells (Bazin et al. (4)). When suppression was stopped, complete regeneration of spleen follicular structures was obtained 5 days later with germinal centre development by day 8. It appeared that the $\mu^+\delta^+$ population was complete before marginal zones had their full contingent of $\mu^+\delta^-$ cells.

In contrast to the above observations of fast regeneration of the B cell compartment are data from Kumararatne et al. (5) who found that, following 5 days of ^{3}H Thymidine infusion, only 25% of recirculating (= corona type) and static (= marginal zone type) B cells of the splenic white pulp of rats were labelled. Surprisingly, the two subpopulations did not seem to differ in labellingindex, what might have supported the notion of two separate lineages.

B CELL SUBSETS in the BONE MARROW

Using a double immunofluorescence technique Opstelten et al. (6) identified cμ^+sμ^- pre-B cells in young adult rat bone marrow to subsequently stain these cells with a panel of monoclonal antibodies. Of these, HIS 11 and HIS 25 detected 50% of the μ^+ cells, mostly the small ones. As these monoclonals stain all peripheral B cells, they may therefore discriminate between the more and less mature pre-B cells. When bone marrow cells after depletion of sIg$^+$ B cells were stained with HIS 24, 90% of HIS 24$^+$ cells were cμ^+sμ^- pre-B cells whereas the HIS 24$^+$ μ^- cells (7%) were very large blastlike cells. When comparing these data with those of Kroese et al. (2) it is worth noting that, in spleen, both germinal centre cells and corona lymphocytes also react with HIS 24, but not marginal zone cells.

Specificity of "markers" for pre-B and B cells.

The HIS 24 Mab, by Opstelten et al. (7) was also found to react with our pre-B cell (LAMA) leukaemia. Apparently, LAMA represents a malignant tranformation in an early stage of B cell differentiation, at the transition of pre-B to B lymphocyte, as witnessed by K gene rearrangement.
Using this pre-B cell leukaemia, Rozing et al. (8) produced a panel of monoclonal antibodies, two of which (ER 14 and ER 1) also reacted with thymocytes (100%), peripheral T cells, as well as normal pre-B cells from rat bone marrow, but not with peripheral B cells. Putting these data together we get the following picture:

	Bone marrow pre-B cells cμ^+		Periphery B cells sμ^+	T cells -
HIS 11	+		+ (100%)	-
HIS 25	+		+	some +
ER 1	+		-	+ (100%)
ER 14	+		-	+ (100%)
HIS 24	+	GC/Ly.Cor.	+	-
		Marg. Zone	-	

Apparently, pre-B cells carry a variety of determinants part of which are lost upon further differentiation and some of which are shared with mature T cells. It is clear that these determinants should only be used as "markers" in strictly defined experimental systems. It would be nice to know whether (any of) these determinants are also carried by prothymocytes.

B LYMPHOCYTE GENESIS *in* the BONE MARROW

Whether B lymphocyte production in the bone marrow is regulated either by endogenous and/or exogenous factors (and how!) is still a matter of discussion.
CBA/HN mice, when homozygous for the X-chromosome linked immunodeficiency (Xid) trait, show abnormalities in their B lymphocyte subpopulations. Witkowski et al. (9) using mice heterogeneous for Xid as well as for the A and B alleles of the X-linked enzyme, phosphoglycerate kinase (PGK-1), showed, that, whatever the mechanism of action of the Xid gene on B lymphocyte development, it does not

affect the bone marrow B lymphocyte populations (including their progenitors) at the cellular level.

From pathology to "normal" physiology, Hooykaas et al. (10), in their studies on germfree mice -fed a chemically defined ultra filtered "antigen free" diet (GF/CD)- vs conventional mice (CV), obtained evidence that the IgM repertoire is established independently of exogenous antigenic and/or mitogenic stimuli. They also concluded that endogenous factors must account for the generation of "spontaneoulsy" occurring ("background") Ig secreting cells.

In addition to this endogenous component of regulation of B lymphocyte production, Osmond et al. (11) discussed the effects and mechanism of action of exogenous stimuli on pre-B cells, B lymphocytes and bone marrow lymphocyte turnover. Following the injection of SRBC in 8-10 wk C_3H/HeJ mice the number of pre-B cells and B lymphocytes per femur increased. This effect was abolished either by the injection of silica (a macrophage poison) or splenectomy. They concluded that SRBC injections stimulate bone marrow B lymphocyte genesis by a silica-sensitive spleen dependent mechanism and suggested that the normal level of B lymphocyte production in the bone marrow may in part reflect the influence of environmental stimuli.

B LYMPHOCYTE GENESIS outside the BONE MARROW

That environmental stimuli not only influence B lymphocyte production in the bone marrow but also affect other organs, systems or structures like the gut associated lymphoid tissue, the Bursa of Fabricius as well as germinal centres was a further topic of interest.

Gut associated lymphoid tissue

Comparing the relative ratio's of IgM, IgA or IgE isotype bearing B lymphocytes in Peijer's patches from germfree vs conventional mice, Durkin et al. (12) observed that, in GF mice, 30% of PP B cells carry sIgE, one half of which were simultaneously positive for sIgA. In CV mice these cells were virtually absent. In addition GF mice contained elevated numbers of $sIgA^+$ cells, decreased $sIgM^+$ cells as well as high numbers of T suppressor cells (20% vs 2% in CV mice) (see also Kroese et al. (1)). By either thymectomy or the use of unautoclaved chow these relative ratio's could be "normalized within 24 hours, indicating that the "normal" situation largely depends on the exposure to gut flora, either directly or through the mediation of T regulatory cells possibly instrumental in isotype switching.

Particular attention was focused on the role and development of Peijer's patches in sheep (Reynolds et al. (13), (14)). The ileocecal part of these PP's (length ± 2.5 meters) making up 90% of total PP mass, develops early during intra-uterine life, apparently independent of external antigen, and by birth the follicles appear fully mature and have a high rate of lymphopoiesis. Ileocecal PP's involute at about one year of age. Although morphologically indistinguishable from normal germinal centres, Reynolds et al., because of the essentially antigen-independent development of PP follicles, concluded that the development of GC and PP follicles involve different processes.

Moreover, Miyasaka et al. (15), who consider the ileocecal PP the primary source of B cells in the sheep, demonstrated that some B cells leaving ileal PP by way of the efferent lymphatics are still immature (large blast cells which are dull sIg^+, dull Ia^+) and require further maturation in the periphery. The latter observation is compatable with previous data by Opstelten et al. (16) obtained for the rabbit appendix. (There is a difference, however, as appendix germinal centres do not develop before birth and are antigen dependent structures).

Considering these data, one cannot escape reminiscing previous theories on the possible role of the GALT as a primary lymphoid organ comparable with the Bursa of Fabricius in chicken.

Bursa of Fabricius

To complicate matters, conflicting data as to a role even for the Bursa of Fabricius as a primary lymphoid organ were presented by Ewert et al. (17) who removed the primordium of the BF from 60-hr chicken embryos by cauterizing the tailbud (Tbx). Their findings indicated that embryonic bursectomy reduces quantitatively the B cell pool but does not prevent differentiation and maturation of B (or T) lymphocytes.

To further complicate matters, Ekino et al. (18) suggested that a major part of the role of the BF depends upon its exposure to gut-derived antigens. Following bursal duct ligation in ovo (day 19) the development of spontaneous germinal centre formation in the spleen and "natural" serum agglutinins against bacteria and hetero-erythrocytes was suppressed. These results suggest that information about environmmental antigens is mainly obtained by way of the antigen trapping mechanism of the BF. Unfortunately, no data were presented on total B cell pool contained in the spleen of these chickens. They concluded that the antigen trapping mechansim of the bursa may play a major role in acceleration and amplification of B cell differentiation pathways for the protection of the newly hatched chick.

B LYMPHOCYTE PRODUCTION by GERMINAL CENTRES

As germinal centres, including GC in the GALT with the possible exception of PP in the sheep, are -by definition- antigen dependent structures, and as the role of the BF at least in part seems to be antigen-mediated one is left to wonder whether the bone marrow is the only "real" primary lymphoid organ associated with autonomous B lymphocyte genesis. What then is the contribution of these antigen dependent processes going on in germinal centres?

Follicular dendritic cells

Follicular dendritic cells have been described in primary follicles as well as in germinal centres and generally are considered to be non phagocytic antigen presenting cells. FDC are a distinct population differing in various aspects from the -phagocytic- marginal zone macrophages. In a rather elegant series of experiments by Humphrey et al. (19) origin and turnover of these two cell types was studied using F_1 -> P bone marrow chimeras between CBA ($H2^k$) and B10 ($H2^b$) mice, thus allowing the distinction between cells of donor or host origin. In spleens of chimeras established up to 1 year, the dendritic processes of FDC exhibited recipient H2 antigen only, in contrast to the associated B cells which were clearly of donor origin, like the newly formed marginal zone macrophages. FDC therefore are not BM derived.

In mice, Radoux et al. (20) at the EM level studied the localisation of immune complexes onto the surface of FDC in draining lymph nodes using BSA-gold coupled to homologous gold-labelled anti-BSA antibody, once more showing the non phagocytic character of this cell type. Moreover, they showed that immune complex binding by FDC occurs mainly via FC receptors.
In a parallel study, on FDC isolated from human tonsils, Heinen et al. (21) could corroborate the above findings, even though in the human some FDC-associated antigens (Leu M_3, HLA-DR) are shared with macrophages. Because these FDC envelope lymphoid cells within their cytoplasmic extensions an important role as accessory cells within germinal centres was postulated.

Finally, the ontogenetic development of FDC, both at the LM as well as at the EM level, was investigated by Dijkstra et al. (22). I.v. injected horse radish peroxidase (HRP)-anti HRP complexes were used as marker for the immune complex retaining FDC. It appeared that these immune complexes were trapped in primary follicles on immature forms of FDC. At the ultrastructural level these cells showed characteristics of fibroblastic reticulum cells of mesenchymal origin. The latter observation corresponds nicely with the above mentioned data by Humphrey et al. i.e. that FDC are not BM derived.

Germinal Centres

In mice, Germinal Centre Cells (GCC) can be identified by the high binding of PNA. By separating PNA-hi cells and PNA-lo cells from stimulated lymph nodes and subsequent transfer into irradiated recipients, Kraal et al. (23) demonstrated that after primary stimulation virtually all memory is carried by the PNA-hi cells, whereas after secondary stimulation memory cells were also present among the PNA-lo Mel 14^+ cells outside germinal centres. Only in the PNA-hi cell population, a relatively high percentage (15-25%) of specific antigen binding cells was identified, by day 7 after stimulation. Most probably these are the cells capable of memory transfer after primary stimulation. Surprisingly, however, little attention is usually paid to the 75-85% of GCC that do <u>not</u> specifically bind with the antigen used for lymph node stimulation. It would be worthwhile if this type of experiment were repeated under germfree conditions to eliminate any "contaminating" background GC activity (for discussion see also Nieuwenhuis and Opstelten (24)).

Above data were largely confirmed by Coico and Thorbecke (25). They also presented some very interesting preliminary results regarding the in vitro properties of memory B cells. At least under certain experimental conditions colony forming B cells are approximately 5-fold higher in PNA^+ than in PNA^- B cell populations. Whether these observations are the hallmark of "in vitro germinal centre formation" only the future can tell. Up till now, germinal centres, as structural elements in highly organised lymphoid tissues, can only be studied in the in vivo situation.

The question of whether or not GC play a role in antibody synthesis was reiterated by McCarty et al. (26). GCC of stimulated lymph nodes were "purified" by depletion of IgD^+ and Thy 1^+ cells. Between day 5 and 13 after priming the GCC population secreted antigen specific antibodies correlating with serum antibody titers. These data, however, are in discordance with the very sophisticated studies by van Rooijen et al. (27) in which specific antibody forming cells were identified in vivo (in situ) in rabbit spleen during a primary or secondary response. Only after secondary stimulation could specific antibody forming cells be detected inside splenic follicles. A possible explanation for this discrepancy might be that the depletion procedure as used by McCarty et al. only results in the depletion of inactivated B and T cells. Thus their "GCC" population would still contain plasmablasts and plasmacells located outside GC but not properly eliminated.

Germinal Centre Precursor Cells

In comparison to the rapidly increasing knowledge on GC and Germinal Centre Derived Cells (GCDC) (including B memory cells) little attention is usually paid to the cell population from which GC (and their progeny) derive i.e. Germinal Centre Precursor Cells (GCPC). From migration studies few and mostly negative data were obtained regarding cell types homing to GC. Recently we could demonstrate that GCC (isolated from rabbit appendix) would migrate to spleen GC when injected i.v. and to lymph node GC only when administered via the afferent lymphatics (Deenen et al. (28)).

Perhaps the only relevant information present available on GCPC derives from experiments as reported by Thorbecke et al. (29), who showed that treatment of mice from birth with anti IgD did not prevent GC (and MZ) to develop, implying that GCPC never express sIgD. Turning back to the first part of this summary we may close the circle and add one more to the two already postulated separate lineages for (i) lymphocyte corona and (ii) marginal zone B cells, i.e. (iii) GCPC and GCC.

<u>Bone marrow</u> | <u>Spleen follicle</u>

pre-B immature B

$c\mu^{+}\ 24^{+} \rightarrow s\mu^{+}\ \delta - 24^{+}$ →
- $s\mu^{+}\ \delta - 24^{-}$ marginal zone
- $s\mu^{+}\ \delta + 24^{+}$ lymphocyte corona
- low $s\mu^{+}\ \delta - 24^{+}$ germinal centre

Future research will certainly focus on establishing precursor-product relationships between B lymphocytopoiesis in the bone marrow and peripheral B cells constituting the lymphoid follicle. At present, especially with regard to these highly visible structures, as Butcher refered to GC, "the mystery deepens" (to quote HG Wells: "The <u>In</u>visible Man").

REFERENCES

1. F.G.M. Kroese, A.S. Wubbena, P. Joling and P. Nieuwenhuis. This volume, p. 443
2. F.G.M. Kroese, D. Opstelten, A.S. Wubbena, G.J. Deenen, J. Aten, E.H. Schwander, L. de Leij and P. Nieuwenhuis. This volume, p. 81
3. I.C.M. MacLennan, H. Bazin, D. Chassoux, D. Gray and J.E. Lortan. This volume, p. 139
4. H. Bazin, B. Platteau, I.C.M. MacLennan, M. Khan and G.D. Johnson. This volume, p. 65
5. D.S. Kumararatne, D. Gray, I.C.M. MacLennan and J. Lortan. This volume, p. 73
6. D. Opstelten, G.J. Deenen, T. Jaarsma and S.V. Hunt. This volume, p. 9
7. D. Opstelten, G.J. Deenen, B. de Jong, V.J.S. Idenburg and S.V. Hunt. This volume, p. 27
8. J. Rozing, P. Joling, P. Nieuwenhuis and L.M.B. Vaessen. This volume, p. 17
9. J. Witkowski, L.M. Forrester, J.D. Ansell and H.S. Micklem. This volume, p. 47
10. H. Hooijkaas, N. Bos, R. Benner, J.R. Pleasants and B.S. Wostman. This volume, p. 131
11. D.G. Osmond, G.M. Fulop, D. Opstelten and C. Pietrangeli. This volume, p. 35
12. H.G. Durkin, S.M. Chice and S.I. Jiang. This volume, p. 127

13. J.D. Reynolds, R. Pabst and G. Bordmann. This volume, p. 101
14. J.D. Reynolds. This volume, p. 111
15. M. Miyasaka, J. Reynolds, L. Dudler, M-F. Beya, W. Leiserson and Z. Trnka. This volume, p. 119
16. D. Opstelten, R. Stikker, D. van der Heijden and P. Nieuwenhuis. Germinal Centres and the B Cell System IV. Functional Characteristics of Rabbit Appendix Germinal Centre (-Derived) Cells. Virch. Arch. B Cell Path. 34: 53 (1980).
17. D.L. Ewert, C. Oats and Ch. Chen. This volume, p. 91
18. S. Ekino, T. Urano, H. Fujii and M. Kotani. This volume, p. 487
19. J.H. Humphrey and V. Sundaram. This volume, p. 167
20. D. Radoux, E. Heinen, C. Kinet-Denoël and L.J. Simar. This volume, p. 185
21. E. Heinen, D. Radoux, D.Y. Mason, H. Stein, C. Kinet-Denoël, C. Lilet-Leclerq and L.J. Simar. This volume, p. 171
22. C.D. Dijkstra, E.A. Döpp and E.W.A. Kamperdijk. This volume, p.161
23. G. Kraal, I.L. Weissman and E.C. Butcher. This volume, p. 145
24. P. Nieuwenhuis and D. Opstelten. Functional anatomy of germinal centres. Am. J. Anat. 170: 421 (1984). 25. R.F.
25. R.F. Coico and G.J. Thorbecke. This volume, p. 729
26. J. McCarty, G. Kraal and T. Leanderson. This volume, p. 621
27. N. van Rooijen, N. Kors and R. van Niewmegen. This volume, p. 153
28. G.J. Deenen, D. Opstelten and P. Nieuwenhuis. Homing of germinal center cells into germinal centers of lymph node via afferent lymphatics - an autoradiographic study on rabbits. Cell Tiss. Res. 236: 1984, in press.
29. G.J. Thorbecke, T.J. Flotte and Y. Baine. Maturity of precursor cells for Germinal Centres. Adv. Exp. Med. Biol. 149: 845 (1982).

SESSION 2

ROLE OF THE THYMUS IN T CELL DIFFERENTIATION AND REPERTOIRE ACQUISITION

CHAIRPERSONS: R. Schwartz
A. Munro

PRECURSORS OF T-CELL MITOGEN REACTIVE CELLS AND THEIR SUPPRESSORS IN THE MOUSE FETAL LIVER

M. Rosenberg*, L. Abel**, T. Kukulansky**, N. Sharon* and A. Globerson**

Departments of *Biophysics and **Cell Biology
The Weizmann Institute of Science
Rehovot, Israel

INTRODUCTION

A considerable volume of information has been accumulated on the development of T cell lineages from the embryonic thymus (Ceredig et al., 1982). However, the original source of the stem cells and the developmental processes preceding the earliest phase in which the thymus becomes lymphoid have not yet been clearly delineated. An intriguing possibility is that the stem cells originate from, and/or temporarily reside in, the fetal liver. It should be noted in this respect that although the fetal liver is recognized as an early B cell organ (Owen et al., 1974), it also contains pre-thymic and post-thymic cells that can develop into T cells (Stutman, 1978). Indeed, chimeras obtained by reconstitution of lethally irradiated mice with 14-day fetal liver cells developed a wide spectrum of liver type cells expressing T cell functions (Rabinowich et al., 1983). By the 14th day of gestation, embryonic liver cells could be induced to elicit a graft-versus-host response (Umiel et al., 1968), suggesting that embryonic liver may contain cells which can contribute to the T cell compartment even before the thymus itself is established as a lymphoid or secretory organ. Yet the ability to produce a response required thymic induction. On the other hand, the embryonic liver is a major source of suppressor cells during ontogeny (Globerson et al., 1975). The early fetal liver thus represents an anatomical site at which processes crucial for the establishment of regulation and manifestation of immunocompetence occur both prior and together with thymic development. Since fetal liver cells at this stage of differentiation do not express membrane markers specific for the T cell lineages, the peanut agglutinin (PNA) receptor on the early appearing suppressor cells is a unique and useful marker (Rabinowich et al., 1979).

The present communication reports on our attempts to determine (a) whether the fetal liver can give rise to any T cells without passage through the thymus, and (b) what is the developmental relationship of the PNA^+ suppressor cells in this tissue to the early potentially reactive cells.

MATERIALS AND METHODS

Mice

C57BL/6j mice were obtained from Jackson Laboratories (Bar-Harbor, Maine) at the age of 5-6 weeks and were used 4 weeks later. Embryonic stage was determined by considering the day of first observed vaginal plug as day 0 of gestation.

Cell separation by PNA

The separation was as previously described by Reisner et al. (1976), except that the separation was on 5% bovine serum albumin. The procedure was repeated twice to reduce cross-contamination and obtain more pure subpopulations.

Culture procedures

Cells were cultured in RPMI-1640 medium, supplemented with 20% fetal calf serum and 20% supernatant of secondary mixed lymphocyte cultures prepared as described by Ryser et al. (1978). Cells were seeded either before or after selective agglutination by PNA, in 24-well Costar culture plates, at a concentration of 5 x 10^5 cells/ml/well. The cultures were further supplemented with media once a week. Mitogen-induced proliferation and suppression of the response were assayed as described by Rabinowich et al. (1979).

RESULTS AND DISCUSSION

Cultured cells were harvested at weekly intervals and tested for responsiveness to T and B cell mitogens. In some experiments the cells were further separated by PNA before the assay. Results show that after 7 days in vitro, fetal liver cells from the 12th, 13th or 14th day of gestation proliferated in response to mitogenic stimulation. Furthermore, the response was manifest mainly in the PNA^- population isolated from an unseparated or from PNA^+ cell cultures just before exposure to mitogens (Figs.1-3).

The possibility that the PNA^+ fraction contains suppressor cells was then examined by testing the ability of cultured cells to suppress concanavalin A-induced response in splenocytes from young adult mice. Suppression was exerted by the PNA^+ fraction after 7 and 21 days in culture (Fig.4). It was interesting to note that at about 14 days the cell cultures seemed to pass through a "crisis" period. This was expressed as a slowdown in cell proliferation as well as a decrease in their suppressive ability. When subpopulations of PNA^+ or PNA^- cells were grown separately in vitro, the

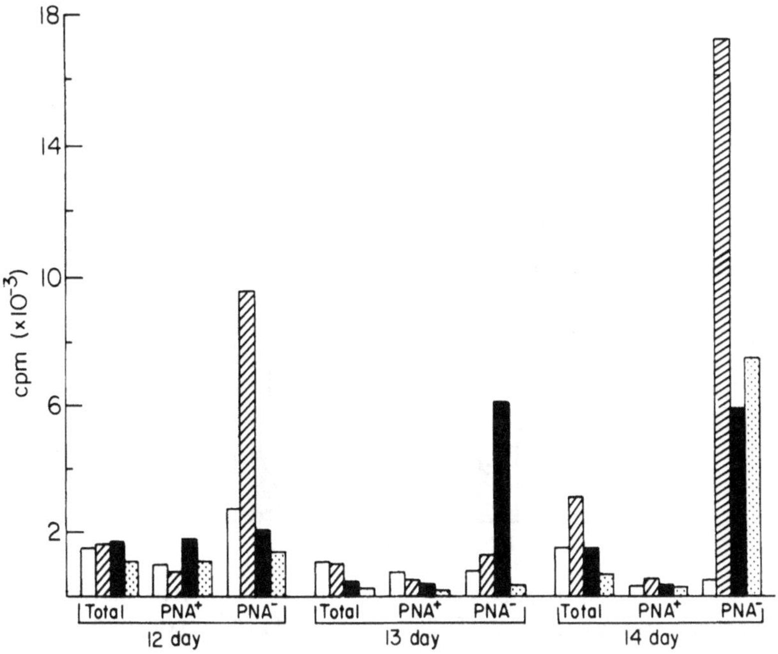

Figure 1. Mitogen-induced proliferative response of fetal liver cells cultured for one week, and tested before and after separation by PNA. Symbols: ☐ , medium alone; ▨ , concanavalin A; ■ , PHA; ▩ , lipopolysaccharide.

suppressive activity was higher in the PNA$^+$ than in the PNA$^-$ fraction. When these subpopulations were subsequently fractionated again by PNA before being tested for suppression, the PNA binding cell population, harvested from either the PNA$^+$ or the PNA$^-$ cultures, were found to be most suppressive (Fig.5).

On the other hand, the liver cells seem to lose during the culture period the ability to develop in vitro hemopoietic colonies (Table 1). It thus appears that the culture conditions employed support development of certain types of T cells, and the depletion of hemopoietic stem cells.

Our results indicate that, from the 12th day of gestation, the fetal liver can give rise to cells which have the potential to react to T cell mitogens without direct passage through the thymus. These reactive cells have no receptors for PNA. The PNA$^+$ suppressor cells residing in the fetal liver can develop and/or further expand in vitro. These cells continue to carry receptors for PNA and still maintain their suppressive capacity after 21 days in culture. The observations are in accord with the idea that these suppressor cells may play a continuous role of surveillance in preventing response of

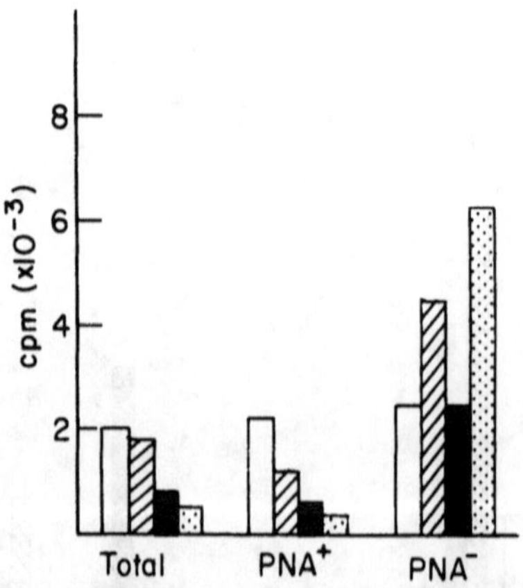

Figure 2. Mitogen-induced proliferative response of cultured subpopulations of 14 day fetal liver cells after one week in culture. Symbols as in Fig.1.

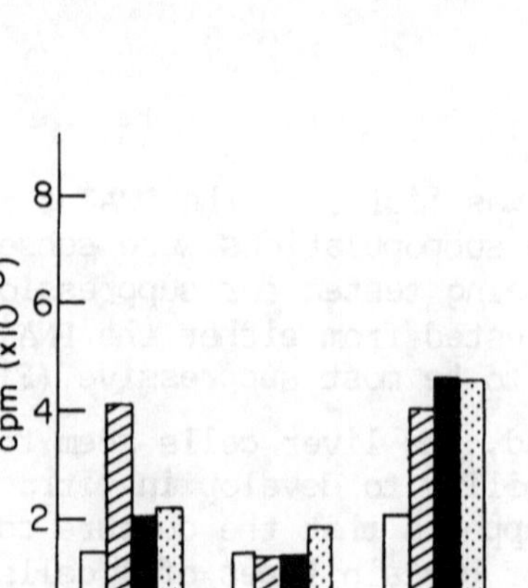

Figure 3. Mitogen-induced proliferative response of subpopulations of cultured PNA+ fetal liver cells after one week in culture. Symbols as in Fig.1.

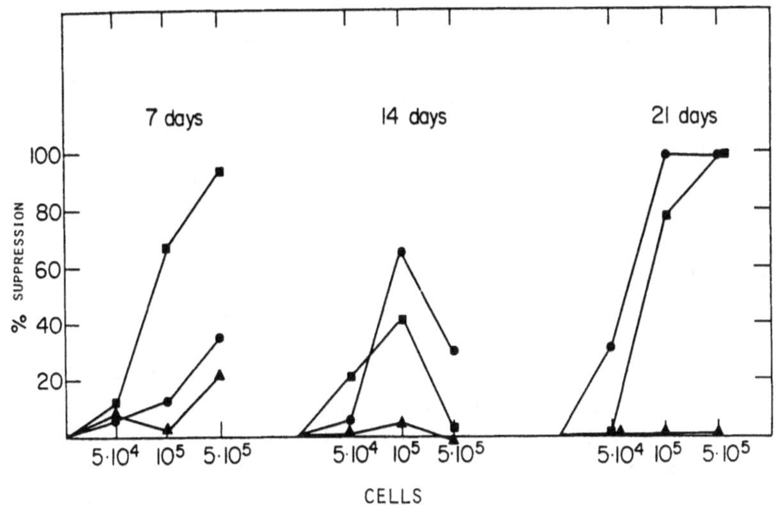

Figure 4. Suppression of the response of adult splenocytes to concanavalin A by cultured fetal liver cells
Symbols: ● , unseparated; ▲ , PNA^+; ■ , PNA^-.

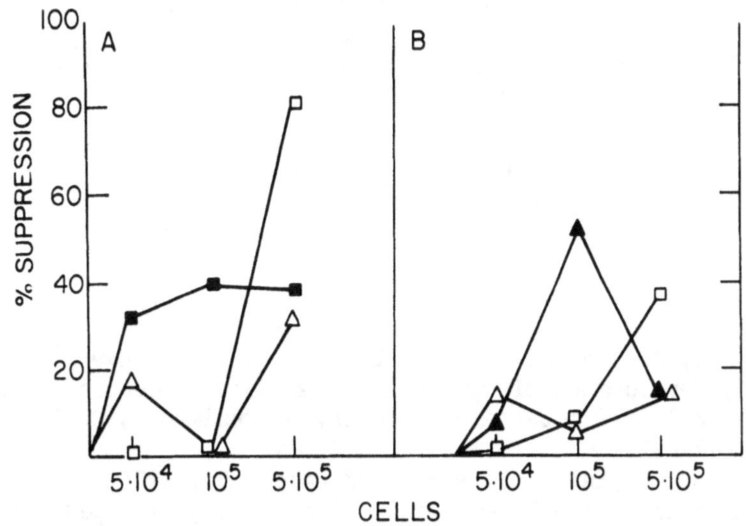

Figure 5. Suppression of the response of adult splenocytes to concanavalin A by subpopulations of fetal liver cells cultured for 7 days.
Symbols for (A): ■ , PNA^+; △ , PNA^+PNA^-; □ , PNA^+PNA^+.
Symbols for (B): ▲ , PNA^-; △ , PNA^-PNA^-; □ , PNA^-PNA^+.

Table 1. In vitro colony-forming cells generated by cultured fetal liver cells

Cells tested	Culture period (days)	Number of colonies per 10^5 cells
12 day fetal liver	1	9-12
12 day fetal liver	8	1-2
13 day fetal liver	45	0
Bone marrow	0	80-100

any emerging autoreactive cells in ontogeny. The observations that the PNA receptor is still present on suppressor cells after one month in culture, and that PNA^-, potentially reactive, cells develop in parallel, suggest that these cell populations may develop independently of each other. However, some PNA^- mitogen-reactive cells were obtained from cultures of PNA^+ cells.

The appearance of cells reactive to T cell mitogens without direct thymic influence seems to represent an additional, or an alternative, pathway of development under conditions of thymic deficiency, including the situation in the nude mice (Hünig, 1983).

The questions remain open whether cells manifesting the T cell properties in culture already express any restricted properties, or whether they represent precomitted T cells which would normally pass through the thymus for further differentiation.

Acknowledgements: This study was supported by the Hermann and Lilly Schilling Stiftung im Stifterverband. AG is incumbant of the Harriett and Harold Brady Chair in Cancer Research.

REFERENCES

Ceredig, Rh., Jenkinson, E.J., MacDonald, H.R., and Owen, J.J.T., 1982, Development of cytolytic T lymphocyte precursors in organ-cultured mouse embryonic thymus rudiments, J. Exp. Med. 155:617-622.

Globerson, A., Zinkernagel, R.M., and Umiel, T., 1975, Immunosuppression by embryonic liver cells, Transplantation 20:480-484.

Hünig, T., 1983, T-cell function and specificity in athymic mice, Immunol. Today 4:84-87.

Owen, J.J.T., Cooper, M.D., and Raff, M.C., 1974, In vitro generation of B lymphocytes in mouse fetal liver, a mammalian 'bursa equivalent', Nature 249:361.

Rabinowich, H., Umiel, T., Reisner, Y., Sharon, N., and Globerson, A., 1979, Characterization of embryonic liver suppressor cells by peanut agglutinin, Cell. Immunol. 47:347-355.

Rabinowich, H., Umiel, T., and Globerson, A., 1983, T cell progenitors in the mouse fetal liver, Transplantation 35:40-48.

Reisner, Y., Linker-Israeli, M., and Sharon, N., 1976, Separation of mouse thymocytes into two subpopulations by the use of peanut agglutinin, Cell. Immunol. 25:129-134.

Ryser, J.-E., Cerottini, J.-C., and Brunner, K.T., 1978, Generation of cytolytic T lymphocytes in vitro. IX. Induction of secondary CTL responses in primary long-term MLC by supernatants from secondary MLC, J. Immunol. 120:370-377.

Stutman, O., 1978, Intrathymic and extrathymic T cell maturation, Immunol. Rev. 42:139-184.

Umiel, T., Globerson, A., and Auerbach, R., 1968, Role of the thymus in the development of immunocompetence of embryonic liver cells in vitro, Exp. Biol. Med. 129:598-600.

THYMUS HOMING CLONOGENIC BONE MARROW CELLS

Sophie Ezine, Irving L. Weissman and Robert V. Rouse

Department of Pathology
Stanford University
Stanford, California 94305

INTRODUCTION

The thymus plays a major role in guiding the maturation, differentiation and specificity of T cells (1-4). In the adult mouse, the thymus receives small numbers of T cell precursors from the bone marrow and exports to the periphery a large variety of cells that differ by their surface antigens and functions. Bone marrow cells homing to the thymus undergo different steps of proliferation, maturation, and selection before they leave the thymus. Each of these steps is characterized by the involvement of various cell types (lymphoid or non-lymphoid) and factors whose sequential participation is still unknown.

Using a rapid *in vivo* assay for thymus homing bone marrow cells, LePault and Weissman (5) showed that 0.1% of the total bone marrow cells are capable of homing to the irradiated thymus when injected IV. The actual number of pre-T cells which enter and proliferate in the normal thymus is not known. In order to study the first steps of thymocyte development, we reconstituted irradiated mice with a limiting number of bone marrow cells from Thy-1 congenic donors. We have shown that distinct clones of reconstituting cells can be identified within the thymus and that at least 2 independent thymus lineages exist: cortex plus medulla and medulla alone.

MATERIALS AND METHODS

The following mouse strains were used: C57Bl/Ka (Thy-1.2) and C57Bl/KaThy-1.1 (BA mice), C57Bl/J (Thy-1.2, Lyt-1.2) and Alpha (Thy-1.2, Lyt-1.1). Mice were irradiated (850 rad) and reconstituted with a mixture of anti-Thy-1 and complement treated bone marrow cells from identical and congenic mice. The animals received intravenously a total cell number of $2.5x10^6$ or $2.5x10^7$ containing a mixture of donor to host type varying from 1:10 to 1:1000. The thymuses were taken from hosts 2-10 weeks post-irradiation. The patterns of thymocyte repopulation were examined as described elsewhere (6) using monoclonal antibodies anti-Thy-1.1 (clone 19XE5, gift of R. Nowinski) and anti-Thy-1.2 (clone 30H12, Becton Dickinson), anti-Lyt-1 (clone 53-7.313. from J. Ledbetter), and anti-Lyt-1.1 (clone 18-2.3 from E.A. Boyse). The present report is based on the analysis of 126 thymus lobes, 80 of which have been previously reported (7).

RESULTS AND DISCUSSION

About half of the lobes examined contained very few or rare donor type cells (non-host type cells); the remaining 61 lobes showed a variety of patterns as described (7). Briefly, the following patterns of donor type Thy-1 bearing cells were observed: 1) scattered cells present either individually or as clusters (2-5 cells each), 2) discrete foci (20-30 cells diameter), 3) complete reconstitution of an entire thymic lobe by cells of one Thy-1 type, and 4) repopulation of the thymic medulla only (Fig. 1).

Thymuses from mice injected with mixtures of bone marrow cells differing for Lyt-1 show similar patterns of focal reconstitution (Fig. 2). This indicates that the focal patterns observed are not merely due to Thy-1 specific recruitment or exclusion.

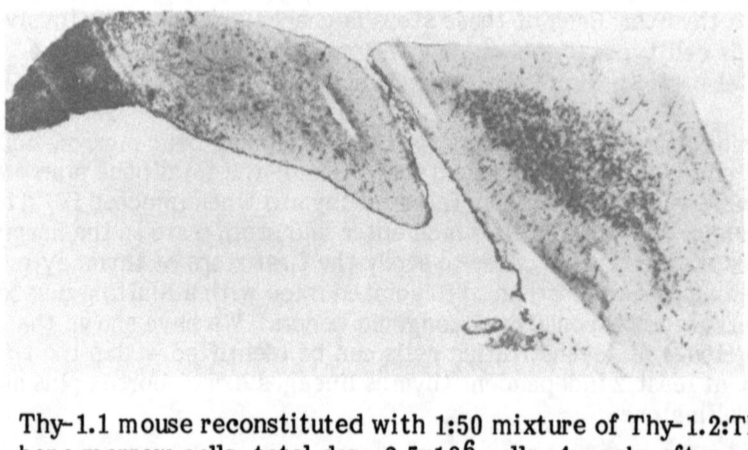

Fig. 1. Thy-1.1 mouse reconstituted with 1:50 mixture of Thy-1.2:Thy-1.1 bone marrow cells, total dose $2.5x10^6$ cells, 4 weeks after reconstitution. Anti-Thy-1.2 stain shows a discrete confluent focus of donor type cells in the left lobe while the right lobe contains large nubmers of donor type cells confined to the medulla. (Reprinted by permission from (7)).

Table 1. Frequency of Clonogenic Repopulation at Four Weeks When $2.5x10^6$ Total Cells are Injected

Number of Donor Type Cells Injected	Ratio D:H	Fraction of Total Lobes Repopulated	Pattern
$2.5x10^3$	1/1000	0/10	
$5x10^3$	1/500	0/6	
$2.5x10^4$	1/100	5/8	foci,clusters
$5.0x10^4$	1/50	11/16	foci,cluster,medulla
$2.5x10^5$	1/10	6/8	foci,clusters

Table II. Comparison of Frequencies of Repopulation at 4W When a Fixed Number of Donor-Type Cells is Coinjected with $2.5x10^6$ vs. $2.5x10^7$ Host-Type Cells

Number of Donor Type Cells Injected	Fraction of Lobes Repopulated Total cells injected	
	$2.5x10^4$	$2.5x10^7$
$2.5x10^4$	5/8	0/8
$5x10^4$	11/16	3/6
$s.5x10^5$	6/8	1/6
$5x10^5$	n.t.*	4/14
$2.5x10^6$	n.t.	4/10

*Not tested

Table III. Frequency of Clonogenic Repopulation as a Function of Time After Injection of $2.5x10^6$ Cells

Number of Donor Type Cells Injected	4W	6W	10W	14-18W
$2.5\text{-}5x10^3$	0/16	n.t.	n.t.	n.t.
$2.5x10^4$	5/8	3/4	n.t.	n.t.
$5x10^4$	11/16	3/4	n.t.	n.t.
$2.5x10^5$	6/8	4/4	7/10	10/12

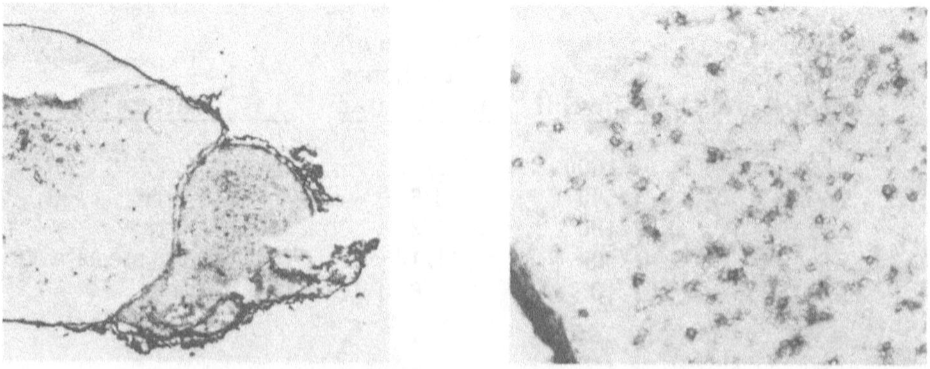

Fig. 2. Thymus from C57Bl/J (Thy-1.2, Lyt-1.2) mouse at 7 weeks after reconstitution with a mixture of bone marrow cells at the ratio Lyt-1.1:Lyt-1.2 = 1:50. The thymus is stained with anti-Lyt-1.1 and shows positive cells throughout the right lobe (see high power) but none in the left lobe.

All thymus lobes taken at 4 weeks from mice receiving a total of 2.5×10^6 cells injected were analyzed for frequency of repopulation (Table I). When 5×10^3 or fewer donor cells were present in the inoculum, no reconstitution by donor type cells was observed. When 2.5×10^4 to 2.5×10^5 donor cells were present in the inoculum, no significant difference in the frequency of repopulation was observed, ranging from 5 of 8 to 6 of 8 lobes. No significant difference in patterns of reconstitution were observed at these various higher doses.

When a fixed number of donor type cells was coinjected with 2.5×10^6 versus 2.5×10^7 host type cells, the number of positive lobes decreased (Table II). We interpret these results as the consequence of competition between donor type cells and increasing numbers of host type cells.

Although only small numbers of lobes have been analyzed at most time points, between 4 and 18 weeks after reconstitution, the number of positive lobes does not change significantly with time (Table III). This suggests that the seeding frequency is determined early and is constant thereafter.

CONCLUSION

These findings support the proposition that more than one lineage of intrathymic cells exists. Whether the diversity of these lineages is determined at the precursor cell level or after entry into the thymus, perhaps by interaction with stromal cells, is not clear. That under some circumstances apparently very few cells can reconstitute an entire thymic lobe (as suggested by Wallis et al (8)) not only has implications relevant to the generation of diversity but also suggests the existence of a strict control over thymocyte precursor entry into the thymus. This method permits study of the kinetics of thymocyte seeding and repopulation.

ACKNOWLEDGMENT

This project was supported by CNRS (S.E.), NIH grant AI 09072 (I.W.) and by the Veterans Administration (R.R.)

REFERENCES

1. Burnet, F.M., 1962, Role of the thymus and related organs in immunity. Br. Med. J. 2:807.
2. Jerne, N.K., 1971, The somatic generation of immune recognition. Eur. J. Immunol. 1:19.
3. Davies, A.J.S., Leuchars, E., Wallis V., Marchant, R., and Elliot, E.V., 1967, The failure of thymus derived cells to produce antibody. Transplantation 5:222.
4. Zinkernagel, R.M., Althage, A., Waterfield, E., Kindred B., Welsh, R.M., Callahan, G. and Pincetl, P., 1980, Restriction specificities, alloreconstituted with H-2 compatible or incompatible thymus grafts. J. Exp. Med. 151:376.
5. Lepault, F. and Weissman, I.L., 1981, An *in vivo* assay for thymus homing bone marrow cells. Nature 293:151.
6. Rouse, R.V., Ledbetter, J.A., and Weissman, I.L., 1982, Mouse lymph node germinal centers contain a selected subset of T cells. J. Immunol. 128:2243.
7. Ezine, S., Weissman, I.L., and Rouse, R.V., 1984, Bone marrow cells give rise to distinct cell clones within the thymus. Nature 309:629.
8. Wallis, V.J., Leuchars, E., Chwalinski, S., and Davies, A.J.S., 1975, On the sparse seeding of bone marrow and thymus in radiation chimeras. Transplantation 19:2.

THE ROLE OF THE THYMIC CORTEX AND MEDULLA IN T CELL DIFFERENTIATION

Roland Scollay[1], Paul Andrews[1], Richard Boyd[2] and Ken Shortman[1]

[1]Walter and Eliza Hall Institute, Melbourne, 3050 Australia, and [2]Department of Pathology and Immunology Monash University, Melbourne, 3168, Australia

INTRODUCTION

The thymus has two main cellular zones, clearly distinguishable histologically; the major outer zone, the cortex, and the smaller central zone, the medulla. Although this morphological subdivision has been known for a long time, we still don't know the respective roles of these two thymic compartments in the generation of peripheral T cells [reviewed in 1,2]. Our ignorance in this area is surprising since we have a lot of information concerning the nature of the lymphoid and non-lymphoid cells in the two compartments. Thus, isolated thymic lymphocytes fall into two major categories: "mature" cells binding low levels of peanut agglutinin, and expressing relatively little Thy 1 but a lot of H-2 antigen, and "immature" cells which are PNA^{hi}, Thy 1^{hi} and $H\text{-}2^{low}$ [reviewed in 3]. The mature cells are located mainly in the medulla and hence we will call them medullary-phenotype cells, while the major population (85%) of immature cells is located in the cortex and hence these are termed cortical-phenotype cells. We use the term cortical- and medullary-phenotype because a small number of cells of either phenotype could exist in the "wrong" region and could thus be, for example, of medullary phenotype but located in the cortex. We should stress, though, that most cells do occur in the "right" region, and exceptions (for which there is little hard evidence, see below) must be few in number [see ref. 3]. The key questions are whether the cells the thymus exports (thymus emigrants) come from the medulla or the cortex, and whether cells move between the cortex and medulla during the process of intrathymic differentiation. One direct approach to these questions is to analyse the emigrants leaving the thymus, to see whether they resemble cortical thymocytes or medullary thymocytes, or neither.

Analysis of Thymus Emigrants

In a number of studies in the last few years we have used the techniques of intrathymic injection of fluorescein isothiocyanate (FITC) into the thymus, followed by analysis of the fluorescent cells appearing in spleen and lymph nodes over the next 3-4 hours, to assess thymus migrants. Details of the technique and a wide variety of controls, as well as discussion of the possible limitations of the method, have been published [4,5]. Migrants are phenotypically exactly like medullary cells, quite different from cortical cells and slightly different from normal peripheral T cells [6,7]. As shown in Table I, they are divided into the mature functional subsets defined by Ly2 and L3T4 [8]. All T cells express one or other of these markers, rarely both. Thymus migrants also express either one or other of these markers and the $Ly2^{+}L3T4^{-}$:$Ly2^{-}L3T4^{+}$ ratio is the same as that of medullary thymocytes or peripheral T cells (i.e. 40:60). Migrants are also immunocompetent [5]. Thus by a wide variety of phenotypic and functional criteria, migrants have a medullary phenotype and hence come either from the medulla or from a small subpopulation of medullary-phenotype cells in the cortex.

Cortical Minorities

There are a number of reasons for looking for cortical minority populations. It seems clear that the majority of cortical thymocytes die in situ and never differentiate further to medullary cells or peripheral cells [4,9, reviewed in 10]. If the cortex is to have any role, it must be to select out a small subgroup of the cells produced. Since emigrant cells make up only 1% of total thymocytes each day [4], a postulated selected cortical population could be this small. This subgroup must either emigrate directly, and so should have a relatively mature phenotype, or go to the medulla first, in which case it could have an intermediate or partially mature phenotype. In addition to these "mature" minorities, it would be of great interest to be able to identify the truly immature, early thymocytes which have not yet become committed to either the typical, "suicidal" cortical pathway, or to the "successful" pathway.

Our own multiparameter analysis has revealed several small populations of probable early blast cells [3,11], such as those lacking the lineage markers Ly1, Ly2 and L3T4. However, we have found little evidence for typical medium-sized medullary-phenotype cells in the cortex, although there do seem to be small numbers of cortical blasts with an "intermediate" phenotype [data in preparation]. On the other hand, there are cells scattered through the

Table I. Ly2 and L3T4 phenotype of recent thymus emigrants

Stain	% of cells positive Spleen migrants	Lymph node migrants
αLy2	31	34
αL3T4	62	63
αLy2 + αL3T4	95	97

Subpopulation	% of migrants expressing phenotype Spleen migrants	Lymph node migrants
$Ly2^+$ $L3T4^+$	0	0
$Ly2^-$ $L3T4^+$	63	63
$Ly2^+$ $L3T4^-$	32	34
$Ly2^-$ $L3T4^-$	5	3

Spleens and lymph nodes were taken from 4 (C57BL/6 x BALB/c)F_1 mice which had been injected intrathymically with FITC 4 hours previously. The cells were counter-stained with biotin conjugated anti-Ly2 or anti-L3T4 or a mixture of the two, followed by avidin conjugated to phycoerythrin (a red emitting fluorochrome). FACS analysis of the fluorescein positive cells gave proportions positive for the two antigens or the mixture (top part of table). From these values, the proportions of the four possible subpopulations can be estimated (bottom part of table). Fluorescein positive cells occur in lymph nodes and spleen with a frequency of about 1 in 1500 nucleated cells. For more details of methods see ref. 6.

cortex which are positive for the antibody MEL14, a monoclonal which binds to the lymphocyte homing receptor and so is presumed to be a marker of mature cells [12 and Reichert, Gallatin, Butcher and Weissman, private communication]. This interesting minority needs more detailed analysis, as this antigen could indicate that these cells are ready to leave (typical cortical cells are low or negative for this antigen, while peripheral T cells are positive).

Another cortical minority population which has received some attention is the one associated with epithelial cells in the lympho-epithelial complex called the thymic nurse cell [13-15]. Although

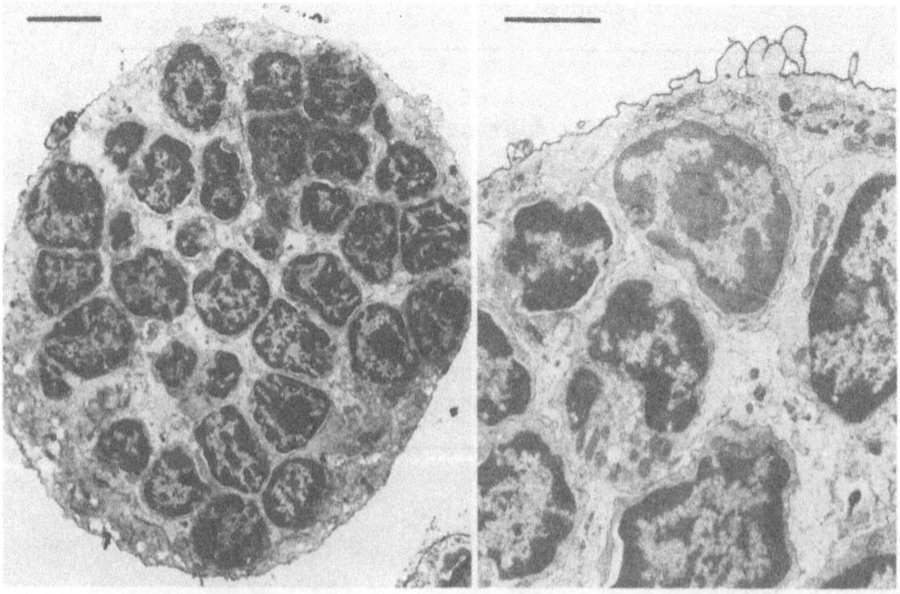

Fig. 1. Thymic nurse cells isolated from alcian blue injected CBA thymuses. The left side shows a low power EM photograph of a single nurse cell. The enclosed lymphocytes can be clearly seen. The high power photo on the right shows the outer nurse cell membrane stained with alcian blue (the dark stain associated with the plasma membrane), but no stain can be seen on any of the internal lymphoid or epithelial cell membranes. Scale bars represent 5 μm (left) and 2 μm (right).

these complexes have been implicated in the process of differentiation in the thymus, little is known of their structure <u>in situ</u>. We have introduced various dyes into the intact thymus and used their differential permeativity to determine whether or not the nurse cell represents an isolated microenvironment <u>in situ</u> [16]. One example is the polycationic dye, alcian blue, which binds to acidic glycoproteins at the cell surface, but is unable to cross the plasma membrane or penetrate sealed membrane junctions [17]. Injection of the dye into the thymus labelled about 25% of free thymocytes and the outer membrane of about 30% of nurse cells. None of the lymphocytes inside nurse cells were labelled, nor was the inner epithelial plasma membrane surrounding the enclosed thymocytes ever found to be labelled (Figure 1). The same result was found with several other dyes. The results indicate that the inside of the nurse cell is not open to the general thymic milieu.

We estimated that there are about $2x10^4$ nurse cells in a thymus, and the average content is about 80 thymocytes. Thus this isolated cortical [15] microenvironment contains about 1% of all thymocytes. Although most of the nurse cell thymocytes are apparently phenotypically immature [14,15], they represent a very interesting population which demands much more detailed study.

CONCLUSION

In considering the relative roles of thymic cortex and medulla in the generation of T cells we must take into account the facts that most cortical thymocytes die *in situ*, and that most exported cells are of medullary phenotype. This means that the cortex contains some small minorities which either migrate to the medulla or leave directly and hence should stand out as mature cells in a predominantly immature region. There are several cortical minorities of interest, including nurse cell lymphocytes, but all of them need to be further analysed before firm conclusions can be drawn.

ACKNOWLEDGEMENTS

This work was supported in part by National Institutes of Health grant no. AI-17310 and by the National Health and Medical Research Council of Australia.

REFERENCES

1. K. Shortman, and R. Scollay, The interrelationship between cortical and medullary thymocytes, *in*: "Recognition and regulation in cell mediated immunity", J. Watson and J. Marbrook, eds., Marcel Dekker, New York (in press) (1984).
2. R. Scollay, Intrathymic events in the differentiation of T lymphocytes: A continuing enigma. *Immunol. Today* 4:282-286 (1983).
3. R. Scollay, and K. Shortman, Thymocyte subpopulations: An experimental review, including flow cytometric cross-correlations between the major murine thymocyte markers, *Thymus* 5:245-295 (1983).
4. R. Scollay, E. Butcher, and I. Weissman, Thymus migration: Quantitative studies on the rate of migration of cells from the thymus to the periphery in mice, *Eur. J. Immunol.* 10:210-218 (1980).
5. R. Scollay, W.-F. Chen, and K. Shortman, The functional capabilities of cells leaving the thymus, *J. Immunol.* 132:25-30 (1984).

6. R. Scollay, Thymus cell migration: Cells migrating from the thymus to the peripheral lymphoid organs have a mature phenotype, J. Immunol. 128:1566-1570 (1982).
7. R. Scollay, A. Wilson, and K. Shortman, Thymus cell migration: Analysis of thymus emigrants with markers which distinguish medullary thymocytes from peripheral T cells, J. Immunol. 132:1089-1094 (1984).
8. D.P. Dialynas, D.B. Wilde, P. Marrack, A. Pierres, K.A. Wall, W. Havran, G. Otten, M. Loken, M. Pierres, J. Kappler, and F. Fitch, Characterisation of the murine antigenic determinant, designated L3T4a, recognised by the monoclonal antibody GK1.5, Immunol. Rev. 74:29 (1983).
9. D. McPhee, J. Pye, and K. Shortman, The differentiation of T lymphocytes: Evidence for intrathymic death of most thymocytes, Thymus 1:151 (1979).
10. R. Scollay, and K. Shortman, Cell traffic in the adult thymus: Cell entry and exit, cell birth and death, in: Recognition and regulation in cell mediated immunity, J. Watson and J. Marbrook, eds., Marcel Dekker, New York (in press) (1984).
11. R. Scollay, K. Shortman, P. Bartlett, D. Dialynas, and F. Fitch, T cell development in the adult murine thymus: Changes in the expression of the surface antigens Ly2, L3T4 and B2A2 during development from early precursor cells to emigrants, Immunol. Rev. (in press) (1984).
12. W.M. Gallatin, I.L. Weissman, and E.C. Butcher, A cell surface molecule involved in organ-specific homing of lymphocytes, Nature 304:30 (1983).
13. H. Wekerle, and U.-P. Ketelson, Thymic nurse cells: Ia-bearing epithelium involved in T lymphocyte differentiation, Nature 283:402 (1980).
14. H. Wekerle, U.-P. Ketelson, and M. Ernst, Thymic nurse cells. Lymphoepithelial cell complexes in murine thymuses: Morphological and serological characterisation, J. Exp. Med. 151:925 (1980).
15. B.A. Kyewski, and H.S. Kaplan, Lymphoepithelial interactions in the mouse thymus: phenotypic and kinetic studies on thymic nurse cells, J. Immunol. 128:2287 (1982).
16. P.A. Andrews, and R.L. Boyd, The murine thymic nurse cell: An isolated thymic microenvironment, Eur. J. Immunol. (in press) (1984).
17. O. Behnke, and T. Zelander, Preservation of intercellular substances by the cationic dye alcian blue in preparative procedures for EM, J. Ultrastruct. Res. 31:424 (1970).

INTRATHYMIC DIFFERENTIATION IN THE RAT

P.Joling, F.J.Tielen, L.M.B.Vaessen, C.J.Hesse and J.Rozing

Dept. of Internal Medicine I, Renal Division
Erasmus University Rotterdam, P.O.Box 1738
Rotterdam, The Netherlands

INTRODUCTION

In view of our interest in the role of various subpopulations of T lymphocytes in the process of immunological tolerance induction or rejection after allogeneic renal transplantation in an experimental rat model, we have started to produce monoclonal antibodies against cell surface determinants of cell types at various stages along the T cell lineage. Such cell types may represent an early phase in T cell differentiation, such as thymocytes or even prothymocytes, or more mature and functionally differentiated stages, such as T helper or T suppressor/cytotoxic cells. This paper describes five new mouse anti rat monoclonal antibodies. Three of these recognise distinct subpopulations in peripheral lymphoid organs. From comparative studies with existing markers in the rat [1], it can be concluded that ER2 reacts with the helper subset, whereas ER3 and ER10 react with the suppressor/cytotoxic subset. ER4 and ER14, on the other hand, do not react with peripheral cells in adult animals. These antibodies recognise different determinants on the 25 KD THY-1 glycoprotein, which are expressed dissimilarly during T cell differentiation. Both are present on almost all thymocytes, but in contrast to ER14 the expression of the ER4 marker on cortical cells exceeded that of medullary cells by manyfold. The recognition pattern of all of these antibodies in the thymus is described, especially in relation to the effect of corticosteroid treatment. Data obtained in such experiments have led to a proposed model of intrathymic T cell differentiation in the rat.

MATERIALS AND METHODS

Animals

PVG rats (7-10 weeks old) were used for analysis. They were purchased from Olac, England. Balb/c mice (10-14 weeks old) were used for immunisation and ascites production. AKR mice (8-12 weeks old) were used for analysis. The mice were bred in our own colony at the Erasmus University Rotterdam.

Immunisation

Balb/c mice were immunised at least twice with various cell types as antigens (see Table 1). Four days after the last booster the spleen of these mice were used for fusion.

Fusion

$2x10^7$ spleen cells and $5x10^7$ SP2/0 cells were fused with PEG 4000 as a fusion reagent using the classical fusion protocol of Köhler and Milstein [2], with some modifications after Fazekas de St. Groth and Scheidegger [3]. After the fusion procedure the cells were plated immediately in HAT medium under limiting dilution conditions [3]. Peritoneal macrophages of Balb/c mice, harvested one day before by flushing the peritoneal cavity with a sucrose solution, were provided as feeder cells. After establishing growing cultures and screening for immunoglobulin production, positive cultures were recloned and used for further analysis.

Antisera

Clones OX7 (anti-THY-1), W3/25 (anti-T helper subset) and OX8 (anti-T suppressor/cytotoxic subset) were kindly provided by Dr.A.F.Williams, Oxford, England.

Conjugation

Monoclonal antibodies were, after DEAE purification, directly conjugated to FITC (Sigma, USA) and TRITC (Nordic, The Netherlands) using standard methods as described by Goding [4].

Immuneprecipitation

The molecular weight of the cell surface antigens, recognised by the various monoclonal antibodies, was determined on thymocytes. Thymocytes were radiolabelled with 125-Iodine with Iodogen and lysed with NP40. Immuneprecipitation was done according to Agthoven et al [5]. Molecular size of the antigens was determined by running the precipitate on a 7-20% gradient SDS-PAGE gel, followed by autoradiography.

Immunefluorescence

Preparation of cellsuspensions of thymus and cercival lymph nodes and membrane fluorescence staining were done as described previously [6]. Fluorescent cells were analysed on a FACS II (Becton and Dickinson, USA), equipped with a ND 100 pulse height analyser (Nuclear Data, USA) interphased with a PDP 11/70 computer (Digital, USA) for further analysis, or on a Zeiss fluorescence microscope with the appropriate filter systems for FITC/TRITC emission.

Immuneperoxidase histology

Specific immune histology of monoclonal antibodies was done on 5 micrometer cryostate sections of the thymus. After incubation with the monoclonal antibody, the sections were washed and then incubated with horse-raddish peroxidase conjugated rabbit-anti-mouse immuneglobulin (Miles, England). After washing the substrate diamino benzidine tetrahydrochloride (DAB) was added. The preparation were mounted finally in depex and examined under a standard Zeiss microscope.

THY-1 binding

Binding of monoclonal antibodies to the THY-1 glycoprotein was studied in three ways. First, crossreactivity of the antibodies was determined in a fluorescence staining test on thymocytes of AKR mice. These mice carry the same THY-1 allele as rats [7]. Second binding of the antibodies was measured in a micro-ELISA system, in which purified rat THY-1 was coated to the bottom of microtiter wells and the presence of bound antibodies was established with a β-galactosidase rabbit-anti-mouse serum in combination with the substrate ortho-phenyl-diamine (OPD). Purified rat THY-1 was kindly provided by Dr.A.F.Williams, Oxford, England. Finally, this purified THY-1 was also used in a blocking assay. Therefore antibodies were preincubated with THY-1 and remaining activity was measured in a fluorescence assay on rat thymocytes.

RESULTS

In a series of fusion experiments a panel of monoclonal antibodies directed against T cell markers in the rat was obtained. Of these, two antibodies, ER4 and ER14 (see Table 1) reacted with a 25 KD molecule on the surface of thymocytes, but not with peripheral T cells. Further analysis of these antibodies indicated, as shown in Table 2, that they recognised the THY-1 glycoprotein. In three binding assays for THY-1, ER4

Table 1. List of monoclonal antibodies.

Antibody	Antigen used for immunisation	MW antibody	Clone
ER2*	PVG lymphoid tumour	52 KD	64.7A3
ER3	PVG lymphoid tumour	95 KD	62.7A6
ER4	LEW thymocytes	25 KD	46.2D3
ER10	LEW thymocytes	62 KD	46.10C1
ER14	(AOxBN)F1 lymphoid tumour	25 KD	37.4D4

* ER stands for Erasmus Rotterdam

and ER14 showed a similar reactivity pattern as the anti THY-1 antibody OX7 [8] did. The anti-52 KD antibody ER2, on the other hand was negative in all three tests. Nevertheless it is obvious, that ER4 and ER14 have also major differences. ER4, for instance, recognised in the bone marrow only small cells, whereas ER14 reacted also with blast cells (data not shown). Also, as shown in Fig.1, a differential recognition pattern for these antibodies was found in the thymus. ER14 gave the classical distribution known for THY-1, with most cells equally stained and a slightly more intensive staining of the subcapsular cells (Fig.1A). ER4, however showed a quite different pattern. This antibody also stained most if not all of the thymocytes, but the

Table 2. Binding of monoclonal antibodies to THY-1.

Antibody	Binding: AKR-thymus[a]	Binding: THY-1[b]	Blocking[c]: THY-1
OX7	+	+	+
ER2	–	–	–
ER4	+	+	+
ER14	+	+	+

a All antibodies were directly conjugated with FITC and analysis was done on a FACS II.

b Binding of antibody to purified THY-1 was measured in a micro-ELISA.

c Blocking with purified THY-1 of the antibody interaction with rat thymocytes.

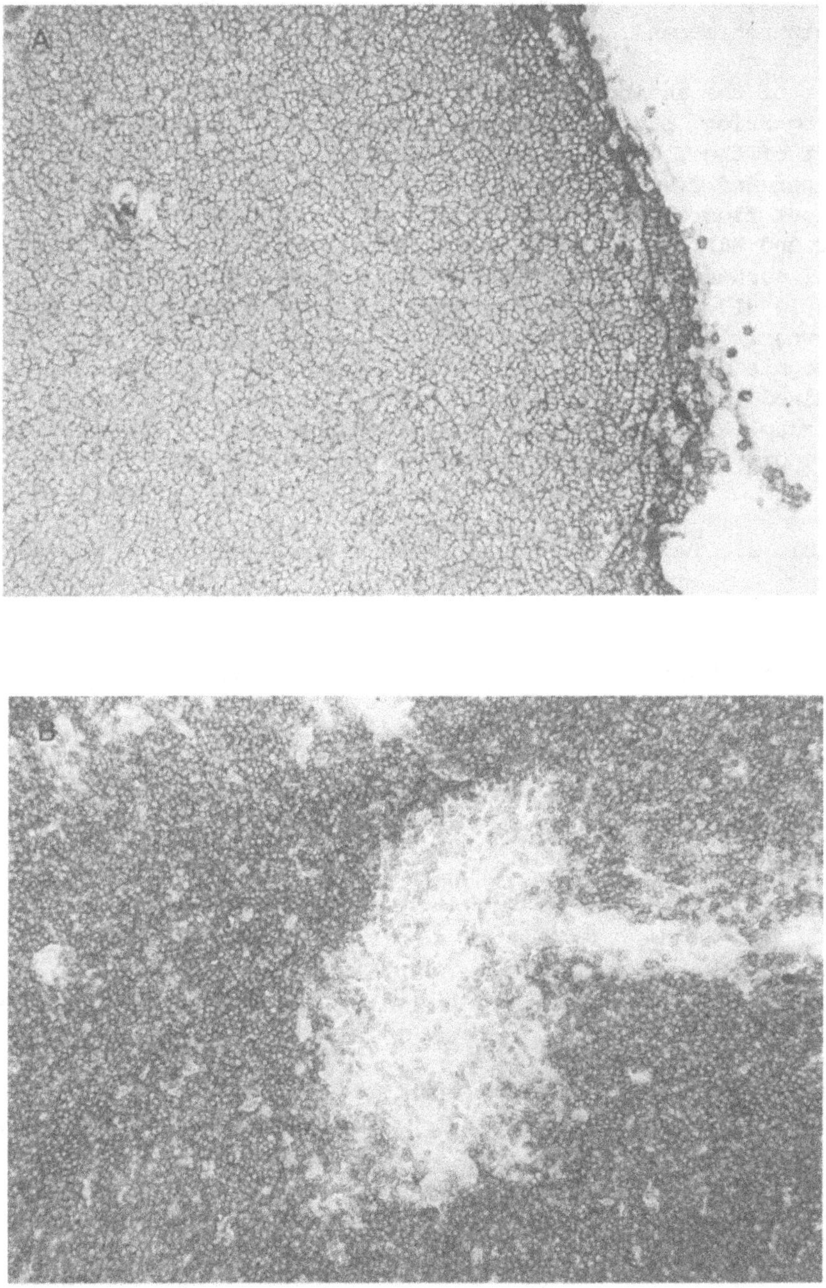

Fig.1. Immune peroxidase histology of monoclonal antibodies ER14 (A) and ER4 (B) on cryostate section of the thymus.

intensity of the cortical cells exceeded that of the medullary cells by manyfold (Fig.1B). In FACS analysis these two populations could actually be seen as two separated populations (data not shown).

Of the antibodies produced, three showed a differential recognition pattern on lymph node cells. They reacted with only part of the T cells. Comparison with existing markers for T helper and T suppressor/cytotoxic subsets in the rat [1] in a two colour fluorescence assay indicated, as shown in Table 3, that ER2 and W3/25 recognised the same cell population in a lymph node cell suspension. ER2 reacted with a 52 KD antigen on these cells (Table 1). In sequential immunoprecipitation studies (data not shown) it could be shown, that this was the same molecule as recognised by W3/25 [9]. ER3 and ER10, on the other hand, stained the same cells as OX8 (Table 3), with ER10 having some overlap with the W3/25 positive population. However, ER3 and ER10 did react with these cells through different

Table 3. Two colour fluorescence analysis of monoclonal antibodies.

Antibody[a]		Positive cells[b]		
FITC	TRITC	Green	Red	Green/red
W3/25	--	41.0	0	0
OX8	--	16.2	0	0
W3/25	OX8	36.3	16.9	1.8
--	ER2	0	42.4	0
W3/25	ER2	0	0	41.8
OX8	ER2	16.2	40.3	1.5
--	ER3	0	17.2	0
W3/25	ER3	39.3	15.7	2.5
OX8	ER3	0	2.1	17.8
--	ER10	0	24.0	0
W3/25	ER10	30.2	19.8	5.2
OX8	ER10	0	6.4	19.2

a Antibodies were directly conjugated with FITC and TRITC.
b Fluorescent cells were analysed under a Zeiss fluorescence microscope and are expressed as a percentage of all nucleated cells in the lymph nodes.

determinants than OX8 did, since OX8 recognised a complex molecule with 35+39+70 KD chains [9], whereas ER3 and ER10 reacted with single chain molecules of 95 and 62 KD, respectively.

All of the antibodies described above stained almost all of the cells in the thymus of normal young adult rats (Table 4). However, they behaved quite dissimilar in the recovery process after corticosteroid treatment. At 3 days after a single dose of 30 mg Dexamethason / kg bodyweight ip, the remaining thymus was very small, with many dead cells. As far as the cell surface phenotypes of the viable cells are concerned, there were many "null" cells and the ones showing cell membrane markers (approximately 40%) showed a non-differential distribution pattern, as found in a normal thymus. At 4 days the recovery had clearly started. The thymus was larger and there were less dead cells and hardly any "null" cells. The markers used showed a differential distribution pattern, with high percentages for ER3 and ER4, lower numbers for ER10 and OX8 and just very low numbers for ER2.

Table 4. Effect of corticosteroid treatment on the thymus.

Antibody[a]	Days after treatment[b]			
	0	3	4	
ER2	91.9	31.5	29.2	
+ ER3				92.2
+ ER4				90.3
+ ER10				85.9
+ OX8				72.5
ER3	97.0	36.8	89.3	
+ ER4				95.2
+ ER10				90.7
+ OX8				89.0
ER4	95.6	32.8	81.0	
+ ER10				88.1
+ OX8				87.2
ER10	94.0	31.0	75.1	
+ OX8				76.3
OX8	90.3	32.1	57.3	

a All antibodies were conjugated to FITC. Binding was analysed on a FACS II. Results of a typical experiment are shown.
b PVG rats were treated with Dexamethason (30mg/kg bodyweight ip)

Double staining experiments indicated, that the populations stained by ER2 and OX8 excluded each other to a great extend. The other antibodies did not show such an almost complete mutual exclusion. The ER10 positive population, actually, was found completely within the ER3 positive population, and the OX8 positive cells again were found completely within the ER10 positive population.

DISCUSSION

Over the years it has become clear from many experiments, that the thymus plays an important role in T cell differentiation. However, as reviewed by Scollay [10], the exact nature of this role remains unclear. One of the unresolved questions is the relationship of cells in the cortex and the medulla of the thymus. There exist a variety of models describing such a relation. They can vary from a simple model in which all cells enter the cortex, develope an immature cortical phenotype and migrate from there to the medulla, getting a mature phenotype, and to the periphery [11], to very complex models. An example of such a model is proposed by Scollay [10] based on previous mouse and recent human data. In this model some cells go to the cortex. These cells are class II reactive and develope there an immature cortical phenotype. They migrate to the medulla, get a mature inducer-type phenotype and leave as such for the periphery. Some other cells go directly from the bone marrow to the medulla, where they obtain the mature phenotype of the cytotoxic/suppressor subset. Then they migrate to the periphery. Also in the rat are now markers available, which define functional subsets [1]. Biochemical data suggests a strong homology with similar human markers [9]. However, not much progress has been made, using these markers, to get more insight on intrathymic differentiation in this species. This paper describes five new monoclonal mouse-anti-rat antibodies, which can contribute probably considerably to such experiments. From the data presented, it is obvious, that ER2 recognises the inducer subset through the same 52 KD molecule as W3/25 does [9]. ER3 and ER10, on the other hand, react with the suppressor/cytotoxic subset, but through a different determinant, then OX8 does [9]. All these markers are normally expressed on almost all cells in the thymus. However, in the recovery phase after corticosteroid treatment, a rather differential pattern is seen. From these data as well as double staining results we calculated the proposed model for intrathymic differentiation in the rat, as shown in Fig.2. The percentages shown are those at 4 days after corticosteroid treatment. The first cell type carries only the ER4 marker. This cell differentiates into an ER3, ER4 positive cell and subsequently also gains the marker ER10. The next marker to be expressed is OX8, which is followed by ER2. This cell type, which expresses the markers for ER4, ER3, ER10,

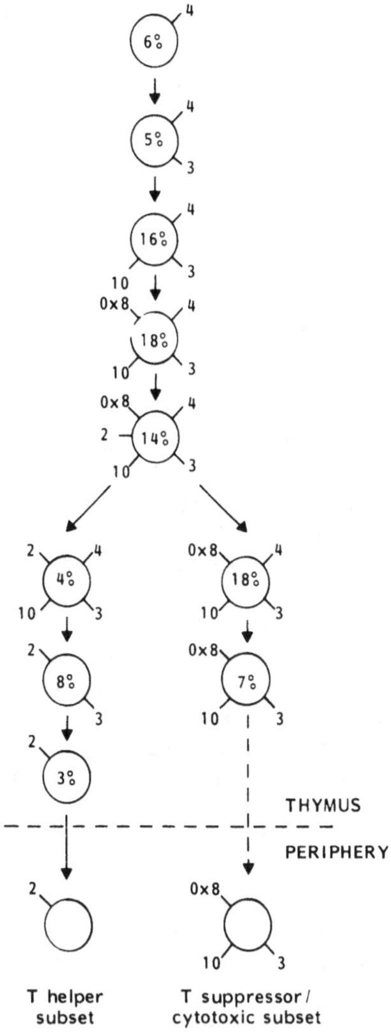

Fig.2.
Model of intrathymic differentiation in the rat. Percentages shown and relations between cell types are calculated at 4 days after treatment of rats with 30 mg Dexamethason / kg bodyweight ip.

OX8 and ER2, is in a normal thymus the dominating cell type. From here on the functionally different cellines may be separated. One line differentiates from a ER2, ER4, ER3 and ER10 positive cell finally into an only ER2$^+$ cell, which then migrates to the periphery. During this process the markers OX8, ER3 and ER10, which are in the periphery specific for the T suppressor/cytotoxic subline, are gradually lost from the helper subline. The suppressor/cytotoxic subline, on the other hand, looses only the markers ER2 and ER4 to get its mature peripheral phenotype. Although one should be careful with the interpretation of data obtained after corticosteroid treatment in respect to the normal situation in the thymus, this model opens new perspectives for the studies of intrathymic differentiation in the rat. Monoclonal antibody ER4, finally, which, through a specific determinant on the THY-1 molecule, enables us to make an almost complete distinction between cortical and medullary cells, will help us to study this process in relation to the specific place in the thymus, where it takes place.

ACKNOWLEDGEMENTS

Part of this investigation has been subsidised by the Dutch Renal Foundation (Grant C245), Amsterdam, The Netherlands. The authors wishes to express their gratitude to Prof.Dr.J.Jongkind for the use of his FACS equipment.

REFERENCES

1. D.W.Mason, R.P.Arthur, M.J. Dallman, J.R.Green, G.P.Spickett and M.L.Thomas, Functions of rat T-lymphocyte subsets isolated by means of monoclonal antibodies, Immunol.Rev. 74:57 (1983)
2. G.Köhler and C.Milstein, Continuous cultures of fused cells secreting antibody of predefined specificity, Nature, 256:495 (1975)
3. S.Fazekas de St.Groth and D.Scheidegger,Production of monoclonal antibody:Strategy and tactics, J.Immunol.Meth., 35:1 (1980)
4. J.W.Goding.,Conjugation of antibodies with fluorochromes: modifications to the standard methods, J.Immunol.Meth., 13:215 (1976)
5. A.Agthoven, C.Terhorst, E.Reinherz and S.F.Schlossman, Characterisation of T cell surface glycoproteins T1 and T3 present on all human peripheral T lymphocytes and functionally mature thymocytes, Eur.J.Immunol., 11,18 (1981)
6. J.Rozing, N.H.C.Brons and R.Benner., B lymphocyte differentiation in lethally irradiated and reconstituted mice. II.Recovery of humoral immune responsiveness, Cell.Immunol. 29:37 (1977)
7. T.C.Douglas.,A rat analog of the mouse theta antigen, Transpl.Proc. 5:79 (1973)
8. D.W.Mason and A.F.Williams, The kinetics of antibody binding to membrane antigens in solution and at the cell surface, Biochemical J. 187:1 (1980)
9. M.L.Thomas and J.R.Green, Molecular nature of the W3/25 and MRC OX-8 marker antigens for rat T lymphocytes: comparison with mouse and human antigens,Eur.J.Immunol. 13:855 (1983)
10. R.Scollay. Intrathymic events in the differentiation of T lymphocytes: a continuing enigma,Immunol.Today. 4:282 (1983)
11. J.J.T.Owen and M.C.Raff, Studies on the differentiation of thymus-derived lymphocytes, J.Exp.Med. 132:1216 (1970)

STROMAL CELL POPULATIONS IN THE DEVELOPING THYMUS OF NORMAL AND NUDE MICE

Eric J. Jenkinson, Rosetta Kingston and John J.T. Owen

Department of Anatomy, Medical School University of Birmingham

INTRODUCTION

Considerable evidence is available to indicate that the thymus plays a vital role in T-cell development (1). Although little is known of the manner in which this is achieved, both direct membrane interaction between stromal cells and lymphoid cells, for example in selecting the receptor repertoire to produce self tolerance and MHC restriction (2) as well as growth factor or hormonal effect may be involved. In this context characterisation of the stromal elements of the thymus is an essential step in elucidating micro environmental influences that may support and direct T-cell development.

Previously, we have investigated the importance of the different thymic components in supporting T-cell development by comparing the ontogeny of MHC antigen expression in the normal thymic rudiment with that in the thymic rudiment of the nude mouse, which is unable to support thymopoiesis. These studies indicated that the Ia antigen positive (Ia^+) epithelium of the normal rudiment was absent or deficient in the nude rudiment (3). We have now extended this comparison to examine two other markers, a GQ ganglioside characteristic of neuro-endocrine cells, recognised by antibody A2B5 and reported to be present in the adult thymus (4) and a determinant associated with the tonofilaments of simple epithelia recognised by antibody Le.61 (5). To determine the ontogeny, distribution and relationship of these markers to the Ia^+ epithelium, these comparisons have been carried out in freshly isolated nude and normal thymus lobes at various stages of gestation, and on organ cultures of intact thymus lobes and thymus lobes depleted of their lymphoid cells to allow the stroma to collapse into more compact cell groupings.

MATERIALS AND METHODS

Animals and Embryonic Material

The following inbred strains were employed. B.10Br ($H-2^k$), B.10Br nu/nu ($H-2^k$), CBA ($H-2^k$), and Balb/c ($H-2^d$). All embryonic material was obtained from timed matings with the day of finding a vaginal plug designated as day 0. Nude embryonic material was obtained from homozygous matings between B.10Br nu/nu male and females or from matings of nu/+ females with nu/nu males. In the latter case nu/nu embryos were easily disinguished from their nu/+ littermates by gross difference in thymus size and morphology.

Culture techniques

Embryonic rudiments were isolated by micro dissection as described previously (3). Organ cultures were carried out by placing explanted thymus lobes on the surface 0.8μ nucleopore filters resting on the surface of gelatin foam sponge (Sterispong No. 3, Allen and Hanbury , London) and in a 90 m.m. petri dish containing Dulbeccos modification of Eagles medium supplemented with 10% foetal calf serum 5×10^{-5} M 2-mercapto-ethanol and additional amino acids. Cultures were maintained for 6-12 days. For deoxyguanosine (dGuo) treatment, thymus rudiments were organ cultured for five days in the above medium containing 1.35mM dGuo (6).

Antibodies and Immunolabelling

All immunolabelling procedures were two-step indirect immunofluorescence assays. For double labelling first step reagents were monoclonal A2B5 (I_gM) (4) and monoclonal anti Ia^k (clone 10-2-6 I_gG_2), (7) and were mixed together to give a final distribution of 1/20 A2B5 ascites fluid and 1/12 10-216 culture supernatant. Second step reagents for these antibodies were anti-mouse I_gG_2-FL (Nordic Labs Maidenhead, U.K.) and anti I_gM(Fc)Rh (Nordic) both at a final dilution of 1/15. For tonofilament labelling monoclonal antibody Le-61 (5) was used at a final dilution of 1/100 of concentrated culture supernatant followed by anti-mouse I_gG Rh at a dilution of 1/15. Frozen sections were prepared and stained as described previously (3) by sequential 30 min incubations at room temperature in humidified boxes. Labelling of monolayer cultures was also carried out by sequential incubations with first and second step reagents with intervening washes as described before (3).

Labelled material was examined on a Zeiss Universal microscope with epi - illumination and standard filter set 9. Photographic recordings were made using Ilford HP-5 film.

RESULTS AND DISCUSSION

Staining of embryonic thymus lobes at various stages of development with monoclonal antibody A2B5 revealed that cells recognised by this antibody are detectable in the thymus by d.13 of gestation. These cells increased in number with development forming irregular patches within which labelling was reticular in pattern, with occasional whorl like clusters of more compact cells. In the pre-natal thymus, A2B5$^+$ cells were found in both the peripheral and central regions of the lobes and did not show preferential localisation to the sub-capsular cortex as described for the adult human thymus (4). Thymus lobes developing in organ culture showed a pattern of A2B5 expression similar to that seen <u>in vivo</u> demonstrating that in this regard the cultures are representative of the <u>in vivo</u> situation, an important consideration in the use of such cultures in the study of T-cell maturation (8).

In contrast to the patchy pattern of A2B5 labelling, by d.16 of gestation, Ia antigens showed a more or less continuous distribution throughout the thymic lobes, as described previously (3). The majority population of Ia$^+$ epithelial cells are, therefore, A2B5$^-$. Conversely, two colour labelling revealed that most, if not all, A2B5$^+$ cells lacked Ia expression. This was most easily seen in thymic lobes treated with deoxyguanosine in vitro (6). This treatment results in the elimination of the lymphoid cells and the collapse of the distended stromal reticulum into more distinguishable cell groups (Fig 1a & b).

Our previous observations have demonstrated that the Ia$^+$ epithelium of the normal thymic rudiment is absent or deficient in the nude mouse (3). Consistent with an epithelial nature for this defect, nude thymic rudiments at various stages of gestation failed to label with antibody LE.61 unlike their normal counterparts where labelling of tonofilament associated determinants in a widespead reticular pattern was seen as early as d.14. Nude thymic rudiments did, however, contain A2B5$^+$ components both as scattered individual cells and as cells present in the walls of the cystic epithelial structures characteristic of the nude thymic rudiment.

Overall, these observations indicate that, in contrast to its effect on the Ia$^+$, LE-61$^+$ epithelium, the nude mutation does not directly affect the thymic A2B5$^+$ population, supporting the conclusion that these populations are distinct and suggesting that they are unlikely to share the same developmental origin. The frequent association of A2B5 labelling with neural crest derivatives (4) may indicate that thymic A2B5$^+$ cells arise from the neural crest input into the thymus (9) although the presence of A2B5$^+$ cell in

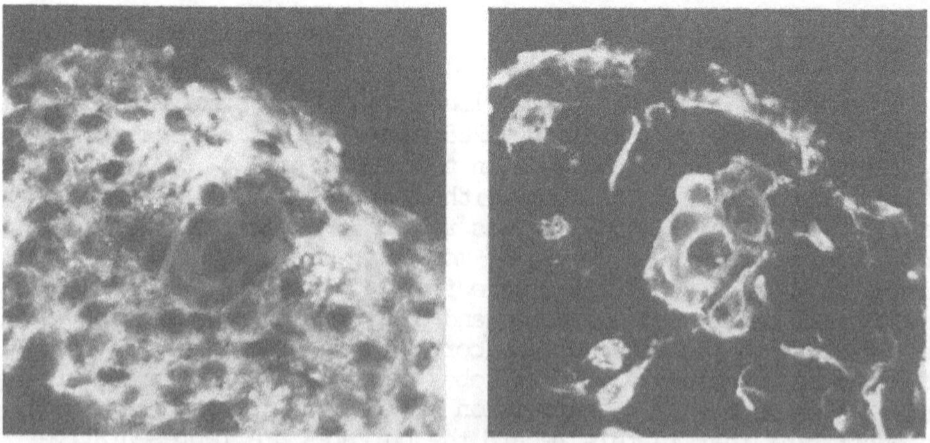

Fig. 1. Deoxyguanosine treated thymic rudiment showing two colour fluorescence labelling with monoclonal anti Ia^k (a) and monoclonal antibody A2B5 (b). The stromal elements have collapsed into a compact mass which shows a confluent distribution of Ia antigens surrounding smaller Ia negative patches (a). In (b) these Ia negative patches are seen to label with antibody A2B5.

the thymus of W/W_v mice, (unpublished observations) which are deficient in a variety of neural crest derivatives (10) argues against this. An alternative possibility is that at least some of the thymic $A2B5^+$ cells are derived from the endodermal component of the thymus. This would be consistent with their presence in the nude thymic rudiment where endodermal derivatives of the third pharyngeal pouch are thought to persist, whilst the ectoderm of the third pharyngeal cleft that may normally give rise to the Ia^+(3), $LE\text{-}61^+$ epithelium fails (11). Interestingly, the epithelium of the developing trachea, also a derivative of the pharyngeal endoderm, is strongly $A2B5^+$ (unpublished observations).

The functional role, if any, of thymic $A2B5^+$ cells also awaits clarification. A role for these cells in the production of thymic growth factors or thymic hormones is possible in view of the association of A2B5 labelling with neuro-endocrine cells (4) and it is noteworthy that antibody recognising the thymic factor FTS also picks out a population of thymic stromal cells distinct from the Ia^+ epithelium (12). As a direct approach to determining whether or not $A2B5^+$ cells play a part in T-cell development, we are currently investigating the effect of eliminating them on the ability of thymys organ cultures to support T-cell differentiation. Whatever the outcome of these studies, however, it is already possible to conclude from the presence of $A2B5^+$ cells in the nonfunctional nude thymic rudiment that alone they are not sufficient to attract T-cell precursors and support their maturation.

REFERENCES

1. Miller, J.F.A.P. Thymus 1:3 (1979)

2. Zinkernagel, R.M. Immunol. Rev. 42:224 (1978)

3. Jenkinson, E.J., van Ewijk, W & Owen J.J.T. J. Exp. Med. 153:280 (1981)

4. Haynes, B.F., Shimizu, K & Eisenbarth, G.S. J.Clin. Invest. 9:14 (1983)

5. Lane, E.B. J. Cell Biol. 92:665 (1982)

6. Jenkinson, E.J., Franchi, L.F., Kingston, R, & Owen J.J.T. Eur. J. Immunol 12:583 (1982)

7. Oi. V.T., Jones, P.P., Goding, J.W., Herzenberg, L.A. Curr. Top. Microbiol, Immunol. 81:115 (1975)

8. Owen, J.J.T., Jenkinson, E.J. & Kingston, R. Ann. Immunol. (Inst. Pasteur) 134D:115 (1983)

9. Le Douarin, N.M. & Jotereau, F. J. Exp. Med. 142:17 (1975)

10. Mayer, T.C., & Green, M.C. Dev. Biol 18:62 (1968)

11. Cordier, A.C. & Haumont, S.M. Am. J. Anat 157:227 (1980)

12. Savino, W., Dardenne, M & Bach, J.F. Clin. Exp. Immunol 52:1 (1983)

NEW SURFACE ANTIGENS ON CELLS IN THE EARLY PHASE OF T-CELL DIFFERENTIATION, DETECTED BY MONOCLONAL ANTIBODIES

L.M.B. Vaessen, P. Joling, F.J. Tielen and J. Rozing

Department of Internal Medicine, Erasmus University Rotterdam, The Netherlands

The precursor of the thymocytes, the pro-thymocyte, is a specialized cells, committed to the T-cell differentiation line and discrete from both precursors of B-cells and the pluripotent stem cell (HSC).[1,2] However, based upon morphological and physical criteria such as buoyant density, sedimentation velocity and surface charge, prothymocytes can not be distinguished from HSC and pre-B cells.[3]

For the rat it has been reported that the 25 kD Thy-1 molecule[4] as well as the 95 kD W3/13 molecule[5] are present on prothymocytes. Both molecules, however, can also be detected on the HSC,[5,6] pre-B cells and plasma cells (this issue, p 17),[7,8] some myeloid cell types and their precursors.[5,7] The W3/13 molecule is also present on all mature T-cells.[8]

Although studies with these markers gave a great deal of information about the early phase of lymphoid differentiation, in rats more specific markers are still needed. In cell fusion experiments we produced monoclonal antibodies (Mab) that recognized part of the cells in the bone marrow (BM), nearly all thymocytes and non or just a few cells in spleen and lymph nodes of adult rats. This distribution pattern prompted us to study the relation between these Mab and the cells in early phase of T-cell differentiation. In this report we describe the partial characterization of the ER-4, ER-5 and ER-6 Mab and the comparison of ER-4 with the Mab ER-14, which recognizes the same determinant as Mab MRC OX-7 on the the Thy-1 molecule (this issue, p 17).[7]

Materials and methods

Animals. The rat strain Lew/Ola (RT-1^l) was obtained from Olac, Bicester, England; the strain RP/EUR (RT-1^p) and the Balb/c mice were bred in our colony at the Erasmus University, Rotterdam. The rats were used at 4 weeks (young) and 10-20 weeks (adult) of age.

Antisera. Cloned hybrid cell lines 46.2D3, 46.5A2 and 46.4C5 producing Mab ER-4, ER-5 and ER-6, respectively, were derived from a fusion between mouse spleen cells and the mouse myeloma cell line Sp2/0-Ag14 according to the methods as described by Fazekas de St. Groth.[9] Spleen cells were obtained from a Balb/c mouse that had been immunized twice with Lewis thymocytes. Mab ER-14 described in detail elsewhere in this issue (p 17) recognizes the Thy-1 molecule. Mab MRC OX-19 was a generous gift of Dr A.F. Williams, MRC, Oxford, England. This antibody binds to all thymocytes and mature peripheral T-cells.[10] An affinity column purified rabbit anti-rat IgM was used for cytoplasmic immunoglobulin (cIg) staining.

All antibodies used for immunofluorescence staining were conjugated with fluorescein isothiocyanate (FITC) or tetramethyl rhodamine isothiocyanate (TRITC). Conjugation was performed as described by Goding.[11]

Terminal deoxynucleotidyl transferase (TdT) assay. The immunofluorescent assay for TdT was carried out with a test-kit from Bethesda Research Lab Inc. under standard conditions. As second step a TRITC-labeled affinity column purified goat anti-rabbit Ig was used.

Determination of the lactate dehydrogenase isoenzyme pattern was kindly done by Dr J. Plum, State University, Gent, Belgium, according to the method described before.[12]

Cell suspensions, fluorescence staining and FACS analysis. BM was collected by flushing the femora and tibiae with RPMI-1640, supplemented with 0.1% BSA or 1% FCS. Spleens, thymuses and lymph nodes were minced with scissors, sqeezed through a 0.003 inch vittalium wire mesh and subsequently brought over a nylon-gauze filter to give a single cell suspension. Erythrocytes were removed from spleen and BM by lysing with 0.16M ammoniumchloride during 5 min on ice. Nucleated cells were counted with a sysmex-TOA model CC-1340 blood cell counter. For staining procedures, cells were washed with 0.1% BSA and 0.1 % azide (S-RPMI). Fifty µl of a cell suspension containing $4x10^7$ cells/ml were incubated with 50 µl diluted antibody (10 µg/ml) for 30 min on ice. After staining, the cells were washed twice with 3 ml S-RPMI and were analysed on a fluorescence activated cell-sorter (FACS II, Becton and Dickinson, USA) with an argon laser (Spectra Physics) at 488 nm using the following setting: laser 200 mW, photomultiplier 560V, fluorescence gain ranging from 2-16. Forward light scatter-windows were used to exclude dead cells and remaining erythrocytes. Scatter and fluorescence signals were analysed and stored on a Nuclear Data ND-100 pulse height analyser interphased with a PDP-11 computer system (Digital, USA).

Table 1. Percentage of positive cells in several lymphoid tissues of juvenile and adult Lewis rats

Antibody	Bone marrow		Thymus		Spleen		Lymph nodes	
	juv[a]	adult[b]	juv	adult	juv	adult	juv	adult
ER-4	35[c]	20	98	97	28	1	18	1
ER-5	25	27	97	97	25	15	16	11
ER-6	23	18	95	83	23	11	14	8
ER-14	35	31	98	96	25	3	16	4
OX-19	2	3	98	98	23	36	44	53

[a] juvenile, 4 weeks old
[b] adult, 10-20 weeks old
[c] percentage positive cells determined on FACS II after staining with direct FITC-conjugated monoclonal antibody.

Results

Cells from several lymphoid organs of adult (10-20 weeks old) and young (4 weeks of age) Lewis rats were incubated with the FITC-conjugated Mabs ER-4, ER-5, or ER-6. Mab ER-14 was used as a marker for early cells in lymphoid and myeloid differentiation, whereas the Mab OX-19 was used as a marker for mature T-cells.

The percentage of labeled cells in BM, spleen, lymph nodes and thymus are shown in TABLE 1. With the exception of ER-6, all antibodies tested labeled virtually all thymocytes in adult as well as young animals. In the spleen of the 4 weeks old rats the Mabs ER-4, ER-5 and ER-6 bound to approximately 25% of the nucleated cells. Double analysis showed that these markers were shared by the same cells (data not shown). In the lymph nodes of young animals 14-18% of the cells were recognized by these antibodies. A striking difference, however, was seen between these three Mabs in the peripheral lymphoid organs of adult rats. ER-4 positive cells could not be detected, whereas ER-5 and ER-6 stained 10 and 20% of the cells. Also in the BM of young rats a different staining pattern could be observed. Whereas ER-4 stained 35% of the nucleated cells, the Mabs ER-5 and ER-6 did bind to approximately 25% of the cells. In BM of adult rats on the other hand ER-5 did recognize the highest percentage of cells (27%). ER-4 and ER-6 bound to approximately 20% of the cells in this organ. Double labeling studies on FACS (data not shown) revealed that all ER-5 and ER-6 positive BM-cells reacted also with the ER4 antibody.

Table 2. Double analysis with monoclonal antibodies

Antibody[a]	Percentage[b]	
	Bone marrow[c]	Spleen[d]
ER-4	21	28
ER-14	31	25
OX-19	2	23
ER-4 + ER-14	39	nd[e]
ER-4 + OX-19	22	52

[a] monoclonal antibodies directly conjugated with FITC
[b] percentage of all viable cells analysed on a FACS II
[c] bone marrow cells from adult rats
[d] spleen cells from 4 weeks old rats
[e] nd, not determined

As can be seen from Table 2, showing the results of double labeling experiments of ER-4 in combination with OX-19 and ER-14 on BM and spleen, the ER-4 positive cells in the spleen of young rats do not carry the mature T-cell marker OX-19. Furthermore, it is obvious from these data, that in the bone marrow of adult rats there is a considerable overlap between the cell populations recognized by ER-4 and ER-14. It can be calculated that about 13% of the nucleated cells in the BM of adult rats carry the $ER\text{-}4^+ER\text{-}14^+$ phenotype.

Distribution of ER-4 positive cells during ontogeny. The distribution pattern of ER-4 positive cells in spleen, lymph nodes and BM of young and adult animals urged us to study the distribution of the ER-4 and ER-14 markers from day 14 of gestation up to 20 weeks after birth. In Fig. 1 it is shown that at the time of birth in both BM and spleen just a low percentage of ER-4 and ER-14 positive cells were present. The highest amount of positive cells in the spleen was found at three weeks after birth. From this point of time onwards the cells, binding to these markers, disappeared from the spleen. In the BM the maximal values for both markers were reached at 4 weeks of age. During the remaining period studied the percentage of BM-cells carrying the antigenic determinant recognized by Mab ER-14 remained at the same level (35%). Cells recognized by the ER-4 antibody, however, disappeared partly from the BM

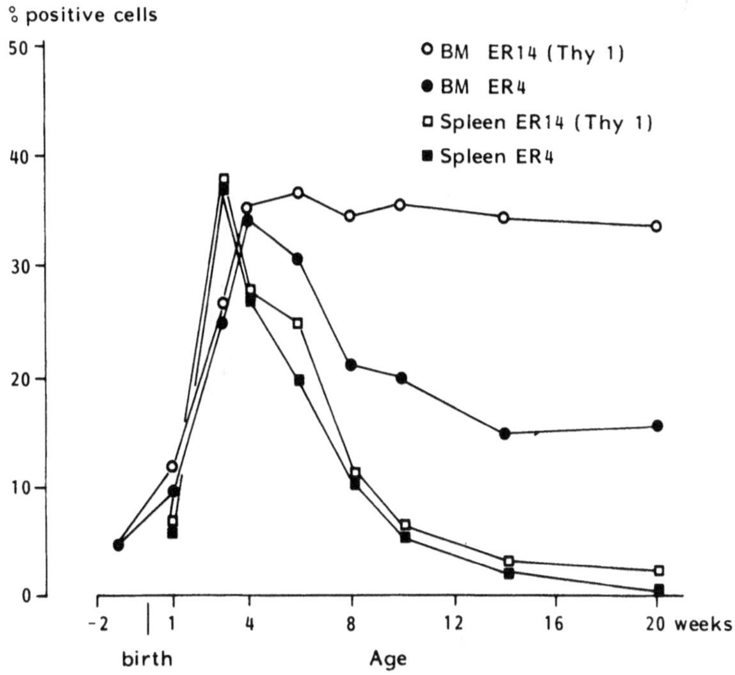

Fig. 1. Percentage ER-4 (closed symbols) and ER-14 (open symbols) positive cells in BM (circles) and spleen (squares) of Lewis rats as a function of age. All values represent the percentage of all viable nucleated cells.

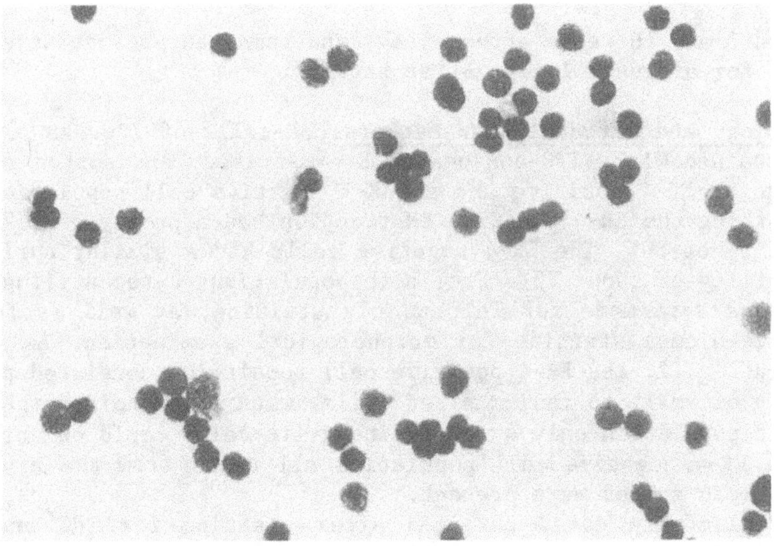

Fig. 2. Cytospin preparation of FACS sorted ER-4 positive BM cells stained with May Grünwald-Giemsa.

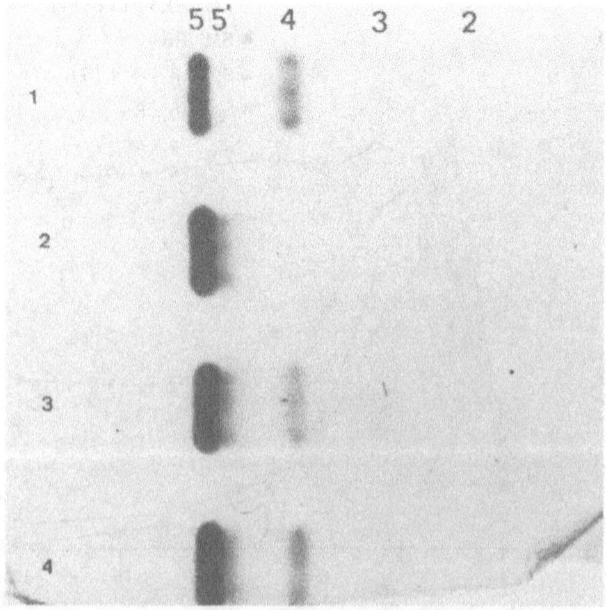

Fig. 3. LDH-isozyme profiles of BM cells from 12 weeks old RP rats. In lane 1: FACS-sorted ER-4 positive cells, lane 2: FACS-sorted ER-4 negative cells, lane 3: unsorted stained cells and in lane 4: unsorted, unstained BM cells. From left to right hand the isozyme bands LDH-5, LDH5', LDH-4, LDH-3 and LDH-2 can be seen.

between 8 and 16 weeks after birth, and remained present at a level of 15% for at least 1 year after birth.

Morphology and intracellular markers. BM-cells of 7 weeks old rats were stained with FITC-conjugated ER-4 antibody and sorted on FACS II into an ER-4 positive and an ER-4 negative cell population. After sorting the ER-4 positive BM fraction had a purity of 95% and a viability of 80%. The ER-4 negative cells had a similar purity and a viability of about 75%. From both populations cytocentrifuge preparations were made for TdT and cIg staining, as well as for May Grünwald-Giemsa staining for morphological examination. As can be seen in Fig. 2, the ER-4 positive cell population consisted predominantly of small to medium sized cells, with a lymphoid morphology. In this population only a few eosinophylic cells could be observed. In the ER-4 negative cell population all cells from the erythroid and myeloid series were present.

Preliminary data, obtained after staining for TdT and cIg, suggest that approximately 5-10% of the ER-4 positive cells were stained for nuclear TdT. Furthermore, also a part of the cIg positive cells were present in this ER-4 positive fraction. A fraction

of the ER4 sorted BM cells was also analysed for the presence of LDH-isozymes. The results are shown in Fig. 3. The most striking result was the absence in the ER-4 positive BM cells of a band that electrophoresed between the LDH-4 and LDH-5 isozyme band. This band was called LDH-5' by Barton et al.[13] This band was clearly present in the ER-4 negative cells and in the unsorted BM. Other differences between the ER-4 positive and ER4 negative cells were the lack of the isozyme bands LDH-2 and LDH-3 and a low LDH-4 band in the ER-4 negative cells as compared to the ER4 positive BM population.

Discussion

In a series of fusion experiments a panel of monoclonal antibodies was obtained directed against cells of the T-cell lineage of the rat. In this report we describe the antibodies ER-4, ER-5 and ER-6. Because of the peculiar staining pattern in BM, spleen, lymph nodes and thymus of both adult and juvenile animals, those antibodies drew our attention as possible markers for prothymocytes. All these antibodies react with a 25 kD cell surface determinant on thymocytes and the same cell population in the bone marrow. However, in view of its specific properties, ER-4 seems especially interesting in relation to pre-T cell differentiation and was therefore selected for further analysis. This antibody showed a unique thymic distribution, with strong staining in the cortex and rather weak staining in the medulla (shown elsewhere in this issue, p 235). This in combination with the fact, that ER-4 does not recognize any cells in the peripheral lymphoid organs of adult rats, indicates that this determinant is lost during T-cell differentiation. Double labeling experiments with MRC OX-19 and ER-4 indicated furthermore, that also in the spleen of young animals, the differentiation antigen recognized by ER-4 is not present on mature T-cells.

The ER-4 antibody has some features in common with the ER-14 antibody. Both antibodies recognize a 25 kD molecule on thymocytes and in binding assays with purified Thy-1, ER-4 showed the same reactivity pattern as Mab ER-14 and OX-7 did. These latter antibodies react with the "classical" Thy-1 determinant (data presented elsewhere in this issue, p 17). However, besides the dissimilarities found for ER-4 and ER-14 in the thymus (this issue, p 235), also striking differences between ER-4 and ER-14 were noticed in the BM. In adult rats ER-4 reacted with a consistently lower percentage of BM cells than ER-14 did. Moreover, as shown in Fig. 2, ER-4 binds only to small and medium sized cells with a lymphoid morphology, whereas it has been reported that the "classical" Thy-1 determinant is also present on blastoid cells and on small cells of the erythroid and myeloid differentiation lineages.[14] The exclusive lymphoid morphology of ER-4 positive BM-cells is also in contrast with the cellular distribution of another known marker for prothymocytes, the W3/13 molecule.[5] This marker is also present on precursors and mature cells of the erythroid and myeloid series.

Also the LDH-isozyme studies on FACS sorted BM cells supported the view, that ER-4 reacts in the BM with cells, which belong to the early phase of T-cell differentiation. As can be seen in Fig. 3, ER-4 positive BM cells represent a LDH-isozym profile resembling that given by Barton et al.[13] for rat thymocytes. Both thymocytes and ER-4 positive BM cells lack the LDH-5' band. The ER-4 negative cells do contain this LDH-5' band, but do not show the LDH-2 and LDH-3 bands, whereas the LDH-4 band is very faint. This ER-4 negative cell population, which gave this profile was very heterogeneous in morphology and consisted of all cells from the myeloid and erythroid series.

Another observation supporting our hypothesis, that ER-4 binds to prothymocytes in the BM, was obtained from an ontogeny experiment. In this study a low percentage of ER-4 positive cells could be demonstrated in both BM and spleen just before and after birth. Shortly after birth the number of ER-4 positive cells started to increase, reaching maximal levels between 3 and 4 weeks after birth. Thereafter, the number of positive cells started to decrease in these organs. At 16 weeks of age hardly any ER-4 positive cells could be detected in the spleen, whereas the percentage of positive cells in the BM was only half of that found at 4 weeks of age. These observations strongly coincides with the data given by Gregoire et al.[15] for the presence of prothymocytes in BM and spleen. In these studies prothymocytes were defined as TdT positive cells. Although the absolute percentages found for ER-4 positive cells in the present study are higher than those demonstrated for TdT, these data seem to indicate, that the prothymocyte is present within the ER-4 positive BM population. The fact that the TdT positive cells indeed are present within the ER-4 positive BM population was clearly demonstrated in the FACS-sorted cell population.

The discrepancy between the percentages TdT and ER-4 positive cells could be due to the fact, that not all prothymocytes are TdT positive.[4,15] Possibly ER-4 recognizes not only the TdT positive, but also the TdT negative prothymocytes.

A similar, but even stronger discrepancy exist between the percentage ER-4 positive BM cells and the number of precursor cells, which seem to migrate from the BM to the thymus, as defined in migration studies by Lepault et al.[16] In their experiments they estimate that less than 0.1% of the BM-cells migrate to the thymus. Although technical aspects may be involved, this number is in striking contrast to the 20% ER-4 positive BM-cells found in the present study. Perhaps not all prothymocytes in the BM do migrate to the thymus. In this respect Wagner[17] recently presented an interesting point of view. He suggested, that the number of prothymocytes in the BM must be rather high, in order to enable a selective and restrictive procedure in T-cell differentiation, which probably already takes place in the BM, even before the cells go to the thymus. According to this point of view, ER-4 might recognize not only the cells which will go to the thymus, but also the cells, which will not.

Nevertheless, ER-4 will probably not react specifically with only cells of the T-cell lineage. As suggested by our preliminary data on cIg staining of ER-4 positive BM-cells, at least part of the precursor B population in the BM seems to be included in the ER-4 positive cell population. More specific definition of the prothymocyte population has to come therefore, from double and perhaps triple marker studies, in which ER-4 is combined with known, but rather aspecific, markers as Thy-1 and W3/13.[4,5]

These type of experiments as well as studies to provide formal prove for a precursor-progeny relationship between ER-4 positive BM cells and mature T-cells are currently in progress.

Acknowledgements

The LDH-isozyme analysis was kindly done by Dr J. Plum, State University of Gent, Belgium. Part of this work was supported by a Grant (C 245) of the Dutch Kidney Foundation. Prof. Dr. J. Jonkind is kindly acknowledged for allowing the use of his FACS-equipment.

References

1. S. Abrahamson, R.G. Miller, and R.A. Philips, The identification in adult bone marrow of pluripotent and restricted stem cells of myeloid and lymphoid systems, J. Exp. Med. 145: 1567 (1977).
2. P. Jones-Villeneuve, and R.E. Philips, Potentials for lymphoid differentiation by cells from long term cultures of bone marrow, Exp. Hemat. 8: 65 (1980).
3. W.J.A. Boersma, R. Dakulsi, and G. van der Westen, Post-irradiation thymocyte regeneration after BM transplantation. II. Physical characteristics of thymocyte progenitors, Cell Tissue Kinet. 14: 197 (1981).
4. D.L. Greiner, I. Goldschneider, and R.W. Barton, Identification of thymocyte progenitors in hemopoietic tissues of the rat. II. Enrichment of functional prothymocytes on the Fluorescence-activated Cell Sorter, J. Exp. Med. 156: 1448 (1982).
5. M.J.S. Dyer, and S.V. Hunt, Committed T-lymphocyte stem cells of rats. Characterization by surface W3/13 antigen and radiosensitivity, J. Exp. Med. 154: 1164 (1981).
6. I. Goldschneider, L.K. Gordon, and R.J. Morris, Demonstration of Thy-1 antigen on pluripotent hemopoietic stem cells in the rat, J. Exp. Med. 148: 1351 (1978).
7. M.A. Ritter, L.K. Gordon, and I. Goldschneider, Distribution and identity of Thy-1-bearing cells during ontogeny in rat hemopoietic and lymphoid tissues, J. Immunol. 121: 2463 (1978).
8. A.F. Williams, G. Galfre, and C. Milstein, Analysis of cell surfaces by xenogeneic myeloma-hybrid antibodies: differentiation antigens of rat lymphocytes, Cell 12: 663 (1977).
9. S. Fazekas deSt. Growth, and D. Scheidegger, Production of monoclonal antibodies; strategies and technics, J. Immunol. Meth. 35: 1 (1980).

10. M.J. Dallman, M.L. Thomas, and J.R. Green, MRC OX-19: a monoclonal antibody that labels rat T-lymphocytes and augments in vitro proliferative responsens, Eur. J. Immunol. 14: 260 (1984).
11. J.W. Goding, Conjugation of antibodies with fluorochromes: modifications to the standard methods, J. Immunol. Meth. 13: 215 (1976).
12. J. Plum, and M. de Smet, Enzymatic relationship between human $T\gamma^{+}$ lymphocytes and thymocytes, Cell. Immunol. 55: 485 (1980).
13. R.W. Barton, F. Tausche, and I. Goldschneider, Evidence for the cellular origin of gross virus-induced leukemia in the rat: description of a unique LDH-isozyme sub-band in leukemic lymphoid cells and lymphohemopoietic precursor cells, J. Immunol. 120: 2299 (1980).
14. M.A. Ritter, The differential expression of Thy-1 antigen on cell surfaces and extracellular matrix during lymphopoiesis, in: "Development and Differentiation of Vertebrate Lymphocytes," J.D. Horton, ed., Elsevier/North Holland Biomedical Press, Amsterdam, p. 75 (1980).
15. K.E. Gregoire, I. Goldschneider, R.W. Barton, and F.J. Bollum, Ontogeny of terminal deoxynucleotidyl transferase-positive cells in lymphohemopoietic tissues of rat and mouse, J. Immunol., 123: 1347 (1979).
16. F.P. Lepault, R.L. Coffman, and I.L. Weissman, Characteristics of thymus-homing bone marrow cells, J. Immunol. 131: 64 (1983).
17. H. Wagner, Where is MHC restriction determined?, Progr. Immunol. 5: 809 (1983).

STUDIES ON THE ROLE OF THE THYMUS IN THE SELECTION AND CLONAL EXPANSION OF T-CELL PRECURSORS

R.K.Jordan, J.H.Robinson, A.L.Bentley, K.C.House and N.A.Hopkinson

Department of Anatomy, Medical School, University of Newcastle upon Tyne, Newcastle upon Tyne, NE2 4HH

INTRODUCTION

Since the observation that neonatal thymectomy in mice causes a severe and definitive impairment of cell mediated immune function[1] there has been little doubt that the thymus plays a central role in the development of the fully competent T-lymphocyte system. While events such as the expression of antigen specific receptors[2], the acquisition of MHC restriction and self tolerance[3] as well as the emergence of phenotypically divergent functional subpopulations[4] are all accepted as features of T-cell ontogeny our understanding of the influence of the thymus upon, and its mechanistic involvement in the differentiation process has progressed little. However it remains attractive to view the thymus as providing the primary site for the selection and amplification of those differentiating T-cell clones physiologically capable of functioning with the self environment of the host. Such a view supposes a series of interactions between the thymic epithelial matrix and the colonizing lymphohaematopoietic cells which include the T progenitors, macrophages and a population of dendritic cells[5-7]. Within this thymic milieu cell surface MHC gene products may act as ligands to provide signals for the selection of thymocytes of relevant specificity while soluble factors such as interleukins may induce precursor proliferation.

To date experimental approaches to the study of the thymic microenvironment have been based largely on the *in vitro* isolation of thymic epithelium and its controlled recombination with lymphohaematopoietic cells in either *in vitro* or *in vivo* engineered chimeras[8-17]. Various organ culture protocols have been employed to first isolate lymphoid free thymic epithelium from embryonic mouse thymus, including culturing under suboptimal conditions[11,12], in the presence of deoxyguanosine[13,17] or at low temperature[10,16]. There is a broad measure of agreement that thymic epithelium isolated by any of these methods can support the differentiation and maturation of exogenously derived T progenitors in a variety of chimeric systems. When transplanted into intact recipient mice thymic epithelial explants become repopulated by host-derived T lineage cells which regenerate an intragraft lymphoid population expressing allotypic and functional markers typical of thymocytes [12,17,18]. Similarly, when associated with fetal liver as a source of stem cells in transfilter experiments *in vitro*, organ culture-derived thymic epithelium again provides the environment for T-cell differentiation[13]. We provide here evidence that (i) lymphoid-free thymic epithelium derived from low temperature organ culture of fetal mouse thymus survives for prolonged periods when transplanted into normal intact histoincompatible recipients, (ii) thymic epithelium has a role in tolerance induction as judged by the lack of MLC responsiveness to graft-type of host derived thymocytes developing within allogeneic epithelial grafts in nude recipients, and (iii) the T-cell deficit in athymic nude mice can be restored, at least in part, by the transplantation of such thymic epithelial explants. Data from the same experimental system are being published elsewhere[19,20].

MATERIALS AND METHODS

Animals

Inbred C57Bl/6, C57Bl/6-Tla^a, C_3H, Balb/c and nu/nu mice with a Balb/c background were used at 6 weeks. Embryonic material was obtained within the inbred strain C57Bl/6. Gestation was calculated counting the day of appearance of a plug as day 0. Pregnant mice were sacrificed at day 14 of gestation and embryos were dissected from the uterus. Individual thymic lobes were dissected free under aseptic conditions using a dissecting microscope.

Organ culture

Intact isolated thymic lobes were explanted on nucleopore filters floating on RPMI 1640 medium supplemented with 10% heat inactivated fetal calf serum. Details of this method have been published elsewhere[9]. To obtain lymphoid-free thymic epithelium, cultures were incubated at 24°C in an atmosphere of 5%CO_2 in air for a period of 7 days. Following low temperature treatment cultures were maintained for a further 7 days under conventional 37°C conditions. Control organotypic cultures were incubated from initiation at 37°C.

Grafting procedure

Intact adult male C57Bl/6, C57Bl/6-Tla^a, C_3H and Balb/c nu/nu mice were used as graft recipients. Organ cultures were grafted beneath the renal capsule using a fine drawn glass probe. In experiments designed to evaluate the origin and cell surface marker expression of the reconstituting intragraft lymphoid population, thymic epithelial grafts from C57Bl/6 embryos were grafted into the congenic partner C57Bl/6-Tla^a.

Analysis of cell surface markers

The following monoclonal antibodies were titrated and subsequently used at optimal cytotoxic titer: anti-mouse Thy1.2 (clone 30-H12, Becton and Dickinson), anti-mouse TL.m6, Lyt-1.2 and Lyt-2.2 (New England Nuclear). In addition to ascertain the origin of intragraft lymphocytes in the nude mouse chimaeric studies, anti-H_2-D^d and anti-H_2-K^b monoclonals were employed.

Marker analysis was performed by complement mediated (Low Tox M rabbit complement, Cedarlane) antibody dependant cytotoxicity using trypan blue dye exclusion as the indicator. Stained and unstained cells (a total of 400) were enumerated with phase microscopy.

Mixed leucocyte culture

In most experiments responding cells were either graft lymphocytes or control young adult thymocytes. In those experiments designed to evaluate the peripheral reconstitution of nude mice, host spleen cells were also used as responders. Spleen cells (depleted of red cells by hypotonic shock) were treated with mitomycin C at 30 ug/ml for 40 minutes at 37°C were used as stimulators. MLC assays were performed in triplicate in microtiter plates with 2×10^5 responding cells per well titrated against doubling dilutions of stimulators over the range 1×10^6 to 6.25×10^4. 2-mercaptoethanol (2×10^{-5}M) was added, and plates were incubated at 37°C in 5% CO_2 in air for 96 hours, 0.2uCi/well tritiated thymidine was added for the final 18 hours.

RESULTS AND DISCUSSION

Low temperature organ culture abolishes lymphopoiesis in fetal thymus

Organ culture of 14 days fetal thymus lobes at 24°C for 7 days abolishes lymphopoiesis and yields a lymphoid free epithelial matrix. Returning such cultures to 37°C culture conditions for an additional 7 days does not restore morphologically detectable lymphopoiesis. Furthermore, flow cytometric analysis of low temperature thymic organ cultures recovered subsequent to a period of conventional post-treatment culture and dissociated by trypsin, failed to demonstrate the presence of any cells bearing the alloantigen markers Thy-1, Lyt-1 or Lyt-2 (unpublished observations). Similar analysis of fresh 14 day fetal thymus and conventionally cultured fetal thymus demonstrated a normal pattern of T lymphocyte marker development. It thus appears that low temperature treatment causes a permanent deletion of all lymphoid cells and their progenitors.

Thymic epithelial grafts exhibit organotypic thymic development

When thymic epithelium, derived from 24°C organ culture is transplanted beneath the renal capsule in normal intact syngeneic mice, the grafts become colonized by lymphocytes to form an heterotopic thymus. The first lymphohaematopoietic cells can be detected within the epithelial grafts by the fifth day post-transplantation and a subsequent progressive expansion in the lymphoid component leads to the attainment of normal thymic morphology by 30 days.

C57Bl/6 and congenic C57Bl/6-Tla^a mice were used to analyse the acquisition of T lineage specific alloantigens by, and host derivation of the developing intragraft lymphoid population. Low temperature treated organ cultures prepared with fetal C57Bl/6 thymus grafted into C57Bl/6-Tla^a recipients were recovered at various times post-grafting and the presence of the specific T lineage alloantigens Tla, Thy-1, Lyt-1 and Lyt-2 was demonstrated by cytotoxic assay using monoclonal antibodies. Results of this study are summarized in Fig.1.

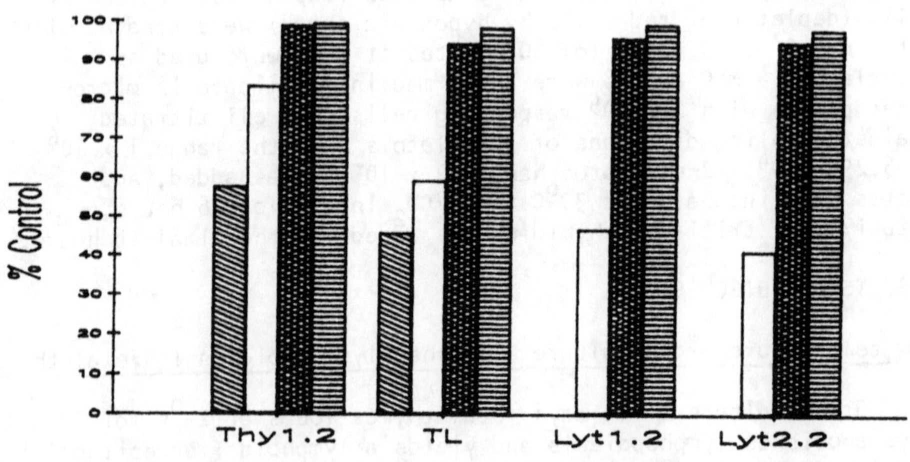

Fig. 1. Temporal expression of T lineage markers by cells recovered from thymic epithelial grafts 7 days (▨), 10 days (□), 23 days (▩) and 35 days (▤) post-transplantation.

The data show that the percentage of graft lymphocytes expressing Tla, Thy-1, Lyt-1 and Lyt-2 increased temporally such that from 25 days onwards normal adult thymic distribution of these markers was attained. As Tla antigens are normally expressed only by lymphoid cells of the thymus, the Tla data indicate that the maturing intragraft lymphocytes may be characterised as thymocytes. Furthermore, as the epithelial grafts were derived from Tla^b donors, in which the Tla antigens are not normally expressed, the presence of Tla antigens on intragraft thymocytes shows their origin to be from the Tla^a host.

The capacity of thymic epithelium to support functional maturation of immigrant T lymphoid progenitors has been demonstrated previously both *in vivo*[17,18] and *in vitro*[13,15]. For example, thymocytes recovered from low temperature derived thymic epithelial grafts include a mature subpopulation, equivalent to that found in normal medullary thymus, which is responsive to alloantigens in MLC (Table 1).

Table 1. Optimal MLC response of Intragraft Thymocytes ($H\text{-}2^b$) from C57Bl/6 ($H\text{-}2^b$) Mice Bearing Low Temperature C57Bl/6 ($H\text{-}2^b$) Thymic Epithelial Grafts.

Days Post Grafting	Mitomycin C treated stimulators	CPM±SEM*		S.I.
10	$H\text{-}2^b$ spleen	185±13		
	$(H\text{-}2^b \times H\text{-}2^k)_{F1}$spleen	963±65	$p<.001$	5.2
23	$H\text{-}2^b$ spleen	798±96		
	$(H\text{-}2^b \times H\text{-}2^k)_{F1}$spleen	4009±124	$p<.001$	5.0
35	$H\text{-}2^b$ spleen	1707±38		
	$(H\text{-}2^b \times H\text{-}2^k)_{F1}$spleen	7853±430	$p<.001$	4.6
Central**	$H\text{-}2^b$ spleen	242±13		
	$(H\text{-}2^b \times H\text{-}2^k)_{F1}$spleen	1498±17	$p<.001$	6.2

* Incorporation of tritiated thymidine by 2×10^5 graft lymphocytes at optimal stimulator to responder ratio.
** Thymocytes of young adult $H\text{-}2^b$ recipients.

Prolonged transplantation across major histocompatability barrier of low temperature organ culture derived thymic epithelium

It has been reported previously that embryonic thymus lobes depleted of their endogenous lymphohaematopoietic cells by deoxyguanosine treatment *in vitro* can be successfully transplanted across major histocompatability barriers, even though the surviving thymic epithelium continues to express class II MHC antigens.[17] Similar prolongation of graft survival is seen when low temperature derived thymic epithelium is transplanted to allogeneic intact recipients, thus C57Bl/6 ($H-2^b$) fetal thymus lobes organ cultured at 24°C for seven days escape acute rejection when transplanted to C_3H ($H-2^k$) recipients, whereas 14 days gestation fetal thymus either untreated or organ cultured for 7 days at 37°C is rejected normally by 15 days post grafting. (Fig.2).

The most likely explanation for this finding is that the low temperature treatment, which we adopted for its efficacy in disrupting thymic lymphopoiesis *in vitro* and which was originally used to deplete passenger leucocytes in studies of allograft survival[21], eliminates not only the thymocytes and their progenitors but also the donor class II MHC antigen expressing dendritic cells of the thymus thus avoiding route I rejection as described by Lechler and Batchelor[22]. Interestingly we found that the survival of thymic epithelial allografts, at least in this strain combination (B6→C_3H), is not indefinite. While 100% of low temperature treated grafts were recovered at 25 days, this proportion fell to 60% by 40 days and to 20% by 50 days.

It is tempting to speculate that such late rejection is triggered by the processing of graft antigens by host derived antigen presenting cells following the alternative late rejection route proposed by Lechler and Batchelor[22].from their renal allograft studies in the rat. Up to 25 days allogeneic epithelial grafts survive and grow in the same way as they do in syngeneic recipients giving a maximal lymphoid cell yield of about 5×10^5 /lobe. Morphologically such grafts develop normally. Furthermore, host derived intragraft lymphocytes contain a functionally mature subpopulation (See Table 2). Thus when allografted the thymic epithelium derived by low temperature organ culture seems to behave in a similar fashion to that obtained following deoxyguanosine treatment[17], except that over the longer time scale reported here low temperature grafts appear to be eventually rejected.

Thymic epithelium and tolerance induction

The prolonged survival of allogeneic thymic epithelial grafts makes possible studies directed at assessing the role of MHC gene products in the putative selective functions of thymic epithelium in *in vivo* thymic chimeras. Thus, it has been reported that while deoxyguanosine treated thymic epithelium continues to express class II MHC products it failed to induce allotolerance when transplanted to intact allogeneic receipients[17]. In this study,

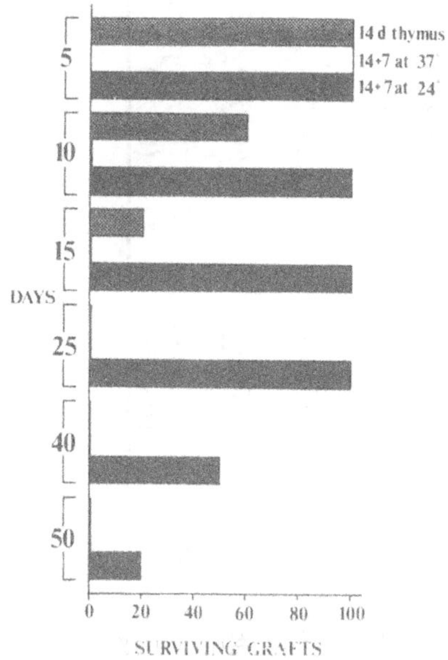

Fig. 2. Effect of organ culture for 7 days at 24°C on the survival of transplanted 14 day fetal thymus lobes as compared with no treatment or conventional culture at 37°C.

lymphocytes of recipient origin repopulating and maturing within donor allogeneic thymic epithelial grafts nevertheless responded to graft haplotype in MLC. However, since the deoxyguanosine treated thymic epithelial graft were always transplanted into normal recipients, it is not possible to conclude that the reactive T cells present within the grafts had in fact matured there. Indeed, it is possible that the reactive T cells had matured in the syngeneic receipient thymus and then migrated into the allogeneic thymic graft. To eliminate this possibility, we have

Table 2. MLC Responses of Spleen and Intragraft Lymphocytes from Mice bearing Low Temperature C57Bl/6 ($H-2^b$) Thymic Epithelial Allografts.

Recipient	Responders	Stimulators* Host ($H-2^k$)	Graft-type ($H-2^b$)	Third party ($H-2^d$)	p-value
Intact normal C_3H ($H-2^k$)	Intragraft lymphocytes	564±92	4815±737	5392±124	<.001; <.001
	Host spleen	16006±1079	38048±311	40673±2007	<.001: <.01
		Host ($H-2^d$)	Graft-type ($H-2^b$)	Third party ($H-2^k$)	
Athymic Balb/c nude ($H-2^d$)	Intragraft Lymphocytes	412±95	326±101	4337±58	N.S.; <.001
	Host spleen	12261±1473	16671±897	25614±2351	N.S.; <.01

*Incorporation of tritiated thymidine(DPM±SEM) by 2×10^5 responding lymphocytes at optimal stimulator to responder ratio.

repeated this study using low temperature organ culture derived thymic epithelium transplanted not only to normal intact allogeneic recipients but also to athymic nude mice, where there is no possibility of host thymic processing.

Intact normal C_3H ($H\text{-}2^k$) mice and athymic Balb/c nu/nu ($H\text{-}2^d$) mice were grafted with low temperature organ culture derived C57Bl/6 ($H\text{-}2^b$) thymic epithelium. We found (Table 2) that in intact normal recipients host lymphocytes ($H\text{-}2^k$) both within the epithelial grafts and within the spleen, can respond to the MHC antigens of the thymic graft type ($H\text{-}2^b$) in MLC, thus confirming the findings of Ready et al.[17] However, in the nude recipients we find that host derived lymphocytes ($H\text{-}2^d$), both within the epithelial grafts and within the spleen, respond to the MHC antigens of a third party ($H\text{-}2^b$) but not this time to the MHC antigens of the epithelial donor ($H\text{-}2^b$). These results suggest that when investigated in a system which precludes the possible involvement of extrinsic mature T cells processed by the syngeneic host thymus, thymic epithelium can induce transplantation tolerance. It appears that class II antigen expression by thymic epithelial cells might therefore be related to a mechanism of negative selection for the deletion of self-reactive T cell clones thus establishing self tolerance.

Reconstitution of nude mice with thymic epithelial grafts

While a number of reports indicate that isolated thymic epithelium retains the functional capacity to form a nidus for organotypic thymic regeneration upon transplantation[8,10,17] and to support the differentiation of host derived intragraft T progenitors[17,18], the attainment of functional restoration within the peripheral lymphon of T deficient mice using such grafts has been at best both partial and variable[11,12]. Indeed previously we have reported our failure to reconstitute adult thymectomised, lethally irradiated, fetal liver reconstituted mice with low temperature organ culture derived thymic epithelial grafts[23]. However, we report here that in the course of our chimaeric studies in nude mice, designed to address the question of tolerance induction, we noted significant reconstitution of the peripheral T cell compartment of the nude recipients of thymic epithelial grafts. In the spleen, the number of Thy-1 positive cells increased significantly and T cell function, as judged by the proliferative responses to the mitogen concanavalin A and to alloantigens in MLC, was restored. In addition, the thymus-dependent areas of the host peripheral lymphoid tissues (spleen and lymph node) were reconstituted (Table 3).

The results reported here, using nude mouse recipients, are in agreement with those of Schuurman et al[24], who have achieved reconstitution of certain T cell functions in nude rats with cultured thymic epithelium, but contrast with our previous findings

Table 3. Cytotoxic Assays for Thy.1 and Optimal Con A Responses of Spleen Lymphocytes Recovered from Nude Mice with and without Thymic Epithelial Grafts.

Recipient	Graft	Thy.1*	Concanavalin A 0ug/ml DPM±SEM	Concanavalin A 2ug/ml DPM±SEM	p-value
nu/+	none (+ve control)	28.2±4.2	2042±149	69629±3990	<0.01
nu/nu	none	.0.1±0.6	1149±81	207±303	N.S.
nu/nu	Isolated thymic epithelium	9.6±1.5	1978±389	19445±1512	<0.01
nu/nu	Untreated fetal thymus (+ve control)	7.6±1.8	1316±322	14451±2099	<0.05

*Mean Cytotoxic Index ± SEM.

in 'B' mice. Of course the exploitation of the difference between these two systems (i.e. the nude mouse versus the irradiation chimaera) may itself prove to be useful in further elucidating the role of the thymus in T cell differentiation.

ACKNOWLEDGEMENTS

These studies were supported by NIH grant AI 15819

REFERENCES

1. J.F.A.P. Miller, and G.F. Mitchell, Transplant Rev. 1:3 (1969).
2. H.von Boehmer, W.Haas, and N.K. Jerne, Proc.Natl.Acad.Sci. 75:2439 (1978).
3. R.M. Zinkernagel, Immunol.Rev. 42:224 (1978).
4. H. Cantor, and E.A. Boyse, Immunol.Rev. 33:105 (1977).
5. A.N. Barclay, and G.Mayrhofer, J.Exp.Med. 153:1666 (1982).
6. J.H. Robinson, J.Immunol. 130:1592 (1983).
7. J.H. Robinson, Cell Immunol. 84:422 (1984).
8. R.K. Jordan, D.A. Crouse, and J.J.T. Owen, J.Reticuloendothel. Soc. 26:373 (1979).
9. J.H. Robinson, and R.K. Jordan, Immunol.Tod. 4:41 (1983).
10. R.K. Jordan, and D.A. Crouse, Developments Immunol. 8:47 (1980).
11. F.Loor, and L.Haag, B.Cell Immunol. 29:200 (1977).
12. R. Hong, H. Schulte-Wissermann, E. Jarrot-Toth, S. Horowitz, andD.D. Manning, J.Exp.Med. 149:398 (1979).
13. E.J. Jenkinson, L.L. Franchi, R. Kingston and J.J.T. Owen, Eur. J. Immunol. 12:583 (1982).
14. J.C. Fontaine-Perus, F.M. Calman, C. Kaplan, and N.M. LeDouarin, J.Immunol. 126:2310 (1981).
15. M.F. Good, K.W. Pyke, and G.J.V. Nossal, Proc.Natl.Acad.Sci. 80:3045 (1983).
16. K.W. Pyke, P.F. Bartlett, and T.E. Mandel, Thymus 5:95 (1983).
17. A.R. Ready, E.J. Jenkinson, R. Kingston, and J.J.T. Owen, Nature 310:231 (1984).
18. R.K. Jordan, J.H. Robinson, A.L. Bentley, and D.A. Crouse, Behring Inst.Mitt. 70:3 (1982).
19. R.K. Jordan, A.L. Bentley, G.A. Perry, and D.A. Crouse. Submitted to J.Immunol.
20. R.K. Jordan, J.H. Robinson, and A.L. Bentley. Submitted to Nature.
21. P.E. Lacy, J.M. Davie, and E.H. Finke, Science 204:312 (1979).
22. R.I. Lechler, and J.R. Batchelor, J.Exp.Med. 155:31 (1982).
23. D.A. Crouse, G.A. Perry, and R.K. Jordan, "Regulation of the Immune Response", A.R. Sercarz, H. Cantor, and L. Chess, eds., Liss. In press.
24. H.J. Schuurman, J.G. Vos, E. van de Brink, and L. Kater, Clin.Immunol.Immunopathol 24:418 (1982).

[illegible] mice. Of course the explanation of the difference between these two systems (i.e., the nude mouse versus the irradiated [illegible] mouse) [illegible] [illegible] [illegible] [illegible] the role of the thymus in T cell differentiation.

ACKNOWLEDGEMENTS

These studies were supported by NIH grant AI 15019.

REFERENCES

1. [illegible] and G.F. Mitchell, [illegible] Rev. [illegible] (1969).
2. [illegible] Wheeler, and R. [illegible], Proc. Natl. Acad. Sci. 75:2439 (1978).
3. R.M. Zinkernagel, Immunol. Rev. 42:224 (1978).
4. [illegible] and [illegible] Royle, Immunol. Rev. 33: [illegible] (1977).
5. [illegible], and G. [illegible], J. Exp. Med. [illegible] (1982).
6. J.B. Robinson, J. Immunol. [illegible] (1983).
7. J.B. Robinson, Cell. Immunol. [illegible] (1984).
8. R.K. Jordan, D.A. [illegible], and J.J.T. Owen, J. Reticuloendothel. Soc. 26:3 (1979).
9. J.B. Robinson and R.K. Jordan, Immunol. Rev. [illegible] (1983).
10. R.K. Jordan and G.A. Crouse, Developments Immunol. 10:47 (1980).
11. [illegible], and [illegible] Hong, Scand. J. Immunol. 28:203 (1977).
12. R. Hong, [illegible], Transplant Proc. [illegible]
13. [illegible], J. Exp. Med. 149:358 (1979).
14. [illegible], [illegible], R. Kingston and J.J.T. Owen, Eur. J. Immunol. 12:484 (1982).
15. [illegible], [illegible], and [illegible], J. Immunol. 130:2340 (1983).
16. [illegible], Proc. Natl. Acad. Sci. 80:[illegible] (1983).
17. [illegible], and [illegible], Thymus 5:[illegible] (1983).
18. [illegible], [illegible] Kingston, and J.J.T. Owen, Nature 310:[illegible] (1984).
19. R.K. Jordan, J.B. Robinson, A.L. Bentley, and G.A. Crouse, [illegible] 70:[illegible] (1982).
20. R.K. Jordan, A.L. Bentley, [illegible], and G.A. Crouse, submitted for publication.
21. R.K. Jordan, J.B. Robinson, and A.L. Bentley, submitted to Nature.
22. [illegible], and [illegible], Science 204:[illegible] (1979).
23. [illegible] and [illegible], J. Exp. Med. 155:[illegible] (1982).
24. G.A. Crouse, [illegible], and R.K. Jordan, "Regulation of the Immune Response", [illegible], eds. (1984), in press.
25. H. [illegible], [illegible] Vos, [illegible] van der Brink, and [illegible], [illegible] Immunol. [illegible] 72:[illegible] (1982).

CULTURE OF MOUSE THYMIC EPITHELIAL CELLS AND STUDIES OF AGE-RELATED CHANGES

Myra Small, Liliane Barr-Nea and Moshe Aronson

Dept. Histology and Cell Biology, Sackler School of Medicine, Tel-Aviv University Ramat Aviv, Israel

Evidence accumulating over the last few years strongly suggests a role for epithelial cells of the thymus in the development of the T cell repertoire (1,2), but relatively little is known of the mechanism of this activity. In addition, we know almost nothing on the role and fate of these epithelial cells in thymic involution, a process which results in decreased T cell activity in advanced age. Both lines of research could be advanced by tissue culture of epithelial cells. In mice this objective has been difficult to achieve as attested by the large number of unsuccessful, or complicated and only partially successful reports (3-9), which appeared before this investigation was undertaken. In recent months other procedures have also been published (10-13), and it is now possible to compare the advantages and shortcomings of each. Difficulties have been described especially with advanced donor age, and we sought a novel approach to facilitate cultures from mice of all ages. Thus, we modified a medium known to support growth of hepatic epithelial cells (14), to bring about simultaneous inhibition of fibroblasts (15) which have plagued other culture systems.

MATERIALS AND METHODS

Cultures: Cultures were set up from inbred CBA/Lac mice, or random-bred ICR mice. Thymuses were removed aseptically and the majority of the lymphocytes dispersed by pressure through a stainless steel grid placed in Dulbecco's modified Eagle medium (EM). The remaining stroma was transferred as a single thymus to each plastic Falcon petri dish and minced to small fragments. The fragments were spread over the dish and allowed to dry a few moments to facilitate adherence. Modified medium was then added and replaced or supplemented twice weekly. Cultures were maintained in a 37° incubator with 10% CO_2. In some instances, cultures were grown on coverslips, or alternatively in Lab-Tek tissue culture chamber/slides. Cultures of kidney stroma were set up in parallel. Control cultures of mouse fibroblasts (from trypsinised embryos) were grown in EM + 10% fetal calf serum (FCS) and antibiotics.

Modified Medium: EM without valine was obtained from Biolab Labs., Jerusalem. To each 100 ml we added 9.4 mg D-valine, 10 mg glutathione, 5 mg ascorbic acid, 0.4 mg insulin, 1 mg NADH, 10 mg kanamycin, 2.5 mg gentamycin, and 10% heat-inactivated FCS (Dutchland Labs., Philadelphia). Hydrocortisone succinate (5 mg) was added unless otherwise indicated below.

Fluorescence Microscopy: Cultures of thymus stroma, or control fibroblasts, were fixed in cold acetone and stained 1 hr at 37° with rabbit anti-human keratin (DAKO lot 092), 15 min with FITC-goat anti-rabbit IgG at 1:40 (Miles Yeda), and 0.1% Evans Blue, with washes in between each step in phosphate-buffered saline (PBS). Frozen sections of thymus were stained in parallel, and all preparations were observed with a Zeiss fluorescent microscope fitted with epi-illumination. Excitation was 450-490 nm and emitted light >520 nm.

Autoradiography: To study division of cells in culture, cells were grown on cover slips, and 2 uCi ^{3}H-thymidine was added to each well in 3 ml medium. After 12 hr the cultures were washed repeatedly in PBS, and fixed in methanol:acetic acid (3:1). The coverslips were glued to slides and dipped in Ilford K-5 emulsion. After 1 week exposure, autoradiographs were developed and stained with Crystal violet.

Electron Microscopy: Cultures were rinsed in PBS, fixed in 3% glutaraldehyde pH 7.2, rinsed in buffer, and post-fixed in 1% osmium tetroxide. After dehydration, Beem capsules were filled with Epon and inverted on marked areas of the culture. After polymerisation the embedded tissues were separated from the dish and processed for electron microscopy with lead citrate and uranyl acetate staining.

Cell Transfer: A vigorous stream (13) of 0.02% EDTA in PBS was found to release all the epithelial cells from the cultures. When the dishes were sprayed vigorously macrophages were released as well, but with moderate treatment macrophages remained adherent. The cells were centrifuged, resuspended in modified medium with FCS and replated.

Transfer of Microareas in the Cultures: Primary cultures were grown on strips of silicone rubber (16). Microareas surrunding individual explants were located microscopically and dissected out with the substrate. The transferred areas were either maintained on the substrate, or released as described above.

Transfer of Cells Differing Morphologically: Cells released from primary cultures as described above were layered on preformed step-gradients of Percoll in EM from 10-55% and centrifuged at 1700 rpm for 20 min, at 5°. The bands obtained were washed with PBS, resuspended in modified medium and grown in Costar Clusters.

RESULTS

The first experiments were performed to check the inhibition of fibroblast growth in medium with D-valine by splitting cultures from mouse embryonic fibroblasts into dishes with Eagles medium , and others with liver medium (14), modified with D-valine (15). As can be seen in Fig. 1, fibroblasts were indeed inhibited. The modified medium did support the growth of cells from thymic explants, and cultures were estab-lished within 1-2 weeks containing a variety of morphological types.

A large proportion of cells in the cultures appeared as carpets of confluent cells resembling paving stones, each attached to neighboring cells all along the length of their membrane (Fig. 2). They contained abundant cytoplasm, without noticeable pro-jections and were often found immediately surrounding the explant tissue. These confluent carpets of cells were stained with anti-keratin antibody (Fig. 2), were attached to neighboring cells by desmosomes (Fig. 3), were relatively more abundant in cultures derived from older mice, and were relatively unaffected by the

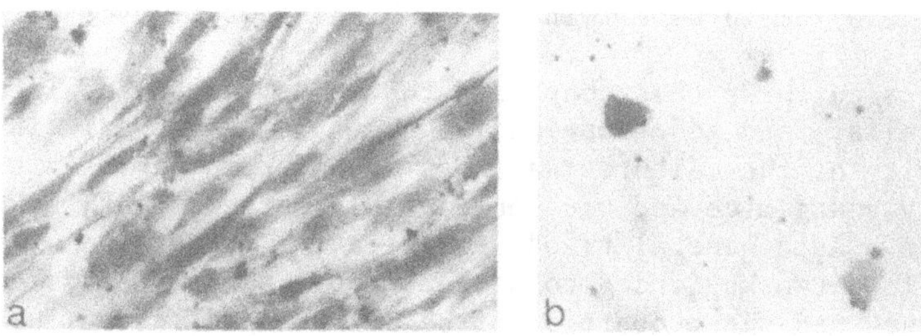

Fig. 1. Inhibition of fibroblast growth by modified medium. Mouse embryonic fibroblasts were passaged into EM with FCS (1-a), or modified medium with FCS (1-b) (X220).

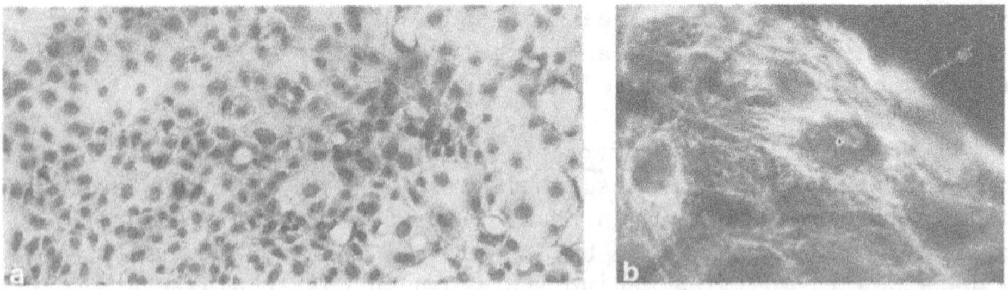

Fig. 2. Confluent carpets of cells growing in culture. a) grown 3½ weeks and stained with crystal violet (X220), and b) grown 12 days and stained with rabbit anti-keratin, and fluorescein-goat anti-rabbit serum.

presence of cortisone in the medium (as described below).

Another large proportion of cells was found to connect to their neighbors by long cytoplasmic projections, and the cells appeared to be organised into networks as shown in Fig. 4. Heterogeneity of the cells growing in networks was apparent, especially in medium without cortisone. In other words, these cells were especially susceptible to the effects of cortisone. Their abundance in culture tended to decrease with the age of the donor mice. They did not stain with the anti-keratin antibody used, but desmosomes were found between adjacent projections (Fig. 3).

There were also a considerable number of dispersed cells characterised by multiple processes. In addition, other morphological types appeared in some cultures in low numbers. Macrophages were often encountered, but as described by Boniver et al. (7), they could be eliminated by high doses of mycostatin.

These cultures thus contained two major cell types identified as epithelial, and additional cells that are as yet unideintified. Cellularity of the cultures was most abundant in thymuses taken from very young mice and was considerably decreased at 6 mo of donor age. As a general trend, the relative proportion of cells growing in networks tended to decrease more sharply as a function of age, and the proportion of cells arranged in confluent carpets was correspondingly increased. It was subsequently found that both of these aspects could be altered by changing the culture medium in two respects. The cellularity of cultures derived from older mice could be increased considerably by adding low doses of EDTA to the medium. It is possible that migration of cells from the explants was facilitated by reducing the cementing action of intercellular calcium. Although this effect seemed also to occur

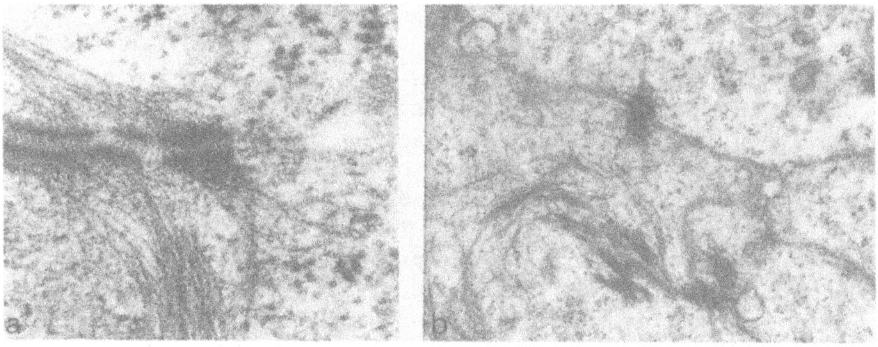

Fig. 3. Electron microscopic identification of thymic epithelial cells. a) Desmosomes between confluent cells in carpets (X40,000). b) Desmosomes linking cytoplasmic processes in network of cells (X33,000).

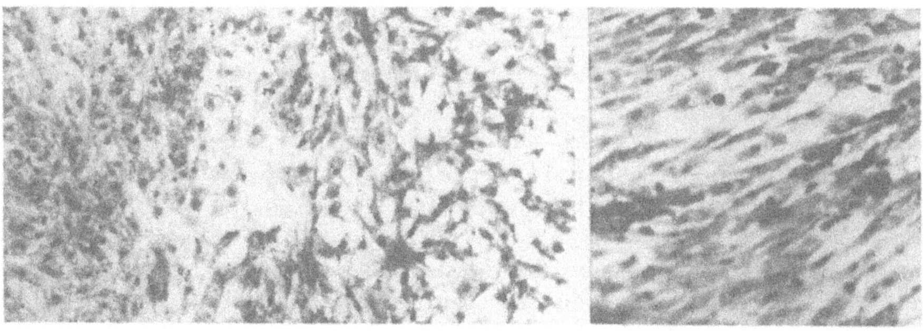

Fig. 4. Networks of cells grown 10 days in culture without cortisone (X220).

Table 1. Effect of EDTA on Growth of Thymic Cultures from 10 Month Old ICR Mice

Conc. EDTA (%):	0	.0002	.001	.01	.02	.1
Relative	±	++	++++	+++	++	-
Cellularity/	±	±	+++	+++	++	-
Culture*	±		++			
	±		++			
	±		+			
	±		±			
	++					

*Pooled results from 3 experiments scored at 2 weeks.

in thymuses from young mice, it was especially useful in stimulating cultures from older donors, as shown in Table 1.

Secondly, we tested whether the high concentration of cortisone in the medium could be detrimental to growth of the cells. When thymic cultures were set up parallel in medium with and without hydrocortisone succinate, cell growth was greatly enhanced in cultures without cortisone (but fibroblasts still failed to grow). As summarised in Table 2, cortisone even at doses considered unphysiological (17), dramatically inhibited growth of cells from thymic stroma, while not affecting cultures set up in parallel from kidney. The inhibitory effect of cortisone was especially apparent on the cells organised as networks, but reduced growth of the confluent carpets was also observed, and these effects were noted at all ages tested. These results raise the interesting possibility that cortisone or other steroid hormones which begin to be secreted around the onset of thymic involution might influence this process by acting on the epithelial framework of the thymus.

A second and possibly related effect of age was noted at the level of DNA synthesis. A decreased rate of DNA synthesis was observed in a portion of the epithelial cells in cultures set up from mice after the onset of thymic involution. Since this process begins after 7 weeks of age in the strain of mice used, cultures of thymus (or kidney as control) were set up at <6 weeks, or >10 weeks (in medium containing cortisone).

A pulse of tritiated thymidine was given for the last 12 hrs and the cultures were processed for autoradiography. We then compared heavily labelled cells to those which were lightly labelled, an indication of as yet incomplete DNA synthesis. Results of

Table 2. Effect of cortisone in the culture on growth of epithelial cells

Cortisone (mg/100ml)	0	$1x10^{-3}$	$1x10^{-2}$	$2x10^{-2}$	5
Relative cellularity/cultures*					
Thymus (5 months)	+++				+
Thymus (4 months)	+++				+
Kidney (4 months)	++				++
Thymus (3 months)	++++	++++	++	+	±
Kidney (3 months)	++				++
Thymus (2 weeks)	++++				+

*Results of 4 separate experiments scored at 9-12 days.

Table 3. Rate of DNA Synthesis in Cultured Epithelial Cells as a Function of Age of Donor Mice

Exp. No.	Age of donor mice: Preinvolution	Age of donor mice: Postinvolution	Tissue	% cells/culture with decreased rate of DNA synthesis	Total labelled cells/culture
1	10 days		Thymus-confluent carpet	12, 12, 18	844, 391, 631
		$2\frac{1}{2}$ months	Thymus-confluent carpet	28, 24, 31	589, 128, 1060
2	1 month		Thymus-confluent carpet	13, 16, 14	130, 339, 547
		5 months	Thymus-confluent carpet	32, 29, 35	207, 104, 410
	1 month		Thymus-network	7, 5	415, 314
		5 months	Thymus-network	22, 24	51, 58
	1 month		Kidney	9	133
		5 months	Kidney	8	203

such experiments are shown in Table 3. A significant increase in the percentage of lightly labelled cells was apparent in thymic epithelium (organised both in confluent carpets and in networks) in the cultures derived from postinvolution mice. While a significant proportion of the cells thus exhibited a decreased rate of DNA synthesis, neighboring cells were heavily labelled. These results raise the possibility that decreased potential for regeneration or turnover in some of the epithelial cells could be expressed gradually in the process of thymic involution. When we repeated these experiments in medium without cortisone, morphological heterogeneity of the cells forming networks was apparent in the abundant cell growth, with a corresponding heterogeneity in degree of labelling. For meaningful comparison with respect to age we must learn to unravel this complexity in terms of the component cells involved.

Our approach to this problem is two-fold. Microareas from the cultures can be isolated and transferred separately (as described in Methods) with the ultimate aim of characterising their surface properties. Since cell division in culture seems far in excess of that expected in vivo, the microareas growing out from individual explants should contain an exaggerated number of identical cells, and thus might facilitate study of their interaction with developing lymphocytes. Secondly, cells differing morphologically can be isolated from the cultures and subcultured (see Methods). This step is essential to determine whether or not all of the various cell types growing in networks are epithelial (work in progess). For those cells found to contain desmosomes the next steps should also be the characterisation of surface properties and then interaction with developing lymphocytes.

DISCUSSION

We have described a simple method for culturing the stromal elements of mouse thymus that can also be used with adult mice, and thus facilitates a study of the changes occurring as a function of age. So far, cultured cells of two main morphologies have been identified as epithelial: confluent carpets of polygonal cells and networks of cells connected by long cytoplasmic processes. The latter category is also heterogeneous in morphology and although desmosomes are a general feature, we are currently testing the epithelial identity of each cell type separately. The heterogeneity observed in these cultures seems to us advantageous for fuller understanding of the multi-faceted processes occurring in the thymus, since epithelial heterogeneity occurs in vivo (18-23). Thus, this technique is not less important than the production of cloned thymic epithelial cells which has been reported in AKR mice (10). Two other recent reports have described success in culture of thymic epithelial cells (11, 12), but only from donors 1 month of age. A more general success has recently been reported (13), using a feeder layer to grow trysinised cell cultures. Although this technique may be useful we see a distinct advantage in culture of explants for producing homogeneous microareas of cells. The conflicting results (12, 13) reported with anti-Ia serum illustrates the point that a variety of culture methods should complement each other and together further an understanding of thymic function.

A primary aim of all these culture systems is to shed light on the role of epithelial and other thymic stromal cells in the development of T lymphocytes. Since surface properties of the epithelial (and nonepithelial) cells are undoubtedly involved in this process, the next step in this direction would seem to be characterisation of surface antigens and surface carbohydrates of each cell type. A second objective of our research is to study the causes and results of the involution process. The observation that cortisone inhibited thymic epithelial growth (incidentally not found with feeder layers (13)) raises the possibility that this hormone or other steroids might exert an inhibitory effect on the thymic framework in vivo. The finding that the rate of DNA synthesis was reduced in some, but not all, epithelial cells from post-involution mice also points to a possible mechanism of the involution process. In this respect, it is interesting to note that as well as an increased incidence of autoimmune disease correlated with aging, at least one autoimmune disease has been linked with deficiency of a particular thymic epithelial type (24).

REFERENCES

1. R.M. Zinkernagel, C.N. Callhan, A. Althage, S. Cooper, P.A. Klein and J. Klein, J. Exp. Med. 147: 882 (1978).
2. M.J. Bevan and P.G. Fink, Immunol. Rev. 42: 3 (1978).
3. J.M. Garland and T.M. Dexter, Amer. J. Reprod. Immunol. 1: 129 (1981).
4. M.V. Berridge, J.C. Nargang, P.T. Loverde and E.S. Golub, J. Immunol. 124: 2738 (1980).
5. R. K. Jordan and D.A. Crouse, J. Reticuloendothel. Soc. 26: 385 (1979).
6. K.H. Jones and R.L. St. Pierre, In Vitro 17: 431 (1981).
7. J. Boniver, A. Decleve, M.O. Dailey, C. Hensik, M. Lieberman and H.S. Kaplan, Thymus 2: 193 (1981).
8. F. Loor, Immunology 37: 157 (1979).
9. M.E. Gershwin, R.M. Ikeda, W.L Kruse, F. Wilson, M. Shiofrine and W. Spangler, J. Immunol. 140: 971 (1978).
10. T.R. Beardsley, M. Piersbacher, G.D. Wetzel and E.F. Hays, PNAS 80: 6005 (1983).
11. M. Sugimoto, Y. Nakaniski, M. Otokawa, N. Vashida, T. Yasuda, H. Sato and Y. Sato, J. Immunol. 130: 2767 (1983).
12. J.J. Jason and C.A. Janeway, Clin. Immunol. and Immunopathol. 30: 214 (1984).

13. T.T. Sun, P. Bonitz and W.H. Burns, Cell. Immunol. 83: 1 (1984).
14. D.C. Anuforo, D. Acosta and R.V. Smith, In Vitro 14: 981 (1978).
15. S.F. Gilbert and B.R. Migeon, Cell 5: 11 (1975).
16. A. Shahar, R. Monzain and Y. Straussman, Tissue & Cell 5: 691 (1973).
17. I.L. Weissman and R. Levy, Israel J. Med. Sci. 11: 884 (1975).
18. R.V. Rouse, R. van Ewijk, P.P. Jones and I.L. Weissman, J. Immunol. 122: 2508 (1979).
19. I.L. Weissman, R.V. Rouse, B.A. Kyeski, F. Lepault, E.C. Butcher, H.S. Kaplan and R.G. Scollay, Behring Inst. Mitteilung 70: 242 (1982).
20. E.J. Jenkinson, W. van Ewijk and J.J.T. Owen, J.Exp. Med 153: 280 (1981).
21. W.S. Hwang, T.Y. Ho, S.C. Luk and G.T. Simon, Lab. Invest. 31: 473 (1974).
22. H. Wekerle, U.P. Ketelsen and M. Ernst, J. Exp. Med. 151: 925 (1980).
23. B. Kyewski and H.S. Kaplan, J. Immunol. 128: 2287 (1982).
24. R.L. Boyd, G. Oberhuber, K. Hala and G. Wick, J. Immunol. 132: 718 (1984).

HETEROGENEITY OF HUMAN THYMUS EPITHELIAL CELLS REVEALED BY MONOCLONAL ANTI-EPITHELIAL CELL ANTIBODIES

Mary A. Ritter, Henk-Jan Schuurman, Wilma A. MacKenzie, Ruud A. de Maagd, Kevin M. Price, Roel Broekhuizen and Louis Kater

Department of Immunology, Royal Postgraduate Medical School, Hammersmith Hospital, London, England, and Division of Immunopathology, Departments of Internal Medicine and Pathology, University Hospital, Utrecht The Netherlands

The epithelial cell matrix of the thymus, which is involved in the generation of T lymphocyte immunocompetence,[1] shows considerable (cytological) heterogeneity. This phenomenon has been well-established in ultrastructural morphology (this issue, p 288),[2] but also in expression of thymic hormone components in the cytoplasm.[3] We started the analysis of the thymic microenvironment for antigenic heterogeneity by raising monoclonal antibodies to thymus stromal cells.

Materials and methods

Thymus tissue was obtained from children undergoing cardiac surgery. The tissue was thoroughly minced. In the London laboratory, a homogenate was used for immunization, and in the Utrecht laboratory preparations enriched for stromal cells were used. These preparations were obtained by collagenase and Dispase enzyme-desintegration, followed by Percoll (d=1.066 g/cm^3) density centrifugation. After repeated immunization of mice, using as adjuvant Complete Freunds Adjuvant (London) or Bordetella Pertussis particles (Utrecht), spleen cells were fused with X63-Ag8.653 non-producer mouse myeloma cells according to standard procedures. Supernatants of hybridomas were screened by immunohistochemistry on frozen tissue sections in standard immunoperoxidase (staining of peroxidase with 3-amino-9-ethylcarbazole)[4] or immunofluorescence.[5] The anti-mouse immunoglobulin conjugates used in these assays were obtained commercially (Dakopatts, Copenhagen, Denmark, and Nordic

Immunological Laboratories, Tilburg, The Netherlands). Relevant hybridomas were cloned three times. The immunoglobulin class of monoclonal antibodies was determined by immunofluorescence on cytocentrifuge preparations of antibody secreting hybridoma cells using subclass specific antibodies (Nordic).

To identify epithelial cells in tissue sections, a polyclonal rabbit anti-human keratin antibody (Dakopatts) was applied. In two-color immunofluorescence the reactivity of monoclonal antibodies to (keratin-positive) epithelial cells was evaluated. In these experiments, tetramethyl rhodamin isothiocyanate conjugated anti-rabbit immunoglobulin (Nordic) and fluorescein isothiocyanate conjugated anti-mouse immunoglobulin (Nordic) were applied in indirect immunofluorescence. Control experiments included ommission of the first antibody (anti-keratin or mouse monoclonal antibody) and replacement of the first antibody by irrelevant antibody; in these experiments no staining was observed.

Results and discussion

In both laboratories hybridomas secreting antibodies were obtained (antibodies with prefix MR were raised in London, all others in Utrecht). Apart from antibodies with reactivity to connective tissue and/or blood vessels, or antibodies which gave staining patterns similar to anti-HLA class I or class II antibodies, we obtained a series of antibodies with 6 main staining patterns of binding to epithelial cells (Table 1, Fig. 1). The staining of epithelial cells, demonstrated in two-color immunofluorescence with anti-keratin antibody, is illustrated in Fig. 2.

Some antibodies (18B5, MR3, MR6) gave a dendritic staining of epithelial cells in cortex and/or medulla, which suggests that the antigen detected is located on the plasma membrane. In addition to this dendritic staining pattern of epithelial cells in the cortex, antibodies MR3 and MR6 weakly stained the cytoplasm of non-epithelial cells in the medulla. Other antibodies (MR10, MR14, 7C5) apparently stained antigen on the plasma membrane and in the cytoplasm. Finally, the antibody 8A2 preferentially stained the cytoplasm of epithelial cells.

Most of the antibodies gave consistent staining patterns in sections of one thymus or of different thymus specimens. However, the antibodies 8A2, MR13 and 4A6 revealed variability in this aspect. In all thymuses studied (n=8, age range of donors 2 months-6 years) antibody 8A2 consistently stained Hassall's corpuscles, but the extent of staining of adjacent cells varied from thymus to thymus. Antibodies MR13 and 4A6 gave variable staining intensity of Hassall's corpuscles even within the same thymus: highly keratinized areas were most intensily stained.

The antibodies were evaluated for reactivity to components in human lymph node, tonsil, spleen, gut, kidney, brain and skin. Antibody 18B5 binds to epithelium of tonsil, skin, gut and kidney, indicating that the antigen detected may be a general epithelial

Table 1. Reactivity of monoclonal antibodies to human thymus epithelial cells[a]

Code	Immunoglobulin class	Reactivity pattern[b]
18B5 (U)	IgM	Epithelial cells in cortex and medulla. Stains only keratin-positive cells.
MR10, MR14 (L)	IgM	Subcapsular epithelial cells and subpopulation of epithelial cells in medulla. Stains only keratin-positive epithelial cells.
MR3, MR6 (L)	IgG1	Epithelial cells in cortex (excluding subcapsular area). Stains keratin-negative cells in medulla.
7C5 (U)	IgG1	Some epithelial cells in cortex (excluding subcapsular area). Stains only keratin-positive cells.
8A2 (U)	IgM	Hassall's corpuscles and adjacent epithelial cells. Stains only keratin-positive cells.
MR13 (L) 4A6 (U)	IgM IgG1	Hassall's corpuscles. Stains scattered keratin-negative cells in cortex, medulla and septa.

[a] Within brackets, laboratory of preparation (London, Utrecht).
[b] Immunohistochemistry on frozen tissue sections, including two-color immunofluorescence with anti-keratin antibody.

cell marker. Tubular epithelium of kidney was found to be MR3, MR6 and 7C5 positive, and epithelium of tonsil and gut stained for 8A2. Antibodies MR3 and MR6 stained non-epithelial cells in tonsil, lymph node and splenic white pulp similar to cells in thymic medulla. Preliminary experiments indicated that these cells are of dendritic/macrophage origin. To a less extent, MR13 and 4A6 stained keratin-negative cells in thymic medulla. Similar cells were found in tonsil, lymph node and spleen, some of which apparently belong to the granulocyte lineage. MR10 and MR14 were completely negative in all control tissues tested.

The different staining patterns of our monoclonal antibodies lead us to conclude heterogeneity within the epithelial meshwork of the human thymus. Our observations extend those of others who studied antibodies to thymic stroma: e.g., the antibody RFD4,[1] A2B5,

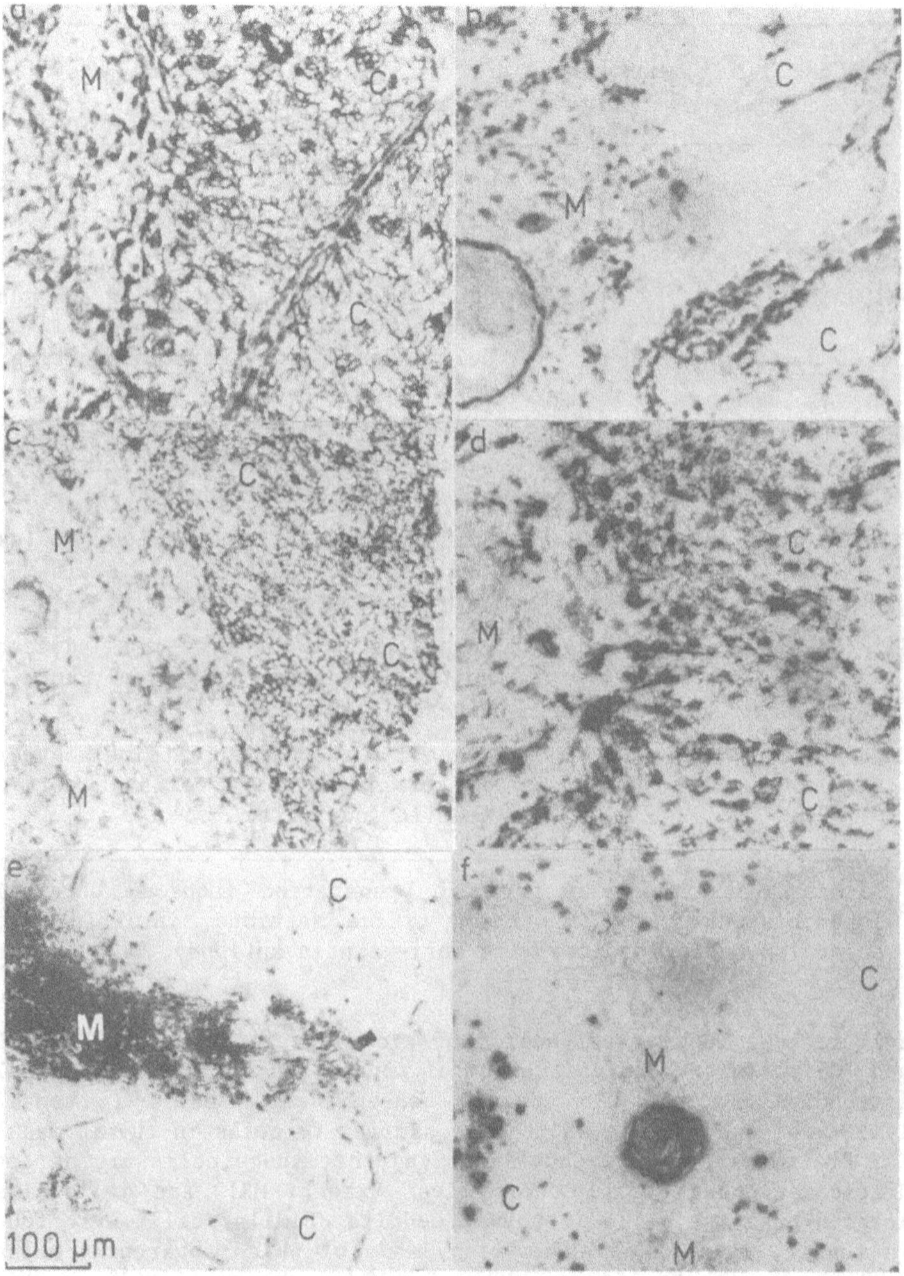

Fig. 1. Immunoperoxidase staining of human thymus tissue sections with monoclonal anti-epithelial cell antibodies. Cortex (C) and medulla (M) are indicated. a, 18B5; b, MR10; c, MR3; d, 7C5; e, 8A2; f, MR13.

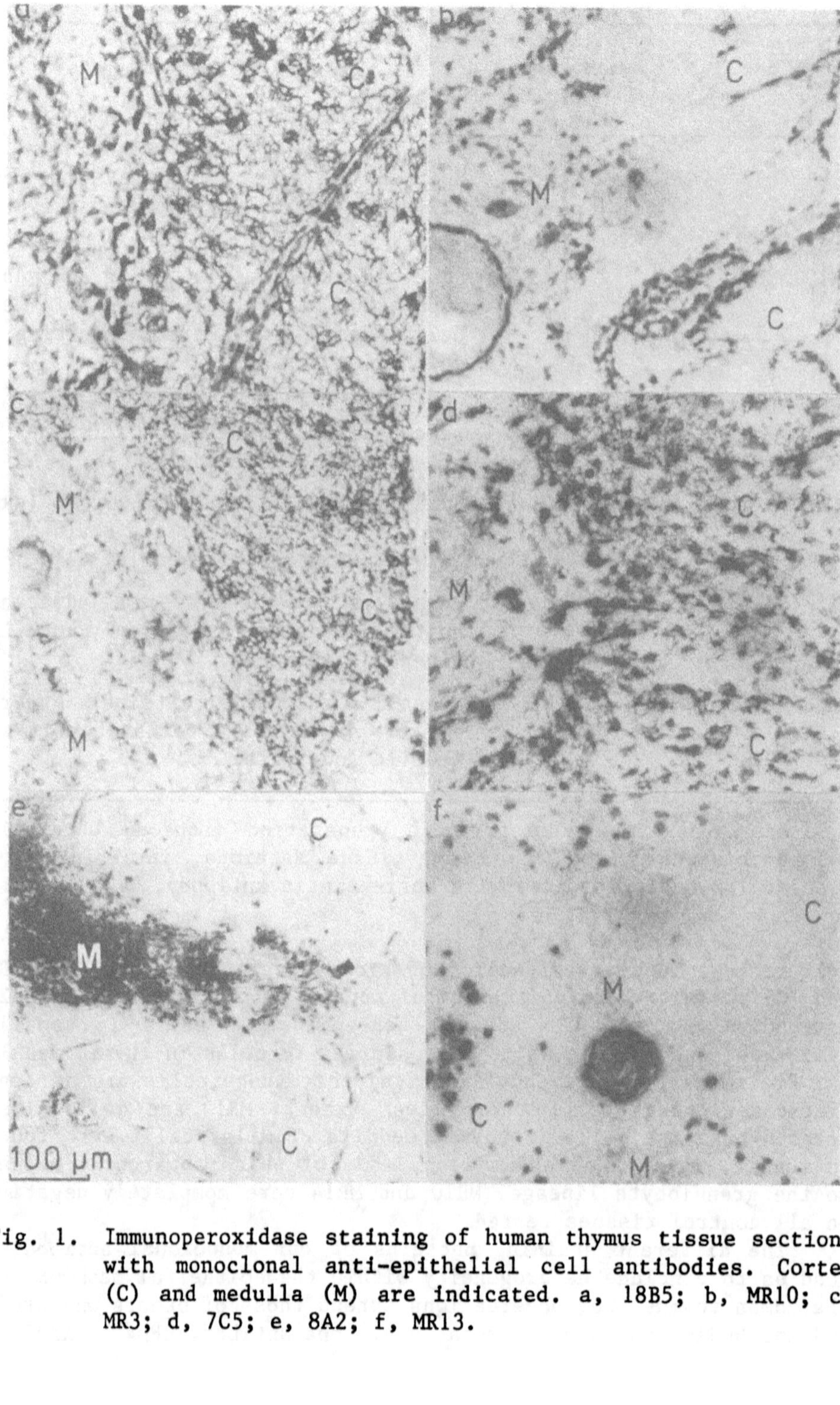

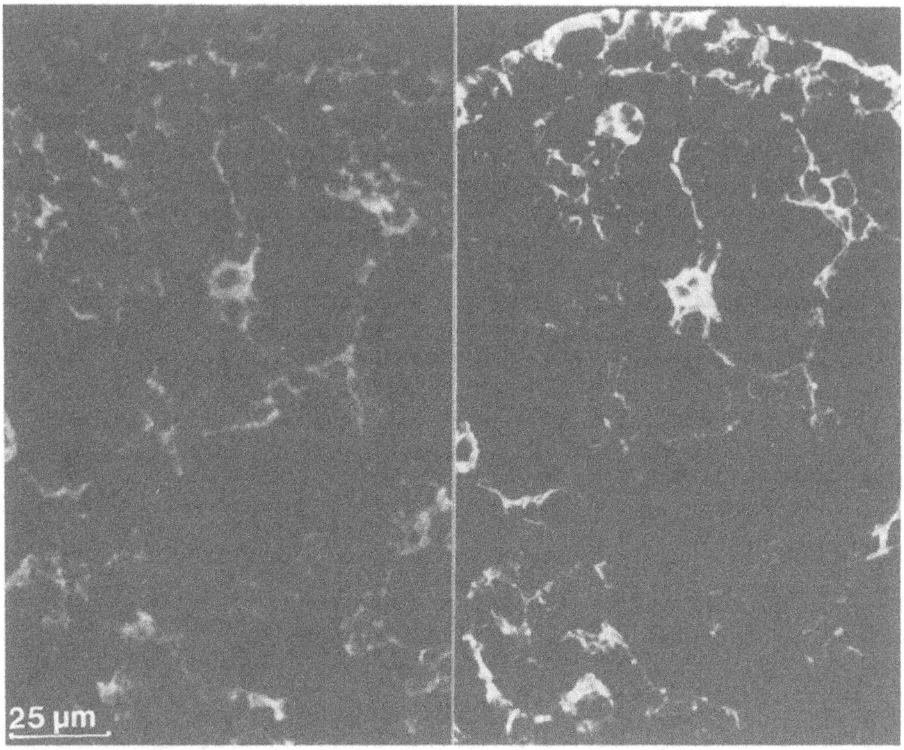

Fig. 2. Two-color immunofluorescence with MR3 mouse monoclonal antibody (left) and rabbit anti-human keratin antibody (right) on thymus tissue section (cortex). MR3 stains keratin-positive epithelial cells in the cortex, but not subcapsular epithelial cells.

anti-p19 and TE7[6] detect subcapsular and medullary epithelium in a similar way to our antibodies MR10 and MR14, although MR10 and MR14 are restricted to only a subpopulation of cells in the medulla. It remains to be established to what extent indentity in reactivity to epithelial cell subpopulations or even antigenic molecules exists for these antibodies. Antibodies with restricted reactivity to (part of) cortex epithelium (MR3, MR6 and 7C5) have not been described before: we elsewhere reported on differences in ultrastructural morphology between epithelial cell types in cortex and medulla (this volume, p 229);[2,4] the restricted antigen expression may be linked with these morphological characteristics.

The detection of antigenic heterogeneity of epithelial cells can be helpful to elucidate the function of the various subpopulations in T lymphocyte maturation.[1] Bone marrow precursor T cells migrate to subcapsular and medullary areas: at these sites epithelial cells reside which can be distinguished from other epithelial cells by positivity for MR10 and MR14. Concerning thymic hormone synthesis, thymosin α_1 positive epithelial cells are present at these sites, and thymosin β_4 positive cells are restricted to the subcapsular area.[3] These thymosin peptides are known to be involved in early intrathymic T lymphocyte maturation. Close contacts between epithelial cells and lymphocytes occur especially in the cortex.[2,4] In this compartment epithelial cells are specifically positive for MR3 and MR6, whereas "thymic nurse cells" in this area were previously found to be Thy-1 positive.[5] The fine heterogeneity of cortical epithelium is further illustrated by 7C5 staining of cells at the cortico-medullary junction, where mature lymphocytes are believed to leave the thymus. The antibodies described in this report may be especially helpful in elucidation of the surface and cytoplasmic molecules of various thymic epithelial cell subpopulations with different biological function in T lymphocyte maturation.

Acknowledgement

Work in the London laboratory was supported by a grant from the Medical Research Council (no: G8210410CA).

References

1. G. Janossy, M. Bofill, L.K. Trejdosiewicz, and M. Chilosi, Cellular differentiation of lymphoid subpopulations and their microenvironments in the human thymus, in: "The thymus, histophysiology and pathology," H.K. Müller-Hermelink, ed., Springer Verlag, Berlin (1984).
2. F.P. van de Wijngaert, M.D. Kendall, H.-J. Schuurman, L.H.P.M. Rademakers, and L. Kater, Heterogeneity of epithelial cells in the human thymus. An ultrastructural study, Cell Tissue Res. 237: 227 (1984).
3. H.J. Schuurman, F.P. van de Wijngaert, L. Delvoye, R. Broekhuizen, J.E. McClure, A.L. Goldstein, and L. Kater, Heterogeneity and age dependency of human thymus reticulo-epithelium in production of thymosin components, Thymus in press.
4. F.P. van de Wijngaert, L.H.P.M. Rademakers, H.J. Schuurman, R.A. de Weger, and L. Kater, Identification and in situ localization of the "Thymic Nurse Cell" in man, J. Immunol. 130: 2348 (1983).
5. M.A. Ritter, C.A. Sauvage, and S.F. Cotmore, The human thymus microenvironment: in vivo identification of thymic nurse cells and other antigenetically distinct subpopulations of epithelial cells, Immunology 44: 439 (1981).
6. B.F. Haynes, R.M. Scearce, D.F. Lobach, and L.L. Hensley, Phenotypic characterization and ontogeny of mesodermal-derived and endocrine epithelial components of the human thymic microenvironment, J. exp. Med. 159: 1149 (1984).

HETEROGENEITY OF THE HUMAN THYMUS EPITHELIAL MICROENVIRONMENT AT THE ULTRASTRUCTURAL LEVEL

Marion D. Kendall, Frank P. van de Wijngaert, Henk-Jan Schuurman, Loek H.P.M. Rademakers and Louis Kater

Department of Anatomy, St. Thomas's Hospital Medical School, London, England, and Division of Immunopathology, University Hospital and Institute for Pathology Utrecht, The Netherlands

The supporting meshwork of the thymus, which is composed of epithelial cells (epithelial-reticular cells), plays a major role in the generation of immunocompetent T lymphocytes. From the characterization and location of lymphocytes in the thymus it can be hypothesized that different microenvironments exist promoting different steps in intrathymic T cell maturation. Our findings on heterogeneity of human thymus epithelial cells synthesizing various thymosin components[1] and heterogeneity in antigen expression revealed by monoclonal anti-epithelial cell antibodies (this issue, p 297) gives support to this hypothesis. Also in electron microscopy, variation in ultrastructural characteristics of epithelial cells has been noted.[2,3] We studied the ultrastructural morphology of epithelial cells in sections of the human thymus to evaluate whether interrelationships exist between epithelial cell subtypes and between morphology and function in the microenvironment.

This study included thymus tissue specimens from 23 children and young adults (age range 3 months - 31 years, average age 4 years) undergoing cardiac surgery. All thymuses showed a normal histology for the respective ages. The processing of tissue for electron microscopy was done according to standard techniques.[4,5]

Results and discussion

In the light microscopic study on 1 µm thick sections of Epon-embedded tissue, 6 main types of epithelial cells could be distinguished on the basis of different affinity for stains and morphology (Fig. 1). In the electron microscope, the ultrastructural characteristics of these cell types were even more pronounced (summa-

Table 1. Morphological characteristics and location of types of epithelial cells in the human thymus

	Morphological characteristics	Location
Type 1 "subcap-sular-perivas-cular"	basal lamina heterochromatic, irregularly shaped nucleus, 5x10 μm long cisterna of RER well-developed Golgi complex micropinocytotic vesicles	continuous sheet beneath capsule around capillaries in cortex and cortico-medullary region
Type 2 "pale"	euchromatic, round to oval nucleus, diameter about 12 μm well-developed Golgi complex short profiles of RER electron-dense granules	scattered in cortex and medulla predominant in outer cortex
Type 3 "interme-diate"	spectrum between types 2 and 4 eu- or hetero-chromatic,irregular-ly shaped nucleus, 5x10 μm dilated cisterna of nucleus and RER	mid cortex, deep cortex, medulla
Type 4 "dark"	heterochromatic, electron-dense, irregularly shaped nucleus, 3x6 μm dilated cisterna of nucleus and RER residual bodies, swollen mitochon-dria, vacuoles	deep cortex scattered in medulla and around Hassall's corpuscles
Type 5 "undif-feren-tiated"	rounded nucleus with some hetero-chromatin, 5x7 μm polyribosomes small bundles of tonofilaments small desmosomes	in groups in cortico-medullary region scattered in medulla
Type 6 "large-medul-lary"	large, eu- or hetero-chromatic, rounded nucleus, diameter 15 μm sometimes RER well-developed abundant tonofilaments	scattered in medulla adjacent to large Hassall's corpuscles part of small Hassall's corpuscles

rized in Table 1, and illustrated for types 1, 2, 4-6 in Figures 2-5). The cells of type 3 comprised a spectrum, with a range in electron density of cytoplasm and nucleus from electron-lucent (like type 2) to electron-dense (like type 4). The more electron-lucent cells showed a Golgi-complex like type 2 cells, but less

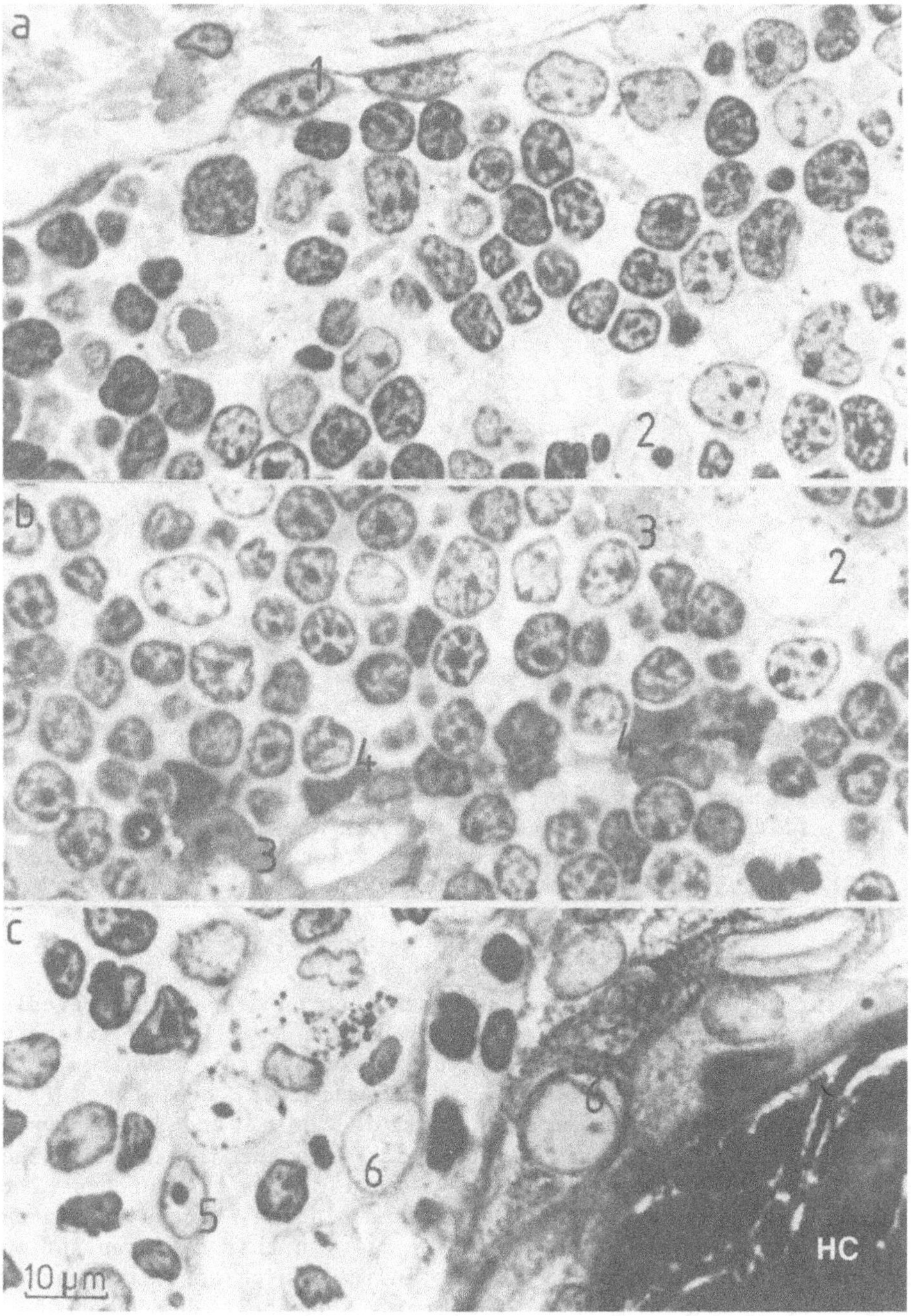

Fig. 1. One µm Epon-embedded section of thymus cortex (a,b) and medulla (c, from ref. 5, with permission) (2-year-old donor) stained with pararosaniline and methylene blue. Epithelial cells of types 1-6 are indicated. HC, Hassall's corpuscle.

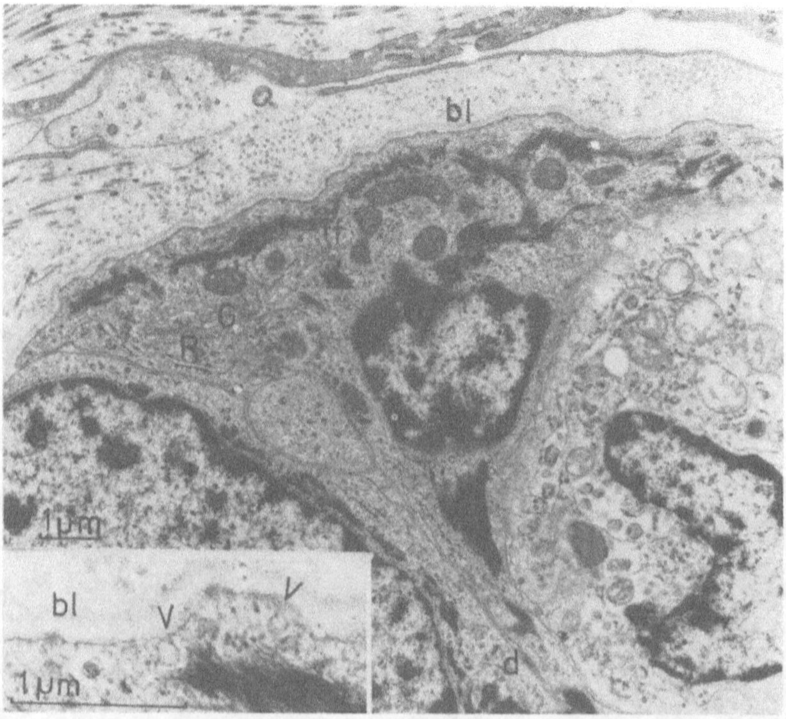

Fig. 2. Electron micrograph of a type 1 epithelial cell (6½-year-old donor). The cell has a nucleus with marginal heterochromatin. Profiles of RER (R), a Golgi complex (G), tonofilaments (tf) and desmosomes (d) are indicated. bl, basal lamina. Inset: higher magnification of part of the cell showing micropinocytotic vesicles (V, about 50-100 nm in diameter) facing the basal lamina.

prominent than type 2 cells. The more electron-dense cells revealed swollen mitochondria, but less so than type 4 cells. An extensive description of the various cell types is presented elsewhere.[5]

This documentation of six (main) epithelial cell types (Fig. 6) adds to previous classifications of "pale" and "dark" variants.[2,3] Type 3 cells have not been described before as a distinct entity. The spectrum in morphology, ranging from that of type 2 to that of type 4 cells, suggest a differentiation process within the epithelial cells from type 2 into type 4. In this differentiation, the characteristics of type 2 cells indicate high activity, whereas type 4 cells show signs of inactivity, possibly degeneration. A similar process may occur within the subcapsular-perivascular cell population, as some of these cells resembled type 2 and others type 4 cells.

Type 5 cells have not been explicitely reported in the literature. The presence of polyribosomes suggests an early stage of differentiation.

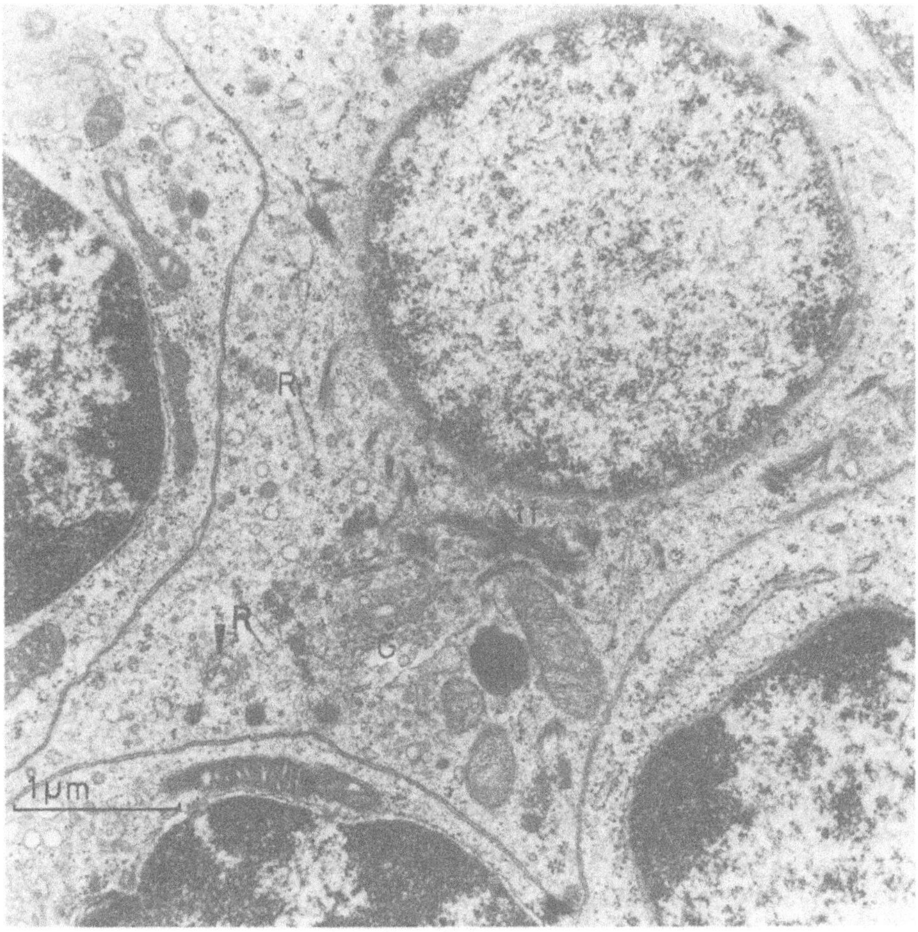

Fig. 3. Electron micrograph of a type 2 epithelial cell (6-year-old donor). For indications, see legend to Fig. 2. ▼, multivesicular body. From ref. 5, with permission.

How do these morphological characteristics relate to the function in the microenvironment? Epithelial cells of types 2 and 6 are likely candidates for thymus humoral factor production as these cells have a pronounced RER (especially type 2) and a well-developed Golgi complex (especially type 6). In the thymus cortex lymphocyte-epithelial cell-cell interactions are thought to be involved in T cell maturation. Epithelial cells with internalized lymphocytes have been observed in cell suspension of thymus (so-called "thymic nurse cells"):[4,6,7] these cells may represent a special type of this interaction. In sections of thymus cortex there are epithelial cells of types 2 and 3 of which the cytoplasmic exten-

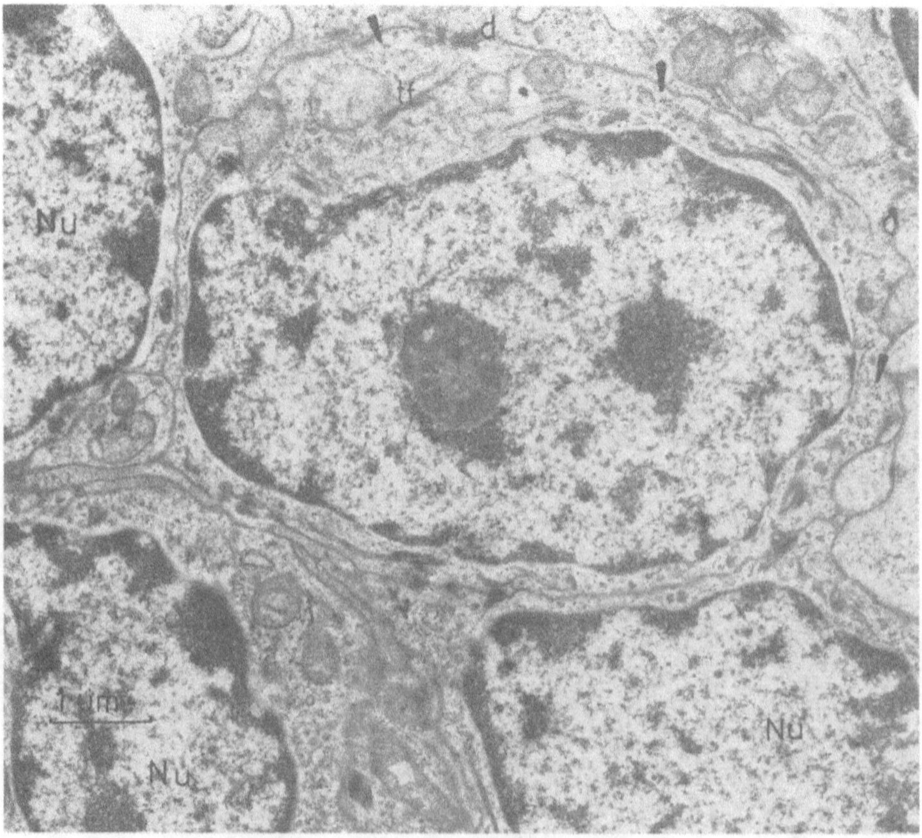

Fig. 4. Electron micrograph of a type 5 epithelial cell (4-year-old donor). For indications, see legend to Fig. 2. ▼, polyribosomes; Nu, nuclei of adjacent type 5 cells.

sions appear to surround lymphocytes completely.[4,5] These cells may be the "thymic nurse cells" found in suspension.

Finally, in type 1 epithelial cells we observed micropinocytotic vesicles facing the basal lamina (Fig. 2). The function of these vesicles may be participation in basal lamina formation, on the other hand these may be involved in secretion of factors with biological activity in T-cell maturation, e.g., chemo-attractants for precursor T cells which are usually found in the subcapsular area. Subcapsular epithelial cells synthesize thymosin α_1 and β_4, components which promote early intrathymic T lymphocyte maturation.[1]

Our study shows that there is a compartmentalization of morphologically distinct epithelial cells in the human thymus (Fig. 6). It is tempting to suggest, that this heterogeneity in the microenvironment is related with the compartmentalization of lymphoid elements in different stages of maturation.

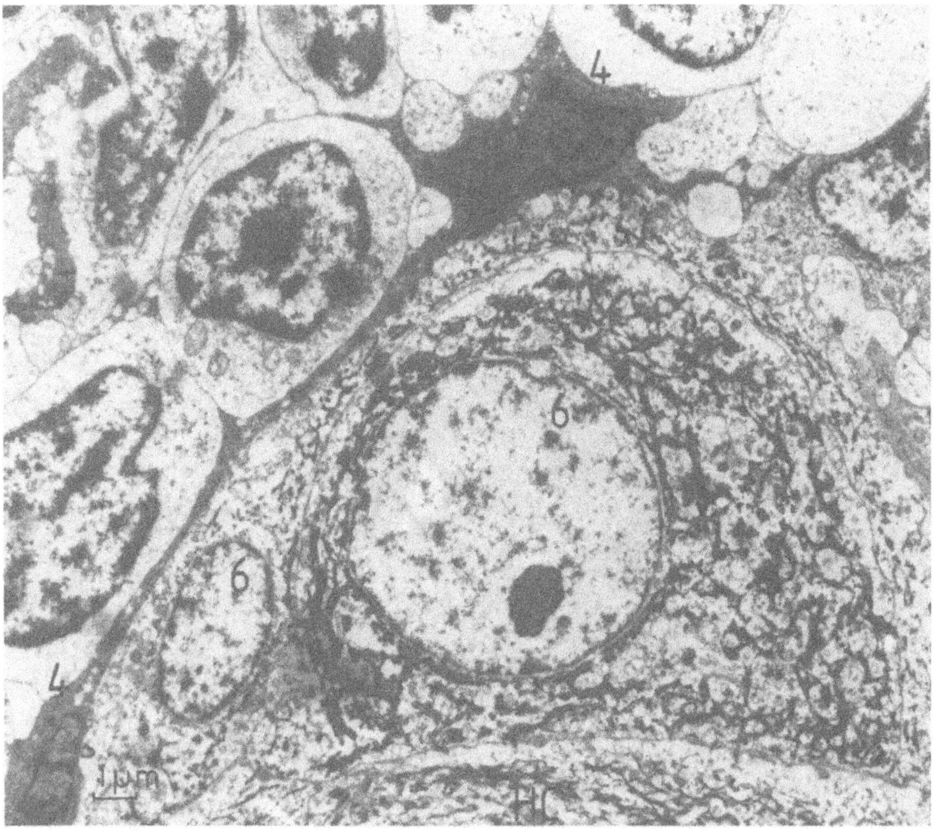

Fig. 5. Electron micrograph of a two type 6 epithelial cells (6) adjacent to a Hassall's corpuscle (HC) and two type 4 epithelial cells (4) (2-year-old donor). Note the slender cytoplasmic extensions of the type 4 cells.

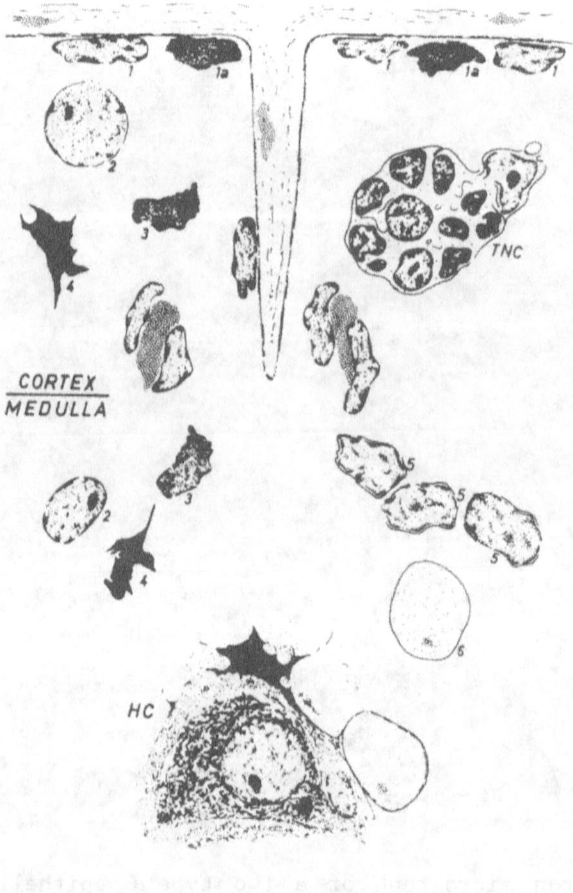

Fig. 6. Diagrammatic representation of types of epithelial cells in the human thymus. TNC, thymic nurse cell; HC, Hassall's corpuscle.

References

1. H.J. Schuurman, F.P. van de Wijngaert, L. Delvoye, R. Broekhuizen, J.E. McClure, A.L. Goldstein, and L. Kater, Heterogeneity and age dependency of human thymus reticulo-epithelium in production of thymosin components, Thymus in press.
2. S. Arya, E.F. Gilbert, R. Hong, and J.M.B. Bloodworth Jr., The thymus, in: "Endocrine pathology," J.M.B. Bloodworth, ed., Williams and Wilkins, Baltimore, chapter 19 (1982).
3. J. Singh, The ultrastructure of epithelial reticular cells, in: "The thymus gland," M.D. Kendall, ed., Academic Press, London, p. 133 (1981).
4. F.P. van de Wijngaert, L.H.P.M. Rademakers, H.-J. Schuurman, R. A. de Weger, and L. Kater, Identification and in situ localization of the "Thymic Nurse Cell" in man, J. Immunol. 130: 2348 (1983).
5. F.P. van de Wijngaert, M.D. Kendall, H.-J. Schuurman, L.H.P.M. Rademakers, and L. Kater, Heterogeneity of epithelial cells in the human thymus. An ultrastructural study, Cell Tissue Res. 237: 227 (1984).
6. H. Wekerle, U.-W. Ketelsen, and M. Ernst, Thymic nurse cells. Lymphoepithelial cell complexes in murine thymuses: morphological and serological characterization. J. exp. Med. 151: 925 (1980).
7. M.A. Ritter, C.A. Sauvage, and S.F. Cotmore, The human thymus microenvironment: in vivo identification of thymic nurse cells and other antigenetically distinct subpopulations of epithelial cells, Immunology 44: 439 (1981).

T CELL DIFFERENTIATION WITHIN TNC RELATED LYMPHO-EPITHELIAL COMPLEXES IN THE THYMUS CORTEX

W. Leene, R. de Waal Malefijt, P.J.M. Roholl* and K.A. Hoeben

Department of Histology and Cell Biology, University of Amsterdam, Academic Medical Center, Meibergdreef 15 1105 AZ Amsterdam and *Department of Pathology State University Utrecht, The Netherlands

INTRODUCTION

The isolation and characterization of thymic nurse cells (TNC) (1,2) has renewed the interest in microenvironmental events in intrathymic T cell maturation, and may contribute to detect the hitherto hidden intrathymic pathways of T cell differentiation for which many mutually exclusive models (for a review see ref.3) have been presented so far.

For a number of reasons thymic nurse cells are attractive candidates to play a specific role in T cell maturation events. They are predominantly located in the thymus cortex (2), they constitute with their caveolae a micro-environment completely separated from the extracellular space (1), the TNC caveolar membranes expose class I and II MHC determinants fit for the selection of MHC-receptor bearing prethymocytes or imposing MHC restriction specificities on developing T cells (4), and finally, the high rate of thymocyte proliferation (1) coupled to an extensive cell death (5) within TNC is suggestive for the existence of a selection mechanism operating in TNC on developing T cells (6).

In the present paper we will first consider the much disputed *in situ* existence of TNC, using a cortisone induced lymphocyte depletion in the cortical thymic tissue and isolating according to Wekerle (1) TNC from the depleted tissue. The results of this approach confirm the relation between TNC and preexisting lympho-epithelial complexes (LEC) *in situ* in the thymus cortex suggested earlier by other authors (2,7,8). Furthermore we will demonstrate that cortical cortisone-resistant thymocytes (CRT)-a thymocyte category (9) receiving not much attention so far - are present within the cortical LEC. Finally we will conclude from investigations on the phenotype of thymocytes within LEC and TNC before and after cortisone treatment that cortisone-resistance and a phenotype shift indicative for T cell maturation is imposed on thymocytes developing within the TNC related LEC in the thymus cortex.

MATERIALS AND METHODS

Female C57Bl/10 mice were used at 6 - 8 weeks. Cortisone treatment was achieved with one single but high dose (125 mg/kg b.w.) i.p. injection of dexamethasone-21-phosphate (Organon-Oss, Holland) at 6-48 hrs before sacrifice. Thymic nurse cells were isolated using the method of Wekerle (1) with some minor modifications (2,7). Monoclonals anti-Ia and anti-H-2 K/D (Camon C 9012 and C 9003), anti-Thy 1.2 (Cedarlaine Cl 8600-A) and anti-Lyt-1 and anti-Lyt-2 (from clones 53-7.3 and 53-6.7 respectively, Ledbetter and Herzenberg, 1979; Immunol. Rev. 47, 63) and the lectin PNA (PNA-Fitc from E & Y Laboratories) were used in labeling experiments. Labeling of TNC caveolar membranes and thymocytes within TNC was realized by perforating lighly fixed (0.05% glutaraldehyde) TNC with 0.04% Triton X-100. Frozen thymus sections briefly fixed with acetone, intact and perforated TNC in wet and cytospin preparations were incubated with proper dilutions of the monoclonals and PNA. Anti-Ia, anti-H-2 K/D and anti-Thy 1.2 were biotinylated and detected with a second layer of Avidin-D-Fitc (Vector), Anti-Lyt-1 and anti-Lyt-2 antibodies were detected with rabbit-anti-rat-IgG-(H+L)-Fitc or - peroxydase in case of cytospin preparations. For light and electron microscopic investigation thymic tissue fragments and isolated TNC were processed as described earlier (9).

RESULTS

In vitro TNC are derived from preexisting in situ LEC

Using the thymic nurse cell (TNC) isolation procedure developed by Wekerle (1) we were able to enrich thymic cell suspensions for lympho-epithelial complexes (LEC) even in dexamethasone (DX) treated animals. Enriched suspensions were processed for electron-microscopical examination to compare *in vitro* LEC with their *in situ* counterparts. LEC isolated from thymus fragments of untreated mice show the characteristics of TNC, fully intact lymphocytes lying in a cytoplasm rich in tonofilament bundles (Fig. 1a); occasionally a different type of TNC is found in such preparations, characterized by a highly vacuolated cytoplasm in which only a few intact lymphocytes are present (Fig. 1b, LC). LEC in tissue sections as well as TNC from thymus fragments 12 h after DX injection contain lymphocytes in various stages of degradation and apoptotic bodies. At 24 h after DX, lymphocyte depletion in LEC in tissue

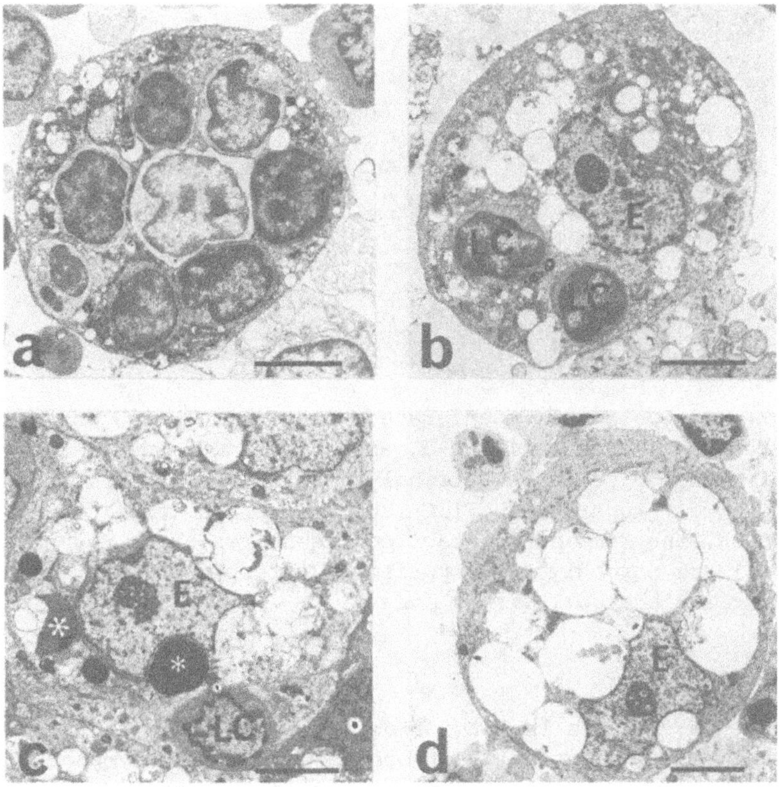

Fig. 1. a. Control TNC isolated before dexamethasone treatment. b. Control TNC partially depleted under physiological conditions, note the intact lymphocytes (LC) in the vacuolated cytoplasm. c. Dexamethasone (24h) induced lymphocyte depletion in a cortical lympho-epithelial complex in tissue section, in the vacuolated cytoplasm remnants of apoptotic bodies (asteriks) and a single intact lymphocyte (LC) are present. d. TNC isolated from thymic tissue 24 h after dexamethasone treatment, lymphocyte depletion is complete, leaving a highly vacuolated cytoplasm. The nuclei of the epithelial cells are easily recognized (E) by their outstanding nucleolus and characteristic heterochromatin distribution. Bar 5 µm.

sections (Fig. 1c) as well as in TNC isolated at this stage (Fig. 1d) is complete, leaving a highly vacuolated cytoplasm around the nuclei of the epithelial cells (Fig. 1c and d, E) (immunofluorescence studies reveal the expression of Ia as well as H-2K/D antigens on the TNC vacuolar membrane system). From these experiments, in which LEC *in situ* are visualized by a DX induced lymphocyte

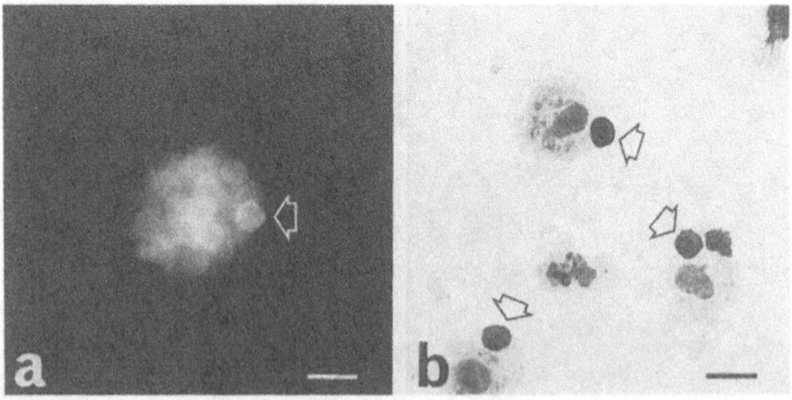

Fig. 2. a. Control TNC perforated with Triton X-100 and stained with the monoclonal anti-Lyt-1, among the predominantly weakly positive thymocytes a few strongly positive cells are present (arrow) (wet preparation, FITC-labeled). b. TNC after 24 h dexamethasone, the remaining intact cortisone-resistant thymocytes (arrows) are strongly Lyt-1 positive (cytospin preparation, peroxydase-DAB-labeled). Bar 20 µm.

depletion and TNC are isolated from such tissue displaying morphological features identical to those of the *in situ* lympho-epithelial complexes, it is concluded that *in vitro* TNC are derived from preexisting *in situ* LEC.

Cortisone-resistant lymphocytes in TNC and LEC

As reported earlier in the rabbit (9), also in the mouse thymus a small number of cortical lymphocytes were found to be resistant to DX, to be indicated as cortisone-resistant thymocytes (CRT). Analysis of samples of semi-thin sections of the thymus cortex at 24-48 h intervals after DX injection revealed that 30-40% of the LEC sections examined contain one or more CRT. Taking into account the probability of hitting a single lymphocyte in a one micron thick section of a lympho-epithelial complex, it is obvious that most LEC contain one or more CRT. An example of a CRT containing LEC is given in Fig. 1c (LC). In cytospin preparations of whole TNC from 24 h DX treated mice one or more CRT within each individual TNC was observed (Fig. 2b, arrows). From these data it is concluded that lymphocytes within LEC or TNC are predominantly cortisone-sensitive, but more interesting, that within the LEC/TNC micro-environment a minority of CRT is invariably present. Most intriguing is the observation that apparently partially depleted TNC can also be derived from control thymic tissue (Fig. 1b), indicating that a selective mechanism may operate also under physiological conditions, although at a much lower level than under the experimental conditions of extremely high cortisone exposition.

Surface phenotype of lymphocytes in TNC and LEC

To investigate the differentiative capacity of LEC and TNC microenvironments, the presence of a number of relevant T cell differentiation markers on the surface of thymocytes within individual LEC and TNC was studied. TNC were perforated with Triton X-100 to give monoclonals and lectin access to the surface of the intra-epithelial thymocytes. Peanut agglutinin (PNA) binding - until recently considered to be the most valuable feature to distinguish cortical and medullary thymocytes in the mouse system - was varying among the lymphocytes in TNC, the majority being strongly to weakly PNA-positive, a minority being PNA-negative.

The presence of Thy 1.2, Lyt-1 and Lyt-2 markers was studied on thymocytes in LEC in frozen sections as well as on thymocytes in isolated TNC. The frozen section data are in agreement with those presented earlier by van Ewijk (10), dexamethasone treatment (24 h) leaving only weakly Thy 1.2-positive, strongly Lyt-1 positive and no Lyt-2-positive lymphocytes in the otherwise depleted cortex. (It was observed that 48 h after DX strongly Thy 1.2-positive lymphocytes reappear in the outer cortex). These Thy 1.2^{+} Lyt-1^{+++} Lyt-2^{-} thymocytes constitute the pool of cortical cortisone-resistant (CRT) thymocytes. In the previous section it was argued that a considerable number of these cortical CRT is present within LEC.

More direct evidence was obtained on the phenotype of thymocytes in TNC. Thy 1.2 expression on thymocytes in individual TNC varied from weakly to strongly positive, Lyt-1 expression from

Table 1. Cortisone-sensitivity and Surface Phenotype of TNC Thymocytes

Marker	Thymocytes within TNC	
	Majority	Minority
Cortisone-sensitivity	Cortisone-sensitive	Cortisone-resistant
PNA receptors	PNA^{+}	PNA^{-}
Thy-1 antigen	Thy-1^{+++}	Thy-1^{+}
Lyt-1 antigen	Lyt-1^{+}	Lyt-1^{+++}
Lyt-2 antigen	Lyt-2^{+}	Lyt-2^{-}

- negative, + positive, +++ strongly positive

positive to strongly positive, viz. only a few thymocytes per TNC showing a high Lyt-1 antigen density on their surface (Fig. 2a, arrow), and Lyt-2 expression was positive with little variation in individual TNC. Depleted TNC (24 h DX) were studied in cytospin preparations using peroxydase-labeling. It is shown that the number of thymocytes within TNC is greatly reduced after dexamethasone treatment and, moreover, that the surviving intraepithelial lymphocytes are strongly positive for the Lyt-1 marker (Fig. 2b, arrows) and negative for Lyt-2 (not shown).

From these data, summarized in Table 1, it is concluded that there is a remarkable heterogeneity in surface phenotypes of lymphocytes in TNC and LEC related to cortisone-sensitivity, indicating that a phenotype shift, if not a functional maturation, may be imposed upon thymocytes residing in the TNC/LEC microenvironment by the stromal (epithelial) element in such complexes.

DISCUSSION

Thymic nurse cells (TNC) are large epithelial cells, filled with high numbers of lymphocytes (1) derived from and related to preexisting <u>in vivo</u> associations between immature thymocytes and cortical epithelial cells (LEC) (2), the precise nature of the preexisting LEC was, however, not known thus far.

With a method based on the 'all or none' effect of glucocorticoids upon lymphocytes (11) a large scale lymphocyte depletion was induced in the thymus cortex. Surprisingly, in most cortical regions examined apoptotic cells (i.e. cells undergoing morphological changes of a selective death) were not randomly distributed but present in circumscribed areas surrounding epithelial cell nuclei. After prolonged glucocorticoid exposition lymphocyte depletion is complete, leaving behind the vacuoles in which the lymphocytes were lodging. We were able to isolate highly vacuolated TNC from the depleted tissue, thus confirming the notion (2,7,8) that TNC are derived from preexisting LEC, which, according to our data are well defined cortical structures.

The presence of cortisone-resistant thymocytes (CRT) in the thymus cortex, the significance of which is hardly understood hitherto, comes into another light, now they appear to be the phenotypically more mature (Thy 1.2^{+} Lyt-1^{+++} Lyt-2^{-}) thymocytes within the LEC from which TNC are derived.

Finally it should be noticed that other authors (2) did not find indications for the existence of TNC subpopulations, which may be the result of overlooking the subpopulation of more mature CRT which constitutes only a small minority in a total population of TNC-derived thymocytes; furthermore it should be noticed that one should be cautious in interpreting any result of phenotype studies on TNC thymocytes considering the susceptibility of some surface antigens on thymocytes towards the enzymatic procedures used in TNC isolation. By studying thymocyte phenotypes in LEC in frozen sections as well as in isolated TNC, we tried to overcome such difficulties.

REFERENCES

1. H. Wekerle, U-P. Ketelsen, and M. Ernst, Thymic Nurse Cells, J. Exp. Med. 151:925 (1980).
2. B.A. Kyewski, and H.S. Kaplan, Lymphoepithelial interactions in the mouse thymus phenotypic and kinetic studies on thymic nurse cells, J. Immunol. 128:2287 (1982).
3. R. Scollay, Intrathymic events in the differentiation of T lymphocytes: A continuing enigma, Immunol. Today 4: 282 (1983).
4. R. de Waal Malefijt, W. Leene, and P.J.M. Roholl, T cell maturation within the TNC microenvironment (submitted).
5. W. Leene, R. de Waal Malefijt, P.J.M. Roholl, and K.A. Hoeben, Lymphocyte depletion in the mouse thymus: A tool to identify *in situ* Thymic Nurse Cells (submitted).
6. N.K. Jerne, The somatic generation of immune recognition, Eur. J. Immunol. 1:1 (1971).
7. M.A. Ritter, C.A. Sauvage, and S.F. Cotmore, The human thymus microenvironment: *in vitro* identification of thymic nurse cells and other antigenically distinct subpopulations of epithelial cells, Immunol. 44:439 (1981).
8. F.P. van de Wijngaert, L.H.P.M. Rademakers, H.J. Schuurman, R.A. de Weger, and L. Kater, Identification and *in situ* localization of the "thymic nurse" cell in man, J. Immunol. 130:2348 (1983).
9. P.J.M. Roholl, W. Leene, and J.W.M. Visser, T cell differentiation in the rabbit I. Cytofluorometric analysis of a T cell antigen on lymphocytes from normal and dexamethasone treated animals, Thymus 5:153 (1983).
10. W. van Ewijk, P.L. van Soest, and G.J. van den Engh, Fluorescence analysis and anatomic distribution of mouse T lymphocyte subsets defined by monoclonal antibodies to the antigens Thy-1, Lyt-1, Lyt-2, and T-200, J. Immunol. 127:2594 (1981).
11. J.J. Cohen, and R.C. Duke, Glucocorticoid activation of a calcium-dependent endonuclease in thymocyte nuclei leads to cell death, J. Immunol. 132:38 (1983).

THE RELATION BETWEEN THYMIC INTERDIGITATING CELLS AND DENDRITIC CELLS

E.W.A. Kamperdijk[1], M. van den Berg[1], M. Kapsenberg[2]
A.M. Duijvestijn[1] and E.C.M. Hoefsmit[1]

[1]Department of Electron Microscopy, Medical Faculty
Free University, Amsterdam, The Netherlands
[2]Laboratory of Histology and Cell Biology, Academic
Medical Center, University of Amsterdam, Amsterdam
The Netherlands

INTRODUCTION

In the thymus histologically and functionally distinct areas can be distinguished. The cortex predominantly comprises closely packed proliferating thymocytes. The medulla is characterized by the presence of immunocompetent T cells intermingled with interdigitating cells (IDC) (1). These IDC sometimes show distinct signs of phagocytic activity suggesting that they belong to the mononuclear phagocytes (2). However, it can not be excluded that they belong to a separate population of dendritic cells (DC) recently isolated from rat thymus (3).*

To study the relation between thymic IDC in situ and DC in suspension we characterize and compare these cells on morphological-, enzyme- and immunohisto(cyto)chemical level.

METHODS

Male Wistar rats, 2 months old and weighing 180-200 g were used.

*These DC function both as accessory cells in mitogenic responses and as stimulators in a mixed leucocyte reaction (3).

IDC In Situ

For light microscopy a combined Ia and acid phosphatase (APh) staining was carried out on cryosections (fixed 10' in aceton). Ia labelling was performed with an indirect immunoperoxidase method, for APh staining the method according to Burnstone (Naphthol-AS-BI phosphate as substrate) (4) was used. For electron microscopy perfusion fixation with 1.5% glutaraldehyde and postfixation with 1% OsO_4 was carried out.

DC In Suspension

The isolation was performed by a modified procedure according to Klinkert et al. (5) using Ficoll gradients. For light microscopy

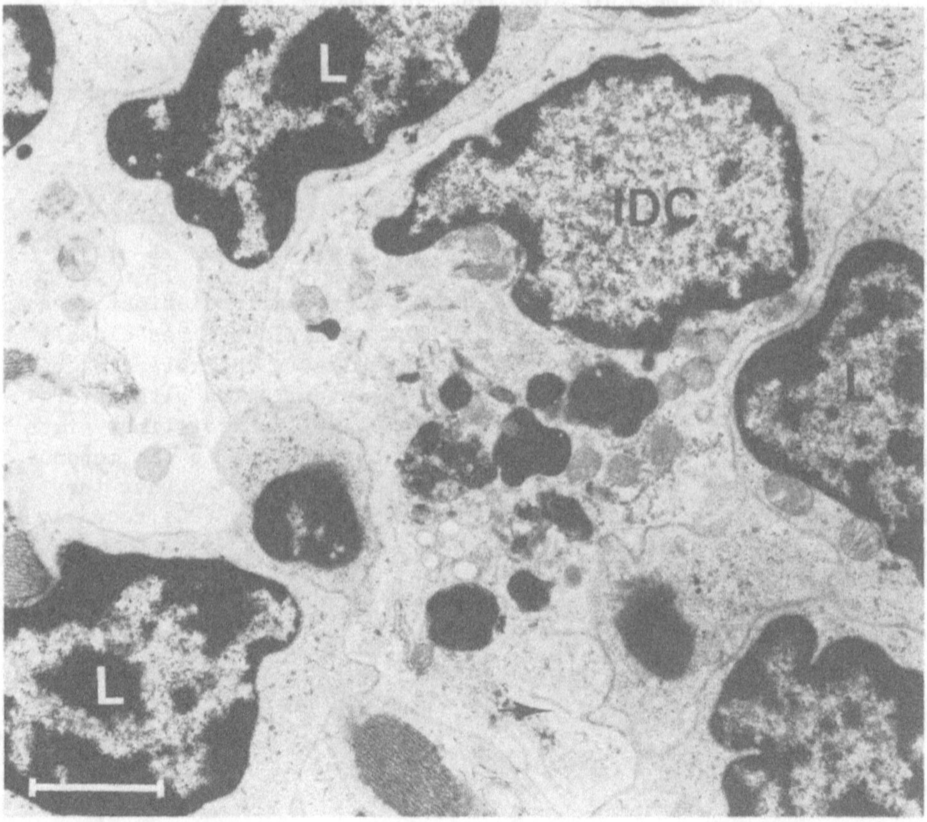

Fig. 1. Phagocytic IDC in thymic medulla. The cytoplasm is electron lucent, the cell membrane shows tubular membrane invaginations (arrow). L = lymphocyte. Bar 2.7 μm

a combined Ia and APh staining was performed on cytocentrifuge preparations. For electron microscopy the cells were fixed in 1.5% glutaraldehyde and postfixed with 1% OsO_4. Ia labelling on ultramicroscopical level was performed with an indirect immunoperoxidase method (the cells were fixed in 0.5% glutaraldehyde).

RESULTS

IDC In Situ

According to Duijvestijn et al. (1) well developed IDC were only present in the well vascularized cortico-medullary region (CMR) and in the medulla. The cells showed APh activity in a spot

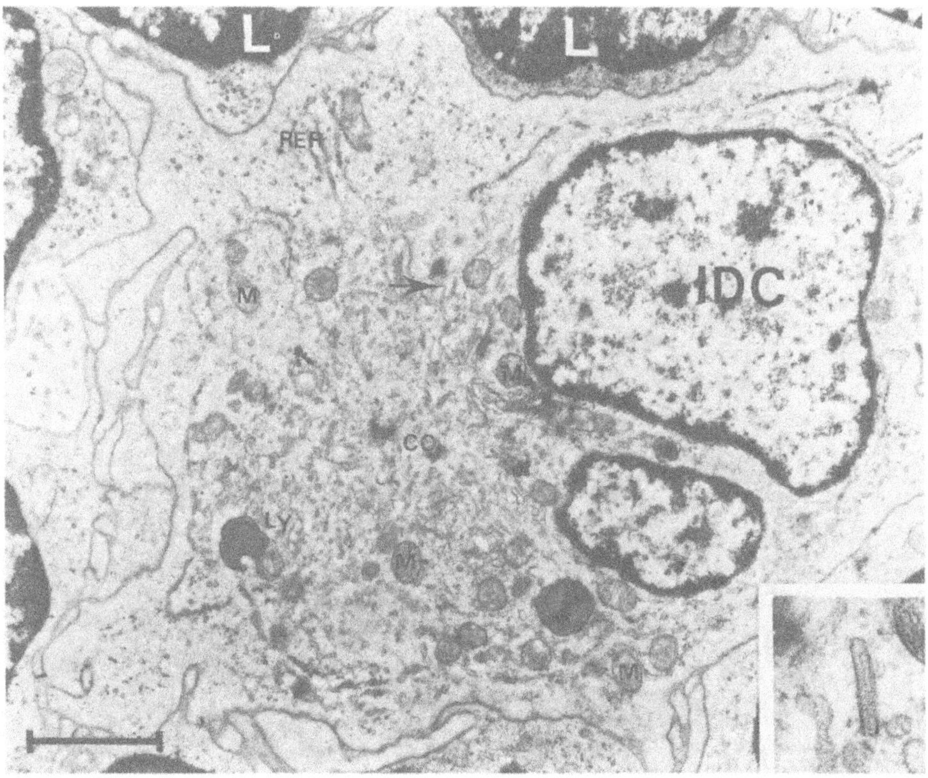

Fig. 2. IDC in the medulla. The electron lucent cytoplasm shows a cytocentre (cc), mitochondria (m), rough endoplasmic reticulum (rer), lysosomes (ly) and Birbeck granules (arrow). L = lymphocyte. Bar 2.1 µm. Inset: detail Birbeck granulum.

near the nuclei. Moreover, they all were positive for Ia on the cell membrane. The cells had a voluminous pale cytoplasm with cell processes interdigitating with surrounding thymocytes. The euchromatic nuclei were frequently indented and eccentrically situated. They usually had a varying number of small- and medium sized phagolysosomes in the central area of the cell. In some cases they showed phagocytic activity as mononuclear phagocytes (Fig. 1). About 60% of the IDC contained Birbeck granules (Fig. 2), the characteristic cell organelles of the epidermal Langerhans cell. These granules had a rod-like profile with rounded ends and a limiting membrane approximately 60 Å thick and an inside cristalloid structure.

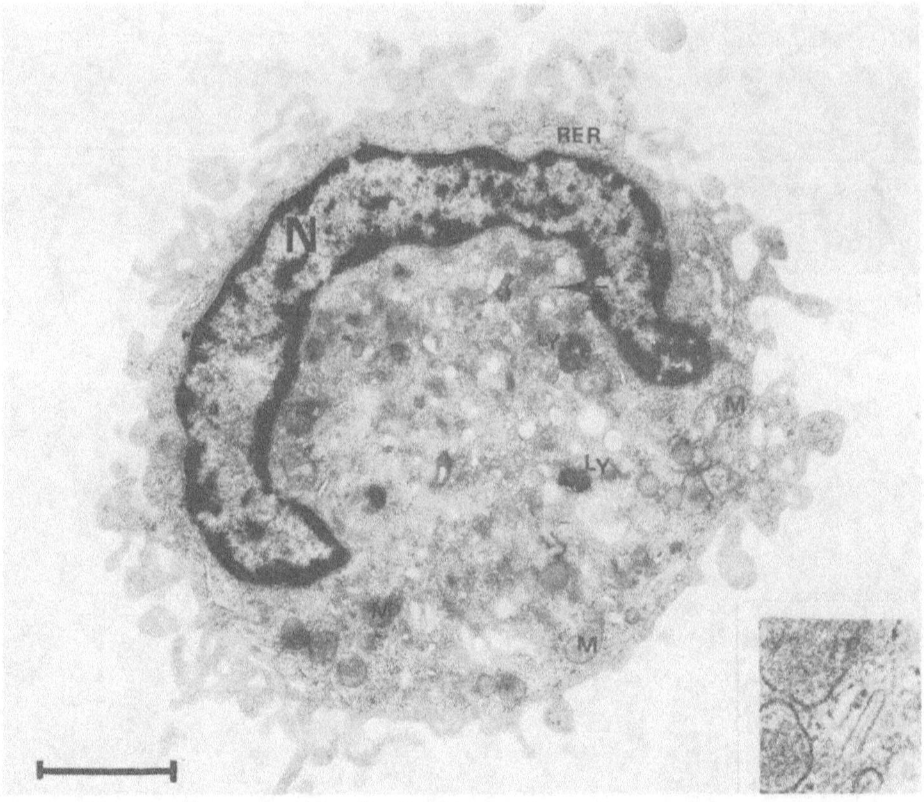

Fig. 3. DC, showing an irregular outline. The nucleus (N) is eccentrically situated, the cytoplasm contains mitochondria (m), rough endoplasmic reticulum (rer), lysosomes (ly) and Birbeck granules (arrow). Bar 3.2 µm. Inset: detail Birbeck granulum.

DC In Suspension

DC, depleted from FcR-positive cells by rozetting, showed many characteristics of the IDC: they had an irregular outline and the nuclei were eccentrically situated (F9g. 3). All of them had APh activity in a central spot and were strongly positive for Ia on the cell membrane (Fig. 4). Moreover, about the same percentage of these cells contained Birbeck granules as IDC in situ (Fig. 3 + 4). However, they never showed any signs of phagocytic activity.

DISCUSSION AND CONCLUSION

Thymic IDC in situ show many characteristics with DC in suspension with respect to morphology, enzyme- and immunohisto(cyto)-chemistry such as a) an irregular outline, b) the presence of an

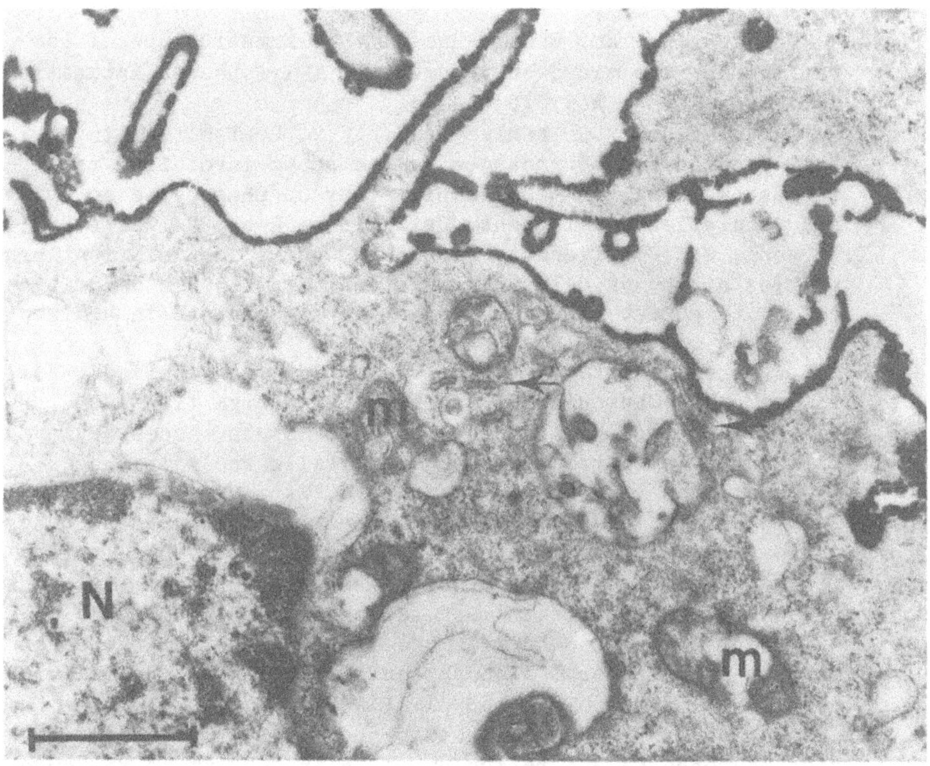

Fig. 4. Part of a DC, the cell membrane is strongly Ia positive. The cytoplasm shows Birbeck granules (arrows). N = nucleus, m = mitochondrium. Bar 1.2 μm

irregularly shaped, eccentrically situated nucleus, c) APh activity in a central spot near the nucleus, d) the presence of Ia antigen on the cell membrane and e) the presence of Birbeck granules.

Moreover, there are strong indications that both IDC (6) and DC (7) originate from precursors in the bone marrow. All these aspects, especially the presence of Birbeck granules, unequivocally show a relation between these cell types. However, it has been demonstrated that DC are non-phagocytic, possibly due to the absence of FcR receptors, and that IDC may show phagocytic activity, especially after sublethal thymus irradiation (2). Phagocytic activity of IDC has also been described in rat lymph nodes after antigenic stimulation. This means that under certain circumstances IDC can phagocytize like mononuclear phagocytes.

From these results it is concluded that DC are the in vitro equivalent of only the FcR-negative non-actively phagocytizing population of IDC.

REFERENCES

1. A.M. Duijvestijn, and E.C.M. Hoefsmit, Ultrastructure of the rat thymus: the micro-environment of T-lymphocyte maturation, Cell Tissue Res. 218:279 (1981).
2. A.M. Duijvestijn, Y.G. Köhler, and E.C.M. Hoefsmit, Interdigitating cells and macrophages in the acute involuting rat thymus. An electron microscopic study on phagocytic activity and population development, Cell Tissue Res. 224:291 (1982).
3. T.W. Wong, W.E.F. Klinkert, and W.E. Bowers, Immunological properties of thymus cell subpopulations: rat thymic dendritic cells are potent accessory cells and stimulators in a mixed leucocyte culture, Immunobiol. 160:413 (1982).
4. A.G.E. Pearse, "Histochemistry: theoretical and applied", Vol. I, 3rd. ed., Churchill Livingstone, Edinburgh (1968).
5. W.E.F. Klinkert, J.H. LaBadie, and W.E. Bowers, Accessory and stimulating properties of dendritic cells and macrophages isolated from various rat tissues, J. Exp. Med. 156:1 (1982).
6. A.M. Duijvestijn, and A.N. Barclay, Identification of the bone marrow derived Ia positive cells in the rat thymus. A morphological and cytochemical study, J. Leucocyte Biol., in press.
7. W.E. Bowers, M. Berkowitz, W.E.F. Klinkert, and J.H. LaBadie, Differentiation of dendritic accessory cells from rat bone marrow and peripheral blood precursors cultured in conditioned medium, Immunobiol. 163:112 (1982).
8. E.W.A. Kamperdijk, J.H.S. de Leeuw, and E.C.M. Hoefsmit, Lymph node macrophages and reticulum cells in the immune response. The secondary response to paratyphoid vaccine, Cell Tissue Res. 227:277 (1982).

"T-CELLS" IN NUDE RATS

L.M.B. Vaessen, R. Broekhuizen, J.G. Vos, H.-J. Schuurman and J. Rozing

Department of Internal Medicine, Erasmus University Rotterdam; Division of Immunopathology, Departments of Internal Medicine and Pathology, University Hospital Utrecht; and Laboratory for Pathology, National Institute for Public Health and Environmental Hygiene, Bilthoven, The Netherlands

There are several reports that congenitally athymic nude mice and rats have cells with T-cell markers in spleen and lymph nodes. For the nude mouse, the numbers of these cells increase with age.[1] In athymic rats we found cells with T-cell markers in T-dependent areas of mesenteric lymph node and Peyer's patches, especially at ages older than 35-40 weeks.[2] This observation prompted us to study in detail the presence of cells with T-cell phenotypes in lymphoid organs of nude rats as a function of age. Lymphoid organs of 8, 12, 20, 36 and 68-week-old nude rats were analysed by both flow cytometry (FACS) on cell suspensions and by immunohistochemistry on tissue sections. A panel of monoclonal antibodies to T-cell subpopulations was applied. The proliferative response after mitogenic stimulation was simultaneously investigated. In these assays athymic nude rats were compared with age- and sex-matched immunocompetent littermates.

Materials and methods

Rats. The *rnu* gene was derived from an athymic hooded rat strain obtained from the MRC Laboratory Animals Centre, Carshalton, England. Homozygous *rnu/rnu* males were mated with heterozygous +/*rnu* females and backcrossed to the WAG/RIV ($RT-1^u$) strain for 15 generations. The animals were bred in SPF conditions and nude animals were maintained after weaning in laminar flow units. At each age point, 4 male and 4 female *rnu/rnu* rats were compared with 4 male and 4 female +/*rnu* immunocompetent littermates. As there were no sex differences in the results of all assays, data from both sexes were combined. For cell suspension analysis by FACS, spleen cells from 2 animals were pooled (at each age point 4 determinations for rnu/rnu rats and 4 determinations for +/rnu controls). Similarly, cervical lymph node suspensions from 2 animals were pooled, and for mesenteric lymph nodes pools from 4 animals were prepared. In immunohistochemistry and mitogen responsiveness each individual rat was analysed separately. The results of cervical and mesenteric lymph nodes did not differ significantly, and results for both lymph node sources are combined in data presentation.

Antisera. The following monoclonal antibodies were applied:

ER-1 recognizes the same molecule as the W3/13 antibody and binds to all T-cells, thymocytes, hematopoietic precursor cells and mature myeloid cells[3] (this issue, p 17),

ER-2 recognizes the same molecule as the W3/25 antibody and binds[4] to T-helper cells, thymocytes and macrophages (this issue, p 235),

MRC OX-8 reacts with T-nonhelper cells (i.e., cytotoxic and suppressor cells)[4] and with Natural Killer (NK) cells,[5]

ER-4 binds to thymocytes, pro-thymocytes and probably a small part of pre-B cells (this issue, p 251),

MRC OX-19 binds to all T-cells and thymocytes.[6]

OX-8 and OX-19 were kindly provided by Dr A.N. Barclay and Dr A.F. Williams, Oxford, England.

FACS analysis and immunohistochemistry. The isolation of cells in suspension, fluorescence staining and FACS analysis are described elsewhere (this issue, p 251). In cell suspension analysis purified monoclonal antibodies conjugated to fluorescein isothiocyanate (FITC) were applied. In immunohistochemistry on frozen tissue sections purified antibodies were applied in an indirect immunoperoxidase technique using horseradish peroxidase conjugated rabbit anti-mouse immunoglobulin (Miles) in the second incubation. Staining of the peroxidase was performed with diamino benzidine tetrahydrochloride and H_2O_2 as substrate. Sections were counterstained with hematoxylin.

Mitogen responsiveness. $5x10^5$ cells were cultured with 2, 4, and 8 µg/ml Phytohemagglutinin (PHA, HA 15, Wellcome, Beckenham, England) or with 2.5, 5 and 10 µg/ml Concanavalin A (Con A, A grade, Calbiochem-Behring, La Jolla, Ca.) for 24 and 48 h. Proliferative responses were measured by ^{3}H-thymidine incorporation; results are expressed in counts per minute (cpm). Data presented are from optimal culture conditions (8 µg/ml PHA, 48 h). The response of unstimulated cultures was negligible.

Results

FACS analysis. In spleen and lymph nodes from +/rnu immunocompetent rats, there was no significant difference in percentages of cells with ER-1, ER-2, OX-8 or OX-19 expression between different age points (Fig 1). For ER-4 expression an age-dependency was ob-

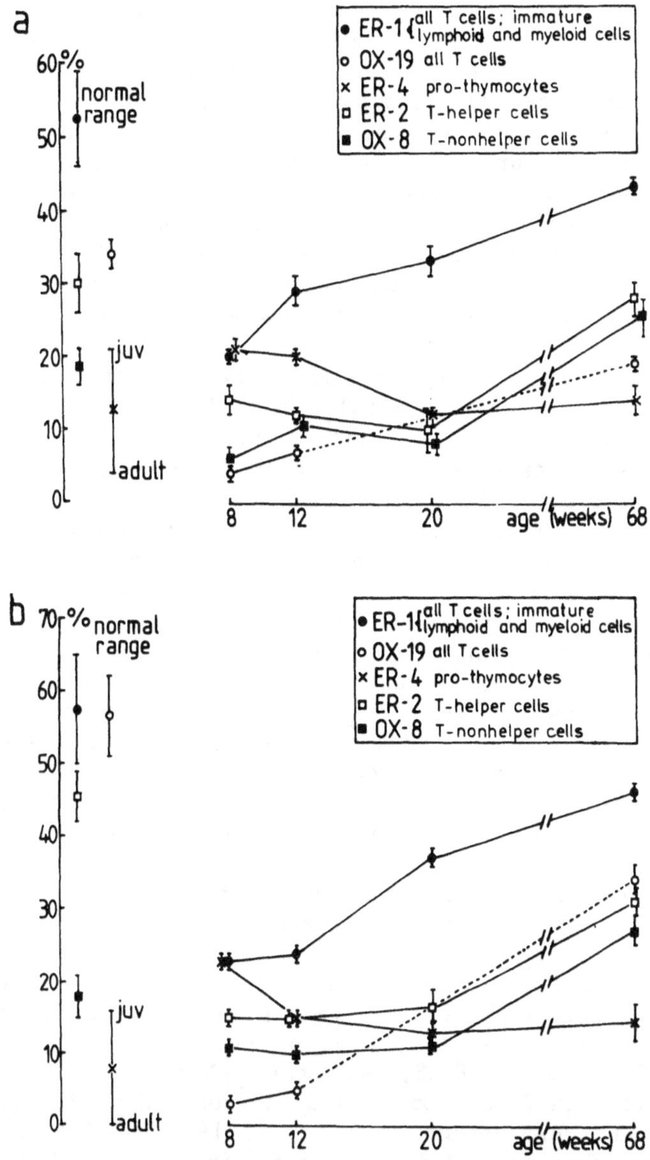

Fig. 1. Phenotype expression on lymphocytes in cell suspension from spleen (a) and lymph nodes (b) of athymic nude rats at various ages. The normal range in +/rnu littermates is presented at the left side. Data figured are the geometric mean percentages ± SEM of all viable nucleated cells (at least 8 experiments for each data point). All monoclonal antibodies were conjugated to FITC.

served. In 8-week-old animals the highest percentages were found (21% in spleen and 16% in lymph node suspensions); lower percentages to a complete absence were seen in older animals. In spleen suspension the percentage of ER-1 positive cells was higher than the percentages positive cells found with the other antibodies. This phenomenon can be ascribed to binding of ER-1 antibody to other cell types than T-cells, such as cells of the myeloid lineage, hematopoietic precursor cells and plasma cells.

In spleen and lymph nodes from rnu/rnu rats the percentages of cells with expression of ER-1 and OX-19 revealed an age-associated increase (Fig. 1). For OX-8 and ER-2 positive cells, the values in animals at ages of 8, 12 and 20 weeks were about the same, but higher percentages were found in 68-week-old animals. The highest percentages of $ER\text{-}4^+$ cells were observed in 8-week-old animals (21% in spleen and 23% in lymph node); in animals of older ages lower values were seen (about 12-15%). This observation is in contrast with the absence of $ER\text{-}4^+$ cells in +/rnu rats at a higher age. Since the ER-4 marker is present on immature T-cells, we evaluated the relation between ER-4 expression and that of other T-cell markers by double immunofluorescence (incubation with a combination of antibodies). The data of spleen cells from 8-week-old and 68-week-old rnu/rnu animals are presented in Table 1. Based on these double immunofluorescence results we calculated the percentages of cells with different composition of phenotypes (Table 1). In 8-week-old animals 5% of spleen cells carried only the ER-4 marker; this cell subpopulation was not found in 68-week-old animals. On the other hand, there were 8% $ER\text{-}4^-OX\text{-}19^+$ cells in suspension from 68-week-old animals; this subpopulation was not present in 8-week-old nude rats. Furthermore, cells positive for ER-4 in combination with OX-19, ER-2 and OX-8 were observed: in 8-week-old animals 3% of spleen cells carried this $ER\text{-}4^+OX\text{-}19^+ER\text{-}2^+OX\text{-}8^+$ phenotype, whereas 8% of spleen cells from old rats showed this phenotype expression.

Immunohistochemistry. In frozen spleen tissue sections from +/rnu rats, periarteriolar lymphocyte sheaths (PALS) showed a high density of cells positive for ER-1, ER-2 and OX-19; in marginal zones and follicles these cells were present in scattered location. OX-8 was present on small groups of cells or on single cells at scattered location in PALS, and some cells in the marginal zones and follicles. In PALS from 8- and 12-week-old rnu/rnu rats only a few cells with expression of ER-1, OX-19 and OX-8 were observed Fig. 2a-f); also in follicles there was occasionally a positive cell. Some cells in the marginal zones and follicles were OX-8 positive. However, there was a higher density of cells expressing ER-2 in the PALS: in marginal zones and follicles rarely an ER-2 positive cell was observed. In the red pulp there was a predominant presence of ER-1 positive cells: the positive cells probably are myeloid cells or hematopoietic precursor cells. In rnu/rnu rats of 36 and 68 weeks of age the PALS showed a higher density of cells with T-cell markers than in 8 and 12 week old animals. The density of positive cells was similar to that of +/rnu littermates. In 20-

Table 1
T-cell marker expression on spleen cells from athymic nude rats

monoclonal antibody*		percentage positive cells** 8-week-old	68-week-old
ER-4	thymocytes, prothymocytes	18	11
OX-19	all T-cells, thymocytes	3	19
ER-2	T-helper cells, thymocytes macrophages	17	24
OX-8	T-nonhelper cells, thymocytes, NK-cells	5	19
ER-4 + OX-19		18	19
ER-4 + OX-8		18	22
ER-4 + ER-2		23	27
OX-8 + ER-2		19	36
phenotype		calculated percentage	
$ER\text{-}4^+$ only		5	0
$ER\text{-}4^+OX\text{-}19^+ER\text{-}2^+OX\text{-}8^+$		3	8
$ER\text{-}4^+ER\text{-}2^+$		9	0
$ER\text{-}4^-ER\text{-}2^+$		0	16
$ER\text{-}4^+OX\text{-}8^+$		2	0
$ER\text{-}4^-OX\text{-}8^+$		0	11
$ER\text{-}4^+OX\text{-}19^+$		0	3
$ER\text{-}4^-OX\text{-}19^+$		0	8
$ER\text{-}2^+$ only		5	?
$OX\text{-}8^+$ only		0	?

* purified antibody conjugated to fluorescein isothiocyanate
** percentage of all viable nucleated cells

and 32-week-old rnu/rnu rats the densities of cells with T-cell markers in PALS were intermediate between those found in younger and in older animals.

Preliminary analysis of lymph node sections and sections of Peyer's patches (Fig. 2g,h) revealed similar results as the immunohistochemistry of spleen. In the paracortical areas of lymph nodes and interfollicular areas of Peyer's patches, there were almost no

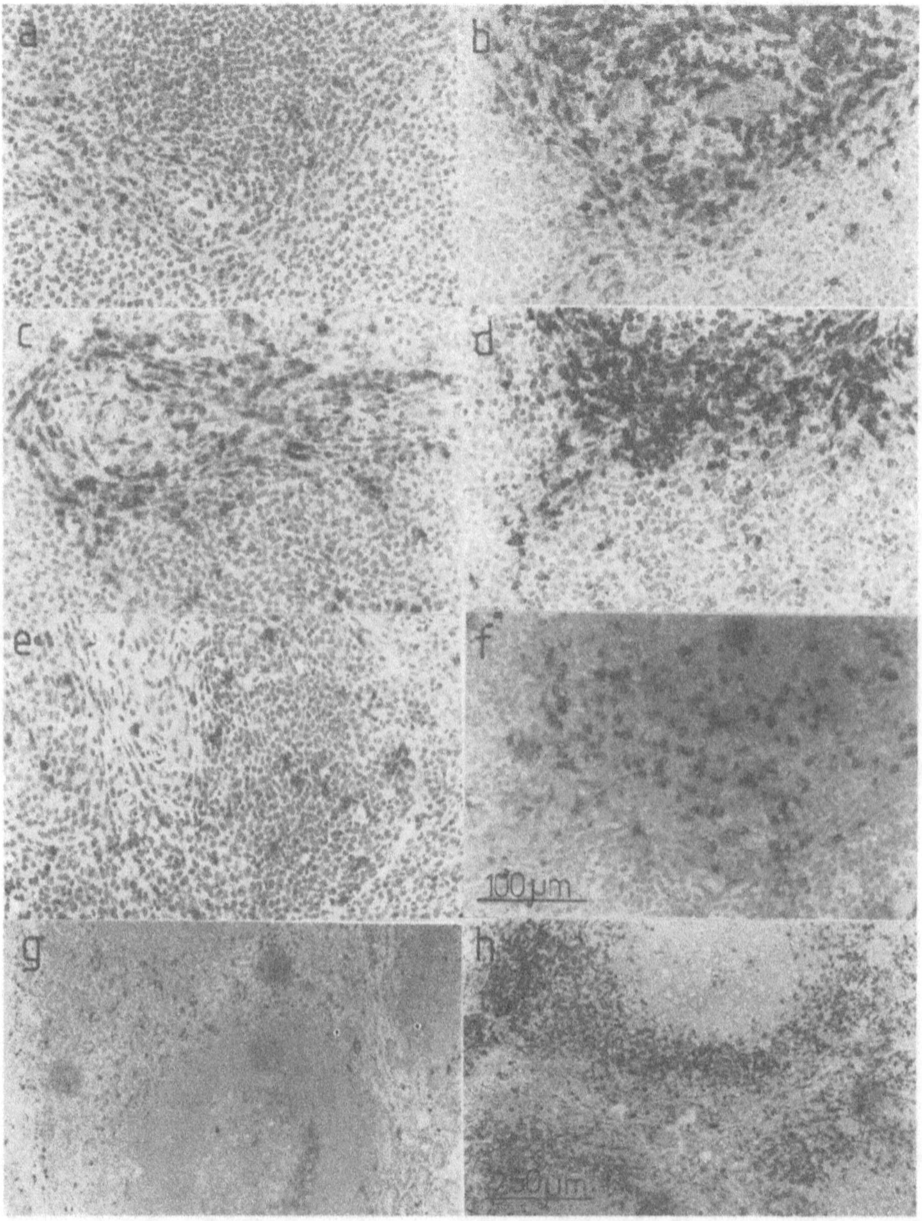

Fig. 2. Immunohistochemistry on frozen tissue sections of spleen (a-f) and Peyer's patches (g,h) from athymic nude rats at 8 (left panel, a,c,e,g) and 68 (right panel, b,d,f,h) weeks of age. Immunoperoxidase staining with monoclonal antibodies OX-19 (a,b,g,h), ER-2 (c,d) and OX-8 (e,f), counterstaining with hematoxylin.

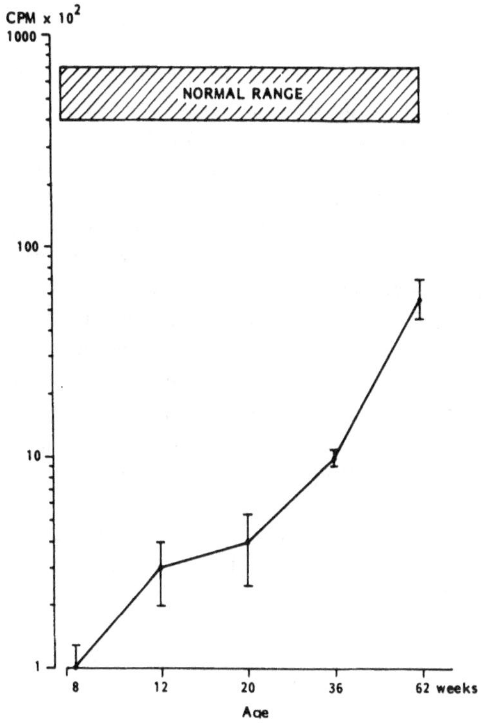

Fig. 3.
Phytohemagglutinin response of spleen cells from athymic nude rats at various ages; proliferative response (counts per minute after ^{3}H-thymidine incorporation) of $3x10^{5}$ cells after 48 h culture with 8 µg/ml PHA). Data presented are geometric mean values ± SD of 8 experiments per data point. The normal range in +/rnu rats is also presented.

cells with T-cell markers in 8- and 12-week-old rnu/rnu rats, whereas the density and phenotype distribution of 68-week-old animals in these areas resembled that of +/rnu rats. In 20- and 36-week-old animals the T-dependent areas showed a positivity intermediate between young and old animals. However, the interfollicular areas of Peyer's patches in 12- and 20-week-old nude rats showed a higher density of OX-19 positive cells than the T-dependent areas in lymph node or spleen. The same phenomenon was observed for ER-1 and ER-2 expression. In T-dependent areas of Peyer's patches and lymph nodes a small population of cells at scattered location were OX-8 positive. The density of positive cells was higher in older animals than in young animals. For lymph nodes figures resembling +/rnu animals were observed in rnu/rnu animals of 68-week-old, for Peyer's patches already in 36-week-old animals.

Mitogen responsiveness. The PHA response of spleen cells from rnu/rnu rats at various age points and the normal range in +/rnu littermates is shown in Fig. 3. In 8-week-old nude rats there was no detectable response. A low but significant response (about 0.5% of the normal range) was observed in animals at 12 and 20 weeks of age. The highest response was found in 68-week-old animals (mounting about 10% of the normal range). Similar results were observed for Con A responsiveness (data not shown). At most age points, mitogen responses in lymph node cells were comparable to those measured for spleen cells; only in 68-week-old nude rats somewhat higher resonses were observed (data not shown).

Discussion

In FACS analysis on cell suspensions and immunohistochemistry on frozen tissue sections we found an age-associated increase in cells with T-cell markers in congenitally athymic nude rats. In young animals at 8 and 12 weeks of age low percentages of cells positive for ER-2, OX-8 and OX-19 were observed in suspension analysis. Tissue sections revealed an almost absence of cells with these markers in T-dependent areas of spleen, lymph node and Peyer's patches. Higher percentages were found in animals at older ages; especially in 68- week-old animals the figures found in FACS and immunohistochemistry resembled those of heterozygous immunocompetent littermates. Both in young and old nude animals there were ER-2 positive cells which did not react with OX-19 antibody (Table 1). Since ER-2 recognizes the same membrane molecule as the W3/25 antibody, which reacts also with macrophages,[7] the ER-2$^+$OX-19$^-$ cells are probably macrophages. The ER-1 antibody, recognizing the same molecule as the W3/13 antibody (this issue, p 17), bound in young nude rats only to cells in splenic red pulp. The positive cells at this location are mainly myeloid cells and hematopoietic precursor cells.[3] The OX-8 positive cells, which were found in the red and white pulp of the spleen of nude rats irrespective of their age, did not all react with OX-19. This indicates that these cells are not T-cells. According to the binding pattern in suspension and tissue localization in follicles, marginal zones and red pulp, these OX-8$^+$OX-19$^-$ cells presumably are NK cells. The presence of NK cells in athymic nude rats in higher numbers than in normal rats,[8] and the expression of OX-8 by these cells,[5] has been documented.

The double labeling experiments revealed that the spleen of young as well as old nude rats contains a cell population with the phenotype ER-4$^+$OX-19$^+$ER-2$^+$OX-8$^+$. This phenotype resembles that of lymphocytes in the thymus cortex. In young rats this population is small, being 3% of all nucleated cells, but in old animals it accounted for 8% of the spleen cells (Table 1). In contrast to the presence of cells positive only for ER-4 in spleen from young animals (5%), there were no cells with this particular phenotype in spleen suspension of old nude rats. In addition, in spleen cell suspension of young animals a hitherto unknown and probably immature T-cell population was observed, which expressed both ER-4 and ER-2 (9% of all nucleated cells). In older nude rats this cell population is absent. In spleen suspensions of old rats 8% of cells carried the phenotype of mature T-cells, being OX-19 positive, ER-4 negative and either ER-2 or OX-8 positive. The exact percentages of ER-2$^+$OX-19$^+$ and OX-8$^+$OX-19$^+$ cells were not determined. The ER-2$^+$-OX-19$^+$ cells are most likely responsible for the Con A response in 68-week-old nude rats.

In nude animals the presence of ER-4 positive pro-thymocytes at 8 weeks of age is compatible with the situation in normal rats at 3-4 weeks of age (this issue, p 251). In normal rats these cells disappear between 8 and 12 weeks of age, but in nude animals they remain at percentages of about 15% of all nucleated cells. The results of ER-4 expression in combination with the markers ER-2, OX-8 and OX-19, seem to indicate an *in situ* differentiation in nude animals from pro-thymocytes, via cells with an ER-4$^+$OX-19$^+$ER-2$^+$OX-8$^+$ phenotype (also found on thymus cortex lymphocytes), to cells with an OX-19$^+$OX-8$^+$ or an OX-19$^+$ER-2$^+$ phenotype (peripheral mature T cell). This latter cell population may be mitogen responsive. As significant mitogen responsiveness becomes measurable at first at higher ages (10% of normal value at 68 weeks of age), this *in situ* differentiation occurs rather slowly, and seems not to be very efficient in the generation of immunocompetent T cells.

Acknowledgements

The authors are grateful to Mrs J.M.A. Huijbregts, Mr F.J. Tielen and Mr C.J.W.M. Brandt for excellent assistance in various experiments. Professor Dr J. Jongkind is gratefully acknowledged for allowing the use of his FACS equipment.

References

1. H.R. McDonald, R.W. Lees, D. Sordat, P. Zaech, J.L. Maryanski, and C. Bron, Age-associated increase in expression of T-cell surface markers Thy-1, Lyt-1, and Lyt-2 in congenitally athymic nu/nu mice: analysis by flow microfluorometry, *J. Immunol.* 126: 865 (1981).
2. H.J. Schuurman, J.G. Vos, R. Broekhuizen, C.J.W.M. Brandt, and L. Kater, *In vivo* biological effect of allogeneic cultured thymic epithelium on thymus-dependent immunity in athymic nude rats, *Scand. J. Immunol.* in press.
3. A.F. Williams, G. Galfre, and C. Milstein, Analysis of cell surface by xenogeneic myeloma-hybrid antibodies: differentiation antigens of rat lymphocytes, *Cell* 12: 663 (1977).
4. D.W. Mason, R.P. Arthur, M.J. Dallman, J.R. Green, G.D. Spickett, and M.L. Thomas, Functions of rat T-lymphocyte subsets isolated by means of monoclonal antibodies. *Imm. Rev.* 74: 57 (1983).
5. D.A. Cantrell, R.A. Robins, and C.G. Brooks, Phenotype of rat natural killer cells defined by monoclonal antibodies marking rat lymphocyte subsets, *Immunology* 45: 97 (1982).
6. M.J. Dallman, M.L. Thomas, and J.R. Green, MRC OX-19: a monoclonal antibody that labels rat T-lymphocytes and augments *in vitro* proliferative responses, *Eur. J. Immunol.* 14: 260 (1984).
7. A.N. Barclay, The localization of populations of lymphocytes defined by monoclonal antibodies in rat lymphoid tissues, *Immunology* 42: 593 (1981).
8. W.H. de Jong, P.A. Steerenberg, P.S. Ursem, A.D.M.E. Osterhaus, J.G. Vos, and E.J. Ruitenberg, The athymic rat: III. Natural cell-mediated cytotoxicity, *Clin. Immunol. Imunopathol.* 17: 163 (1981).

T-CELLS AND T-CELL SUBSETS IN RAT BRONCHUS ASSOCIATED LYMPHOID TISSUE (BALT) IN SITU AND IN SUSPENSION

Gerda J. van der Brugge-Gamelkoorn and Taede Sminia

Department of Histology, Medical Faculty, Free University
P.O. Box 7161, 1007 MC AMSTERDAM, The Netherlands

INTRODUCTION

Mucosa associated lymphoid tissue (MALT) comprises two main elements, viz. bronchus associated lymphoid tissue (BALT) and gut associated lymphoid tissue (GALT), which show morphological and functional similarities[1-4]. The GALT will be exemplified by Peyer's patch, the lymphoid organ to which the morphological similarities refer[1].

One phenomenon which supports the similarity between BALT and Peyer's patch is the lack of cells which contain immunoglobulin G (IgG), IgA or IgM in their cytoplasm, suggesting that local production of these immunoglobulins does not occur[1]. However, Sminia and Plesch[5] presented evidence that Peyer's patch does contain cytoplasmic IgG (cIgG), cIgA and cIgM cells indicating that local synthesis can occur in Peyer's patch. In BALT these cells are scarcely present (unpublished own observations). As to the lymphocyte with surface Ig (sIg) the percentages of sIgG, sIgA, sIgM and sIgE are comparable in BALT and Peyer's patch[5,6].

With regard to functional similarities it appears that both BALT and Peyer's patch contain precursor cells, which are destined to synthesize and secrete IgA in the mucosae[3,7].

IgA responses are T cell dependent[8,9]. Although T cells play a prominent part in the onset of the IgA synthesis, no data concerning T cell and T cell subsets in BALT are available.

The aim of the present investigation was to study T cells in BALT, in particular to determine the in situ localization of T cell subsets (T helper and T suppressor/cytotoxic cells) and to qualify the percentages of the subsets in suspension. The results will be compared to data available on the percentages of T helper and T suppressor/cytotoxic cells in Peyer's patch.

Because the lung is regarded as an immunologic organ[10], attention has also been paid to the in situ localization of T helper and T suppressor/cytotoxic cells in the lung parenchyma.

MATERIALS AND METHODS

Animals

Young adult Wistar rats and nude rats, Sd/Gbnu, were obtained from TNO (Zeist, The Netherlands).

Histology

Cryostat and sublimate formaldehyde fixed paraffin sections were used for routine histology and for immunocytochemistry.

Cell preparation

Five to seven BALT's were removed from the upper part of the lung under a dissecting microscope. Cell suspensions of these BALT's were obtained by nylon gauze filtration. From the suspensions cytocentrifuge preparations were made.

Immunocytochemistry in situ and in suspension

The aceton fixed cytocentrifuge preparations and cryostat sections were incubated for 1 h with each of the following monoclonal antisera: mouse-anti-rat Ia (Seralab), mouse-anti-rat-T cell (W3/13), mouse-anti-rat-T-helper cells (W3/25) and mouse-anti-rat T suppressor/cytotoxic cell (Ox8). Thereupon the slides were incubated for 30 min with the conjugate rabbit-anti-mouse-IgG peroxidase (Miles). The sections were stained for peroxidase activity with 3-3'-diaminobenzidine-tetrahydrochloride (DAB, Sigma) in 0.5 mg/ml Tris-HCl (pH 7.6) containing 0.01% H_2O_2 for 10 min. Next the sections were counterstained with hematoxylin, dehydrated and covered.

Quantitation

From four rats the number of positive cells per 2000-5000 cells present in the cell suspensions was determined in duplo. The standard errors were calculated with the Student's-t-test.

RESULTS

BALT in the rat consists of lymphoid aggregates which are mostly situated between a bronchus and an artery (Fig. 1). It contains lymphoid and non-lymphoid structures characteristic for a peripheral lymphoid organ[6,11]. Although BALT is comprised of discrete T and B cell areas, T and B cells occupy no predestinated areas, so they have no regular place within BALT. That is why the

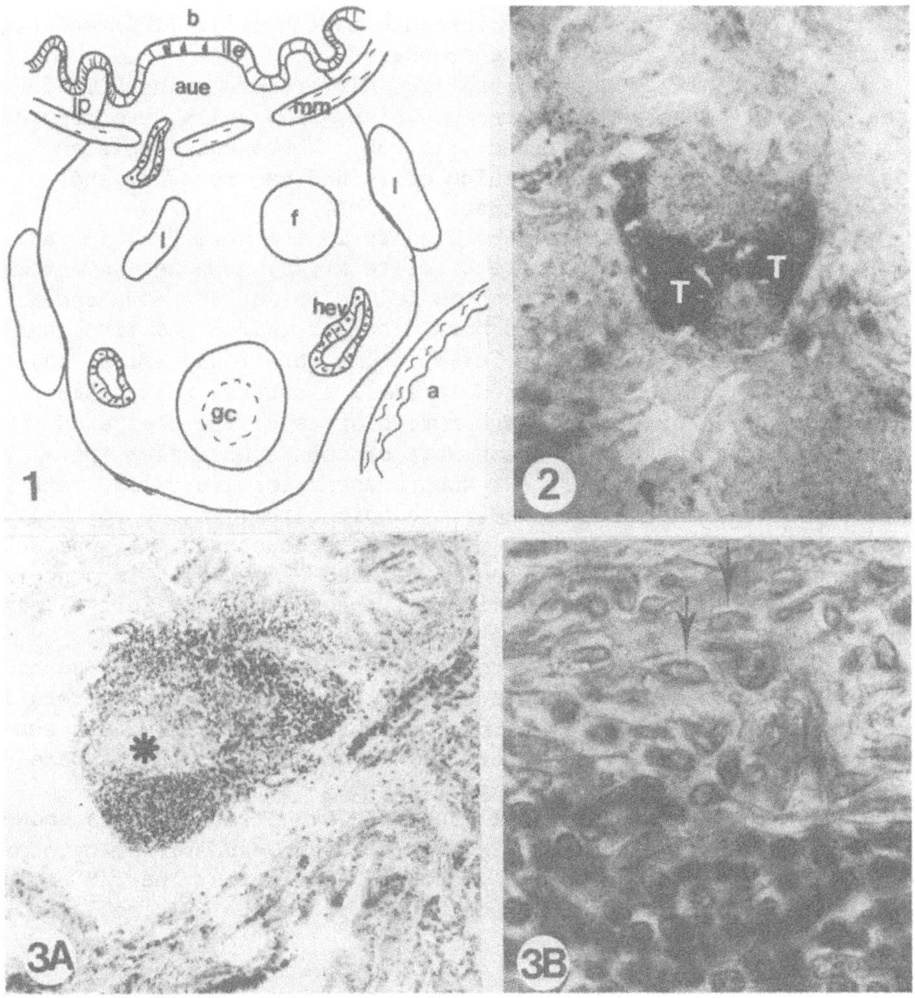

Fig. 1. Schematic figure of rat BALT. a: artery; aue: area under epithelium; b: bronchus; f: follicle; gc: germinal center; hev: high endothelial venule; l: lymph vessel; le: lymphoepithelium; lp: lamina propria; mm: muscularis mucosae.

Fig. 2. Anti-W3/13 showing T cell area (T). 60x.

Fig. 3A. BALT of nude rat with cell depleted area (asterisk). 80x.
Fig. 3B. Enlargement of the boundary between the cell depleted T cell area and B cell area. Reticulum cells (arrows) are visible in the T cell area. 800x.

T-cell areas were made visible by staining with anti-W3/13, a mouse-anti-rat T cell monoclonal (Fig. 2). In these areas characteristic structures, notably Ia-positive interdigitating cells (IDC) and high endothelial venules (HEV) were encountered.

Examining BALT of the nude rat two differences with BALT of normal rat were observed, viz. no positive W3/13 cells were present and cell-depleted areas occurred (Fig. 3A). These cell-depleted areas contained IDC, HEV, reticulum cells and macrophages, and, hardly any lymphocytes were present (Fig. 3B).

Although the majority of the T cells in the normal adult rat BALT was confined to one or more discrete areas, some were scattered throughout non-T cell areas, such as follicles and the area under the epithelium. In T cell areas both T-helper (W3/25 positive) and T-suppressor/cytotoxic (Ox8 positive) cells were found (Figs. 4A, 4B). Most T-helper cells occurred in small clusters, while the T-suppressor/cytotoxic cells were more dispersed. The scattered T lymphocytes in the B cell areas were of both subclasses; the T helper cells outnumbered the T suppressor/cytotoxic cells, both in the follicle and in the area under the epithelium. In the epithelium covering unstimulated BALT no T cells were encountered. This presents a contrast to epithelium covering stimulated BALT, which is penetrated by lymphocytes, thus forming a lymphoepithelium. In the lamina propria and the rest of the lung parenchyma both T helper and T-suppressor/cytotoxic cells were present. On account of the sections examined no pronunciation can be done on which of the two subsets is most represented. However, the total number of cells from both subsets in the lamina propria is very little compared to the numbers present in the rest of the lung parenchyma.

Quantification of the T cell subsets in BALT suspensions showed that T helper cells contribute 28.9 $\pm$ 2.3% and T-suppressor/cytotoxic cells 10.8 $\pm$ 2.0% to the total lymphocyte population. The T/B ratio in BALT is 0.7, so for every T cell there is about 1.5 B cell.

DISCUSSION

The aim of the present study was to examine and describe T cell areas and T cell subsets in BALT of the rat. T cell areas are distinct compartments within the peripheral lymphoid organs, which are characterized by a special type of non-lymphoid cell[12,13] and which become lymphocyte depleted after neonatal thymectomy[14,15]. According to the literature BALT showed no areas of cell depletion up to two months after neonatal thymectomy[2,16]. Recently it was found that some BALT's of thymectomized animals, not much older than two months, did show areas of cell depletion be it that the areas were not totally free of lymphocytes (unpublished own observations).

The majority of BALT's present in the nude rat contained clear areas of cell-depletion. In these areas high endothelial venules (HEV) and interdigitating cells, the non-lymphoid cells characteristic for T cell areas[12,13] are present. So it does seem that BALT contains predisposed T cell areas, but they probably appear late in the development.

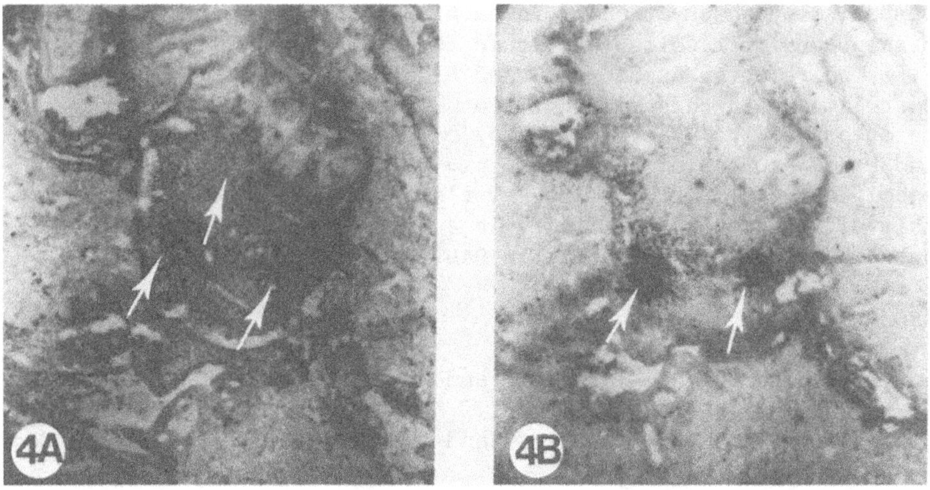

Fig. 4A. Anti-W3/25 showing T helper cells (arrows). 60x.
Fig. 4B. Anti-Ox8 showing T suppressor/cytotoxic cells (arrows). 60x.

The T cell areas have a different location from one BALT to another in rat. That this is species dependent is clearly shown by the more organized structure of the guinea pig BALT[17].

The T cell area in the BALT is composed of T helper and T suppressor/cytotoxic cells. In the B cell areas, such as the area under the epithelium and the follicles, there are also T cells present, which are mainly of the T helper cell phenotype. This phenomenon has also been described for the follicles in human lymph node[18], rat lymph node[19] and for the germinal centers in the mouse lymph node[20]. Nude mice, which lack germinal centers, develop these after thymus cells are injected[21]. The effector cells for germinal center development are probably of the T helper phenotype[20]. Normal adult rat BALT contains no germinal centers, only six weeks after intratracheally stimulation germinal centers can be encountered[22]. The epithelium covering unstimulated BALT contains no T helper or T suppressor/cytotoxic cells. Infiltration of the epithelium by lymphocytes only happens after stimulation[22]. The few lymphocytes present in the epithelium covering unstimulated BALT are Ia-positive[6]. No data are yet available on the phenotype of the lymphocytes penetrating the epithelium after antigenic stimulation.

The quantitative observations on cell suspensions showed that BALT comprises two times as much T helper as T suppressor/cytotoxic cells. The fraction of W3/13 positive (T) cells in BALT and Peyer's patches is not equal: BALT contains a higher percentage of T cells than does Peyer's patch. In rat BALT for every T cell there is about

1.5 B cell, while in rat Peyer's patch this is about 2.7 B cell (unpublished observations). In rabbit, on the other hand, equal percentages of T cells were found for BALT and Peyer's patch[22].

The relative number of T and B cells present in lymphoid organ correlates with the number of T and B cells entering these organs through the HEV[24]. A recent study on the HEV specificity in BALT and Peyer's patch of the guinea pig points to a different binding specificity for BALT HEV and Peyer's patch HEV[17]; most probably this difference in specificity is reflected in the different percentages of T and B cells in BALT and GALT.

REFERENCES

1. J. Bienenstock, N. Johnston, and D.Y.E. Perey. Lab. Invest. 28: 686 (1973).
2. J. Bienenstock, N. Johnston, and D.Y.E. Perey. Lab. Invest. 28: 693 (1973).
3. R. Rudzik, R.L. Clancy, D.Y.E. Perey, and J. Bienenstock. J. Immunol. 114:1599 (1975).
4. M.R. McDermott, and J. Bienenstock. J. Immunol. 122:1892 (1979).
5. T. Sminia, and B.E.C. Plesch. Virch. Archiv 40:181 (1982).
6. B.E.C. Plesch. In: "In vivo immunology", P. Nieuwenhuis, A.A. van den Broek, and M.G. Hanna, eds., Plenum Press, New York (1982).
7. S.W. Graig, and J.J. Cebra. J. Exp. Med. 134:188 (1971).
8. J.J. Cebra, P.J. Gearhart, R. Kamat, S.M. Robertson, and J. Tseng. Cold Spring Harbor Symposium on quantitative biology 41:201 (1977).
9. C.O. Elson, J.A. Heck, and W. Strober. J. Exp. Med. 149:632 (1979).
10. J. Bienenstock. Ann. Rev. Med. 35:49 (1984).
11. G.J. van der Brugge-Gamelkoorn, T. Sminia, and M.B. van de Ende. Submitted.
12. A.J.P. Veerman, and W. van Ewijk. Cell Tissue Res. 136:417 (1975).
13. E.W.A. Kamperdijk, E.M. Raaymakers, J.H.S. de Leeuw, and E.C.M. Hoefsmit. Cell Tissue Res. 192:1 (1978).
14. B.H. Waksman, B.S. Arnason, and B.D. Jonkovic. J. Exp. Med. 116: 187 (1962).
15. D.M.V. Parrott, M.A.B. de Sousa, and J. East. J. Exp. Med. 123: 191 (1966).
16. B.E.C. Plesch, G.J. Gamelkoorn, and M.B. van de Ende. Develop. Comp. Immunol. 7:179 (1983).
17. G.J. van der Brugge-Gamelkoorn, and G. Kraal. Manuscript in preparation.
18. S. Poppema, A.K. Bhan, E.L. Reinherz, R.T. McCluskey, and S.F. Schlossman. J. Exp. Med. 153:30 (1981).
19. A.M. Barclay. Immunology 42:593 (1981).
20. R.V. Rouse, J.A. Ledbetter, and I.L. Weissman. J. Immunol. 128: 2243 (1982).
21. E.B. Jacobson, L.H. Caporale, and G.J. Thorbecke. Cell. Immunol. 13:416 (1974).

22. G.J. van der Brugge-Gamelkoorn, B.E.C. Plesch, T. Sminia, and H.L. Langevoort. Submitted.
23. O. Rudzik, R.L. Clancy, D.Y.E. Perey, J. Bienenstock, and D.P. Singal. J. Immunol. 114:1 (1975).
24. E.C. Butcher, S.K. Stevens, R.A. Reichert, R.G. Scollay, and I.L. Weissman. In: "Recent advances in mucosal immunology", W. Strober, L.A. Hanson, and K.W. Sell, eds., Raven Press, New York (1982).

K CELL ACTIVITY IS INDEPENDENT OF BOTH THYMUS AND SPLEEN

Danielle M. Chassoux*, J.P. Kolb* and I.C.M. MacLennan+

*Institut de Recherche Scientifique sur le Cancer CNRS, Villejuif
+Department of Immunology, University of Birmingham (Birmingham), U.K.

K cells, mediating antibody-dependent cellular cytotoxicity of nucleated dividing target cells, are highly efficient effectors in the presence of minute amounts of antibody in vitro (MacLennan,1972). Their distinct Fc receptor may now be identified by a monoclonal antibody, B73.1 (Perussia et al.,1983a). The role they may play in vivo, in normal and/or pathological situations remains unknown. K cells may be involved in the lysis of virus-infected cells (Kohl and Moore,1983), and of established tumour cell-lines as Chang cells - Also, sera from patients with chronic active liver disease (Kawanishi and McDermott,1979) rheumatoid arthritis (MacLennan,1969) may mediate ADCC of relevant target cells. Recently, a role of K cells in the regulation of immune responses has been considered. Anti-idiotypes have been shown to be able to mediate ADCC against mouse hybridoma cells (Caraux and Weigle,1983) or plasmocytoma cells (Chassoux and Stanislawski, in preparation). The origin and conditions of maturation of K cells are not known. K/NK cells do not appear to be B cells (Chassoux et al.,1983a). Their relationships with other lymphoid cell lineages has received attention with the development of monoclonal antibodies. Some authors have described common antigens on T and K/NK cells and therefore consider that they are part of the same lineage (Fast, Hansen and Newman,1981): other reagents, however, may separate T from K/NK cells (Beverley and Callard,1981). The separation of K and NK cells by surface markers is not clear at the present time, although functional data often suggest that these activities may well be mediated by different cells (Koren and Williams,1978; Dokhelar et al.,1981, Perussia et al.,1983b). Another approach is the use of an in vivo situation where a spontaneous mutation has led to the functional absence of one cell lineage. This has been possible since the development of

Nude rat (rnu/rnu) colonies - Rats are chosen, as opposed to mice, since rats display adequate K cell activity. In such animals, T cell functions are absent and T cell areas in lymphoid organs are devoid of lymphocytes (Salomon and Fogh, 1982).

In this paper, we describe the kinetics of appearance of K and NK cell cytotoxic activities in rats congenitally deprived of a thymus, in comparison with their normal littermates. Moreover, a possible role of the spleen in providing a micro-environment necessary for the maturation of K cells has been considered and newborn splenectomised rats have been used to measure K and NK cytotoxic activities.

Materials and Methods

1 - Neonatal splenectomy

Rats were operated on within 24 hours of birth. They were cooled on ice and when they had stopped moving, a small incision was made with a scalpel blade just below the left costal grid. The spleen was removed after suturing blood vessels and the skin wound was closed by one or two stitches. Usually no major bleeding occurred. Sham operated rats were treated in the same way but the spleen was not removed. Animals were left to warm up in an incubator at 37°C before being returned to their mothers. No deaths were noted after that stage.

2 - Preparation of effector cells from peripheral blood

Animals were used when 4-6 months old. They were bled, under anaesthetic, from the retro-orbital plexus. Blood samples (5 ml) were collected in heparinised tubes. Aliquots were kept for mononuclear white blood cell counts. Blood samples were processed individually for the separation of mononuclear cells. This was done by layering blood, diluted 1:2 in heparinised MEM, onto 5ml of Lymphoprep and spinning at 1500 rpm for 30 min at 20°C. Cells in the upper layer were collected and washed three times in MEM + 2% fetal calf serum. Cells were then counted and adjusted to $5X10^6$/ml in RPMI 1640 + 10% fetal calf serum. When fewer cells were collected, they were used at the concentration obtained under the minimum volume necessary for the assay.

3 - Preparation of effector cells from spleens

Animals were killed by ether and their spleens removed. Spleen cell suspensions were performed individually. Cells were forced through a stainless steel mesh, washed twice in MEM and adjusted to $10X10^6$/ml in RPMI 1640 + 10% fetal calf serum.

4 - Cytotoxicity assays

K cell activity was measured by using 51-Cr labelled Chang cells sensitised with a rat antibody (rat anti-serum used at 1:10 000) in a six hour assay. NK cell activity was measured by using 51-Cr labelled YAC-1 cells in a four hour assay. For both tests, effector cells were used at three different ratios against 10^4 target cells. Results are expressed as lytic units per 10^6 effector cells (LU/106) One lytic unit is defined as the number of cells required to achieve 30% specific cytotoxicity of the targets. When total numbers of cells per organ (spleen) or per volume unit (blood) are taken into account, results are then expressed as LU/spleen or LU/ml. Statistics were performed according to the non-parametric test of Mann & Whitney.

Results

1 - Development of K and NK cytotoxic activities in the spleen of young male Nude rats and their controls

Results are shown in Fig.1 (K) and Fig.2 (NK). Neither cytotoxic activity was measurable for the first ten days of life. They were less consistently found in rnu/rnu rats than in rnu/+ rats at the time when cytotoxicity was first detected (day 12-13 for K, day 13-14 for NK). Only on day 14 did all animals tested show K cell activity. NK cell activity, however, awaited day 20-21. One exceptional rat had high cytoxic activity on day 14. Clearly, cytotoxic activities were not higher in rnu/rnu than in rnu/+ before day 29 for K cell activity and day 35 for NK. When total spleen cell counts were considered, K and NK cytotoxic activities were not higher in rnu/rnu than in rnu/+, at any time. (Data not shown). Total spleen cell counts became consistently higher in rnu/+, compared with rnu/rnu, at the time at which cytotoxic activities ($LU/10^6$) became higher in rnu/rnu, indicating that the increase in activity in rnu/rnu reflect a dilution of effectors by T cells in rnu/+ spleen cells.

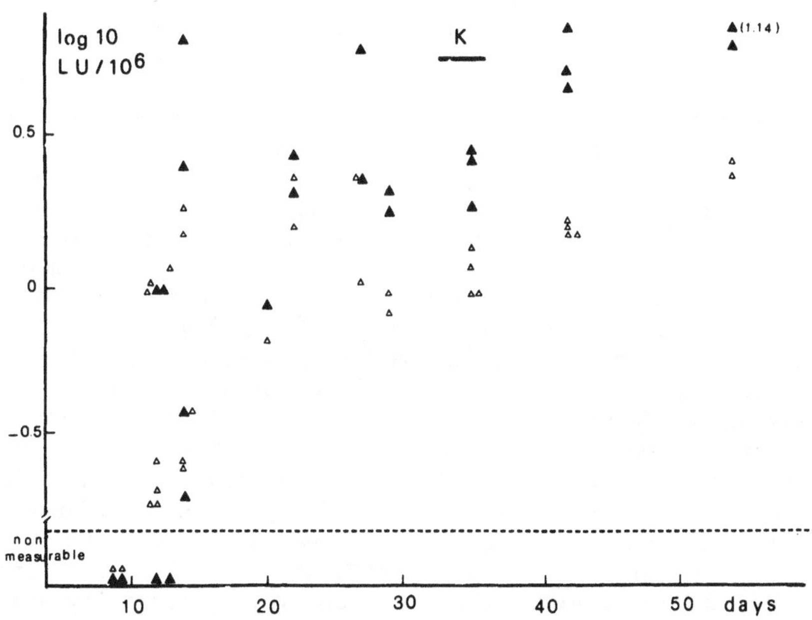

Fig. 1. K cell cytotoxicity of spleen cells in relation to the age of male rnu/rnu (▲) and rnu/+ (Δ) rats.

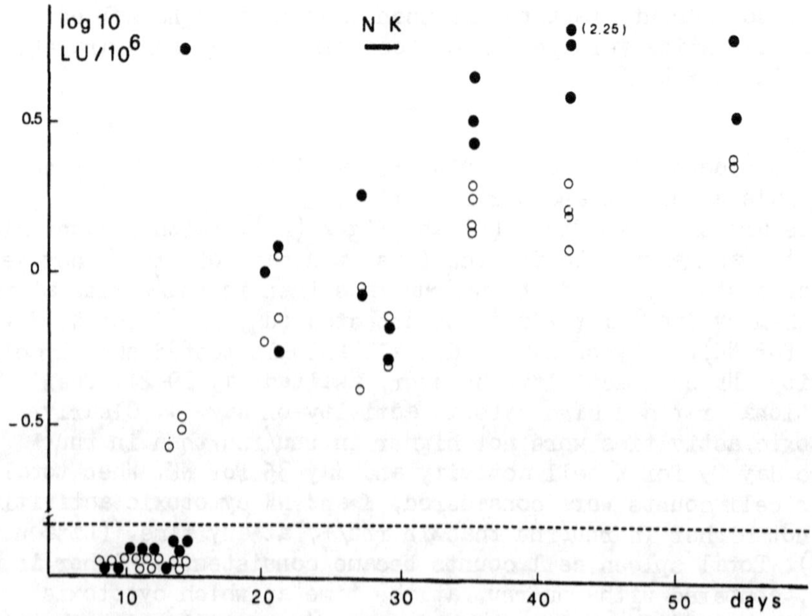

Fig. 2 NK cell cytotoxicity in the same rnu/rnu (●) and rnu /+(O) animals as Fig.1.

2 - Assessment of K and NK cytotoxic activities in neonatally splenectomised male rnu/rnu, rnu/+ and their controls

Animals splenectomised at birth were tested when they reached four, five or six months of age. Independent experiments always included the sham-operated littermates as controls. K cell activity (Fig.3) was found to be lower in the blood of splenectomised animals than in controls, although this is not statistically significant: p=0.02 for rnu/rnu, p=0.14 for rnu/+, due to the wide range of activity from one individual to another. NK activity, on the other hand, was at similar levels for all groups (Fig.4). Mononuclear cell counts per milliliter of blood were strikingly higher in splenectomised rats, in rnu/rnu (6.73 ± 0.1 vs 6.33 ± 0.11 for controls, p<0.01) as well as in rnu/+ (6.78 ± 0.045 versus 6.5 ± 0.06 for controls, p<0.01). An increase in B cells was found to be only partly responsible for the increase in peripheral blood mononuclear cells (D. Gray and D. Chassoux, unpublished). High cell counts were noticed at two months of age in splenectomised animals, but were more pronounced at four months and later. As indicated in Table 1, K cell activity per unit volume of blood (a) was not different in splenectomised and control animals, whereas NK activity (b) was higher, but without statistical significance (p=0.1). In several instances, and particularly in K cell assays, cytotoxicity was present, but results could not be included as an increase in effectors did not produce an increase in cytotoxicity and calculation of lytic units could not be done precisely.

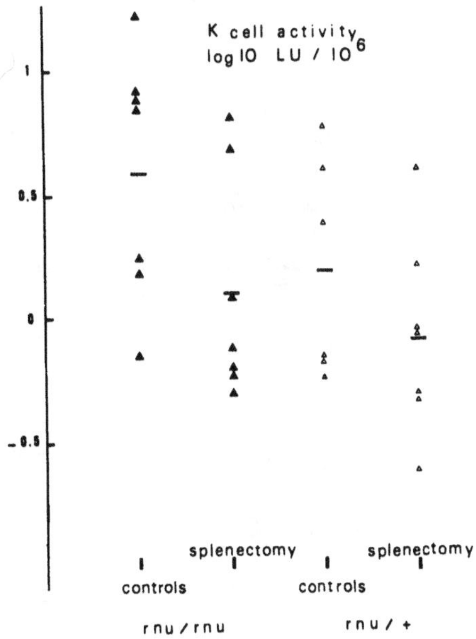

Fig. 3 K cell activity of peripheral blood mononuclear cells of splenectomised rnu/rnu (▲), rnu/+ (Δ) and their controls.

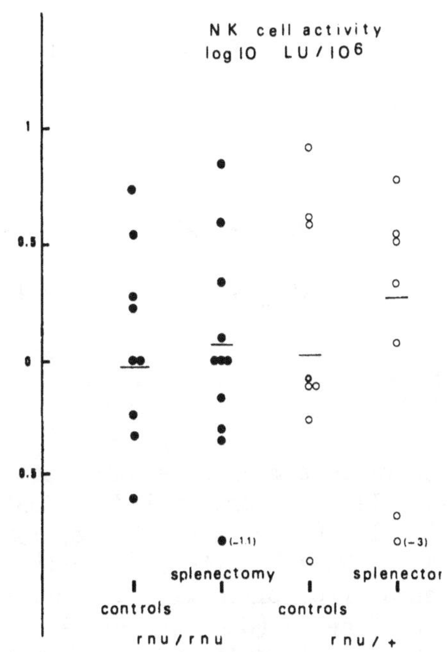

Fig. 4 NK cell activity of peripheral blood mononuclear cells of splenectomised rnu/rnu (●) rnu/+ (○) and their controls.

Table 1

(a) K cell activity

rnu/+	LU/106	LU/ml	n
Splenectomised	-0.07±0.15	0.77±0.14	7
Controls	0.2±0.18	0.72±0.15	6
	p=0.02	NS	
rnu/rnu			
Splenectomised	0.11±0.17	0.8±0.12	7
Controls	0.59±0.19	0.95±0.18	7
	p=0.147	NS	

(b) NK cell activity

rnu/+	LU/106	LU/ml	n
Splenectomised	0.26±0.24	1.04±0.23	6
Controls	-0.03±0.33	0.48±0.36	8
	NS	NS	
rnu/rnu			
Splenectomised	-0.006±0.27	0.73±0.25	11
Controls	0.07±0.2	0.4±0.27	9
	NS	NS	

Discussion

There is now a considerable number of experimental arguments indicating that K/NK cells may form an independent cell lineage from that of T and B cells (Horwitz and Bakke, 1984).
We have shown that K and NK cell cytotoxic activities were not impaired in rats treated from birth with an anti-mu sera, to prevent development of B cells (Chassoux et al., 1983a). We have also reported previously that adult nude rats have K/NK cytotoxic activities which are not different from rnu/+ normal littermates (Chassoux et al., 1983a; Chassoux et al., 1983b). High cytotoxic activities obtained with nude effector cells are due to a relative

concentration of effectors per given number of mononuclear cells, consequently to the absence of T cells. In young rats, however, cytotoxic activities per given number of effectors were not higher in rnu/rnu than in rnu/+, as reported in the present paper. This does not agree with K/NK cells being early T cells accumulating in an absence of functional thymus, as this ought to have taken place before animals were about one month of age. This is the time at which rats start responding to various antigenic stimuli. It seems that expression of K/NK cytotoxic functions develops together with the occurrence of normal immune responses. Data in favour of a regulatory role of these cells are now accumulating (Abruzzo and Rowley, 1983; Brieva, Targan and Stevens, 1984; Caraux and Weigle, 1983).

Appearance of NK cytotoxicity was delayed compared with K cytotoxicity, although of a few days only. This is reminiscent of the delay in NK recovery compared with K recovery in bone-marrow transplant patients (Dokhelar et al., 1981) and indicates again that K and NK cytotoxities may be mediated by different cells (Koren and Williams, 1978). Also, it may be noted that individuals, in this study, with high NK values were not always the ones with high K cytotoxicity.

Cells recognised by reagents against K/NK surface membrane antigens have been identified in germinal centres (Ritchie, James and Micklem, 1983). In the absence of the spleen, main tissue involved in antibody production, those cells might be either absent, if they need the spleen to develop, or be in greater number in the circulation, if they accumulate there. Results of neonatally splenectomised rats show that K/NK cells may exist despite the absence of splenic tissue from birth onwards and that they do not accumulate in the blood. It may be that an alternative site of maturation is used, as suggested for static B cells (Gray, Chassoux, Bazin and MacLennan, 1984, submitted for publication) after neonatal splenectomy. K/NK activities are minimal in the bone-marrow and were not increased after splenectomy (data not shown), indicating that there was not an accumulation of effector cells at that site. There was an apparent decrease of K cell cytotoxicity, as opposed to NK cell cytotoxicity in the peripheral blood of adult rats splenectomised at birth. It may be that K, but not NK, inhibitory factors were present (MacLennan, 1972). It may also be considered that part of the mononuclear cells found in greater numbers in splenectomised animals are immature K cells. Combined use of monoclonal antibodies as HNKI and B73.1 (Horwitz and Bakke, 1984) may help to clarify this point.

K and NK cells have a number of features in common as morphology (large granular lymphocytes), stimulation by interferon (Herberman, Ortaldo and Bonnard, 1979), inhibition by placental factors (Kolb, Chaouat and Chassoux, 1984) or monoclonal antibodies as MHM23 (Chassoux and Mac Michael, unpublished), but they may now be differentiated by a monoclonal antibody against the Fc receptor, that inhibits K but not NK cytotoxicity and does not recognise B cells (Perussia et al., 1983b). Such reagents will be useful for the purification of K cells and the preparation of normal cell lines having K cell activity may now be considered. This should be feasible if K cells indeed differ from NK cells as well as from T and B cells and would permit the study of the role of these cells and their mechanism ofaaction.

Acknowledgements

We would like to thank Dr J.C. Salomon for laboratory facilities and Viviane Lascaux for expert technical assistance.
This work was supported by INSERM grant P.R.C. number 1270 20.

References

Abruzzo, L.V. and Rowley, O.A., 1983, Science, 222:581.

Beverley, P.C.L. and Callard, R.E., 1981, Distinctive functional characteristics of human 'T' lymphocytes defined by E-rosetting or a monoclonal anti-T cell antibody, Eur. J. Immunol., 11:329.

Brieva, J.A., Targan, S. and Stevens, R.H., 1984, NK and T cell subsets regulate antibody production by human in vitro antigen-induced lymphoblastoid B cells, J. Immunol., 132:611.

Caraux, J. and Weigle, W.O., 1983, Anti-idiotype antibody-dependent cell mediated cytotoxicity (ADCC) against idiotype-bearing cells, Cell. Immunol., 78:23.

Chassoux, D., Kolb, J.P., Bazin, H. and MacLennan, I.C.M, 1983a, Antibody-dependent cellular cytotoxicity (K) and natural killing (NK) in B-suppressed germ-free nude rats, Immunology, 50:327.

Chassoux, D., Dokhelar, M.C., Tursz, T. and Salomon, J.C., 1983b, Antibody-dependent cellular cytotoxicity and natural killing in nude rats : quantitative study according to age and sex, Ann. Immunol., 134D:309.

Dokhelar, M.C., Wiels, J., Lipinski, M., Tetaud, C., Devergie, A., Gluckman, E. and Tursz, T., 1981, Natural killer cell activity in human bone marrow recipients, Transplant, 31:61.

Fast, L.D., Hansen, J.A. and Newman, W., 1981, Evidence for T cell nature and heterogeneity within natural killer (NK) and antibody-dependent cellular cytotoxicity (ADCC) effectors : a comparison with cytolytic T lymphocytes (CTL), J. Immunol., 127:448.

Herberman, R.B., Ortaldo, J.R. and Bonnard, G.D., 1979, Augmentation by interferon of human natural and antibody-dependent cell-mediated cytotoxicity, Nature, 277:221.

Horwitz, D.A. and Bakke, A.C., 1984, An Fc receptor-bearing, third population of human mononuclear cells with cytotoxic and regulatory function, Immunology Today, 5:118.

Kawanishi, H., and MacDermott, R.P., 1979, K cell-mediated antibody-dependent cellular cytotoxicity in chronic active liver disease, Gastroenterology, 76:151.

Kohl, S. and Moore, C.M., 1983, Human antibody-dependent cellular cytotoxicity and natural killer cytotoxicity to herpes simplex virus-infected autologous and allogeneic cells, Immunology, 48:187.

Kohren, H.S. and Williams, M.S., 1978, Natural killing and antibody-dependent cellular cytotoxicity are mediated by different mechanisms and by different cells, J. Immunol., 121:1956.

Kolb, J.P., Chaouat, G. and Chassoux, D., 1984, Immunoactive products of placenta III. Suppression of natural killing activity, J. Immunol., 132:2305.

MacLennan, I.C.M., Loewi, G. and Howard, A., 1969, A human serum immunoglobulin with specificity for certain homologous target cells which induce target cell damage by normal human lymphocytes, Immunology, 17:897.

MacLennan, I.C.M., 1972, Antibody in the induction and inhibition of lymphocyte cytotoxicity, Transplant, Rev., 13:67.

Perussia, B, Starr, S., Abraham, S., Fanning, V. and Trinchieri, G., 1983a, Human natural killer cells analysed by B73.1, amonoclonal antibody blocking Fc receptor functions. I. Characterization of the lymphocyte subset reactive with B73.1, J. Immunol., 130:2133.

Perussia, B., Acuto, O., Terhorst, C., Faust, J., Lazarus, R., Fanning, V. and Trinchieri, G., 1983b, Human natural killer cells analysed by B73.1, a monoclonal antibody blocking Fc receptor functions II studies on B73.1 antibody-antigen interaction on lymphocyte membrane, J. Immunol., 130:2142.

Ritchie, A.W.S., James, K. and Micklem, H.S., 1983, The distribution and potential significance of cells identified in human lymphoid tissue by the monoclonal HNK.1., Clin. Exp. Immunol., 51:43.

Salomon, J.C. and Fogh, J., 1982, The nude rat. In: The nude mouse in experimental and clinical research II (Fogh, J. and Giovanella, B., eds.), p.229, Acad. Press, N.Y.

USE OF TRANSGENIC MICE WITH HUMAN MHC CLASS 1 GENE (HLA-A2)

Takeshi Matsunaga+#, Orly Cabilly-Horesh+, David Terz+, Alex Lee+, Lawrence Christian#, Susumu Ohno#, and Andrew P. Biro*

+Department of General and Oncologic Surgery and #Beckman Research Institute of City of Hope, City of Hope National Medical Center, Duarte, California 91010,USA *Department of Tumor Virology, DANA Farber Cancer Institute, Boston Massachusetts 02115, USA

Studies of MHC restriction in T cell antigen recognition (1) suggest that the two sets of multigene families, i.e. MHC class 1 and 2 genes on the one hand and T cell antigen receptor genes on the other, evolved in parallel. The notion of Jerne (2) that germ-line v-genes encode combining sites of antibodies or T cell receptors which have specificities for all MHC alleles of the species almost predicted the MHC restriction found several years later (1).

The basis for MHC restriction would be a complementarity present between MHC and T cell receptor molecules which would allow mutual recognition in presence of foreign antigens, such as viruses. Therefore, relationship between MHC and T cell receptors is interdependent, i.e. if one of the two components is missing or defective, the outcome of immune response is affected.

Thus, holes exist in T cell recognition repertoire, known as Ir gene controlled unresponsiveness, due to defect of T cell clone (2) and/or appropriate MHC alleles for antigen presentation (3). In its extreme, when all the MHC's are absent in individuals, total malfunction of the immune system seems to have resulted (4).

Although human cytotoxic T cells can recognize mouse H-2 polymorphic determinants (5), such alloMHC-recognition across species's barrier has not been always observed (6). Apparently, more systematic studies are needed.

Transgenic mice (7) which can express MHC antigens from foreign species might be useful to study the Jerne's hypothesis (2) and other aspects of MHC functions (e.g. HLA-linked diseases (8)), since in transgenic mice foreign MHC genes introduced at 1-cell stage may be expressed early in ontogeny and recognized as self MHC antigens. One can ask whether such foreign MHC functions as restricting elements in anti-virus Tc responses. We have produced transgenic mice by microinjecting human MHC class 1 gene (HLA-A2) into fertilized eggs. Preliminary results with regard to chromosomal integration and gene expression are reported in this article.

Materials and Methods

1. Production of Transgenic Mice with HLA-A2 Gene

The HLA-A2 genomic DNA clone was originally isolated (9) from Eco R1 genomic library of human lymphoid cell line, JY, (HLA-A2/A2, B7/B7). Then about 6.7 kb Eco R1 fragment of HLA-A2 gene was subcloned in plasmid pBR 328 and used for microinjection. This 6.7 kb fragment contains the complete HLA-A2 gene flanked by about 500 base-long 5' upstream region and also 3' untranslated region. Transfected mouse L-cells with this clone express HLA-A2 determinants (9).

Supercoiled plasmid (about 10^4 copies in a few picoliter) was microinjected directly into one of the two pronuclei of 1-cell stage of fertilized eggs (C57BL/6xC56BL/6, Swiss Albino x Swiss Albino) using Leitz micromanipulators (10). Eggs that survived physical damage were transferred to the opening of oviduct of a pseudopregnant recipient female mouse. All mice born were delivered by Caesarean section and transferred to foster Swiss Albino mothers.

2. Southern Blot Hybridization

The Southern blot hybridization (11) was used to identify mice that integrated injected HLA genes. When born mice became more than 6 weeks old, half of the tail was cut off and high molecular weight DNA was extracted (12). 10μg of DNA from individual mice was digested by Eco R1 endonucleases (37°C, overnight), electrophoresed in 1% agarose gel, transferred to nitrocellulose filters after alkaline denaturation and baked in vacuum oven.

The hybridization probe we used was an about 600 base-long DNA fragment which contains regions of a part of C2 domain, membrane and cytoplasmic regions as well as untranslated 3' region (a gift from Dr. Arnot). This probe detects only a few hybridization bands in the JY DNA indicating that it is fairly specific to the HLA-A2 gene and does not cross-hybridize with other members of HLA or mouse H-2 genes. The probe DNA was labelled with ^{32}P by nick-translation to a specific activity of 10^8 cpm/μg DNA. Hybridization was done at

65°C for 16 hours (6xSSC, 5xDenhardt solution, 50μg/ml salmon sperm DNA, 0.5% SDS, probe $5x10^6$ cpm per ml). The filters were washed 3 times (65°C, 3xSSC) and once (65°C, 0.3xSSC) before exposed to x-ray film (XAR-5) for autoradiography (several hours to overnight at -70°C with two intensifier-plus screens). DNA from human JY cells was also extracted and used as positive controls for hybridization.

3. Immunofluorescence

A simple method to examine HLA-gene expression is to use antibodies specific to HLA antigens. We have obtained mouse monoclonal antibody reagents, W6/32, (13) which bind a common determinant of HLA-A, B, C proteins. This antigenic determinant is preserved when HLA heavy chains associate with mouse β2 microglobulins (13) and therefore useful for our experiment. FITC conjugated antibodies were incubated with frozen sections or live spleen cell suspension without erythrocytes (1% BSA in PBS, 1 hour, room temperature), washed, placed in 0.1% phenylenediamine, 90% glycerol in PBS, mounted on slide glasses and sealed for observation by a fluorescent microscope (Olympus, HB-2).

Results

1. Production of Transgenic Mice

In early stage of this experiment, HLA genes have been microinjected into cytoplasm of mouse eggs rather than into pronucleus. About 4 dozen of mice were born, but none of them showed evidence for chromosomal integration of injected HLA genes. Although there is a report (14) that cytoplasmically injected retrovirus genome was integrated in the host chromosome, efficiency for integration may be much lower than pronucleus injection. Subsequently, only pronucleus injection was employed in later stage. About 300 eggs injected with HLA-A2 genes were transferred to 22 recipients. Fourteen mice were born and most survived to adulthood. However, a dramatic decline in the pregnancy rate was observed (15%; pronucleus injection versus 50%; cytoplasmic injection). The reason for this difference is not clear.

2. Southern Blot Hybridization

The figure 1 shows the result of one hybridization experiment using tail DNA from several mice that derived from pronucleus-injected eggs.

In the lane a to e, two major bands can be seen. Each band corresponds to the size of HLA-A2 insert (6.7 kb) and pBR 328 (4.8 kb) with the 4.8 kb band more intense. As the probe used in this hybridization was contained in PUC 9 plasmid vectors which constitute major portion of DNA fragment and cross-hybridize extensively with pBR 322, stronger signals may have been generated in the 4.8 kb plasmid band. When the 600 base-long DNA was separated from PUC 9 vector in polyacrylamide gel and used as a hybridization probe with the same filter from which the first probe was washed off (95°C, water, 10 minutes), only upper bands (6.7 kb) were observed supporting the above interpretation.

There is a notable variation in intensity of hybridization signals among different individual mice. This indicates that the integration of multiple copies of injected HLA genes occurred in some mice. Comparing with the signal intensity of the third band (6.7 kb) of JY DNA (lane f), two mice (i.e. TG 69; lane c, TG 73; lane e) might have as many as one hundred copies per haploid DNA. The low molecular weight 4th band in JY DNA is an artifact. The 6.7 kb band contains a fragment of the HLA-A2 genes from which micro-injected DNA clone derived.

There is an extra-band (indicated by an arrow) in higher molecular weight position in the lane d (TG 70). In other individuals (TG 71 and TG 72) we found other extra bands in lower molecular weight positions. These bands may represent boundary fragments for integration sites.

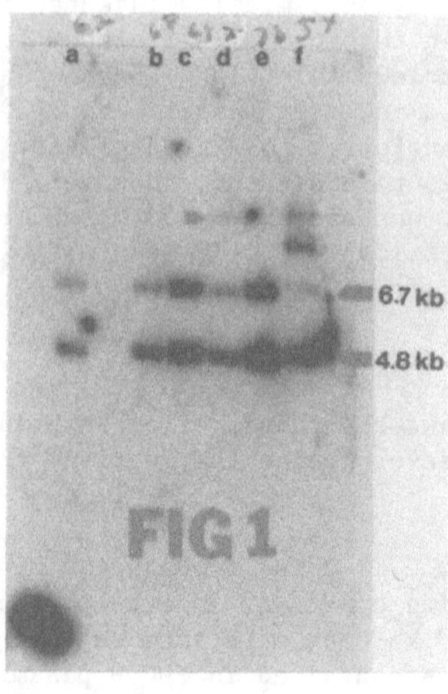

Figure 1

These hybridization bands are remarkably similar to the published reports (10, 15, 16, 17) and are compatible with the interpretation that head-to-tail multiple concatemers of injected DNA are integrated in mouse chromosomes. Six out of 14 mice were positive for HLA-A2 sequence judged by Southern blot hybridization.

3. Immunofluorescence and a Case of Immunodeficiency

Two transgenic mice died postnatally. The first (TG 73, C57BL/6) died at about 6 weeks of age. This mouse looked sick from the beginning and did not grow normally. The second mouse (TG 72, C57BL/6) died at about 4 months of age. The TG 72 did not grow normally either (approximately 2/3 of normal body size) and one eye was not well opened. A few hours after his death was noticed, various organs were inspected. A very small spleen (1/20 of normal size) and only vestigial thymus-like glands were found. The size of all other organs in this mouse was normal. Histological examination of frozen section of the spleen of TG 72 showed clear evidence of massive atrophy of lymphoid cells. It should be mentioned that both TG 72 and TG 73 mice had multiple copies (about 20 copies; TG 72, about 100 copies; TG 73) of HLA-A2 genes in their genomes.

Immunofluorescent staining of spleen and liver sections did not show clear difference between the TG 72 and normal control mouse. However, when photographic pictures were compared, the cytoplasmic staining of the liver of TG 72 was noticed suggesting that the HLA-A2 genes may have been expressed weakly.

Spleen cell suspensions from several other transgenic mice were prepared by partial splenectomy and stained with FITC-conjugated anti-HLA antibodies. Cells from one transgenic mouse (TG 71, C57/BL6, female) seems to be stained producing caps and patches of surface fluoresence. This mouse has not produced offsprings so far. The result is promising but needs further confirmation at this moment.

4. Discussion

Expression of microinjected genes in transgenic mice seems to depend on several ill-defined variables, such as effects of chromosomal position (10). On the other hand, injected genes with appropriate regulatory elements (eg. enhancer) lead to expression in appropriate tissues (16). Furthermore, exogeneous strong promoters can be used to enhance gene expression (15).

The HLA-A2 genes used in our microinjection experiments can be expressed in transfected mouse L-cells ((9) and unpublished observation). This might indicate that regulatory DNA sequences needed for gene expression may be contained in the 6.8 kb fragment. Such regulatory sequences may be recognized by mouse factors but a species' barrier may lead to weak expression, although this is by no means a general rule (17). It is possible that two of our transgenic mice (TG 71, TG 72) may express HLA-A2 antigens as suggested by immunofluorescent staining. This finding should be further confirmed using other method (eg. Northern blot hybridization).

The reason for the atrophy of the immune tissues in TG 72 mouse is not clear. One possibility is that the presence of expressed HLA antigens triggered the process of autoimmune diseases and auto-antibodies or cytotoxic T cells may have wiped out lymphoid cells. Alternatively, the presence of HLA antigens or genes caused blockade in the development of the immune system.

In summary, we have shown that HLA genes microinjected into mouse pronucleus can be integrated in the host chromosomes and some of these transgenic mice may express the HLA genes.

Acknowledgement

We thank Drs. A. McMahon and F. Costantini for advices for microinjection, Dr. D. Arnot for the hybridization probe, Mrs. Sharon Van de Velde for preparing tissue sections of a transgenic mouse, Dr. R. Teplitz for examining some of the sections, Dr. P. Salvaterra for making fluorescent microscope available to us, and Drs. J. J. Terz and C. Mittman for support.

5. References

1. Zinkernagel, R.M., Doherty, P.C. (1974) Nature (London) 251, 547
2. Jerne, N.K. (1971) Eur. J. Immunol, 1, 1
3. Matsunaga, T., Simpson, E. (1978) Proc. Natl. Acad. Sci. USA 75, 6207
4. Touraine, J.L., Betual, H., Souillet, G., Jeune, M. (1978) J. Pediatr. 93, 47
5. Lindahl, K.F., Bach, F.H. (1975) Nature, 254, 609
6. Lafferty, K.J., Jones, M.A.S. (1969) Aust. J. Exp. Biol. Med. Sci. 47, 17
7. Gordon, J.W., Ruddle, F.H. (1981) Science 214, 1244
8. Ryder, L.P., Svejgaard, A., Dausset, J. (1981) Ann. Rev. Genet. 15, 169
9. Barbosa, J.A., Kamarck, M.E., Biro, P.A., Weissman, S.M., Ruddle, F.H. (1982) Proc. Natl. Acad. Sci. USA 79, 6327
10. Lacy, E., Roberts, S., Evans, E., Burtenshaw, M.D., Costantini, F.D. (1983) Cell 34, 343
11. Southern, E.M. (1975) J. Mol. Biol. 98, 503
12. Blin, N., Stafford, D.W. (1976) Nucleic Acid Res 3, 2303
13. Brodsky, F.M., Parham, P. (1982) J. Immunol. 128, 129
14. Harbers, K., Jahner, D., Jaenisch, R. (1981) Nature 293, 540.

15. Palmiter, R.D., Brinster, R.L., Hammer, R.E., Trambauer, M.E., Rosenfeld, M.G., Birnberg, N.C., Evans, R.M. (1982) Nature 300, 611
16. Brinster, R.L., Ritchie, K.A., Hammer, R.E., O'Brien, R.E., Arp, B., Storb, U. (1983) Nature 306, 332
17. McNight, G.S., Hammer, R.E., Kuenzel, E.A., Brinster, R.E. (1983) Cell 134, 335

15. Palmiter, R.D., Brinster, R.L., Hammer, R.E., Trambauer, M.E., Rosenfeld, M.G., Birnberg, N.C., Evans, R.M. (1982) Nature 300, 611
16. Brinster, R.L., Ritchie, K.A., Hammer, R.E., O'Brien, R.E., Arp, B., Storb, U. (1983) Nature 306, 332
17. McNight, G.S., Hammer, R.E., Kuenzel, E.A., Brinster, R.E. (1983) Cell 134, 335

. Palmiter, R. [illegible], Brinster, R.L., Hammer, R.E., Trumbauer, M.E., Rosenfeld, M.G., Birnberg, N.C., [illegible] (1982) Nature 300, [illegible]

14. [illegible] Palmiter, R.D., [illegible] Hammer, R.E., [illegible] (1983) Nature 306, [illegible]

15. [illegible] Hammer, R.E., [illegible] Brinster, R.L. [illegible] (1985) Cell [illegible]

EFFECT OF SYMPATHECTOMY ON THE MATURATION OF FETAL THYMOCYTES GROWN WITHIN THE ANTERIOR EYE CHAMBERS IN MICE

Upendra Singh

Department of Anatomy
University of Utah School of Medicine
Salt Lake City, Utah 84132 USA

INTRODUCTION

Generation of an immunocompetent pool of peripheral lymophocytes requires an intrathymic differentiation of their precursor cells which involves some interactions between these cells and the stroma of the thymic microenvironment or its secretory products such as the thymic hormones and interlukins (1,2). The factors which regulate these intrathymic interactions are still far from clear. Recent reports indicate that surgical and chemical sympathectomy affect the immunoreactivity of T lymphocytes (3,4). Furthermore, autonomic nerve fibers have been observed within the parenchyma of various lymphoid tissues including the adult and the fetal thymus (5). In view of these studies it is likely that the autonomic nervous system may modulate the thymic microenvironment and thus affect the development and differentiation of T lymphocytes. This study, therefore, was initiated to study the role of the sympathetic neuronal input in the maturation of thymocytes during ontogeny.

MATERIAL AND METHODS

1. Mice - BALB/c albino mice and black NZB mice were obtained from Jackson laboratories, Ann Harbour. The mice (BALB/c) were mated in trios (2 female, 1 male) and the day of the vaginal plug was counted as day 0. The mice were maintained in the vivarium of the University of Utah.
2. Chemicals - General chemicals were obtained from Sigma Chemical Co., St. Louis, ^{3}H-thymidine from New England

Nuclear, Boston and media from Gibco, Grand Island, New York.

3. Organ cultures - The thymic rudiments were aseptically removed at day 13/14 of gestation, and were organ cultured in the presence of 1.3 mM of deoxyguanosine for 5 days/or grown at 24°C for 2 weeks to eliminate the lymphoid cells (6,7).
4. Sympathecetomy - Sympathetic cervical ganglion along with part of the sympathetic chain was surgically removed under tribromoethenol general anaesthesia, in the adult mice of both sexes, the ganglia were removed on the right side only while the left side was sham operated. The success of this operation was confirmed by the appearance of drooping of the eyelid (ptosis) and constriction of the pupil on the operated side after the operation and the histological examination of the tissue removed.
5. Anterior chamber implantation - After one week of sympathectomy organ cultured thymic lobes were implanted within the anterior eye chambers by making an incision into the cornea under general anesthesia. The animals were observed carefully for any signs of infection following the operations. The thymic grafts were allowed to grow within the eye for up to 21 days when the animals were sacrificed and the thymic lobes removed from the eyes to isolate the thymocytes.
6. Chemical sympathectomy: Animals were chemically sympathectomised by injecting 100 mg/kg of 6-hydroxy-dopamine intraperitoneally daily for 5 days followed by one weekly injection for another 2 weeks. Such animals received thymic grafts on both sides into their eyes in groups of 3.
7. Compliment cytoxicity tests were performed as described earlier (8).
8. Histofluorescence microscopy. To examine the presence of catecholaminergic nerves tissues obtained from the eyes (thymic grafts) were immediately frozen at -70°C and sections cut by cryostat and examined under a fluorescent microscope after treating with galoxylic acid according to the method of De La Tore (9).
9. Mitogen responses: Thymocytes removed from the grafts were isolated in serum free (RPM1-1640) media containing glutamine and 10 mg/ml of streptomycin, 100 units/ml of penicillin and 10 mg/ml of myecostatin. 2x10^{5} cells were then treated with 5 ug/ml of con-A for 72 hours and the incorporation of ^{3}H-thymidine into DNA in the last 18 hours was measured by scintillation counting as previously described (10).
10. Mixed lymphocyte reaction - 2 x 10^{5} thymocytes were mixed with 2 x 10^{5} mitomycine -C treated (25 ug/ml for 1 hour)

spleen cells from black NZB mice and kept in suspension cultures for 90 hours at 37°C in the presence of 5% CO_2 in air. Incorporation of ^{3}H-thymidine was measured in last 18 hours by scintillation counting. Differences between the stimulated and unstimulated cells were calculated for the stimulation index. A difference of about 2½ times between the stimulated and unstimulated cultures were considered positive.

RESULTS

Morphological observation: Examination of the grafts under dissecting microscope indicated that the grafts had increased in size in the duration of 3 weeks and had received blood vessls from the iris. (Fig 1).

Histology: Examination of the tissues removed from the eye indicated that the graft had lymphoid cells when stained by toluidine blue histologically and they had increased in number as compared at the time of grafting. (Fig. 2-3). Histological examination by haematoxilin and eosin showed that the tissues removed from the neck were sympathetic ganglia showing large ganglion cells. (Fig. 4).

Histofluorescence microscopy: Examination of thymic grafts removed from the anterior eye chambers revealed that on the sham operated side catechotamenergic nerve fibers were visible within the thymic graft which had grown into the graft along the blood vessels from the iris. On the operated side (Rt) there was total absence of fluorescent fibers both in the iris and in the thymic graft (Fig. 5, 6).

Cell yield: Thymic lobes removed from the operated side had higher cell yeild ($6.3 \pm 1.8 \times 10^6$)cells per lobe as compared to ($3.2 \pm 1.3 \times 10^6$) cells from the sham operated side.

Surface antigens: Compliment cytotoxicity tests using anti Thy 1.2 and guniae pig compliment showed that 90 to 96% of cells removed from the grafts were thy 1.+ in both sets.

Immunoreactivity: Thymocytes recovered from the operated side had higher incorporation of ^{3}H-Thymidine when stimulated by con-A or in mixed lymphocyte reaction with the mitomycin treated NZB spleen cells as compared to the cells recovered from the sham operated side. (Table 1-2).

DISCUSSION

It is well established that the generation of immunocompetent T lymphocytes depends upon an interaction between the prethymic progenitor cells and the thymic microenvironment or its secretory products. However the precise mechanism of this interaction is

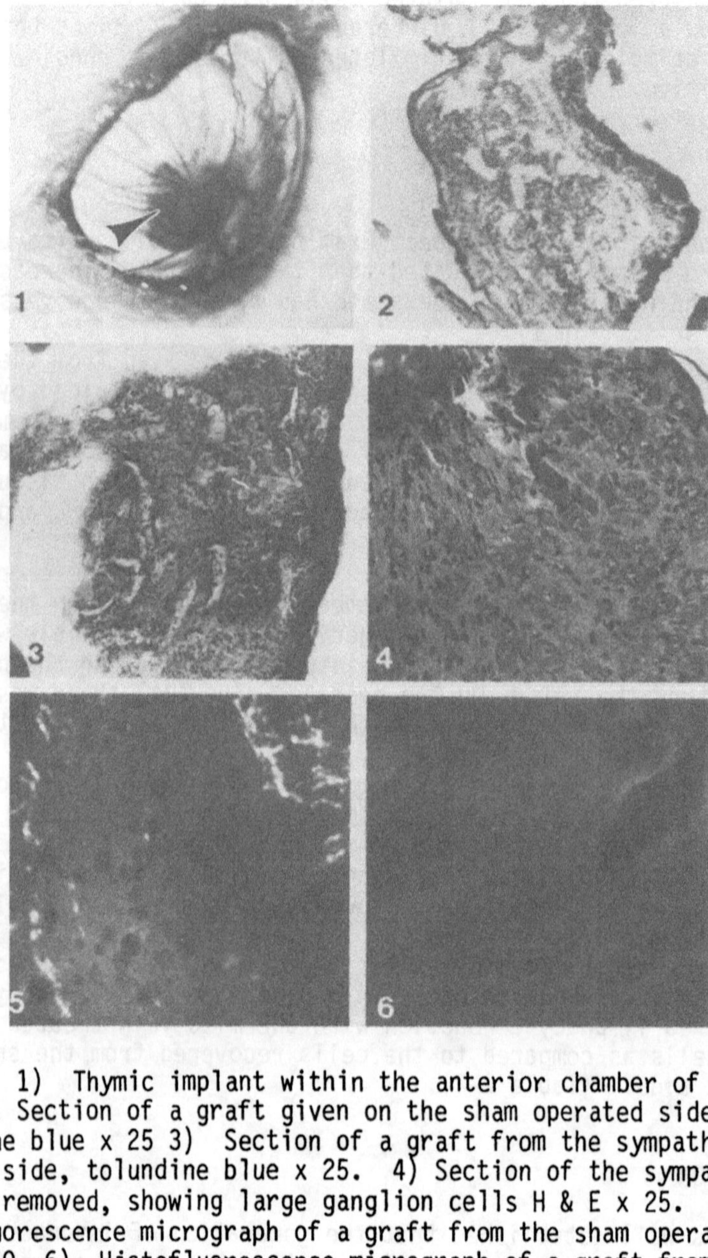

FIGURES 1) Thymic implant within the anterior chamber of the eye. 2) Section of a graft given on the sham operated side. Tolundine blue x 25 3) Section of a graft from the sympathectomised side, tolundine blue x 25. 4) Section of the sympathetic ganglia removed, showing large ganglion cells H & E x 25. 5) Histofluorescence micrograph of a graft from the sham operated side x 40. 6) Histofluorescence micrograph of a graft from the sympathetomised side x 40.

Table 1. Incorporation of ^{3}H-Thymidine by thymocytes (counts per minute) following treatment with 5ug/ml of con-A at 37°C in the presence of 5% CO_2 in air. Last 18 hours of a 72 hour culture period was pulsed with ^{3}H-thymidine. Data form 6 experiments.

Cells	Unstimulated	Con-A Stimulated	P.value
A	165 +- 63	362 +- 74	>.1
B	447+-212	72622+-1281	<.002
C	269+-101	2818+-462	<.002
D	265+-86	41251+-912	<.002

A = 14 day thymocytes
B = Adult thymocytes
C = Thymocytes from the sham operated side
D = Thymocytes from the operated side
$\pm$ = Standard deviation of 4 to 6 experiments
$\bar{p}$ = Student's t test

Table 2. Intercorporation of ^{3}H-Thymidine by BALB/C thymocytes (counts per minute) following co-culture with spleen cells from young adult NZB/black mice (treated with mitomycin - C(25 μg/ml for 1 hr), for a period of 90 hours last 18 hrs of that period the cells were pulsed with ^{3}H-thymidine (three experiments).

Cells	Unstimulated	Stimulated	P.value
A	212+-91	341+-87	>.1
B	319+-79	5762+-918	<.002
C	198+-32	687+-175	<.05
D	251+-69	1128+-1375	<.02

A = 14 day thymocytes
B = Adult thymocytes
C = Thymocytes from the sham operated side
D = Thymocytes from the operated side
$\pm$ = Standard deviation of 4 to 6 experiments
$\bar{p}$ = Student's t test

far from clear. This study indicates that the sympathetic nervous system may have some regulatory role in the development and maturation of thymocytes and further substantiates our earlier preliminary work (10). Histofluoresence studies clearly indicate that following cervical sympathectomy there is an absence of fluorescent nerve fibers (catecholamenergic) within the iris and the implanted graft, where as these fibers are present on the sham operated side. These fibers innervate the graft along with the blood vessels which also vascularise the graft in about 4-5 days from the day of implantation. The exact nature of the influence, these fibers have in modulating the thymic environment, however, is not clear from this study. It may be that the nerves affect the lymphoid cells directly through an interaction between the neurotransmitters and the beta adnenoceptors present on these cells (11) or they may affect the release of various factors from the non lymphoid cells of the thymus, thereby affecting the interaction between the progenitor cells and various thymic hormones and factors such as interlukin which have been shown to affect the maturation and proliferation of thymocytes (2,12). Such an involvement of the autonomic nervous system in the development of the thymus is possible due to the anatomical connections between the hypothalamic areas and the sympathetic neurons in the spinal cord, directly and indirectly through the reticular formation. Sympathetic neurons in turn have been observed to innervate not only the thymus but other lymphoid tissues as well (3). Additionally, hypothalamic neurons have been shown to be affected by the antigenic challenge and that hypopituatary animals have been shown to be deficient in cell mediated immunity (13, 14) which could be restored by the administration of the growth hormone (15, 16, 17). Based upon the experiments in which some neuroendocrine peptide hormones were shown to regulate the production of lymphhokines it has recently been suggested that the neuroendocrine and immune systems might share common peptide signals and receptors (13). In light of these observations it is quite likely that the autonomic nervous system is capable of modulating the thymic microenvironment directly or through another secretory factors released within the thymus. It is possible that the increased immunoreactivity seen in the thymocytes from the sympathectomized side could be due to the unopposed action of the parasympathic nerves which may have also innervated the graft. This is supported to some extent by our previous study where an increase in the cell yield and mitogen reactivity in thymocytes from the organ cultures treated with cholinergic agents was observed (10). Whether similar phenomenon may have operated in the eye grafts is unclear at present since we failed to observe acetyl choline exterase positive fibers within the grafts on any side, in this study. However parasympathetic nerves have been reported within the thymic parenchyma (5). All these observations suggest the possibility of a neuronal modulation of the thymic microenvironment.

It should be pointed out, however, that the difference in cellularity and the immunoreactivity between the two groups of thymocytes could also have resulted due to the differences in the vascularization and the pattern of repopulation of the non-lymphoid thymic rudiments following their grafting into the anterior eye chambers. The pattern of repopulation could be affected by the change in the permiability of the blood vessels following sympathectomy, and therefore raises various important aspects for further studies to note the differences in the repopulation of the thymic grafts using marker cell systems. Whatever the source of the cells developing within the graft, this study demonstrates an inhibitory role of the sympathetic nervous system in the development of thymocytes during ontogeny. This experimental model provides an excellent approach to study the regulatory role of the automonic nervous system in the development of thymocytes and further studies are in progress to ascertain the source of the repopulating lymphoid cells and the patterns of repopulation of the grafts following sympathectomy both in the normal and in the nude mice using a marker system to identify the origin of cells developing within the grafts.

Acknowledgements: The author thanks Karen Evans and June Stephenson for typing the manuscript and Lubena Dhoon for the technical assistance. The work was supported by a grant from NIH.

REFERENCES

1. Stutman, O. 1974, Inability to restore immune function in nude mice with humoral thymic factors. Fed. Proc. 33:736.
2. Trainin, N., Small, M., and Kook, A.I. 1977. In B and T cells in immune recoginition. Ed. Foor F. and Roelants, G.E. John Wiley and Sons. London 6.83-102.
3. Williams, J.M. and Felton, D.L. 1981. Sympathetic Innervation of Murine Thymus and Spleen. Anat. Rec. 199:531-592.
4. Besedovsky, M.O., Rey, A., Sorkin, E., Prada, M.D.A. and Keller. H.H. 1979. Immunoregeneration mediated by the sympathetic nervous system. Cell. Immunol. 48:346-355.
5. Bullock, K. and Moore, R.Y. 1980. Nucleus Ambiguous projections to the thymus gland. Anat. Rec. 196:25.
6. Jenkinson, E.J., Franchi, L.L., Kingston, R., and Owen, J.J.T. Effect of deoxygguanosine on lymphopoiesis in the developing thymic rudiment in vitro: application in the

production of chimeric thymic rudiments. 1982. Eur. J. Immol. 12:582-587.

7. Jordon, R.K., and Crouse, D.A. 1980, In: Development and Differentiation of vertebrate lymphocytes, ed. Horton J. D., Elsevier, North-Holland Biomedical press, p. 47-54.
8. Singh, U. and Owen, J.J.T., 1976. The effects of catcholamines, histamine and various peptide hormones on the expression of T cell alloantigens. Eur. J. Immunol. 6:59-63.
9. DeLaTorre, J.C., 1979. Standardization of the histofluorenscence method for monoamine transmitters. Soc. Neuroscience Abstr. 5:333.
10. Singh, U. 1979. Effect of catecholamines on lymphopoiesis in fetal mouse thymic explants. J. Anat. 129:279-292.
11. Singh, U., Millson, D.S., Smith, P.A. and Owen J.J.T., 1979. Identification of B adrenoceptors during thymocyte ontogeny in mice. Eur. J. Immunol. 9:31-35.
12. Papiernick, M. and Homo-Delarche, F. 1983. Thymic reticulum in mice. III. Phagocytic cells of the thymic reticulum in culture secrete both prostagland in E_2 and interlukin 1 which regulate thymocyte proliferation. Eur. J. Immol. 13:689-692.
13. Blalock, J.E. and Stanton, J.D., 1980. Common pathways of interferon and hormonal action. Nature 283:406-408.
14. Cross, R.J., Marksberry, W.R., Brooks, W.H. and Roszman, T.L., 1980. Hypothalamic-immune interactions. I. Brain Res. 196:79-87.
15. Brooks, W.K., Cross, R.J. Roszman, T.L. and Marksberry W.R., 1982, Neuroimmunomodulation neural anatomical basis for impavement and facilitation. Ann Neurol. 12:56-61.
16. Blalock, J.E., 1984. The Immune system as a sensory organ. J. Immunol. 132:1667-1670.
17. Payan, D.G., Levine, J.D. and Goetzl, E.J., 1984. Modulation of immunity and hypersensitivity by sensory neumopeptides. J. Immol. 132:1601-1604.

THE EFFECT OF DIOCTYLTIN DICHLORIDE ON THE THYMUS AND T-CELL DIFFERENTIATION IN THE RAT

S. Nicklin, M.P. Scott, J. Evans and K. Miller

Dept. Immunotoxicology, The British Industrial Biological Research Association
Carshalton, Surrey, U.K.

Various industrial and environmental chemicals have been shown to have a profound effect on thymus and thymus dependent functions (Seinen and Willems, 1976; Vos, 1977; Miller, 1983). Recently, particular attention has been paid to the selective effect of dioctyltin compounds on the thymus. These materials are used extensively in the plastics industry, in particular, during the heat/light stabilisation process in the manufacture of PVC items intended to come into contact with drugs and food stuffs. In 1976 Seinen and Willems reported that dioctyltin dichloride (DOTC) caused a severe reduction of thymic weight in weanling rats after exposure to diets containing 50 or 150 ppm DOTC. Subsequent work in our laboratories indicated that 6-8 week old PVG strain rats were equally sensitive to DOTC. Animals exposed to 75 ppm DOTC via their diet developed overt thymic atrophy after 2 wks exposure (Fig. 1), almost complete atrophy was evident after 4 wks exposure and by 8 wks the thymic remnant consisted almost entirely of brown fat. Body weight gain and food intake were not influenced by the feeding regime and animals showed no signs of abnormal behaviour or appearance. Examination of plastic embedded thymic sections from DOTC treated rats revealed a marked depletion of cortical thymocytes, a concurrent loss of the cortico-medullary junction and an apparent increase in the number of vacuolated reticulo-epithelial cells (RECs). Tissue sections of spleen, liver, prostate and seminal vesicles from treated rats revealed no abnormal histological features or pathological lesions attributable to treatment, indicating the selective thymic toxicity of DOTC.

Thymic involution was also shown to be associated with a decreased number of peripheral blood leucocytes. Table 1

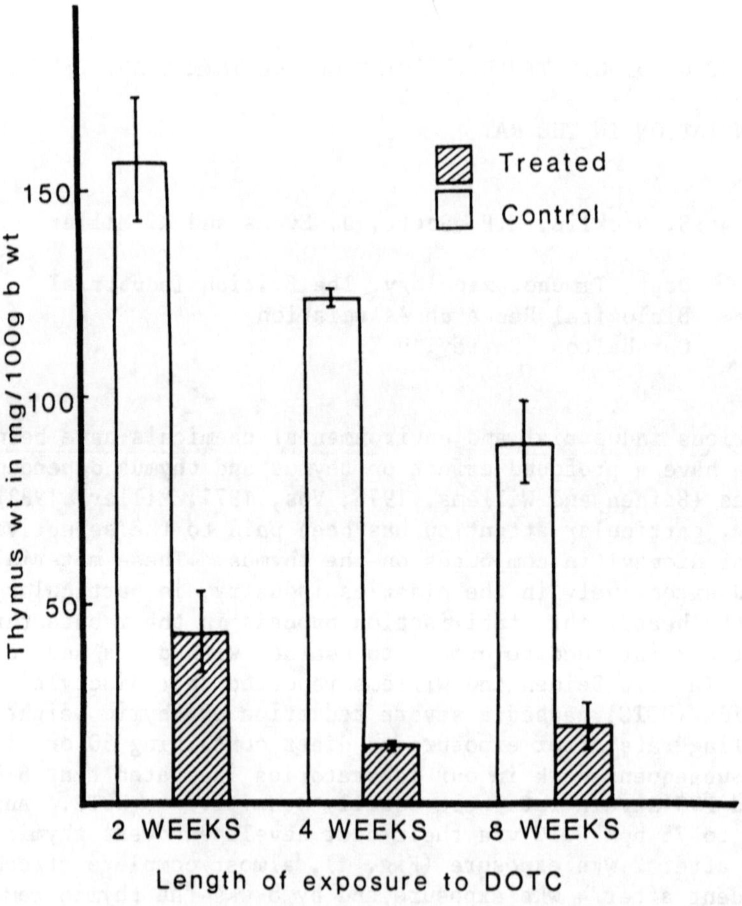

Fig. 1 Effect of exposure to diet containing 75 PPM DOTC on relative thymus weight of PVG rats.

Table 1. Effect of dietary exposure to 75 ppm DOTC on the number of circulating white blood cells

	WBC	% Lymphocyte	Absolute no. of lymphocytes	% Neutrophils	Absolute no. of neutrophils	No. of rats
8 weeks exposure						
Test	9.9 ± 0.22	78.2 ± 0.35	7.69 ± 0.31	18.82 ± 1.31	1.83 ± 0.12	10
Control	13.27 ± 0.48	85.1 ± 1.0	10.76 ± 0.37	11.4 ± 1.07	1.45 ± 0.14	11
	***	**	***	***	NS	
12 weeks exposure						
Test	10.4 ± 1.2	77.0 ± 1.15	7.98 ± 0.30	20.08 ± 1.13	2.03 ± 0.12	13
Control	13.8 ± 1.5	82.0 ± 1.27	11.48 ± 0.32	13.46 ± 0.87	1.84 ± 0.14	14
	***	**	***	***	NS	

NS = Not significant
** = Significant at 1% level
*** = Significant at 0.1% level
Mean ± standard error
Absolute no. of cells = Cells x 10^{-3}/ mm^3 blood

indicates that the peripheral leucocyte count was suppressed at both 8 and 12 weeks of treatment. Differential cell counts revealed that the percentage of lymphocytes in the blood was decreased whilst the percentage of neutrophils had increased. However, when these figures were corrected for the lowered white blood cell count the decrease in lymphocytes became more apparent whereas the increase in neutrophils was no longer significant.

Monoclonal antibody analysis of the T cell sub-populations making up the circulating lymphocyte pool, demonstrated a reduction in the number of cells expressing the W3/25 phenotype (helper/inducer T cell sub-set, Webb, Mason and Williams 1979) in DOTC treated rats (Table 2). The "non-helper", suppressor/effector T cell sub-set identified by the MRC monoclonal antibody OX 8 (Brideau et al. 1980) were not influenced by the DOTC regime. As W3/25 + ve T cells are required for helper/

Table 2. Effect of dietary exposure to 75 ppm DOTC on % T cell subsets

	Abs W3/25	Abs MRC OX/8	No. of rats
8 weeks			
Control	7.001 ± 0.665	2.336 ± 0.222	10
Test	4.420 ± 0.293	1.990 ± 0.126	11
P	**	NS	
12 weeks			
Control	6.184 ± 0.290	3.133 ± 0.186	8
Test	4.351 ± 0.319	2.561 ± 0.250	8
P	***	NS	

$$\text{Abs Nos. subsets} = \frac{\text{Abs Lymphocyte No} \times \%\ \text{subset}}{100}$$

NS = Not significant
** = Significant at 1% level
*** = Significant at 0.1% level
Mean ± standard error
Absolute no. of cells = cells x $10^{-3}/mm^3$ blood

Lymphocyte subsets were determined by immunofluorescence using monoclonal antisera

alloantigen recognition responses we performed experiments to compare lymphocytes, obtained from control and DOTC treated donors, in terms of their ability to react either specifically against alloantigens, as assessed by the mixed lymphocyte response (MLR), or non-specifically to mitogenic stimulation, as assessed by PHA induced blastogenesis. Table 3 indicates the effect of 8 wks dietary exposure to 75 ppm DOTC on mitogen induced and alloantigen induced lymphocyte proliferation. It was apparent that the capacity of lymphocytes to respond to PHA was reduced by the DOTC treatment, significant suppression of blastogenesis was however only apparent at sub-optimal doses of PHA. This effect is similar to the results from investigations into lymphocyte reactivity in patients suffering from immunodeficiency syndromes, where depressed lymphocyte responses were also only evident at sub-optimal levels of PHA exposure (Oppenheim, Blaese & Waldman, 1970; Ziegler, Hansen & Penny, 1975). These observations considered in association with

Table 3. Effect of dietary exposure to 75 ppm DOTC on Mitogen-induced and alloantigen-induced proliferation of blood lymphocytes

		Mitogen (PHA) induced proliferation	
		Control	DOTC treated
PHA Conc in µg/ml	.0	883 ± 208	859 ± 192
	1.25	17177 ± 5032	4596 ± 1000
	2.5	55733 ± 12533	23463 ± 3578
	3.75	73429 ± 16559	46342 ± 10362
N.		10	7
		Alloantigen induced proliferation (MLR)	
Stimulator cells		Control	DOTC treated
Syngeneic (PVG)		13677 ± 612	9863 ± 446
Allogeneic (PVG x DA)F1		36357 ± 1671	23570 ± 583
N.		15	16

Results are mean DPM ± standard error

reduced alloantigen reactivity in the MLR are consistent with the reduced numbers of circulating W3/25 + ve T cells.

These results clearly suggest that DOTC can lead to altered immune function, however, other experiments from this laboratory indicate that the overall immune competence of 6-8 wk old animals exposed to 75 ppm DOTC for up to 12 wks is not significantly different from untreated control animals (Miller, Scott, & Foster, 1984; Miller & Scott, 1985). These observations indicate that if DOTC induced thymic atrophy occurs after peripheral lymphoid tissues are seeded with T cells then although peripheral blood lymphocyte traffic/reactivity is altered, overt immune competence is maintained (at least for the life span of the peripheral T cells). Nevertheless the fact that immune suppression may occur when DOTC is given prior to weaning (Seinen et al. 1977) indicates that DOTC induced neonatal "chemical thymectomy" could compromise the developing immune system. However, the precise interaction that occurs between DOTC and the cellular components of the thymus still remains to be determined. As the thymus consists almost entirely of thymocytes and reticulo-epithelial cells (RECs) the question arose as to whether DOTC induced thymic atrophy may be mediated via an effect on the RECs. These cells are believed to be the source of thymic humoral factors, and as such are considered vital for T cell maturation (Stutman, 1978; Bach et al. 1979). There is also considerable evidence in laboratory animals of thymic involution during pregnancy, followed by subsequent regeneration in the post partum period (Millar, Mills and Baines, 1973; McLean, Mosley and Gibbs, 1974; Bauminger and Peleg, 1978). It was therefore decided to investigate the effects of DOTC and/or pregnancy on thymic ultrastructure with particular reference to the effects on the RECs. In this study syngenetically pregnant PVG rats were exposed to control diet or diet containing 60 ppm DOTC from day 5 of pregnancy and killed immediately after parturition (3 wks later). Virgin rats fed identical diets and killed at the same time served as negative controls. At post mortem, thymuses from pregnant rats maintained on control diets showed slight thymic involution; however the thymuses from both virgin and pregnant DOTC treated rats were markedly atrophied. Tissue sections of thymuses from post partum control animals apeared normal, with a clear distinction between cortex and medulla. Reticulo-epithelial cells (RECs) were present in both the cortex and medulla and often contained large numbers of non-osmiophillic vacuoles. Virgin animals exposed to DOTC for the same period showed massive thymocyte depletion in the cortical region, the number of thymocytes was also slightly reduced in the medullary region. There was no marked increase in the degree of vacuolation in the RECs. Pregnant animals given DOTC during pregnancy and killed immediately post partum presented a similar picture in terms of thymocyte depletion, however, the degree of vacuolation in the RECs was greatly increased. It would therefore appear that the changes in the thymus occurring during pregnancy are qualitatively, but not quantitatively similar to those observed during DOTC-exposure. In both instances there was evidence of vacuolation in RECs however, the degree of vacuolation observed as well as the number of vacuolated cells was greater in rats exposed during pregnancy. It therefore appears that the changes in endocrine control that occur during pregnancy may exacerbate the effects of DOTC.

Additional studies investigating the organ distribution of orally administered DOTC also focuses attention on the possibility that endocrine factors may play a role in DOTC induced thymic atrophy. To date very few studies have examined the tissue distribution and possible metabolism of alkyl tin compounds. Using atomic absorption spectrophotometry and HPLC Iwai, Wata and Arakawa (1981), found tributyltin dichloride and its metabolites to be present only in the liver and brain of rats after a single oral dose of 40 mg/kg. Another study (Yu and Arakawa, 1983) found dibutyltin distributed to every organ, with levels in the liver and kidney very much higher than in other organs, but very little was localised within the thymus.

In view of the lack of information concerning DOTC organ distribution, it was decided to examine the tissue distribution of ^{14}C-labelled DOTC with a view to determining whether the *in vivo* effects of DOTC result from specific DOTC accumulation into the thymus or whether other organs may also be involved.

PVG male rats weighing 150g were orally dosed with 100 μCi ^{14}C-labelled DOTC in corn oil and housed individually in closed metabolic units. Animals were killed at 6, 24 and 48 hrs later and thymus, pituitary, spleen, liver, mesenteric lymph nodes, urine and faeces taken for sample oxidation and scintillation counting. It was apparent that almost all of the radioactivity was excreted in the faeces by 48 hrs. Nevertheless the presence of radioactivity in the liver and urine indicated that a small proportion of the administered compound was absorbed and possibly metabolised. Furthermore although only a minute amount of label was found to be present in the thymus significantly more was found in the pituitary (Table 4). This supports the earlier observations of Seidler, Härtig and Woggon (1971) who likewise demonstrated slight uptake of ^{14}C-DOTC into the thymus and pituitary.

Despite the very specific action DOTC has on rat thymic tissues only a trace of radioactivity was found associated with the organ, it therefore seems unlikely that thymic atrophy is the result of a local DOTC-mediated toxic event. The

Table 4. Distribution of ^{14}C-labelled material in rat after oral administration of ^{14}C-DOTC expressed as percentage of amount administered

	6 hr	24 hr	48 hr
Liver	0.24%	0.40%	0.39%
Spleen	0.002%	0.002%	0.002%
Mesenteric lymph nodes	0.008%	0.006%	0.006%
Thymus	0.003%	0.0005%	0.001%
Pituitary	0.001%	0.019%	0.013%

observation that label was also present in the pituitary lends further support to the postulate that endocrine factors may play a role in DOTC induced thymic atrophy.

ACKNOWLEDGEMENTS

These studies form part of a research project funded by the U.K. Ministry of Agriculture, Fisheries and Food.

REFERENCES

Bach, J.F., Bach, M.A., Charriere, J., Dardenne, M., and Pleau, J.M. 1979, The mode of action of thymic hormones. Ann. N.Y. Acad. Sci. 332:23.

Bauminger, S., and Peleg, S., 1978, Changes in immunological activity of rat lymphoid organs during pregnancy, Clin. Exp. Immunol. 32:179.

Brideau, R.J., Carter, P.B., McMaster, W.R., Mason, D.W., and Williams, A.F., 1980. Two sub-sets of rat lymphocytes defined with monoclonal antibodies, Eur. J. Immunol. 10:609.

Iwai, H., Wata, O., and Arakawa, Y., 1981, Determination of tri-, di- and monobutyl tin and inorganic tin in biological materials and some aspects of their metabolism in rats, J. Anal. Tox., 5:300.

McLean, J.M., Mosley, J.G., and Gibbs, A.C.C., 1974, Changes in the thymus, spleen and lymph nodes during pregnancy and lactation in the rat, J. Anat., 118:223.

Millar, K.G., Mills, P., and Baines, M.G., 1973, A Study on the influence of pregnancy on the thymus gland of the mouse, Am. J obstet. Gynec. 117:913.

Miller, K., 1983, Various mechanisms in Chemically-Induced thymic injury, in: "Immunotoxicology" G.Gordon Gibson, R. Hubbard, and D.V. Parke, eds. Academic Press, London, U.K.

Miller, K., Scott, M.P., and Foster, J.R., 1984, Thymic involution in rats given diets containing dioctyltindichloride, Clin. Immunol. Immuno-path., 30:62.

Miller, K., and Scott, M.P., 1985, Immunological consequences of dioctyltin dichloride (DOTC) induced thymic injury, Toxicol. Appl. Pharmacol. (in press).
Oppenheim, J.J., Blaese, R.M. and Waldman, J.A., 1970, Defective lymphocyte transformation and delayed hypersensitivity in Wiskott-Aldrich Syndrome, J. Immunol., 104:835.
Seidler, H., Härtig, M., and Woggon, H., 1971, Das Verhalten von ^{14}C-markiertem Di-n-octylzinndithioglykolsäure-2-äthylhexylester und Di-n-octylzinnchlorid Sowie ^{35}S-markierstem Thiotglykolsäure-2-äthylhexylester in Rattenoranismus, Z Ges Hyg 17:603.
Seinen, W., and Willems, M.I. 1976, Toxicity of organotin compounds I. Atrophy of thymus and thymus-dependent lymphoid tissue in rats fed di-n-octyltin dichloride, Toxicol. Appl. Pharmacol. 35:63.
Seinen, W., Vos, J.G., Van Krieken, R., Penninks, A., Brands, R., and Hooykaas, H., 1977, Toxicity of organotin compounds III. Suppression of thymus-dependent immunity in rats by Di-n-Butyltindichloride and Di-n-Octyltindichloride, Toxicol. Appl. Pharmacol. 42:1.
Stutman, O., 1978, Intrathymic and extrathymic T cell maturation, Immunological Rev., 42:138.
Vos, J., 1977, Immune suppression as related to toxicology, CRC Critical Reviews in toxicology, 5:67.
Webb, M., Mason, D.W. and Williams, A.F., 1979, Inhibition of mixed lymphocyte response by a monoclonal antibody specific for a rat T lymphocyte sub-set, Nature, 282:841.
Yu, T.H., and Arakawa, Y., 1983, HPLC determination of dialkyltin homologues using fluorescence detection. J. Chromatog., 258:189.
Ziegler, J.B., Hansen, P., and Penny, R., 1975, Intrinsic lymphocyte defect in Hodgkin's disease: analysis of the phytohaemagglutinin dose-response, Clin. Immunol. Immunopath., 3:451.

SESSION 3

STRUCTURAL ASPECTS OF IMMUNE INDUCTION

CHAIRPERSONS: R.M. Steinman
D.M. Mason

ANTIBODY RESPONSES TO T-DEPENDENT ANTIGENS:

CONTRIBUTIONS OF DENDRITIC CELLS AND HELPER T LYMPHOCYTES

Kayo Inaba and Ralph M. Steinman

The Rockefeller University and Irvington House Institute
1230 York Ave., New York, N.Y. 10021, U.S.A.

INTRODUCTION

The standard system for the study of antigen-specific, thymus dependent, antibody responses utilizes mouse spleen. B and T lymphocytes (Sephadex G10 nonadherent spleen) are cultured with accessory cells (plastic adherent spleen) and antigen (red cells or TNP/DNP-KLH*) for 4-5 days to generate specific IgM PFC. The accessory cells likely induce functioning helper T cells, and the latter cooperate with B lymphocytes during antibody formation.

TABLE I

Properties of mouse spleen dendritic cells and macrophages

Property	Dendritic cell	Macrophage
Origin	marrow	marrow
Adherence properties	adherent,then not	adherent
Endocytosis:		
phagocytosis	-/weak	active
pinocytosis	-/weak	active
Surface markers:		
leukocyte antigen	+	+
Ig	-	+/-
thy-1, lyt-1, lyt-2	-	-
33D1, α-DC	+	-
F4/80, α-MO	-	+
2.4G2, α-FcR	-	+
Ia	++	++/-

*Abbreviations: DC, dendritic cell; PFC, plaque forming cell; KLH, keyhole limpet hemocyanin; MLR, mixed leukocyte reaction; MHC, major histocompatibility complex; DNP/TNP, di/trinitrophenyl.

Studies on the need for adherent accessory cells in 1° antibody responses led to the discovery of dendritic cells (DC) several years ago (1,2). The early work established a large number of differences between DC and adherent macrophages (Table I and ref 3-5 for illustrative material). DC are Ia rich and Fc receptor negative, and becomes nonadherent after overnight culture. Macrophages are $Ia^{+ \text{ or } -}$, Fc^{+} and retain the capacity to adhere to glass or plastic. In mouse spleen, DC and macrophages usually are separated on the basis of their distinct adherence properties (3-5). Following separation, the macrophage and DC-rich fractions exhibit stable differences in morphology, phenotype and function over 2-3 days in culture. This article summarizes our recent work on the role of DC in antibody responses. Emphasis will be placed on the induction and function of T helper cells.

DC ARE REQUIRED FOR 1° T-DEPENDENT ANTIBODY RESPONSES

Once methods had been developed for selectively enriching and depleting DC, it became possible to analyze the accessory cell requirement for 1° antibody responses in culture. Enriched populations of DC reconstituted the response to control levels at doses of 0.3-1.0% of total cultured cells (6). Selective depletion of DC with the 33D1 monoclonal antibody and complement ablated accessory activity in heterogenous populations of spleen adherent cells and whole spleen (6). When spleen adherent cells were separated into macrophage and DC-rich components, accessory function was largely confined to the DC-rich fraction (Table II).

Although macrophages do not independently stimulate 1° antibody responses, these cells still could make many contributions. Macrophages could make IL-1 and thereby amplify the response. At this time there is no information on the production of IL-1 by either DC or macrophages <u>during</u> 1° T-dependent responses. Macrophages could also "process" particulate antigens. However, the cultures we have studied contained very few macrophages (<1 in 2000 cells), and there is as yet no data that heterologous red cells or hapten carrier conjugates need to be processed, especially by the "stimulator cells". Macrophages could also inhibit antibody formation (e.g., Table II), and this is especially evident in human blood cell cultures (6). DC-mediated inhibition of antibody responses has not been observed, e.g., as one increases the dose of DC one gets a larger and/or more rapid response until a plateau is reached at ~3% DC.

B cells can act as "antigen presenting" cells (7,8). Yet DC are still required to <u>induce</u> antibody responses to T-dependent antigens, i.e., the B cells in the Sephadex G10 nonadherent fraction do not substitute for DC (e.g., Table II). Most likely, B cells present antigens to helper T cells, but this only occurs after the helper cells are sensitized by DC (see below).

TABLE II

Accessory cells for T-dependent antibody responses

Accessory cells for 4×10^6 B/T	α-sRBC plaque forming cells day 4 (+/- RBC)	α-sRBC plaque forming cells day 5 (+/- RBC)
None	4/0	8/0
Spleen adherents, MO-rich		
1.5×10^5	76/44	308/8
5×10^4	30/0	26/0
1.5×10^4	25/16	22/4
Spleen adherents, DC-rich		
1.5×10^5	1244/96	1794/112
5×10^4	886/116	1440/120
1.5×10^4	230/20	602/4
Mix of MO and DC		
3×10^5	624/N.T.	376/N.T.
1×10^5	388/N.T.	572/N.T.
3×10^4	48/N.T.	152/N.T.
MO-rich, treat with		
No Ab, plus C	16/N.T.	2/N.T.
α-DC, "	8	6
α-Ia, "	8	0
α-lyt 1, "	14	26
DC-rich, treat with		
No Ab, plus C	1174/N.T.	1986/N.T.
α-DC, "	144	344
α-Ia, "	162	622
α-lyt 1, "	1134	1692

Spleen adherent cells were separated into macrophage (MO) and DC-rich components by readherence to plastic (3-5). The readherent fraction was >95% MO, while the nonadherent fraction was > 50% DC and < 5% MO. Other aliquots of spleen adherent cells were allowed to readhere in the presence of rabbit complement +/- monoclonals (33D1, α-DC; 10.216, α-Ia^k; and C3PO, α- lyt-1.2) to provide readherent MO and nonadherent DC that had been depleted of specific cell types. G10 nonadherent spleen cells were then added to the accessory cells and cultured 4 or 5 days +/- sheep red cells as antigen.

PRODUCTION OF B CELL HELPER (STIMULATING) FACTORS:

One function of the helper cell is to generate B cell stimulating factors, which probably include at least two distinct class of mediators termed B cell growth and differentiation factors (reviewed in 9,10). Typically these factors are obtained from media conditioned by T cell lines, lectin-stimulated spleen cells, or secondary mixed leukocyte reactions (MLR's).

We have recently shown that DC stimulate T cells to release B cell helper factors in the syngeneic MLR (11). The latter is a weak T cell proliferative response that occurs whenever DC are cocultured with T cells from mouse peripheral lymphoid organs (5). The helper assay is to add conditioned medium to B cells (DC and T-depleted) plus heterologous red cells. The factor-mediated response is antigen dependent and specific, but MHC nonrestricted (Table III and ref 11). We noted (11) that: a) DC were the principal stimulators for helper factor release in the syn MLR; b) the helper factors were released from aggregates of dendritic and T cells, which accounted for about 5% of the total culture; and c) factor release occurred by an IL-2 dependent pathway in which DC first activated syngeneic T cells to respond to IL-2, and then IL-2 induced factor release.

TABLE III

Helper factor production in the sygeneic and allogeneic MLR

Exp't #	B cells: Strain	Helper factors MLR	Helper factors Conc.	PFC from B cells plus no RBC	PFC from B cells plus sheep RBC	PFC from B cells plus horse RBC
1*	CxD2 F_1	Syn (CxD2)	50%	40/0	6117/17	60/5867
			25	60/3	4153/0	10/1890
		Allo (B6 α-AxB6)	50%	160/33	8587/37	83/5823
2	CxD2 F_1	Syn (CXD2)	40%	7	350	
		Allo (CXD2 α-H-2s)	40	3	883	
3	B6.H-2k	Syn (CXD2)	50%	N.T.	1644	
			25	N.T.	500	
			12.5	N.T.	68	
		Allo (CXD2 α-H-2s)	50%	N.T.	3272	
			25	N.T.	2440	
			12.5	N.T.	1064	

* In Exp't 1, PFC were determined on sheep and horse red cell layers (sheep/horse). PFC responses without MLR media were always <4.

Conditioned medium from the syn MLR is about 1/4 - 1/2 as active as the allogeneic MLR (Table III). Both the syn and alloMLR are considered to be examples in which DC trigger resting T cells to release helper factors. However it is still unclear if the synMLR is a true 1° reaction to self, or if some unidentified antigen is contributing. We are now studying the capacity of DC to induce the formation of helper cells during primary responses to protein antigens.

DIRECT HELPER T - B CELL INTERACTION DURING 1° RESPONSES:

A second function of the helper cell is to interact directly with <u>resting</u> B cells. As a consequence, the B cells develop into antibody secreting cells and/or become responsive to stimulating factors. This kind of helper T cell function seems to be essential for responses to certain hapten-carrier conjugates like TNP/DNP-KLH; i.e., antibody responses do not occur unless intact T cells are present in the culture e.g., ref 12.

We recently studied cell aggregates which develop in cultured spleen suspensions (13). The clusters appear to be the sites in which B cells become activated to helper factors. As antigens, we used heterologous red cells or DNP/TNP-KLH. Clusters were isolated at 20 and 40 hrs and proved to be the principal site in which antibody secreting cells (PFC) developed. B, T, and dendritic cells were present in the clusters. The B cells entered the cluster <u>after</u> exposure to antigen in vitro (for at least 8 hrs) and in an MHC-restricted fashion. Thus when DC and T cells were syngeneic, only syngeneic B cells clustered (Table IV and ref 13). When DC were P_1 and T cells P_2, then P_1 B cells became PFC within clusters.

The critical observation was that the B cells in the clusters had become responsive to helper factors (13). If T cells were eliminated from the clusters with α-thy-1 or α-lyt-1 and complement, T cell function could be restored with T cell replacing factors. In contrast TNP-specific responses did not occur when resting B cells were exposed to antigen and factors. It will be important to test if DC are inducing antigen specific helper cells, since our published experiments did not distinguish if helper cells in the clusters were carrier specific or "syn MLR reactive."

We also evaluated the role of DC during antibody responses in clusters. If we used antibody and complement to remove DC from clusters at day 1, prior to transformation of the responding T cells, then PFC development ceased; but if we removed DC at day 2, then PFC were generated at 50-100% of the levels seen in control cultures. This implied that DC were no longer needed by day 2, and that T cells were interacting directly with B cells to induce responsiveness to helper factors. To pursue this notion further, we studied the system of "allogeneic" help in which alloreactive T blasts could be recovered in high yield and purity (14).

TABLE IV
Clustering of dendritic, T, and B cells during α-TNP responses

Rx prior to PFC	PFC developing from clusters which were formed with B cells that were: Syn: H-2b	Allo: H-2k	H-2b and H-2k
No Ab/No C	992	148	512
α-K^b/No C	918	130	560
α-K^k/No C	980	134	602
α-D^b/No C	968	120	584
α-D^k/No C	966	134	568
No Ab/C	946	146	590
α-K^b/C	16	98	16
α-K^k/C	976	22	574
α-D^b/C	42	146	24
α-D^k/C	952	34	588

2.5×10^4 B6 (H-2b) dendritic cells, 2×10^6 B6 T cells and 3×10^6 B6 or B6.H-2k B cells (or 1.5×10^6 of each) were cultured 2 days with 5 ug TNP-KLH. Clusters were isolated (2.5×10^5/culture) and cultured 3 more days. α-TNP plaques were enumerated but prior to the assay, the cells were exposed to various alloantibodies +/- complement to phenotype the PFC. Elsewhere (14) it has been shown that the cultures must be exposed to antigen for specific B cells to enter clusters at day 2. It is evident that B6 (syngeneic) cells preferentially develop into PFC in clusters, but that allogeneic H-2k B cells do not suppress the B6 cells in the "mix" experiment. In the absence of T cells, PFC responses were <25. The positive control was that all cultures made similar responses (~1300 PFC) to TNP-Ficoll.

INTERACTION OF SENSITIZED T CELLS WITH DIFFERENT PRESENTING CELLS

If resting T cells were cocultured with allogeneic B cells, then there was no 1° MLR and no B cell help (13,14). However when enriched populations of lyt-2^- alloreactive blasts were retrieved from a 1° MLR, the blasts interacted effectively with B cells both large and small. The helper blasts would quickly aggregate with allogeneic small B lymphocytes, whereas resting T cells did not even after 5 days of culture. The interaction between small B cells and T blasts was shown to be bidirectional. The T blasts quickly resumed growth and lymphokine production when cultured with mitomycin C-treated or irradiated (750 rads) B cells, apparently in the absence of DC. Furthermore, the B cells began to proliferate and differentiate into antibody secreting cells when cultured with irradiated helper T blasts.

Therefore, there are really two stages in the life history of helper T cells (Table V). Resting cells need DC to be stimulated; once sensitized, the blast interacts effectively with other Ia$^+$ cells including B cells and macrophages (14).

TABLE V
Two stages in the life history of a helper T cell

Parameter	Resting T	Activated T
Source	Nylon nonadherent	DC-T cell clusters
Phenotype	Ia$^-$, lyt-2$^-$	Ia$^-$, lyt-2$^-$
Size	Small	Large
IL-2 responsiveness	No	Yes
Responds to:	Dendritic cells	DC, MO, B cells

SUMMARY

Dendritic cells (DC) are important accessory cells for T-dependent antibody formation. Other leukocytes -- such as macrophages and B cells -- do not independently initiate antibody responses. DC induce the two principal functions of helper T cells, i.e., direct activation of small, antigen-specific B cells, and release of B cell stimulating factors. These two helper functions can operate in tandem during the 1° response to hapten-carrier conjugates. Once the helper cell is sensitized in concert with DC, it interacts directly with B cells apparently in the absence of DC. These functions of helper cells all occur in discrete aggregates which contain DC and the appropriate responding lymphocytes. The DC/lymphocyte aggregates likely represent the appropriate structural unit for the study of cell-cell interactions during antibody responses.

ACKNOWLEDGMENT

The authors acknowledge the collaboration of Drs. Z.A. Cohn, S. Muramatsu, A. Granelli-Piperno, and M. Witmer. Dr. Inaba is a Visting Investigator from Kyoto Univ., Kyoto 606, Japan. Dr. Steinman is an Established Investigator of the American Heart Association. Funds supporting this work were provided by the N.I.H . grants AI 13013 and CA 30198.

REFERENCES

1. Steinman R.M., and Cohn Z.A. 1973. Identification of a novel cell type in peripheral lymphoid organs of mice. I. Morphology, quantitation, tissue distribution. *J. Exp. Med.* 137: 1142-1162.
2. Steinman R.M., and Cohn Z.A. 1974. Identification of a novel cell type in peripheral lymphoid organs of mice. II. Functional properties *in vitro*. *J. Exp. Med.* 139: 380-397.
3. Nussenzweig M.C., and Steinman R.M. 1980. Contribution of dendritic cells to stimulation of the murine syngeneic mixed leukocyte reaction. *J. Exp. Med.* *151*:1196-1212.
4. Nussenzweig M.C., Steinman R.M., Unkeless J.C., Witmer M.D., Gutchinov B. and Cohn Z.A. 1981. Studies of the cell surface of mouse dendritic cells and other leukocytes. *J. Exp. Med.* *154*:168-187.
5. Steinman R.M., Gutchinov B., Witmer M.D., and Nussenzweig M.C. 1983. Dendritic cells are the principal stimulators of the primary mixed leukocyte reaction in mice. *J. Exp. Med.* 157:613-627.
6. Inaba K., Steinman R.M., Van Voorhis W.C., and Muramatsu S. 1983. Dendritic cells are critical accessory cells for thymus-dependent antibody responses in mouse and in man. *Proc. Natl. Acad. Sci.* U.S.A. 80: 6041-6045.
7. Chesnut R.W. and Grey H.M. 1981. Studies on the capacity of B cells to serve as antigen-presenting cells. *J. Immunol.* 126: 1075-1079.
8. Ashwell J., DeFranco A., Paul W., and Schwartz R. 1984. Antigen presentation by resting B cells: radiosensitivity of the antigen presentation function and two distinct pathways of T cell activation. *J. Exp. Med.* 159: 881-905.
9. Melchers F. and Andersson J. 1984. B cell activation: three steps and their variations. *Cell* 37: 715-720.
10. Finkelman F.D. and Vitetta E.S. 1984. Role of surface immunoglobulin in B lymphocyte activation. *Fed. Proc.* 43: 2624-2632.
11. Inaba K., Granelli-Piperno A., and Steinman R.M. 1983. Dendritic cells induce T lymphocytes to release B cell stimulating factors by an interleukin 2 dependent mechanism. *J. Exp. Med.* 158: 2040-2057.
12. Roehm N.W., Marrack P., and Kappler J.W. 1982. Antigen specific H-2 restricted helper T cell hybridomas. *J. Exp. Med.* 156: 191-204.
13. Inaba K., Witmer M.D., and Steinman R.M. 1984. The clustering of dendritic cells, helper T lymphocytes, and histocompatible B cells during primary antibody responses *in vitro*. *J. Exp. Med.* In press.
14. Inaba K., and Steinman R.M. 1984. Resting and sensitized T lymphocytes exhibit distinct stimulatory requirements for growth and lymphokine releases. *J. Exp. Med.* In Press.

DEVELOPMENT OF DENDRITIC CELLS FROM RAT BONE MARROW

William E. Bowers and Mary R. Berkowitz

The Mary Imogene Bassett Hospital
Cooperstown, N.Y. 13326

INTRODUCTION

The dendritic cell is a recently recognized new cell type having important immunological functions. It provides accessory activity required for responses of T lymphocytes, acts as a potent stimulator of a mixed leukocyte reaction and, owing to its richness in Ia antigens, serves as a critical "passenger cell" that elicits rejection of transplanted tissues. In rat, the dendritic cell is the principal accessory cell required for proliferation of Tx lymphocytes induced by treatment with sodium periodate. As measured by ^{3}H-thymidine incorporation, proliferation shows a linear dose-response with the addition of dendritic cells, which provided a useful assay.

Based on this assay, we developed a procedure for purifying dendritic cells from lymph nodes. This procedure took advantage of the fact that rat dendritic cells have a low density, are non-adherent, are insensitive to irradiation (1,000 rads), and lack Fc receptors. As a result it was possible to purify rat dendritic cells some 1,200-fold over the starting lymph node cell preparation and to recover >70% of the dendritic cells in a fraction that contained >80% dendritic cells (Klinkert et. al., 1980).

We applied the same fractionation procedure to cell suspensions of other rat tissues in an effort to characterize and enrich dendritic cells. In all tissues studied (spleen, thymus, liver, skin, peritoneal exudate cells, and thoracic duct lymph), the final preparation was greatly enriched in dendritic cells, but the extent of contamination varied. Once dendritic cells were sufficiently purified, the measured accessory activity in all tissues reflected the content of dendritic cells.

The similarity of morphological features and functional properties suggested that dendritic cells isolated from all these tissues belong to the same family (Klinkert et. al., 1982).

Bone marrow was included in the group of tissues surveyed. Unlike cell preparations from other tissues, however, bone marrow cells exhibited unique features that led us to examine this tissue in more detail. The main findings of our study are presented here.

RESULTS AND DISCUSSION

After fractionation of a fresh preparation of bone marrow cells, extremely low levels of accessory activity were measured and recognizable dendritic cells were not present. If, however, bone marrow cells were cultured for five days, significant levels of both accessory activity and dendritic cells resulted (Table 1). Irradiation of freshly prepared bone marrow cells abolished the production of accessory activity and dendritic cells after five days of culture (Table 1). As was the case with dendritic cells prepared from other tissues, those that appeared in bone marrow cultures were recovered in a low density fraction. They were non-adherent, Ia-positive, Fc receptor-negative, and were morphologically indistinguishable from dendritic cells isolated from other tissues. Moreover, they possessed an accessory activity per cell identical to that of dendritic cells purified from lymph nodes. Thus, it appears that dendritic cells arise from radio-sensitive precursors in cultures of bone marrow cells.

TABLE 1 Accessory activity of irradiated and unirradiated bone marrow cells cultured for five days

No. of viable cells tested ($x 10^{-3}$)	^{3}H-Thymidine incorporation[b]	
	Irradiated[a]	Unirradiated[a]
0	385 ± 105	385 ± 105
7	------	957 ± 106
15	384 ± 79	1557 ± 504
30	619 ± 245	4253 ± 747
60	378 ± 111	9334 ± 1253
120	713 ± 126	16019 ± 444
240	497 ± 58	-------

[a]Bone marrow cells were irradiated prior to being tested for accessory activity with 6 x 10^5 periodate-treated high-density lymph node cells.

[b]^{3}H-Thymidine was measured during the last four hours near the peak of the response (48 hours).

Some of the properties of the precursors were explored. We found that virtually all (>90%) of the dendritic cells produced in culture derived from a minor population of low density cells that comprised only 4.38 ± 1.18% of the total unfractionated bone marrow cells (N=29). We therefore used this low density fraction as the starting preparation for all subsequent experiments.

The effect of irradiation on dendritic cell precursors was examined by irradiating one culture of low density bone marrow cells each day over a five-day period and determining the accessory activity and number of dendritic cells produced at the end of five days. The results, presented in Table 2, show that both accessory activity and dendritic cells increase daily through the first four days and then reach a plateau. These findings suggest that the production of dendritic cells results from cell division of precursors, a process impaired by irradiation.

A similar experiment was performed to compare lymph node dendritic cells cultured under the same conditions as low density bone marrow cells. In this experiment, the number of dendritic cells was determined the day after irradiation. The results for low density bone marrow cells, recorded in Table 3, are similar to those of Table 2; they exclude the possibility that the number of dendritic

Table 2 Effect of Irradiation of Low Density Bone Marrow Cells on Successive Days on the Development of Accessory Activity and Number of Dendritic Cells Produced

	Experiment 1		Experiment 2	
Day of Irradiation[a]	Units of Accessory Activity[b]	Number of Dendritic Cells	Units of Accessory Activity	Number of Dendritic Cells
0	59[c]	9,800[c]	8	2,700
1	369	39,400	---	---
2	825	74,200	668	29,600
3	1276	106,400	634	40,900
4	1300	173,300	900	60,200
5	1485	161,000	988	64,200

[a]Six cultures of low density bone marrow cells were established. Starting on day 0, one was removed daily, irradiated, and returned to the incubator. After 5 days, accessory activity was determined and the number of dendritic cells counted.
[b]The unit of accessory activity is defined as the number of cells producing a response 2,000 cpm above the control response.
[c]Units of accessory activity and number of dendritic cells are both given per 10^7 low density bone marrow cells initially cultured.

Table 3 Number of Dendritic Cells in Cultures of Low Density Bone Marrow Cells and Lymph Node Dendritic Cells Irradiated on Successive Days and Counted the Following Day

Day Irradiated	Number of dendritic cells in culture of: LD-BMC	Lymph node dendritic cells
0	14,400	107,400
1	45,000	116,000
2	93,900	111,000
3	117,800	72,800
4	139,400	65,000
5	148,300	67,200

cells in each culture was initially the same, but that the longer the time elapsing between irradiation and counting, the greater the loss of dendritic cells, thereby resulting in an apparent increase in dendritic cells over the five day culture period. In contrast to the increase in the number of dendritic cells in the bone marrow cell cultures, the number of lymph node dendritic cells did not change during the first few days and then declined (Table 3). Thus, the data suggest that a dividing precursor in bone marrow gives rise to dendritic cells, but once mature, as exemplified by lymph node dendritic cells, it does not divide under the same conditions.

The dendritic cell expresses abundant amounts of Ia, and it is possible to remove these cells by an indirect panning method that utilizes anti-mouse Ig-coated dishes to select cells pre-incubated with mouse anti-rat Ia monoclonal Ig. We performed six experiments in which low density bone marrow cells were either pre-incubated with anti-Ia Ig or not, and then panned over rabbit anti-mouse Ig-F(ab')$_2$-coated plates. As a control, cells were also panned over uncoated dishes. The resulting preparations were then cultured for five days, at which time the number of dendritic cells and accessory activity were determined. The results are given in Table 4. They show that panning for Ia-positive cells has little effect on the production of dendritic cells by five days and no influence on the measured accessory activity.

A number of controls were performed to assure that the panning procedure was effective. Lymph node cells or low density lymph node cells were panned at the same time. The ability of these preparations to respond to periodate or to function as accessory cells was decreased by more the 90%. Immunofluorescence for Ia-positive cells was performed before and after panning of low density bone marrow cells. Virtually all (2-5%) of the Ia-positive cells were removed by specific anti-Ia and panning, whereas the low density bone marrow cells panned on Ig-coated dishes or blank dishes showed no difference in their content of Ia-positive cells.

These results indicate that the precursor of the dendritic

Table 4 Effect of Panning for Ia-positive Cells in Low Density Bone Marrow Cells on Development of Dendritic Cells and Accessory Activity

Panning	Number of Dendritic Cells ($x 10^{-3}$)	Units of Accessory Activity[b]
Blank dish	118.5 ± 47[a]	1452 ± 566[a]
Ig-coated dish	121.0 ± 49	1906 ± 698
Anti Ia + Ig-coated dish	96 ± 35	1968 ± 604

[a]The values given are means ± standard deviations for six experiments and are based on 10^7 low density bone marrow cells initially cultured.
[b]The unit of accessory activity is defined in the legend to Table 2.

cell either lacks Ia or expresses insufficient Ia on its surface to be removed in the panning procedure. In either case, there is a difference between the precursor and the dendritic cell.

CONCLUSIONS

We draw several conclusions from this work. Unlike the other tissues studied, bone marrow contains cells that are precursors of dendritic cells. Based on their sensitivity to irradiation, precursors most likely divide to produce dendritic cells. Although they possess a low density, as do mature dendritic cells, the precursors differ in terms of their low content, or lack, of Ia on the cell surface. Mature dendritic cells possess abundant amounts of Ia.

This work was supported in part by NIH grant AI-17887 and the Stephen Carlton Clark Fund of The Mary Imogene Bassett Hospital.

REFERENCES

Klinkert, W.E.F., LaBadie, J.H., O'Brien, J.P., Beyer, C.F., and Bowers, W.E., 1980, Rat Dendritic Cells Function as Accessory Cells and Control the Production of a Soluble Factor Required for Mitogenic Responses of T Lymphocytes, Proc. Natl. Acad. Sci, USA, 77:5414.

Klinkert, W.E.F., LaBadie, J.H., and Bowers, W.E., 1982, Accessory and Stimulating Properties of Dendritic Cells and Macrophages Isolated From Various Rat Tissues, J. Exp. Med., 156:1

THE RELATION BETWEEN RAT LYMPH NODE INTERDIGITATING CELLS IN SITU AND DENDRITIC CELLS IN SUSPENSION

E.W.A. Kamperdijk[1], M. van den Berg[1], M. Kapsenberg[2] and E.C.M. Hoefsmit[1]

[1]Department of Electron Microscopy, Medical Faculty Free University, Amsterdam, The Netherlands
[2]Laboratory of Histology and Cell Biology, Academic Medical Center, University of Amsterdam, Amsterdam The Netherlands

THE NORMAL LYMPH NODE

Interdigitating cells (IDC) are present in the thymus dependent areas of peripheral lymphoid organs such as lymph nodes. They resemble the epidermal Langerhans cells in many aspects: both cell types form the same characteristic interdigitations with the surrounding cells, having the pale cytoplasm and the irregularly shaped, most eccentrically situated nucleus. However, in non-stimulated lymph nodes Birbeck granules, which are the characteristic cell organelles of the Langerhans cells, are usually not present in IDC (1). IDC have a central spot of acid phosphatase (APh) activity and the plasma membrane is strongly Ia positive. These cells may function as accessory cells in antigen presentation to immunocompetent T cells (2).

Dendritic cells (DC) isolated from non-experimentally stimulated lymph nodes are large cells with long cytoplasmic extensions similar to IDC, the nuclei of these FcR-negative cells are eccentrically situated and the cell membrane is strongly Ia positive. They function as stimulators of mixed leucocyte reactions and as accessory cells (antigen presenting cells) in immune responses in vitro (3).

To study the relationship between IDC in situ and DC in suspension we injected paratyphoid vaccine (PTV) in rat footpads and characterized these cells on light- and electron microscopical level in the different stages of the provoked immune response in the draining (skin associated) lymph nodes.

IDC IN STIMULATED LYMPH NODES

A series of male Wistar rats was given 0.1 ml PTV containing 5 x 10^9 formol killed micro-organisms per ml, in one footpad. This vaccine causes thymus-dependent and thymus-independent humoral immune responses (1).

After 1 day (induction phase) and 2 weeks (memory phase) a combined Ia and acid phosphatase (APh) staining was carried out on lymph node cryosections. Ia labelling was performed with an indirect immunoperoxidase method, for APh staining the method according to Burnstone (4) (Naphthol-AS-BI phosphate as substrate) was used.

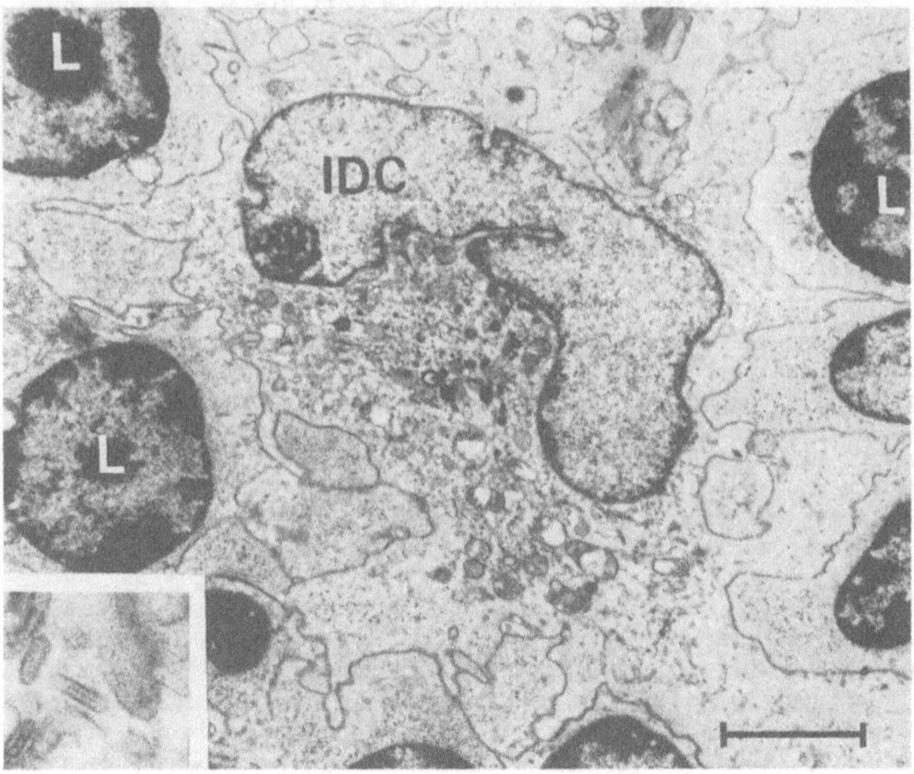

FIG. 1. IDC in the paracortex of a popliteal lymph node, 1 day after PTV (induction phase). The cytoplasm contains Birbeck granules (inset). L = lymphocyte. Bar 2.3 µm

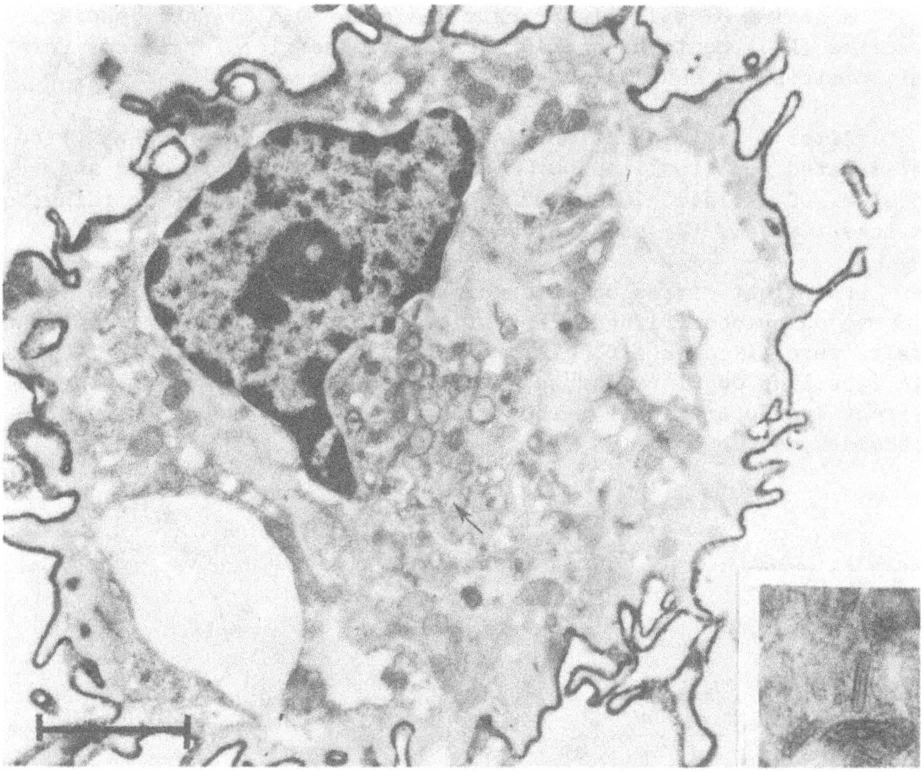

Fig. 2. DC, 1 day after PTV (induction phase). The cell has a strongly Ia positive cell membrane. Bar 2.6 μm. Inset: magnification of a Birbeck granule.

For electron microscopy perfusion fixation with 1.5% glutaraldehyde and postfixation with 1% OsO_4 was carried out.

In both phases of the provoked immune response on light microscopical level IDC could easily be distinguished by their eccentrically situated nucleus, the centrally positioned APh activity and the Ia staining of the cell membrane. Electron microscopical observations revealed that in the induction phase many IDC contained Birbeck granules (Fig. 1), by which they were very similar to the epidermal Langerhans cells. On this moment IDC did not show a distinct phagocytic activity. In the memory phase these cells never contained Birbeck granules and some of them were actively phagocytizing necrotic lymphocytes.

DC FROM STIMULATED LYMPH NODES

A series of male Wistar rats was given 0.4 ml paratyphoid vaccine (PTV) containing $5x10^9$ formol-killed micro-organisms per ml, distributed equally in all four footpads.

After 1 day (induction phase) and 2 weeks (memory phase) the stimulated popliteal- and axillary lymph nodes were removed and pooled. DC isolation occurred by a modified procedure according to Klinkert et al. (3) using Ficoll gradients.

For light microscopy a combined Ia and APh staining was performed on cytocentrifuge preparations. For electron microscopy the cells were fixed in 1.5% glutaraldehyde and postfixed with 1% OsO_4. Ia labelling on ultramicroscopical level was performed with an indirect immunoperoxidase method (the cells were fixed in 0.5% glutaraldehyde).

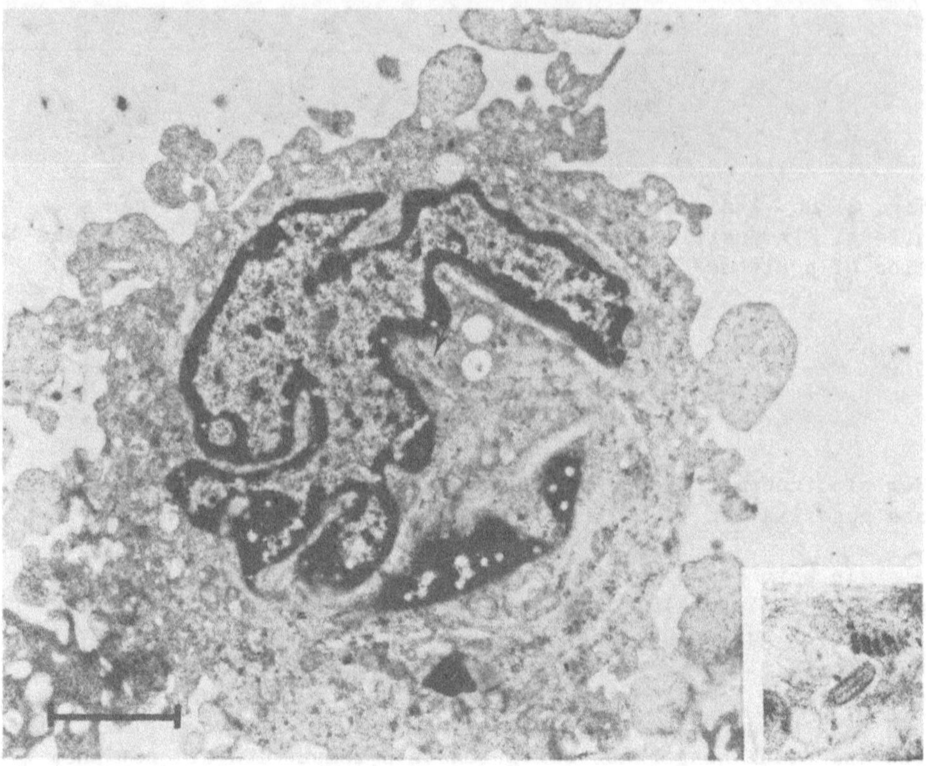

Fig. 3. DC, 1 day after PTV (induction phase). The cytoplasm contains Birbeck granules (inset). Bar 2.7 µm

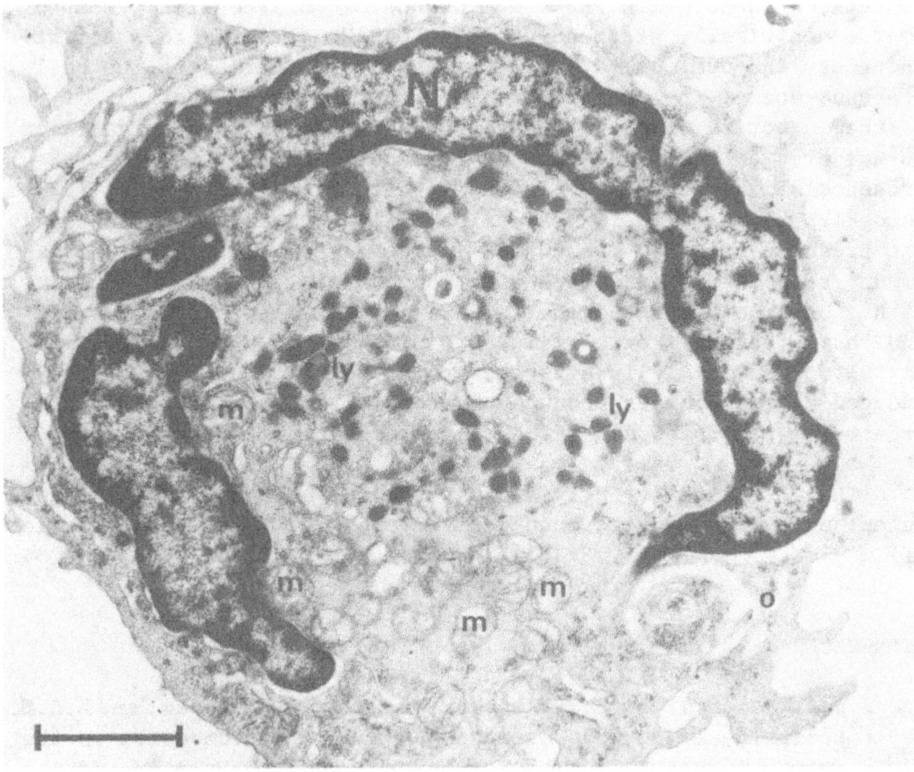

Fig. 4. DC, 2 weeks after PTV (memory phase). The nucleus (N) is eccentrically situated and the cytoplasm contains mitochondria (m) and lysosomes (ly). Bar 2.4 μm. Birbeck granules are not present.

All isolated DC had an irregular outline, APh activity in a spot near the eccentrically situated, irregularly shaped nucleus and were strongly Ia positive (Fig. 2). DC isolated from lymph nodes in the induction phase of the response frequently contained Birbeck granules (Fig. 3), which were not seen in cells isolated from lymph nodes in the memory phase (Fig. 4). These organelles did not show any relationship with other cell organelles or with the cell membrane. In all experiments these DC in suspension never showed distinct signs of phagocytic activity.

DISCUSSION AND CONCLUSION

Our experiments clearly show that lymph node IDC in the different phases of an immune response resemble DC isolated from nodes in the corresponding phase, with respect to morphology, enzyme- and immunohisto(cyto)chemistry. This appears from a) the dendritic aspects of these cells, b) the presence of an irregularly shaped most eccentrically situated nucleus, c) APh activity in a central spot near the nucleus, d) the presence of Ia antigen on the cell membrane and e) the presence of Birbeck granules in both cell types in the induction phase of the response. Although these organelles do not play an important role in lymphocyte stimulation (5), the presence of these granules in IDC in situ gives the possibility to recognize these cells in suspension. All these aspects, especially the presence of Birbeck granules in the same phase of the provoked immune response, indicates a relation between IDC and DC. However, it has been demonstrated that DC are non-phagocytic (absence FcR?) and that after antigenic stimulation IDC may display phagocytic activity, especially in a secondary immune response (6) possibly induced by lymphokines produced by sensitized T cells, or after treatment with Tilorone, which causes depletion of T cells (7).

These results strongly suggest that DC are the in vitro equivalent of a FcR-negative, non-actively phagocytizing subpopulation of IDC.

REFERENCES

1. E.W.A. Kamperdijk, E.M. Raaymakers, J.H.S. de Leeuw, and E.C.M. Hoefsmit, Lymph node macrophages and reticulum cells in the immune response. I. The primary response to Paratyphoid vaccine, Cell Tissue Res. 192:1 (1978).
2. E.C.M. Hoefsmit, A.M. Duijvestijn, and E.W.A. Kamperdijk, Relation between Langerhans cells, veiled cells and interdigitating cells, Immunobiol. 161:255 (1982).
3. W.E.F. Klinkert, J.H. LaBadie, and W.E. Bowers, Accessory and stimulating properties of dendritic cells and macrophages isolated from various rat tissues, J. Exp. Med. 156:1 (1982).
4. A.G.E. Pearse, "Histochemistry: theoretical and applied", Vol. I, 3rd. ed., Churchill Livingstone, Edinburgh (1968).
5. E.W.A. Kamperdijk, M. van den Berg, and E.C.M. Hoefsmit, The immunological significance of Birbeck granule containing cells, in: "In Vivo Immunology", P. Nieuwenhuis, A.A. van der Broek, and M.G. Hanna Jr., eds., Plenum Publishing Corporation, New York, p. 415 (1982).
6. E.W.A. Kamperdijk, J.H.S. de Leeuw, and E.C.M. Hoefsmit, Lymph node macrophages and reticulum cells in the immune response. The secondary response to paratyphoid vaccine. Cell Tissue Res. 227:277 (1982).
7. S. Levine, P. Cacho, R.A. Sowinski, and W.L. Thelmo, Depletion of lymphoid tissues by Tilorone. Ultrastructure and synergistic interaction with anti lymphocyte serum, Amer. J. Pathol. 111:174 (1983).

DENDRITIC CELLS AND MACROPHAGES AS ANTIGEN PRESENTING CELLS

IN OVALBUMIN-INDUCED T LYMPHOCYTE PROLIFERATION IN VITRO

Martien L. Kapsenberg, Frank E.M. Stiekema and
Marcel B.M. Teunissen

Laboratory of Histology and Cell Biology
University of Amsterdam, Academic Medical Centre
Amsterdam, The Netherlands

INTRODUCTION

A variety of cells has been reported as having the capacity to act as antigen presenting cells in antigen specific activation of helper T cells. They include dendritic cells, macrophages and B cells. Dendritic cells (DC) in particular were shown to be potent antigen presenting cells in the proliferative T cell response to soluble antigens (1). In this volume we (2) have shown that DC most likely represent an in vitro subpopulation of interdigitating cells. Interdigitating cells are localized in T cell areas of peripheral lymphoid organs. Previous experiments have suggested that these cells play a role as inducers of T cells in responses to subcutaneously administered paratyphus vaccin (3). Since both in vitro and in vivo observations stress that DC are obvious antigen presenting cells, we have studied the role of DC as antigen presenting cells in the secondary response to ovalbumin in more detail.

Previous studies on the mechanisms of antigen presentation by macrophages (MØ) and B tumor cells have demonstrated that lysosomal degradation of antigen is essential for conversion of native antigen into immunogeneic fragments (4,5). When compared to classical macrophages (FcR^+), DC (FcR^-) are poorly endocytizing cells (6). In a comparative investigation on antigen presentation by DC and MØ we have therefore paid special attention to the question of how these cells present antigen.

MATERIALS AND METHODS

Mice

Female C57Bl/10ScSn mice were obtained from the Central Institute for the Breeding of Laboratory Animals TNO (Zeist, The Netherlands).

Antigens and immunization

Ovalbumin (OVA) grade V was obtained from Sigma (St. Louis, Mo. USA). OVA was coupled to poly acroleine latex beads (Polyscience, Warrington, U.K.) by incubation of 25 mg/ml OVA with 5 mg of these beads for 2 h followed by a fivefold washing. OVA in saline was emulsified in an equal volume of complete Freund's adjuvant. Animals were injected subcutaneously at both sides of the tailbase and in the hind footpads with 30 μl emulsion containing 25 μl OVA.

Cells.

10-16 days after priming, T cells were isolated from a mixed cell suspension of popliteal, inguinal and paraaortic lymph nodes by nylon wool filtration. Antigen presenting cells were thoroughly depleted from the T cells by removal of Ia^+ cells using an immuno-selective rosette method followed by density gradient centrifugation. T cells consisted of 91% Thy 1.2^+ cells, less than 0.5% sIg^+ cells and less than 0.5% Ia^+ cells, as determined with the appropriate antisera in the fluorescence microscope. Peritoneal macrophages were obtained by lavage of the peritoneal cavity. They consisted of 70-85% macrophages as determined by staining for Mac 1 antigen and for staining of acid phosphatase activity (a strong activity throughout the whole cytoplasm). Dendritic cells were isolated according to the method of Steinman (6). The non-EA-rosetting, transiently adherent cells could be killed for 84% by treatment with anti-dendritic cell Mab 33D1 (a generous gift of Dr. R.M. Steinman) and complement. 93% were Ia^+, 2% sIg^+, 5% Thy 1^+ and less than 1% Mac 1^+. DC showed a acid phosphatase staining that is concentrated in a discrete area near the nucleus. Antigen presenting cells were pulsed in vitro by incubating 1.10^5 cells with 1 mg/ml OVA for 2 h. Where indicated in the text 0.1 mM chloroquine was added during the pulsing. In that case the cells were additionally incubated with chloroquine prior to the antigen pulsing.

Functional assays

Cells were cultured in IMDM supplemented with 3.2 mg/ml albumin, 32 μg/ml transferrin, 10 μg/ml insulin, 1.2 ng/ml zinc chloride, 0.1% fetal calf serum, 5.10^{-5} M 2-ME and antibiotics. $3x10^5$ T cells were cocultured with antigen presenting cells in 200 μl medium with or without 100 μg/ml OVA for 5 days. Proliferation was measured by the incorporation of tritiated thymidin.

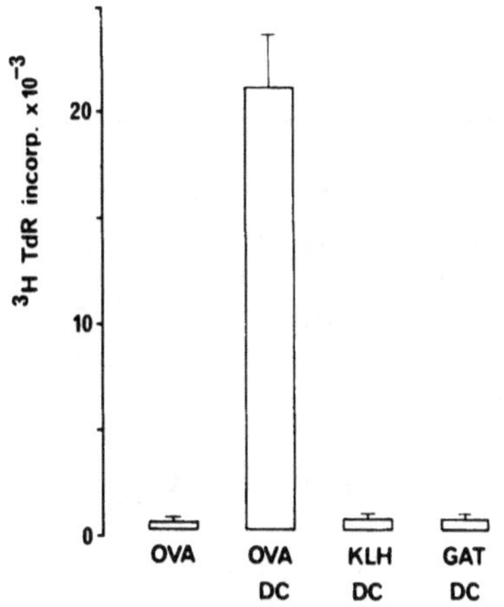

Fig. 1. Proliferation of $3x10^5$ T cells in the presence of OVA, KLH and GAT in presence or absence of dendritic cells (DC).

RESULTS AND DISCUSSION

Comparison of antigen presenting cell function of DC and MØ

In vitro proliferation of primed lymph node T cells in the secondary response of OVA required the presence of antigen presenting cells (Fig. 1). Since the presence of antigens like KLH or GAT did

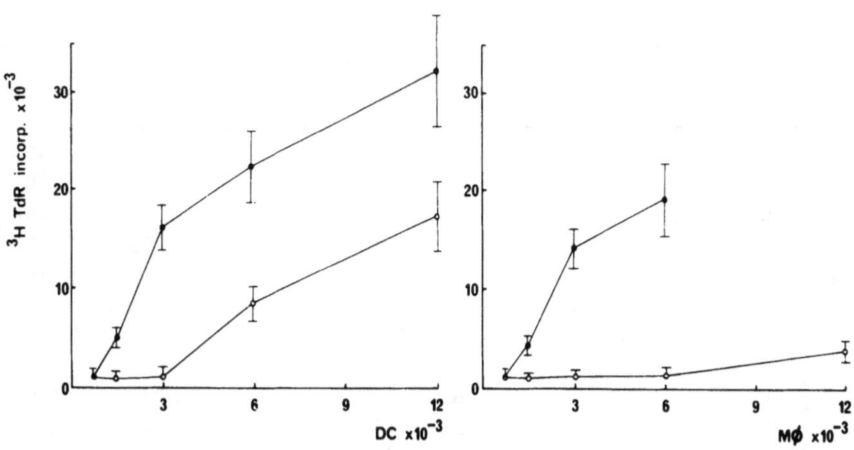

Fig. 2. Proliferation of purified T cells in the presence of a varying number of dendritic cells (a) or macrophages (b) with (0-0) or without (0-0) 100 μg/ml OVA.

not lead to T cell proliferation, the response to OVA is antigen specific. We have compared the antigen presenting cell function of splenic DC with that of peritoneal MØ. Addition of a varying number of DC and MØ to purified T cells resulted to comparable levels of proliferation (Fig. 2). However, the use of relatively high numbers of DC induced T cell proliferation in absence of antigen. This response known as a syngeneic mixed leucocyte reaction is evident when more than $3x10^3$ DC are added to $3x10^5$ T cells and should be avoided in studies on antigen-specific responses. In subsequent experiments we used no more than $3x10^3$ per well.

Synergism in presentation of particulate OVA by DC and MØ

DC and MØ can present OVA synergistically, as appeared from experiments using mixtures of these cells. Addition of a small number of MØ to DC or a small number of DC to MØ resulted in T cell proliferations that were significantly higher than the proliferations induced by these antigen presenting cells individually (Table 1)

However, a far stronger synergism is found when particulate OVA instead of soluble OVA was used. OVA coupled to poly acrolein latex beads is readily presented by macrophages. The latex beads are phagocytized by these cells and presentation will follow after digestion of OVA molecules from the beads. T cell proliferation in response to particulate OVA in the presence of DC was rather low (Table 1). The poor antigen presenting capacity of particulate OVA by DC probably can be explained by the fact that DC are poorly endocytizing cells. Addition of a small number of MØ - too small to induce high T cell proliferations - to DC resulted in T cell

Table 1. Synergisms in antigen presentation of DC and MØ in response to soluble and particulate OVA.

Antigen presenting cells	Antigen	T cell proliferation
-	OVA	170 ± 20
$0.75x10^3$ DC	OVA	1.820 ± 100
$3x10^3$ DC	OVA	16.100 ± 1.800
$0.75x10^3$ MØ	OVA	3.400 ± 200
$3x10^3$ MØ	OVA	11.800 ± 1.200
$3x10^3$ DC + $0.75x10^3$ MØ	OVA	24.500 ± 3.700
$3x10^3$ MØ + $0.75x10^3$ DC	OVA	15.300 ± 2.700
$3x10^3$ DC	latex-OVA	2.800 ± 1.100
$0.75x10^3$ MØ	latex-OVA	2.100 ± 400
$3x10^3$ MØ	latex-OVA	14.400 ± 2.700
$3x10^3$ DC + $0.75x10^3$ MØ	latex-OVA	14.700 ± 2.500

Table 2. Effect of chloroquine on presentation of OVA by DC and MØ

Antigen presenting cells	Treatment	T cell proliferation cpm x 10^{-3}	
		no antigen	OVA
-	-	0.1 ± 0	0.6 ± 0
$3x10^3$ DC-OVA	-	29 ± 4.4	n.d.
$3x10^3$ DC-OVA	chloroquine	26 ± 5.1	n.d.
$6x10^3$ MØ-OVA	-	21 ± 2.0	24 ± 0.8
$6x10^3$ MØ-OVA	chloroquine	1.4 ± 0.4	12.5 ± 3.2

proliferations that were nearly as high as observed in T cells that had OVA presented by MØ. It is tempting to explain this observation by the fact that DC present OVA or fragments of OVA delivered by MØ.

Different mechanisms of antigen presentation by DC and MØ

An obvious difference in the mechanism of antigen presentation of DC and MØ seems to exist. We have studied the role of lysosomal enzymes in antigen handling by these cells using the lysosomotropic drug chroloquine. Chloroquine selectively inhibits lysosomal enzymes by raising the intracellular pH. In these experiments T cell proliferation was induced by OVA-pulsed antigen presenting cells. Table 2 shows that addition of 0.1 mM chloroquine during antigen pulsing markedly reduced the antigen presenting capacity of MØ, whereas the antigen presenting capacity of DC was unaffected. In another series of experiments similar results were obtained with KLH as antigen in the response of KLH-primed T cells (data not shown).

These observations strongly suggest that DC possess the unique capacity to present native unprocessed antigen in an immunogeneic form. Since these experiments do not prove whether DC possess an additional chloroquine-sensitive pathway for antigen processing it is possible that these cells also present fragmented antigen. The possibility that dendritic cells present native unprocessed antigen is interesting as it may imply that DC and MØ activate different T cell repertoires. Previous experiments (7,8) on cross reactivity in T cell proliferation in response to native and fragmented antigen are difficult to interpret, when it is considered that native antigen is presented in both forms when various antigen presenting cells are present.

REFERENCES

1. G.H. Sunshine, D.R. Katz, and M. Feldmann, Dendritic cells induce T cell proliferation under Ir gene control, J. Exp. Med. 152:817 (1980).
2. E.A.W. Kamperdijk, M. van den Berg, M.L. Kapsenberg, Relation between lymph node interdigitating cells in situ and dendritic cells in suspension, In: this volume.
3. E.C.M. Hoefsmit, A.M. Duyvestijn, and E.W.A. Kamperdijk, Relationship between Langerhans cells, veiled cells and interdigitating cells, Immunobiology. 161:255 (1982).
4. H.K. Ziegler, and E.R. Unanue, Decrease in macrophage antigen catabolism caused by ammonia and chloroquine is associated with inhibition of antigen presentation to T cells, Proc. Natl. Acad. Sci. USA. 79:175 (1982).
5. R.W. Chesnut, S.M. Colon, and H.M. Grey, Requirements for the processing of antigens by antigen presenting B cells, J. Immunol. 129:2382 (1982).
6. R.M. Steinman, G. Kaplan, M. Witmer, and Z.A. Cohn, Identification of a novel cell type in the peripheral lymphoid organs of mice, J. Exp. Med. 149:1 (1979).
7. R.W. Chesnut, R.O. Endres, and H.M. Grey, Antigen recognition by T cells and B cells: recognition of cross reactivity between native and denaturated forms of globular antigens, Clin. Exp. Immunopath. 15:397 (1980).
8. A.J. Infante, M.Z. Atassi, and C.G. Fathman, T cell clones reactive with sperm whale myoglobin, J. Exp. Med. 154:1342 (1981).

VEILED CELL LYMPHOCYTE INTERACTION

B.M. Balfour, L. Buttifant, J. O'Brien, J. Clarke and S.C. Knight

Clinical Research Centre, Harrow, England

INTRODUCTION AND METHODS

Interactions between veiled cells and lymphocytes have been studied by time-lapse cinematography. For this purpose rabbits were immunized with 10mgs Keyhole Limpet Haemocyanin (KLH) or Human Gamma Globulin (HgG) in Freunds' complete adjuvant and later boosted by an injection of 20mgs antigen alone. At various times after the second injection, veiled cells were prepared from the afferent lymph and lymphocytes from the peripheral blood (PBL). The veiled cells were pulsed with the appropriate antigen 0.4mgs KLH/ml or 1.5mgs HgG/ml for ½ hour, washed and added to autologous PBL at a level of 1%. The cells were filmed in an inverted microscope for up to 6 hours after mixing and after a delay of 10-16 hours.

RESULTS

KLH immunized rabbits

Veiled cells were prepared from immunized animals at 3 and 5 days after boosting, the cells were pulsed with KLH and mixed with autologous PBL. During the first 6 hours after mixing, interactions between veiled cells and a single large or small lymphocyte were frequently seen. They were of short duration, about 10-15 minutes and there was no change in the appearance, or behaviour of the veiled cells after such contacts. A more permanent type of interaction was also seen, but less frequently. A group of 3 small lymphocytes, already attached to each other, made contact with a veiled cell. One of the group assumed a rounded shape and remained immobile and close to the veiled cell, while the other two cells appeared to be attached to this cell by narrow uropod-like structures. The two satellite cells made vigorous movements away from the veiled cell, but peripolesis was not observed. The body of the veiled cell itself was contracted and the veils were hyperextended and continued to make characteristic movements. This type of interaction remained unchanged for several hours.

When these preparations were observed after a delay of 14 hours a number of larger aggregates were seen, containing one veiled cell and up to 10 lymphocytes. At least two of these were round and immobile and the remainder were making swimming movements away from the veiled cell; again peripolesis was not observed and the veiled cell continued to make active movements. When mixtures of cells were made 6 days after boosting no interactions were seen, but on the 7th day after boosting lymphocytes were seen to attach to veiled cells and perform peripolesis, but the veiled cells continued to make movements and seemed to be unaffected by this treatment.

HgG immunized rabbits

Veiled cells were pulsed with antigen and mixed with autologous PBL from animals immunized with HgG. In one preparation made 3 days after boosting a small lymphocyte was seen to attach to a veiled cell and immediately start to perform peripolesis. Shortly afterwards two small lymphocytes, already attached to each other, began to swim towards the veiled cell and quickly attached themselves to the first lymphocyte. One of the trio then became round and passive and remained close to the veiled cell, while the other two seemed to be attached to the round cell; again the satellite cells made swimming movements away from the veiled cell and did not perform peripolesis. In other preparations a single attached cell continued to perform peripolesis for long periods, but the veiled cell was not damaged by this activity.

When cells were taken 5 days after boosting with HgG the lymphocyte fraction contained a proportion of large cells with well-developed uropods. These cells were extremely active and frequently made contact with the veiled cells, aligning themselves very close to the veiled cell, but only for short periods. Some permanent interactions were also seen, but trio formation with one passive and two active cells was not observed. On the 6th day after boosting there were increased numbers of large lymphocytes with uropods, they appeared even more active, making short attacks on the veiled cells, after one of which the veiled cell was seen to die. No permanent interactions were observed at this time. On the 8th day after boosting large lymphocytes were still present and although they approached the veiled cells they did not actually attack them. However it is possible that these lymphocytes were secreting a toxic substance (1), since many of the veiled cells showed signs of degeneration, with the appearance of long spiky processes in place of the veils.

When untreated veiled cells were mixed with autologous PBL taken on the 3rd day after boosting, a few permanent interactions were seen, usually with a single small lymphocyte which remained round and passive, but peripolesis was not seen. On the 5th day after boosting large lymphocytes were present and a number of brief interactions were observed, but here the lymphocytes became attached by their uropods and did not align themselves close to the veiled cell. Peripolesis was not seen.

DISCUSSION

The interactions observed between lymphocytes and KLH-pulsed veiled cells were very similar to those described by Werdelin (2) in which immune lymphocytes became attached to antigen-pulsed macrophages. The round passive cell probably corresponds to the lymphocyte which attaches directly to the macrophage. This bond was shown to be antigen-specific and MHC-restricted and the satellite cells were all attached to this cell by their uropods, but these bonds were not antigen-specific or MHC-restricted (3). Veiled cells, unlike macrophages, can accommodate more than one group or trio of cells, perhaps because more of the cell surface is available, moreover electron micrographs of cultures containing antigen-pulsed veiled cells and immune lymphocytes show that the cell aggregates contain a veiled cell which is in contact with several of the lymphocytes which surround it. (Figure I) In KLH immunized animals peripolesis of veiled cells by lymphocytes was not observed until the 7th day after boosting, nor were there large killer-type cells present in the peripheral blood. There was no evidence of damage to the veiled cells at any stage of the reaction.

In HgG immunized animals the interactions seen on the 3rd day were quite similar to those observed in KLH immunized animals. The same trio of cells was seen, but here a single attached cell performed peripolesis until two other cells became attached and if other cells failed to appear peripolesis continued for many hours. By the 5th day large killer-type cells were present in the blood but lethal attacks on veiled cells were not seen until 6 days after boosting. On the 8th day the killer cells did not actually attack the veiled cells but may have been secreting a toxic substance. Peripolesis by immune lymphocytes seemed to be dependent on pulsing the veiled cell with the antigen and it is likely that killing is also dependent on pulsing, but this has not yet been established.

There is a clear difference between the response generated by KLH and that generated by HgG. The fact that one of the antigens is an immunoglobulin may be of some significance.

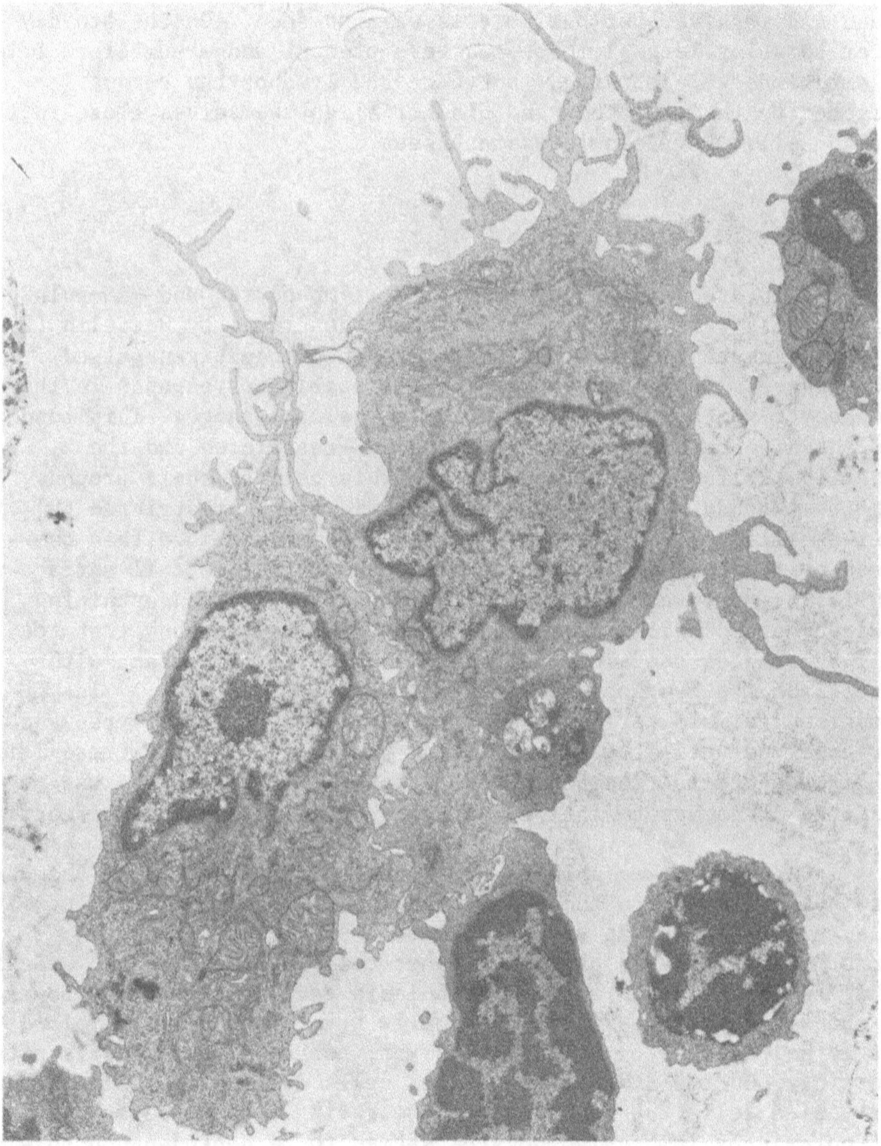

Fig. 1 Electron micrograph of hanging drop 24 hour culture containing antigen-pulsed veiled cells and autologous lymphocytes from a rabbit immunized with KLH. x 9000.

REFERENCES

1. J. C. Hiserodt, L., Britvan and S. R. Targan, Characterization of the cytolytic reaction mechanism of the human natural killer (NK) lymphocyte : resolution into binding, programming and killer cell independent steps. J. Immunol. 129:1782-1787 (1982).
2. J. Petri, O. Braenstrup and O. Werdelin, Macrophage-lymphocyte clusters in the immune response to soluble protein antigen in vitro VIII Cinematographic Studies. Scand. J. Immunol. 10:493-498 (1979).
3. O. Braendstrup, O. Werdelin, E. M. Shevach and A. S. Rosenthal, Macrophage-lyymphocyte clusters in th immune response to soluble protein antigens in vitro VII Genetically Restricted and Nonrestricted Interactions. J. Immunol. 79:1608-1613 (1979).

A COMPARISON BETWEEN ANTIGEN-LADEN CELLS AND DENDRITIC CELLS IN AFFERENT AND EFFERENT LYMPH

M.T. Drayson, D.R. Canning and E.B. Bell

Department of Immunology
Manchester University Medical School
Manchester, U.K.

INTRODUCTION

The importance of antigen presenting cells in the initiation of an immune response has focused attention on two contrasting cell types - the antigen-laden (Ag-L) cell (a macrophage-like cell) and the dendritic cell (DC) alias the "veiled" cell or interdigitating cell of the lymph node (LN) paracortex. Whereas the DC is recognized by a high surface density of Ia molecules and can be found in afferent lymph,[1-4] the Ag-L cell is found in low frequency (<0.01%) in the efferent lymph of the thoracic duct,[5] lacks Ia[6] and appears to transport injected antigens.[7] Functionally, the DC stimulates T cell proliferation[3,8,9] while the Ag-L cell is active in triggering specific memory cells into antibody synthesis on adoptive transfer.[5,6]

Many of the properties of Ag-L cells and DC were published previously.[4,10-12] In the present report we have investigated the immediate origin of those Ag-L cells which gain entrance to the thoracic duct. We have also looked for uptake of thymus dependant antigens, thymus independant antigens and allogeneic cells by Ag-L cells and DC.

The studies made use of mesenteric-lymphadenectomized (MLA-X) rats in which the mesenteric lymph node (MLN) chain was surgically removed. On healing, the afferent lymphatics which drain the gut and Peyer's patches reanastimose directly with the efferent lymphatics.[13] Subsequent cannulation of the thoracic duct yields a cell population that is representative of afferent lymph and is particularly useful for its enriched content of DC.[3,4] Whether Ag-L cells could be recovered from MLA-X rats was not previously known.

MATERIALS AND METHODS

The procedures used were described in detail elsewhere.[5-7] Briefly, Ag-L cells were obtained from thoracic duct lymph of inbred rats from a surgically implanted indwelling cannula. Ag-L cells appeared in lymph within a few hours of antigen injection into the peritoneal cavity and were collected for periods up to 3 days. Thymus-dependant and thymus-independant antigens were conjugated with fluorescien (FITC) or rhodamine (RHOD).

RESULTS

Ag-L cells and DC in MLA-X rats

Because the MLN contributes the bulk of the cells normally found in thoracic duct lymph, we asked whether removal of this LN chain influenced the output of Ag-L cells. Thoracic duct-cannulated, MLA-X rats were injected ip with rhodamine labelled, alum precipitated human gamma globulin (RHOD-apHGG). Thoracic duct cells were examined on slides for the presence of rhodamine bearing cells. The mean output of Ag-L cells from 3 MLA-X rats is given in Fig 1. During the first 16 h collection, 1-3% of the cells in lymph were carrying RHOD-apHGG. However, in the absence of further injections the number of Ag-L cells

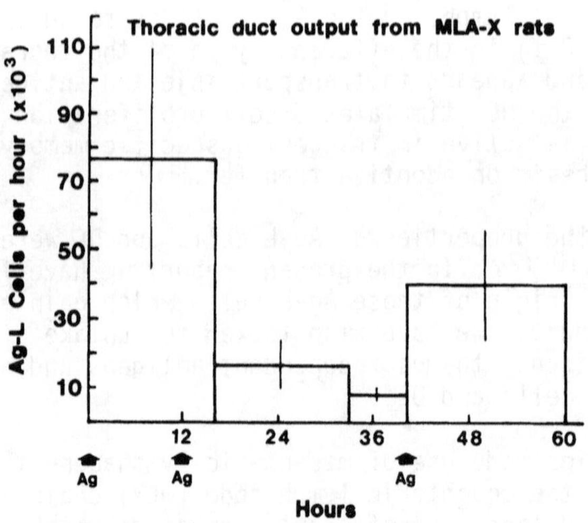

Fig. 1. The output of Ag-L cells in thoracic duct lymph from MLA-X rats. The arrows indicate the ip injection of RHOD-apHGG. The results are means ±SE of 3 rats.

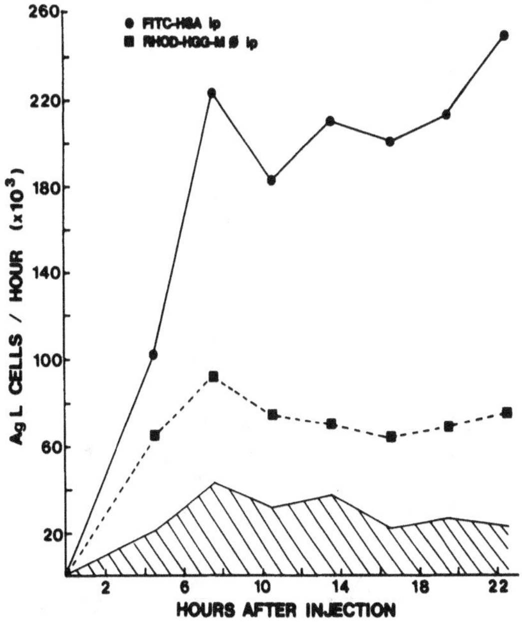

Fig 2. The migration into the thoracic duct of Ag-L cells from the perioteneal cavity of intact AO rats. Macrophages (MØ) bearing RHOD-apHGG were introduced into the periotenal cavity of thoracic-duct cannulated syngeneic hosts followed immediately by ip injection of FITC-apHSA. The hatched area represents cells carrying both RHOD and FITC antigens.

fell to low levels. A further injection of antigen at 40 h again markedly increased the Ag-L cell output.

Since MLA-X rats provided an enriched source of DC, we looked for DC that also bound RHOD-apHGG. DC were identified by morphology and by evidence of strong surface staining for Ia. Ia was detected with the mouse-anti-rat monoclonal OX6 [14], using an immunopurified FITC-sheep anti-mouse IgG. All OX6 positive DC were negative for RHOD-apHGG. Conversely, RHOD positive Ag-L cells were void of Ia staining.

Origin of Ag-L cells

The presence of RHOD-apHGG positive cells in thoracic duct lymph of MLA-X rats suggested that Ag-L cells originated other than from the MLN. Several possibilities were considered. They could have originated directly from the population of macrophages in the peritoneal cavity. Alternatively, they could have migrated from the gut or emerged from the retroperitoneal LN that empty into the thoracic duct.[15] The first possibility was tested.

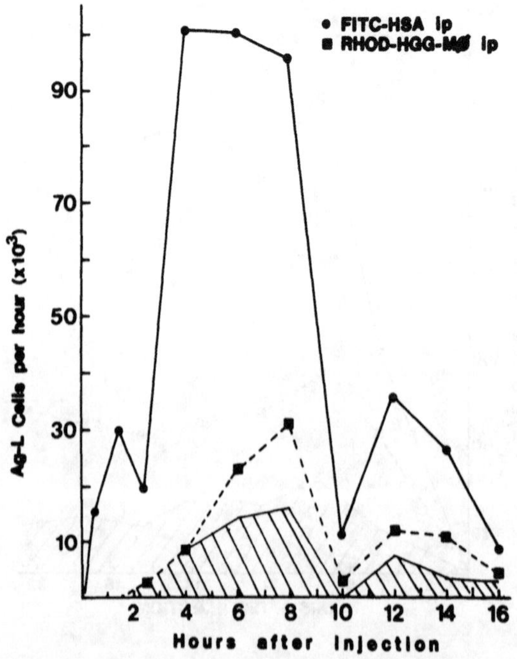

Fig 3. The migration into the thoracic duct of Ag-L cells from the peritoneal cavity of MLA-X AO rats. Details as for Fig 2.

Donor peritoneal macrophages were pre-labelled by injecting RHOD-apHGG ip. Two hours later the peritoneal contents were harvested, washed and transferred ip into thoracic duct-cannulated, MLA-X or intact hosts. The recipients also received an ip injection of FITC-apHSA (human serum albumin). Lymph was collected at 1.5 to 3 h intervals on a fraction collector and assessed for the presence of FITC-labelled, RHOD-labelled or double-labelled cells.

RHOD-labelled donor macrophages originating from the peritoneal cavity were present in the thoracic duct lymph of both intact (Fig 2) and MLA-X (Fig 3) recipients. Approximately one third of the Ag-L cells were of donor origin. The kinetics of the donor macrophage-migration paralleled that of FITC-labelled host cells. The identification of double-labelled Ag-L cells suggested that donor macrophages were able to phagocytose FITC-apHSA within the peritoneal cavity before reaching the thoracic duct.

These experiments also demonstrated some important differences in the migration of Ag-L cells between intact and MLA-X rats. Firstly, the total number of Ag-L cells in MLA-X rats was greatly reduced compared with intact rats (cf Fig 2 and 3).

Table 1. The mean output of Ag-L cells carrying the thymus independant antigen hydroxyethyl starch (OES) into the thoracic duct of intact and MLA-X rats.

Time interval	Intact	MLA-X
1.5 to 3 h	15.5	6.1
to 6	29.3	5.3
to 9	58.0	22.8
to 12	102.6	7.0
to 15	122.1	8.4
to 18	125.5	8.8
to 21	108.7	7.8
to 24	136.3	7.6

Values represent Ag-L cells x 10^{-3} per h following the ip injection of FITC-OES (3 intact, 3 MLA-X) or RHOD-OES (1 intact, 1 MLA-X). The frequency of Ag-L cells was determined at each time interval for each rat by counting 5000-25,000 cells.

Secondly, the number of Ag-L cells reached an early peak (4-8 h) in MLA-X rats and fell to low levels by 16 h. In contrast, the tempo of migration was slower in the intact rat; the output of Ag-L cells failed to reach a plateau before 8 h but thereafter remained elevated for at least 24 h.

Antigen uptake by Ag-L cells and DC

We determined whether thymus independant antigens were taken up by Ag-L cells to the same extent as were thymus dependant antigens. Rats with established indwelling thoracic duct cannulae were injected ip with one of the following antigens kindly provided by Prof. John Humphrey: FITC- or RHOD-conjugated hydroxyethyl starch (OES) and RHOD-conjugated pneumococcal polysaccharide Type III (SIII). The pooled results from 8 rats are shown in Table 1. Both intact and MLA-X rats gave the expected pattern and number of OES bearing Ag-L cells in thoracic duct lymph. Preliminary studies have also demonstrated that Ag-L cells will take up RHOD-SIII. In general therefore the results using these two thymus independant antigens were comparable to studies using thymus dependant antigens.

Recent studies by Fossum and colleagues[16] showed that resident interdigitating cells of LN (the believed homologue of the migrating DC), engulfed allogeneic lymphocytes that were injected iv. We have looked for the uptake of ^{51}Cr-labelled allogeneic and semi-allogeneic lymphocytes by DC and by Ag-L cells in thoracic duct lymph of MLA-X rats. Thoracic duct-cannulated rats were injected ip or iv with ^{51}Cr-labelled cells along with FITC-apHSA. The cells from collections of lymph were examined using combined autoradiography and fluorescence microscopy. We found no evidence for the uptake of ^{51}Cr-labelled lymphocytes by DC or by Ag-L cells.

DISCUSSION

We have compared the output of Ag-L cells in MLA-X and intact rats. Removal of the MLN chain not only reduced the total number of Ag-L cells in thoracic duct lymph but also altered the kinetics. Ag-L cells appeared sooner in the thoracic duct of MLA-X rats, reached a rapid peak (4-8 h) and declined to undetectable levels by 18 h. In the intact rat, 2-3 times as many Ag-L cells were found during a 24 h period. Their appearance in the thoracic duct was delayed but the output was maintained for up to 48 h. A high level of Ag-L cells could be re-established in MLA-X rats by additional ip injections of antigen.

The origin of Ag-L cells in thoracic duct lymph after an ip injection of antigen has been a puzzle. Drainage from the peritoneal space is primarily through the diaphram;[15] retroperitoneal drainage is not extensive.[15] Furthermore, LN are interposed between the afferent and efferent lymphatics which link the peritoneal cavity to the thoracic duct. We wanted to establish whether antigen was phagocytosed by Ag-L cells locally within the peritoneal space or alternatively by cells already within LN. It was clear from the present study that donor macrophages, pre-labelled with a fluorescent antigen, were able to escape from the peritoneal cavity and enter the thoracic duct coincident with host macrophages. The migration kinetics of donor and host Ag-L cells in MLA-X rats was similar. Likewise in intact rats, the donor and host Ag-L cells followed a comparable time course. We concluded that many if not all Ag-L cells in thoracic duct lymph took up antigen within the peritoneal space. The precise route from the peritonium could not be determined. If Ag-L cells indeed enter afferent lymphatics, they must traverse the LN matrix quickly, judging from their rapid appearance in thoracic duct lymph. It is interesting to contrast the peritoneal origin of the Ag-L cells with the intestinal origin of the DC.[3,4]

In addition to the uptake of a number of thymus dependant antigens by Ag-L cells,[12] we now add two contrasting thymus independant antigens to the list. SIII was shown to be phagocytosed indiscriminately by macrophages from various sites.[17] However, Humphrey and Grennan found that OES was engulfed by cells in the marginal zone of the spleen and not by red pulp macrophages.[17] We have recently found that Ag-L cells carrying HSA and injected into the blood stream, pass quickly through the marginal zone to localize almost exclusively in the red pulp.[7]

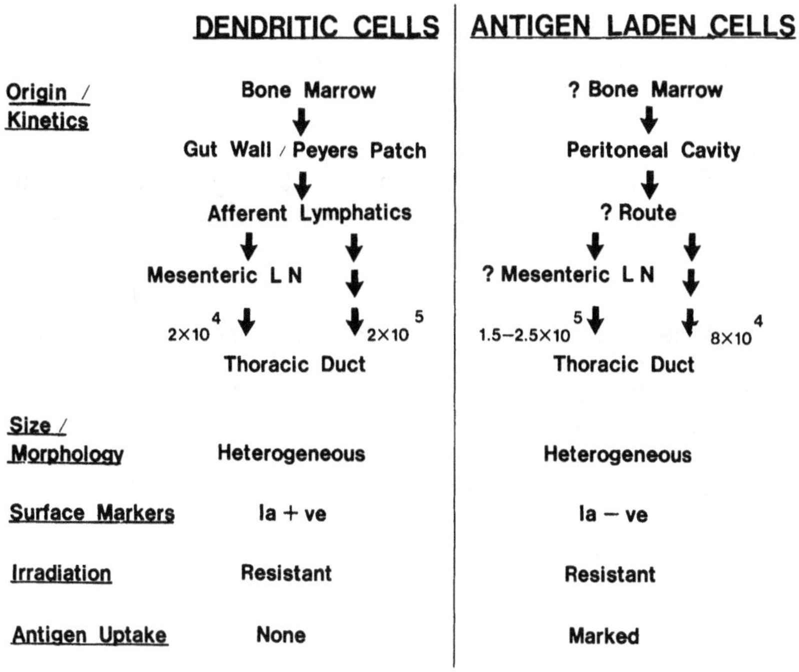

Fig 4. A comparison of DC and Ag-L cells.

It will be interesting to test whether Ag-L cells carrying OES have a selective migration to the marginal zone.

In terms of antigen uptake, the DC was the antithesis of the Ag-L cell. In most laboratories, phagocytosis or even surface binding of antigen was not a distinguishing feature of the DC. Recently, however, cells with a dendritic morphology in afferent lymph of sheep were shown to bind complexes of antigen/antibody.[18] Furthermore, in the rat LN, Ia positive, resident interdigitating cells were found to have phagocytosed allogeneic lymphocytes.[16] Although we failed to find engulfment of allogeneic lymphocytes or other antigens by DC free within lymph, perhaps DC must be in a stationary mode or at a later stage of differentiation before they will handle antigen.

A summary of the major features that characterize DC and Ag-L cells is illustrated in Fig 4. The numbers refer to the thoracic duct output of each cell type per hour in the presence (left) or absence (right) of the MLN. Note that removal of the MLN increased the output of DC tenfold but reduced Ag-L cells by one half to one third. Note also the difference in expression of surface Ia.

If, as the evidence suggests, both Ag-L cells and DC present antigen, one can only surmise that two very different mechanisms are being served by these contrasting cell types.

Acknowledgements. We wish to acknowledge Tamar Aslan for technical assistance and Shirley John for typing the manuscript. The work was supported by the British Medical Research Council.

REFERENCES

1. H A Drexhage, H Mullink, J de Groot, J Clarke & B M Balfour. Cell Tissue Res. 202:407 (1979).
2. R H Kelly, B M Balfour, J A Armstrong & S Griffiths. Anat. Rec. 190:5 (1978).
3. D W Mason, C W Pugh & M Webb. Immunology 44: 75 (1981).
4. C W Pugh, G G MacPherson & H W Steer. J. Exp. Med. 157:1758 (1983).
5. E B Bell. Immunology 38:797 (1979).
6. E B Bell & J Botham. Immunology 47: 477 (1982).
7. E B Bell. Cell. Immun. 88 (in press) (1984).
8. R M Steinman & M D Witmer. Proc. Natn. Acad. Sci. 75:5132 (1978).
9. W E F Klinkert, J H LaBadie, J P O'Brien, C F Beyer & W E Bowers. Proc. Natn. Acad. Sci. 77: 5414 (1980).
10. R M Steinman & Z A Cohn. J. Exp. Med. 137: 1142 (1973).
11. R M Steinman & Z A Cohn. J. Exp. Med. 139:380 (1974).
12. E B Bell. Immunobiol. (in press) (1984).
13. A G Sanders & H W Florey. Brit. J. Exp. Path. 21: 275 (1940).14. W R McMaster & A F Williams. Eur. J. Immun. 9:426 (1979).
15. T Olin & T Saldeen. Cancer Res. 24:1700 (1964).
16. S Fossum, B Rolstad & W L Ford. Immunobiol. (in press) (1984).17. J H Humphrey & D Grennan. Eur. J. Immun. 11:221 (1981).
18. J G Hall & D Robertson. Int. Archs. Allergy Appl. Immun. 73:155 (1984).

THE HETEROGENEITY OF MONONUCLEAR PHAGOCYTES IN LYMPHOID ORGANS: DISTINCT MACROPHAGE SUBPOPULATIONS IN RAT RECOGNIZED BY MONOCLONAL ANTIBODIES ED1, ED2 AND ED3

C.D. Dijkstra, E.A. Döpp, P. Joling[1], and G. Kraal

Department of Histology, Free University, Amsterdam
[1] Department of Internal Medicine I, Erasmus University Rotterdam, The Netherlands

INTRODUCTION

The precursors of macrophages, originating in the bone marrow, differentiate into circulating monocytes, which, after leaving the peripheral blood for the tissue compartments, differentiate into macrophages. The macrophages form a heterogeneous population, as expressed by differences in their characteristics, like morphology, density, enzyme activity, cell surface properties and functional capacities. Together these characteristics determine the phenotypical appearance of macrophages, which differs with different environmental circumstances, like the state of activity[3,6,20,32], the developmental stage of the individual macrophage[7], the localization in different tissues[12,14,26] and compartments[15] and often a combination of these factors. Whether these environmental conditions determine the phenotypical appearance of the macrophages[24] or the origin of the heterogeneity is already determined at bone marrow level[5] is still a subject of discussion.

In the present study we describe a set of three monoclonal antibodies, each of which recognizes exclusively cells of the mononuclear phagocyte line in the rat. The tissue distribution of each of the three monoclonals is determined by immunoenzymehistochemistry. Results show that ED1 recognizes a cytoplasmic antigen in monocytes and in most macrophages, free and fixed, whereas ED2 and ED3 recognize membrane antigens of tissue macrophages, discriminating between subsets of macrophages, each with a characteristic localization in the compartments of the lymphoid organs.

MATERIALS AND METHODS

Animals

Wistar rats and BALB/c mice were obtained from Centraal Proefdierenbedrijf TNO (Zeist, The Netherlands) and kept under routine laboratory conditions.

Preparation of the monoclonal antibodies

BALB/c mice were primed intraperitoneally with rat spleen cells and Freund's complete adjuvant and boosted i.p. after 4 weeks with rat spleen cells and Freund's incomplete adjuvant. Four days after the last immunization two mice were killed, their spleen cells were pooled and fused with Sp2/0-ag14 myeloma cells according to the method of Köhler and Milstein[19], modified after Fazekas de St. Groth and Scheidegger[11]. The fusion mixture was plated in 96 wells culture trays (Costar) with BALB/c peritoneal exudate cells as feeder cells. Hybrids surviving in hypoxanthin, aminopterin and thymidin (HAT) selective RPMI-1640 medium, supplemented with 15% foetal calf serum (FCS, High Clone, Greiner, USA) were tested. Screening of the supernatants for monoclonal antibodies was performed on cryostat sections of rat spleen by a two-step immunoperoxidase method using peroxidase conjugated rabbit anti mouse Ig (Dako, Denmark) as second antibody[9]. The procedure is described under "Immuno- and enzymehistochemistry". Clones which reacted with putative macrophages in the spleen sections were selected and cloned. Ascites fluid was obtained by peritoneal injection of 3×10^6 hybrid cells into BALB/c mice.

Immuno- and enzymehistochemistry

The following organs were frozen in liquid nitrogen and stored at -20°C: spleen, thymus, mesenterial lymph node, Peyer's patches, lung (including BALT) and liver. Cryostat sections of 8-12 µm were picked up on slides and air-dried. Sections were fixed in acetone for 10 min and air-dried for at least 30 min. Sections were then incubated with culture supernatant for 60 min or with ascites fluid, diluted 1:3,000. After washing in 0.01 M phosphate buffered saline (PBS, pH 7.4), slides were covered with a 1:200 dilution of rabbit anti mouse Ig peroxidase (Dako, Denmark) in PBS with 0.2% BSA and 1% normal rat serum. After washing in PBS the slides were stained for peroxidase activity with 3,3'-diaminobenzidine-tetrahydrochloride (DAB, Sigma) (0.5 mg/ml Tris-HCl buffer, pH 7.6, containing 0.01% H_2O_2). Control slides were incubated in the same way omitting the first step. Where desired, endogenous peroxidase activity was eliminated by incubation in methanol with 0.03% H_2O_2[25].

Acid phosphatase activity was demonstrated according to Burnstone[22] with naphthol AS-BI phosphate (Sigma) as substrate and hexazotized pararosaniline as diazonium salt. This procedure was carried out on the cryostat sections following the immunoperoxidase procedure.

Uptake of particles

Rats were injected intravenously with 0.5 ml of a 1% solution of Dow Latex particles (particle Ø 1.091 μm, Serva, Hannover, FRG). After 24 h the animals were killed and their spleens were frozen in liquid nitrogen. Cryostat sections were treated as described above.

Isotype determination

The isotype of the monoclonal antibodies was determined in a two step immunoperoxidase method on cryostat sections of rat spleen, using peroxidase conjugates, specific for each of the mouse immunoglobulin isotypes (Serotec) as second step.

RESULTS

ED1, ED2 and ED3 are specific for macrophages

The ED1, ED2 and ED3 monoclonal antibodies all reacted with large, round oval cells or cells with a dendritic appearance by their slender cell processes. These cells always showed acid phosphatase activity in every organ tested, as demonstrated by combination of the immunoperoxidase procedure with the demonstration of acid phosphatase activity in the same cryostat section. In the splenic red pulp as well as in the white pulp, cells positive for each of the three monoclonal antibodies, contained latex particles 24 h after intravenous injection (e.g. Fig. 1D). These findings, together with the localization pattern, indicate that these monoclonal antibodies recognize macrophages. They did not recognize other cells, like granulocytes, lymphocytes, follicular dendritic cells, vascular endothelium, mucosal or bronchial epithelium or liver parenchymal cells.

The distribution and staining pattern of macrophages recognized by each of the three monoclonals is summarized in Table 1. The ED1 monoclonal antibody showed a granular staining pattern within the cytoplasm of cells. ED2 and ED3 showed a more diffuse staining pattern, not only on the cell bodies but also on the blunt or dendritic cell processes (Figs. 1D, 2B). This staining pattern is comparable to that of anti-Ia antigen[9] and suggestive for membrane staining. The heavy chain class of ED1 is IgG1, of ED2 and ED3 IgG2a. Some relevant findings in different lymphoid organs are described below.

Table 1 Distribution and staining pattern of macrophages identified by ED1, ED2 and ED3 in various organs

Monoclonal antibody	ED1	ED2	ED3
Isotype	IgG1	IgG2a	IgG2a
Staining pattern	granular, patchy, cyto-plasmic	diffuse membrane	diffuse membrane
Spleen			
White pulp			
inner PALS	++	-	+ weakly
outer PALS	++	+	+ weakly
follicle	-	-	-
marg. metalloph.	-/+ weakly	-	+++ dendritic
marginal zone	+ weakly	-	+++ dendritic
Red pulp	+++	+++	+++ weakly
Lymph node			
Cortex			
outer cortex	+ weakly	-	+++ subsinus-oidal dendr.
paracortical area	++	+	-
follicles	-/+	-	-
Medulla	+++	+ 10-20%	+++
Capsule	+	+	-
Peyer's patches			
Interfollicular area	+++	++	+ small groups 3-4 dendr. cells
Dome	+	-	-
Follicle	-	-	-
Villi	+++ apex basis	++ apex basis	-
Lung			
BALT	++	only at perip. of BALT	++ dendritic
Perivascular/ peribronchial	+	+++	+
Alveolar	+++	-	-
Thymus			
Cortex	++	++ dendritic	-/+ weakly
Medulla	++	-	-/+ weakly
Corticomedullar area	+++	+++	-/+ weakly
Capsule	+	++	+
Liver	+++ dendritic	+++ dendritic	++ dendritic

+++ = (almost) all acid phosphatase positive cells stain with the monoclonal antibody;
++ = a considerable number;
+ = few
-/+ = very few or none

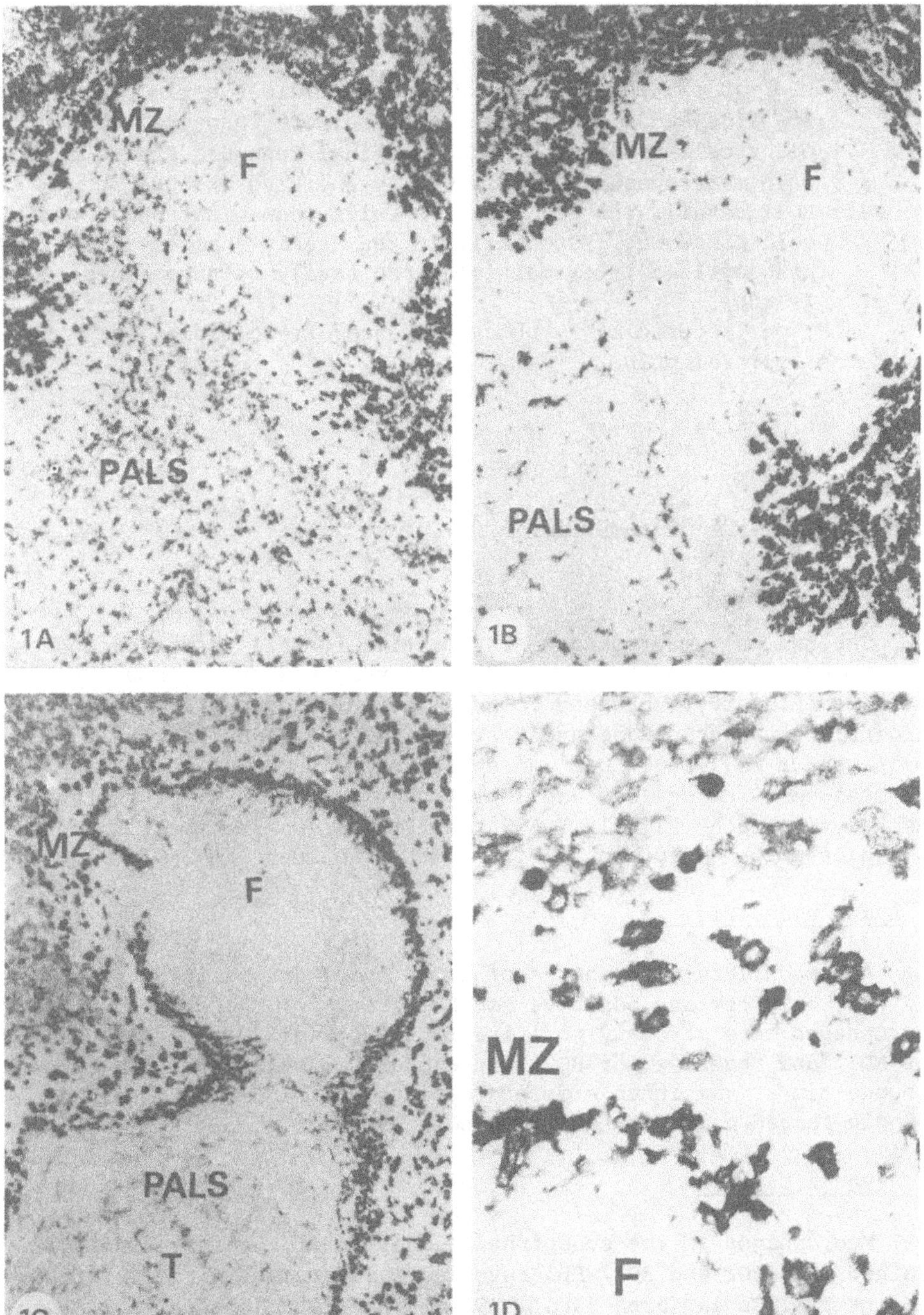

Fig. 1. Rat spleen, cryostat sections stained for either of the three monoclonal antibodies in an immunoperoxidase procedure.

(Continued)

Fig. 1A: ED1: a patchy staining pattern is found in PALS, follicle and marginal zone. The red pulp macrophages show heavy staining for ED1 (x 100).

1B: ED2: all red pulp macrophages are stained as well as a few cells in the outer PALS (x 100).

1C: ED3: at the periphery of PALS and follicle a rim of strongly positive cells are present. Furthermore large cells with blunt cell processes in the marginal zone are stained. Red pulp macrophages are only weakly positive (x 100).

1D: ED3: detail. The ED3 positive cells show blunt and dendritic cell processes. Especially in the weakly stained cells the phagocytized latex particles can easily be recognized (x 400).

PALS: periarteriolar lymphocyte sheath; f: follicle; mz: marginal zone; rp: red pulp.

Staining patterns and localization

A. Spleen (Fig. 1)

The red pulp macrophages are positive for all three antibodies. ED3, however, stained the red pulp macrophages weakly, though the staining intensity increased towards the white pulp (Fig. 1C). The marginal zone macrophages and the marginal metallophils were strongly positive for ED3 (Fig. 1D), and negative for ED2. In the white pulp ED2 stained a few cells or cell processes in the outer periarteriolar lymphocyte sheath (Fig. 1B). A patchy staining pattern throughout the white pulp, except in the follicles, was seen with ED1 (Fig. 1A).

B. Lymph node (Fig. 2)

All medullary macrophages of lymph nodes are positive for ED1 and ED3, only few are positive for ED2 (Fig. 2). The subsinusoidal macrophages were strongly positive for ED3 (Fig. 2B), weakly positive for ED1, and negative for ED2. These branched cells formed a continuous rim at the inner side of the marginal sinus. ED2 stained spindle shaped cells in the lymph node capsule.

C. Peyer's patches

Macrophages in the subepithelial tissue of the intestinal villi stained with ED1 and ED2. ED1 revealed a patchy staining pattern in the interfollicular area (IFA), ED2 stained dendritic cells in the IFA and a few cells in the dome area, ED3 stained only small groups of 1-4 cells at the base of the IFA. At the mesothelial side a rim of spindle shaped cells stained with ED1 and ED2.

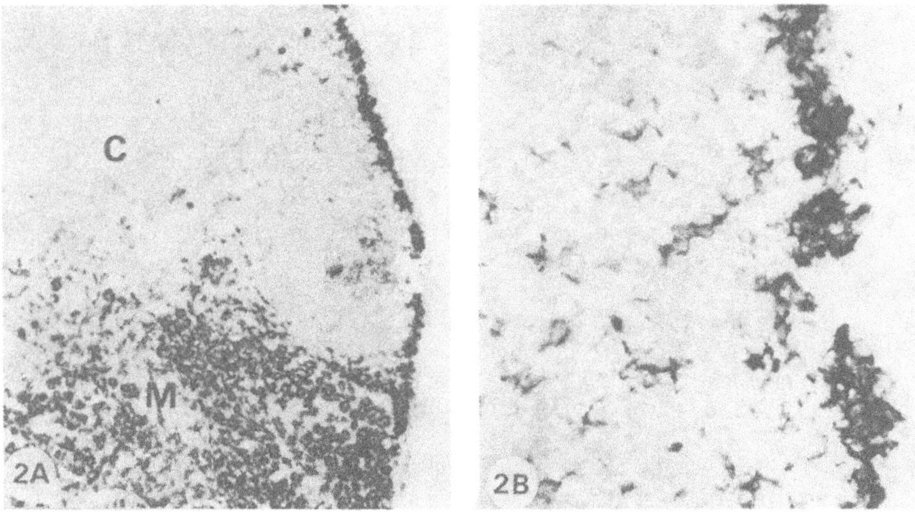

Fig. 2. Lymph node

A: ED3: all medullary macrophages are strongly positive for ED3. Furthermore a rim of dendritic cells is present at the inner border of the marginal sinus (x 100).

B: ED3: detail of the corticomedullary border, dendritic positive cells are visible at the inner border of the marginal sinus and in the medulla of the lymph node (x 400).

C: cortex; M: medulla.

D. Thymus (Fig. 3)

The most conspicious staining pattern in the thymus was observed after staining with ED2 (Fig. 3A). In the cortex large dendritic cells were found in intimate contacts with the surrounding lymphocytes (Fig. 3B). These cells showed strong acid phosphatase activity, and only very few cells with acid phosphatase activity in the cortex did not stain with ED2. The medulla was negative for ED2. ED1 showed a patchy staining pattern in cortex and medulla, the highest concentration of patches was found in the cortico-medullary region. ED3 stained only few cells in the thymic medulla very weakly.

E. Lung

In the bronchus associated tissue (BALT) large, branched cells were stained by ED3. In the perivascular and peribronchial tissue as well as at the periphery of the BALT, dendritic cells, strongly positive for ED2, were found. The alveolar macrophages were strongly

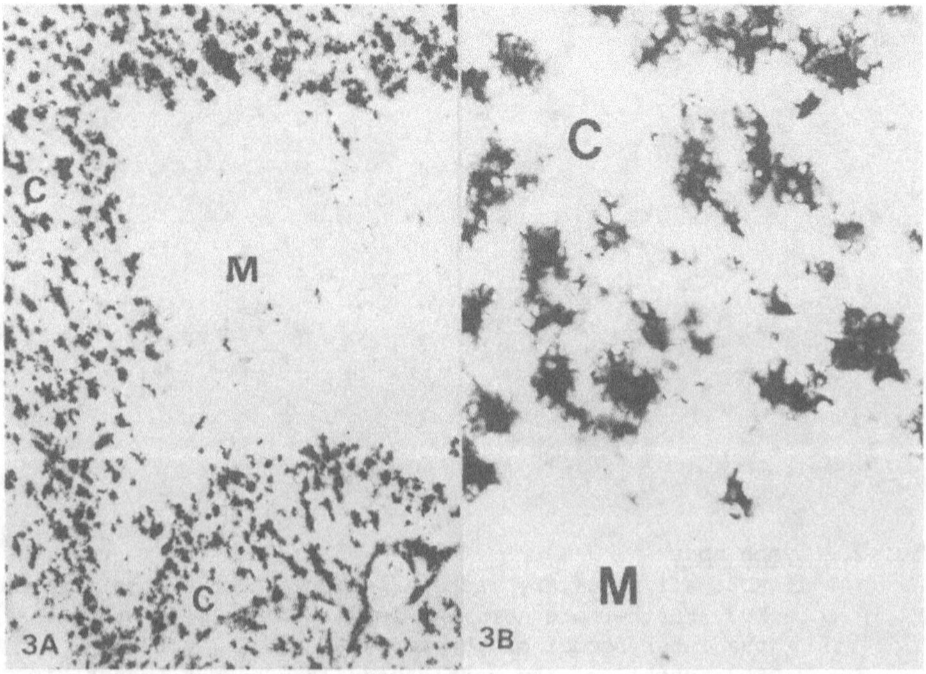

Fig. 3. Thymus

A: ED2: positive dendritic cells are present in the thymic cortex, the medulla does not contain ED positive cells (x 100).

B: ED2: detail, the ED2$^+$ cells in the thymic cortex show extensive dendritic cell processes in intimate contact with the surrounding lymphocytes (x 400).

C: cortex; M: medulla.

positive for ED1, and negative for ED2 and ED3.

F. Liver

ED1 and ED2 recognize dendritic cells in the sinuses, probably the cells of Von Kupffer. A smaller number of these dendritic cells was recognized by ED3.

DISCUSSION

In this study three monoclonal antibodies have been described that recognize different antigenic determinants of rat tissue macrophages. That these antibodies recognize macrophages is based on morphological criteria, the tissue distribution and the correlation with acid phosphatase activity. In addition it has been demonstrated that the spleen cells, positive for any of the three antibodies, are capable of phagocytosis. The monoclonal antibodies did not react with other cell types or tissue structures.

In the compartments of the lymphoid organs distinct macrophage subpopulations can be recognized exclusively by use of the monoclonal antibodies. ED1 showed a patchy staining pattern in all compartments of the organs investigated, and recognized most macrophage subpopulations. However, the staining intensity differed per localization: the red pulp macrophages and the alveolar macrophages showed heavy staining, whereas in the white pulp only small patches were seen. ED2 and ED3 revealed distinct staining patterns in the different compartments. An $ED2^-$ and $ED3^+$ subpopulation is found in the spleen, consisting of both the marginal zone macrophages[18] and the marginal metallophils[23]. Also in the lymph node, an $ED2^-$ and $ED3^+$ subpopulation was found, comprising the subsinusoidal macrophages and the major portion of the medullary macrophages. This distribution resembles the localization of FITC ficoll after injection by which a distinct macrophage subpopulation can be distinguished in mice[18], namely the marginal zone macrophages in spleen and the subsinusoidal macrophages in the lymph node. ED3, however, recognizes in addition the marginal metallophils in spleen and the medullary macrophages in lymph node.

A second distinct subpopulation is found in the thymic cortex, where large dendritic $ED2^+$ $ED3^-$ cells are found, with extensive cell processes between the surrounding lymphocytes. These cells showed strong acid phosphatase activity and thus represent the cortical macrophages. In the medulla no $ED2^+$ cells are found. The extensive cell processes in intimate contact with the surrounding lymphocytes and the exclusive localization in the cortex may indicate a possible role in the maturation of cortical thymocytes as suggested for thymic macrophages[4].

Taken together these results present evidence that the phenotype of macrophages differs depending on the organ localization. ED1 recognizes a cytoplasmic antigen, distributed more or less randomly in all compartments of the lymphoid organs. ED2 and ED3 probably recognize membrane antigens, by which separate subpopulations of macrophages can be distinguished in lymphoid compartments. On the other hand there exists a considerable overlap between the distribution of these two antigens. These results suggest that ED2 and ED3 may recognize differentiation antigens of macrophages.

The tissue distribution of ED1, ED2 or ED3 does not correlate with that of any of the macrophage-antigens described in mice. Neither of the Mac-1, Mac-2 and Mac-3 antigens is located exclusively on macrophages[12]. By the localization pattern of Mac-1, Mac-2 and Mac-3 also subpopulations of macrophages can be distinguished[12], but none of these correlates exactly with one of the subpopulations as recognized by ED1, ED2 or ED3. F4/80 antigen is located exclusively on macrophages in mice, with a distribution pattern comparable with ED2, but in contrast with ED2 F4/80 is present in both thymic medulla and cortex[17]. W3/25 monoclonal antibody recognizes macrophages in rat, but T-helper cells are also positive[1].

Many of the cells recognized by ED2 and ED3 have a dendritic appearance. This makes it necessary to compare them with the other dendritic cells described in lymphoid organs i.e. the follicular dendritic cell[21] and the interdigitating cell. None of the antibodies react with the FDC, which is not surprising since there exists considerable evidence that the FDC originates from a reticulum cell[10,29,31] and not from a member of the mononuclear phagocyte system. There are, however, studies that describe macrophage-related antigens on FDC's[13]. The interdigitating cell (IDC), occur in the T-cell areas of lymphoid organs and can be recognized on ultrastructural level[28,30] or by staining for Ia antigen[2,9,16]. Staining with anti-Ia antibody shows a considerable number of dendritic strongly positive cells (IDC) in all T-cell compartments of lymphoid organs[2,9,16]. ED3 showed in all T cell areas investigated a patchy staining pattern, the distribution of which was similar to that of Ia positive cells. Both Ia positive cells and ED3 positive cells in T-cell areas showed acid phosphatase activity. The staining pattern of ED2 and ED3 in T-cell areas does not correlate with that of Ia positive cells. Therefore we must conclude that all IDC's might be recognized by ED1, but that the distribution of ED2 and ED3 on IDC is variable, if any.

The three monoclonal antibodies, described in this study, have considerable value in the identification of rat macrophages. In further studies the ED2 and ED3 antibodies will be used to elucidate the functional aspects and differentiation pathways of the macrophage subpopulations recognized by these unique antibodies.

REFERENCES

1. A.N. Barclay. Immunology 42:593 (1981).
2. A.N. Barclay. Immunology 44:727 (1981).
3. R.H.J. Beelen, D.M. Broekhuis-Fluitsma, and E.C.M. Hoefsmit. Ultramicroscopy 3:129 (1978).
4. D.I. Beller, and E.R. Unanue. J. Immunol. 121:1861 (1978).
5. I. Bursuker, and R. Goldman. J. Reticuloendothel. Soc. 33:207 (1983).
6. Z.A. Cohn. J. Immunol. 121:813 (1978).
7. Z.A. Cohn, and B. Benson. J. Exp. Med. 121:153 (1964).
8. A.M. Duijvestijn, T. Sminia, Y.G. Köhler, E.M. Janse, and E.C.M. Hoefsmit. In: In vivo immunology (Nieuwenhuis, P., Van den Broek, A.A. and Hanna, M.G. Jr. eds.), Plenum Press, New York, p. 441 (1982).
9. C.D. Dijkstra. J. Reticuloendothel. Soc. 32:167 (1982).
10. C.D. Dijkstra, E.A. Döpp, and E.W.A. Kamperdijk. Cell Tissue Res. 236:203 (1984).
11. S. Fazekas de St. Groth, and D. Scheidegger. J. Immunol. Meth. 35:1 (1980).

12. T.J. Flotte, T.A. Springer, and G.J. Thorbecke. Am. J. Pathol. 111:112 (1982).
13. J. Gerdes, H. Stein, D.Y. Mason, and A. Ziegler. Virch. Arch. B 42:161 (1983).
14. K.A. Haines, T.J. Flotte, T.A. Springer, I. Gigli, and G.J. Thorbecke. Proc. Natl. Acad. Sci. USA 80:3448 (1983).
15. E.C.M. Hoefsmit, E.W.A. Kamperdijk, H.R. Hendriks, R.H.J. Beelen, and B.M. Balfour. In: Lymph node macrophages (Carr, and Daems, eds.), Plenum Publishing Corporation, p. 417 (1980).
16. G. Hoffman-Fezer, D. Gotze, H. Rodt, and S. Thierfelder. Immunogenetics 6:367 (1978).
17. D.A. Hume, A.P. Robinson, G.G. MacPherson, and S. Gordon. J. Exp. Med. 158:1522 (1983).
18. J.H. Humphrey, and D. Grennan. Eur. J. Immunol. 11:221 (1981).
19. G. Köhler, and C. Milstein. Nature 256:495 (1975).
20. P.S. Morahan. J. Reticuloendothel. Soc. 27:223 (1980).
21. G.J.V. Nossal, A. Abbot, J. Mitchell, and Z. Lummus. J. Exp. Med. 127:277 (1968).
22. A.G.E. Pearse. Histochemistry. Theoretical and Applied. Vol. 1, 3rd ed. Edinburgh, Churchill Livingstone (1968).
23. T. Snook. Anat. Rec. 148:149 (1964).
24. C. Sorg. Mol. Immunol. 19:1275 (1982).
25. J.G. Streefkerk, and M. van der Ploeg. Histochemistry 40:105 (1974).
26. J.J. Twomey, A. Laughter, and M.F. Brown. Clin. Exp. Immunol. 52:449 (1983).
27. R. van Furth, Z.A. Cohn, J.G. Hirsch, W.G. Spector, and H.L. Langevoort. Bull WHO 46:845 (1972).
28. A.J.P. Veerman. Cell Tissue Res. 148:247 (1974).
29. A.J.P. Veerman. Z. Immun. Forsch. 150:45 (1975).
30. J.E. Veldman. Acad. Thesis, Groningen, The Netherlands (1970).
31. A. Villena, A. Zapata, J.M. Rivèra-Pomar, M.G. Barrutia, and J. Fonfria. Cell Tissue Res. 229:219 (1983).
32. W.S. Walker. J. Reticuloendothel. Soc. 20:57 (1976).

MACROPHAGES MIGRATE FROM THE MARGINAL ZONE INTO THE GERMINAL CENTRE OF THE RODENT SPLEEN

Kenjiro Matsuno, Kazuhisa Miyakawa and Masahiko Kotani

Department of Anatomy, Kumamoto University Medical School, Kumamoto 860, Japan

INTRODUCTION

The marginal zone (MZ) is the outer compartment of the white pulp (WP) of the spleen. Intravenously injected antigens flood out first into MZ and then localize in the lymphoid follicles for a long period (Nossal and Ada, 1971). However, the transport mechanisms of antigens or immune complexes from MZ to the lymph follicles are still not precisely understood. In our previous studies (Kotani et al., 1977, 1979, 1982), macrophages (MP), after phagocytosis of colloidal carbon, were seen to reach the regional lymph nodes via afferent lymphatics and then accumulate into germinal centres (GC) of their lymph follicles. This was a experimental model of extrinsic MP migration into lymph nodes via lymphatics. The present study was carried out to examine whether or not carbon-laden MP which were either coming from blood stream or resident in the spleen could really enter GC as well as in lymph nodes.

MATERIALS AND METHODS

Fate of Intravenously Injected Carbon in the Mouse Spleen (Experiment 1)

Male C3H/He mice (8 weeks old) were injected slowly into the tail vein with 0.3 ml of Pelikan ink (16mg/ml of carbon; Guenther Wagner, Hanover). At various intervals from 10 minutes to 180 days after injection, the spleens of 3 mice were removed, fixed in Carnoy's fluid, embedded in paraffin wax and were cut at 6 μm. The serial sections were stained with methyl green-pyronin.

Adoptive Transfer of Rat Peritoneal Macrophages: Retrograde Injection into the Hepatic Artery (Experiment 2)

Adult (PvG/c x DA)F1 male rats were injected intraperitoneally with the mixture of 10 ml of 5% sodium caseinate plus 10 ml of 0.25 mg/ml aggregate-free Pelikan ink. Four days later, peritoneal washes were collected and 1-2 $x10^7$ carbon-laden MP in 0.25 ml were injected into the hepatic artery of syngeneic male recipient under ether anaesthesia. For the injection, the gastric artery was temporarily ligated with a haemoclip and pulled outward. Cells were injected retrograde into the hepatic artery at the point just after a trifurcation of the coeliac trunk, so that the inoculum flowed back to the trifurcation and then into the splenic artery (SA). In this manner, a local concentration of transferred cells were increased with a minor influence on the splenic circulation. Rats were sacrificed 5 minutes, 6,12,24 hours, 3 and 5 days after the cell transfer. The excised spleens were processed similarly as in experiment 1.

Adoptive Transfer of Rat Peritoneal Macrophages: Injection into the Splenic Artery (Experiment 3)

Experiment 3 was performed as same as experiment 2, except that SA was selected as a route of injection. SA was located and injected slowly at a point 5 mm after a trifurcation of coeliac trunk.

RESULTS

Experiment 1

Intravenously injected carbon was taken up quickly within 10 to 20 minutes by large number of cells with big oval or round nuclei and wide cytoplasm full of carbon particles located in MZ. Therefore, they seemed to be a resident MP in the MZ. Morphologically they were very similar to so-called MZ-type MP described by Humphrey and Grennan (1981). These MP appeared to migrate from MZ either to the red pulp (RP) or to WP. The former accumulated particularly outside MZ and RP and fused with each other. The latter moved into the deeper WP diffusely from the periphery of WP or into the periarterial lymphatic sheaths (PALS) surrounding the central artery and its branches. Diffuse migration of MP across WP began by 6 hours after carbon injection and selective migration into PALS 24 hours later. They came together favorably around GC and entered the majority of GC on a large scale by 10 days. They stayed in the active portion of GC, where large pyroninophilic cells were accumulated. The number of MP in GC increased with time after carbon injection. They clustered in GC and frequently fused with each other which were continuously observed until the end of this experiment, 180 days. Since the migration of MP towards RP and WP was so massive, the number of these cells in MZ decreased markedly within a few days after carbon injection. However, the same type of carbon-laden MP could be seen persistently in MZ, and migration of them by two pathways towards GC was also observed persistently.

Experiment 2

Immediately after transfer into the hepatic artery, carbon-laden MP were seen only in MZ. At 6 hour they scattered in MZ and RP, and by 12 to 24 hours a few of them were occasionally found also in WP, i.e., outer area of the PALS, near the central artery, lymphocyte corona and in GC. Incidence in the number of GC with migrating MP was low ranging from 5 to 10% of GC examined (6 rats). Carbon-laden MP stayed in RP cord or MZ until the end of the experiment (5 days).

Experiment 3

Results obtained from the injection into the splenic artery were fundamentally the same as those in experiment 2. However, a migration of carbon-laden MP into GC at 12 to 24 hours was more obvious (Figs.1, 2,3). The incidence of MP entry in GC varied ranging 10 to 40% of GC examined (4 rats). The number of carbon-laden MP in one GC was also variable. These variations seemed to reflect an actual number of transferred cells which entered the splenic tissue investigated, i.e., carbon-laden MP in the lymph follicle were obvious where neighbouring MZ and RP were occupied by a lot of transferred cells.

DISCUSSION

The results of the first experiment suggested that resident MP in MZ were mobile and entered into GC at least after ingestion of carbon particles, though slow turning over of MZ-MP was reported (Humphrey and Grennan, 1981). Since persistent migration of carbon-laden MP from MZ into WP was observed, some of them might arise from the blood stream.

It has been reported that when labeled MP were transferred systemically, they scarcely localized in the WP and not in GC of the spleen (Roser, 1965, Roser, 1970, Unanue & Cerottini, 1971). This might be explained as such that trapping of transferred cells in the spleen was not enough, thus, if the number of labeled MP in the spleen was much more increased, their traffic pattern within WP would become more easily visualized. The results of adoptive MP transfer by local injection performed in the present study may support this hypothesis. The rate of carbon-laden MP entrance into

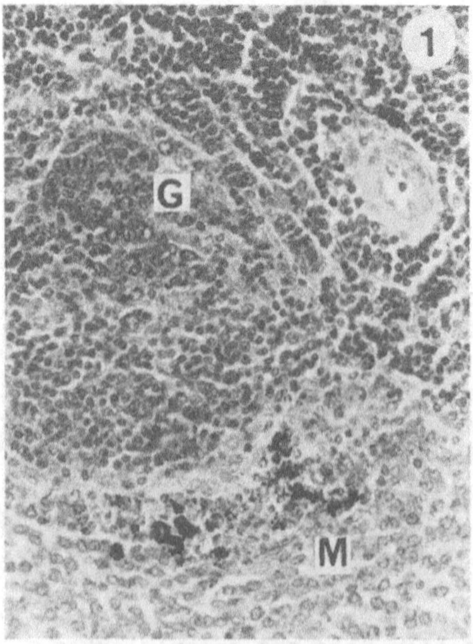

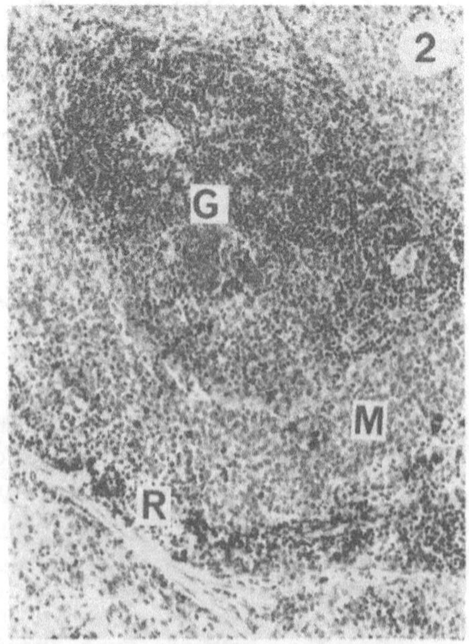

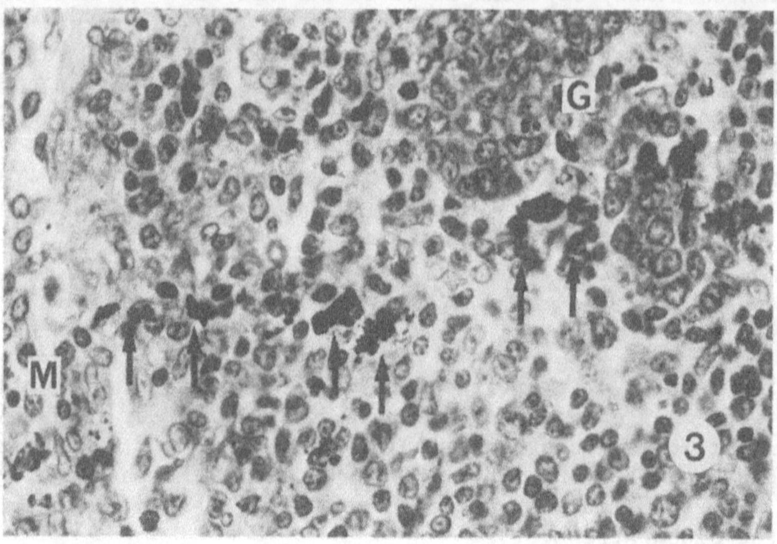

Fig. 1. Immediately after injection of carbon-laden macrophages into the splenic artery. They were found only in the marginal zone. x 240.

Fig. 2. Twelve hours after injection of carbon-laden macrophges into the splenic artery. x 120.

Fig. 3. A higher magnification of a part of Fig. 2. Large carbon-laden macrophages can be seen in a row (arrow) from the marginal zone to the germinal centre. x 600.

G: germinal centre, M: marginal zone, R: red pulp.

GC appeared to be parallel to the absolute number of transferred cells found in MZ and RP. Splenic arterial (SA) injection was more effective in sequestrating carbon cells within spleen than the hepatic arterial injection. This might account for definite but unimpressive migration of carbon-laden MP into lymph follicles in experiment 2. Alternatively, GC entrance of MP might be produced by an injection artefact due to unappropriately high infusion pressure or an interrupted circulation of SA by direct handling. Since result of experiment 3 is comparable with that of experiment 2 where splenic circulation was protected intact, and a structural integrity of the MP-containing GC after SA injection was preserved, this artefact might liest account for the MP migration.

As a conclusion, possible migration of carbon-laden MP of both intrinsic and extrinsic origin into the splenic GC from MZ was demonstrated. These MP may participate in carrying antigens or immune complexes from MZ to GC in the spleen.

SUMMARY

In the first experiment, histological observation of the mouse spleen was carried out after intravenous carbon injection. Large carbon-laden macrophages were found in the marginal zone soon after injection and then they appeared to migrate into the white or red pulp. Macrophages migrating deeply into the white pulp came together favorably around the germinal centre and entered the majority of the germinal centres by 10 days.

In the second and third experiments, syngeneic transfer of rat peritoneal macrophages labeled with carbon were performed. They were injected either retrograde via hepatic artery or directly into the splenic artery. Carbon-laden macrophages were found in the germinal centres in the spleen 12 to 24 hours after injection though in a limited number. Possible migration of intrinsic as well as extrinsic macrophage from the marginal zone into the germinal centre was demonstrated.

REFERENCES

Humphrey, J.H., and Grennan, D., 1981, Different macrophage populations distinguished by means of fluorescent polysaccharides. Recognition and properties of marginal zone macrophages, Eur. J. Immunol., 11:221.

Kotani, M., Okada, K ., Fujii, H., Tsuchiya, H., Matsuno, K., Ekino, S., and Fukuda, S., 1977, Lymph macrophages enter the germinal center of lymph nodes of guinea pigs, Acta anat., 99:391.

Kotani, M., Ezaki, T., Fujii, H., Matsuno, K., Ekino, S., and Harada, S., 1979, Peritoneal macrophages introduced into mouse foot pads enter the germinal center of regional lymph nodes nonspecifically, Acta anat., 104:406.

Kotani, M., Ezaki, T., Ekino, S., Matsuno, K., Fujii, H., and Nawa, Y., 1982, Lymph macrophages enter the germinal center of regional lymph nodes, in:"In Vivo Immunology," P.Nieuwenhuis, A.A.van den Broek, and M.G.Hanna,Jr., eds., Plenum Press, New York, pp.837.

Nossal, G.J.V., Ada, G.L., 1971, "Antigens, Lymphoid Cells and the Immune Response," Academic Press, London, pp. 121.

Roser, B.,1965, The distribution of intravenously injected peritoneal macrophages in the mouse, Aust. J. exp. Biol. med. Sci., 43:553.

Roser, B., 1970, The migration of macrophages in vivo, in: "Mononuclear Phagocytes," R.van Furth, ed.,Blackwell Scientific Publications, Oxford, pp. 166.

Unanue, E.R., and Cerottini, J.C., 1971, Fate and immunogenicity of macrophage associated hemocyanins, in:"Developmental Aspects of Antibody Formation and Structure," J.Sterzl, and M.Riha, eds., vol. 2., Publ. House Czech. Acad. Sci., Prague.

SPLENIC DEPENDENCE OF THE ANTIBODY RESPONSE TO THE TI-2 ANTIGEN, DNP-FICOLL: EFFECT OF THE FICOLL CARRIER ON THE TISSUE DISTRIBUTION OF 125-I-FICOLL CONJUGATES

P.L. Amlot*, D. Grennan** and J.H. Humphrey**

Departments of Medicine, Guy's Hospital Medical School* and Royal Postgraduate Medical School Hammersmith Hospital**, London U.K.

INTRODUCTION

In two situations the spleen is the major, if not only, site of the antibody response. Firstly, particulate antigens, usually sheep red blood cells (SRBC), are readily trapped in the splenic architecture when administered intravenously (i.v.) and in this situation SRBC antigens are concentrated and presented selectively to splenic B-cells. In this case splenectomy may impair the antibody response to such antigens following i.v. but not when other routes of immunisation are used. [1,2,3] Secondly, there is increasing evidence that the spleen responds more readily than other lymphoid tissues to polysaccharide antigens, irrespective of the route of immunisation. Examples of this are that splenectomised (Sx) mice respond poorly to the pneumococcal polysaccharide SIII[4] and thymus-independent (TI) antigens induce plaque-forming cells in the spleen earlier and more easily than in the bone marrow[5]. There are two, possibly complementary, features that might explain why the spleen responds more sensitively to polysaccharide antigens. First, it has been reported that certain polysaccharides, particularly TI type 2 (TI-2) antigens, localise preferentially in the macrophages of the splenic marginal zone (MZM)[6] and secondly this area contains a distinctive B-cell population[7] providing a putatively specialised milieu for the response to polysaccharide antigens.

We report here the splenic dependence of the antibody response to the TI-2 antigen, DNP-Ficoll, and examine conditions affecting localisation of radiolabelled Ficoll conjugates to the MZM of the spleen and lymph nodes.

MATERIALS AND METHODS

Mice used were CBA aged 10-12 weeks (Olac, UK) and Charles River OUtbred CD1 aged 12-16 weeks (Guy's Hospital Medical School). Splenectomy (Sx) and sham-splenectomy (Sh) were performed via a left subcostal incision, under chloral hydrate anaesthesia. Experiments were performed at least 18 days later.

Antigens. The number of residues as moles/10^5 daltons of the carrier are shown in subscript. Ficoll (FIC) was coupled as $Tyramine_{2.5}$-FIC, $DNP_{3 \text{ or } 16}$-lys-tyr-FIC and DNP_9-lys-FIC[6], Lipopolysaccharide as TNP_{15}-LPS[8] and keyhole limpet haemocyanin as $TNP_{9.5}$-KLH[9]. Immunisation with DNP_9-lys-FIC was i.p. and ^{125}I-FIC conjugates were given i.v. Unconjugated FIC loads were either given concurrently (i.v.) or 18 hours prior to (i.p.) the ^{125}I-FIC.

Antibody assay for anti-TNP antibodies was by haemagglutination of TNP sensitised horse RBC[10].

Tissue distribution of ^{125}I-FIC conjugates was assessed 2 days after injection. Blood, spleen, lymph nodes, liver, heart, lungs, bone marrow, intestines, kidneys, carcase and skin were counted for retained radioactivity. The relative concentration of ^{125}I-FIC in tissues was given by the

$$\text{Tissue Index} - \frac{\text{dpm per mg. of tissue or organ}}{\text{dpm per mg of the whole body}}$$

Cyclosphosphamide (CY) was given as a single dose of 300mg/kg i.v. and localisation experiments started 4 days later. Values shown in graphs and histograms are the mean $\pm$ SE of 3-6 mice each.

RESULTS

The effect of splenectomy (Sx) upon the antibody response to the TI-2 antigen, DNP-FIC, is shown in Fig. 1. This is representative of the response made by neonatally and adult Sx mice, with doses from 0.5-100μg DNP-FIC and with mice immunised from 18-70 days post-splenectomy. This severe and long-lasting impariment of the response to DNP-FIC was not seen with the TI-1 antigen, TNP-LPS, nor with the thymus-dependent (TD) antigen TNP-KLH. Furthermore the response to DNP-FIC could not be reconstituted with normal spleen cells (Fig.1) even though these were given in sufficient number to reconstitute the response to DNP-FIC by irradiated (750cGy) but not Sx mice.

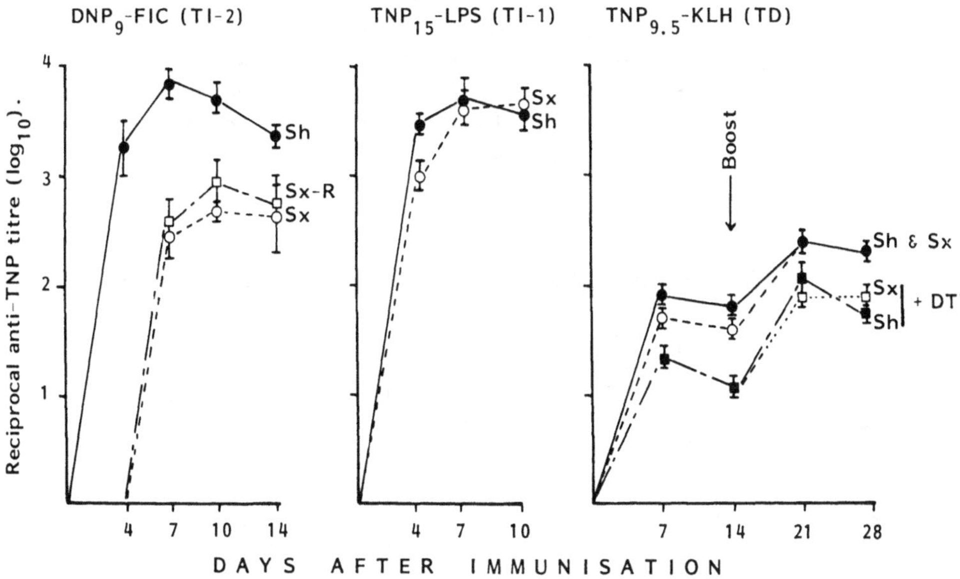

FIG. 1 The effect of splenectomy on the antibody response to TI-1, TI-2 and TD antigens in CBA mice.

The statistical difference between Sh splenectomised mice and Sx and Sx-R mice are as follows:-

<u>DNP-FIC</u> Sh, Sx and Sx reconstituted (Sx-R) mice received 1µ DNP-FIC. Sx-R mice were given 2.5 - 5 x 10^7 normal spleen cells i.v. 24 hours earlier.
<u>Day 4 and 7</u>: Sx, Sx-R $p<0.001$;
<u>Day 10</u>: Sx $p<0.001$, Sx-R $p<0,02$
<u>Day 14</u>: Sx is NS, Sx-R $p<0.01$
NS difference between Sx and Sx-R at any time.

<u>TNP-LPS</u> Immunisation with 1µg TNP-LPS.
<u>Day 4</u>: $p<0.02$; <u>Day 7 and 10</u>: NS

<u>TNP-KLH</u> Immunised with 100µg TNP-KLH on day 0 and boosted day 14 Total and 0.01M dithiothreitol (DTT) sensitive titres showed no significant difference between Sh and Sx at any time.

These findings were consistent with a requirement for presentation of DNP-FIC in immunogenic form by the radio-resistant MZM to splenic B-cells. In an attempt to displace DNP-FIC from MZM excess unconjugated FIC was given to mice either prior to or concurrently with DNP-FIC and we expected to see a suppression of antibody formation. Unexpectedly, antibody formation was not affected (Fig.2) even when a 5 million fold excess of unconjugated FIC was used.

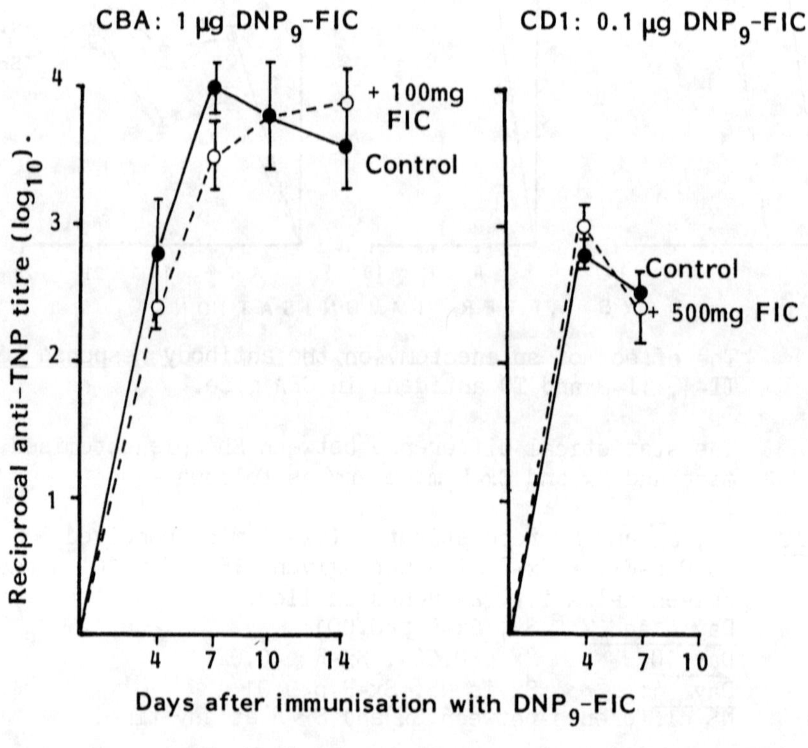

FIG. 2 The effect of unconjugated-FIC on the antibody response to DNP_9-FIC

Quantitative studies were therefore carried out to examine the efficacy of unconjugated FIC in suppressing localisation of ^{125}I-FIC conjugates to the spleen and other tissues. We found that selective localisation of ^{125}I-FIC (Tissue Indices> 1) only occurred regularly in the spleen, lymph nodes (LN) and liver. The results are shown in Fig. 3 and 4 and are summarised below.

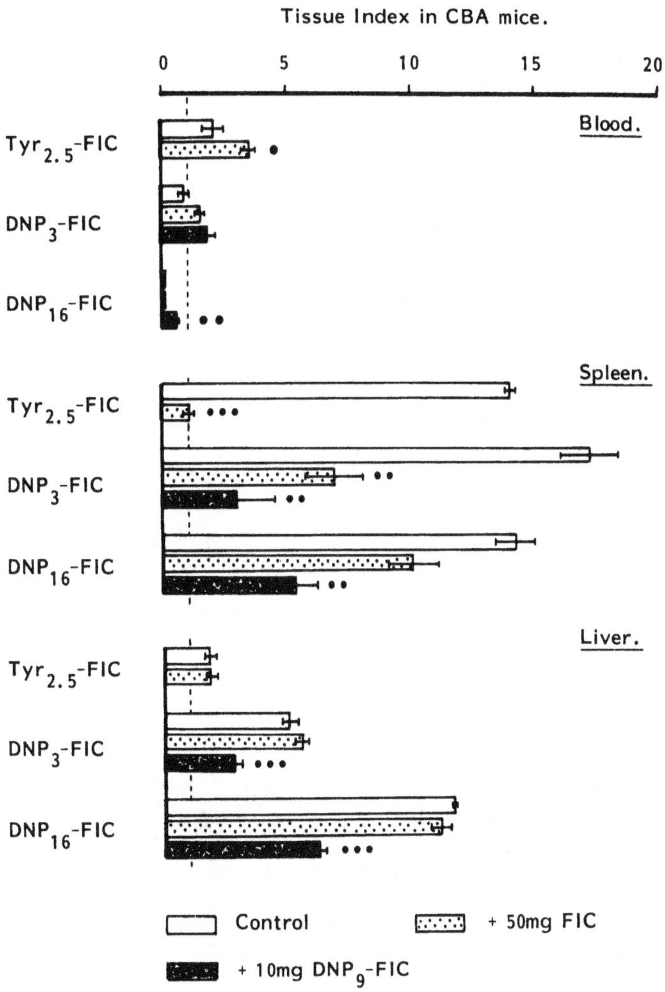

FIG. 3 Effect of a Ficoll load on the tissue indices of ^{125}I-FIC conjugates in CBA mice at 48 hours.

Dotted line shows a tissue index of 1 - the mean radioactivity throughout the body compared with controls: • $p<0.05$; ••
••• $p<0.002$

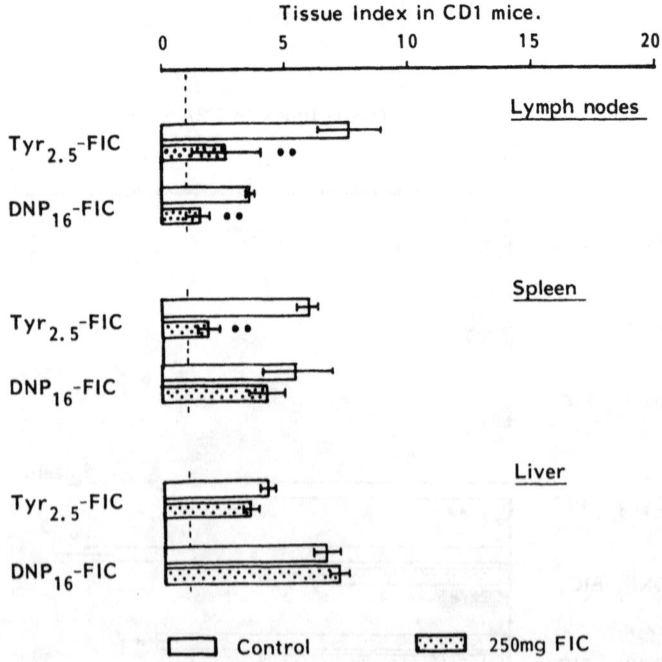

FIG. 4 Effect of a Ficoll load on the tissue indices of ^{125}I-FIC conjugates in CD1 mice at 48 hours.

Legend as in Fig. 3.

(1) Spleen and lymph nodes. Tyr-$_{2.5}$-FIC and DNP$_{3\ or\ 16}$-FIC accumulated to the same extent in both these tissues. However, although excess unconjugated FIC suppressed localisation of both types of conjugate, in the spleen it was much less effective at suppressing DNP$_{3\ or\ 16}$-FIC than Tyr$_{2.5}$-FIC. On the other hand, an excess of unlabelled DNP$_9$-FIC was much more effective at suppressing localisation of DNP$_{3\ or\ 16}$-FIC than an excess of unconjugated FIC.

(2) Liver.Here localisation differed from spleen and lymph nodes. DNP$_{3\ or\ 16}$-FIC accumulated to much greater extent in the liver than Tyr$_{2.5}$-FIC. Excess unconjugated FIC had no effect on the hepatic uptake of ^{125}I-FIC conjugates but unlabelled excess DNP$_9$-FIC suppressed localisation of DNP$_{3\ or\ 16}$-FIC

(3) Cyclophosphamide (CY) treatment was used to deplete murine B-cells and Table 1 shows its effects upon splenic localisation of ^{125}I-FIC conjugates. There was no diminution in uptake of either Tyr$_{2.5}$-FIC or DNP$_{3\ or\ 16}$-FIC, indeed in CBA mice the splenic uptake of DNP$_3$-FIC was increased. In CD1 mice uptake of DNP$_{16}$-FIC was more effectively suppressed by a FIC load in CY-treated than in untreated mice.

TABLE 1 The Effect of Unconjugated Ficoll and Cyclophosphamide (CY) Treatment on the Splenic Localisation of 125-I-Ficoll Conjugates

MOUSE STRAIN	^{125}I-FICOLL CONJUGATE	FICOLL LOAD	MEAN $\pm$ SE OF MOLS (x 10^3) PER MZM PER μg*:	
			CONTROL	CY-TREATED
CBA	$Tyr_{2.5}$-FIC	-	41.4$\pm$12	54.4$\pm$2
	$Tyr_{2.5}$-FIC	150mg	7.3$\pm$0.6	12.9$\pm$7
	DNP_3-FIC	-	69.2$\pm$11	110 $\pm$7.8**
	DNP_3-FIC	150mg	17.6$\pm$0.4	11.4$\pm$4
CD1	DNP_{16}-FIC	-	22.3$\pm$4	18 $\pm$6
	DNP_{16}-FIC	250mg	16.6$\pm$0.2	7.2$\pm$2**

* Calculation of mols/MZM/μg of ^{125}I-FIC injected by the formula:

$$\frac{s - b}{t} \times \frac{A}{M \times Z}$$

where:

s = splenic dpm; b = background dpm;
t = total dpm; A = Avogadro's constant;
M = M.W. of Ficoll (4×10^5); Z = 10^6 MZM/spleen

** p = <0.05 cfd with control

DISCUSSION

Both spleen and lymph nodes contain macrophages in the marginal zone (MZM) which can now be distinguished by a monoclonal antibody[11] as well as by their ability to take up ^{125}I-FIC or FITC-FIC conjugates[6]. These MZM take up polysaccharides (FIC, hydroxyethyl-starch or dextran) which are TI-2 antigens either intrinsically (dextran) or because of attached epitopes (DNP-FIC). The DNP-epitope on the FIC carrier does not contribute quantitatively to binding of DNP-FIC by MZM but does make a qualitative difference. Splenic uptake of DNP-FIC conjugates was <u>less</u> easily suppressed than $Tyr_{2.5}$-FIC or FITC-FIC by unconjugated FIC and <u>more</u> easily suppressed by unlabelled DNP-FIC than by unconjugated FIC. The influence of the DNP-epitope on the binding of DNP-FIC may be explained by a different population of cells in the splenic marginal zone namely B-cells. Firstly because DNP-binding B-cells occur frequently in murine spleens (1-20 per 10^3 spleen cells)[12] and secondly because excess unconjugated FIC can suppress the splenic uptake of DNP-FIC more easily in B-cell depleted than in normal mice.

We conclude that presentation of DNP-FIC by MZM to splenic B-cells is still likely to be essential for the antibody response to this TI-2 antigen. The reason why this could not be demonstrated definitively by displacement of immunogenic DNP-FIC from MZM by carrier FIC is attributable to the different binding properties of DNP-FIC compared to the carrier FIC alone.

REFERENCES

1. D.A. Rowley, Effect of splenectomy on formation of circulating antibody in adult male albino rat, J.Immunol. 64: 289-295 (1950)

2. F.L. Adler, Studies on mouse antibodies I. The response to sheep red cell, J. Immunol. 95: 26-38 (1965)

3. R. Benner and A. van Oudenaren, Antibody formation in mouse bone marrow 1V. The influence of splenectomy on the bone-marrow plaque-forming cell response to sheep red cells, Cell Immunol 19: 167-182 (1975)

4. J.M. Jones, D.F. Amsbaugh and B. Prescott, Kinetics of the antibody response to type 11 pneumococcal polysaccharide 11. Factors influencing the serum antibody levels after immunisation of an optimally immunogenic dose of antigen. J.Immunol 116: 52-64 (1976)

5. G. Koch, B.D., Lok, A. van Oudenaren and R.J. Benner, The capacity and mechanism of bone marrow antibody formation bt TI-antigens. J. Immunol. 128: 1497-1501 (1982)

6. J.H. Humphrey, Tolerogenic or immunogenic activity of hapten-conjugated polysaccharides correlated with cellular localisation Eur. J. Immunol. 11: 212-220 (1981)

7. D.S. Kumaratne, H. Bazin, and I.C.M. MacLennan, Marginal zones : the major B-cell compartment of rat spleens, Eur. J. Immunol 11: 858-864 (1981)

8. J. Fidler, In Vivo immune response to TNP hapten coupled to thymus-independent carrier lipopolysaccharide, Cell. Immunol 16:223-236 (1975)

9. M.B. Rittenberg and A.A. Amkraut, Immunogenicity of trinitrophenyl-haemocyanin: production of primary and secondary anti-hapten precipitins, J.Immunol 97: 421-430 (1966)

10. M.B. Rittenberg and K.L. Pratt, Antitrinitrophenyl (TNP) plaque assay. Primary response of Balb/c mice to soluble and particulate immunogen, Proc. Soc. Exp. Biol. Med. 132 : 575-581 (1969)

11. E. Van Vliet, M. Melis and W. van Ewijk. Marginal zone macrophages in the mouse spleen identified by a monoclonal antibody I. Anatomical correlation with a B-cell sub-populations, Histochem. Cytochem. Submitted for publication.

12. N.H. Sigal and N.R. Klinman, The B-cell clonotype repertoire, Adv.Immunol. 26: 255-387 (1978)

EVIDENCE THAT STATIC BUT NOT RECIRCULATING B CELLS ARE RESPONSIBLE FOR ANTIBODY PRODUCTION AGAINST DINITROPHENOL ON NEUTRAL POLYSACCHARIDE, A TI-2 ANTIGEN

D. Gray[1], I.C.M. MacLennan[1], B. Platteau[2], H. Bazin[2], J. Lortan[1] and G.D. Johnson[1]

[1]Dept. Immunology, University of Birmingham Medical School, Birmingham B15 2TJ
[2]Experimental Immunology Unit, University of Louvain, 1200 Brussels, Belgium

INTRODUCTION

The peripheral B cell pool of the rat can be divided into a subpopulation that recirculates between follicles in secondary lymphoid organs and one that is static (1). A large component of the non-recirculating population is found in the marginal zones (MZ) of the spleen. MZ B cells differ from recirculating follicular (RF) B cells in lacking surface membrane (Sm) IgD. Both express Sm IgM (2). Recently, several lines of indirect evdence have emerged that infer the involvement of MZ B cells in responses to thymus-independent type-2 (TI-2) antigens: i) Rats suppressed from birth with anti-IgD antibodies develop intact MZ, but not RF B cell populations (3). ii) This treatment results in an elevation of of serum IgG2c with loss of IgG2a (3). IgG2c is the predominant isotype elicited in response to TI-2 antigens, while IgG2a is dominant in T cell-dependent responses (4). iii) Approximately 20% of MZ B cells express Sm IgG2c (3). iv) Dendritic cells within the marginal zone selectively localise neutral polysacccharides which are TI-2 antigens (5,6). v) Adult splenectomy causes a profound depression of the response to the the TI-2 antigen dinitrophenylated hydroxyethyl starch (DNP-HES), but not to DNP-haemocyanin, a thymus-dependent antigen (7). In this communication we provide evidence of a more direct nature to associate MZ B cells with responses to TI-2 antigens. This is derived from experiments using congenic strains of rats differing in their kappa light chain allotype. Parabiosis of congenic pairs for 10 days leads to an exchange of RF B cells, but only a low level of MZ B cell chimaerism is achieved. Consequently, follicles in these animals consist of a mixture of recipient and donor B cells, while the MZ remains almost wholly of recipient allotype. In the experiments described here, animals were separated after 10 days parabiosis. After a further 5 days each rat was immunised with DNP-HES. The degree of chimaerism in follicles and MZ and the allotype of the response to DNP on a HES carrier were assessed 10 days after immunisation.

MATERIALS AND METHODS

Animals

LOU/C strain of rats bear the Igkla allotype. A congenic strain bearing the OKA Igklb allotype on a LOU/C background has been derived at the Experimental Immunology Unit, University of Louvain, Brussels (8). Parabiosis experiments used pairs of these congenic rats.

Parabiosis

The procedure for joining animals was similar to that described in (9). Parabiosed pairs were bandaged together and the dressing changed daily. Rats 6-8 wks of age were used in these experiments. Previous time-course experiments demonstrated that cross-circulation and extensive exchange between recirculating lymphocytes occurred within 10 days.

Antibodies

Polyclonal rabbit antibodies to rat IgM and IgD and monoclonal antibodies to rat kappa (MARK-1)(10), and to rat kappa la allotype (MARK-3) were produced in the Experimental Immunology Unit, Brussels. A polyclonal rat anti-rat kappa lb was raised in the Dept. of Immunology in Birmingham.

Immunofluorescence

Frozen sections of spleen were cut and fixed in acetone at $4^{o}C$ for 20 min. Pairs of antibodies labelled with either fluorescein or rhodamine were used simultaneously to stain for IgM + donor allotype, or IgD + donor allotype. Antibodies to IgM, IgD and kappa la allotype (MARK-3) were directly labelled with fluorescent dye, while anti-kappa lb antibody was biotinylated and identified using fluorescein-avidin. The double immunofluorescence staining procedure is described by Johnson and Holborow (11).

Radioimmunoassay

Standard solid-phase RIA techniques were employed to detect

anti-DNP antibody in serum samples. Polystyrene mictrotitre plates were coated (in 0.1M carbonate buffer, pH 9.5) with DNP-ovalbumin (25 ug/well). Dilutions of sera were incubated for 2 hr on these plates. Detection of total kappa and kappa la antibodies to DNP was carried out using ^{125}I-labelled MARK-1 and MARK-3 antibodies, respectively. Detection of kappa lb antibodies emp-loyed a three-layer technique: the anti-kappa lb was biotinylated and detected using ^{125}I-avidin.

RESULTS AND DISCUSSION

Double immunofluorescence staining of spleen sections from pairs of parabiosed (and separated) congenic rats revealed the degree of recipient/donor B cell chimaerism in the MZ and follicles (Figs. la and lb). Fig. 2 shows that in follicles the IgD+ve B cells were 20-30% of donor allotype. In contrast, IgD-ve B cells of donor allotype comprised less than 5% of the MZ B cell population in all 12 animals. The distinction is made between Sm IgD+ve and Sm IgD-ve cells in the MZ, as IgD+ve cells in this region have been shown to be recirculating B cells in transit to follicles (2,12).

The serum anti-DNP response was assessed in a RIA, firstly to detect antibodies of donor kappa allotype, and secondly all kappa antibodies (recipient and donor). The donor allotype response expressed as a proportion of the total kappa response to DNP-HES in the rats separated after parabiosis is shown in Fig. 2. This varies from 0.4 to 16%. This is well below the level of donor RF B cell chimaerism and correlates more closely with the level of non-recirculating B cell chimaerism. If the response originated from RF B cells, 20 to 30% chimaerism in this population might be expected to elicit a much higher proportion of antibody of donor allotype.

The results are consistent with the hypothesis that static B cells are involved in the antibody response to DNP-HES. This may reflect inherent differences in the responsiveness of MZ and RF B cells to TI-2 antigens. However, the response of MZ B cells is likely, in part, to reflect the localisation of DNP-HES on MZ dendritic cells. Although RF B cells migrate through the MZ, they may not be activated by DNP at that site because of lack of T cell help. The cells localising neutral polysaccharides have been reported to lack Class II MHC antigens, and so may not be able to activate T helper cells. It is more likely that T cell-dependent antigens activate RF B cells as they pass interdigitating cells in the T zones of secondary lymphoid organs. Experiments are currently underway to determine whether the response of rats to DNP-haemocyanin, a thymus-dependent antigen, is induced in static or recirculating B cells.

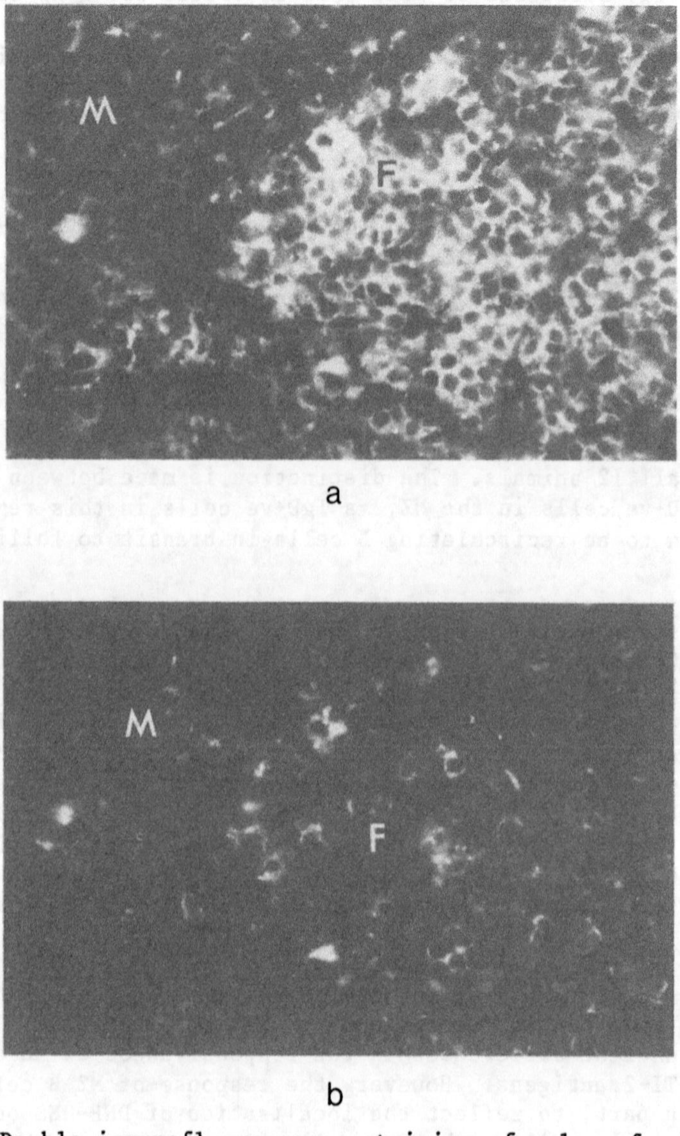

Fig. 1. Double immunofluorescence staining of spleen from a rat parabiosed for 10 days, separated and killed after a further 15 days. 1a. rhodamine-anti-IgD staining shows a follicle (F), containing +ve cells and marginal zone (M), containing IgD-ve cells. 1b. Fluorescein-anti-donor k allotype shows that while a significant proportion of follicular B cells are of donor allotype, the marginal zone contains only very few of these cells.

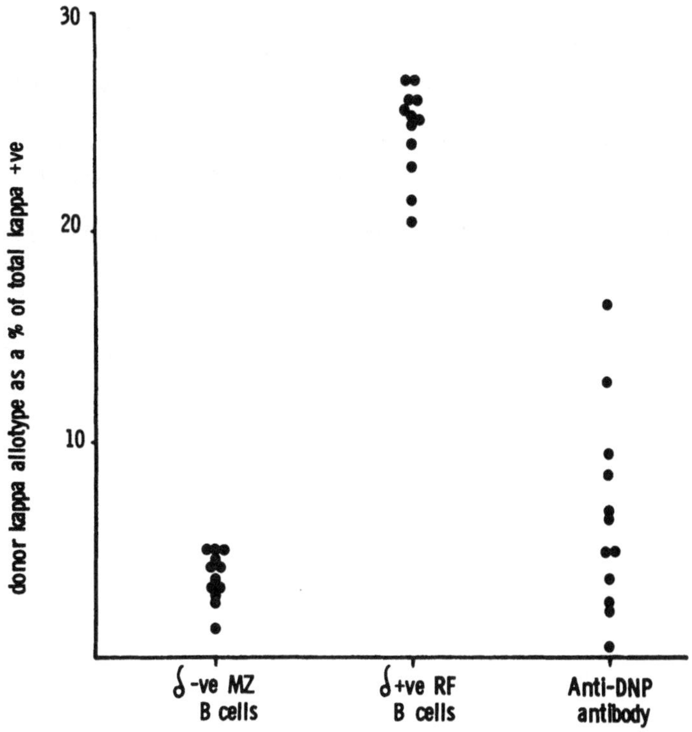

Fig. 2. Donor B cell chimaerism in the marginal zone (MZ), and the follicular compartments compared with the proportion of donor kappa allotype detected in the serum antibody response to DNP-HES.

REFERENCES

1. D.S. Kumararatne, I.C.M. MacLennan and H. Bazin. Eur. J. Immunol. 11: 858 (1981).
2. D. Gray, I.C.M. MacLennan, H. Bazin and M.Khan. Eur. J. Immunol. 12: 564 (1982).
3. H. Bazin, D. Gray, B. Platteau and I.C.M. MacLennan. Ann. N.Y. Acad. Sci. 399: 157 (1982).
4. G.P. der Balian, J. Slack, B.L. Clevinger, H. Bazin and J.M. Davie. J. Exp. Med. 152: 209 (1980).
5. J.H. Humphrey and D. Grennan. Eur. J. Immunol. 11: 221 (1981).
6. D. Fray, I. McConnell, D.S. Kumararatne, I.C.M. MacLennan, J.H. Humphrey and H. Bazin. Eur. J. Immunol. 14: 47 (1984).
7. D. Gray, D. Chassoux, I.C.M. MacLennan and H. Bazin. (In Press).
8. H. Bazin, F. Cormont and L. DeClercq. J. Immunol. Methods 71: 9 (1984).

9. E. Bunster and R.K. Meyer. Anat. Rec. 57: 339 (1933).
10. H. Bazin, L.M. Xhurdebise, G. Burtonboy, A.M. Lebacq, L. Declercq and F. Cormont. J. Immunol. Methods 66: 261 (1984).
11. G.D. Johnson and E.J. Holborow in: "Handbook of Experimental Immunology", Blackwell Scientific Publications, Oxford (1984).
12. P. Nieuwenhuis, and W.L. Ford. Cell. Immunol. 23: 254 (1976).

T-LYMPHOCYTES IN RAT LYMPHOID FOLLICLES ARE A SUBSET OF T HELPER CELLS

F.G.M. Kroese, A.S. Wubbena, P.Joling* and P.Nieuwenhuis

Dept. of Histology, State University, Oostersingel 69/1 Groningen and *Dept. of Internal Medicine I, Erasmus University, Rotterdam, The Netherlands

INTRODUCTION

Germinal centres are well defined morphological structures found in peripheral lymphoid organs upon stimulation with antigen (for review see 1). Although their role in the generation of immunological memory seems to be well established, the mechanism of germinal centre reactivity is hardly understood. However, it is clear that T-lymphocytes play an important role in their generation, as indicated by the absence of germinal centres in nude (thymusless) animals (2). Moreover, reconstitution of lethally irradiated mice and rats requires both B and T-lymphocytes for germinal centres to develop (3) and T-independent antigens are poor inducers of germinal centre activity (4).

The way T-lymphocytes are involved in the induction and/or continuation of germinal centres is obscure. In this context, the presence of T-lymphocytes in lymphoid follicles as demonstrated in mouse (5) and man (6), may be of functional importance. Using monoclonal antibodies (Mab's) directed against T-lymphocyte subpopulations, it was found that in secondary follicles these T-cells belong predominantly to the T-helper subset (6-8). In the rat, data on the presence of T-lymphocytes in lymphoid follicles are conflicting (cf. 9-11) and incomplete. In this study the occurrence of T-lymphocytes and their subpopulations in splenic lymphoid follicles of rats was investigated with the aid of a panel of Mab's.

MATERIALS AND METHODS

Rats and tissues

To study primary follicles, splenic tissue of germfree AO and Sprague Dawley rats was used.
To study secondary follicles, conventionally bred AO rats were immunized intravenously with 1×10^9 sheep red blood cells. These rats were killed 7 days after injection, when spleen samples were taken out. Fragments of splenic tissue were deep frozen with liquid freon and stored at -70°C.

Immunoperoxidase staining

To demonstrate the presence of T-lymphocytes in tissue sections, the indirect immunoperoxidase technique was applied. Acetone fixed serial cryostat sections of spleen were overlayed with Mab in the appropriate dilution in phosphate buffered saline (PBS). Table 1 shows the specificities of the Mab's. Control sections were incubated with an irrelevant antibody (2G12), directed against human melanoma cells (gift of Dr. L. de Leij). After washing PBS (3x, 10'), the sections were incubated with peroxidase conjugated rabbit-anti-mouse Ig antiserum (DAKO, Denmark), diluted 1:40 in PBS, and supplemented with 1% normal rat serum. The sections were rinsed in PBS (3x, 10') and stained for peroxidase activity with 3.3'-diaminobenzidine tetra HCl (Sigma, St. Louis, U.S.A.), 0.5 mg/ml in 0.05 M Tris/HCl buffer (pH 7.6), containing 0.01% hydrogen peroxide. The reaction product was intensified by incubation for 5 minutes with 0.5% w/v $CuSO_4$ in 0.9% w/v NaCl. Sections were then counterstained with Hematoxylin according to Mayer.

Monoclonal antibody HIS 17

The Mab HIS 17 was obtained by fusion of spleen cells of a Balb/c mouse, immunized with lymphocytes of rat Peyer's patches, with the murine myeloma celline X 63-Ag8 (12). This Mab of the IgG1 subclass, reacted with all thymocytes as well as virtually all lymphoid cells in T cell areas of peripheral lymphoid organs (Fig. 2). In bone marrow, HIS 17 reacts with the surface of cells of the myeloid and lymphoid lineages, including plasmacells. This staining pattern closely resembles that of the Mab W3/13 (anti-rat T-lymphocytes; 13).

RESULTS

T-lymphocytes in splenic primary follicles of germfree rats

To study the presence of T-lymphocytes in primary follicles, spleens of germfree (GF) rats were used. Immunoperoxidase staining of serial sections of splenic tissue of both GF AO rats and GF Sprague Dawley rats revealed the presence of low numbers W3/13, ER- 1, HIS 17 and MRC OX-19 positive cells (all pan-T Mab's) in primary follicles (Fig. 1). These cells were distributed randomly throughout the follicles. Staining with Mab's against T-lymphocyte subsets showed the presence of both W3/25$^+$ and ER-2$^+$ (T helper subset, Th) lymphocyte as well as MRC OX-8$^+$ and ER-3$^+$ cells (T suppressor/cytotoxic subset, Ts/c) with the former in slightly excess (not shown).

T lymphocytes in splenic secondary follicles of immunized rats

Secondary follicles were studied in the spleens of antigen (SRBC) stimulated rats. W3/13, ER-1, HIS 17 and MRC OX-19 positive cells (pan-T Mab's) were detected in both lymphocyte corona and germinal centre of secondary follicles (Fig. 2 and 3a). Within germinal centres, most positive cells were found in the light zone, with only a few cells in the dark zone. In the corona they were found scattered throughout. The percentage of positive cells in the corona and in the light zone of the germinal centre of secondary follicles was much higher than found in primary follicles of GF rats (compare Fig. 1 and Fig. 3a). Staining of serial sections showed an almost identical staining pattern with W3/25 (or ER-2) (Th subset) (Fig. 3b). However, only few cells in corona and germinal centre expressed the MRC OX-8 determinant (Ts/c subset) (Fig. 3c). The Mab ER-3 recognizes all MRC OX-8 positive cells and a small proportion (5%) of the W3/25 subset (16). Surprisingly, in secondary follicles, the number of ER-3$^+$ cells exceeded the number of OX-8$^+$ cells, approaching the number of W3/25$^+$ cell (Fig. 3d) as judged by similar staining patterns.

DISCUSSION

The presence of lymphocytes bearing T cell antigens in follicles of rat spleen was unequivocally demonstrated by immunoperoxidase staining, using a panel of Mab's. These putative T-lymphocytes were present both in primary and in secondary follicles, although in different relative numbers. In primary follicles of germfree rats, the few T-lymphocytes were distributed randomly throughout the follicle. In secondary follicles of antigen stimulated rats, T-lymphocytes were found throughout the corona, and in the light zone of germinal centres. Few T-lymphocytes were observed in the dark zone of the germinal centre. This distribution pattern is in

TABLE 1: Monoclonal antibodies used to detect T lymphocytes and their subsets in rat spleen.

MAB	TARGET	SOURCE/REFERENCE
W3/13	All T-lymphocytes	Seralab, (13, 14)
ER-1	All T-lymphocytes (W3/13 like)	EUR*, (15)
HIS 17	All T-lymphocytes (W3/13 like)	RUG, see text
MRC OX-19	All T-lymphocytes	MRC, (14)
W3/25	T-helper cells	Seralab, (14)
ER-2	T-helper cells (W3/25 like)	EUR, (15, 16)
MRC OX-8	T-suppressor/cytotoxic cells	MRC, (14)
ER-3	T-suppressor/cytotoxic cells and 5% of W3/25^{+} cells	EUR, (15, 16)

* EUR: Erasmus University Rotterdam, The Netherlands
RUG: State University Groningen, The Netherlands
MRC: Gift of the MRC Cellular Immunology Unit, Oxford, U.K.

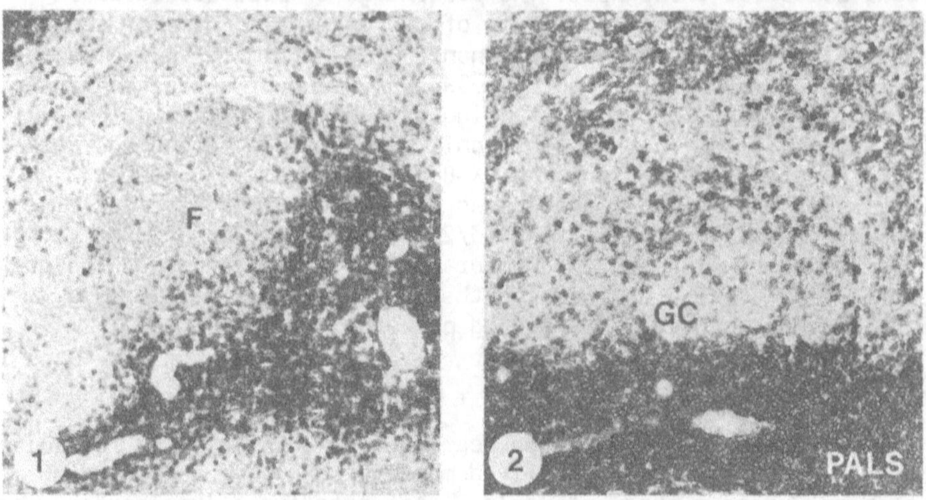

Fig. 1 and 2: Immunoperoxidase-staining pattern of frozen sections of spleen. (1): Spleen of germfree rat stained with MRC OX-19. (F = primary follicle). (2): Immunized rat spleen, stained with HIS 17. (PALS = periarteriolar lymphocyte sheath; GC = germinal centre.

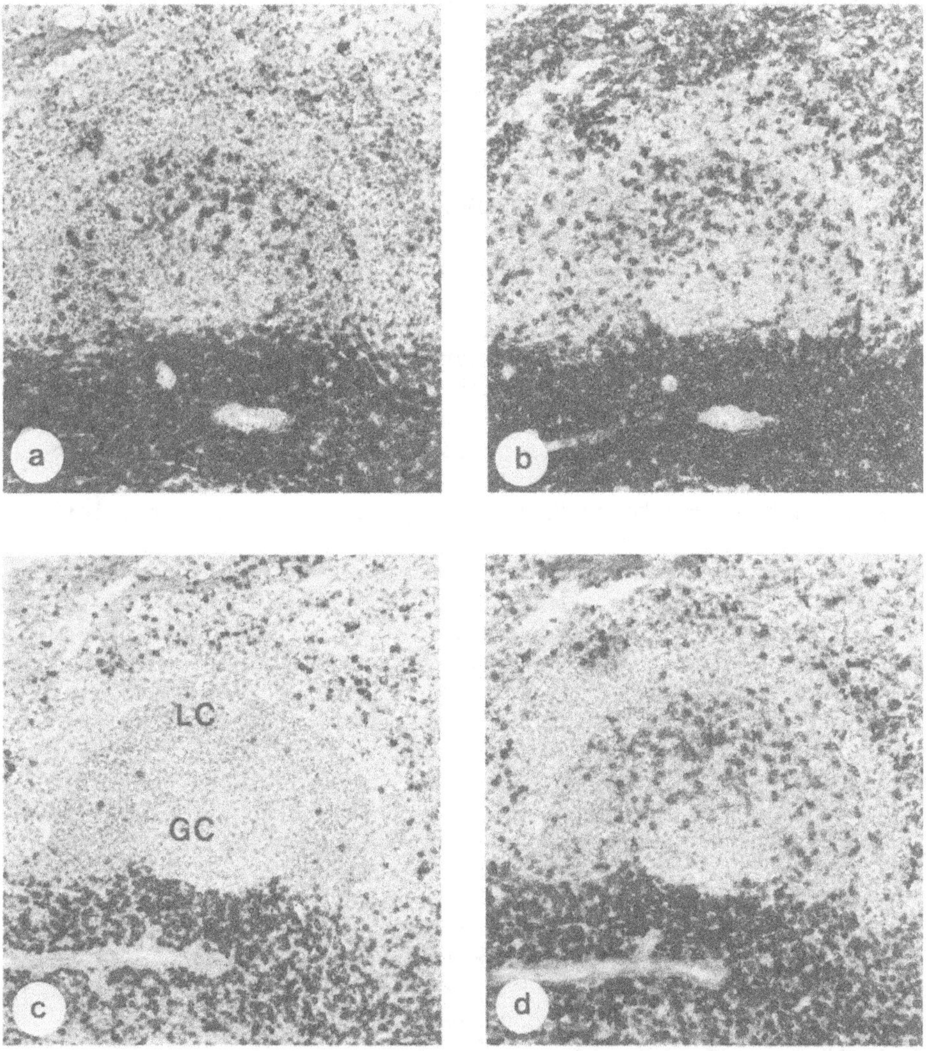

Fig. 3: Immunoperoxidase-staining pattern of serial frozen sections of immunized rat spleen incubated with different Mab's showing a secondary follicle with lymphoycte corona (LC) and germinal centre (GC). The sections were incubated with a) MRC OX-19, b) ER-2, c) MRC OX-8 and d) ER-3.

accordance with that described for follicles of mouse (4; 7) and human (6-8) peripheral lymphoid organs.

Analysis of the subsets of T-lymphocytes present in lymphoid follicles demonstrated that in primary follicles both W3/25$^+$ (Th subset) and MRC OX-8$^+$ (Ts/c subset) cells were found, with W3/25$^+$ cells in only slight excess, as observed in mice (7). On the other hand, in secondary follicles of rat spleen the vast majority of T-lymphocytes appeared to belong to the Th subset, because of their staining with W3/25 and ER-2, but not with MRC OX-8. A similar predominance of Th cells in secondary follicles has been reported for both mouse (7) and human (6-8). Our most striking finding was the reactivity of the majority of T-lymphocytes in rat splenic follicles with Mab ER-3. This Mab recognizes only 5% of splenic W3/25$^+$ cells (Th subset), and it therefore appears that the majority of T-lymphocytes in rat splenic follicles seems to belong to a particular subset of Th cells, characterized by the presence of the ER-3 determinant. However, definite proof awaits double staining procedures. Whether this putative W3/25$^+$ ER-3$^+$ cell population is unique to secondary follicles or is also present in the T cell area (periarteriolar lymphocyte sheath) of the spleen could not be established, due to confluence of positive cells in the latter area.

In secondary follicles of human peripheral lymphoid organs the presence of cells bearing the Leu 7 (HNK-1) determinant has recently been described by several authors (e.g. 17, 18). This determinant present on natural killer (NK) cells. Double stained sections with anti-T-lymphocyte Mab's and anti-Leu 7 revealed that in follicles the Leu 7 determinant is borne by the majority of the T(helper) cells (17; 18). Thus also in human lymphoid follicles most of the follicular T-lymphoyctes appear to be a special subset of Th cells (17). In this context, it should be emphasized that the ER-3 antigen -analogous to Leu 7- is also expressed by NK cells (19). Possibly, these cells may exert both Th cell and NK cell functions, involved in the regulation of germinal centre reactivity. In addition to T cell regulation, recently NK cell regulation of B-lymphocyte responses has been described (20). Further studies will therefore aim at the isolation of W3/25$^+$ ER-3$^+$ cells from rat lymphoid tissues and to study their possible role(s) in germinal centre activity.

ACKNOWLEDGEMENTS

The authors wish to thank Dr. D. Opstelten for discussions, Miss Aukje Zandstra for typing the manuscript and Mr. Peter van der Sijde for the photographs.

REFERENCES

1. P. Nieuwenhuis and D. Opstelten. Am. J. Anat. 170: 421 (1984).
2. E.B. Jacobson, L.H. Caporale and G.J. Thorbecke. Cell. Immunol. 13: 416 (1974).
3. N.A. Gastkemper, A.S. Wubbena, F.J.H. Gimbrère, A. de Graaff and P. Nieuwenhuis. Cell. Tiss. Res. 219: 281 (1981).
4. I.L. Weissman, G.A. Gutman, S.H. Friedberg and L. Jerabek. Adv. Exp. Med. Biol. 66: 229 (1976).
5. G.A. Gutman and I.L.Weissman. Immunology 23: 465 (1972).
6. S. Poppema, A.K. Bhan, E.L. Reinhertz, R.T. McCluskey and S.F. Schlossman. J. Exp. Med. 153: 30 (1981).
7. R.V. Rouse, I.L. Weissman, J.A. Ledbetter and R.A. Warnke. Adv. Exp. Med. Biol. 149: 751 (1982).
8. D.J. Bailey, M. Rainey and J.A. Habeshaw. Adv. Exp. Med. Biol. 149: 809 (1982).
9. J. Goldschneider and D.D. McGregor. J. Exp. Med. 138: 1443 (1973).
10. A.N. Barclay. Immunology 42: 593 (1981).
11. G. Mayrhofer, C.W. Pugh and A.N. Barclay. Eur. J. Immunol. 13: 112 (1983).
12. F.G.M. Kroese, D. Opstelten, A.S. Wubbena, G.J. Deenen, E.H. Schwander, L. de Leij and P. Nieuwenhuis (this volume).
13. A.F. Williams, G. Galfré and C. Milstein. Cell 12: 663 (1977).
14. D.W. Mason, R.P. Arthur, M.J. Dallman, J.R. Green, G.P. Spickett and M.L. Thomas. Immunological Rev. 74: 57 (1983).
15. P. Joling, L.M.B. Vaessen, F.J. Thielen and J. Rozing. In: "Modern Trends in Clinical Immunosuppression". Eds.: W. Weimar, R.L. Marquet, A.B. Bijnen and R.J. Ploeg, Rotterdam, The Netherlands, p. 201 (1983).
16. P. Joling, F.J. Thielen, L.M.B. Vaessen, C.J. Hesse and J. Rozing (this volume).
17. S. Poppema, L. Visser and L. de Leij. Clin. Exp. Immunol. 54: 834 (1983)
18. A. Porwit-Ksiazek, T. Ksiazek and P. Biberfeld. Scand. J. Immunol. 18: 485 (1983).
19. J. Rozing, personal communication.
20. J.A. Brieve, S. Targan and R.H. Stevens. J. Immunol. 132: 611 (1984).

FUNCTIONAL DIFFERENCES ASSOCIATED WITH QUANTITATIVE DISTRIBUTION OF Ia ANTIGENS ON MOUSE B CELLS

E.M. Andrew, N.M. Mackenzie and
R.M.E. Parkhouse

National Institute for Medical Research
London NW7 1AA, UK

Cellular interaction, the sine qua non of the immune system, is based on ligand-receptor systems expressed on the surface of the interacting cell types. Our understanding of the cellular functioning of the immune system has been made possible largely through the use of antibodies to surface molecules which are selectively expressed on subpopulations of lymphocytes at different stages of their development. In many studies the emphasis has been on the association of cell function with the presence or absence of a particular surface marker[1-3]. In other words, the function of the cell has been correlated with operationally qualitative differences in marker expression.

One could envisage, however, in such a system which relies on receptor-ligand interactions, that cell function could be regulated by variations in the amount of the interacting molecules. In this case, quantitative differences in the expression of the surface molecules would be the crucial controlling influence.

One problem in using this approach to study B cell interactions is that the function of many of the surface markers is poorly understood. This is not so for the Ia antigens, however. The biochemistry of these molecules is very well described, both at the protein and nucleic acid levels[4]. They function as restriction elements for T cell recognition[5] and are present on the majority of, if not all, adult B cells where they show quantitative differences of expression on the cells as a group. In principle, therefore, B cell subpopulations could be isolated with few or many surface Ia molecules. By looking at the ability of these subpopulations to respond to T-dependent antigens, one could ask whether the quantitative expression of Ia on the surface of the B cells is involved, perhaps through T cell interaction, in regulating the B cell response to the antigen.

Two studies have already been carried out which give some information in this direction. In one, subpopulations of B cells bearing many or few Ia molecules were shown to have similar precursor frequencies for TNP presented on a T-dependent carrier[6]. In another study to a different antigen (PC), however, these two subpopulations appeared not to generate similar numbers of plaque-forming cells (PFC). In this case, after killing the highest Ia-bearing cells (with antibody and complement) the low Ia-expressing cells made far fewer PFC than the original population[7].

One way to interpret this apparently conflicting data would be to suggest that although both groups of cells are equally capable of making T-dependent responses, the group with more surface Ia interacts more effectively with T helper cells and consequently generates more PFC. The fact that different antigens and different experimental systems were used in these two studies should, of course, also be taken into consideration when interpreting these findings.

In order to test this idea further we used the fluorescence-activated cell sorter (FACS) to fractionate splenic B cells into groups bearing high or low levels of Ia and looked at their ability to make T-dependent responses, or to respond to a B cell mitogen, LPS. Briefly, (CBA x Balb/c) F_1 B cells (spleen cells treated with anti-Thy 1 and guinea pig complement) were stained with a monoclonal rat anti-mouse Ia antibody, NIM-R4, (made in this laboratory and to be described elsewhere) followed by a fluoresceinated monoclonal mouse anti-rat κ chain antibody, FITC-OX12, given by Dr. Alan Williams, University of Oxford. They were then divided by flow cytometry into two subpopulations expressing either a low (the dullest 40%) or a high (the brightest 40%) amount of the marker. After overnight culture to allow loss of the staining antibodies, three different types of B cell function were assessed: (1) primary responses, by measuring the number of PFC generated in vitro to sheep red blood cells (SRBC), (2) secondary responses to dinitrophenylated ovalbumin (DNP-OA), using an adoptive transfer system, (3) LPS responses, by measuring proliferation after culture with LPS in vitro.

The results are summarized in Fig.1. B cells which had greater amounts of surface Ia made more PFC in both the primary and secondary responses than the cells which had less, and this finding, which is similar to those reported previously[6,7], fits in with the idea that the cells bearing more Ia receive better T cell help than those which bear less. In contrast, both groups of cells were equally able to proliferate in response to LPS (Fig.1(c)). This is not altogether surprising in terms of the hypothesis, since LPS is a

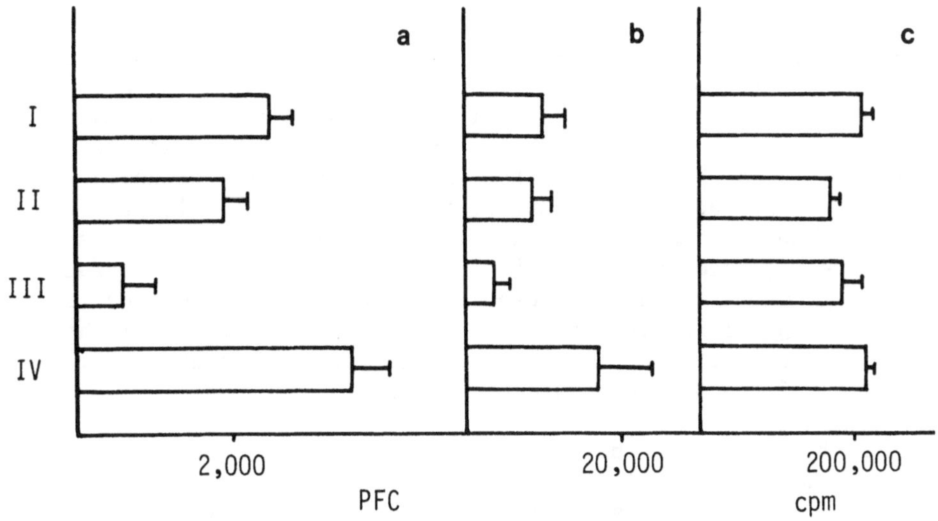

Fig. 1. Responses of B cells selected for Ia expression and modulated overnight to allow loss of staining reagents. The function of unsorted (I), stained unsorted (II), low Ia^+ (III) and high Ia^+ (IV) cells was assessed in a) primary response in vitro to SRBC, b) adoptive secondary response in vivo to TNP and c) proliferative response in vitro to LPS. Details of the procedures are in the text and see ref. 2.

polyclonal B cell mitogen which is thought to act largely independently of T cells[8].

An inherent limitation in these experiments is the use of antibodies in the selection procedure, and their possible effects upon subsequent responses. However, in other experiments to be reported elsewhere, B cells which were selected for the quantitative expression of membrane immunoglobulin (mIg) or Fc receptors (FcR) behaved differently from those selected for the expression of Ia. In this case, the groups bearing high or low levels of these two markers made similar primary responses to SRBC. On the other hand, cells with low densities of mIg or FcR were clearly more effective responders than those with a high expression of these two molecules. Surface-bound antibody therefore does not produce a consistent 'non-specific' effect on B cell responses.

In order to minimise any unforseen effects of surface-bound antibody, we cultured the sorted cells overnight to allow loss of the surface-bound reagents prior to challenge with antigen. Parallel experiments without the overnight incubation were also done for comparison (Fig.2). Although the primary response was more erratic without the overnight 'modulation' the high Ia-bearing cells still generated more PFC than the low ones. However, in the secondary response there was a considerable difference, the unmodulated cells making extremely low numbers of PFC (Fig. 2(b)). In this context it is worth noting that different experimental systems were used to generate primary and secondary responses. For example, there could be problems associated with the migration of cells to the spleen after adoptive transfer, whereas cells in culture obviously cannot escape from the culture vessel. Thus when the cells were injected into mice directly after sorting, the surface-bound antibody may have interfered with their ability to localise in the spleen. This may explain the lower yield of PFC from the unmodulated cells, as opposed to those cultured overnight to allow modulation of the surface staining reagents.

Finally, although these studies show that quantitative differences in the expression of Ia on B cells can be associated with cell function, and one can postulate a mechnistic link in terms of T cell help between the two, there is no formal evidence for this at present. The quantitative expression of surface receptors may either play a role in determining cell function, or may be a consequence of it, and further studies will be necessary to show which possibility is correct. Whatever the underlying mechanisms, however, it is clear that subpopulations of B cells which have been selected on the basis of the quantitative expression of surface Ia can also be distinguished by their functional properties.

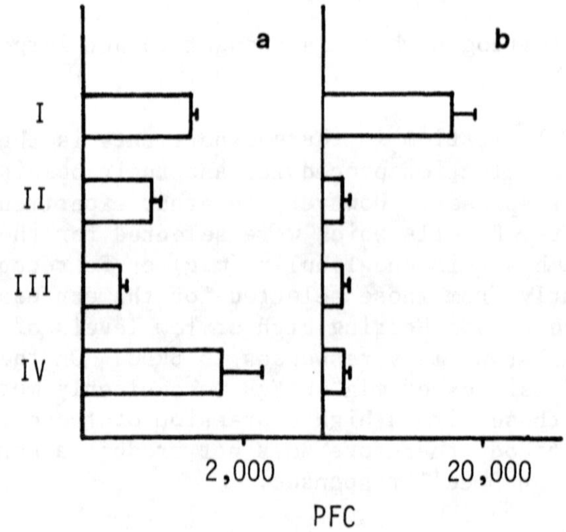

Fig. 2. Responses of B cells selected for Ia expression. Unstained (I), stained unsorted (II) low Ia^+ (III) and high Ia^+ (IV) cells were put directly from the FACS without prior overnight modulation into a) primary response in vitro to SRBC and b) adoptive secondary response in vivo to TNP.

ACKNOWLEDGEMENTS

We wish to thank Keith Keeler for assisting with the FACS operations and Nick Clark and Graham Preece for their technical help.

REFERENCES

1. B. Subbarao, D.E. Mosier, A. Ahmed, J.J. Mond, I. Scher and W.E. Paul, Role of a nonimmunoglobulin cell surface determinant in the activation of B lymphocytes by thymus-independent antigens, J. Exp. Med. 149: 495 (1979).
2. S. Marshall-Clarke, A. Chayen and R.M.E. Parkhouse, Monoclonal antibody NIM-R2 shows differential reactivity with virgin and memory B cells, Eur. J. Immunol. 12: 733 (1982).
3. K. Hayakawa, R.R. Hardy, M. Honda, L.A. Herzenberg, A.D. Steinberg and L.A. Herzenberg, Ly-1 B cells: Functionally distinct lymphocytes that secrete IgM autoantibodies, Proc. Natl. Acad. Sci. USA. 81: 2494 (1984).
4. L. Hood, M. Steinmetz and B. Malissen, Genes of the major histocompatibility complex of the mouse, Ann. Rev. Immunol. 1: 529 (1983).
5. Z.A. Nagy, C.N. Baxevanis, N. Ishii and J. Klein, Ia antigens as restriction molecules in Ir-gene controlled T-cell proliferation, Immunol. Rev. 60: 59 (1981).
6. J.L. Greenstein, E.M. Lord, P. Horan, J.W. Kappler and P. Marrack, Functional subsets of B cells defined by quantitative differences in surface I-A, J. Immunol. 126: 2419 (1981).
7. K. Bottomly, B. Jones, J. Kaye and F. Jones, Subpopulations of B cells distinguished by cell surface expression of Ia antigens. Correlation of Ia and idiotype during activation by cloned Ia-restricted T cells, J. Exp. Med. 158: 265 (1983).
8. A. Coutinho, L. Forni and B. Blomberg, Shared antigenic determinants by mitogen receptors and antibody molecules to the same thymus-independent antigen, J. Exp. Med. 148: 862 (1978).

EVIDENCE FOR A CLONAL DEVELOPMENT OF SPECIFIC ANTIBODY FORMING CELLS IN THE SPLEEN

N. van Rooijen, N. Kors and R. van Nieuwmegen

Immunocytochemistry Unit, Medical Faculty, Dept. of Histology, Free University, P.O. Box 7161, 1007 MC AMSTERDAM, The Netherlands

INTRODUCTION

The clonal development of antigen triggered lymphoid cells has been studied extensively in vitro. No technique was available until now to distinguish between two different antibody producing cells in vivo. It could not be investigated for that reason whether or not, after simultaneous administration of two different antigens, specific antibody producing cells develop in the lymphoid organs of the responding animal randomly. When, after contact with the antigens, the antigen reactive cells and/or the specific antibody producing cells that develop during the immune response against the antigen, migrate not at all or only slowly, specific antibody producing cells arising from one precursor will remain close together forming a group of cells, all producing the same antibodies. When, in the contrary, they show a relatively strong migration, specific antibody producing cells arising from different precursors will be distributed randomly. It is shown in a series of recent experiments, reviewed in the present contribution, that after intravenous injection of two different antigens, clones of similar specific antibody producing cells develop in the spleen of rabbits. Double immunocytochemical staining, using both horseradish peroxidase (HRP) and alkaline phosphatase (AP) was performed for the simultaneous detection of two or three different specific antibody producing cells, after injection of two different protein antigens[1-3].

MATERIALS AND METHODS

Experimental design

Young male chinchilla rabbits were primed either simultaneously with the antigens human serum albumin (HSA) and bovine gamma globulin (BGG) or simultaneously with the antigens human gamma globulin (HGG) and BGG. Two till ten mg of the antigens was administered intravenously in saline. Spleen biopsies were taken at different time intervals after priming. After 1-2 months a booster injection with the same antigens was given, and again a piece of spleen was removed at different time intervals after the booster. Specific antibody containing cells were demonstrated by incubation of cryostat sections with antigen-enzyme conjugates followed by enzyme histochemistry[1-3].

Antigens and chemicals

Human gamma globulin (HGG) and human serum albumin (HSA) were obtained from Miles Laboratories Inc. Bridgend, Glamorgan, South Wales, U.K. Bovine gamma globulin (BGG) was obtained from Pentex, Kankakee, Ill, U.S.A. Horseradish peroxidase (HRP, Grad I) was obtained from Boehringer, Mannheim, Germany. 3-amino-9-ethylcarbazole (AEC) was obtained from K + K Labs, Plainville, N.Y., U.S.A. Alkaline phosphatase (P-6774, type VII-T, 1020 units per mg protein) was obtained from Sigma Chem. Comp. St. Louis, Missouri, U.S.A.

BGG-AP conjugate

BGG-AP conjugate was prepared using the one step glutaraldehyde method[4]. 2.5 Mg BGG was mixed with 5 mg alkaline phosphatase in 1 ml PBS and 0.1 ml 1% glutaraldehyde was added. After 2 h the mixture was dialysed overnight against PBS. The conjugate was chromatographed on Ultrogel AcA 34 and the relevant fractions containing the conjugate were pooled.

HSA-HRP conjugate and HGG-HRP conjugate

Four mg HRP in 1 ml distilled water was oxidized by 0.2 ml 0.1 M sodium periodate[5] for 20 min. Five drops ethylene glycol were added and after 5 min the reaction mixture was filtered on Ultrogel AcA 202[6]. The activated HRP was concentrated to 1 ml and 4 mg HSA or HGG in 1 ml 0.15 M NaCl was added. Immediately the pH was raised to 9.0 - 9.5 by the addition of sodium carbonate buffer pH 9.5. After 2 h at room temperature the reaction was stabilized by the addition of 0.1 ml freshly prepared sodium borohydride (4 mg/ml distilled water). After an additional 2 h at 4°C the conjugate was purified on Ultrogel AcA 44[7].

Detection of anti-BGG antibody producing cells and anti-HSA or anti-HGG antibody producing cells in the same spleen section

Pieces of spleen were frozen in liquid nitrogen, and afterwards cryostat sections of 8 µm thickness were cut. Sections were air-dried overnight. They were fixed for 10 min at 20°C with acetone (pro analyse) to which H_2O_2 was added (0.02% final concentration). They were rinsed 3x in PBS (pH 7.4). The sections were incubated overnight horizontally at 4°C with a mixture of HSA-HRP or HGG-HRP conjugate in PBS + 0.05% BSA (final antigen concentration 10 µg/ml) and BGG-AP conjugate (final antigen concentration 5 µg/ml). They were washed again 3 times with PBS. They were treated for alkaline phosphatase using Fast blue BB salt (no. F 0250, Sigma Chem. Comp.) for 45 min at 37°C according to Burnstone (see Pearse, ref. 8). Afterwards they were washed with PBS and treated for 8 min with AEC + H_2O_2[9]. They were briefly counterstained with haematoxiline and mounted in glycerine-gelatine.

RESULTS

Localization of specific antibody producing cells in the spleen during the primary and secondary immune response

The development and disappearance of specific antibody producing cells in the spleen of rabbits during the primary and secondary immune response have been described in earlier studies[10,11] and have also been reviewed in this volume. In short, the bulk of specific antibody producing cells were found in the coaxial sheaths of lymphoid tissue surrounding the terminal arterioles in the spleen and in the periphery of the periarteriolar lymphocyte sheaths (PALS). To a lesser extent specific antibody producing cells were found in the follicles, in the inner parts of the PALS and in the red pulp. Especially during the primary immune response specific antibody producing cells were not evenly divided throughout these areas of the spleen. Rather, they occurred in defined small groups, each group consisting of several to many specific antibody producing cells located closely together. In the areas between these groups only very few individual antibody producing cells were found.

Simultaneous detection of different specific antibody producing cells in the spleen during the primary and secondary immune response

In some groups of specific antibody producing cells, found during the primary immune response that was elicited by simultaneous injection of HSA and BGG, both anti-HSA producing cells and anti-BGG producing cells were detected. Most groups, however, consisted of either specifically anti-HSA producing cells or specifically anti-BGG producing cells (Figs. 1, 2). Positive cells were stained blue with the BGG-AP conjugate or red with the HSA-HRP conjugate. Double staining cells were not found, indicating the absence of cells

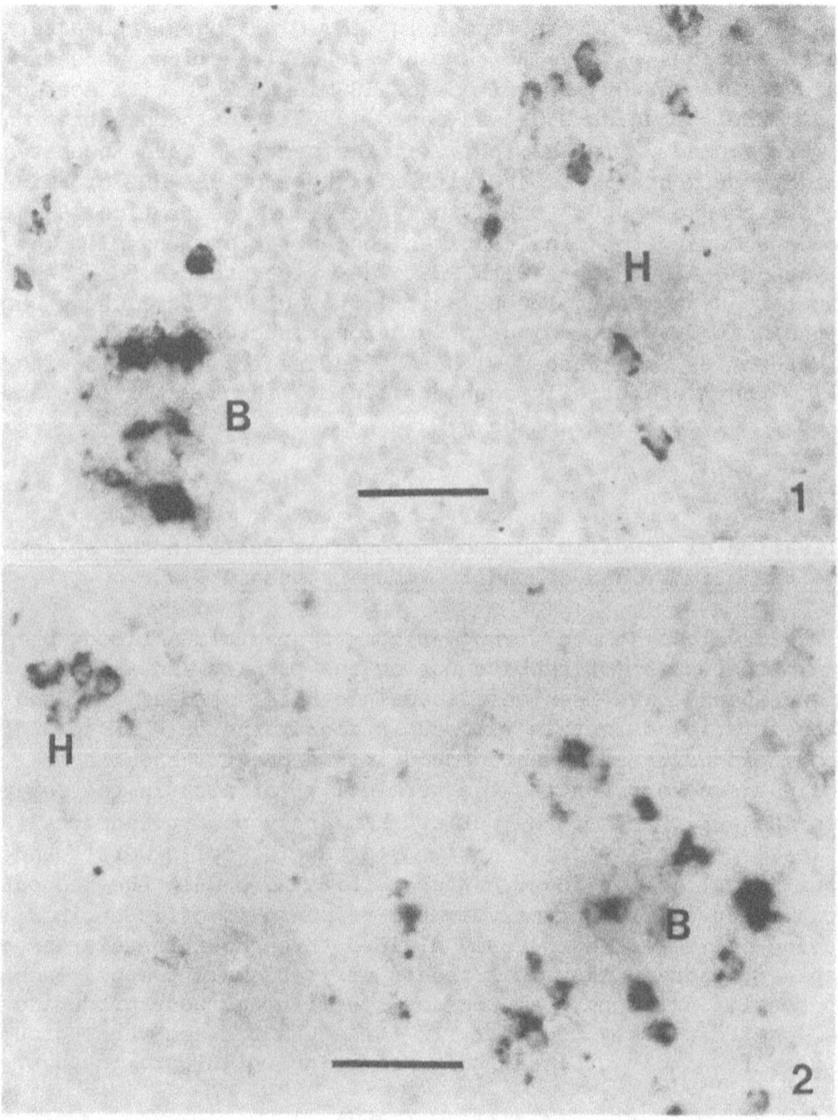

Figs. 1 and 2
Both figures show a group of anti-HSA producing cells (H) and a group of anti-BGG producing cells (B) in coaxial sheaths of lymphoid tissue surrounding the terminal arterioles in the spleen of a rabbit. The rabbit had been given a single intravenous injection with both the antigens HSA and BGG at 6 days before taking the spleen biopsy. Blue stained cells are represented by black spots and red stained cells are reproduced in shades of gray in the black and white photographs (color photographs of these figures are shown in ref. 2). Bar = 100 μm.

producing antibodies that cross-react with both antigens. Three types of positive cells were found in the spleen after simultaneous injection with the antigens HGG and BGG: 1) cells with a red-stained cytoplasm had bound the HGG-HRP conjugate indicating these cells contained specific anti-HGG antibodies; 2) cells with a blue-stained cytoplasm had bound the BGG-AP conjugate indicating these cells contained specific anti-BGG antibodies; 3) cells with a violet-stained cytoplasm had bound both the HGG-HRP conjugate and the BGG-AP conjugate indicating these cells contained antibodies cross-reacting with both HGG and BGG. Also in these spleens, similarly stained cells were frequently found in groups.

Groups of specific antibody producing cells were also found in the spleen of rabbits during the secondary immune response. However, since many more specific antibody producing cells were found during the secondary immune response, compared to the primary response, groups of specific antibody producing cells were overlapping in most cases. This explains why larger concentrations with a mixed population of e.g. anti-HSA producing cells and anti-BGG producing cells were often found after a simultaneous booster injection with HSA and BGG. However, specific antibody producing cells were not randomly mixed. Within the larger concentrations, small groups of anti-HSA antibody producing cells and small groups of anti-BGG antibody producing cells could often be distinguished.

DISCUSSION

It is shown that specific antibody-producing cells develop in defined groups in the spleen during a primary immune response. Different specific antibody producing cells were not randomly distributed in these groups after induction of a simultaneous immune response against two antigens. During the primary immune response against simultaneously injected HSA and BGG, most groups of specific antibody producing cells were either anti-HSA specific or anti-BGG specific. Similar results were obtained after simultaneous injection of HGG and BGG. Although specific antibody producing cells showed an overlapping localization pattern during the secondary immune response, different specific antibody producing cells were not distributed randomly. After a simultaneous booster injection with HSA and BGG, small groups of densely packed cells, consisting of only anti-HSA or only anti-BGG antibody producing cells, could often be distinguished within larger less densely packed fields of antibody producing cells.

The most probable explanation for these observations is that lymphoid cells, once triggered to proliferation by a particular antigen, show a clonal development in the spleen. During their proliferation and successive antibody formation, they migrate only slowly so that they remain close together for that time. This also explains why only few groups of antibody producing cells have a mixed population of different specific antibody producing cells.

When e.g. an anti-HSA reactive precursor and an anti-BGG reactive precursor localize close together in an unstimulated spleen, they will form a mixed concentration of anti-HSA producing cells and anti-BGG producing cells after simultaneous stimulation with the antigens HSA and BGG. The more precursors of both specific antibody producing cells are present, the more increases the chance that two different precursors are localized close together. After priming, much more precursors of both specific antibody producing cells are present so that, compared to the primary response, anti-HSA precursors and anti-BGG precursors (B-memory cells) may often localize close together. This explains the overlapping localization pattern of two different specific antibody producing cells during the secondary immune response, after a simultaneous booster injection with two antigens.

The 'in vitro' clonal development of antigen specific lymphoid cells has been studied and discussed extensively, but 'in situ' studies on specific antibody producing cells were not performed, since no technique was available to distinguish between two different antibody producing cells[1-3].

REFERENCES

1. N. van Rooijen, N. Kors, D.M. Boorsma, P. de Haan, and R. van Nieuwmegen. Immunology 51:417 (1984).
2. N. van Rooijen, N. Kors, and R. van Nieuwmegen. Anat. Rec. 209:385 (1984).
3. N. van Rooijen, N. Kors, and R. van Nieuwmegen. J. Histochem. Cytochem. 32 (1984).
4. S. Avrameas. Immunochemistry 6:43 (1969).
5. M.B. Wilson, and P.K. Nakane. In: "Immunofluorescence and related staining techniques", W. Knapp, K. Holubar, G. Wick, ed., Elsevier, Amsterdam, p. 215 (1978).
6. D.M. Boorsma, and J.G. Streefkerk. J. Immunol. Meth. 30:245 (1979).
7. D.M. Boorsma, and J.G. Streefkerk. J. Histochem. Cytochem. 24: 481 (1976).
8. A.G.E. Pearse. Histochemistry. Theoretical and Applied. Vols. I and II, Churchill Livingstone, Edinburgh, U.K.
9. R.C. Graham, U. Lundholm, and N.J. Karnovsky. J. Histochem. Cytochem. 13:150 (1965).
10. N. van Rooijen, N. Kors, R. van Nieuwmegen, and P. Eikelenboom. Anat. Rec. 206:189 (1983).
11. N. van Rooijen, R. van Nieuwmegen, N. Kors, and P. Eikelenboom. Anat. Rec. 208:579 (1984).

MODIFICATION OF IMMUNE INDUCTION BY ADJUVANT

A.T.J. Bianchi, Ph.J van der Heijden and B.A. Bokhout

Department of Immunology, Central Veterinary Institute
P.O. Box 65, 8200 AB Lelystad, The Netherlands

SUMMARY

The role of adjuvanticity in the modification of the immune induction was studied by separate application of antigen and adjuvant in mice. An antigen specific plaque forming cell (PFC) test and a delayed type hypersensitivity (DTH) assay were used for these studies. Antigen (SRBC) and adjuvant were applicated separately in time and place. A stable well defined water-in-oil emulsion without antigen was used as adjuvant. Our studies have shown that intraperitoneal (i.p.) injection of this adjuvant stimulates the intravenous (i.v.) induced PFC response, while the DTH response is suppressed. This effect can be demonstrated even when adjuvant has been injected i.p. 5 days prior to i.v. antigen injection. The stimulation of the PFC response depends on the dosage of i.v. injected antigen.

INTRODUCTION

In an earlier study (1) we described the observation that an intraperitoneal (i.p.) injection of a stable water-in-oil emulsion (2), an adjuvant without antigen, shortly before weaning of piglets led to the prevention of postweaning diarrhoea and mortality due to infectious diseases of the gut. This benificial effect has been observed on a number of farms, where untill now more than 100.000 piglets have been treated by i.p. adjuvant application. The observed protection by i.p. adjuvant treatment is probably due to modification of the immune response to antigens presented via the gut wall.

The first indication in mouse studies (3), that adjuvant application influences at least part of the immune system, was derived from the comparison of the weight of the spleens of adjuvant treated and control mice. The spleens of the adjuvant treated mice were significantly heavier than those of the control mice. In addition activation and an increase in number of peritoneal macrophages in adjuvant treated mice was observed.

Other studies (4, 5) using microorganisms or products of microorganisms as immunogenic immunomodulators showed that administration of this type of immunomodulator in combination with an injection of an unrelated antigen influences the immune response against the antigen.

In this paper we investigated whether application of antigen and adjuvant (a stable non-immunogenic water-in-oil (W/O) emulsion) separated in time and place also modifies the immune response against the injected antigen.

MATERIALS AND METHODS

Mice

Female C3H/He mice were obtained from OLAC Ltd., Bicester, United Kingdom, and used at an age of 10-14 weeks.

Antigen

Sheep red blood cells (SRBC) from a single donor were used. They were stored in Alsever's solution and washed 3 times in physiological saline before intravenous (i.v.) immunization.

Adjuvant

As adjuvant a stable W/O emulsion was used. The procedure for the preparation of this W/O emulsion has been described in a previous paper (2). The adjuvant was injected i.p. before or simultaneously with the i.v. immunization with SRBC.

Cell suspensions

Mice were killed with carbon dioxide. Immediately after killing the spleens were removed, placed in balanced salt solution (BSS) and squeezed through a nylon gauze filter to provide a single-cell suspension.

Plaque assay

For enumeration of IgM secreting anti-SRBC PFC the Jerne-type plaque assay according to the description of Benner e.a. (6) was used.

Assay for DTH

DTH reactions were determined by measuring the difference in thickness of the hind feet 24 hr after s.c. injection of 1×10^8 SRBC into the dorsum of the right hind foot. As a control for background DTH reactivity, naive mice were used which only received the challenge dose. The specific DTH response was calculated as the relative increase in foot thickness of the immune mice minus the relative increase in foot thickness of the control mice. The swelling in the control mice ranged between 15 and 20%.

RESULTS

The effect of separate application of antigen and adjuvant on the PFC response was investigated. C3H/He mice were i.v. immunized with 1 x 10^6 or 1 x 10^8 SRBC. Half of the mice were simultaneously injected i.p. with 100 μl of the stable W/O emulsion, the other half were left untreated. Subsequently the SRBC specific PFC response in the spleen was determined on day 2, 3, 4, 5 and 8. The results (Fig. 1A and C) show that adjuvant application causes a clear stimulation of the IgM PFC response in mice immunized with a low dosage of 1 x 10^6 SRBC. While the IgM PFC response in mice immunized with a high dosage of 1 x 10^8 SRBC is hardly affected.

The same results are obtained with mice injected i.p. with adjuvant 5 days before the antigen injection (Fig. 1B and D). The stimulatory effect of the adjuvant application appears earlier in mice injected with adjuvant on day–5 (Fig. 1B) than in mice injected with adjuvant on day 0 (Fig. 1A).

Comparable experiments were carried out to investigate the effect of separate application of antigen and adjuvant on the DTH response. C3H/He mice were injected i.p. with 100 μl of the stable W/O emulsion on day –5, on day 0 or were left untreated. On day 0 all mice were immunized i.v. with 1 x 10^6 SRBC and tested for DTH reactivity on day 3, 4, 5 and 6 after immunization. The results (Fig. 2) show that in contrast to the stimulatory effect on the PFC response, the DTH response is suppressed by i.p. adjuvant application.

The strongest suppression of the DTH response is found when the adjuvant was injected i.p. simultaneously with the i.v. immunization with SRBC.

DISCUSSION

This paper demonstrates that an i.p. injection of a W/O-emulsion, an adjuvant without antigen, stimulates the i.v. induced PFC response (Fig. 1), while the i.v. induced DTH response in suppressed (Fig. 2). Both effects are also found when the adjuvant has been injected 5 days before the i.v. antigen injection.

The combination of a high PFC response with a suppressed DTH response has been described by Lagrange *et al.* (7). These authors showed that high dosage of i.v. injected SRBC induce clear splenic PFC responses, while the DTH reactivity in these mice is suppressed. Passive transfer experiments showed that DTH reactive cells were still present in the spleens of these mice. The authors explain the suppression of DTH reactivity by a selective trapping of DTH mediating cells in the spleen under influence of serum factors. Other studies (8) have shown that adjuvant application causes trapping of antigen reactive lymphocytes in the lymphnodes near the place of adjuvant injection leading to an enhanced local immune response.

Low dosages of SRBC induce PFC as well as DTH responses. Our results suggests that treatment with our adjuvant in combination with low dosages of SRBC mimics the effect obtained after a high dosage of i.v. injected SRBC (7), i.e. the splenic

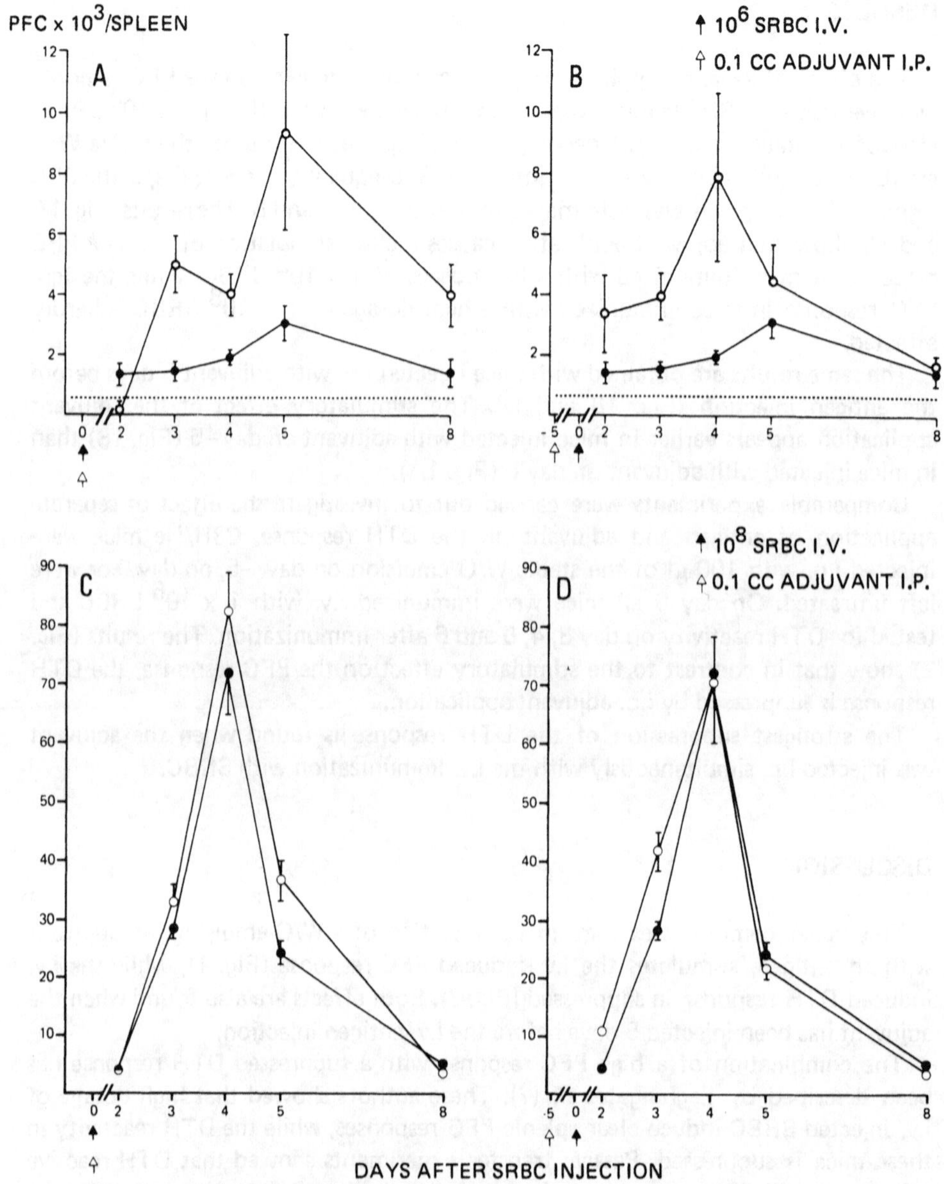

Fig. 1. **Comparison of IgM specific PFC anti SRBC response in spleens of adjuvant treated (○—○) and control (●—●) C3H/He mice. A and B: Mice were i.v. immunized with 1 x 10^6 SRBC. C and D: Mice were i.v. immunized with 1 x 10^8 SRBC. The moment of i.p. adjuvant application is indicated with ⇧ . PFC responses were measured on day 2, 3, 4, 5 and 8 after i.v. immunization. Each experimental group represents the arithmic mean ± SE (n=5).**

PFC response is increased, while the DTH reactive T-cells are trapped in the spleen leading to a suppressed DTH response.

In other studies (4), dealing with immune modulation by separate application of adjuvant and antigen, microbial agents are used as adjuvants. Most of these microbial adjuvants have strong intrinsic adjuvant capacities due to their antigenic

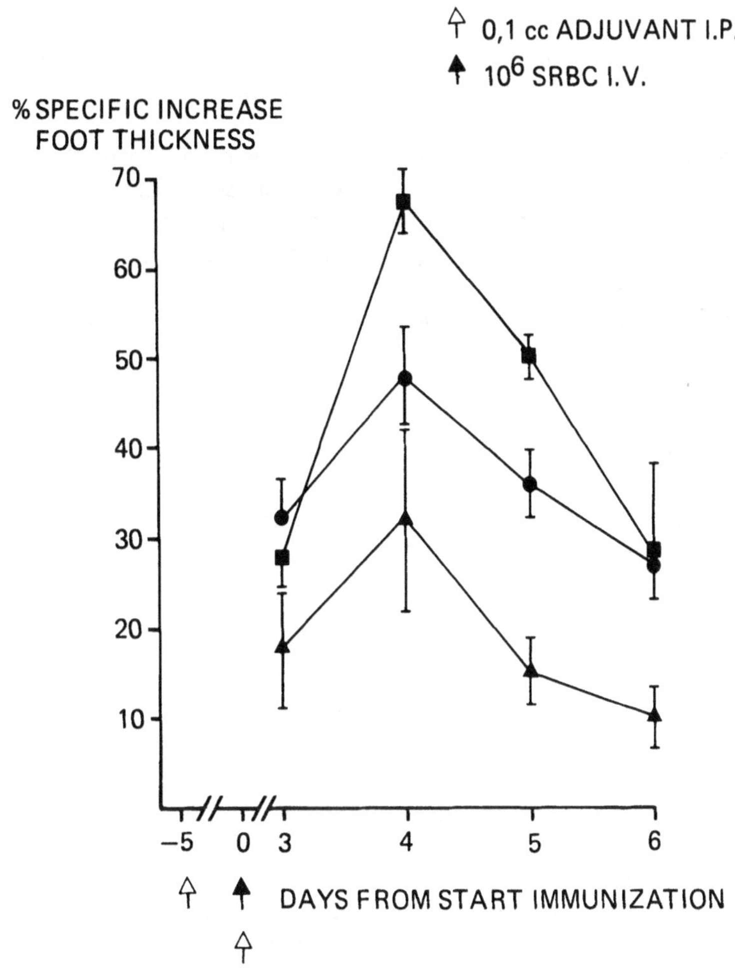

Fig. 2. Comparison of the SRBC specific DTH response of adjuvant treated (●–● day −5, ▲–▲ day 0) and control (■–■) C3H/He mice. Mice were i.v. immunized with 1×10^6 SRBC on day 0 (↑). The moments of adjuvant application are indicated with (⇧). DTH reactivity was measured on day 3, 4, 5 and 6 after i.v. immunization. Each experimental group represents the arithmic mean ± SE (n=6).

(9) or mitogenic (10) nature. A W/O-emulsion has none of these properties. However, we show that this adjuvant is a potent immunomodulator and not a mere vehicle for antigens, which provides a long-lasting antigen stimulating effect a.o. by depot-formation. The immunomodulation obtained by our adjuvant probably stems from the induction of lymphocyte trapping as described for such products (8).

In these studies the same effects are obtained when the W/O emulsion is injected

5 days before or simultaneously with the i.v. injected antigen. This indicates that the modulatory effect is less time dependent than the effect obtained with microbial adjuvants where the effect can be either stimulating or suppressive in relation to the moment of adjuvant application.

It is unknown in piglets when the infectious agent will penetrate. Therefore the moment of adjuvant treatment to prevent the disease cannot be timed exactly. For that reason it is more profitable to use an immunomodulator which induces the same effects over a relatively long period without the risk of inducing opposite effects.

We therefore think that a stable W/O emulsion can be beneficially used in the prevention of diseases.

ACKNOWLEDGEMENTS

We thank J.W. Scholten and W. Stok for excellent technical assistance, and C.D. Propsma for typing the manuscript.

REFERENCES

1. M.J.A. Nabuurs, B.A. Bokhout, and Ph.J. van der Heijden. Intraperitoneal injection of an adjuvant for the prevention of post-weaning diarrhoea and oedema disease in piglets: A field study. *Prev. Vet. Med.* **1**: 65 (1982).
2. B.A. Bokhout, C. van Gaalen, and Ph.J. van der Heijden. A selected water-in-oil emulsion: Composition and usefulness as an immunological adjuvant. *Vet. Immun. Immunopath.* **2**: 491 (1981).
3. B.A. Bokhout, Ph.J. van der Heijden, M.J.A. Nabuurs, J.W. Pals, J.W. Scholten and W. Stok. Adjuvant application without antigen for prevention of diseases in piglets. Effect obtained by manipulation of the immune system. E.C. meeting on: Adjuvants and non-specific immunity (1983). EUR report (in press).
4. J.H. Schwab. Suppression of the immune response by microorganisms. *Bact. Rev.* **39**: 121 (1975).
5. D.W. Dresser and J.M. Philips. T and B adjuvants. In: Immunopotentiation. Ciba Foundation Symposium 18 (new series) Elsevier, Excerpta Medica, North Holland, Amsterdam, London, New York 3 (1973).
6. R. Benner, F. Meima, G.M. van der Meulen, and W.B. van Muiswinkel. Antibody formation in the mouse bone marrow. I. Evidence for the development of plaque-forming cells in situ. *Immunology* **26**: 247 (1974).
7. P.H. Lagrange and G.B. Mackaness. Site of action of serumfactors that block delayed-type hypersensitivity in mice. *J. Exp. Med.* **148**: 235 (1978).
8. P. Frost and E.M. Lance. The relation of lymphocyte trapping to the mode of action of adjuvants. In: Immunopotentiation. Ciba Foundation Symposium 18 (new series). Elsevier, Excerpta Medica, North Holland, Amsterdam, London, New York 3 (1973).

9. R. Benner, A.M. Rijnbeek, A. van Oudenaren and A. Coutinho. Quantitative aspects of the nonspecific humoral immune response to sheep erythrocytes. *Adv. Exp. Med. Biol.* **149**: 651 (1982).
10. M. Björklund and A. Coutinho. Isotype commitment in the *in vivo* immune responses. II. Polyclonal plaque-forming cell responses to lipopolysaccharide in the spleen and bone marrow. *Eur. J. Immunol.* **13**: 44 (1983).

FOOTPAD RESPONSES IN NUDE MICE

IMMUNIZED WITH CHICKEN ERYTHROCYTES

John D. Jackson, Greg A. Perry
and J.G. Sharp

Department of Anatomy
University of Nebraska Medical Center
Omaha, Nebraska 68105 USA

INTRODUCTION

It is not yet clear whether the precursors of helper T lymphocytes and cytotoxic/suppressor T lymphocytes have common or independent origins and pathways of intrathymic maturation[1-3]. Evidence has been presented for a differential timing of the appearance in the thymus of the progeny of these precursors[1]. In addition, there is increasing evidence for alternate extrathymic pathways of T-cell differentiation and maturation[4,5]. Shimamoto et al.[6], reported that T-cells responsible for delayed type hypersensitivity (DTH) to chicken red blood cells (CRBC) were less thymic dependent than T-cells involved in cytotoxicity or in helper functions for antibody production. We have attempted to clarify the degree of thymus dependency of such DTH responses. We employed young adult athymic (nude) mice, which do not possess significant numbers of T-cells as shown by flow cytometric analysis. We appeared to be able to demonstrate a "DTH" response to chicken erythrocytes in such mice. However we believe these observations are most easily explained as a potent non-specific immune reaction.

METHODS AND MATERIALS

Animals: Breeding pairs of C57B1/10ScN nude (nu/nu), Type III nude (nu/nu, bg/bg, xid/xid), Swiss nude (nu/nu), and the haired littermates (nu/+) were obtained from Dr. Carl Hansen (Small Animal Section, Veterinary Resources Branch, NIH, Bethesda, MD). The animals were housed on laminar flow stations in sterile cages containing sterilized bedding and given sterile food and water

TABLE 1: FOOTPAD SWELLING (CRBC:CRBC)[a]

Group	Time after challenge (hr.)		
	24	48	72
Swiss			
(nu/nu)	198±34[b]	91±26	93±30
(nu/+)	244±32	75±27	85±13
C57Bl/10			
(nu/nu)	159±24	122±91	44±29
(nu/+)	131±50	50±23	31±1
Type III			
(nu/nu)	192±38	129±18	118±1
(nu/+)	169±44	73±28	62±24

[a] Mice were immunized I.P. with CRBC and challenged 6 days later with CRBC. Values represent the mean difference in thickness of the PBS injected (right) and the CRBC injected (left) footpads and are expressed in $1/10^4$ inch units.
[b] Mean ± SEM

ad libitum. Normal, intact control Swiss mice (+/+) were obtained from a breeding colony in the Eppley Institute for Research on Cancer and Allied Diseases, Omaha, NE and were given food and acidified water (pH 2.2) ad libitum.

Delayed Type Hypersensitivity (DTH) assay: The DTH assay was performed according to the method described by Shimamoto et al.[6]. Briefly, mice were injected intraperitoneally (IP) with either $1x10^8$ chicken red blood cells (CRBC) or in a control experiment with phosphate buffered saline (PBS). Six days later the mice were challenged by injecting $1x10^8$ CRBC in 50 µl of PBS into the left hind footpad. The right hind footpad was injected with 50 µl of PBS to serve as a control. Twenty-four, 48, and 72 hours later the footpad thickness was measured using a micrometer. Data were expressed as the difference between the thickness of the CRBC injected and the PBS injected footpads.

RESULTS

As can be seen in Table 1, all groups of nude mice responded to the CRBC challenge. Except for one case (Swiss nu/nu at 24 hr.

TABLE 2: FOOTPAD SWELLING (PBS:CRBC)[a]

Group	Time after challenge (hr.)	
	24	48
<u>Swiss</u>		
(nu/nu)	122±26[b]	53±22
(+/+)	104±28	21±23

[a] Mice were immunized with PBS and challenged 6 days later with CRBC. Values represent the mean difference in thickness of the PBS injected (right) and the CRBC injected (left) footpads and are expressed in $1/10^4$ inch units.
[b] Mean ± SEM

after challenge), the nu/nu mice had numerically greater, but not significantly greater, responses to the CRBC than their haired littermates. In order to assure that the footpad responses to the CRBC were true DTH responses, two other groups of mice (Swiss nu/nu and +/+) were immunized with PBS followed by a CRBC challenge. This data is presented in Table 2. Although the responses were slightly reduced compared to the mice immunized with CRBC, both the nu/nu and the +/+ Swiss mice responded to the CRBC challenge by an increase in the footpad thickness even though the mice were immunized with PBS. The nu/nu also had a slightly larger increase in the swelling of the footpad than the +/+ mice.

DISCUSSION

A DTH reaction is believed to be a helper T-cell dependent response mediated by a $Ly\text{-}1^+$ effector cell[7-9]. The peak response is usually seen at 24 and 48 hours following antigenic challenge[8,10]. Shimamoto et al.[6], and Thompson et al.[11], have utilized a method of injecting foreign erythrocytes into the footpad of erythrocyte immunized mice to elicit a DTH response as determined by the increase in thickness of the footpad after 24 and 48 hours[6]. We elected to use the method as described by Shimamoto et al.[6], to test the thymus dependency of the DTH effector cell in the athymic nude mouse. All three nude mouse strains exhibited footpad responses to the CRBC challenge. The nu/nu mice tended to have numerically greater footpad responses than their nu/+ littermates. In addition, the Swiss nu/nu mice and the normal (+/+) Swiss mice responded to a "challenge" by CRBC even though the mice were naive to the CRBC antigen. We believe that these observations suggest that the footpad swelling response seen in mice challenged with CRBC is largely a non-specific inflammatory response and not a DTH reaction. This suggests that the large number ($1x10^8$) of CRBC injected into the footpad may have caused localized tissue damage resulting in local inflammation. However, the slightly greater response of the mice immunized with CRBC might indicate a minor specific component to the response.

These results emphasize the obvious but sometime overlooked necessity of performing appropriate control experiments. These studies have permitted us to compare the method of eliciting a DTH response using cells as the sensitizing agent, versus a chemical sensitizer which we have employed previously[12]. We believe that the latter is more easily controlled and provides fewer complications of interpretation than the former. As might be anticipated, the lack of DTH responses induced by chemical sensitizers does not appear to have been studied in the nude mouse. Although young nude mice lack significant numbers of T-cells as demonstrated by flow cytometry, our results demonstrate that they are very capable of showing a footpad response. In several instances this response appears to be greater than that of their intact littermates. An investigation of the effector mechanisms of this response might be instructive. Presumably this response does not involve T-cell derived lymphokines unless immature T-cell precursors can produce such factors in significant quantities or unless such precursors mature to functional cells at sites of local immunization. This response might involve neutrophils which are observed early in most DTH responses and are especially prominent in the mouse[7]. We cannot exclude a role for basophils/mast cells as in cutaneous basophil hypersensitivity[7,13] (Jones-Mote reaction), however since the mouse has few if any basophils[14] we think this is unlikely.

On the basis of these results we feel that any conclusions about thymus dependency of DTH cells determined using CRBC as the sensitizing agent must be interpreted with extreme caution. Although the extent of DTH responses might be altered by the availibility of suitable numbers of T-cells and other accessory cells, it is unlikely that such cells are less thymus dependent than cytotoxic T-cells. Indeed recent data suggest that T-cells can have a common precursor[3] and that cytotoxic T-cells can be generated extrathymically in situations where the appropriate humoral stimulator is provided[5]. Thus the thymus dependancy of helper T-cells and cytotoxic T-cells appears similar under normal circumstances and if anything, there is better evidence for the extrathymic generation of cytotoxic T-cells than there is for the generation of helper cells.

We conclude that the reponse of nude mice to CRBC by footpad swelling does not appear to be a specific DTH response. We suspect that it is largely a non-specific inflammatory response to cellular materials and/or tissue disruption perhaps involving neutrophils, although a role for basophils/mast cells similar to that of a Jones-Mote reaction cannot be excluded. Whatever the mechanism, we do not feel that studies performed using this assay system provide convincing evidence suggesting a partial lack of thymus dependency of the DTH response.

ACKNOWLEDGEMENTS

The authors would like to thank Mrs. Sally Mann and Mrs. Barbara Murphy for their expert technical assistance. In addition we would like to thank Dr. David Crouse for his helpful discussion during these experiments. This work was supported by NIH grants AM 30419 and AI 15819.

REFERENCES

1) B.J. Mathieson, S.O. Sharrow, P.S. Cambell and R. Asofsky, An Lyt differentiated thymocyte subpopulation detected by flow microfluorometry, Nature 277:478 (1979).
2) R. Scollay, Intrathymic events in the differentiation of T lymphocytes: a continuing enigma, Immunol. Today 4:282 (1983).
3) S. Ezine, I.L. Weissman and R. Rouse, Bone marrow cells give rise to distinct cell clones within the thymus, Nature 309:629 (1984).
4) J. Reimann and R.G. Miller, Differentiation from precursors in athymic nude mouse bone marrow of unusual spontaneously cytolytic cells showing anti-self-H-2 specificity and bearing T cell markers, J. Exp. Med. 158:1672 (1983).
5) V. Duprez, R. Maziarz, O. Weinberger and S.J. Burakoff, Thymectomized, irradiated, and bone marrow-reconstituted chimeras have normal cytolytic T lymphocyte precursors but a defect in lymphokine production, J. Immunol. 132:2185 (1984).
6) Y. Shimamoto, K. Taniguchi, C. Kubo and K. Nomoto, Differences in thymus-dependency among various T-cell functions, Immunology 41:167 (1980).
7) S. Cohen, Lymphokines in delayed hypersensitivity, in:"Immunology 80: Progress in Immunology IV", M. Fougereau and J. Dausset, eds., Academic Press, London (1980).
8) A.T.J. Bianchi, H. Hooijkaas, R. Benner, R. Tees, A.A. Nordin and M.H. Schreier, Clones of helper T cells mediate antigen specific, H-2 restricted DTH, Nature 290:62 (1981).
9) H. Cantor and E.A. Boyse, Cooperation between subclasses of Lyl^+ cells in the generation of killer activity, J. Exp. Med. 141:1390 (1975).

10) H. Bril, Th.W. van den Akker, B.D. Molendijk-Lok, A.T.J. Bianchi and R. Benner, Influence of 2'-deoxyguanosine upon the development of DTH effector T cells and suppressor T cells _in vivo_, _J. Immunol._ 132:599 (1984).
11) C.H. Thompson, T.A. Potter, I.F.C. McKenzie and C.R. Parish, The surface phenotype of a suppressor cell of delayed-type hypersensitivity in the mouse, _Immunology_ 40:87 (1980).
12) E.D. Prentice, H. Lipscomb, W.K. Metcalf and J.G. Sharp, Effects of hypophysectomy on DNCB-induced contact sensitivity in rats, _Scand. J. Immunol._ 5:955 (1975).
13) P.W. Askenase, Cutaneous basophil hypersensitivity in contact-sensitized guinea pigs, _J. Exp. Med._ 138:1144 (1973).
14) S. Schermer, _in_: "The Blood Morphology of Laboratory Animals," F.A. Davis, Philadelphia (1967).

CELLULAR CHARACTERISTICS OF THE ROSETTE FORMING CELLS IN THE MOUSE SPLEEN

Hun-Taeg Chung and Tai-You Ha

Departments of Immunology, Won-Kwang University Medical School and Chonbuk University Medical School
Chollabuk-Do, Republic of Korea

INTRODUCTION

Isolated spleen cells of mouse have been used for the investigation of antibodies against sheep red blood cells(SRBC) either in the hemolytic plaque technique described by Jerne et al[1] or the rosette formation test described by Zaalberg.[2] However, rosette formation requires only the presence of receptor on the surface of the rosette forming cell(RFC), in contrast to the fact that the plaque formation requires the secretion and diffusion of antibody.[3] So the RFC can be not only B cell but also T cell, but the plaque forming cell is always B cell. There have been many controversial reports about the nature and number of mouse spleen RFCs due to the variations in the affinities of the receptors of RFCs.

In the present study, we determined the relative proportions of T and B lymphocytes in the rosette forming spleen cells of the mouse and tried to characterize the RFCs by using glass wool, anti-thymocyte serum(ATS), transfer factor(TF), prostaglandin $F_{2\alpha}$(PG $F_{2\alpha}$), cyclophosphamide(CY), epinephrine and levamisole.

MATERIALS AND METHODS

Animals. ICR mice were bred locally in the Won-Kwang University Medical School Animal Room and used when 2-3 months old.

Antigen. SRBCs were obtained from the same sheep weekly. The cells were stored in Alsevers solution at 4°C. They were washed three times with sterile phosphate buffered saline(PBS) and suspended to give a desired concentration by hemocytometer.

Immunization. Immunization was carried out by intravenous injection of 0.2ml of SRBC suspension($5X10^{8}$ cells/ml). Four days after immunization, the spleen was removed from the mouse.

<u>Drugs</u>. PG $F_{2\alpha}$ (Soya Pharmacological Co., Japan) had been injected intramuscularly for 3 days before immunization with SRBCs in the daily amount of 2 g per mouse. CY(Endoxan; Astawerke, Brackwede, Germany) was dissolved in sterile saline immediately before being used and injected 3 days before immunization with SRBCs through intraperitoneal route as a single dose in the amount of 200mg kg body weight. Epinephrine(Wyeth Laboratories INC., USA) was diluted with minimun essential medium for suspension culture(MEMS) and added to the mixtures for rosette forming assay to final concentrations varying between 10^{-5} mg/ml and 10^{-1} mg/ml. Levamisole(Jansen Pharmaceutica, Belgium) was dissolved in MEMS and added to the mixtures for rosette formation assay to final concentrations varying between 10^{-8} mg/ml and 10^{-5} mg/ml.

<u>Preparation of cell suspension</u>. We harvested spleen cells by mincing spleen, gently squeezing the fragments between the frosted ends of two sterile microscope slides, expelling them several times through a 23 gauge needle, filtering them through nylon mesh, and finally washing them three times with cold PBS. In most experiments, spleens from each mouse in a group(5 mice) were pooled and spleen cells were isolated by the gradient centrifugation for rosette assay using lymphoprep(Nyegaard & Co., Oslo, Norway).

<u>Separation of glass wool-nonadherent cells.</u> Glass wool(Wako Pure Chemical Industries, LTD) was packed to 5ml in 10ml syringes. Three ml of cell suspension($5X10^7$ cells/ml) was incubated in the column at 37°C for 30 min, then eluted with 30ml of warm PBS and washed.

<u>Treatment with ATS and complement.</u> Splenic lymphoid cells were suspended in MEMS at the concentration of $5X10^6$ cells/ml. ATS(1:1024) that was obtained from the rabbit immunized three times with mouse thymocytes was added and the cells were incubated for 30 min at 4°C. The cells were washed once, resuspended in MEMS to $5X10^6$ cells/ml, and fresh guinea pig serum was added. The cells were incubated for one hour at 37°C and then washed two times with cold PBS.

<u>Treatment with lymphocyte dialysates(Transfer factor)</u>. Lymphocyte dialysates were prepared by the method of Holzman et al[4] and Nekam et al[5]. Spleen cells isolated from the immunized mice by gradient centrifugation were suspended in sterile distilled water. The cells were frozen at 20°C and thawed at 25°C for 10 cycles. The lysate was dialysed against 5ml of MEMS for 18hr at 4°C. The dialysate treatment was carried out before the spleen cells being mixed with SRBCs. Fifty microliter of dialysates was added to $5X10^6$ spleen cells in MEMS, and incubated at 37°C for 15 min. Thereafter the cells were washed 3 times with MEMS and rosette assay was carried out.

<u>Rosette assay.</u> A modification of the method of Zaalberg[2] was employed. Spleen cells were suspended in one ml of MEMS at the concentration of $5X10^6$ cells/ml. One ml of SRBC suspension($5X10^7$ cells/ml) were added to the spleen cell suspension and the mixtures were incubated for 1 hour at 37°C. The mixtures were then transferred to a 4°C refrigerator, held overnight, and RFC were enumerated the following day after resuspending the mixture using a pasteur pipette.

The number of RFC was calculated in an improved Neubauer hemocytometer at X400 magnification. Nucleated cells binding more than three SRBCs were regarded as a RFC.

RESULTS

Effect of the immunization on the number of mouse spleen RFCs. As shown in Fig.1, the number of RFCs per 10^6 spleen cells in the immune mice to SRBCs was 5917, in contrast to 680 in the non-immune mice. The RFCs of the immune mice spleens were more than eight times those of the non-immune mice spleens.

RFCs in the glass wool-nonadherent spleen cells. Whole spleen cells were separated by glass wool into glass wool-adherent and -nonadherent cells. As shown in Fig.2, almost all of the RFCs of both non-immune and immune mice to SRBCs were glass wool-adherent.

Effect of ATS on the mouse spleen RFCs. Spleen cells from non-immune and immune mice to SRBCs were treated with ATS and complement and were assayed for RFCs that were resistant to ATS cytotoxicity. As shown in Fig.3, RFCs that were not resistant to ATS and complement were less than one third the RFCs resistant to them in the spleens of non-immune and immune mice.

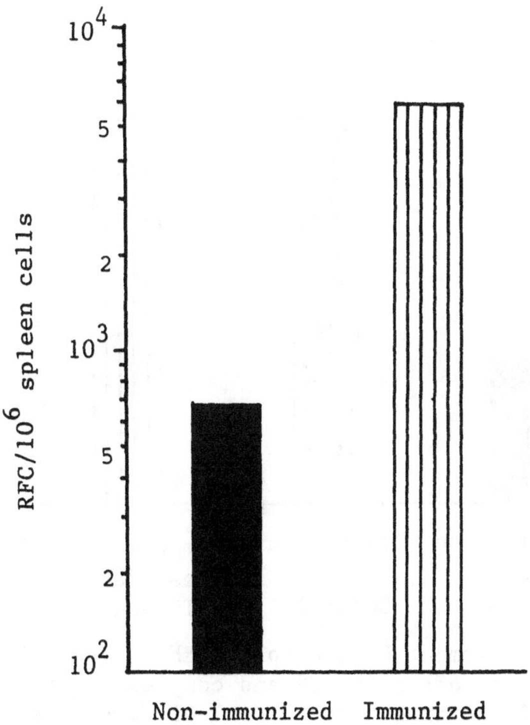

Fig.1. The effect of immunization on the spleen RFC. Mice were immunized with SRBC and spleens were assayed for RFC 4 days after immunization Non-immunized mice were injected with PBS. Data represent the RFC obtained from a pool of five spleens.

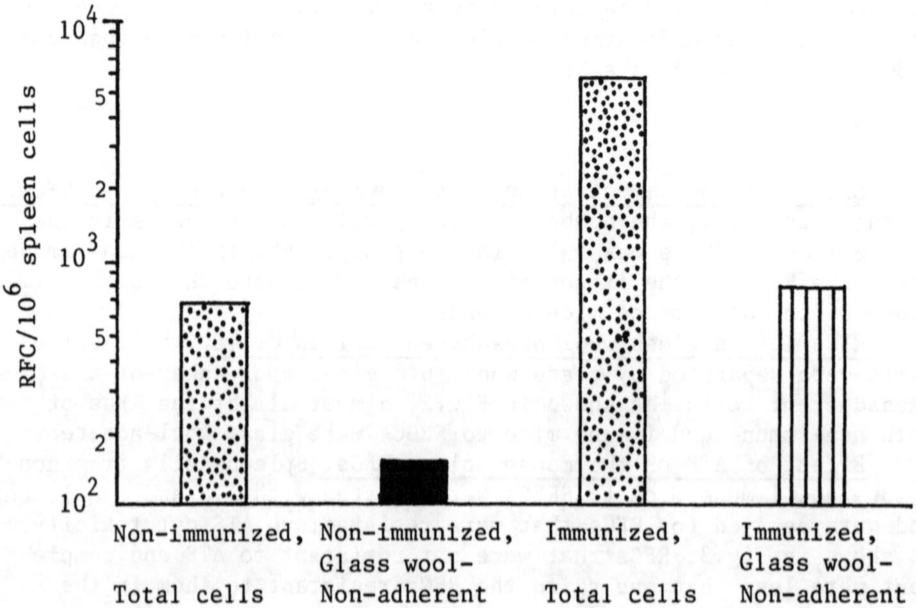

Fig.2. The RFC of glass wool-nonadherent spleen cells. Total cells means glass wool-adherent cells plus -nonadherent cells. Data represent the RFCs obtained from a pool of five spleens.

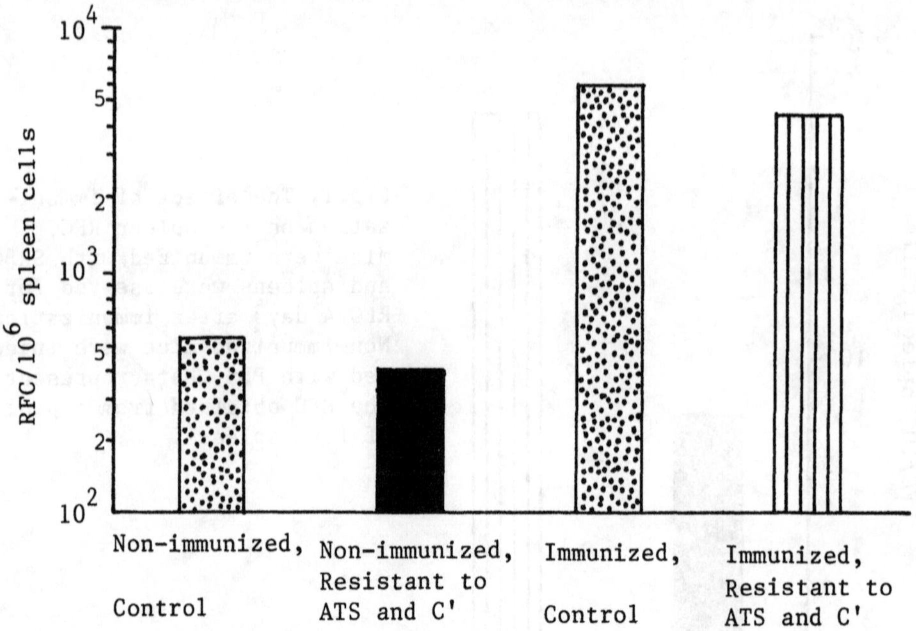

Fig.3. The effect of ATS and complement on the spleen RFCs. The spleen RFCs resistant to the treatment with ATS and complement were assayed. Data represent the RFCs obtained from a pool of five mice spleens.

Effect of the TF on the mouse spleen RFCs. Spleen cells from non-immune mice were incubated with the dialysates from the immune mice to SRBCs. As shown in Fig.4, the dialysates increased the number of RFCs of the non-immune mice to SRBCs to more than two times of it.

Effect of PG $F_{2\alpha}$ treatment on the mouse spleen RFCs. The effect of PG $F_{2\alpha}$ treatment on the RFC of non-immune mice was slight as shown in Fig.5. However, the pretreatment of PG $F_{2\alpha}$ for three days before immunization affected profoundly the rosette formation of the immune mice. The numbers of RFCs of the pretreated immune mice were halves the numbers of RFCs of the non-treated immune mice.

Effect of CY treatment on the mouse spleen RFCs. The effect of CY treatments on the spleen RFCs was severe both in the non-immune and immune mice. As shown in Fig.6, the numbers of RFCs of the CY treated mice were only one fifth of those of the control mice.

Effect of epinephrine on the mouse spleen RFCs. Spleen cells from immune mice were incubated with epinephrine solutions of various concentrations. As shown in Table 1, rosette formations of immune mice spleen cells were strikingly inhibited by epinephrine especially at the concentration of 10^{-4} mg/ml. The degree of decrement was 23%.

Effect of levamisole on the mouse spleen RFCs. Spleen cells of immune mice were incubated with levamisole solutions of various concentrations. As shown in Table 2, levamisole especially at the concentration of 10^{-6} mg/ml profoundly increased the number of rosettes of immune mice spleen cells. The degree of increment was 180%.

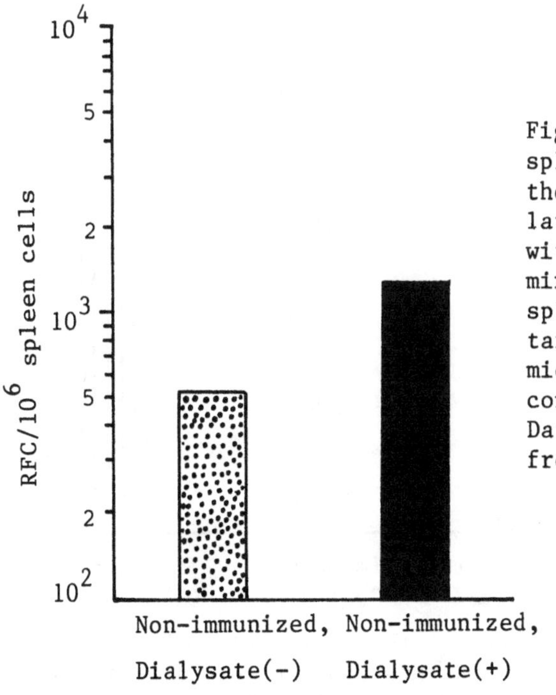

Fig.4. The effect of TF on the spleen RFC. The dialysates of the mouse spleen cells, isolated from the mice immunized with SRBC, were added to the mixtures of SRBC and mouse spleen cells which were obtained from the non-immune mice. MEMS was added to the controls instead of dialysates. Data represent the RFC obtained from a pool of five spleens.

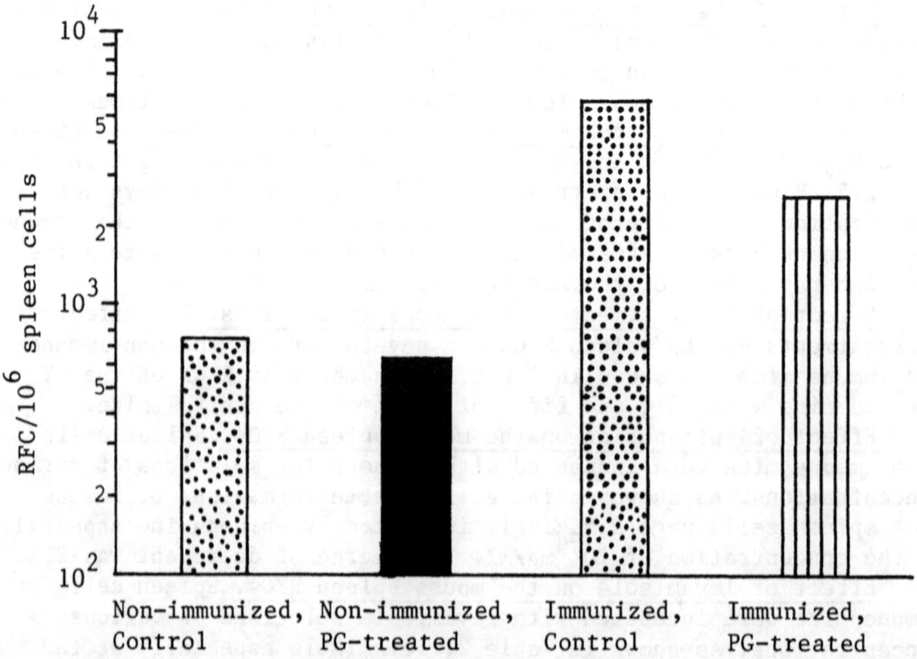

Fig.5. The effect of PG $F_{2\alpha}$ treatment on the mouse spleen RFCs. PG $F_{2\alpha}$ had been injected before immunization for three days. Data represent the RFCs obtained from a pool of five spleens.

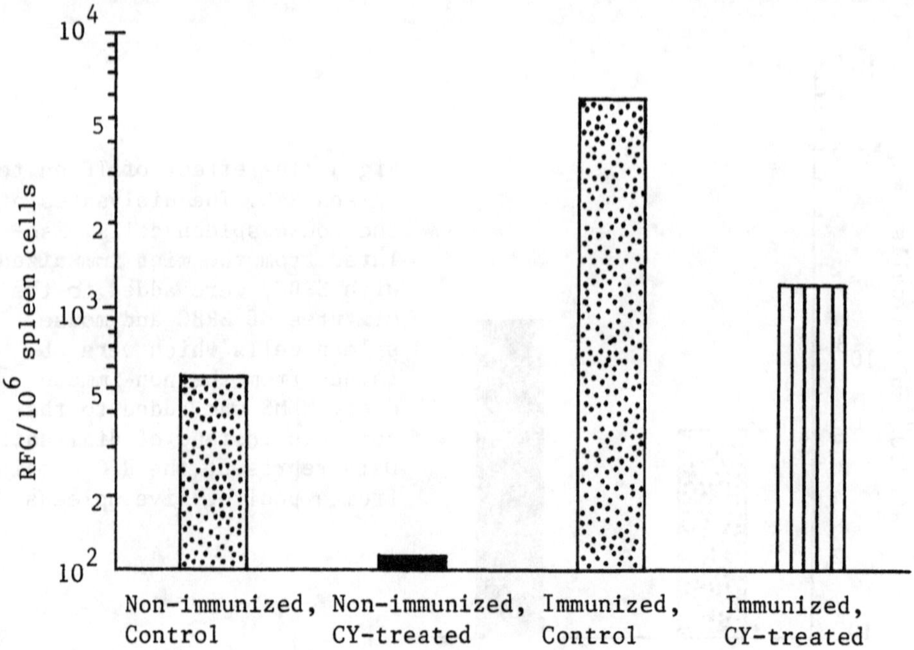

Fig.6. The effect of CY treatment on the mouse spleen RFCs. CY (200mg/kg body weight) was treated three days before immunization. Data represent the RFCs obtained from a pool of five spleens.

Table 1. Effect of Epinephrine on the RFC of Mouse Spleen[a]

Concentration (mg/ml)	0	10^{-5}	10^{-4}	10^{-3}	10^{-2}	10^{-1}
Total RFC/ 10^6 cells [b]	5911±131	5007±557	1365±796*	5372±1242	5462±553	5117±561

[a]The spleen cells of the mice, isolated from the mice immunized with SRBC, were incubated in MEMS containing 0.25% SRBC plus various concentrations of epinephrine for 12 hours at 4°C.
[b]An average of 16 hemocytometer chamber counts from three incubations involving two different cellular pools, each of which is comprised of spleen cells from five mice, is used for all data presented in this table. Data are expressed as mean±s.d. Statistical analysis was done using the Student s t-test.
*$P < 0.05$ compared to control.

Table 2. Effect of Levamisole on the RFC of Mouse Spleen[a]

Concentration (mg/ml)	0	10^{-8}	10^{-7}	10^{-6}	10^{-5}
Total RFC/ 10^6 cells [b]	5863±114	5931±424	5971±345	10848±623*	9862±475*

[a]The spleen cells of the mice, isolated from the mice immunized with SRBC, were incubated in MEMS containing 0.25% SRBC plus various concentrations of levamisole for 12 hours at 4°C.
[b]An average of 16 hemocytometer chamber counts from three incubations involving two different cellular pools, each of which is comprised of spleen cells from five mice, is used for all data presented in this table. Data are expressed as mean±s.d. Statistical nanlysis was done using the Student s t-test.
*$P < 0.05$ compared to control.

DISCUSSION

Rosette technique had been widely used for the detection of antibody formation in vitro by isolated spleen cells. However, it is known that both B and T lymphocytes interact with antigens in a highly specific fashion.[6] The most direct demonstration of lymphocyte recognition of antigen has been the in vitro demonstration of antigen binding cells obtained from immune and non-immune individuals.

In this study, we showed that immunization increased the RFCs greatly(Fig.1), and almost all of the RFCs of both non-immune and immune mice spleens were glass wool-adherent. This glass wool-adherent character can be used for the separation of RFCs of mouse spleen cells.

However, while antigen binding B cells can be demonstrated readily, direct visualization of antigen binding T cells has been more problematic.[7] It is thought that the difficulty of the binding of soluble antigens by T cells is due to the low site density of antigen-specific receptors in the T cell membrane[8,9] and the difficulty of the demonstration of the binding of soluble antigens by T cells is due to particular temperature and antigen concentration requirements requisite for antigen binding to occur.[10,11] The demonstrations of antigen binding T cells in immune and non-immune individuals have been more successfully achieved by making T cells bind antigens in solid phase; either soluble antigens linked to fibers[12] or cell membrane associated antigens.[13,14] However, immune T cells which bind heterologous erythrocytes differ from their counterparts in the fact that they are highly unstable and dissociate rapidly in vitro.[15] These observations suggest that the avidity of T cell antigen receptors is lower than that of B cell receptors and the physiology of the lymphocyte-antigen interaction is different for T and B cells. According to the rosette forming method used in this study, 80% of the RFCs of both the non-immune and immune mice were ATS resistant B cells and only 20% of the RFCs were ATS susceptable T cells (Fig.3). These results are in agreement with those of Elliot et al.[3]

TF, prepared from dialysates of human leukocytes, confers on its recipients the specific cell-mediated cutaneous hypersensitivities of the donor.[16] An increasing body of evidences indicates that TF posseses some antigen-independent non-specific 'stimulatory' effects as well.[17-19] Our TF preparations from immunized mice spleen cells enhanced the number of RFCs of non-immune spleen cells to two times of it(Fig.4). This result is correspondent with the results of other researchers.[16-19]

In the previous study,[20,21] we reported that PG $F_{2\alpha}$ could greatly enhance the carcinogenic capacity of 3-methylcholanthrene and suggested that PG F_2 might act as a suppressor factor of immune response of an individual. In this study, the treatment of the mice with PG $F_{2\alpha}$ decreased severely the RFCs only in the immune mice(Fig.5). This result suggests that PG $F_{2\alpha}$ inhibits the antigen committed lymphocytes.

We showed that CY affected B cells severely and T cells mildly in the other studies.[22,23] In this study, treatment of mice with CY profoundly decreased the RFCs of mice spleens(Fig.6).

It is well known that adrenergic receptors are present on the lymphocytes,[24] and the catecholamines, epinephrine and norepinephrine, have been implicated in modulating various immune effector functions. Mitogenic lymphocyte stimulation[26] and active human T cell rosettes are inhibited by -adrenergic stimulants.[27] In this study, epinephrine inhibited the rosette formation in the immune mice spleens(Table 1), and further study is required to elucidate the mechanism.

Levamisole, an antihelmintic drug, has been shown to potentiate and restore both in vitro and in vivo immune responses in animals and human.[28] Levamisole can influence on the response of normal human lymphocytes to mitogens[29] and change the human lymphocyte surface markers.[30] In this study, we demonstrated that levamisole also enhanced the rosette forming ability of immune mice spleen cells(Table2). However, we do not know whether the increment of RFCs was due to the increased stability of T cell antigen receptors or increased proliferation of the receptor-positive B cells.

Further study about the functional significance of the rosette forming spleen cells is thought to be required.

REFERENCES

1. N.K.Jerne, A.Nordin,and C.Henry, Cell bound antibodies, B.Amos, and H.Koprowski ed., The Wistar Institute Press, Philadelphia (1963).
2. O.B.Zaalberg, A simple method for detecting single antibody forming cells, Nature(Lond.), 202:1231(1964).
3. B.E.Elliott, J.S.Haskill, Characteristics of thymus-derived and bone marrow-derived rosette forming lymphocytes, Eur.J.Immunol. 3:68(1973).
4. R.S.Holzman,and H.S.Lawrence, In vitro augmentation of lymphocyte sheep cell rosette formation by leukocyte dialysates, J. Immunol.118:1672(1977).
5. Nekam, L.Kalmar, P.Gergely, G.Kelemen, B.Fekete, I.Lang, J.Levai and GY.Petrayi, In vitro effect of transfer factor on active rosettes and leukocyte migration of patients with cancer, Clin. exp.Immunol. 27:416(1977).
6. J.F.Bach and M.Dardenne, Antigen recognition by T lymphocytes, I. Thymus and marrow dependence of spontaneous rosette forming cells in the mouse, Cell.Immunol. 3:1(1971).
7. N.L.Warner, Membrane immunoglobulins and antigen receptors on B and T lymphocytes, Adv.Immunol. 19:67(1974).
8. G.L.Ada.and P.L.Ey, Lymphocyte receptors for antigens, in The Antigens, M.Sela,ed., Academic Press Inc., New York(1975).
9. G.J.Hammerling, T.Masuda,and H.McDevitt, Genetic control of immune response. Frequency and characteristics of antigen-binding cells in high and low responder mice, J.Exp.Med. 137: 1180(1973).
10. G.E.Roelants,and A.Ryden, Dose dependence of antigen binding to B and T lymphocytes, Nature(lond.), 247:104(1974).
11. S.J.Black, G.H.Hammerling, C.Berek, K.Rajewsky,and K.Eichman, Idiotypic analysis of lymphocytes in vitro. I. Specificity and heterogeheity of B and T lymphocytes reactive with anti-idiotype antibody, J.Eëp.Med.143:846(1976).
12. C.S.Henney, On the mechanism of T cell mediated cytolysis, Transplant.Rev. 17:37(1973).
13. N.M.Hogg,and M.F.Graves, Antigen binding thymus-derived lymphocytes. I. Rapid methods for isolation of θ-positive antigen stimulated cells, Immunology 22:959(1972).
14. R.F.Ashman,and M.C.Raff, Direct demonstration of θ-positive antigen binding cells with antigen-induced movement of thymus-dependent cell receptors, J.Exp.Med. 137:69(1973).

15. J.Kettman, Delayed hypersensitivity: Is the same population of thymus-derived cells responsible for cellular immunity reactions and the carrier effect? Immunol.Commun. 1:289(1972).
16. H.S.Lawrence, Transfer factor in cellular immunity, Harvey lectures series 68:293, Academic Press, New York.
17. B.Dupont, M.Ballow, J.A.Hansen, C.Quick, E.J.Yunis,and R.A.Good, Effect of transfer factor therapy on mixed lymphocyte culture reactivity, Proc.Nat.Acad.Sci.(Wash.) 71:867(1974).
18. N.F.Mendes, P.J.Saraiva,and O.B.O.Santos, Restorative effect of normal human serum, transfer factor, and thymosin on the ability of heated human lymphocytes to form rosettes with sheep erythrocytes, Cell.Immunol. 17:560(1975).
19. A.S.Hamblin, R.N.Maini,and D.C.Dumonde, Human transfer factor in vitro. I. Augmentation of lymphocyte transformation to tuberculin PPD, Clin.exp.Immunol. 23:290(1976).
20. H.T.Chung,and T.Y.Ha, Immune response to sheep red blood cells in mice bearing tumor induced by 3-methylcholanthrene, Kor.J. Immunol. 1:35(1979).
21. T.Y.Ha,and H.T.Chung, Potentiation of carcinogenic effect of 3-methylcholanthrene by prostaglandin $F_{2\alpha}$, J.Kor.Med.Assoc.23: 785(1980).
22. T.Y.Ha,and H.T.Chung, Effect of cyclophosphamide on humoral and cellular immune responses in mice, J.Kor.Med.Assoc.20:985(1977).
23. T.Y.Ha,and H.T.Chung, Effects of immune serum on the immune responses in the mice pretreated with cyclophosphamide, Jeonbug Univ.Med.J. 3:11(1979).
24. N.H.Bishopric, H.J.Cohen,and R.J.Lefkowitz, β-adrenergic receptors in lymphocyte subpopulations, J.Allergy Clin.Immunol. 65: 29(1980).
25. H.D.Basedovsky, A.del Ray, E.Sorkin, M.Da Prada,and H.H.Keller, Immunomodulation mediated by the sympathetic nervous system, Cell.Immunol. 48:346(1979).
26. B.Crary, M.Borysenko, D.C.Sutherland,and I.Kutz, J.Z.Borysenko, and H.Benson, Decrease in mitogen responsiveness of mononuclear cells from peripheral blood after epinephrine administration in humans, J.Immunol. 130:694(1983).
27. S.P.Galant,and R.A.Remo, β-Adrenergic inhibition of human T lymphocyte rosettes, J.Immunol. 114:512(1975).
28. D.Sampson,and A.Lui, The effect of levamisole on cell-mediated immunity and suppressor cell function, Cancer Res. 36:952 (1976).
29. J.W.Hadden, R.G.Coffey, E.M.Hadden, E.Lopez-Corrale,and G.H.Sunshine, Effects of levamisole and imidazole on lymphocyte proliferation and cyclic nucleotide levels, Cell.Immunol. 20:98 (1975).
30. J.Wybran,and A.Govaerts, Levamisole and human lymphocyte surface markers, Clin.exp.Immunol. 27:319(1977).

THE BURSA OF FABRICIUS AS A TRAPPING MECHANISM FOR ENVIRONMENTAL ANTIGENS

S. Ekino*, T. Urano**, H. Fujii* and M. Kotani*

Department of Anatomy*, Labo. Animal Res. Center**
Kumamoto University Medical School, 2-2-1 Honjo
Kumamoto City Japan

SUMMARY

The bursa of Fabricius was isolated from gut-derived antigens by ligating the bursal duct on the 18th day of incubation. Ligation of the bursal duct (BDL) suppressed the bursal development, spontaneous germinal centre formation in the spleen and the development of serum "natural" agglutinins for bacteria or heteroerythrocytes. Moreover, administration of sterilized ceacal contents into the bursal lumen at BDL led to normal level of serum "natural" agglutinins. These findings strongly suggest that the bursa of Fabricius possesses an antigen-dependent process which modulates the development of immune system. Antigen-trapping mechanism of the bursa may supply information about environmental antigens for this process.

INTRODUCTION

The development of immune system is regulated by antigen-dependent and independent processes. It is well known that, in avian species, the B cell differentiation in the bursa of Fabricius is regulated by antigen-independent processes (1). On the other hand, the bursa of Fabricius has microflora (2) and takes up gut-derived antigens through the bursal duct by a sucking function and antigenic materials in the bursal lumen are transported into the bursal follicle by the follicle-associated epitherium (3,4). These facts suggest that the bursa of Fabricius may possess an antigen-dependent process which modulates the development of immune system. This problem has been examined in germfree chicken (1,5). However, even in germ-free condition, the bursa is exposed to non-bacterial conponents such as digested foods and/or macromolecular materials from the mother (3). We, therefore, ligated the bursal duct on the 18th day of incubation so as to arrest the entrance of gut-derived antigens into the bursal lumen and examined the bursal development, spontaneous germinal centre formation in the spleen and the development serum "natural" agglutinins for bacteria or heteroerythrocytes after hatching.

MATERIALS AND METHODS

Chickens

White Leghorn chickens were housed together in conventional condition.

In ovo ligation of the bursal duct

The bursal duct was ligated on the 18th day of incubation without any damage to blood and lymphatic vessels (6,7). Formaline-sterilized ceacal contents (8μl) which were washed three times in sterile saline were administered into the bursal lumen just before the bursal duct was ligated.

Experimental design

Chicken were bled and killed at 4 and 11 weeks old.The bursal weight, number of germinal centre in a cross-section of the spleen and titer of serum "natural" agglutinins for bacteria (indigeneous E.coli or Brucella abortus) or human erythrocytes (ARBC, BRBC or ORBC) were measured. In BDL chicken, the bursal lumen was tested for sterility at autopsy. The spleen was fixed in Carnoy's fluid and embedded in paraffin. Cross section (8μ) was cut in the thickest point in the spleen and stained with methyl green pyronin. The number of germinal centre per cross section was recorded. E.coli(B) was isolated from the bursal lumen and E.coli(C) was isolated from ceacal contents. They were killed by 2 % folmalin. B. abortus which was killed by phenol was obtained from KAKETSUKEN. Human erythrocytes were obtained from healthy men. Serum "natural" agglutinins for them were titrated by using microtitration technique after sera were heated to 56°C for 30 min.

RESULTS

Development of the bursa

As Fig. 1 shows, The bursal weight of BDL chickens was smaller than that of normal controls (NC) in 4 (A) ($P<0.001$) and 11 weeks old (B) ($P<0.001$).

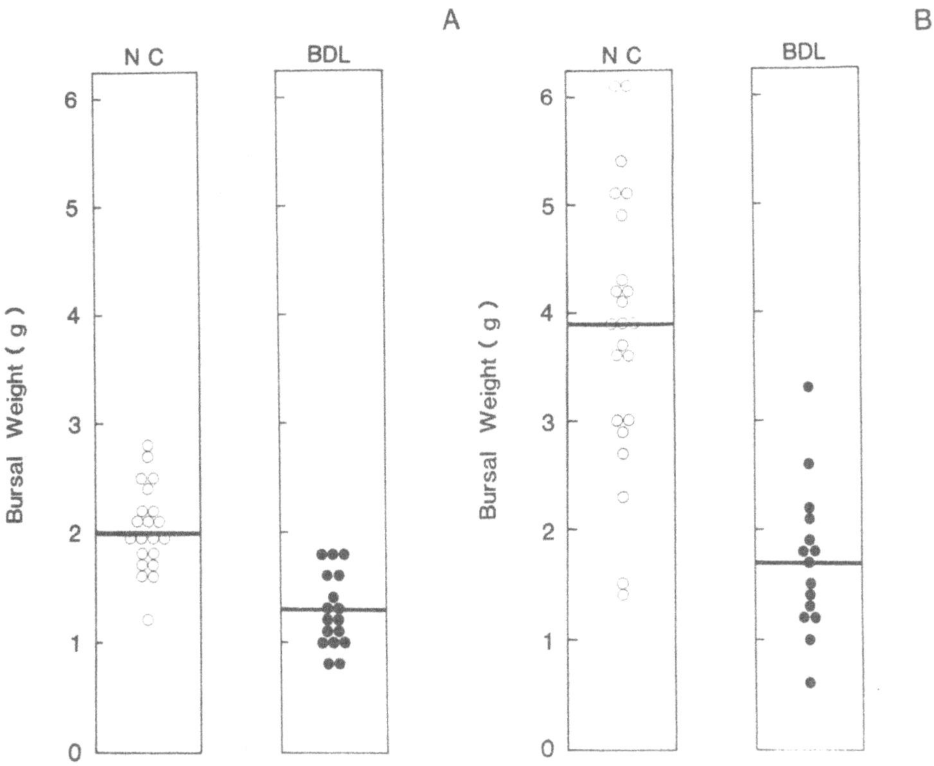

Fig. 1. The bursal weight of normal controls (○) and BDL chickens (●) 4 (A) and 11 weeks old (B). The mean is indicated by a horizontal heavy line.

During the fourth to eleventh week, the bursal weight of normal controls increased 1.91-fold, while that of BDL chickens increased 1.35-fold.

<u>Development of spontaneous germinal centre formation in the spleen</u>

As Fig. 2 shows, although normal controls (NC) were not immunized with any antigens, they showed a wide range of the number of germinal centre at both age and the number of germinal centre increased with age. In BDL chicken, spontaneous germinal formation in the spleen was suppressed at age of 4 (A) ($P<0.01$) and 11 weeks (B) ($P<0.05$), although the number of germinal centre increased with age.

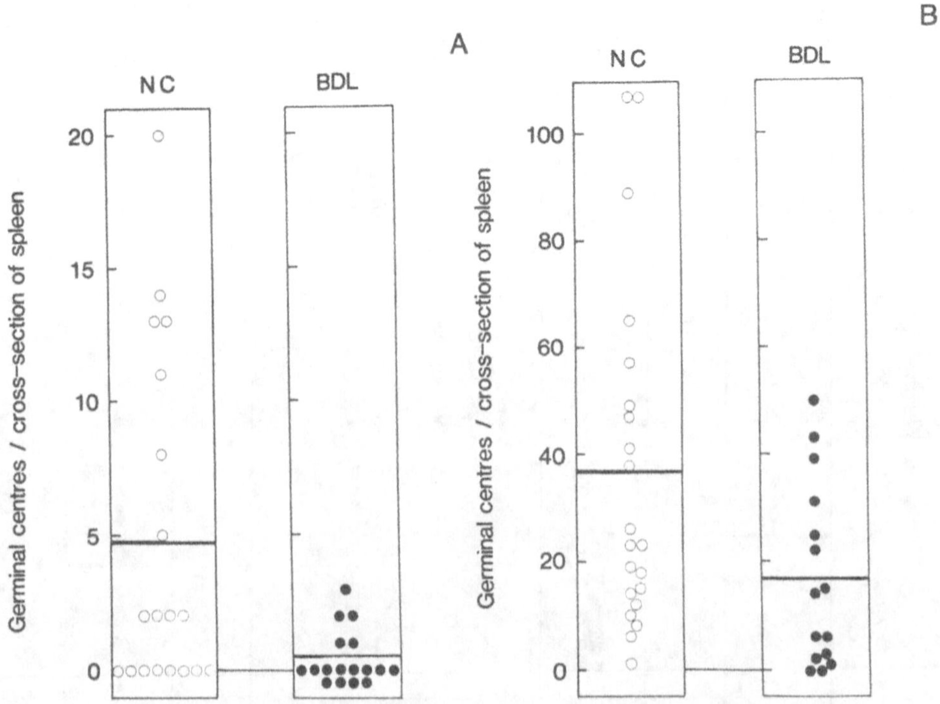

Fig. 2. The number of spontaneously formed germinal centre in the spleen of normal controls (○) and BDL chickens (●) 4 (A) and 11 weeks old (B). Number of germinal centre per cross section was counted in the thickest point. The mean is indicated by a horizontal heavy line.

Development of serum "natural" agglutinin for bacteria or human erythrocytes

As Fig. 3 shows, sera of normal controls (NC) showed high titer for E.coli(B) or E.coli(C) and low titer for B.abortus or human erythrocytes. In BDL chicken, the development of serum "natural" agglutinins was significantly suppressed for any tested antigens (α - E.coli(B) ; $P<0.001$, α - E.coli(C) ; $P<0.001$, α - B. abortus ; $P<0.02$, α - ARBC ; $P<0.01$, α - BRBC ; $P<0.01$, α - ORBC ; $P<0.01$). It is probably that normal chickens showed high titer for E.coli(B) and E.coli(C) because these E.coli were inhabitant in their gut. In BDL chicken, we could not observe positve titer for E.coli(B) but 74% of them showed positve titer for E.coli(C). Although B. abortus is pathogenic bacterium and not inhabitant in the gut, 57% of normal controls showed positive titers. But , in BDL chicken , only 7% of them showed positive titers for B. abortus.

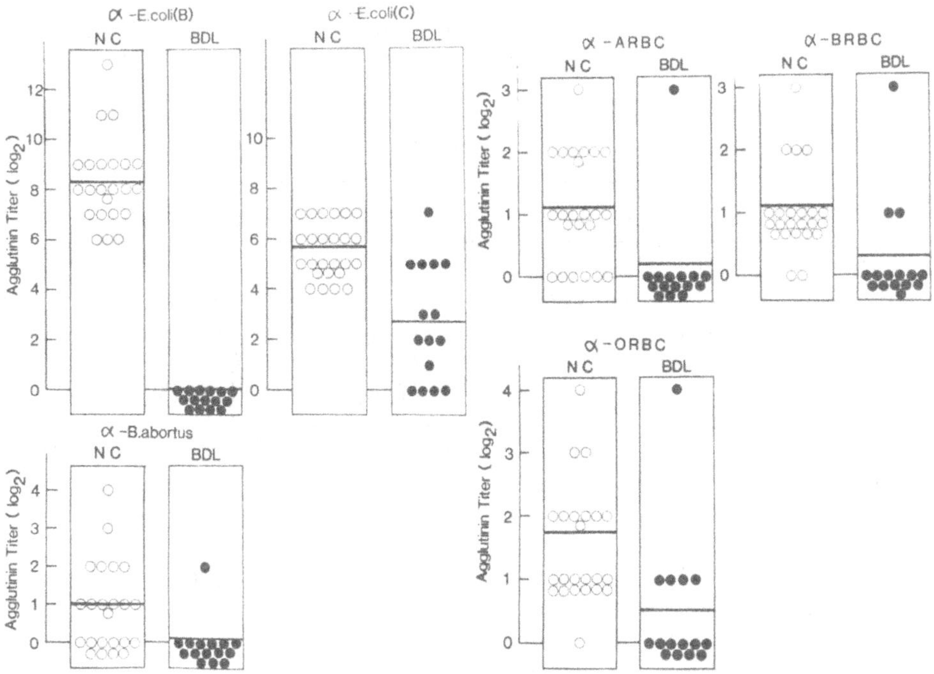

Fig. 3. Titer of serum "natural" agglutinins for bacteria or human erythrocytes in normal controls (O) and BDL chickens(●) 11 weeks. The mean is indicated by a horizontal heavy line.

<u>Recovary of serum "natural" agglutinin by administration of ceacal contents into the bursal lumen at BDL</u>

As Table 1 shows, administration of sterilized ceacal contents into the bursal lumen at BDL led to normal level of serum "natural" agglutinins for any tested antigens.

DISCUSSION

Present results showed that when the bursa of Fabricius was isolated from environmental stimuli derived from the gut by BDL, the bursal development, spontaneous germinal centre formation and the development of serum "natural" agglutinins were markedly suppressed. Moreover, administration of sterilized ceacal contents into the bursal lumen at BDL could lead to normal level of serum "natural" agglutinins. This result suggests that surgical procedures of BDL did not directly damage the bursa. Previously, Thorbecke reported that the bursa developed normally in germfree condition. However, even in germfree condition,

Table 1. Effects of administration of sterilized ceacal contents (c.c.) into the bursal lumen on serum "natural" agglutinins in 11 week-old chickens

Group	Agglutinin titer ($\log_2$)		
	α-E.coli(B)	α-E.coli(C)	α-B R B C
Control	8.2±0.3* (23/23)**	5.7±0.2 (23/23)	1.1±0.1 (21/23)
BDL(saline) a)	0 (0/15)	2.7±0.2 (11/15)	0.3±0.2 (3/15)
BDL(c.c.) b)	6.7±1.4 (6/7)	5.1±0.5 (7/7)	1.6±0.7 (5/7)

a), b) ; Sterile saline (a)) or sterilized ceacal contents (b)) were administrated into the bursal lumen at BDL.
* ; Each value represents mean±S.D.
** ; Figure in parentheses represents the number of positive titer / number of chicken examined.

the bursal lumen may be exposed to various gut-derived antigens except microflora (3). Our results suggest that the bursal development may be partly modulated by environmental antigens derived from the gut.

In BDL chicken, the number of spontaneously formed germinal centre was reduced especially at young age, but it increased with age. These results suggest that, in young chicken, the bursa may play a major role in induction of spontaneous germinal centre formation in the spleen. On the other hand, in 11 week-old chickens, non-bursal sites may effect partly induction of spontaneous germinal centre formation in the spleen. Moreover, non-bursal sites may play an important role in induction of positive titers for E.coli(C), B. abortus, ARBC, BRBC or ORBC in BDL chicken 11 weeks old.

Spontaneous germinal centre formation in the spleen and serum "natural" agglutinins are thought to arise as a result of stimulation by environmental antigens. Our results showed that isolation of the bursa from environmental stimuli by BDL markedly suppressed development of them and administration of sterilized ceacal contents into the bursal lumen at BDL led normal level of serum "natural" agglutinins for bacteria or human erythrocytes.

These results and our previous results (7,8) strongly suggest that the bursa of Fabricius possesses an antigen-dependent process which modulates the development of immune system. It is probably that antigen-trapping mechanism of the bursa supplies information about environmental antigens for this process. We hypothesize that stimuli of environmental antigens trapped by the bursa of Fabricius may accelerate or amplify the development of immune system.

REFERENCES

1. P.M. Lydyard, C.E. Grossi, and M.D. Cooper, Ontogeny of B cells in the chicken, J. Exp. Med. 144:79 (1976)
2. R. Fuller, Differences in the microflora of the intestine and the bursa of Fabricius as illustrated by the Coliform / Lactobacillus ratio in the two sites, Br. Pout. Sci. 14:221 (1974)
3. T. Schaffner, M.W. Hess, and H. Cottier, A reappraisal of bursal function, Ser. Haemt. 7:568 (1974)
4. R. Sorvari and T.E. Sorvari, Bursa of Fabricius as a peripheral lymphoid organ, Immunology 32:499 (1977)
5. G.J. Thorbecke, H.A. Gordon, B. Wostman, M. Wagner, and J.A. Reyniers, Lymphoid tissue and serum gamma globulin in young germfree chickens, J. Infect. Dis. 101:237 (1957)
6. S. Ekino, K. Matsuno, and M. Kotani, Distribution and role of lymph vessels of the bursa Fabricii, Lymphology 12:247 (1979)
7. S. Ekino, Y. Nawa, K. Tanaka, K. Matsuno, H. Fujii, and M. Kotani, Suppression of immune response by isolation of the bursa of Fabricius from environmental stimuli, Aust. J. exp. Biol. med. Sci. 58:289 (1980)
8. S. Ekino, K. Matsuno, S. Harada, H. Fujii, Y. Nawa, and M. Kotani, Amplification of plaque-forming cells in the spleen after intracloacal antigen stimulation in neonatal chicken, Immunology 37:811 (1979)

MATERNAL AUTOIMMUNE DISEASE INFLUENCES SELF-TOLERANCE IN OFFSPRING

J. Gibson, K. L. Wong and A. Basten

The Clinical Immunology Research Centre
University of Sydney
Sydney N.S.W. 2006 Australia

INTRODUCTION AND AIM

It is now recognised that the placenta is not an inpenetrable barrier separating foetus from mother. Not only can IgG antibodies cross but lymphocyte traffic has recently been reported as well (Adinolfi, 1978). The potential therefore exists for maternal influences (cells, antibodies, antigen or non-specific factors) to affect self-tolerance in offspring. To investigate this possibility, autoimmune disease was induced in prospective murine mothers and subsequent offspring tested for their susceptibility to develop the same autoimmune process.

MATERIALS AND METHODS

Autoimmune haemolytic anaemia (AIHA) was induced in prospective CBA/Ca/T6 mothers by three intraperitoneal (IP) injections of 2×10^8 washed rat erythrocytes (RRBC) at weekly intervals (Playfair and Marshall-Clarke, 1973). Control mice of the same strain were immunised with sheep erythrocytes using an identical protocol. Mating with normal CBA males took place immediately after the last immunisation. Offspring from each group were then tested for the presence of autoantibodies to erythrocytes at six weeks of age. Half of the test group were found to have passively acquired anti-erythrocyte autoantibodies which always disappeared by eight weeks of age whilst in none of the control group could such autoantibodies be demonstrated.

At eight weeks of age both test and control groups (minimum of six animals) were immunised twice IP with 2×10^8 washed RRBC a week apart. Animals were then bled at regular intervals for the measurement of anti-erythrocyte autoantibodies by direct antiglobulin test (DAGT) and anti-RRBC antibodies by passive haemagglutination.

Assay for Suppression: To test for the presence of suppressor cells, spleen cells from test and control offspring were transferred into naive non-irradiated recipients which were subsequently immunised twice with 2×10^8 RRBC a week apart.

Neonatal Injections: Groups of normal neonates were injected IP within 12 hours of birth with 10^7 RRBC, normal MRBC and DAGT positive MRBC.

Cell Traffic Analysis: Pregnant mice were injected IV on days 14 to 18 of gestation with either 75×10^6 TRITC labelled syngeneic spleen cells (test groups) or 75×10^6 unlabelled spleen cells (control groups). Offspring were sacrificed on the day of birth and splenic lymphocyte populations labelled with monoclonal antibodies to Thy-1, Ly-2 and L_3T_4 cell surface antigens followed by appropriate FITC conjugated second antibodies. Two colour FACS analysis was then undertaken to determine the percentage of cells of maternal origin in the foetal spleen and their relative subset distribution.

RESULTS

Effect of Active Autoimmune Disease During Pregnancy on Self-Tolerance in Offspring

Mean DAGT titres (± SEM) in offspring of mothers with an active autoimmune disease (solid bars) were found to be significantly less ($\rho < 0.05$) than those in offspring born to control mothers (open bars) both at two and four weeks after immunisation (Figure 1). The effect of maternal autoimmune disease was specific for the autoantibody response since the mean levels of anti-RRBC antibodies were comparable in test and control groups. A similar trend was observed when the percentage of mice with positive DAGT titres in test (solid bars) and control (open bars) groups was compared (Figure 2). No difference was seen in the level of autoantibody response in those test animals which were DAGT positive at six weeks of age compared to those which were DAGT negative at this time.

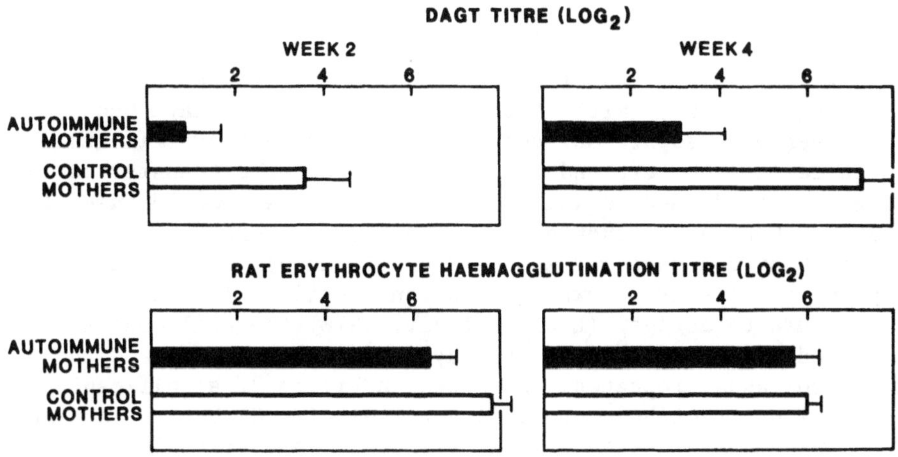

FIGURE 1
Mean DAGT and rat erythrocyte haemagglutination titres ($\log_2 \pm$ SEM) of offspring from mothers with AIHA and normal mothers two and four weeks after immunisation.

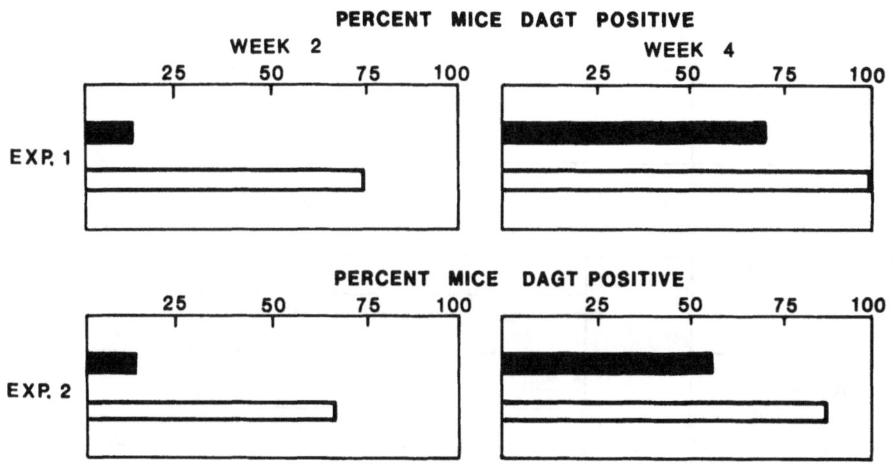

FIGURE 2
Percentage of offspring with positive DAGT and rat erythrocyte haemagglutination titres from mothers with AIHA and normal mothers at two and four weeks after immunisation.

Demonstration of Suppressor Cells

It is possible that the reduced anti-self response in offspring of mothers with autoimmune disease could have been due to the presence of suppressor cells. In preliminary experiments spleen cells from test and control offspring were transferred into naive non-irradiated recipients which were given two IP injections of RRBC. However no suppression of either anti-RRBC or autoantibody response was seen (data not shown). Since memory suppressor cells are known to be silent until activated by re-exposure to antigen (Loblay et al., 1978) the cell donors were immunised with 2 x 10^8 RRBC ten days prior to transfer and the suppressor assay repeated. As shown in Figure 3, significant suppression of antibody production to self (MRBC) but not non-self (RRBC) was obtained with spleen cells from test offspring. By contrast the response in recipients of cells from the control group to both self and non-self was unaffected.

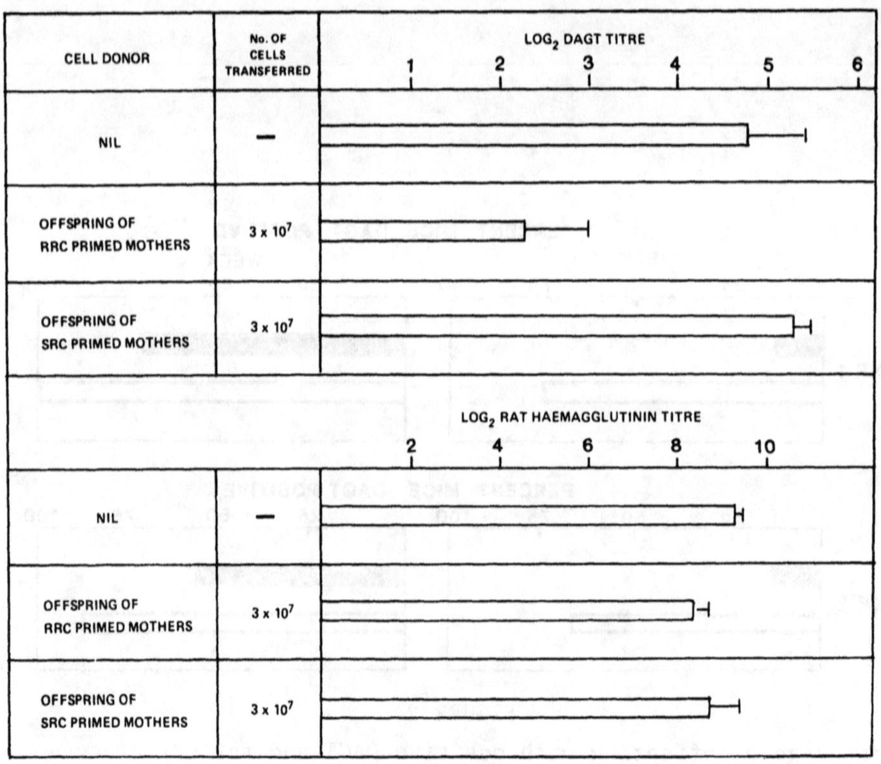

FIGURE 3

Self-reactive suppressor cells in offspring of mothers with AIHA.

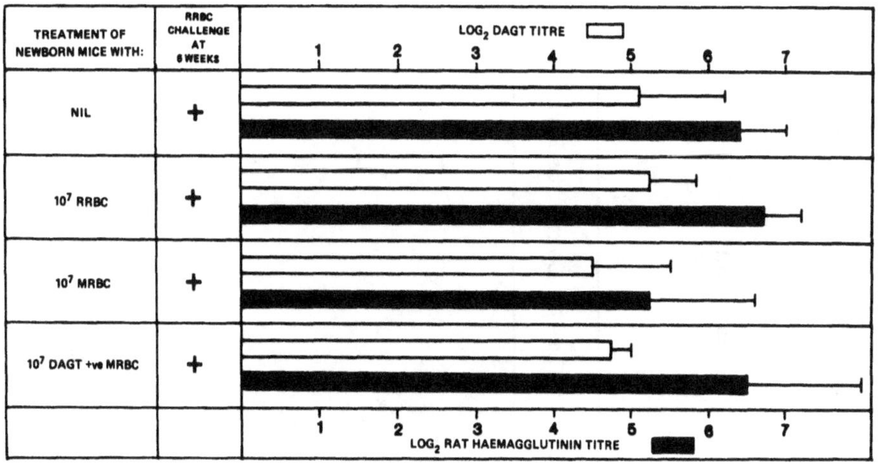

FIGURE 4
Failure of neonatal erythrocyte injection to inhibit subsequent autoantibody response.

Effect of Neonatal Injection of Red Cell Preparations

Such suppressor cells could in theory have been generated by exposure to maternal influences (e.g., antibody, antigen or cells) antenatally or postnatally (via colostrum). In an attempt to distinguish between these possibilities normal mice were injected with RRBC, normal MRBC and DAGT positive MRBC and examined for their ability to generate an autoantibody response. All animals were found to be DAGT negative at six weeks of age and as shown in Figure 4 all groups developed similar anti-self and anti-nonself responses after two RRBC immunisations thus demonstrating the importance of ante-natal influences compared to post-natal in this system.

Demonstration of Materno-Foetal Cell Traffic

In an attempt to more precisely define maternal influences on the developing foetus, two colour FACS analysis was performed, as described, to demonstrate and phenotype materno-foetal cell traffic. Preliminary analysis of TRITC-labelled cells remaining in the maternal spleens demonstrated expected proportions of Thy-1, Ly-2 and L_3T_4 bearing cells. When foetal spleen mononuclears were examined 6% were found to be of maternal origin (i.e., TRITC positive). Of this population 30% bore the surface antigen (Ly2) that characterises suppressor T cells whilst only 1% were helper in phenotype (L_3T_4 positive). It is interesting to note that only approximately 50% of Ly2 positive cells were Thy-1 positive.

TABLE 1

Two colour FACS Analysis of Phenotype of Maternal Lymphocytes Present in Spleens of Syngeneic Neonates and of TRITC labelled cells remaining in the maternal spleen.

Maternal TRITC positive cells* labelled with:	% Cells in Neonatal Spleen		% Cells in Maternal Spleen	
	Absolute Value	Corrected Value Δ	Absolute Value	Corrected Value Δ
Normal Mouse Serum + FITC-SαMIg#	19	-	57	-
Normal Mouse Serum + FITC-RαRIgθ	1	-	1	-
Anti-Ly 2 + FITC-SαMIg	48	29	63	6
Anti-Thy 1.2 + FITC-RαRIg	13	12	25	24
Anti-L_3T_4 + FITC-RαRIg	2	1	22	21

* Constituted 6% of small lymphocytes of neonatal spleen cells.
Δ Corrected for anti-Ig positive (B) cells and non-specific binding of RαRIg.
\# FITC conjugated sheep anti-mouse immunoglobulin.
θ FITC conjugated rabbit anti-rat immunoglobulin.
Note: 39% TRITC positive cells were negative with all labels used.

DISCUSSION

Active autoimmune disease during pregnacy was found to result in a decreased susceptibility to development of a similar autoimmune process in the offspring as demonstrated by both a lower mean DAGT titre (Figure 1) and a smaller percentage of mice with a positive DAGT in test versus control groups (Figure 2). The characteristics of this effect may be summarised as follows:

1. It is specific for the response to self antigens since comparable levels of anti-RRBC antibodies were observed in test and control groups (Figure 1).

2. The events responsible for enhancement of self tolerance probably take place _in utero_ and not post-natally (Figure 4).

3. It appears to be mediated by memory suppressor cells with specificity for self rather than maternal antibody with possible anti-idiotype specificity (Figure 3).

In theory the suppressor cells could have originated in the offspring during embroyonic life due to transplacental passage of antigen (RRBC) carrying cross-reactive murine antigens. However their specificity for self argues against this possibility. Alternatively materno-foetal traffic of specific memory suppressor cells from mothers with active autoimmune disease could have occurred. The finding, in the foetus, of significant numbers of maternal cells which displayed the suppressor phenotype is consistent with this possibility as is the formal demonstration, in another system of transfer of functionally active suppressor cells from mothers to offspring (Wong, Loblay and Basten; unpublished observations). Experiments are currently in progress to distinguish between these two alternatives by determining whether cells mediating suppression in the adoptive transfer system carry maternal or F_1 antigens on their surface.

ACKNOWLEGEMENTS

The following people generously supplied the cell lines producing the monoclonal antibodies used in this project: Ms J. Gamble, Walter and Eliza Hall Institute, Melbourne, Australia for anti-Thyl-2; Dr I. F. McKenzie, Research Centre for Cancer and Transplantaion, University of Melbourne, Melbourne, Australia for anti-Ly-2 and Professor F. W. Fitch, University of Chicargo, Chicargo, Illinois, U.S.A. for anti-L_3T_4. This work was supported by grants from the National Health and Medical Research Council of Australia and the Australia Development Assistance Bureau. We are greatful to Ms M. Pritchard for assistance with manuscript preparation.

REFERENCES

Adinolfi, M., 1982, Two Aspects of the Materno-Foetal Relation: The Immunosuppressive Role of Alpha-Foetoprotein and the Transfer of Lymphocytes Across the Placenta, in: "Serno Symposium No. 45. Immunological Factors in Human Reproduction", S. Shulman., F. Donero., and M. Nicotra, Ed's., Academic Press, London.

Loblay, R. H. L., Basten, A., and Pritchard-Briscoe, H, 1978, Suppressor T Cell Memory. Nature 272:620.

Playfair, J. H. L., and Marshall-Clarke, S, 1973, Induction of Red Cell Autoantibodies in Normal Mice. Nature (New Biol.) 243:213.

THE ROLE OF SUPPRESSOR T LYMPHOCYTES IN THE MAINTENANCE OF TOLERANCE TO ALLOANTIGENS IN BONE MARROW CHIMERAS PREPARED WITH TOTAL LYMPHOID IRRADIATION (TLI)

S. Slavin, M. Weigensberg, S. Bar, L. Weiss and S. Morecki

The Bone Marrow Transplantation and Cryopreservation Unit and the Immunobiology Research Laboratory Department of Medicine A, Hadassah University Hospital, Jerusalem, Israel

INTRODUCTION

Treatment of BALB/c mice with high dose fractionated total lymphoid irradiation (TLI) allows for the engraftment of allogeneic bone marrow (BM) without the subsequent development of graft vs. host disease (GVHD) (1-3). The resultant chimeras are specifically tolerant to tissues of the marrow donor strain, and reject third-party allografts (1-3). The mechanism and cellular basis of the induction and maintenance of tolerance and lack of GVHD in the chimeras is not clear. However, studies of mice treated with TLI and tolerized to bovine serum albumin (BSA) or bovine gamma globulin (BGG) show that specific suppressor T cells are involved in the maintenance of tolerance to the heterologous proteins (4).

The presence of nonspecific suppressor cells immediately following TLI, and specific suppressor cells in well established chimeras was previously demonstrated by blocking mixed lymphocyte reactions (MLR) using co-culturing experimental systems (4-8). Although it is tempting to suggest that suppressor cells may play a role in establishing and/or maintaining the bilateral tolerance of host versus graft and graft versus host following TLI, no conclusive data could be generated on the biological role of suppressor cells induced after TLI in stable BM chimeras. Data presented in this report supports the concept that specific unresponsiveness in stable chimeras may be the result of operational clonal deletion rather than active suppression by classical suppressor T cells. Nevertheless, the possibility that suppressor cells, not necessarily T lymphocytes, may play a role in the induction of the unresponsive state or in maintaining the

the state of tolerance by reducing or maintaining a low frequency of alloreactive precursors cannot be excluded.

MATERIALS AND METHODS

Animals. Inbred BALB/c, C57BL/6, (BALB/c x C57BL/6)F_1 (F_1), C3H/He, A/Crgl (A) mice were used in all experiments. Mice were kept in conventional animal facilities.

Induction of chimerism by conditioning with TLI. F_1 mice were anesthetized for proper irradiation of the major lymphoid organs as previously described (1, 2). TLI consisted of 8 fractions of 200 rad per day x 6/week, to a total dose of 1600 rad, delivered from a Phillips X-ray unit (250 kv, 20 mA) at a rate of 52 rad per minute. Whole body irradiation (WBI) was delivered using the same source with a single exposure to 200 rad for induction of semiallogeneic or 400 rad for induction of completely allogeneic chimeras. Aliquots of 0.25 ml medium containing 30 x 10^6 nucleated cells were injected intravenously into recipient mice one day following irradiation. Chimerism was confirmed by microcytotoxicity using specific antibodies and rabbit complement. Allogeneic chimeras received donor type skin allografts to confirm tolerance to donor alloantigens.

Mixed lymphocyte reaction (MLR). 1 x 10^6 responding cells were cultured in flat bottom microculture wells with 1 x 10^6 stimulator lymphocytes inactivated by a single exposure to 3000 rad. MLR suppression was assayed by co-culturing 1 x 10^6 spleen cells from experimental chimeras. Co-culturing of equal numbers of spleen cells obtained from normal animals was carried out in each experiment, as previously described in detail (5).

Depletion of LyT-2 cells by flow cytometry. Spleen cells obtained from chimeras were stained with anti-LyT-2 antibody (clone 53.6.72). Cells were labeled with FITC-F(ab')$_2$-rabbit anti-mouse antibody to detect LyT-2^+ cells. Analysis and sorting were carried out by fluorescence activated cell sorter (FACS), using asceptic conditions.

Graft versus host disease (GVHD) assays. Adoptive recipients were irradiated by WBI 550 rad 24 h prior to intravenous inoculation of cells used for challenge. Survival was observed daily.

RESULTS

Stable chimerism was documented in mice conditioned with a short TLI protocol consisting of 1600 rad and WBI 200 rad in semiallogeneic combinations (C57BL/6 → F_1). Chimerism with completely allogeneic BM cells (A → F_1) was established following pre-treatment of recipients

Table 1. Specific unresponsiveness of spleen cells obtained from C57BL/6→F_1 and A→F_1 in MLR

Responders ($1x10^6$ spleen)	Stimulators ($1x10^6$ lymph node)	Mean ^{3}H-TdR uptake allogeneic / Mean ^{3}H-TdR uptake syngeneic
C57BL/6 → F_1	C57BL/6	1.2
C57BL/6 → F_1	BALB/c	1.3
C57BL/6 → F_1	C3H/He	7.1
A → F_1	A	1.0
A → F_1	F_1	0.9
A → F_1	ATH	6.1

with TLI 1600 rad and WBI 400 rad. Percent donor type cells in the spleen was >90% by microcytotoxicity assays. Allogeneic chimeras carried A skin allografts for an observation period of >100 days. No signs of overt GVHD were apparent in chimeras throughout all experiments.

Tolerance to alloantigens in chimeras. Tolerance to host-type alloantigens was documented in vivo by the lack of GVHD-associated mortality. In vitro studies using MLR assays indicated that donor-type cells present in chimeras were tolerant of host alloantigens (Table 1).

Unresponsiveness to host-type alloantigens was further demonstrated by in vivo experiments using adoptive transfer systems (Table 2). Specific unresponsiveness was documented by lack of GVHD alloreactivity of spleen cells obtained from tolerant chimeras against sublethally irradiated naive recipients of the host-type (Table 2).

Role of T cells and T cell subsets in suppression of alloreactivity in vitro. In order to test whether suppression of MLR in vitro was mediated by LyT-2^+ suppressor cells spleen cells obtained from stable A→F chimeras were sorted for LyT-2^+ cells using FACS. As shown in Table 3, spleen cells obtained from tolerant chimeras were capable of blocking in vitro reactivity of donor reacting against host-type cells. Elimination of LyT-2^+ cells did not result in abrogation of MLR suppressive capacity. Likewise, addition of LyT-2^+ cells into MLR mixtures did not result in any suppressive activity. Data shown in Table 3 suggest that the specific suppressor cells present in the spleen of tolerant chimeras are not classical LyT-2^+ suppressor cells nor Thy 1.2^+ cells.

The role of LyT-2^+ cells in the maintenance of unresponsiveness to host-type alloantigens by in vivo experiments. Spleen cells obtained from stable C57BL/6→F_1 chimeras were stained for LyT-2 and

Table 2. Documentation of specific unresponsiveness by adoptive transfer experiments by lack of alloreactivity in GVHD of spleen cells obtained from C57BL/6→BALB/c chimeras

Adoptive recipients prepared with WBI 550 rads	Cell transfer $25x10^6$ spleen cells	No. of mice	Survival in days
BALB/c	C57BL/6	12	9-13
C3H/He	C57BL/6	12	8-14
BALB/c	C57BL/6→BALB/c	12	>180
C3H/He	C57BL/6→BALB/c	6	12-16

Table 3. The role of Thy 1.2 and LyT-2 cells obtained from A→F_1 chimeras in the suppression of MLR*

Design of experimental system	Co-culture cells	% suppression
Suppression of GVH reactivity	Unseparated chimera spleen cells	30
Suppression of graft vs. unrelated host reactivity	Unseparated chimera spleen cells	4
Suppression of GVH reactivity	LyT-2 depleted chimera spleen cells	52
Suppression of GVH reactivity	$LyT-2^+$ cells from chimera spleen	0
Suppression of GVH reactivity	Unseparated chimera spleen cells	60
Suppression of GVH reactivity	Thy 1.2 depleted chimera spleen cells	82

* Spleen cells obtained from well established chimeras were cultured in one way MLR and results are expressed as percentages of ^{3}H-TdR as compared to that obtained when normal donor-type spleen cells were co-cultured under similar experimental conditions

sorted by FACS in order to test their role in the maintenance of unresponsiveness to host-type alloantigens. As shown in Table 4, although normal C57BL/6 spleen cells were capable of killing sublethally treated BALB/c recipients, equal numbers of spleen cells obtained from tolerant chimeras (also consisting of C57BL/6 type cells) were incapable of provoking lethal GVHD in naive recipients. Injection of LyT-2-depleted spleen cells obtained from tolerant chimeras failed to induce GVHD in adoptive recipients. Likewise, addition of $1x10^6$ $LyT-2^+$ cells to $5x10^6$ normal C57BL/6 spleen cells failed to abrogate

Table 4. The role of LyT-2 cells in the maintenance of unresponsiveness to host-type alloantigens

Spleen cells used for GVHD	Median survival-BALB/c recipients
$5x10^6$ normal C57BL/6	9 days
$5x10^6$ C57BL/6 F_1→chimeras	>60 days
$5x10^6$ $LyT\text{-}2^-$ C57BL/6→F_1 chimeras	>60 days
$5x10^6$ normal C57BL/6 + $1x10^6$ $LyT\text{-}2^+$ obtained from C57BL/6→F_1 chimeras	12 days

lethal GVHD. The data suggest that LyT-2 cells do not seem to play a major role in the maintenance of unresponsiveness to BALB/c alloantigens in stable C57BL/6→F_1 chimeras.

DISCUSSION

Although the induction of transplantation tolerance in adult recipients of semiallogeneic and allogeneic BM cells following conditioning with TLI was previously reported by various investigators, the details of the mechanisms involved in induction and maintenance of the specific unresponsiveness in chimeras is not well understood. Suppressor cells of MLR were previously demonstrated in mice following TLI as well as in chimeras prepared by prior conditioning with TLI (5-8). Since tolerance acquired after prior TLI conditioning is antigen specific it was previously assumed that it is mediated by suppressor T cells (9). However, subsequent studies indicated that TLI-induced suppressor cells do not bear known T cell surface markers and, in fact, they are probably not lymphoid cells at all (10, 5). These suppressor cells appear to be large immature cells abundant in the spleen of TLI-treated recipients for up to 6 weeks following radiotherapy. Our present report suggests that the specific suppressor cells observed in stable chimeras prepared with TLI are not classical LyT-2 suppressor cells and most probably do not bear the Thy 1.2 cell surface phenotype. This conclusion is based on our failure to abrogate MLR suppressive capacity by elimination of $LyT\text{-}2^+$ cells using FACS. Likewise, recovery of alloreactivity in GVHD assays in vivo could not be restored by depletion of tolerant spleen cells of LyT-2 cells, again suggesting that the maintenance of unresponsiveness against host-type cells is probably not maintained by classical $LyT\text{-}2^+$ suppressor cells. In addition, $LyT\text{-}2^+$ cells sorted from the spleens of stable chimeras fail to suppress GVHD capacity of normal spleen cells, again pointing to the same conclusion. It appears therefore, that specific unresponsiveness to host-type cells is mediated by operational clonal deletion rather than by active suppression of alloreactive clones present in chimeras. TLI-induced suppressor cells may nevertheless play an important role in the induction of the state of tolerance by blocking alloreactive clones thus resulting in functional clonal deletion as may be suggested by some of our ongoing studies. Though functional clonal deletion seems to be the final common pathway of the mechanism operating towards manipulation of the regenerating immune system towards specific unresponsiveness in chimeras as also suggested by limiting dilution assays (Morecki, et al, submitted), it cannot be ruled out that suppressor cells present in chimeras do play a role in maintaining the functional clonal deletion by inhibiting expansion of potentially alloreactive clones. Nonetheless, if such a mechanism operates, our study suggests that it may not be mediated by classical LyT-2 suppressor cells. Be the mechanism as it may, the fact that tolerance to alloantigens can be induced by introducing BM cells derived from an adult donor suggests that tolerance to a new host may be reinduced even in adult recipients - a conclusion that may be relevant towards establishing transplantation tolerance to BM and organ allografts in man.

ACKNOWLEDGMENTS

This work was supported by NIH grants 15387, 3030013; Israel's Ministry of Health, Chief Scientist's Office; The Israel Academy of Sciences and Humanities; Lady Tata Memorial Trust.

REFERENCES

1. S. Slavin, S. Strober, Z. Fuks and H.S. Kaplan, Long term survival of skin allografts in mice given fractionated total lymphoid irradiation, Science 193:1252 (1976).
2. S. Slavin, S. Strober, Z. Fuks and H.S. Kaplan, Induction of specific tissue transplantation tolerance using fractionated total lymphoid irradiation in adult mice: long-term survival of allogeneic bone marrow and skin grafts, J. Exp. Med. 146:34 (1977).
3. S Slavin, Z. Fuks, H.S. Kaplan and S. Strober, Transplantation of allogeneic bone marrow without graft vs. host disease using total lymphoid irradiation, J. Exp. Med. 147:963 (1978).
4. I. Zan-Bar, S. Slavin and S. Strober, Induction and mechanism of tolerance to bovine serum albumin after total lymphoid irradiation, J. Immunol. 124:1400 (1978).
5. M. Weigensbergm S. Morecki, L. Weiss, Z. Fuks and S. Slavin, Suppression of cell mediated immune responses following total lymphoid irradiation. I Characterization of suppressor cell of the mixed lymphocyte reaction, J. Immunol. 132:971 (1984).
6. S. Slavin and S. Strober, Induction of allograft tolerance after total lymphoid irradiation: Development of suppressor cells of the mixed leucocyte reaction, J. Immunol. 123:942 (1979).
7. S. Strober, S. Slavin, M. Gottlieb, et al, Allograft tolerance after total lymphoid irradiation, Immunol. Rev. 46:87 (1979).

8. S. Strober, S. Okada and A. Oseroff, Role of natural suppressor cells in allograft tolerance, Fed. Proc. 43:263 (1984).
9. D.P. King, S. Strober and H.S. Kaplan, Suppression of the mixed leucocyte response and of graft vs. host disease by spleen cells following total lymphoid irradiation (TLI), J. Immunol. 126: 1140 (1980).
10. S. Morecki, M. Weigensberg and S. Slavin, Suppression of cell-mediated immune responses following total lymphoid irradiation (TLI). II Lectin separation of nonlymphoid suppressor cells induced by TLI. Eur. J. Immunol. in press.

THE ROLE OF MEMORY SUPPRESSOR T CELLS IN SELF TOLERANCE: INDUCTION *IN UTERO* AND IN ATHYMIC MICE

Basten, A., Gibson, J., Loblay, R.H., Wong, K.L. and Fazekas de St Groth, B.

Clinical Immunology Research Centre, University of Sydney
Sydney, 2006 Australia

INTRODUCTION

Exposure of mice in adult life to foreign antigens such as human IgG (HGG) in tolerogenic form has been shown to result in generation of long-lived memory bearing suppressor T cells (Ts) as well as short-lived effector Ts (Loblay et al., 1983). In contrast to effector Ts, memory Ts are functionally silent until reactivated by secondary challenge of tolerant hosts with antigen which results in the rapid appearance of another wave of effector cells capable of preemptively inhibiting antibody production. A detailed study of these two suppressor cell populations has revealed that a key role can be attributed to memory (but not effector) Ts in maintenance of tolerance to HGG. Furthermore, when foetal mice are exposed to HGG via the placenta so that the antigen is handled by the developing immune system as if it were self, postnatal tolerance ensues and is associated with the parallel development of memory Ts (Fazekas de St Groth et al., 1984). Thus memory Ts may be important in mediating tolerance to self as well as nonself. This hypothesis is supported by the recent demonstration of Ts with specificity for self antigens such as C5 (Harris et al., 1982), myelin basic protein (Ada et al., 1977) and autologous erythrocytes (Cooke et al., 1978). On the other hand it remains unclear whether suppression subserves a failsafe function or is directly involved in induction of self tolerance. The aim of the current experiments was to seek evidence in favour of the latter possibility by testing whether (a) memory Ts were preferentially generated in athymic mice i.e. early in ontogeny and (b) elimination of Ts during embryonic life resulted in abrogation of self tolerance.

Induction of HGG Specific Ts In Tolerant Athymic Mice

The possibility that functional Ts may arise early in ontogeny was tested in nude mice using the HGG system described previously (Loblay et al., 1983; Fazekas de St Groth et al., 1984). Adult nude mice on a BALB/cJ background were given a single intravenous injection of 3 mg deaggregated (d)HGG. As expected this resulted in complete tolerance on subsequent challenge with HGG in immunogenic form (antibodies measured by a sensitive haemagglutination technique). Suppression was assayed by transferring spleen cells from nu/nu murine donors boosted with HGG into irradiated recipients together with HGG primed (Th) cells, FITC-KLH primed (B) cells, FITC-HGG and a non crossreacting antigen, horse red cells (HRC). Plaque forming cells (PFC) were then measured 7 days later. As shown in Figure 1, significant antigen specific suppression was observed with tolerant nu/nu spleen cells which was comparable to that obtained with cells from normal donors given a single dose of dHGG. The cells responsible for suppression carried the phenotype $Ly\text{-}2^+$, $Ia(J)^+$ (data not shown) and presumably represent the progeny of memory Ts generated by previous exposure to antigen. By contrast, minimal help was induced under the appropriate experimental conditions which indicates that antigenic exposure during ontogeny leads to activation of Ts in preference to other lymphocyte subsets. The previous demonstration of anti-idiotype specific Ts (Kim and Hopkins, 1978) and nonspecific Ts (Smith and Elson, 1976) in nude mice provides further evidence in favour of this concept.

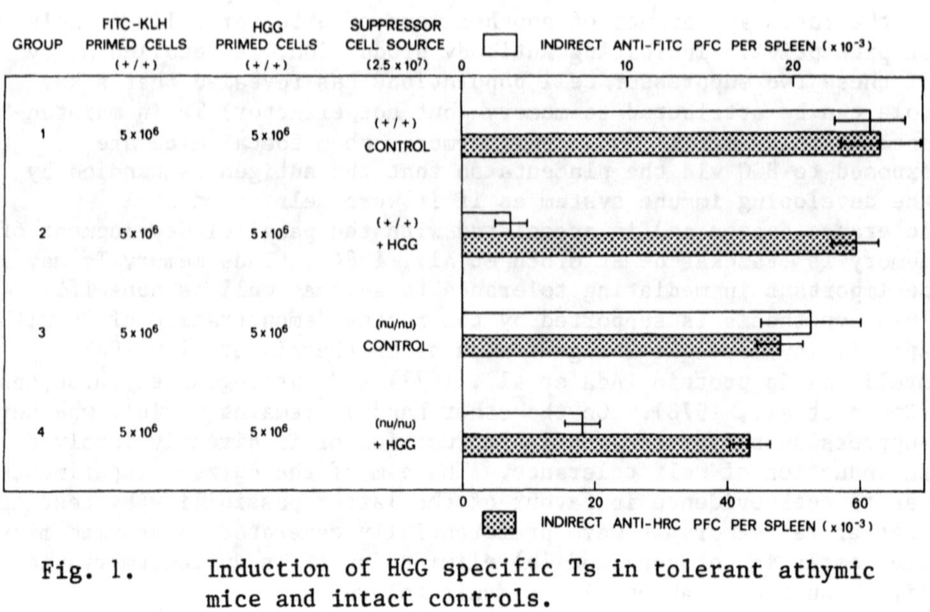

Fig. 1. Induction of HGG specific Ts in tolerant athymic mice and intact controls.

Model for Elimination of Ts During Embryonic Life

The data presented in the previous section and published previously (Fazekas de St Groth et al., 1984) point to the preferential induction of memory Ts by exposure of the developing immune system to antigen, but do not establish whether they are important in mediating self tolerance. This possibility was examined by setting up a model in which Ts were depleted during embryonic life. The model was based on the fact that Ts carry the serologically and genetically defined I-J determinant (Hayes et al., 1984) and can therefore be selectively removed by antibodies with anti-I-J specificity (Fig. 2). B10.A(3R)$(I\text{-}J)^b$ females were hyperimmunised with 6 injections of 10^7 cells from B10.A(5R) $(I\text{-}J^k)$ female donors at weekly intervals in order to generate circulating IgG anti-$I\text{-}J^k$ antibodies. Immediately after the final injection the B10.A(3R) females were mated with CBA $(I\text{-}J^k)$ males, thus exposing the F1 progeny to anti-$I\text{-}J^k$ antibodies *in utero* and during lactation. The efficacy of this immunisation regimen was confirmed by showing that maternal serum contained sufficient anti-I-J activity to eliminate $I\text{-}J^+$ Ts in our standard assay for suppression (Loblay et al., 1983). Between six and eight weeks of age F1 offspring born to anti-I-J antibody positive mothers and controls comprising normal F1 mice were assayed for: [a] antenatally induced tolerance to HGG and the presence of HGG specific Ts; and [b] their susceptibility to production of autoantibodies to mouse erythrocytes.

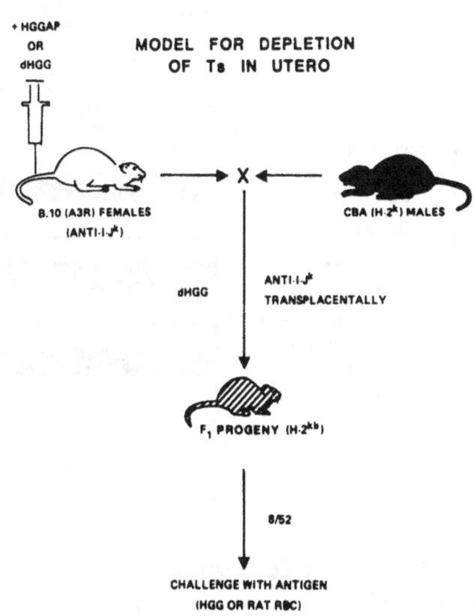

Fig. 2. Experimental model for depletion of Ts *in utero*.

Effect of Depletion of I-J^+ Ts In Utero On Antenatal Tolerance and Suppression To HGG

As mentioned previously exposure of mice _in utero_ to HGG results in induction of tolerance postnatally and generation of specific memory Ts (Fazekas de St Groth et al., 1984). To test whether Ts might be important in mediating this form of 'self' tolerance, anti-I-J antibody positive female B10.A(3R) mice or normal controls were crossed with CBA (H-2^k) males and injected on day 10 of gestation with 5 mg dHGG, thereby exposing the F1 progeny to HGG as well as anti-I-J^k antibodies _in utero_ (Fig. 2). When assayed in adult life, both tolerance and primary and secondary (memory) suppression to HGG (measured as described in 2) were abrogated (Fig. 3). Thus when a foreign antigen like HGG is handled as if it were self, induction of tolerance appears to depend on the presence of I-J^+ Ts early in ontogeny.

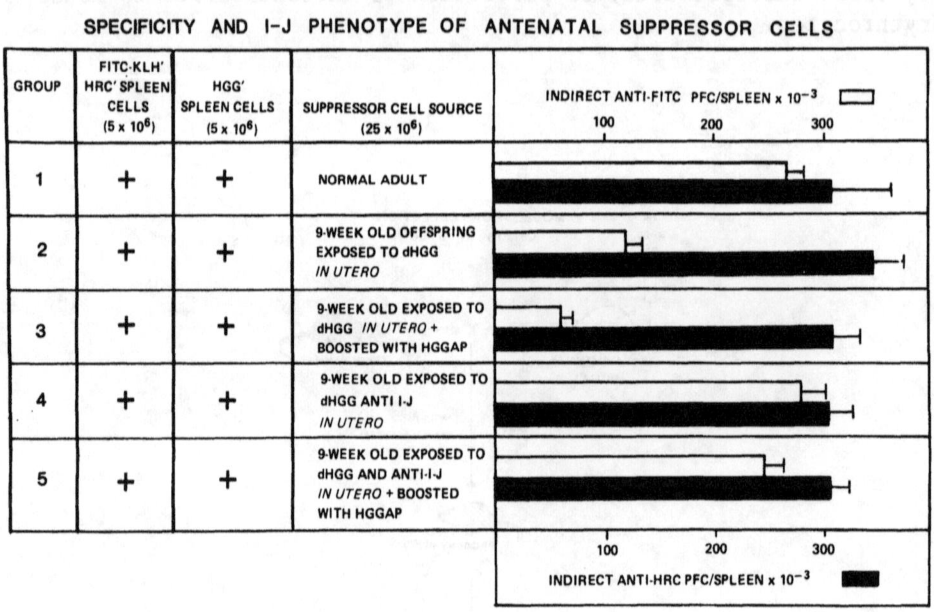

Fig. 3. Effect of depletion of I-J^+ Ts in utero on antenatal tolerance and suppression to HGG

Effect of Depletion of I-J^+ Ts In Utero on Anti-Red Cell Autoantibody Production

The same system was then applied to the issue of whether I-J^+ memory Ts play a role in tolerance induction to defined self antigens. For this purpose, use was made of the well described rat erythrocyte model of autoimmune haemolytic anaemia (Cooke, A. et al., 1978) which allows the simultaneous measurement of antibodies to self (mouse erythrocytes) and nonself (rat erythrocytes). Antibody responses to mouse RBC and rat RBC were measured by the direct anti-globulin test (DAGT) and passive haemagglutination respectively. The F1 progeny of anti-I-J^k antibody positive mothers (Fig. 2) and controls were given a single injection of 2 x 10^8 rat RBC six weeks postnatally and antibody production to autologous and rat RBC measured at intervals thereafter. A highly significant autoantibody (DAGT) response occurred in the group exposed to anti-I-J^k _in utero_ whereas that of the controls was negligible (Fig. 4).

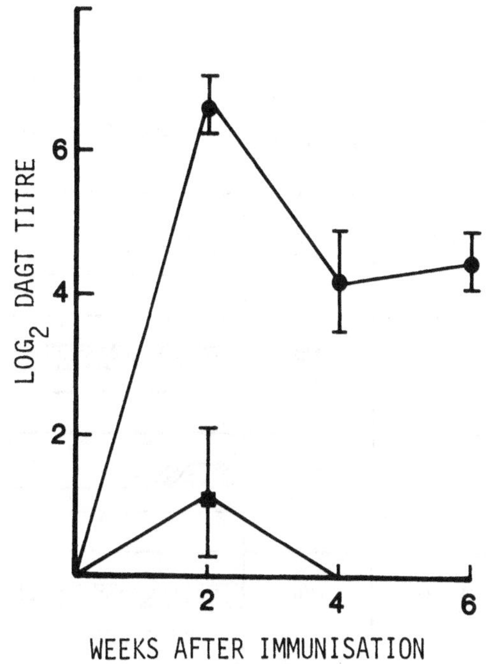

Fig. 4. Effect of depletion of I-J^+ Ts in utero on antibody production to mouse erythrocytes.

By contrast when antibody production to rat RBC was assayed in parallel in another experiment, the responses were comparable in test and control groups (Fig. 5). The selectivity of I-J$^+$ Ts depletion _in utero_ on the anti-self response is important for two reasons: first it excludes the remote possibility that the enhanced autoantibody response observed in anti-I-J antibody exposed offspring was due to a nonspecific effect on Ly-1$^+$I-J$^+$ helper/inducer cells rather than to elimination of self reactive memory Ts. Secondly, it suggests that self reactive Ts act early during ontogeny. If the enhanced autoantibody response had been caused by residual circulating anti-I-J antibodies at the time of immunization a higher anti-rat response would have been expected in the test group, since both self and non-self specific Ts should have been equally depleted. This possibility was in fact formally excluded by the failure of excess anti-I-J^k antibodies given postnatally to influence the autoantibody response (data not shown).

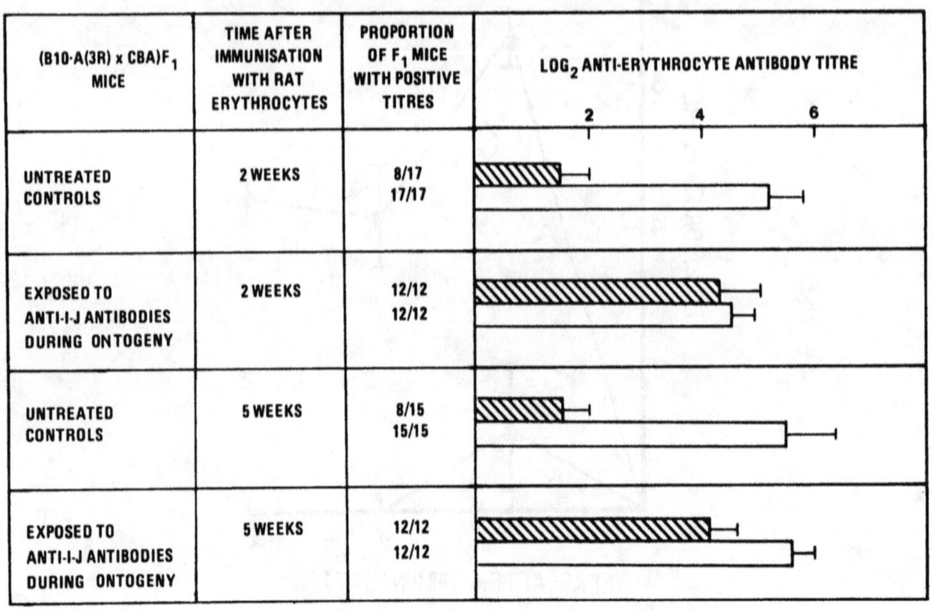

Fig. 5. Effect of depletion of I-J$^+$ Ts in utero on antibody production to mouse (self) and rat (non self) erythrocytes.

Effect of Depletion of I-J^{+} Ts In Utero On Frequency Of Anti-Red Cell Antibody Producing Hybrids

The findings to date are consistent with a role for Ts in induction of self tolerance. If true it can be predicted that the loss of self tolerance following depletion of Ts *in utero* should be accompanied by the appearance of B cells secreting anti- mouse RBC autoantibodies. This was formally tested by carrying out a series of fusions (Kohler and Milstein, 1975) between P3-NS1-1-Ag4-1 and rat RBC immunized spleen cells from normal F1 donors or F1 progeny exposed to anti-I-J^{k} antibody *in utero*, and then quantitating the proportion of hybrids secreting antibodies to rat RBC, mouse RBC or both (i.e. to crossreactive determinants) by cellular RIA. When donor mice were primed for suppression (to self) by two injections of rat RBC plus two subsequent boosters eight weeks later the great majority of hybrids produced antibodies to rat (non self) RBC as expected (Table 1). When they were primed for help (to self) by five weekly injections, antibodies from most hybrids recognised cross reacting determinants on both mouse and rat RBC. In neither situation were any hybrids with true self specificity obtained. By contrast, the majority of hybrids derived from spleen cells from donors exposed to anti-I-J^{k} *in utero* and primed for help (to self) produced antibodies which reacted exclusively with mouse (self) RBC.

Table 1. Frequency of Anti-Erythrocyte Antibody Producing Hybrids From Spleens of Normal and Ts Depleted Donors

Pre-Treatment Of Donors	Immunisation Schedule	Percentage of Clones * With Specificity For: Rat RBC	Mouse RBC	Both (X reactive)
Nil	Primed for Suppression	92	0	8
Nil	Primed for Help	11	0	89
Anti-I-J^{k} *in utero*	Primed for Help	20	75	5

* 40-55 per cent hybrids secreted anti-RBC antibody

DISCUSSION AND CONCLUSIONS

Memory Ts have been shown to play a key role in maintenance of tolerance to foreign antigen in adult life (Loblay et al., 1983). The recent finding of Ts with specificity for self antigens (Harris et al., 1982; Ada et al., 1977; Cooke et al., 1978) has raised the possibility that suppression may be important in self tolerance as well. On the other hand, it remains unclear whether Ts merely subserve a failsafe function or are ipso facto the principal mediators of self tolerance. For the latter to be true, two essential prerequisites need to be met. First, the appearance of long-lived memory Ts should precede that of other lymphocyte subsets in ontogeny; secondly, elimination of Ts during embryonic life should result in abrogation of self tolerance. The experiments described here provide evidence in favour of both these prerequisites. Thus memory Ts to HGG were readily demonstrable in nude mice (Fig. 1) which suggests that they may normally arise very early in development of the immune system, even before the thymus is fully developed. Furthermore, depletion of Ts _in utero_ by exposure to maternal anti-I-J antibodies led to abrogation of tolerance as well as suppression to HGG when injected during pregnancy (Fig. 3) and reversed tolerance to the defined self antigen, mouse erythrocytes (Figs 4,5). Similar results have been obtained by us in the case of a second autoantigen to DNA, and by others in models of neonatal tolerance to MHC antigens following exposure to anti-I-J antibody at birth (Nossal, 1983). Indirect support for the second prerequisite also carries from the demonstration of potent nonspecific Ts in the lymphoid tissues of neonates (Mosier and Johnson, 1975).

Taken together these findings are consistent with a role for Ts in induction as well as maintenance of self tolerance. However, this concept needs to be reconciled with the finding that immature B cells can be rendered tolerant on exposure _in vitro_ to physiological concentrations of antigen without overt participation of T cells (Pike and Nossal, 1979). Since I-J$^+$ Ts are generated early in ontogeny (Fig. 1), they could be responsible in the _in vivo_ setting for suppression of developing B cells. If the latter are capable of storing a negative signal they would appear to be 'anergic' on subsequent testing _in vitro_. The finding here of a dramatic increase in mouse RBC (self) reactive B cell hybrids following elimination of Ts _in utero_ (Table 1) provides evidence in support of this concept. Furthermore, the capacity of Ts to bind antigen directly in significant amounts (Basten et al., 1977) means that they could serve as an efficient mechanism for focusing small amounts of self antigens onto B cell precursors at a time when their surface receptors are first expressed.

ACKNOWLEDGEMENTS

This work was supported by a grant from the National Health and Medical Research Council of Australia. KLW and JG were in receipt of Scholarships from ADAB and NH&MRC respectively.

REFERENCES

Ada, D.H., Bernard, E.P., Depieds, R. (1977) Evidence for suppressor T cells in Lewis rats' experimental allergic encephalomyelitis. Eur. J. Immunol., 7:620 .

Basten, A. Loblay, R.H., Chia, E., Callard, R. and Pritchard-Briscoe, H. (1977) Suppressor T cells in tolerance to self and non self. Cold Spring Harb. Symp. Quant. Biol., 41:93.

Cooke, A., Hutchings, P.R. and Playfair, J.H.L. (1978) Suppressor T cells in experimental autoimmune haemolytic anaemia. Nature, 273:154.

Fazekas de St Groth, B., Basten, A. and Loblay, R.H. (1984) Induction of memory and effector suppressor T cells by perinatal exposure to antigen. Eur. J. Immunol., in press.

Harris, D.B., Cairns, L., Rosen, F.S. and Borel, Y.(1982) A natural model of immunologic tolerance: tolerance to murine C5 is mediated by T cells and antigen is required to maintain unresponsiveness. J. exp. Med. 156:567.

Hayes, C.E., Klyczek, K.K., Krum, D.P., Whitcomb, R.M., Hullett, D.A., and Cantor, H. (1984) Chromosome 4 Jt gene controls murine T cell surface I-J expression. Science, 223:559.

Kim, B.S. and Hopkins, W.J. (1978) Tolerance by neonatal treatment with anti-idiotypic antibodies: induction and maintenance in athymic mice. Cell Immunol., 35:460.

Kohler, G. and Milstein, C. Continuous cultures of fused cells secreting antibody of predefined specificity. (1975) Nature, 256:495.

Loblay, R.H., Pritchard-Briscoe, H., Fazekas de St Groth, B. and Basten, A. (1983) Suppressor T cell memory. II. The role of memory suppressor T cells in tolerance to human gammaglobulin. J. exp. Med., 157:957.

Mosier, D.E. and Johnson, B.M. (1975) Origin of mouse lymphocyte function II. Development of the ability to produce antibody is modulated by T lymphocytes. J. exp. Med. 141:216.

Nossal, G.J.V. (1983) Cellular mechanisms of immunological tolerance. Ann. Rev. Immunol. 1:33.

Pike, B.L. and Nossal, G.J.V. (1979) Mechanisms of clonal abortion tolerogenesis. III. Antigen abrogates functional maturation of surface Ig-negative adult bone marrow lymphocytes. Eur. J. Immunol. 9:708.

Smith, J.B. and Eison, G.J. (1976) Suppressor cells in spleens from 'nude' mice: their effect on the mitogenic response of B lymphocytes. J. Immunol. 117:319.

THE LONG-LASTING STATE OF SPECIFIC NONRESPONSIVENESS INDUCED BY INTRAVENOUS IMMUNIZATION WITH ALLOANTIGENS IS DUE TO THE GENERATION OF RECIRCULATING SUPPRESSOR T CELLS

A.T.J. Bianchi, L.M. Hussaarts-Odijk, H. Bril, H. de Ruiter and R. Benner

Dept. of Cell Biology and Genetics, Erasmus University Rotterdam and Dept. of Immunology, Central Veterinary Institute, Lelystad, The Netherlands

SUMMARY

We investigated whether the long-lasting state of nonresponsiveness that is induced by intravenous immunization with alloantigens is mediated by suppressor T cells, or is caused by inactivation or deletion of the relevant alloreactive T cell clones. The data from parabiosis and thoracic duct drainage experiments suggest that the state of nonresponsiveness depends on recirculating non-proliferating Ts memory cells.

INTRODUCTION

Suppressor T cells that suppress delayed type hypersensitivity (DTH) against haptens (1), heterologous erythrocytes (2,3) and alloantigens (4,5) can be induced by iv immunization of mice with the relevant antigens. In contrast to the cascades of several Lyt subtypes of suppressor T cells (Ts) described for DTH against haptens (1), we have shown in previous studies (6,7) that suppression of both the afferent and the efferent limb of alloantigen-specific DTH depend on Lyt-1+2+ Ts cells. Transfer of the state of nonresponsiveness by such Ts cells isolated from spleen and lymph nodes of iv suppressed mice is effective during the first few days after the iv immunization only (6). In the suppressed mice themselves, on the other hand, the state of nonresponsiveness is long-lasting (4,6). Similar results have been described in studies dealing with suppression of DTH against other antigens (2,8-10). The authors of these studies suggested that iv induced nonresponsiveness of DTH is based on clone inactivation as well as active suppression. The clone inactivation is supposed to be a long-lived phenomenon and non-transferable, while transferable Ts cells only occur during a short period after their induction (8-10).

The present paper describes experiments aimed to investigate the nature of the long-lasting state of nonresponsiveness to alloantigens that is induced by the protocol originally developed by our laboratory (4).

MATERIALS AND METHODS

Mice

(C57BL/Rij x CBA/Rij)F1 (H-$2^{b/q}$) and BALB/c (H-2^{d}) mice were purchased from the Laboratory Animals Centre of the Erasmus University, Rotterdam, The Netherlands. DBA/2 (H-2^{d}) mice were purchased from the Radiobiological Institute TNO, Rijswijk, The Netherlands. The age of the responder mice varied between 10 and 24 weeks. Only female mice were used.

Immunization

The preparation of cell suspensions, cell counting and irradiation of spleen cells have been described in a previous paper (4).

Induction of nonresponsiveness to certain alloantigens was done by iv injection of 5 x 10^7 of the appropriate irradiated (20 Gy) allogeneic spleen cells. DTH was induced by subcutaneous (sc) immunization with 1 x 10^7 of the appropriate allogeneic spleen cells.

Parabiotic mice

Parabionts were joined unilaterally under Avertin (Merck-Schuchardt, Hohenbrunn bei München, F.R.G.) anaesthesia. The mice were united by suturing the peritoneum and abdominal wall muscles, the connective tissue along the thorax and the scapulae. The skin was closed with linen thread from behind the ears to the femora. Parabiotic mice were separated from each other under Avertin anaesthesia 3 weeks after joining. Again the skin was closed with linen thread.

Thoracic duct drainage

Thoracic duct fistulas were established according to techniques previously described (11). The surgery was done under Avertin (Merck-Schuchardt) anaesthesia. During drainage, the mice were infused with saline at a flow rate of 0.5 ml/hr. Thoracic duct lymphocytes (TDL) were collected in 10 ml of Dulbecco's balanced salt solution that was supplemented with 50 to 100 IU preservative-free heparin (Hoffmann-La Roche, Basel, Switzerland).

The mice were cannulated for 24 hr, and lymph was collected at 4°C. The yield of TDL after 24 hr drainage of BALB/c mice varied between 5 and 10 x 10^7; 72% of these cells were Thy-1.2 positive (12). The viability of the collected cells was usually 95%.

Vinblastine treatment

Vinblastine sulphate (Eli-Lilli and Co, Indianapolis, Ind. U.S.A.) was iv injected in a dose of 100 µg/recipient, dissolved in a volume of 0.2 ml sterile phosphate buffered balanced salt solution (PBS).

Assay for DTH

DTH reactions were determined by measuring the difference in thickness of the hind feet 24 hr after sc injection of 2 x 10^7 of the appropriate allogeneic spleen cells into the dorsum of the right hind leg. As a control for background DTH reactivity, naive responder mice were used, which only received the challenge dose. DTH responses were measured 24 hr after challenge. This was done using a foot thickness meter with an accuracy of 0.05 mm. The specific DTH response was calculated as the relative increase in foot thickness of the immune mice minus the relative increase in foot thickness of the control mice. The swelling of the control mice ranged between 12 and 20%.

RESULTS

Transfer of nonresponsiveness by parabiosis

The involvement of suppressor cells in the long-lasting state of nonresponsiveness was studied by parabiosis of naive mice and nonresponsive mice, which had been iv immunized with alloantigens several weeks earlier. After joining, an equilibrium of potentially circulating cells between both partners can be expected within three weeks of parabiosis (13). If the long-lasting state of nonresponsiveness would be due to putative suppressor cells, these cells can be supposed to equilibrate as well.

(C57BL x CBA)F1 mice, that had been iv immunized four weeks earlier with H-2 and non-H-2 incompatible DBA/2 spleen cells, were joined for parabiosis with untreated (C57BL x CBA)F1 mice. As controls, nonresponsive and untreated mice were joined with nonresponsive and untreated mice, respectively. Three weeks after establishment of parabiosis, the parabionts were separated. Three weeks later all mice were sc immunized with DBA/2 spleen cells and tested for DTH reactivity another six days later. The results (Fig. 1) show that parabiosis of nonresponsive mice and untreated mice resulted in an equally suppressed DTH reactivity, while parabiosis of nonresponsive mice with each other did not affect this state of nonresponsiveness.

Transfer of nonresponsiveness by thoracic duct lymphocytes

Next we examined the occurrence of suppressor cells in the thoracic duct lymph of nonresponsive mice at various intervals after the iv suppressive injection. Therefore we used BALB/c responder mice which were iv immunized with H-2 compatible, but non-H-2 incompatible DBA/2 spleen cells. At various intervals after the induction of nonresponsiveness thoracic duct fistulas were made and TDL were collected. Thereafter, the TDL were transferred iv to naive BALB/c recipients which were sc immunized with DBA/2 spleen cells a few hours later. Six days later all mice were challenged with similar spleen cells and the subsequent DTH response was measured. As can be seen in Table 1, transfer of TDL caused a significant suppression of DTH reactivity till 80 days after iv immunization. This suppressive effect mediated by TDL could be abolished by treatment with anti-Thy-1.2 antibodies and complement *in vitro* (data not shown).

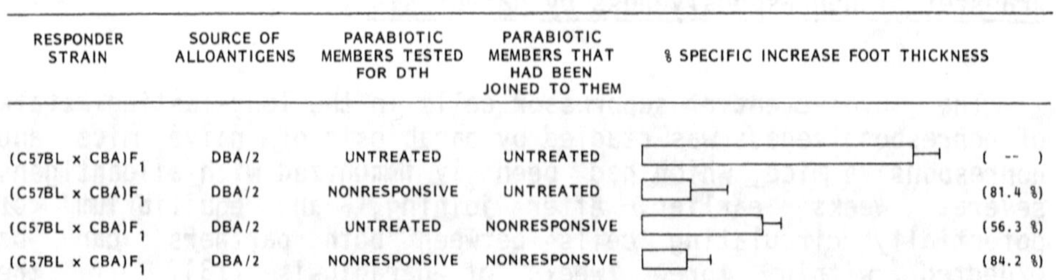

Fig. 1. Transfer of nonresponsiveness to naive mice by parabiosis. (C57BL x CBA)F1 mice were iv injected with 5×10^7 irradiated allogeneic DBA/2 spleen cells, rested for 4 weeks and joined for parabiosis with either similar responsive mice or untreated mice. A control group of untreated pairs was included. Three weeks later all pairs were separated and after another three weeks sc immunized with DBA/2 spleen cells in order to elicit a state of DTH. Six days later all mice were challenged and the subsequent DTH responses were measured after 24 hr. The horizontal bars represent the arithmetic mean $\pm$ 1 SEM (n = 6). The percentage suppression is given in parentheses.

Effect of an antimitotic agent during the early and late phase of the state of nonresponsiveness

(C57BL x CBA)F1 responder mice were iv immunized with 5 x 10^7 irradiated DBA/2 spleen cells. Two days and forty days after the iv immunization, one group of mice was iv injected with 100 μg vinblastine while a control group was iv injected with PBS. Control groups of mice that had not been iv immunized were also treated with vinblastine or PBS. Six days after vinblastine or PBS injection all groups of mice were sc immunized with DBA/2 spleen cells and tested for DTH reactivity another six days later. The results show that vinblastine abolished the state of nonresponsiveness, when it was given during the early phase of the state of nonresponsiveness (Fig. 2, first line). Injection of vinblastine 40 days after the induction of the state of nonresponsiveness had no effect (Fig 2, second line).

Table 1. Transfer of nonresponsiveness to naive mice by thoracic duct lymphocytes.

Interval after iv immunization (days)	Donors of TDL	Specific increase foot thickness (%)	Suppression
4	Nonresponsive	14.1 ± 2.3	50%
	Untreated	28.1 ± 3.8	
11	Nonresponsive	9.6 ± 3.5	68%
	Untreated	30.0 ± 3.2	
40	Nonresponsive	11.2 ± 4.5	54%
	Untreated	24.2 ± 3.9	
80	Nonresponsive	24.0 ± 4.7	47%
	Untreated	45.5 ± 4.2	

Naive BALB/c recipient mice were iv injected with 5 x 10^7 viable 'nonresponsive' or 'untreated' TDL. 'Nonresponsive' TDL were obtained from BALB/c mice which had been iv immunized with 5 x 10^7 irradiated DBA/2 spleen cells at various intervals before drainage. 'Untreated' TDL were obtained from naive BALB/c control mice. Within 2 hr after infusion of the TDL, the recipient mice were sc immunized with 1 x 10^7 DBA/2 spleen cells and challenged with similar spleen cells 6 days later. Figures represent mean DTH responses ± 1 SEM (n = 6).

DISCUSSION

In previous studies dealing with suppression of Host-versus-Graft (4,6,7) and Graft-versus-Host reactions (14), we have shown that suppression of DTH to alloantigens can be transferred by spleen and lymph node cells during the first week after the induction of suppression only. Although this Ts cell-dependent, transferable state of nonresponsiveness is short-lasting (6), the state of nonresponsiveness in the suppressed mice themselves lasts for more than 70 days (6).

Several authors describing suppression of DTH against haptens (8,9) and heterologous erythrocytes (2,10) explain the long-lasting nonresponsiveness by a deletion of the reactive T cell clones. On the other hand, this phenomenon can also be explained by postulating the existence of different types of Ts cells or, alternatively, that Ts cells in a different state of activation are involved in the early and long-lasting state of nonresponsiveness.

Our present study indicates that, in addition to early phase Ts cells that localize in spleen and lymph nodes, also long-lived recirculating Ts cells occur (Fig. 1 and Table 1). These long-lived, recirculating Ts cells probably account for the long-lasting state of nonresponsiveness to alloantigens induced by

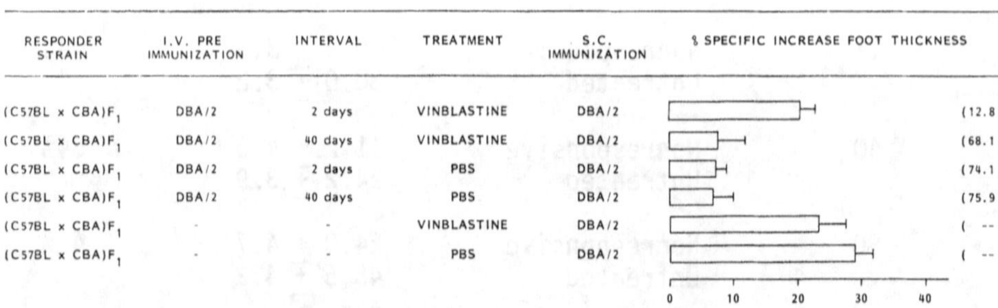

Fig. 2. Effect of vinblastine treatment during the early and late phase of the state of nonresponsiveness. (C57BL x CBA)F1 mice were iv injected with 5 x 10^7 irradiated allogeneic DBA/2 spleen cells. Two and 40 days later, groups of these mice were iv injected with vinblastine (100 µg in 0.2 ml PBS) or with PBS (0.2 ml) only. Control groups of untreated mice were also injected with vinblastine or PBS. Six days later all groups of mice were sc immunized with DBA/2 spleen cells and tested for DTH reactivity another six days later. The horizontal bars represent the arithmetic mean + 1 SEM (n = 6). The percentage suppression is given in parentheses.

iv immunization. However, the long-lasting occurrence of Ts cells does not exclude the simultaneous occurrence of clonal deletion of specific alloreactive T cells. Green et al. (15) have shown that Ts cells caused a deletion of the functional activity of two sets of T lymphocytes, namely the specific helper T cells for antibody formation and the specific helper T cells that had induced the Lyt-1+ Ts cells. This clonal deletion could even be demonstrated after that the Ts cells had been removed.

Our results are compatible with those of Polak and Rinck (16) and Liew and Howard (3). These investigators have shown that tolerance to dinitrobenzenesulphonic acid in guinea pigs and iv induced suppression to heterologous erythrocytes are transferable by means of parabiosis several weeks after induction of suppression. The latter authors (3) demonstrated that these putative Ts memory cells could be recalled by reexposure to the antigen in a dosage/route combination which is inappropriate for the induction of Ts cells in naive mice. Our experiments show that the long-lived Ts cells display the same characteristics as the well documented DTH reactive memory T cells which are involved in secondary type DTH reactions: both cell types are transferable by parabiosis (Fig. 1) (17) and TDL (Table 1) (18) and both are resistant to antimitotic drugs (Fig. 2) (17). That Ts memory cells really exist has been shown by Loblay et al. (19,20). They demonstrated secondary type suppressor T cell activity generated in a system in which suppression of the anti-DNP plaque-forming cell response was assessed. The secondary Ts cell response was characterized by an accelerated onset, a heightened peak and a prolonged duration as compared with primary suppression. We are currently investigating whether the long-lived Ts cells described here, also fulfill these criteria.

ACKNOWLEDGEMENTS

We gratefully acknowledge Mrs. M. Stout for typing the manuscript and the Dutch Kidney Foundation, Bussum, The Netherlands, for its financial support.

REFERENCES

1. M.I. Greene and B. Benacerraf, Studies on hapten specific T cell immunity and suppression. Immunol. Rev. 50:163 (1980).
2. I.A. Ramshaw, P.A. Bretscher, and C.R. Parish, Regulation of the immune response. I. Suppression of delayed-type-hypersensitivity by T cells from mice expressing humoral immunity. Eur. J. Immunol. 6:674 (1976).
3. F.Y. Liew and J.G. Howard, Regulation of delayed-type-hypersensitivity. V. Suppressor cell memory in antigen-specific suppression of delayed-type hypersensitivity to sheep

erythrocytes. Eur. J. Immunol. 10:937 (1980).
4. Th.H. van der Kwast, A.T.J. Bianchi, H. Bril, and R. Benner, Suppression of antigraft immunity by preimmunization. I. Kinetic aspects and specificity. Transplantation 31:79 (1981).
5. F.Y. Liew, Regulation of delayed-type hypersensitivity. VI. Antigen-specific suppressor T cells and suppressor factor for delayed-type hypersensitivity to histocompatibility antigens. Transplantation 33:69 (1982).
6. A.T.J. Bianchi, L.M. Hussaarts-Odijk, Th.H. van der Kwast, H. Bril, and R. Benner, Suppression of antigraft immunity by preimmunization. II. Characterization of the suppressor cells. Transplantation 37:490 (1984).
7. A.T.J. Bianchi, L.M. Hussaarts-Odijk, E.A.J. Wolters, and R. Benner, Suppression of antigraft immunity by preimmunization. III. Characterization of the suppressor T cells involved in suppression of the efferent phase of DTH against alloantigens. Submitted for publication.
8. M.H. Dietz, M.S. Sy, B. Benacerraf, A. Nisonoff, M.I. Greene, and R.N. Germain, Antigen- and receptor-driven regulatory mechanisms. VII. H-2 restricted anti-idiotypic suppressor factor from efferent suppressor T cells. J. Exp. Med. 153: 450 (1981).
9. J.W. Moorhead, Tolerance and contact sensitivity to DNFB in mice. VI. Inhibition of afferent sensitivity by suppressor T cells in adoptive tolerance. J. Immunol. 117:802 (1976).
10. S.H.E. Kaufmann, J.S. Ahmed, and H. Hahn. Transferable suppression and intrinsic unresponsiveness in delayed-type hypersensitivity to sheep red blood cells of mice: two distinct mechanisms? Immunobiol. 157:331 (1980).
11. J. Sprent, Circulating T and B lymphocytes in the mouse. I. Migration properties. Cell. Immunol. 7:10 (1973).
12. W. van Ewijk, P.L. van Soest, and G.J. van den Engh, Fluorescence analysis and anatomic distribution of mouse T lymphocyte subsets defined by monoclonal antibodies to the antigens Thy-1, Lyt-1, Lyt-2, and T-200. J. Immunol. 127: 2594 (1981).
13. R. Benner, A. van Oudenaren, and H. de Ruiter, Antibody formation in mouse bone marrow. VII. Evidence against the migration of plaque-forming cells as the underlying cause for bone marrow plaque-forming cell activity: a study with parabiotic mice. Cell. Immunol. 29:28 (1977).
14. H. Bril, B.D. Molendijk-Lok, L.M. Hussaarts-Odijk, and R. Benner, Characterization of suppressor T cells in graft-versus-host reactions. Immunobiol., in press.
15. D.R. Green, R.K. Gershon, and D.D. Eardley, Functional deletion of different Ly-1 T-cell-inducer subset activities by Ly-2 suppressor T lymphocytes. Proc. Natl. Acad. Sci. USA 78:3819 (1981).
16. L. Polak and C. Rinck, Persistent activity of suppressor

cells in T cell-mediated immune response. J. Immunol. 121: 762 (1978).

17. Th.H. van der Kwast, J.G. Olthof, H. de Ruiter, and R. Benner, Secondary delayed type hypersensitivity to sheep red blood cells in mice: dependence on long-lived memory cells. Cell. Immunol. 43:94 (1979).

18. A.T.J. Bianchi, H. de Ruiter, L.M. Hussaarts-Odijk, and R. Benner, Secondary delayed-type hypersensitivity to sheep red blood cells and minor histocompatibility antigens in mice: transfer of memory by recirculating thoracic duct lymphocytes. Immunobiol. 165:200 (1983).

19. R.H. Loblay, H. Pritchard-Briscoe, and A. Basten, Suppressor T-cell memory. Nature 272:620 (1978).

20. R.H. Loblay, B. Fazekas de St.Groth, H. Pritchard-Briscoe, and A. Basten, Suppressor T cell memory. II. The role of memory suppressor T cells in tolerance to human gamma globulin. J. Exp. Med. 157:957 (1983).

cells and cell-mediated immune response. J. Immunol. 121: [illegible] (1978).

17. Th.H. van der Kwast, J.G. Olthof, H. de Ruiter, and R. Benner, Secondary delayed type hypersensitivity to sheep red blood cells in mice: dependence on long-lived memory cells. Cell. Immunol. 45:84 (1979).

18. A.T.J. Bianchi, H. de Ruiter, L.M. Hussaarts-Odijk, and R. Benner, Secondary delayed type hypersensitivity to sheep red blood cells and minor histocompatibility antigens in mice: transfer of memory by recirculating thoracic duct lymphocytes. Immunobiol. 165:200 (1983).

19. R.H. Loblay, H. Pritchard-Briscoe, and A. Basten, Suppressor T-cell memory. Nature 272:620 (1978).

20. R.H. Loblay, B. Fazekas de St. Groth, H. Pritchard-Briscoe, and A. Basten, Suppressor T cell memory. II. The role of memory suppressor T cells in tolerance to human gamma globulin. J. Exp. Med. 157:957 (1983).

PATHOGENESIS OF THE INTESTINAL PHASE OF THE GRAFT-VERSUS-HOST REACTION IN F_1 HYBRID MICE

Allan McI. Mowat, Annette Borland & Delphine M.V.Parrott

Department of Bacteriology and Immunology

Western Infirmary, Glasgow G11 6NT, Scotland

INTRODUCTION

Graft-versus-host disease (GvHD) is the most important complication of allogeneic bone marrow transplantation (BMT) in humans and the associated intestinal damage is a major cause of morbidity and mortality in these patients[1]. If the intestinal consequences of GvHD are to be prevented or treated, it will be necessary to elucidate the pathogenic mechanisms which are responsible and to identify the nature of the alloantigenic stimuli. Although it has usually been assumed that cytotoxic T lymphocytes (CTL) cause the tissue damage during allograft rejection and in GvHD, recent evidence indicates that CTL may not be necessary for these phenomena[2,3].

In previous studies, we have shown that the intestinal phase of the graft-versus-host reaction (GvHR) in unirradiated F_1 hybrid mice is not associated with detectable CTL[4], but is characterised by increases in crypt cell production rate (CCPR), crypt length and by an increased number of intraepithelial lymphocytes (IEL). Using these indices, we have obtained indirect evidence that the mucosal pathology is caused by a local delayed-type hypersensitivity (DTH) reaction[5,6]. We have now examined this hypothesis directly by investigating the phenotype of the T cells responsible for the intestinal alterations and the region of the major histocompatibility complex (MHC) which stimulates them. In addition, we have examined whether tissue or bone-marrow derived cells provide the stimulus for the local GvHR.

MATERIALS AND METHODS

Mice: CBA/Ca, BALB/c and C57Bl/10 mice were obtained from departmental stocks while B10.A, B10.A3R, B10.A4R, B10.A5R, B10.AKM, B10.AQR, B10.BR, A.TH and A.TL mice were obtained from Olac Limited, (Bicester, England). F_1 hybrids were bred and maintained in the department.

Induction of Graft-versus-host reactions: Adult F_1 hybrid mice, aged 6-8 weeks, were injected with 6×10^7 spleen cells intraperitoneally, while neonatal mice were first used at 6-8 days old and injected with 2.5×10^7 parental spleen cells i.p. Control mice received medium alone.

Depletion of T cell subsets: Parental spleen cells were depleted of individual T cell subsets by treatment with monoclonal anti-Lyt 1.1 or anti-Lyt 2.1 (NEN Clones NEI 003 and 004) plus rabbit complement (C') (Low Tox M - Cedarlane). Cells were washed x3 before use.

Assessment of Intestinal GvHR: Intraepithelial lymphocyte counts were performed on H&E stained sections of jejunum as described previously[5,6] and expressed as IEL/100 epithelial cells. Measurements of CCPR, villus and crypt length were performed by a micro-dissection technique employing colchicine to block crypt cell mitosis in metaphase[5,6].

Preparation of Bone Marrow Chimeras: Adult mice were lethally irradiated with 900 rad and reconstituted immediately with an intravenous injection of 1.6×10^7 BM cells which had been treated previously with monoclonal anti-Thy 1.2+C'. 4 months later, these mice were completely chimeric for donor cells and a GvHR was induced as above.

Statistics: Groups of means + standard deviations were compared by Student's t test, while crypt cell production rates were compared by covariance analysis.

RESULTS

T cell subsets mediating the intestinal phase of GvHR. Neonatal (CBAxBALB/c)F_1 mice were injected with 2.5×10^7 CBA cells i.p. and sacrificed at the peak of the GvHR, 9 days later. As before[5,6] inoculation with intact parental spleen cells produced significant increases in IEL count (Fig.1) and in CCPR and crypt length (Fig.2), compared with controls. Depletion of Lyt 1^+ T cells abolished the increase in IEL count, while depletion of Lyt 2^+ T cells had no effect on this aspect of intestinal GvHR. Significantly, addition of Lyt 2^- T cells restored the ability of Lyt 1^- T cells to produce an increased IEL count. Thus Lyt 1+2- T cells are alone sufficient to induce increased lymphocytic infiltration of the epithelium in GvHR.

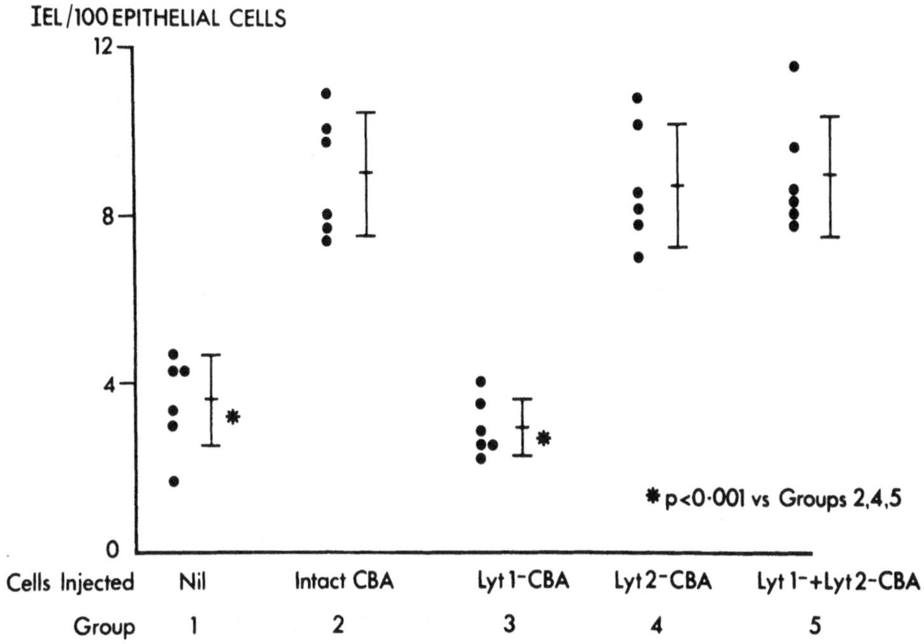

Fig. 1. Phenotype of T cells mediating intestinal GvHR. Intra-epithelial lymphocyte counts in jejunum of (CBAxBALB/c)F_1 mice 9 days after injection of 2.5×10^7 CBA spleen cells i.p. at 6-8 days of age. Parental spleen cells were depleted with either monoclocal anti Lyt 1.1 or anti-Lyt 2.1+C' or with C' alone. Results shown are mean + 1 standard deviation.

Depletion of Lyt 1^+ T cells also abolished the increases in CCPR and crypt length during the GvHR (Fig.2) but in this case, mice receiving Lyt 2^- cells also had alterations in CCPR and crypt length which were significantly less than those observed with intact cells. However, these parameters were still significantly greater than those in control mice. Addition of Lyt 2^- cells to Lyt 1^- cells did not restore the full changes in mucosal architecture and we interpret the results to show that Lyt 1^+2^- T cells alone can produce significant alterations in crypt cell turnover, but that these require to interact with Lyt 1^+2^+ T cells to induce the full picture of intestinal GvHR. Our findings are, therefore, similar to those obtained by Loveland and McKenzie studying allograft rejection by unprimed mouse T cells which were considered to reflect cooperation between T_{DTH} and their precursors[7]. No alterations in villous length were found

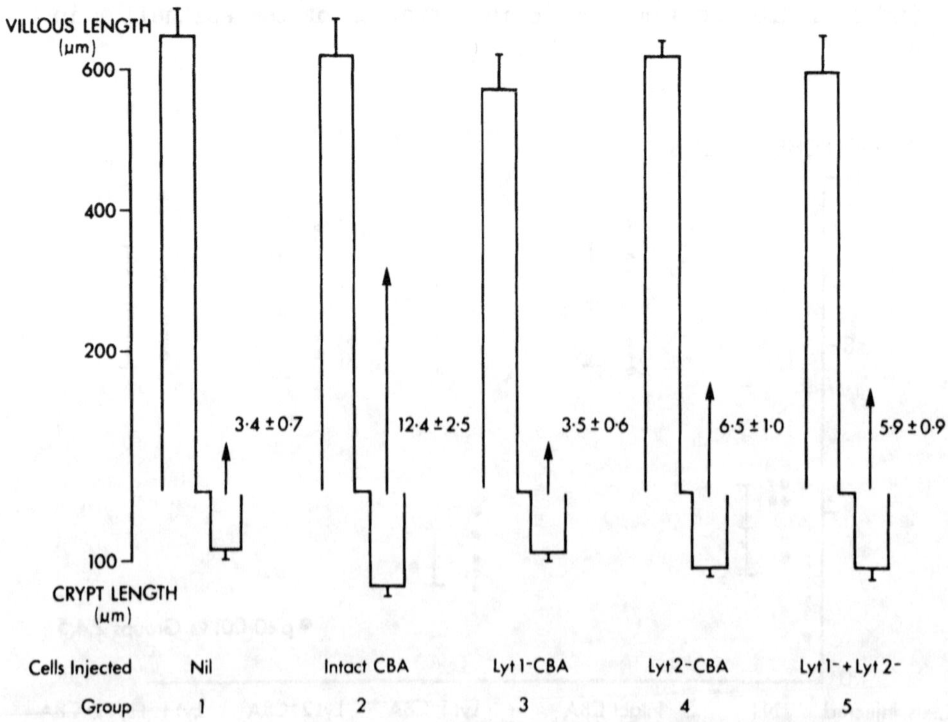

Fig. 2. Phenotype of T cells mediating intestinal GvHR. Villus length, crypt length and CCPR (arrows) during GvHR in neonatal (CBAxBALB/C)F_1 mice 9 days after injection of intact or Lyt-depleted CBA spleen cells i.p. Results shown are means $\pm$1standard deviation for 6 mice/group. Mice in Groups 2, 4 and 5 had significant increases in crypt length ($p<0.001$) and in CCPR ($p<0.02$) compared with controls (Group 1). Crypt lengths and CCPR were lower in Groups 4 and 5 than in recipients of intact cells ($p<0.001$ and <0.02 respectively).

<u>MHC alloantigens providing the stimuls for intestinal GvHR:</u> Adult F_1 hybrid mice were bred from parental strains differing at specified MHC loci and used as hosts for a GvHR induced by $6x10^7$ parental spleen cells. 9 days later, the mice were sacrificed and their jejunum examined for alterations in IEL count and mucosal architecture. Significant increases in IEL count (Fig.3) crypt length and CCPR (results not shown) were observed only when the GvHR was induced by the whole H-2 complex [B10⟶(B10xB10BR)], the I region [A.TH⟶(A.THxA.TL)] or by H-2K+I-A[B10⟶(B10xB10.A4R)]. These increases were not induced by H-2K or H-2D alone or by a difference at J-D: indeed, significant suppression of the mucosal indices was occasionally induced by these incompatibilities while mice with a GvHR across the I-J region always had significantly lower IEL counts, crypt lengths and CCPR than controls. Thus,

incompatibility at I-A is alone sufficient and necessary for the intestinal phase of GvHR in unirradiated F_1 mice, while Class I MHC and the I-J region may induce allosuppressor T cells which can also regulate intestinal cell renewal. Once again, villous atrophy was not observed in any GvHR mice.

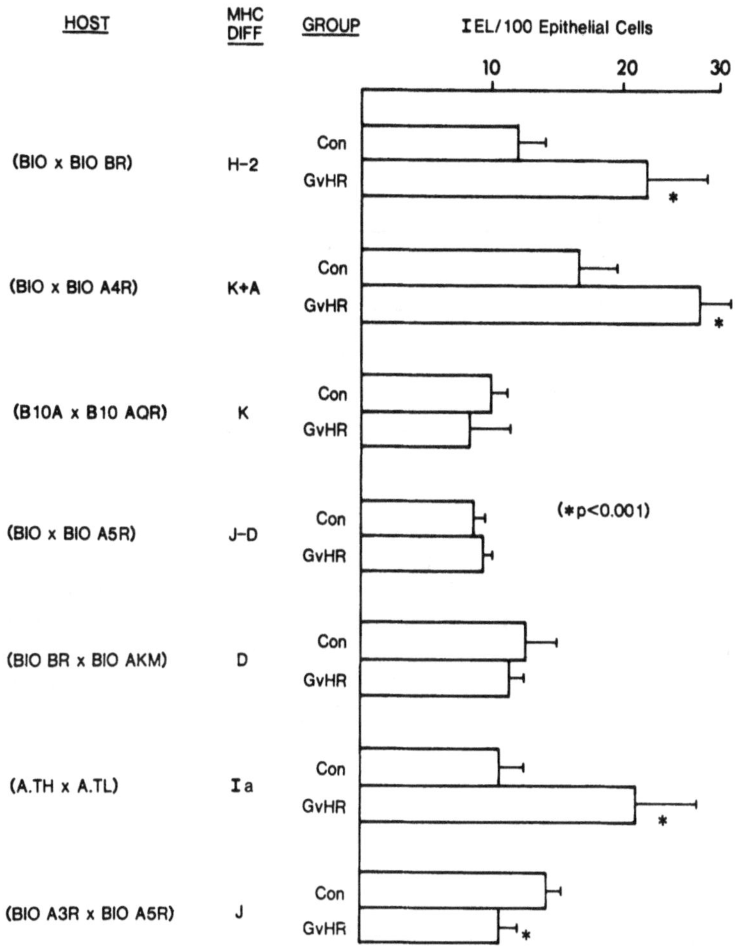

Fig. 3. MHC antigens stimulating intestinal GvHR. Intraepithelial lymphocyte counts in jejunum of adult F_1 hybrid mice 9 days after injection of $6x10^7$ parental spleen cells i.p. Mice received spleen cells from the first parent named in the F_1 hybrid and results shown are means + 1 standard deviation of 5-7 mice/group.

Role of bone marrow derived cells in intestinal GvHR: To examine the relative roles of bone marrow and tissue derived cells in producing the intestinal pathology in GvHR, a GvHR was induced by injecting $6x10^7$ CBA spleen cells into adult (CBAxBALB/c)F_1——>CBA or CBA——>F_1 bone marrow chimeras. 12 days later F_1——>CBA mice with GvHR had significant increases in IEL count, crypt length and

CCPR compared with $F_1 \longrightarrow$ CBA controls (Figs. 4 and 5). No such alterations were observed in CBA $\longrightarrow F_1$ mice after transfer of CBA cells (results not shown). Thus, bone marrow derived cells provide the stimulus for intestinal GvHR and the presence of alloantigens on intestinal cells is of no consequence.

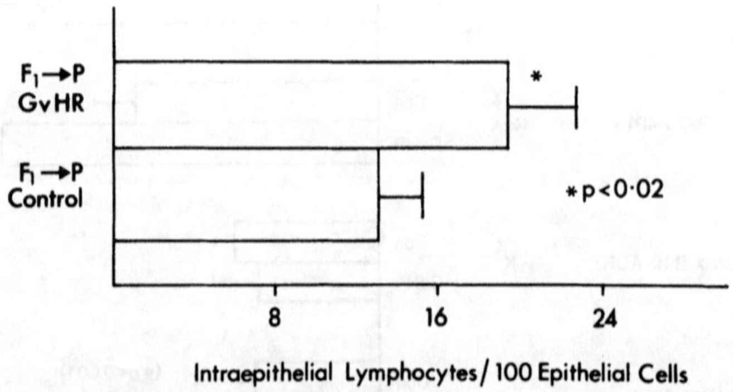

Fig. 4. Intestinal GvHR in (CBAxBALB/c)$F_1 \longrightarrow$CBA bone marrow chimeras. Intraepithelial lymphocyte counts in jejunum of $F_1 \longrightarrow$CBA mice 12 days after injection of 6×10^7 CBA spleen cells i.p. Chimeric mice were used 4 months after lethal irradiation and transfer of anti-Thy treated CBA-BM. Results shown are means $\pm$ 1 standard deviation for 5-6 mice/group.

CONCLUSIONS

This study indicates that the proliferative changes in intestinal lymphoid populations and in epithelial cell turnover which characterise the GvHR in unirradiated mice are due to recognition of bone marrow derived I-A alloantigens by Lyt 1^+ T cells. Thus, the findings support the hypothesis that the intestinal changes are due to soluble factors released by a local DTH response in the mucosa[5]. We propose that alloreactive donor T cells recirculate randomly through the intestinal mucosa where they are activated by

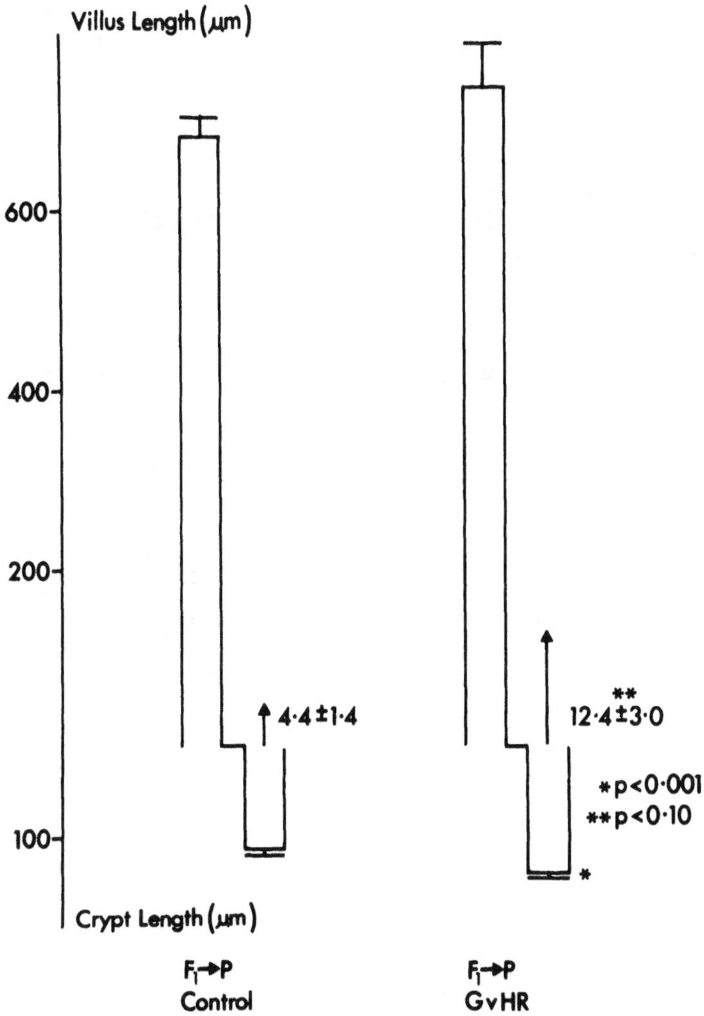

Fig. 5. Intestinal GvHR in (CBAxBALB/C)F_1——>CBA bone marrow chimeras. Villus lengths, crypt lengths and CCPR (arrows) in jejunum 12 days after injection of $6x10^7$ CBA spleen cells. Results are means + 1 standard deviation for 5-6 mice/group.

host bone marrow derived, I-A$^+$ cells and release enteropathic lymphokines which modify the structure and function of the small intestine. An additional possibility is that I-A+ non-lymphoid cells in Peyer's Patches[8] provide the stimulus for the activation of donor T cells which then recirculate specifically to the intestinal wall as lymphokine-producing effector T_{DTH}. These concepts are the subject of continuing study.

ACKNOWLEDGEMENTS

Supported by MRC Grant No: 8224559

Annette Borland was in receipt of a Vacational Scholarship from the Wellcome Trust.

REFERENCES

1. R. E. Slavin and G. W. Santos, Clin. Immunol. Immunopathol. 1:472 (1973).
2. D. W. Mason, Transplantation, 32:222 (1981).
3. M. R. Jadus & A. B. Peck, Transplantation, 36:281 (1983)
4. A. Borland, A. McI. Mowat and D. M. V. Parrott. Transplantation, 36:513 (1983).
5. A. McI. Mowat and A. Ferguson, Transplantation, 32:238 (1981).
6. A. McI. Mowat and A. Ferguson, Gastroenterology, 83:417 (1982)
7. B. E. Loveland and I. F. C. McKenzie, Transplantation, 33:407 (1982).
8. L. K. Richman, A. S. Graeff and W. S. Strober, Cell Immunol., 62:110 (1981).

GRAFT VERSUS HOST REACTION IN THE RAT: EFFECT OF CYCLOSPORIN A AND CYCLOPHOSPHAMIDE ON LYMPHOCYTE SUBSETS INVOLVED IN GVHR*

A.A. van den Broek, R.J.M. Stet, G.G. van Eden,
J. Kuiper, B. Meedendorp and C. Thomas

Department of Histology, State University of Groningen
Oostersingel 69/1, 9713 EZ Groningen, The Netherlands

Immunocompetent cells proved to be sensitive to both Cyclosporin A (CyA) and Cyclophosphamide (Cy). CyA would affect proliferating T-cells (1) or would interfere with the formation of receptors for interleukins on T-cells (2). Only B-lymphocytes, which react to the so called "TI2" antigens should be affected by this agent (3). In rats and rabbits we had found that precursors T-helper cells and both activated helper- and cytotoxic T-cells were sensitive to CyA. On B-cells and plasmacells no effect could be demonstrated. Histologically no sign of cell damage was found (4). Cy on the other hand affected immature plasmacells or cells involved in germinal centre reactions. With regard to T-cells only those T-cells, which are actively transforming into helper cells, were affected by Cy. No effect was obtained on other T-cell activity (5).

The experiments presented were designed to investigate the effect of a single injection of CyA or Cy on both regional and systemic Graft versus Host Reactions. Parental spleen cells were transfered to F_1 hybrid rats. Either parental or F_1 hybrid animals were treated with CyA or Cy. Lymphocyte subpopulations of the injected spleen cell suspensions were determined using monoclonal antibodies against different subsets of lymphocytes.

---

* This investigation was supported by a grant of the IRS Leiden. Cyclosporin A was provided by Sandoz, Basel and ER-2 by Dr. J. Rozing, Erasmus University, Rotterdam.
The skillfull help of Ms. A. Zandstra and Mr. P. van der Sijde is gratefully acknowledged.

MATERIALS AND METHODS

Highly inbred rats AO/G ($RT1^u$), BN/G ($RT1^n$) and their F_1 hybrids (AOxBN) were used. Regional Graft versus Host Reactivity (GvHR) was tested using a Popliteal Lymph Node Assay (6). Parental spleen cells (2×10^7 cells) were injected s.c. into the right footpad (R in figures). The left footpad (L in figures) served as control and was injected with medium only. Lymph node weights were determined 7 days later. A systemic GvHR was induced by i.v. injection of parental spleen cells (1×10^8 or 5×10^7 cells) into F_1 hybrids. This type of GvHR was monitored by assesement of clinical signs (weight loss and mortality). Cyclosporin A (CyA; Sandoz) was dissolved in Arachis oil and given as a single intramuscular injection (25 mg/kg BW). Cyclophosphamide (Cy; Asta Werke) was dissolved in aqua bidest and given as a single intravenous injection (100 mg/kg BW). The drugs were administered to either parental spleen cell donors or to F_1 hybrid recipients. Monoclonal antibodies used are: MRC OX-8; MRC OX-19; MRC OX-22 and ER-2 (W3/25 like) (7). Donor spleen lymphocytes, isolated on Percoll (Pharmacia), cryostat sections or parental spleens were stained with monoclonal antibodies using an immunoperoxydase technique.

RESULTS

Effect of CyA and Cy on Regional Graft versus Host Reaction. To analyse the effect of CyA or Cy on regional GvHR each drug was administered to either parental spleen cell donors or the F_1 hybrid recipients. The time of drug treatment is indicated in figure 1. CyA treatment of F_1 hybrids results in a reduction of severity of GvH reaction. In comparison to stimulated controls increase in lymph node weight is reduced (fig. 1a, 2 and 4). However, a complete suppression of the GvHR is obtained, when F_1 hybrids are treated with Cy (fig. 1b). No significant increase in weight is found if Cy is administered uptill 4 or 5 days after spleen cell transfer. Treatment of parental donor animals with CyA, 1 day before cell transfer has no effect on GvHR (fig. 1a; -1). Cy, on the other hand, given at the same time, results in a complete absence of a GvHR (fig. 1c). A slightly reducing effect is only found when Cy is given 4 days before cell transfer.

Effect of Cy on a systemic Graft versus Host Reaction. Treatment of parental donor with Cy 1, 2, 4 or 6 days before cell transfer, results in a spleen cell population, which can not induce systemic GvHR in F_1 hybrids. Transfer of spleen cell of untreated animals induces a GvHR, as assessed by weight loss over a period of 30 days. This weight loss is not observed by the Cy treated groups.

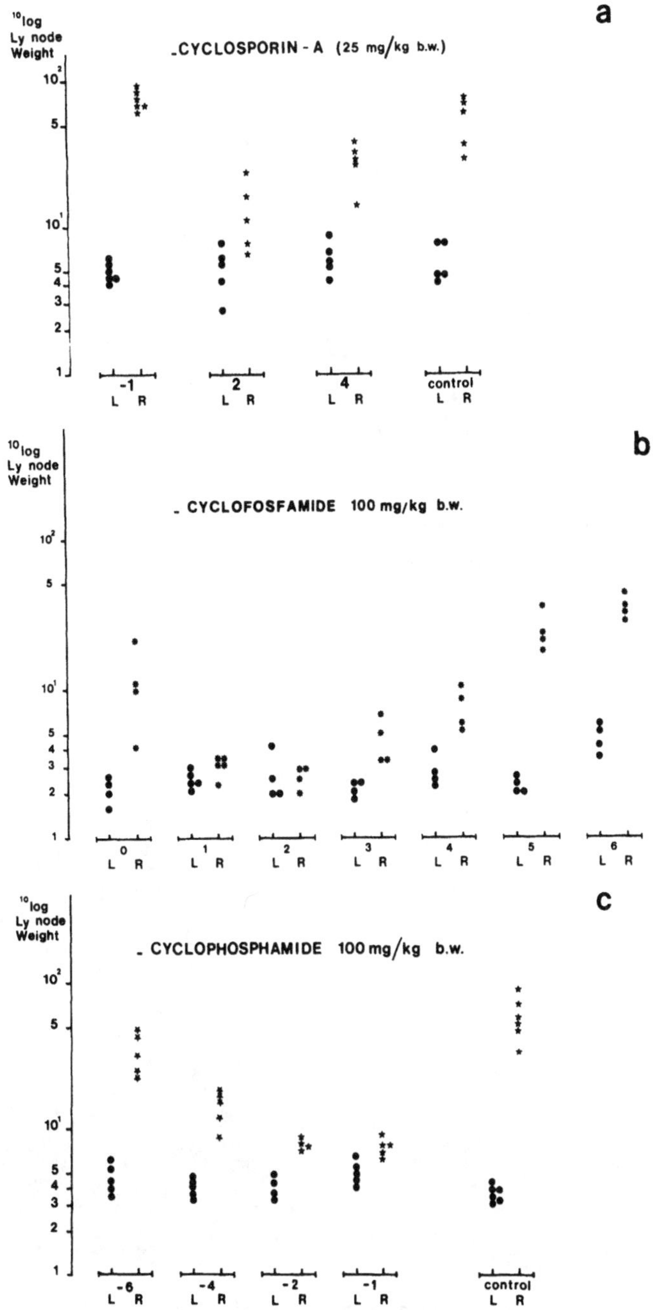

Fig. 1. Regional GvHR; a. CyA treatment; b and c. Cy treatment; -ve numbers donor treated; +ve numbers recipient treated.

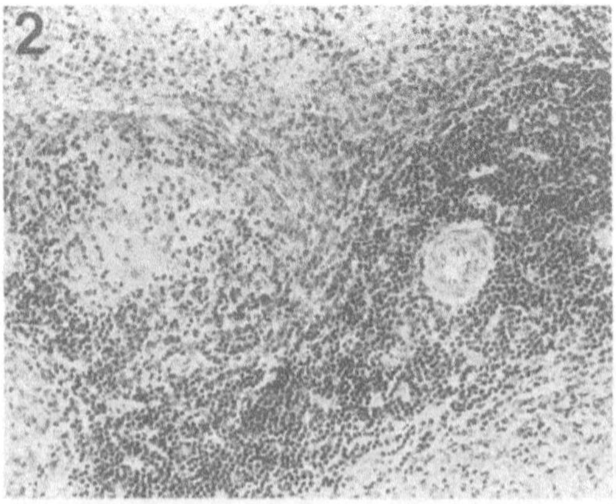

Fig. 2. Spleen section; 24 hours after Cy treatment (Methylgreen-Pyronin) (x120)

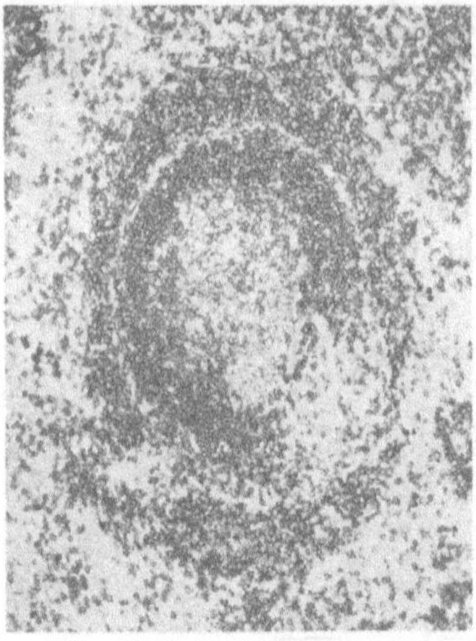

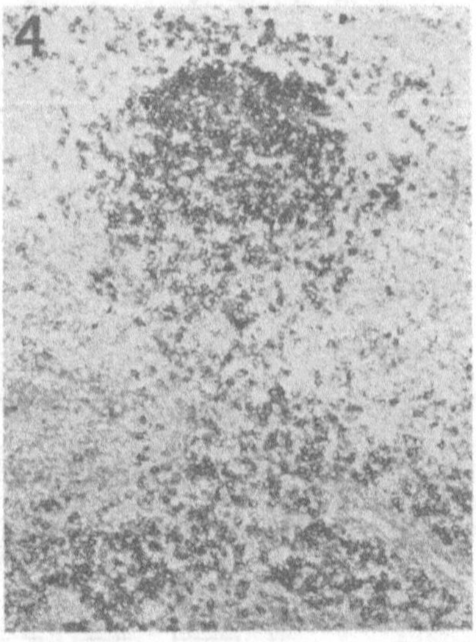

Spleen sections, immunoperoxidase staining with MRC OX-22 (x120)

Fig. 3. Untreated rat

Fig. 4. 24 hours after Cy treatment (GC largely composed with stained cell debris).

Effect of CyA and Cy on histology of the spleen. Histologically no effect of CyA on spleen morphology is observed. In the white pulp all T- or B-cell areas appear to be normal. After treatment with Cy, on the other hand, an atrophy of lymphoid follicles and marginal zones, is seen. The periarteriolar lymphocyte sheath seems to remain intact (fig. 2).

Effect of Cy on lymphocyte subsets in the spleen. After staining of both cytospots of spleen lymphoid cells and cryostat sections, it is found that MRC OX-22 +ve cells are significantly reduced (table 1, fig. 3a and 4). The relative percentages of ER-2, MRC OX-8 and MRC OX-19 +ve cells seem to remain unchanged.

Table 1

Total cell number/spleen		Untreated 4.7×10^8	Cy treated 1.7×10^8
ER-T2 (W3/25)	T helper cells	32.3	28.9
MRC OX-8	Supp/cyt. T cells	32.6	28.4
MRC OX-19	all T cells	46.2	49.1
MRC OX-22	GvH cells	65.8	37.4

DISCUSSION AND CONCLUSION

The results of the experiments indicated that CyA affects GvH reactions when the drug is given to F_1 hybrid recipients 2 or 4 days after transfer of normal parental spleen cell donors. A complete absence of GvH reactivity was only seen when either parental or F_1 hybrid rats were treated with Cy. Histologically no effect on lymphoid organs was observed after CyA administration. On the other hand Cy treatment causes a nearly complete atrophy of follicles and marginal zones. It remains unclear whether the T-cell areas are affected at all. After staining of both cytospots and cryostat sections with monoclonal antibodies a profound effect of Cy on MRC OX-22 +ve cells situated in marginal zones and follicles, was found. Ford et al. (8) showed that during GvH reaction the proliferative reaction in the popliteal lymph node is initiated uniquely by W3/25 +ve "helper cell" subset of T lymphocytes. This initiation is followed by a secondary phase of host cell proliferation probably involving both B- and T-lymphocytes with a bias towards the activation of B-cells (8). The latter may be activated by a form of non-specific allo-activation. The observed reduction of B-cells in F_1 hybrid spleens after Cy treatment accounts for the absence of increase in popliteal lymph node weight. If Cy also has a cytostatic effect on donor T-cells involved in initiating activation of recipient B-and T-cells remains to be answered.

Spickett et al. (9) demonstrated that in rats MRC OX-22 antigen is expressed besides on all B-cells, on 70% of W3/25 +ve cells and all MRC OX-8 +ve cell. W3/25 +ve, MRC OX-22 +ve cells are potently allo-reactive both in MLR, regional and systemic GvHR. On the other hand, a systemic GvHR can also be induced by MRC OX-22 +ve and MRC OX-8 +ve cells. W3/25 +ve, MRC OX-22 -ve cells play a role in providing help for B cells (7). These cells seem to be Cy unsensitive (5). The above described data show that treatment of spleen cell donors with Cy has a pronounced effect on GvHR and on the number of MRC OX-22 +ve cells. It appears that apart from earlier described data, Cy has a specific effect on W3/25 +ve, MRC OX-22 +ve cells, which would be responsible for both regional and systemic GvHR. Regarding the unability to induce a sytemic GvHR with Cy treated spleen cell donors, an observed difference between spleen cells and thoracic duct lymphocytes has to be considered. It was found that both cell populations differed with respect to GvHR inducing capacity giving different kinetics of clinical signs and mortality. Further experiments with double staining, involving both MRC OX-22 and W3/25 are in progress to test the hypothesis.

REFERENCES

1. Borel, J.F. Transplant. Proc. 13: 344 (1981).
2. Farrar, J.J., Benjamin, W.R., Hilfikes, M.L., Howard, M., Farrar, W.L. and Tullar Farrar, J. Immunol. Rev 63: 129 (1982).
3. Dongworth, D.W. and Klaus, G.G.B. Eur. J. Immunol. 12: 1018 (1982).
4. Broek, A.A. van den and Meedendorp, B. Int. J. Immunopharm. 4: 333 (1982).
5. Broek, A.A. van den and Keuning, F.J. In: Dev.in Biol. Standard. Eds. A.H. Griffith and R.H. Regamey, S. Karger, 38: 167 (1978).
6. Ford, W.L., Burr, W. and Simonson, M. Transplantation 10: 2 (1970).
7. Mason, D.W., Arthur, R.P., Dallman, M.J., Grein, J.R., Spickett, G.P. and Thomas, M.L. Immunol. Rev. 74: 57 (1983).
8. Ford, W.L., Rolstad, B., Fossum, S., Hunt, S.V., Smith, M.E. and Sparshott, S.M. Scand. J. Immunol. 14: 705 (1981).
9. Spickett, G.P., Brandon, M.R., Mason, D.W., Williams, A.F. and Woolett, G.R. J. Exp. med. 158: 795 (1983).

GRAFT VERSUS HOST DISEASE IN THE RAT:
Ia EXPRESSION ON TARGET TISSUES

R.J.M. Stet, C. Thomas and P. Nieuwenhuis

Department of Histology, University of Groningen
Groningen, The Netherlands

INTRODUCTION

Class II (Ia) antigens, in the rat encoded for in the RT1B subregion of the MHC, are potent stimulators of a Graft versus Host Reaction (GvHR) (1). However, the distribution pattern and quantity of Ia antigens expressed on host tissues during a GvHR have not yet been fully established.
Normally, Ia antigens are expressed on i) B lymphocytes, some macrophages, Langerhans cells, interdigitating cells and dendritic cells (2-5) and ii) on epithelial cells of thymus, gut and kidney (6-8).
Following the induction of a GvHR, studies demonstrated an increase in the expression of Ia antigens on epithelial cells of the intestine (9), certain renal tubuli (10) and a de novo expression on epidermal cells (11). Since a.o. skin and intestine are tissues severely damaged during a GvHR (12), and since increased Ia expression on different cell types has been reported in a variety of experimental and disease conditions (13), a relationship between increased Ia expression and specific damage to these particular tissues may be hypothesized.
As to a possible mechanism, in the case of a GvHR induced over a full MHC barrier, it has been suggested that Ia alloantigens cause activation of T helper cells. The latter in turn may activate class I reactive T suppressor/cytotoxic cells (14). Activated T cells produce a factor with -interferon (γ -IFN) activity (15), capable of induction of Ia antigens on a wide variety of cells (16). Newly expressed Ia antigens may then function as new targets for T helper cells, eventually causing destruction of these tissues due to T suppressor/cytotoxic cells.

Therefore, the purpose of this study was to analyse extensively the relationship between Ia antigen distribution pattern and quantity in tissues known to be damaged (targets) during a GvHR. In addition, the nature of the immunologically reactive cells infiltrating these tissues was studied, in particular with regard to lymphocyte phenotypes.

MATERIALS AND METHODS

Rats: Twelve to sixteen weeks old male inbred AO/G ($RT1^u$) and BN/G ($RT1^n$) and their F_1 hybrid AO/GxBN/G ($RT1^{u/n}$) rats were obtained from stocks held at the Department of Histology, Groningen.
Induction of systemic Graft versus Host Reaction: Thoracic Duct Lymphocytes (TDL) were obtained as a first overnight collection from cannulated AO/G donor rats. GvHR was induced by injecting 10^8 TDL per 200 g body weight intravenously into unirradiated F_1 hybrids.
Tissue sampling: Tissues were sampled from TDL recipient rats 7, 14 and 21 days after induction of GvHR. As controls age and sex-matched F_1 hybrids were used.
Monoclonal antibodies: Mouse monoclonal antibodies used are depicted in Table I. For staining with monoclonal antibodies, tissue samples were snap-frozen in freon. Cryostat sections were cut and transferred to gelatinized slides, air dried and fixed in acetone. Sections were incubated with the monoclonal antibody followed by incubation with peroxidase-conjugated rabbit immunoglobulins to mouse Ig (DakoPatts a/s, Copenhagen, Denmark), containing 1% normal rat serum to block non-specific staining. The sections were developed by incubation with diaminobenzidine and H_2O_2. Sections were counterstained with haematoxylin and mounted in Depex.

Table I

Monoclonal antibody	Lymphocyte subset or antigen labeled	Reference
MRC OX-19	All T cells	17
MRC OX-8	Suppr./cytotox. T cells	17
MRC OX-22	All B cells/subset T cells	17
HIS 14	All B cells	18
HIS 17	All T cells (W3/13 like)	18
ER-2	Helper T cells/subset MO (W3/25-like)	19
ER-13	Ia (common part)	20
F 17-23-2	Ia (polymorphic; $RT1^u$:-ve; $RT1^n$:+ve)	21

As control antibody 2G12 was used, directed against human melanoma antigen and not reacting with rat tissues.

RESULTS

Ia expression on target tissues

Spleen

In normal spleen, Ia antigens, as detected with ER-13, were expressed on B cells in germinal centres (GC), corona and marginal zone (MZ), on interdigitating cells (IDC) in the periarteriolar lymphocyte sheath (PALS) and on some macrophages in the red pulp.
On day 7 after induction of GvHR, apart from IDC, an increase of Ia antigens, both on lymphocytes and on an unidentified cell type, seen as a confluent pattern, was observed in the PALS. GC had disappeared, while corona and MZ were normal as to distribution and quantity of Ia expression.
On day 14 changes were dramatic. Corona had disappeared and only small clusters of Ia^+ marginal zone cells remained. The area occupied by the PALS had increased considerably and consisted of lymphocytes, some of which were Ia^+, and irregulary shaped Ia^+ cells (Fig. 1a).
On day 21 an overall decrease of Ia^+ cells was observed, as well as a condensation of the area occupied by the PALS.

Thymus

Normal thymus showed a distinct pattern of Ia expression as detected with ER-13. In the cortex a reticular and in the medulla a confluent pattern of Ia^+ cells was observed.
On day 7 after induction of GvHR the first changes in Ia epxression were observed as a loss of Ia^+ cortical epithelial cells at the cortico-medullary juntion. Also the confluent Ia^+ staining pattern in the medulla was now interrupted by Ia^- stained areas.
On day 14 the reticular staining pattern of the cortex was largely replaced by a confluent pattern resembing that of the medulla (Fig. 1b). There were no changes observed in the medulla compared to day 7.
At day 21 the changes described for the cortex had even more progressed.

Tongue

In the tongue of normal rats Ia antigens were detected on dendritic cells in the lamina propria and the interstitium of the muscle layer.
On day 7 after induction of GvHR there was an increase of Ia^+ cells, mainly in the intersitium of the muscle layer but also in the lamina propria. Epithelial cells of the mucosa now showed a weak expression of Ia antigens and in some locations Ia^+ cells of a lymphoid morphology had penetrated the basal layer of the epithelium.

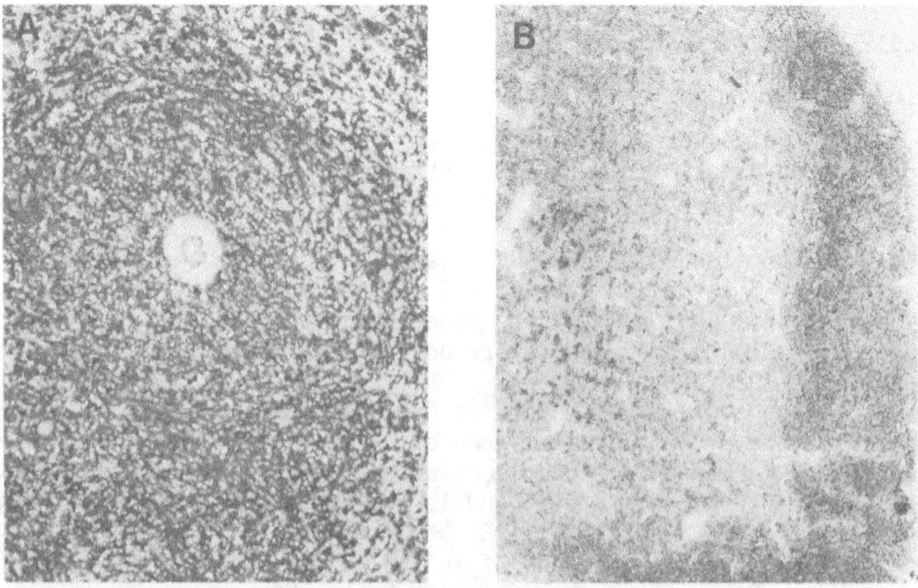

Fig. 1. Detection of Ia antigens using ER 13 in frozen sections of spleen (a.x75) and thymus (b.x40) 14 days after induction of GvHR.

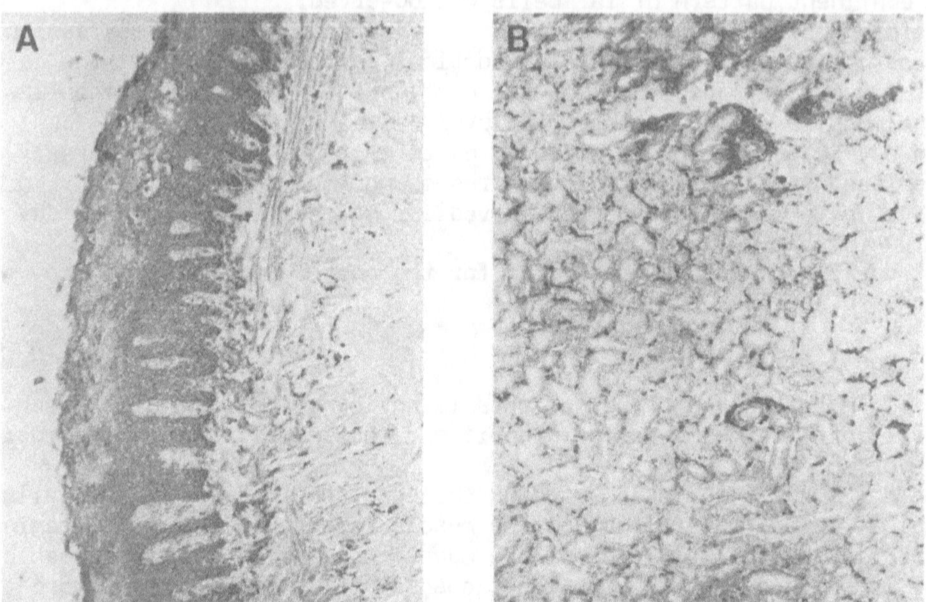

Fig. 2. Detection of Ia antigens using ER 13 in frozen sections of tongue (a.x40) and kidney (b.x40) 14 days after induction of GvHR.

On day 14 a large increase in Ia antigens was seen on both infiltrated cells, epithelial cells of the mucosa and endothelial cells of arterioles. The most prominent increase in Ia antigens was found in the lamina propria, probably on lymphocytes and dendritic cells. The now Ia^+ epithelium was also infiltrated by Ia^+ cells (Fig. 2a). The epithelium contained numerous vacuoles, filled with Ia^+ cells and debris.
On day 21 the changes described had even more progressed.

Kidney

In normal kidneys, Ia, as detected with ER-13, was expressed on certain tubuli and scattered dendritic cells.
7 Days after induction of a GvHR an increased proportion of cortical tubuli expressed Ia antigens, whereas in the interstitium in both cortex and medulla Ia^+ cells (probably dendritic cells and lymphocytes) had also increased.
On day 14 the proportion of Ia^+ tubuli had even further increased (Fig. 2b). This increase coincided with an increase in number of interstitial Ia^+ cells. The epithelial cells of the renal calyx were strongly Ia^+ and were heavily infiltrated with Ia^+ lymphocytes. The endothelial cells of the arterioles were now also Ia^+.
After 21 days an even more abundant Ia expression was observed showing the same distribution pattern.

Salivary gland

Normally salivary gland contained only very few Ia^+ dendritic cells.
On day 7 after induction of a GvHR, Ia expression was restricted to epithelial cells of the major extra-lobular ducts in the serous part of the gland; in most instances associated with peritubular Ia^+ cells.
On day 14 all intra- and extra-lobular ducts of the serous part now expressed Ia antigens and were surrounded by interstitial Ia^+ lymphocytes and dendritic cells (Fig. 3a). Acini were Ia^-, but surrounded by Ia^+ lymphocytes, some of which had penetrated into the acini. In the mucous part of the gland only the large extra- lobular ducts expressed Ia antigens in association with interstitial Ia^+ cells.
On day 21 the serous part showed a further increase of Ia expression, while the mucous part remained unchanged.

Pancreas

The pancreas of normal rats, like the salivary gland, only contained few Ia^+ dendritic cells.
On day 7 after induction a slight increase of Ia^+ cells was observed mainly in the interstitium around bloodvessels and extra-lobular ducts.
A sharp increase was observed on day 14, again of interstitial

cells. Now also the epithelial cells of the extra- and intra-lobular ducts expressed Ia antigens. The Ia^- islets of Langerhans were surrounded by Ia^+ cells. In the exocrine part Ia^+ cells were present in large numbers in the interstitium, but acini remained Ia^-.
On day 21 a further increase of Ia^+ cells was observed showing the same distribution as described, with the addition of Ia^+ cells now also present within the islets of Langerhans.

Liver

In normal liver only dendritic cells in the portal interstitium and a few Kupffer cells expressed Ia antigens.
On day 7 after induction, Ia^+ infiltrates were observed containing lymphocytes and macrophage-like cells. Also a slight increase of Ia^+ Kupffer cells in the liver parenchyma was detected.
On day 14 portal infiltrates spread into surrounding parenchyma and in the sinusoids a sharp increase in Ia^+ cells was seen (Fig. 3b).
On day 21 no further changes compared to day 14 were observed.

Is the newly expressed Ia of recipient and/or donor type?

In all tissues screened the Ia expressed on epithelial and endothelial cells during a GvHR was of recipient type as detected with F 17-23-2. However, using F 17-23-2, considerably less Ia^+ infiltrating cells were detected as compared to ER-13, suggesting that these cells were of predominantly donor origin.

Lymphocyte subset analysis in target tissues

Spleen

Using MRC OX-19 as a marker for all T cells, considerable changes were observed in spleen of rats undergoing a GvHR.
At day 7 the proportion of MRC $OX\text{-}19^+$ cells in B cell areas (GC and corona) was increased (Fig. 4a); in T cell areas (PALS) the staining pattern of MRC OX-19 was weaker than normal, possibly indicating loss of marker due to activation and/or proliferation.
On day 21, MRC $OX\text{-}19^+$ cells were depleted in the PALS. Similar observations were made with HIS 17 (all T cells), although staining was more intensely compared to MRC OX-19. Both ER-2 and MRC OX-8 (T_H and $T_{S/C}$ resp.) detected on day 7 increased numbers of +ve cells in GC and corona. On day 14 and 21 both subsets equally participated in the changes observed in the PALS.
The B cell population in spleen, as detected with MRC OX-22 and HIS 14, was drastically reduced: at day 7 GC and at day 14 corona and most of the MZ were devoid of B cells. On day 21 no B cells could be detected.

Thymus

During an ongoing GvHR the number of cortical MRC OX-19$^+$ cells was reduced, resulting in a negatively stained area at the cortico-medullary border. The distribution of MRC OX-19$^+$ cells in the medulla remained unchanged. In contrast, all thymocytes remained HIS 17$^+$ during the observed changes. Staining with MRC OX-8 showed a similar pattern as seen with MRC OX-19, whereas ER-2 stained all thymocytes similar to HIS 17. No changes were observed with MRC OX-22, which stained only a few scattered cells in the cortico-medullary area.

Non-lymphoid target tissues

In all other target tissues screened, an increasing number of infiltrating donor derived HIS 17$^+$ cells was observed during the GvHR (Fig. 4b). Surprisingly only a few of these cells were MRC OX-19$^+$. Both ER-2$^+$ and MRC OX-8 lymphocytes were present, with ER-2$^+$ cells in slight excess. Staining with HIS 14 indicated that only a few B cells were present in these tissues. Thus, infiltrating lymphocytes were predominantly of T cell phenotype.

DISCUSSION

Screening for class II (Ia) antigen during a GvHR clearly reveals that in all tissues studied Ia expression is modulated during an ongoing GvHR. The first signs can be observed in lymphoid tissues. Notably the spleen, as the major lymphoid organ through which donor derived recirculating lymphocytes pass (22), shows the first changes. Ia alloantigens are potent stimulators for a GvHR (1), therefore these changes are likely to involve recipient Ia$^+$ cells (either IDC or B cells). This is indicated by disappearance of follicular structures and detectable IDC. These Ia$^+$ cells besides being initiators, also seem to be target. Whether the loss of spleen B cells is due to an effect of donor T cells on bone marrow cells as indicated by decreasing number of pre-B cells during a GvHR (Xenocostas, pers. comm.) or a direct effect on the population itself, is not clear. The latter can be substantiated by the presence of increasing numbers of T cells in follicular structures as shown in Fig. 4a. In the thymus, Ia$^+$ cortical epithelial cells gradually disappear during an ongoing GvHR. Consequently thymocytes present in the affected cortical area lose the MRC OX-8 and MRC OX-19 determinants. This loss of markers may be indicative for the observed T cell functional defect described for mice undergoing a GvHR (23).

The changes observed in these lymphoid organs are clearly different form those described for the non-lymphoid tissues viz. these containing Ia$^-$ epithelial structures and only Ia$^+$ dendritic cells like tongue, salivary gland and pancreas. These tissues, from day 7 on, show a gradual increase in Ia expression on epithelial cells coinciding with an increase of Ia$^+$ lymphocytes and in-

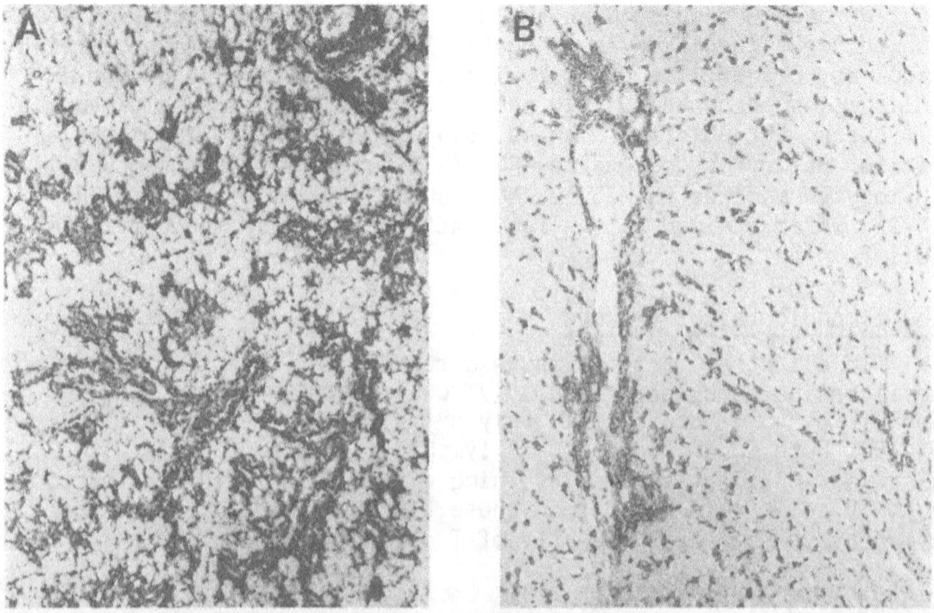

Fig. 3. Detection of Ia antigens using ER 13 in frozen sections of salivary gland (a.x40) and liver (b.x40) 14 days after induction of GvHR.

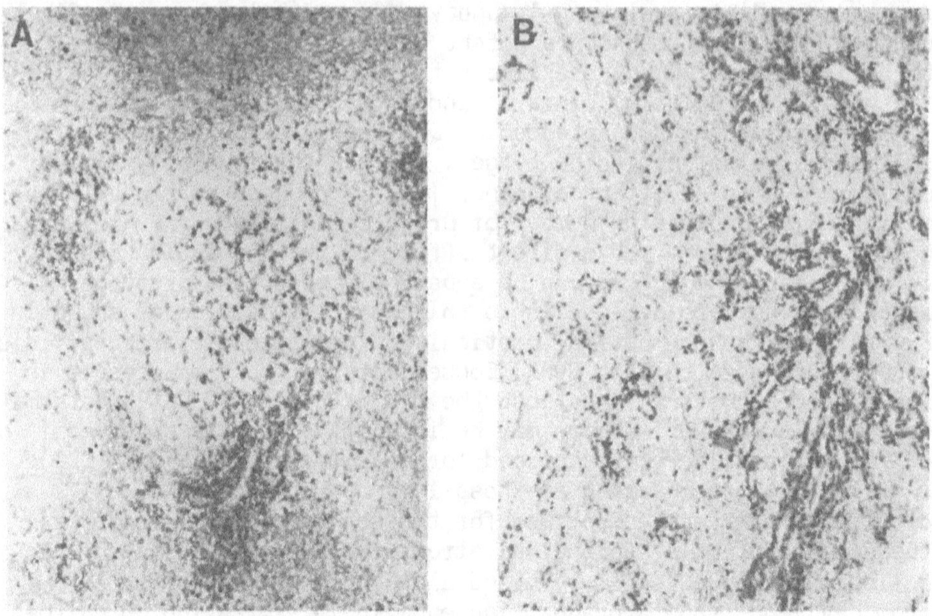

Fig. 4a. (x40). Detection of MRC OX-19$^+$ cells in follicular structures of spleen 7 days after induction of a GvHR.

4b. (x40). Detection of HIS 17$^+$ cells in salivary gland 14 days after induction of GvHR.

creasing number of dendritic cells. This increase during a GvHR has also been described for skin and intestine (9, 11). It has been suggested that Ia epxression on epithelial cells might well be consistent with a role in antigen presentation, because these epithelial structures an directly exposed to the external environments. However, our data clearly show that during a GvHR epithelial cells like those of salivary gland and pancreas also becomes Ia^+ These epithelial cells are not in direct contact with the external environment. Therefore the observed Ia expression on these latter organs possibly has a different significance.

We like to postulate the following sequence of events which might lead to the observed Ia expression in these tissues. After injection of parental T cells, delayed type hypersensitivity T cells are generated from T_H cells due to local (e.g. in the spleen) recognition of recipient Ia^+ cells (24). These cells, upon migration encounter Ia^+ dendritic cells present in almost every tissue (5). As a consequence of this "peripheral" recognition of these Ia^+ cells by migrating T_H cells soluble fraction are produced (γ-IFN), which in turn can induce recipient type Ia epxression on a number of cells types (16). This sequence eventually induces new targets for alloreactive T cells (25) by presenting new allo MHC antigens. The first peripheral signs of a GvHR are always detected in sites where normally Ia^+ dendritic cells are already present viz. perivasculary and peritubulary. The cytotoxic effect on other Ia^- cell types (e.g. acini) can be attributed to class I reactive $T_{S/C}$, activated by T_H cells (14). This has been substantiated by our demonstration of increasing numbers of either subset in these tissues.

In normal kidney Ia is already expressed on a number of proximal tubuli. This would make this organ an immediate target. However, kidney shows the same pattern of increase in Ia expression coinciding with an infiltration of Ia^+ lymphocytes as seen in other organs. The reason why this organ is not immediately affected is probably due to the fact that it is not a lymphoid organ to which lymphocytes have immediate access. The liver on the other hand though readily accessible to donor lymphocytes, is not immediately affected, because neither endothelial cells nor hepatocytes express Ia antigen. Therefore damage is at first restricted to the portal triads where Ia^+ recipient cells are present. The large increase in Ia^+ Kupffer cells and lymphocytes will be discussed in detail elsewhere (Stet et al., in preparation).

The question remains whether damage to target tissues as a result of a GvHR is only depended on the presence, either initial or de novo, of Ia antigens, or also involves changes in the expression of class I antigens.

REFERENCES

1. D.V. Cramer, B.K. Davis, J.W. Shonnard and T.J. Gill, III. Transplantation 23: 498 (1977).
2. A.N. Barclay. Immunol. 44: 727 (1981).
3. R.M. Steinman and M.C. Nussenzweig. Immunol. Rev. 53: 127 (1980).
4. I. Silberberg-Sinakin, I. Gigli, R.L. Bear and G.J. Thorbecke. Immunol. Rev. 53: 203 (1980).
5. D.N.J. Hart and J.W. Fabre. J. Exp. Med. 153: 347 (1981).
6. G. Mayrhofer, C.W. Pugh and A.N. Barclay. Eur. J. Immunol. 13: 112 (1983).
7. A.N. Barclay and G. Mayrhofer. J. Exp. Med. 153: 1666 (1981).
8. D.N.J. Hart and J.W. Fabre. J. Immunol. 126: 2109 (1981).
9. A.N. Barclay and D.W. Mason. J. Exp. Med. 156: 1665 (1982).
10. A. Wagdymar, J. Urmson, R. Baumal and P. Halloran. J. Immunol. 132: 1826 (1984).
11. I.A. Lampert, A.J. Suitters and P.M. Cnisholm. Nature 293: 149 (1981).
12. W.E. Beschroner, P.J. Tutschka and G.W. Santos. Clin. Immunol. and Immunopath. 22: 203 (1982).
13. A.N. Barclay, D.W. Mason and G. Mayrhofer. Transplant. Proc. 15: 1587 (1983).
14. S.T. Pals, T. Radaszkiewicz and E. Gleichman. J. Immunol. 132: 1669 (1984).
15. N. Cerf-Bensussan, A. Quaroni, J.T. Kurnick and A.K. Bhan. J. Immunol. 132: 2244 (1984).
16. G.H.W. Wong, I. Clark-Lewis, A.W. Harris and J.W. Schrader. Eur. J. Immunol. 14: 52 (1984).
17. D.W. Mason, R.P. Arthur, M.J. Dallman, J.R. Green, G.P. Spicket and M.L. Thomas. Immunol. Rev. 74: 57 (1983).
18. F.G.M. Kroese, D. Opstelten, A.S. Wubbena, G.J. Deenen, J. Aten, E.H. Schwander, L. de Leij and P. Nieuwenhuis. This volume.
19. P. Joling, F.J. Tielen, L.M.B. Vaessen, C.J. Hesse and J. Rozing. This volume.
20. R.J.M. Stet, J. Rzoing, G.D. Majoor, F.G.M. Kroese, D. Opstelten and P. Nieuwenhuis. Transplant. Proc. In press (1985).
21. D.N.J Hart and J.W. Fabre. J. Immunol. 126: 2109 (1981).
22. J.L. Gowans. Ann. NY Ac. Sci. 99: 432 (1969).
23. M. Seddik, T.A. Seemayer and W.S. Lapp. Transplantation 29: 61 (1980).
24. E.A.J. Wolters, Th.H. van der Kwast, L.M. Odijk and R. Benner. Cell. Immunol. 57: 389 (1981).
25. M.J. Dallman and D.W. Mason. Transplantation 36: 222 (1983).

INDUCTION OF CLASS I AND CLASS II ANTIGENS DURING REJECTION OF ALLOGRAFTED HUMAN HEARTS

M.L.Rose, M.I.Coles, R.J.Griffin and M.H.Yacoub

Thoracic and Cardiac Surgical Unit
Harefield Hospital, Harefield, Middlesex, England

INTRODUCTION

The distribution of class II (HLA-DR) antigens in various organs and tissues has been well documented in animals and man (1). Tissue distribution is restricted compared to class I antigens (see below) but is nevertheless present on a wide variety of cells, including macrophages and dendritic cells. It is known that on macrophages and dendritic cells the class II antigens are involved in antigen presentation and are an essential part of inducing an immune response (2). The expression of class II antigens can be induced both in *vivo* (3) and in *vitro* (4) and it is thought such changes in class II expression may regulate the immune response (5). The distribution of class II antigens in transplant organs, both before and after transplantation, is therefore of interest as it may shed light on the cellular mechanisms of graft rejection.

The distribution of class I (HLA-ABC) antigen in all organs and tissues is less well documented. Until recently it was thought that all cells except erythrocytes, expressed class I antigens (6). Such studies were based on use of cell suspensions and hetero-antisera or allo-antisera as typing reagents. The use of tissue sections and immunoperoxidase techniques is much better able to detect the distribution of class I antigen in *situ*. Here we demonstrate that the monomorphic class I determinant detected by monoclonal antibody (mab) W6/32 is not detectable on normal myocardium but is expressed on myocardium during a rejection episode.

MATERIALS AND METHODS

Patients and Tissues

Twelve patients had received orthotopic cardiac transplants 1-6 months previously. All were receiving Cyclosporine, azathioprine and low dose steroid immunosuppression. Pieces of right ventricular myocardium were obtained by transvenous endomyocardial biopsy (7). Pieces of biopsy approximately $2mm^2$ were snap-frozen in liquid nitrogen and 6μm cryostat sections were cut and fixed as previously described (8). Pieces of non-transplanted heart were either from the donor heart at the time of transplantation or were biopsies from patients undergoing assessment for transplantation.

Monoclonal antibodies and immunoperoxidase technique

Details of the monoclonal antibodies are described in Table I. A triple layer immunoperoxidase technique was utilised as previously described (8). Briefly, mouse mab was followed by biotinylated goat anti-mouse lgM and lgG, followed by avidin-biotin-peroxidase complex. The sites of peroxidase fixation were identified using diaminobenzidine tetrahydrochloride and hydrogen peroxide.

TABLE I

Source of Specificity of Monoclonal Antibodies

Designation	Specificity	Dilution*	Reference
HLA-DR	Common determinant class II antigens	1:10	Clone L243 Becton-Dickinson
NFK-1	Common determinant to DR, DC and SB	1:100	Fuggle et al[9]
Tu22	Monomorphic DC	1:100	Wernet et al[10]
10	Endothelial cells	neat (supernatant)	N.Hogg & Y.Selevenden. ICRF London
W6/32	Monomorphic class I common determinant	neat (supernatant)	Barnstaple et al[11]

* diluted in phosphate buffered saline for use

Class II antigens

Ten myocardial biopsies from 10 non-transplanted hearts showed an even and consistent distribution of interstitial structures staining with HLA-DR (Figure 1). NFK-1 gave similar results. By contrast Tu22 showed very little staining of interstitial structures in normal heart, myocardium was negative with all three mabs. Some of the HLA-DR + structures were obviously endothelium of blood vessels by morphological criteria. To facilitate indentification of small capillaries the mab 10 was used (Figure 2). This mab produced a very similar picture to that seen using HLA-DR and NFK-1 mabs. Biopsies taken from transplanted heart during rejection episodes showed an increase in the numbers of interstitial structures binding mabs against class II antigens. The interstitial cells could be identified as large blood vessels, capillaries and the infiltrating mononuclear cells which are characteristic of rejection. The increased staining of mabs against class II antigens was particularly noticable using Tu22 where vascular endothelium may become strongly positive and many interstitial cells are positive (Figures 3 and 4). Myocardial membranes and intercalating discs remained negative.

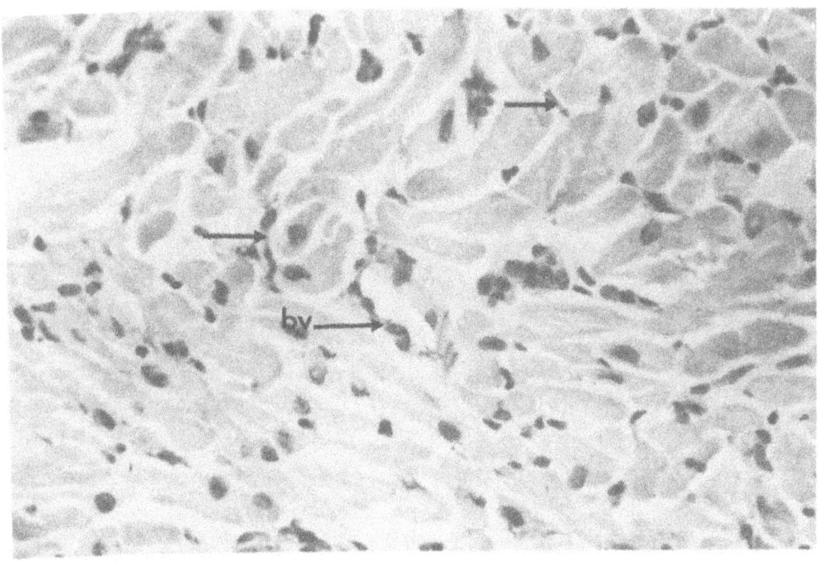

Fig.1: Non-transplanted heart stained with HLA-DR.
Arrow points to positive blood vessels (bv) and positive interstitial structures.

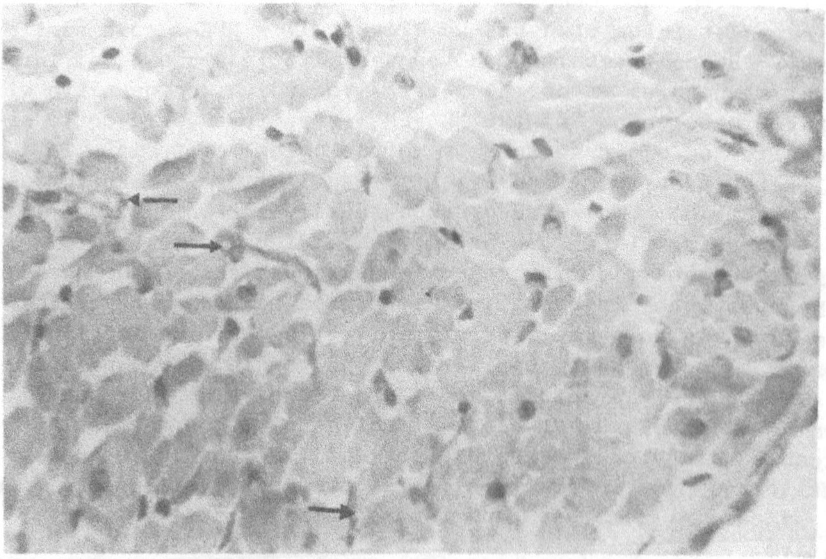

Fig.2: Non-transplanted heart stained with 10. Arrow points to positive blood vessels

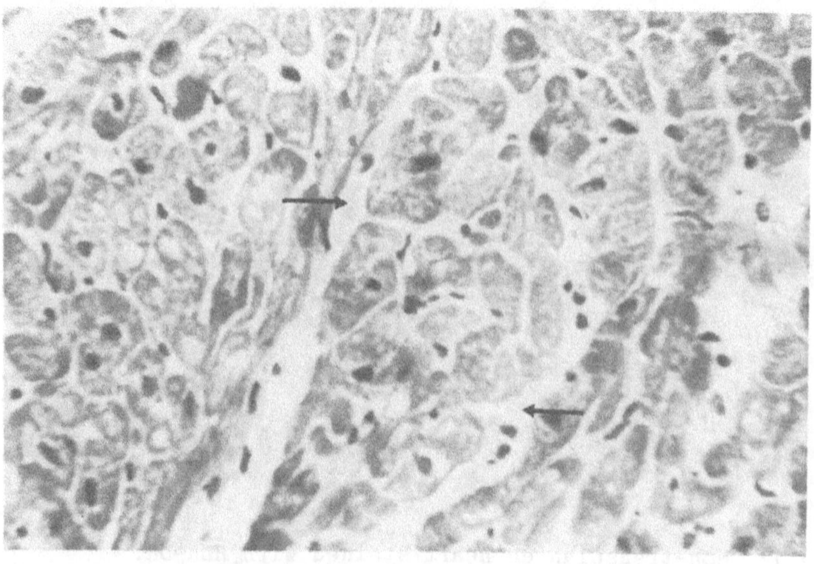

Fig.3: Non-transplanted heart stained with Tu22. Arrow points to negative blood vessels.

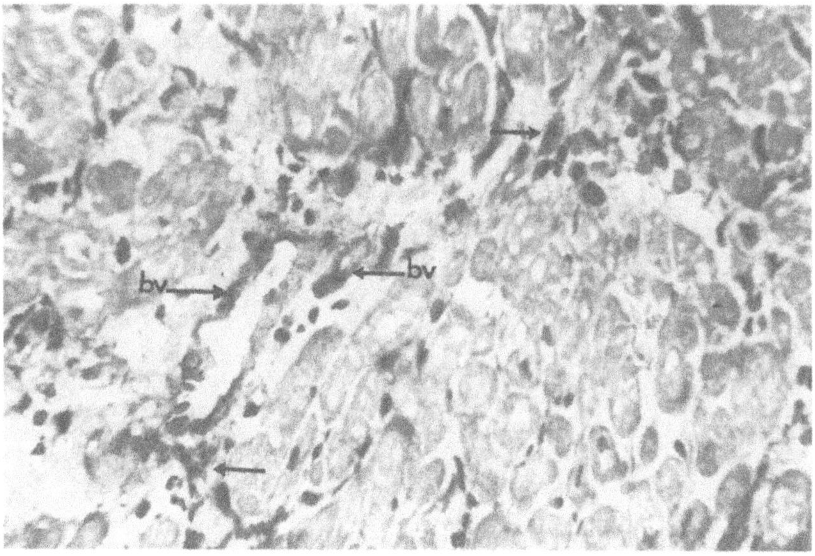

Fig.4: Transplanted heart (rejecting) stained with Tu22. Arrows point to positive blood vessels (bv) and to positive infiltrating cells.

Class I Expression

Eleven biopsies from 11 pieces of non-transplanted heart were stained with W6/32. The myocardium showed virtually no staining but the interstitial structures were heavily stained. (Figure 5). A further 30 biopsies from transplanted hearts were investigated and of these 20 showed positive staining of W6/32 on the myocardium (Figure 6). Where myocardium was cut longitudially, intercalating discs could be seen to be binding W6/32.

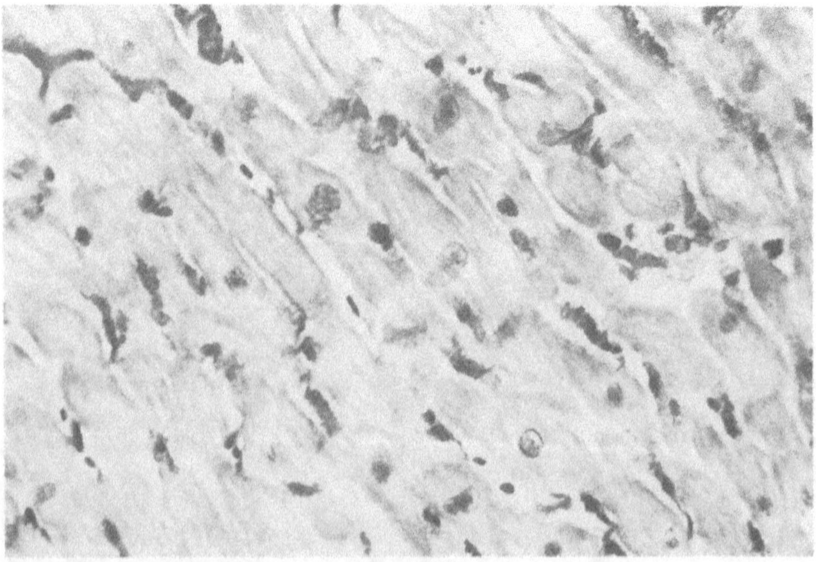

Fig.5: Non-transplanted heart stained with W6/32. Interstitial cells are positive, Myocardium is negative.

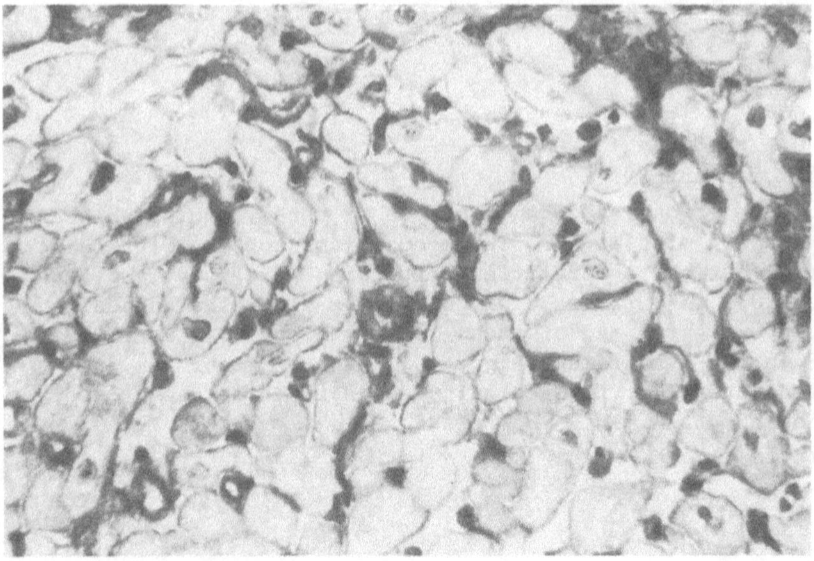

Fig.6: Transplanted heart (rejecting) stained with W6/32. Interstitial cells are positive, Myocardium is positive.

DISCUSSION

This paper briefly described the distribution of class I and class II antigens on normal heart and the induced expression of such antigens after allogeneic transplantation.

In normal (non-transplanted) heart mabs against class II antigens bind to numerous interstitial structures. Class II antigens are expressed on the surface of T cells, monocytes, macrophages, dendritic cells and vascular endothelium. The latter cell types all bear the leucocyte common antigen (12) except for vascular endothelium. Studies in this laboratory have shown that few cells bearing the leucocyte common antigen are present in normal heart, thus rendering it more likely that most of the class II positive cells in normal heart are endothelial cells. It was found that the pattern of binding of the anti-endothelial mab 10 closely corresponded to that using HLA-DR and NFK-1, strongly suggesting that the vast majority of class II positive structures in normal heart are endothelial cells of small capillaries. This observation raises the question whether endothelial cells are important for the induction of allogeneic immune responses, as has been suggested (13) and whether class II antigens on endothelial cells are modified during an immune response. The mab Tu22 showed little staining on normal heart, it was used at a concentration of 1/100 whereas in cytotoxic tests it is used at 1/10,000. It may be that this determinant is expressed on the cell surface at lower density than determinants to DC, DR and SB. The other possibility is that DC is a class II determinant which is differentially expressed and is linked to cell maturation (14).

The increase expression of class II antigens in transplanted hearts accords well with the increased expression of these antigens in experimental models of graft rejection (15) or graft versus host disease (3). DC determinants were found to be expressed on structures of donor origin (endothelial cells) as well as on the host infiltrating cells which are characteristic of mild rejection. Myocardium was not found to bind mabs against class II antigens in these studies.

The absence of binding of W6/32 to myocardium of non-transplanted heart was unexpected in view of the earlier studies that all nucleated cells express class I antigens (6). This finding has been corroborated by a recent study (16) using mab PA2.6 which has also shown that expression of class I determinants on human myocardium is poor or absent. Of the 30 biopsies from transplanted hearts used in this study, 15 were showing histological signs of rejection, all 15 showed heavy staining with W6/32 on the myocardium. In addition, 5 biopsies thought not to be showing signs of rejection also stained heavily with W6/32 on the myocardium. There are in vitro precedents for induction of class I antigens (17) but as yet no other in vivo examples of this phenomenon.

Biopsies showing histological signs of rejection contain T cells (8) and acid phosphatase positive macrophages (unpublished). It is possible that γ interferon released from activated T cells causes the induction of class I antigens on myocardium but the significance of this event to the host-versus-graft response is as yet unknown.

REFERENCES

1. P.G.Natali, C.DeMartino, V.Quaranta, M.R.Nicotra, F.Frezza, M.A.Pellegrino, S.Ferrone; Expression of Ia-like antigens in normal human non-lymphoid tissues. Transplantation 31:75 (1981)

2. J.Klein, V.Hauptfeld; Ia antigens their serology, molecular relationships and their role in allograft reactions. Transplant.Rev.30:83 (1976).

3. I.A.Lampert, A.J.Suitters, P.M.Chisholm; Expression of Ia antigens on epidermal keratinocytes in graft-versus-host disease. Nature (Lond)293:14 (1981).

4. J.S.Pober, M.A.Gimbrone, R.S.Cottran, C.S.Reiss, S.J.Burakoff, W.Fiers, K.A.Ault; Ia expression by vascular endothelium is inducable by activated T cells and by human γ interferon. J.Exp.Med.157:1339 (1983).

5. P.S.Steeg, R.N.Moore, J.J.Oppenheim; Regulation of murine macrophage Ia-antigen expression by products of activated spleen cells. J.Exp.Med.152:1734 (1980).

6. M.Berah, J.Hors, J.Dausset; A study of HLA antigens in human organs. Transplantation 9:183, (1970).

7. P.K.Caves, W.P.Schulz, E.Dong, E.B.Stinson, N.E.Shumway; New instrument for transvenous cardiac biopsy. Am.J.Cardiol. 33:264 (1974).

8. M.L.Rose, J.A.Gracie, A.Fraser, P.M.Chisholm, M.H.Yacoub; Use of monoclonal antibodies to quantitate T lymphocyte subpopulations in human cardiac allografts. Transplantation (in press, 1984).

9 S.V.Fuggle, P.Errasti, A.S.Daar, J.W.Fabre, A.Ting, P.J.Morris; Localisation of major histocompatibility complex (HLA-ABC and DR) antigens in 46 kidneys. Transplantation 35:385 (1983).

10 P.Wernet, A.Ziegler, J.Shaw, N.Blaurock, G.Pawelec; Monoclonal antibodies against Ia like antigens inhibiting HLA-D and or SB directed secondary proliferative responses. Leukaemia markers (1983).

11 C.J.Barnstable, W.F.Bodmar, G.Brown, G.Galfre, C.Milstein, A.F.Williams, A.Ziegler; Production of monoclonal antibodies to group A erythrocytes, HLA and other human cell surface antigens - new tools for genetic analysis. Cell 14:9 (1978).

12 R.Dalchau, J.Kirkley, J.W.Fabre; Monoclonal antibody to human leucocyte specific membrane glycoproteins probably homologous to the leucocyte common (L-C) antigens of the rat. Eur.J. Immunol.10:737 (1980).

13 H.Hirschberg, L.R.Braathen, E.Thorsby; Antigen presentation by vascular endothelial cells and epidermal largerhans cells: the role of HLA-DR. Immun.Revs.65:57 (1982).

14 R.A.Newman, Domenico Delia, M.F.Greaves, C.Navarette, L.Fairboim H.Festenstein; Differential expression of HLA-DR and DR-linked determinants as human leukemias and lymphoid cells. Euro.J.Immuno.13:172 (1983).

15 M.J.Dallman, D.W.Masion; Induction of Ia antigens on murine epidermal cells during the rejection of skin allografts. Transplantation 36:222 (1983).

16 A.S.Daar, S.V.Fuggle, J.W.Fabre, A.Ting, P.J.Morris; The detailed description of HLA-ABC antigens in normal human organs. Transplantation (in press).

17 M.Fellous, M.Kamoun, I.Gresserand, R.Bona; Enhanced expression of HLA antigens and B_2-microglobin on interferon treated human lymphoid cells. Euro.J.Immunol.9:446 (1979).

ENHANCED REGENERATION OF TRANSPLANTED SPLENIC TISSUE BY INCREASED WORK LOAD

Reinhard Pabst, Robert Hafke, and Jürgen Hillebrand

Centre of Anatomy, Medical School of Hannover
D-3000 Hannover 61, West Germany

INTRODUCTION

In splenectomized patients there is an increased risk of overwhelming septicaemia (1). To prevent this, either splenic tissue should be salvaged by partial splenectomy,splenorrhaphy or haemostyptic agents,or splenic fragments should be autotransplanted into the greater omentum (2). In experimental animals and in patients transplanted splenic fragments become necrotic, but later they regenerate with an almost normal histological structure. A critical mass of splenic tissue seems to be essential in preventing an overwhelming sepsis (3,4,5), but the size and the clearance function of the regenerated splenic tissue only reach a small fraction of normal values (6). The aim of the present study was to enhance splenic regeneration by stimulating either the white pulp, the red pulp or the reticuloendothelial parts of the spleen in rats.

MATERIAL AND METHODS

Male Lewis rats were housed under SPF conditions. All operations were performed under clean conditions under ether and ketamin hydrochloride anaesthesia. Normal rats (n=42) were treated as follows: A. Intravenous injections of red blood cells (2×10^8 per dose), twice pig, twice human, twice sheep red blood cells. B. Phenylhydrazine hydrochloride (1mg/100g body weight) was given every fortnight i.p. C. A 2.5% methylcellulose solution was injected i.p. twice a week with a dose increasing from 1ml to 2ml for a total of 12 weeks. In a second series 48 rats were splenectomized and half of the spleen transplanted in fragments into the greater omentum. Three weeks after the operation groups of rats were treated with the following; A. xenogeneic red blood cells i.v. to stimulate the white pulp, B. phenylhydrazine to damage red cells and increase the work load to the red pulp, C. red blood cells and phenylhydrazine and D. methylcellulose. In a third series rats were treated with xenogeneic red cells and phenylhydrazine and the size of the spleen was studied either at the end of that treatment or 11 weeks later without any further injections. This was done to find out whether the greater splenic mass persists after the end of the increased work load to the transplant. The spleens or transplants were weighed (mean ± SEM) and methacrylate semi-thin sections prepared.

RESULTS

In normal rats the splenic weight was 0.50 ± 0.01g. After multiple injections of xenogeneic red blood cells only an insignificant increase to 0.56 ± 0.04g was found. Phenylhydrazine induced larger spleens of a mean weight of 1.06 ± 0.08g and methylcellulose treatment resulted in an enormous splenomegaly (mean weight 5.82 ± 1.13g), which is about 12 times the normal splenic weight.

Table 1. Relative weight (weight of the regenerated tissue as a percentage of the implanted mass) of splenic transplants after different treatments of the rats

Treatment	No. rats	Relative splenic weight (%) ± SEM	Significance		
NaCl	5	20.7 ± 3.5	$p < 0.01$		
xenogeneic red blood cells	15	46.9 ± 5.8			$p < 0.005$
			$p < 0.005$		
phenylhydrazine	17	102.0 ± 9.7			
xenogeneic red blood cells + phenylhydrazine	9	118.1 ± 15.1	$p < 0.005$		
methylcellulose	11	205.7 ± 38.0			

When a comparable protocol was used in rats after trans planting splenic fragments all treatments resulted in significantly more splenic tissue (Table 1). To study the duration of the stimulatory effect, the weight of the regenerated splenic tissue after stimulation with xenogeneic red blood cells and phenylhydrazine was studied one and eleven weeks after the end of the stimulation period. Just after the last injection the mean weight was 118.1% of the implanted mass and after a delay of 11 weeks 63.9 ± 12.0%. Thus, the splenic mass diminished when the splenic tissue was no longer stimulated, but even after 11 weeks there was still significantly more regenerated tissue than in the control rats treated with NaCl.

The histological sections revealed obvious differences (Fig. 1). In transplants treated with xenogeneic red blood cells many active germinal centres and a relatively normal marginal zone could be seen. The phenylhydrazine treatment resulted in splenic tissue consisting almost exclusively of red pulp with foci of haemopoesis. Transplants treated with methylcellulose showed hardly any normal splenic structure. Many activated, partly foamy-looking macrophages were seen in an atypical red pulp and areas of white pulp were very rare.

DISCUSSION AND CONCLUSIONS

Fatal septicaemia has been described in patients despite splenotic regeneration or accessory spleens (summarized in 6) and animal experiments have shown that about 30 to 50% of the normal splenic tissue is essential for protection against septicaemia (3,4,5). In rodents (7,8), pigs (6) and humans (2,9) the mass of regenerated splenic tissue after heterotopical autotransplantation only reached a small proportion of the normal weight. In addition, the blood flow to the splenules was much lower than to the normal spleen (6,10,11), and this had an obvious effect on the clearance function of the spleen. Our concept was to increase the work load to the different compartments of the spleen and thus stimulate regeneration. Phenylhydrazine (12) clumps the haemoglobin in erythrocytes, which are destroyed when they are squeezed through the cords of the red pulp. Particulate antigen injected intravenously stimulates the white pulp. Methylcellulose produces hypersplenism in rodents (13). Each treatment resulted in an increase in the regeneration of splenic tissue. It has yet to be shown whether the blood flow increases in a parallel fashion and whether the clearance function is more effective in such stimulated splenic transplants. The factors which regulate the size and functional capacity of a lymphoid organ, such as transplanted splenic fragments, are still unknown but of great clinical relevance.

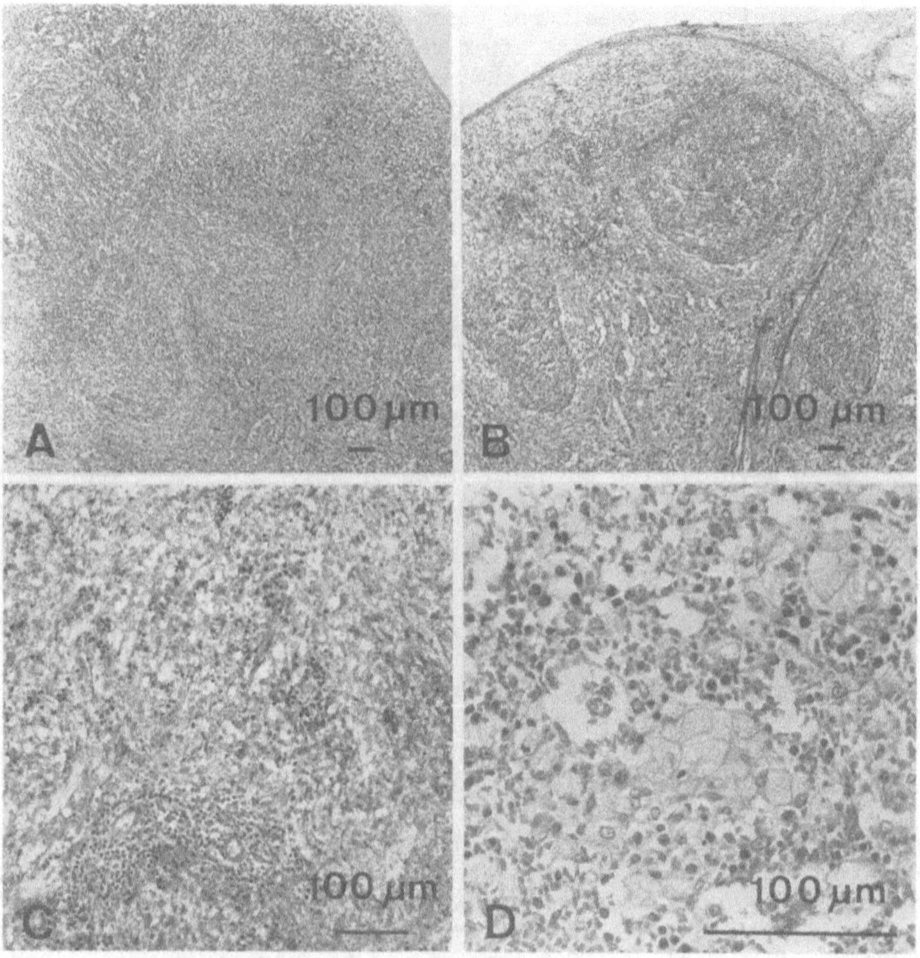

Fig. 1. Histological sections of regenerated splenic tissue after transplantation into the greater omentum.

A. No stimulation. Normal structure with white and red pulp.

B. Multiple injection of xenogeneic red cells. Many active germinal centres and a normal marginal zone.

C. Rats treated with phenylhydrazine. The spleen consists almost exclusively of red pulp.

D. Rats injected with methylcellulose. No typical histological splenic structure, but many activated macrophages can be seen.

ACKNOWLEDGEMENTS

This study was supported by the Deutsche Forschungsgemeinschaft SFB 146, A5a. We would like to thank I. Dressendörfer and M. Kaatz for their skilful technical assistance and S. Fryk for correcting the English of the manuscript.

REFERENCES

1. A.S. Leonard, G.S. Giebink, T.J. Baesl and W. Krivit. World J. Surg. 4:423 (1980).
2. J. Patel, J.S. Williams, B. Shmigel and J.R. Hinshaw. Surgery 90:683 (1981).
3. P.H. Bradshaw and C.G. Thomas. J. Surg. Res. 32:173 (1982).
4. J.A. Coil, J.D. Dickerman, S.R. Horner and B.J. Chalmer. J. Surg. Res. 28:18 (1980).
5. D.B. van Wyck, M.H. Witte, C.L. Witte and A.C. Thies. J. Surg. Res. 28:14 (1980).
6. R. Pabst and D. Kamran. J. Pediatr. Surg., in press (1984).
7. A. Alwmark, S. Bengmark, P.Gullstrand, I. Idvall and C. Schalen. Eur. Surg. Res. 15:217 (1983).
8. J.L. Nielsen, H.M.K. Andersen, K.B. Hansen, P. Sachsø, E.S. Kristensen and F.H. Sørensen. Scand. J. Haemat. 30:367 (1983).
9. R.M. Seufert, W. Böttcher, D. Munz and U. Heusermann. Chirurg 52:525 (1981).
10. R. Pabst and D. Kamran. Amer. J. Surg. 147:382 (1984).
11. J. Horton, M.E. Ogden, S. Williams and D. Colne. Ann. Surg. 195:172 (1982).
12. L.T. Chen. Blood 56:737 (1980).
13. A. Alwmark, S. Bengmark, B. Börjesson and P. Gullstrand. Eur. Surg. Res. 14:322 (1982).

RETINOID EFFECTS ON MURINE IMMUNE RESPONSES

M.Drzymala, D.R.Katz, J.A.Turton and R.M.Hicks

Pathology Department, The Middlesex Hospital Medical School, London W1P 7PN

INTRODUCTION

There is considerable literature concerning the immunomodulatory effects of vitamin A acetate and other retinoids, but several aspects of these effects remain poorly understood. An example of this is the well-known enlargement of lymphomedullary organs seen in animals fed on either vitamin A acetate (VAA) or 13-cis retinoic acid (CRA) enriched diets (Hunt et al, 1983). Amongst the questions raised by this observation are queries about the cellular nature of the enlargement, whether the enlargement is generalised or is associated with shifts in cell phenotype within the tissues; and what the functional significance of the increased peripheral lymphomedullary pool may be in relationship to the overall host immune response.

Questions such as these have assumed increased significance as evidence has accumulated which suggests that the tissue micro-environment may be an important determinant in the type of immune response elicited (e.g. Humphrey and Grennan, 1980; Klaus et al, 1981; Feldmann et al, 1984). A possible explanation for the retinoid effects could be that lymph node enlargement is associated with microenvironmental changes and that this is the underlying mechanism for some of the long term effects of these compounds.

In this presentation we describe our recent observations in a murine system. Our findings suggest that the lymph node enlargement is not a random event. Both our structural and our functional observations suggest that the interaction between T cells and non-lymphoid accessory cells (AC) may be the microenvironmental locus for retinoid effects.

MATERIALS AND METHODS

Animal Handling

Six-week-old female (bxd) haptotype F_1 mice (Olac) were used in this study. Allogeneic (k) haptotype mice were obtained from the Immunology Department, The Middlesex Hospital Medical School.

All animals were fed on a basic ground diet. Randomised animals were fed on a diet supplemented with VAA or CRA. Both the VAA and the CRA were the gift of Roche Products Ltd (Dr. N R Rogers). The additives were blended into the ground diet using a commercial mixing machine (Hobart Manufacturing). VAA was used at a concentration of 328 mg/kg diet and CRA at a concentration of 540 mg/kg diet. No significant differences in either amount of diet consumed or weight gain were observed. Animals were sacrificed at different time points up to twelve weeks and their lymph nodes and spleen removed under sterile conditions.

Tissue Handling

(i) Morphology

a) For histological studies, tissue was fixed immediately in 0.25% glutaraldehyde in phosphate buffered saline and processed routinely for haematoxylin and eosin sections.

b) For immunohistology tissue was snap frozen and sections were cut. The sections were exposed to a panel of monoclonal antisera. Reagents used included an anti Ia antibody (against $I\text{-}A^b$ determinants) (Dr G Hammerling), antibodies against Lyl and Ly2 determinants (Dr G Hammerling) and an antibody against the Qa4 antigen (Dr L Flaherty). The slides were exposed to the antisera for one hour at 4°C, post-fixed in methanol with 0.02% hydrogen peroxide and then kept at 4°C for one hour with peroxidase conjugated anti-mouse IgG (Tago). After repeated washing in Tris buffer the slides were exposed to diaminobenzidine and hydrogen peroxide, washed and then lightly counter stained. Control serum used included a non-cross reactive anti Ia antiserum and the second label antibody alone.

(ii) A single cell suspension was prepared from the lymph nodes and spleens of the test animals. Cells from part of this preparation were spun on a discontinuous bovine serum albumin gradient (8000g; 4°C; 30 minutes) and the resultant low density cells harvested. Dendritic cells (DC) and macrophages (MO) were prepared from these cells by a standard protocol (Sunshine et al 1982). The remainder of the cells, and spleen and lymph node cells from allogeneic mice, were used to make T cell enriched responder populations for subsequent functional studies.

In Vitro Functional Studies

Mixed leucocyte responses (MLR): DC and MO from the three groups were used as stimulators in a 96-hour mixed leucocyte response. The stimulator cells were irradiated (1200 rads from a 60 Co source) and different stimulator cell concentrations were co-cultured in triplicate with responder T cells. All cultures were performed in 200 ml volume under optimum conditions (37°C; humidified; 5% CO_2 in air). Medium used was RPMI (Gibco) supplemented with 5% heat inactivated foetal calf serum (Gibco) 2mM L-glutamine; 10mM HEPES buffer; 50 M mercaptoethanol and antibiotics. For the last six hours of the incubation period the cultures were pulsed with 0.5 Ci 125 Iododeoxyuridine in 10 1 volume. Radioactive macromolecular DNA was assessed by gamma counting and the results expressed as mean upon standard deviation was less than 15%.

RESULTS AND DISCUSSION

Our histological studies on enlargement of lymphomedullary organs in animals fed on VAA and CRA showed that this is not a random event. As illustrated in Fig 1 the enlargement is selective for the paracortical region which is the particular zone in the tissues at which the TS-interaction between T cells and accessory cells is considered to occur. A feature of both spleen and lymph nodes examined is that there are so few germinal centres present in these animals; that this is not a function of aging alone is illustrated by the presence of these structures in the control animals. Following these observations our standard density gradient method was used to see if we could identify an increase in the accessory cell population. In every experiment performed using animals fed on VAA there was a minimum ten fold increase in the number of low density non-adherent cells present in the mixed cell pool; with CRA fed mice the increase was less marked but equally consistent.

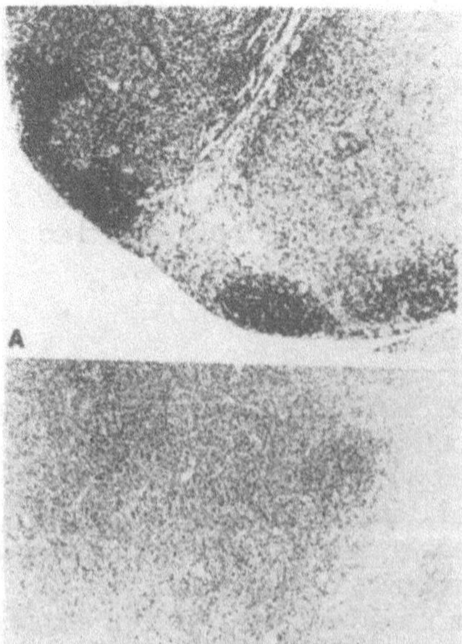

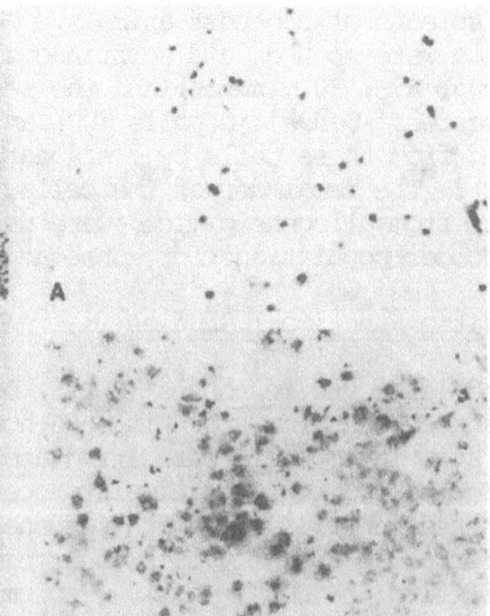

FIGURE 1. (H & E x 250). Lymph node from 12 week control (A) and VAA fed mice (B). Note the striking paracortical expansion in the test animal.

FIGURE 2. (Immunoperoxidase x 400). Paracortical region of lymph node from 12 week control (A) and VAA fed (B) mice. Note the increase in Ia expression in the test animal.

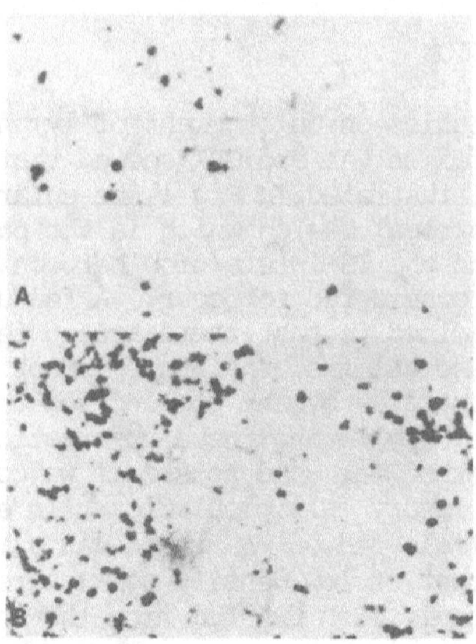

FIGURE 3. (Immunoperoxidase x 400). Paracortical region of lymph node from 12 week control (A) and VAA fed (B) mice. Note the increase in Ly1 expression in the test animal.

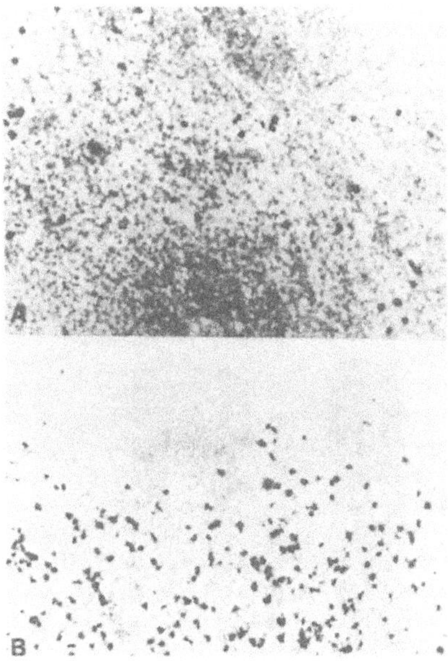

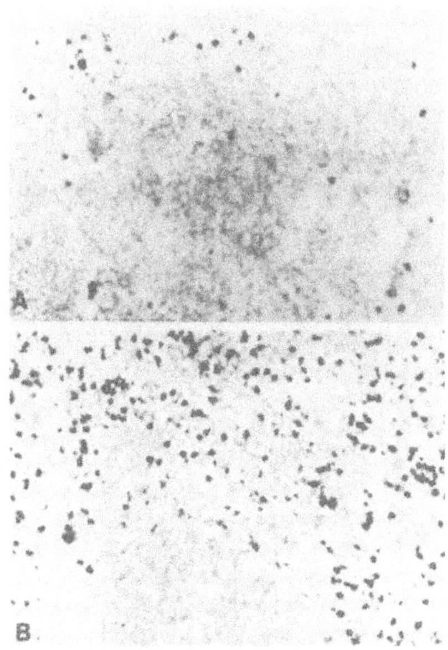

FIGURE 4. (Immunoperoxidase x 400). Paracortical region of lymph node from 12 week control (A) and VAA fed (B) mice. Note the increase in Ly2 expression in the test animal.
FIGURE 5. (Immunoperoxidase x 400). Paracortical region of lymph node from 12 week control (A) and VAA fed (B) mice. Note the increase in Qa4 expression in the test animal.

In view of both the increase in low density cells and the relative lack of B cell activation as judged by morphology, we used a panel of monoclonal antibodies to examine the in situ phenotype of the increase in paracortical cells. There was an obvious increase in the number of cells which express Ia (Fig 2). There was also an increase in the number of cells which express both the Ly markers (Fig 3 and Fig 4) as well as the Qa4 antigen (Fig 5); in contrast the number of cells which express receptors for the mouse immunoglobulin was not increased.

Taken together with the morphology and cell counts these results re-emphasise that the effect of the dietary manipulation is to increase the potential for accessory cell - T cell interaction as well as to increase the cells which express the markers linked to this interaction.

Therefore our subsequent studies were directed at this question. Results shown in Fig 6a illustrate that after seven weeks of enriched diet DC derived from the VAA group were more potent inducers of both syngeneic and allogeneic responses than DC from

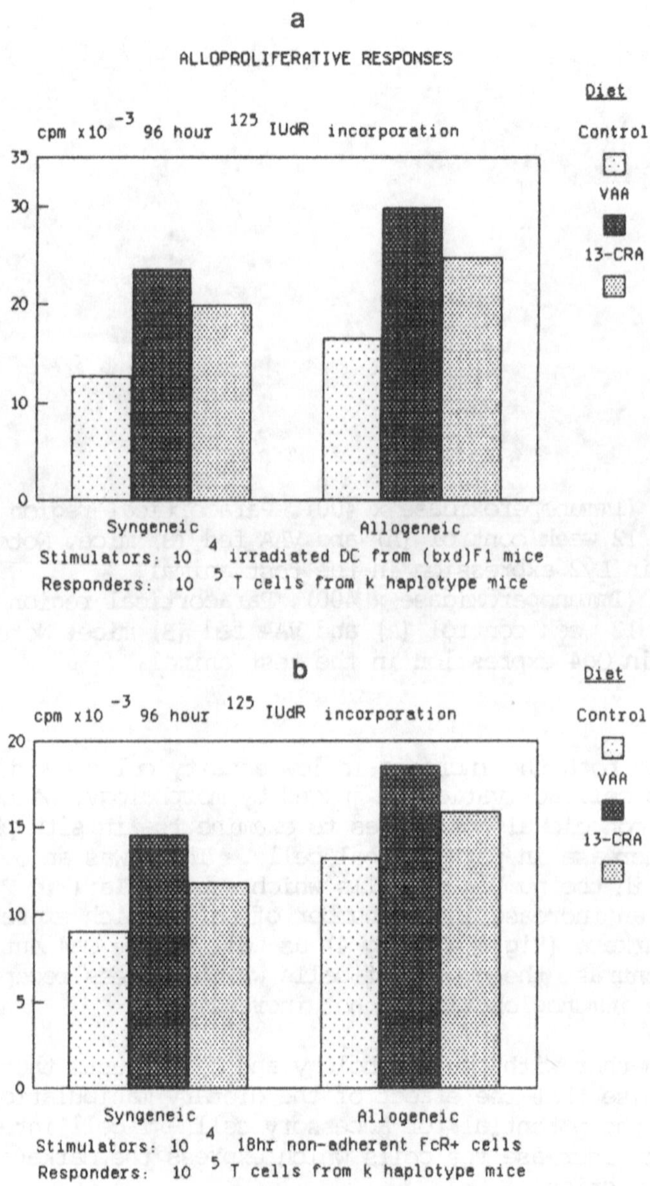
a
ALLOPROLIFERATIVE RESPONSES
cpm x10 -3 96 hour 125 IUdR incorporation
Diet
Control
VAA
13-CRA
35
30
20
10
0
Syngeneic
Allogeneic
Stimulators: 10 4 irradiated DC from (bxd)F1 mice
Responders: 10 5 T cells from k haplotype mice
b
cpm x10 -3 96 hour 125 IUdR incorporation
Diet
Control
VAA
13-CRA
20
15
10
5
0
Syngeneic
Allogeneic
Stimulators: 10 4 18hr non-adherent FcR+ cells
Responders: 10 5 T cells from k haplotype mice

Fig 6.

control animals; CRA fed mice showed a similar but less striking increase in inducer capacity. In parallel with this we also examined the MO induced response; the results are shown in Fig 6b. As expected, there is less proliferation induced than is seen with DC, but the effect of both VAA and CRA on the MO is similar to that on DC.

Thus our results indicate that there is both a quantitative and qualitative difference in the accessory cells isolated from the mice fed on these diets, i.e. that there are both more cells capable of inducer activity and more induction on a single cell basis.

From the retinoid vantage point these findings show that the underlying retinoid induced changes are clearly demonstrable at the accessory cell level. In this type of model the effector T cells which mediate the enhancement (Malkowsky et al, 1983) could either act via a mechanism such as interferon release to induce more accessory cell Ia, or could be a responder population following increased accessory cell activation and localisation to the peripheri. From the accessory cell vantage point, the apparent capacity of these compounds to act as 'activators' of both DC and MO function appears to be a unique property and is currently under further investigation.

REFERENCES

1. M Feldmann, G H Sunshine and D R Katz in The Reticulo-endothelial System vol 7A, Plenum Press New York 1984

2. J H Humphrey and D Grennan, Eur J. Immunol, 11, 221, 1981.

3. R Hunt, L Palmer, P B Medawar, J A Turton, J Gwynne and R M Hicks, Proc Nutrition Soc, 42, 13a, (1983) (abstract).

4. G G B Klaus, J H Humphrey, A Kunkl, and D W Dongworth, Immunol Rev, 53, 3, 1980.

5. M Malkovsky, A J Edwards, R Hunt, L Palmer and P B Medawar, Nature 302, 338, (1983).

6. G H Sunshine, D R Katz and A A Czitrom, Eur J Immunol, 12, 9, (1982).

EFFECT OF AVRIDINE ON ENTERIC ANTIGEN UPTAKE AND MUCOSAL IMMUNITY TO REOVIRUS (1/Lang)

A.O. Anderson and D.H. Rubin

USAMRIID, Fort Detrick, Frederick, MD 21701, USA and
Department of Medicine, The U.of Pennsylvania
Philadelphia, PA 19104, USA

INTRODUCTION

Antigen priming by a mucosal route has been shown to be superior to parenteral inoculation in eliciting a protective secretory IgA response [1]. Indeed, parenteral priming with non-replicating antigens often depresses or prevents specific IgA responses when that antigen is subsequently exposed to the mucosa[2]. Conversely, mucosal priming is often associated with induction of peripheral tolerance or reduced IgG levels and cell-mediated immunity to parenterally-administered antigen[3-5]. In addition, inactivated, attenuated, or toxoided antigens often fail to retain the essential immunogenic properties needed for an IgA response, and unmodified pathogens and toxins carry with them unacceptable toxicity. Therefore, studies focused on development of immunological adjuvants that can facilitate mucosal priming may simplify these difficulties, especially in cases where it is desirable to use polypeptide subunit vaccines produced by gene cloning. A number of studies have proposed that the ability of an antigen to elicit a secretory response depends upon its uptake and processing by mucosal lymphatic tissues. A number of hypotheses have been proposed, including one concerning specific receptors that facilitate binding of the antigen to mucosal surfaces, and the ability of the antigen to colonize the bowel lumen or invade the mucosa[6]. It is instructive that gastrointestinal pathogens, such as Salmonella , enteropathogenic E. coli, and Cholera toxin, normally have these properties, and convalescence often is associated with the presence of specific IgA in the secretions[7]. The present study presents new data indicating that the immunological adjuvant Avridine enhances secretory immunity when delivered in combination with live and inactivated reovirus antigen by facilitating the uptake and retention of antigen in mucosal lymphatic tissues.

MATERIALS AND METHODS

Animals

Adult female Balb/c mice (NIH) were used for the immunization and radiolabel uptake studies. $A/_{J}$, Balb/c(Jax), $C_{57}BL/_{Ka}$, DBA/2, CD_{F1}, and C_3H/Hej strains were used for analysis of susceptibility to development of accessory Peyer's patches. All animals were housed in Biogard laminar flow units and fed conventional mouse chow.

Antigens

Reovirus serotype 1/Lang were propagated on mouse L cells and purified as described previously[8]. The final concentration of viral particles was determined by optical density measurement at 260 nm. Infectious titers on mouse L cells were approximately 100-fold less than particle counts. Each mouse received 1×10^{10} particles of either live or UV-inactivated virus. Purified *Cholera* toxin was lot 09-72, prepared by R. A. Finkelstein and provided by Juliett Fuhrman of the University of Pennsylvania. Each mouse received a 10-microgram dose. Avridine (previously called CP20-961) was provided by Dr. Ivan Otterness (Pfizer Central Research, Groton, Conn.). A 6-mg/ml emulsion was prepared by dissolving 42 mg Avridine in 0.3 ml absolute ethanol and 0.03 ml Tween 80. The mixture was agitated with 6.67 ml of Intralipid. Avridine was mixed 1:1 with antigen and sonicated prior to inoculation.

Immunization procedures

Mice were anesthetized with methoxyflurane for direct intraduodenal (ID) injection via laparotomy[9]. Virus or *Cholera* toxin was mixed with Avridine in soybean oil emulsion (Intralipid; Cutter Laboratories, Berkley, Calif.) and sonicated for 10 seconds. The total injection volume was 0.1 ml. Groups of mice were sacrificed at 1, 3, 7, and 14 days. The entire small bowel, from duodenum through ileocecal valve, was removed, flushed with ice cold saline, and threaded onto a black glass rod for rapid assessment of the size and number of Peyer's patches. In other experiments, the Peyer's patches were individually excised and weighed.

The small bowel was opened along the mesenteric border and scraped with a scalpel blade to remove the mucus membrane. This material, containing epithelial cells, mucus, and other secretions, was sonicated and frozen for antibody measurements. Excised Peyer's patches, mesenteric lymph nodes, and spleens were prepared for histological examination.

These data have been corrected for differences in the DPM/mg of ^{3}H-reovirus compared to ^{3}H-Avridine. Total tissue recovery of labelled adjuvant was less than 10.7% of the injected activity compared to 24.9% for reovirus.

Radiolabel uptake

Reovirus 1/Lang were grown in the presence of 10mCi/ml of ^{3}H-uridine to a specific activity of 600,000 DPM/10^{10} virions. Tritiated Avridine (specific activity 11.79 mCi/mg) was obtained as a gift from Drs. Keith Jensen and Sanford Figdor of Pfizer Central Research . ^{3}H-reovirus, with or without unlabelled Avridine, was injected ID, and mice were sacrificed; the intestines, Peyer's patches, mesenteric lymph nodes, liver, spleen, and lungs were removed, weighed, and dissolved in scintillation cocktail. A Beckman scintillation counter was used to determine the DPM.

Immunoassays

Antibody was measured in serum and intestinal secretions by enzyme-linked immunosorbent assay (ELISA)[10] and hemagglutination[8]. Enzyme-labelled, affinity-purified rabbit antisera specific for alpha or gamma heavy chains were used to score the response. Values are expressed as the optical density reading at 405 nm (x 10^{-3}) of samples diluted 1/20, since OD correlated directly with titer.

RESULTS

Eighty to ninety percent of the radioactivity associated with either virus or adjuvant was present in the feces collected between 6 and 10 hours after intraduodenal dosing. Association of ^{3}H-virus or ^{3}H-adjuvant with tissues reached a peak at 2 hours, through 48 hours, and gradually declined to barely detectable levels by 7 to 10 days after ID inoculation. The specific activities for ^{3}H-virus, ^{3}H-virus plus Avridine, and ^{3}H-Avridine alone at 2 hours are representative of the uptake pattern found through 48 hours (table 1). These data have been corrected for differences in the DPM/mg of ^{3}H-reovirus compared to ^{3}H-Avridine. Total tissue activity compared to 24.9% for reovirus.

There was virtually no difference between the DPM/mg tissue for ^{3}H-virus and ^{3}H-virus plus Avridine with respect to uptake by small bowel minus Peyer's patches. In contrast, uptake of virus in Peyer's patch was significantly enhanced when virus was associated with Avridine. Since comparable specific activities were obtained with labelled Avridine alone in Peyer's patch, and there was no difference between the values for intestine and Peyer's patch in the ^{3}H-virus group, it is concluded that Avridine enhances uptake and retention of antigen in mucosal lymphatic tissue. Very little free Avridine passes beyond the intestinal cell wall. By the seventh day after inoculation, specific activities of 24.7 and 30.3 DPM/mg, respectively, were found in Peyer's patches of mice receiving ^{3}H-virus plus Avridine or ^{3}H-Avridine alone. In that time period some of the activity had migrated to the mesenteric lymph node, but no further. Radioautographs of Peyer's patches indicated that most of the retained Avridine was sequestered in mononuclear cells under the follicle-associated epithelium.

Intraduodenal immunization with reovirus, live or UV-inactivated, and Cholera toxin resulted in measureable antibody by 7 days after a single inoculation for each antigen except inactivated reovirus. Presence of Avridine in the ID-injected inoculum enhanced the reovirus and Cholera toxin secretory immune responses 2- to 4-fold in this time period (table 2). The immunopotentiating effect of Avridine depended upon its being administered with the antigen, and Avridine alone failed to prime for any of the above antigens. In contrast, reovirus-specific hemagglutinin titers were reduced when the intraduodenal inoculum contained Avridine (table 3).

TABLE 1

UPTAKE AND DISTRIBUTION OF ^{3}H-U LABELED REOVIRUS
Specific activity (DPM/mg tissue $\pm$ SEM)

Organ	^{3}H-reovirus	^{3}H-reovirus+Avridine	^{3}H-Avridine
Intestine	75.0 $\pm$ 9	75.1 $\pm$ 6	111.7 $\pm$ 7
Peyer's Patch	78.5 $\pm$ 6	219.1 $\pm$ 11	294.0 $\pm$ 14
M Lymph Node	68.9 $\pm$ 6	45.7 $\pm$ 2	6.2 $\pm$ 3
Liver	162.0 $\pm$ 13	80.6 $\pm$ 5	7.2 $\pm$ 2
Lung	75.8 $\pm$ 8	73.6 $\pm$ 4	1.6 $\pm$ 0
Spleen	101.1 $\pm$ 8	88.8 $\pm$ 4	2.9 $\pm$ 1

Table 2
SPECIFIC ANTIBODY IN INTESTINAL SECRETIONS
FOLLOWING INTRADUODENAL IMMUNIZATION

	Reovirus antibody ELISA units*			
	IgA		IgG	
vaccine composition	Day 7	Day 14	Day 7	Day 14
Reovirus 10^8 PFU	588	234	0	15
Reovirus 10^8 PFU + Avridine	1352	849	78	0
UV-ReoV 10^{10} paarticles (p)	0	67	0	100
UV-ReoV 10^{10}p + Avridine	160	220	0	100
Avridine alone	0	0	0	0
Cholera toxin	0	0	0	0
	Cholera toxin antibody			
	Day 7	Day 14	Day 7	Day 14
Cholera toxin 10 mcg	41	187	15	0
C. Toxin 10 mcg + Avridine	165	754	9	12
Avridine alone	0	0	0	0
Reovirus 10^8 PFU	0	0	0	0

*As determined by ELISA using alpha and gamma heavy chain specific antisera. Values expressed as the mean optical density reading at 405 nm (x 10^{-3}) of duplicate tests per sample diluted by 1/20.

TABLE 3

REDUCTION IN SERUM HA TITERS AFTER INTRADUODENAL PRIMING WITH
LIVE REOVIRUS (1/LANG) AND AVRIDINE

Vaccine composition ID inoculation (n=3)	Serum hemagglutination Geometric mean titers ($x10^{-1}$)		
	day 7	day 14	day 28
Reovirus (10^{10} P)	30	1210	3840
Reovirus (10^{10} P) + Avridine (0.3 mg)	38	762	605
Avridine alone (0.3 mg)	<10	<10	<10
Untreated	<10	<10	<10

The response of mucosa-associated lymphatic tissues to enteric immunization with reovirus, reovirus and Avridine, or Avridine alone is shown in table 4. Surprisingly, there were no changes in the mesenteric lymph nodes. In contrast, the spleen and Peyer's patches were the major sites of tissue reaction. All tissues were maintained at uniform hydration in buffered 10% formalin. There were three age-matched Balb/c(NIH) mice per group. The spleens enlarged reproducibly by the third day after priming, when reovirus or reovirus plus Avridine was given. Very slight splenic enlargement was seen later in the response when Avridine was given alone. Morphometric analysis of the splenic enlargement indicated that the periarteriolar lymphatic sheath (PALS) increased 4 and 6 times its mass between the first and third days following ID inoculation with reovirus and Avridine (table 5). By day 7 the PALS of both the reovirus + Avridine- and reovirus-treated mice were of comparable size, 2.5 times the untreated controls. Germinal centers showed progressive increase in mass till 7 and 14 days after ID priming. On day 7 the follicles in the reovirus + Avridine-treated spleens were 2.5 times the mass of the follicles in the reovirus control. A surprising finding was that red pulp compartments, where extramedullary hematopoiesis (EMH) occurs, were markedly enlarged when Avridine was given ID without antigen, but showed no appreciable changes if antigen was included. All tissues were maintained at uniform hydration in buffered 10% formalin. There were three age-matched Balb/c(NIH) mice per group.

TABLE 4

RESPONSE OF LYMPHATIC TISSUES TO ID IMMUNIZATION

		Wet weight in mg + -1SD			
Treatment	Day	Spleen	Lymph node	Peyer's Patches	Number PP
Untreated	0	97.3 ± 9	53.7 ± 1	43.7 ± 1	7.3 ± 1
ReoV + Avridine	3	164.1 ± 3	57.9 ± 1	68.3 ± 3	8.7 ± 1
ReoV + Avridine	7	126.1 ± 1	61.1 ± 9	74.2 ± 2	9.7 ± 0
ReoV + Avridine	14	130.0 ± 2	60.0 ± 7	67.8 ± 4	12.3 ± 1
ReoV alone	3	148.6 ± 3	53.9 ± 3	58.1 ± 5	8.0 ± 2
ReoV alone	7	123.5 ± 5	55.9 ± 2	53.9 ± 6	8.3 ± 1
ReoV alone	14	113.8 ± 9	50.1 ± 5	75.8 ± 9	11.3 ± 0
Avridine alone	3	112.3 ± 9	63.1 ± 3	61.4 ± 7	7.5 ± 1
Avridine alone	7	126.6 ± 8	61.9 ± 6	54.6 ± 1	6.3 ± 1
Avridine alone	14	120.8 ± 5	79.9 ± 8	52.7 ± 9	6.8 ± 1

TABLE 5

MORPHOMETRY OF SPLENIC LYMPHOID COMPARTMENTS

Mass of Lymphatic tissue (mg)					
Treatment ID	Day	GC	PALS	MZ	EMH
Untreated	0	1.94	14.69	14.68	6.81
ReoV + Avridine	1	2.09	58.47	24.74	2.50
ReoV + Avridine	3	4.43	86.49	37.25	4.10
ReoV + Avridine	7	8.87	39.53	28.76	10.13
ReoV + Avridine	14	8.71	42.78	27.44	9.90
ReoV alone	7	3.58	35.46	35.95	10.37
Avridine alone	7	3.67	21.02	19.25	31.66

The mass of lymphoid tissue was calculated from the wet weights using percent volumes determined on histological sections via planimetry. GC=Germinal Centers, PALS=Periarteriolar lymphatic sheath, MZ=Marginal zone, and EMH=Extramedullary hematopoietic compartments in red pulp.

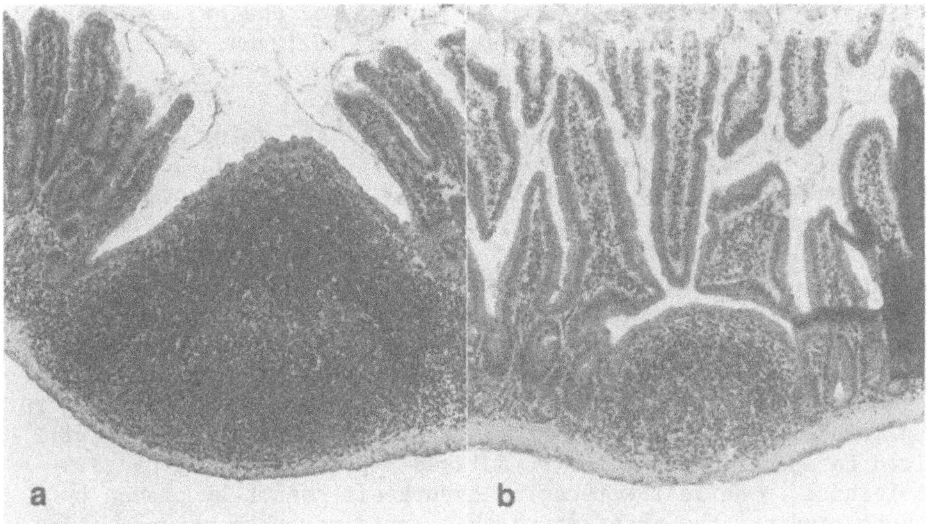

Figure 1. Accessory Peyer's patches are more than just isolated mucosal follicles since they possess specialized follicle associated epithelium (arrows) including structures that resemble M-cells, Germinal Centers (GC), and interfollicular areas containing high endothelial venules (TDA).

DISCUSSION

Secretory IgA plays an important role in host resistance against many intestinal diseases by impairing toxin binding, bacterial colonization of the mucus membranes, and by preventing reinfection by the same virus. The efficacy of this process has been well established by the use of oral immunization with live-attenuated viruses, such as the Sabin poliovaccine, which prevents both reinfection and establishment of a carrier state[1]. With the exception of intraperitoneal priming, inoculation of non-replicating protein antigens by other parenteral routes may cause depression of subsequent secretory IgA responses[2]. In addition, inactivated or toxoided antigens may not only fail to induce protective mucosal immunity but can also result in suppression of peripheral immunity following oral priming[3-5]. This is an especially important concern with regard to potential use of genetically-engineered polypeptide or subunit vaccines, since replicating/invasive viral or microbial antigens effectively cross-prime for mucosal and peripheral immunity[1,6-8], but inactivated, synthetic or toxoided antigens do not cross-prime.

An important feature of effective mucosal immunogen is the ability of these agents to bind to mucosal epithelium or at least the M-cells of Peyer's patches[11-14]. Others are examining the utility of bacterial pili or the B subunit of Cholera toxin that have a natural ability to bind to mucosal surfaces, as carriers for mucosally-applied antigens.

In the present study we used the lipoidal amine Avridine as a candidate mucosal adjuvant because previous studies suggested that Avridine might insert in epithelial cell membranes, possibly via 18 carbon lipid tails, while electrostatically binding antigen to the hydrophilic propanediamine end[15,16]. In effect, Avridine might substitute for an antigen receptor on the M-cell. A receptor-like interaction between reovirus (1/Lang) and Peyer's patch M-cells has been observed using live virus[13], but the ability of UV-inactivated reovirus to bind these specialized cells has never been demonstrated. UV-inactived reovirus is ineffective in priming for an intestinal IgA response after ID or oral inoculation but will lead to specific reduction of delayed hypersensitivity or tolerance[5]. While the receptor hypothesis cannot be proven by these studies, the observation that Avridine enhances uptake and retention of labelled antigen in Peyer's patches lends some support to the idea. We have recently reported that an IgA-isotypic indirect plaque-forming-cell response can be initiated following a single ID inoculation with SRBC and Avridine[14]. A specific receptor for SRBC is not likely to exist on follicle-associated epithelium since chronic exposure to SRBC in drinking water is required for an Iga-SRBC response in the absence of Avridine. This would imply that Avridine might substitute for the absent receptor.

Concentration of antigen in a tissue that clearly has a functional role in generating a clonal repertoire of IgA-committed B-cell precursors which react with environmental microbial antigens[9,17], certainly must contribute to the selective enhancement of antigen-specific IgA responses over other isotypes when Avridine is administered ID. Since mucosal lymphatic tissues are also centers of priming for peripheral tolerance[3-5] it is not surprising that the use of Avridine with ID-administered antigen results in enhancement of the peripheral depression of IgG simultaneous to enhancement of a secretory IgA response.

Avridine was originally developed as a topical interferon inducer[15] but was more potent as an adjuvant when given parenterally[16]. However, the ability of Avridine to stimulate interferon secretion may be responsible for some of the adjuvant effects, since interferons effect surface expression of important molecules on accessory cells such as immune response-associated (Ia) antigens[18,19]. Avridine accelerates lymphoid cell traffic into lymph nodes while slowing the traffic through the paracortex[16]. It stimulates the proliferation and release into efferent lymph of monocytes and Ia-antigen positive dendritic cells[20] which present antigens (in the context of self), thereby causing helper T-cells to proliferate[21].

By itself, Avridine is not a polyclonal B-cell activator; but, in the presence of antigen, proliferation of clonally-selected B-cells may be enhanced. Studies by Pierce and Sacci[22] indicate that Avridine enhances mucosal B-cell memory in rats. The histological nature of Pierce's antibody-forming cell assay would favor such an interpretation, since a secondary response would be required before a significant increase in the number of visually-scored specific IgA-containing B-cells could be seen in the intestine[23]. The enhancement of specific IgA secretion we observed following a single priming inoculation may have resulted from either increased committment of B-cells to specific IgA secretion or increased secretion by fewer cells. In addition, there is likely to be specific s-IgA present in intestinal secretions that had been contributed by plasma cells located in other tissues. It is now well-recognized that s-IgA is scavenged from the circulating blood in the liver and re-secreted into bile that flows into the proximal small intestine[24]. Clonal expansion of IgA-committed B-cells in Peyer's patches may result in migration and lodging of antigen-specific IgA-plasma cells in mucosal sites all over the body[25].

The ability of Avridine to enlarge the lymphoid compartments of the spleen and Peyer's patches, and to induce development of accessory mucosal lymphatic tissue in Balb/c(NIH) mice, indicates that local sequestration of adjuvant results in systemic effects that are trophic for mucosal lymphatic tissue. Development of additional germinal follicles in the intestine might be related to the enhancement of immunologic memory, since germinal follicles function to maintain B-cell memory. Although the devlopment of accessory Peyer's patches in Balb/c(NIH) mice may be a genetic peculiarity, the phenomenon correlates with the susceptibility of this strain to develop IgA-secreting plasmacytomas after IP inoculation with mineral oil. Balb/c(NIH) mice are susceptible and Balb/c(JAX) mice are resistant to plasmacytoma induction. A more thorough genetic analysis will be necessary before this phenomenon is explained.

In conclusion, Avridine enhances mucosal immunity to reovirus antigens following intraduodenal inoculation while simultaneously enhancing suppression of serum hemagglutination titers. Mucosal priming with Avridine and antigen has a trophic effect on lymphatic tissues associated with sites of initiation and maturation of mucosal immune responses; e.g, it has been demonstrated that B-cells that have been primed in mucosal lymphatic tissues must lodge in the spleen for up to 7 days before they are mature enough to lodge in the elamina propria of the intestine and differentiate into IgA secreting plasma cells[9,17]. These effects correlate with the ability of Avridine to enhance the uptake and retention of ID-inoculated antigens in mucosal lymphatic tissues, and indicate a possible mechanism by which Avridine might be useful in enhancing secretory immune responses to a variety of enteric vaccines.

REFERENCES

1. P. L. Ogra, D. T. Karzon, F. Righthand, and M. MacGillivray, Immunoglobulin response in serum and secretions after immunization with live and inactivated polio-vaccine and natural infection, New Engl. J. Med. 279:894 (1968).

2. F. T. Koster, and N. F. Pierce, Parenteral immunization causes antigen-specific cell-mediated suppression of an intestinal IgA response, J. Immunol. 131:115 (1983).

3. M. W. Chase, Inhibition of experimental drug allergy by prior feeding of the sensitizing agent, Proc. Soc. Exp. Biol. Med. 61:257 (1946).

4. J. A. Mattingly, and B. H. Waksman, Immunologic suppression after oral administration of antigen. I. Specific suppressor cells formed in rat Peyer's patches after oral administration of sheep erythrocytes and their systemic migration, J. Immunol. 25:941 (1978).

5. D. H. Rubin, H. L. Weiner, B. N. Fields, and M. I. Greene, Immunologic tolerance after oral administration of reovirus: requirement for two viral gene products for tolerance induction. J. Immunol. 127:1697 (1981).

6. P. B. Carter and F. M. Collins, Peyer's patch responsiveness to Salmonella in mice, J. Reticuloendothel. Soc. 17:38 (1975).

7. J. J. Cebra, P. J. Gearhart, J. F. Halsey, J. L. Hurwitz, and R. D. Shahin, Role of environmental antigen in the ontogeny of the secretory immune response, J. Reticuloendothel. Soc. 28:61(Suppl.) (1980).

8. D. H. Rubin, and B. N. Fields, Molecular basis of reovirus virulence: role of the M2 gene, J. Exp. Med. 152:853 (1980).

9. J. A. Fuhrman, and J. J. Cebra, Special features of the priming process for a secretory IgA response. B-cell priming with Cholera toxin, J. Exp. Med. 153:534 (1981).

10. D. H. Rubin, A. O. Anderson, and D. Lucis, Potentiation of the secretory IgA response by oral and enteric administration of CP-20,961 Ann. N. Y. Acad. Sci. 409:866 (1983).

11. D. E. Bockman and M. D. Cooper, Pinocytosis by epithelium associated with lymphoid follicles in the bursa of Fabricius, appendix, and Peyer's patches: an electron microscopic study, Am. J. Anat. 136:455 (1973).

12. R. L. Owen, and A. L Jones, Epithelial cell specialization within human Peyer's patches: an ultrastructural study of intestinal lymphoid follicles, Gastroenterology 66:189 (1974).

13. J. L. Wolf, D. H. Rubin, R. Finberg, R. S. Kaufman, A. H. Sharpe, J. S. Trier, and B. N. Fields, Intestinal M cells: a pathway for entry of reovirus into the host, Science 212:471 (1981).

14. A. O. Anderson, T. T. MacDonald, and D. H. Rubin, Effect of orally administered Avridine on enteric antigen uptake and mucosal immunity, Internatl. J. Tiss. React. (in press).

15. J. F. Niblack, I. G. Otterness, G. R. Hemsworth, J.S. Wolff, W. W. Hoffman, III, and A. R. Kraska, CP-20,961: a structurally novel, syntetic adjuvant, J. Reticuloendothel. Soc. 26:655 (Suppl) (1979).

16. A. O. Anderson and J. A. Reynolds, Adjuvant effects of the lipid amine CP-20, 961, J. Reticuloendothel. Soc. 26:667 (Suppl) (1979).

17. S. W. Craig and J. J. Cebra, Peyer's patches: an enriched source of precursors for IgA producing immunocytes in the rabbit. J. Exp. Med. 134:188 (1971).

18. P. S. Steeg, R. N. Moore, and J. J. Oppenheim, Regulation or murine macrophage Ia antigen expression by products of activated spleen cells, J. Exp. Med. 152:1734 (1980).

19. M. Nakamura, T. Manser, G. D. N. Pearson, M. J. Daley, and M. L Gefter, Effect of IFN-gamma on the immune response in vivo and on gene expression in vitro, Nature 307:381 (1984).

20. A. O. Anderson and J. T. Warren, Effect of immunological adjuvants on the appearance of small monocytes and dendritic cells in rat thoracic duct lymph, Adv. Exp. Med. Biol.149:791 (1982).

21. R. M. Steinman, and M. D. Witmer, Lymphoid dendritic cells are potent stimulators of the primary mixed leukocyte reaction in mice, Proc. Natl. Acad. Sci. 75:5132 (1978).

22. N. Pierce and J. B. Sacci, Jr., Enhanced mucosal priming by Cholera toxin and Pocholeragenoid using a lipoidal amine adjuvant (Avridine) delivered in Liposomes, Infect. Immun. 44:469 (1984).

23. A. J. Husband and J. L. Gowans, The origin and antigen-dependent distribution of IgA-containing cells in the intestine. J. Exp. Med. 148:1146 (1978).

24. E. Orlans, J. Peppard, J. Reynolds, and J. Hall, Rapid active transport of immunoglobulin A from blood to bile, J. Exp. Med. 147:588 (1978).

25. M. R. McDermott, and J. Bienenstock, Evidence for a common mucosal immunologic system: I. migration of B immunoblasts into intestinal, respiratory and genital tissues, J. Immunol. 122:1892 (1979).

SESSION 4

LYMPHOID CELL TRAFFIC PATTERNS

CHAIRPERSONS: I. McConnell
I.C.M. MacLennan

REGULATION OF THE SIZE OF THE RECIRCULATING B CELL POOL OF ADULT RATS

J. Lortan[1], D. Gray[1], D. S. Kumararatne[1], B. Platteau[2], H. Bazin[2], and I.C.M. MacLennan[1]

[1]Dept. of Immunology, University of Birmingham Medical School, Birmingham B15 2TJ. [2]Experimental Immunology Unit, University of Louvain, 1200 Brussels, Belgium

INTRODUCTION

Transfer of syngeneic B lymphocytes into immunologically-intact animals results in rapid elimination of these cells (1). This has been interpreted to indicate a relatively short half life of mature B-lymphocytes. In contrast, the transfer of lymphocytes into lymphocyte-depleted animals (2), allows for prolonged survival of these cells.

We have investigated the transfer of kappa-allotype-marked B cells into normal and cyclophosphamide-treated recipients in an attempt to establish a stable, long term, B-cell-chimaeric state. Parabiosis between congenic animals, differing only in kappa-light-chain immunoglobulin allotype, was undertaken with the same aim.

MATERIALS AND METHODS

Rats

Adult, male, PVG/cKla and PVG/cKlb congenic pairs which differ in their kappa immunoglobulin-light-chain allotype (3), were used in the cell transfer experiments. PVG/c Kla rats were used as donors and PVG/c Klb rats as recipients. In the parabiosis experiments, age,and sex-matched, congenic LOU/CK1a and LOU/CKlb rats were used (4).

Cyclophosphamide

Cyclophosphamide was given intraperitoneally, as a single bolus of 500 mg/m^2 (5) 48 hours prior to cell transfer.

Thoracic Duct Lymphocytes

Thoracic duct lymphocytes were collected into RPMI 1640, supplemented with antibiotics and heparin 20 units/ml (6).

Spleen Lymphocytes

Spleen lymphocytes were suspended in RPMI 1640 and used within one hour.

Cells Used for Transfer

Cells used for transfer were maintained at 0-4°C, suspended in 1 ml of sterile normal saline and injected intravenously via the dorsal vein of the penis. Each recipient animal received 2 x 10^8 thoracic duct lymphocytes or all the lymphocytes obtained from one donor spleen.

Parabiosis

Parabiosis was performed according to the method of Bunster and Meyer (7) between LOU/CKla and LOU/CKlb animals.

Spleens

Spleens were snap frozen in liquid nitrogen and stored at -70°C.

Antisera

Polyclonal rabbit anti-rat IgM, rabbit anti-rat IgD and monoclonal mouse anti-rat kappa la Mark-3[4] antisera were produced in the Experimental Immunology Unit, Brussels. The antisera were directly conjugated to fluorochromes by Dr. G. D. Johnson, Department of Immunology, University of Birmingham.

Immunohistology

5 µm frozen sections of spleen were stained for the presence of surface membrane IgM, IgD and kappa la immunoglobulin-light-chain allotype by the peroxidase-anti-peroxidase (PAP) (8) and double immunofluorescence techniques (9).

RESULTS

The Effect of Lymphocyte Depletion on the Level of B-Cell Chimaerism Achieved by Passive Transfer of Lymphocytes from K Allotype-Marked, Congenic Donors

a) Thoracic-duct-lymphocyte transfer. Two groups of PVG/c Klb rats were selected. One group was treated by I.P. injection of 500 mgs/m^2 of cyclophosphamide. Two days later all rats in both the groups were given 2×10^8 thoracic duct lymphocytes from congenic,PvG/c Kla donors. The level of chimaerism achieved was assessed immunohistologically 7 days later. Only occasional donor B-cells were detected in sections of the spleens from intact (non-drug-treated) rats (Fig. la, b, c). By contrast cyclophosphamide-treated rats showed a high level of donor B cell chimaerism. Most of the donor B cells were located in the follicles and were SmIgM+ve IgD+ve (Fig. 2a, b).

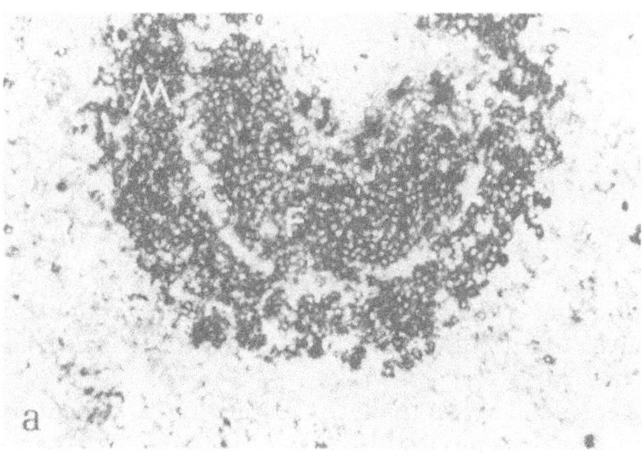

Fig. la. Spleen of normal rat following,cell-transfer. Stained for SmIgM (PAP). F = follicle, M = marginal zone.

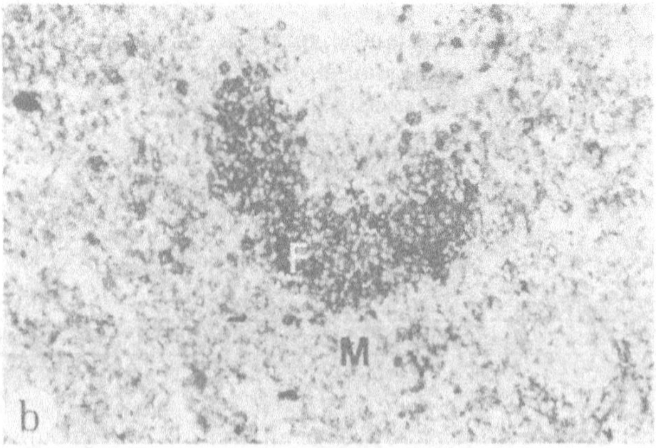

Fig. 1b

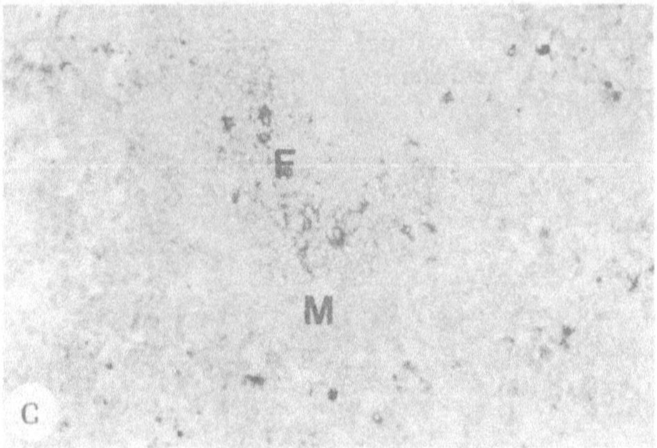

Fig. 1c

Fig. 1b. Spleen of normal rat following cell transfer. Stained for SmIgD (PAP). F = follicle, M = marginal zone.

Fig. 1c. Spleen of normal rat following cell transfer. Stained for K-allotype (PAP). F = follicle, M = marginal zone. 7 days after cell transfer very few PVG/c Kla B-cells are detected in PVG/c Klb spleen.

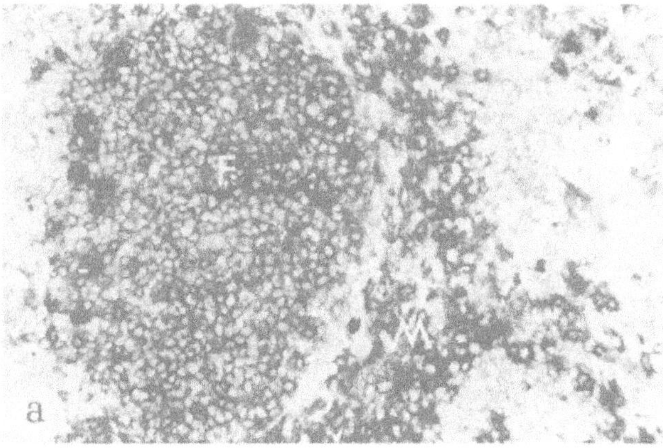

Fig. 2a

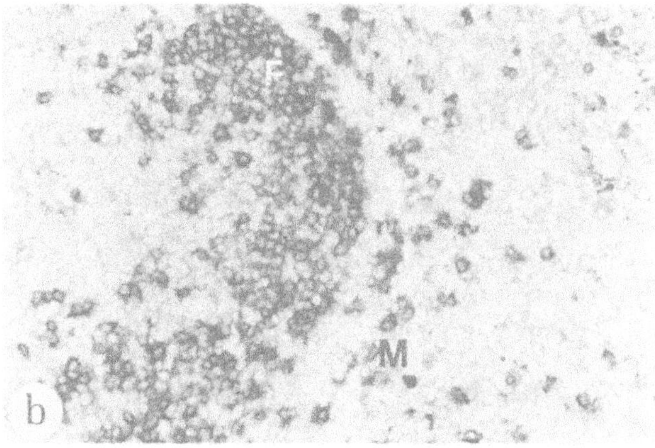

Fig. 2b

Fig. 2a. Spleen of cyclophosphamide-treated rat following thoracic-duct-lymphocyte transfer. Stained for SmIgM (PAP).

Fig. 2b. Spleen of cyclophosphamide-treated rat following thoracic-duct-lymphocyte transfer. Stained for K-allotype (PAP). F = follicle, M = marginal zone. 7 days after cell transfer extensive follicular and minimal marginal zone B cell chimaerism is seen.

b) Spleen cell transfer. The above experiment was repeated but with the substitution of spleen cells for thoracic duct lymphocytes. Each rat received the entire cell yield from a PVG/c Kla spleen. Again intact recipients failed to achieve more than minimal chimaerism. The rats pre-treated with cyclophosphamide were seen to have donor B cells in both the follicles and marginal zones (Fig. 3a, b).

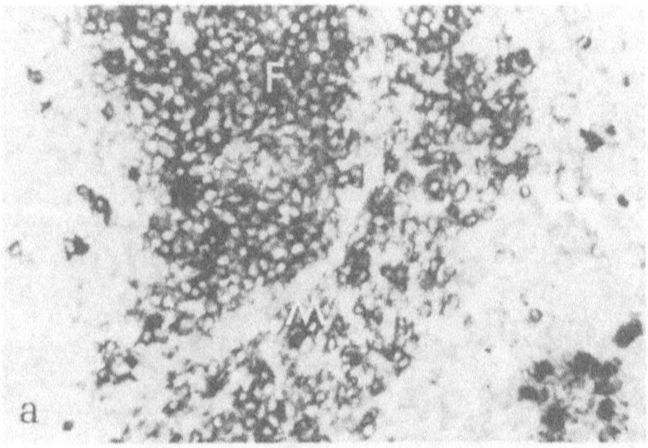

Fig. 3a

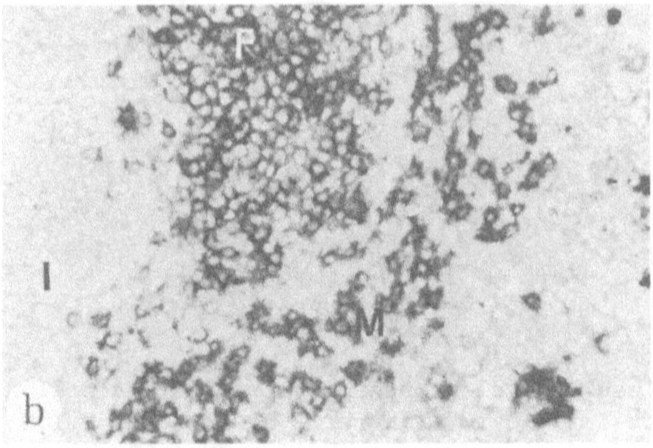

Fig. 3b

Fig. 3a. Spleen of cyclophosphamide-treated rat following spleen cell transfer. Stained for SmIgM (PAP).

Fig. 3b. Spleen of cyclophosphamide-treated rat following spleen cell transfer. Stained for K-allotype (PAP). 7 days after cell transfer both follicular and marginal zone B cell chimaerism has occurred.

Evidence That Intact Donor Resistance to Allotype-Marked Congenic B Cells is not Due to Rejection Directed Specifically Against Allo-Antigens

To test the resistance of intact Klb rats to congenic Kla B cells LOU/CKla rats were parabiosed to LOU/CKlb rats. The cross circulation achieved allows the exchange of recirculating, follicular B cells. As the combined B cell pool of the parabiosed pair would remain unaltered during recirculating B cell exchange, chimaerism would be anticipated, unless donor cells were immunologically rejected. Ten days after establishing parabiosis some rats were sacrificed for immunohistological assessment of their lymphoid organs. The remaining pairs were separated and were subsequently killed 15 days after separation. The level of chimaerism achieved in SmIgD-positive B cells in follicles at 10 days was 25.8% donor cells ± S.E. 1.3% (Fig. 4a, 4b). Fifteen days after separation this was 24.7% ± S.E. 0.6%. Marginal zone SmIgD-ve B cell chimaerism at 10 days was minimal at 1.3% ± S.E. 0.2%. These data provide no evidence of immunological rejection of B cells between these congenic strains.

DISCUSSION

The hypothesis that rapid elimination of mature B lymphocytes transferred into intact animals is a reflection of physiological rapid turnover of these cells (1) is not compatible with the data presented here. Chimaerism was achieved by two means: destruction of host B cells by cyclophosphamide and balanced B lymphocyte exchange. In both cases the chimaerism achieved was stable. These results are in accord with those of other experiments which have shown labelling indices in the order of 25% in follicles and marginal zones following 5-day tritiated-thymidine infusion into rats (11).

Our data are compatible with the hypothesis that both recirculating follicular and marginal zone B cell numbers are closely regulated, excess B cells in either pool being eliminated. The mechanism by which this regulation of B cell pool size is achieved is obscure. It seems unlikely that this is controlled by assessment of the rate of traffic through lymphoid organs, for donor cells must have traversed high endothelial venules at supra-normal rates following thoracic duct lymphocyte transfer into cyclophosphamide-treated rats. It seems likely that a separate mechanism exists for regulating the number of newly-formed virgin B cells into the mature pool. Osmond and co-workers (12) have estimated that B cell production in adult mouse bone marrow is sufficient to replenism the mature B cell pool within a week. However, the data presented here and those discussed above do not indicate that such rates of mature B cell turnover occur.

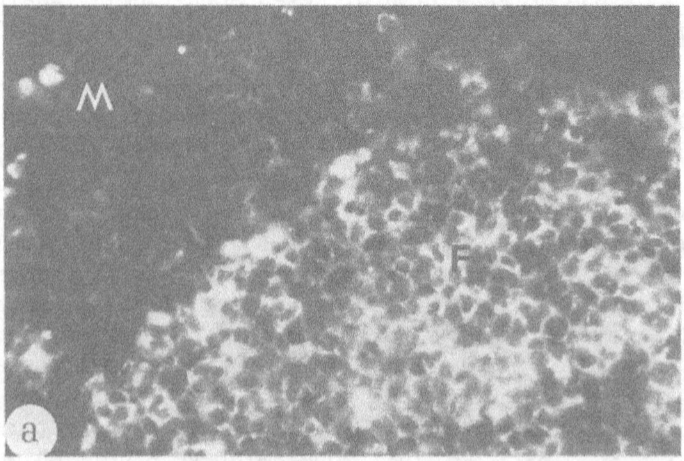

Fig. 4a

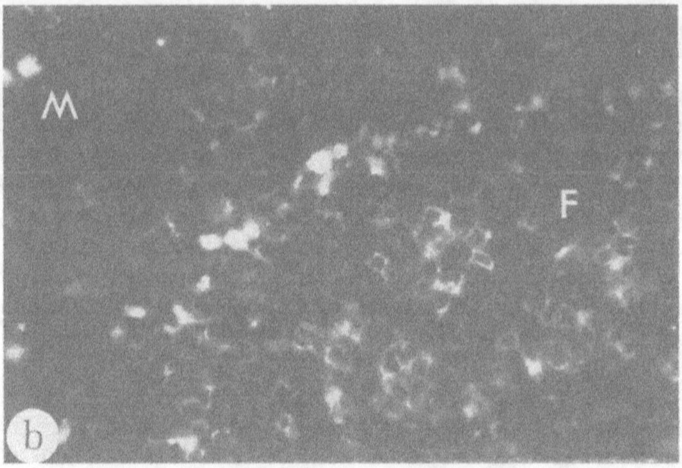

Fig. 4b

Fig. 4a. Spleen of LOU/ CKlb rat following parabiosis. Stained for SmIgM (Immunofluorescence).

Fig. 4b. Spleen of LOU/Cklb rat following parabiosis. Stained for Kla allotype (Immunofluorescence). 10 days after parabiosis follicular B cell chimaerism has been achieved.

It is of interest to speculate on the effect of mature B-cell-pool-regulation in neoplasias of recirculating B cells such as chronic lymphocytic leukaemia and prolymphocytic leukaemia. On the one hand these mechanisms may tend to reduce neoplastic B cell numbers. Conversely the same mechanisms might result in a markedly reduced number of non-neoplastic B lymphocytes. This may be one mechanism for the secondary hypogammaglobulinaemia seen in these conditions. Clearly it is necessary to investigate the sites of elimination of excess mature B cells. The use of congenic strains of rats differing at kappa allotype and immunohistology should prove a valuable tool in such studies.

REFERENCES

1. A. A. de Freitas, A. Coutinho. J. Exp. Med. 154:994 (1981).
2. R. F. Gagnon, I. C. M. MacLennan. Immunology 45:283 (1982).
3. S. V. Hunt, M. H. Fowler. Cell tissue Kinetics 14:445 (1981).
4. H. Bazin, F. Cormont and L. De Clerq. J. Immunol. Methods 71:9-16 (1984).
5. D. S. Kumararatne, R. F. Gagnon, Y. Smart. Immunol. 40:123 (1980).
6. W. L. Ford, _in_: "Handbook of Experimental Immunology", D. M. Weir, ed., Blackwell Scientific Publications, 3rd Edition, Oxford (1978).
7. E. Bunster, R. K. Meyer. The Anatomical Record 57(4):339 (1933).
8. G. Hoffman-Fezer, H. Rodr, M. Eulitz, S. J. Thierfelder. J. Immunol. Methods 13:261 (1976).
9. G. D. Johnson, E. J. Holborow, J. Dorling, _in_: "Handbook of Experimental Immunology", D. M. Weir, ed., Blackwell Scientific Publications, 4th Edition, Oxford (In print).
10. D. Gray, I. C. M. MacLennan, H. Bazin, M. Khan. Eur. J. Immunol. 12:564 (1982).
11. D. S. Kumararatne and I. C. M. MacLennan. Eur. J. Immunol. 11.865 (1981).
12. D. G. Osmond, M. T. E. Fahlman, G. M. Fulop and D. M. Rahal. CIBA Foundation Symposium 84:68 (1981).

THE INFLUENCE OF COMPLEMENT RECEPTORS ON THE LOCALIZATION OF LYMPHOCYTES IN VIVO

G. Kraal and I. van Hoogstraten

Department of Histology, Medical Faculty
Free University, Amsterdam
The Netherlands

INTRODUCTION

T and B lymphocytes entering the spleen leave the blood stream in the sinuses of the marginal zone and will migrate from there into the white pulp of the spleen[1]. It has been demonstrated that there is a preference of B cells to localize in the spleen[2-4] which is comparable to the localization patterns of lymphocytes in Peyer's patches[2]. The splenic immigration of T cell subsets, however, is more comparable with peripheral lymph nodes[5]. The selective entrance of T and B cells in the spleen is therefore different from that in the high endothelial venule (HEV)-bearing lymphoid organs such as lymph nodes and Peyer's patches. How this selection in the spleen is accomplished is not known. In the marginal zone various macrophages can be identified[6,7] and it has been suggested that these cells play a role in lymphocyte traffic[8-10]. Macrophages can be influenced by complement factors[11] and we therefore approached the question of selective lymphocyte migration in the spleen by looking at the effects of complement depletion by Cobra Venom (CVF) administration.

MATERIALS AND METHODS

In vivo homing studies

Spleen of lymph node cells from C3D2 mice (G.O. Bomholtgärd Ltd., Denmark) were labeled with TRITC as described[4] and injected into syngeneic recipients treated with CVF 24 h prior to transfer. Two and 24 h after cell transfer suspensions of axillary and brachial lymph nodes, spleen and Peyer's patches were stained for fluorescein labeled Thy-1. Using a fluorescence microscope, donor-derived cells were detected using filters for rhodamine and by switching to fluorescein filters their phenotype was determined.

Cobra Venom Factor

Cobra Venom Factor (CVF) (Naja Naja; Sigma) was purified as described by Ballow and Cochrane[12]. Mice received two intraperitoneal injections of 1 unit CVF in saline. Unit determination and determination of serum C3 depletion of treated mice has been described[13].

Preparation of EAC and EAC rosette test

The preparation of EAC has been described[14]. The rosette test was performed according to Bianco et al.[15]. Nucleated cells surrounded by at least three erythrocytes were scored as rosettes.

Preparation of C3 deficient serum in vitro

Depletion of serum C3 in vitro was performed according to Jessen[16] using methylamine. The serum was dialyzed extensively with PBS prior to cell incubation.

RESULTS

When control lymph node cells were transferred into CVF-treated recipients the B/T cell ratio of lymphocytes localizing in the spleen had decreased (Fig. 1). This low B/T ratio was found at 2 h as well as at 24 h after cell transfer. The localization of donor lymphocytes in HEV-bearing lymph nodes and Peyer's patches was completely comparable to the control situation (Fig. 1). When cells from CVF-treated animals were transferred into control animals no different localization patterns could be observed compared with the transfer of control cells. After treatment with CVF no changes were seen in the ratio's of B and T cells that were present in the lymphoid organs. These results suggest that the B/T localization shift after CVF treatment results from direct interference of CVF with the selective entrance of lymphocytes into the spleen and has no influence on the distribution of lymphocyte subsets within the organs.

Serum factors

CVF administration results in a depletion of intact C3 and high levels of C3b in serum. C3b, interacting with B cells via their CR1 receptors, could be involved in the observed disturbances. To test what the effects of CR1 receptor blockade would be on migration, lymphocytes were treated in vitro with CVF alone, or in the presence of normal mouse serum. Blockade of CR1 receptors was determined by the ability to form EAC-rosettes. The number of EAC-rosette forming cells was not influenced by incubation with CVF alone but a strong reduction was seen after incubation with CVF plus serum as well as

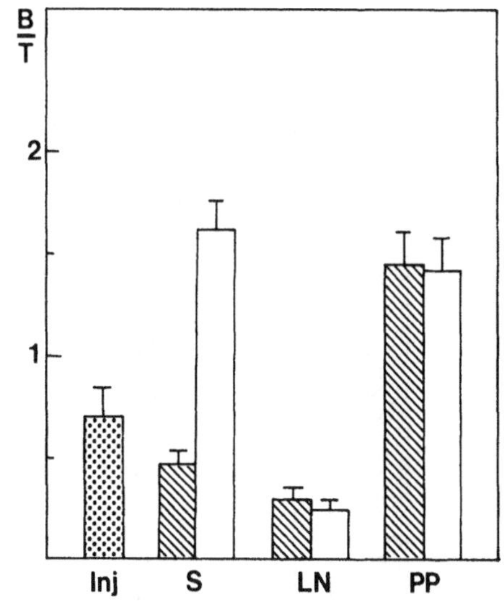

Fig. 1. Localization of B and T cells after CVF-treatment. Normal lymph node cells were rhodaminated and transferred into recipients that had been treated with CVF 24 h previously ($50x10^6$ cells per recipient). Two hours after transfer spleen, axillary, and brachial lymph nodes and Peyer's patches were removed and cell suspensions were stained for Thy-1. Using double immunofluorescence the ratio of B over T cells of the immigrated cells was determined. Per organ the phenotype of 300-400 immigrant cells was determined. Results are expressed as ratio B/T (± S.D.) of a representative experiment. The dotted bar represents the ratio of the injected donor population. Striped bars are the ratio's found in the spleen (S), lymph nodes (LN) and Peyer's patches (PP) of CVF-treated animals. Open bars represent the ratio's found in control animals.

after incubation in serum only (Table 1). The incubation in serum was performed at 37°C. At this temperature the complement cascade resulting in C3b formation takes place thereby mimicing the in vivo situation after CVF administration.

CR1 receptor blockade by serum incubation of the cells before transfer into normal recipients resulted in a B/T localization shift in the spleen comparable to the one found after CVF-treatment. Prior blockade of CR1 receptors, however, as in the case of CVF-treatment in vivo, had no influence on the localization patterns via HEV's (Fig. 2).

Further proof that the observed shifts were induced by CR1 blockade was obtained by transferring cells after in vitro incubation

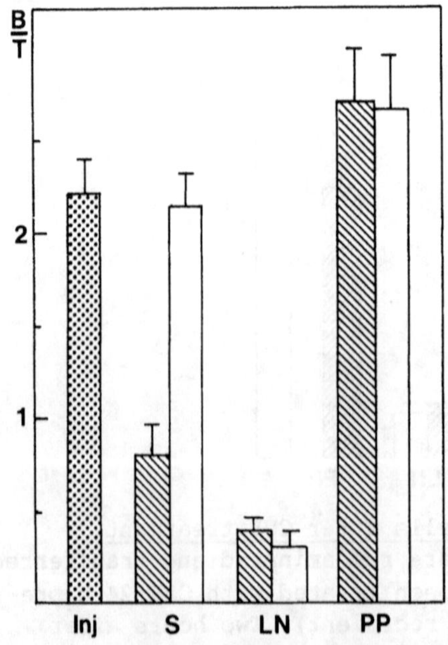

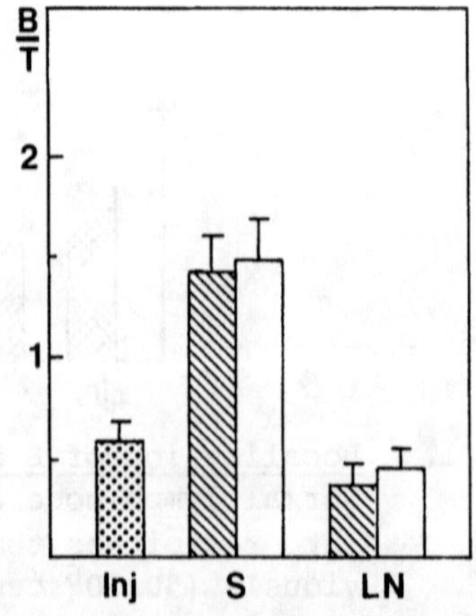

Fig. 2. The influence of serum incubation prior to cell transfer. Lymph node cells were rhodaminated and 250 µl aliquots at $3x10^8$ cells/ml were incubated for 30 min at 37°C in normal mouse serum (50% v.v.). The cells were washed twice and transferred in normal recipients. Two hours after cell transfer spleen (S), lymph nodes (LN) and Peyer's patches (PP) were removed and single cell suspensions were stained for Thy-1. Using double immunofluorescence the ratio of B/T cells of the immigrant cells was determined. Per organ the phenotype of 300-400 immigrant cells was determined. The dotted bar represents the ratio of the injected donor population. Striped bars are the ratio's found after serum incubation of the transferred population. Open bars represent the values without prior serum incubation. Results are expressed as the ratio B over T (± S.D.) of a representative experiment.

Fig. 3. Cell transfer after incubation in C3-deficient serum. The experiments are similar to those described in Fig. 2, except that spleen cells were used as donor cells and that serum incubation was performed in serum depleted of C3 using the methyl-amine method. See legend Fig. 2 for data analysis.

Table 1. In vitro incubation of spleen cells with CVF and serum

Incubation	EAC-rosette forming cells % of control
medium	100
CVF	87.4
CVF + NMS 25% v/v	18.7
CVF + NMS 50% v/v	3.5
NMS 25%	27.1
NMS 50%	12.5
C3 depleted serum	98.2

Spleen cells were incubated at 37°C for 30 min at $2x10^6$/ml in the various incubation mixtures. Concentration of CVF was 1 unit/ml. The cells were washed twice and the number of EAC-rosettes was determined. Results are expressed as percentage of control (incubation in medium). Per incubation at least 200 viable cells were counted.

in serum that had been depleted of C3 previously. Unchanged localization patterns (Fig. 3) and no reduction in EAC-rosette forming cells was observed (Table 1), illustrating that blockade of CR1 receptors by C3 cleavage products interferes with the ability of B cells to localize in the spleen.

DISCUSSION

The correlation between the blockade of CR1 receptors by serum factors and the observed shift in B/T localization ratio in the spleen points to a role of the CR1 receptor in B cell retention in the spleen. Occupation of CR1 receptors might result in a disturbed ability of these B cells to migrate into the splenic lymphoid compartment leading to a faster transit time for B cells. This explains why after CVF treatment in vivo as well as after serum incubation in vitro similar localization disturbances can be observed. It is not clear how important the CR1 receptor is at splenic B cell migration or retention under physiological conditions and if so, how this is accomplished. However, the fact that the receptor blockade has no influence on the localization in lymph nodes and Peyer's patches clearly demonstrates that the blockade does not interfere with major physiological processes of the B cell. It is an event specific for the spleen and illustrates the differences in localization mechanisms between the spleen and HEV-bearing lymphoid organs.

REFERENCES

1. P. Nieuwenhuis, and W.L. Ford. Cell. Immunol. 23:254 (1976).
2. G.A. Gutman, and I.L. Weissman. Transplantation 16:621 (1973).
3. J. Sprent. Cell. Immunol. 7:10 (1973).
4. S.K. Stevens, I.L. Weissman, and E.C. Butcher. J. Immunol. 128: 844 (1982).
5. G. Kraal, I.L. Weissman, and E.C. Butcher. J. Immunol. 130:1097 (1983).
6. J.H. Humphrey, and D. Grennan. Eur. J. Immunol. 11:221 (1981).
7. C.D. Dijkstra, E. van Vliet, E.A. Döpp, G. v.d. Lelij, and G. Kraal. Immunology, in press.
8. W.L. Ford. Progr. Allergy 19:1 (1975).
9. A.A. Freitas, and M. de Sousa. Eur. J. Immunol. 6:629 (1976).
10. R. Brelinska, and Ch. Pilgrim. Cell Tissue Res. 233:671 (1983).
11. B. Rutherford, and H.A. Schenkein. J. Immunol. 130:874 (1983).
12. M. Ballow, and Ch.G. Cochrane. J. Immunol. 103:944 (1969).
13. H. van Dijk, P.M. Rademaker, and J.M.N. Willers. J. Immunol. Meth. 36:29 (1980).
14. H. van Dijk, D. van Heuven-Nolsen, P.M. Rademaker, N. Bloksma, and J.M.N. Willers. Cell. Immunol. 51:402 (1980).
15. C. Bianco, R. Patrick, and V. Nussenzweig. J. Exp. Med. 132:702 (1970).
16. T.E. Jessen, V. Barkholt, and K.L. Welinder. J. Immunol. Meth. 60:89 (1983).

INTERACTION OF HIGH ENDOTHELIAL VENULES WITH T AND B CELLS AFTER ANTIGENIC STIMULATION

G. Kraal and A. Twisk

Department of Histology, Medical Faculty, Free University Amsterdam, The Netherlands

INTRODUCTION

The migration of lymphocytes into lymph nodes and Peyer's patches is controlled by interaction with the specialized endothelium of postcapillary, high endothelial venules (HEV). This selective interaction was studied by Butcher et al. using short-term *in vivo* homing studies and *in vitro* HEV adherence assays[1,2]. It was suggested that the limited organ preferences shown by most T and B cells as well as by T cell subsets can be explained by expression of different proportions of peripheral node and Peyer's patch specific receptors[1-4]. Further indications for such dual receptor specificity have come from the development of a monoclonal antibody, MEL-14, specific for the lymphocyte receptor mediating the recognition of peripheral node HEV's[5].

It is well documented that blast cells arising after antigenic stimulation of lymph nodes localize specifically in lymph nodes, whereas gut-derived lymphoblasts show remarkable specificity for Peyer's patches and the gut wall after i.v. injection[6,7]. Because of such almost absolute specificity for one of the two HEV receptor types antigen-reactive cells will preferentially return to sites where they are likely to meet similar antigens.

Very shortly after antigenic stimulation (e.g. 3 days) we have observed a striking change in the ratio of T and B cells in peripheral lymph nodes[8]. After stimulation the percentage of B cells in lymph nodes can increase from 25% up to 75%. These rapidly occurring changes cannot be explained on the basis of cell proliferation only, but may be the result of specific retention of especially B cells. Another explanation could be that the specificity of the high endothelial venules will be influenced by the physiological changes that occur in the lymph nodes after antigenic stimulation, contributing to the observed changes in T and B cell distribution. Therefore in this study the specificity of HEV's was compared in stimulated vs non-stimulated lymph nodes in short-term homing experiments using double immunofluorescence.

MATERIALS AND METHODS

In vivo homing

BALB/c spleen cells were labeled with TRITC as described[3] and injected into syngeneic recipients ($5x10^7$ cells/animal). The lymph nodes of the axillary region of the recipients had been challenged by injecting antigens in the front footpads or, in the case of oxazolone-treatment by painting the axillary region. Two and 24 h after transfer the recipients were killed and suspensions of the axillary and brachial lymph nodes were stained for fluorescein labeld Thy-1. Using a fluorescence microscope donor derived cells were detected using filters for rhodamine and by switching to fluorescein filters their phenotype was determined.

In vitro lymphocyte binding

This technique has been described[2]. Briefly, lymphocytes were incubated at 4°C on unfixed frozen sections of either non-stimulated or oxazolon-stimulated peripheral lymph nodes. The lymphocytes were a mixture of rhodaminated mesenteric lymph node and Peyer's patch lymphocytes as standard population and a sample population of unlabeled peripheral lymph node lymphocytes. The slides were agitated throughout incubation. After 30 min the sections were fixed and examined under darkfield illumination to identify HEV-adherent cells and after a change to UV epi-illumination the numbers of rhodaminated and unlabeled cells on a given HEV were scored and tabulated as data pairs. The ratio of labeled to unlabeled cells and its standard error were calculated using the delta method[9].

RESULTS

Localization in vivo

In control lymph nodes as well as in lymph nodes undergoing an immune response a clear preference of T cells to localize was demonstrated at 2 h after cell transfer (Fig. 1). This 2 h time point was chosen to exclude excessive recirculation so that the localization of donor cells will reflect the initial localization as determined by the high endothelial venules. In Fig. 1 the results after KLH stimulation are shown. Similar results were found after oxazolon treatment and injection of allogeneic spleen cells.

When the localization of T and non-T donor cells was determined in the two types of lymph nodes 24 h after injection marked differences were observed between lymph nodes of immunized versus

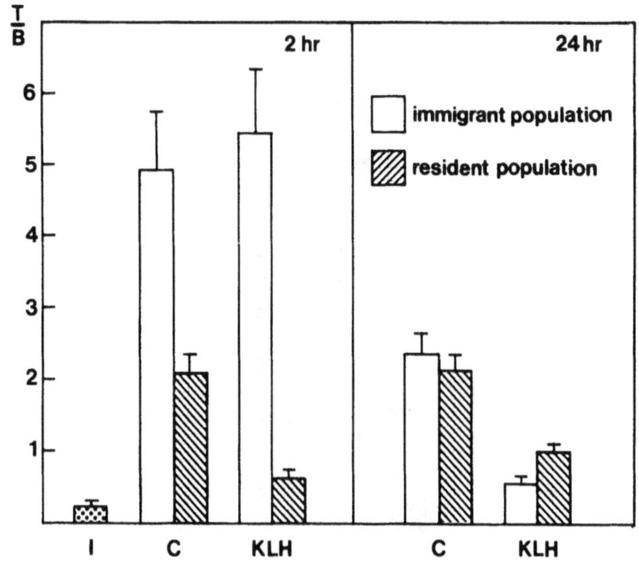

Fig. 1. BALB/c mice were immunized in the front footpad with KLH. Four days later syngeneic rhodaminated spleen cells were injected and at 2 h and 24 h axillary and brachial lymph nodes were removed, suspended and stained for T cells. With double immunofluorescence the ratio of T over non-T (B) cells was determined of the rhodaminated immigrant (open bars) and resident (striped bars) population. The dotted bar (I) represents the ratio of the injected spleen cells. C = control. The results are expressed as ratio ± standard error from a representative experiment. Per sample 300-400 cells were counted.

non-immunized animals. In control animals the ratio of T over non-T donor cells that had localized in the lymph nodes was still high but in immunized animals this ratio had dropped at 24 h. In control as well as in immunized lymph nodes the ratio of T over non-T immigrant cells at 24 h was a reflection of the actual distribution of the lymphocytes in the organs (Fig. 1).

Adherence in vitro

The comparable initial localization at 2 h in stimulated and non-stimulated lymph nodes clearly suggests that the specificity of the HEV had not changed after antigenic stimulation at the time point tested (4-5 days after stimulation). Also when later or earlier time points after antigenic stimulation were studied similar results were obtained. To prove definitively that antigenic stimulation does not change the receptor-specificity an in vitro binding assay was performed using sections of stimulated and non-stimulated lymph nodes. The relative adherence of two different

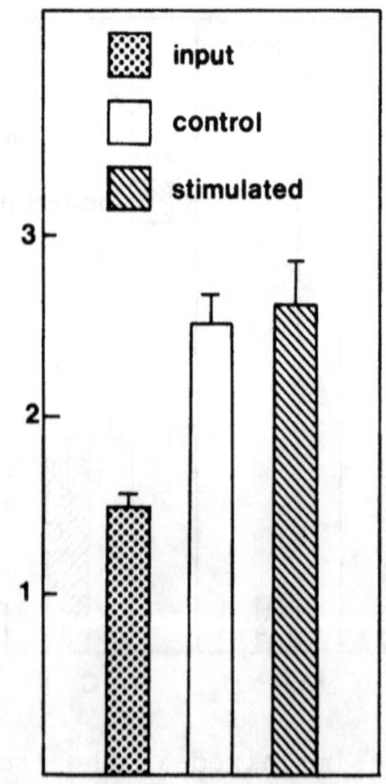

Fig. 2. Peripheral lymph node cells (sample cells) were mixed with rhodaminated cells from Peyer's patches and mesenteric lymph nodes and incubated on frozen sections of control lymph nodes (striped bar) or antigenically stimulated (oxazolon) lymph nodes (vertically striped bars). The open bar represents the input ratio of sample cells/standard cells. The results are from a representative experiment and expressed as ratio ± standard error.

populations (peripheral lymph node cells over mesenteric lymph node cells) was compared. No statistically significant difference of the adherence ratio of both types of HEV's could be found ($p < 0.01$), demonstrating that the specificity of the high endothelial venules had not changed after antigenic stimulation (Fig. 2).

DISCUSSION

B cells bind better than T cells to Peyer's patch HEV and form the major compartment of these organs, whereas T cells demonstrate a relative preference for binding to HEV of peripheral nodes in which they are the predominant population. The preference of the lymphocyte populations to adhere to the different types of HEV is a direct reflection of the actual distribution of these populations in the organs involved. If we assume that the specificity of the HEV is not organ-specific but merely dependent on the state of activation of the organ, the differences in specificity of the HEV can be explained by the continuous antigenic stimulation at mucosal sites versus the inactivated state of peripheral lymph nodes.

Activation of the peripheral lymph nodes resulting in an increase of the percentage of B cells comparable to Peyer's patches will then lead to a change in receptor specificity of the peripheral lymph node HEV.

The results presented in this paper, however, are in favor of a fixed specificity of the HEV which will not be changed during an ongoing immune response.

Both in control as well as in stimulated lymph nodes a preferential localization of T cells was found two h after cell transfer, in accordance with the findings of Steven et al.[3]. This initial localization is directly based on the selective interaction of the HEV and lymphocytes, since the influence of recirculation on the composition of the immigrant cells at this time point can be neglected[10].

The 24 h time point allows us to study other selective mechanisms that interact with the recirculating cells inside the lymph node besides the endothelial cells of the HEV. The greater percentage of donor B cells in the stimulated lymph nodes compared to the control situation at this time point clearly suggests that B cells will be retained specifically in the stimulated nodes so that, although the influx of T and B cells as determined by the HEV is similar for both types of lymph node, the efflux has drastically changed after antigenic stimulation. These results suggest a very precise regulation of the efflux of lymphocytes from lymph nodes. The nature of these regulatory mechanisms remains speculative but interesting candidates may be found in the non-lymphoid cells characteristic for the specific T and B cell areas, the interdigitating cell[11] and the follicular dendritic cell[12, 13], respectively.

Due to their fine elongations and close contact with lymphocytes they may very well determine the size of the T and B cell area, expanding their surface area after stimulation and thereby directly influencing the efflux of lymphocyte subpopulations from the lymph node. These cells may be responsible for the selective recruitment of antigen specific cells from the circulation. In addition to the recruitment of antigen specific B cells there also must be a substantial retention of non-specific B cells, to account for the observed increase of B cells in our immigrant populations 24 h after cell transfer (Fig. 1), since the number of antigen specific cells in normal, non-immunized lymphocyte populations is very low as determined by antigen binding[14]. Antigenic stimulation brings about many important changes in the lymph nodes: the occurrence of germinal centers, predominantly populated by non-migrating B cells[15] as well as an increase of the circulation of lymphocytes[16,17] are factors that will directly influence the retention of lymphocyte subsets.

This increase of circulation through the nodes may be the result of activation of the endothelial cells in the HEV facilitating the entry of lymphocytes into the lymph nodes. It has been suggested that this activation is the result of mediators produced by macrophages inside the lymph node[18], an attractive explanation which can also account for the formation of HEV-like vessels at sites of inflammation. However, it is clear from our results that such increase of activity of the endothelial cells of the HEV after antigenic stimulation does not alter the selective recognition and specific interaction of these cells with circulating lymphocytes.

REFERENCES

1. E.C. Butcher, R.G. Scollay, and I.L. Weissman. J. Immunol. 123: 1996 (1979).
2. E.C. Butcher, R.G. Scollay, and I.L. Weissman. Eur. J. Immunol. 10:556 (1980).
3. S. Stevens, I.L. Weissman, and E.C. Butcher. J. Immunol. 128: 844 (1982).
4. G. Kraal, I.L. Weissman, and E.C. Butcher. J. Immunol. 130:1097 (1983).
5. W.M. Gallatin, I.L. Weissman, and E.C. Butcher. Nature 304:30 (1983).
6. D. Guy-Grand, C. Griscelli, and P. Vassali. Eur. J. Immunol. 4:435 (1974).
7. M.E. Smith, A.F. Martin, and W.L. Ford. Monogr. Allergy 16:203 (1979).
8. G. Kraal, I.L. Weissman, and E.C. Butcher. This volume.
9. C.R. Rao. Linear statistical inference and its applications. Wiley, New York, p. 319 (1965).
10. M.E. Smith, and W.L. Ford. Immunology 49:83 (1983).
11. A.J.P. Veerman, and W. van Ewijk. Cell Tissue Res. 156:417 (1975).
12. G.J.V. Nossal, A. Abbot, J. Mitchell, and Z. Lummus. J. Exp. Med. 127:277 (1968).
13. G.G.B. Klaus, J.H. Humphrey, A. Kunkel, and D.W. Dongworth. Immunol. Rev. 53:3 (1980).
14. J.E. Merrill, and R.F. Ashman. J. Immunol. 123:434 (1979).
15. R. Reichtert, W.M. Gallatin, I.L. Weissman, and E.C. Butcher. J. Exp. Med. 157:813 (1983).
16. J.B. Hay, and B.B. Hobbs. J. Exp. Med. 145:31 (1977).
17. R.N.P. Cahill, H. Frost, and Z. Trnka. J. Exp. Med. 143:870 (1976).
18. H.R. Hendriks, and I.L. Eestermans. Eur. J. Immunol. 13:663 (1983).

HUMAN LYMPHOCYTE-HIGH ENDOTHELIAL VENULE INTERACTION: FUNCTIONAL AND MOLECULAR CHARACTERIZATION

Sirpa Jalkanen, Robert Bargatze, Lynne Herron,
and Eugene C. Butcher

Department of Pathology
Stanford University Medical Center
Stanford, California 94305, and
The Veterans Administration Medical Center
Palo Alto, California 94304

INTRODUCTION

Mature lymphocytes continuously recirculate between blood and lymphoid organs (1). They leave the blood by selectively binding to specialized high endothelial venules (HEV) in lymph nodes or in the mucosa-associated lymphoid organs (appendix and Peyer's patches). The capacity of lymphoid cells to recognize HEV in murine systems can be assessed *in vitro* by incubating lymphocytes on frozen sections of lymph nodes and Peyer's patches (2,3), and the binding of lymphocytes to HEV in this assay correlates well with their ability to recirculate and home to lymphoid organs *in vivo*. Using this *in vitro* assay, in conjunction with *in vivo* homing studies, at least two independent lymphocyte-endothelial cell recognition systems have been found, one mediating lymphocyte traffic through lymph nodes and another directing traffic through Peyer's patches. Recently, the existence of a homing receptor molecule responsible for organ specific HEV recognition has been proved by producing a monoclonal antibody, MEL-14, against the mouse homing receptor for lymph node HEV (4-see also Woodruff and Chin, this volume).

In this paper we have adapted the *in vitro* assay of HEV-binding to human cells and lymph nodes, and we describe a monoclonal antibody, LSRE-1, which recognizes determinants on or in close association with the human homing receptor.

MATERIALS AND METHODS

Human Cells and Tissues

Human lymph nodes, thymuses, tonsils and bone marrows were from pathologically normal biopsy or surgical specimens. Peripheral blood lymphocytes (PBL) and bone marrow cells were isolated using Ficoll-hypaque and T cells were isolated from peripheral blood lymphocytes by passage through nylon wood, followed by centrifugation on a Percoll gradient.

In Vitro HEV Binding Assay

3×10^6 cells in 100 μl medium were incubated with agitation on frozen sections of human lymph nodes as described earlier (3,5). After 30 minutes, bound lymphocytes were fixed to the sections in 1% glutaraldehyde in cold PBS. Slides were gently rinsed with PBS to remove unbound cells. To facilitate quantitative comparisons, an internal standard population of fluorescein isothiocyanate labeled lymphocytes were mixed with each sample population before incubation. The adherence of different human cell types were compared to that of a control population of PBL.

Immunoperoxidase Tissue Section Staining

Immunoperoxidase staining of frozen sections was carried out by using standard techniques (6). The second stage was peroxidase-conjugated rabbit anti-rat Ig (Dako).

Radioiodination and Immunoprecipitation.

Ficoll purified human PBL and mouse peripheral lymph node cells were labeled with ^{125}I using the lactoperoxidase method, and the cells were lysed with 1% Triton X-100. Affigel-10 beads coupled with Mel-14 or LSRE-1 antibody were used for overnight immunoprecipitation at +4°C.

RESULTS

Production and initial characterization of LSRE-1 antibody

The monoclonal antibody LSRE-1 was produced by immunizing rats with human tonsillar and peripheral blood lymphocytes, using Sp 2/0 cells as a fusion partner. It was selected as a possible antibody against the lymphocyte surface receptor for HEV on the basis of its staining pattern on normal human tonsillar lymphocyte populations, which resembles the staining pattern of MEL-14 in mouse lymphatic tissues. Thus germinal center cells and HEV cells are LSRE-1 negative, whereas T cells in interfollicular zones and B cells around germinal centers stain intensely with LSRE-1 (Figure 1a). Like the MEL-14 antigen, LSRE-1 is expressed at high levels only by a small subpopulation of cortical thymocytes, but unlike the weakly staining thymic medulla in mice, the human medulla stains intensely with LSRE-1 (Figure 1b).

The molecular weight analysis revealed that the human antigen molecule identified by LSRE-1 is remarkably similar to the MEL-14 defined mouse homing receptor: in SDS-PAGE both molecules run with apparent molecular weight of about 80-90 kd under non-reducing conditions, but migrate more slowly (5-10 kd apparent shift) under reducing conditions (data not shown, manuscript in preparation) suggesting the presence of internal disulfide linkages.

HEV-Binding of Various Human Cell Types and Expression of LSRE-1

To determine if the migratory properties of human lymphocytes and the expression of LSRE-1 antigen correlate with each other, we compared the antigen expression and HEV-binding abilities of various human lymphocyte populations representing distinct stages in lymphocyte differentiation. As shown in Table 1 and Figure 2, mature peripheral T lymphocytes, PBL and tonsillar lymphocytes, which participate in the circulating lymphocyte pool, bind efficiently to HEV, whereas immature T cells from thymus and bone marrow cells, consisting largely of non-circulating cell types, bind poorly. Expression of LSRE-1 parallels the HEV-binding capacity of these cell types, consistent with the proposal that the antigenic determinant recognized by LSRE-1 is on or at least is expressed in close association with the human homing receptor.

Table 1. Correlation of HEV-binding ability of various human cell types with their expression of the LSRE-1 antigen.

Cell Type	Binding to HEV	LSRE-1 Binding[a]
Peripheral blood lymphocytes	++++	++++
Peripheral blood T cells[b]	+++++	++++
Tonsil cells	++++	++++
Thymocytes	+	+
Bone marrow cells[c]	+	+

a) by FACS analysis
b) Nylon wool and Percoll gradient purified
c) Ficoll separated bone marrow cells

DISCUSSION

The fact that the expression of the LSRE-1 antigen correlates extremely well with the HEV-binding properties of different human cell types, in combination with its similarity to the mouse receptor for lymph

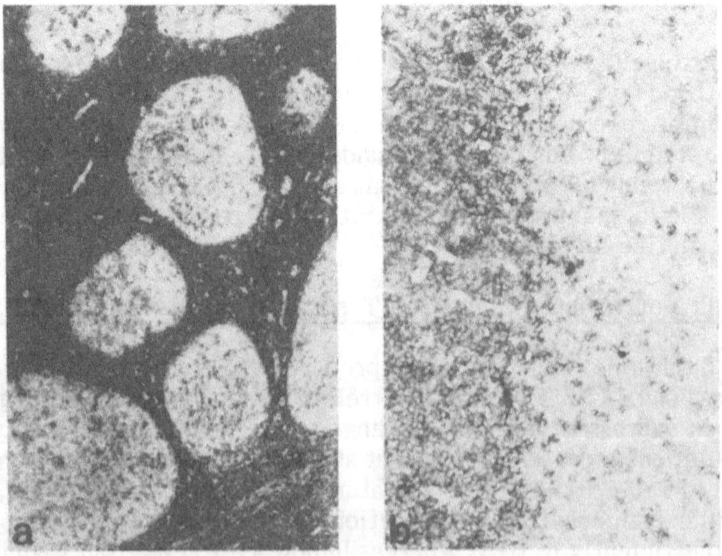

Fig. 1 Immunohistologic (immunoperoxidase) localization of LSRE-1 expressing lymphocytes. (a) Human tonsil stained with LSRE-1. Germinal centers are largely negative, but surrounding B and T cell areas are positive. (b) Human thymus stained with LSRE-1. A small subpopulation of cortical cells is positive, but the medullary region stains intensely.

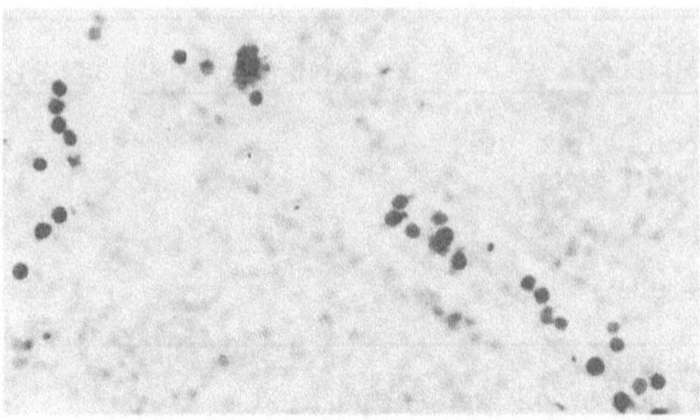

Fig. 2 Human peripheral blood lymphocytes binding selectively to HEV in a lymph node frozen section. Bound lymphocytes are round dark circles lying in focus above the plane of the section. A distinct, poorly stained basement membrane delineates the outline of the HEV. (Thionine stain, x 125).

node HEV, strongly suggests that the structures recognized by LSRE-1 antibody may play a role in lymphocyte-endothelial cell recognition in man.

The difference in the staining of thymic medulla between mouse and human may relate to the hormonal status of the human donors (young patients undergoing cardiac surgery), because mouse medullary thymocytes become MEL-14 positive following hydrocortisone treatment (7). Alternatively, the staining of medullary thymocytes in humans may reflect a true species difference; or may imply cross-reaction of LSRE-1 with a related receptor system (perhaps responsible at some level for initial homing of thymocytes to the medullary region, but not involved in lymphocyte recognition of lymph node HEV).

While in the murine system MEL-14 can directly inhibit lymphocyte binding to lymph node HEV, precoating of human lymphocytes with LSRE-1 does not directly block their binding to HEV, suggesting that the antigenic determinant is not in the recognition domain of the homing receptor. However, in preliminary studies the LSRE-1 antigen has been linked to HEV recognition. Conventional rat antiserum against human lymphocytes inhibits HEV-binding of lymphocytes, and the blocking activity in this antiserum can be selectively removed by absorption with purified receptor bound to an LSRE-1 affinity column. Direct proof of the role of the LSRE-1 antigen in lymphocyte-endothelial cell recognition must await the isolation of that molecule in sufficient quantities to measure its capacity for specific binding to lymph node and Peyer's patch HEV.

ACKNOWLEDGMENTS

We thank M. Beers for excellent secretarial assistant and R. Coffin for photographic help. This work was supported by grants CA 37233, CA34709 and AI 19957 from the USPHS. E.B. is a scholar of the Leukemia Society of America, and S. J. is a postdoctoral fellow of Jane Coffin Childs Memorial Fund for Medical Research and a recipient of a grant from the Finnish Cancer Union.

REFERENCES

1. J.L. Gowans and W.J. Knight, The route of recirculation of lymphocytes in the rat, Proc. Roy. Ser. B. 159:257 (1964).
2. H.B. Stamper, Jr. and J.J. Woodruff, Lymphocyte homing into lymph nodes: in vitro demonstration of the selective affinity of recirculating lymphocytes for high endothelial venules, J. Exp. Med. 144:828 (1976).
3. E.C. Butcer, R.G. Scollay, and I.L. Weissman, Lymphocyte adherence to high endothelial venules: characterization of a modified in vitro assay, and examination of the binding of syngeneic and allogeneic lymphocyte population, J. Immunol. 123:1996 (1979).

4. W.M. Gallatin, I.L. Weissman, and E.C. Butcher, A cell surface molecule involved in organ specific homing of lymphocytes, Nature 303:30 (1983).
5. S.T. Jalkanen and E.C. Butcher, In vitro analysis of the homing properties of human lymphocytes: acquisition of functional receptors for high endothelial venules during lymphocyte development, (submitted)
6. R.V. Rouse and R.A. Warnke, Special applications of tissue section immunologic staining in the characterization of monoclonal antibodies and in the study of normal and neoplastic tissues, in: "Handbook of Experimental Immunology," D. Weir, ed., in press.
7. R.A. Reichert, W.M. Gallatin, E.C. Butcher, and I.L. Weissman, A homing receptor-bearing cortical thymocyte subset: implications of thymus cell migration and the nature of cortisone-resistant thymocytes, Cell 38:89 (1984).

T CELL CLONES: A MODEL OF A NON-RECIRCULATING PHASE OF T CELL DIFFERENTIATION

Morris O. Dailey, W.M. Gallatin, Georg Kraal, Eugene Butcher, and Irving Weissman

Department of Pathology
Stanford University School of Medicine
Stanford, California 94305

Cloned lines of T lymphocytes have been valuable in investigating many of the in vitro properties of thymus-derived lymphocytes. When such clones have been tested in vivo, however, they have generally shown much less activity than they have in vitro. Thus, cytolytic T cell (T_C) clones that are highly active in lysing tumor target cells in vitro are poorly effective in causing rejection of the same cells in vivo, unless they are injected into the same site as the tumor cells[1,2]. Similarly, in vitro T cell clones efficiently mediate delayed hypersensitivity only when injected at the same site as the relevant antigen[3].

Both T and B lymphoyctes recirculate from blood, through lymphoid tissue to efferent lymph, and then back to the blood stream. The migration and specific tissue localization of lymphocytes is important in maintaining the normal host immune responsiveness. The entry of lymphocytes into the peripheral lymphoid tissues lymph nodes (LN) and Peyer's patches (PP) is dependent upon the interaction of the lymphocytes with the walls of specialized blood vessels, high endothelial venules (HEV), in these organs. This interaction, mediated by specific cell surface receptors on the lymphocytes, leads to the binding of the lymphocyte to the vessel followed by passage through its wall and into the parenchyma of the lymphoid organ. Cells which lack this receptor, such as most thymocytes, do not home to LN or PP.

While attempting to use clones to study the in vivo function of well defined homogeneous T cells, it was found that such clones do not migrate to peripheral lymphoid tissues as do most normal T lymphoyctes[4]. The lack of homing of these cells to LN and PP, and their inability to recognize HEV in a frozen section binding assay, suggested that these cells lack the lymphocyte surface receptors for HEV. The results reviewed in this report demonstrate that T cell clones, of both helper and cytolytic phenotype, do in fact lack this receptor, and that this loss is apparently irreversible. Furthermore, the changes in lymphoycte surface phenotype are identical to those seen in heterogeneous mixed lymphocyte culture (MLC) - activated blast cells. T cell clones appear to represent lymphocytes that are permanently locked in a stage of differentiation representative of large, activated, nonhoming blast cells.

MATERIALS AND METHODS

Cell Lines. Several independently derived T cell clones were used in these studies, including soluble antigen-reactive and alloreactive helper clones, allospecific cytolytic, and H-2-restricted tumor-specific cytolytic clones, as was previously described[4].

Lymphocyte Activation. Murine spleen cells were stimulated in allogeneic mixed lymphocyte cultures using 7.5 x 10^7 responders with an equal number of irradiated stimulator cells in 36 ml of medium.

Immunofluorescence. Five x 10^5 cells were stained with rat monoclonal antibodies and rabbit anti-rat-FITC or with PNA-FITC, fixed, and analyzed on the fluorescence activated cell sorter (FACS II) using logarithmic amplification for fluorescence. The MEL-14 monoclonal antibody, specific for the mouse LN HEV receptor, was described previously[5].

In Vitro HEV Binding Assay. The recognition of HEV by lymphocytes was measured by their ability to adhere to HEV in frozen sections of lymph nodes and Peyer's patches, as previously described[6].

RESULTS

T Cell Clones Do Not Home to LN or PP. The pattern of migration of T cell clones was assessed by tracing ^{51}Cr-labeled cells in vivo. At 2 and 24 hours after intravenous injection neither helper nor cytolytic clone cells could be found in recipients lymph nodes or Peyer's patches (localization < 5% that of normal lymphocytes). This finding was true for all clones tested, including alloreactive cytotoxic, H-2 restricted antiviral T_C, and alloreactive and soluble antigen-specific helper clones. Although there was some increased localization of cells in the lungs and liver, the number of cells trapped in these organs was

TABLE 1. Summary of Homing and Fluorescence Experiments

Cells	Immunofluorescence MEL-14	PNA	HEV Binding	Homing to Lymph Nodes
Normal T Cells	++	+	++	++
T Cell Clones	-	+++	-	-
MLC Cells:				
Large	-	+++	-	-
Small	++	+	++	++

insufficient to explain the lack of homing to LN or PP. The localization of clones was also assessed by injecting unlabeled cells into Thy-1 congeneic recipients. Immunofluorescence staining of recipient organs with monoclonal anti-Thy-1.2 revealed virtually no donor cells in the lymph nodes, confirming the results obtained with radiolabeled cells.

T Cell Clones Lack Surface Homing Receptors. Previous results showed that these T cell clone cells did not recognize LN HEV's, since they did not bind to HEV in the frozen section adherence assay (relative adherence $\leq$ 2% that of normal cells). This suggests that these cells might have lost the specific lymphocyte surface receptor that mediates the binding of lymphocytes to HEV, resulting in their inability to migrate through HEV into lymph nodes in vivo. This possibility was tested by examining these cells for MEL-14, a monoclonal antibody specific for this lymphocyte surface receptor. When clones were analyzed by immunofluorescence using the fluorescence activated cell sorter, no MEL-14 antigen could be detected on any of these cells (results summarized in Table 1). This is in contrast to normal peripheral T lymphocytes, most of which are brightly MEL-14$^+$. In addition, all clones stained brightly with the lectin PNA, demonstrating a change in surface glycosylation reflecting a decrease in terminal sialic acid content. The significance of this result will become apparent later.

The lack of fluorescence staining by MEL-14 could conceivably be explained by the burying of the antigenic determinant in the membrane even though significant levels of the protein could be present. A cytolytic clone was therefore surface labeled with ^{125}I, lysed, and extracted with appropriate detergents. SDS gel electrophorectic analysis of control lysates (from MEL-14$^+$ cells) showed the 80-90 kd stand typical of the MEL-14 antigen, while T cell clone lysates showed no such band. Thus, the surface of these cells do in fact lack the MEL-14 homing receptor antigen.

The MEL-14 Non-Homing Phenotype of T Cell Clones is Identical to That of Primary Activated T Cells. Most resting T and B lymphocytes express large amounts of the MEL-14 antigen on their surface. In order to see if this expression varies after activation, cells were stimulated by alloantigen in mixed lymphocyte cultures and subsequently tested for the presence of the MEL-14 antigen and for their ability to recognize HEV (Table 1). On day five of a primary C57Bl/6 anti-BALB/c MLC the cells were examined by FACS analysis using MEL-14 and PNA. A representative dot plot (Figure 1) shows that these cultures consist of two major populations. The large (activated) cells express little or no MEL-14 antigen, while the small, non-responding cells in the same culture retained the normal degree of receptor expression. Furthermore, all of the large, MEL-14^- cells stained brightly with PNA, and the small cells remained PNA-low. When the cells were separated by size, using elutriator centrifugation, only the small, unactivated cells were able to bind to HEV in the frozen section adherence assay, while the large cells were unable to bind (relative adherence < 2%). It is apparent that the T cell clones are identical in phenotype to the large, activated PNA^{++} MEL-14^- blast cells generated in the MLC's.

Irreversability of the Non-Homing Phenotype of T Cell Clones. It is is possible that lack of the MEL-14 homing receptor reflected the loss of a "luxury" function in rapidly proliferating cells which might reverse if the cells were allowed to cease replicating. Clones 1E4 (cytotoxic) and 4 (helper) were rested in a nonproliferating state without stimulation by antigen or exogenous growth

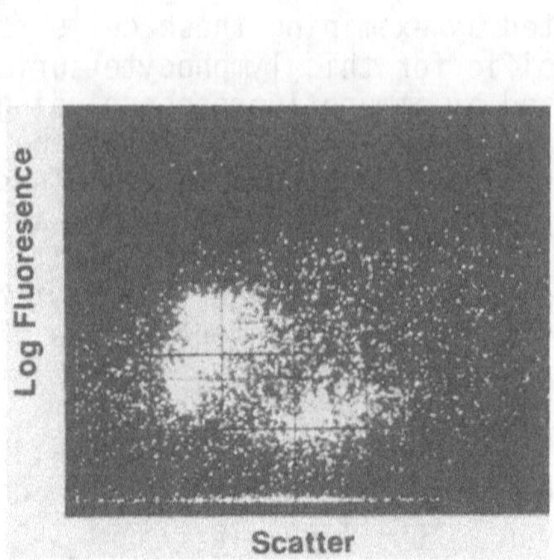

Figure 1. Immunofluorescence analysis of day 5 C57Bl/6 anti-BALB/c MLC cells. Fluorescence intensity is on a log scale, while scatter, a correlate of cell size, is linear.

factors for up to one month. When examined by immunofluorescence, they remained MEL-14$^-$ and strongly PNA$^+$. In addition, they did not recognize HEV in the in vitro adherence assay or home to LN or PP in vivo. Thus, even in nonproliferating cells, the loss of the lymphocyte homing receptor is irreversible, at least in vitro and for short terms in vivo (24 hrs).

DISCUSSION

The studies summarized in this report demonstrate that murine T cell clones, of several types, and isolated independently, do not migrate to peripheral lymphoid tissue as do most normal lymphocytes. After intravenous injection, these cells are not found in LN or PP. The inability of these cells to gain access to LN appears to be due to their inability to recognize HEV's in these organs, due to the loss of the surface HEV or "homing" receptor, as shown by experiments with the MEL-14 antibody. The lack of homing of these cells to PP in vivo and their nonrecognition of PP HEV in vitro indicates that the Peyer's patch-specific HEV receptor is also absent from these cells. The absence of lymphocyte homing receptors is not merely due to the highly active, rapidly proliferating state of these cells, but rather seems to be an irreversible feature of murine T cell clones.

The homing properties and surface phenotype of T cell clones is identical to that of primary T cells activated by alloantigen in vitro. Both cell types are large blast cells which are PNA^{++} MEL-14$^-$, do not bind to HEV in vitro, and do not localize in LN or PP in vivo. The loss of the HEV receptor from activated lymphocytes is not just an artifact of in vitro culture conditions, since unactivated (small) cells in the same cultures remain MEL-14$^+$. Substantiating this view is the finding of MEL-14$^-$ PNA^{++} T cells in LN draining skin allografts in vivo (M.O.D., preliminary results). We therefore propose that the loss of MEL-14 homing receptor expression and the change in cell surface glycosylation leading to PNA positivity represent a stage of differentiation through which T cells pass after interaction with specific antigen. This would likely be a transient stage, with the cells later regaining the potential for recirculation[7]. T cell clones would then be cells that are permanently arrested in this stage, perhaps because of being repetitively activated through many cycles of antigen and IL-2 stimulation.

The passage of activated T cells through a MEL-14$^-$ nonhoming stage might facilitate a sessile phase of development during which lymphocytes remain localized in a microenvironment that optimizes the cell-to-cell interactions necessary for the immune response. This would be identical to the case of germinal center B cells, antigen-activated B cell blasts which are also MEL-14$^-$ PNA^{++} and do not home to LN or PP[8]. Spleen marginal zone B cells, another example of stationary nonrecirculating lymphocytes, are also MEL-14⁻.[9] Thus, although it is probably not involved in controlling the egress of lymphocytes from LN, lack of the MEL-14 homing receptor may be a necessary feature of sessile lymphocytes in vivo.

There is an alternative, and not necessarily exclusive, explanation for the loss of homing receptors from activated lymphocytes and T cell clones. In this hypothesis, T cell clones would represent a normal in vivo lineage of effector T cells which permanently remain MEL-14$^-$, PNA-high, and do not recirculate. This lineage could be distinct from that giving rise to recirculating memory T cells and perhaps from that of other end-stage effector cells. By not expressing HEV receptors, effector cells in the blood stream would not be cleared during their circulation through LN or PP, and thus would be made more likely to reach peripheral sites of antigen deposition, such as a skin graft or DTH inflammatory lesion. Loss of homing receptors from these cells would therefore be a mechanism of increasing the traffic of lymphocytes through peripheral non-lymphoid tissues in the host.

In what way would these changes in homing patterns affect the in vivo activity of T cell clones? If loss of the PP and LN-specific HEV receptors represent a generalized defect in lymphoycte migration, then their reduced efficacy could be do to the inability of these cells to home to antigenic sites, such as a tumor graft, as was previously suggested[2]. Alternatively, other mechanisms might be responsible for their poor activity. A single homogeneous T_c clone may require the simultaneous action of another, perhaps Ly1$^+$2$^-$, cell which is able to enter the graft and initiate the inflammatory reaction. Further study will be necessary to determine the in vivo physiologic significance of the homing phenotype displayed by activated T cells and these T cell clones.

REFERENCES

1. Georgi, J.V. and N. L. Warner. Continuous cytotoxic T cell lines reactive against murine plasmacytoma tumor-associated antigens. J. Immunol. 126:322 (1981).
2. Dailey, M.O., E. Pillemer, and I.L. Weissman. Protection against syngeneic lymphoma by a long-lived cytotoxic T cell clone. Proc. Natl. Acad. Sci. 79:5384 (1982).
3. Bianchi, A.T.J., H. Hooikjkaas, R. Benner, R. Tees, A.A. Nordin and M.H. Schreier. Clones of helper T cells mediate antigen-specific, H-2 restricted DTH. Nature 290:62 (1981).
4. Dailey, M.O., C.G. Fathman, E.C. Butcher, E. Pillemer, and I. Weissman. Abnormal migration of T lymphocyte clones. J. Immunol. 128:2134 (1982).
5. Gallatin, W.M., I.L. Weissman, and E.C. Butcher. A cell surface molecule involved in organ-specific homing of lymphocytes. Nature 303:30 (1983).

6. Butcher, E.C., R.G. Scollay, and I.L. Weissman. Lymphocyte adherence to high endothelial venules: characterization of a modified in vitro assay, and examination of the binding of syngeneic and allogeneic lymphocyte populations. J. Immunol. 123:1996 (1979).
7. Sprent, J., and J.F.A.P. Miller. Fate of H-2-activated T lymphocytes in syngeneic hosts. II. Residence in recirculating lymphocyte pool and capacity to migrate to allografts. Cell. Immunol. 21:303 (1976).
8. Reichert, R.A., W.M. Gallatin, I.L. Weissman, and E.C. Butcher. Germinal center B cells lack homing receptors necessary for normal lymphocyte recirculation. J. Exp. Med. 157:813 (1983).
9. Grey, D., I.C.M. MacLennon, H. Bazin, and M. Khan. Migrant $\mu+\delta+$ and static $\mu+\delta-$ B lymphocyte subsets. Eur. J. Immunol. 12:564 (1982).

RADIOSENSITIVITY OF THE CIRCULATION OF DTH EFFECTOR T LYMPHOCYTES : AN ACCURATE ESTIMATE USING A QUANTITATIVE ASSAY

Geneviève Milon and Gilles Marchal

Unité d'Immunophysiologie cellulaire
Institut Pasteur
25, rue du Dr. Roux, 75724 Paris Cedex 15 (France)

INTRODUCTION

The effects of ionizing radiation on the immune response have been extensively investigated during the last decades (reviewed by Anderson and Warner, 1976). Many of these studies were performed before the cooperative relationships among the various subpopulations of immunocompetent cells as well as their circulation began to be investigated. Though the circulation of sensitized lymphoid cells and their precursors is critical to immune responses *in vivo*, it remains poorly analyzed in quantitative terms.

To further explore this critical function of lymphoid cells and its radiosensitivity, we have chosen to study the delayed-type hypersensitivity reaction (DTH). Indeed, DTH skin reactions are mediated by sensitized T cells that circulate and enter site of antigen challenge. The ability of these T cells to interact specifically with antigen leads to recruitment of bone marrow derived effector leukocytes which comprise more than 95% of the infiltrate at the DTH reaction. The studies on the effects of radiation on such cooperative responses have to take into account each function of each cell type. Using qualitative adoptive transfer experiments, Kettman and Mathews (1975) have already documented the high radioresistance of DTH-mediating T cells when assayed after local transfer in contrast to their radiosensitivity when assayed after iv. transfer. Using a quantitative assay of DTH-mediating T cells, we also demonstrated that the induction of DTH reaction was very sensitive (D37 of approximately 0.5 Gray) while its expression was very radioresistant (D37 of approximately 20 Gray (Marchal, Milon and Hurtrel, 1979). In these data the high radioresistance of the DTH reaction was related to the fact that irradiated DTH-mediating cells were transferred in concert with the antigen to the site where the evaluation was done. Direct studies of radiosensitivity of DTH reaction in sensitized animal are much more complex to interpret since radiation may have affected either clonal expansion and / or circulation of DTH effector T lymphocytes. In order to approach this difficult problem, we report here additional data upon effect of radiation on circulation of DTH-mediating T lymphocytes.

MATERIALS and METHODS

Mice

Specific pathogen free male or female C57Bl/6 mice 6 to 8 weeks of age were reared in our animal facility.

Immunization

Sheep red blood cells (SRBC) were collected and stored in Alsever's solution at 4°C. SRBC were washed three times before use and diluted to the appropriate concentration in sterile pyrogen free 0.9% NaCl or Hanks' medium. For immunization 0.5 ml of SRBC suspensions were injected iv. into the tail vein.

Assay for DTH reaction

Four days after immunization mice were challenged by injection of 5×10^8 or 2.5×10^7 SRBC suspended in 50 µl of pyrogen free Hanks' medium into one hind footpad. Footpad swelling was measured 18 to 24 h later with a dial gauge caliper. The difference in thickness between feet that received SRBC and those that did not was used as a measure of the DTH reaction.

Assay for local transfer of DTH reaction

Sensitized blood cells were obtained from immunized mice or from intermediate recipients which have been treated with heparin (50 IU ip.), 30 min before blood collection from the retroorbital sinus. Fifty µl of diluted blood to which were added 5×10^8 SRBC were injected in one hind footpad of syngeneic naive mice. Footpad swelling was measured 14 to 18 h later and the results expressed as the difference between feet that received sensitized cells and SRBC and those that did not.

Limiting dilution analysis

The method used has been previously described (Marchal et al., 1982). Briefly, the criterium for negative transfer was established by estimating the 95 percentile of the distribution of footpad thickness increases after local injection of unsensitized cells mixed with 5×10^8 SRBC into 20 to 25 naive mice. The 95 percentile upper value for negative transfer was calculated by adding the mean (m) value of footpad thickness increases to the product SD x $t_{0.05}$ in which SD represents the standard deviation and t is the value of Student's test for $p = 0.05$. The negative transfer criteria were always less than 0.2 mm, provided the mice were kept on clean litter. When injected in excess, cells to be titrated transferred a DTH reaction with values of footpad swelling ranging from 0.3 to 0.9 mm. Thus the values of positive transfers were delineated from values of negative ones allowing for a clear cut delineation of positive and negative transfers when limiting numbers of sampled cells were assayed. Starting from these data titration curves were drawn by fitting the Poisson distribution. The percentage (y) of negative transfers per group and the number (x) of cells per recipient were plotted on a semi-log plot (Lefkovitz and Waldmann, 1979).

Irradiation of mice and cells

Sensitized mice received 0.4 - 0.8 - 1.2 Gy of ^{137}Cs whole body irradiation immediately before challenge with antigen. They were exposed in lucite containers at a target distance of 50 cm (dose rate 0.4 Gy/min). Heparinized blood containing DTH-mediating T lymphocytes was irradiated in syringues under the same conditions.

RESULTS and DISCUSSION

The development of a typical DTH reaction involves an initial non-specific inflammatory reaction followed by an antigen specific step which results from interactions between antigen and sensitized T lymphocytes (Marchal and Milon, 1984). The initial non specific inflammatory reaction is critical because it creates optimal conditions for recruitment of specific effector lymphocytes as well as non-specific phagocytic cells at the site of the reaction. With respect with DTH to SRBC for example, we have demonstrated previously that the number of DTH-mediating T cells which can be detected upon local transfer depends on the amount of antigen injected. With 5 x 10^8 SRBC a single T cell was sufficient to evoke maximal DTH while when the dose was lowered to 2.5 x 10^7 at least 10^4 T cells were required for a positive reaction (Marchal et al., 1982).

When immunized mice were irradiated 30 min before challenge, the suppressive effects on the DTH reaction were more clearly detectable at low doses of challenging antigen (Figure 1). From these experiments alone we cannot distinguish radiation effects on (1) the traffic of T cells (2) their effector function of recruitment or (3) the recruitment of phagocytic cells. The effector function of T cells present in the challenging site has been shown to be remarkably radioresistant (Kettman and Mathews, 1975 ; Marchal et al., 1979). Likewise the recruitment of

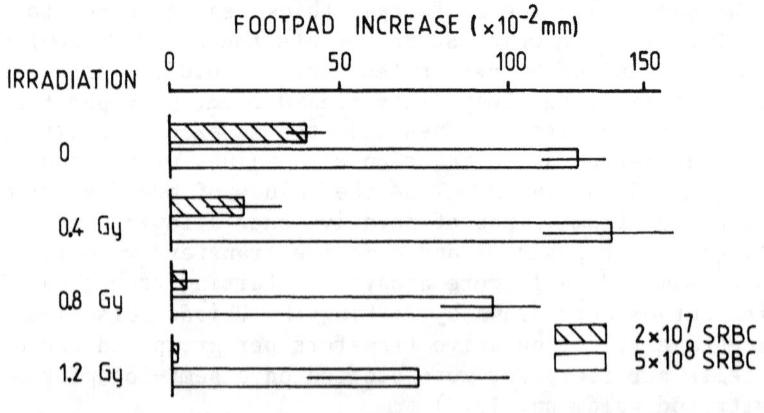

Fig. 1. Effects of irradiation on DTH responses to two doses of sheep erythrocytes.

phagocytic cells was relatively unaffected by irradiation (10 Gy) given 30 min to 2 h before local transfer of sensitized lymphocytes and antigen (Milon and Marchal, 1984). In contrast, we are left with the first critical function of traffic. It will be shown here that the circulation of sensitized T cells is in fact very radiosensitive.

These studies suggest that the inflammatory reaction found in sensitized mice irradiated 30 min before challenge with the high dose of SRBC results from few T cells which have remained unaffected by irradiation. Thus, the diminished DTH reaction in sensitized mice given low dose irradiation just before challenge might be explained by the known capacity of such radiation to alter lymphocyte traffic. To approach quantitatively this hypothesis we have used the experimental design summarized in figure 2.

Blood containing DTH-mediating lymphocytes was collected from mice 4 days after iv. injection of 5 x 10^5 SRBC. The blood was diluted in an equal volume of Hanks' medium and injected iv. (200 µl) in a tail vein of syngeneic untreated mouse. A direct assay of DTH transfer was performed 30 min after iv. transfer by challenge with 5 x 10^8 SRBC injected in a hind footpad of the recipients. Measurement of the clearance of DTH-mediating lymphocytes was performed on blood samples obtained 15 min after iv. injection in intermediate recipients. Different blood dilutions were mixed with antigen and injected in the footpads of naive recipients (limiting dilution assay of DTH-mediating cells). The blood was exposed to different radiation doses 30 min before transfer in intermediate recipients.

The experiments shown on Table 1 indicated that irradiation of blood containing sensitized T lymphocytes significantly reduced

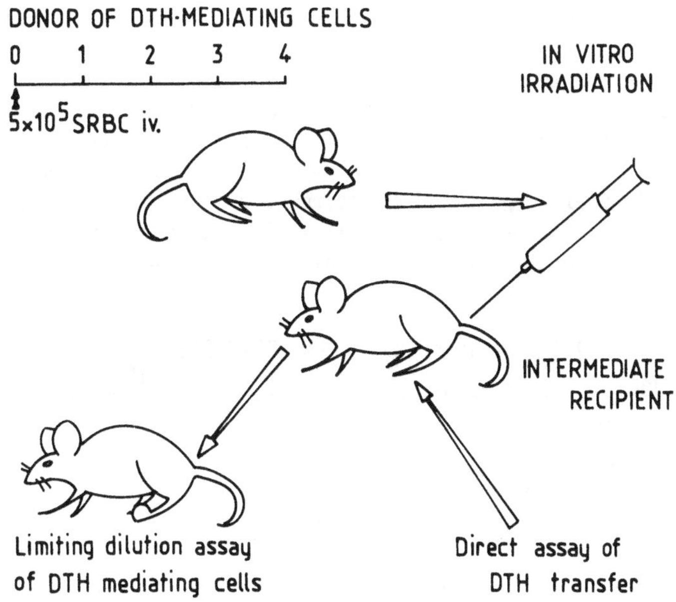

Fig. 2. Experimental design for quantitating DTH-mediating cells.

their capacity upon iv. injection to adoptively transfer the DTH reaction to recipient mice. The table also shows that the standard error's values increase after irradiation of the donor cells, suggesting that only few T cells have migrated to the site of antigen injection.

Table 1.

Irradiation of cells before iv. transfer	Footpad increase after iv. transfer of sensitized cells (x 0.01 mm)
0	55 ± 17
0.4 Gy	36 ± 35
1.2 Gy	26 ± 25

Since the irradiated lymphocytes directly transferred to the footpads did elicit DTH, these experiments support that irradiation reduced the ability of the T cells to move from the blood compartment to the footpad.

In order to quantitate precisely this inability of irradiated sensitized T cells to circulate we titrated remaining specific T lymphocytes present in blood of intermediate recipients 15 min after their iv. injection. The background of spontaneously reactive DTH-mediating cells present in the blood of intermediate recipients was also determined and found to be 1 per 80000 lymphocytes (control in figure 3). Therefore, for the estimate of the D37 of surviving fraction after irradiation, the value due to spontaneously reactive cells had to be substracted from the value found in mice that

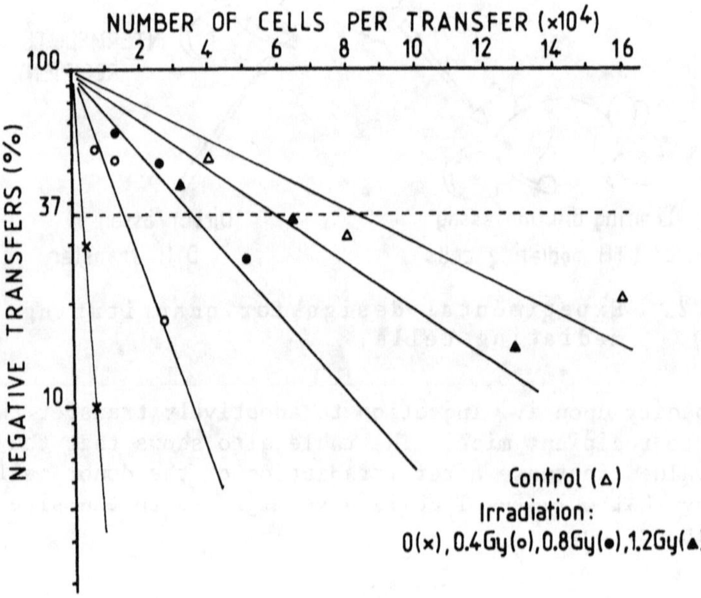

Fig. 3. Frequency of DTH-mediating cells in the blood of intermediate recipients.

received irradiated sensitized lymphocytes. Figure 3 displays the frequency of DTH-mediating cells present in blood of intermediate recipients. Blood of mice receiving unirradiated sensitized cells contained approximately 1 DTH T cell per 2500 lymphocytes. After 0.4, 0.8 and 1.2 Gy in vitro irradiation the frequencies were respectively 1 per 8000, 1 per 35000, 1 per 60000 lymphocytes. Substraction of background value from these results allowed to calculate the frequency of transferred cells persisting in blood after their in vitro irradiation. Taking the frequency for unirradiated cells as one (better surviving factor) we estimated the surviving fraction after the different doses of radiation. A semi-log plot of the surviving fraction against dose of radiation allowed the radiosensitivity of the circulation of DTH-mediating

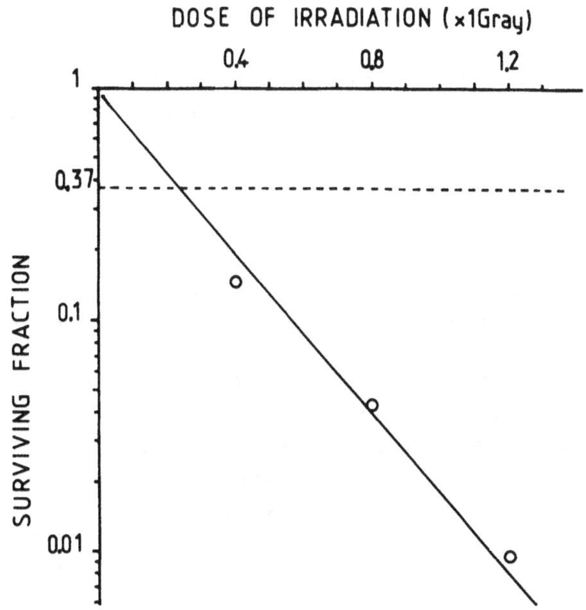

Fig. 4. Surviving fraction of irradiated DTH-mediating cells.

cells to be determined. The D37 was approximately 0.2 Gray (figure 4).

"Irradiation greatly modifies the capacity of lymphocytes to traffic in normal fashion" (Anderson and Warner, 1976). This statement was based upon numerous observations in rodents but also in human subjects. Most frequently found was a drop in lymphocyte numbers in the blood or in thoracic duct lymph after irradiation. These observations were at best semi-quantitative. Anderson et al. (1974) introduced quantitative data on the effect of radiation on lymphocyte traffic. They reported that low dose irradiation of labeled lymph node cells led to marked change in their organ redistribution in recipient mice between 4 or 24 h after irradiation. The redistribution from spleen to lymph nodes was markedly inhibited by as little as 50 r (0.5 Gy).

Recently we demonstrate that the DTH reaction observed after a low dose of antigen is mediated by helper T cells (Milon et al., 1983). The present data together with our earlier results may account for the remarkable radiosensitivity of the inductive phase of antibody response. In addition to the radiosensitivity of the proliferative phase of T and B cells, altered circulation and recruitment of antigen reactive cells may also represent a critical step affected by irradiation.

This work was supported by grants from Institut Pasteur, INSERM (CRE n° 83.3009), MRI (n° 83.C.0859), and ARC (n° 6200). The authors express their appreciation to Dr. M. Rabinovitch for critical reviewing of the manuscript. They also thank Mrs C. Maczuka for preparing the manuscript.

REFERENCES

Anderson, R.E., Sprent, J., and Miller, J.F.A.P., 1974, Radiosensitivity of T and B lymphocytes. I. Effect of irradiation on cell migration, Eur.J.Immunol., 4 : 139.

Anderson, R.E., and Warner, N.L., 1976, Ionizing radiation and the immune response, Adv.Immunol., 24 : 215.

Kettman, J., and Mathews, M.C., 1975, Radioresistance of cells responsible for delayed-type hypersensitivity reaction in the mouse, J.Immunol., 115 : 606.

Lefkovitz, I., and Waldmann, H., 1979, Limiting dilution analysis of cells in the immune system, Cambridge University Press, Boston.

Marchal, G., Milon, G., and Hurtrel, B., 1979, High radiosensitivity of induction in contrast to expression of cells mediating delayed-type hypersensitivity during response to sheep red blood cells in mice. Immunology, 37 : 273.

Marchal, G., Seman, M., Milon, G., Truffa-Bachi, P., and Zilberfarb, V., 1982, Local adoptive transfer of skin delayed-type hypersensitivity initiated by a single T lymphocyte, J.Immunol., 129 : 954.

Marchal, G., and Milon, G., 1984, Numeration of DTH-mediating T lymphocytes in mice under optimal titration conditions, Ann.Immunol.(Inst.Pasteur), 135C : 345.

Milon, G., Goossens, P., and Marchal, G., 1984, Role of T lymphocytes during murine BCG infection, Proceedings of Leiden's Conference.

Milon, G., Marchal, G., Seman, M., Truffa-Bachi, P., and Zilberfarb, V., 1983, Is the delayed-type hypersensitivity observed after a low dose of antigen mediated by helper T cells ? J.Immunol., 130 : 1103.

Turk, J.L., 1980, Delayed hypersensitivity, 3rd ed., Research Monographs in Immunology, vol. 1, Elsevier / North-Holland Biomedical Press, Amsterdam.

EVIDENCE FOR LOCAL NEUROMODULATION OF T CELL MIGRATION IN VIVO

C. A. Ottaway

Department of Medicine, St. Michael's Hospital
University of Toronto
Toronto, Canada

Evidence is accumulating rapidly to support the concept that specific receptors for certain neuropeptides on lymphocytes may provide a means through which neurophysiological signals can influence immunological processes. Human lymphoid cells have been reported to have receptors for beta-endorphin (1), somatostatin (2), substance P (3), and vasoactive intestinal peptide (4), and a variety of investigations suggest that each of these neuropeptides can affect the behaviour of human lymphocytes in vitro. It has also been proposed that neurophysiological factors could have an important impact on the immune system in vivo by regulating the migration of lymphocytes through lymphoid tissue (5,6). This paper concerns experiments that were performed to test the hypothesis that receptors for the neuropeptide vasoactive intestinal peptide (VIP) influence the migration of T cells.

Previous work from this laboratory has demonstrated that mouse T cells can recognize and respond to VIP (7). This 28 amino acid neuropeptide is found within neurons of the gut and other mucosal tissues, and is well represented in the brain and the peripheral fibers of the autonomic nervous system (8). If the the ability of T cells to interact with VIP were involved in the migratory behaviour of these cells in vivo, one might hope to disturb their migration by disrupting VIP-containing nerves. At present, there is no known means by which VIPergic nerves can be disturbed selectively, and so an alternate approach was used.

Induced alterations either in the number of available receptors or the receptor binding affinity of a target cell can influence the biological responses elicited by polypeptide messengers (9). Furthermore, a variety of polypeptide hormones and biogenic amines have been shown to be able to regulate their own receptor numbers in a variety of target cells through a mechanism known as "down regulation", whereby exposure of the cells to the ligand leads to a dose-dependent decrease in the binding capacity of the cells and an alteration of the dose-response of the cells to the ligand (9,10). In the present studies, incubation of T cells in vitro in the presence of VIP led to a decrease in the number of receptors for VIP expressed by the cells and this was accompanied by a decrease in the ability of the treated cells to localize in vivo in the mesenteric lymph nodes and Peyer's patches.

Lymphocytes were obtained from the mesenteric lymph nodes (MLN) of 6-8 week old female Balb/c mice and single cell suspensions of cells enriched for T cells were prepared using nylon wool columns (7). These cells were incubated in the presence or absence of VIP for 18h and then washed extensively under conditions that lead to the rapid dissociation of cell bound VIP (7). The ability of these cells to bind VIP was investigated using radio-iodinated VIP as previously described (4,7) and their ability to localize in vivo was assessed by examining their distribution in various organs after intravenous cell transfer to syngeneic recipients. The methods used for cell isotopic labelling (sodium chromate) and organ retrieval are described elsewhere (11).

When T cells were incubated in the presence of 10^{-7} M VIP for 18h there was a decrease in the ability of these cells to specifically bind radioiodinated VIP compared with T cells incubated in the absence of added peptide. This altered binding was present throughout a wide range of competing VIP concentrations and Scatchard analyses of the tracer dilution profiles indicated that there was a decrease in the density of VIP binding sites in the T cell suspensions that had been treated with the neuropeptide (Table I). In contrast, there was no significant difference either in the affinity of the differently treated cells for VIP or in the expression of Thy-1.2, Lyt-1.2, or Lyt-2.2 T cell surface markers by the differently treated cells (Table I). The expression of these markers was tested because B and T cells are well known to differ in their migration, and it has been shown that subpopulations of T cells defined according to their expression of Lyt-1 (12) and Lyt-2 (13) determinants can display different migratory properties.

When suspensions of T cells that had been incubated in the presence or absence of VIP were labelled and transferred to recipients, the differently treated cells displayed a selective alteration in their organ distribution (Table II). There was a marked decrease in the recovery of VIP-pretreated cells in the mesenteric lymph node and Peyer's patches , but there was no difference either in the availability of VIP treated or control treated cells in the blood or in the recovery of the differently treated cell populations in other organs. (Table II).

Table I. Properties of T cells incubated with or without VIP*.

Property	Incubation	
	no VIP	10^{-7} M VIP
VIP Binding		
binding sites/cell	5200 (970)	2810 (520)
K_D (10^{-10} M)	3 (1)	4 (2)
Surface Markers (%)		
Thy-1.2	94 (3)	95 (4)
Lyt-1.2	84 (4)	85 (3)
Lyt-2.2	10 (3)	12 (4)

* Nylon wool effluent T cells were incubated for 18h in the presence or absence of VIP, harvested, washed and examined for the ability to specifically bind radiolabelled VIP, or to be specifically killed in cytotoxicity assays. The results are the means (S.D.) measured in four experiments.

A concentration of 10^{-7} M VIP was used in the experiments here. In other experiments, it was found that T cells incubated in the presence of different concentrations of VIP (0.1 nM to 100 nM) showed a dose-dependent decrease in their ability to specifically bind VIP. Similarly, T cells incubated with varying concentrations of VIP showed a dose-dependent decrease in their localization in both mesenteric lymph nodes and Peyer's patches at 1h after cell transfer without any accompanying disturbance in their distribution in the blood or other major organs. These data have been presented elsewhere (11).

To investigate further the nature of this altered migration displayed by T cells after exposure to the neuropeptide, the early time course of accumulation of T cells treated with or without VIP was examined (Table III). The accumulation of injected radioactivity in the spleen was equivalent with both of the treated cell populations, but there was a substantial decrease in the apparent rate at which the VIP pretreated cells accumulated in the Peyer's

Table II. Tissue distribution 1h after cell transfer of T cells incubated with or without VIP*.

Tissue	Treatment (±) VIP	Experiment 1	2	3
Blood/ml	-	1.7 ± 0.2	1.4 ± 0.1	1.5 ± 0.5
	+	1.6 ± 0.9	1.1 ± 0.1	1.4 ± 0.2
Mesenteric	-	106.9 ± 7.5	101.3 ± 8.6	140.2 ±24.3
node/g	+	80.3 ±16.3	67.9 ± 9.5	81.0 ±22.8
Peyer's	-	53.6 ±11.3	54.0 ± 8.2	73.1 ±18.1
patch/g	+	34.2 ± 3.2	30.4 ± 6.9	31.3 ± 5.2
Spleen	-	29.2 ± 4.4	35.2 ± 2.8	32.6 ± 1.2
	+	28.1 ± 3.5	34.9 ± 2.9	34.2 ± 3.6
Liver	-	14.5 ± 1.5	11.0 ± 0.5	12.5 ± 0.3
	+	15.0 ± 1.6	10.1 ± 0.2	14.9 ± 2.0
Lung	-	7.3 ± 1.2	7.1 ± 1.1	6.7 ± 0.9
	+	10.7 ± 1.2	6.1 ± 1.1	7.4 ± 1.8
Small	-	1.6 ± 0.5	0.9 ± 0.2	1.5 ± 0.2
Intestine	+	1.7 ± 0.7	1.2 ± 0.4	1.5 ± 0.5
Large	-	2.6 ± 0.7	1.5 ± 0.3	2.3 ± 0.4
Intestine	+	2.6 ± 0.7	1.2 ± 0.5	2.2 ± 0.7

* Nylon wool effluent T cells were incubated for 18h in the presence or absence of VIP, harvested, washed and labelled with Cr and transferred to recipients.Results are the means ± S.D. of % injected dose for four or five animals in each group.

patches and mesenteric nodes of the recipients throughout the first 50 minutes after cell transfer (Table III).

These alterations in localization of the peptide pretreated cells are consistent with the notion that perturbation of the T cell receptors for VIP selectively disturbs events contributing to the entry of T cells into Peyer's patches and mesenteric lymph nodes. Previous experiments have implicated the cell concentration in the blood stream, the delivery of blood flow to tissues, and the interactions of cells with the vascular endothelium as the principal factors controlling the entry of lymphoid cells into tissues (14-16). One way in which to explore the relationship of the early accumulation of transferred cells in a particular tissue to those in the blood is to evaluate an experimentally determined function V(t), defined by Equation 1.

$$V_i(t) = P_i(t) \left[1/t \times \int_0^t P_b(t)dt \right]^{-1} \qquad (1)$$

where at time t,

$P_i(t)$ = % injected dose of cells/g in the tissue, and

$P_b(t)$ = % injected dose of cells/ml of blood.

The dimensions of V(t) are thus ml/g, and V(t) represents a measure of the tissue accumulation of the transferred cells at a given time in terms of the mean value of the availability of cells in the blood up to that time. For convenience, I will refer to V(t) as the volume transform of the tissue observation.

In these experiments there was no difference in the availability of either the peptide-treated or control-treated cells in the blood (Table II) and in the time course experiment the amounts of radioactivity in the blood fell rapidly and equivalently with both types of cell preparations. Therefore, the observed differences in tissue localization cannot be accounted for by differences in the presence of the cells in the blood.

Table III. Early time course of the accumulation of T cells incubated with or without VIP*.

Tissue	Treatment (±)VIP	Time (minutes after cell transfer) 10	20	30	40	50
Peyer's patch/g	-	11.3(1.3)	24.7(5.0)	46.2(9.5)	58.7(2.2)	68.5(7.2)
	+	8.2(2.4)	16.2(2.6)	30.2(5.5)	38.2(5.7)	42.5(4.8)
Mesenteric node/g	-	17.7(2.6)	28.6(2.6)	53.0(6.1)	65.3(6.7)	88.1(5.4)
	+	13.3(3.2)	21.6(2.8)	45.0(5.5)	59.8(6.7)	67.7(8.4)
Spleen/organ	-	4.9(0.7)	7.2(1.7)	12.3(1.8)	16.6(3.4)	18.6(1.8)
	+	3.8(0.4)	7.8(1.7)	12.6(4.8)	17.9(1.2)	19.3(1.3)

* Nylon wool effluent T cells were incubated for 18h in the presence or absence of VIP. The results are the means (S.D) for three or four animals at each time point.

The time courses of the percent of the injected dose in the blood for both cell populations, however, were found to be satisfactorily described (r=0.98) by the exponential function,

$$P_b(t) = 13.5\, e^{-0.025t} \qquad (2)$$

Using this relationship, the integral in the denominator of Eqn. 1 can be explicitly calculated and the volume transform of the tissue observations derived. When this was done for the time course of the accumulation of the differently treated cells in different lymphoid tissues (Table III), V(t) was found, with a high degree of confidence, to be linear with time (Fig. 1, Table IV). The slopes of these volume transforms were determined by regression analysis and provide estimates of the rate at which the cells are cleared from the blood by these tissues (Table IV).

These alterations in the clearance of the peptide treated T cells by Peyer's patches and mesenteric nodes are unlikely to have arisen from changes in the blood flow to these tissues for several reasons. First, the conditions that were used to wash the treated cells lead to the rapid and efficient dissociation of cell bound VIP (7,11). It is possible that the peptide treated cells could contain small amounts of internalized VIP subsequent to their incubation with the peptide, but it is unlikely that this would disturb local blood flow. Infusion of large doses of VIP can produce transient increases of splanchnic blood flow in experimental animals (17-18), but if increased blood flow resulted one would expect enhanced localization of cells in the lymphoid tissue (14,19).

The decreased clearance of the peptide treated cells by Peyer's patches and mesenteric node, therefore, most likely represents an alteration in the way in which the "down regulated" T cells interact with the endothelium of these tissues. Lymphocytes enter lymph nodes and Peyer's patches via the specialized high endothelium of postcapillary venules. The interaction of lymphocytes with these structures is extremely efficient and quite similar in different species of experimental animals (15). The neuroanatomy of lymphoid tissues is incompletely understood, but nerve fibers have been identified within the medulla and internodular areas (5) and in close proximity to both the arterio-venous communications and the postcapillary venules of rat lymph nodes (20). So far, these nerves have not been examined for the presence of peptidergic neurons, but VIP-containing fibers have been demonstrated to be associated with the small vessels of other tissues (21). It is possible, therefore, that VIP may be locally available from VIPergic nerve endings within the microvasculature of lymphoid tissues.

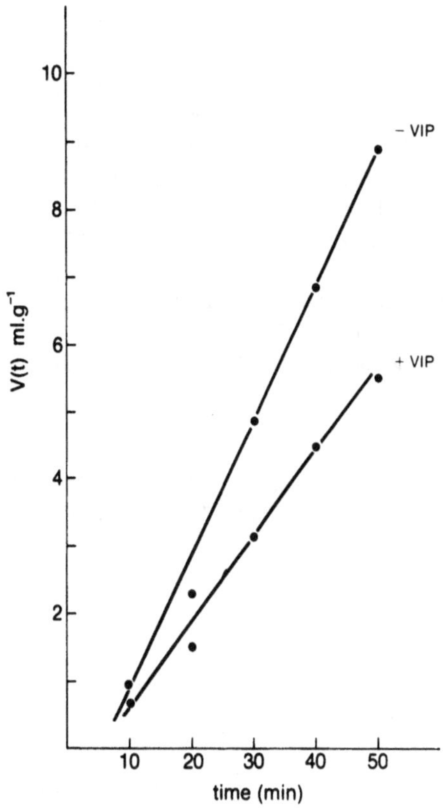

Fig. 1 Volume transforms of the Peyer's patches accumulation of T cells incubated with or without VIP.

Table IV. Apparent clearance rates determined from volume transforms of the early accumulation of T cells.

Tissue	Treatment (±) VIP	Correlation Coefficient	Clearance ml/g/min.
Peyer's	-	0.99	0.20 ± 0.01
patch	+	0.99	0.13 ± 0.01
Mesenteric	-	0.99	0.25 ± 0.02
node	+	0.99	0.20 ± 0.01
Spleen	-	0.99	0.53 ± 0.04
	+	0.99	0.57 ± 0.03

A plausible explanation for the observations presented here is that local occupancy of the VIP receptors of T cells by the neuropeptide facilitates the interaction of T cells with the specialized endothelium of these tissues. This could arise through peptide induced alterations in the state of activation of the T cells. Alternatively, binding of VIP to the T cells may lead to rearrangements of other surface components on the cells that participate in their interaction with the endothelium. The precise details remain to be examined, but the present findings clearly support the concept that the ability of T cells to recognize and respond to a specific neurophysiological signal contributes to the ability of T cells to migrate into Peyer's patches and mesenteric lymph nodes in vivo.

Acknowledgments: This work was supported by the Medical Research Council of Canada. C.A. Ottaway is a Scholar of the Medical Research Council of Canada.

References:

1. E. Hazum, K.J. Chang and P. Cautrecasas. Specific nonopiate receptors for beta-endorphins. Science (Wash. D.C.) 205:1033 (1979).
2. S.J. Bhathena, J. Louie, G.P. Schecter, R. Redman, L. Wahl and L. Recant. Identification of human mononuclear leukocytes bearing receptors for somatosatin and glucagon. Diabetes 30:127 (1981).
3. D.G. Payan, D.R. Brewster and E.J. Goetzl. Binding properties of substance P neuropeptide receptors on human lymphoblastoid cells. Fed. Proc. 132:1759 (1984).
4. C.A. Ottaway, C. Bernaerts, B.Chan and G.R. Greenberg. Specific binding of vasoactive intestinal peptide to human circulating mononuclear cells. Can. J. Physiol. Pharmacol. 61:664 (1983).
5. L.T. Giron, K.A. Crutcher and J.N. Davis. Lymph nodes: a possible site for sympathetic neuronal regulation of immune responses. Ann. Neurol. 8:520 (1980).
6. A.J. Cunningham. Mind, body and immune responses, p609, in: "Psychoneuroimmunology", R. Ader, ed., Academic Press, New York (1981).
7. C.A. Ottaway and G.R. Greenberg. Interaction of vasoactive intestinal peptide with mouse lymphocytes: specific binding and the modulation of mitogen responses. J. Immunol. 131:417 (1984).
8. J.M. Polak and S.R. Bloom. Distribution and tissue localization of VIP in the central nervous system and seven peripheral organs, p107, in, "Vasoactive Intestinal Peptide", S.I. Said, ed., Raven Press, New York (1982).

9. M.D. Ruggere and B.I. Posner. Polypeptide hormone receptors: their role in the pathogenesis of human disease. Clin. Invest. Med. 5:69 (1982).
10. G.P. Tell, F. Haour and J.M. Saez. Hormonal regulation of membrane receptors and cell responsiveness: a review. Metabolism 27:1566 (1978).
11. C.A. Ottaway. In vitro alteration of receptors for vasoactive intestinal peptide changes the in vivo localization of mouse T cells. J. Exp. Med. in press, (1984).
12. A.M. Carroll, Y. Reisner and M. deSousa. Lyt phenotype and lectin binding properties of mouse lymphocytes which enter lymph nodes. Adv. Exp. Med. Biol. 149:161 (1982).
13. G. Kraal, I.L. Weissman and E.C. Butcher. Differences in in vivo distribution and homing of T cell subsets to mucosal vs. nonmucosal lymphoid organs. J. Immunol. 130:1097 (1983).
14. C.A. Ottaway and D.M.V. Parrott. A method for the quantitative analysis of lymphoid cell migration experiments. Immunol. Lett. 2:283 (1981).
15. C.A. Ottaway. The efficiency of entry of lymphoid cells into lymphoid and non-lymphoid tissues. Adv. Exp. Med. Biol. 149:219 (1982).
16. C.A. Ottaway, R.G. Bruce and D.M.V. Parrott. The in vivo kinetics of lymphoblast localization in the small intestine. Immunology 49:641 (1983).
17. Y. Matsuo and A. Seki. The coordination of gastrointestinal hormones and the autonomic nervous system. Am. J. Gastroenterol. 69:21 (1978).
18. C. Holliger, M.Radzyner and M. Knoblauch. Effects of glucagon, vasoactive intestinal peptide and vasopressin on villous microcirculation and superior mesenteric artery blood flow of the rat. Gastroenterology 85:1036 (1983).
19. C.A. Ottaway and D.M.V. Parrott. Regional blood flow and its relationship to lymphocyte and lymphoblast traffic during a primary immune reaction. J. Exp. Med. 150:218 (1979).
20. A.O. Anderson and N.D. Anderson. Studies on the structure and permeability of the microvasculature in normal rat lymph nodes. Am. J. Pathol. 80:387 (1975).
21. R. Uddman, J. Alumets, L. Edvinsonn, R. Hakanson and F. Sundler. VIP nerve fibers around peripheral blood vessels. Acta Physiol. Scand. 112:65 (1981).

LYMPHOCYTE MIGRATION FROM PEYER'S PATCHES BY DIAPEDESIS THROUGH M CELLS INTO THE INTESTINAL LUMEN

Robert L. Owen and Martin F. Heyworth

Intestinal Immunology Research Center, Department of Medicine, University of California, San Francisco and Cell Biology and Aging Section, Veterans Administration Medical Center, San Francisco, CA 94121

INTRODUCTION

Lymphocytes and other leukocytes are found in the intestinal lumen in a variety of pathological conditions, and also in the absence of disease. Although Mitchison (1) found few lymphocytes in luminal washings from rabbit intestinal loops, Kotani et al. (2) found large numbers of lymphocytes in the lumen of rat intestinal loops, and calculated that a number of lymphocytes equal to the blood lymphocyte count entered the intestinal lumen five times per day. In jejunal biopsies from patients with giardiasis, and in mice infected with *Giardia muris*, we have observed intraluminal lymphocytes in close association with *Giardia* trophozoites (3,4). These lymphocytes are numerous over Peyer's patch lymphoid follicles. Leukocytes have also been found over Peyer's patches and appendix lymphoid follicles in the rabbit intestinal lumen (5). In mice Abe and Ito found twice as many intraepithelial lymphocytes over Peyer's patch follicle domes as over villi (6). Over half of these dome intraepithelial lymphocytes lay above the level of epithelial cell nuclei suggesting migration into the lumen, whereas only 7% of intraepithelial lymphocytes over villi lay above epithelial nuclei. After hydrocortisone administration intraepithelial lymphocytes dropped to one third but the proportion of lymphocytes above the level of epithelial nuclei remained unchanged, consistent with continued movement toward the lumen (6). Using monoclonal antibodies which label different subsets of lymphocytes, we have found that there is a preponderance of specific subsets within the mouse intestinal lumen, consistent with selective migration through the epithelium (7). Suggested pathways for lymphocyte entry into the lumen have included exudation through eroded portions of epithelium, passive release along with desquamated epithelial cells at villus tips, and migration between epithelial cells. In an attempt to determine the mechanisms by which lymphocytes enter the intestinal lumen, we examined the epithelium over Peyer's patch lymphoid follicles obtained from rabbits and mice after experimental manipulations.

METHODS

Ileal loops of adult male rabbits were inoculated with cholera vibrios. After 90 minutes, these loops were excised, and Peyer's patches in the loops were fixed with glutaraldehyde and paraformaldehyde and prepared for transmission electron microscopy (see Owen et al. (8) for further details). Follicle surfaces in the patches were examined for evidence of lymphocyte migration into the lumen. Male BALB/c mice were injected subcutaneously with 5 mg hydrocortisone, on each of two successive days, and on the third day colloidal gold was introduced into loops of small intestine. After 120 minutes, Peyer's patches from the loops were fixed as above, and examined by transmission electron microscopy.

RESULTS

Lymphocytes with morphologically intact external membranes and cytoplasmic organelles were observed migrating through the epithelium in both species of animal. Lymphocytes in cavities within M cells extended past M cells into the lumen (Fig. 1). Membranes of M cells were intact, and formed channels around migrating lymphocytes. The M cell membranes were closely apposed to the surfaces of these lymphocytes. In addition, lymphocytes were observed migrating through opened junctional complexes between epithelial cells over lymphoid follicles (Fig. 2). Migration of lymphocytes through the epithelium was most easily seen in rabbits, in which M cells comprise half the follicle surface. In both animal species, the "hand mirror" configuration of migrating lymphocytes (nucleus followed by cytoplasm) indicated that the direction of lymphocyte movement was into the lumen (Fig. 1). Groups of lymphocytes streaming through channels past M cells formed sheets over follicle surfaces (Fig. 3). Lymphocytes containing ingested bacteria were observed in the intestinal lumen of rabbits (Fig. 4). In mice, lymphocytes in the lumen were seen to be pinocytosing particles of colloidal gold (Fig. 2). We found no instances in which the structural configuration of epithelial lymphocytes indicated that they were migrating from the lumen back into the mucosa.

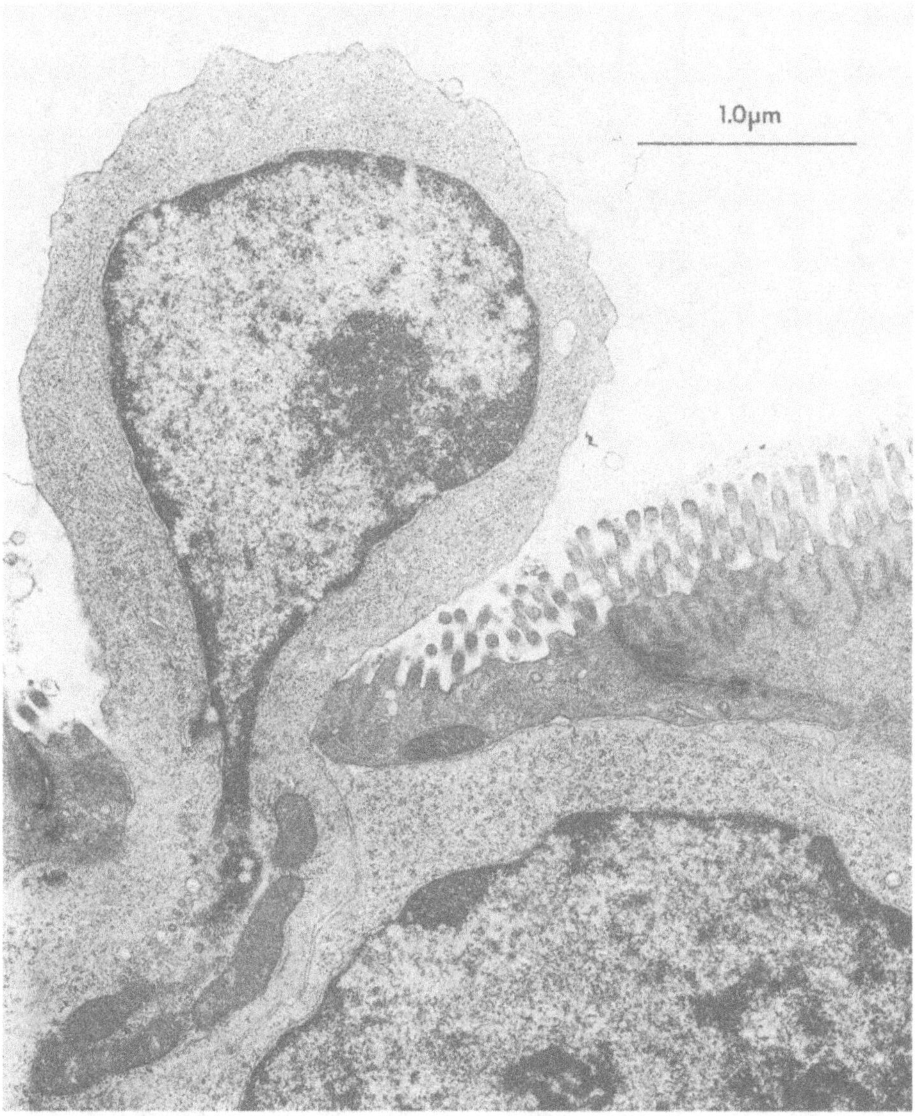

Figure 1. A lymphocyte with a hand mirror configuration is migrating through an M cell into the ileal lumen of a rabbit. The M cell forms tight junctions with adjacent cells but no evidence of a disrupted tight junction is apparent along the channel through which the lymphocyte is passing.

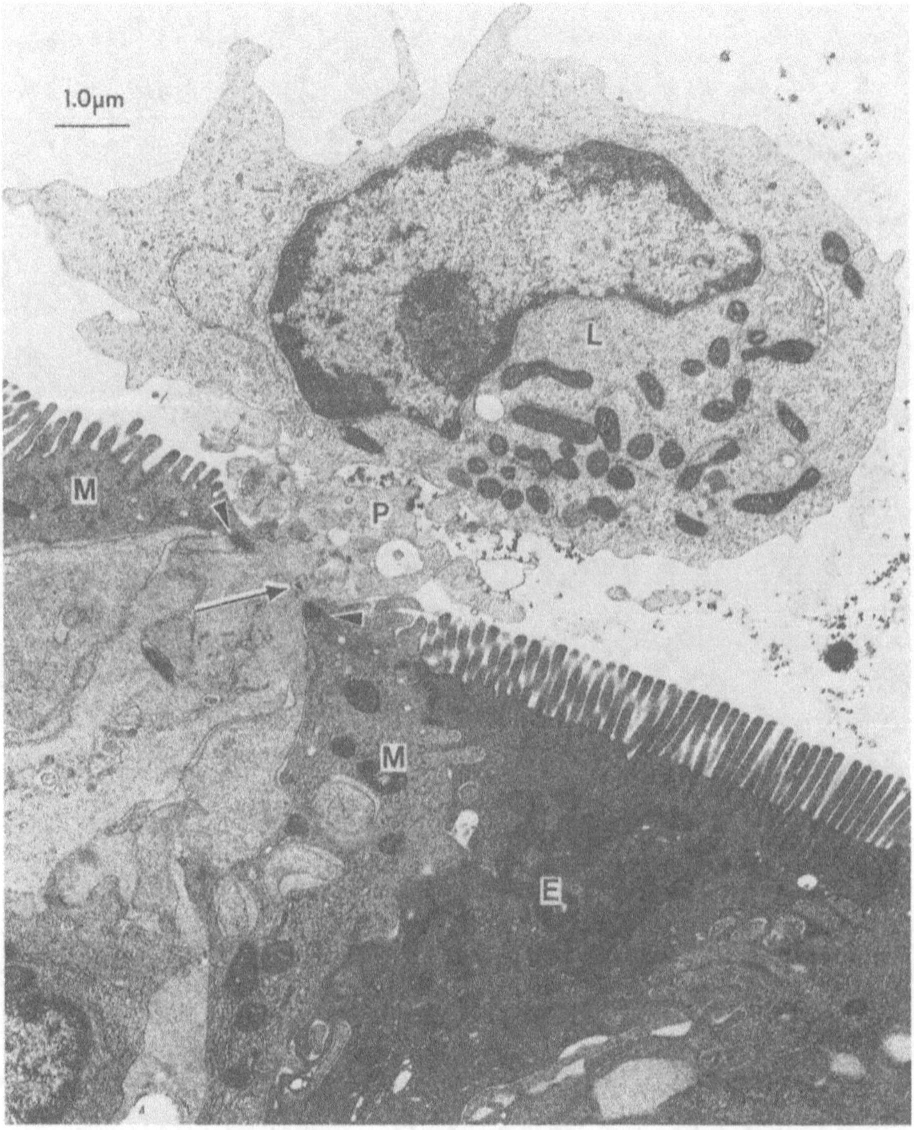

Figure 2. In a hydrocortisone-treated mouse, a lymphocyte (L) lies in the intestinal lumen. The pseudopod (P) of an additional lymphocyte is protruding into the lumen past an intact M cell (M) attached on each side to enterocytes (E). Portions of a disrupted tight junction are seen on each side of the pseudopod (arrowheads). Colloidal gold particles are visible in a vesicle within the pseudopod (arrow).

DISCUSSION

Structural evidence of lymphocyte diapedesis through Peyer's patch follicle epithelium is consistent with our previous finding that specific lymphocyte subsets populate the intestinal lumen (7). As Abe and Ito previously inferred from the position of lymphocytes above the level of epithelial cell nuclei over Peyer's patch domes (6), we have confirmed that hydrocortisone administration does not prevent lymphocyte migration into the lumen. Our observations that lymphocytes pass between M cells and enterocytes, between adjacent M cells, and through new pores in these cells, suggest that M cells may constitute a physiologically important site for lymphocyte entry into the lumen. Unlike columnar epithelial cells, M cells lack a developed terminal web and their cytoplasm appears to be readily deformed by pressure from lymphocytes. This deformability may permit lymphocytes to cross the intestinal epithelium more readily in the vicinity of M cells than in epithelial regions where M cells are absent. Our observations suggest the existence of at least two mechanisms for lymphocyte migration into the lumen, namely undoing of junctional complexes by migrating lymphocytes, and induction of channels through the cytoplasm of M cells. The suggestion that migrating lymphocytes form channels through M cells is based on electron micrographs in which intact portions of an individual M cell are seen on either side of a migrating lymphocyte, with no evidence of residual junctional complexes (Fig 1).

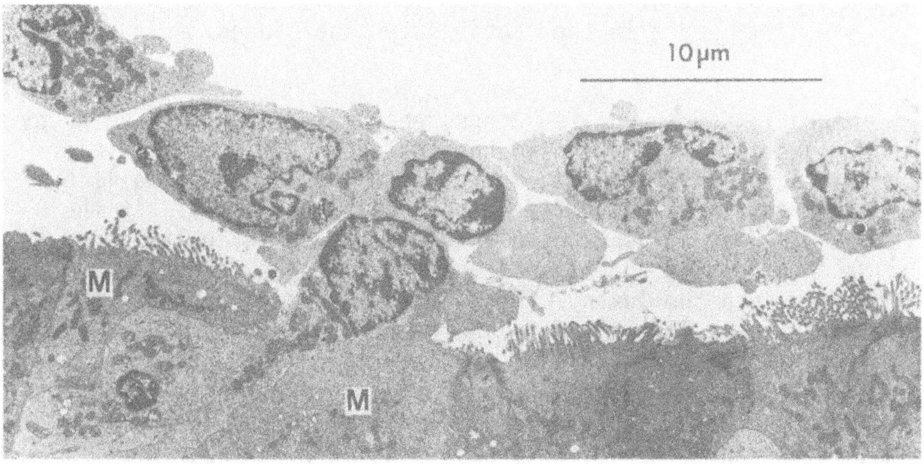

Figure 3. Numerous intact lymphocytes lie in the intestinal lumen over a rabbit Peyer's patch follicle. Additional lymphocytes are migrating through an M cell (M).

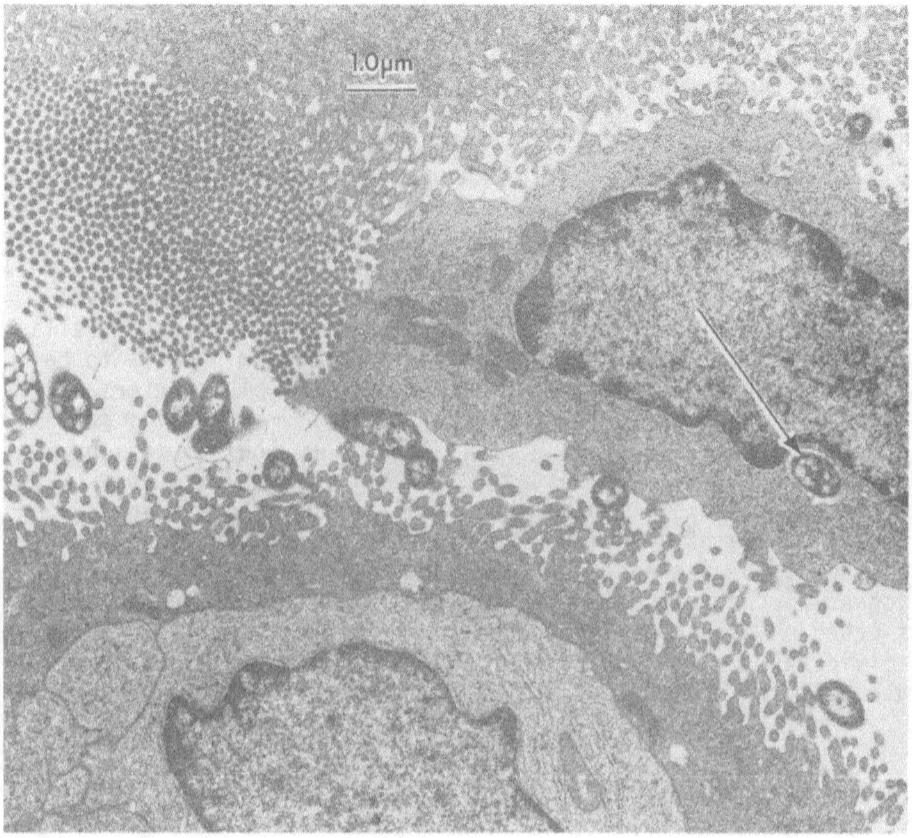

Figure 4. A noninvasive cholera vibrio (arrow) is seen within a lymphocyte in the rabbit intestinal lumen.

Migration of leukocytes across epithelia has been studied by other investigators. When lymphocytes enter lymphoid follicles from post-capillary venules, the cells traverse venule walls by opening tight junctions between endothelial cells (9). In the trachea of guinea pigs exposed to cigarette smoke, polymorphonuclear neutrophils and other leukocytes migrate across the epithelium into the lumen. These leukocytes pass through well-defined channels in the epithelium rather than migrating randomly between epithelial cells (10). In the stomach of patients with gastric ulceration, polymorphonuclear leukocytes and lymphocytes pass between gastric epithelial cells to enter the lumen (11). Lymphocytes can be harvested from the lumen of the bursa of Fabricius in chicks (12). There is evidence that most of these lymphocytes enter the bursal lumen by crossing epithelium in the dorsal region of the bursa.

Different types of leukocyte are found in the intestinal lumen under different circumstances. In calves infected with *Escherichia coli*, neutrophils enter the intestinal lumen, particularly above Peyer's patch follicles (13). Neutrophils also migrate into the intestinal lumen of pigs sensitized to bovine serum albumin (BSA), after exposure of the small intestinal mucosa to BSA (14). In baby pigs which have not been deliberately sensitized to any particular antigen, lymphocytes have been demonstrated in the small intestinal lumen (15). Lymphocytes and macrophages have been harvested from the lumen of the rabbit appendix (5). There is evidence that local antigenic stimulation with sheep erythrocytes leads to increased entry of mononuclear leukocytes into the rabbit appendix lumen (5).

At present, the physiological role of leukocytes in the intestinal lumen is uncertain. There is some evidence that luminal macrophages phagocytose local microorganisms. Macrophages containing bacteria have been harvested from the appendiceal lumen of rabbits (5), and we have observed small numbers of macrophages adhering to *Giardia muris* trophozoites obtained from the intestinal lumen of *Giardia*-infected mice. It is not known whether luminal macrophages play a significant part in the clearance of intestinal infections. There is little evidence that intraluminal lymphocytes have a cytotoxic effect on local pathogenic organisms. We have previously observed intraluminal leukocytes in contact with *Giardia* trophozoites, but have not identified any lymphocyte subset which attaches to the parasites (16). Whether intraluminal lymphocytes ever become sensitized to antigens in the lumen of the intestine, and then migrate back into the mucosa to initiate an immune response is not known.

ACKNOWLEDGEMENTS

Supported by the U.S. Veterans Administration and NIH Grants AM 33004 and AM 33930. The technical contributions of R.T. Apple to this study are gratefully acknowledged.

REFERENCES

1. N.A. Mitchison, The passage of antibodies into the intestine in rabbits, *Quart. J. Exp. Physiol.* 38:139 (1953).
2. M. Kotani, A. Yamashita, F. Rai, K. Seiki, and I. Horii, Reutilization of DNA breakdown products from lymphocytes in lumen of intestine, *Blood* 29:616 (1967).
3. R.L. Owen, The immune response in clinical and experimental giardiasis, *Trans. R. Soc. Trop. Med. Hyg.* 74:443 (1980).

4. R.L. Owen, P.C. Nemanic, and D.P. Stevens, Ultrastructural observations on giardiasis in a murine model: I. Intestinal distribution, attachment, and relationship to the immune system of Giardia muris, Gastroenterology 76:757 (1979).
5. R.V. Heatley and J. Bienenstock, Luminal lymphoid cells in the rabbit intestine, Gastroenterology 82:268 (1982)
6. K. Abe and T. Ito, Qualitative and quantitative morphologic study of Peyer's patches of the mouse after neonatal thymectomy and hydrocortisone injection. Am. J. Anat. 151:227 (1978).
7. M.F. Heyworth and R.L. Owen, Intestinal leukocyte subpopulations in Giardia muris-infected immunocompetent and immunodeficient mice (Abstract), Immunobiology, in press (1984).
8. R.L. Owen, N.F. Pierce, R.T. Apple, and W.C. Cray, Jr., M cells transport Vibrio cholerae from intestinal lumen into Peyer's patches, J. Infec. Dis., in press.
9. R.L. Owen and D.K. Bhalla, Lympho-epithelial organs and lymph nodes, in "Biomedical Research Applications of Scanning Electron Microscopy", Vol. 3, G.H. Hodges and K.E. Carr, eds., Academic Press, London (1983).
10. W.C. Hulbert, D.C. Walker, and J.C. Hogg, The site of leukocyte migration through the tracheal mucosa in the guinea pig, Am. Rev. Respir. Dis. 124:310 (1981).
11. H.W. Steer, Ultrastructure of cell migration through the gastric epithelium and its relationship to bacteria, J. Clin. Path. 28:639 (1975).
12. S. Odend'hal and J.E. Breazile, Luminal lymphoid cells of the cloacal bursa, Am. J. Vet. Res. 40:1015 (1979).
13. J.E.C. Bellamy and S.D. Acres, Enterotoxigenic colibacillosis in colostrum-fed calves: Pathologic changes, Am. J. Vet. Res. 40:1391 (1979).
14. J.E.C. Bellamy and N.O. Nielsen, Immune-mediated emigration of neutrophils into the lumen of the small intestine, Infect. Immun. 9:615 (1974).
15. R.M. Chu, R.D. Glock, R.F. Ross, and D.F. Cox, Lymphoid tissues of the small intestine of swine from birth to one month of age, Am. J. Vet. Res. 40:1713 (1979).
16. M.F. Heyworth, R.L. Owen, W.E. Seaman, F.W. Schaefer, and A.L. Jones, Harvesting of leukocytes from intestinal lumen in murine giardiasis and preliminary characterization of these cells, Dig. Dis. Sci., in press (1984).

EMIGRATION OF NEWLY FORMED LYMPHOCYTES FROM ILEAL PEYER'S PATCHES IN LAMBS

Reinhard Pabst[1] and John D. Reynolds[2]

[1]Centre of Anatomy, Medical School of Hannover
D-3000 Hannover 61, West Germany
[2]Department Medical Physiology, University of Calgary
Calgary, Canada T2N 4N1

INTRODUCTION

In lambs there is a long continuous Peyer's patch of about 2m length at the end of the ileum. This patch has several peculiarities which are comparable to the primary lymphoid organ for B cells in chickens, the Bursa of Fabricius: it is well developed before birth (1), it is the largest lymphoid organ in lambs in the first few months of life, later it almost completely disappears, the lymphocyte production is not influenced by antigen (2), nearly all lymphoid cells in ileal Peyer's patches are B lymphocytes (3).

In the present study the lymphocyte production in the Peyer's patches of lambs was quantified, and the proportion of newly formed lymphocytes which leave Peyer's patches and migrate to other lymphoid organs was estimated. In addition the emigrants were classified by cytological criteria. Different parts of the small intestine were isolated temporarily in young lambs and perfused with medium containing a radioactive precursor of DNA. The lymphocyte production was determined on autoradiographs.

MATERIAL AND METHODS

In seven lambs the mesenteric lymph nodes along the ileum were removed at 1-2 weeks of age. The operation and the selective perfusion of parts of the gut containing either ileal Peyer's patches (ilPP) or jejunal Peyer's patches (jejPP) have already been described (4). The perfusion system has been used for in vivo

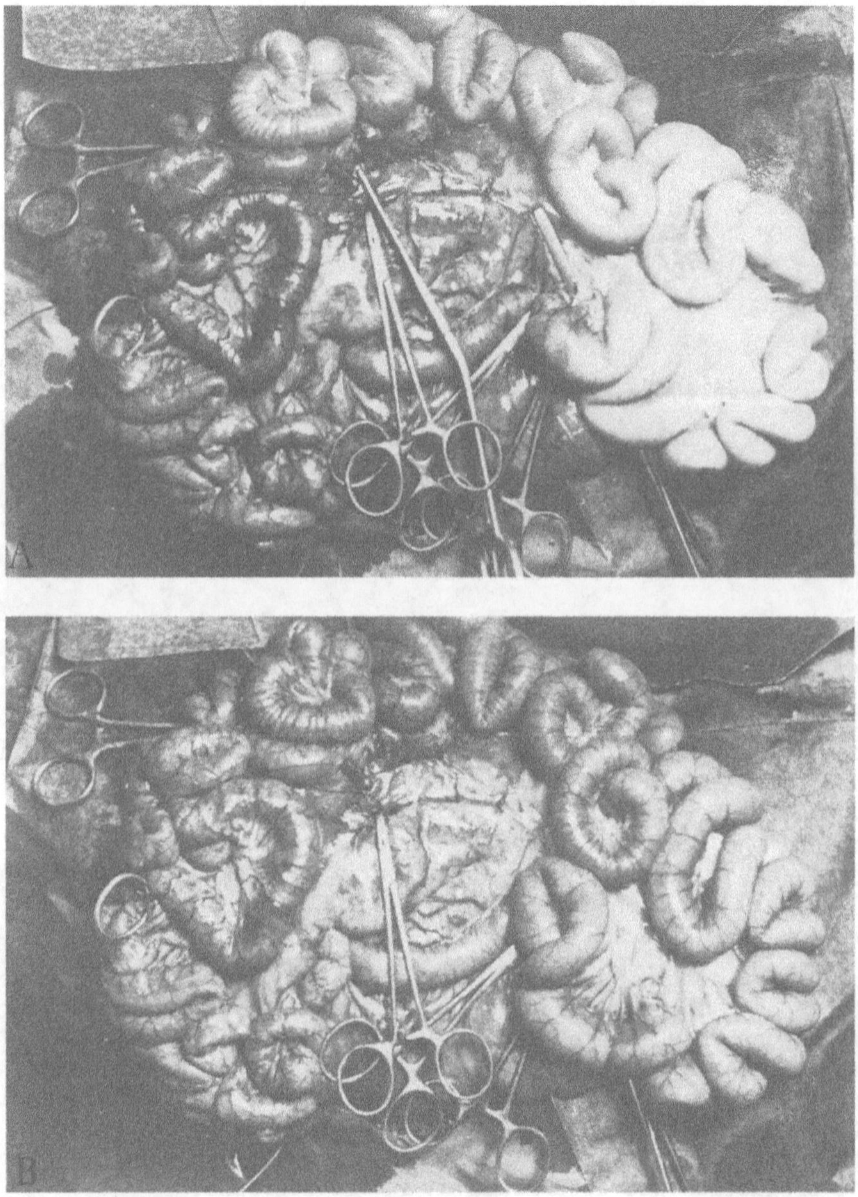

Fig. 1. A. During the extracorporeal perfusion of the terminal ileum the perfused part (right side) is almost free of red cells and the border of the ilPP can be seen.
B. Shortly after opening the clamps on the main intestinal vessels in the perfused gut the colour returns.

labelling of pig spleens (5). In brief, the side branch of the superior mesenteric artery and vein was cannulated and connected to an extracorporeal perfusion system, the main vessels to the gut and the gut itself clamped and the isolated part of the small intestine perfused with 300ml tissue culture medium, containing 1-3mCi of tritiated thymidine, for 15 min (Fig. 1). Then the non-incorporated label was flushed from the vascular system with fresh unlabelled medium. After this 30 min perfusion period the gut was once again perfused by the lamb's own blood. One or three days later the labelling index of lymphoid cells in ilPP, jejPP and other lymphoid organs was determined on smears after autoradiography. Four lambs were studied 1 day after perfusion, in two of these the intestinal lymph duct was cannulated and the lymph drained. Three lambs were examined after 3 days. In two lambs the jejunum, only containing a few jejPP, was perfused and studied 1 day later.

RESULTS

The isolated perfusion of a particular region of the gut restricts the label to that specific part of the gut (4) while preserving the structure and function of the gut, as demonstrated by normal numbers of mitotic figures in the crypts (Fig. 2).

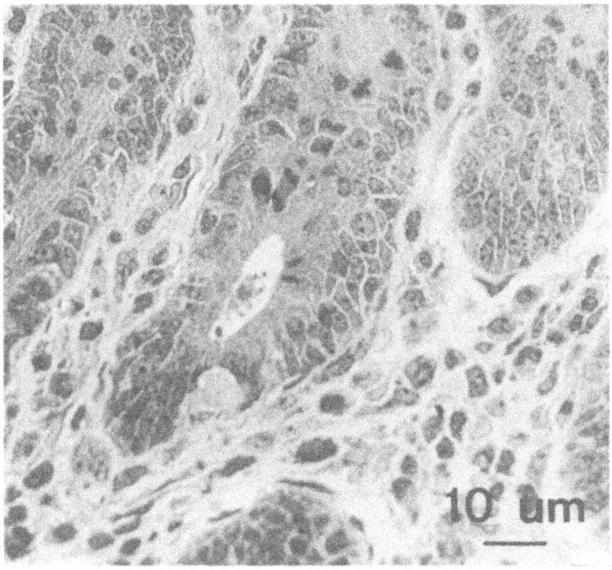

Fig. 2. Histological section of intestinal crypts one day after extracorporeal perfusion of this part of the gut. A normal structure and many mitotic figures can be seen.

Thus, proliferating lymphoid cells incorporate the label in their normal microenvironment under almost physiological conditions. Histoautoradiographs showed densely packed labelled cells in Peyer's patches follicles with far less in the interfollicular region and the dome area. Smears of perfused Peyer's patches (after one day) revealed obvious differences between ilPP and jejPP, 63.7% and 26.1% respectively. The labelling index of lymphocytes in other lymphoid organs is given in Table 1. When the intestinal lymph duct was drained, far less lymphocytes derived from Peyer's patches were found in other organs. The labelling index of blood lymphocytes was always higher than in lymph nodes, jejPP and spleen, these three showing labelled cells in comparable numbers. A few labelled lymphocytes were identified in the thymus and hardly any in the bone marrow. In all experiments far fewer newly formed lymphocytes migrated to non-perfused parts of the ilPP than to the jejPP.

Table 1. Labelling index of lymphoid cells after selective perfusion of parts of the continuous Peyer's patch at the terminal ileum or parts of the jejunum with few Peyer's patches in lambs, in which the ileal mesenteric lymph nodes were excised previously.

Organ	Labelling index			
	day 1 ilPP perfused + intest. lymph cannula (n=2)	day 1 ilPP perfused (n=2)	day 3 ilPP perfused (n=3)	day 1 jejPP perfused (n=2)
perfused ilPP	65.7%	55.5%	84.9%	26.1%
non-perfused PP	0.007%	0.05%	0.08%	0.02%
jejPP	0.04%	0.24%	0.39%	0.08%
mesenteric LN	0.05%	0.24%	0.30%	0.13%
prescapular LN	0.03%	0.10%	0.19%	0.05%
spleen	0.05%	0.06%	0.30%	0.06%
thymus	0.001%	0.007%	0.01%	0.003%
blood	0.06%	1.1%	1.6%	0.5%

The total number of lymphocytes in the lymphoid organs of 5-8 week old lambs are: Peyer's patches 17 x 10^{10}, thymus 18 x 10^{10}, lymph nodes 5 x 10^{10}, spleen 1.6 x 10^{10} and blood 0,27 x 10^{10} (Reynolds, unpublished). Based on the labelling index and the absolute numbers, the total number of newly formed lymphocytes could be extrapolated. One day after selectively labelling ilPP about 108 x 10^9 newly formed lymphocytes were still in the perfused organ and only 0.18 x 10^9 in all other lymphoid organs studied. That means about only 0.2% of all newly produced lymphocytes left ilPP within 1 day after labelling.

The newly formed lymphocytes which emigrated from ilPP were classified in mesenteric and prescapular lymph nodes, spleen and jejPP. There were no obvious differences in these four organs and therefore the data were pooled. More than 90% of the emigrated newly formed cells were small and medium sized lymphocytes, about 6% large lymphocytes and hardly any labelled mature plasma cells were identified (Fig. 3). In some experiments histological sections of unperfused areas in the small intestine were studied after autoradiography. Up to 50% of labelled cells were plasma cells in the lamina propria three days after labelling.

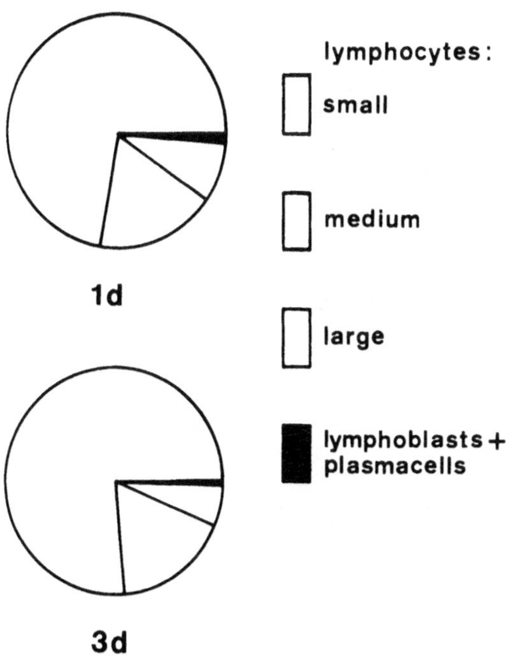

Fig. 3. Newly formed lymphoid cells from ilPP were classified by cytological criteria in mesenteric and prescapular lymph nodes, spleen and jejPP. Only a minority of immigrants from ilPP are large lymphoid cells or lymphoblasts one or three days after selective perfusion of ilPP.

DISCUSSION AND CONCLUSIONS

The technique of a temporary extracorporeal perfusion of parts of the gut for labelling proliferating lymphocytes in their normal microenvironment is useful in studying the amount of lymphocyte production and emigration. Many more lymphocytes proliferate in ilPP than in jejPP. Only a very small proportion of all newly formed cells leave Peyer's patches. This is comparable to the primary lymphoid organ for T cells, the thymus (6). Most emigrating cells are small to medium sized lymphocytes, presumably B lymphocytes. The majority migrates via the lymphatics. These data add further evidence that the long continuous Peyer's patch at the terminal ileum can be thought of as an equivalent in lambs to the Bursa of Fabricius (7).

ACKNOWLEDGEMENTS

Most of the animal experiments were done at the Basel Institute for Immunology,which was founded and supported by Hoffmann La Roche Ltd., Basel, Switzerland.

REFERENCES

1. J.D. Reynolds and B. Morris. Eur. J. Immunol. 13:627 (1983).
2. J.D. Reynolds and B. Morris. Eur. J. Immunol. 14:1 (1984).
3. M. Miyasaka. Immunology, in press (1984).
4. J.D. Reynolds and R. Pabst. Eur. J. Immunol. 14:7 (1984).
5. R. Pabst and K. Pötschick. Immunology 50:281 (1983).
6. R. Scollay and K. Shortman. Thymus 5:245 (1983).
7. J.D. Reynolds. Curr. Top. Microbiol. Immunol., in press (1984).

PATTERNS OF MIGRATION OF LABELLED BLOOD LYMPHOCYTE SUBPOPULATIONS: EVIDENCE FOR TWO TYPES OF PEYER'S PATCH IN THE YOUNG PIG

R.M.Binns and S.T.Licence

Immunology Department
AFRC Institute of Animal Physiology
Babraham, Cambridge CB2 4AT, UK

INTRODUCTION

Although in many ways apparently conventional, the lymphoid system of the pig differs from that of other species in some important respects: The lymph node is structurally inverted with a peculiar medulla and its efferent lymph is almost devoid of lymphocytes since recirculating lymphocytes emigrate directly into veins. Moreover a large proportion of blood lymphocytes (~40% and 4000/mm^3) in young pigs are Null T-dependent cells which apparently do not recirculate[1,2]. Although there have been careful studies of the incidence of lymphocytes in different lymphoid organs (see [2]), quantitative data are lacking on the migration of labelled blood lymphocytes to the lymphoid tissues and their changes with time. This paper using ^{51}Cr- and FITC-labelling provides evidence of major heterogeneity of tissue and lymphocyte population behaviour in blood lymphocyte migration including evidence for two different types of Peyer's patch.

MATERIALS AND METHODS

These studies used blood lymphocytes from 3-8 month old female donor pigs of the Babraham MHC-homozygous line labelled at 10^8/ml with 10-20μCi $Na_2\,^{51}CrO_4$ for 30 minutes or 50μg FITC for 15 minutes, washed and injected intravenously into histocompatible 4-6 week normal immunocompetent recipients. Lymphocyte isolation, FITC-labelling, enumeration and study of subset type are as detailed previously[1,2,3]. Studies were also made of the distribution of lymphocytes from thymectomized and control sham-thymectomized donors (see [5]) in normal recipients and of normal donor lymphocytes in immunocompetent antigenically-isolated pig foetuses 3 weeks before birth (94 days) following injection into a tarsal vein. In the ^{51}Cr-label experiments blood and some 30 lymphoid and non-lymphoid tissues were taken after exsanguination at varying times between 5 minutes and 329 hours post-injection, cleaned, weighed and their gamma-emission counted whole or using a representative sample. The factor of labelled cell concentration in each tissue was computed from the cpm/gm tissue by dividing it by the mean body cpm/gm (i.e. total injected cpm, adjusted for isotope decay, divided by gms body-weight). This "cell concentration factor" (which would be 1 if lymphocytes were uniformly distributed throughout the body) varied from ~75 for postnatal spleen at 30 minutes and ~100 for foetal lymph nodes at day 1 down to ~0.02 for postnatal subcutaneous fat at 3 hours and the eyeball at day 1. At individual times, mean data from 2-7 pigs are shown (see Fig.1), frequently using more than one donor and several experiments, but nevertheless yielding very consistent data. Early data showed that transferred histocompatible lymphocytes yielded essentially the same results as autologous lymphocytes, confirming similar experiments using FITC-labelling[3]. The following tissues taken are shown in all text illustrations as signified in bold capital type: Lymph nodes:- Superficial Inguinal, PreScapular, PreCrural, SpLeNic MaNDibular, RETrosternal, ParaAortic, BROnchial, MESenteric and PORtal. Other tissues:- THYmus, palatine TONsils, papillae of ToNGue dorsum, EPiglottis, one inch sections of OEsophagus, DUOdenum, Upper ILeum, Lower ILeum by the ileocaecal junction and proximal spiral COLon, isolated jejunal Discrete Peyer's Patches and upper ileal Continuous Peyer's Patch, SPLeen, femoral Bone Marrow, BLood, LuNG, LIVer, KIDney, PANcreas, submandibular SALivary gland, THyroiD, HeaRT, hamstring MuScLe, shoulder SKiN and subcutaneous FAT.

RESULTS AND DISCUSSION

Distribution of ^{51}Cr-labelled isogeneic blood lymphocytes

Injected isogeneic labelled lymphocytes, traced by whole tissue gamma counting, show a highly reproducible series of changes in distribution summarised in Fig.1A & B. Lymphocyte concentration in lung, blood and non-lymphoid tissues (except liver and kidney which are sites of lymphocyte death and ^{51}Cr excretion) fell rapidly over the first few hours after injection from levels at 5-30 minutes which seem to broadly correlate with the vascularity of the organ (e.g. heart >> leg muscle; thyroid, pancreas, salivary gland >> skin, fat, muscle). The initial high lymphocyte entry within 5 minutes into lung and spleen is continued in the spleen at 30 minutes when lymphocytes are emerging from the lung and large numbers have entered the bone marrow. Then marrow shows a consistent fall in counts at one hour, returning to a higher level by 3 hours and falling slowly thereafter. Other lymphoid tissues show differing initial rates of lymphocyte entry but their labelled cell contents increase for ~1-3 days (Fig.1B). Thus different lymph nodes showed consistent differences in the rate of entry and final maximum labelled cell concentration, e.g. portal - slow entry, high level; prescapular - slow entry, low level; mesenteric - fast entry, long consistent intermediate plateau concentration from ~1-90 hours - differences which presumably reflect their normal physiological cell entry from afferent lymph and blood. Mesenteric nodes also seemed to differ in their faster rate of loss of cells between 90 and 329 hours, raising the possibility of differences in average lymphocyte life-span in different classes of lymph node. Though at a lower level, migration to the thymus shows a similar time pattern to most lymph nodes. Taken together, these data suggest the existence of major differences in lymphocyte migration behaviour both through different lymphoid tissues and between different lymphocyte populations and are consistent with the accepted dogma of a fast major recirculation through spleen and a slower one through lymph nodes.

The suggestion of lymphocyte heterogeneity in recirculation receives support from the observation (Fig.2) that injected FITC-labelled isogeneic blood lymphocytes show marked rapid changes in the first 24 hours in the proportion of lymphocyte subpopulations superimposed upon the changes in cell number seen during their equilibration with the tissue lymphocyte pool[4]. In the spleen, Null cells predominate early and B cells at day 1, while lymph nodes show enrichment of strongly-rosetting T cells. The changes in Null, B and T cells (strong and weak rosetting) in blood reflect these differing

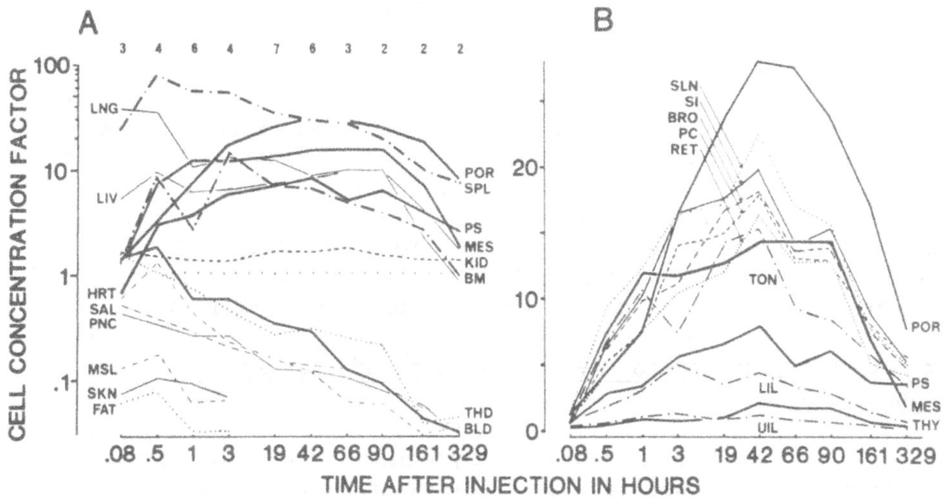

Fig. 1. The mean cell concentration factor in 25 tissues at various times after intravenous injection of ^{51}Cr-labelled histocompatible blood lymphocytes. 1A. Lymphoid and non-lymphoid tissues (log scale; also shows numbers of recipients per time). 1B. Lymph nodes, thymus and alimentary lymphoid tissue (Abbreviations see M and M).

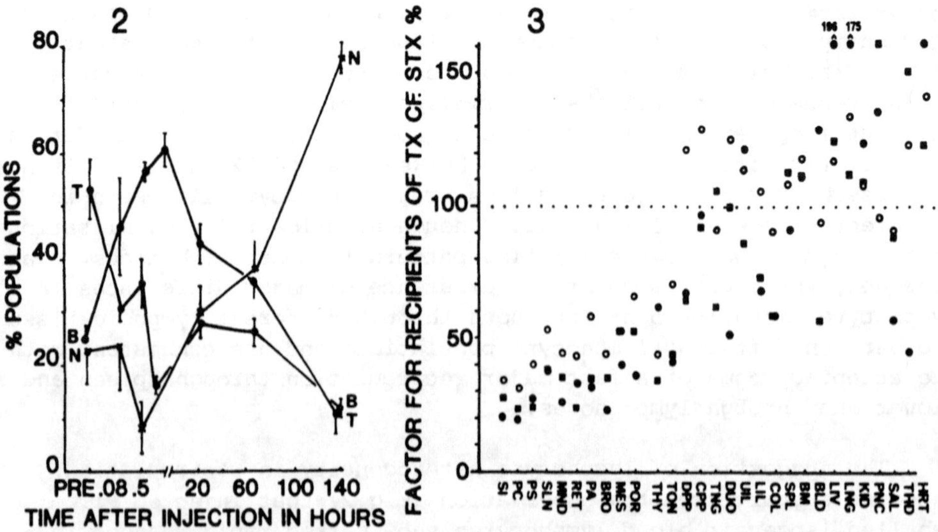

Fig. 2. Changes in proportions of FITC-labelled T, B and Null blood lymphocytes in the blood after intravenous injection.

Fig. 3. Percent reduction in the cell concentration factor in lymph nodes, thymus and tonsils of recipients of thymectomized (Tx) compared to sham-thymectomized (STx) donor ^{51}Cr-labelled blood lymphocytes at 3 hours (●), 1 day (○) and 4 days (■) (Abbreviations see M and M).

phases of tissue entry. After this, the proportion of labelled Null cells shows a progressive increase. There is no reason to assume that such changes are peculiar to FITC-labelled lymphocytes since ^{51}Cr-labelled blood lymphocytes also labelled with FITC show the same tissue distribution in detail as those only labelled with ^{51}Cr (data not shown). Though incomplete these data indicate marked differences in tissue migration behaviour of different pig lymphocyte populations and are consistent with our previous data suggesting that Null cells apparently do not recirculate through spleen or lymph nodes[2].

Effect of thymectomy on the distribution of ^{51}Cr-labelled lymphocytes

Using littermate neonatally thymectomized and sham thymectomized female pigs as lymphocyte donors for transfer to normal young pigs, it is clear that thymectomy results in inhibition of the capacity of blood lymphocytes to enter lymph nodes and thymus (Fig. 3) but not significantly other tissues (e.g. spleen, bone marrow, liver and the non-lymphoreticular tissues) with the possible exception of tonsil. Moreover this deficiency, evident at 3 hours and 1 day has not been 'repaired' significantly by circulation for 4 days in the normal environment of the recipients. These data on difference in lymph node entry obtained using ^{51}Cr-labelling were confirmed with FITC labelling. The deficiency in lymph node entry correlates with several other functional deficits and with a selective depletion of lymphocytes from the lymph node paracortex[6] and is also consistent with the existence of as-yet-unrecognised lymphocyte subsets.

Lymphocyte migration in the antigen-free immunocompetent foetus

The differences seen between tissues in lymphocyte migration behaviour would appear to be largely an inherent property rather than a function of antigenic experience, since normal donor lymphocytes show broadly the same distribution in foetuses 3 weeks before birth. Thus Figure 4 shows that at day 1 post injection, donor lymphocytes show an even higher concentration in foetal than in postnatal lymph nodes and some other tissues, and that although there are differences in the detail of distribution, presumably due to differences in the physiological blood flow, maturity and antigen-driven development of different tissues, the broad pattern is the same, such that for instance the difference in lymphocyte entry between portal, precrural and prescapular lymph nodes and between upper (UIL) and lower (LIL) ileum is already evident.

Lymphocyte migration to alimentary lymphoid tissues

Initial observations of the difference in lymphocyte entry in the Peyer's patch of upper (UIL) and lower (LIL) ileum (Fig. 1B) was

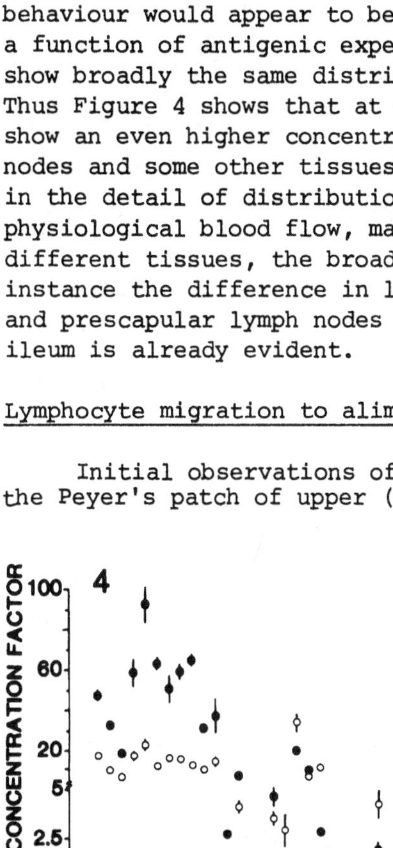

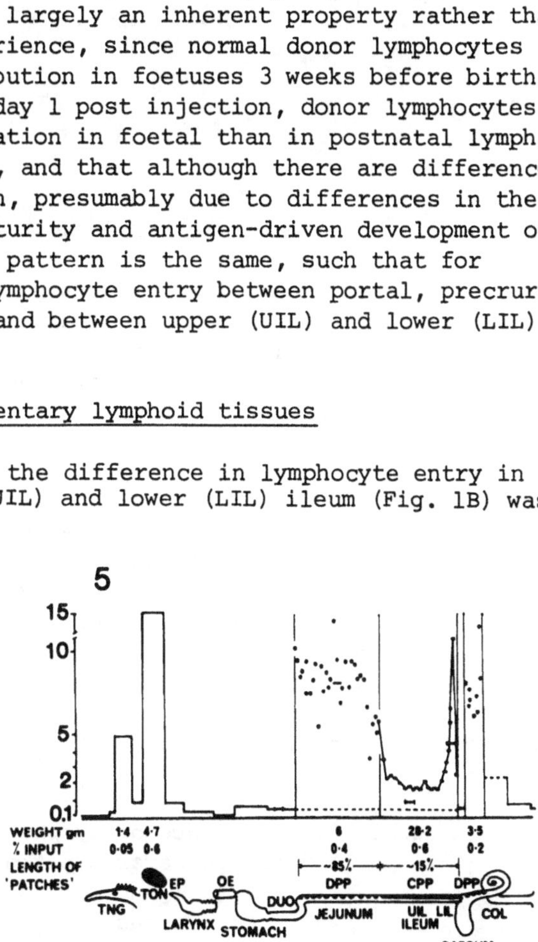

Fig. 4. Cell concentration factor in the tissues of 94 day old foetal (●) and postnatal (○) recipients of normal postnatal ^{51}Cr-labelled blood lymphocytes at day 1. Standard errors are shown if larger than the symbols.

Fig. 5. Cell concentration factor of entry of ^{51}Cr-labelled blood lymphocytes in the alimentary tissues (open histograms) and isolated lymphoid "patches" (solid symbols) in a typical young pig (7 weeks old and weighing 12.5kg) 42 hours after intravenous injection (Abbreviations see M & M).

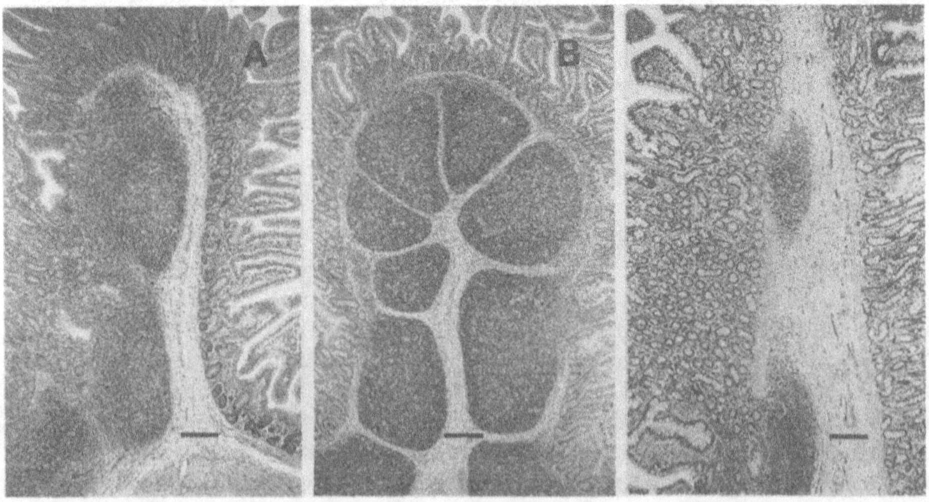

Fig. 6. Pig small intestinal Peyer's patches. A. Jejunal discrete patch-DPP and B. Upper ileal continuous (UIL) Peyer's patch in a young pig (1 month old). C. Regressed upper ileal continuous patch in a 4 year old adult pig. (Magnification = x 27.5. Scale bar = 200μm. H & E stained 5μ paraffin sections).

later shown to be due to a consistently low intake of lymphocytes in almost the whole continuous ileal patch (except for the last ~20%) when compared with the discrete patches of duodenum, jejunum and proximal colon and with the tonsils and focal lymphoid tissue of the filiform papillae of the tongue dorsum (see Fig.5). This difference in blood lymphocyte entry in continuous and discrete patches correlated with a lack in the former of interfollicular tissue (Fig. 6A & B) and of T cells (Table 1) and with its tendency to regress in adulthood (Fig. 6C). Strikingly the last~20% of the continuous patch

Table 1. Incidence of lymphocyte subpopulations in pig Peyer's patches.

	Continuous			Discrete		
	T	B	Null	T	B	Null
Pig 1	7	26	67	34	31	35
2	10	46	44	31	48	21

Tonsils showed population indices similar to those of discrete patches. T (SRBC rosettes in dextran), B (sIg^+ by DARR test) and Null cells were identified exactly as described previously[2].

shows a progressive change from the predominantly follicular low entry type to the mixed type characteristic of discrete patches and the ileocaecal junction and that this very last section in the foetal pig also shows the difference in lymphocyte uptake from the more proximal adjacent section of the patch.

Conclusion

These early studies of pig lymphocyte migration reveal major differences in the behaviour both of lymphocyte subpopulations and of lymphoreticular tissues in the pig which suggest the existence of important functional heterogeneity of lymphocytes - the subject of continuing active research. They also reveal both the existence in young pigs of two types of Peyer's patches, reminiscent of the findings of Reynolds and Morris[7] in lambs, and of a significant migration of blood lymphocytes to the dorsal tongue papillae, a potentially important supplement to the palatine tonsils in the buccal immune system.

Acknowledgements

We are grateful to Mr. Ian King for histology, Mr. Peter Taylor for photography and Mrs. Jenny Richards for typing the manuscript.

REFERENCES

1. Binns, R.M. 1982, Organization of the lymphoreticular system and lymphocyte markers in the pig. Vet. Immunol. Immunopathol. 3: 95.
2. Binns, R.M., Pabst, R. and Licence, S.T. 1984, Lymphocyte emigration from lymph nodes, by blood in the pig and efferent lymph in the sheep. Immunology in press.
3. Binns, R.M., Blakeley, D. and Licence, S.T. 1981, Migration of fluoresceinated pig lymphocytes in vivo: technical aspects and use in studies of autologous and homologous cell survival for up to three weeks. Int. Archs. Allergy appl. Immun. 66: 341.
4. Binns, R.M. and Licence, S.T. 1983, Long-term survival of MHC-compatible lymphocytes and two-phase clearance of allogeneic incompatible lymphocytes in the young pig. Immunology 49: 727.
5. Binns, R.M., Pallares, V., Symons, D.B.A. and Sibbons, P. 1977, Effect of thymectomy on lymphocyte subpopulations in the pig. Int. Archs. Allergy appl. Immun. 55: 96.
6. Binns, R.M., McFarlin, D.E. and Sugar, J.R. 1972, Lymphoid depletion and immunosuppression after thymectomy in the young pig. Nature New Biol. 238: 181.
7. Reynolds, J.D. and Morris, B. 1983, The evolution and involution of Peyer's patches in fetal and postnatal sheep. Eur. J. Immunol. 13: 627.

8. Chu, R.M. and Liu, C.H. 1984, Morphological and functional comparisons of Peyer's patches in different parts of the swine small intestine. Vet. Immunol. Immunopathol. 6: 391.

Note added after completion of manuscript: Chu and Liu[8] have just reported their failure to detect morphological differences between ileal and jejunal patches in 8 week old pigs, a disparity with our data which might be explainable by failure to study the central region of the continuous ileal patch.

INTERACTION MECHANISM OF CLUSTERING LYMPHOCYTES: SPECIFICITY MOLECULAR PROPERTIES AND ITS RELATION TO MIGRATION

Alf Hamann, Dorothee Jablonski-Westrich and Heinz-Günter Thiele

Abteilung für Immunologie
I. Medizinische Universitätsklinik
2000 Hamburg 20, FRG

INTRODUCTION

Specific receptors on lymphocytes recognise and bind to structures on high endothelial cells. This is the first, now well-known event in the migration of lymphocytes though lymph nodes or Peyer's patches. Little is known, however, about the mechanisms regulating the further traffic within lymphoid organs including phenomena such as the antigen-induced trapping of specific lymphocytes. Soluble factors or other signals influencing the locomotion of the lymphocytes are one possibility. Another (or additional) mechanism could involve differential adhesion between the cells. We have studied a contact intact interaction (adhesion) mechanism diplayed by activated lymphocytes which may play a role in cellular traffic.

RESULTS AND DISCUSSION

If lymphocytes are stimulated in allogeneic or autologous mixed lymphocyte cultures, or by mitogens, clusters of cells are formed within a short time (Fig. 1). The question arises whether this is only a sign of increased stickiness of the cells, or whether specific cell interaction sites are displayed by these cluster-forming cells. To answer this, mixing experiments were carried out, in which FITC-labelled, dissociated clustering cells of periodate-stimulated cultures were allowed to reaggregate in the presence of different other cell types.

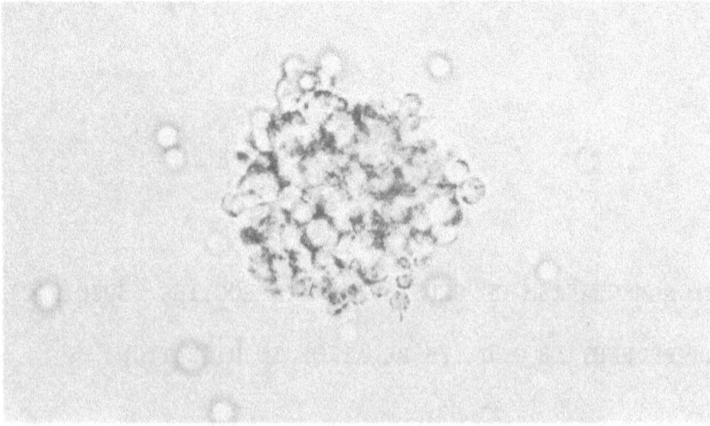

Fig. 1. Cluster of activated cells in a 3-day autologous culture of rat lymph node lymphocytes (depleted of adherent cells)

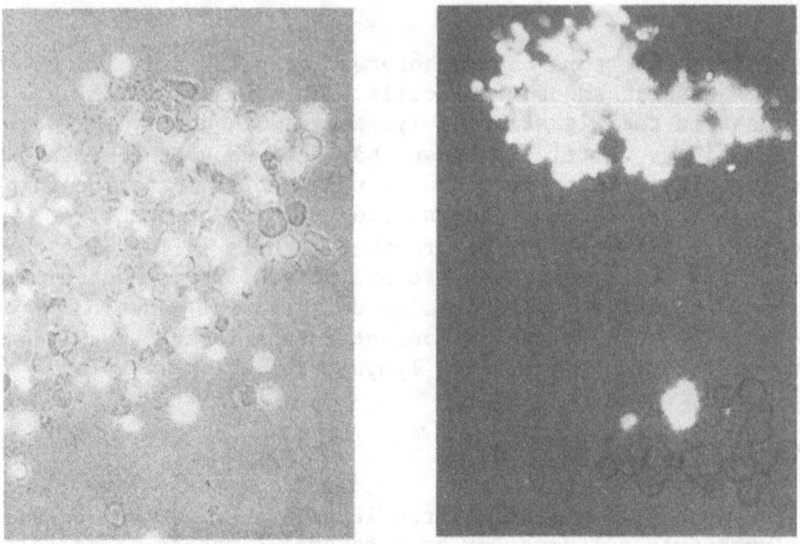

Fig. 2 Coaggregation (left) and sorting-out (right) of different types of cluster forming cells. Activated mouse lymphocytes were FITC-labelled (bright cells), mixed with unlabelled rat lymphocytes (activated, left panel), or P3-X63 myeloma cells (right panel), and were allowed to reform clusters after previous dissection

Fig. 2 shows an example of coaggregation (formation of mixed aggregates) between activated rat and mouse lymphocytes (cluster cell fraction), and of sorting out, respectively (formation of separate clusters of activated mouse lymph node lymphocytes and of myeloma cells P3-X63). Table 1 summarises the results of several experiments (1). It can be seen that activated lymphocytes of different strains of mice, or of rat and mouse, form perfectly mixed aggregates, whereas cells from more distant species (rat/human) had a restricted capacity to cluster together. Among a panel of different cell lines only one, i.e. the BCL1, coaggregates well, whereas the others, including cells of different tissue origin, are completely or partially sorted out. This is also the case for the majority fresh lymphocytes. Thus, it may be concluded that the mutual binding of activated lymphocytes is a selective recognition event.

The appropriate receptors are obviously newly expressed, or activated upon stimulation, since fresh lymphocytes do not bind, and the cluster cell fraction contains mainly the activated cell populations of B as well as T cells (1), as evidenced by several criteria (Table 2).

The contact sites (or part of them) are trypsin-sensitive, magnesium-dependent, and are inhibited by monoclonal antibodies against LFA-1 (Table 3), a presumed cell interaction molecule involved in the adhesion step of the interaction of cytotoxic T cells with their targets (2,3).

Table 1. Results of mixing experiments: coaggregation (inclusion) versus sorting out

Activated Lymphocytes of:	Degree of coaggregation (inclusion)
$H-2^k$ mouse/$H-2^d$ mouse	+++
mouse/rat	+++
mouse/human	±
Act. ly./hepatoma (rat/rat)	0
Act. ly./Ehrlich ascites (mouse/mouse)	0
Act. ly./HeLa cell (rat/human)	0
Act. ly./fresh ly. (mouse/mouse)	±
Act. ly./myeloma (P3X63)	0
Act. ly/L1210 (T cell line: mouse/mouse)	+
Act. ly/BCL1 (B cell line: mouse/mouse)	+++

Table 2. Activation State of the Cluster-Forming Fraction of Lymphocytes Evidenced by:

- O Ultrastructure (Blast-like)
- O Thymidine incorporation (3-50 X higher than in single cell fraction)
- O Immunoglobulin synthesis (4-30 X higher plaque-forming cells/cell in cluster fraction)
- O Mitogen stimulation (periodate stimulation increases clustering drastically)

This is an interesting finding, since it is an indication for the possible existence of a universal adhesion mechanism of lymphocytes, working in different kinds of functional interactions (but not including adhesion of lymphocytes to high endothelial cells (4)).

With regard to the migration behviour of the cluster-forming activated lymphocytes, so far only the first step, i.e. the binding of lymphocytes to high endothelial venules has been tested. As seen in Table 4, the binding of the cluster cell fraction of periodate stimulated lymphocytes to high endothelial cells does not differ

Table 3. Inhibition of Clustering of Activated Lymphocytes by Monoclonal Antibodies against LFA-1. Reaggregation of Dissociated Cluster Cells in the Presence or Absence (Control) of Antibody.

Antibody:		Reaggregation (% of Control):
H-35.89.9[a]	(10 ug/ml)	36%
	(100 ug/ml)	21%
M-17/5.2[b]	(Supernatant)	30%

[a]Provided by M. Pierres, Marseilles
[b]Provided by T. Springer, Boston

Table 4. Binding of Periodate-Stimulated Lymphocytes to Postcapillary Venules of Peripheral Lymph Nodes. Relative Binding to High Endothelial Venules on Frozen Sections.

(Lymph Node Lymphocytes = 100%) (Unstimulated)	
Clustering Cells:	110 $\pm$ 25%
Single Cells:	82 $\pm$ 17%

significantly from that of fresh lymphocytes and also the respective receptor, reactive with the monoclonal antibody Mel-14 (5), is expressed on the majority of these cells. This result is in some contrast to findings with lymphocytes activated by a MLR or Con A, which show greatly reduced binding to high endothelial cells, and expression of Mel-14 (6). Future work may clarify whether the adhesive properties of activated lymphocytes have a function for cell traffic, by immobilizing the activated population within the lymph node or elsewhere, where the activation has occurred.

Acknowledgment: This work was supported by the Deutsche Forschungsgemeinschaft.

REFERENCES

1. Hamann, A., Jablonski-Westrich, D., Raedler, A., Thiele, H.G. Lymphocytes express specific antigen independent contact interaction sites upon activation. Cell. Immunol. 86: 14, 1984.
2. Davignon, D., Martz, E., Reynolds T., Kurzinger, K. and Springer, T.A. Monoclonal antibody to a novel lymphocyte function associated antigen (LFA-1): mechanism of blockade of T lymphocyte-mediated killing and effects on other T and B lymphocyte functions. J. Immunol. 127: 590, 1981.
3. Pierres, M., Goridis, C. and Golstein, P. Inhibition of murine T cell-mediated cytolysis and T cell proliferation by a rat monoclonal antibody immunoprecipitating two lymphoid cell surface polypeptides of 94,000 and 180,000 molecular weight. Eur.J. Immunol. 12: 60, 1982.
4. Hamann, A., Butcher, E.C., Duijvestijn, A. and Thiele, H.G. (unpublished).

5. Gallatin, W.M., Weissman, I.L., Butcher, E.C. A cell surface molecule involved in organ-specific homing of lymphocytes. Nature 303: 30, 1983.
6. Dailey M.O., Gallatin, W.M., Weissman, I.L. and Butcher, E.C. Surface phenotype and migration properties of activated lymphocytes and T cell clones. In: Intercellular Communication in Leukocyte Function. Ed. J.W. Parker and R.L. O'Brion, p. 641. J. Wiley & Sons Ltd., London, 1983.

EXPLOITING LYMPHOCYTE TRAFFIC TO DELIVER RADIOACTIVITY OR RICIN TO LYMPHATIC TISSUES

W.L. Ford, A.J.S. Davies, M. Birch, J.A. Forrester
D. McIntosh, H. Sharma, S. Sparshott, and C. Wood

Departments of Immunology, University of Manchester and
Immunobiology, Institute for Cancer Research, London

INTRODUCTION

After injection into the bloodstream lymphocytes localize in the traffic areas of the spleen and lymph nodes where they reach a concentration approximately 500 times greater than the concentration averaged over the whole body and about 25,000 times the concentration in most non-lymphatic tissues such as muscle. This selective localization develops rapidly so that under physiological conditions over 90% of lymphocytes leave the blood stream to enter lymphatic tissues within one hour[1]. Can this remarkable process be exploited in order to target drugs to certain sites? Two current projects in the Manchester Immunology department have the common objective of examining the practicability of using lymphocytes in order to carry, deposit and concentrate materials in lymphatic tissues. For this principle to be effective three conditions require to be fulfilled - 1) the material must be suitable for "loading" into lymphocytes in vitro; 2) loaded lymphocytes must remain capable of their normal migratory behaviour for at least 30-60 min so that they enter the lymphatic tissues after intravenous injection; 3) when the loaded cells have reached the organ they must release the material in such a form that it affects neighbouring cells. It may be taken up by macrophages and so retained in the tissue. In this preliminary report we claim that two materials - Indium-114^{m} and ricin-fulfil these conditions with certain reservations and can be concentrated in lymphatic tissues with the predicted biological effects. The following is a brief overview of two series of experiments that will be published in detail.

A. INDIUM-114 ^{m}OXINE

Continuous irradiation of the spleen by attaching a ^{32}P-impregnated strip to the surface of the rat spleen[2] or by intrasplenic or intranodal injection of colloidal ^{185}W-tungsten trioxide[3] produced a profound deficit of lymphocytes in healthy rats. The field of radiation produced in the lymphatic organ destroyed its content of recirculating lymphocytes and lymphocytes that had been elsewhere continued to recirculate into the irradiated organ so that recirculating lymphocytes were destroyed over a long period. The "Mark-III" version of this principle was suggested by the finding that 7 days after rats had been injected with lymphocytes that had been trace labelled with Indium-111-oxine most of the radioactivity was fixed in the spleen and lymph nodes having been transferred from lymphocytes to sessile cells probably macrophages .

Indium-114^{m} is an energetic β-emitter (Emax=1.988 Kev) with a 50 day half life. When chelated with oxine it is avidly taken up by lymphocytes in vitro as expected from studies with Indium-111[4]. . A series of adult female rats of the AO strain were injected intravenously with approximately 10^8 syngeneic thoracic duct lymphocytes which had been loaded in vitro with 20μCi of Indium-114m.

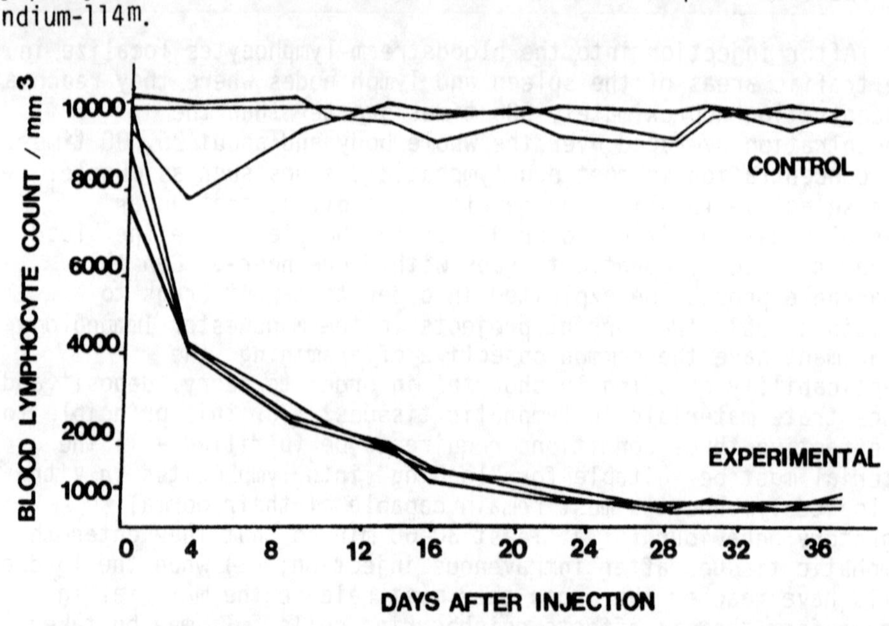

Fig 1 shows that the blood lymphocyte count fell to 3-5% of control levels and remained low for at least 5 weeks. The rats were healthy throughout this period, gained weight normally and no evidence of damage to the liver or bone marrow was found. The same dose of free Indium-114^{m} oxine injected intravenously did not significantly affect the blood count. Autoradiography and whole

(Continued)

Having established that Indium-114^m loaded lymphocytes produced profound lymphocytopenia presumably by irradiation of recirculating lymphocytes as they pass through the traffic areas the effect of this procedure a) on immune responses and b) on an experimental T cell leukemia (the Roser leukemia[5]) was investigated. Indium-114^m had only a marginal effect on skin graft survival but almost totally suppressed the antibody response to an i.v. injection of 10^7 sheep erythrocytes. However Indium-114^m lymphocytes had a marked effect in delaying the onset of the Roser leukemia when injected 24 hours after an inoculum of 10^6 leukemic cells. Control rats injected with 10^6 Roser leukemia cells developed a high leucocytosis after 8 days and died in 11-12 days. Single doses of lymphocytes loaded with Indium-114^m progressively delayed the onset of the leukemia as the dose of Indium was increased so that after a dose of 250μCi leucocytosis did not begin until 20 days followed by death in 23 days. Weekly doses of 150μCi of Indium-114^mlymphocytes delayed the onset of the leukemia apparently indefinitely. In another laboratory similarly encouraging results have been obtained by injecting Indium-114^m loaded lymphocytes into mice with B cell lymphomas. (L.Cobb, MRC Radiobiology Unit, Harwell, personal communication).

B. RICIN

This powerful plant toxin binds to galactose residues on the cell surface. Some is internalized and after a latent period kills cells by enzymatic inactivation of ribosomes. When cells were incubated with ricin in vitro, exhaustively washed and then added to a surplus of untreated cells the toxic effect was clearly transferred to the untreated cells in vitro[6].

Thoracic duct lymphocytes (TDL) from AO rats were labelled in vitro with ^{51}Cr[7] and passaged from blood to lymph in syngeneic rats. They were collected for up to 3 hours at room temperature so that their ability to enter lymph nodes from the blood was high as has been expounded previously[1]. At this stage the TDL were incubated with ricin (A & B chains) at concentrations varying from 10μg to 100μg/10^8 cells/ml. After two centrifugation cycles they were injected intravenously into syngeneic rats.

---

organ counting showed that most of the radioactivity in the spleen was localized in the vicinity of the marginal zone. The amount of radioactivity in the spleen and lymph nodes remained remarkably constant (allowing for radioactive decay) between 24 hours and six weeks after injection.

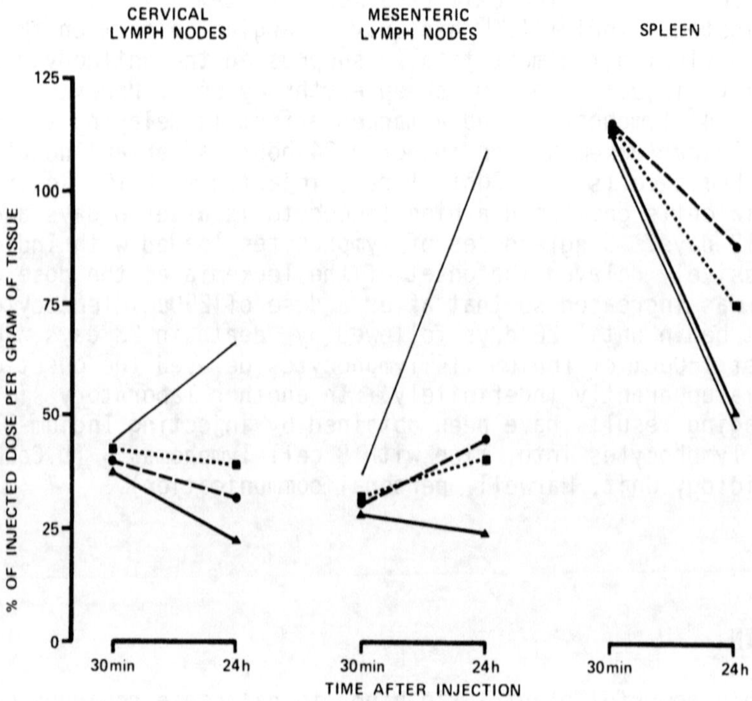

▲——▲ incubated with ricin at 100μg/10 cells/ml.
■······■ " " " " 10μg/10 cells/ml.
●– – –● " " " " 100μg/10 cells/ml in the presence of 100mM lactose
——— incubated without ricin.

Fig 2 shows that at 30 minutes after injection the ricin loaded lymphocytes had entered the spleen just as efficiently as had the control cells. Entry into the cervical and mesenteric lymph nodes was not impaired by ricin loading at 10μg/ml and was only slightly affected at 100μg/ml. Measurement of radioactivity in other tissues including the lungs and liver confirmed that for the first 30 minutes after injection the ricin loaded-lymphocytes were distributed very similarly to control lymphocytes; by contrast at 24 hours after injection the distribution of ^{51}Cr differed greatly according to whether ricin-loaded or control lymphocytes had been injected.

Ricin-loaded lymphocytes were also injected into recipients with thoracic duct fistulae from which timed fractions of lymph were collected. Only a very few labelled lymphocytes were recovered in the first 6 hours after injection and none thereafter. The conclusion is clear: ricin loaded lymphocytes reached the tissues after i.v. injection and did not return to the blood. This no doubt reflects the latency of ricin toxicity which in the case of intact cells extends over several hours.

The histology of the spleen, lymph and other tissues of rats injected 24 hours previously with ricin-loaded lymphocytes, typically 200×10^6 incubated at $25 \mu g/10^8$ cells/ml, has been examined comprehensively. The traffic areas of the spleen (periarteriolar lymphatic sheath) and the lymph nodes (paracortex) were purged of intact lymphocytes although the architecture of the tissues was preserved. The high endothelial venules in the lymph nodes were severely damaged but usually remained intact so that there was not a great deal of haemorrhage into the tissues. The marginal zone of the spleen also sustained moderate damage on morphological criteria. Probably as a result of damage to the portals of lymphocyte entry there was a redistribution of the surviving lymphocytes so that they accumulated in the red pulp of the spleen which was much more cellular than normal. Paradoxically the number of lymphocytes in the blood almost doubled no doubt because lymphocytes could no longer migrate out of the blood to the lymph nodes and white pulp. Unfortunately most rats eventually died when given 200×10^6 TDL loaded with ricin at 25µg/ml and sublethal doses produced less complete cell destruction in the traffic areas.

An equivalent dose of free ricin injected i.v. did cause some diffuse damage to the spleen but the histological pattern was entirely different from that following the injection of ricin laden cells in that the cellular destruction was less intense and not so restricted to the traffic areas.

CONCLUSION

These experiments clearly showed that lymphocytes can be exploited to carry Indium-114^m or ricin into lymphatic tissues. Both of these agents cause the destruction of "host" lymphocytes in the traffic areas but beyond that the consequences of concentrating ricin or Indium-114^m in lymphatic tissues were starkly contrasting. Indium-114^m produced a gradual but profound depletion of the recirculating lymphocyte pool because the material was held in tissue macrophages for prolonged periods. Ricin purged the traffic areas of lymphocytes within 24 hours but only about 50-60% of recirculating lymphocytes in the body were destroyed after a single injection.

REFERENCES

1. Smith M.E. and Ford W.L. (1983). The recirculating lymphocyte pool of the rat: a systematic description of the migratory behaviour of recirculating lymphocytes. Immunology 49 83.
2. Ford W.L. (1968). the mechanism of lymphopenia produced by chronic irradiation of the rat spleen Br. J. exp. Path. 49 502.
3. Roser B. and Ford W.L. (1972). Prolonged lymphocytopenia in the rat. The depletion of blood and thoracic duct lymphocyte populations following injection of β-emitting colloids into the spleen or lymph nodes. Aust. J. exp. Biol. med. Sci. 50 165.
4. Sparshott S.M., Sharma H., Kelly J.D. & Ford W.L. (1981). Factors influencing the fate of 111-Indium labelled lymphocytes after transfer to syngeneic rats. J. immun. Meth. 41 303.
5 Dibley M., Dorsch S. and Roser B. (1975). T cell leukemia in the rat: the pathophysiology. Pathology 7 219.
6. McIntosh D.P., Edwards D.C. and Davies A.J.S. (1984). Transfer of ricin toxicity by spleen cells. Toxicon 22 293.
7. Ford W.L. (1977). The preparation and labelling of lymphocytes. Chap 23. in Handbook of Experimental Immunology. ed. D.M. Weir. 3rd edn. Blackwell Edinburgh.

SESSION 5

REGULATION OF IMMUNE RESPONSES BY LYMPHOID CELL PRODUCTS

CHAIRPERSON: I.M. Roitt

TMF, A THYMIC EPITHELIAL GLYCOPROTEIN:

CHEMOTAXIS OF HEMATOPOIETIC PRECURSOR CELLS

E.F. Potworowski[1], K.W. Pyke[2], Teresa Brodniewicz-Proba[1] and Maria Zelechowska[1]

[1] Institut Armand-Frappier, Laval, QC, Canada and
[2] Ludwig Institute for Cancer Research, Melbourne, Australia

INTRODUCTION

T-cell precursors which migrate to the thymus have been described in fetal liver (Lepault and Weissman, 1981) and in adult bone-marrow (Weissman et al, 1977). The mechanism of entry of these cells into the thymus, as well as of subsequent intra-thymic differentiation events are largely unknown, although thymic stromal elements, particularly epithelial cells, have been singled out as playing a major role both through their secretions and by cell-to-cell contact.

In the present work, we describe a thymic microenvironmental factor, TMF, distinct from thymic epithelial culture supernatant (TES). We show that TMF is a glycoprotein, that it is associated with the membranes of intracytoplasmic vacuoles of thymic epithelial cells, and that it is capable to attract, in a migration-under-agar assay (Pyke and Bach, 1979), hematopoietic precursor cells which can colonize epithelial thymus.

MATERIAL AND METHODS

Cell suspensions

Fetal livers were collected aseptically on day 12 of gestation, care being taken to avoid contamination with maternal blood. The cells were dissociated by vigorous pipetting with a sterile pasteur pipette in RPMI-1640 medium containing 10 % human serum (RPMI-HS). The suspension was then allowed to stand at room temperature until the large aggregates had settled to the bottom. The remainder of the suspension was centrifuged over Ficoll-Paque (Pharmacia) to eliminate dead cells, erythrocytes and granulocytes. The cells at the interface were then washed in medium and their concentration adjusted to 10^8/ml.

Preparation of TMF and control fractions

This technique was basically the same as that described previously (Brodniewicz-Proba et al, 1983). Thymic stroma was obtained from young adult C57BL/6Ka mice by X-irradiating the animals (750R) and collecting the thymic remnants aseptically four days later; these were thoroughly homogenized in ice-cold PBS with a tight fitting Potter homogenizer and were washed 5 times in 100 volumes of cold PBS. After centrifugation at 10,000 g for 30 min, the pellet was stored at -90°C when not used immediately. Macrophage digestion of a pellet from 50 thymuses was done by mixing it with 10^8 peritoneal macrophages in 2 ml of RPMI-HS medium and incubating the mixture for 48 h at 37°C in a 5 % CO_2 atmosphere. Cultures were then spun at 6,000 g for 10 min; the supernatant, centrifuged again for 1 h at 100,000 g, constituted TMF. Control preparations included similarly prepared spleen extracts as well as the supernatant of peritoneal macrophages cultured alone in RPMI-HS.

In order to establish its glycoprotein nature, TMF was applied to a Sepharose-lentil lectin column (Pharmacia) and eluted with 0.5 M glucose. This fraction is referred to as TMF-L. TMF was also eluted at pH 2.0 from an anti TMF Ig - CNBr activated Sepharose 4B column. This fraction is referred to as TMF-A.

For the preparation of thymic epithelial supernatant, fresh thymuses were minced in Hank's balanced salt solution, washed twice, incubated with collagenase (1000 u/ml) for 1 h, washed again and resuspended in HEPES buffered Minimal Essential Medium with 10 % fetal calf serum (Flow Laboratories). The cells were then dissociated and cultured in P16 (75 cm^2) flasks (Corning). The medium was changed two days later. The thymus epithelial supernatant (TES) was collected five days later. The activity of TES was assayed by its ability to induce T-cells with a low surface concentration of the Thy 1 antigen (Kruisbeeck and Astaldi, 1979; Kowalczyk et al, 1981).

Migration-under-agar assay

This assay was described in detail elsewhere (Pyke and Bach 1979, 1981). Essentially, 1.5 ml of 1 % agar (Gibco) in RPMI-HS was poured into sterile plastic petri dishes (35 mm diameter) and allowed to gel. A row of three wells (2.5 mm diameter, 1.5 mm apart) was cut in the gel. The gels were then equilibrated with a few drops of RPMI-HS for 1 h at 37°C in 5 % CO_2 in air, and excess medium was carefully removed. To the center well, 5 µl of the cell suspension were placed while the test substance was placed in one of the outer wells and RPMI-HS in the other. The gels were then incubated in a fully humidified atmosphere of 5 % CO_2 in air. After 24 h, the cells migrating from the central well under the agar were counted with an inverted phase contrast microscope. Specific migration of the cells towards the test substance was expressed as the migration index (the number of cells migrating towards the test substance/the number of cells migrating towards the control well). In each case, the tests were set up at least in triplicates, and the results of several experiments pooled. Results were expressed as mean $\pm$ s.d.

In competition experiments, the competing substance was added to the central well with the cells, whereas in neutralization experiments the neutralizing antibody was added to the attractant well.

Preparation of anti-TMF serum

Anti-TMF serum was prepared in two steps, using essentially the technique described by Kelly and Levy (1979) for the production of antibodies to tumor antigens. A first set of rabbits was immunized against a "contaminating antigen cocktail". This cocktail consisted of non TMF tissue components likely to be present in TMF preparations. These included connective tissue, thymic epithelial supernatant, thymocytes, and macrophages. The antiserum resulting from this immunization was incubated with TMF and the precipitating antigen-antibody complex removed by centrifugation, leaving purified TMF in solution for injection into a second set of rabbits. This approach had the advantage of giving high titer, high specificity antibodies requiring little subsequent absorption.

Colonization studies

Third pharyngeal pouches (TPP) from 10-day fetuses were grown on Millipore filters, as previously described (Bartlett and Pyke, 1982; Good et al, 1983). After 24-48 h at 37°C in 10 % CO_2 in air, the filters were trimmed to leave the TPP surrounded by a small rim of filter. This was placed, filter side down, on the surface of the attractant and control wells of migration plates that had been incubated for 24 h. After an over-night incubation, the filters with TPP were removed and placed in organ culture, for 3 weeks, when they were fixed, paraffin-embedded and sectioned at 5 um, according to the method of Sainte-Marie (1962). The migrating cells were of Thy 1.1 phenotype and the TPP of Thy 1.2 phenotype. Sections of tissue were stained with monoclonal antibodies to Thy 1.1 and Thy 1.2 followed by FITC-labeled sheep anti-mouse IgG (Bartlett and Pyke, 1982).

Immunocytochemistry

The ultrastructural localization of TMF was done using the protein A-gold technique introduced by Romano et al. (1977). The same rabbit anti TMF serum was used as in the affinity chromatography and in the neutralization experiments. It is worth noting that in this technique, the gluteraldehyde-fixed tissue blocks cannot be post-fixed in OsO_4, hence the poor definition of ultrastructural elements.

RESULTS

Migration of fetal liver cells

In the migration-under-agar assay, fetal liver cells migrated towards TMF with an index of 2.15 $\pm$ .40 (Table 1). This was comparable to the previously reported migration towards a thymic anlage (Pyke and Bach, 1979). Thymic epithelial supernatant, however did not induce specific migration (index = 0.94 $\pm$ .12). Spleen extract, prepared in the same way as TMF, and supernatant from macrophages cultured alone, used as controls did not attract fetal liver cells either.

TMF after application to a lentil lectin column, and elution with 0.5 M glucose (TMF-L) was as active as the original fraction, if not more (migration index 3.2 $\pm$ .03) indicating that the activity of TMF was in the glycoprotein fraction. The glucose-eluted peak from a lentil lectin column to which a spleen extract was applied did not attract fetal liver cells. Macrophage medium applied to such a column did not give a detectable glycoprotein peak.

In order to establish whether the antigenicity of TMF and its biological activity were on the same molecule, affinity chromatography was carried out using protein A purified anti TMF Ig bound to a CNBr activated Separose 4B column. TMF was applied to such a column, and a peak could be eluted at pH 2. The material in this peak when tested in the migration assay proved to attract fetal liver cells (index 2.2 $\pm$ .4). Macrophage medium eluate gave no appreciable peak and did not attract fetal liver cells.

Competition

In order to determine whether TMF is the same element which attracts thymocyte precursors to the thymic primordium, a competition experiment was set up where thymic anlagen were competing with TMF for the migration of fetal liver cells (Table 2). When TMF was placed both in an outer well and together with the fetal liver cells in the central well, migration to the outer well was abolished (index 1.03 $\pm$ .14) indicating that the TMF in the central well successfully competed with that in the outer well.

Migration to thymic anlage in the outer well was also abolished by TMF in the central well (index 1.00 $\pm$ .12). Similarly migration towards TMF was successfully abolished by thymic anlage in the central well (0.86 $\pm$ .26).

Thymic Epithelial Supernatant, previously shown to be unable to attract cells, was equally unable to compete with TMF. Spleen extract, on the other hand showed a degree of migration inhibition, although this is likely to be a non-specific effect, since it inhibits other cells from non-specific migration (data not shown).

Table 1. Migration of fetal liver cells to TMF fractions

Attractant	Migration index
TMF[a]	2.15 ± .40
14 day thymic anlage	1.71 ± .18
TES	0.94 ± .12
Spleen extract	1.01 ± .18
Macrophage medium	1.03 ± .17
TMF-L[b]	3.2 ± .03
Macrophage medium (lentil lectin)	no eluate
TMF-A[c]	2.2 ± .4
Macrophage medium (antibody column)	0.9 ± .26

a) Crude TMF, before affinity chromatography
b) Glucose eluate of TMF applied to a lentil lectin column
c) pH 2 eluate of TMF applied to anti TMF-Sepharose column.

Table 2. Migration competition

Attractant	Competitor[a)]	Migration index
TMF	TMF	1.03 ± .14
Thymic anlage	TMF	1.00 ± .12
TMF	Thymic anlage	0.84 ± .26
TMF	TES	2.06 ± .8
TMF	Spleen extract	1.57 ± .21

a) Competitor is added to the fetal liver cells in the central well.

Neutralization

Confirmation that the antigenicity and biological activity of TMF are on the same molecule was sought in the following experiment in which TMF was diluted in anti-TMF antibody before being placed in the attractant well (Table 3). The activity of TMF appeared inhibited by this procedure (index 0.97 ± .17). Anti-TES was unable to achieve this neutralization as was the case with normal rabbit serum.

Table 3. Neutralization

Attractant	Neutralizing serum[a]	Migration index
TMF	Anti-TMF	.97 ± .17
TMF	anti-TES	2.18 ± .28
TMF	normal serum	2.12 ± .28

a) Neutralizing serum is used to dilute TMF, placed in the attractant well.

Localization of TMF

Protein A-gold binding to anti TMF antibody gave a pattern of electron-dense granules found almost exclusively in the cytoplasm of thymic epithelial cells (Fig. 1). The plasma membrane was seldom positive except at points of contact between epithelial cells and thymocytes. In the immediate vicinity of such contact points, the cytoplasm of thymocytes also had some granules. Cytoplasmic vacuoles of epithelial cells were invariably devoid of granulations, but high concentrations of granules were often found at the periphery of such vacuoles.

Colonization of epithelial thymus by TMF-attracted cells

Having shown that TMF and thymic anlage could compete for the migration of the same cells, the question addressed here is whether the cells attracted by TMF can colonize an epithelial thymus.

A standard migration experiment was therefore set up with TMF on one side of AKR/J fetal liver cells and medium on the other. After 24 h, third pharyngeal pouches from 10 day old BALB/c fetuses were placed either in the TMF well or in the medium well for a further 24 h, then removed and cultured for three weeks. They were then examined histologically to see whether lymphoid colonization had occurred and whether the Thy 1 phenotype of the lymphocytes was that of the fetal liver or that of the thymus. All the thymic primordia placed in TMF wells were colonized (Table 4); those from the contralateral side were entirely epithelial. Furthermore, the lymphocytes present in the lymphoid thymuses were positive for Thy 1.1, showing that they derived from the fetal liver.

DISCUSSION

The data presented here show that TMF is a biologically active and antigenically specific entity, distinct from thymic hormone-containing

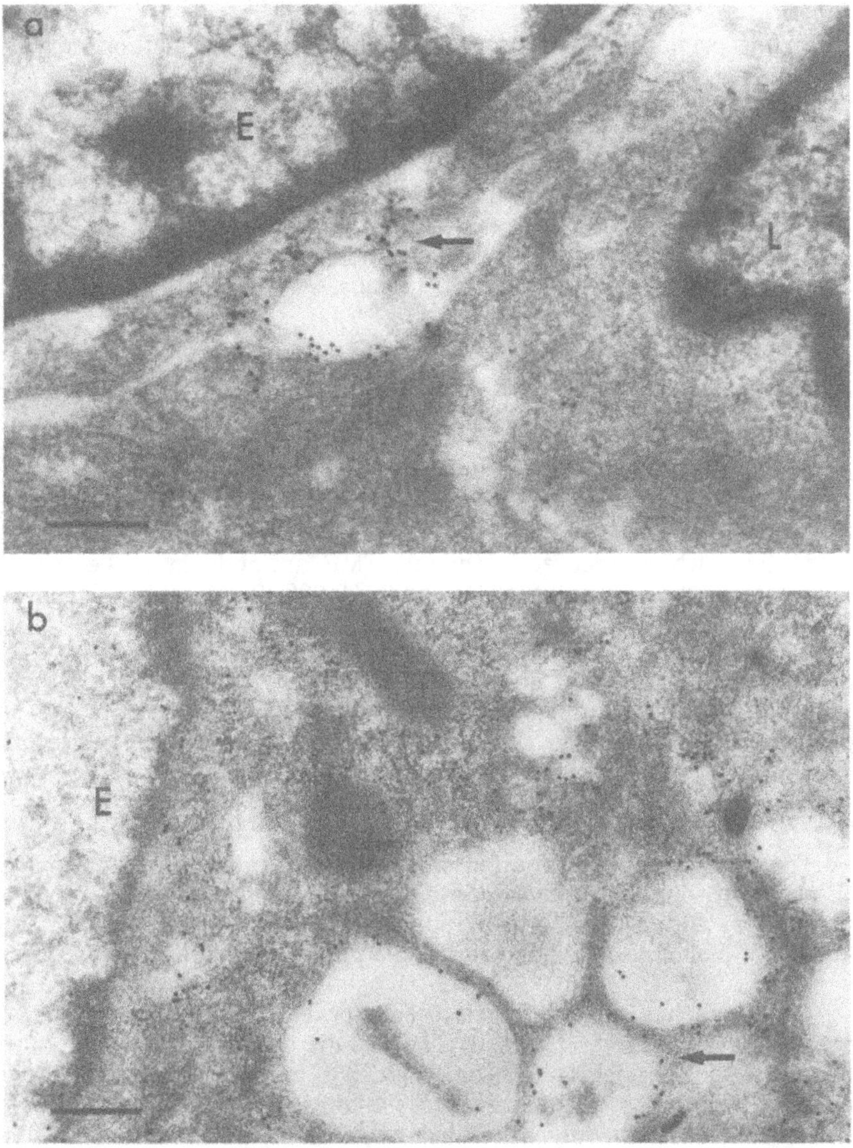

Fig. 1 Electron micrographs of thymus, treated with anti TMF antibody and Protein A-gold.

a: Epithelial-lymphocyte contact point; arrow indicates granules on the lymphocyte (L) as well as on the epithelial (E) side.

b: Cytoplasmic vacuoles in an epithelial cell. Note the concentration of granules (arrow) around the vacuoles. Bar indicates 0.5 μm.

Table 4. Colonization of Epithelial Thymus

Thymus primordia	Lymphoid thymus	T cell phenotype
From TMF well	4/4	Thy 1.1[a)]
From medium well	0/4	-

a) Phenotype of fetal liver cells indicating that these cells did in fact colonize the epithelial thymus.

TES, but associated with thymic epithelial cells. In previous work, we had shown that TMF injected intraperitoneally to mice stimulated the migration of bone marrow cells to the thymus, and that the molecular weight of the fractions which reacted with anti TMF antibody in a Western blot assay, were in the ranges of 55,000 and 140,000 (Brodniewicz-Proba et al, 1983).

One of the important questions that is yet unanswered is whether the _in vitro_ migration of fetal liver cells towards TMF represents the same phenomenon as the _in vivo_ stimulation of bone marrow cell migration to the thymus, and whether those two phenomena are related in any way to the physiological function of endogenous TMF.

The preliminary information on ultrastructural localization indicates TMF concentration at points of contact between epithelial cells and thymic lymphocytes, and around intracytoplasmic vacuoles, which would argue against TMF being secreted and forming an _in vivo_ concentration gradient.

Other mechanisms of T cell precursor entry into the thymus must therefore be envisaged, if indeed TMF is the factor responsible for this activity. One possibility is the selective entry of a precursor cell population having appropriate receptors and circulating at random. Even if such were the case, the migration-under-agar assay still retains its value as a convenient way of monitoring TMF purification.

REFERENCES

Bartlett, P. F., and Pyke, K. W., 1982, Evidence for extrinsic origin of Ia-positive cells in embryonic murine thymus. In: "_In vivo_ immunology: Histophysiology of the lymphoid system". M. G. Hanna Jr. ed. Plenum Press, New York, p. 375.

Brodniewicz-Proba, T., Perreault, C., and Potworowski, E. F., 1983, Thymic microenvironmental factor: solubilization and preliminary characterization. Molecular Immunol., 20:137.

Good, M. F., Pyke, K. W., and Nossal, G. J. V., 1983, Functional clonal deletion of cytotoxic T-lymphocyte precursors in chimeric thymus produced in vitro from embryonic anlagen. Proc. Nat. Acad. Sci., (USA) 80:3045.

Kowalczyk, D., Fournier, M., and Potworowski, E. F., 1981, Age-dependent accumulation of thymic hormone-sensitive cells. Thymus, 3:335.

Kruisbeeck, A. M., and Astaldi, G. C. B., 1979, Distinct effects of thymic epithelial culture supernatant on T cell properties of mouse thymocytes separated by the use of peanut agglutinin. J. Immunol., 123:984.

Kelly, B., and Levy, J. G., 1977, Evidence for a common tumour-associated antigen in extracts of human bronchogenic carcinoma. Brit. J. Cancer, 35:828.

Lepault, F., and Weissman, I. L., 1981, An in vivo assay for thymus-homing bone marrow cells. Nature, (London) 293:151.

Pyke, K. W., and Bach, J. F., 1979, The in vivo migration of murine fetal liver cells to thymic rudiments. Eur. J. Immunol., 9:317.

Pyke, K. W., and Bach, J. F., 1981, In vitro migration of potential hemopoietic precursors from the murine fetus. Thymus, 3:1.

Romano, E. L., and Romano, M., 1977, Staphylococcal protein A bound to colloidal gold: a useful reagent to label antigen-antibody sites in electron microscopy. Immunochemistry, 14:74.

Sainte-Marie, G., 1962, A paraffin embedding technique for studies employing immunofluorescence. J. Histochem. Cytochem., 10:250.

Weissman, I. L., Baird, S., Gardner, R. L., Papaiouannou, V. E., and Raschlke, W., 1977, Normal and neoplastic maturation of T-lineage lymphocytes. Cold Spring Harbor Symp. Quant. Biol., 41:9.

ACKNOWLEDGEMENTS

Part of this work was done at the Walter and Eliza Hall Institute, Melbourne, during the tenure by EFP of an Eleanor Roosevelt International Cancer Fellowship. Supported by grants of the National Health and Medical Research Council of Australia, the Medical Research Council of Canada and the National Cancer Institute of Canada.

CHARACTERIZATION OF T CELL GROWTH FACTOR RESPONSIVE THYMOCYTES IN GUINEA PIGS*

Olof Söder, Göran Sandberg, Ulf Ernström,
and Jan Löthwall

Department of Histology
Karolinska Institutet, P.O.Box 60 400
S-10401 Stockholm, Sweden

INTRODUCTION

Thymus ontogeny and thymocyte subpopulations are best studied in mice and rats.[1,2] In these species, conflicting results have been reported on the role of T cell growth factor (TCGF) in the regulation of thymocyte proliferation. Thus, it has been claimed that TCGF, in addition to its effect on peripheral T cells, also stimulates the proliferation of immature thymocytes,[3] and that this may be an important mechanism behind the intense growth of immature thymocytes in the young thymus.[4,5] In contrast, other authors have reported that the rapid proliferation of immature thymocytes during thymus ontogeny starts before the appearance of TCGF producing cells,[6] and that only mature medullary thymocytes are TCGF responsive.[7]

In this study, thymocytes from guinea pigs have been separated into subpopulations based on their differences in buoyant density and ability to bind and form rosettes with rabbit erythrocytes. The different subpopulations were tested as regards their responsiveness to PHA and guinea pig TCGF. In addition, rosette separated lymph node cells were studied as representative of peripheral T lymphocytes in guinea pigs.

---

*Supported by grants from the Swedish Society of Medical Sciences and Karolinska Institutet.

METHODS

Cell Culture

Genetically heterogeneous guinea pigs were used as cell donors. Cultures (100 µl, $5x10^5$ cells) were set up in α-modified Eagle's Minimum Essential Medium in Linbro plates as described previously.[8]

Mitogen Stimulation

Cultures were stimulated with PHA-P in optimal (5 µg/ml) or suboptimal (0.2 µg/ml) concentration and harvested at 48 h after a 2-h pulse of ^3H-thymidine. Incorporated radioactivity was determined by liquid scintillation counting.

Preparation and Assay of Guinea Pig TCGF

The cell-free supernatant of cultures of guinea pig spleen cells incubated for 24 h in the presence of concanavalin A-Sepharose beads (3% v/v) was precipitated with ammoniumsulfate (80% saturation), dialysed and supplemented with α-methylmannoside (0.2 M). TCGF preparations were tested for comitogenic effect in combination with PHA (0.2 and 5 µg/ml) on unseparated guinea pig thymocytes cultured for 48 h as described above. In a maintenance assay, splenic lymphoblasts obtained at 72 h after stimulation with Con A (1 µg/ml) were cultured in the presence of serial dilutions of TCGF and harvested after an additional 48 h and a 2-h pulse of ^3H-thymidine.

Density Gradient Centrifugation with Percoll

Thymocytes were mixed with isotonic Percoll and separated according to buoyant density during centrifugation at 15,000 g for 11 min. Population Ia was recovered from the top 2 ml, Ib from the next 3.5 ml, and population II from the remaining 5.5 ml. For details see Sandberg et al.[9]

Separation of Rosetting and Nonrosetting Cells

The rosette separation procedure has recently been published.[10] Briefly, thymocytes or lymph node cells were mixed with an equal volume of 2% rabbit erythrocytes, incubated for 15 min at 37° C, centrifugated and incubated for 1 h at 0° C, before estimation of rosette frequency. The cells were then layered over Percoll (1.09 g/ml) and centrifugated for 30 min at 400 g. Rosette forming cells were enriched in the sediment, and nonrosetting cells recovered from the supernatant.

RESULTS

Activity of Guinea Pig TCGF

Con A-Sepharose-induced and α-methylmannoside-supplemented TCGF was comitogenic in combination with PHA, but lacked direct mitogenic effect, in a test with unseparated guinea pig thymocytes (Fig. 1). In a dose dependent way, this TCGF sustained the proliferation of Con A-induced splenic lymphoblasts (Fig. 2).

Effect of PHA and TCGF on Thymocytes Separated According to Density

In the three Percoll separated populations Ia, Ib and II the cell number (% of recovered cells) and mean cell volume (fl) were 20 and 160, 21 and 130, and 59 and 100, respectively. The low density population Ia contained all mitogen reactive and TCGF responsive thymocytes (Fig. 3).

Effect of PHA and TCGF on Thymocytes Separated by Rosette Formation

Approximately 78% of guinea pig thymocytes formed rosettes with rabbit erythrocytes. After separation, nonrosetting thymocytes constituted 9 (5-14)% (mean and range of 4 exp.) of recovered cells, and were markedly enriched in PHA reactive cells and cells responsive to

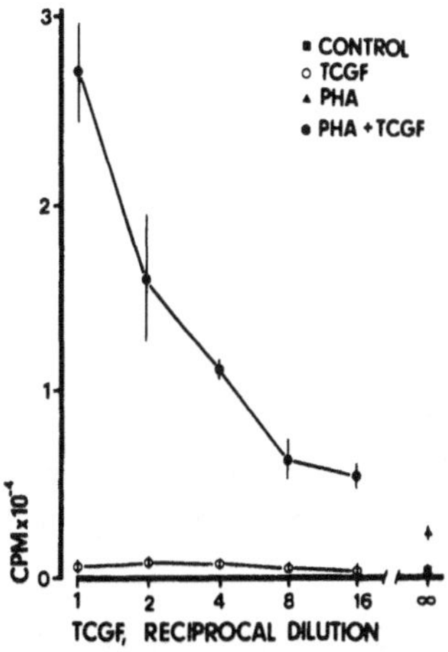

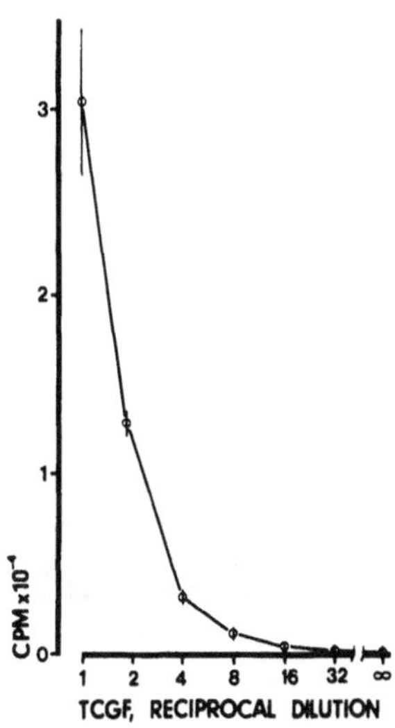

Fig. 1. (left) Comitogenic assay of guinea pig TCGF with unseparated guinea pig thymocytes.
Fig. 2. (right) Effect of guinea pig TCGF on the proliferation of Con A-induced splenic lymphoblasts from guinea pigs. Mean cpm and range of duplicate cultures.

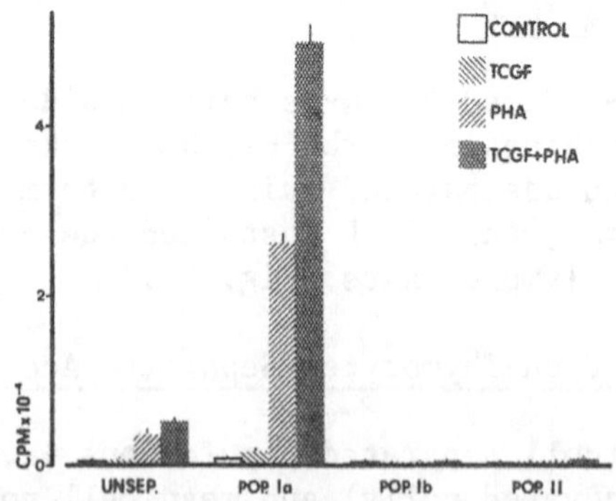

Fig. 3. Effect of PHA and guinea pig TCGF on the proliferation of guinea pig thymocyte populations separated on a density gradient of Percoll (mean cpm and range of duplicate cultures).

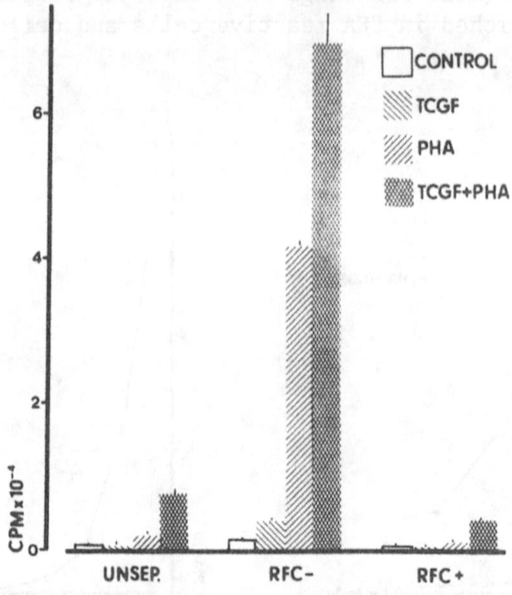

Fig. 4. Effect of PHA and guinea pig TCGF on the proliferation of guinea pig thymocytes separated on the basis of rosette formation with rabbit erythrocytes (mean cpm and range of duplicate cultures). RFC-, nonrosetting cells; RFC+, rosette forming cells.

the comitogenic effect of TCGF. The nonrosetting cells also displayed a slight but reproducible (in 4 of 4 exp.) mitogenic response to TCGF alone (Fig. 4). Identical reactivity was observed with suboptimal concentration of PHA, although the magnitude of the response was lower (data not shown).

Effect of PHA and TCGF on Rosette Separated Lymph Node Lymphocytes

Rosette forming lymph node lymphocytes constituted 24 (7-44)% (mean and range of 7 exp.) of total cells. After separation, 54% of the sedimented and 0.5% of the nonsedimented cells were rosette forming. Cells responsive to an optimal concentration of PHA were slightly enriched among rosette forming lymphocytes (Fig. 5). Nonrosetting lymph node cells were slightly but reproducibly (in 7 of 7 exp.) stimulated by TCGF alone (Fig. 5), and markedly enriched in cells responding to suboptimal concentration of PHA, although the magnitude of the response was ten-fold lower than with optimal dose of PHA (Fig. 6). No significant comitogenic effect of TCGF in combination with PHA was found with guinea pig lymph node lymphocytes (Fig. 5 and 6).

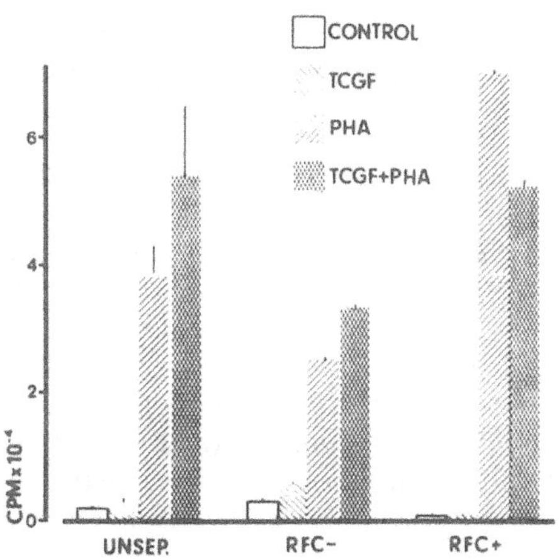

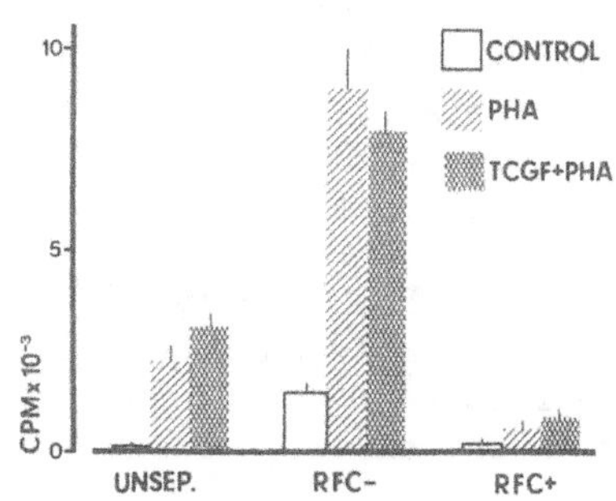

Fig. 5. (left) Effect of PHA (optimal concentration) and guinea pig TCGF on the proliferation of guinea pig lymph node cells separated on the basis of rosette formation with rabbit erythrocytes.

Fig. 6. (right) Effect of PHA (suboptimal concentration) and the combination with guinea pig TCGF on the proliferation of guinea pig lymph node cells separated on the basis of rosette formation with rabbit erythrocytes.
Mean cpm and range of duplicate cultures. RFC-, nonrosetting cells; RFC+, rosette forming cells.

DISCUSSION

In guinea pigs, rosette formation with rabbit erythrocytes has been used as a marker for T lymphocytes,[11] similar to the binding of sheep erythrocytes to human T cells.[12] We here show in the guinea pig, that 78% of thymocytes and 24% of lymph node cells, and have previously found that 46% of splenic lymphocytes and 42% of peripheral blood lymphocytes form rosettes with rabbit erythrocytes.[13] After subtraction of the frequency of B lymphocytes in the different organs,[13] it is evident from our data that a substantial proportion of peripheral T cells in guinea pigs is nonrosetting. The functional relationship between the rabbit erythrocyte binding and nonbinding peripheral T cells is not known. Recently, we have found that only nonrosetting thymocytes are mitogen responsive, whereas the target cells of a thymocyte growth factor with specific effect on cycling immature thymocytes are recovered among rosetting cells.[10] We have also demonstrated that low density guinea pig thymocytes (population Ia) include 56% peanut agglutinin binding cells, which are enriched in DNA synthesizing thymic blasts and target cells of the thymocyte growth factor. Peanut agglutinin nonbinding thymocytes in population Ia (44%), on the other hand, include no thymic blasts, but are enriched in mitogen responsive cells.[9]

In the present study, we find that TCGF responsive thymocytes in guinea pigs are enriched among rabbit erythrocyte nonbinding cells of low buoyant density, constituting less than 10% of total thymocytes, and including all PHA reactive cells. Thus, it can be concluded that a distinct subpopulation of guinea pig thymocytes with characteristics of mature medullary cells contains all TCGF responsive cells. There was no indication that TCGF could affect the proliferation of immature thymocytes in guinea pigs, which is in agreement with the results of Ceredig et al.[6] and Chen et al.[7] in mice. The present findings support the conclusion that separate mechanisms for growth control operate in different thymocyte compartments. TCGF obviously regulates the growth of immunocompetent medullary thymocytes, whereas our previously described thymocyte growth factor[14,15] influences the proliferation of cycling precursor cells in the thymic cortex.

As regards peripheral T lymphocytes, both rosetting and nonrosetting lymph node cells responded to PHA, but no significant augmenting effect of TCGF was observed. The nonrosetting lymph node cells were reactive to a suboptimal concentration of PHA, and, similar to nonrosetting thymocytes, slightly, but reproducibly, stimulated by TCGF in the absence of mitogen. Although variation in the participation of accessory cells might explain some of the different reactivities observed, it is tempting to speculate that nonrosetting thymocytes emigrate and can be traced in the periphery as nonrosetting T lymphocytes. Rosette forming peripheral T cells might be the result of postthymic maturation, or represent a separate subset of T lymphocytes in guinea pigs.

REFERENCES

1. R. Scollay and K. Shortman, Thymocyte subpopulations: an experimental review, including flow cytometric cross-correlations between the major murine thymocyte markers, Thymus 5:245 (1983).
2. Centre Diagram, Immunol. Today 4, n:o 8 (1983).
3. P. Dräber and P. Kisielow, Identification and characterization of immature thymocytes responsive to T cell growth factor, Eur. J. Immunol. 11:1 (1981).
4. H. Wagner, C. Hardt, K. Heeg, K. Pfizenmaier, W. Solbach, R. Bartlett, H. Stockinger, and M. Rollinghoff, T-T cell interactions during cytolytic T lymphocyte (CTL) responses: T cell derived helper factor (Interleukin 2) as a probe to analyze CTL progenitors, Immunol. Rev. 51:215 (1980).
5. R. Jordan and J. Robinson, T lymphocyte differentiation, in: The thymus gland, M. Kendall, ed., Academic Press, London (1981).
6. R. Ceredig, D. Dialynas, F. Fitch, and H. MacDonald, Precursors of T cell growth factor producing cells in the thymus: ontogeny, frequency, and quantitative recovery in a subpopulation of phenotypically mature thymocytes defined by antibody GK-1.5, J. Exp. Med. 158: 1654 (1983).
7. W.-F. Chen, R. Scollay, and K. Shortman, The functional capacity of thymus subpopulations: limit-dilution analysis of all precursors of cytolytic lymphocytes and of all T cells capable of proliferation in subpopulations separated by the use of peanut agglutinin, J. Immunol. 129:18 (1982).
8. O. Söder and U. Ernström, Chemical and biological properties of a thymocyte specific growth factor from calf thymus, Thymus 3:289 (1981).
9. G. Sandberg, O. Söder, S. Kölare, and U. Ernström, Studies on thymocyte subpopulations in guinea pigs. III. Physical and functional characterization of six subpopulations separated by density gradient centrifugation and PNA binding, Exp. Cell Biol. 51:257 (1983).
10. G. Sandberg and O. Söder, Studies on thymocyte subpopulations in guinea pigs. 5. Rosette formation as a tool to separate thymocyte growth factor responsive and mitogen reactive cells, Int. Archs Allergy appl. Immunol. 74:297 (1984).
11. A. B. Wilson and R. R. A. Coombs, Rosette formation between guinea pig lymphoid cells and rabbit erythrocytes - a possible T cell marker, Int. Archs Allergy appl. Immunol. 44:544 (1973).
12. M. Jondal, G. Holm, and H. Wigzell, Surface markers on human T and B lymphocytes. I. A large population of lymphocytes forming non-immune rosettes with sheep red blood cells. J. Exp. Med. 136:207 (1972).

13. G. Sandberg, O. Söder, and U. Ernström, Quantitation of B and T lymphocytes in guinea pigs with evidence for a release of both cell types from the spleen into the blood, Int. Archs Allergy appl. Immunol, 50:374 (1976).
14. O. Söder and U. Ernström, Characterization of the target cell for a thymocyte specific growth factor in guinea pigs, Thymus 5:141 (1983).
15. O. Söder and U. Ernström, Purification to apparent homogeneity of a thymocyte specific growth factor from calf thymus, Biochem. Biophys. Res. Commun. 120:420 (1984).

LYMPHOKINE PRODUCTION BY T LYMPHOCYTES ISOLATED FROM MURINE SMALL INTESTINAL EPITHELIUM

Susan B. Dillon and Thomas T. MacDonald*

Dept. of Microbiology, Thomas Jefferson University, Philadelphia, Pa. 19107 U.S.A.

Mouse small intestinal intraepithelial lymphocytes (IEL) are predominantly Lyt 2^+ (suppressor/cytotoxic phenotype) T cells. IEL T lymphocytes display immunocompetence *in vitro* insofar as they respond to Concanavalin A (Con A) and alloantigens (1). There is also some evidence from histologic studies that IEL respond to local stimuli *in vivo*, because their numbers can be markedly increased (parasite infection; food allergy) or decreased (germ free animals) and many of them have the characteristics of immunoblasts (2). However it is not known whether T cells in the intestinal epithelium have the potential to function in the regulation of gut immune responses as suppressor cells, reflecting their predominantly Lyt 2^+ phenotype, or whether they act in the afferent limb of cell mediated immunity by secreting lymphokines.

In the present study, IEL were tested for regulatory T cell activities by analyzing Con A activated IEL for suppressor cell function and for production of the lymphokines Interleukin 2 (IL-2) and Interleukin 3 (IL-3, or mast cell growth cell factor - see ref. 3). In addition, IEL isolated from *Trichinella spiralis* (*T. spiralis*) infected mice were tested for IL-3 production, since a thymus dependent intestinal mastocytosis occurs during the enteric phase of this infection (4).

---

*Present address: Dept. of Immunoparasitology, Merck, Sharp and Dohme Research Laboratories, Rahway, N.J. 07065

METHODS

Lymphocyte Isolation

IEL were isolated exactly as described previously (1) from normal (C57BL/6 x DBA/2)F_1 (BDF_1) male mice, 6-16 weeks of age, purchased from the Jackson Laboratories, Bar Harbor, Maine. For some experiments, IEL were isolated from mice infected with *T. spiralis* (see below). Spleens (SPL) were removed asceptically and gently teased into suspension in sterile medium.

T. Spiralis Infection

T. Spiralis muscle larvae were prepared from infected mice (provided by Dr. Joye F. Jones, Dept. of Microbiology, Thomas Jefferson University) by pepsin HCL digestion. Mice were infected orally with 100 *T. spiralis* larvae using a syringe fitted with a feeding needle (Harvard Bioscience, S. Natick, Massachusetts).

Isolation and Assay of Con A Activated Suppressor T Cells

The procedure described by Rich and Pierce (5) was modified as follows: 10^6 IEL or SPL per ml were cultured in $16mm^2$ tissue culture wells in Mishell and Dutton medium (6) supplemented with 5×10^5 2-mercaptoethanol (2-ME) (Calbiochem, San Diego, California). Con A (Sigma; C-2010) was added at a final concentration of 750 ng/ml. Cultures were incubated in a gassed box on a rocking platform (6). After 48 hours the cells were harvested, washed three times, and graded doses were added to flat bottomed microtiter wells containing 10^6 freshly isolated SPL cells and 5×10^5 sheep red blood cells (SRBC) in a final volume of 225 ul. On day 5 of culture, triplicate wells for each experimental group were pooled and assayed for direct anti SRBC plaque forming cells (PFC) by the slide modification of the Jerne PFC assay (6).

IL-2 and IL-3 Production by IEL

IEL isolated from normal or *T. spiralis* infected mice were cultured in completed RPMI-1640 medium containing 2.8 mM L-glutamine, 1.0 mM L-asparagine, 10mM HEPES, 5×10^5 2-ME, and 10% heat inactivated fetal calf serum (Reheis Chemical Co., Phoenix, Arizona). To test for

IL-2 production, Con A (5 ug/ml) was added to IEL cultured at various cell densities as described in the text. After 2-4 days, supernatants from triplicate wells were pooled and assayed on the IL-2 dependent cell line CTLL-16 (the gift of Dr. Delores Byrne, Dept. of Biochemistry, Thomas Jefferson University). Units per ml (U/ml) IL-2 activity were calculated as described by Farrar, et. al. (7). Partially purified human IL-2 (Hu IL-2) (Electronucleonics, Inc., Silver Spring, Maryland; lot #1464-81), prediluted 1:160 before serial titration to yield less than maximum stimulation of CTLL proliferation, was assigned the value of 100 U/ml. To test for IL-3 production, IEL from normal, uninfected or 21 day *T. spiralis* infected mice were cultured at 2.5 x 10^5 cells per well in the presence of either Con A, Hu IL-2 (1:200 final dilution), or Con A plus Hu IL-2. Supernatants were assayed for IL-3 activity on the IL-3 dependent cell line, FDC-P2, which does not proliferate in the presence of IL-2 (8). The Hu IL-2 preparation used in this study also failed to stimulate FDC-P2 proliferation (Fig 3). The FDC-P2 cell line was the gift of Dr. John C. Lee (Smith, Kline and French Research Laboratories, King of Prussia, Pa).

Mast cell characterization

IEL were lysed in 2% perchloric acid for measuring the total histamine content by an automated fluorometric method (9). Histamine analyses were performed by Dr. E.S. Shulman, Dept. of Medicine, Thomas Jefferson University. For morphologic studies, cytocentrifuged cells were fixed in Carnoy's solution and stained in 0.5% toluidine blue in McIlvaines citrate buffer.

RESULTS

Con A Activated IEL Do Not Suppress A Splenic Primary Antibody Response to SRBC

IEL and SPL cells were cultured with Con A as described in 'Methods'. Viable cell recoveries at 48 hours were 20% and 60% of the original cell number cultured for IEL and SPL respectively. Many of the recovered cells resembled blasts. When graded doses of Con A activated SPL cells were added to 10^6 freshly isolated SPL cells plus SRBC as antigen, the day 5 PFC response was markedly reduced (83-95% suppression) when compared with the control response made by 10^6 SPL

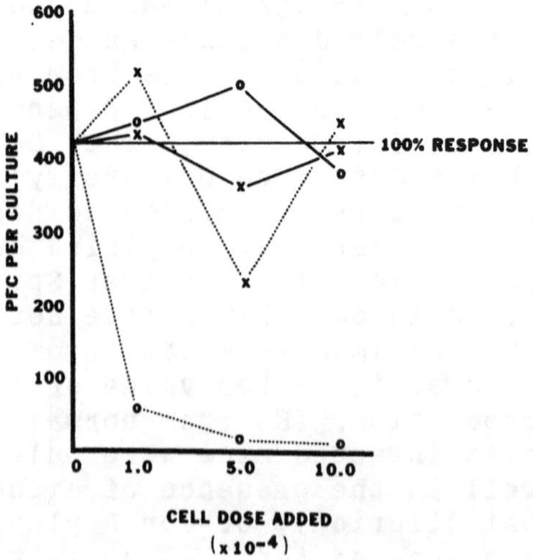

Fig. 1 Con A activated IEL do not suppress a splenic primary antibody response to SRBC. Graded doses of SPL (O) or IEL (X) that had been pre-cultured in the presence (hatched lines) or absence (solid lines) of 750 ng/ml Con A were added to 10^6 freshly isolated SPL cells plus SRBC in microtiter wells. Results are shown as direct anti-SRBC PFC per well on day 5 of culture.

cells cultured with antigen only (Fig. 1). In contrast, when SRBC stimulated SPL cells were co-cultured with Con A activated IEL, no suppression was seen at the highest (10^5) or lowest(10^4) number of IEL added. Although an intermediate dose of Con A activated IEL (5 x 10^4) reduced the splenic PFC response by 46% in the experiment shown in Fig 1; this cell dose did not have any suppressive effect in several other experiments (not shown), and therefore the reduced PFC seen here is not considered to be significant. When either SPL or IEL were pre-cultured in medium alone for 48 hours before addition to cultures of SPL cells plus SRBC, there was no suppression seen at any cell dose in either group (Fig. 1). Because we have found that the peak IEL proliferative response to Con A occurs later than that for SPL cells, (Dillon and MacDonald, unpublished), we also tested 4 day Con A activated IEL for suppressor cell activity, but none was detected.

Production of IL-2 By Con A Activated IEL

The results in Table 1 show that after 48 hours incubation, Con A activated IEL produced low levels of IL-2 in comparison to Con A activated SPL cells. When IEL were cultured with Con A for longer periods (3-4 days), the IL-2 activity detected in culture supernatants remained in the same range (9.5-15.1 U/ml) (Table 1). The proliferative response of IEL stimulated with Con A is also shown to demonstrate that although the IL-2 activity detected in IEL supernatants was relatively low, the concentration of IL-2 generated in these cultures was sufficient to support cell division. No IL-2 activity was detected in supernatants of IEL cultured in medium alone for 2,3, or 4 days.

Table 1. IL-2 Activity in Supernatants of Con A Activated IEL

Supernatant source[a] Cells/ml + Con A	IL-2 Activity[b] Day	Units/ml	Proliferation[c] CPM	(SI)
Exp. 1				
5×10^6 IEL	2	9.5	4,565	(39)
5×10^6 SPL	2	69.5	23,424	(31)
Exp. 2				
2×10^6 IEL	4	9.5	1,295	(14)
Exp. 3				
1×10^6 IEL	3	15.1	17,542	(153)

[a] Cells cultured in $16mm^2$ wells 1 ml in Exp. 1; and in in microtiter wells (in 0.2 ml) in Exp. 2 and 3.

[b] Measured on 4000 CTLL/ml in a 48 hour proliferation assay and compared to a Hu IL-2 standard as described in 'Methods'.

[c] One microcurie ^{3}H-thymidine was added per well 18 hours before harvest on day 4. Stimulation index (SI) was calculated by dividing the mean cpm in cultures with Con A by the mean cpm in cultures without Con A.

Production of IL-3 in Cultures of IEL Isolated From Normal and T. spiralis Infected Mice

At 21 days after infection with 100 *T. spiralis* larvae, the small intestinal architecture, assesed on histologic sections, had returned to normal and the cell yields of IEL from infected mice were comparable to those from uninfected controls. To test mitogen activated IEL for IL-3 production, IEL from *T. spiralis* infected and normal mice were cultured with Con A. IEL from both groups of mice were also cultured with Con A plus Hu IL-2, since our previous work has shown that adding an exogenous source of IL-2 to Con A stimulated IEL enhances the proliferative response (1). Supernatants were collected from each experimental group on day 3 of culture and tested for IL-3 activity as described in 'Methods'. Normal IEL stimulated with either Con A alone or with both Con A and Hu IL-2 produced low levels of IL-3 (Fig. 2). Con A stimulated IEL from *T. spiralis* infected mice produced only slightly higher levels of IL-3 in comparison to normal IEL cultured under the same conditions. By contrast, relatively high levels of IL-3 were found in culture supernatants of IEL from infected mice cultured in the presence of Con A plus Hu IL-2 or IL-2 alone (Fig. 2). It is important to note here that the Hu IL-2 used in this study contained no detectable IL-3 activity as measured on the FDC-P2 cell line (Fig. 3). In addition, the Hu IL-2 preparation was not able to induce mast cell differentiation in cultures of anti - Thy 1.2 plus complement treated bone marrow (not shown).

Appearance of Mucosal Mast Cells in Cultures of IEL Isolated from T. Spiralis infected mice

In addition to T cells, IEL from normal mice also contain undifferentiated mast cell precursors (1) (or 'P' cells; see ref. 10). To determine whether the IL-3 detected in IEL culture supernatants could promote the differentiation of mucosal mast cell precursors into identifiable mast cells, the IEL cultured in the experiment described in Fig. 2 were reincubated after supernatant collection (day 3). On day 8 of culture, the wells were harvested and cytocentrifuged preparations were Toluidine Blue stained. The viable cell recoveries were greatest in wells that contained IEL isolated from *T. spiralis* infected mice cultured with Con A plus Hu IL-2. The majority (93%) of these cells resembled mast cells morphologically and contained histamine (Table 2).

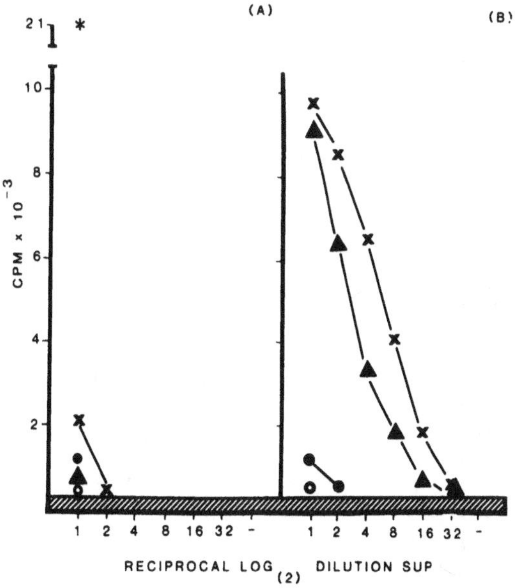

Fig. 2 Supernatants from 2.5 x 10^5 IEL from uninfected (A) or 21 day *T. spiralis* infected (B) mice cultured with medium alone (O); 5 ug/ml Con A (●); Hu IL-2 (1:200) (▲) or Con A + Hu IL-2 (X) were collected on day 3 of culture and assayed for IL-3 activity. Results are shown as the mean cpm of 4000 FDC-P2 cells after 48 hours culture in serially diluted supernatants. Conditioned medium from the Wehi 3B tumor line, (which constituatively produces IL-3) (final dilution = 1:20) is shown for reference (*). All points above the dashed line were greater than 3 standard deviations above the mean background cpm.

The results in Fig. 2 and Table 2 therefore show that IEL isolated from *T. spiralis* infected mice produced factor(s) in response to *in vitro* stimulation with Con A plus Hu IL-2 that mediated the differentiation of mast cells from precursors also present in the IEL population.

DISCUSSION

The major finding reported here is that although both IL-2 and IL-3 activity (Table 1 and Fig. 2) were present in low amounts in supernatants from normal or *T. spiralis* infected IEL cultured with Con A; *T. spiralis* IEL cultured with Con A and Hu IL-2 produced markedly increased amounts of IL-3 (Fig. 2). These results therefore show that whereas T cells from the gut epithelium of worm infected mice have the potential to

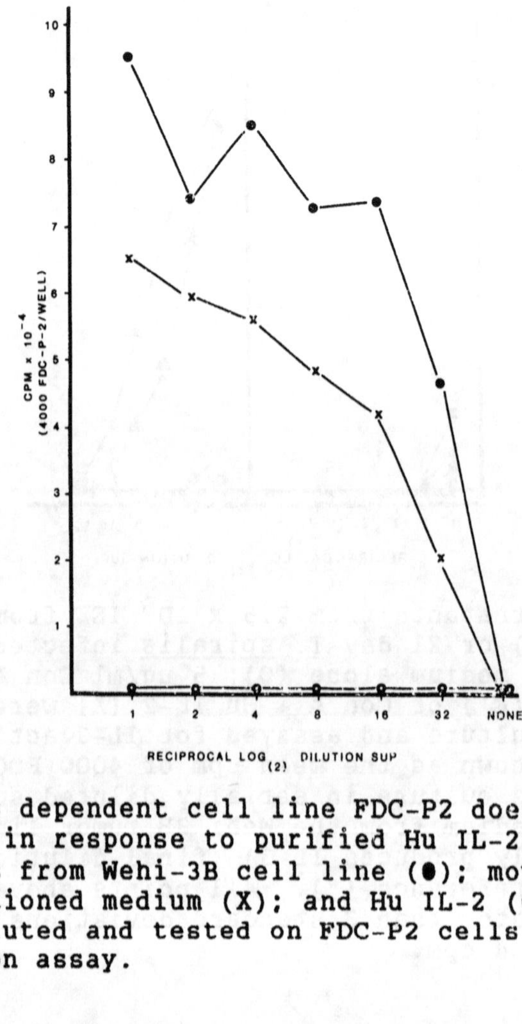

Fig. 3 IL-3 dependent cell line FDC-P2 does not proliferate in response to purified Hu IL-2. Supernatants from Wehi-3B cell line (●); mouse spleen Con A conditioned medium (X); and Hu IL-2 (O) were serially diluted and tested on FDC-P2 cells in a 48 hour proliferation assay.

favor the differentiation of mucosal mast cells via the production of IL-3, (which acts as a mast cell growth factor), this potential is only minimally expressed in IEL isolated from normal mice. Since these results reflect the relative numbers of intestinal mast cells found in normal vs. T. spiralis infected mice (4) it is probable that one function of intraepithelial T lymphocytes is to regulate intestinal mast cell differentiation in situ.

The results in this study also raise some questions as to the signals required for activation of IEL T cells. It is reasonable to assume that in vivo worm antigens act to stimulate IEL; but it is not known if IL-2 is also involved. It is particularly difficult to assess the role of IL-2 in vivo in view of the fact that IEL do not produce high levels of this lymphokine *in vitro*. In a recent report by Guy Grand et. al. (11), gut mucosal lymphocytes were isolated from *Nippostrongylus brasiliensis* infected mice by collagenase digestion of small intestinal pieces after removal of the villus epithelium. The cells isolated, representing a mixture of lamina propria lymphocytes and crypt IEL, produced mucosal mast cell growth factor(s) when cultured in the presence of worm antigen without the exogenous addition of IL-2.

Another point for consideration is the fact that significant mucosal mast cell differentiation was only promoted in cultures containing IEL stimulated with Con A plus Hu IL-2, in spite of the comparable levels of IL-3 found in supernatants from *T. spiralis* infected IEL stimulated with IL-2 alone. This suggests that another factor(s) in addition to IL-3 may be necessary to allow the differentiation of mast cells from precursors present in the IEL population. However, in the case of mucosal mast cells derived from bone marrow precursors, purified homogeneous IL-3 supported both the differentiation and growth of cultured mast cells (12). Finally, it is

Table 2. Generation of Mucosal Mast Cells in Cultures of IEL From 21 Day T. spiralis Infected Mice.

Cells Cultured(Day 0)[a]	Cells Recovered (Day 8)	
IEL (Infected) +	Total Live ($x\ 10^5$)	Toluidine Blue (% Mast Cells)[b]
Medium	0.0	-
Con A	0.2	100
Hu IL-2	0.2	33
Con A + Hu IL-2	1.4[c]	93
IEL (Normal) +		
Con A + IL-2	0.2	50

[a]2.5×10^5 IEL per well; 3 wells per group.

[b]Toluidine blue stained cytocentrifuged slides IEL do

[c]0.13 pg histamine per cell.

interesting to note that in several studies in which Lyt 2^+ spleen cells were cultured at limiting dilutions with either Con A or allogeneic cells, IL-2 was found to enhance the production of gamma interferon or macrophage activating factor, but did not enhance the production of IL-3 (13-15).

In conclusion, IEL from T. spiralis infected mice contain both mast cell precursors and T cells capable of producing a factor (IL-3) that acts to promote mast cell differentiation. Although IEL isolated from normal uninfected mice also contain mast cell precursors (1, 10,11) significant numbers of these do not undergo differentiation in culture unless a source of IL-3 is exogenously added. In addition, we have shown that although Lyt 2^+ cells from the gut epithelium respond to Con A (Dillon and MacDonald, manuscript in preparation), these cells are unlike Con A activated Lyt 2^+ spleen cells (16) in that they do not function as non-specific suppressor cells. Therefore, the subset of Lyt 2^+ cells from the gut epithelium may be qualitatively different from the Lyt 2^+ cells found in peripheral lymphoid organs.

REFERENCES

1. S.B. Dillon and T.T. MacDonald. Immunology 52:501 (1984).
2. A. Ferguson, Gut 18:921 (1977).
3. J.N. Ihle, J. Keller, S. Oroszlan, L.E. Henderson, T.D. Copeland, F.Fitch, M.B. PRystowsky, E. Goldwaser, J.W. Schrader, E. Palaszyski, M. Dy and B. Lebel. J. Immunol. 131:282 (1983).
4. E.J. Ruitenberg and A. Elgersma. Nature 264:258 (1976).
5. R.R. Rich and C.W. Pierce. J. Exp. Med. 131:649 (1973).
6. R.I. Mishell and R.W. Dutton. J.Exp.Med. 126:423 (1967).
7. J.J. Farrar, S.B. Mizel, J. Fuller-Farrar, W.L. Farrar, and M.L. Hilfiker. J. Immunol. 125:793 (1980).
8. J.N. Ihle, L. Rebar, J. Keller, J.C. Lee, and A. Hapel. Immunol. Rev. 63:5 (1982).
9. R.P. Siraganian, R.P. and W.A. Hook in "Manual of Clinical Immunology", N.R. Rose and H. Friedman, eds. Am. Soc. for Microbiol., Washington, D.C. (1980). p.808.

10. J.W. Shrader, R. Scollay and F. Battye. J. Immunol. 130;558 (1983).
11. D. Guy-Grand, M. Dy, G. Luffau and P. Vassalli. J. Exp. Med. 160:12 (1984).
12. E. Razin, J.N. Ihle, D. Seldin, J.M. Mencia-Huerta, H.R. Katz, P.A. LeBlanc, A. Hein, J.P. Caulfield, K F. Austen, and R.L. Stevens. J. Immunol. 132:1479 (1984).
13. R.A. Miller and O. Stutman. J. Immunol. 130:1749 (1983).
14. A. Kelso, H.R. MacDonald, K.A. Smith, J. Cerottini, and K.T. Brunner. J. Immunol. 132:2932 (1984).
15. P.Guerne, P. Piguet and P. Vassalli. J. Immunol. 132:1869 (1984).
16. J. Jandinski, H. Cantor, T. Tadakuma, P.L. Peavy and C.W. Pierce. J. Exp. Med. 143:1382 (1976).

RESISTANCE TO THE INDUCTION OF T CELL-MEDIATED AUTOIMMUNITY CORRELATES WITH LOWER IL 2 PRODUCTION

Marija Mostarica-Stojković, Lota Ejdus-Konstantinović, Mirjana Kostić and Miodrag L. Lukić

Institute of Microbiology and Immunology, School of Medicine, University of Belgrade and Institute for Biological Research, Belgrade, Yugoslavia

INTRODUCTION

According to the generally accepted model, IL 2 generated by ligand activated T cells regulates the clonal expansion of antigen-stimulated T cells in a hormonal manner by binding to specific surface receptors on such T cells (1). Because of this central role believed to be played by IL 2 in regulating T cell responses and in view of the importance of IL 2 in the generation of effector cells in experimental allergic encephalomyelitis (EAE) (2) we have analyzed in detail patterns of IL 2 production and consumption by T cells of Albino Oxford (AO) and Dark August (DA) rats previously shown to differ in their susceptibility to EAE (3).

MATERIALS AND METHODS

<u>Animals and immunizations</u>. Inbred DA and AO rats, when 12-16 wekks old, were used in the experiments. Rats were immunized with rat myelin basic protein (BP) or bovine retinal tissue (BRT) according to the previously described procedures (3,4).

<u>Culture medium</u>. Medium 199 (Torlak) supplemented with 5% heat-inactivated fetal calf serum (FCS,GIBCO),20mM HEPES (Calbiochem)

5×10^{-5}M 2-mercaptoethanol (2-ME,Galenika) was used throughout.

Cell preparation. Rats were sacrificed under ether anesthesia and popliteal lymph nodes and spleens were aseptically removed. Single cell suspension was prepared by gently teasing the tissue and passing the cells through stainless steel wire mesh screen.

IL 2 production. Cells were resuspended in culture medium containing 2.5, 5 or 10 ug Concanavalin A (Con A, Pharmacia) per ml, at different cell densities as indicated in Results. Volumes of 2.5 ml were added to 35 x 10 mm plastic Petri dishes (Falcon 1008). Cultures were harvested after 24, 48 or 72 hr and supernatants obtained by centrifugation at 1500 rpm for 15 min were supplemented with 0.3M 1-methyl- -D-glycopyranoside (MAG, Serva), filter sterilized and stored at -20°C until assayed.

Assay for IL 2. IL 2 activity in supernatants was assayed according to the method of Andersson et al. (5). T cell blasts were prepared by culturing 1×10^7 spleen cells/ml for 48 hr in culture medium supplemented with Con A (2.5 ug/ml). The harvested cells were washed 3 times with medium containing 0.3M MAG and thereafter cultured in standardized IL 2 supplemented medium (50% v/v) for 24 hr before use. The proliferative response of these T cell blasts cultured at 3×10^5/ml in 0.2 ml aliquots in flat-bottomed microculture plates (Falcon,Microtest II) containing 50% of tested IL 2 preparation was assessed on day 3 after 16 hours pulse with 1 uCi/ well of ^{3}H-thymidine (NEN, spec.act. 6.7 Ci/mM). The levels of IL 2 activity in the tested supernatants were expressed as the mean cpm observed in triplicate cultures ± SE.

IL 2 Absorption. Cells used for absorption of IL 2 were Con A induced T cell blasts. The cells were washed three times and then resuspended in 1 ml of standard IL 2 preparation. This suspension was incubated for 2 hr at 37°C with occasional shaking. The supernatant obtained by centrifugation at 1500 rpm for 15 min was assayed for residual IL 2 activity in the assay described above.

RESULTS

Production of IL 2 by antigen primed lymphocytes. We have shown previously that AO rats, in contrast to DA rats exibited complete resistance to the induction of EAE with BP+CFA (3) and experimental allergic uveitis (EAU) with BRT+CFA (4). As the initial phase of the immune response induced by i.d. injection of antigen emulsified

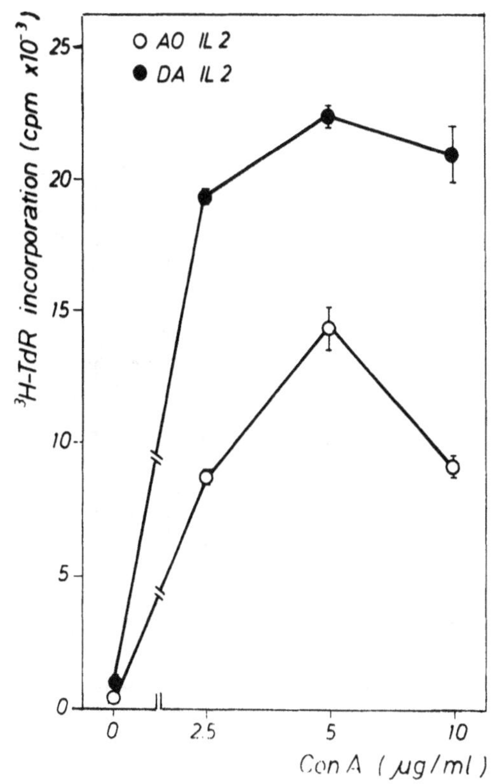

Fig. 1. IL-2 activity in Con A culture supernatants of lymph node cells of rats immunized with BP+CFA

in CFA occurs at the level of the lymph node draining the site of inoculation, we first examined the levels of IL 2 produced by Con A stimulated lymph node cells of AO and DA rats previously injected with BP+CFA. As shown in Fig. 1. culture supernatants of antigen primed AO lymphocytes contained markedly lower levels of IL 2 activity in comparison to supernatants obtained by culturing DA lymphocytes under the same conditions.

Production of IL 2 by spleen lymphocytes from nonimmunized rats. In an attempt to delineate whether the low IL 2 activity observed in the supernatants of AO lymph node cells represents their inherent capacity or reflects the outcome of a complex set of regulatory events initiated by strong antigenic stimulation, we have performed further experiments in which Con A induce IL 2 generation by spleen cells from normal nonimmunised rats was examined. The levels of IL 2 in supernatants of Con A stimulated spleen cells derived from

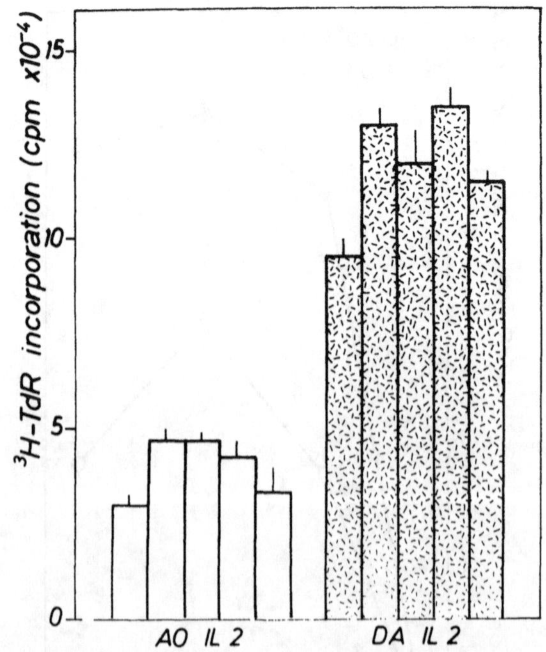

Fig. 2. IL 2 activity in Con A culture supernatants of rat spleen cells

individual AO and DA rats are shown in Fig. 2. The IL 2 generation within each of tested strains is highly uniform, but it is evident that culture supernatants of AO cells contained markedly lower levels of IL 2 activity.

Kinetics of IL 2 production. The kinetics of IL 2 production by spleen cells from AO and DA rats is shown in Fig. 3. These results are in agreement with previous data (6) showing IL 2 activity in spleen cell cultures decreases with time. Despite similar kinetics, AO spleen cells expressed significantly lower IL 2 generation than DA cells when cultured 24, 48 or 72 hr with Con A. This indicated that lower IL 2 activity demonstrated in the AO spleen cell supernatant is not due to altered kinetics of IL 2 generation in this strain.

Effects of culture conditions on IL 2 production in AO and DA rats. One possible explanation for this strain difference in IL 2 activity was simply inadequate culture conditions e.g. suboptimal concentration of responding cells or Con A. However, results of several

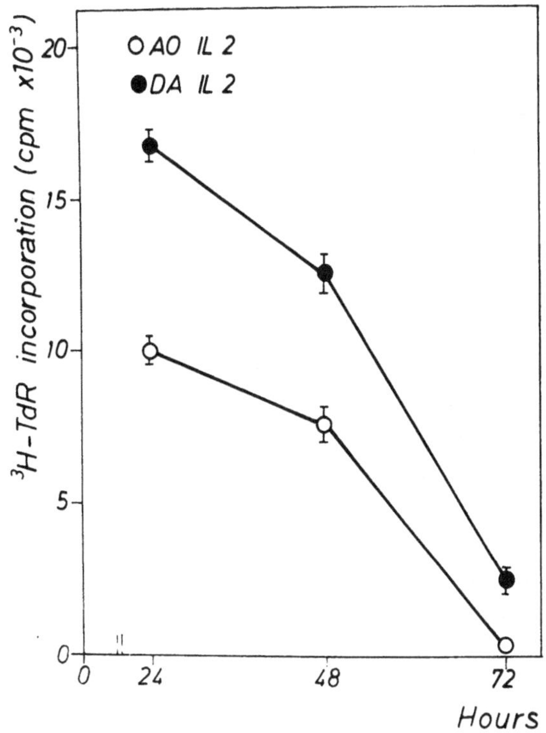

Fig. 3. Kinetics of IL 2 production

experiments revealed that the defect in IL 2 production by AO lymphocytes was consistent across a wide range of mitogen dose (2.5-10 ug/ml) and cell (2.5-10 x 10^6/ml) concentrations (data not shown).

<u>IL 2 Absorption Studies</u>. The low level of IL 2 activity demonstrated in supernatants of AO cultures might reflect enhanced passive absorption of once produced IL 2 by lymphocytes that express receptors for IL 2. In order to test this possibility, we compared standard IL 2 preparations for their residual activity after absorption with AO and DA cells. These experiments indicated that both fresh and Con A activated lymphocytes from AO rats had not increased capacity to absorb IL 2 (Fig. 4) when compared with DA cells.

<u>Responsiveness of AO and DA blasts to IL 2</u>. To determine whether the responsiveness of T cell blasts correlates with IL 2 production standard preparations of IL 2 were assayed for their growth promo-

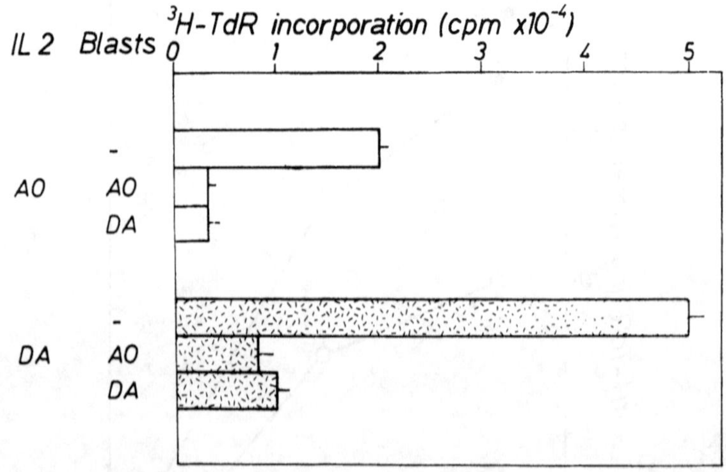

Fig. 4. Residual IL 2 activity in Con A culture supernatants after absorbtion with T cell blasts

ting activity on blasts obtained from AO or DA rats. No statistically significant differences were observed between the responsiveness of AO and DA blasts to IL 2 (data not shown). This is in agreement with the results of the absorbtion studies indicating that similar level of IL 2 receptors is expressed on Con A activated cells from both strains.

DISCUSSION

Differences in immune response may play an important role in susceptibility to a variety of diseases including autoimmune diseases in both animal and man. This was demonstrated for EAE in several species. Differences in susceptibility to EAE in DA and AO rats are genetically determined. We have previously obtained, in DA x AO combination, the evidence suggesting the existance of the dominant autosomal gene(s) not linked to the MHC which controls the decision making event in the inductive phase of EAE (7) and EAU (to be published). The results presented in this paper demonstrate the striking strain difference in IL 2 produced by Con A stimulated lymphocytes of AO and DA rats. Supernatants obtained from cultured

AO lymphocytes revealed significantly lower IL 2 activity when compared with preparations derived from DA cells. Thus, IL 2 activity correlates with the susceptibility to the experimentaly induced autoimmunity. The defect in the production and/or response to IL 2 have recently been demonstrated in some human disorders (8,9,10) as well as in the animal models (11,12) of T cell incompetence. On the other hand, the importance of IL 2 in generating the autoreactive effector cells in the model of EAE has been clearly established (2). Taken together, these data strongly argue that IL 2 plays an important role in regulating the appropriate and integrated responses of the immune system to internal and external signals.

Diminished IL 2 activity in culture supernatants of AO lymphocytes is not due to the altered kinetics of production, inadequate culture conditions or increased absorption of IL 2 by T cells. It would reflect changes in the number of IL 2 producing cells, alterations in the amount of IL 2 generated by each individual responding cell or multicellular interactions that suppressed IL 2 production. "Deficient" production of IL 2 in AO rats must be considered in relation to the previous report (3) showing that relative resistance to the induction of EAE in low susceptible AO strain could be overcome by the procedures which eliminate the precurssors of suppressor T cells. Assuming that suppressor cells regulate IL 2 production *in vivo* as it has been shown *in vitro* (13,14), it would be resonable to accept that failure of AO rats to generate T cell mediated autoimmune diseases might result from the genetically determined inbalance between their T helper IL 2 production and concomitant activity of suppressor cells.

REFERENCES

1. Watson,J and Moshizuki,D.: Immunol.Rev. 51:257 (1980)
2. Ortuz-Ortiz,L. and Weigle,W.O.: J.Immunol. 128:1545 (1982)
3. Mostarica-Stojković,M. and Lukić,M.L.: Clin.exp.Immunol. 9:581 (1979)
4. Mostarica-Stojković,M.,Nikolić,M.,Petrović,M. and Lukić,M.L.: Period.Biol. 85:63 (1984)
5. Andersson,J.,Grönvik,K.-O.,Larsson,E.-L. and Couthinho,A.: Eur.J.Immunol. 9:581 (1979)
6. Gillis,S.,Ferm,M.M.,Ou,W. and Smith,K.A.: J.Immunol. 120:2027 (1978)
7. Lalić,N.M.,Latković,Z.V.,Mostarica-Stojković,M. and Lukić, M.L.: Period.Biol. 85:77 (1984)

8. Linker-Israeli,M.,Bakke,A.C.,Kitridou,R.C.,Gender,S.,Gillis,S. and Horwitz,D.A.: J.Immunol. 130:2651 (1983)
9. Warred,H.S.,Atkinson,K.,Pembrey,R.G. and Bigs,J.C.: J.Immunol. 131:1771 (1983)
10. López-Botet,M.,Fontán,G.,Garcia Rodrigez,M.C. and Landázuri, M.O.: J.Immunol. 128:679 (1982)
11. Reiner,N.E. and Finke,J.H.: J.Immunol. 131:1487 (1983)
12. Altman,A.,Theophilopoulos,A.N.,Weiner,R.,Katz,D.H. and Dixon,F.J.: J.exp.Med. 154:791 (1981)
13. Palacios,R., ans Möller,G.: J.exp.Med. 153:1360 (1981)
14. Gullberg,M. and Larsson,E.-L.: J.Immunol. 128:746 (1982)

DETECTION OF ANTIBODY-INDUCED FEEDBACK SUPPRESSION IN SPLEEN CELL POPULATIONS OF PRIMED AND UNPRIMED MICE

Annette L. Rothermel and Catherine E. Calkins

Department of Microbiology
Thomas Jefferson University
1020 Locust Street, Philadelphia, PA 19107 USA

INTRODUCTION

Feedback suppression has been proposed as a natural mechanism of regulating immune responses. Both activated T and B cells, and their products have been shown to induce feedback regulatory circuits by interacting with T cells from normal, unprimed mice (1-4). If feedback suppression is a regulatory mechanism that occurs *in vivo* to down-regulate antibody responses, it would require that the antigen-primed host be able to generate feedback suppression in the absence of the exogenous unprimed T cells used to detect suppression in the experimental models. This study reports the results of experiments designed to investigate whether all the cells participating in the feedback suppression circuit are present in the antigen-primed host.

We have described a system of antibody-induced feedback suppression which is detected when unprimed spleen cells are added to cultures of SRBC-primed spleen cells stimulated with SRBC (1). This suppression requires the presence of PC-1$^+$ B cells from SRBC-primed mice (5) or anti-SRBC specific antibody produced by these B cells (6) to provide a feedback induction signal. This signal then activates unprimed Lyt123$^+$, Qa-1$^+$ T cells to mediate suppression of further anti-SRBC responses (7). In the experiments reported here T cell populations from mice primed with SRBC were tested to determine whether, like unprimed T cells, they could be activated by feedback signals to mediate suppression. To test this, T cells from primed mice were added to cultures of SRBC-primed spleen cells which provide the induction signal from primed B cells and also serve as the control response.

MATERIALS AND METHODS

In these experiments $4x10^6$ spleen cells taken from BDF_1 mice immunized ip 8 days earlier with $2x10^8$ sheep erythrocytes (SRBC) plus pertussis vaccine were cultured with SRBC in standard Mishell and Dutton cultures (8) as described previously (7). Various T cell subpopulations from unprimed mice, or from mice primed _in vivo_ with SRBC, horse erythrocytes (HRBC), or 1.0 mg of heat aggregated DNP-BGG, were added to the SRBC-primed spleen cell cultures to test their ability to mediate feedback suppression. Duplicate cultures were assayed after 3 through 6 days of culture for direct and indirect anti-SRBC antibody-forming cells by the slide modification of the Jerne plaque assay (8). Results obtained from the peak day of the anti-SRBC PFC response are shown.

T cells were obtained by passage through nylon wool columns (9) followed by further enrichment on rabbit anti-mouse Ig-coated dishes (10). $Lyt2^+$ and $Lyt2^-$ T cell subsets were then obtained as adherent and non-adherent cell fractions, respectively, by indirect panning of T cells pre-incubated with rat anti-Lyt2 on dishes coated with anti-rat IgG (11). $Lyt1^-23^+$ T cells were obtained as the non-adherent fraction by indirect panning of T cells pre-incubated with rat anti-Lyt1, followed by indirect panning for $Lyt2^+$ T cells. $Lyt123^+$ T cells were obtained by sequential indirect panning for $Lyt1^+$ and then for $Lyt2^+$ T cells after removing the anti-Lyt1 antibody with a series of 6 fifteen minute incubations in fresh 37° medium.

RESULTS

Loss of Feedback Suppressor T Cell Precursors Following Specific Antigen-Priming

Initial experiments were performed to determine if splenic T cells from SRBC-primed mice were able to participate in the generation of feedback suppression in the same manner as splenic T cells from unprimed mice. T cells from unprimed mice, or from mice primed with SRBC,

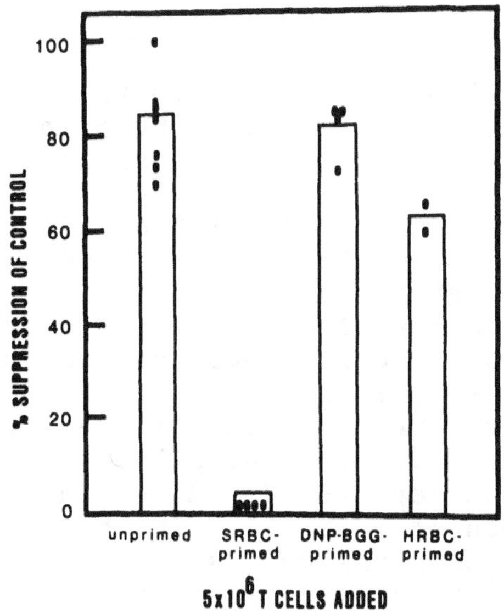

Fig. 1. Loss of feedback suppressor T cell precursors following specific antigen priming. 5x10^6 T cells from unprimed, SRBC-, DNP-BGG-, or HRBC-primed mice were added to cultures of 4x10^6 SRBC-primed spleen cells stimulated with SRBC. Results are reported as % suppression of control anti-SRBC IgG PFC response of SRBC-primed spleen cells cultured without additional T cells (4,700 to 24,313 IgG PFC/Culture).

were added to cultures of unfractionated SRBC-primed spleen cells challenged in culture with SRBC (Fig. 1). Addition of unprimed T cells to the culture resulted in greater than 80% suppression of the anti-SRBC PFC response, while addition of SRBC-primed T cells did not produce detectable suppression. It appeared, therefore, from these experiments that feedback suppressor T cell activity was lost after antigen priming.

Further experiments were performed to determine whether the loss of feedback suppression for an anti-SRBC response occured only after priming with SRBC, or if the loss occured non-specifically after priming with other antigens. T cells taken from mice immunized with either DNP-BGG or HRBC were added to cultures of SRBC-primed spleen cells to test their ability to suppress the anti-SRBC PFC response (Fig.1). Both DNP-BGG- and HRBC-primed T cells suppressed the anti-SRBC similarly to unprimed T cells. Therefore, the loss of feedback suppression for an anti-SRBC response occurs after specific priming with SRBC, and not after priming with other antigens.

Detection of Lyt2+ Suppressor T Cells in SRBC-Primed Mice

One explanation to account for the loss of feedback suppressor T cell activity following priming is that feedback suppressor T cell precursors are present but their activity is masked by the helper T cell activity that is generated after SRBC-priming. This was tested by separating T cells from SRBC-primed mice into Lyt2$^+$ and Lyt2$^-$ subpopulations by indirect panning. (In similar separations of unprimed T cells, we have found that the feedback suppressor T cell precursors are present in the Lyt2$^+$, but not the Lyt2$^-$ subpopulation.) Following removal of helper T cells present in the Lyt2$^-$ T cell fraction, the Lyt2$^+$ T cell population from SRBC-primed mice suppressed the anti-SRBC PFC response in a dose-dependent manner (Fig. 2). In this and four other experiments, 80% suppression was routinely obtained with addition of 5x10^6 Lyt2$^+$ T cells to the culture. Lyt2$^-$ T cells from primed mice enhanced the anti-SRBC PFC response. Unseparated T cells were not suppressive up to the 5x10^6 T cells tested in this experiment. These results indicate that Lyt2$^+$ suppressor T cells are present in the primed spleen, but their activity can only be detected upon removal of the Lyt2$^-$ population.

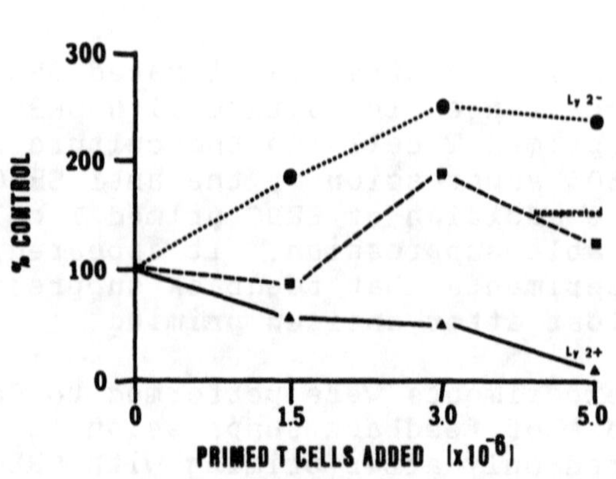

Fig. 2. Detection of Lyt2$^+$ suppressor T cells in the SRBC-primed T cell population. Graded numbers of SRBC-primed T cells (■ unselected T cells, ▲ Lyt2$^+$ T cells, or ● Lyt2$^-$ T cells) were added to cultures of 4x10^6 SRBC-primed spleen cells stimulated with SRBC. Results are reported as % control IgG PFC/culture of SRBC-primed spleen cells cultured without additional T cells.

The $Lyt2^+$ population contains both $Lyt123^+$ T cells, the phenotype of feedback suppressor cell precursors, as well as $Lyt1^-23^+$ T cells (those cells with a low amount of surface Lyt1) the characteristic phenotype of activated suppressor effector cells. The objects of the present study were the suppressor cell precursors that require induction by primed B cells in order to become suppressive. The $Lyt2^+$ T cells were therefore tested to determine whether they required a feedback induction signal in order to mediate suppression. Preliminary experiments indicated that $Lyt2^+$ T cells from primed mice were suppressive in the absence of the primed B cell induction signal. Thus at least part of the suppression observed by the primed $Lyt2^+$ T cells was due to suppressor T cells that are "pre-activated" during in vivo priming. The next experiments were designed to determine if feedback suppressor T cell precursors were present along with the "pre-activated" suppressor T cells in the spleens of SRBC-primed mice.

Demonstration of Feedback Suppressor T Cell Precursors within the SRBC-Primed Spleen Cell Population

Further T cell separations were performed to separate the suppressive primed $Lyt2^+$ T cell population into $Lyt123^+$ and $Lyt1^-23^+$ T cell subpopulations in order to separate potential feedback suppressor T cell precursors from the "pre-activated" suppressor T cells. T cell fractions separated by sequential indirect panning were tested for their ability to suppress in the presence or absence of the feedback induction signal. For this they were added to SRBC-stimulated cultures containing either SRBC-primed cells which provide the primed B cell inducing signal (Fig. 3A) or unprimed spleen cells which do not provide the induction signal (Fig. 3B). Primed $Lyt123^+$ T cells were only suppressive when added to cultures of primed spleen cells. They suppressed both the IgM and IgG anti-SRBC PFC response in these cultures. They did not suppress the response of unprimed spleen cells indicating a requirement for an inducing signal not present in the cultures of unprimed cells. In contrast, primed $Lyt1^-23^+$ T cells were highly suppressive when added to cultures of unprimed spleen cells plus SRBC, accounting for the suppression seen previously in the preliminary experiments using the mixed population of $Lyt2^+$ T cells. Surprisingly, the $Lyt1^-23^+$ T cell population did not suppress the secondary IgG anti-SRBC PFC response. The IgM response

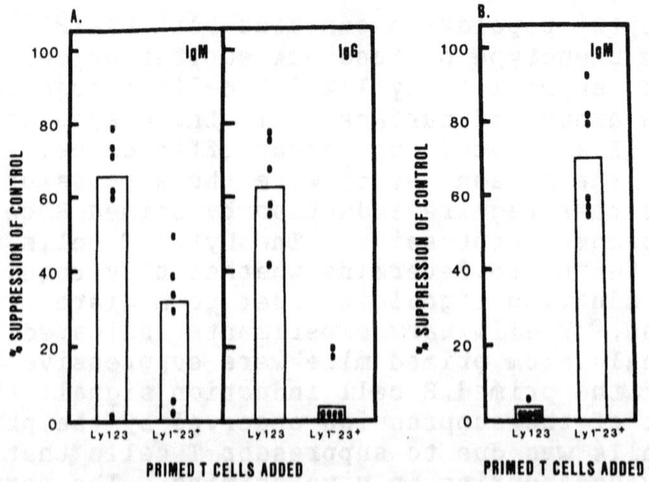

Fig. 3. $3x10^6$ Lyt123$^+$ or Lyt1$^-$23$^+$ T cells were added to cultures of $4x10^6$ primed spleen cells or 10^7 unprimed spleen cells. Results are reported as % suppression of the PFC response of primed spleen cells (255 to 4,712 IgM and 1,025 to 26,913 IgG PFC/culture) or unprimed spleen cells (870 to 3700 IgM PFC/culture)cultured without additional T cells.

of the SRBC-primed cells was suppressed to varying degrees by the Lyt1$^-$23$^+$ T cells.

DISCUSSION

Two findings stand out from the results reported here. Firstly, the studies were originally undertaken to determine whether antigen-primed mice possessed a counterpart to the feedback suppressor T cell precursors shown to be present in the spleens of unprimed mice. Similar precursors are indeed demonstrable in the spleens of antigen-primed mice, but their activity is masked by the Lyt2$^-$ T cell subpopulation. Whether this masking by the Lyt2$^-$ T cells is due to helper or contrasuppressor activity is as yet unclear. The suppressor cell precursors are found in the Lyt123$^+$ subpopulation, as are the feedback suppressor T cell precursors present in the unprimed spleen. These feedback suppressor T cell precursors require an induction signal from SRBC-primed B cells in order to mediate suppression. However, once activated, they suppress secondary IgM and IgG anti-SRBC responses.

Secondly, a suppressor cell population with a different function was also identified in the T cell population from SRBC-primed mice. This suppressor cell is an $Lyt1^{-}23^{+}$ T cell which suppresses in the absence of the induction signal provided by primed cells and is therefore probably "pre-activated" in the SRBC-primed mice. These "pre-activated" cells differ from feedback induced suppressor T cell precursors in that they suppress only primary responses and appear to have no effect on secondary IgG responses. It is of interest that these "pre-activated" suppressor cells are also masked in the unselected primed T cell population. These results demonstrate that there are at least two coexistent suppressor T cell pathways that regulate antibody responses following antigen priming. As these two suppressor T cells clearly differ in their phenotypes, their activation requirements, and their subsequent suppressive activities, it is also likely that they play different roles in regulating the immune response.

REFERENCES

1. C. E. Calkins, S. Orbach-Arboys, O. Stutman,and R. Gershon. J. Exp. Med. 143:1421 (1976).
2. T. Tada. Prog. Immunol. 3:369 (1977).
3. D. D. Eardley,J. Hugenberger, L. McVay-Boudreau, F. W. Shen, R.K. Gershon, and H. Cantor. J. Exp. Med. 147:1104 (1978).
4. J. L'age-Stehr, Teichmann, R. K. Gershon, and H. Cantor. Eur. J. Immunol. 10:21 (1980).
5. C. E. Calkins. Eur. J. Immunol. 12:70 (1982).
6. C. E. Calkins. J. Molec. Biochem. Suppl. 8A (1984).
7. C. E. Calkins, T.A. Stanton, and O. Stutman. Eur J. Immunol. 10:449 (1980).
8. R. Mishell and R. Dutton. J. Exp. Med. 126:423 (1967).
9. M.H. Julius, E. Simpson and L.A. Herzenberg, Eur. Immunol. 3:645 (1973).
10. M.G. Mage, L.L. McHugh and T.L. Rothstein. J. Immunol Methods. 15:47 (1977).
11. M. Mage, B. Mathieson, S. Sharrow, L. McHugh, U. Hammerling, C. Kanellopoulos-Langevin, D. Brideau Jr., and C.A. Thomas III. Eur. J. Immunol. 11:228 (1981).

IgD-INDUCED ENHANCEMENT OF THE IMMUNE RESPONSE

Richard F. Coico, David Wallace, Bin Xue,
Judith Rosen, Ben Pernis, G. Jeanette Thorbecke,
and Gregory W. Siskind

Dept. of Pathology, NYU School of Medicine, NY 10016
Dept. of Medicine, Cornell U. Medical College, NY 10021
and Dept. of Microbiology, Columbia U. College of
Physicians and Surgeons, NY 10032

INTRODUCTION

Mice bearing IgD-producing myelomas have greatly enhanced primary and secondary immune responses [1]. This enhancement appears to be related to the presence of high levels of serum IgD, since pretreatment of BALB/c mice with IgD-containing ascites fluid or with isolated IgD also causes a marked increase in their primary responses to intravenously injected trinitrophenylated hemocyanin (TNP-KLH) [1]. Ascites fluid from mice bearing myelomas producing other Ig isotypes appears to cause an inhibition rather than an enhancement of immune responses [1,2]. These data suggest an immunoregulatory effect of IgD itself, but do not provide insight into the mechanism of this effect.

The present studies were undertaken to determine whether the immune enhancement effect of IgD-containing ascites fluid is exerted directly on B cells or indirectly via accessory and/or T cells. It will be shown that both primary and secondary responses can be enhanced by pretreatment with IgD-containing ascites fluid in normal, but not in athymic mice and that the tendency to give enhanced responses after such pretreatment can be transferred to normal recipients by T cells. Moreover, pretreatment with anti-IgD enhances primary, but not secondary responses in athymic mice.

MATERIALS AND METHODS

<u>Mice</u>. BALB/c mice were obtained from Charles River Breeding Lab-

oratories, Inc., Wilmington, MA. Athymic BALB/c mice were supplied by the Division of Cancer Treatment Animal Program, NCI, Bethesda, MD. BALB.K mice were bred from a stock originally obtained from Dr. D. Meruelo (NYU School of Medicine).

Antigens. TNP-conjugated KLH (Sigma Chemical Co., St. Louis, MO) and TNP-conjugated *Brucella abortus* (BA) (U.S. Department of Agriculture, Ames, IA) were prepared as described by Little and Eisen [3].

Immunizations. Mice were injected intravenously at 6-8 weeks of age (Day 0) with either 100 μg TNP-KLH, or 80 μg TNP-BA. Unless otherwise indicated, for secondary responses, mice were boosted 4 days after priming and their responses were determined 5 days after challenge.

Plasmacytomas. IgD-secreting plasmacytoma, TEPC 1017, was maintained *in vivo* as previously described [1]. Ascites fluid was examined for the presence of IgD by double diffusion in agar using affinity purified goat anti-mouse IgD.

Purification of IgD. TEPC 1017 ascites was passed over a column of rabbit anti-mouse IgD-coupled Sepharose 4B (Pharmacia Fine Chemicals). After extensive washing with phosphate-buffered saline, the IgD was eluted with 4.0 M $MgCl_2$. The eluate was dialyzed and reconstituted to the same volume as the original ascites fluid.

Spleen Cell Fractionation. Adherent cells were removed by incubation of spleen cells in tissue culture grade petri dishes (Nunclon 1400-1, Roskilde, Denmark) at 37°C for 1 hr. T cells were eliminated by complement- (C) mediated cytolysis using a monoclonal antibody to Thy 1.2 (6.80) provided by Dr. U. Hammerling (Sloan-Kettering Institute, New York). Splenic T cells were purified by negative selection with the use of rabbit anti-mouse Ig-coated petri dishes as described [4].

Cell Transfer. Spleen cells, suspended in RPMI 1630, were transferred iv into recipients that had received 100R whole body γ-irradiation one day earlier by using a ^{137}Cs source (Isomedix, Parsippany, NJ).

Plaque-Forming Cell (PFC) Assay. Anti-TNP and anti-sheep erythrocyte (SE) PFC were assayed by the slide modification of the technique of Jerne *et al.* [5,6]. TNP-SE were prepared using the method described by Rittenberg and Pratt [7]. IgG-producing cells were developed with rabbit anti-mouse Ig in the complement and goat anti-μ in the agar [8].

RESULTS AND DISCUSSION

The enhancing effect of pretreatment with IgD-containing ascites fluid prior to the primary injection of TNP-KLH is shown, in three experiments each, for both the primary response (Fig. 1) and the "secondary" response induced with a 4-day interval between primary and secondary injections (Fig. 2). It can be seen that the enhancing effect on the early secondary response is

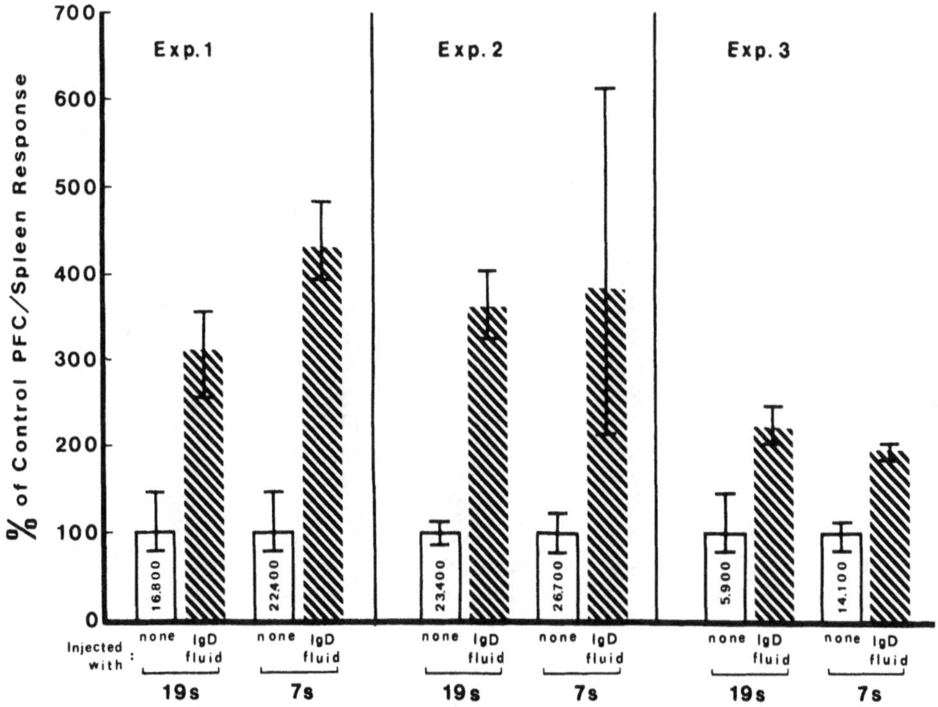

Fig. 1. Effect Of Pretreatment With IgD-Containing Ascites Fluid On The Primary Immune Response Of BALB/c Mice To TNP-KLH. Pretreatment consisted of ip injections of 0.5 ml ascites fluid [IgD fluid] on days -8, -7, -6 and -5. Antigen (100 μg TNP-KLH) was injected iv on day 0. Anti-TNP PFC/spleen were determined on day 5. Results are presented as °/° of control responses $\overset{\times}{\div}$ SE. Control geometric mean PFC/spleen are given in the open bars. (n=5).

much greater than that on the primary response. These findings are in agreement with our previous observations using IgD myeloma-bearing mice. Both 19S and 7S responses are equally enhanced (up to 40-fold), suggesting that at least a large part of the effect is exerted on B cells prior to isotype switching. The rationale for using such a short interval between primary and secondary injections is that B cells are known to undergo a progressive reduction of their surface IgD (sIgD) after activation [9]. Thus, an effect mediated via sIgD might not be obtained later after priming. The results in Fig. 3 show, however, that enhanced secondary responses can still be induced by treatment with IgD containing ascites begun 10 days after priming. It is of interest to note that in this case the relative effect on the IgM response is much greater (3.4-fold) than on the IgG response (1.7-fold).

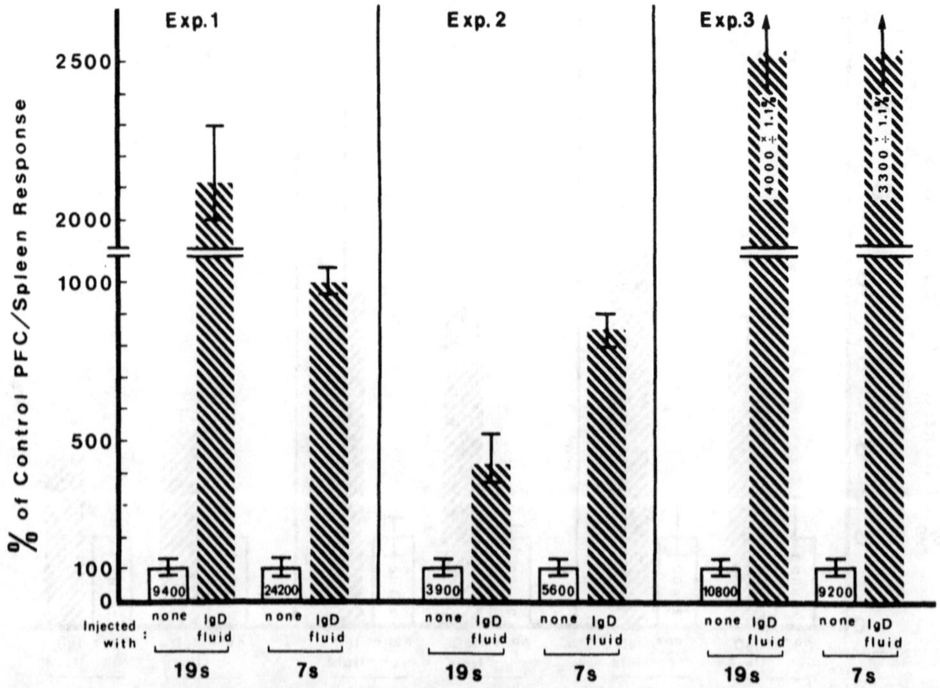

Fig. 2. Effect Of Pretreatment With IgD-Containing Ascites Fluid On The Secondary Immune Response To TNP-KLH.
Pretreatment consisted of ip injections of 0.5 ml ascites fluid [IgD fluid] on days -8, -7, -6, -5, and, in expt. 2, on day +3. Antigen (100 μg TNP-KLH) was given on days 0 and +4. Anti-TNP PFC/spleen were determined on day +9. Expts. 1 and 3 were performed in BALB/c mice, expt. 2 in BALB.K mice. Results are presented as °/° of control responses $\overset{\times}{\div}$ SE. Control geometric mean PFC/spleen are given in the open bars. (n=5).

In order to obtain insight into the cellular basis for the enhancing effect of IgD we determined whether the effect of pretreatment with IgD containing ascites could be transferred to untreated recipients. Spleen cells from IgD pretreated donors were injected simultaneously with TNP-KLH into 100R γ-irradiated recipients. As seen in Table 1 the anti-TNP PFC responses in recipients of IgD pretreated donor cells are significantly higher than in recipients of control cells. This effect is clearly due to the IgD in the ascites fluid, since isolated IgD causes the same effect (Expt. 2). In addition, the results of transfers of splenic T and B cell fractions are also presented. They show that T cells from IgD pretreated donors confer the immunostimulatory effect (Expt. 4), whereas B cells from similar donors do not (Expt. 3, Table 1).

TABLE 1

ABILITY OF SPLENIC T CELLS TO TRANSFER THE IMMUNE ENHANCING EFFECT OF IgD ASCITES PRETREATMENT

DONOR CELLS	Geom. Mean Direct PFC/Spleen*			
	Expt. 1+	Expt. 2	Expt. 3**	Expt. 4***
Spleen from untreated mice	2,420	3,400	4,420	2,210
Spleen from IgD-ascites treated	5,710	14,760[a]	10,680	4,090
Spleen from purified IgD treated		11,060[b]		
	($p<.0005$)	($p<.005^a$)	($p<.002$)	($p<.02$)
		($p<.01^b$)		
B cells from untreated mice			3,840	
B cells from IgD-ascites treated			4,370	
			($p=.5$)	
T cells from untreated mice				2,670
T cells from IgD-ascites treated				4,410
				($p<.05$)

* Recipients were irradiated 100R on day -1, received 3 x 10^7 cells iv on day 0 and their spleens were assayed on day 5. Only results for 19S PFC are recorded.

\+ In expt. 1 individual donor - recipient pairs were used.

** B cells prepared by anti-Thy 1.2 + C treatment, control cells were treated with C alone.

*** T cells prepared by panning on anti-Ig coated dishes at 4^oC, after removal of plastic-adherent cells at 37^oC. Control cells were simultaneously incubated on uncoated dishes.

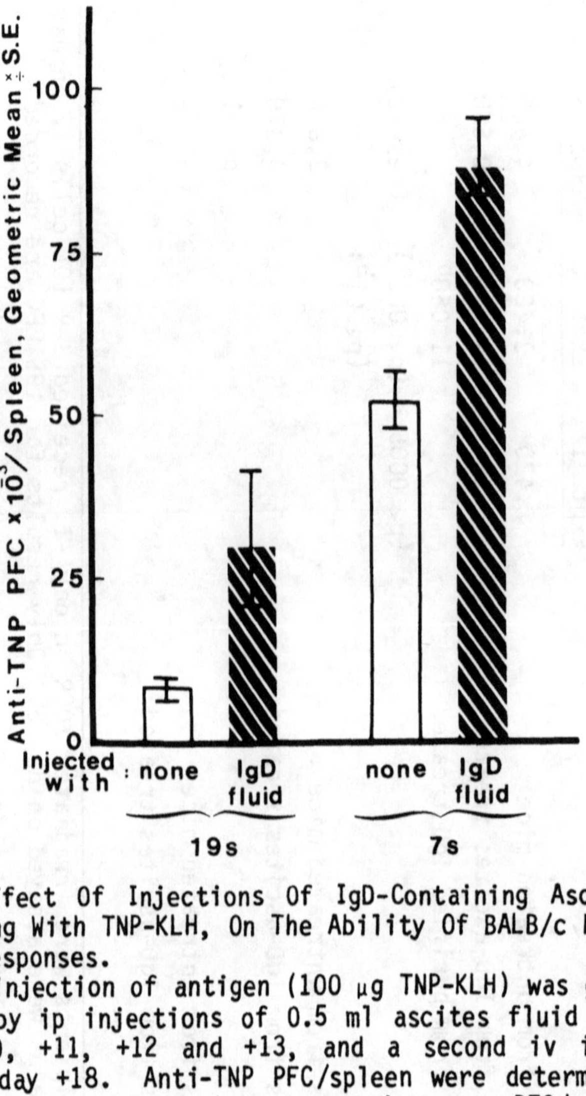

Fig. 3. Effect Of Injections Of IgD-Containing Ascites Fluid, After Priming With TNP-KLH, On The Ability Of BALB/c Mice To Give Secondary Responses.
Primary iv injection of antigen (100 μg TNP-KLH) was given on day 0 followed by ip injections of 0.5 ml ascites fluid [IgD fluid] on days +10, +11, +12 and +13, and a second iv injection of TNP-KLH on day +18. Anti-TNP PFC/spleen were determined on day +23. Results are expressed as geometric means PFC/spleen ×÷ SE. (n=5).

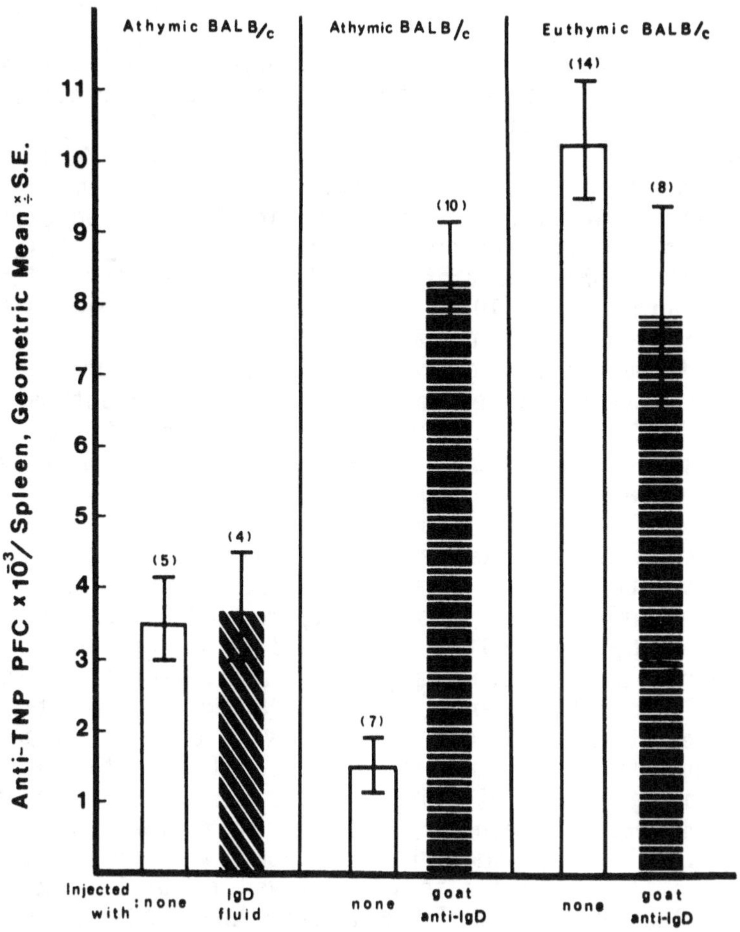

Fig. 4. Effect Of Pretreatment Of Athymic (nu/nu) BALB/c Mice Either With IgD-Containing Ascites Fluid (left-hand panel) Or With Anti-IgD On Primary Response To TNP-KLH.
Pretreatment with 0.5 ml ascites fluid ip was on days -8, -7, -6 and -5. Pretreatment with goat anti-IgD ip was on days -10 (3.5 mg), -7, -3 and 0 (1.4 mg each). Antigen (100 µg TNP-KLH) was injected iv on day 0. PFC/spleen were determined on day 5. The effect of anti-IgD on the response of euthymic BALB/c mice is shown in the right-hand panel. Results are presented as geometric means of anti-TNP PFC/spleen ×÷ SE (n is given in parentheses).

If the effect of IgD is mediated via T cells, it should be absent in athymic mice. The results in Fig. 4 indeed show this to be the case. In our previous studies [1] we hypothesized that T cells with receptors for IgD might down-regulate IgD expression, particularly in response to sIgD cross-linking on B cells. Furthermore, we suggested that this decrease in IgD transcription might promote the tendency of the affected B cells to differentiate, resulting in enhanced secretion of IgM and/or IgG. Since treatment with anti-IgD can also down-regulate sIgD expression on B cells, it might be expected that such a pretreatment would enhance antibody production. In agreement with this postulate we have previously shown that pretreatment with anti-IgD from birth causes a marked enhancement of antibody production to iv injected antigen in normal mice [10]. The present results show that even a 2 week pretreatment with goat anti-IgD may cause a marked enhancement of the primary IgM response to TNP-KLH in athymic mice, causing the response to become as high as that in euthymic mice with or without anti-IgD pretreatment. Thus, the effect of sIgD down-regulation in this system is exerted directly on B cells whereas the effect of IgD is mediated by T cells.

In previous studies we have shown that treatment with goat anti-IgD in euthymic mice causes an increase in the number and size of germinal centers starting approximately 5 days after injection of the anti-IgD [11]. No effect on germinal centers was seen prior to this time, suggesting that, like the effect of antigen on germinal centers [12], this effect is mediated by T cells rather than a direct effect on B cells. Since goat anti-IgD appeared to improve the response of mice to the thymus dependent antigen TNP-KLH (Fig. 4), it seemed of interest to determine whether the secondary response to a thymus independent antigen, e.g. TNP-BA, which is much lower in athymic than in control mice [13], could also be improved by pretreatment with anti-IgD. The results in Table 2 show that this is not the case. Thus, pretreatment with goat anti-IgD may mimic the effect of T cells in the generation of a primary, but not a secondary response.

In summary, the data presented show that IgD causes a T cell-mediated enhancement of the primary response as well as enhanced priming for a secondary response. In contrast, prolonged treatment with anti-IgD, but not a single anti-IgD injection (data not shown), causes a T cell independent enhancement of the primary response only.

Supported by PHS Grant AG-04980 and NIH Cancer Research Training Grant CA-09161, DHHS, and New York State Department of Health AIDS Institute Grant A0190.

TABLE 2

LACK OF EFFECT OF PRETREATMENT WITH ANTI-IgD ON THE SECONDARY IMMUNE RESPONSE OF ATHYMIC BALB/c MICE TO TNP-B. ABORTUS

			Geom. Mean PFC/Spleen $\stackrel{\times}{\div}$ SE (n)		
Expt.	Mice	Pretreatment	19S	7S	
1	Athymic	None	27,500 $\stackrel{\times}{\div}$ 1.7	<500	(3)
		Goat anti-IgD	22,600 $\stackrel{\times}{\div}$ 1.1	<500	(3)
	Euthymic	None	69,200 $\stackrel{\times}{\div}$ 1.1	66,100 $\stackrel{\times}{\div}$ 1.1	(3)
			19S + 7S		
2	Athymic	Normal goat Ig	9,900 $\stackrel{\times}{\div}$ 1.1 (6)		
		Goat anti-IgD	7,500 $\stackrel{\times}{\div}$ 1.2 (6)		
	Euthymic	None	34,200 $\stackrel{\times}{\div}$ 1.1 (6)		

TNP-B. abortus (80 μg, iv) was injected on days 0 and +10; anti-TNP PFC/spleen were determined on day +15.

Goat anti-IgD or normal goat Ig was injected ip on days -10 (3.5 mg), -7, -3, 0, +4, +7 and +11 (1.4 mg each).

REFERENCES

1. B. Xue, R. Coico, D. Wallace, G. W. Siskind, B. Pernis, and G. J. Thorbecke, Physiology of IgD. IV. Enhancement of antibody production in mice bearing IgD-secreting plasmacytomas, J. Exp. Med. 159:103 (1984).
2. S. Zolla-Pazner, D. Noar, and T. Tonapatchaiyapong, Cellular basis of immunodepression in mice with plasmacytomas, J. Immunol. 112:2068 (1974).
3. J. R. Little, and H. N. Eisen, Preparation of immunogenic 2,4-dinitrophenyl and 2,4,6-trinitrophenyl proteins, Methods Immunol. Immunochem. 1:128 (1967).
4. L. J. Wysocki, and V. L. Sato, Panning for lymphocytes: a method for cell selection, Proc. Natl. Acad Sci. USA 75:2844 (1978).
5. M. B. Rittenberg, and K. L. Pratt, Antitrinitrophenyl plaque assay. Primary response of BALB/c mice to soluble and particulate immunogen, Proc. Soc. Exp. Biol. Med. 132:575 (1969).

6. R. Mishell, and R. W. Dutton, Immunization of dissociated cell cultures from normal mice, J. Exp. Med. 126:423 (1967).
7. N. K. Jerne, A. A. Nordin, and C. Henry, The agar plaque technique for recognizing the antibody-producing cells, In: "Cell Bound Antibody", B. Amos and H. Koprowsky, eds., Wistar Institute Press, Philadelphia, p. 109, (1963).
8. E. A. Goidl, T. J. Romano, G. W. Siskind, and G. J. Thorbecke, Changes in affinity of 19S and 7S antibodies at the cellular level in responses to hapten conjugates of varying T dependency, Cell. Immunol. 35:2311 (1978).
9. S. Kanowith-Klein, E. S. Vitetta, E. L. Korn, and R. F. Ashman, Antigen-induced changes in the proportion of antigen-binding cells expressing IgM, IgG, and IgD receptors, J. Immunol. 122:2349 (1979).
10. E. B. Jacobson, Y. Baine, Y. -W. Chen, T. Flotte, M. J. O'Neil, B. Pernis, G. W. Siskind, G. J. Thorbecke, and P. Tonda, Physiology of IgD. I. Compensatory phenomena in B lymphocyte activation in mice treated with anti-IgD antibodies, J. Exp. Med. 154:318 (1981).
11. T. J. Flotte, F. D. Finkelman, and G. J. Thorbecke, Polyclonal activation of the murine immune system by an antibody to IgD. V. Effect in germinal centers, Eur. J. Immunol. (in press, 1984).
12. G. J. Thorbecke, and S. P. Lerman, In: "The Reticuloendothelial System in Health and Disease", Reichard, Escobar, and Friedman, eds., Plenum Press, NY, Adv. Exp. Med. Biol. 73A:83 (1976).
13. E. B. Jacobson, L. H. Caporale, and G. J. Thorbecke, Effect of thymus cell injections on germinal center formation in lymphoid tissues of nude (thymusless) mice, Cell. Immunol. 13:416 (1974).

REGULATION OF ISOTYPE EXPRESSION IN THE IMMUNE RESPONSE TO TYPE-2 THYMUS INDEPENDENT DEXTRAN ANTIGENS

Kerstin Struckmann, Walter Schuler and Eckehart Kölsch

Department of Immunology
University of Münster
Domagkstr. 10, D-4400 Münster, FRG

The immune responses of mice to bacterial dextrans have been classified as being "thymus independent" (TI) since there is no evidence that T helper cells are involved in their regulation. One of the two most commonly investigated dextrans is dextran B 1355S [α(1->3)dextran] towards which BALB/c (Igh^a) mice mount a strong antibody response which is allotype linked and λ light chain associated (high responders). C57BL/6 (Igh^b) mice fail to respond to Dex 1355 (1). The other dextran is Dex B 512F [α(1->6)dextran] the response to which seems to be under a less pronounced allotype-linked genetic control (2,3). The immune response to Dex 1355 in BALB/c mice is characterized by several unexpected features which prompted us to investigate a possible regulatory role of T lymphocytes in so called type-2 TI immune responses (4,5). The response is associated with the appearance of a major germline associated idiotype and in most conventionally raised BALB/c mice it remains restricted to the production of IgM antibodies. This isotype restriction is however not an intrinsic property of the B lymphocyte population committed to produce these antibodies but is the outcome of regulatory interactions of T lymphocytes with the Dex 1355-specific B lymphocyte pool.

The majority of conventionally raised euthymic BALB/c or BALB/c nu/+ mice produce only IgM anti Dex 1355 antibodies. Together with the lack of an anamnestic response these are features commonly associated with type-2 TI responses. Germfree raised euthymic BALB/c as well as conventionally raised athymic BALB/c nu/nu mice produce upon primary immunization IgG anti Dex 1355 antibodies in addition to IgM. The idiotype restriction found in the IgM response is retained. The isotype shift in serum antibodies is controlled at the level of B lymphocytes by T suppressor (Ts) lymphocytes which in conventionally raised mice are apparently pre- or perinatally induced by the environmental antigen Dex 1355. These Ts cells not only prevent an isotype shift in antibodies appearing upon immunization but apparently play an important role in the differentiation of Dex 1355-specific B lymphocytes. It was possible to demonstrate Dex 1355-specific Bγ memory cells (i.e. resting B cells committed to production of IgG antibodies) upon only a single immunization, although no enhanced secondary IgG can be elicited even after repeated immunization. Dex 1355-specific Bγ memory cells are not found in BALB/c nu/nu

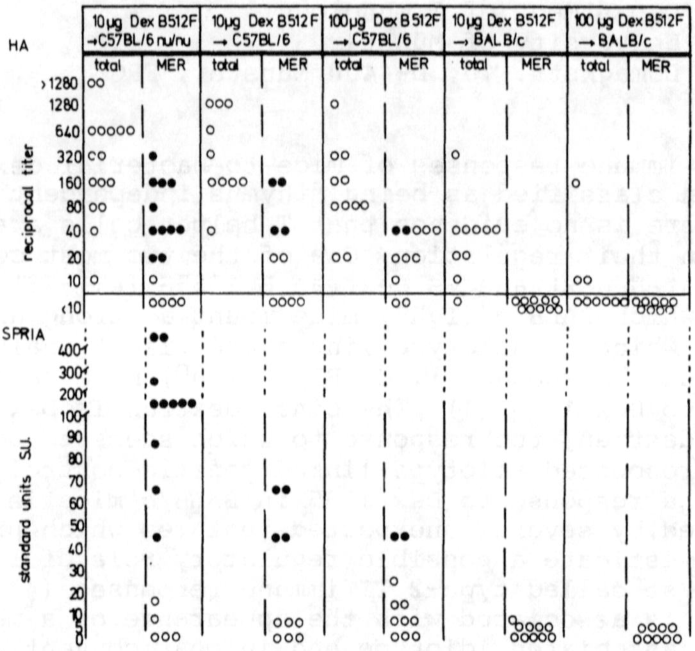

Fig. 1. The immune response to Dex 512. BALB/c, C57BL/6 and C57BL/6 nu/nu mice were immunized by a single injection of Dex 512 and bled 14 days later. Total and mercaptoethanol resistant (MER) Dex 512-specific hemagglutination titers (HA) of individual mice are given. Antibodies were also assayed in a solid phase radioimmune assay (SPRIA) employing protein A. In this assay only IgG antibodies can be detected. IgG responders are marked by closed circles.

mice although they do produce IgG anti Dex 1355. Reconstitution of BALB/c nu/nu mice with T cells suppresses IgG anti Dex 1355 production upon immunization but restores the capacity to develop Dex 1355-specific Bγ memory cells. These results led to the view that Ts cells not only block terminal differentiation to IgG secreting plasma cells but might play an important positive regulatory role in the differentiation pathway of Bγ memory cells (5,6).

The present report shows that similar to the anti Dex 1355 response in BALB/c mice the immune response of C57BL/6 mice to Dex 512 seems also to be under the control of Ts cells. It furthermore gives a description of the anti Dex 1355 and anti Dex 512 responses in (BALB/c xC57BL/6)F_1 hybrid ($CB6F_1$) mice.

MATERIALS AND METHODS

Dex 1355 and Dex 512 have been obtained from Dr. M.E. Slodki, USDA, Peoria, Il. USA. C57BL/6 and $CB6F_1$ mice were purchased from Jackson Laboratory, Bar Harbor, Maine, USA. C57BL/6 nu/nu mice were obtained from Olac, Bicester, England. All other materials and methods have been described (4,5).

RESULTS AND DISCUSSION

Extension of the concept of the regulation of type-2 TI responses by Ts cells

Figure 1 shows that the immune response of C57BL/6 mice to Dex 512 is largely confined to IgM antibodies, whereas a considerable number of C57BL/6 nu/nu mice produce IgG anti Dex 512 antibodies. Up to now a total of 61 C57BL/6 mice have been tested, out of which only 14 (=23%) show a weak IgG response, whereas 11 out of 16 C57BL/6 nu/nu mice (=69%) make high titers of IgG anti Dex 512 antibodies. Concerning the isotype distribution these results are comparable to the data obtained upon immunization of BALB/c mice with Dex 1355. The data also show that BALB/c mice are low responders to Dex 512. Showing a difference between C57BL/6 and C57BL/6 nu/nu mice in the response to Dex 512 does not, of course, imply a regulatory influence of Ts cells. But one should keep in mind that demonstration of the nu-gene influence on the anti Dex 1355 response in BALB/c mice was a key experiment leading to the demonstration of Ts regulatory functions in type 2 TI responses.

Dominance of IgG isotype suppression in $CB6F_1$ mice

In Figure 2 the responses of $CB6F_1$ animals to Dex 1355 and Dex 512 are compared with the respective parental immune reactions. Of the $CB6F_1$ animals 28% make IgG anti Dex 1355, 11% produce IgG anti Dex 512 antibodies. This shows that the F_1 responses are quantitatively similar to the parental ones with respect to IgM isotype dominance.

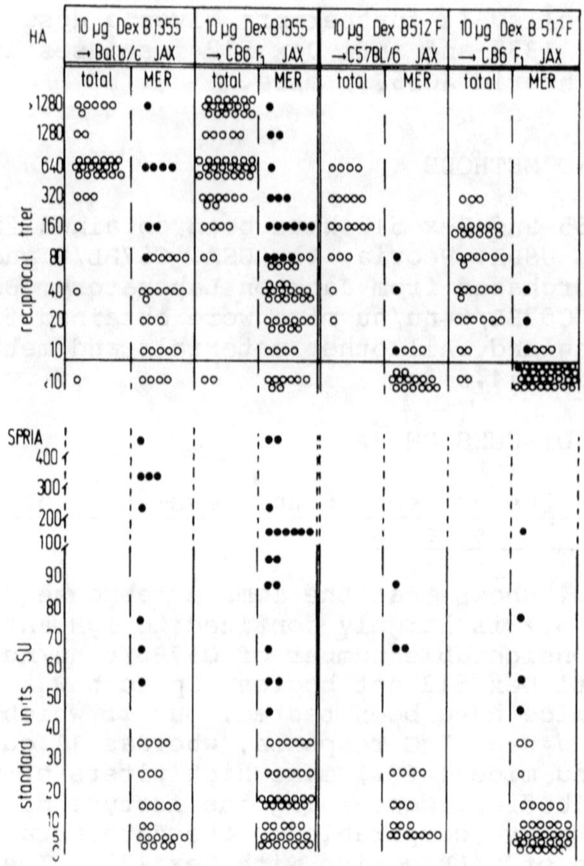

Fig. 2. The anti Dex 1355 and anti Dex 512 immune response of $CB6F_1$ mice. Animals, between 9-11 weeks of age at beginning of the experiment, were immunized by a single injection. The responses are compared to those of the parental strains (BALB/c anti Dex 1355 and C57BL/6 anti Dex 512). Legends as in Figure 1.

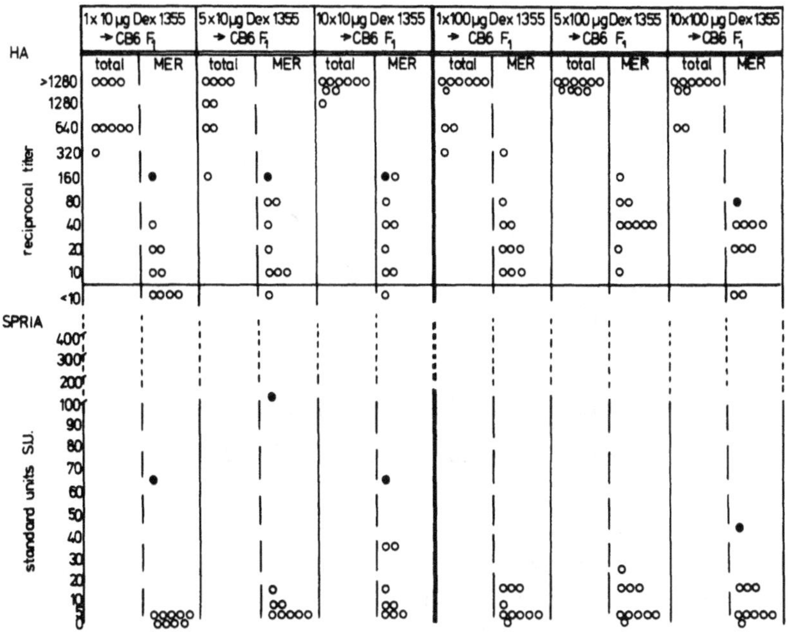

Fig. 3. The effect of repeated immunization of $CB6F_1$ mice with Dex 1355. Legends as in Figure 1.

Next it was tested whether repeated injections of antigen would change the response pattern. When eleven week old $CB6F_1$ animals were repeatedly immunized with Dex 1355 (Fig. 3) it turned out that even upon this repeated immunization it was virtually impossible to elicit an IgG anti Dex 1355 response. Similar results have been obtained upon immunization of 11 week old $CB6F_1$ with Dex 512 (data not shown).

When a group of older animals (23 weeks old at beginning of the experiment) was immunized by repeated injections with Dex 1355 a marked IgG anti Dex 1355 was found in the majority of animals tested. Nevertheless there were also individuals which could not mount an IgG response (Fig. 4). As far as the anti Dex 512 response of $CB6F_1$ mice is concerned no difference between young and old animals was found upon repeated immunizations. Four out of 28 $CB6F_1$ mice showed a low titer IgG response, the rest produced only IgM antibodies (data not shown). The isoelectric focussing data for an IgG producing individual $CB6F_1$ mouse are shown from different subsequent bleedings, demonstrating both preservation of the BALB/c clonotype through the time course of immunization as well as appearance of greater hetero-

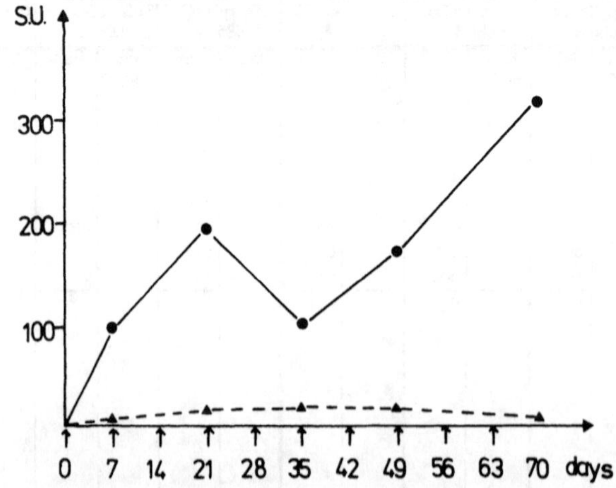

Fig. 4. Kinetics of the IgG anti Dex 1355 response in two individual $CB6F_1$ mice, 23 weeks of age at the beginning of the experiment. The animals were injected with 100µg Dex 1355 for 10 weeks in weekly intervals (arrows). Sera were taken at the times given by circles and triangles. 60-90% of animals in this experimental group show high titers of IgG as does mouse "circle". Most of them showed IgM anti Dex 1355 antibodies in their preimmune serum. No anti Dex 1355 activity was observed in the preimmune serum of mouse "triangle", which could not be boosted to IgG production.

geneity (Fig. 5). The isotype heterogeneity in the anti Dex 1355 response pattern of old $CB6F_1$ mice is probably due to an age dependent uncontrollable exposure to Dex 1355 in a thymus-dependent hapten-carrier complex. This interpretation is supported by an experiment challenging with Dex 1355 BALB/c mice primed with Dex 1355 coupled to sheep red blood cells (Tab. 1). An analogous situation was never observed in $CB6F_1$ animals immunized with Dex 512. For unknown reasons natural exposure to a hapten-carrier complex seems to be more likely for haptenic Dex 1355 than for Dex 512.

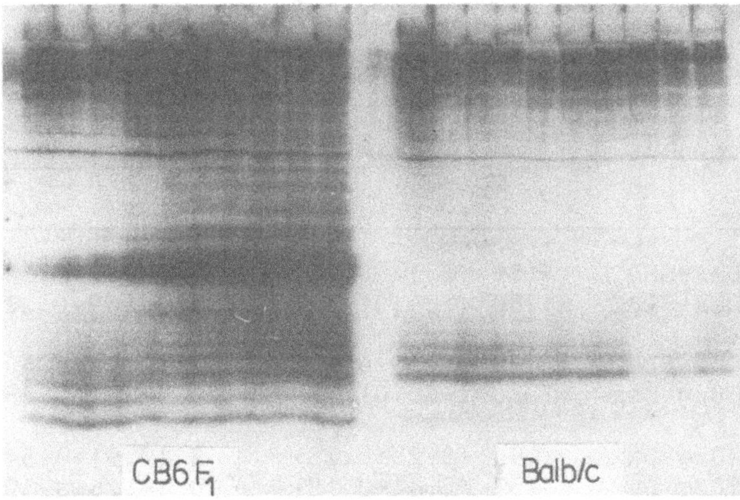

Fig. 5. Isoelectric focussing pattern of Dex 1355-specific antibodies from a $CB6F_1$ mouse which produced high titers of IgG upon repeated immunization. Sera from day 0 until 70 were tested (left). For comparison the pattern of the sera from a repeatedly injected BALB/c mouse are given (right). Details of the methods see reference (4).

Despite the fact that the allotype-linked control of the immune response to Dex 512 seems to be less stringent (1,2) than the one of BALB/c mice to Dex 1355 there are similiarities in the regulation of isotype expression of the two type-2 TI antigens. The direct involvement of Ts cells in the down regulation of IgG antibody production has sofar only been demonstrated in the BALB/c - Dex 1355 system. But the facilitation of an IgG anti Dex 512 response in C57BL/6 nu/nu mice as compared to the one in euthymic C57BL/6 mice suggests that the phenomenon observed is not restricted to one strain and one antigen yet seems to indicate a more general involvement of Ts cells in type-2 TI responses.

In this context it was of special interest to compare the immune response to the two type-2 TI antigens in F_1 hybrids of the two parental strains. In general the $CB6F_1$ hybrid mice respond to each of the two antigens in question as do animals of the parental strains. It

Table 1. The effect of preimmunization of BALB/c mice with Dex 1355 in a "thymusdependent" form upon challenge with Dex 1355 in TI form.[a]

Antigen	Anti-Dex HA-Titer[b]	
	Total	MER
1° Dex-SRBC	139 (2.14±0.19)	7 (0.82±0.59)
2° Dex-SRBC	>1280 (>3.10)	485 (2.69±0.22)
1° Dex-SRBC	99 (1.99±0.20)	3 (0.43±0.56)
2° 10μg Dex	>1280 (>3.10)	106 (2.04±0.25)
1° 10μg Dex	194 (2.29±0.23)	8 (0.91±0.54)
2° 10μg Dex	257 (2.41±0.18)	18 (1.26±0.52)

[a] Dextran coupled sheep red blood cells, single injection of $1.5x10^8$ Dex SRBC/mouse. Challenge two weeks later. Bleeding 2 weeks after antigen injection.

[b] Geometric mean of 10-19 mice within each group. Log 10 of the geometric means ± standard error given in parentheses.

can be stated that suppression of the IgG antibody response to type-2 TI antigens as observed in conventionally raised euthymic parental strains is dominant in the F_1 hybrids. One deviation from this rule is found in the response of old $CB6F_1$ animals to Dex 1355, where, always associated with an individual trait, some mice can mount a high IgG anti Dex 1355 response. This is most likely due to a previous exposure of such individuals to Dex 1355 in a hapten-like form linked to a carrier molecule and thus to a response no longer being truely type-2 TI. It is however in itself interesting that B lymphocytes primed in this fashion can be challenged by Dex 1355 lacking the carrier moiety.

The Ts mechanism regulating the isotype expression in the BALB/c anti Dex 1355 response remains operating in F_1 animals where half of the immunological repertoire of the individual comes from a different genomic background. This is an important finding because it shows that the potential for the Ts-B clonotype interactions is genetically determined and not disturbed even when the idiotype spectrum of the lymphocyte population of an individual is altered.

Acknowledgement

This work was supported by the Deutsche Forschungsgemeinschaft through SFB 104.

REFERENCES

1. B. Blomberg, W.R. Geckeler, and M. Weigert, Genetics of the antibody response to dextran in mice, Science 177:178 (1972).
2. C. Fernandez, R. Lieberman, and G. Möller, The immune response to the alpha 1-6 epitope of dextran is determined by a gene linked to the IgCH locus, Scand. J. Immunol. 10:77 (1979).
3. F. Ivars, D. Holmberg, and A. Continho, The immune response to bacterial dextrans. I. Genetic control of reponsiveness, Scand. J. Immunol. 17:419 (1983).
4. W. Schuler, G. Lehle, E. Weiler, and E. Kölsch, Immune response against the T-independent antigen α(1->3) dextran. I. Demonstration of an unexpected IgG response of athymic and germ-free-raised euthymic BALB/c mice, Europ.J. Immunol. 12:120 (1982).
5. W. Schuler, A. Schuler, and E. Kölsch, Immune response against the T-independent antigen α(1->3)dextran. II. Occurence of Bγ memory cells in the course of immunization with the native polysaccharide is T cell dependent, Europ. J. Immunol. 14:578 (1984).
6. E. Kölsch, Interaction of suppressor and helper antigenic determinants in the dominance of either tolerance or immunity, Scand. J. Immunol. 19:387 (1984).

ORGAN-SELECTIVE IMPAIRMENT OF PHAGOCYTOSIS BY CADMIUM

D.L. Vredevoe and L. Levy

UCLA School of Nursing and School of Medicine
Los Angeles, California 90024

INTRODUCTION

Cadmium is an environmental pollutant from industrial and mining sources[1]. It has an effect on delayed-type hypersensitivity and humoral immunity[2,3].

Studies presented here focus on the effects on phagocytosis to determine whether organ uptake in the liver and spleen is affected differently upon continued exposure to cadmium and later withdrawal. Previous work from this laboratory showed that chronic ingestion of cadmium chloride in drinking water by mice caused a decrease in the rate of clearance from the circulation of ^{51}Cr labelled sheep red blood cells (E) and E coated with immunoglobulin (IgG). This was shown to be the result of a decrease in localization of the E and E-IgG in the liver. The effect could be partially reversed upon removal of the cadmium from drinking water[4]. Another study showed reduced clearance of soluble aggregated IgG by mice ingesting 300 ppm of cadmium in their drinking water for at least 90 days. This appeared to have been a selective defect in that clearance of both aggregated bovine serum albumin and colloidal carbon were not affected by ingestion of cadmium.[5]

This study analyzes phagocytosis of E and E-IgG in the liver and spleen, in parallel, during and after treatment of drinking water of CBA/H mice with cadmium chloride. Data indicate that 1) there is a decreasing ability of the liver to remove E and E-IgG, during cadmium exposure; at the same time, there is an increasing ability of the spleen to remove these materials; 2) removal of cadmium from the drinking water consistently causes a partial reversal of these effects in both liver and spleen with return towards control levels.

This reversal occurs despite a continuing organ burden of cadmium which is particularly high in the liver.

METHODS

Mice

Inbred male CBA/H mice, age 2-3 months at the initiation of the experiments, were maintained in a temperature and light controlled environment, fed a standard diet of mouse pellets (Wayne Lab Blox) and given free access to water. Cadmium was administered to the experimental group of mice by dissolving 0.6 g $CdCl_2$, (Aldrich Chemical Corp., #20,829-9) per liter (300 ppm) in their drinking water available in standard drinking bottles in cages.

Preparation of Sheep Red Blood Cells (E)

E (Mission Laboratory Supply Assoc., Rosemead, California), kept in Alsever's Solution and aged for two weeks before use, were centrifuged and washed 3 times in sterile saline. They were then suspended in phosphate buffered saline (PBS, ph 7.3) to make a 50% concentration of washed E. The E suspension was then labeled by adding $Na^{51}CrO_4$ (Amersham Corp., Arlington Heights, Illinois) to final concentration of 50 uCi/ml. This mixture was incubated at room temperature for 30 minutes. After incubation, the ^{51}Cr-E were centrifuged and washed 3 times with PBS and reconstituted with PBS to make a 10% suspension of washed ^{51}Cr-E.

Equal volumes of 5% ^{51}Cr-E and 7S fraction of rabbit anti-E diluted in PBS (at the maximal subagglutinating dilution) were mixed and incubated at 37° for 30 minutes. The ^{51}Cr-E-IgG was then washed 7 times in PBS and resuspended in PBS to make a 10% suspension.

Accumulation of ^{51}Cr-E or E-IgG

A 10% ^{51}Cr-E or ^{51}Cr-E-IgG suspension (Specific Activity approximately 10^7 cpm/ml), 0.1 ml per 10 grams of body weight, was injected I.V. in the tail vein. After 15 minutes the animals were sacrificed by cervical dislocation. The spleen and liver were removed, rinsed, weighed, and the radioactivity in each organ was assessed in a Beckman gamma counter.

Measurement of Cadmium

The method of Alexander and McAnulty[6] was used for elemental analysis of the tissues using optical emission spectroscopy.

Statistics

Students t test was used to test for significant differences between the randomized control and experimental groups with $p \leq 0.05$ being considered significantly different.

RESULTS

Animals receiving cadmium chloride in drinking water for various periods of time were studied for uptake of E or E-IgG in liver and spleen (Table 1). As cadmium intake increased, uptake of both E and E-IgG decreased in liver and increased in the spleen. This uptake could be reversed (Tables 2,3).

Table 1. Percent of ^{51}Cr-E or E-IgG Uptake by Liver and Spleen of Cadmium-treated Mice Relative to Controls (no Cadmium)

Duration of Cd Exposure (Days)	Liver		Spleen	
	E	E-IgG	E	E-IgG
8	118	103	84	82
22	67	99	123	115
48	78	51	115	112
105	35	33	154	98
365	24	43	136	213
Slope K	-0.19	-0.13	+0.09	+0.33
Correl. Coef.	-0.75	-0.59	+0.49	+0.94

n =9/assay

Table 2. Percent of ^{51}Cr-E Uptake (Relative to no Cd Controls) by Liver and Spleen

Organ Assayed	Cadmium Treatment					
	Continuing Regimen			Withdrawal Regimen		
Days on Cd	115	135	362	109	116	355
Days off Cd	0	0	0	6	19	7
Liver	53	48	24	85	65	39
Spleen	89	125	140	75	104	123

Table 3. Percent of ^{51}Cr-E-IgG Uptake (Relative to no Cd Controls) by Liver and Spleen

Organ Assayed	Cadmium Treatment					
	Continuing Regimen			Withdrawal Regimen		
Days on Cd	115	135	362	109	116	355
Days off Cd	0	0	0	6	19	7
Liver	38	33	28	67	52	36
Spleen	103	119	260	93	90	247

When the organ burden of cadmium was assayed by optical emission spectroscopy, cadmium levels in liver remained high after withdrawal of cadmium in the drinking water. Levels in the spleen declined significantly ($p \leq 0.05$) after 6 weeks of withdrawal of cadmium. (Table 4).

Table 4. Cadmium Levels[a] in Organs of Mice that have been given Cadmium Chloride in Drinking Water (300 ppm) for One Year and Effects of Withdrawal of Cadmium.

Organ Assayed	One Year Tap Water (Control)		One Year Cadmium Water		One Year Cadmium 6 Weeks Tap Water	
Liver	12[b]	0	3	602[c] ± 113	8	631[c] ± 36
Spleen	12	0	3	8.7[c] ± 0.5	8	5.9[c] ± 0.4

[a]Optical emission spectroscopy. All values expressed as mean parts per million ± SEM.
[b]Number of mice.
[c]$p \leq 0.05$ relative to control

When uptake of E and E-IgG were analyzed in normal liver and spleen of CBA/H mice, the uptake was considerably greater, per mg organ weight, for spleen than for liver [liver uptake of E:85 ± 2 (SEM) cpm/mg, n = 18; spleen uptake of E:248 ± 7 (SEM) cpm/mg, n = 18; liver uptake of E-IgG: 94 ± 4 (SEM) cpm/mg, n = 18; spleen uptake of E-IgG: 215 ± 28 (SEM) cpm/mg].

DISCUSSION

Data presented here indicate that as the duration of exposure to cadmium chloride in drinking water (300 ppm) increases, the accumulation of E and E-IgG by the liver decreases significantly. In the spleen, in marked contrast, as the duration of exposure to cadmium increases, the rate of clearance of E and E-IgG increases relative to controls. These alterations in clearances are reversible in the liver despite continuing high organ burdens of cadmium.

Cadmium seems to have an organ-selective effect. The relative increased uptake of E and E-IgG seen in the spleen of mice on cadmium drinking water may be the result of the decreased ability of the liver to remove E and E-IgG resulting in prolonged circulation times of free E and E-IgG. The spleen may then function as a secondary phagocytic organ for the circulating E and E-IgG exceeding the phagocytic activity of the controls.

Optical emission spectroscopy of cadmium burdens in the spleen and liver after one year of treatment of animals with cadmium chloride water indicated that there was significantly less cadmium in the spleen than in the liver. This indicates an organ-selectivity for cadmium. This may well be related to the fact that the liver is more active in phagocytosis of some foreign materials than is the spleen. In normal animals, the spleen accumulated more E per mg of tissue than did the liver. During cadmium treatment the liver accumulated more cadmium than the spleen. This indicated the different selectivity for tissue association of substances in the spleen and liver.

The data on withdrawal of cadmium drinking water show that a high organ burden of cadmium remains constant in the liver six weeks after replacing the animals on tap water, whereas there is a significant drop in cadmium levels in the spleen. This observation may indicate a different binding site for cadmium in each organ. Despite the remaining high cadmium burden in the liver the organ begins to regain its phagocytic function after withdrawal of the cadmium. The spleen function also returns toward normal. Thus, it appears that, as the liver resumes its normal phagocytic capabilities, the spleen is able to reduce its increased phagocytosis to normal levels.

There is some indication that humoral antibody production may not be affected by cadmium as significantly as is cellular immunity[2,3]. This may be the result of this organ-selective effect on phagocytosis. However, many other factors would have to be explored as well.

ACKNOWLEDGMENTS

This research was supported by research grant ES-02410 from the National Institute of Environmental Health Sciences. We are grateful for the technical assistance of Mrs. Peggy Cohen.

REFERENCES

1. D.W. Fassett, Cadmium, *in*: "Metals in the Environment," H.A. Waldron, ed., Academic Press, London (1980).

2. L.D. Koller, Immunosuppression produced by lead, cadmium and mercury. Am. J. Vet. Res. 34:1457 (1973)
3. L.D. Koller, Review/commentary immunotoxicology of heavy metals. Int. J. Immunopharm. 2:269 (1980).
4. L. Levy, D.L. Vredevoe, G. Cook and P. Cohen. In vivo effect of chronic cadmium dosage in phagocytosis of sheep red blood cells. Proc. West. Pharmacol. Soc. 27:465 (1984).
5. D.W. Knutson, D.L. Vredevoe, K.R. Aoki, E.F. Hays and L. Levy. Cadmium and the reticuloendothelial system (RES). A specific defect in blood clearance of soluble aggregates of IgG by the liver in mice given cadmium. Immunology 40:17 (1980)
6. G.V. Alexander and L.T. McAnulty. Multi-element analysis of plant-related tissues and fluids by optical emission spectrometry. J. Plant Nutrition. 3:51 (1981).

QUANTITATION OF MURINE IgE IN AN AUTOMATIC ELISA SYSTEM

H.F.J. Savelkoul, P.W.C. Soeting, R. Benner and *J. Radl

Department of Cell Biology and Genetics, Erasmus University, Rotterdam and *Institute for Experimental Gerontology TNO, Rijswijk, The Netherlands

We describe a two-site micro ELISA system for rapid, automatic and quantitative measurement of total IgE in mouse serum, that is based upon a hybridoma IgE standard and heterologous anti-IgE antibodies. With this method we determined the absolute total IgE concentration in the serum of aging athymic nude mice and their heterozygous litter mates.

INTRODUCTION

Until recently, quantitation of murine IgE has been hampered by the lack of suitable reagents. The absence of murine IgE-producing myelomas and the very low concentration of IgE in mouse serum made it extremely difficult to prepare anti-IgE antibodies (1,2). With the recent development of murine antigen-specific IgE-secreting hybridomas it became feasible to purify IgE-antibodies and to generate heterologous anti-IgE antibodies (3-7). With these reagents, detection of IgE-antibodies is possible in sensitive radio (RIA) or enzyme immunoassays (ELISA). Quantitation in these assays is based upon a standard curve of highly purified hybridoma IgE (6-10).

Recently, our laboratory introduced a highly sensitive ELISA in an automatically scanning microtest system (11). We adapted this system for quantitating total murine IgE utilizing affinity a pool of four different affinity purified hybridoma IgE as an absolute standard and heterologous antibodies raised against the hybridoma product. With this ELISA we determined absolute IgE levels in the sera of athymic nude mice and their heterozygous littermates at 6, 9 and 12 months of age.

MATERIALS AND METHODS

Mice

Female BALB/c mice, at least 10 weeks of age, were obtained from Bomholtgard, Ry, Denmark. Female and male (B10 x B10.D2)F1 mice of various ages were purchased from OLAC Ltd., Bicester, United Kingdom. Female BALB/c nude mice (nu/nu) and their heterozygous littermates (nu/+) were bred and maintained at the Radiobiological Institute TNO, Rijswijk, The Netherlands. At the beginning of the experiments, all nu/nu and nu/+ mice were specific pathogen free and barrier maintained. These mice were transferred to conventional facilities as described previously (12). Blood samples of about 200 µl were serially taken at various ages and stored at -20°C for later use.

Murine hybridoma IgE

As a source of murine IgE we used ascitic fluids of four TNP-specific mouse IgE secreting hybridomas of the a and b allotypes (IGEL a4 6.3, IGEL a4 6.5, IGEL b4 1, IGEL b4 14.2). These cell lines (5) were generous gifts of Dr. M. Wabl. Ascitic fluids were obtained by tapping pristane-primed (Janssen Chimica, Beerse, Belgium) mice that had been intraperitoneally injected with 10^6 cultured hybridoma cells 10 days earlier. The hybridomas of the a allotype were injected into BALB/c mice whereas the hybridomas of the b allotype were injected into (B10 x B10.D2)F1 mice. Supernatant of an OVA-specific IgE secreting hybridoma (3) was obtained from Sera Lab (Sussex, United Kingdom), and ascitic fluid of a DNP-specific hybridoma (4) was a gift from Dr. Z. Eshhar.

Heterologous antisera against murine IgE

The anti-IgE antisera were raised in rabbits (RAM/IgE) and goats (GAM/IgE) by repeated injections of mixtures of purified IgE from all four above mentioned hybridomas from Dr. Wabl and the hybridoma of Dr. Eshhar. Details about the isolation and purification of the absolute IgE standard and the heterologous anti-IgE antisera are described elsewhere (13).

ELISA for murine IgE

The assay was performed essentially as described elsewhere (13). Briefly, standard 60-well Terasaki plates (Greiner type 653180, Nurtingen, F.R.G.) were used with wells filled with 5µl during all incubation steps. Plates were coated with the GAM/IgE antiserum (0.5 µg per well) in phosphate buffered saline (PBS) by overnight incubation at 4°C. Nonspecific binding was prevented by incubation with PBS containing 0.02% gelatin (UCB Biochemicals, Bruxelles, Belgium) (PBS-gel). The sample containing murine IgE was allowed to react for 1 hour at room temperature (RT). Plates were then extensively washed with PBS-gel containing 0.05% Tween-20 (Merck, Darmstadt, F.R.G.) (PBS-Tw-gel) to remove unbound material. The IgE bound was sandwiched (1 hour, RT) by addition of the non-crossreacting specific RAM/IgE antiserum (1.6 μg per

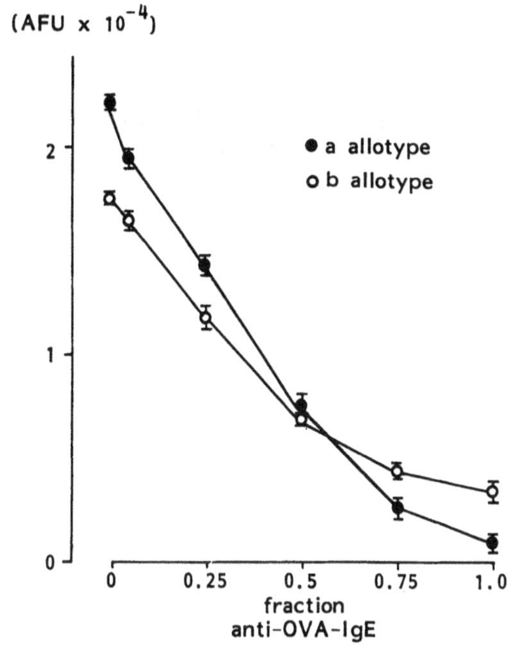

Fig. 1. Specificity of the RAM/IgE antiserum for murine IgE as determined in ELISA. Wells were coated with 1 μg TNP-specific IgE from the IGEL a4. 6.5 hybridoma of the a allotype (closed circles) or of the IGEL b4 1 hybridoma of the b allotype (open circles). Subsequently, the wells were incubated with mixtures of a constant concentration (1.6 μg/well) of the RAM/IgE antiserum and increasing amounts (0-10 μg/well) of OVA-specific IgE. The fluorescence intensity of the wells is expressed in arbitrary fluorescence units (AFU). Each experimental point represents the arithmetic mean ± 1 SEM of 6 wells.

well). Subsequently, plates were incubated (1 hour, RT) with a conjugate of protein-A linked to β-galactosidase (Amersham International, Amersham, United Kingdom) and after extensive washing with PBS-Tw-gel, the substrate 4-methylumbelliferyl-β-D-galactoside (MUF-G) (Sigma, St. Louis, U.S.A.) was applicated. After 1 hour at 37°C, the reaction was stopped by adding 5 μl per well of 0.5 M carbonate buffer (pH 10.3). Detection was based upon the fluorescence intensity of the reaction product 4-methylumbelliferone (MUF), which was measured by an automatically scanning inverted microfluorometer (Leitz MPV1, Wetzlar, F.R.G.) equipped with a scanning stage, automatic shutter control and a photomultiplier connected to a digital voltmeter and a desk-top calculator (11,14). Plates were analysed within 120 s and results are expressed as arbitrary fluorescence units (AFU).

RESULTS

The GAM/IgE and RAM/IgE antisera used in our ELISA system were found to be specific for mouse IgE and contained no anti-idiotypic or anti-allotypic activity as determined by double radial

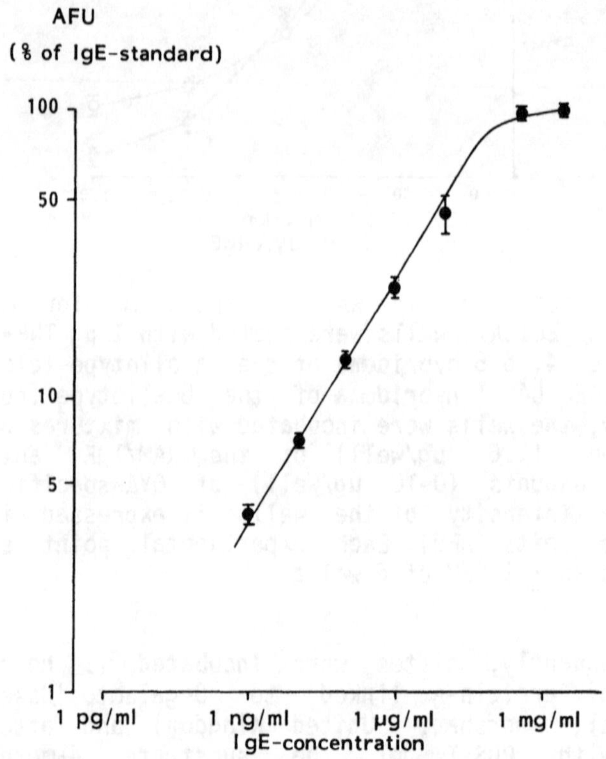

Fig. 2. Standard curve of murine total IgE as determined by ELISA. Fluorescence (AFU) of the sample corrected for negative control is expressed as percentage of the absolute IgE standard. Each experimental point represents the arithmetic mean ± 1 SEM of 6 wells.

immunodiffusion in PEG 6000 and immunoelectrophoresis (data not shown). The same holds when tested in the ELISA system itself, since on plates coated with TNP-specific IgE from one of the four IGEL hybridomas, the binding of the RAM/IgE antiserum could be completely inhibited by addition of OVA-specific IgE (Fig. 1). This is true for hybridoma IgE of both allotypes used for raising the RAM/IgE antiserum. Essentially the same results were found in similar experiments involving the GAM/IgE antiserum (data not shown). Thus, both antisera did not contain anti-idiotypic or anti-allotypic activity as far as could be detected in the ELISA system they were produced for.

In order to be able to use the ELISA system for quantitative measurement of IgE, a standard curve of a pool of the four highly purified hybridoma IgEs was prepared (Fig. 2). By calculating the percentage of the sample AFU of the positive control (absolute IgE standard) and by using this standard curve, the absolute IgE concentration in the sample can be found. This ELISA system has a detection limit of 1 ng/ml and is most sensitive for IgE concentrations ranging from 1 ng/ml to 100 µg/ml, which is sufficient to determine the low levels of total IgE in mouse serum.

We applied the above ELISA system for measurement of the IgE levels in the sera of BALB/c nude (nu/nu) mice and their heterozygous (nu/+) littermates at various ages. It was found that 6-months-old nu/nu and nu/+ mice have 9.24 and 5.76 µg/ml IgE in their serum, respectively (Table I). After 6 months of age there was a gradual increase in the serum IgE level, which was greater in nu/+ than in nu/nu mice. In the serum of 12-months-old mice, 9.79 (nu/nu) and 7.22 (nu/+) µg/ml IgE were found.

The IgG1 levels, determined in a similar type ELISA, did not differ between nu/nu and nu/+ mice at the different ages studied. The total Ig level, on the other hand, showed an age-related increase. At 12 months of age, the Ig level of nu/+ mice was significantly higher ($p < 0.01$) than in nu/nu mice (Table I).

DISCUSSION

So far, all ELISA's for detecting murine IgE are performed in microtiterplates using 100 to 200 µl samples and making use of chromogenic substrates (8-10). In the present paper we present a method which is able to detect IgE in only 5 µl samples using a fluorogenic substrate, giving highly reproducible results with the same sensitivity and efficiency as the methods mentioned above. By making use of a two-site (sandwich) ELISA system in which the antigen IgE is characterized by two geometrically and molecular distinct binding sites for the two specific antibodies employed, a much higher degree of specificity to the overall system is rendered (16).

Table I. Immunoglobulin concentration in the serum of BALB/c nude mice (nu/nu) and their heterozygous (nu/+) littermates as determined by ELISA.

age	IgE (µg/ml)		IgG1 (mg/ml)		Total Ig (mg/ml)	
(months)	nu/nu	nu/+	nu/nu	nu/+	nu/nu	nu/+
6	9.24 ± 0.11	5.76 ± 0.44	3.21 ± 1.35	2.50 ± 0.20	6.45 ± 0.19	7.67 ± 1.30
9	9.44 ± 0.14	6.38 ± 0.25	2.63 ± 0.71	2.13 ± 0.36	6.98 ± 1.22	6.09 ± 0.13
12	9.79 ± 0.18	7.22 ± 0.38	3.84 ± 0.66	2.67 ± 0.20	8.21 ± 0.18	11.49 ± 0.89

IgE concentrations were determined in the IgE ELISA. IgG1 and total Ig concentrations were determined in a similar type ELISA using specific antisera for IgG1 and total Ig, respectively. The absolute standards for IgG and total Ig were the same as previously described (15). Figures represent the arithmetic mean ± 1 SEM of the concentrations of serum levels determined in 12-15 animals per group.

For measurement of extremely small quantities in ELISA, conjugates of the enzyme ß-galactosidase and the use of a fluorogenic substrate are superior compared to other enzymes and substrates (17). This is caused by their superb dose-response relationship, while the reproducibility and the sensitivity is equal to that achieved by RIA. Methods based upon fluorescence intensity determination are particularly favourable to adapt for rapid and automatic measuring procedures.

The specificity (Fig. 1) and the sensitivity (Fig. 2) of the assay system developed are sufficient to allow its use for quantitation of total IgE in mouse serum (Table I). The absolute IgE levels in the serum of BALB/c nu/nu and nu/+ mice as determined by our ELISA system (Table I) are in agreement with the IgE concentrations measured by others in mice (8,18) and other species like man (19) and rat (20).

It is well known that the IgE synthesis is highly regulated by antigen-specific suppressor and helper T cells and the factors they produce. Moreover, antigen non-specific T cells and factors may also be involved (21). IgE expression during clonal outgrowth of B cell clones from Peyer's patches cultured in vitro with cholera toxin-primed T cells, showed that a hapten-linked regulatory mechanism is acting on the IgE secretion (22). The increased IgE levels in nude mice are probably caused by the absence of functional suppressor T cells and their lymphokines and the occurrence of remnant precursor or helper T cells and their lymphokines (18).

The total Ig and IgG1 concentrations as determined in this ELISA system are 1000 times higher than the IgE levels and contrast to the IgE levels, is greatly increased. The same result was found by other groups using single radial immunodiffusion or solid phase RIA (18,23). Since most of the nude mice were infected with hepatitis virus, this infection may in part account for the heterogeneity in the individual serum IgG1 levels (15). The mean serum IgG1 levels, on the other hand, seem to depend on the age and the genetic background of the nude mice studied (15,18).

ACKNOWLEDGEMENTS

We gratefully acknowledge Mr. P.L. van Soest and Mr. J.E. de Josselin de Jong for technical advise and assistance with regard to the development of the ELISA for quantitative measurements, and Ms. M. Stout for typing the manuscript. This investigation was supported by the Netherlands Foundation for Medical Research (FUNGO) and the Netherlands Asthma Foundation.

REFERENCES

1. G.M. Lang, W.Y. Lee, B.G. Carter, and A.H. Sehon, Production of goat, rat, and guinea pig antisera to murine IgE, J. Immunol. 116:881 (1976).
2. S.B. Lehrer, Isolation and immunochemical properties of mouse IgE, Immunochemistry 13:837 (1976).
3. I. Bottcher, G. Hammerling and J.-F. Kapp, Continuous production of monoclonal mouse IgE antibodies with known specificities by a hybrid cell line, Nature 275:761 (1978).
4. Z. Eshhar, M. Ofarim, and T. Waks, Generation of hybridomas secreting murine reaginic antibodies of anti-DNP specificity, J. Immunol. 124:775 (1980).
5. A.K. Rudolph, P.D. Burrows, and M.R. Wabl, Thirteen hybridomas secreting hapten-specific immunoglobulin E from mice with Ig^a or Ig^b heavy chain haplotype, Eur. J. Immunol. 11:527 (1981).
6. A. Bohn and W. Konig, Generation of monoclonal DNP-specific IgM and IgE murine antibodies on the efficacy of hybridization, Molecular Immunol. 19:193 (1982).
7. F.-T. Liu, J.W. Bohn, E.L. Ferry, H. Yamamoto, C.A. Molinaro, L.A. Sherman, N.R. Klinman, and D.H. Katz, Monoclonal dinitrophenyl-specific murine IgE : antibody-preparation, isolation and characterization, J. Immunol. 124:2728 (1980).
8. B.E. Bozelka, M.L. McCants, J.E. Salvaggio, and S.B. Lehrer, IgE isotype suppression in anti- -treated mice, Immunology 48:527 (1982).
9. A. Giallongo, L. Kochoumian, and T.P. King, Enzyme and radioimmunoassays for specific murine IgE and IgG with different solid-phase immunosorbents, J. Immunol. Methods 52:379 (1982).
10. P.N. Hill and F.-T. Liu, A sensitive enzyme-linked immunosorbent assay for the quantification of antigen-specific murine immunoglobulin E, J. Immunol. Meth. 45: 51 (1981).
11. P.L. van Soest, J.E. de Josselin de Jong, P.M. Lansdorp, and W. van Ewijk, An automatic fluorescence micro-ELISA system for quantitative screening of hybridoma supernatants using a protein-A-ß-galactosidase conjugate, Histochem. J. 16: 21 (1984).
12. J. Radl, J.G. Mink, P. van den Berg, M.J. van Zwieten, and R. Benner, Increased frequency of homogeneous immunoglobulins in the sera of nude athymic mice with age, Clin. Immunol. Immunopath. 17:469 (1980).
13. H.F.J. Savelkoul, P.W.C. Soeting, P.L. van Soest, J.E. de Josselin de Jong, R. Benner, and J. Radl, Quantitative determination of antigen-specific and total IgE in mice using a highly sensitive ELISA, Submitted for publication.
14. J.E. de Josselin de Jong, J.F. Jongkind, and H.R. Ywema, A

scanning inverted microfluormeter with electronic shutter control for automatic measurements in micro-test plates, Analyt. Biochem. 102:120 (1980).

15. J.G. Mink, J. Radl, P. van den Berg, J.J. Haaijman, M.J. van Zwieten, and R. Benner, Serum immunoglobulins in nude mice and their heterozygous littermates during aging, Immunology 40:539 (1980).
16. R. Ekins, Merits and disadvantages of different labels and methods of immunoassay, in: Immunoassays for the 80s, A. Voller, A. Bartlett and D. Bidwell eds., MTP Press Ltd, Falcon House, United Kingdom 5 (1981).
17. S. Avrameas, Heterogeneous enzyme immunoassays, ibid., 85.
18. K. Ito, T. Ogita, M. Suko, M. Mori, K. Kudo, T. Hayakawa, H. Okudaira, and Y. Horiuchi, IgE levels in nude mice, Int. Archs. Allergy appl. Immunol. 58:474 (1979).
19. R.H. Buckley and S.A. Fiscus, Serum IgD and IgE concentrations in immunodeficiency diseases, J. Clin. Invest. 55:157 (1975).
20. T. Ishizaka, J.F. Urban, and K. Ishizaka, IgE formation in the rat following infection with Nippostrongylus brasiliensis. I. Proliferation and differentiation of IgE bearing cells, Cell. Immunol. 22:248 (1976).
21. T. Kishimoto, IgE class-specific suppressor T cells and regulation of the IgE response, Prog. Allergy, 32:265 (1982).
22. J.J. Cebra, E.R. Cebra, E.R. Clough, J.A. Fuhrman, and P.A. Schweitzer, Relationships and regulation of IgG, IgA and IgE expression during clonal outgrowth of primed B cells, in: Regulation of the immune response. 8th Int. Convoc. Immunol. Karger, Basel, 107 (1983).
23. G.C. Bosma, J. Owen, G. Eaton, G. Marshall, C. Dewitt, and M.J. Bosma, Concentration of IgG1 and IgG2a allotypes in serum of nude and normal allotype-congenic mice, J. Immunol. 124:879 (1980).

DOUBLE IMMUNOENZYMATIC STAINING EMPLOYING RAT AND MOUSE MONOCLONAL ANTIBODIES

S. Poppema, A. Boes, L. de Ley, T. Rozeboom, W. Timens
E. Schwander, and L. Visser

Departments of Pathology and Immunology
University of Groningen, Oostersingel 63
Groningen, 9713 EZ, The Netherlands

INTRODUCTION

Double immunoenzymatic labelling has been described by Nakane, employing two different substrates for peroxidase staining[1] and by Mason, employing alkaline phosphatase labelled antibodies[2].
In the procedure described by Nakane, no mixed staining of cells could be obtained when diaminobenzidin was used as a primary chromogen[3,4]. The method described by Mason, allows mixed staining of cells containing two different antigens.
The availability of monoclonal antibodies directed against subsets of cells, especially lymphocytes, has provided the opportunity to analyze the localization of such subsets in tissue sections[5].
Double labelling of cells in suspensions is usually performed with immunofluorescence techniques, since each of the fluorescent labels can be examined alone, without interference of the other[6].
Double immunoenzymatic labelling allows for better tissue localization and also for simultaneous visualization of the two antigens.
Immunoenzymatic techniques usually demand the use of a two or more step procedure to obtain sufficient sensitivity. Therefore, monoclonal mouse antibodies, when of the same immunoglobulin subclasses, are not usefull in such procedures. In addition, when monoclonal antibodies of different subclasses are available, the second step antibodies, directed against mouse immunoglobulin subclasses are generally neither very specific nor sensitive. Therefore, we have chosen for an approach employing monoclonal antibodies derived from rats in combination with the already available mouse monoclonal antibodies.

Here we will describe the production of a rat anti-human IgM monoclonal antibody and a double immunoenzymatic staining procedure employing this rat anti-IgM antibody in combination with mouse anti-T cell antibodies and mouse anti-IgD antibodies respectively.

MATERIALS AND METHODS

Preparation of tissue

Tissue samples of lymph nodes and tonsils were snap frozen in phreon and stored at -80°C. Cryostat sections were prepared on a Bright microtome, air dried with a hair dryer during 20 minutes and fixed in acetone at room temperature during 10 minutes.

Antibodies

Mouse monoclonal antibodies directed against T lymphocytes were a gift by Dr. Goldstein, Ortho Pharmaceutical Company (OKT11) and by Dr. Matthijs van de Rijn, Sidney Farber Institute, Boston (66II G5). OKT11 is directed against an antigen with a molecular weight of 50 kD, associated with the sheep red blood cell receptor on T lymphocytes. 66II G5 is directed against a 40 kD antigen present on the majority of immature and mature T lymphocytes, except for a small population of mature T lymphocytes and is equivalent to the 3A1 antibody of Haynes. An admixture of OKT11 and 66II G5 was used to obtain a maximum specific staining of T lymphocytes. Mouse monoclonal antibodies directed against IgD and IgM were a gift by Dr. Lee Nadler, Sidney Farber Institute, Boston and by Dr. Joost Haayman, RIV.

Preparation of rat anti-human IgM monoclonal antibody

A Wistar rat was given two intraperitoneal injections, 14 days apart, of monoclonal IgM, emulsified in Freund's complete adjuvant in the first and Freund's complete adjuvant in the second injection and obtained from a patient with Waldenström's macroglobulinaemia. This antigen was kindly provided by Dr. Theo Ockhuizen, department of Internal Medicine, University of Groningen. Two weeks after the second immunization, the serum of this rat was tested by immunodiffusion. This was followed by a final intravenous immunization with IgM in phosphate buffered saline. Four days there after the rat was sacrified and the spleen removed aseptically and a spleen suspension was prepared. Fusion of the spleen cells with myeloma was done according to standard techniques, using polyethyleneglycol (PEG) as a fusion reagent. The fused cells were suspended in HAT medium and distributed in six 96-well tissue culture plates. After incubation for two weeks almost all wells contained growing clones and 200 well were tested for anti-IgM activity by incubating frozen sections of human tonsil with the supernatants for 30 minutes followed by a wash in phosphate buffered saline and a further incubation with peroxidase conjugated rabbit anti-rat Ig antiserum. After another wash the sections were incubated with aminoethylcarbazole and H_2O_2. Four strongly positive cultures were recloned

Table I

PRIMARY ANTIBODY	SECUNDARY ANTIBODY	ENZYME STAINING
OKT11 + 66II G5 (mouse) (2 μg/ml)	goat anti-mouse Ig *A.Ph. (Sigma, diluted 1/20) + 1% human serum	naphthol AS-MX - Fast Blue BB
anti-IgD (mouse) (2 μg/ml)	goat anti-mouse Ig *A.Ph. (Sigma, diluted 1/20) + 1% human serum	naphthol AS-MX + Fast Blue BB
anti-IgM (rat) (2 μg/ml)	rabbit anti-rat Ig *P.O. (Dakopatts, diluted 1/20) + 1% human serum	aminoethylcarbazole + H_2O_2 (7 minutes)
OKT11 + 66II G5 (mouse) and anti-IgM (rat)	goat anti-mouse Ig *A.Ph. diluted 1/10 and rabbit anti-rat Ig *P.O. diluted 1/10 + 1% human serum	firstly: aminoethyl-carbazole + H_2O_2 secondly: naphthol AS-MX + Fast Blue BB
anti-IgD (mouse) and anti-IgM (rat)	goat anti-mouse Ig *A.Ph. diluted 1/10 and rabbit anti-rat Ig *P.O. diluted 1/10 + 1% human serum	firstly: aminoethyl-carbazole + H_2O_2 secondly: naphthol AS-MX + Fast Blue BB

and retested by the same procedures. Three cultures appeared to be unstable, but one showed continuous specific antibody production. This culture was transferred to tissue culture flasks and larger volumes of supernatant were prepared. This supernatant showed strong reactivity with IgM in tissue sections in a 1:20 dilution and was also reactive with human IgM in immunodiffusion.

Immunohistological staining procedures

Acetone fixed slides were transferred to phosphate buffered saline (PBS) and incubated with the monoclonal mouse and/or rat antibodies during 30 minutes. After a 5 minute wash in PBS, the sections were incubated with alkaline phosphatase conjugated goat anti-mouse Ig antibodies and/or peroxidase conjugated rabbit anti-rat Ig antibodies, supplemented with 1% human serum, during 15 minutes. The enzymehistochemical stainings were performed with the aminoethylcarbazole method as described by Karnovsky[7] for peroxidase activity during 7 minutes at pH 4.9 with Naphthol AS-MX Fast Blue BB as described by Mason[2] for alkaline phosphatase activity during 15 minutes. The conditions used for the double immunoenzymatic staining procedure and the controls are summarized in table I.

Inhibition of endogenous enzyme activities

Endogenous peroxidase is not a significant problem when aminoethylcarbazole is used as a substrate, since in contrast to diaminoethylcarbazole appears to be much more sensitive for horse radish peroxidase than for endogenous peroxidase in the conditions used (pH 4.9, staining duration less than 10 minutes, acetone fixation). Therefore, no or only weakly brownish endogenous peroxidase staining is observed, clearly contrasting with the bright reddish brown staining for horse radish peroxidase.
Endogenous alkaline phosphatase can be inhibited completely by the addition of 1 mM levamisole to the substrate solution.

RESULTS

A) Single staining

Staining with OKT11 + 66II G5 resulted in intense specific staining of T lymphocytes in interfollicular areas and also in secundary follicles, mainly in the light zones of germinal centers. Staining with anti-IgD resulted in strong specific staining of small lymphocytes in primary follicles and in the mantle zones of secundary follicles. The interfollicular areas contained only scattered positive cells, but in tonsils also several positive staining plasma cells with intracytoplasmic IgD were observed. Germinal centers of secundary follicles contained no IgD positive cells. Staining with the rat anti-IgM monoclonal antibodies resulted in strong staining of small lymphocytes in primary follicles and the mantle zones of secundary follicles. The interfollicular areas contained variable amounts of IgM positive cells, partly clearly with intracytoplasmic staining, especially in tonsils. In contrast to staining with anti-IgD, anti-IgM also resulted in intense staining of germinal centers of secundary follicles. The staining was similar to that obtained with polyclonal rabbit or goat anti-human IgM antibodies (Dakopatts, Kallestad) and also similar to staining with monoclonal mouse anti-human IgM antibodies. Most of this staining appeared to be confined to the surface of dendritic reticulum cells in germinal centers, but also a variable number of positive lymphoid cells could be observed.

B) Double staining

Double staining with OKT11 + 66II G5 versus anti-IgM resulted in two non overlapping cell populations with a clear distribution pattern (fig. 1A,B). The brownish-red staining for IgM was found in the follicles, similar to the results with single staining, and bluish staining T lymphocytes were mainly found in interfollicular areas, but could also clearly be identified inside follicles, in between the mesh work of dendritic reticulum cell staining of germinal centers.

Double staining with anti-IgD versus anti-IgM resulted in two largely overlapping cell populations (fig. 2A,B). The small lymphocytes of mantle zones and primary follicles nearly all exhibited mixed staining, and only a minority of red staining IgM only cells was found. These cells were predominantly localized in the interfollicular areas. Cells with intracytoplasmic Ig had no mixed staining pattern; IgM positive reddish staining, as well as IgD positive bluish staining plasma cells were observed in the tonsils. The dendritic mesh work in germinal centers only stained reddish brown.

Controls, designed to detect unwanted reactivity of the second step antibodies with human tissue antigens, with the other primary antibodies, or with each other, all resulted in absence of staining.

DISCUSSION

The results reported above indicate that double immunoenzymatic staining on tissue sections can be performed employing mouse and rat monoclonal antibodies as primary reagens. Since a broad range of mouse monoclonal antibodies is available, we decided to produce rat monoclonal antibodies. Therefore, rats were immunized with human IgM, and the rat splenocytes were fused with the mouse myeloma X63. This resulted in a stable rat-mouse hybridoma producing monoclonal rat anti-human IgM antibodies with strong reactivity on tissue sections. Culture supernatant could be used at a 1:20 dilution.

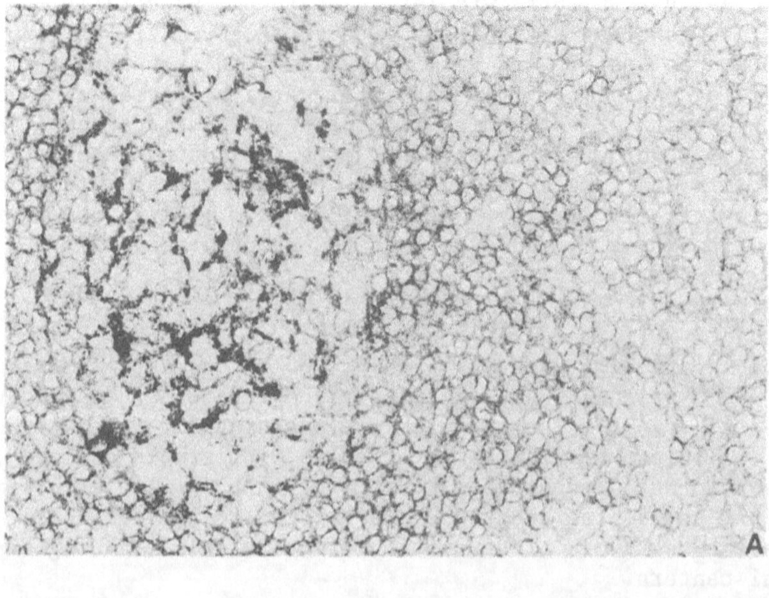

fig. 1A,B. Double immunoenzymatic staining with anti IgM and anti T11. Attempt to demonstrate two colour staining in black and white with filters. Fig. 1A: Anti IgM (red) with blue filter to diminish T cell staining (blue). Fig. 1B: Anti T cell (blue) with orange filter to diminish IgM staining (red). Note the presence of T cells outside the follicle (right), but also in the germinal center (left).

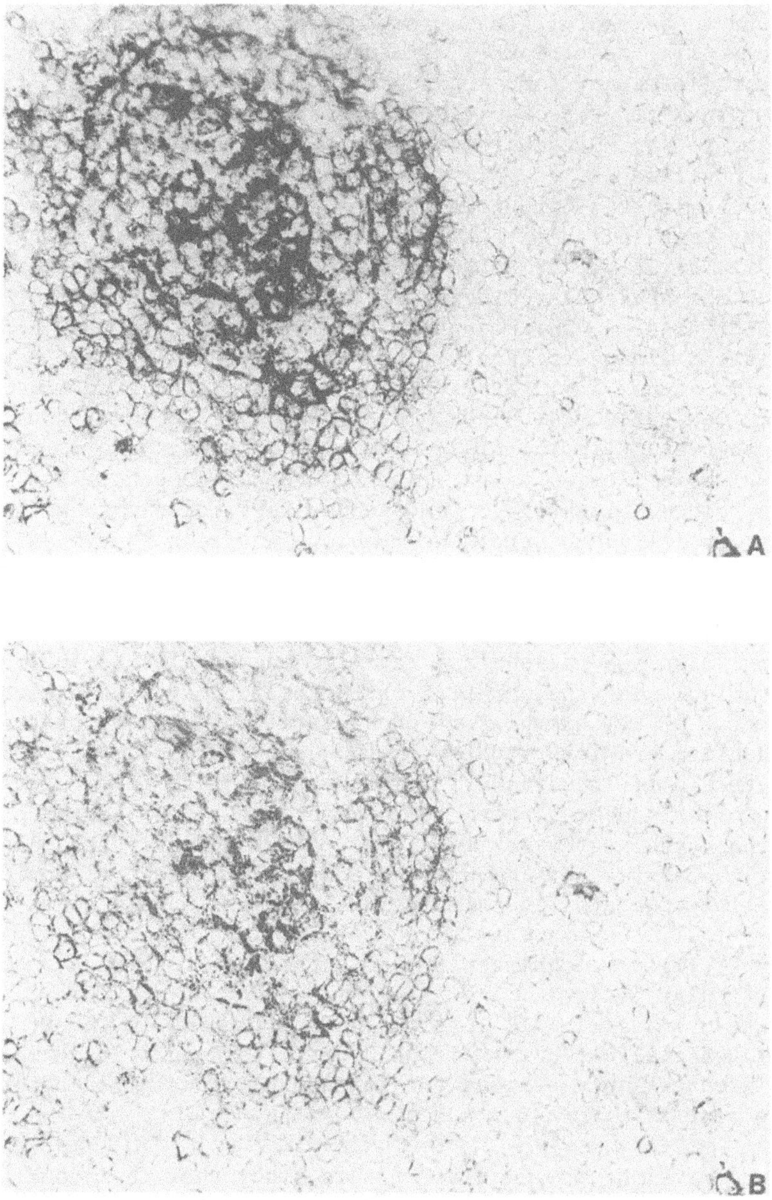

fig. 2A,B. Attempt to demonstrate double staining for IgM and IgD in black and white. Fig. 2A: Strong staining for IgM in mantle zone and germinal center (red) with blue filter. Fig. 2B: Strong staining for IgD in mantle zone (blue) with orange filter. The reactivity seen in the germinal center results from the red immunostaining for IgM, which could not entirely be obscured by filtering.

As second step reagents peroxidase conjugated rabbit anti-rat Ig and alkaline phosphatase conjugated goat anti-mouse antibodies were found to be useful. These reagents could be stained enzyme-histochemically, resulting in a reddish reaction product (amino-ethylcarbazole method for peroxidase) and a bluish product (fast Blue BB for alkaline phosphatase). When both antigens, and therefore both reagents were present on the same cells, a mixed purple staining could be observed.

The use of aminoethylcarbazole for peroxidase staining has some advantages which have not generally been appreciated in the literature. Firstly, the aminoethylcarbazole reaction product does allow for the subsequent binding of other reagents, such in contrast to diaminobenzidine, which appears to block subsequent antibody binding and staining. Secondly, when the aminoethylcarbazole method is performed at pH 4.9, this staining appears to be much more sensitive for horse radish peroxidase than for the endogenous peroxidase present in the tissue. Such in contrast to diaminobenzidine which is leading to very strong endogenous peroxidase staining. Since "blocking" of endogenous peroxidase with methanol and H_2O_2 generally denatures the membrane antigens in frozen tissue sections, such endogenous peroxidase activity can be a serious problem in situations where many eosinophils are present like in Hodgkin's disease.

Endogenous alkaline phosphatase activity can be blocked sufficiently with levamisol when the duration of the staining is not exceeding 30 minutes. The two examples of double staining chosen for this report confirm previous studies on the distribution of IgM positive lymphocytes in follicles and T lymphocytes in interfollicular areas, and also clearly demonstrates the presence of a significant number of T lymphocytes in the germinal centers of secondary follicles. The double staining for IgM and IgD confirms the general notion that small lymphocytes in primary follicles and mantle zones of secondary follicles express IgM as well as IgD. Cells staining for IgM only are mainly localized on the outside of follicles and in interfollicular areas.

Cells with cytoplasmic IgD were a frequent finding in interfollicular areas of tonsils, but not in lymph nodes. The dendritic network in germinal centers only stained for IgM and not for IgD in accordance with the results of other studies[8,9].

In conclusion, the results of this study show that double immunoenzymatic staining employing mouse and rat monoclonal antibodies is feasible with relatively simple procedures. This fact should be a stimulus to prepare rat monoclonal antibodies in addition to the already available mouse monoclonal antibodies, since the simultaneous demonstration of two antigens may provide a better insight in the phenotype of cells and also in the possible interrelations of different cell types.

REFERENCES

1. P. K. Nakane, Simultaneous localization of multiple tissue antigens using the peroxidase-labeled antibody method: A study on pituitary glands of the rat, J. Histochem. Cytochem. 16:557 (1968).
2. D. Y. Mason and R. Sammons, Alkaline phosphatase and peroxidase for double immunoenzymatic labelling of cellular constituents, J. Clin. Pathol. 31:454 (1978).
3. S. Poppema, J. D. Elema and M. R. Halie, The significance of intracytoplasmic proteins in Reed-Sternberg cells, Cancer 42:1793 (1978).
4. L. A. Sternberger and S. A. Joseph, The unlabeled antibody method. Contrasting color staining of paired pituitary hormones without antibody removal, J. Histochem. Cytochem. 27:1424 (1979).
5. S. Poppema, A. K. Bhan, E. L. Reinherz, R. T. McCluskey and S. F. Schlossman, Distribution of T cell subsets in human lymph nodes, J. Exp. Med. 153:30 (1981).
6. K. B. Pryzwanski, Applications of double label immunofluorescence, in: Techniques in Immunocytochemistry, G. R. Bullock and P. Petrusz, ed., Academic Press, London, (1982).
7. R. C. Graham, U. Lundholm and M. J. Karnovsky, Cytochemical demonstration of peroxidase activity with 3-amino-9-ethylcarbazole, J. Histochem. Cytochem. 13:150 (1965).
8. H. Stein, A. Bonk, G. Tolksdorf, K. Lennert, H. Rodt and J. Gerdes, Immunohistologic analysis of the organization of normal lymphoid tissue and non Hodgkin's lymphomas, J. Histochem. Cytochem. 28:746 (1980).
9. N. L. Harris and R. E. Data, The distribution of neoplastic and normal B-lymphoid cells in nodular lymphomas, Hum. Pathol. 13:610 (1982).

SESSION 6

ANTIBODIES AS THERAPEUTIC TOOLS

CHAIRPERSONS: G.F. Rowland
H. Waldmann

MONOCLONAL ANTIBODIES AS THERAPEUTIC TOOLS IN VIVO

G.F. Rowland

Lilly Research Centre Limited
Eli Lilly & Company
Erl Wood Manor
Windlesham, Surrey GU20 6PH

INTRODUCTION

Antibodies, obtained from human or animal sera have played a small role in therapy for many years. They are used to neutralise venoms, some bacterial toxins and viral antigens. More recently antilymphocyte globulin has been used as an adjunct to the treatment of allograft rejection. Attempts to use antibodies more widely *in vivo* have been limited by lack of specificity, by inhomogeneity and variability of preparations and by adverse side-effects, especially associated with the early widespread use of horse serum as a source of antibodies.

In recent years the development of hybridoma technology has provided scientists and clinicians with monoclonal antibodies that may avoid the limitations set by polyclonal antisera. This has given a major impetus to the re-examination of the therapeutic potential of antibodies; particularly in the field of cancer. Studies are now being undertaken to evaluate the potential of monoclonal antibodies as therapeutic tools in their own right and also as target site specific delivery agents.

MONOCLONAL ANTIBODIES AS INHERENT THERAPEUTIC AGENTS

Antibody molecules may act therapeutically by simple interaction with antigen or by a combination of antigen binding and a functional interaction of the Fc portion with cells or complement. In situations where antibodies are used to neutralize circulating soluble antigens, simple interaction is probably sufficient and the choice of immunoglobulin species used, of little importance. In the treatment of disease, however, antibody action generally involves the Fc portion of the molecule and here the choice of species and isotype may be crucial. The hybridoma technique allows these factors to be examined with greater precision since monoclonal antibodies are homogeneous with respect to immunochemical composition.

Treatment of kidney allograft rejection

Anti-lymphocyte and anti-thymocyte globulin derived from polyclonal antisera has found a place in transplantation not, as originally hoped, to prevent rejection but as part of the treatment to reverse rejection episodes (Streem et al., 1983). These antibodies however are not without side-effects. Chills, fevers and thrombocytopenia are sometimes seen, possibly associated with the generalised lymphocyte destruction obtained. A more specific approach has now been described using a murine IgM monoclonal antibody designated CBL 1 (Takahashi et al., 1983). This antibody recognises only the T-lymphoblast sub-population of human lymphocytes and when administered at 5 mg daily for 9 successive days was able to reverse kidney allograft rejection in 17 out of 19 patients without side-effects. The lack of effect on peripheral blood lymphocyte counts supports the view that the antibody destroys only a small population of active lymphoblasts.

In this study, no evidence was provided for the mechanisms of action. Since the antibody was of the IgM isotype, the possibility exists that human complement was involved. In general, murine IgG monoclonals, even those of the appropriate isotype are not able to utilise human complement. By contrast, rat monoclonal antibodies, particularly of the IgG2b isotype, are capable of utilising human complement to produce lysis of target cells (Clark et al., 1983). The phenomenon of synergistic lysis, in which more than one rat monoclonal antibody, each recognizing a different epitope on the same molecule is required to produce complement mediated cytolysis, has also been found to require IgG2b as one of the antibody isotype partners (Hugh-Jones et al., 1983). Appropriate rat monoclonal antibodies recognising T-cell antigens may be useful for treatment of allograft rejection in man.

Cancer therapy

The significance of immunoglobulin isotype is also observed in the use of monoclonal antibodies for cancer therapy. Studies using mouse antibodies suggest that IgG2a monoclonals (such as the anti-colorectal carcinoma antibody 17-(A), Herlyn and Koprowski, 1982) are more likely to produce anti-target cell effects than other isotypes. With IgG2a mouse monoclonals, the mechanism does not involve complement but probably results from activation of macrophage/monocyte cells. It should be noted, however, that the isotype specificity for this mechanism is not absolute; not all IgG2a monoclonals will act in this way, and a range of different isotypes recognizing a given antigen can display a wide spectrum of activities in macrophage arming.

The clinical therapeutic applications of monoclonals that recognize human tumour-associated are now being tested in phase I and II clinical trials. Use of the 17-1A antibody is an early example of such studies and some limited data has recently been published (Sears et al., 1984). It appears from these studies that 17-1A can produce responses in patients with colorectal carcinoma, perhaps by a similar mechanism to that observed in the tumour xenograft studies (Herlyn & Koprowski, 1982).

A different approach to cancer therapy with monoclonals was adopted by Miller et al., (1982). The target antigen chosen was the surface immunoglobulin secreted by human B-lymphoma cells. The idiotypic nature of the antigen calls for the preparation of antibodies that are "tailor-made" to patient specificity.

Since the original publication, based on treatment of a single patient (Miller et al., 1982), work has continued and several patients have been treated, some with more than 1000 mg amounts of patient-specific anti-idiotype antibody. (Levy, communicated at "International Meeting on Monoclonal Antibodies in Oncology", Nantes, France, June 1984). Although responses were observed in more than 50% of patients, none of the anti-tumour effects were long-lasting. Tumour cells removed from the circulation by opsonisation after antibody binding are presumably replaced by cells resulting from the proliferation and differentiation of antigen free precursor cells.

Other therapeutic applications of antibody

Haber (1983) has described a number of monoclonal antibodies specific for digoxin. These could be used *in vivo* to reverse digitalis intoxication resulting from accidental or suicidal overdose. The use of Fab fragments of a polyclonal sheep anti-digoxin antibody has been described to treat 26 of patients with life-threatening doses of digoxin or digitoxin (Haber, 1983). Complete recovery was obtained with 21 patients. It appears probable that the monoclonals described would eventually replace the sheep antibody for this application.

A quite different therapeutic application of antibodies, in which monoclonals are now being considered, is in the treatment of rhesus D blood group incompatibility in pregnancy. Presently the only source of antibody considered appropriate for use is from male human volunteers immunised with the antigen. A human anti-D monoclonal antibody has now been prepared by the strategy of EB virus transformation of human antibody-producing cells in culture (Crawford et al., 1983). Concerns that viral contaminants may prove a safety hazard in the use of such an antibody have been countered by the claim that the purification of antibody from the culture supernatants will effectively eliminate viral contaminants (Crawford et al., 1983).

MONOCLONAL ANTIBODIES AS TARGETING AGENTS

Therapeutic attempts with antibodies that rely on both antigen binding activity and functional properties of the Fc portion may not always succeed. Limitations may include failure of the antibody to bind complement, failure to bind effector cells or simply a shortage of available effector cells. To circumvent these limitations, considerable effort is being applied to the use of the antigen binding properties of antibodies for target specific delivery. At present this work is almost entirely confined to the cancer field, the aim being to improve the tumour selectivity of anticancer agents, thereby rendering them more effective and/or less generally toxic. Three major types of delivered agent are under investigation; those involving radiation toxicity, those involving anticancer drugs and those using the highly potent polypeptide toxins, so-called immunotoxins.

Radiation

A major thrust in the clinical application of monoclonal antibodies is to attach a radiolabel for use in immunoscintigraphy to detect tumour metastases. The concept of delivering much higher and, hence cytotoxic, levels of radioactivity by the antibodies, is not new but has had to wait until suitable monoclonals were available before being tested. A recent report from the Hammersmith Oncology Group (Epenetos, 1984) illustrates the potential of this approach. Three patients were given monoclonal antibody HMFG2 labelled with ^{131}I to treat accessible tumours by intracavitary routes. This method was chosen to avoid removal or destruction of the labelled antibodies by blood or reticuloendothelial system. The results demonstrated that much higher concentrations of radioactivity were delivered to the malignant sites than to normal organs, and that this appeared to produce tumour cell destruction.

An different approach to the application of antibodies for induction of cytotoxic radiation is the potential use of boron-10 to absorb thermal neutrons and release α-particles. This idea was recently examined in a model system using carcino-embryonic antigen (CEA) in a human colorectal xenograft (Goldenberg et al., 1984). The studies showed that boron conjugated anti-CEA would retain the ability to localise selectivity in the tumour. No data on the effectiveness of neutron capture was provided.

Drugs

The use of monoclonal antibodies recognising human tumours to target anti-cancer drugs has been reviewed recently (Rowland, 1983). The drugs studied are those which lend themselves to chemical modifications that can lead to covalent coupling to suitable residues on the antibody molecule. The most commonly used sites on antibody for coupling are the amino side chains of lysine residues. Most laboratories with continuity in this field tend to restrict themselves to one or two anticancer agents for study. Thus daunomycin and adriamycin have been examined intensively by the group at the Weizmann Institute (Arnon & Sela, 1982); chlorambucil has been studied by the group at Halifax, Nova Scotia (Guclu et al., 1976) and methotrexate has been investigated by most workers in this field at some time (Burstein and Knapp, 1977; Latif et al., 1980; Kulkarni et al., 1981; Garnett et al., 1983). Despite numerous studies, most reports continue to describe *in vitro* effects only. Where *in vivo* effects are described, they are generally based on non-human tumour models. In an attempt to overcome these limitations, studies were undertaken to couple the anti-cancer alkaloid vindesine to monoclonal antibodies recognising human tumour-associated antigens. Conjugates against melanoma, osteogenic sarcoma and tumours bearing CEA were prepared and have been shown to be effective both in vitro and in vivo in tumour xenograft studies (Rowland et al., 1983; Rowland et al., 1984).

In most of the studies involving anti-cancer drug antibody conjugates, there is no information about the mechanism of drug action. It is generally assumed that conjugates act by the same mechanism as free drug, perhaps following release of free drug at the target cell surface or internally.

Immunotoxins

Whereas the drug-antibody studies are generally pragamatic and have avoided questions of mechanism, the work using immunotoxins stems from a belief that the mechanism of polypeptide toxin action is understood and can be made target-specific by antibody conjugation. The plant toxin ricin is composed of two chains, A and B, linked by a disulphide bond. The A chain is an enzymic cellular protein synthesis inhibitor, acting after ricin is internalised through interaction between the B chain and cell surface galactose-containing glycoproteins. Immunotoxin strategy has been of two general types. In the holotoxin approach, intact ricin is coupled to antibody. This provides highly potent conjugates with target selectivity _in vitro_ but with, as yet, little evidence that these agents can act _in vivo_ on human tumours (Moolten et al., 1982). The use of the extremely potent ricin holotoxin also produces immunotoxins that have considerable _in vivo_ non-specific toxicity. One elegant approach to reduce this, is to block the galactose binding site of the B chain without interfering with the other functional properties of the immunotoxin (Thorpe and Ross, 1982).

The second immunotoxin strategy is to dispense with the B chain and to couple A chain to antibody. This approach has produced immunotoxins with a remarkable degree of target cell specificity _in vitro_ (Jansen et al., 1982). The overall potency however may be considerably lower than when using holotoxin. One method for improving potency is to reduce intracellular lysosomal breakdown of the immunotoxin. Ammonium chloride acts in this manner and can be shown to increase potency (Jansen et al., 1982).

A novel strategy to improve the efficiency of A-chain immunotoxins is to provide a separate B-chain antibody conjugate. Synergy between these conjugates has recently been described in the killing of human B-cell tumour cells _in vitro_ (Vitetta et al., 1983).

The wealth of evidence for selective cytotoxicity when using immunotoxins _in vitro_ is as yet not matched by _in vivo_ data. Problems of reticulo-endothelial system damage, rapid _in vivo_ elimination, non-specific toxicity and high immunogenicity (Jansen et al., 1982) provide major hurdles to the clinical application of immunotoxins _in vivo_. However their use in the removal of tumour cells from autologous bone marrow grafts is not limited by such considerations and is currently being evaluated clinically.

CONCLUSIONS

The availability of a technique that can provide homogeneous preparations of antibodies of fine specificity, high purity and quantity has led to enormous growth in antibody research. Much of this research is now being directed at therapeutic applications and when success is achieved in some areas, it is safe to predict that new applications will appear. In support of the research, considerable development work on scale-up, analysis and toxicology will be required before extensive clinical trials can begin. It is now clear that this support is being rapidly generated and that many of the therapeutic applications of monoclonal antibodies described above will soon be widely tested.

REFERENCES

Arnon, R. and Sela M., 1982, In vitro and in vivo efficacy of conjugates of daunomycin with anti-tumour antibodies, Immunol Rev., 62: 5.

Burstein, S. and Knapp, R., 1977, Chemotherapy of murine ovarian carcinoma by methotrexate-antibody conjugates, J. Med. Chem., 20: 950.

Clark, M., Cobbold, S., Hale, G. and Waldmann, H., 1983, Advantages of rat monoclonal antibodies, Immunol. Today, 4: 100.

Crawford, D.H., Huehns, E.R. and Epstein, M.A., 1983, Therapeutic use of human monoclonal antibodies, Lancet 1: 1040.

Epenetos, A.A., 1984, Antibody-guided irradiation of malignant lesions: three cases illustrating a new method of treatment, Lancet 1:1441.

Garnett, M.C., Embleton, M.J., Jacobs, E. and Baldwin, R.W., 1983, Preparation and properties of a drug-carrier-antibody conjugate showing selective antibody-directed cytotoxicity in vitro, Int. J. Cancer 31: 661.

Goldenberg, D.M., Sharkey, R.M., Primus, F.J., Mizusawa, E. and Hawthorne, M.F., 1984, Neutron-capture therapy of human cancer: In vivo results on tumor localization of boron-10-labelled antibodies to carcinoembroyonic antigen in the GW-39 tumor model system, Proc. Natl. Acad. Sci 81: 560.

Guclu, A., Ghose, T., Tai, J. and Mammen, M., 1976, Binding of chlorambucil with antitumor globulins and its effect on drug and antibody activities, Eur J. Cancer 12: 95.

Haber, E., 1983. Antibodies as models for rational drug design, Biochem Pharmacol 32 : 1967

Herlyn, D. and Koprowski, H., 1982, IgG2a monoclonal antibodies inhibit human tumour growth through interaction with effector cells, Proc. Natl. Acad. Sci., 79: 4761.

Hugh-Jones, N.C., Gorick, B.D. and Howard, J.C., 1983, The mechanism of synergistic complement-mediated lysis of rat red cells by monoclonal IgG antibodies, Eur. J. Immunol., 13: 635

Jansen, F.K., Blythman, H.E., Carriere, D., Casellas, P., Gros, O., Gros, P., Laurent, J.C., Paolucci, F., Pau, B., Poncelet, P., Richer, G., Vidal, H. and Voisin, G.A., 1982, Immunotoxins: hybrid molecules combining high specifity and potent cytotoxicity, Immunol Rev. 62: 185.

Kulkarni, P.N., Blair, A.H. and Ghose, T.I., 1981, Covalent binding of methotrexate to immunoglobulins and the effect of antibody-linked drug on tumour growth in vivo, Cancer Res., 41: 2700

Latif, Z.A., Lozzio, B.B., Wust, C.J., 1980, Evaluation of drug-antibody conjugates in the treatment of human myelosarcomas transplanted in nude mice, Cancer, 45: 1326.

Moolten, F.L., Schreiber, B.M. and Zajdel, S.H., 1982, Antibodies conjugated to potent cytotoxins as specific antitumour agents, Immunol Rev., 62: 47.

Miller, R.A., Maloney, D.G., Warnke, R. and Levy, R., 1982. Treatment of B cell lymphoma with monoclonal anti-idiotype antibody, N. Engl. J. Med., 306: 517.

Rowland, G.F., 1983, Use of antibodies to target drugs to tumour cells, Clinics in Immunol and Allergy, 3 : 235.

Rowland, G.F., Simmonds, R.G., Corvalan, J.R.F., Baldwin, R.W., Brown, J.P., Embleton, M.J., Ford, C.H.J., Hellstrom, K.E., Hellstrom I., Kemshead, J.T., Newman, C.F. and Woodhouse, C.S., 1983, Monoclonal antibodies for targeted therapy with vindesine. Protides of the Biol. Fluids, 30: 375.

Rowland, G.F., Corvalan, J.R.F., Axton, C.A., Gore, V.A., Marsden, C.H., Smith, W. and Simmonds, R.G., 1984. Suppression of growth of a human colorectal tumour in nude mice by vindesine-monoclonal anti-CEA conjugates, Protides of the Biol Fluids, 31: 783.

Sears, H.F., Herlyn, D., Steplewski, Z. and Koprowski, H., 1984. Effects of monoclonal antibody immunotherapy on patients with gastrointestinal adenocarcinoma, J. Biol. Resp. Modifiers, 3: 138.

Streem, S.B., Novick, A.C., Braun, W.E. and Steinmuller, D., 1983. Antilymphocyte and antithymocyte globulines, <u>Transplant. Proc.</u> 15 : 590.

Takahashi, H., Terasaki P.I., Kinukawa, T., Chia, D., Miura, K., Okazaki, H., Iwaki, Y., Taguchi, Y., Hardiwidjaja, S., Ishizaki, M. and Billing R., 1983, Reversal of transplant rejection by antiblast antibody, <u>Lancet</u> II: 1155.

Thorpe, P.E. and Ross, W.C.J., 1982, The preparation and cytotoxic properties of antibody-toxin conjugates, <u>Immunol. Rev</u>, 62: 119.

Vitetta, E.S., Cushley, W. and Uhr, J.W., 1983, Synergy of ricin A chain-containing immunotoxins and ricin B chain-containing immunotoxins in in vitro killing of neoplastic human B cells, <u>Proc. Natl. Acad. Sci.</u>, 80 : 6332.

IMMUNOSUPPRESSION WITH MONOCLONAL ANTIBODIES - RULES FOR EFFECTIVE SEROTHERAPY

S.P.Cobbold, G.Martin, P.E.Lovat, and H.Waldmann

Division of Immunology, Dept. of Pathology
Cambridge University

INTRODUCTION

The potential advantages of monoclonal antibodies (Kohler & Milstein, 1975) over anti- lymphocyte globulin (ALG, Monaco et al., 1966a) for immunosuppression in organ grafting (Hunt and Stinson, 1981; Groth et al., 1981) are generally accepted. However, this potential has not yet been fully realised due to a number of problems (Ritz and Schlossman, 1982), some of which stem from a lack of information about the factors which lead to antibody- mediated cell destruction *in vivo* and the contribution of different T cell subsets in graft rejection. A model of immunosuppression using rat monoclonal antibodies to mouse T cell antigens was used to determine the ground rules for effective serotherapy. It was found that rat IgG2b monoclonal antibodies could be highly immunosuppressive when given intra-venously, and that target T cells were substantially depleted (for periods up to 12 months in adult thymectomised animals). Treatment of mice receiving fully allogeneic skin grafts showed that rat IgG2b anti-Thy-1 or anti-L3/T4 gave considerable delays in rejection, while anti-Lyt-2 was ineffective. These results suggest that the L3/T4$^+$ 'helper' T cells should represent a major target for immunosuppression of allograft rejection.

MATERIALS AND METHODS

Animals

CBA and BALB/c mice, and (DAxLOU)F_1 rats were bred and

maintained in the animal facility of the Dept. Pathology, University of Cambridge. Mice were thymectomised when 5-7 weeks old by the method of Monaco et al. (1966b) and allowed to recover 3 weeks before antibody administration. Tail skin grafts were placed on the lateral thoracic wall, and covered with Vaseline impregnated gauze and plaster bandages. The dressings were removed on day 10 and the grafts inspected every day for rejection.

Antibody responses

Thymectomised CBA mice were given two iv. injections of monoclonal antibodies as ascitic fluid (approx. 0.5mg active antibody) one week apart. After three weeks, 0.2ml 10% sheep red blood cells (SRBC) were given ip. and sera collected from the tail vein on day 10. The IgG agglutination titre against SRBC was tested in the presence of 0.1M 2-mercaptoethanol.

Skin grafted mice which had been given a two week course of antibody (0.2 ml ascitic fluid iv. and ip.; figure 1) were bled from the tail vein 10 days after the last injection of monoclonal antibody. Sera were tested for antiglobulin responses using a solid-phase enzyme-linked immunosorbent assay with irrelevant rat IgG2b coated plates, biotin rat monoclonal anti-mouse kappa light chain, and avidin-peroxidase. The assay was standardised by comparing the results with dilutions of a mixture of mouse anti-rat Ig monoclonal ascitic fluids.

Monoclonal Antibodies

Rat monoclonal antibodies to mouse T cells were prepared (Cobbold, 1983) using the Y3/Ag1.2.3 rat myeloma line (Galfre et al., 1979). Some of these were shown (Cobbold et al., 1983) to react with purified mouse brain Thy-1 (four IgMs, IgG2a, IgG2b and IgG2c), others (IgG2bs) showed cross- inhibition with the antibodies to Lyt-1 or Lyt-2 of Ledbetter and Herzenberg (1979), and YTS 191.1 reacted with a 55K protein of Lyt-2 negative peripheral T cells designated L3/T4 (Dialynas et al., 1983). Ascitic fluids were produced by *in vivo* passage of the hybrid myelomas in pristane-primed $(DAxLOU)F_1$ rats, and had titres of binding between 10^{-5} and 10^{-6} on mouse thymocytes.

Immunohistology

Frozen sections of spleen taken on the day after skin graft rejection were stained by the immunoperoxidase method of Aqel et al. (1984), using biotin- coupled monoclonal anti-rat IgG2b and avidin-peroxidase.

Table 1
Immunosuppression by monoclonal rat IgG2b antibodies to T cells

Antibody treatment [1]	Anti-SRBC response [2]	% Depletion [3]
None (pre-immune)	<1	- [4]
None (immunised)	6 ± 0	0
IgM anti-Thy-1 (four)	4 ± 2	-
IgG2a anti-Thy-1	4 ± 2	-
IgG2a anti-T200	6 ± 2	-
IgG2c anti-Thy-1	3 ± 1	-
IgG2b anti-Lyt-2 (two)	6 ± 1	>96
IgG2b anti-Thy-1	* <1 *	>90
IgG2b anti-Lyt-1	* <1 *	>82
IgG2b anti-L3/T4	* <1 *	>86
IgG2b anti-T200	* 2 ± 1 *	-

1 groups of 3-8 adult thymectomised CBA mice, 2 iv injections
2 2-mercaptoethanol resistant agglutination (log_2 titre)
3 % depletion of antigen positive cells by cytofluorograph as measured 5 weeks post-antibody injection. 4 not determined

RESULTS AND DISCUSSION

Previous studies have suggested that the rat IgG2b isotype might be effective for serotherapy in mice (Ledbetter & Seaman, 1982; Cobbold et al., 1983). Table 1 shows that, of a series of rat monoclonal antibodies to mouse T cells, the IgG2b sub- class was always the most effective at suppressing the *in vivo* antibody response to sheep red blood cells (SRBC). All the rat IgG2b reagents tested (when administered iv.) gave a substantial and long term depletion (up to 12 months in thymectomised mice) of the specific target T cells. Such effective monoclonal antibodies with specificity for the 'helper' ($L3/T4^+$) or 'cytotoxic' ($Lyt\text{-}2^+$) T cell subsets allowed the development of a mouse model for serotherapy to control allograft rejection.

Fully allogeneic (BALB/c) skin was grafted onto CBA mice receiving a two week course of monoclonal antibodies to different T cell subsets (figure 1). Antibody to L3/T4 and Thy-1 greatly increased graft survival, while anti-Lyt-1 was less effective (possibly because elimination of either T cell subset was less

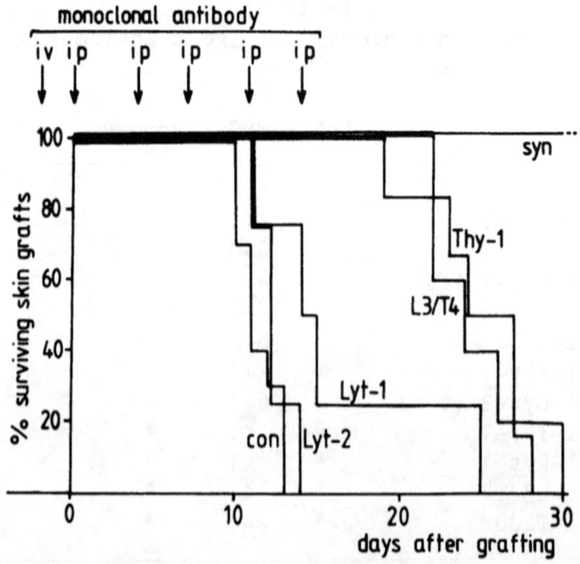

Fig. 1. Suppression of skin graft rejection (BALB/c to CBA) with monoclonal antibodies to T cell subpopulations in vivo. con = no antibody, syn = syngeneic skin grafts

complete than with the other reagents, figure 2). Serotherapy against Lyt-2$^+$ cells did not cause any delay in rejection, even though immunoperoxidase staining of spleen sections (taken on the day after rejection) confirmed that target cells were markedly, and specifically, depleted in all the treated groups (figure 2). These results suggest that the L3/T4$^+$ 'helper' T cells are the major promotors, and therefore the important targets for serotherapy, of skin graft rejection. However, anti-L3/T4 was less effective if given after day 4 of grafting (not shown), so L3/T4$^-$ cells may also contribute to rejection if the helper T cell subset is present.

There are two main factors which can limit serotherapy with monoclonal antibodies: antigenic modulation (Ritz et al., 1980) and a response to the administered antibody (Hale et al., 1983; Ritz and Schlossman, 1982). The first problem of modulation may be reduced by using monovalent antibodies (Glennie & Stevenson, 1982; Cobbold and Waldmann, 1984), and it should be noted that all the monoclonal reagents used in this paper contain both bivalent and monovalent immunoglobulin due to the Y3/Ag1.2.3 light chain. The second problem of antiglobulin responses may be solved by simply using reagents

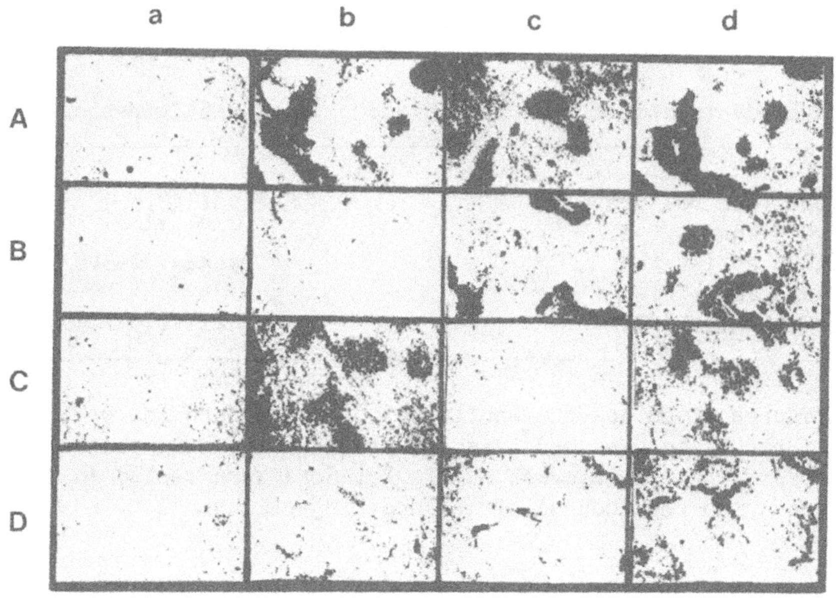

Fig. 2. Immunoperoxidase staining of spleen sections confirms the depletion of T cell subsets in immunosuppressed mice. Sections from mice depleted with antibody to; A control, B L3/T4, C Lyt-2, D Lyt-1 were stained for; a diluent, b L3/T4, c Lyt-2, d Lyt-1.

which are sufficiently immunosuppressive. Table 2 shows that anti-Ig responses were made by all the treated mice in the skin graft model, except in the case of anti-L3/T4 where no response was detectable. This demonstrates that rat IgG2b antibodies to the L3/T4 antigen can be used to control allograft rejection without the limitations imposed by responses to the xenogeneic protein.

The results described above may be applicable to serotherapy in man, because rat IgG2b is effective with human effector systems (Clark et al., 1983; Clark, Hale & Waldmann, this volume) and the mouse L3/T4 is directly homologous to the human OKT4 antigen (Dialynas et al., 1983).

We wish to thank Dr T.Prospero for help with the cytofluorograf. This work was supported by grants from the Medical Research Council.

Table 2
Anti-Ig responses after monoclonal antibody immunosuppression

Antibody treatment	Number/group	Anti-rat Ig response[1]
None (control group)	3	all<0.5
IgG2b anti-Lyt-2		
IgG2b anti-Lyt-1	8	range= 10-500
IgG2b anti-Thy-1		
IgG2b anti-L3/T4	4	all<0.5

[1] Figures given are the equivalent of a standard (nl/well) anti-rat Ig as measured by solid-phase ELISA using plates coated with an irrelevant rat IgG2b. Sera were tested 10 days after a two week course of antibody injections.

REFERENCES

Aqel,N.M., Clark,M., Cobbold,S.P., & Waldmann,H., 1984, Immunohistological screening in the selection of monoclonal antibodies: the use of isotype-specific antiglobulins, J. Immunol. Meth., 69: 207.

Clark,M., Cobbold,S., Hale,G., & Waldmann,H., 1983, Advantages of rat monoclonal antibodies, Immunol. Today, 4: 100.

Cobbold,S.P., & Waldmann,H., 1984, Therapeutic potential of monovalent monoclonal antibodies., Nature, 308: 460..

Cobbold,S.P., 1983, Monoclonal antibodies against differentiation antigens of the mouse haemopoetic system, PhD thesis, Cambridge: 53.

Cobbold,S.P., Thierfelder,S., & Waldmann,H., 1983, Immunosuppression with monoclonal antibodies: A model to determine the rules for effective serotherapy, Mol. Biol. Med., 1: 285.

Dialynas,D.P., Quan,Z.S., Wall,K.A., Pierres,A., Quintans,J., Loken,M.R., Pierres,M., & Fitch,F.W., 1983, Characterization of the murine T cell surface molecule designated L3T4, identified by monoclonal antibody GK1.5: similarity of L3T4 to the human Leu3/T4 molecule, J. Immunol., 131: 2445.

Galfre,G., Milstein,C., & Wright,B., 1979, Rat x rat hybrid myelomas

and a monoclonal anti-Fd portion of mouse IgG, Nature, 277: 131.
Glennie,M.J., & Stevenson,G.T., 1982, Univalent antibodies kill tumour cells in vitro and in vivo, Nature, 295: 712.
Groth,C.G., 1981, The value of ALG and ATG in clinical renal transplantation: recent European experience, Transplant. Proc., 13: 460.
Hale,G., Swirsky,D.M., Hayhoe,F.G.J., & Waldmann,H., 1983, Effects of monoclonal anti-lymphocyte antibodies in vivo in monkeys and humans, Mol. Biol. Med., 1: 321.
Hunt,S.A., & Stinson.E.B., 1981, Cardiac transplantation, Annu. Rev. Med., 32: 213.
Kohler,G., & Milstein,C., 1975, Continuous cultures of fused cells secreting antibody of predefined specificity, Nature, 256: 495.
Ledbetter,J.A., & Herzenberg,, 1979, Xenogeneic monoclonal antibodies to mouse lymphoid differentiation antigens, Immunol. Rev., 47: 63.
Ledbetter,J.A., & Seaman,W.E., 1982, The Lyt-2, Lyt-3 macromolecules: structural and functional studies, Immunol. Rev., 68: 197.
Monaco,A.P., Wood,M.L., Gray,J.G., & Russell,P.S., 1966a, Studies on heterologous anti-lymphocyte serum in mice II Effect on the immune response, J. Immunol., 96: 229.
Monaco,A.P., Wood,M.L., & Russel,P.S., 1966b, Studies on heterologous anti-lymphocyte serum in mice. III Immunologic tolerance and chimerism produced across the H-2 locus with adult thymectomy and anti-lymphocyte serum, Ann. N.Y. Acad. Sci, 129: 190.
Ritz,J., & Schlossman,S.F., 1982, Utilization of monoclonal antibodies in the treatment of leukemia and lymphoma, Blood, 59: 1.
Ritz,J., Pesando,J.M., Notis-McConarty,J., Schlossman,S.F., 1980, Modulation of human acute lymphoblastic leukaemia antigen induced by monoclonal antibody, J. Immunol., 125: 1506.

INTERACTION OF RAT MONOCLONAL ANTIBODIES WITH HUMAN KILLER CELLS

Mike Clark, Geoffrey Hale and Herman Waldmann

Department of Pathology
University of Cambridge
Cambridge CB2 1QP, U.K.

INTRODUCTION

The advent of monoclonal antibody technology has renewed interest in serotherapy for immunosuppression and for treating neoplastic disease because potentially unlimited quantities of reagents are now available and their specificity can be assured. A variety of antibodies have now been tested in vivo with limited success (Ritz and Schlossman, 1982). In addition to antibody specificity it is necessary to define the rules for effective ways of using antibodies to deplete cells and it is not yet clear which effector mechanisms are the most significant for clearance of antibody-coated cells in vivo. Some experiments have suggested that both complement-mediated lysis and antibody-dependent cell-mediated cytotoxicity (ADCC) may be of importance (Bernstein et al.,1981), but others have suggested that phagocytosis by macrophages is of more significance (Herlyn and Koprowski, 1982).

We have chosen to study rat monoclonal antibodies because of certain technical advantages in their production (Clark et al., 1983) and have prepared a large number of rat monoclonal antibodies against a variety of human leucocyte differentiation antigens and have tested them to see whether they can participate in ADCC with human mononuclear cells as effectors. The isotypes of these reagents have been determined and although we have never obtained IgG1 monoclonal antibodies against cell surface antigens, we have been able to test the other isotypes and here we show that only IgG2b antibodies are effective in promoting ADCC with human K cells.

MATERIALS AND METHODS

Cells. Blood from laboratory donors was collected by venepuncture and defibrinated with glass beads. Mononuclear cells were isolated by centrifugation on Ficoll-Hypaque as described before (Hale et al., 1983). Mononuclear cells were depleted of monocytes by incubation with carbonyl iron (15mg/ml of blood) in medium containing 10% (v/v) autologous serum for 45min at 37°C. The iron particles and phagocytic cells were removed with a magnet. Mononuclear cells were depleted of K cells by incubation with the mouse monoclonal antibody HNK-1 (Becton-Dickerson, Sunnyvale, CA) plus 20% (v/v) rabbit complement (Buxted Rabbit Co. Ltd., Buxted, UK) for 1h at 37°C.

Monoclonal antibodies. Rat monoclonal antibodies were made by fusion of the kappa-secreting myeloma line Y3/Ag1.2.3 (Galfre et al., 1981) with spleen cells from DA rats immunized intravenously with human or mouse cells. The antibodies described here were from rats immunized with normal human lymphocytes (YTH), or cells from patients with acute myeloid leukemia (YAML) or a human pre-B cell line NALM-1 (YHB)

Isotyping of monoclonal antibodies. The sizes of the heavy and light chains were determined by SDS-PAGE of internally radiolabelled antibody (Galfre and Milstein, 1981). The subclasses of the IgG antibodies were established by Ouchterlony double diffusion of the radiolabelled antibodies against subclass specific antisera to rat immunoglobulin (Miles Laboratories, Stoke Poges, U.K.) For each monoclonal antibody only one antiserum gave a radioactive precipitin line corresponding with the subclass of that antibody. Isotypes were also determined by reverse passive haemagglutination of human group O or protease-treated sheep cells coupled to mouse monoclonal anti-(rat Ig) antibodies. The mouse monoclonal antibodies used were MAR18.5 anti-(rat kappa) (Lanier et al., 1982), RG7/1.7 anti-(rat IgG2a) (Springer et al., 1982), NORIG 7.16 anti-(rat IgG2b) and NORIG 31.12 anti-(rat IgG2c) (Aqel et al., 1984) These antibodies were affinity purified and coupled to the red cells as described for MAR18.5 (Clark et al., 1984). For isotyping, five-fold dilutions of the culture supernatants were made in 50µl PBS and 50µl of a 1% (v/v) suspension of appropriate red cells were added. When the cells had settled, agglutination was scored visually. Protease-treated sheep red cells were found to be more sensitive, but suffered from autoagglutination when coupled with NORIG 31.12, therefore human group O cells were used for routine isotype analysis.

Antibody-dependant cell-mediated cytotoxicity. Target cells , mononuclear cells (mainly T cells), were incubated at 37°C for 1h in 0.3ml medium containing 150µCi sodium [^{51}Cr]chromate. The labelled cells were washed and resuspended at $2x10^5$/ml. Aliquots (50μl)of the suspension were added to U-well microtitre plates containing 50μl of monoclonal antibody (usually 1-10μg/ml). After incubation at 20°C for 30min, the plates were centrifuged at 100g for 2min and the supernatant was removed with great care to avoid cross-contamination of adjacent wells. The cells were resuspended in 200μl of medium containing unlabelled autologous mononuclear cells as effectors (usually $2x10^6$ to $5x10^6$/ml) and centrifuged again to pellet the cells. After 4h at 37°C, 100μl of the supernatants was collected for measurement of radioactivity. Controls with no antibody or no effectors were used to measure the spontaneous radioactivity release and samples of the labelled cells were taken to measure the total radioactivity for calculation of the specific release in the test samples.

RESULTS AND DISCUSSION

It has already been established in several systems that IgG antibodies, but not IgM, can promote ADCC by K cells (Perlmann et al., 1972; McLennan, 1972) although there have been recent reports that monoclonal IgM antibodies are active with mouse effectors (Zoller et al., 1983). MacLennan & Howard (1972) showed that polyclonal rat IgG but not IgM antibodies can induce ADCC. Most studies on the IgG subclass specificity of the relevant Fc receptors have employed aggregated myeloma proteins in high concentration to inhibit ADCC with polyclonal antiserum and thus it was shown that human IgG1, IgG2, IgG3 and possibly IgG4 could inhibit ADCC (Larsson et al., 1973). Now that monoclonal antibodies of defined specificity are available, it is possible to examine the isotype specificity direct and to test whether the nature of the antigen influences killing. There have been no reports of systematic studies of mouse monoclonal antibodies in ADCC with human cells. The antibodies which have been tested clinically have mostly been mouse IgG2a and were generally ineffective in ADCC in vitro (eg. Nadler et al., 1980; Dillman et al., 1982).

Rat monoclonal antibodies which bound to human lymphocytes were selected by solid-phase enzyme-linked binding assays (Cobbold and Waldmann, 1981). The fraction of target cells recognized by each antibody was measured by flow cytometry and in some case the antigens were characterized by affinity purification and/or cross-inhibition of binding of other previously-defined monoclonal antibodies. The set of antibodies which recognize human B and T cells and monocytes (CAMPATH-1 like) have been described in detail before (Hale et al., 1983; Hale et al., 1983a), the antibody specificities CAMPATH-2, CAMPATH-3 and CAMPATH-4 are described elsewhere this volume (H. Waldmann et al.) and the set of antibodies against the human leucocyte common antigen (LCA) are described elsewhere in this volume (Bindon, Hale and Waldmann).

Each antibody was tested for its ability to promote cell-mediated killing with human peripheral mononuclear cells as effectors. The radiolabelled target cells were incubated with the antibodies in the form of culture supernatants, diluted 10 or 100 times. Rabbit anti-lymphocyte globulin was used as a positive control. After removing excess antibody the effector cells were added at a ratio of 10:1 to 50:1. The results are shown in Table 1. Only the IgG2b antibodies were effective.

The antibody YTH 54.12 (anti-leucocyte common antigen) was tested at a range of different concentrations and compared with the polyclonal anti-lymphocyte globulin (data not shown).The optimal concentration of IgG2b antibody required for ADCC was about 0.1μg/ml which is very low compared with that required for complement-mediated lysis (about 10μg/ml; (Hale et al., 1983)), therefore if ADCC is a mechanism which can be exploited in vivo, only relatively small amounts might be required for effective serotherapy. The IgM, IgG2a and IgG2c antibodies were tested at concentrations up to 10μg/ml but gave no killing at any level.

Most experiments were carried out using unseparated peripheral mononuclear cells, i.e. lymphocytes plus monocytes. In order to characterize the effector cell, titrations were carried out of mononuclear cells depleted either of monocytes by treatment with carbonyl iron or of K cells by treatment with HNK-1 plus complement (Abo and Balch, 1981). Depletion of K cells virtually abolished effector activity but removal of the monocytes had no effect . This is in agreement with other published results which also indicate that antibody-dependent killing of nucleated cells is mainly mediated by lymphocytes (Perlmann et al., 1972; Trinchieri et al., 1975).

By titration of the effector cells it was shown that lymphocytes from some donors were much more effective than others at mediating cell lysis. Cells from nine healthy donors were compared as effectors, using targets coated with YTH 66.9 (IgM), YTH 34.5 (IgG2a), YTH 54.12 (IgG2b), YTH 86.1 (IgG2c) or anti-lymphocyte globulin. The effector to target ratio was approx. 30:1. In each case only the target cells coated with YTH 54.12 or the polyclonal antibody were killed; however the extent of lysis varied from one donor to another (varying from 8% to 51% specific release with YTH 54.12). It has been suggested that human monocytes show a polymorphism in their Fc receptors for mouse IgG1 (Tax et al., 1983), but our experiments did not reveal any comparable polymorphism in the lymphocyte Fc receptors, at least as far as rat immunoglobulin was concerned.

In addition to the data on ADCC described here the rat IgG2b isotype has been shown to be the most effective rat IgG subclass at fixing human C1q (Hughes-Jones et al., 1983). Also rat IgG2b

Table 1. ADCC with rat monoclonal antibodies

Clone number	Isotype	Antigen	% Specific release
YTH 14.20	IgM	CAMPATH-1	1
YTH 62.7	IgM	CAMPATH-1	0
YTH 66.9	IgM	CAMPATH-1	0
YTH 73.4	IgM	CAMPATH-1	1
YTH 75.5	IgM		0
YTH 81.111	IgM		0
YTH 85.9	IgM		0
YHB 66.25	IgM	LCA	0
YTH 10.2	IgG2a		0
YTH 16.7	IgG2a		0
YTH 19.9	IgG2a		1
YTH 23.11	IgG2a		1
YTH 25.2	IgG2a		0
YTH 27.7	IgG2a		1
YTH 34.5	IgG2a	CAMPATH-1	2
YTH 36.11	IgG2a	CAMPATH-4	0
YTH 37.5	IgG2a		0
YTH 56.8	IgG2a		0
YTH 58.1	IgG2a		1
YTH 62.1	IgG2a		0
YTH 67.1	IgG2a	CAMPATH-4	1
YTH 76.3	IgG2a	HLA-B	1
YTH 81.5	IgG2a		1
YTH 91.4	IgG2a		2
YAML 501.4	IgG2a	LCA	0
YAML 568.4	IgG2a		0
YTH 3.2	IgG2b	CAMPATH-2	36
YTH 5.3	IgG2b	LCA	48
YTH 12.5	IgG2b	CAMPATH-3	22
YTH 24.5	IgG2b	LCA	43
YTH 26.5	IgG2b	CAMPATH-4	19
YTH 53.1	IgG2b		19
YTH 54.12	IgG2b	LCA	51
YTH 65.3	IgG2b	LCA	35
YTH 68.2	IgG2b	LCA	23
YTH 79.6	IgG2b	LCA	17
YTH 86.10	IgG2b	LCA	37
YTH 95.2	IgG2b		22
YTH 86.1	IgG2c	CAMPATH-1	1
YHB 143.3	IgG2c	LCA	0
Rabbit ATG	polyclonal		64

antibodies against mouse lymphocytes are able to eliminate cells in vivo (Cobbold et al., 1983; and elsewhere this volume, Cobbold, Martin, Lovat and Waldmann) There is now a need to evaluate the role of rat IgG2b antibodies for serotherapy in man.

ACKNOWLEDGMENTS

We are grateful to Dr. T. Prospero for help with the cyto-fluorograf and to Jill Free and Penny Lovat for expert technical assistance. We also thank Drs J. Howard and G. Butcher, ARC Institute of Animal Physiology, Babraham, for telling us their unpublished results in which they demonstrated ADCC with rat IgG2b antibodies.

REFERENCES

Abo,T., & Balch,C.M., 1981, A differential antigen of human NK and K cells identified by a monoclonal antibody (HNK-1), J. Immunol., 127:1024

Aqel,N.M., Clark,M., Cobbold,S.P., & Waldmann,H., 1984, Immunological screening in the selection of monoclonal antibodies: the use of isotype-specific antiglobulins, J. Immunol. Meth., 69:207

Bernstein,I.D., Nowinski,R.C., Tam,M.R., McMaster,B., Houston,L.I., Clark,E.A., 1981, Monoclonal antibody therapy of mouse leukemia, in: "Monoclonal antibodies: hybridomas: a new dimension in biological analysis", Kennet, McKearn & Bechtel, eds Plenum NY.

Clark,M., Cobbold,S., Hale,G., & Waldmann,H., 1983, Advantages of rat monoclonal antibodies, Immunol. Today, 4:100

Clark,M.R., Cobbold,S.P., Waldmann,H., & Coombs.R.R.A., 1984, Detection of monoclonal antibodies against cell surface antigens: the use of antiglobulins coupled to red blood cells, J. Immunol. Meth., 66:81

Cobbold,S.P., & Waldmann,H., 1981, A rapid solid-phase enzyme-linked binding assay for screening monoclonal antibodies to cell surface antigens, J. Immunol. Meth., 44:125

Cobbold,S.P., Thierfelder,S., & Waldmann,H., 1983, Immunosuppression with monoclonal antibodies: A model to determine the rules for effective serotherapy, Mol. Biol. Med., 1:285

Dillman,R.O., Shawler,D.L, Sobol,R.E., Collins,H.A., Beauregard,J.C., Wormsley,S.B., & Royston,I, 1982, Murine monoclonal antibody therapy in two patients with chronic lymphocytic leukemia, Blood, 59:1036

Galfre,G., & Milstein,C., 1981, Preparation of monoclonal antibodies: strategies and procedures, Methods Enzymol., 73:3

Hale,G., Bright,S., Chumbley,G., Hoang,T., Metcalf,D., Munro,A.J., Waldmann,H., 1983, Removal of T cells from bone marrow for transplantation:a monoclonal antilymphocyte antibody that fixes human complement, Blood, 62:873

Hale,G., Hoang,T., Prospero,T., Watt,S.M., & Waldmann,H., 1983a, Removal of T cells from bone marrow for transplantation: comparison of rat monoclonal anti-lymphocyte antibodies of different isotypes, Mol. Biol. Med., 1:305

Herlyn,D., & Koprowski,H., 1982, IgG2a monoclonal antibodies inhibit human tumour growth through interaction with effector cells, Proc. Nat. Acad. Sci. USA, 79:4761

Hughes-Jones,N.C., Gorick,B.D., & Howard,J.C., 1983, The mechanism of synergistic complement-mediated lysis of rat red cells by monoclonal IgG antibodies, Eur. J. Immunol., 13:635

Lanier,L.L., Gutman,G.A., Lewis,D.E., Griswold,S.T., & Warner,N.L., 1982, Monoclonal antibodies against rat immunoglobulin kappa chains, Hybridoma, 1:125

Larsson,A., Perlmann,P., & Natvig,J.B., 1973, Cytotoxicity of human lymphocytes induced by rabbit antibodies to chicken erythrocytes, Immunology, 25:675

MacLennan, I.C.M., 1972, Antibody in the induction and inhibition of lymphocyte cytotoxicity, Transplant. Rev., 13:67

MacLennan,I.C.M., & Howard,A., 1972, Evidence for correlation between the antigenic specificity and charge of human IgG, Immunol., 22:1043

Nadler,L.M., Stashenko,D., Hardy,R., Kaplan,W.D., Button,L.N., Kuke,D.W., Antman,K.H., & Schlossman,S.F., 1980, Serotherapy of a patient with a monoclonal antibody directed against a human lymphoma-associated antigen, Cancer Res., 40:3147

Perlmann,P., Perlmann,H., & Wigzell,H., 1972, Lymphocyte mediated cytotoxicity in vitro: Induction and inhibition by humoral antibody and nature of effector cells, Transplant. Rev., 13:91

Ritz,J., & Schlossman,S.F., 1982, Utilization of monoclonal antibodies in the treatment of leukemia and lymphoma, Blood, 59:1

Springer,T.A., Bhattacharya,A., Cardoza,J.T., & Sanchez-Madrid,F., 1982, Monoclonal antibodies specific for rat IgG1, IgG2a and IgG2b subclasses and kappa chain monotypic and allotypic determinants: reagents for use with rat monoclonal antibodies, Hybridoma, 1:257

Tax,W.J.M., Willems,H.W., Reekers,P.P.M., Capel,P.J.A., & Koene,R.A.P., 1983, Polymorphism in mitogenic effect of IgG1 monoclonal antibodies against T3 antigen on human T cells, Nature, 304:445

Trinchieri,G., Baumann,P., de March,M., & Tokes,Z., 1975, Antibody-dependent cell-mediated cytotoxicity in humans 1 Characterization of the effector cell, J. Immunol., 115:249

THERAPEUTIC POTENTIAL OF MONOCLONAL ANTIBODIES TO THE LEUCOCYTE COMMON ANTIGEN

Carol Bindon, Geoff Hale and Herman Waldmann

University of Cambridge Department of Pathology
Tennis Court Road
Cambridge. CB2 1QP. U.K.

INTRODUCTION

The leucocyte common (LC) antigen is a major glycoprotein family of leucocyte membranes. Monoclonal antibodies against the rat, mouse and human antigens have been described (Sunderland et al., 1979; Trowbridge, 1978; Dalchau et al., 1980). Its cell distribution and molecular weight are very similar in the three species. It is widely distributed on leucocytes, being found on T and B lymphocytes, thymocytes, macrophages and granulocytes, but is absent from other tissues. Its apparent molecular weight by SDS-PAGE is about 200 kilodaltons, but up to four species with slightly different mobilities have been observed according to the cell type studied (Dalchau and Fabre, 1981; Banga et al., 1984).

Anti-LC antibodies might be useful for immunosuppression and for conditioning of patients to receive bone marrow grafts. The effectiveness of cell destruction *in vivo* will presumably depend on their ability to activate appropriate effector mechanisms such as complement-mediated lysis, antibody dependent cell-mediated lysis (ADCC) or phagocytosis. However, ADCC and phagocytic mechanisms could become self-limiting *in vivo*, since the effector cells also express the LC antigen and might themselves be depleted. Complement-mediated lysis would not suffer from this drawback, but anti-LC antibodies are generally poor at fixing complement. Here we show that antibodies which recognize different epitopes on the molecule can synergize in complement-mediated lysis, as has been described in other systems (Elliot et al., 1978; Hughes-Jones et al., 1983), and thus the basis for a rational cocktail of antibodies for therapy is proposed. We will principally describe our data with rat antibodies to the human LC antigen, but very similar results were also obtained with analogous antibodies to the mouse LC antigen, and these will be useful for model experiments _in vivo_.

MATERIALS AND METHODS

Monoclonal Antibodies

Rat monoclonal antibodies were made by fusion of the myeloma line Y3/Ag 1.2.3 (Galfre et al., 1979) with spleen cells from DA rats immunized with human T cells (YTH) or acute myeloid leukaemia cells (YAML). Isotypes were determined by standard methods as described elsewhere in this volume (M.R.Clark, G.Hale and H.Waldmann).

Solid-Phase Enzyme Linked Binding Assays

Binding to fixed cells or to purified LC antigen was carried out in microtitre plates as described by Cobbold and Waldmann (1981). LC antigen was purified from human tonsil cell membranes extracted with Tween 40 (Standring and Williams, 1978) by affinity chromatography on YTH 54.12 coupled to Sepharose. To examine whether antibodies were mutually competitive for antigen binding, they were concentrated 10-fold by ammonium-sulphate precipitation. The first (concentrated) antibody was pre-incubated on the antigen-coated plate for 45min. The second (unconcentrated) antibody of a different Ig subclass was then added, without washing out the first, and incubated for a further 45min. Binding of the second antibody was measured using B-galactosidase coupled to the appropriate isotype-specific mouse monoclonal anti-(rat Ig), either RG7/1.7 (anti-IgG2a, Springer et al., 1982) or NORIG 7.16 (anti-IgG2b, Aqel et al., 1984).

Characterization of Antibodies by Immunoprecipitation and Flow Cytometry

Lymphocytes were labelled with ^{125}I using lactoperoxidase (Kessler, 1975) and solubilized with detergent. The lysate was absorbed on Sepharose coupled to human Ig to remove non-specific binding activity, then affinity purified on Sepharose coupled to anti-(rat Ig), which had been preincubated with the test monoclonal antibody. Material specifically bound by the second column was eluted and analysed by SDS-PAGE.

For flow cytometry human mononuclear cells were incubated with monoclonal antibody, and then stained with fluorescent mouse monoclonal anti-(rat kappa light chain) and analysed in a Cytofluorograf model 50-H (Ortho, Westwood, MA) as previously described (Hale et al., 1983).

Complement-Mediated Lysis

Peripheral blood mononuclear cells were labelled with ^{51}Cr and treated with monoclonal antibodies in the presence of autologous or rabbit complement as described before (Hale et al., 1983).

RESULTS

Identification and Characterization of Antibodies

Monoclonal anti-LC antibodies were first identified by their ability to bind specifically to purified antigen in a solid-phase enzyme-linked binding assay. They were further characterized by immunoprecipitation and cell distribution. Each of the anti-LC antibodies but not control antibodies precipitated four components of approximate molecular weight 200kd. These probably correspond to those described by others (Banga et al., 1984). Indirect fluorescence analysis showed that all of the antibodies labelled 96-97% of peripheral blood mononuclear cells, suggesting that they recognized a determinant of the LC molecule expressed on all lymphocytes unlike some anti-LC antibodies which recognize a determinant expressed predominantly on B cells (Dalchau and Fabre, 1981).

Competitive Binding Assays

Each IgG2b antibody was tested for its ability to inhibit binding of the two IgG2a antibodies YAML 501.4 and YAML 568.4 (Fig 1) and conversely, the inhibition of IgG2b antibodies by the two IgG2a antibodies was also tested. Two groups of mutually competitive antibodies were defined. Group 1 contained YAML 501.4 and YTH 54.12. Group 2 contained YAML 568.4 and YTH 5.3, YTH 24.5, YTH 65.3, YTH 68.2, YTH 72.1 YTH 75.6 YTH 79.6 and YTH 86.10. These results suggest that the two groups of antibodies recognize at least two distinct epitopes on the human LC molecule. It is of course possible that antibodies within a group could recognise distinct but adjacent sites, such that binding of one prevented a second antibody binding to a nearby site.

Complement-Mediated Lysis

Each antibody was titrated to establish a saturating concentration for complement-mediated lysis. IgG2b antibodies were generally better at causing complement-mediated lysis, which is consistent with other observations for rat antibodies (Fust et al., 1980; Hughes-Jones et al., 1983). A comparison of the use of autologous human complement or rabbit complement for the lysis of human cells confirmed other reports that heterologous complement was more effective (e.g. Hansch et al., 1981).

All possible pairs of antibodies were tested at saturation to examine whether synergistic lysis was possible between antibodies binding to different epitopes (Table 1). Synergistic lysis occurred only between YTH 54.12 and another IgG2b antibody. This was consistent with the results of the competitive binding experiments, and

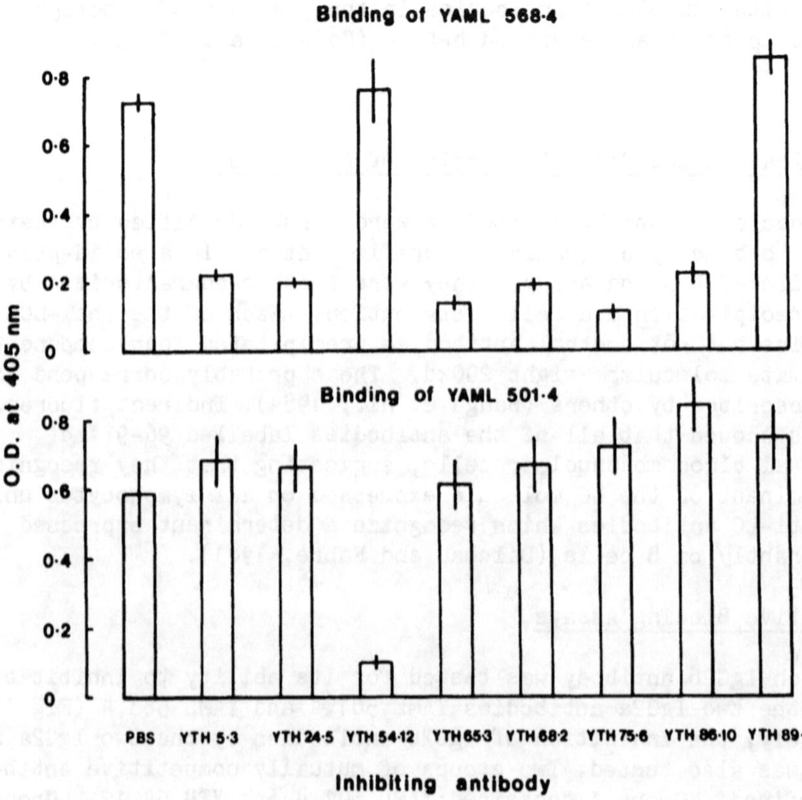

Fig.1. Binding of the two IgG2a antibodies YAML501.4 and YAML568.4 to purified LC antigen in the presence of 10x concentrated IgG2b anti-LC antibodies or YTH89.1, specific for red cell glycophorin. Mean ± one standard error is shown.

confirmed the hypothesis that YTH 54.12 recognized a different epitope on the LC molecule from the remainder of the IgG2b antibodies. An optimum combination of two IgG2b antibodies gave a marked increase in killing, being about 4-6 times the sum of that obtained with each antibody alone. The maximum lysis corresponded to about 75% of the Cr-release given by CAMPATH-1, a control IgM antibody which is known to kill all lymphocytes under these conditions (Hale et al., 1983). A combination of an anti-LC antibody and a control IgG2b anti-T cell antibody gave a slight increase in lysis, but much less than the optimum combination of two anti-LC antibodies. The same pattern of lysis was seen in several experiments, although as expected, the precise extent of killing was dependent on the complement concentration and source. Under conditions where one antibody alone gave significant lysis (using rabbit complement), this was reduced by combination with a second, non-lytic antibody binding to the same epitope.

Table 1: Complement-Mediated Lysis Using Pairs of Anti-Human LC Antibodies.

	YTH 5.3	YTH 24.5	YTH 54.12	YTH 65.3	YTH 68.5	YTH 72.1	YTH 75.6	YTH 86.10	YAML 501.4	YAML 568.4	YTH 12.5	YTH 3.2	Medium
IgG subclass:	2b	2b	2b	2b	2b	2b	2b	2b	2a	2a	2b	2b	
YTH 5.3	0												
YTH 24.5	0	7											
YTH 54.12	51	64	6										
YTH 65.3	0	4	56	2									
YTH 68.5	0	3	51	1	0								
YTH 72.1	0	6	63	3	2	4							
YTH 75.6	0	8	66	9	3	9	10						
YTH 86.10	0	0	18	0	0	0	0	0					
YAML 501.4	0	0	0	0	0	0	1	0	0				
YAML 568.4	0	0	4	0	0	0	3	0	0	0			
YTH 12.5	0	9	11	3	4	9	11	1	0	2	0		
YTH 3.2	8	21	27	15	15	18	21	5	3	3	11	5	
Medium	0	7	8	1	3	3	8	0	0	0	2	5	0

Antibodies were specific for LC or for a T cell marker (YTH 12.5 and YTH 3.2). Autologous human serum was used as a complement source at a final concentration of 20%. The mean % specific chromium release of duplicate points is shown.

DISCUSSION

Compared with conventional antisera, monoclonal antibodies often have a reduced ability to cause complement-mediated lysis and this could limit their therapeutic usefulness. The use of cocktails of monoclonal antibodies recognizing different determinants on a cell-surface molecule may provide a means of restoring lytic capability, while retaining the advantages of specificity and unlimited supply associated with monoclonal antibodies. The data presented here suggest that this might be a feasable approach in the case of one potentially useful specificity.

Several explanations might account for the increased lysis seen with combinations of antibodies. Initiation of the complement cascade requires C1q attachment by more than one head (Metzger, 1978) so that large numbers of IgG molecules bound to a cell are normally required for lysis and therefore simply doubling the amount of antibody on the cell might give a significant improvement (Borsos and Rapp, 1965; Hughes-Jones and Ghosh, 1981). However, in the special case of two antibodies binding different epitopes of the same antigen molecule there is the potential for a greatly increased efficiency of lysis, since every antigen molecule can become a C1q binding site, through the attachment of two heads of C1q. Convincing data in favour of this possibility have been reported for monoclonal antibodies against rat major histocompatability antigens (Hughes-Jones et al., 1983). From the data presented here it is impossible to distinguish the two mechanisms.

Whether the antibodies will be therapeutically effective remains unknown, and they need to be tested _in vivo_. Some were found to weakly react with lymphocytes from baboons, rhesus and cynomolgus monkeys but it would probably be better to use the appropriate species-specific antibodies for model experiments. Therefore we have identified similar antibodies which recognize mouse LC antigen and give synergistic lysis (C. Bindon, S.P.Cobbold and H. Waldmann, unpublished work). These can now be tested _in vivo_ alone and in combination, using normal mice and mice depleted of complement using cobra venom factor. It is also possible that the principle of synergy will apply to other physiological effectors e.g. ADCC (Elliott et al., 1977). The immunosuppressive ability of anti-LC monoclonal antibodies has in fact been examined before, when they were found to be unable (Fabre et al., 1980) or only partially able (Ledbetter and Seaman, 1982) to prevent graft rejection. Nevertheless, a polyclonal anti-LC antiserum did give a significant degree of immunosuppression (Fabre et al., 1980), which might be mimicked by a synergizing cocktail of monoclonal antibodies. Indeed the observation that antibodies of non-lytic subclasses were able to inhibit lysis by other lytic antibodies of the same group suggests that such a cocktail of antibodies of the most appropriate isotype may well be more effective than a polyclonal antiserum since the latter might contain a mixture of lytic and non-lytic antibodies competing for the same binding sites.

Complement-mediated lysis using combinations of the anti-human LC antibodies and autologous complement was still less than killing by CAMPATH 1, but it might be further increased if antibodies recognising more epitopes on the LC molecule could be found, e.g. by using a different immunization or screening procedure. However, one mechanism which could prevent complete lysis from occuring, even in a synergistic system, might be redistribution of antigen-antibody complexes on the cell surface, a process known as antigenic modulation (Boyse and Old, 1969). The preparation of a monovalent form (Glennie and Stevenson, 1982; Cobbold and Waldmann, 1984) of the synergizing antibodies could be a means of overcoming this problem.

REFERENCES

Aqel,N.M., Clark,M., Cobbold,S.P. and Waldmann,H., 1984, Immunohistochemical screening in the selection of monoclonal antibodies. The use of isotype-specific antiglobulins, J. Immunol. Meth., in press.

Banga,J.P., Guarnotta,G., Harte,A., Pryce,G., Campbell,M.A., Quartey-Papafio,R., Lydyard,P.M., and Roitt,I.M., 1984, A common epitope identified by a monoclonal antibody, MID2, present on all leukocytes and associated with a group of high molecular weight glycopeptides, Scand. J. Immunol., 19:11.

Borsos,T., and Rapp,H.J., 1965, Complement fixation on cell surfaces by 19S and 7S antibodies, Science, 150:505.

Boyse,E.A., and Old,L.J., 1969, Some aspects of normal and abnormal cell surface genetics, Ann. Rev. Genetics, 3:269.

Cobbold,S.P., and Waldmann,H., 1981, A rapid solid-phase enzyme-linked binding assay for screening monoclonal antibodies to cell surface antigens, J. Immunol. Meth., 44:125.

Cobbold,S.P., and Waldmann,H., 1984, Therapeutic potential of monovalent monoclonal antibodies., Nature, 308:460.

Dalchau,R., Kirkley,J., and Fabre,J.W., 1980, Monoclonal antibody to a human leukocyte specific membrane glycoprotein probably homologous to the leukocyte common antigen of the rat, Eur. J. Immunol., 10:737.

Dalchau,R., and Fabre,J.W., 1981, Identification with a monoclonal antibody of a predominantly B lymphocyte-specific determinant of the human leukocyte common antigen. Evidence for structural and possible functional diversity of the human leukocyte common molecule, J. Exp. Med., 153:753.

Elliot,E.V., Pinder,A., Stevenson,F.K., and Stevenson, G.T., 1978, Synergistic cytotoxic effects of antibodies directed against different cell surface determinants., Immunology, 34:405.

Fabre,J.W., Sunderland,C.A., and Williams,A.F., 1980, The immunosuppressive properties of rabbit antibodies against a major glycoprotein restricted to rat leukocyte membranes, Transplantation, 23:349.

Fust,G., Medgyesi,G.A., Bazin,H., and Gergely,J., 1980, Differences in the ability of rat IgG subclasses to consume complement in homologous and heterologous serum, Immunol. Lett., 1:249.

Galfre,G., Milstein,C., and Wright,B., 1979, Rat x rat hybrid myelomas and a monoclonal anti-Fd portion of mouse IgG, Nature, 277:131.

Glennie,M.J., and Stevenson,G.T., 1982, Univalent antibodies kill tumour cells in vitro and in vivo, Nature, 295:712.

Hale,G., Bright,S., Chumbley,G., Hoang,T., Metcalf,D., Munro,A.J., Waldmann,H., 1983, Removal of T cells from bone marrow for transplantation: a monoclonal antilymphocyte antibody that fixes human complement, Blood, 62:873.

Hansch,G.M., Betz,M., and Shin,M., 1984, Cytolysis of nucleated cells by complement. Inhibition of membrane transmethylation enhances cell death by C5b-9, J. Immunol., 132:1440.

Hughes-Jones,N.C., and Ghosh,S., 1981, Anti-D-coated Rh positive red cells will bind the first component of the complement pathway, C1q, FEBS Letts., 128(2):318.

Hughes-Jones,N.C., Gorick,B.D., and Howard,J.C., 1983, The mechanism of synergistic complement-mediated lysis of rat red cells by monoclonal IgG antibodies, Eur. J. Immunol., 13:635.

Kessler,S.W., 1975, Rapid isolation of antigens from cells with a staphlococcal protein A-antibody adsorbent: Parameters of the interaction of antibody-antigen complexes with protein A., J. Immunol., 115:1617.

Ledbetter,J.A., and Seaman,W.E., 1982, The Lyt-2, Lyt-3 macromolecules: structural and functional studies, Immunol. Rev., 68:197.

Metzger,H., 1978, The effect of antigen on antibodies: Recent studies, Contemp. Top. Mol. Immunol., 7:119

Springer,T.A., Bhattacharya,A., Cardoza,J.T., and Sanchez-Madrid,F., 1982, Monoclonal antibodies specific for rat IgG1, IgG2a and IgG2b subclasses and kappa chain monotypic and allotypic determinants: reagents for use with rat monoclonal antibodies, Hybridoma, 1:257.

Standring,R., and Williams,A.F., 1978, Glycoproteins and antigens of membranes prepared from rat thymocytes after lysis by shearing or with the detergent Tween-40, Biochim. Biophys. Acta, 508:85.

Sunderland,C.A., McMaster,W.R., and Williams,A.F., 1979, Purification with monoclonal antbody of a predominant leukocyte common antigen and glycoprotein from rat thymocytes, Eur. J. Immunol., 9:155.

Trowbridge,I.S., 1978, Interspecies spleen-myeloma hybrid producing monoclonal antibodies against mouse lymphocyte surface glycoprotein, T200., J. Exp. Med., 148:313.

ELIMINATION OF GRAFT VERSUS HOST DISEASE IN MATCHED ALLOGENEIC LEUKEMIC TRANSPLANT RECIPIENTS USING CAMPATH-1

S. Slavin[1], R. Or[1], L. Weiss[1], D. Steiner-Zalz[2], G. Cividalli[3], C. Brautbar[4], Z. Weshler[5], E. Galun[1], S. Samuels[1], D. Galili[6], M. Ephros[7], M. Shapiro[7], G. Hale[8] and H. Waldmann[8]

The Bone Marrow Transplantation and Cryopreservation Unit and The Immunobiology Research Laboratory, Departments of Medicine A[1], Hematology[2], Pediatrics[3] Tissue Typing Unit[4], Radiation Therapy[5], Hospital Oral Medicine Service[6] and Clinical Microbiology[7] Hadassah University Hospital, Jerusalem, Israel and The Department of Pathology[8], University of Cambridge, Cambridge, United Kingdom

INTRODUCTION

Bone marrow transplantation (BMT) offers one of the major advances in the treatment of certain malignant hematological disorders for which no specific effective alternative therapeutic modality exists, particularly leukemias. Although results of BMT in man have improved over the last few years, success rates have been limited mainly due to the reactivity of the graft against the host, a disease called graft versus host disease (GVHD). None of the treatment modalities studied thus far seems to be effective against established GVHD. Therefore, it appears that prevention of GVHD is the best approach to overcome the obstacle of GVHD. Since experimental data and recent pilot clinical studies suggest that acute GVHD results from donor type mature T cells present in the marrow aspirate (1-9), it would seem that the safest way to prevent GVHD is by elimination of T cells from the marrow allograft prior to transplantation. The present report summarizes our initial results in trying to prevent GVHD in patients undergoing BMT for leukemia using a monoclonal anti-lymphocyte antibody, CAMPATH-1 (10). Two major advantages of this novel antibody are: 1) its ability to deplete all lymphoid cells, including T cells, without damaging the stem cell compartment; and, 2) the unique characteristic of this antibody to bind human complement. Therefore, CAMPATH-1 can be used for efficient depletion of all mature lymphoid cells without exposing the unbound stem cells to heterologous complement sources.

PATIENTS AND METHODS

Ten patients with HLA identical sibling donors have entered our pilot clinical study thus far. Essential patient characteristics are described in Table 1. Three patients were in overt relapse upon admission for transplantation (1 patient with drug-resistant Burkitt's disease with CNS involvement, and 2 patients with ALL). The protocol used for conditioning of these patients consisted mainly of cyclophosphamide 60 mg/kg x 2, followed by 6 fractions of 200 rads whole body irradiation given over 3 days (the only exception was the patient with Burkitt's leukemia that was resistant to cyclophosphamide and therefore conditioned with ARA-C 3 g/m^2 x 12 over 6 days). All patients received prophylactic treatment against CNS involvement consisting of 2 injections of methotrexate 12 mg intrathecally. No post-transplant immunosuppression was used. Elimination of immunocompetent lymphocytes was carried out as follows: Anticoagulated bone marrow cells were mixed with hetastarch (Volex, Idrossoetilco, McGaw, Laboratori Don Baxter, Trieste, Italy) at a final concentration of 0.66% and placed in plastic transfusion bags for about 1 h at room temperature. Sedimenting red blood cells were discarded and mononuclear cell-enriched plasma was washed once in phosphate buffered saline (PBS) without calcium and magnesium, supplemented with penicillin 100 U/ml and gentamycin 10 µg/ml. Cells were resuspended at 50-100 x 10^6 cells/ml in PBS containing calcium chloride ($CaCl_2 \cdot 6H_2O$) and magnesium chloride ($MgCl_2 \cdot 6H_2O$) at 0.1 g/l each. CAMPATH-1 was added to a final concentration of 80-100 µg/ml for 30 min at room temperature. Fresh autologous serum was added for an additional 30-45 min at 37°C and the cells were stirred. Large aggregates were removed and the residual cells were washed once and resuspended in saline supplemented with 0.5% human serum albumin. Prior to infusion of bone marrow cells, patients received hydrocortisone 100 mg intravenously for prevention of untoward hypersensitivity reactions. Marrow cells were administered over a 15-30 min period. The constitution of various lymphocyte subsets, as well as in vitro immune functions, were studied at different time intervals following transplantation and will be reported separately.

RESULTS

Efficient removal of lymphocytes capable of forming E-rosettes with sheep erythrocytes was documented in all patients. No rosettes could be documented in marrow samples of 8 out of 10 CAMPATH-1 treated marrow samples. The proportion of rosettes documented in only the remaining 2 out of 10 treated marrow samples was extremely

Table 1. Characteristics of leukemia patients treated by lymphocyte depleted HLA-idenitical bone marrow transplantation, using CAMPATH-1 and autologous serum as a source of complement

Diagnosis	Sex/Age	Donor sex	No. of cells aspirated/kgx10^8	% E-RFC*	No. of cells infused/kgx10^8
ALL, 3rd CR prior CNS disease	F/7	M	11.30	5.5	1.11
Resistant Burkitt's ALL CNS & systemic relapse	F/10	F	6.70	8.3	1.56
ALL, 2nd CR	F/11	M	7.50	6.1	1.29
ALL, 1st relapse	M/8	M	5.33	10.0	1.29
CML, chr. phase	F/35	F	6.43	18.1	2.75
CML, chr. phase	M/9	F	8.48	3.6	3.81
ALL, resistant in relapse	F/5	F	8.26	5.7	7.45
AML, 1st CR	M/23	M	5.0	1.6	3.1
AML, 3rd CR	M/8	F	1.7	1.7	1.2
AML, 1st CR	F/35	M	2.0	1.4	1.4

* Cells forming rosettes with sheep red blood cells (T lymphocytes)

low: 0.04% as compared to an initial 8.3% in one patient, and 0.18% as compared to 10.0% in another patient. Prompt marrow engraftment occurred in all patients and none have shown any clinical signs of GVHD up to the time of this report, with the longest observation period being 13 months. Polymorphonuclear counts of $\geq 500/mm^3$ were evident on days 12-24 (median, day 15), and polymorphonuclear counts of $\geq 1000/mm^3$ were evident on days 19-28 (median, day 22). Patients were off platelet transfusions ($>20000/mm^3$ without bleeding diathesis) on days 10-29 (median, day 15). The immediate post-transplantation period was uneventful with the exception of mild mucositis and febrile episodes in most patients. Six patients are alive and well. Three patients out of the first 7 patients have normal blood counts and maintain normal daily activities for observations of 4-13 months post-transplantation. Two patients died due to relapse: the patient with stage IV Burkitt's leukemia with CNS involvement died of relapse (in fact she was never in remission), at day 48 post-transplantation with completely normal marrow function. The second patient was an 8 year old boy transplanted in relapse who died 7 months post-transplantation with normal marrow function until relapse occurred. The most worrisome problem was death of the 2 patients with chronic phase CML due to graft rejection on days 63 and 126 post-transplantation. Both had normal peripheral blood counts following bone marrow transplantation, so the development of late graft failure is compatible with graft rejection. Attempts to reconstitute the marrow of these patients with retransplantation, using non-T cell depleted marrow failed, despite the use of cyclosporin A post-transplantation. The last 3 patients transplanted with the same standard conditioning, augmented with total lymphoid irradiation (TLI) 600 rads given as 4 fractions of 150 rads twice daily, are doing fine, but the observation period is too short for a definite statement.

DISCUSSION

Our pilot trial, in which we have succeeded in depleting all immunocompetent lymphocytes from the marrow prior to transplantation using a monoclonal anti-lymphocyte antibody (CAMPATH-1), indicates that such treatment does not impair engraftment efficiency post-transplantation and that GVHD can be completely abolished without using post-transplant immunosuppression. Reconstitution of hemopoietic function post-transplantation seems to be fast and adequate and therefore the hospitalization post-transplantation is markedly shortened. Nevertheless, the main problem that seems to be a direct result of T cell depletion was rejection of the marrow in 2 patients. Increased incidence of marrow rejection was also observed in a parallel study in which we used two other approaches for depletion of immunocompetent T cells from the marrow prior to transplantation consisting of E-rosette depletion of T cells with and without prior agglutination with soybean agglutinin (results not shown). Taken together, our ongoing studies suggest that increased chances of graft rejection occur in patients transplanted with T cell depleted marrow, regardless of the type of protocol used to deplete the T cells. Based on these observations, we conclude that better immunosuppressive protocols must be determined in order to abrogate what appears to be graft rejection. Two potential approaches are legitimate: 1) to increase immunosuppression pre-transplantation; and, 2) to add post-transplant immunosuppression. The approach that we have chosen is to increase the immunosuppression pre-transplantation by adding TLI, using 4 fractions of 150 rads each. The observation period is too short for conclusions, but the 3 patients transplanted thus far appear to have tolerated the conditioning regimen reasonably well. Further observations are required to assess the long-term effects of T cell depletion, particularly in relation to the incidence of relapse which might be expected to increase due to the known reciprocal interrelationship between GVHD and leukemic relapse post-transplantation (11). Nonetheless, despite the risks associated with using T cell-depleted marrow allografts in leukemia, it appears that the complete elimination of GVHD justifies further clinical research into improving conditioning in order to ensure engraftment without the untoward sequelae of GVHD.

ACKNOWLEDGMENTS

S. Slavin acknowledges grant support from the Israel Cancer Research Fund, Israel Cancer Association and the Ruth Estrin Goldberg Memorial for Cancer Research.

H. Waldmann and G. Hale acknowledge grant support from the Medical Research Council and Leukaemia Research Fund of Great Britain.

REFERENCES

1. K. A. Dicke, J. Van Hooft, and D. W. Van Bekkum, The selective elimination of immunologically competent cells from bone marrow and lymphatic cell mixtures, Transplantation 6:562 (1968).
2. M. L. Tyan, Modification of severe graft vs host disease with antisera to the theta antigen or to whole serum, Transplantation 15:601 (1973).
3. H. V. Rodt, S. Thierfelder, and M. Eulitz, Anti-lymphocytic antibodies and marrow transplantation. III Effect of heterologous anti-brain antibodies on acute secondary disease in mice, Eur. J. Immunol. 4:25 (1974).
4. W. Muller Ruchholtz, H.U. Wottge, H. K. Muller-Hermelink, Bone marrow transplantation in rats across strong histocompatibility barriers by selective elimination of lymphoid cells in donor marrow, Transplant. Proc. 8:537 (1976).

5. Y. Reisner, L. Itzicovitch, A. Meshorer, and N. Sharon, Hemapoietic stem cell transplantation using mouse bone marrow and spleen cells fractionated by lectins, Proc. Natl. Acad. Sci. USA 75:2933 (1978).
6. D. A. Vallera, A. H. Filipovich, C. C. B. Soderling, and J. H. Kersey, Bone marrow transplantation across major histocompatibility barriers in mice, Clin. Immunol. Immunopathol. 23:437 (1982).
7. R. Korngold, and J. Sprent, Surface markers of T cells causing lethal graft-versus-host disease in mice, in "Recent Advances in Bone Marrow Transplantation," R. P. Gale, ed., Alan R. Liss, New York (1983).
8. Y. Reisner, N. Kapoor, Kirkpatrick, et al, Transplantation for severe combined immunodeficiency with HLA-A, B, D, DR incompatible parental marrow cells fractionated by soybean agglutinin and sheep red blood cells, Blood 61:341 (1983).
9. H. G. Prentice, H. A. Blacklock, G. Janossy, M. J. M. L. Gilmore, et al, Depletion of thymocytes in donor marrow prevents significant graft versus host disease in matched allogeneic leukaemic marrow transplant recipients, Lancet 1:472 (1984).
10. G. Hale, S. Bright, G. Chumbley, et al, Removal of T cells from bone marrow for transplantation: a monoclonal antilymphocyte antibody that fixes human complement, Blood 62:873 (1983).
11. P. L. Weiden, N. Flournoy, E. D. Thomas, R. Prentice, A. Fefer, C. D. Buckner, and R. Storb, Antileukemic effect of graft-versus-host disease in human recipients of allogeneic-marrow grafts, N. Eng. J. Med. 300:1068 (1979).

THE IMMUNOTHERAPEUTIC EFFECT OF ANTI LYT-1 ANTIBODIES ON LOCAL AND METASTATIC TUMOR GROWTH

Nurit Hollander

Department of Human Microbiology
Tel-Aviv University
Tel-Aviv 69978
Israel

INTRODUCTION

Immunotherapy of malignant diseases by passively administered antibodies may be accomplished by two main approaches: 1) The use of antibodies against tumor antigens, which mediate direct effects on the tumor cells. 2) The use of antibodies against components of the immune system, which exert immunoregulatory effects and hence modify the host's immune response to the tumor.

We have recently demonstrated[1,2] that anti Lyt-1 antibodies enhance (in the absence of complement) antigen-dependent T cell proliferation, induction of cytotoxic T cells and production of lymphokines such as interleukin-2, lymphotoxin, macrophage activating factor and interferon. Since T cell-mediated immune responses play an important role in tumor rejection, we explored whether anti Lyt-1 antibodies can enhance immune responses in vivo and thus potentiate tumor rejection. This report demonstrates the therapeutic effect of anti Lyt-1 monoclonal antibodies on the progression of a syngeneic metastatic tumor.

MATERIALS AND METHODS

Mice

C57Bl/6 male mice at the age of 8-16 weeks were used.

Antibodies

Hybridoma cell lines producing rat anti Lyt-1 (53-7.3) and anti Lyt-2 (53-6.7) were previously described[3]. They were maintained in Dulbecco's modified Eagle's medium supplemented with 10% fetal calf serum. Monoclonal antibodies were used as culture supernatants of cells incubated for 24 h in medium devoid of serum. Antibodies were precipitated from serum free culture supernatants by 50% ammonium sulfate. For direct conjugation with fluorescein, monoclonal antibodies were purified from supernatants on goat anti rat Ig affinity columns.

Tumor

The 3LL Lewis lung carcinoma is a malignant tumor which metastasizes to the lungs. It originated spontaneously in a C57Bl/6 mouse and was maintained by subcutaneous transfer of 1 x 10^6 tumor cells in C57Bl/6 mice. Single cell suspensions were prepared from solid tumors by treatment of minced tumor tissue with a solution of 0.3% trypsin.

Assay of metastases

Metastatic growth was determined by lung weight and by incorporation of ^{125}IUdR into lung cells. 3LL cells (1 x 10^5) were inoculated into the hind right footpad of C57Bl/6 mice. At different periods after tumor inoculation, established tumors were removed by amputating the leg after ligation above the knee joint. At 13 days following amputation, mice were injected intraperitoneally with 25 g of FUdR and 30 min later with 1 Ci of ^{125}IUdR. One day later, lungs were removed and weighed, and the degree of ^{125}IUdR incorporation was measured by a gamma spectrometer.

RESULTS

C57Bl/6 mice were inoculated subcutaneously with 3LL tumor cells. Anti Lyt-1 or anti Lyt-2 antibodies were injected intraperitoneally to tumor inoculated mice. Antibody injections commenced at time of tumor inoculation and were repeated at intervals of 4 days for 20 days. As shown in Figure 1, anti Lyt-1 antibodies had a significant therapeutic effect on mice inoculated with 2 x 10^4 tumor cells. In contrast to the therapeutic effect of anti Lyt-1, anti Lyt-2 antibodies of the same Ig class (IgG_{2A}) slightly enhanced tumor growth and thus reduced the survival time of tumor-bearing mice.

A possible mechanism by which anti Lyt-1 antibodies directly effect tumor cells is unlikely, because Lyt antigens are not expressed on 3LL cells[4]. It therefore seemed that the antibodies exert their effect via the host immune system. In order to test this possibility, the immunotherapeutic schedule was repeated in B mice (adult thymectomized, lethally irradiated and reconstituted with fetal liver cells). As shown in Figure 1, depletion of T cells from tumor inoculated mice abrogated the curative effect of passively administered anti Lyt-1 antibodies. These results suggest that host

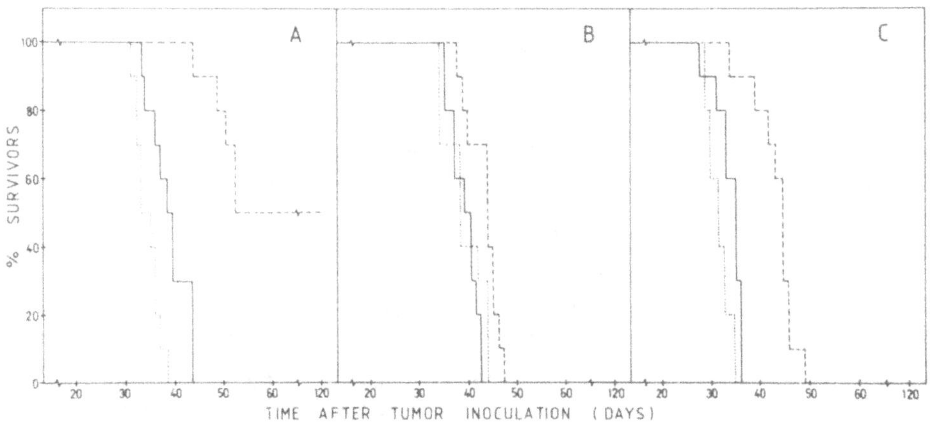

Fig. 1. The therapeutic effect of anti Lyt-1 antibodies on 3LL tumor. Normal mice (A,C) or B mice (B) were inoculated subcutaneously with 2 x 10^4 (A,B) or 1 x 10^5 (C) 3LL tumor cells. 0.2 mg anti Lyt-1 (broken line), 0.2 mg Lyt-2 (dotted line), or phosphate buffered saline (solid line) were injected intraperitoneally at the time of tumor inoculation and thereafter at intervals of 4 days for 20 days. Each experimental group consisted of 10 mice.

Lyt-1^+ lymphocytes are involved in immunotherapy.

Anti Lyt-1 antibodies could effect the host immune system either by depletion of Lyt-1^+ cells or by direct alteration of cell function. To investigate the mechanism of antibody-mediated effects on the host immune system, lymphocyte distribution and activity was analyzed in anti Lyt-1-treated mice. Table 1 shows that antibody treatment did not alter the frequency of Lyt-1^+ cells in spleen and lymph nodes. A significant proportion of cells carried the injected anti Lyt-1 antibodies, as demonstrated by the ability of anti rat Ig antibodies to stain cells from injected mice (Table 1). However. the intensity of staining with anti rat Ig antibodies (not shown) and the percentage of cells stained by anti Lyt-1 and by anti Thy-1 antibodies revealed that the cells bound with anti Lyt-1 were not saturated by the antibodies.

The results indicate that the antibody amounts used for immunotherapy were not sufficient to deplete Lyt-1^+ target cells. Moreover, the administered Lyt-1 antibodies did not reduce the proliferative and cytolytic activity of spleen cells. As shown in Table 2, spleen cells of anti Lyt-1-treated mice manifested an augmented activity in mixed lymphocyte cultures, as compared to normal spleen cells. Thus, administration of anti Lyt-1 antibodies did not cause depletion of Lyt-1^+ cells from lymphoid organs, but most probably alter Lyt-1^+ cell functions.

TABLE 1 FLUORESCENCE ANALYSIS OF LYMPHOCYTES FROM MICE TREATED WITH ANTI LYT-1 ANTIBODIES

In vivo treatment of mice[a]	Percent positive cells by direct immunofluorescence[b]					
	Spleen			Lymph nodes		
	Lyt-1[c]	Thy-1	rat Ig	Lyt-1	Thy-1	rat Ig
-	28	28	1	60	62	1
anti Lyt-1	27	30	13	61	63	20

[a]0.2 mg of anti Lyt-1 were injected intraperitoneally at intervals of 96 h. Analysis of spleen and mesenteric lymph node cells was performed 24 h after the third injection.

[b]Cells were treated with directly fluorescein-conjugated antibodies and analyzed by a fluorescence-activated cell sorter.

[c] :anti.

TABLE 2 ALLOSPECIFIC RESPONSE OF SPLEEN CELLS FROM MICE TREATED WITH ANTI LYT-1 ANTIBODIES [a]

In vivo treatment[b]	In vitro treatment	^{3}H-thymidine uptake (CPM)	Cytolytic activity (% ^{51}Cr release)
-	-	59,472	58
anti Lyt-1	-	85,055	74
-	anti Lyt-1 + complement	10,891	12
-	complement	56,222	57

[a]Spleen cells of anti Lyt-1 treated C57Bl/6 mice were incubated for 96 h with mitomycin C - treated BALB/C spleen cells. Proliferation of responding cells was then determined by ^{3}H-thymidine uptake, as previously described[5]. Cytolytic activity was determined by ^{51}Cr release from P815 target cells at effector cell to target cell ratio of 5:1[5].

[b]0.2 mg of anti Lyt-1 were injected intraperitoneally at intervals of 96 h. Mixed lymphocyte reaction was initiated 24 h after the third injection.

The tumor mass at the time of anti Lyt-1 administration was a critical factor for successful tumor therapy. Mice inoculated with 2 x 10^4 tumor cells were cured by anti Lyt-1 antibodies, whereas anti Lyt-1 treatment of mice inoculated with 1 x 10^5 tumor cells resulted in prolonged survival time with no long-term cure (Fig. 1). The fact that a high tumor burden present at the time of therapy, is a limiting factor, implies that anti Lyt-1 antibodies may not be applicable for curing established primary tumors. However, they may be efficient for controlling the growth of residual tumor cells following removal of primary tumors. We therefore studied the effects of anti Lyt-1 antibodies on development of 3LL lung metastases following excision of primary tumors. Table 3 demonstrates the development of lung metastases in antibody-treated and untreated mice. It indicates that: a) anti Lyt-1 antibodies, in contrast to anti Lyt-2 antibodies, suppress lung metastasis, and b) the efficacy of antibody treatment correlated with the size of tumor at the time of local tumor excision. Thus, anti Lyt-1 antibodies arrested metastases development when local tumors were removed at early stages. However, when local tumors reached large sizes before removal, anti Lyt-1 antibodies attenuated metastases growth, but could not mediate long-term cure.

DISCUSSION

The objective of the present study was to establish an immunotherapy model in which antibodies effect growth of transplanted tumors by influencing tumor-host interactions. It demonstrates the utility of monoclonal anti Lyt-1 antibodies for the treatment of a metastatic tumor, which does not express the Lyt-1 antigen. Although anti Lyt-1 antibodies cannot mediate regression of established tumors, they may be efficient for controlling the growth of small number of cells, such as residual cells following removal of primary tumors. The efficacy of treatment with anti Lyt-1 correlated with the tumor mass at the time of therapy commencement. Antibody treatment which commenced subsequent to removal of small tumors, resulted in 50% survival, compared to 100% mortality in control mice, which were not subjected to immunotherapy. However, excision of large tumors, followed by anti Lyt-1 treatment resulted in attenuated metastases development, but not in long-term cure.

A prerequisite for immunotherapy with monoclonal anti Lyt-1 antibodies is the use of antibody levels that would not deplete target cells. The role of T cell subsets in tumor rejection is controversial, but there is considerable evidence that Lyt-1^+2^- cells mediate tumor graft rejection by delayed hypersensitivity-like mechanisms[6-8]. Depletion of Lyt-1^+ cells may therefore impair resistance to tumors, whereas enhancement or other alteration of Lyt-1^+ cell functions by antibodies, which do not deplete the cells, may augment resistance to tumors. Indeed, this report demonstrated that treatment of tumor bearing mice with anti Lyt-1 antibody doses, which did not effect the frequency of Lyt-1^+ cells, or their in vitro reactivity to allogeneic antigens, resulted in prolonged survival of the treated mice. Our results do not agree with the results of Michaelides et al[9], who demonstrated that administration

TABLE 3 THE EFFECT OF ANTI LYT-1 ANTIBODIES ON PROGRESSION OF 3LL LUNG METASTASES [a]

Diameter of amputated tumor (mm)	In vivo treatment of mice	Weight of lungs (mg)	^{125}IUdR uptake (CPM)	Mean survival time following excision of local tumor (days)	Percent mortality
3-4	-	275 ± 30	1,498 ± 137	25.7 ± 2.2	100
	anti Lyt-1	203 ± 27	706 ± 101	38.0 ± 4.8[b]	50
	anti Lyt-2	374 ± 28	2,781 ± 122	21.1 ± 2.0	100
5-6	-	428 ± 84	3,993 ± 301	21.4 ± 1.9	100
	anti Lyt-1	260 ± 31	1,010 ± 212	33.1 ± 3.5[b]	80
	anti Lyt-2	583 ± 85	4,994 ± 280	17.3 ± 2.7	100
7-8	-	734 ± 109	8,010 ± 632	18.9 ± 1.1	100
	anti Lyt-1	431 ± 49	4,260 ± 710	26.8 ± 3.0	100
	anti Lyt-2	869 ± 80	11,523 ± 998	16.1 ± 2.3	100
9-10	-	908 ± 112	20,870 ± 1279	17.0 ± 1.0	100
	anti Lyt-1	634 ± 70	17,900 ± 1344	19.0 ± 1.9	100
	anti Lyt-2	989 ± 168	21,446 ± 611	16.3 ± 1.2	100
Normal mice		192 ± 25	573 ± 52		

[a] 1×10^5 3LL cells were inoculated into the right hind footpad of C57Bl/6 mice. Nine to seventeen days later mice were divided into groups according to the diameter of their tumors. The local tumors were removed by leg amputation. Antibody treatment commenced at time of tumor excision and was repeated with doses of 0.2 mg at intervals of 72 h for 12 days. Thirteen days following amputation, ten mice of each experimental group were injected with 25 g of FUdR and 30 min later with 1 Ci of ^{125}IUdR. The next day mice were killed, and the weights and radioactivity of lungs were determined. The remaining ten mice of each experimental group were followed up for mean survival time.

[b] Excluding survivor mice.

of small amounts of anti Lyt-1 monoclonal antibodies resulted in depletion of Lyt-1$^+$ cells from lymphoid organs and in inhibition of various immune responses, including tumor rejection. This difference in effects of anti Lyt-1 antibodies might be due to a difference in the abilities of the individual antibodies to cooperate with the host clearance mechanisms.

The anti Lyt-1 antibodies, used by us for treatment of syngeneic tumors, were previously used by Ledbetter and Seaman[10] for treatment of allogeneic tumor grafts. Their studies, resulting in enhancement of allogeneic tumor rejection, agree with our results for syngeneic tumors and hence support a possible therapeutic value of this antibody. We have recently demonstrated[1] that anti Lyt-1 antibodies augment (in the absence of complement) in vitro production of interleukin-2, which results in enhanced cell proliferation and generation of cytotoxic T lymphocytes. Anti Lyt-1 antibodies also augment production of other lymphokines, such as lymphotoxin, macrophage activating factor and immune interferon[2]. These lymphokines may play an important role in the control of tumor growth. It is therefore suggested that the antibodies operate in vivo by the same mechanisms. However, anti Lyt-1 antibodies may as well alter cell functions or cell-cell interactions which are yet unknown.

The effectiveness of antibodies against T cell differentiation antigens for the treatment of metastatic tumors is of considerable clinical importance, because human T cells express differentiation antigens that are analogous to those of mice. The T1 (Leu 1) antigen of human T cells was postulated to be the homologue of the murine Lyt-1 antigen[11]. It has been recently reported[12] that OKT1 antibodies markedly enhanced helper activity of human OKT4$^+$ cells, probably by increased production of helper molecules. It therefore seems that Lyt-1 and T1 have a common functional role in murine and human immune responses. Antibodies directed to the human homologue of mouse Lyt-1 antigen may therefore be useful for tumor immunotherapy.

SUMMARY

Monoclonal antibodies against the T cell differentiation antigen Lyt-1 were effective in the therapy of the murine Lewis lung carcinoma (3LL). A possible mechanism whereby anti Lyt-1 antibodies directly effect the tumor cells was unlikely because 3LL cells lack the Lyt-1 antigen. In addition, the curative effect of anti Lyt-1 antibodies was abrogated in mice depleted of T cells, supporting a mechanism whereby host Lyt-1$^+$ cells were involved in tumor therapy. Treatment with anti Lyt-1 antibodies was not accompanied by depletion of Lyt-1$^+$ cells from lymphoid organs, indicating that the administered antibodies altered Lyt-1$^+$ cell functions without effecting their frequency. In view of the in vitro enhancing effects of anti Lyt-1 antibodies on responses, it is suggested that the potentiation of Lyt-1$^+$ cell activity by passively administered anti Lyt-1 antibodies results in tumor rejection.

ACKNOWLEDGEMENT

This work was supported by a grant from the Israel Cancer Association.

REFERENCES

1. N. Hollander, Effects of anti Lyt antibodies on T cell functions, Immunol. Rev. 68:43 (1982).
2. N. Hollander and J. Shepshalovich, The functional association of Lyt-1 antigen with lymphokine production, Immunology, in press.
3. J. A. Ledbetter and L. A. Herzenberg, Xenogeneic monoclonal antibodies to mouse lymphoid differentiation antigens, Immunol. Rev. 47:63 (1979).
4. N. Hollander, Immunotherapy of lymphoid and non-lymphoid tumors with monoclonal anti Lyt-1 antibodies, J. Immunol., in press.
5. N. Hollander, E. Pillemer and I. L. Weissman, Blocking effect of Lyt-2 antibodies on T cell functions, J. Exp. Med. 152:674 (1980).
6. P. D. Greenberg, M. A. Cheever and A. Fefer, Eradication of disseminated murine leukemia by chemoimmunotherapy with cyclophosphamide and adoptively transferred immune syngeneic Lyt-1^+2^- lymphocytes, J. Exp. Med. 154:952 (1981).
7. A. K. Bhan, L. L. Perry, H. Cantor, R. T. McCluskey, B. Benacerraf and M. I. Greene, The role of T cell sets in the rejection of a methylcholanthrene-induced sarcoma (S150g a) in syngeneic mice, Am. J. Pathol. 102:20 (1981).
8. B. E. Loveland and I. F. C. McKenzie, Cells mediating graft rejection in the mouse. II. The Ly phenotypes of cells producing tumor allograft rejection, Transplantation 33:174 (1982).
9. M. Michaelides, P. M. Hogarth and I. F. C. McKenzie, The immunosuppressive effect of monoclonal anti Lyt-1.1 antibodies in vivo, Eur. J. Immunol. 11:1005 (1981).
10. J. A. Ledbetter and W. E. Seaman, The Lyt-2, Lyt-3 macromolecules: structural and functional studies, Immunol. Rev. 68:197 (1982).
11. J. A. Ledbetter, R. L. Evans, M. Lipinski, C. Cunningham-Rundles, R. A. Good and L. A. Herzenberg, Evolutionary conservation of surface molecules that distinguish T lymphocyte helper/inducer and cytotoxic/suppressor subpopulations in mouse and man, J. Exp. Med. 153:310 (1981).
12. Y. Thomas, L. Rogozinski and L. Chess, Relationship between human T cell functional heterogeneity and human T cell surface molecules, Immunol. Rev. 74:113 (1983).

CHEMOTACTIC ANTI-TUMOUR ANTIBODIES: IN VITRO RESULTS WITH FOUR DIFFERENT ANTIBODY PREPARATIONS

R. Obrist, J.P. Obrecht, R.C. Bast*, Jr. and A.L. Sandberg+

Div. of Oncology, Dept. of Internal Medicine of the University, Kantonsspital, CH-4031 Basel, Switzerland
*Dana Farber Cancer Center, Boston, MA 02115, USA
+Lab. of Microbiology and Immunology, NIDR, NIH Bethesda, MD 20205, USA

INTRODUCTION

The biological behaviour and prognosis of a tumour may be related to the number of intratumour inflammatory cells. However, based on retrospective analysis, tumour infiltrating inflammatory cells have been associated with suppression of metastasis[1], or, to the contrary, enhancement of tumour growth[2]. Recently, we demonstrated that prospective enhancement of tumour infiltrating macrophage numbers in vivo is possible by the administration of anti-tumour heteroantibodies covalently coupled to the chemotactic peptide formyl-methionyl-leucyl-phenylalanine (fMLP)[3]. Using mouse monoclonal antibodies (moab), which have the advantages of being of one molecular species only and of virtually unlimited supply, a more strict evaluation of the biological role of tumour infiltrating macrophages might therefore become feasible. We already demonstrated in vitro chemotactic activity and antigen recognition function of moab-fMLP conjugates similar to the hetero-antibody-fMLP conjugate[4,5]. Whereas a similar average number of binding sites can be assumed for different polyclonal antibodies, conjugation of fMLP to moab is limited by the available number of binding sites on the given moab molecule, which might differ from one moab species to another. In this paper, the in vitro chemotactic activity of polyclonal IgM and IgG conjugates and of three different moab-conjugates is compared. In addition, the number of fMLP molecules bound to one moab molecule is estimated based on the conjugation of fMLP containing trace amounts of ^{3}H-fMLP.

METHODS

Antibody purification, fMLP-conjugation procedure and in vitro chemotaxis assays have all been previously described in detail[3,4,5]. Briefly, rabbit IgM was purified by S200 column chromatography and moab by salt precipitation and S200 chromatography or Staph A protein affinity chromatography. The fMLP-antibody conjugation was done under approppriate conditions with 1-cyclohexyl-3-(2-morpholinoethyl) carbodiimide metho-p-toluenesulfonate (CDI) and human and guinea pig monocyte/macrophage chemotaxis was determined in a modified Boyden chamber system. ^{3}H-fMLP obtained from New England Nuclear was used as trace label in cold fMLP. Molar binding ratios were calculated based on the specific activity of ^{3}H-fMLP, dilution, counter efficiency and Lowry protein determinations. Moab OC 125 and OC 133 were originally produced against ovarian tumour cell lines[6] and moab 96.5, which is directed against a p97 melanoma associated antigen[7], was generously supplied by Hybritech, Inc., San Diego, California.

RESULTS

Chemotactic activity of a rabbit polyclonal IgM-fMLP conjugate is shown in direct comparison to IgG-fMLP in Table 1. In this experiment guinea pig glycogen-induced peritoneal exsudate cells were used. In Figure 1, the dose-response curves of human peripheral blood monocytes to three different moab-fMLP preparations are shown. Data for OC 125 and OC 133 are taken from (5). Chemotaxis results to the moab 96.5- fMLP conjugate are given for two different experiments, in order to demonstrate that even with different reactivity of the monocytes used in a particular experiment (i.e. different spontaneous migration and maximal migration) the shape of the curve remains a constant straight line. This is in contrast to the sigmoid response curves for OC 125 and OC 133, which do descend again after reaching a maximum at protein concentrations below 1000 ug/ml [not shown, see references (4) and (5)].

The differently shaped dose-response curves of different moab conjugates might be due to differences in molar coupling ratios of fMLP to IgG. We have shown earlier that 3-5 molecules of fMLP get conjugated to each molecule of polyclonal rabbit antibody under specified coupling conditions. This, however, might not be the case for every moab species. Therefore fMLP was conjugated to an IgG2a moab with trace amounts of ^{3}H-fMLP added. The OD. 280 nm elution profile of this conjugate from a S200 column is shown together with

Table 1. Chemotaxis of guinea pig peritoneal macrophages to polyclonal rabbit IgM and IgG alone or conjugated to fMLP.

protein ug/ml	IgM[a]	IgM-fMLP	IgG	IgG-fMLP
25	6.8 ± 2.1	3.6 ± 1.0	9.2 ± 0.7	6.2 ± 2.0
50	5.4 ± 0.6	7.8 ± 0.9	8.2 ± 1.2	12.8 ± 3.4
100	6.6 ± 0.6	16.2 ± 2.1	7.2 ± 1.9	15.2 ± 5.9
200	8.2 ± 1.8	49.6 ± 1.6	5.4 ± 1.8	26.8 ± 1.1
400	13.6 ± 0.8	99.8 ± 15.6	13.6 ± 3.3	35.4 ± 4.9

[a]Cells per high power field (hpf), mean ± SEM of triplicate determinations.

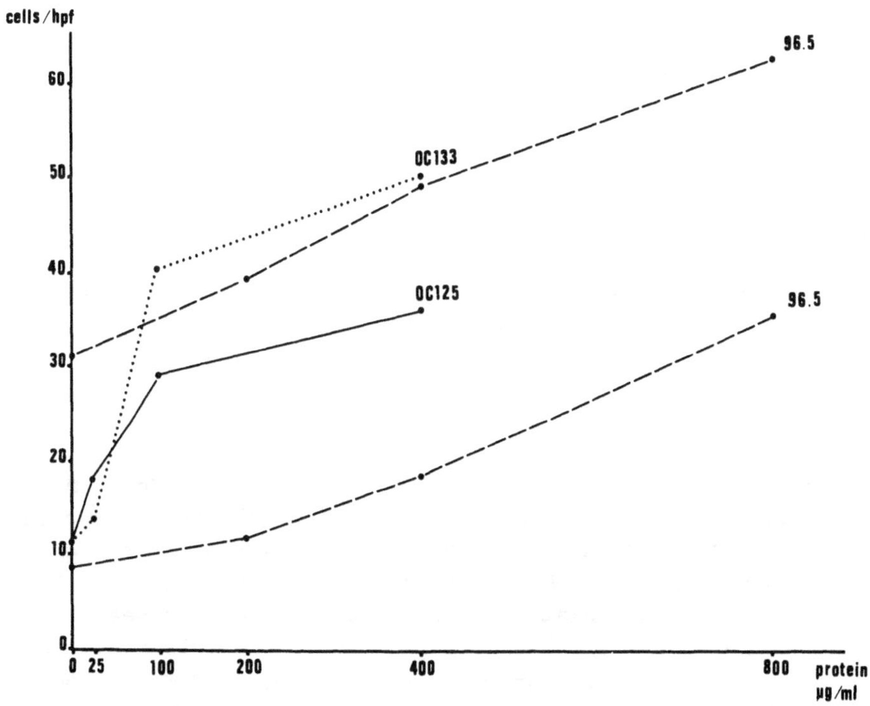

Fig. 1. Chemotaxis of human peripheral blood monocytes to different concentrations of fMLP conjugated with the moab OC 125, OC 133 and 96.5 (two different experiments). Mean of triplicate determinations.

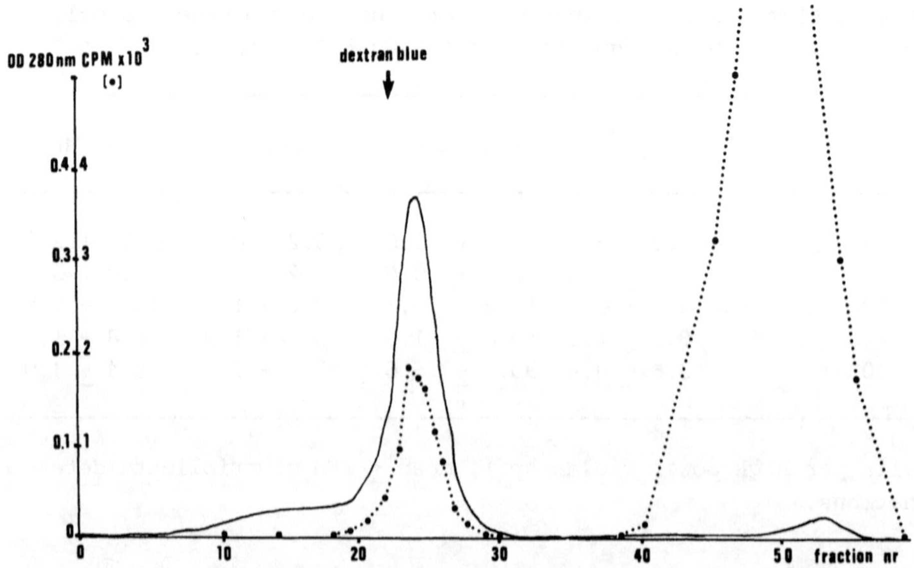

Fig. 2. Elution profile of an IgG2a-fMLP moab conjugate with ^{3}H-fMLP traces from a Sephacryl S200 column. Optical density at 280 nm (OD 280nm) is given as a continuous line, counts per minute of a 300ul aliquot (CPMx10^{3}) as a dotted line.

the number of radioactive counts/300 ul aliquot represented by a heavy dot. The co-elution of the radioactivity with the IgG peak demonstrates coupling of fMLP to IgG. The broad low molecular weight peak represents free fMLP. From these data a molar coupling ratio of only 0.05 to 0.1 for this given antibody has been calculated.

DISCUSSION

Because of the star-like structure of pentameric IgM, fMLP bound to IgM monomers away from the contacting cell surface would remain chemotactically inactive. Therefore a lower activity per protein concentration of IgM-fMLP conjugates in comparison to IgG conjugates could be postulated. This, however, does not seem to be the case, as shown in Table 1. Depending on the intrinsic reactivity of the cells used, IgG-conjugates have been shown to induce chemotaxis at concentrations as low as 12 ug/ml[4]. In this particular experiment conjugate concentrations of 200 ug/ml were needed to obtain a clearcut chemotactic response. Interestingly, IgM-fMLP induces higher responses than the IgG-conjugate. Speculatively this could be explained by either a higher coupling ratio of fMLP to monomeric IgM or by the fMLP of more than one monomer of pentameric IgM reacting with cell surface fMLP-receptors, thus multiplying effective fMLP concentrations. Whether this is sterically possible is unknown. Polyclonal IgG and IgM alone do not show any significant chemotactic activity.

A comparison of chemotactic dose-responses for the three different moab-conjugates is not directly possible from Figure 1., because the data are taken from different experiments. Nevertheless it is evident, that the OC 125 and OC 133 conjugates show sigmoid dose-response curves, whereas chemotaxis induced by the 96.5 conjugate increases linearly with increasing protein concentration. This linearity is reproducible in different experiments with widely different spontaneous migration of the monocytes used, as is the slope of the curve (Fig. 1.). Speculatively, several fMLP molecules present on the carrier protein might react simultaneously with different surface fMLP-receptors until e.g. steric hindrance leads to a flatter shape of the curve. At a very low binding ratio of fMLP to the IgG, reactions with the fMLP receptor would be relatively rare single hit events directly proportional to the concentration of the carrier protein and therefore the dose response curve linear.

In order to get some information on the coupling ratio obtained when using moab, experiments as shown in Figure 2. were done. Based on these data a molar ratio of only 0.05 to 0.1 was calculated, which is well in accordance with the hypothesis discussed and much lower than the molar binding ratio of 3-5 obtained with polyclonal rabbit IgG. We do not know at this point, whether this low binding ratio is due to the structure of this antibody species only, or whether whole groups, e.g. the IgG2a subtype generally, behave in this way. Experiments aimed at clarifiying this question are currently under way. Furthermore the in vivo significance of such differences is clear, but possibly important, if this model is to be used to further clarify the biological role of tumour infiltrating macrophages.

This work was supported in part by the Swiss National Science Foundation and the Swiss Cancer League.

REFERENCES

1. S.A. Eccles, and P. Alexander, Macrophage Content of Tumours in Relation to Metastatic Spread and Host Immune Reaction, Nature 250: 667 (1974).
2. R.T. Prehn, Immunostimulation of the Lymphodependent Phase of Neoplastic Growth, J. Natl. Cancer Inst. 59: 1043 (1977).
3. R. Obrist, and A.L. Sandberg, Enhancement of Macrophage Invasion of Tumors by Administration of Chemotactic Factor - Antitumor Antibody Conjugates, Cellular Immunology 81: 169 (1983).
4. R. Obrist, and A.L. Sandberg, In Vitro Effects of Antitumor Antibody - Chemotactic Factor Complexes, Clin. Immunol. Immunopathol. 25: 91 (1982).
5. R. Obrist, R. Reilly, T. Leavitt, R.C. Knapp, and R.C. Bast, Jr., Monocyte Chemotaxis Mediated by Formyl-Methionyl-Leucyl-Phenylalanine Conjugated with Monoclonal Antibodies Against Human Ovarian Carcinoma, Int. J. Immunopharmac. 5: 307 (1983).
6. R.C. Bast, M. Feeney, M. Lazarus, L.M. Nadler, R.B. Colvin, and R.C. Knapp, Reactivity of a Monoclonal Antibody with Human Ovarian Carcinoma, J. Clin. Invest. 68: 1331 (1981).
7. J.P. Brown, K. Nishiyama, I. Hellstroem, K.E. Hellstroem, Structural Characterization of Human Melanoma - Associated Antigen p97 Using Monoclonal Antibodies, J. Immunol. 127: 539 (1981).
8. R. Obrist, Monoclonal Antibodies as Drug Carriers in Oncology, Trends in Pharmacological Sciences 4: 375 (1983).

A MONOCLONAL ANTIBODY WITH SELECTIVITY FOR HUMAN *KAPPA* MYELOMA AND LYMPHOMA CELLS WHICH HAS POTENTIAL AS A THERAPEUTIC AGENT

Walker, K. Z., Boux, H. A., Hayden, G. E., Goodnow, C. C., and Raison, R. L.

The Clinical Immunology Research Centre
University of Sydney
Sydney N.S.W. 2006 Australia

ABSTRACT

K-1-21 is a monoclonal antibody which binds to human free *kappa* light chains and recognises a determinant, KMA selectively expressed on *kappa* myeloma and lymphoma cells. KMA is absent on plasma cells and resting B cells from normal adults but can be detected on some foetal B cells and a small proportion of activated B cells. Expression of KMA is greatest on cycling cells. K-1-21 is an IgG_1 antibody that elicits ADCC but will not cap the KMA determinant unless a second ligand is present. K-1-21 has potential for practical application.

INTRODUCTION

K-1-21 is a murine IgG_1 monoclonal antibody raised against a human *kappa* Bence Jones protein. It binds to free monomers and dimers of human *kappa* light chains but does not bind to intact *kappa* positive immunoglobulin molecules (Boux and Raison et al., 1983). K-1-21 has also been found to react with an epitope designated KMA (*kappa* myeloma antigen) selectively expressed on the surface of *kappa* myeloma and lymphoma cells. Since K-1-21, in contrast to anti-idiotypic antibodies is not individual patient specific, the antibody has clinical potential. Studies have been made of factors which might affect this practical application.

MATERIALS AND METHODS

Lymphoid Cell Populations and Human Cell Lines

Mononuclear cell (MNC) suspensions from peripheral blood or bone marrow obtained from patients with myeloma and other malignancies were prepared by separation on Ficoll-Hypaque gradients. MNC suspensions were also prepared from normal peripheral blood or bone marrow, from electively removed tonsil, from spleens removed after traumatic rupture or in cases of idiopathic thrombocytopenic purpura, from cord blood, and from foetal liver and spleen. MW 28, an EBV transformed line was obtained from Dr J. Gibson, (Clinical Immunology Research Centre, Sydney) while other cell lines came from the Ludwig Institute of Cancer Research, Sydney and London.

Antibody-dependent Cell Mediated Cytotoxicity (ADCC)

LICR LON/HMy2 (HMy2) cells were stained with Rose Bengal (O'Neill and Parish, 1983) and allowed to bind K-1-21 antibody which had been preincubated with an excess of either *kappa* or *lambda* Bence Jones protein (BJP). Effector cells were adherent cell depleted normal peripheral blood MNC. Effector cells were incubated with antibody-coated target cells for 1 h at 37°C. After centrifugation, the cell supernatant was removed and optical density at 560nm was measured to give an estimate of the dye release from lysed target cells.

Capping of the KMA:K-1-21 complex

To determine whether KMA would be capped from the cell surface after binding by K-1-21, HMy2 cells were incubated with K-1-21 at 37°C for up to 30 h. The cells were then stained at 4°C with FITC-conjugated sheep anti-mouse immunoglobulin (FITC-SaMIg) to detect the KMA:K-1-21 complex. To determine the effect of adding a second ligand, HMy2 cells were incubated at 4°C with K-1-21 followed by incubation with FITC-SaMIg at 37°C. This incubation was also carried out in the presence of capping inhibitors 10^{-4}M dibucaine, 5×10^{-6}M calcium ionophore or 10^{-2}M sodium azide. Stained cells were analysed by flow cytometry.

Elutriation

HMy2 cells were fractionated into enriched growth phases by centrifugal elutriation as described elsewhere (Boux et al., 1984). Separated cells were then analysed by flow cytometry for DNA content and for surface immunofluorescence after staining with either K-1-21 and FITC-SaMIg or with a polyvalent FITC conjugated goat anti-human *kappa* reagent (FITC-GaHK).

TABLE I

Binding of K-1-21 to the surface of cells from patients with myeloma and other immunological disorders and to a panel of human continuous cell line.

Cell Type	Source	Reactivity with K-1-21	
		No.Positive/ No.Samples	% Cells Positive (Range)
Bone Marrow Plasma Cells	Non-Myelomatous*	- (0/8)	-
	Kappa Myeloma	+ (8/11)	5 - 89
	Kappa Macroglobulinemia	+ (2/2)	0.5 - 45
	Lambda Myeloma	- (0/5)	-
Peripheral Blood MNC	*Kappa* B CLL	- (0/6)	-
	Lambda B CLL	- (0/3)	-
	Kappa Myeloma	+ (4/8)	0.5 - 3
	Lambda Myeloma	- (0/7)	-
	Kappa Lymphoma	+ (6/7)	1.0 - 8.5
	Lambda Lymphoma	- (0/3)	-
	Kappa Macroglobulinemia	- (0/4)	-
	Lambda Macroglobulinemia	- (0/2)	-
	Sezary Syndrome	- (0/4)	-
	Multiple Sclerosis	- (0/2)	-
	T CLL	- (0/3)	-
	Hairy Cell Leukemia	- (0/1)	-
	T ALL	- (0/1)	-
	Benign Monoclonal Gammopathy	- (0/1)	-
Human Continuous Cell Lines	B ALL(LAZ-007,RDG,WIL)	- (0/3)	-
	Burkitt Lymphoma (Daudi)	- (0/1)	-
	B Lymphoblastoid (HMy2,MW28)	+ (2/2)	62 - 80
	Melanoma (PMC-22B)	- (0/1)	-
	T ALL (CCRF-HSB,F2/F7 MOLT-4,CCRF-CEM,MPB ALL)	- (0/5)	-
	T Leukemia (Jurkat)	- (0/1)	-

Cells were stained with K-1-21 or a control monoclonal antibody followed by FITC-SaMIg.
* These included amyloid, benign paraproteinemia, SLE, B CLL, AML and lymphoma. All contained *Kappa* positive cells as detected with an FITC-goat anti-human *Kappa* reagent.

In Vitro Stimulation of Human B cells

MNC from human tonsil or spleen were subjected to double rosetting with AET-treated SRBC to yield a B cell enriched (E^-) population which was cultured in RPMI-10%FCS at 10^6 cells ml^{-1} in the presence or absence of formalin fixed Staphylococcus aureus Cowan I strain (SAC) (Muraguchi and Fauci, 1982). On day three, half the culture medium was removed and replaced with fresh medium with 20% MLA 144 supernatant as a source of TRF. On day six, cells were stained for KMA with K-1-21 preincubated with kappa or lambda BJP, followed by FITC-SaMIg. Cell staining was analysed by flow cytometry.

RESULTS

The data given in Table 1 indicate that KMA can be detected by surface immunofluorescence on the bone marrow plasma cells from patients with kappa myeloma and kappa macroglobulinemia and on peripheral blood MNC from patients with kappa myeloma and kappa

TABLE II

Binding of K-1-21 to the surface of normal adult tissues, foetal tissues, or to cells stimulated in vitro.

"Normal" MNC Source	Surface Immunofluorescence	
	K-1-21 (No.positive/No.tested)	GaHK (No.positive/No.tested)
Peripheral Blood	- (0/9)	+ (9/9)
Tonsil	- (0/6)	+ (6/6)
SAC activated Tonsil	+ (3/4)*	ND
Spleen	- (0/5)	+ (5/5)
Bone Marrow	- (0/5)	+ (5/5)
Foetal Liver (14-20wk)	- (0/6)	+ (6/6)
Foetal Spleen (16wk)	+ (3/3)**	+ (3/3)
Cord Blood	- (0/15)	+ (15/15)

* 3.0 - 17.0% cells positive
** 1.0 - 17.8% cells positive
Cells were stained with K-1-21 preincubated with Kappa or Lambda PBL followed by FITC-SaMIg; or with a polyvalent FITC-GaHK reagent.

lymphoma. KMA was not detected on equivalent *lambda* expressing malignancies nor on cells from patients with B CLL or with a

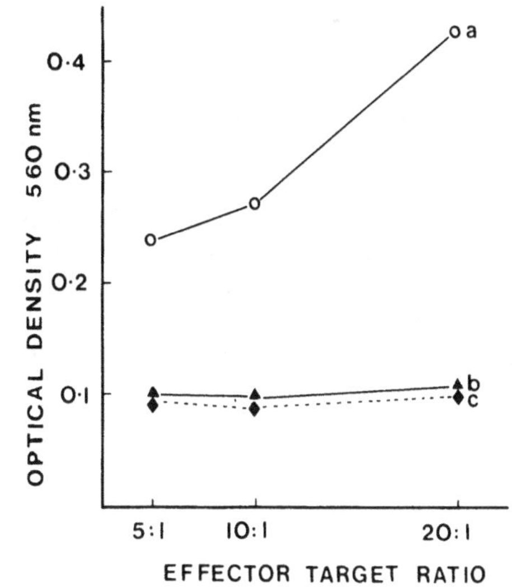

FIGURE 1

K-1-21 (a and c) or K-1-21 preincubated with excess free *kappa* light chain (b) was bound to HMy2 target cells stained with Rose Bengal. Cell lysis, estimated by dye release, was measured after 1 h incubation of the target cells with adherent cell depleted MNC (a and b) or with unstained HMy2 cells (c).

variety of other immunological disorders including one case of benign *kappa* paraproteinemia. KMA is also expressed on the surface of cells from two lymphoblastoid human cell lines (LICR/LON HMy2 and MW28) but was absent on T cell derived lines and on lines derived from B ALL or Burkitt's lymphoma. Table 2 indicates that KMA could not be detected on normal cells from adult peripheral blood, tonsil, spleen or bone marrow, although the determinant was expressed on some foetal B cell and some activated adult B cells. It was found on cells from 16 week foetal spleen where 1.0 - 17.8% of cells were positive but was absent from foetal liver and from cord blood. KMA was also found on a small proportion (4 - 17%) of normal adult tonsillar B cells after stimulation with *Staph. aureus* for six days *in vitro* culture.

The KMA positive human lymphoblastoid cell line LICR LON/HMy2 (HMy2) has been used in studies on the cytotoxic efficiacy of K-1-21 and on the membrane mobility of KMA and its expression during the cell cycle, properties which are relevant to the potential value of K-1-21 as a therapeutic agent. It was found that: (a) K-1-21, which is a monoclonal antibody of the IgG_1 subclass, did not mediate complement dependent cytotoxicity of HMy2 cells but will elicit ADCC (Figure 1); (b) K-1-21 alone does not cap KMA. Even after a prolonged incubation of up to 30 h at 37°C, no capping occurred (Table 3). However, the addition of a second ligand FITC-SaMIg, induced rapid capping which was inhibited by NaN_3 but not by calcium ionophore or dibucaine. The capping is thus of Type II rather than Type I origin (Loor et al., 1981); and (c) the expression of KMA on HMy2 cells is dependent on the stage of the cell cycle (Table 4). Flow cytometric analysis of K-1-21 stained HMy2 cells from elutriated fractions showed that cells enriched for the G_1 growth phase contain the lowest number of cells staining with K-1-21 or GaHK or conversly that KMA is expressed at greatest frequency on cycling cells.

TABLE III

Capping of KMA from the surface of HMy2 with K-1-21 and sheep anti-mouse immunoglobulin.

Capping Antibodies	Inhibitor	Type of capping inhibited	Capping at 90 min
K-1-21	None		-
K-1-21 + FITC-SaMIg	None		+++
K-1-21 + FITC-SaMIg	Calcium ionophore	Type I	+++
K-1-21 + FITC-SaMIg	Dibucaine	Type I	+++
K-1-21 + FITC-SaMIg	NaN_3	Type I and II	-

- indicates that no capping was observed even after 30 h incubation.

+++ indicates that after 90 min incubation all cells were capped.

TABLE IV

Elutriation analysis of KMA on the surface of HMy2 cells at different stages of the cell cycle.

Elutriation fraction	Cell cycle Kinetic parameters			% stained cells*	
	% G_1	%S	% G_2 + M	GaHK	K-1-21
unseparated	40	38	3	ND	89
G_1 enriched	89	14	3	77	64
S and G_2 + M enriched	25	52	22	92	85
G_2 + M enriched	23	36	40	96	89

* Cells were stained with FITC-GaHk or with K-1-21 and FITC-SaMIg.

DISCUSSION

KMA may be regarded in an operational sense as a tumour-associated marker with selectivity for human kappa myeloma, macroglobulinemia and lymphoma cells. However, KMA is not confined to neoplastic tissues. The determinant has been detected on some foetal B cells and a small proportion of in vitro activated B cells although it is absent from resting B cells and plasma cells from freshly isolated lymphoid tissues and from bone marrow. These findings suggest that KMA may be a differentiation antigen expressed at a limited stage or B cell development but represented in an expanded form by the tumour cell population (Figure 2). Thus, KMA, while absent from the surface of resting mature B cells, may be transiently expressed in activated B cells, but lost in normal lymphoblasts or plasmacytes. Malignancies arising at early stages of B cell development (ALL, B CLL) do not exhibit KMA while malignancies arising at or after the stage of B cell activation express this determinant. Relation of KMA to the expression of other B cell differentiation markers such as IgD and the antigens revealed by monoclonals PCA-1 and 4F2 remains to be determined.

The K-1-21 antibody appeared to have potential for practical application as a therapeutic agent. Factors which might affect the

KMA AS A B CELL DIFFERENTIATION ANTIGEN

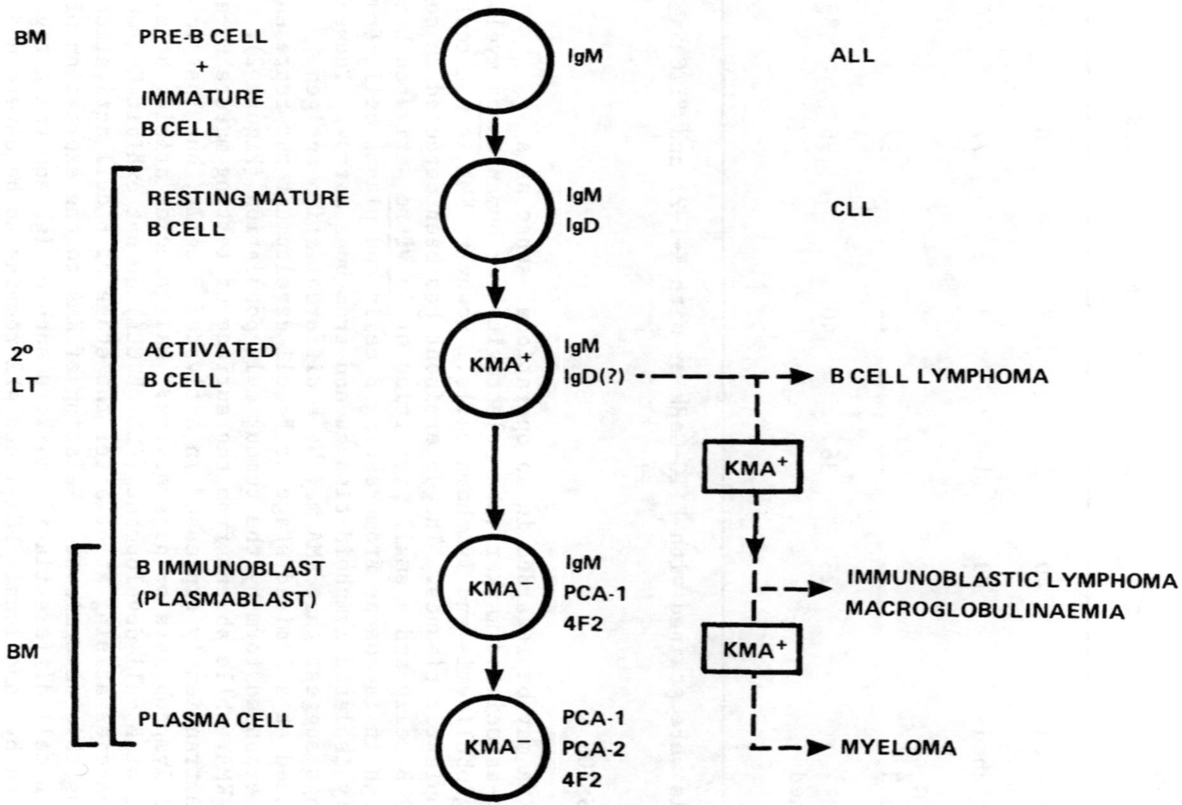

FIGURE 2

KMA as a differentiation antigen expressed transiently on normal B cells and on malignant B cells after the stage of B cell activation.

use of the antibody in a clinical situation have been examined using the KMA positive cell line HMy2. While K-1-21 did not mediate complement dependent cytotoxicity, it did elicit ADCC which might thus be the mechanism of lysis if K-1-21 were used in serotherapy. The fact that K-1-21 alone does not cause capping of KMA suggests that binding of the antibody is unlikely to lead to modulation of antigenic determinants with consequent escape from immunodetection. But if K-1-21 is coupled to a 'cytotoxin', the agent should not require internalisation for its effect. Experiments are therefore underway to couple K-1-21 to the surface active drug adriamycin (Tritton and Yee, 1982). The expression of K-1-21 was shown to vary with phases of the cell cycle, being greatest in S and G_2 + M and least in G_1. This highlights the importance of giving K-1-21 at a time of maximal expansion of the tumour growth fraction such as occurs after administration of conventional chemotherapy. Experiments are in progress to test the efficacy of K-1-21 in a xenograft model using nude mice and its homing potential to the marrow in patients with kappa myeloma. K-1-21 has additional practical applications unlike conventional anti-kappa reagents. First it can be used for precise quantitation of free kappa light chains in serum and urine. Secondly, it has diagnostic potential in discriminating between benign monoclonal gammopathy and true myeloma.

REFERENCES

Boux, H. A., Raison, R. L., Walker, K. Z., Hayden, G. E., and Basten, A., 1983, A tumor-associated antigen specific for human kappa myeloma cells, J. Exp. Med., 158:1769.

Boux, H. A., Raison, R. L., Walker, K. Z., Musgrove, E., and Basten, A., 1984, The surface expression of a tumour-associated antigen on human kappa myeloma cells, Eur. J. Immunol., 14:216.

Loor, F., Martin-Pelissard, C., and Angman, L., 1981, Capping revisited II. Low sensitivity to sulfydryl reagents, Cell. Immunol. 57:73.

Muraguchi, A., and Fauci, A. S., 1982, Proliferative responses of normal human B lymphocytes. Development of an assay system for human B cell growth factor (BCGF), J.Immunol. 129:1104.

O'Neill, H. C., and Parish, C. R., 1983, A rapid automated colorimetric assay for measuring antibody binding to cell surface antigens, J. Immunol. Meth. 64:257.

Tritton T. R., and Yee, G., 1982, The anticancer agent adriamycin can be actively cytotoxic without entering cells, Science (N.Y.) 217:248.

DERIVATIVES OF ANTI-IDIOTYPE ANTIBODIES IN THE TREATMENT OF B-CELL LYMPHOMA

M.J. Glennie and G.T. Stevenson

Lymphoma Research Unit
Tenovus Laboratory
General Hospital
Southampton, SO9 4XY

INTRODUCTION

The widespread availability of monoclonal antibody technology has prompted a renewed interest in antibody as a therapeutic tool for the treatment of neoplastic disease. Unfortunately many of the basic problems which have limited the usefulness of polyclonal reagents still apply to their monoclonal equivalents. One such limitation is that of antigenic modulation. Boyse and Old (1969) were the first to demonstrate that antibody reacting with the surface of a tumour cell could induce changes which rendered the cell resistant to the action of antibody and complement. This work employed antibodies directed towards the thymus-leukaemia antigen of mouse, but the same process has now been described for many antigens on a wide variety of cell types (Chatenoud and Bach, 1984). The ability of cells to avoid destruction has been related to the process of patching, capping and endocytosis commonly observed when fluorescently labelled antibody is cleared from the surface of metabolically active cells. Such a direct correlation may be misleading, since in some systems cells can become resistant to lysis while antibody remains at the cell surface, apparently requiring only minimal reorganisation of the antibody-antigen complexes on the plasma membrane to protect the target cell (Gordon et al, 1982). Antibody-dependent cellular cytotoxicity using K cells or macrophages as effectors is also sensitive to antigenic modulation (Stevenson et al, 1982).

For some time we have been interested in idiotypic determinants of surface immunoglobulin (Ig) found on neoplastic B lymphocytes as targets for antibody attack. Our understanding of Ig structure and the restricted tissue distribution of idiotypes has offered a unique opportunity to study the therapeutic potential of antibody. Throughout this work antigenic modulation has proved a major mechanism whereby tumour cells escape destruction, both in man (Stevenson and Stevenson, 1983) and in the guinea pig lymphoblastic leukaemia L_2C (Stevenson et al, 1975). This was clearly demonstrated by the work of Stevenson and Gordon (1983) who showed that guinea pigs actively immunized against the idiotype of L_2C, to the extent that their serum was directly cytotoxic to tumour cells in vitro, eventually succumbed to the disease and that emerging tumour cells had escaped destruction by modulation. Studies in vitro showed that emerging cells expressed reduced levels of surface Ig, a cellular phenotype which was maintained in the guinea pigs only as long as circulating antibody could be detected. When the tumour bulk was sufficient to utilize all the antibody then surface Ig returned to normal levels.

Investigations in this and other laboratories have shown that tumour cells can only be protected through modulation when the antibody they encounter is at least bivalent (Gordon et al, 1982). This allows neighbouring Ig molecules in the plasma membrane to be cross-linked and form immune complexes which are redistributed and eventually removed in a process which requires both metabolic energy and an intact cell cytoskeleton. One possible means of circumventing antigenic modulation is the use of univalent antibody. It is well known that the univalent antibody fragment Fab'γ, derived by peptic digestion of antibody, will bind to cell surfaces without perturbing antigen expression. Unfortunately the absence of a Fc zone means that such molecules are ineffectual at activating effector systems such as complement. We have now developed a series of antibody derivatives which are univalent, but still retain an intact Fc region (Stevenson et al, 1982). The simplest of these is the Fab/c molecule prepared by enzymatic removal of one Fab arm from rabbit IgG. This molecule, in failing to modulate has proved superior to the parent antibody both in vitro and in immunotherapy of leukaemia in guinea pigs (Glennie and Stevenson, 1982). It has not been possible to evaluate Fab/c therapeutically in man because of logistic problems. However with this in mind we have engineered chimeric univalent derivatives. In these Fab'γ from antibody is linked to normal Ig of the antibody recipient. Figure I shows one such molecule FabIgG in which the antibody Fab'γ is linked by thioether bonds to IgG. In vitro this reagent has performed in a similar manner to Fab/c, but they offer the advantage of being available in large quantities, being less immunogenic, and, because they present the recipient's Fc, should prove efficient recruiters of host effector systems (Stevenson et al, 1984b).

We now describe the production and characterisation of monoclonal antibodies which will be used to supply Fab'γ fragments for the construction of chimeric derivatives. These reagents are specific for the idiotypic determinants found on the L_2C leukaemia of strain 2 guinea pigs. They have behaved in a similar manner to their polyclonal counterparts, both in the clearing of surface Ig from tumour cell

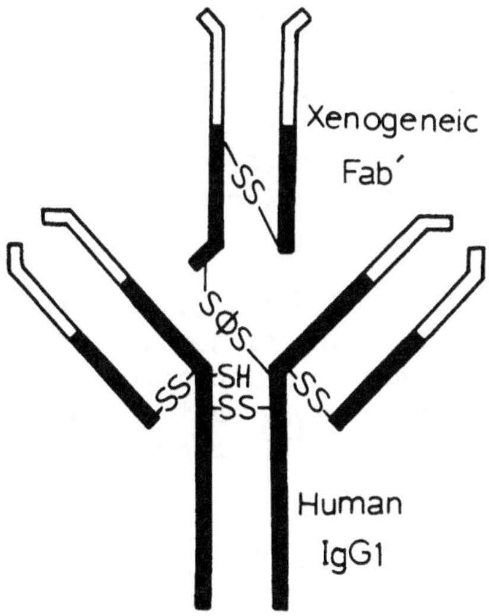

Fig. 1 One of the varieties of FabIgG prepared in our laboratory. It has been formed between a species of Fab'γ (rabbit) with only one hinge region half-cystine, and human IgG1 with two inter-γ SS bonds.

during culture and in immunotherapy. The availability of these reagents in large quantity has also presented an opportunity to investigate the maximum therapeutic effect which might be expected by giving regular daily treatment with unmodified antibody.

MATERIALS AND METHODS

The L_2C lymphoblastic leukaemia of strain 2 guinea pigs was used in the present study (Stevenson et al, 1975). These cells express surface IgM and in culture secrete free λ chains together with a small amount of idiotypic IgM. Idiotypic material to serve as antigen was prepared from the serum of guinea pigs in the terminal phases of the leukaemia by the method of Stevenson et al (1980).

Monoclonal antibodies specific for this idiotype were produced by the method of Kohler and Milstein (1975) as described previously (Stevenson et al, 1984a). Briefly, spleen cells from BALB/c mice which had been immunized with idiotypic IgM were fused with the NS-1 (P3-NS-1/1Ag 4.1) mouse myeloma line at a ratio of 2:1 using 50% polyethylene glycol 4000. Hybrid cells were suspended directly in supplemented Dulbecco's MEM to which HAT (10^{-4}M hypoxanthine, 4 x 10^{-7}M aminopterin and 1.6 x 10^{-5}M thymidine) had been added and distributed into 96-well microculture plates. Colonies were visible in 8-10 days and culture supernatant was screened for antibody activity. Wells showing useful antibody activity were cloned by limiting dilution onto a feeder layer of mouse thymocytes. Larger amounts of antibody was prepared from ascitic fluid by precipitation with ammonium sulphate followed by chromatography on columns of DEAE-Trisacryl and Ultragel AcA34 (LKB Produkter, Sweden).

Screening of antibody activity was carried out by the enzyme-linked immunosorbent assay (ELISA) (Engvall and Perlmann, 1972). Antigen in the form of purified guinea pig Ig was bound to 96-well microplates at 20 ng/ml in carbonate buffer. After washing, monoclonal antibodies were added and the bound mouse Ig detected by exposure to horse-radish peroxidase-labelled rabbit anti-mouse Ig (Nordic, U.K.).

Estimates of the relative quantity of monoclonal antibody bound to L_2C cells were performed indirectly by labelling with radioactive anti-mouse Ig. Cells (2×10^7/ml) were exposed to monoclonal antibody at 0° for 15 min. Then after washing and the various manipulations described below mouse antibody was detected using ^{125}I conjugated rabbit anti-mouse Ig.

Immunotherapy experiments were performed on matched groups of strain 2 guinea pigs. Animals received an intraperitoneal injection of 10^5 viable L_2C cells, followed 24 hours later by a single treatment (unless stated otherwise) of monoclonal anti-idiotype or a non-reactive monoclonal antibody, also given via the peritoneum. Survival of animals was then monitored daily.

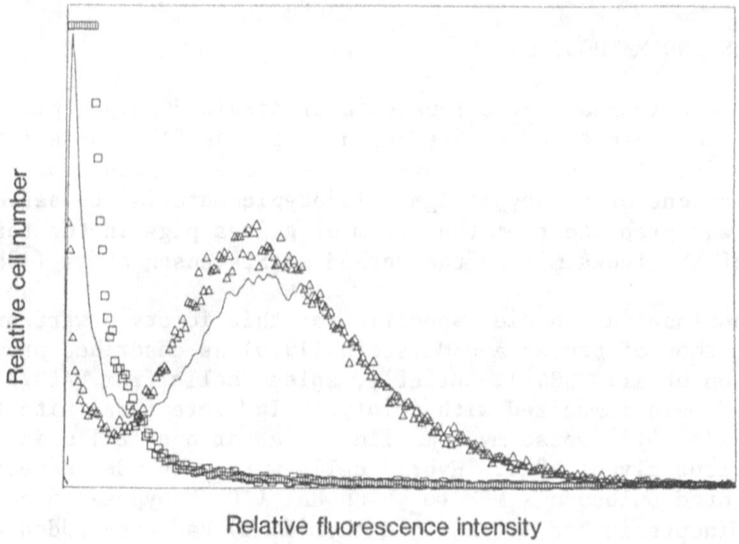

Fig. 2 Reactivity of L_2C leukaemic cells with anti-idiotype antibody. Cells were exposed to monoclonal antibody (5μg/ml) which had been pre-incubated for 15 min at 37°C with an equal volume of phosphate-buffered saline (——), normal guinea pig serum (Δ) or serum obtained from guinea pigs in the terminal phase of the L_2C leukaemia (□).

RESULTS AND DISCUSSION

Monoclonal antibodies raised to L_2C idiotypic IgM were assessed by the ELISA technique against a range of guinea pig Ig antigens coated onto microplates. No reactivity was seen against normal pentameric IgM, normal K or tumour derived λ light chains or normal IgG at concentrations of Ab which gave optical densities of >1.0 in the assay using idiotypic IgM as antigen. The absence of reactivity with light chains produced by the tumour, and recovered from the urine of leukaemic guinea pigs, indicates that the idiotypic determinants depend on either heavy chain or heavy plus light chain combination, as has been found for other idiotypes (Capra 1977).

Antibody specificity was confirmed by the reactivity with L_2C cells in cytofluorimetry using the FACS III. As shown in figure 2 strong fluorescence was seen using antibody at 5μg/ml and such staining was not affected by pre-incubation of the antibody with normal guinea pig serum. In contrast serum obtained from guinea pigs in the terminal phase of L_2C leukaemia or purified idiotypic IgM at 4 μg/ml (data not shown) completely blocked reactivity with the cells.

Three specific anti-idiotype antibodies have been selected during this investigation. However for the present presentation unless stated otherwise we will concentrate on just one. This reagent is of IgG1 isotype with a k light chain and has been designated M6/3D10.

Previous investigations with cultured L_2C cells have shown that surface Ig is rapidly cleared after interaction with a polyclonal anti-idiotype antibody (Gordon et al, 1982). We can now show that an equivalent monoclonal antibody can also perturb surface Ig, this is despite the fact that the monoclonal reagent must rely on a single antigenic determinant for cross-linking surface Ig. The clearance of antibody from the surface of L_2C cells has been monitored at specific intervals by detecting mouse antibody with ^{125}I conjugated rabbit anti-mouse IgG. Two monoclonal antibodies have been investigated, one reactive with the constant region of the light chain (anti-Cλ) and one specific for the Ig idiotype. During the production of anti-idiotype antibodies inevitably a number of anti-constant region antibodies are produced, the anti-Cλ used in this experiment is one such reagent. Figure 3 shows that at saturating levels the two antibodies differ in their ability to clear surface Ig and that both are less efficient in this respect than a polyclonal anti-idiotype serum. Interestingly when both monoclonal antibodies interact with the L_2C cell at the same time then the rate and extent to which Ig is cleared increases and mimicks closely that seen with the polyclonal reagent. This synergistic effect probably reflects the enhanced cross-linking activity of two monoclonal antibodies when recognising different determinants on the same antigen. Evidence that antibody is internalized during clearance rather than being shed into the supernatant is presented in figure 4. Cells treated with ^{125}I

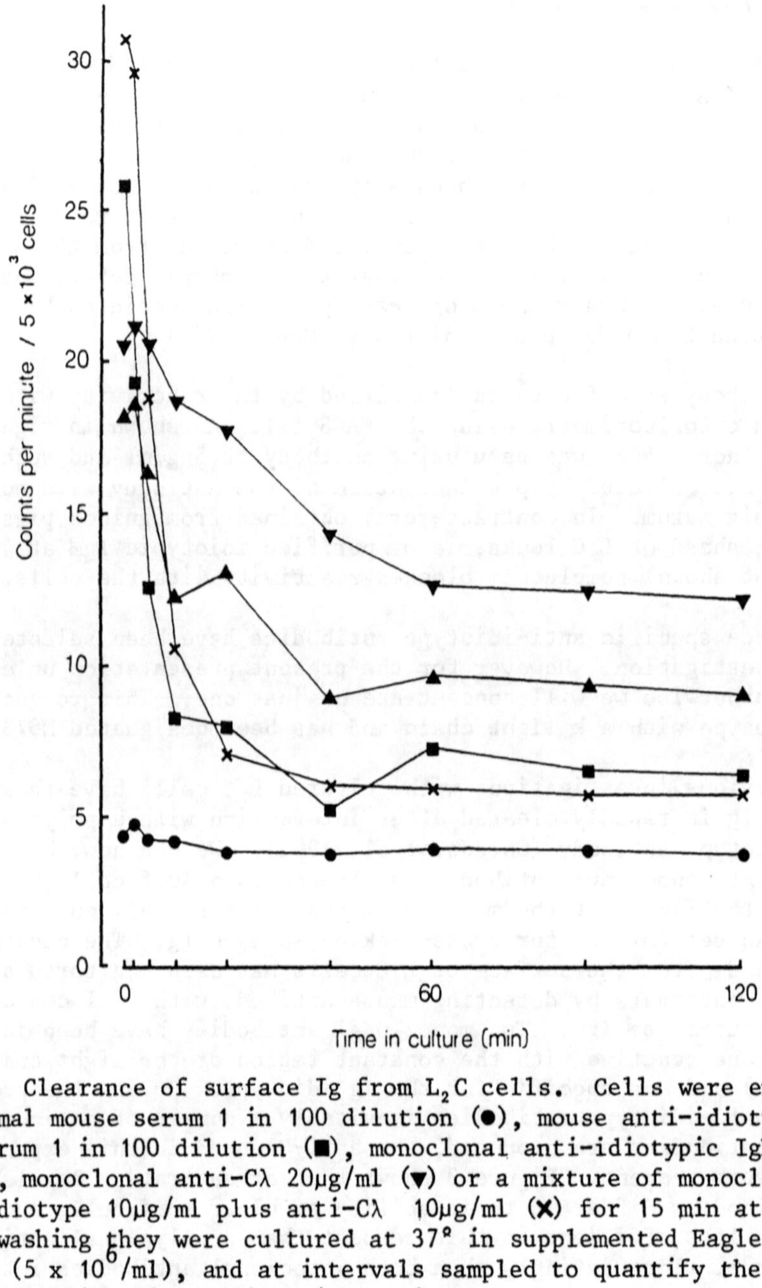

Fig. 3 Clearance of surface Ig from L_2C cells. Cells were exposed to normal mouse serum 1 in 100 dilution (●), mouse anti-idiotypic IgM serum 1 in 100 dilution (■), monoclonal anti-idiotypic IgM 20μg/ml (▲), monoclonal anti-Cλ 20μg/ml (▼) or a mixture of monoclonal anti-idiotype 10μg/ml plus anti-Cλ 10μg/ml (✕) for 15 min at 0°C. After washing they were cultured at 37° in supplemented Eagle's medium (5×10^6/ml), and at intervals sampled to quantify the surface Ig by labelling with radioactive anti-Ig.

conjugated monoclonal anti-idiotype retained radioactivity for several hours in culture with only a gradual depletion. In contrast univalent Fab'γ fragments prepared by peptic digestion of this reagent showed significant shedding when first placed into culture. Such a loss probably occurs during the rapid temperature change experienced by

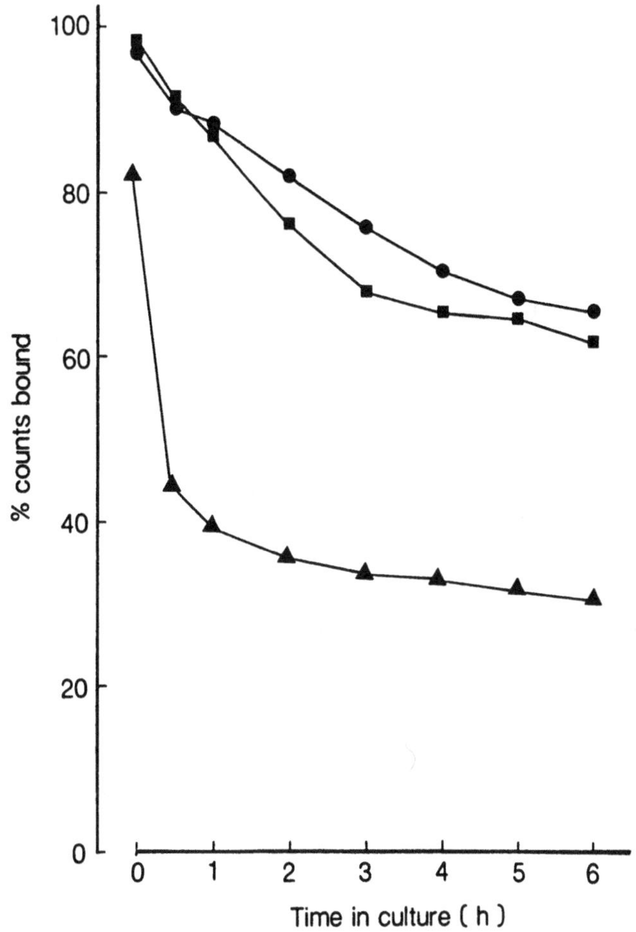

Fig. 4 Retention of monoclonal anti-idiotype and its fragments by L_2C cells during culture. Cells were exposed to radioactive IgG (●), $F(ab')_2$ (■) or Fab'γ (▲) at 20μg/ml for 15 min at 0°. After washing cells were cultured in supplemented Eagle's medium (5×10^6/ml), and at intervals sampled to determine bound radioactivity.

cells at this time, but is not manifest in the parent antibody due to the 'multiple binding bonus' provided by bivalent binding. This result has important implications for the selection of monoclonal antibodies destined to supply the Fab'γ fragments of chimeric derivatives as discussed above. Previously we have always prepared Fab'γ fragments from high affinity polyclonal antibody, thus ensuring a useful proportion of univalent derivatives which bind firmly to the target cells. If monoclonal antibodies are to perform in the same way then it is essential that we select only those with the highest affinity.

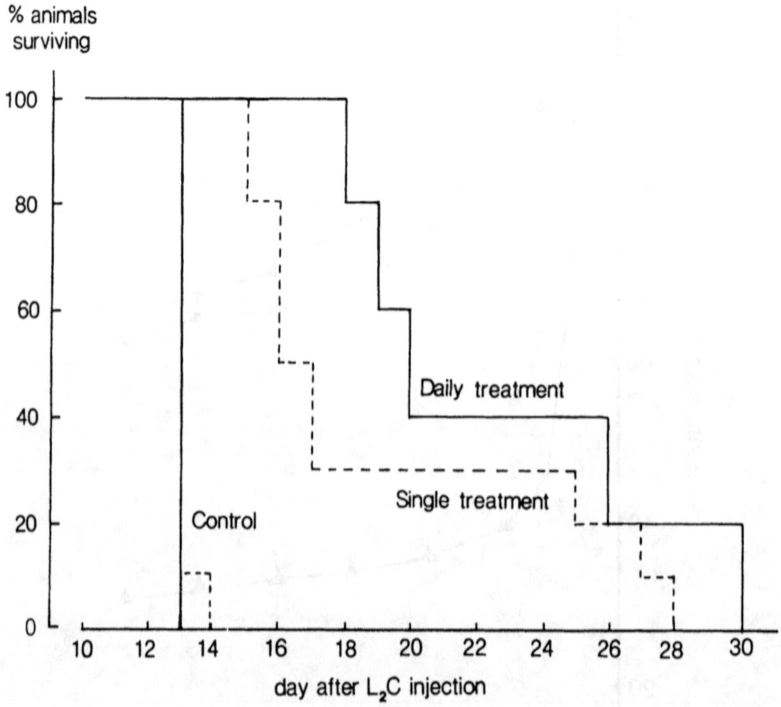

Fig. 5 Survival of leukaemic guinea pigs.
Animals received an intraperitoneal injection of 10^5 viable L_2C cells followed 24 hours later by a single (----) or daily (——) treatment with 0.5 ml of monoclonal anti-idiotype ascites. Control groups were treated similarly but with a non-reactive monoclonal ascites.

In immunotherapy experiments the present monoclonal antibodies have performed in a similar manner to their polyclonal counterparts (Stevenson et al, 1977): a single dose of antibody prolonging the median survival of L_2C leukaemic guinea pigs by about 3 days. This therapeutic effect can only be described as modest although, based on a tumour doubling time of about 19 hours, it would suggest that some 90% of the inoculated cells were killed as a result of antibody treatment.

An attempt to improve this survival curve through intensive treatment with unmodified monoclonal antibody is shown in figure 5. Commencing 24 hours after receiving 10^5 L_2C cells guinea pigs were given a daily dose of anti-idiotype for as long as they survived. Despite such treatment tumour development was only slightly suppressed compared with that seen after a single injection of antibody, and animals succumbed to the leukaemia while still receiving regular treatment. Apparently there is little to be gained from prolonged administration of antibody at least in this system.

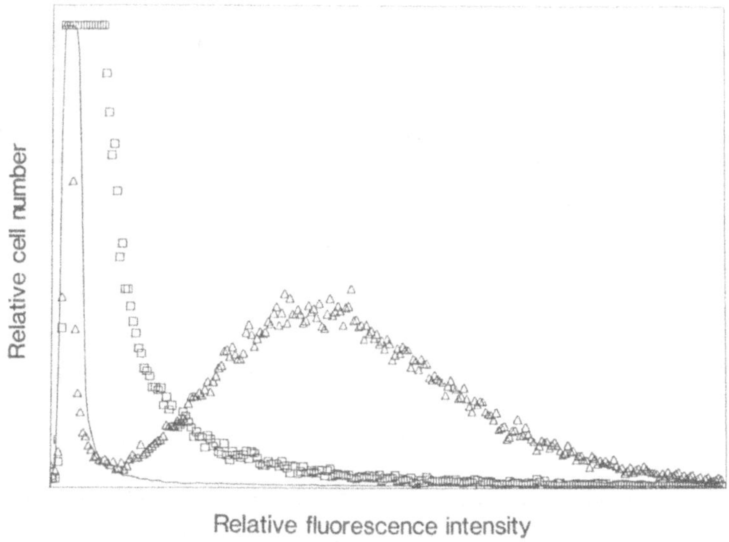

Fig. 6 Density of surface Ig on L_2C cells during immunotherapy. Circulating lymphocytes recovered from leukaemic animals receiving daily treatment with a non-reactive antibody (Δ) or anti-idiotype (□), were stained with monoclonal anti-idiotype followed by fluorescent rabbit anti-mouse Ig.

Circulating lymphocytes recovered from these animals during the terminal phases of the disease exhibit complete modulation. Figure 6 shows that the surface Ig has been reduced to almost undetectable levels by prolonged exposure to antibody in vivo. This is in contrast to the situation in vitro (figure 3) where culture (37°) of L_2C cells with a single monoclonal antibody failed to remove all the surface Ig. It is not known why antibody can perturb a surface antigen more effectively in vivo than in vitro. This phenomenon has been described for several systems using both polyclonal (Boyse et al, 1969) and monoclonal (Chatenoud et al, 1982) reagents in animals and man. Perhaps prolonged exposure to antibody in combination with favourable metabolic conditions for the tumour cells are sufficient explanation. However it is always possible that exogenous factors other than antibody are available in vivo to assist modulation (Stackpole et al, 1978).

The possibility that an Ig-negative variant of L_2C had been selected by antibody treatment was excluded by short-term culture of these cells in the absence of antibody. During this time surface Ig was restored (data not shown), thus confirming the role of antibody in maintaining their negative phenotype during therapy.

The present work strongly confirms that the therapeutic potential of unmodified antibody is severely limited due to modulation and that this limitation is unlikely to be solved simply by more intensive treatment. Circumventing this problem with univalent derivatives as discussed above seem a logical solution and we now look forward to preparing chimeric reagents from these monoclonal antibodies for a direct comparison of therapeutic performance.

REFERENCES

Boyse, E.A. and Old, L.J., 1969, Some aspects of normal and abnormal cell surface genetics, Ann Rev Genet., 3:269

Capra, J.D., 1977, Towards a chemical definition of idiotypy, Fed Proc, 36:204

Chatenoud, L. and Bach, J.F., 1984, Antigenic modulation - a major mechanism of antibody action, Immunol Today, 5:20

Chatenoud, L., Baudrihaye, M.F., Kreis, H., Goldstein, G., Schindler, J. and Bach, J.F., 1982, Human in vivo antigenic modulation induced by the anti-T cell OKT3 monoclonal antibody, Eur J Immunol, 12:979

Engvall, E. and Perlmann, P., 1972, Enzyme-linked immunosorbent assay, ELISA. III. Quantitation of specific antibodies by enzyme-labelled anti-immunoglobulin in antigen-coated tubes, J Immunol, 109:129

Glennie, M.J. and Stevenson, G.T., 1982, Univalent antibodies kill tumour cells in vitro and in vivo, Nature, 295:712

Gordon, J., Anderson, V.A. and Stevenson, G.T., 1982, Loss of surface bound antibody accompanying the anti-complementary modulation of leukaemic B cell immunoglobulin: contrasting effects of antibodies directed against idiotypic and constant regions, J Immunol, 192:2763

Kohler, G. and Milstein, C., 1975, Continuous cultures of fused cells secreting antibody of predefined specificity, Nature, 256:495

Stackpole, C.W., Jacobson, J.B. and Galuska, S., 1978, Antigenic modulation in vitro. II. Modulation of thymus-leukaemia antigenicity requires complement component C3, J Immunol, 120:188

Stevenson, F.K., Glennie, M.J., Johnston, D.M.M., Tutt, A.L. and Stevenson, G.T., 1984a, Consumption of monoclonal anti-idiotypic antibody by neoplastic B lymphocytes: a guide for immunotherapy, Br J Cancer, 50: in press

Stevenson, F.K., and Gordon, J., 1983, Immunisation with idiotypic immunoglobulin protects against development of B lymphocytic leukaemia, but emerging tumour cells can evade antibody attack by modulation, J Immunol, 130:970

Stevenson, F.K., Morris, D. and Stevenson, G.T., 1980, Immunoglobulin produced by guinea-pig leukaemic B lymphocytes: its source and use as a monitor of tumour load, Immunol, 41:313

Stevenson, G.T., Cole, V.M., Summerton, J., Watts, H.F., 1984b, Chimeric univalent antibodies for treating lymphoid malignancies, Medical Oncology and Tumour Pharmacotherapy, in press.
Stevenson, G.T., Eady, R.P., Hough, D.W., Jurd, R.D. and Stevenson, F.K., 1975, Surface immunoglobulin of guinea-pig leukaemic lymphocytes, Immunol, 28:807
Stevenson, G.T., Elliott, E.V. and Stevenson, F.K., 1977, Idiotypic determinants on the surface immunoglobulin of neoplastic lymphocytes: a therapeutic target, Fed Proc, 36:2268
Stevenson, G.T., Glennie, M.J. and Gordon, J., 1982, The killing of lymphoma cells by univalent derivatives of tumour-specific antibody, UCLA Symp Molec Cell Biol, 24:459
Stevenson, G.T. and Stevenson, F.K., 1983, Treatment of lymphoid tumours with anti-idiotype antibodies, Springer Semin Immunopathol, 6:99

PROTECTIVE EFFECTS OF MONOCLONAL ANTIBODIES AND IMMUNOPURIFIED GLYCOPROTEINS IN PRIMARY HSV-1 INFECTION

M. L. Lukic, F. Y. Liew* and W. L. Chan

Imperial Cancer Research Fund Tumor Immunology Unit
University College London, London, and
*Biomedical Research Division
Wellcome Research Laboratories, Beckenham, U.K.

INTRODUCTION

Herpes simplex virus (HSV) glycoproteins are structural components of the virion envelope and have been implicated in virus-induced alterations of mammalian cells (1). These glycoproteins are expressed on infected cell plasma membranes and act as major antigenic stimuli for the cellular and humoral responses of the host (2). It is pertinent to the development of a subunit vaccine effective in controlling primary HSV infection to understand the role of these individual glycoproteins in the induction of the various components of immune reactivity. As a first step, we compared the ability of monoclonal antibodies directed against different individual glycoproteins to protect susceptible mice against lethal HSV-1 challenge. Secondly, glycoproteins immunopurified by monoclonal antibody were examined for their protective capacity, and the immunological mechanisms underlying their protectivity analysed.

CHARACTERIZATION OF MONOCLONAL ANTIBODIES

Three hybridoma cell lines producing monoclonal antibodies to the envelope glycoproteins of herpes simplex virus type 1 were established as described previously (3). Indirect immunofluorescence assay of HSV-1 infected BHK 21/C13 cells showed that G8D1, C2D2 and TI57 monoclonal antibodies reacted specifically with viral antigens located in the plasma membranes of virus-infected and not in mock-infected cells (3, 4). Radioimmune precipitations were carried out with (^{14}C)-glucosamine labelled extracts from infected or mock-infected BHK 21/C13 cells. Fluorography revealed three different patterns of electrophoretically separable glycoproteins of apparent molecular weights of 123,000 (G8D1), 63,000 (C2D2) and 110,000 (TI57). The tentative designations of the glycoproteins precipitated by each monoclonal antibody, as proposed by Spear and Roizman (5) are: gB, gD and gAB (molecular weights 123,000, 63,000 and 110,000, respectively). However, due to the current confusion in HSV-1 antigen terminology we preferred to refer to the detected glycoprotein (gp) species as: 123K gp, 63K gp and 110K gp.

Monoclonal antibodies in culture supernatant were isotyped using rabbit anti-mouse immunoglobulin isotype specific antisera. Each monoclonal antibody gave a single positive precipitin reaction against class-specific or subclass specific antiserum demonstrating that they belong to IgG2a (G8D1), IgM (C2D2) or IgG2b (TI57) (Table 1). Experiments were also performed to test the capacity of the three monoclonal antibodies to neutralize HSV-1 in the presence of complement. Ascites fluids were diluted serially in microtitre plates and mixed with 200 PFU of HSV-1. The virus-antibody mixtures were incubated for 1 hr before being added to BHK 21/C13 cell monolayers. G8D1 and C2D2 antibodies reduced the number of plaques, with neutralizing titres of 1/1280. Monoclonal antibody TI57, however, has no detectable neutralising activity (Table 1).

EFFECT OF PASSIVELY TRANSFERRED MONOCLONAL ANTIBODIES ON HOST RESISTANCE TO HSV-1 INFECTION

It has been shown previously that passive transfer of monoclonal antibodies directed against virion envelope gps protected mice from challenge by a lethal dose of HSV-1 (6, 7). It was also reported that the capacity of glycoprotein specific monoclonal antibodies to neutralize HSV-1, to promote complement-mediated cell lysis and antibody-dependent cellular cytotoxicity (ADCC) _in vitro_ did not correlate with the protective activity of antibody _in vivo_ (8). In fact, one monoclonal antibody devoid of all the three immunologic reactions _in vitro_ was highly effective in promoting recovery from HSV-1 induced ocular disease (8). Therefore we tested the capacity of all three monoclonal antibodies to protect susceptible mice from HSV-1 infection.

Ascitic fluid (0.3ml) was injected i.p. into nine week old male CBA mice 24 hrs after footpad or i.p. infection with 10 x LD_{50} HSV-1. Mice injected with monoclonal antibodies G8D1 or C2D2 were completely protected against the HSV-1 infection, whereas antibody TI57 was ineffective in conferring passive protection (Table 1). Such ineffectiveness was not improved by increasing the amounts of ascitic fluid transferred to 0.6ml (data not shown).

These results are in agreement with previous reports that neutralizing monoclonal antibodies to the type-common epitope of gD and the type specific epitope of gC were protective when given after subcutaneous virus infection (6). They also demonstrate a good correlation between neutralizing activity and protective capacity of antibodies specific for glycoproteins.

TABLE 1 Passive Protectivity of Monoclonal Antibodies to HSV-1 Glycoproteins

MAB (Isotype)	Glycoproteins (MW)	Neutralising+ Titres	Survivors* (Day 21)
G8D1 (G2a)	123K	>1/1280	20/20
C2D2 (M)	63K	>1/1280	19/20
TI57 (G2b)	110K	<1/40	0/10
Normal Mouse Serum	-	<1/40	2/20

+50% neutralization of infective virus by antibodies.

*Mice were challenged in the footpad with 10 x LD_{50} HSV-1. Similar results with i.p. challenge.

PROTECTIVE EFFECTS OF IMMUNOPURIFIED HSV-1 SPECIFIC GLYCOPROTEINS

The demonstration of passive protection clearly indicates that individual viral glycoproteins are important in the generation of protective immunity. Differences in biological activities of monoclonal antibodies *in vivo* may be related to the topographical location or temporal expression of viral glycoproteins on infected cells against which a particular monoclonal antibody binds. Therefore it was of interest to compare the effectiveness of 123K gp, 63K gp and 110K gp to induce immune responses capable of interfering with HSV-1 infection *in vivo*. Viral gps were purified by a modified procedure of Secher and Burke (9) as described previously (3). Initial experiments have shown that 5 x 10^6 PFU heat-inactivated HSV-1 injected s.c. in four sites completely protected CBA mice from the footpad challenge with 10 x LD_{50} dose of live virus. A similar protocol was used to investigate the ability of HSV-1 specific gps to induce protection against primary HSV-1 infection. Mice were immunized s.c. with 30 µg of 123K gp, 63K gp or 110K gp. Control mice were injected with saline or 5 x 10^6 PFU heat-killed HSV-1. Seven days later all animals were challenged with a lethal dose of HSV-1.

TABLE 2 Protective Immunity Induced by s.c. Injection of Specific HSV-1 Glycoproteins

Glycoproteins	Antibody Titre (RIA)[+] Pre-challenge	Antibody Titre (RIA)[+] Post-challenge	Survivors (Day 21)
123K	20	20	4/8
63K	20	20	3/8
110K	320	2560	8/8
Inactivated HSV-1	320	5120	8/8
Saline	<20	<20	1/8

[+]RIA against HSV-1 virus. Immunization, d -6; prechallenge Ab, d -1; challenge, d 0; post-challenge Ab, d 3.

Mice immunized with heat-killed virus and 110K gp were completely protected while 126K gp and 63K gp induced only partial protection (Table 2). Thus, it appears that 110K gp, possessing epitope(s) which induces antibodies without the capacity to neutralize virus infectivity *in vitro* or the ability to interfere with virus spread *in vivo*, is nevertheless by itself very efficient in inducing protective immunity. We therefore attempted to analyse the immunological mechanisms underlying the effectiveness of 110K gp as a subunit HSV-1 vaccine.

ANTIBODY RESPONSE AND PROTECTION

Although TI57 monoclonal antibody was neither neutralizing nor passively protective it was still possible that the protective effects of immunopurified 110K gp is due to humoral antibody directed against other epitopes on the same molecule. Antibody titres to HSV-1 in mice immunized with gps were measured by radioimmune assay because recent work has indicated that protective effectiveness of antibody does not necessarily correlate with virus neutralization, complement mediated lysis or ADCC *in vitro* (8). As shown in Table 2, antibody titres to HSV-1 prior to the infection were significantly elevated only in mice immunized with inactivated virus and 110K gp, but not in those injected with 123K gp or 63K gp. Similarly early postinfection antibody response was seen only in mice immunized with HSV-1 and 110K gp. Thus, the prophylactic capacity of 110K gp correlates to an early induction of specific antibody which become even more evident three days after virus challenge. It is possible, therefore, that 110K gp stimulates the production of protective antibody and that in this respect is superior to 123K gp or 63K gp in inducing protective immunity.

To test this possibility, we examined the protective capacity of the polyclonal antibodies induced by 110K gp. CBA mice were injected i.p. with 10μg of 110K gp in saline at weekly intervals. Five days after the third injection mice were bled and the sera pooled. These sera were used as a source of anti-110K gp antibodies while anti-HSV-1 sera and normal CBA mouse sera were used as controls. Sera (0.35ml) were injected i.p. into syngeneic recipients 24 hrs after a lethal i.p. challenge with HSV-1.

In contrast to the whole inactivated virus, 110K gp did not induce antibody with substantial neutralizing activity or protective capacity (Table 3). Thus, early antibody response alone is probably not responsible for the prophylactic effect of 110K gp.

TABLE 3 Passive Transfer of Polyclonal Antibodies against 110K Glycoprotein

Antibodies	Neutralisation Titre	Survivors (Day 21)
110K	<1/20	0/10
HSV-1	>1/640	9/10
NMS	<1/20	0/10

DELAYED-TYPE HYPERSENSITIVITY (DTH) RESPONSE AND PROTECTION

DTH to HSV-1 has been previously shown in mice (10). Conventionally-purified glycoproteins, gC and gD, although isolated under denaturing conditions were able to induce and elicit DTH response to HSV (11). Mice immunized with gC developed DTH reactivity and were protected from lethal challenge with HSV-1. However, serum neutralizing or precipitating antibodies were not detected prior to the HSV-1 challenge. It was suggested that the protective effect of gC may be ascribed to DTH response (11). We therefore examined the DTH reactivity in animals immunized with immunopurified gps. At first, mice were immunized s.c. with 30µg of gp. Five days after sensitization, they were tested for DTH with challenge injection of antigen (10µg) into the right hind footpad. An equal volume (50µl) of saline was injected into the left hind footpad. Footpad swelling was measured 24 and 48 hrs after injection. DTH was expressed as the difference in thickness of right versus left hind footpad. In repeated experiments none of the gps induced a significant DTH response (data not shown). In the second series of experiments mice were immunized intradermally with HSV-1 or 10µg of gp. Some groups of mice received 150mg/kg body weight of cyclophosphamide (Cy) i.p. two days prior to immunization. Again, gps alone induced only low level of DTH. Pretreatment of mice with Cy significantly enhanced the DTH responses. The Cy pretreatment has been shown to drastically reduce B cells and suppressor T cell precursors (12). Under this regimen, 110K gp was clearly the most efficient gp in the induction of DTH (Table 4). DTH reactivity induced by 110K gp was comparable to that induced by the whole inactivated HSV-1. However, in contrast to HSV-1, 110K gp under such immunising conditions was unable to confer effective protection against a lethal challenge of HSV-1 (Table 4). It thus appears that the cell-mediated immunity alone induced by 110K gp is insufficient for protection.

TABLE 4 DTH and Protective Immunity induced by Specific HSV-1 Glycoproteins*

Immunisation	24 hr footpad reaction to HSV-1 ($\times 10^{-2}$ mm)	Survivors (Day 21)
123K	22 ± 3	3/9
63K	8 ± 1	1/8
110K	38 ± 4	1/8
HSV-1	41 ± 5	7/8
-	4 ± 2	0/8

*Groups of mice were injected i.p. with 150mg/kg Cy and immunised intradermally two days later with 10µg of glycoproteins or 5×10^6 PFU of heat-inactivated HSV-1. Five days after immunisation groups of 5 mice were tested for DTH reactivity to HSV-1 in the footpad while the rest of the mice (8-9 per group) were challenged i.p. with $10 \times LD_{50}$ of HSV-1.

CONCLUDING COMMENTS

Three monoclonal antibodies directed against three different HSV-1 gps differ in their passive protective capacity. Antibodies recognizing 123K gp and 63K gp are protective whereas antibody against 110K gp is ineffective (Table 1). However, immunopurified 110K gp is superior to either the 123K gp or the 63K gp and induced complete protection against a lethal HSV-1 infection (Table 2). The prophylactic capacity of 110K gp is correlated to an early induction of antibody in intact mice and strong DTH response in cyclophosphamide pretreated mice. However, neither the polyclonal 110K specific antibodies (Table 3) nor the DTH alone (Table 4) is sufficient for protective effect. The results therefore suggest that: (a) a monoclonal antibody though by itself not virus-neutralising nor passively protective, is nevertheless capable of affinity-purifying antigen with strong prophylactic capacity; (b) the 110K gp, though not possessing epitopes for inducing virus-neutralising antibodies, contains antigenic determinants recognisable by other B cells and capable of activating T cells which may be responsible for the glycoprotein's prophylactic capacity via a carrier-hapten effect - 110K gp acting as a potent carrier for virus neutralising epitopes on the whole virus upon infectious challenge. Indeed, results to be published elsewhere demonstrate that 110K gp-primed lymph node cells show strong in vitro proliferative response to HSV-1, and the prophylactic effect of 110K gp is adoptively transferrable by purified T cells.

REFERENCES

1. B. Norrild. Curr. Top. Microbiol. Immunol. 90:67 (1980)
2. B. Norrild, S. L. Shore and A. J. Nahmias. J. Virol. 32:741 (1979).
3 W. L. Chan. Immunology 49:343 (1983).
4. N. B. La Thangue, W. L. Chan and J. D. Almeida. J. Med. Virol. 13:227 (1984).
5. P.G. Spear and B. Roizman. In J. Tooze (ed) Molecular Biology of Human Viruses. Cold Spring Harbour Laboratories. Cold Spring Harbour, N.Y. p615-745 (1980).
6. R. D. Dix, L. Pereira and J. R. Baringer. Inf. Immun. 34:192 (1981).
7. N. Balachandran, S. Bacchetti and W. E. Rawls. Inf. Immun. 37:1132 (1982).
8. J. T. Rector, R. N. Lausch and J. E. Oakes. Inf. Immun. 38:168 (1982).
9. D. S. Secher and D. C. Burke. Nature 285:446 (1980).
10. A. A. Nash, H. J. Field and R. Quartey-Papafio. J. Gen. Virol. 48:351 (1980).
11. R. D. Schrier, L. I. Pizer and J. W. Moorhead. J. Immunol. 130:1413 (1983).
12. F. L. Shand and F. Y. Liew. Eur. J. Immunol. 10: 480 (1980).

A NEW INDICATOR OF HUMAN MALIGNANT TUMOUR FOUND ON ERYTHROCYTES

Su Metcalfe

Department of Surgery
E9
Addenbrooke's Hospital, Cambridge, England

INTRODUCTION

A new tumour marker has been discovered which is proving valuable as an aid in the diagnosis and monitoring of a wide range of common tumour types. The tumour marker is detected by the monoclonal antibody "B5" and has the unusual feature of being located on the surface of erythrocytes. Using a simple haemagglutination test it has been found that about 80% of patients with cancer are positive for B5 haemagglutination irrespective of tumour type, whereas B5 positivity is comparatively rare in control groups (20% B5-positive)[1,2]. This distribution of B5 positivity implies that people who are normally B5-negative often become B5-positive if they develop malignant tumour.

MATERIALS AND METHODS

The B5 monoclonal antibody is a rat IgG and its preparation has been detailed elsewhere[2]. The haemagglutination test uses 1 ml of blood (in anticoagulant) washed free of serum and diluted to 1% erythrocytes. Washing buffer is phosphate buffered saline (PBS), and the final diluent is 4% foetal calf serum in PBS. Culture supernatant containing the B5 antibody (25 µl) is mixed with 25 µl of the diluted erythrocytes and left at room temperature for 2 hours. Haemagglutination is scored using an inverted microscope to view the pellet directly under low power. Each sample is then scored in more detail after resuspension and transfer to a glass slide; here, estimation of the percentage free cells and the size of the haemagglutinated clumps is recorded.

RESULTS

Over 570 cancer patients have now been tested with B5, including 100 patients from the Raigemore Hospital, Inverness (in collaboration with Dr. Elia and Jack Urquehart) and 40 patients from the Queen Elizabeth Hospital, Birmingham (in collaboration with Fiona MacDonald). The Inverness and Birmingham samples have been tested both locally and by ourselves: the two sets of results are in very good agreement. Over 40 leukaemia patients in this hospital are being monitored by Jo Milner, and patients with Hodgkin's Disease and Non-Hodgkin's Lymphoma are being monitored by Dr. B. Hancock and Lesley Bruce at the Hallamshire Hospital, Sheffield. The remaining patients included in Table 1 are either surgical or radiotherapy patients in this hospital. These results show that the incidence of B5 positivity is high in most individual tumour types and this is reflected in the overall result of 86% B5-positive individuals in the group of 579 tumour patients.

The preliminary results of the extended survey shown in Table 1 led to the question, is the degree of B5 positivity related to tumour size? This was addressed in a collaborative study with Mr. N.V. Jamieson[3]. Patients with bladder cancer undergoing routine cystoscopy for visual assessment of tumour status were simultaneously assayed for B5 status. The results are summarised in Table 2. Overall, 80% of patients with bladder tumour (74) were B5-positive. Some patients with very small recurrences were B5-positive suggesting that B5 is a sensitive indicator a) of tumour presence, or b) of premalignant changes associated with an unstable urothelium. The strength of B5 haemagglutination seemed to be independent of tumour size.

A second point in question was, does the B5 status of a patient alter when the tumour is successfully removed? We had found that the incidence of B5 positivity in long term follow-up patients was low[2], giving indirect evidence that a switch from B5-positive to B5-negative does occur. More recently this switch has been observed in individual patients monitored over a prolonged period[4], including children on three years' treatment for acute lymphoblastic leukaemia (Jo Milner, in preparation). In a separate, collaborative study with Dr. K. Sikora, teratoma and seminoma patients are being monitored for B5 status and the preliminary results are summarised in Table 3. Figure 1 gives a profile of the B5 status of 3 patients monitored up to 20 months after surgical removal of primary tumour. Interestingly, the B5 status was indicative of tumour status for both seminoma and teratoma patients, whereas raised α-fetoprotein and β-HCG is found only in teratoma patients.

Table 1. Incidence of B5 Positivity in Cancer Patients and in Control individuals.

Type of Tumour	Number individuals tested	Number +ve	%+ve
Breast	146	121	86%
Bladder	90	76	84%
Leukaemia	44	43	97%
Non-Hodgkin's Lymphoma	32	25	78%
Hodgkin's Disease	25	22	88%
Bronchus and Lung	40	37	92%
Ovary	29	28	96%
Teratoma	27	27	100%
Seminoma	11	11	100%
Gastric	33	26	79%
Colorectal	40	33	82%
Prostate	9	8	
Sarcoma	9	7	
Cervix	5	4	
Thyroid	8	6	
Wilm's	4	4	
Uterus	3	3	
Pituitary	3	3	
Larynx	3	1	
Astrocytoma	1	1	
Glioma	2	2	
Melanoma	3	1	
Mesothelioma	1	1	
Oesophagus	4	3	
Pancreas	7	5	
	Total 579	498	86%
Controls	Total 451	100	22%

Blood samples were assayed for B5 positivity by haemagglutination. Erythrocyte clumps of 20 or more cells were scored positive. The controls represent blood donors, haemodialysis patients (44), pregnant women (40) and non-tumour patients (57). Many of the cancer patients were tested prior to treatment, showing that B5 positivity was not caused by therapy. There was no relationship between B5 and any of the major blood groups or other tumour markers normally present in serum.

Table 2. B5 Results from Patients Undergoing Cystoscopy for Diagnosis or Follow-up of Bladder Tumour

Cystoscopy Result	B5 Result +ve	-ve	% B5 positive
Positive			
a) new tumour	18	2	90%
b) recurrence	41	13	76%
Negative			
no recurrence*	11	13	46%

*6 patients in this group who were B5 positive had had tumour within the past 9 months (total 9 patients), whereas, of those patients who had been tumour free for over 9 months, only 5/15 were B5 positive.

Table 3. B5 Status of Teratoma and Seminoma Patients Serially Recorded over 15 Months.

		B5 Status	
		decreasing +ve	constant +ve
Teratoma:	AD*	0	5
	NAD*	13	4
Seminoma:	NAD	7	3
	Total NAD	20 (74%)	7 (26%)

* AD: active disease
NAD: no abnormality detected.

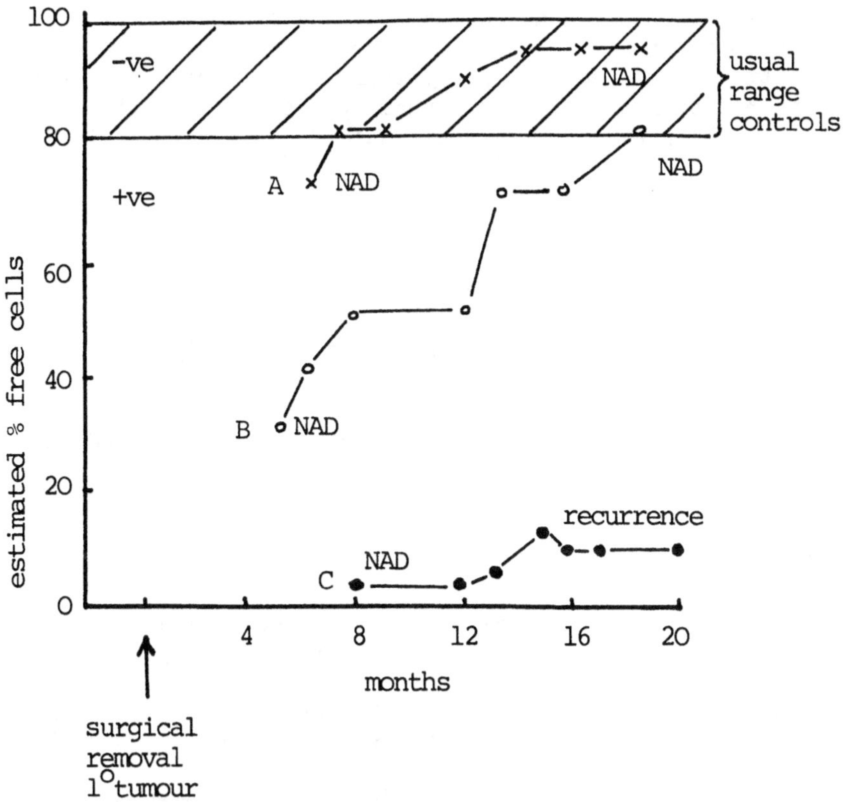

Figure 1. Percentage of non-agglutinable erythrocytes as an Index of B5 Status in three Patients following Removal of Primary Tumour.

Patients A and B were clinically NAD throughout the period shown and remain so. Patient C showed no abnormality by other tests until 17 months after surgery, when tumour recurrence was found.

DISCUSSION

B5 provides a very simple, non-invasive test applicable to a wide variety of malignancies, the only requirement being a small blood sample. Since the B5 marker is on erythrocytes, the recovery of the B5-negative status following tumour removal is prolonged, due to the erythrocyte life span of some 120 days. It is becoming apparent that this characteristic might have its own advantage, since a constant B5 positive status may alert the physician to the possibility of recurrent tumour before other abnormalities are detected. Early indication of recurrent tumour by B5 has now been observed in several patients with different tumour types; for example see patient 'C' in Figure 1.

ACKNOWLEDGEMENTS

Many studies summarised here have involved collaborative work as indicated in the test. In addition I am indebted to the staff of Addenbrooke's Hospital and the Cambridge Regional Blood Transfusion Centre for their help in obtaining samples, and to Rosamund Svvennsen and Teresa Snell for their technical assistance. This work is supported by grants from the Medical Research Council and the Cancer Research Campaign.

REFERENCES

1. J. Milner and S.M. Metcalfe, "A new monoclonal antibody which agglutinates erythrocytes from cancer patients", Lancet, ii, 1100 (1982).
2. S. Metcalfe, J. Milner and R.J. Svvennsen, "A new indicator of human malignant tumour", Br. J. Cancer, 49, 337-342 (1984).
3. S. Metcalfe and N.V. Jamieson, "A new tumour marker tested in 98 patients with bladder carcinoma", J. Roy. Coll. Surgeons, in press, (1984).
4. Su Metcalfe and Jo Milner, "B5, a monoclonal antibody against a new tumour marker on human erythrocytes" in "Protides of Biological Fluids, XXXII" edit. Dr. H. Peeters, in press.

RAT MONOCLONAL ANTIBODIES FOR BONE MARROW TRANSPLANTATION - THE CAMPATH SERIES

H. Waldmann, S. Cobbold, A. Wilson, M. Clark,
S. Watt, G. Hale, and H. Tighe

Division of Immunology
Department of Pathology
University of Cambridge
Cambridge, CB2 1QP, U.K.

Successful allogeneic marrow grafting requires that T-cell mediated responses of both donor and recipient be prevented (Korngold and Sprent, 1978; van Bekkum, 1984). Marrow transplantation is unique in the sense that there is an interplay between Graft versus Host (GvHD) and Host versus Graft (HvG) processes. Subclinical GvHD probably contributes to the immunosuppression which guarantees stem cell engraftment. The corollary of this is that success at purging T cells from donor marrow risks the prospect of marrow rejection by the recipient (see van Bekkum, 1984; Vallera et al, 1982; Vriesendorp et al, 1981a&b; Miller et al, 1983, Wagemaker et al, 1981). The compensating factors which might prevent rejection in these circumstances would be a larger stem cell dose and improved recipient conditioning (immunosuppression). This means that for routine clinical use one needs methods for T cell purging from marrow which do not result in excessive non-specific cell loss. The marrow dose available in any one harvest is often close to the absolute requirement and there is little room for wastage.

In this situation the ideal way of exploiting monoclonal antibodies would be to add them to the unadulterated marrow and to then rely on recipient effector mechanisms to remove coated T cells. This has been tried (unsuccessfully) in a number of centres with mouse monoclonal antibodies such as OKT3 (Filipovitch et al, 1982; Prentice et al, 1982). Three factors may have contributed to the lack of success in these studies. First, the antigens (and therefore the antibodies) may have modulated (Ritz et al, 1980) prior to access to recipient effector systems. Second, the pretransplant conditioning may have interfered with the effector capacity, and third, mouse antibodies may not be efficient in exploiting these human effector mechanisms.

Currently, T-cell depletion is performed wholly in vitro by any of lectin separation (Reisner et al, 1980), toxin-antibody conjugates (Filipovitch et al, 1984) or by complement lysis (Prentice et al, 1984; Slavin et al, this volume). Although remarkably effective the lectin separation method is time consuming and laborious. The use of rabbit complement is now widely adopted but requires mononuclear cell fractions and depends on very careful batch selection to exclude those which are toxic to stem cells or poorly lytic. On the other hand, the development of a potent human complement fixing anti-lymphocyte antibody CAMPATH-1 (Hale et al, 1983a) opens the way to reconsider the simple idea of marrow coating with reliance solely on the recipient complement system. The CAMPATH-1 antibody is lympholytic in vivo (Hale et al, 1983b).

Using CAMPATH-1 as the cornerstone for development of new therapeutic monoclonal antibodies we have directed our attention to producing (a) a marrow coating cocktail which in conjunction with CAMPATH-1 may provide a failsafe to account for all the T cells, and thus minimise GvHD as a problem; and (b) reagents likely to be effective in vivo to contribute to the conditioning required to reduce host resistance to grafting. The three new reagents developed for this purpose are rat monoclonal antibodies (CAMPATH-2,3&4) all of the IgG2b subclass. The rationale underlying selection of the IgG2b isotype is as follows:

IgG2b antibodies can be used to coat mouse T-cells and abrogate GvHD across major histocompatibility barriers in the complete absence of extraneous complement (Thierfelder, Cobbold & Waldmann, in preparation). Furthermore, these same anti-T cell antibodies are potentially immunosuppressive in vivo and small doses can completely ablate peripheral T cell populations or subpopulations in the mouse (see Cobbold et al, this volume; Cobbold et al, 1983); Rat IgG2b seems to be the only subclass of rat antibody recognised by human effector systems such as ADCC (Clark et al, this volume); IgG2b is the most efficient of the IgG subclasses at fixation of C1q and for synergistic complement lysis (Hughes-Jones et al, 1983; Bindon et al, this volume).

The reactivity of CAMPATH-2,3,&4 with human peripheral blood lymphocytes is schematically shown in Figure 1. Whereas CAMPATH-1 reacts with all lymphocytes, CAMPATH-2,3&4 react with virtually all T cells and variable numbers of granular lymphocytes as defined by the VEP13 antibody (Rumpold et al, 1982) and by direct visualisation of rosetted cytosmear preparations. We are still uncertain of the nature of the antigens recognised by CAMPATH-2,3&4 and how they correspond to specificities already well documented with mouse antibodies. None interfere with the E rosette receptor, but one,

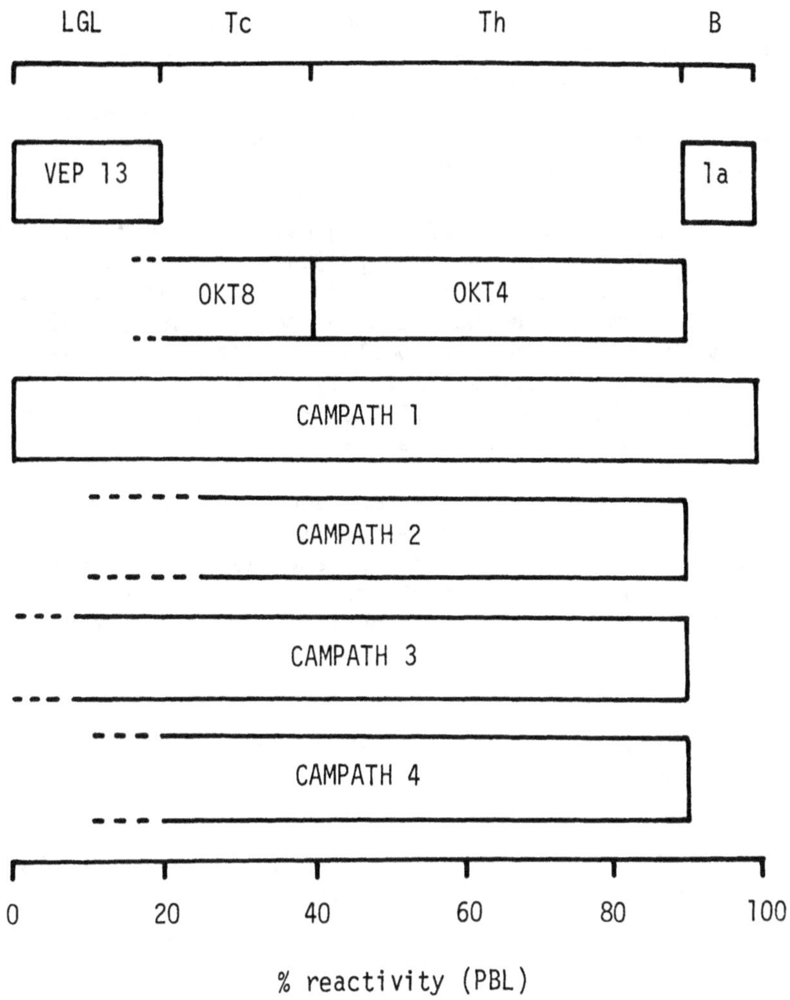

Fig. 1 Expression of the CAMPATH series on peripheral blood lymphocytes.

Schematic representation of the reactivity of the CAMPATH-1 series of rat monoclonal antibodies with human peripheral blood lymphocytes. The data derived from indirect immunofluorescence on the cytofluorograph, or by a direct rosetting method using monoclonal antibodies coupled to ox red cells with chromic chloride (modified from Wilson et al., 1979). The VEP13 antibody (Rumpold, 1982) was used to enumerate the granular lymphocytes and an anti-Ia antibody (YCL 6.8.22, unpublished) was used to provide an estimate of B cells. Broken lines indicate variable reactivity with the granular lymphocytes. The CAMPATH series are all derived from fusions of DA spleen cells with the Y3 myeloma by the method of Galfre and Milstein (1979).

CAMPATH-3, exhibits reciprocal blocking of binding with OKT3 (Kung et al, 1979) and UCHT1 (Beverly and Callard, 1981). CAMPATH-3 differs however in binding to human platelets, a feature which may preclude its use in vivo but not in vitro. None of the antibodies binds to haemopoietic progenitor cells (Table 1) and are therefore unlikely to damage stem cells necessary for engraftment.

Two of the antibodies, CAMPATH-2 & 3 have been produced in the monovalent form (Cobbold and Waldmann, 1984). In this state they exhibit a reduced capacity for modulation and an enhanced capacity for lysis with human complement (Glennie and Stevenson, 1982) and in this form may bypass problems of modulation. It may be however that the rat IgG2b antibodies are sufficiently effective to guarantee total marrow depletion when used with CAMPATH-1, in which case no modification to monovalency will be needed.

Given that the problem of GvHD may soon disappear, we are now faced with the next major challenge of allogeneic bone marrow transplantation. How do we avoid resistance especially in mismatches and sensitised patients? Can we exploit monoclonal antibodies as an alternative to increasing current condition regimens particularly radiation doses?

Table 1. Cell sorting of human bone marrow cells with CAMPATH 2,3&4 monoclonal antibodies

SAMPLE	Day 14 colony forming cells per 10^5 nucleated cells sorted for:-		
	CAMPATH-2	CAMPATH-3	CAMPATH-4
Positive	0.7	0.8	0
Negative	243	73	14
Labelled	225	56	ND
Unlabelled	183	32	84

Cells from bone marrow were stained with CAMPATH-2,3&4 and fluorescent anti-rat Ig. Sorted positive and negative fractions and controls were compared for their ability to produce colonies (>50 cells) after 14 days of culture in methylcellulose. Colony numbers represent pooled data on erythroid, granulocyte and macrophage colonies. In all cases progenitors were negative for the CAMPATH-2,3 and 4 antigens.

It is already clear that CAMPATH-1 and CAMPATH-2 can be administered in large doses intravenously (Hale et al, 1983b; and our own unpublished data) without any harmful sequelae. In preliminary studies both have been administered to patients with steroid resistant kidney rejection episodes with rapid reversal of rejection being observed (unpublished data). If this finding can be confirmed in many more such cases then these reagents may merit assessment for their ability to reduce host resistance to marrow engraftment. It may, however, be that standard conditioning regimes used in marrow grafting may interfere with the effector cells needed to execute the effects of such antibodies. If this turns out to be the case then we are left with the complement system as the only natural effector system that could be exploited. In this context a combination of CAMPATH-1 and a synergistic cocktail of rat IgG2b antibodies to the Leucocyte Common Antigen (on all nucleated blood cells) may be more appropriate (see Bindon, Hale and Waldmann, this volume) with the bonus that the same antibodies may aid in the elimination of residual leukaemia cells, especially of the lymphoid series where the LCA and CAMPATH-1 are well expressed.

AKNOWLEDGEMENTS

This work was supported by a grant from the Medical Research Council. We thank Jill Free, Penny Lovat and Gill Martin for expert technical assistance.

REFERENCES

van Bekkum,D.W., 1984, Conditioning regimens for marrow grafting, Seminars in Haematology, 21:81

Beverly,P.C.L., Callard,R.E., 1981, Destructive functional characterisation of human "T" lymphocytes defined by E-rosetting or a monoclonal anti-T cell antibody, Eur. J. Immunol., 11:329

Cobbold,S.P., Thierfelder,S., & Waldmann,H, 1983, Immunosuppression with monoclonal antibodies: A model to determine rules for effective serotherapy, Mol. Biol. Med., 1:285

Cobbold,S.P. & Waldmann,H., 1984, Therapeutic potential of monovalent monoclonal antibodies, Nature, 308:460

Filipovich,A.H., Ramsay,N.K.C., Warkentin,P. et al, 1982, Pretreatment of donor bone marrow with monoclonal antibody OKT3 for prevention of acute graft versus host disease in allogeneic histocompatible bone marrow transplantation, Lancet, i:1266

Filipovitch,A.H., Vallera,D.A., Youle,R.J., Quinones,R.R., Neville,D.M., Kersey,J.H., 1984, Ex vivo treatment of donor bone marrow with anti T cell immunotoxins for prevention of graft versus host disease, Lancet, i:468

Galfre,G., Milstein,C., & Wright,B., 1979, Rat x rat hybrid myelomas and a monoclonal anti-Fd portion of mouse IgG, Nature, 177:131

Glennie,M.J. & Stevenson,G.T., 1982, Univalent antibodies kill tumour cells in vitro and in vivo, Nature, 195:712

Hale,G., Bright,S., Chumbley,G., Hoang,T., Metcalf,D., Munro,A.J. & Waldmann,H., 1983, Removal of T cells from bone marrow: A monoclonal antilymphocyte antibody that fixes human complement, Blood, 62:873

Hale,G., Swirsky,D.M., Hayhoe,F.G.J. & Waldmann,H., 1983b, Effects of monoclonal anti-lymphocyte antibodies in vivo in monkeys and humans, Mol. Biol. Med., 1:321

Hughes-Jones,N.C., Gorick,B.D. & Howard,J.C., 1983, The mechanism of synergistic complement mediated lysis of rat red cells by monoclonal IgG antibodies, Eur. J. Immunol., 13:365

Korngold,R. & Sprent,J., 1978, Lethal graft versus host disease after bone marrow transplantation across minor histocompatibility barriers in mice, J. exp. Med., 148:1687.

Kung,P.C., Goldstein,G., Reinherz,E.L., Schlossman,S.I., 1979, Monclonal antibodies defining distinctive human T cell surface antigens, Science, 206:347

Miller,A.M., Walma, E.P., Klapwijk,W., et al., 1983, The effect of cyclosporine on host versus graft disease in canine bone marrow transplantation, Transplant Proc., 15:3046

Prentice,H.G., Blacklock,H.A., Janossy,G., Bradstock,K.F., Skeggs,D., Goldstein,G., Hoffbrand,A.V., 1982, Use of anti-T cells monoclonal antibody OKT3 to prevent acute graft-versus-host disease in allogeneic bone marrow transplantat9on for acute leukaemias, Lancet, i:700

Prentice,H.G., Blacklock,H.A., Janossy,G., Gilmore,M.J.M.L., Price-Jones, E. et al, 1984, Depletion of T lymphocytes in donor marrow prevents significant graft versus host disease in matched allogeneic leukaemic marrow transplant recipients, Lancet, i:472

Reisner,Y.,O'Reilly,R., Kapoor,N. et al., 1980, Allogeneic bone marrow transplantation using stem cells fractionated by lectins. VI. In vitro analysis of human and monkey bone marrow cells fractionated by sheep red blood cells and soybean agglutinin, Lancet, i:1320

Ritz,J., Pesando,J.M., Notis-McConarty,J., Schlossman,S.F., 1980, Modulation of human acute lymphoblastic leukaemia antigen induced by monoclonal antibody, J.Immunol., 125:1506

Rumpold,H., Kraft,D., Obexer,G., Bock,G. & Gebhart,G., 1982, A monoclonal antibody against a surface antigen shared by human large granular lymphocytes and granulocytes, J.Immunol., 129:1450

Vallera,D., Soderling,C., Carlson,G. et al., 1982, bone marrow transplantation across major histocompatibility barriers in mice. II. T cell requirement for engraftment in TL1 conditioned recipients, Transplantation, 33:243

Vriesendorp,H.M., Wagemaker,G., Bekkum, D.W.van., 1981a, Engraftment of allogeneic bone marrow, Transplant Proc., 13:643

Vriesendorp,H.M., Klapwijk,W.M., Kessel,A.C.M.van, et al, 1981b, Lasting engraftment of histoincompatible bone marrow in dogs, Transplantation, 32:347

Wagemaker,G.,Vriesendorp,H.M., Bekkum,D.W.van, 1981, Successful bone marrow transplantation across major histocompatibility barriers in rhesus monkeys, Transplant Proc., 13:875

Wilson,A.B., Prichard-Thomas,S., Coombs,R.R.A., 1979, Receptors for activated C3 on thymus dependent (T) lymphocytes of normal guinea pigs, Immunology, 37:377

SESSION 7

HUMAN LYMPHOCYTE SUBPOPULATIONS IN HEALTH AND DISEASE

CHAIRPERSONS: G. Janossy
D.Y. Mason

CHANGES OF LYMPHOID MICROENVIRONMENTS IN HUMAN DISEASES - A SHORT REVIEW

G.Janossy and M.Bofill

Department of Immunology
Royal Free Hospital School of Medicine
London NW3 2QG

INTRODUCTION

During the last two years it has become clear that monoclonal antibodies to the accessory cells of the immune system help to define cell interactions and also indicate that important changes occur in various diseases. The main accessory cell types in the lymphoid organs are as follows. Selective Mc Ab-s have been found which react with the three main types of epithelial cells (subcapsular, cortical and medullary) in the normal human thymus (Fig.1; ref.1-4). These regions of thymus contain different lymphoid elements (reviewed in ref.1). The bone marrow derived interdigitating cells (dominant at the cortico-medullary junction) are complemented by acid phosphatase positive, esterase positive macrophages. Recent evidence in rat shows that these latter cell types show different phenotypes in the cortex and medulla[5].

In the lymph node the accessory cells of the T zone are recognized by their shape and anti-HLA-DR antibodies but additional good reagents are still needed. One candidate is the RFD1 antibody but this is a weak reagent[6], while S100 reacts only with a subpopulation of these cells[7]. The study of macrophage heterogeneity is progressing well but appears to show a complicated pattern at the moment[8]. On the other hand, excellent reagents are available for analysing follicular dendritic cells[9,10]. The microenvironment of the normal human germinal centre has been well described in recent reviews[1,9,11] (Fig.2).

The other important point emerging is that the functional lymphoid subsets are better characterized by reagent combinations than with a single antibody alone. Two relevant examples are as

Fig.1

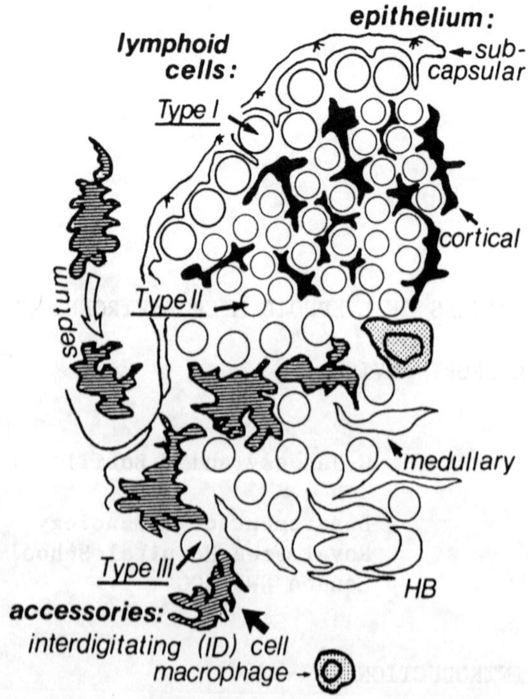

Main cell types in the human thymus. Epithelial cells are selectively labelled with various Mc Ab-s. The subcapsular epithelium is positive with Leu-7 antibody (in fetus and adults; ref. 2,4) and with A2B5 and RFD4 in infants. Cortical epithelium is particularly strongly HLA-DR positive and reacts with MR3 and MR6[3], while medullary epithelium is strongly positive with A2B5[4] and RFD4 but appears to be essentially HLA-DR negative or weak[2]. These Ab-s are representative examples (see also ref.3). The lymphoid elements surrounding these epithelial cells are also fairly distinct and represent large subcapsular blasts, cortical thymocytes and medullary thymocytes, respectively (reviewed in ref.1,11,12). Mesenchymal bone marrow derived elements are interdigitating cells in the medulla (extremely strongly HLA-DR positive and RFD1 positive; ref.2,12), and the macrophage populations of the cortex and medulla (see ref.5).

follows. Firstly, the antibody to a special T and natural killer subset (HNK-1) reacts with a variety of cells including a $T8^-$, $T4^-$ subset (efficient NK cells), a $T8^+$ subset, and a peculiar $T4^+$ subset[14,15]. The latter population is rare in the blood but represents a substantial subset in the germinal centre (ref.15; Fig.2). Kroese et al. have now also shown in the rat that a similar T cell subset exists in the germinal centre. These cells also exhibit a peculiar phenotype ($W3/25^+$,$ER3^+$) which is rare in blood[16]. The function of this T cell type is an enigma but may be rather important for understanding the T cell dependence of germinal centre development and Ig isotype switch. Secondly, a peculiar $T1^+$ B lymphoid population has been identified in the normal lymph node and tonsil[17]. T1 antigen (67k) is normally T cell associated. We searched for this $T1^+$ B cell population in the fetal tissues.

FETAL DEVELOPMENT OF B CELLS

Campana et al. (18) have found that when analysed by a panel of seven Ab-s the earliest identifiable B cells (TdT^+, pre-B and

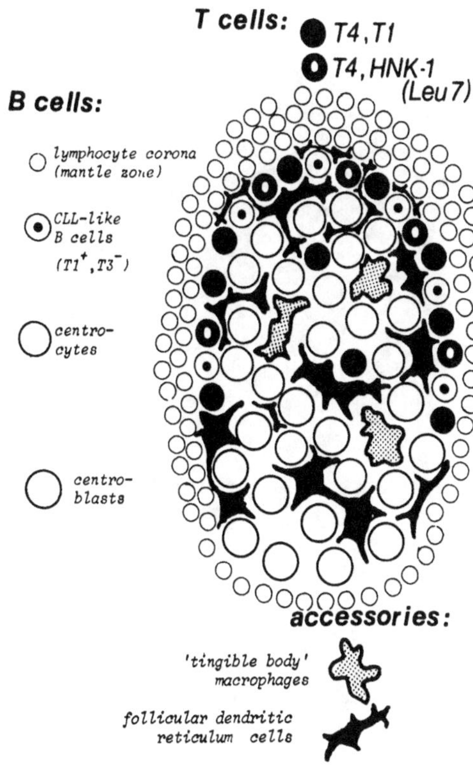

Fig.2

The main cell types in the germinal centres of peripheral lymph nodes. Strong Ab-s are available which react with follicular dendritic cells (R4/23, RFD3; ref.2,9,10). The CLL-like T1$^+$ B cells are a minor population but this is the sole cell type in the fetal primary nodules (see Fig.3).[19] *The HNK-1$^+$,T4$^+$ cells are a special germinal-centre seeking T cell subset.*[15]

IgM$^+$,IgD$^-$) in the normal bone marrow had a different phenotype than the IgD$^+$ BM B cells and the vast majority of lymphocytes in peripheral lymphoid organs (peripheral B cell type; for details see ref.18). Further analysis in the human fetus has shown that the 'early' B cell types have been seen both in the liver and BM. In contrast blood, lymph node and spleen contained the IgM$^+$,IgD$^+$ peripheral B cell type[19]. In the lymph node these peripheral B cells showed heterogeneity. Diffusely distributed B cells were IgM$^+$,IgD$^+$ without T cell associated markers, but after the 19-20th gestational weeks around the easily identifiable RFD3$^+$ follicular dendritic cells a strongly T1$^+$,IgM$^+$, IgD$^+$,HLA-DR$^+$ B cell population could be found (Fig.3). These cells are very similar, by phenotypic criteria, to the malignant B cells seen in CLL and centrocytic lymphoma[9,20]. These T1$^+$,IgM$^+$ cells may also be involved in autoimmune disorders such as rheumatoid arthritis[21]. The equivalent B cell population (Ly-1$^+$ B cells) in the mouse is also involved in both B lymphomas[22] and the disease which resembles systemic lupus erythematosus in the NZB mouse[23]. Furthermore, in this experimental model it is demonstrated that the Ly-1$^+$ B cells are activated and spontaneously secrete autoantibodies of IgM Class[23].

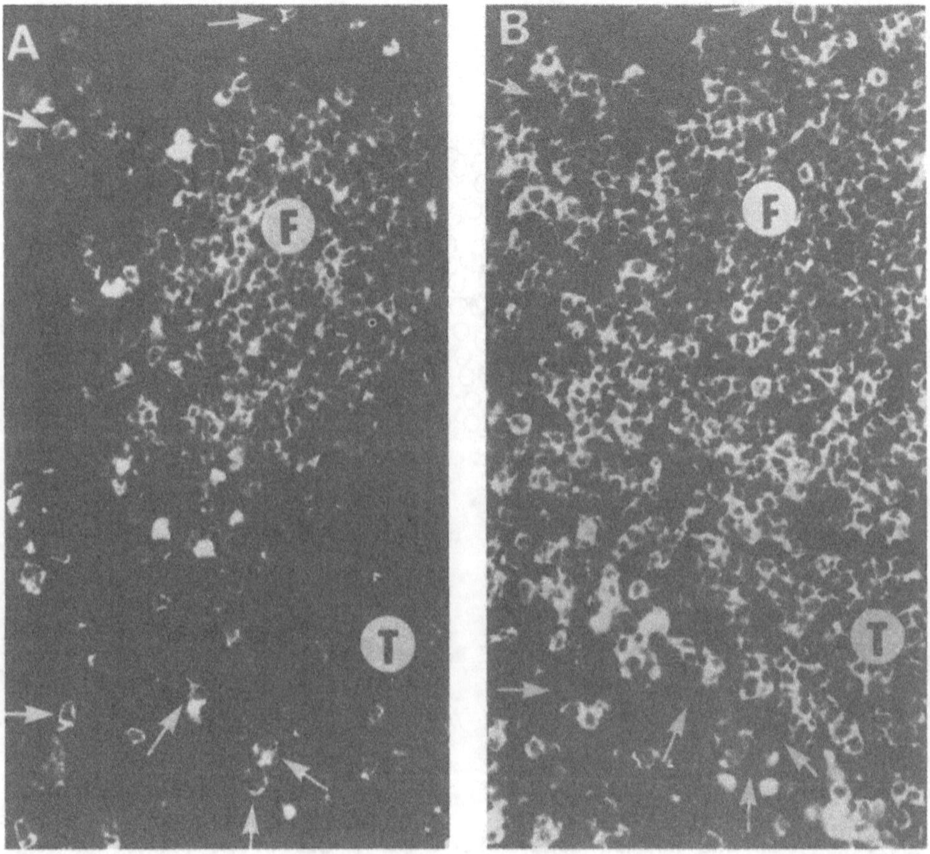

Fig.3

$T1^+$ B lymphocytes in the primary follicles (F) of fetal lymph nodes (22 week old fetus). A: staining for IgM; B: staining for T1 (67k T cell associated antigen, equivalent to murine Ly-1). IgM^+ B cells in the T area are $T1^-$ (arrows).

DESTRUCTION OF GERMINAL CENTRES

In lymphoid tissues stimulated by antigens the $T1^+$ B cells are rare (Fig.2) as new populations of centroblasts and centrocytes emerge inside the germinal centre (GC). This different antigen stimulated population seems to be the normal equivalent cells of follicular lymphomas of centroblastic/centrocytic type[9]. The well organized structure of GC is destroyed in certain diseases, such as Hodgkins disease[13,23] and in a proportion of male homosexuals with persistent generalized lymphadenopathy (PGL). In a recent collaborative study with St.Mary's Hospital[24] 5 out of 11 patients showed explosive follicular hyperplasia with extensive destruction of the $RFD3^+$ network (Fig.4; see also ref.25). This is a peculiar histological picture which is not seen in other

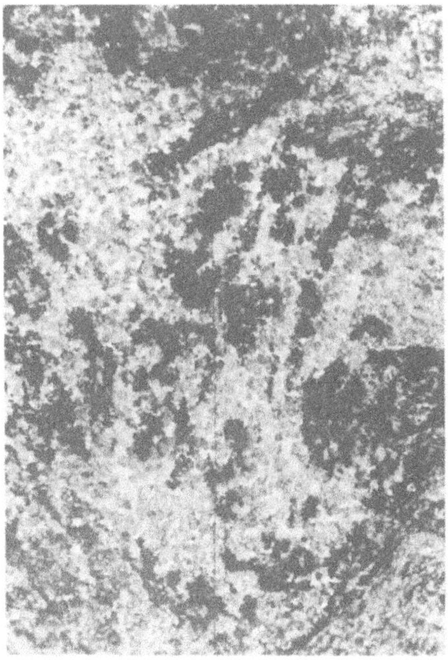

Fig.4

The destruction of germinal centre in a homosexual male who shows persistent generalized lymphadenopathy. The lymph node section has been stained for follicular dendritic cells using RFD3 antibody and immunoperoxidase. The destruction of the regular RFD3$^+$ network is seen only in a proportion of individuals and long term follow-up is required to see whether these patients belong to a poor prognostic group (from ref.24). Many intruding lymphocytes are seen which are a mixture of T8$^+$ and T4$^+$ cells.

conditions, and the labelling with Ab-s to follicular dendritic cells is helpful to visualise the destruction. None of these patients had a fully developed acquired immune deficiency syndrome (AIDS) with opportunistic infections. Nevertheless it will be interesting to follow these individuals for a longer period of time because in this group other "prodromal" symptoms such as anergy in delayed type hypersensitivity, weight loss, oral candidiasis have also been frequently seen. In the other 6 cases of PGL the regular network of dendritic cells was maintained, in spite of explosive follicular hyperplasia - and other prodromal signs were rare. In all cases T8$^+$ cells invaded the GC-s although T4$^+$ cells were also frequently still present[26]. Interestingly in 4 out of 11 cases the depletion of T4$^+$ cells in the lymph node paracortical area was not as severe as expected on the basis of results seen in the peripheral blood. The further phenotypic characterization of the different T4$^+$ and T8$^+$ subpopulations is now in progress in order to investigate why such a strong overactivation of the B lymphoid system can be observed when the HTLV-III virus preferentially destroys T4$^+$ cells of helper phenotype.

THE HISTOLOGY OF MYASTHENIC THYMUS

In the light of the recent developments of characterizing the accessory cells of human thymus it is interesting to study the changes of thymic centres in myasthenia gravis. Previous studies showed no major changes of T cell subpopulations[27], and the frequent involvement of GC development has already been well

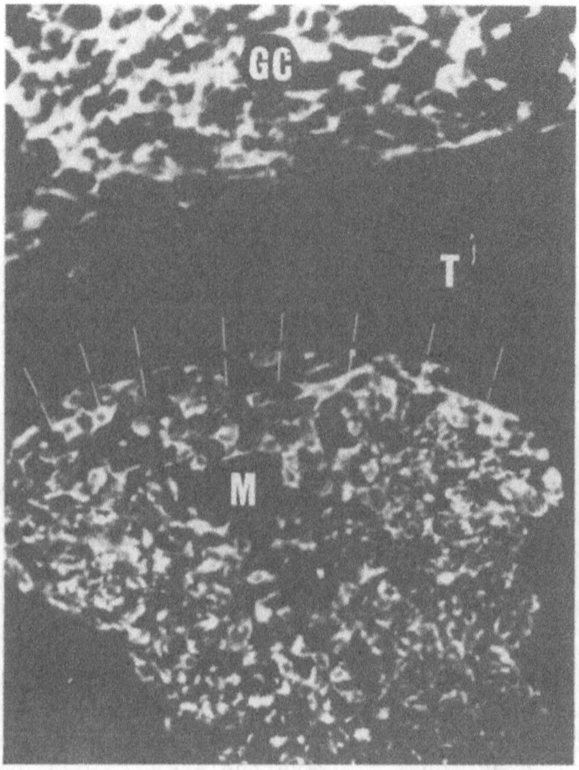

Fig. 5

The altered microenvironment in myasthenia gravis (from ref.2). The follicular dendritic cells of the germinal centre are labelled with RFD-3 Ab (GC). The T cell area is unlabelled but contains a well developed fibronectin network (not shown) which is absent in normal thymus. The medullary epithelium, labelled with RFD-4 Ab (M) is markedly hypertrophic and surrounded by a laminin layer which breaks up (not shown) where the medulla approaches the GC (broken lines; ref.2).

documented[28]. It is also clear, however, that the presence of GC-s is not obligatory and may not be an early event. Thus we have studied the epithelial and mesenchymal elements as well as the extracellular matrix[2].

The most interesting alteration in MG has been the clear development of proper T cell zones with heavy deposition of a fibronectin network and many interdigitating cells. Plenty of high endothelial venules were seen (see also ref.21). In the mouse it has been shown that the normal medulla contains few if any recirculating T cells (Mel-14$^+$). It will be most interesting to investigate the probable increase of recirculating human T cells in myasthenia gravis, when the Ab to the exact human equivalent of Mel-14 will become available[29]. These intruding T cell areas have been observed even in the two cases where proper intramedullay GC development was not seen. Nevertheless, in the fully developed picture of myasthenia gravis the "intrusion" of lymph node features (T zone + GC; Fig.6) is clear.

In addition, in the same tissues the bands which contain hypertrophic medullary epithelial cells are also observed. These areas are "sealed off" from the T zones by a laminin containing

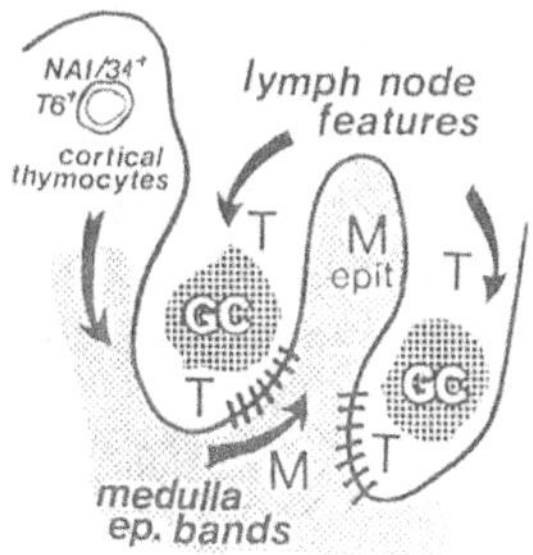

Fig.6

Changes of micro-environments in the myasthenic thymus (see text; ref.2).

basal layer. At specific points near the tips of the GC-s the laminin positive boundaries between lymph node-like areas and the medullary areas break up (Fig.5 and 6). This microenvironment is unique to the myasthenic thymus, and the cell types in the near vicinity include: medullary epithelium, interdigitating cells, follicular dendritic cells and both T cells (including a few immature thymocytes) and various B cell types. Functional assays are in progress (Willcox and Newsom-Davis) to see which of these cell types play a part in the autosensitization process against acetylcholin receptors[30].

CONCLUSIONS

With the help of immunohistology precise questions are being formulated for investigating cell-cell interactions between the relevant in vitro cells. It is therefore most important that the techniques for separating accessory cell types from lymphoid tissues are also progressing[30,31]. The reconstitution of the proper microenvironments in cultures may soon become feasible, in order to establish the cellular basis of various diseases.

Acknowledgement. We gratefully acknowledge the collaboration of Dr.D.Campana (RFH), Dr.A.Pinching (St.Mary's Hospital), Dr.N.Willcox and Prof.J.Newsom-Davis (RFH), Prof.E.Kelemen and Dr.M.Janossa (Budapest) in the original work quoted in the references.

REFERENCES

1. G.Janossy. Progr.Immunol. 5:1085 (1983).
2. M.Bofill, G.Janossy, N.Willcox, M.Chilosi, L.K.Trejdosiewicz, M.Bagott & J.Newsom-Davis. Am.J.Pathology. Submitted (1984).
3. M.A.Ritter, H.J.Schuurman, W.McKenzie, R.A.de Maagd, K.Price, R.Broekhuizen & L.Kater. This volume.
4. B.F.Haynes, K.Shimizu, G.S.Eisenbarth. J.Clin.Invest. 71:9 (1983)
5. C.D.Dijkstra, E.A.Dopp, P.Joling & G.Kraal. This volume.
6. M.J.Raftery, L.W.Poulter, G.Janossy, P.Sweny, O.N.Fernando & J.F.Moorhead. J.Clin.Pathol. 36:734 (1983).
7. K.Takahashi, H.Yamaguchi, P.Ishizeki & S.Watanabe. Wirchows Arch.Cell Pathol. 37:125 (1981).
8. W.W.Hancock, H.Zola & R.C.Atkins. Blood 62:1271 (1983).
9. H.Stein, J.Gerdes & D.Y.Mason. Clin.Haematol. 11:531 (1982).
10. M.Chilosi, G.Pizzolo, L.Fiore-Donati, M.Bofill & G.Janossy. Br.J.Cancer 48:763 (1983).
11. G.Janossy & H.G.Prentice. Clin.Haematol. 11:631 (1982).
12. G.Janossy, M.Bofill, L.K.Trejdosiewicz, L.W.Poulter, M.Chilosi & J.Newsom-Davis. In: The Thymus: Histopathology & Pathology (H.K.Muller-Hermelink, ed), Springer-Verlag, Berlin (1984)
13. Z.Abdulaziz, D.Y.Mason, H.Stein, K.C.Gatter & J.R.G.Nash. Histopathology 8:in press (1984).
14. L.L.Lanier, A.M.Le, J.H.Phillips, N.L.Warner & L.A.Herzenberg. J.Immunol. 131:1789 (1983).
15. G.Pizzolo, G.Semenzato, M.Chilosi, L.Morittu, A.Ambrosetti, N.Warner, M.Bofill & G.Janossy. Clin.Exp.Immunol. 57:195 (1984).
16. F.G.M.Kroese, A.S.Wubbena, P.Joling & P.Nieuwenhuis. This volume.
17. F.Caligaris-Cappio, M.Gobbi, M.Bofill & G.Janossy. J.Exp.Med. 155:623 (1982).
18. D.Campana, G.Janossy, M.Bofill, L.K.Trejdosiewicz, D.Ma, A.V.Hoffbrand, D.Y.Mason, A-M.Lebacq & H.Forster. J.Immunol. Submitted (1984).
19. M.Bofill, G.Janossy, M.Janossa, G.D.Burford, G.J.Seymour & E.Kelemen. J.Immunol. Submitted (1984).
20. F.Caligaris-Cappio, L.Bergui, G.Pizzolo, M.Chilosi, L.Tesio, D.Campana, G.Semenzato, F.Malavasi & M.Gobbi. This volume.
21. C.Plater-Zyberk, K.Lam, T.D.Kennedy, R.N.Maini & G.Janossy. This volume.
22. L.L.Lanier, N.L.Warner, J.A.Ledbetter & L.A.Herzenberg. J.Exp.Med. 153:998 (1981).
23. K.Hayakawa, R.P.Hardy, D.R.Parks & L.A.Herzenberg. J.Exp.Med. 157:202 (1982).
24. G.Janossy, A.J.Pinching, M.Bofill, J.Weber, J.E.McLaughlin, M.Ornstein, K.Ivory, J.R.W.Harris, M.Favrot & D.C.MacDonald-Burns. Clin.Exp.Immunol. In press (1984).
25. P.Racz, A.C.Feller, B.Luther, K.Tenner-Racz, K.Lennert, M.Dietrich & P.Kern. This volume.
26. R.L.Modlin, F.M.Hofman & P.R.Meyer. Lancet ii:768 (1983)
27. J.A.Thomas, H.N.A.Willcox & J.Newsom-Davis. J.Neuroimmunol. 3:319 (1982).

28. G.K.Scadding, A.Vincent, J.Newsom-Davis & K.Henry. Neurology 31:935 (1981).
29. E.C.Butcher, S.Jalkanen & R.Bargatze. This volume.
30. N.Willcox, M.Schluep, M.Bofill & J.Newsom-Davis. This volume.
31. E.Heinen, D.Radoux, D.Y.Mason, H.Stein, C.Kinet-Denoel, C.Lilet-Leclercq & L.J.Simar. This volume.

THE SMALL SUBPOPULATION OF T CELL MARKER$^+$/TdT$^+$ CELLS IN THE HUMAN BONE MARROW MAY REPRESENT PROTHYMOCYTES

J.J.M. van Dongen[1], H. Hooijkaas[1], W.M. Comans-Bitter[1], K. Hählen[2] and G.E. van Zanen[2]

[1]Dept. Cell Biology and Genetics and [2]Dept. Paediatrics Subdivision Paediatric Oncology of the University Hospital/Sophia Children's Hospital, Erasmus University Rotterdam, The Netherlands

INTRODUCTION

Most normal haematopoietic differentiation stages have malignant counterparts in the form of a leukaemia or non-Hodgkin lymphoma (NHL) (Foon et al., 1982). The acute lymphoblastic leukaemias (ALL) may be regarded as the malignant counterparts of normal immature lymphoid cells (Janossy et al., 1980, 1981). These ALL can be characterized immunologically by use of several markers such as: terminal deoxynucleotidyl transferase (TdT), the HLA-DR non-polymorphic antigen, the common ALL antigen (cALL), cytoplasmic immunoglobulin μ heavy chain (cIgM), surface membrane Ig (SmIg) and the T cell markers Tp41, T11, T1, T6, T4, T8 and T3 (Foon et al., 1982; Vodinelich et al., 1983). At least five different immunological types of ALL are distinguishable by these markers: null ALL (HLA-DR$^+$/TdT$^+$), common ALL (cALL$^+$/HLA-DR$^+$/TdT$^+$), pre-B-ALL (cIgM$^+$/cALL$^+$/HLA-DR$^+$/TdT$^+$), B-ALL (SmIg$^+$/cALL$^+$/HLA-DR$^+$/TdT$^-$) and T-ALL (T cell marker$^+$/HLA-DR$^-$/TdT$^+$) (Foon et al., 1982). The T-ALL can be subdivided into several subtypes depending on their positivity for the various T cell markers (Greaves et al., 1981; Catovsky et al., 1982; Roper et al., 1983). Recent studies have demonstrated that null ALL, common ALL and pre-B-ALL have rearranged Ig heavy chain genes or rearranged Ig heavy and light chain genes and are therefore "precursor-B-ALL" (Ford et al., 1983; Korsmeyer et al., 1983). T-ALL, on the other hand, proved to have germline Ig genes (Korsmeyer et al., 1983). Janossy et al. (1980) demonstrated that the normal counterparts of the precursor-B-ALL are present in the normal and regenerating bone marrow (BM) (Janossy et al., 1979, 1980). The normal counterparts of T-ALL (T cell marker$^+$/TdT$^+$ cells) could be detected in the thymus only and not in the BM (Janossy et al., 1981; Neudorf et al., 1984; Tax et al., 1984). Since T cell marker$^+$/TdT$^+$ cells in extrathymic locations were supposed to be malignant cells (Janossy et al., 1980; Bradstock et al., 1981), the use of T cell marker$^+$/TdT$^+$ double immunofluorescence (IF) staining is indicated for the detection of low numbers of malignant cells in the BM and peripheral blood (PB) of patients with T-ALL during follow-up. We currently use this double IF staining for the detection of minimal residual disease and early detection of relapse in patients with T-ALL and TdT positive T cell non-Hodgkin lymphoma (T-NHL) and for staging of TdT positive T-NHL at diagnosis (van Dongen et al., 1984). The detection limit of this double IF staining was determined in limiting dilution experiments and proved to be one T cell marker$^+$/TdT$^+$ cell in 10,000 or even 100,000 normal haematopoietic cells (van Dongen et al., 1984). To obtain control values, we determined whether Tp41$^+$/TdT$^+$, T1$^+$/TdT$^+$ and T6$^+$/TdT$^+$ cells were present in the BM from six patients with non-T-ALL (i.e. precursor-B-ALL) in complete remission but still under maintenance therapy (van Dongen et al., 1984). It appeared that neither T1$^+$/TdT$^+$ nor T6$^+$/TdT$^+$ cells were detectable, which is consistent with the literature (Janossy et al., 1980, 1981; Neudorf et al., 1984). However, Tp41$^+$/TdT$^+$ cells were detected in all six BM samples tested; the frequency ranged from 0.01% to 0.17% (mean 0.08%) (van Dongen et al., 1984). Since the Tp41$^+$ antigen is one of the most consistent markers for T cells (Vodinelich et al., 1983; Tax et al., 1984), these results suggested that the Tp41$^+$/TdT$^+$ cells in the BM are precursor T cells. Therefore we examined the BM samples from another 7 patients with precursor-B-ALL in complete remission under maintenance therapy and 3 patients with precursor-B-ALL in complete remission, 6 weeks after termination of maintenance therapy, to establish the precise phenotype of this putative precursor T cell, by double and triple immunological stainings.

MATERIALS AND METHODS

Cell sampling

BM samples were obtained from seven children with precursor-B-ALL in complete remission under maintenance therapy and from three children with precursor-B-ALL in complete remission 6 weeks after termination of maintenance therapy. Mononuclear cells (MNC) were isolated by Ficoll Paque (d: 1.077 g/cm^3; Pharmacia, Uppsala, Sweden) density centrifugation (Böyum, 1968) and washed. All washings were performed in phosphate buffered saline (PBS) with 1% bovine serum albumin (BSA), pH 7.8. Remaining red blood cells were removed by NH4Cl lysis (Roos and Loos, 1970).

Table 1. Immunological markers used for the immunological characterization of T cells and their malignant counterparts

Antibodies	Antigen recognized	References
WT1[a]	Tp41 antigen	Tax et al., 1984, Hansen et al., 1984
3A1[b]	Tp41 antigen	Haynes et al., 1980, Hansen et al., 1984
E rosette	sheep erythrocyte receptor (E)	Moore and Zusman, 1978
OKT11[c]	T11 antigen (sheep erythrocyte receptor)	Verbi et al., 1982
Leu-1[d]	T1 antigen	Martin et al., 1981
66 III E5[e]	T6 antigen	Lerch et al., 1983
Leu-2[d]	T8 antigen	Engleman et al., 1981
Leu-3[d]	T4 antigen	Engleman et al., 1981
Leu-4[d]	T3 antigen	Borst et al., 1982
anti-TdT[f]	TdT	Goldschneider et al., 1977
anti-HLA-DR[d]	HLA-DR non-polymorphic antigen	Lampson and Levy, 1980

a. Rijksinstituut Volksgezondheid, Bilthoven, The Netherlands.
b. American Type Culture Collection, Rockville, MD, USA.
c. Ortho Diagnostic Systems Inc., Raritan, NJ, USA.
d. Becton Dickinson, Sunnyvale, CA, USA.
e. Dr. M. van de Rijn, Amsterdam, The Netherlands.
f. Bethesda Research Laboratory, Gaithersburg, MD, USA.

Double IF staining

For the Tp41/TdT, T11/TdT, T1/TdT and T6/TdT double IF staining, the MNC were successively incubated with one of the optimally titrated relevant anti-T cell antigen McAb (Table 1) and a tetramethylrhodamine isothiocyanate (TRITC) conjugated goat-anti-mouse-Ig antiserum. Subsequently, cytocentrifuge preparations of these labelled cells were subjected to indirect IF staining with a rabbit-anti-TdT antiserum and a fluorescein isothiocyanate (FITC) conjugated goat-anti-rabbit-Ig antiserum (Bethesda Research Laboratory, Gaithersbrug, MD, USA). The slides were mounted in glycerol/PBS (9:1) containing 1 mg phenylenediamine per ml to prevent fading of the fluorescence (van Dongen et al., 1984). Both the McAb WT1 and 3A1 were used for the detection of the Tp41 antigen in the double IF staining.

Triple immunological staining

For Tp41/HLA-DR/TdT triple immunological staining, the MNC were successively incubated with the McAb 3A1 and a goat-anti-mouse-

Ig antiserum, conjugated with colloidal gold particles of about 30 nm (Janssen Pharmaceutica, Beerse, Belgium). Subsequently the cells were incubated with normal mouse serum (diluted 1 in 100) to block free antigen binding sites of the goat-anti-mouse-Ig antiserum. Finally the cells were incubated with a TRITC conjugated anti-HLA-DR McAb (Becton Dickinson, Sunnyvale, CA, USA). The cells were centrifuged on slides and subjected to the indirect IF staining with rabbit-anti-TdT antiserum and FITC conjugated goat-anti-rabbit-Ig antiserum as described above.

Microscopes

Zeiss and Leitz microscopes were used for the evaluation of the immunofluorescence and the immunogold staining. Both microscopes were equipped with HBO mercury lamps (Osram, Berlin, FRG), phase contrast facilities and filter combinations for the selective visualization of FITC and TRITC. The immunogold staining was evaluated by epiillumination with polarized light (De Mey, 1983).

In both the double and triple labelling experiments at least 50,000 cells from two cytocentrifuge preparations (each cytocentrifuge preparation containing 25,000 to 35,000 cells) were evaluated.

Table 2. Occurrence of T cell marker$^+$/TdT$^+$ cells in BM

	Children with precursor-B-ALL in complete remission under maintenance therapy (n=7)	Children with precursor-B-ALL in complete remission 6 weeks after termination of maintenance therapy (n=3)
% TdT$^+$	0.3-10.8[a] (4.3)	0.5-5.2 (3.3)
% Tp41$^+$/TdT$^+$	0.01-0.13 (0.07)	-[b]
% T11$^+$/TdT$^+$	0.01-0.12 (0.07)	-
% T1$^+$/TdT$^+$	-	-
% T6$^+$/TdT$^+$	-	-

a. Figures represent the ranges of the percentages of positive cells; in the brackets the arithmetic mean is given.
b. No T cell marker$^+$/TdT$^+$ cell in 50,000 cells analyzed.

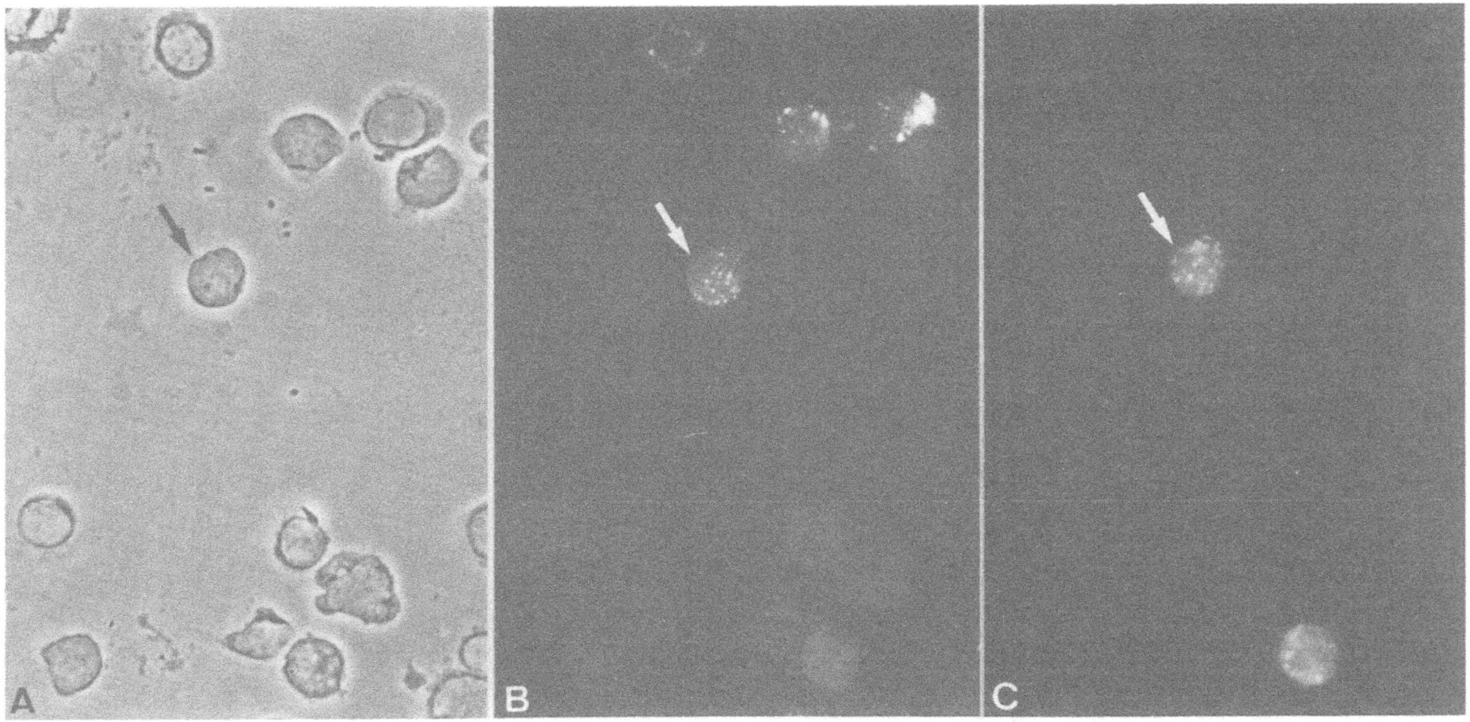

Fig. 1. Double IF staining for Tp41 and TdT. A: phase contrast morphology; B: Tp41 positive cells (TRITC labelled); C: TdT positive cells (FITC labelled). One cell is positive for Tp41 and TdT and probably represents a precursor T cell. Some other cells are positive for Tp41 and not for TdT and probably represent normal T lymphocytes.

RESULTS

Double IF staining

In the BM samples of seven children with precursor-B-ALL in complete remission under maintenance therapy, $Tp41^+/TdT^+$ and $T11^+/TdT^+$ cells were detected in low frequencies (Fig. 1), whereas no $T1^+/TdT^+$ or $T6^+/TdT^+$ cells were observed (Table 2). In the BM of the three children with precursor-B-ALL in complete remission 6 weeks after termination of the maintenance therapy, no T cell $marker^+/TdT^+$ cells could be detected (Table 2).

Triple immunological staining

In the BM samples of four children with precursor-B-ALL in complete remission under maintenance therapy, all $Tp41^+/TdT^+$ cells proved to be positive for HLA-DR.

DISCUSSION

The precursors of the T cells, the so-called 'prothymocytes' are generated in haematopoietic organs, such as BM and foetal liver (Boersma, 1983a; Scollay, 1983). It was suggested that prothymocytes, which occur in a low frequency in murine BM, are already committed to thymocyte differentiation and discrete from precursor-B-cells as well as pluripotent haemopoietic stem cells (Basch et al., 1978; Boersma, 1983b). Some studies suggest that rodent prothymocytes are positive for TdT and a T cell marker (Silverstone et al., 1976; Basch et al., 1978; Goldschneider et al., 1981; Greiner et al., 1982). The human prothymocyte has not been identified as yet. Janossy and Prentice (1982) suggested that the human precursor T cell may belong to the TdT positive cell population in the BM, although T cell $marker^+/TdT^+$ cells were not detected in the human BM (Janossy et al., 1981; Neudorf et al., 1984; Tax et al., 1984).

Table 3. Frequency of the prothymocyte in murine BM

A. Normal murine BM		
- Kadish and Basch (1976):		0.015-0.02%
- Lepault and Weissman (1981):		≤0.1%
- Boersma et al. (1981):		0.03%
B. Regenerating murine BM, after irradiation		
- Kadish and Basch (1976):		
	11 days:	± 10 % of normal
- Boersma et al. (1981):		
	1 week:	1 % of normal
	3 weeks:	2.5% of normal
	8 weeks:	15 % of normal

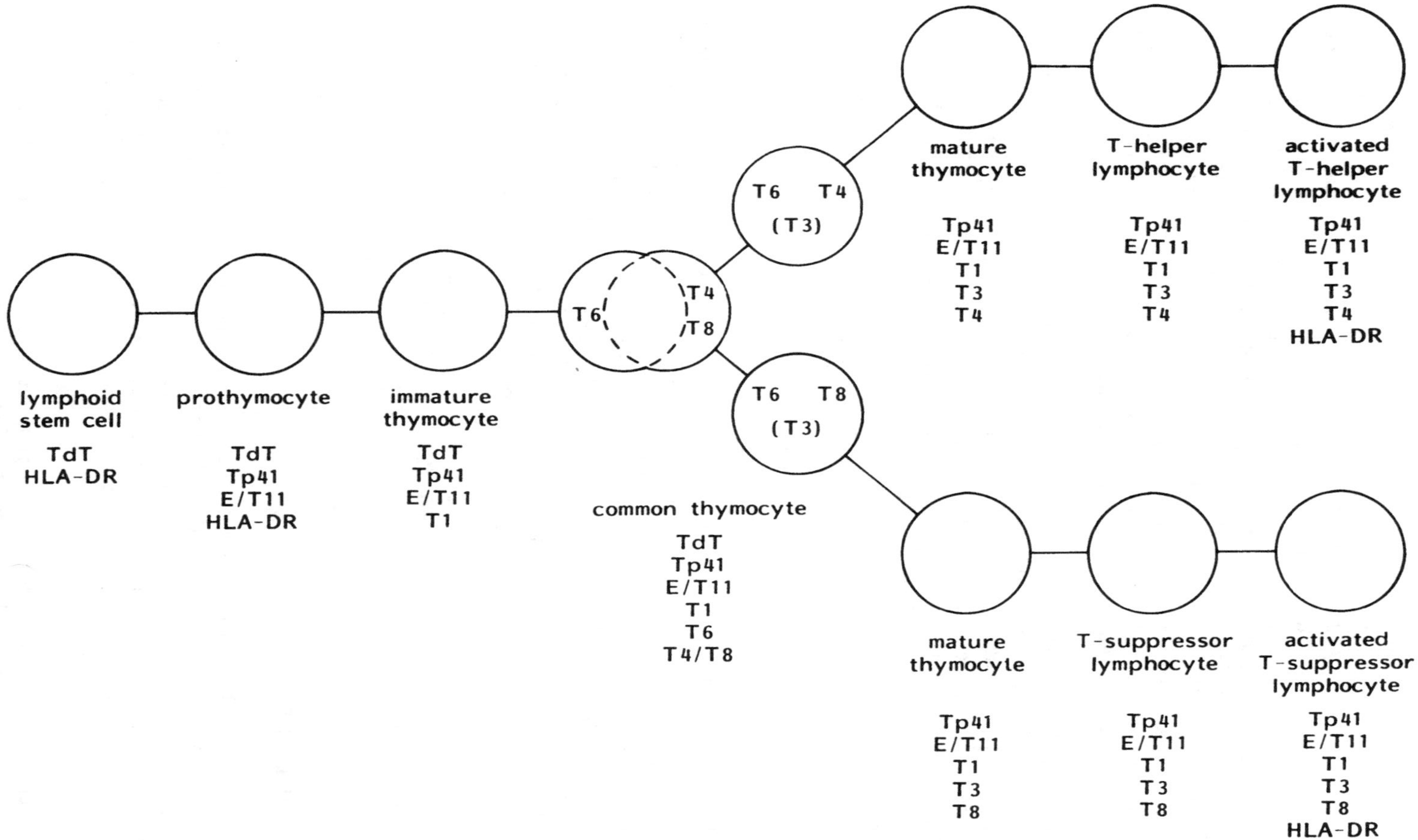

Fig. 2. Hypothetical differentiation pathway of the T cell lineage. The markers used are listed in Table 1.

We did, however, detect T cell marker^{+}/TdT^{+} cells in human BM. Since the Tp41 and T11 antigens are the most consistent markers for T cells, the Tp41^{+}/TdT^{+} and T11^{+}/TdT^{+} cells in the human BM may be precursor T cells. The frequency of these cells (about 0.05%) is similar to the frequency of the prothymocyte in murine BM (Table 3). It was described that the frequency of the murine prothymocyte is strongly reduced in regenerating BM (c.f., Table 3). The absence of Tp41^{+}/TdT^{+} and T11^{+}/TdT^{+} cells in the BM samples of patients with precursor-B-ALL in complete remission, 6 weeks after termination of maintenance therapy supports the above supposition that the Tp41^{+}/TdT^{+} cells in the human BM may represent prothymocytes.

Although the majority of the TdT positive cells in the human BM are positive for HLA-DR, always at least a small percentage is HLA-DR negative (Janossy et al., 1980). To determine whether the T cell marker^{+}/TdT^{+} cells in the BM belong to the HLA-DR^{+}/TdT^{+} cell population, a triple immunological staining was performed, which demonstrated that all Tp41^{+}/TdT^{+} cells were also positive for HLA-DR. Since haematopoietic precursor cells are HLA-DR positive (Falkenburg et al., 1984), the HLA-DR positivity of the Tp41^{+}/TdT^{+} cells in human BM provides additional evidence that this cell is a precursor of T cells, the prothymocyte (Fig. 2).

B cells have rearranged Ig genes (Tonegawa, 1983) and T cells probably have rearranged T cell receptor genes (Chien et al., 1984; Saito et al., 1984). Since precursor B cells have rearranged Ig heavy chain genes, precursor T cells may have rearranged T cell receptor genes. Therefore it would be interesting to analyze the DNA of Tp41^{+}/HLA-DR^{+}/TdT^{+} cells to investigate whether they have germline Ig genes and rearranged T cell receptor genes. However, at least 10^{6} cells are needed for the DNA preparation, which number of cells cannot be obtained by purification procedures, since the Tp41^{+}/HLA-DR^{+}/TdT^{+} cell occurs in too low frequencies in normal BM. This problem can be overcome by analyzing the DNA of the malignant counterpart of the Tp41^{+}/HLA-DR^{+}/TdT^{+} cells. Such studies are in progress in our laboratory.

ACKNOWLEDGMENTS

We gratefully acknowledge Prof.Dr. R. Benner for his continuous support, Mrs. K. Benne, E.G.M. Cristen, A.A. van der Linde-Preesman, and I.L.M. Tettero for their skilfull technical assistance, Mr. J. Fengler for his photographic assistance, Mr. C. Enkelaar from Zeiss Nederland for his technical support and Mrs. C.J.M. Meijerink-Clerkx for typing this manuscript.

This work was financially supported by the Netherlands Cancer Foundation (Koningin Wilhelmina Fonds).

REFERENCES

Basch, R.S., Kadish, J.L., and Goldstein, G., 1978, Hematopoetic thymocyte precursors. IV. Enrichment of the precursors and evidence for heterogeneity, J. Exp. Med., 146:1843.

Boersma, W., Betel, I., Daculsi, R., and Van der Westen, G., 1981, Post-irradiation thymocyte regeneration after bone marrow transplantation. I. Regeneration and quantification of thymocyte progenitor cells in the bone marrow, Cell Tissue Kinet., 14:179.

Boersma, W.J.A., 1983a, Prothymocytes in mouse fetal liver, Thymus, 5:419.

Boersma, W.J.A., 1983b, Radiation sensitivity and cycling status of mouse bone marrow prothymocytes and day 8 colony forming units spleen (CFUs), Exp. Hematol., 11:922.

Borst, J., Prendiville, M.A., Terhorst, C., 1982, Complexity of the human T lymphocyte-specific cell surface antigen T3, J. Immunol., 128:1560.

Böyum, A., 1968, Isolation of mononuclear cells and granulocytes from human blood, Scand. J. Clin. Lab. Invest., 21:77.

Bradstock, K.F., Janossy, G., Tidman, N., Papageorgiou, E.S., Prentice, H.G., Willoughby, M., and Hoffbrand, A.V., 1981, Immunological monitoring of residual disease in treated thymic acute lymphoblastic leukaemia, Leuk. Res., 5:301.

Catovsky, D., Linch, D.C., and Beverley, P.C.L., 1982, T cell disorders in haematological diseases, Clin. Haematol., 11:661.

Chien, Y., Gascoigne, N.R.J., Kavaler, J., Lee, N.E., Davis, M.M., 1984, Somatic recombination in a murine T-cell receptor gene, Nature, 309:322.

De Mey, J., 1983, Colloidal gold probes in immunocytochemistry, in: "Immunocytochemistry. Practical applications in Pathology and Biology", J.M. Polak and S. van Noorden, eds., Wright, Bristol,p.82.

van Dongen, J.J.M., Hooijkaas, H., Hählen, K., Benne, K., Bitter, W.M., van der Linde-Preesman, A.A., Tettero, I.L.M., van de Rijn, M., Hilgers, J., Van Zanen, G.E., and Hagemeijer, A., 1984, Detection of minimal residual disease in TdT positive T cell malignancies by double immunofluorescence staining, in: "Minimal residual disease in acute leukemia", B. Löwenberg and A. Hagenbeek, eds., Martinus Nijhoff Publishers B.V., The Hague, The Netherlands, p. 67.

Engleman, E.G., Benike, C.J., Glickman, E., Evans, R.L., 1981, Antibodies to membrane structures that distinguish suppressor/cytotoxic and helper T lymphocyte subpopulations block the mixed leukocyte reaction in man, J. Exp. Med., 153:193.

Falkenburg, J.H.F., Jansen, J., Van der Vaart-Duinkerken, N., Veenhof, W.F.J., Blotkamp, J., Goselink, H.M., Parlevliet, J., van Rood, J.J., 1984, Polymorphic and monomorphic HLA-DR determinants on human hematopoietic progenitor cells, Blood, 63:1125.

Foon, K.A., Schroff, R.W., and Gale, R.P., 1982, Surface markers on leukemia and lymphoma cells: Recent advances, Blood, 60:1.

Ford, A.M., Molgaard, H.V., Greaves, M.F., and Gould, H.J., 1983,

Immunoglobulin gene organisation and expression in haemopoietic stem cell leukaemia, EMBO J., 2:997.
Goldschneider, I., Gregoire, K.E., Barton, R.W., and Bollum, F.J., 1977, Demonstration of terminal deoxynucleotidyl transferase in thymocytes by immunofluorescence, Proc. Natl. Acad. Sci. USA, 74:734.
Goldschneider, I., Ahmed, A., Bollum, F.J., and Goldstein, A.L., 1981, Induction of terminal deoxynucleotidyl transferase and Lyt antigens with thymosin: Identification of multiple subsets of prothymocytes in mouse bone marrow and spleen, Proc. Natl. Acad. Sci. USA, 78:2469.
Greaves, M.F., Rao, J., Hariri, G., Verbi, W., Catovsky, D., Kung, P., and Goldstein, G., 1981, Phenotypic heterogeneity and cellular origins of T cell malignancies, Leuk. Res., 5:281.
Greiner, D.L., Goldschneider, I., and Barton, R.W., 1982, Identification of thymocyte progenitors in hemopoietic tissues of the rat. II. Enrichment of functional prothymocytes on the fluorescence-activated cell sorter, J. Exp. Med., 156:1448.
Hansen, J.A., Martin, P.J., Beatty, P.G., Clark, E.A., and Ledbetter, J.A., 1984, Human T lymphocyte cell surface molecules defined by the workshop monoclonal antibodies ("T cell protocol"), in "Leucocyte Typing", A. Bernard, L. Boumsell, J. Dausset, C. Milstein, and S.F. Schlossman, eds., Springer Verlag, Berlin, FRG, p. 195.
Haynes, B.F., Mann, D.L., Hemler, M.E., Schroer, J.A., Shelhamer, J.H., Eisenbarth, G.S., Strominger, J.L., Thomas, C.A., Mostowski, H.S., and Fauci, A.S., 1980, Characterization of a monoclonal antibody that defines an immunoregulatory T cell subset for immunoglobulin synthesis in humans, Proc. Natl. Acad. Sci. USA, 77:2914.
Janossy, G., Bollum, F.J., Bradstock, K.F., McMichael, A., Rapson, N., and Greaves, M.F., 1979, Terminal transferase-positive human bone marrow cells exhibit the antigenic phenotype of common acute lymphoblastic leukemia, J. Immunol., 123:1525.
Janossy, G., Bollum, F.J., Bradstock, K.F., and Ashley, J., 1980, Cellular phenotypes of normal and leukemic hemopoietic cells determined by analysis with selected antibody combinations, Blood, 56:430.
Janossy, G., Tidman, N., Papageorgiou, E.S., Kung, P.C., and Goldstein, G., 1981, Distribution of T lymphocyte subsets in the human bone marrow and thymus: An analysis with monoclonal antibodies, J. Immunol., 126:1608.
Janossy, G., and Prentice, H.G., 1982, T cell subpopulations, monoclonal antibodies and their therapeutic applications, Clin. Haematol., 11:631.
Kadish, J.L., and Basch, R.S., 1976, Hematopoietic thymocyte precursors. I. Assay and kinetics of the appearance of progeny, J. Exp. Med., 143:1082.
Korsmeyer, S.J., Arnold, A., Bakhshi, A., Ravetch, J.V., Siebenlist, U., Hieter, P.A., Sharrow, S.O., LeBien, T.W., Kersey, J.H., Poplack, D.G., Leder, P., and Waldmann, T.A., 1983, Immunoglobulin gene rearrangement and cell surface antigen expression in

acute lymphocytic leukemias of T cell and B cell precursor origins, J. Clin. Invest., 71:301.

Lampson, L.A., and Levy, R., 1980, Two populations of Ia-like molecules on a human B cell line, J. Immunol., 125:293.

Lepault, F., and Weissman, I.L., 1981, An in vivo assay for thymus-homing bone marrow cells, Nature, 293:151.

Lerch, P.G., van de Rijn, M., Schrier, P., and Terhorst, C., 1983, Biochemical comparison of the T6 antigen and HLA-A, B antigens, Hum. Immunol., 6:13.

Martin, P.J., Hansen, J.A., Siadak, A.W., Nowinski, R.C., 1981, Monoclonal antibodies recognizing normal human T lymphocytes and malignant human B lymphocytes: A comparative study, J. Immunol., 127:1920.

Moore, A.L., and Zusman, J., 1978, Conditions enhancing the percentage and size of E rosettes formed by T lymphocytes, J. Immunol. Methods, 23:275.

Neudorf, S.M.L., Le Bien, T.W., Bollum, F., and Kersey, J.H., 1984, Double fluorochrome analysis of human bone marrow lymphoid cells: Studies with terminal transferase and anti-T-cell monoclonal antibodies, Exp. Hematol., 12:69.

Roos, D., and Loos, J.A., 1970, Changes in the carbohydrate metabolism of mitogenically stimulated human peripheral lymphocytes. I. Stimulation by phytohaemagglutinin, Biochim. Biophys. Acta, 222:565.

Roper, M., Crist, W.M., Metzgar, R., Ragab, A.H., Smith, S., Starling, K., Pullen, J., Leventhal, B., Bartolucci, A.A., and Cooper, M.D., 1983, Monoclonal antibody characterization of surface antigens in childhood T-cell lymphoid malignancies, Blood, 61:830.

Saito, H., Kranz, D.M., Takagaki, Y., Hayday, A.C., Eisen, H.N., and Tonegawa, S., 1984, Complete primary structure of a heterodimeric T-cell receptor deduced from cDNA sequences, Nature, 309:757.

Scollay, R., 1983, Intrathymic events in the differentiation of T lymphocytes: a continuing enigma, Immunol. Today, 4:282.

Silverstone, A.E., Cantor, H., Goldstein, G., and Baltimore, D., 1976, Terminal deoxynucleotidyl transferase is found in prothymocytes, J. Exp. Med., 144:543.

Tax, W.J.M., Tidman, N., Janossy, G., Trejdosiewicz, L., Willems, R., Leeuwenberg, J., De Witte, T.J.M., Capel, P.J.A., and Koene, R.A.P., 1984, Monoclonal antibody (WT1) directed against a T cell surface glycoprotein: characteristics and immunosuppressive activity, Clin. Exp. Immunol., 55:427.

Tonegawa, S., 1983, Somatic generation of antibody diversity, Nature, 302:575.

Verbi, W., Greaves, M.F., Schneider, C., Koubek, K., Janossy, G., Stein, H., Kung, P., and Goldstein, G., 1982, Monoclonal antibodies OKT11 and OKT11A have pan-T reactivity and block sheep erythrocyte "receptors", Eur. J. Immunol., 12:81.

Vodinelich, L., Tax, W., Bai, Y., Pegram, S., Capel, P., and Greaves, M.F., 1983, A monoclonal antibody (WT1) for detecting leukemias of T-cell precursors (T-ALL), Blood, 62:1108.

M^+ LYMPHOCYTES OF THE HUMAN FOETAL LIVER ARE PRECURSORS OF T LYMPHOCYTES*

B. D. Janković, Katarina Isaković and Tatjana Matejić

Immunology Research Center, Vojvode Stepe 458
11221 Belgrade, Yugoslavia

INTRODUCTION

M receptors are widely considered to be associated with B cells and not with T cells[1-3]. Recent studies on suspensions from the peripheral blood, tonsils and lymph nodes selectively enriched in T cells revealed that M receptors are present on some of the T cells[4]. We reported[5] that a significant number of human foetal thymocytes were M^+, a finding which indicated that M receptor might be a differentiation marker of T lymphocytes. The aim of this study is to define M receptor-bearing cells in the human foetal liver, and the relationship between M receptors and maturation, differentiation and phenotypic expression of T lymphocytes.

MATERIALS AND METHODS

Liver lymphocytes. Livers were obtained from 8.5-21 weeks old foetuses after prostaglandin-induced abortions or Caesarian sections, and gestational age was determined by crown-rump and foot length measurements[6]. Suspensions of liver lymphocytes containing $>90\%$ of viable cells were used in the study. Enriched M^+ cell population (60-66% of M^+ cells) was also employed[7].

Detection of membrane markers, and immunoglobulins. The following tests were employed: rosetting with sheep erythrocytes[8] and mouse erythrocytes[1]; immunofluorescent detection of T and B specific antigens using highly specific heterologous anti-T and anti-B serum respectively; detection of cytoplasmic Ig by means of FITC-labelled anti-IgM, and membrane Ig with FITC-conjugated polyvalent anti-Ig

*Supported by the Republic of Serbia Research Fund, Belgrade.

serum; and immunofluorescence with OKT3, OKT4, OKT6, OKT8 and OKTIa monoclonal antibodies. E and M rosettes were combined with immunofluorescence in double-marker assays for detection of E and M markers, and T and B antigens on the same cell. The percentage of cells stained for membrane antigens and Ig was determined by counting at least 200 cells/suspension.

Regression analysis. The possibility of coexistence of two or more markers on the same lymphocyte was tested by a mathematical probability method. The relationship between different stages of lymphocyte differentiation was evaluated by the 1st, 2nd and 3rd polynomial degree[9] using a DECLAB PDP 11/03 computer.

RESULTS

Membrane markers of foetal liver lymphocytes. During the gestation period between 8.5 and 21 weeks lymphocytes containing M receptors were most numerous and reached the highest level between 14-15.5 weeks (Fig. 1).

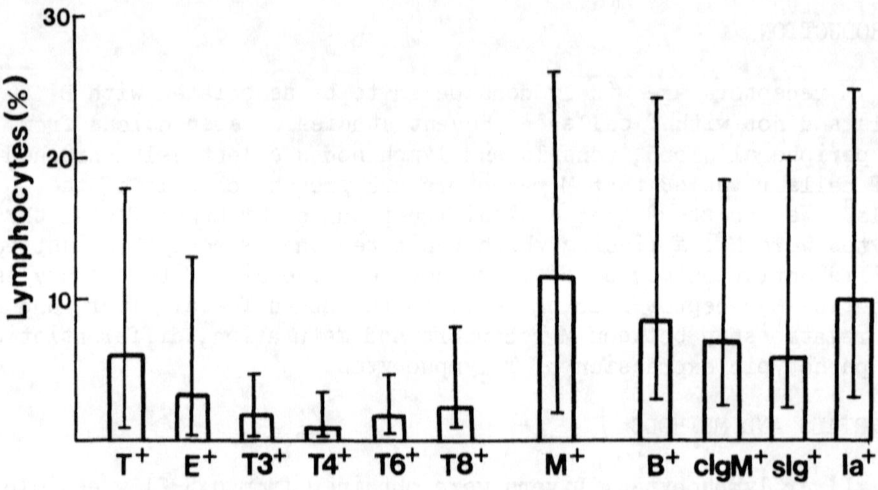

Fig. 1. Membrane markers of foetal liver lymphocytes. Pooled data from 18 foetuses (8.5-21-week-old). Vertical lines indicate range of values.

There were no E^+ lymphocytes in the liver of 8.5-13.5 weeks old foetuses. The first E^+ cells were detected between 14 and 15.5 weeks. On the other hand, T^+ lymphocytes were regularly present in 8.5-13.5 weeks old livers, and the number of these cells increased moderately with development.

The appearance of T3, T4, T6 and T8 antigens was dependent on the gestational age (Table 1). T8 antigen was the first to appear

Table 1. Occurrence of T Antigen-Bearing Cells in the Foetal Liver, as Defined by Monoclonal Antibodies

Gestational Age (weeks)	% of Cells (individual values)			
	$T3^+$	$T4^+$	$T6^+$	$T8^+$
13.0	0	0	0	0
14.5	0	0	0	1.0
15.0	0	0.1	0.5	2.1
17.5	4.9	2.5	3.7	7.7
18.5	5.5	4.9	4.4	10.0

before the 15th week. T3, T4 and T6 made their appearance simultaneously about the 17th week of gestation.

B^+ cells increased steadily during the whole investigated period. The number of $cIgM^+$ and mIg^+ cells was small in very young foetuses but approached a level of 10% of the total liver lymphocytes at 14 weeks.

Subsets of M^+ lymphocytes. Marker analysis (Table 2) of foetal liver M^+ cells revealed that M receptor was associated with T, T3, T6, T8, B and Ia antigens and mIg, but not with T4 antigen.

Table 2. Marker Analysis of M^+ Cells in the Foetal Liver

Incidence of Markers on M^+ Cells (% and Range)[a]							
T	T3	T4	T6	T8	B	Ia	mIg
12-59	0-38	0	0-30	0-67	11-63	25-40	0-30

[a]Pooled data from 12 foetuses.

The wide range of T^+ cells (12-59%) and B^+ cells (11-63%) containing M receptor may be taken as an indication of the presence of cells of the $M^+T^+B^+$ phenotype in the foetal liver. However, analysis of cells of the M^+ population failed to detect $M^+T^+B^+$ cells in the liver. On the other hand, marker analysis showed that the M^+ population is heterogenous and composed of $M^+T^-B^-$, $M^+T^+B^-$ and $M^+T^-B^+$ lymphocytes (Fig. 2).

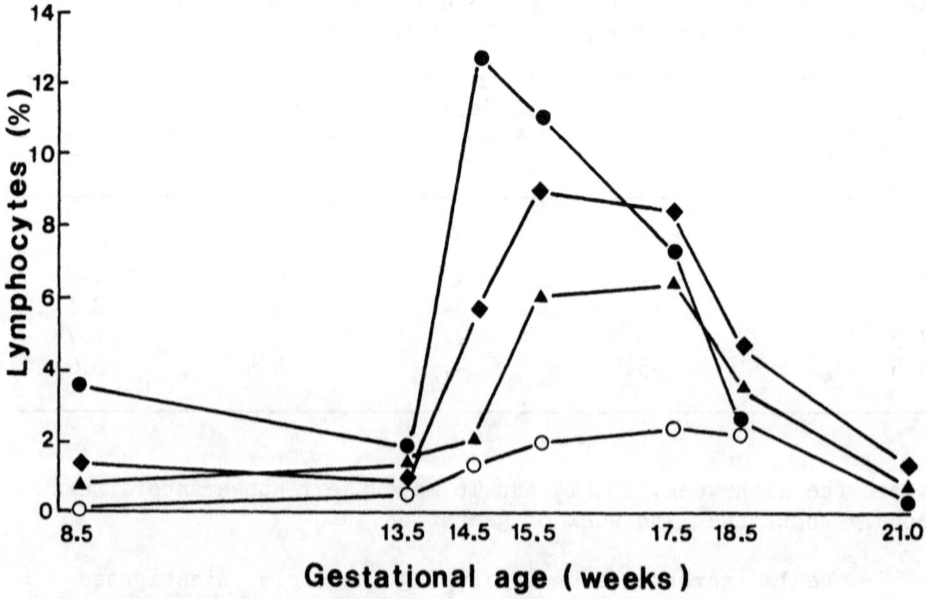

Fig. 2. Lymphocyte populations in the foetal liver detected by double-marker assays. These assays were performed when the number of M^+ cells in the liver was higher than 10%. $M^+T^-B^-$ (●), $M^+T^+B^-$ (▲), $M^+T^-B^+$ (◆), and $M^-T^+B^-$ (○).

Cells of those four phenotypes exhibited similar developmental patterns between 8.5 and 21 weeks of gestation. The $M^+T^-B^-$ population was the largest and the $M^-T^+B^-$ the smallest.

Possible pathways of T and B cell differentiation in the foetal liver. On the basis of the above mentioned marker analysis, the following pathways of lymphocyte differentiation in the human foetal liver are possible (Fig. 3). M^+ lymphocytes lacking T and B antigens ($M^+T^-B^-$ cells) may be regarded as common precursors of the foetal liver T and B lymphocytes. In that context, $M^+T^+B^-$ and $M^+T^-B^+$ lymphocytes represent early differentiation stages of T and B cells respectively. More mature T and B liver lymphocytes were unable to

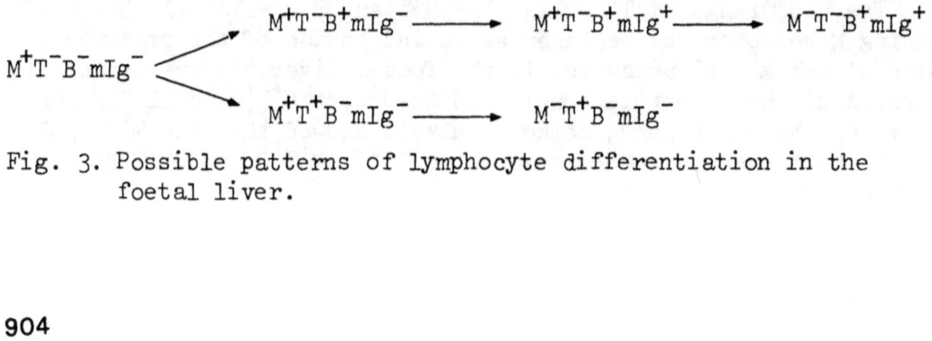

Fig. 3. Possible patterns of lymphocyte differentiation in the foetal liver.

form rosettes with mouse erythrocytes. In the B cell line, it seems that mIg occurred before the disappearance of M receptors from M^{+} cells.

Mathematical analysis. The hypothetical pathways of T and B cell differentiation were tested mathematically by regression analysis[9]. For this purpose, the number of cells at each stage of differentiation represents the function (Y) of the number of cells at a previous stage of differentiation as an independent variable (X). Fig. 4 shows that the best corresponding functions were obtained by the least square method. Those functions were approximated by a polynomial of degree n ($n = 1, 2, 3, \ldots k$).

Regression analysis (Fig. 4) showed that functions representing the interdependence of differentiation stages in the T cell line ($M^{+}T^{-}B^{-}mIg^{-}/M^{+}T^{+}B^{-}mIg^{-}$ and $M^{+}T^{+}B^{-}mIg^{-}/M^{-}T^{+}B^{-}mIg^{-}$) have the same shape (Fig. 4, a and b). The shape of these functions differed from those of the B cell line: $M^{+}T^{-}B^{-}mIg^{-}/M^{+}T^{-}B^{+}mIg^{-}$; $M^{+}T^{-}B^{+}mIg^{-}/M^{+}T^{-}B^{+}mIg^{+}$; and $M^{+}T^{-}B^{+}mIg^{+}/M^{-}T^{-}B^{+}mIg^{+}$ (Fig. 4, c, d and e).

The validity of approximation obtained in Fig. 4 was tested by coefficients of approximation representing the sum of squares of deviations from regression curves (Table 3). Coefficients of approximation showed that the dependence in the T cell differentiation could not be approximated by linear function while the dependence in the B cell differentiation could be approximated as a linear function. The curves that fitted the data satisfactorily and the best corresponding functions were those of the 3rd degree.

DISCUSSION

This study provides evidence that 8.5-21 weeks old human foetal livers contain lymphocytes with membrane properties of T and B cells. However, the central finding of this work concerns T lymphocytes bearing M receptors. These receptors were regarded as a specific marker of B cells in the liver[10] and the bone marrow[2]. We report here that the M^{+} population of human foetal liver is composed of at least three phenotypes: $M^{+}T^{-}B^{-}$, $M^{+}T^{+}B^{-}$ and $M^{+}T^{-}B^{+}$. Since $M^{+}T^{-}B^{-}$ cells occurred earlier than $M^{+}T^{+}B^{-}$ and $M^{+}T^{-}B^{+}$ cells, and since $M^{+}T^{-}B^{-}$ cells represent the largest population in 8.5-week-old foetal liver, i.e., at the time when the first embryonic lymphocytes enter the thymus[11,12], it is possible that the $M^{+}T^{-}B^{-}$ cells are the common progenitors of T and B lymphocytes. Accordingly, M receptor is expressed not only on the membrane of early precursors of B cells[13,14] but also on the membrane of early precursors of T cells.

Whether $M^{+}T^{+}B^{-}$ lymphocytes continue to differentiate in the foetal liver or migrate to the thymus for differentiation into $M^{-}T^{+}B^{-}$ cells remains to be elucidated. These $M^{+}T^{+}B^{-}$ cells are

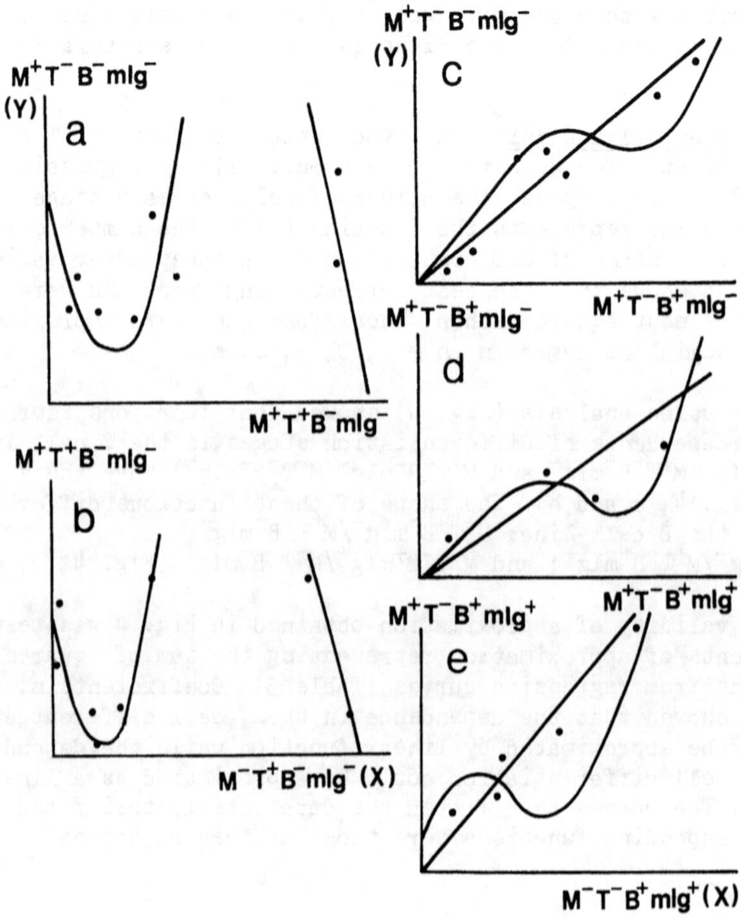

Fig. 4. Hypothetical pathways of T and B liver lymphocyte differentiation as defined by regression analysis. Differentional stages in the T cell line: a and b., and in the B cell line: c, d and e. Difference between successive stages of differentiation are approximated by the third degree function.

certainly the largest M^+ population of the foetal and infant thymus[4]. It appears that the precursors of T cells in the mouse foetal liver are prothymocytes which differentiate into mature T cells exclusively in the thymus[15]. Also, some cells within CALLA-positive foetal liver lymphocytes may belong to T cell precursors with as yet undefined surface markers[16].

Table 3. Coefficients of Approximation Quality by a Polynomial of Degree 1st, 2nd and 3rd in the Foetal Liver

Differentional Stages	Polynomial Degree 1	Polynomial Degree 2	Polynomial Degree 3
$M^+T^-B^-mIg^-/M^+T^+B^-mIg^-$	0.50	0.53	0.59
$M^+T^+B^-mIg^-/M^-T^+B^-mIg^-$	0.18	0.47	0.51
$M^+T^-B^-mIg^-/M^+T^-B^+mIg^-$	0.83	0.83	0.83
$M^+T^-B^+mIg^-/M^+T^-B^+mIg^+$	0.89	0.97	0.98
$M^+T^-B^+mIg^+/M^-T^-B^+mIg^+$	0.88	0.88	0.89

The coefficient of approximation is ranged from 0 to 1.

As for E receptor, and T3, T4, T6 and T8 antigens, it is not clear whether the expression of these membrane macromolecules takes place in the foetal liver or lymphocytes bearing these markers enter the liver from circulation. Under in vitro conditions, it seems that the thymus is needed for the expression of E receptors but probably not for T antigens identified by OKT monoclonal antibodies. The next paper of this series[17] provides further information on that subject.

CONCLUSIONS

The human foetal liver is an important source of T and B lymphocytes in the early gestation. Lymphocytes bearing receptors for mouse erythrocytes (M^+ cells) represent the most numerous population of liver lymphocytes between 14.5 and 17.5 weeks of the foetal development. M^+ population is composed of cells of at least three phenotypes: $M^+T^-B^-$, $M^+T^+B^-$ and $M^+T^-B^+$. Cells of the $M^+T^-B^-$ phenotype are probably very early progenitors of T and B lymphocytes. M receptor is associated with both T and B cell differentiation in the human foetal liver.

REFERENCES

1. S. Gupta and M. H. Griece, Rosette formation with mouse erythrocytes: probable marker for human B lymphocytes, Int.Arch. Allergy Appl.Immunol., 49:734 (1975).
2. W. A. Kagan, F. P. Siegal, S. Gupta and R. A. Good, Early stages of human marrow lymphocyte differentiation: Induction in vitro by thymopoetin and ubiquitin, J.Immunol., 122:686 (1979).

3. J. Forbes, D. Zalewski, L. Valentine and D. Gee, Two maturation-associated mouse erythrocyte receptors of human B cells. I. Identification of four human B-cell subsets, Clin.Exp.Immunol., 47:396 (1982).
4. O. Eremin and R. M. Binns, Mouse red blood cell rosettes: human B and some T lymphocytes express receptors for mouse erythrocytes in the presence of Ficoll, Immunology, 47:321 (1982).
5. B. D. Janković, K. Isaković and T. Matejić, M receptor: a differentiation marker of human fetal and infant thymocytes, Immunobiology, 162:368 (1982).
6. E. L. Potter, "Pathology of the Human Fetus and Infant," 2nd Ed., Year Book, Chicago (1962).
7. F. Caligaris-Cappio, M. Gobbi, M. Bofill and G. Janossy, Infrequent normal B lymphocytes express features of B-chronic lymphocytic leukemia, J.Exp.Med., 155:623 (1982).
8. M. Jondal, G. Holm and H. Wigzell, Surface markers of human T and B lymphocytes. I. A large population of lymphocytes forming non-immune rosettes with sheep red blood cells, J.Exp.Med., 136:207 (1972).
9. W. W. Cooley and P. R. Lohnes, "Multivariant Procedures for the Behavioral Sciences," John Wiley and Sons, New York (1962).
10. S. Gupta, R. A. Good and F. P. Siegal, Rosette-formation with mouse erythrocytes. II. A marker for human B and non-T lymphocytes, Clin.Exp.Immunol., 23:382 (1976).
11. J. B. Solomon, "Fetal and Neonatal Immunology," Elsevier, New York (1971).
12. B. von Gaudecker and H. K. Müller-Hermelink, Ontogenic differentiation of epithelial and non-epithelial cells in the human thymus, in: "Function and Structure of the Immune System," W. Müller-Ruchholtz and H. K. Müller-Hermelink, eds., Plenum Press, New York (1979).
13. A. R. Lawton, K. S. Self, S. Royal and M. D. Cooper, Ontogeny of B lymphocytes in human fetus, Clin.Immunol.Immunopathol., 1:84 (1972).
14. M. L. Raff, M. Megson, J. J. T. Owen and M. D. Cooper, Early production of intracellular IgM by B-lymphocyte precursors in mouse, Nature, 259:224 (1976).
15. H. Rabinowich, T. Umiel and A. Globerson, T cell progenitors in the mouse fetal liver, Transplantation, 35:40 (1983).
16. P. Hokland, P. Rosenthal, J. D. Griffin, L. M. Nadler, J. Daley, M. Hokland, S. F. Schlossman and J. Ritz, Purification and characterization of fetal haemopoietic stem cells that express the common acute lymphoblastic leukemia antigen (CALLA), J.Exp.Med., 157:114 (1983).
17. K. Isaković, T. Matejić and B. D. Janković, M receptor and T-cell differentiation in the human foetal liver: an in vitro study, (this book).

M RECEPTOR AND T-CELL DIFFERENTIATION IN THE HUMAN FOETAL LIVER: AN *IN VITRO* STUDY*

Katarina Isaković, Tatjana Matejić and B. D. Janković

Immunology Research Center, Vojvode Stepe 458
11221 Belgrade, Yugoslavia

INTRODUCTION

The foetal liver is an active site of B lymphocytopoiesis[1,2]. In the preceding paper[3] we reported that the human foetal liver contains a large M^+ population of cells from which common precursors of T and B lymphocytes are derived. Those $M^+T^-B^-$ precursors differentiate into pre-B cells of the $M^+T^-B^+$ phenotype and pre-T cells of the $M^+T^+B^-$ phenotype. Thus, M receptor may be considered as an early differentiation marker of both B and T lymphocytes, and the foetal liver is probably an important source of T cell progenitors. The present paper deals with *in vitro* T cell differentiation of liver lymphocytes from 13-18.5 weeks old human foetuses.

MATERIALS AND MATHODS

Mononuclear liver cells. Suspensions of mononuclear cells from the liver of 13, 15, 17.5 and 18.5 weeks old human foetuses were prepared as described earlier[3]. Briefly, after passing liver tissue through a stainless steel mesh, cells were centrifuged on Ficoll-Isopaque density gradient and mononuclear cells were harvested from the interphase. May-Gruenwald and Giemsa staining revealed that those cells exhibited all the characteristics of lymphocytes at different stages of maturation.

Culture of mononuclear liver cells. 5×10^6 viable cells were suspended in 3 ml of culture medium supplemented with human AB serum (10%), penicillin (100 IU/ml), streptomycin (50 µg/ml) and L-glutamine (2 mM). Cultures were performed in duplicate and cells

*Supported by the Republic of Serbia Research Fund, Belgrade.

were harvested 2 and 5 days after incubation and used in marker analyses.

Assessment of T and B cell membrane markers. The following markers were evaluated: receptors for sheep (E) and mouse (M) red blood cells (rosette formation tests); T- and B-specific membrane antigens (indirect immunofluorescence with highly specific anti-T and anti-B sera produced in rabbits); membrane Ig (direct immunofluorescence using $F(ab)_2$ fragments of a polyvalent anti-human Ig FITC-conjugated serum); and T3, T4, T6 and T8 antigens (immunofluorescent staining with monoclonal antibodies of the OKT series). At least 300 cells were inspected in E and M rosette assays, and percentages of fluorescing cells in other tests were determined by estimating more than 200 cells.

RESULTS

Membrane markers (M, T, E, T3, T4, T6, T8, B and mIg) of cultured lymphocytes from the liver of 13, 15, 17.5 and 18.5 weeks old human foetuses are shown in Fig. 1. In cultured liver from a 13-week-old foetus, the initial decrease in the expression of markers on day 2 of incubation was followed by an increase in the number of cells bearing different membrane markers, and mIg, T, T6 and T8 in particular. It should be emphasized that E receptor was not present on cells of the T cell lineage.

Cultures of liver lymphocytes from a 15-week-old foetus were characterized by vast majority of the M^+ population on day 0 of cultivation. However, M^+ cells disappeared within 5 days of incubation. E^+ cells showed a similar depletion. On the other hand, there was an increase in the number of cells bearing T, T3, T4, T6, T8, B and mIg markers. mIg^+ cells reached the highest level.

In cultures of liver lymphocytes from 17.5-week-old foetuses there was a general and continuous reduction in the number of cells bearing different membrane markers, so that on day 5 of incubation no marker-positive cells could be detected. The only exception were mIg^+ cells the level of which remained about the same throughout the incubation period.

Marker analysis of cells from a 18.5-week-old foetal liver showed that there was no increase in the expression of markers of T and B cell lines.

DISCUSSION

In a previous paper[3] we reported on the M^+ precursors of T and B lymphocytes in the human foetal liver. These cells of the $M^+T^-B^-$ phenotype were considered to be common precursors of the lymphocyte population in the liver. Other cells of the $M^+T^+B^-$ phenotype,

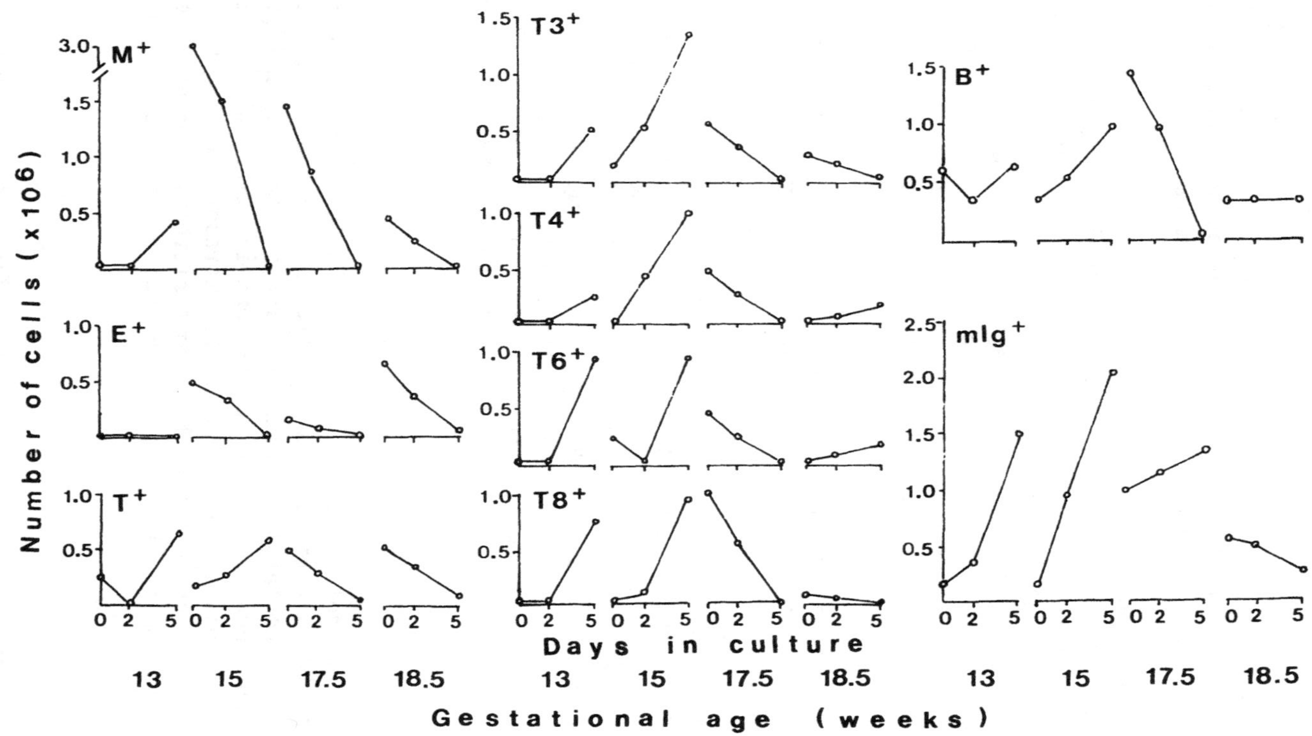

Fig. 1. Occurrence of lymphocyte markers in cultured human foetal livers.

designated as pre-T lymphocytes, were recognized by their ability to form rosettes with mouse erythrocytes and capacity to react with highly specific anti-T serum. The present study showed, however, that M^+ cells did not proliferate under *in vitro* conditions. This finding may suggest that cell inocula from 13-18.5 weeks old foetal livers did not contain precursors of M^+ lymphocytes, or if those precursors were present on day 0 of incubation they were not able to proliferate and differentiate in tissue culture.

The kinetic pattern of M^+ cells in tissue cultures exhibited two characteristics: a relationship between the disappearance of M^+ cells and increase in the number of T and B cell markers on cells from the liver of 13-15 weeks old foetuses, and a steady decrease in the number of cells bearing M receptors and other membrane markers, with the exception of mIg^+ cells, when cultured cells were taken from 17.5-18.5 weeks old foetuses. It appears, therefore, that M^+ lymphocytes may exhibit properties of early precursors of T and B cells only between 13-15 weeks of gestation.

The present experiment suggests that differentiation which includes T3, T4, T6 and T8 antigens was partially inducible *in vitro* since the number of $T3^+$, $T4^+$, $T6^+$ and $T8^+$ cells increased on day 5 of incubation (cultures of cells from 13-15 weeks old foetuses). This may be taken as evidence that the liver T cells from 13-15 weeks old foetuses did not require a contact with the thymic epithelial cells and thymic hormonal factors[4,5]. However, after that age of gestation it seems that further T cell differentiation is thymus-dependent since $T3^+$, $T4^+$, $T6^+$ and $T8^+$ lymphocytes disappeared completely in 5-day-old cultures of liver cells from 17.5-18.5 weeks old foetuses. In other words, the liver-dependent phase of T cell differentiation is replaced or interwoven with a thymus-dependent phase.

CONCLUSIONS

Mononuclear cells from the liver of 13, 15, 17.5 and 18.5 weeks old human foetuses were studied *in vitro*. At 13-15 weeks of gestation, the liver contains M^+ cells with T-specific antigen ("prothymocytes"), but those cells are not present in livers at 17.5-18.5 weeks of gestation. It appears that the expression of T3, T4, T6 and T8 antigens may occur outside the thymus microenvironment at early stages of development (13-15 weeks) but not after that period of gestation.

REFERENCES

1. J. J. T. Owen, M. D. Cooper and M. C. Raff, In vitro generation of B lymphocytes in mouse fetal liver: a mammalian "bursa equivalent", *Nature*, 249:361 (1974).

2. M. D. Cooper, Pre-B cells: normal and abnormal development, J.Clin.Immunol., 1:81 (1981).
3. B. D. Janković, K. Isaković and T. Matejić, M^+ lymphocytes in the human foetal liver are precursors of T lymphocytes (this book).
4. J. L. Touraine, G. S. Incefy, F. Touraine, Y. M. Rho and R. A. Good, Differentiation of human bone marrow cells into T lymphocytes by in vitro incubation with thymic extracts, Clin.Exp.Immunol., 17:151 (1974).
5. G. E. Ranges, G. Goldstein, E. A. Boyse and M. R. Schield, T cell development in normal and thymopoietin-treated nude mice, J.Exp.Med., 156:1057 (1982).

IS Ia ANTIGEN EXPRESSION BY THYMIC EPITHELIAL CELLS CONSTITUTIVE?

Robert V. Rouse

Department of Pathology
Stanford University Department of Pathology
Stanford, California 94305

INTRODUCTION

The thymus plays a special role as a principal site of T cell differentiation and maturation (Cantor and Weissman, 1976). The framework of the thymus is formed principally by thymic epithelial cells (TEC), with which thymocytes are in intimate contact (Rouse and Weissman, 1981). A distinctive feature of TEC is expression of Ia (class II major histocompatibility complex) antigens, antigens which are generally observed on other epithelia only after stimulation (Barclay and Mason, 1982; Scheynius and Tjerlung, 1984; Cerf-Bensussan et al, 1984). In a few situations this apparently distinctive feature of TEC is not observed. These exceptions may shed light upon the role of TEC Ia antigen expression in thymocyte differentiation.

Thymic Epithelial Cells Express Ia Antigens in the Normal Thymus

Frozen sections of normal mouse and human thymuses, when incubated with antibodies specific for class II major histocompatibility complex (MHC) antigens exhibit confluent medullary and dendritic cortical staining patterns (Rouse et al, 1979; van Ewijk et al, 1980; Rouse et al, 1982). Immunoelectron microscopy demonstrates that this cortical pattern is caused by Ia positive cortical TEC and that Ia positive medullary TEC contribute to the medullary pattern (van Ewijk et al, 1980; Rouse et al, 1982; Farr and Nakane, 1983), although a population of Ia negative TEC has been described (Janossy et al, 1984, Savino et al, 1983). Study of bone marrow chimeric mice demonstrates that TEC continue to express their original type Ia antigens while bone marrow derived Ia antigens are expressed only by medullary dendritic cells and not at all in the cortex (Hoffmann-Fezer et al, 1978; Barclay and Mayrhofer, 1981; Rouse et al, 1984).

Such observations appeared to confirm the suggestions based on functional studies of chimeric mice that TEC borne Ia antigens play an important role in the acquisition of MHC specificity by developing T cells in the thymus (Zinkernagel, 1978; Bevan and Fink, 1978). The observations that bone marrow derived cells (Longo and Schwartz, 1980) or the extrathymic environment in some cases (Zinkernagel, 1978; Katz et al, 1980) can determine the MHC specificity of T cells raise some question about this interpretation, or at the least indicate more than one mechanism of such maturation may exist.

Some Human Thymic Epithelial Cell Neoplasms Fail to Express Ia Antigen

In the human, thymomas are neoplasms of TEC which range from spindled, purely epithelial neoplasms to lymphoid rich tumors histologically similar to the normal thymic cortex (Levine and Rosai, 1978). In general, lymphocytes in thymomas express thymocyte phenotype (Lauriola et al, 1981; Mokhtar et al, 1984; Chilosi et al, 1984; Rouse, 1985) indicating that these neoplastic TEC can receive thymocyte precursors and allow or induce their differentiation into thymocytes. Thymomas may thus be useful as models of normal epithelial/lymphoid interactions.

Frozen sections of 14 thymomas stained with monoclonal antibodies specific for T cell differentiation antigens and MHC antigens yields the following results (Rouse, 1985). The TEC of only 3 of these 14 thymomas lacked detectable Ia antigen. Of these 3, one was purely epithelial (containing virtually no lymphocytes) and the other 2 contained only moderate numbers of lymphocytes. Of the 11 thymomas with Ia antigen positive epithelial cells, only 1 was purely epithelial, 2 contained only moderate numbers of lymphocytes and 8 were predominantly lymphoid. These findings suggest an association, although not absolute, between the lack of Ia antigen on neoplastic TEC and decreased numbers of associated lymphocytes.

In Early Development and in some Abnormal Immune States, TEC Fail to Express Ia Antigens

The studies of Jenkinson et al (1979) of the developing mouse thymus have demonstrated that Ia antigen is not expressed on the epithelial cells of the thymic rudiment until day 14 of gestation, concurrent with the beginning of its colonization with lymphoid cells. Jenkinson et al (1980) have also demonstrated that the alymphoid thymic epithelial rudiment of the nude mouse also fails to express Ia antigen even into adulthood. In one human immunodeficiency state, the "bare lymphocyte" syndrome, peripheral blood lymphocytes fail to express detectable class I or II MHC antigens. Schuurman et al (1985) have demonstrated in two such patients that although the thymus is fully developed as a lymphoid organ with an epithelial framework, the epithelial cells fail to express Ia antigen.

Thymic Epithelial Cells in Culture Generally Do Not Express Ia Antigens

When placed in culture, TEC lose both their lymphoid component and their expression of Ia antigen (Sun et al, 1984; M.O. Dailey, personal communication; M. Small, personal communication; S. Berrih, personal communication), although they do retain their ability to form cell-cell contacts and to produce thymic hormones (Berrih et al, 1984).

SPECULATION

This frequent association between a lack of TEC Ia antigen expression and a decrease or lack of associated lymphocytes in a variety of situations can be interpreted in more than one way. A conventional interpretation is that TEC Ia antigen expression is constitutive and that it plays some role in thymocyte recruitment and/or differentiation. Absence of TEC Ia antigen expression might then lead to an alymphoid or paucilymphoid condition. This interpretation is consistent with most of the above observations, however it does not explain the fully lymphoid thymus seen in the "bare lymphocyte" syndrome.

An alternate, less conventional interpretation is that Ia antigen expression by TEC is only secondary to stimulation by thymic lymphocytes or factors they produce. Such an interpretation is consistent with each of the examples given above: Those thymomas which for some other, independent, reason are unable to attract thymocyte precursors or to induce them to proliferate might not receive the necessary signal for Ia antigen expression by their neoplastic epithelial cells. In late gestation, the expression of TEC Ia antigen might be secondary to the concurrent beginning of thymopoiesis rather than causative. The nude mouse rudiment might be Ia antigen negative because it lacks a lymphoid component. Cultured TEC might lose Ia antigens because they are no longer capable of supporting thymocyte proliferation *in vitro*. The inability of the TEC in the "bare lymphocyte" syndrome to express Ia antigen in spite of their abundant thymocyte population might be merely another expression of the same defect which prevents Ia antigen expression by peripheral blood lymphocytes in the same patients.

Other possibilities do exist, such as selective loss or absence of Ia antigen expressing TEC with retention of an Ia antigen negative subpopulation in those special cases noted above. Finally, it is also possible that TEC Ia antigen expression and the lymphoid population of the thymus have no direct relationship. These various interpretations are not necessarily mutually exclusive.

SUMMARY

Thymic epithelial cells express class II MHC antigens (Ia) in the normal thymus. In a variety of special circumstances, generally associated with decreased or absent thymocyte populations, TEC fail to express Ia antigens. Although such observations might be interpreted based on the hypothesis that TEC Ia antigens have an effect on the differentiation of thymocytes, it is also possible that instead, the thymocytes have an effect on TEC Ia antigen expression.

ACKNOWLEDGMENT

Portions of this work were supported by the Veterans Administration.

REFERENCES

Barclay, A.N. and Mayrhofer, G., 1981, Bone marrow origin of Ia-positive cells in the medulla of rat thymus, J. Exp. Med., 153:1666.

Barclay, A.N., and Mason, D.W., 1982, Induction of Ia antigen in rat epidermal cells and gut epithelium by immunological stimuli, J. Exp. Med., 156:1665.

Bevan, M.J., and Fink, P.J., 1978, The influence of thymus H-2 antigens on the specificity of maturing killer and helper cells, Immunol. Rev. 42:3.

Cantor, H., and Weissman, I.L., 1976, Development and function of subpopulations of thymocytes and T lymphocytes, Prog. Allergy, 20:1.

Cerf-Bensussan, N., Quaroni, A., Kurnick, J.T., and Bhan, A.T., 1984, Intraepithelial lymphocytes modulate Ia expression by intestinal epithelial cells, J. Immunol., 132:2244.

Chilosi, M., Iannucci, A.M., Pizzolo, G., Menestrina, F., Fiore-Donati, L., and Janossy, G., 1984, Immunohistochemical analysis of thymoma, evidence for medullary origin of epithelial cells, Am. J. Surg. Path., 8:309.

Farr, A.G., and Nankane, P.K., 1983, Cells bearing Ia antigens in the murine thymus, Am. J. Pathol., 111:88.

Hoffmann-Fezer, G., Gotze, D., Rodt, H., and Thierfelder, S., 1978, Immunohistochemical localization of xenogenic antibodies against Ia^k lymphocytes on B cells and reticular cells, Immunogenetics, 6:367.

Janossy, G., Bofill, M., Trejdosiewicz, L.K., Poulter, L.W., Chilosi, M., and Newsom-David, J., 1984, Cellular differentiation of lymphoid populations and their microenvironment in the human thymus, In The Thymus: Histopathology and Pathology. Current Topics in Pathology. H.K. Muller-Hermelink, ed., Springer-Verlag, Berlin, Heidelberg, New York.

Jenkinson, E.J., Owen, J.J.T., and Aspinall, R., 1980, Lymphocyte differentiation and major histocompatibility complex antigen expression in the embryonic thymus, Nature, 284:177.

Jenkinson, E.J., van Ewijk, W., and Owen, J.J.T., 1981, Major histocompatibility complex antigen expression on the epithelium of the developing thymus in normal and nude mice, J. Exp. Med., 153:280.

Katz, D.H., Katz, L.A., Bogowitz, C.A., and Bargatze, R.F., 1980, The major influence on helper T cell cooperative partner cell preferences is exerted by the extrathymic environment, J. Immunol. 124:1750.

Lauriola, L., Maggiano, N., Marino, M., Carbone, A., Piantelli, M., and Musiani, P., 1981, Human thymoma: immunologic characteristics of the lymphoid component, Cancer, 48:1992.

Levine, G., and Rosai, J., 1978, Thymic hyperplasia and neoplasia: a review of current concepts, Hum. Pathol. 9:495.

Longo, D.L., and Schwartz, R.H., 1980, T cell specificity for H-2 and Ir gene phenotype correlates with the phenotype of thymic antigen-presenting cells, Nature, 287:44.

Mokhtar, N., Hsu, S.-M., Lad, R.P., Haynes, B.F., and Jaffe, E.S., 1984, Thymoma: lymphoid and epithelial components mirror the phenotype of normal thymus, Hum. Pathol. 15:378.

Rouse, R.V., 1985, Human thymomas: immunologically defined microenvironmental differentiation, submitted for publication.

Rouse, R.V., van Ewijk, W., Jones, P.P., and Weissman, I.L., 1979, Expression of MHC antigens by mouse thymic dendritic cells, J. Immunol. 122:2508.

Rouse, R.V., and Weissman, I.L., 1981, Microanatomy of the thymus -its relationship to T cell differentiation, In: Microenvironments in hematopoietic and lymphoid differentiation, Pitman, London (Ciba Foundation Symposium 84).

Rouse, R.V., Parham, P., Grumet, F.C., and Weissman, I.L., 1982, Expression of HLA antigens by human thymic epithelial cells, Human Immunol., 5:21.

Rouse, R.V., Ezine, S., and Weissman, I.L., 1984, Expression of major histocompatibility complex antigens in the thymuses of chimeric mice, submitted for publication.

Savino, W., Dardenne, M., and Bach, J.F., 1983, Thymic hormone containing cells. II. Evolution of cells containing the serum thymic factor (FTS or thymulin) in normal and autoimmune mice, as revealed by anti-FTS monoclonal antibodies. Relationship with Ia bearing cells, Clin. Exp. Immunol. 52:1.

Scheynius, A., and Tjernlund, U., 1984, Human keratinocytes express HLA-DR antigens in the tuberculin reaction, Scand. J. Immunol., 19:141.

Schuurman, H.-J., van de Winjngaert, F.P., Huber, J., Zegers, B.J.M., Schuurman, R.K.B., Roord, J.J., and Kater, L., 1985, The thymus in "bare lymphocyte" syndrome, this volume.

Sun, T.-T., Bonitz, P., and Burns, W.H., 1984, Cell culture of mammalian thymic epithelial cells: growth, structural, and antigenic properties, Cellular Immunol. 83:1.

van Ewijk, W., Rouse, R.V., and Weissman, I.L., 1980, Distribution of H-2 microenvironments in the mouse thymus. Immunoelectron microscopic identification of I-A and H-2K bearing cells, J. Histochem. Cytochem., 28:1089.

Zinkernagel, R.M., 1978, Thymus and lymphohematopoietic cells: Their role in T cell maturation in selection of T cells' H-2 restriction-specificty and in H-2 linked Ir gene control, Immunol. Rev., 42:224.

THE THYMUS IN "BARE LYMPHOCYTE" SYNDROME

H.-J. Schuurman, F.P. van de Wijngaert, J. Huber, B.J. M. Zegers, R.K.B. Schuurman, J.J. Roord and L. Kater

Division of Immunopathology, Department of Internal Medicine and Institute for Pathology, University Hospital, Utrecht; Department of Immunology, University Children's Hospital "Het Wilhelmina Kinderziekenhuis", Utrecht; and Department of Immunohematology, University Medical Center, Leiden, The Netherlands

In patients with various forms of combined immunodeficiency or T lymphocyte deficiency, the histology of the thymus is generally abnormal. During the last years patients with combined immunodeficiency associated with defective expression of HLA of class I and in some cases also of class II on blood mononuclear cells (so-called "bare lymphocyte" syndrome) have been described.[1-3] We were able to study thymus biopsies from two of these cases. This study included histopathology on formalin-fixed, paraffin-embedded tissue, electron microscopy,[4,5] immunohistochemistry on frozen tissue sections with monoclonal antibodies to T lymphocyte subpopulations, monoclonal antibodies to HLA and antisera to thymosin components,[6] as well as cell surface marker tests on thymocytes in suspension and *in vitro* culture of epithelial cells.[7]

Patients

Both patients (KD[1] and TF[2,3]) have been described previously. They were diagnosed as combined immunodeficiency, with absence of *in vivo* and *in vitro* responsiveness to antigenic stimulation. There was hypogammaglobulinemia. The number of blood lymphocytes was normal or lowered, and the percentages of T and B cells were normal. *In vitro* proliferative responses to mitogenic stimulation were present, but *in vitro* differentiation into cytoplasmic immunoglobulin positive cells after pokeweed mitogen stimulation was absent. In allogeneic stimulation, cells from KD showed a response; for TF variable responses were noted. HLA-A,B,C was not detectable on peripheral blood mononuclear cells by routine assays, but for TF FACS analysis revealed expression in very low density. HLA-DR was not detectable on cells from TF, not even by FACS: for KD HLA-DR typing gave inconclusive results, with two out of 6 anti-DR antisera positivity was found. A thymus biopsy was obtained from KD at the age of two years and from TF at the age of 7 months. Before analysis in this study, frozen tissue was stored for six years in liquid nitrogen (KD) or for two years at -70°C (TF).

Results and discussion

In light microscopy both thymus biopsies showed a normal lobulated architecture with distinct cortex and medulla, numerous lymphocytes, well-differentiated epithelium and Hassall's corpuscles (Fig. 1a). In 1 μm sections of Epon-embedded material, clusters of lymphocytes were present in the cortical areas, with a pale nucleus apparently belonging to an epithelial cell (Fig. 1b). These clusters resembled "thymic nurse cells" described by us previously.[4] In electron microscopy subtypes of epithelial cells with characteristic ultrastructural morphology were found (this issue, p);[5] the most common type in the cortex was the so-called "pale" type with an euchromatic irregularly-shaped nucleus, often with a prominent nucleolus. In these aspects the thymus biopsies of both patients did not differ from the normal aspect for age.

In immunohistochemistry, antisera to T lymphocyte antigens of the Leu (Leu 1, 2, 3 and 5, Becton Dickinson, Mountain View, Calif.) and OKT series (OKT 3, 4, 6 and 8, Ortho Diagnostic Systems Inc., Raritan; NJ) gave staining patterns of binding to thymocytes similar to those in normal thymuses. E.g., OKT 6 stained thymocytes in the cortex and OKT 3 stained lymphocytes in the medulla. For TF, this confirmed results of marker tests on suspension of thymocytes reported earlier.[2] In this suspension, 51% of cells showed binding capacity for peanut agglutinin, and in sections these cells were located in the cortex. The percentage of cells positive for nuclear terminal deoxynucleotidyl transferase was 51%. From these data we conclude that intrathymic T cell maturation as judged from T cell phenotype expression was normal in both cases.

In immunohistochemistry for epithelial cell antigen expression, anti-human keratin antibody (Dakopatts, Copenhagen, Denmark) gave staining of epithelial cells throughout the thymus tissue section, being most pronounced in the subcapsular area and in the medulla (including Hassall's corpuscles). With antisera to thymosin components (provided by A.L. Goldstein, Washington, DC), thymosin fractions 5 and 6 were found in epithelial cells throughout the cortex and medulla. In the subcapsular area, cells were positive for thymosin α_1 and β_4, and in the medulla there were scattered thymosin α_1 cells. Thymosin α_7 was observed in Hassall's corpuscles and adjacent cells. These findings do not differ from those in normal thymus.[6]

Part of the biopsy from TF was processed to an epithelial cell

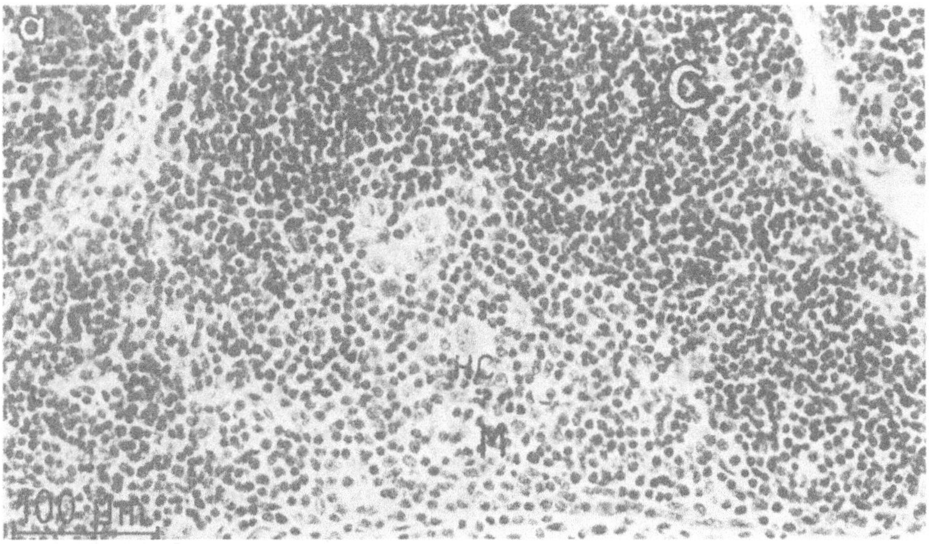

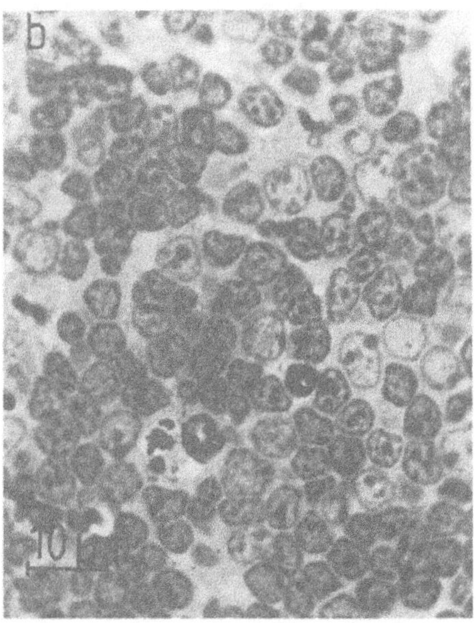

Fig. 1. Light microscopy of thymus biopsy (patient TF). There are distinct cortex (C) and medulla (M) areas with numerous lymphocytes, well differentiated epithelium and Hassall's corpuscles (HC) (a, formalin-fixed, paraffin-embedded tissue, stained with hematoxylin and eosin).
In 1 μm sections of the cortex, there are clusters of lymphocytes with a pale nucleus apparently belonging to an epithelial cell, which resemble "thymic nurse cells" (b, Epon-embedded tissue stained with pararosaniline and methylene blue).

culture.[7] After about 10-12 days of culture, there was an outgrowth of epithelial cells into a confluent monolayer; the supernatant was harvested after 12, 16 and 21 days of culture. In phytohemagglutinin stimulation of mouse thymocytes, these supernatants enhanced the proliferative response 4- to 16-fold, similar to pooled normal thymus epithelial cell culture supernatant.[7]

<u>Expression of HLA</u>. Monoclonal antibodies to <u>HLA class I</u> were obtained commercially (clones W6/32 HLK and 34/28 HLK, Sera-Lab,

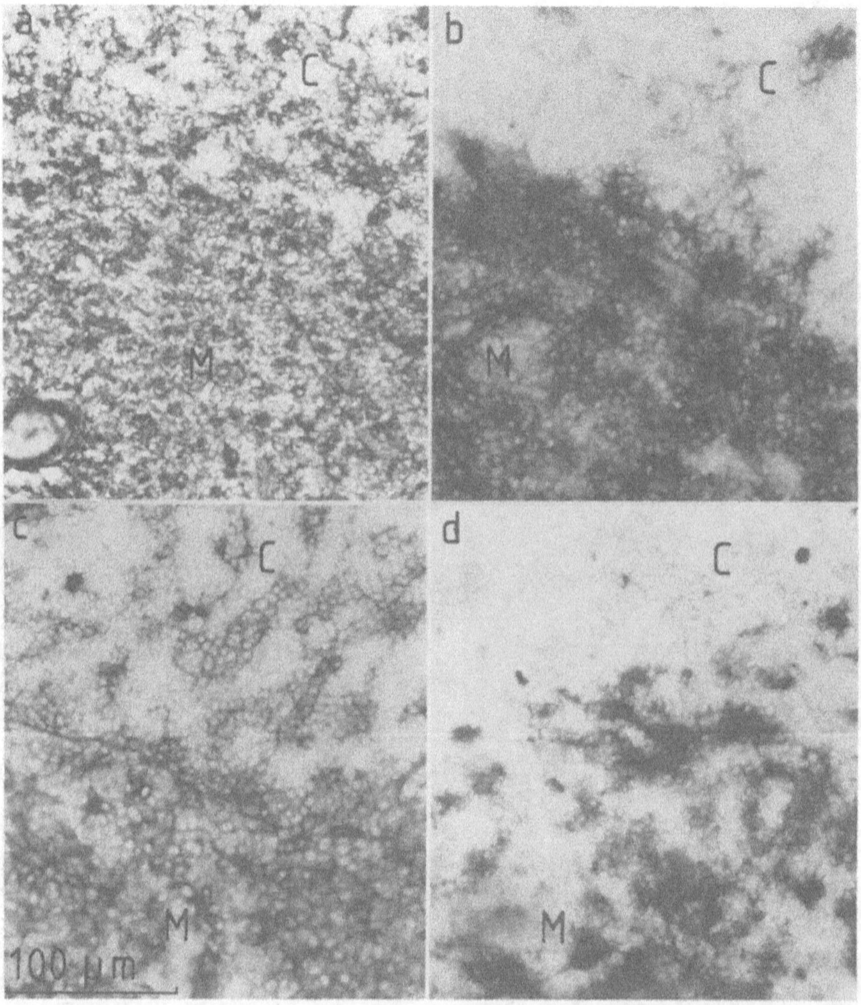

Fig. 2. HLA expression in thymus (immunoperoxidase method on frozen tissue sections). Figures a and b show HLA class I expression (monoclonal antibody B9.12.1). In normal thymus, this is present on stromal cells in cortex (C, dendritic staining pattern) and medulla (M, confluent staining pattern); lymphocytes in the medulla are also positive (a). In thymus from KD, there is only staining of lymphocytes and stromal cells in the medulla (b).
Figures c and d show HLA-DR expression (monoclonal antibody 8.11.2). In normal thymus, this is present on stromal cells in cortex (dendritic staining) and medulla (confluent staining) (c). In thymus from KD there is only staining of stromal components in medulla (d).

Crawley Down, England) or provided by Dr B. Malissem, Marseille, France (B9.12.1). In normal thymus tissue sections, these antisera gave in the cortex a dendritic staining pattern of binding to stromal elements; in the medulla there was a confluent staining of stromal cells and staining of lymphocytes (Fig. 2a). We performed two-color immunofluorescence, using rabbit anti-human keratin and tetramethyl rhodamin isothiocyanate conjugated goat-anti rabbit immunoglobulin on one hand, and mouse anti-HLA and fluorescein isothiocyanate labeled goat anti-mouse immunoglobulin on the other (second layer antibodies were obtained from Nordic Immunological Laboratories, Tilburg, The Netherlands; reactivity of anti-rabbit antibody to mouse immunoglobulin and of anti-mouse immunoglobulin to rabbit immunoglobulin was not detectable; controls including ommission of the first layer were negative). In these experiments, keratin-positive epithelial cells in cortex and medulla were HLA-A,B,C positive. For both patients, HLA-A,B,C was not detectable on stromal elements in the cortex, except for faint staining of some blood vessels (antibody B9.12.1). In the medulla lymphocytes and part of keratin-positive epithelial cells were HLA-A,B,C positive (Fig. 2b). We reported earlier that 19% of cells in the thymocyte suspension from TF were positive for HLA-A and 24% for HLA-B; these cells are probably derived from the medulla.

This HLA class I expression on medullary thymocytes of both patients is in discordance with the absence or presence in very low density (TF, FACS analysis) on blood lymphocytes. This higher density of HLA-A,B,C on medullary thymocytes was also found for β_2-microglobulin expression (data not shown). For patient TF, blood lymphocyte investigations showed another abnormality, viz. a partial deficiency of OKT3 and OKT4 expression (KD could not be evaluated in these assays). These differences between blood lymphocytes and (medullary) intrathymic lymphocytes indicate that somewhere along the line of maturation within the thymus <u>or</u> during or after export from the thymus defective membrane antigen expression occurs. It is tempting to speculate that the thymus provides a microenvironment allowing for normal maturation of lymphocytes within a protective milieu interieur.

Antisera to <u>HLA class II</u> were obtained commercially (anti-HLA-DR, Becton Dickinson; anti-human DR, Bethesda Research Laboratories, Gaithersburg, MD; OKIal, Ortho) or provided by Dr F. Koning, Leiden, The Netherlands (clones 7.5.10.1, anti-DR,DC,SB,MT, and 8.11.2, anti-HLA-DR), Dr H. Spits, Amsterdam, The Netherlands (clone SPV-L3, anti-HLA-DC/DS) and Dr G. Janossy, London, England (fluorescein isothiocyanate conjugated anti-HLA-DR). In the normal thymus, stromal cells in the cortex gave a dendritic staining pattern and stromal elements in the medulla gave a confluent pattern of staining (Fig. 2c). There was no apparent staining of lymphocytes. In two-color immunofluorescence, keratin-positive epithelial cells in the cortex and subcapsular area carried HLA class II, whereas the major part of keratin-positive cells in the medulla was negative. The major part of HLA class II cells in the medulla

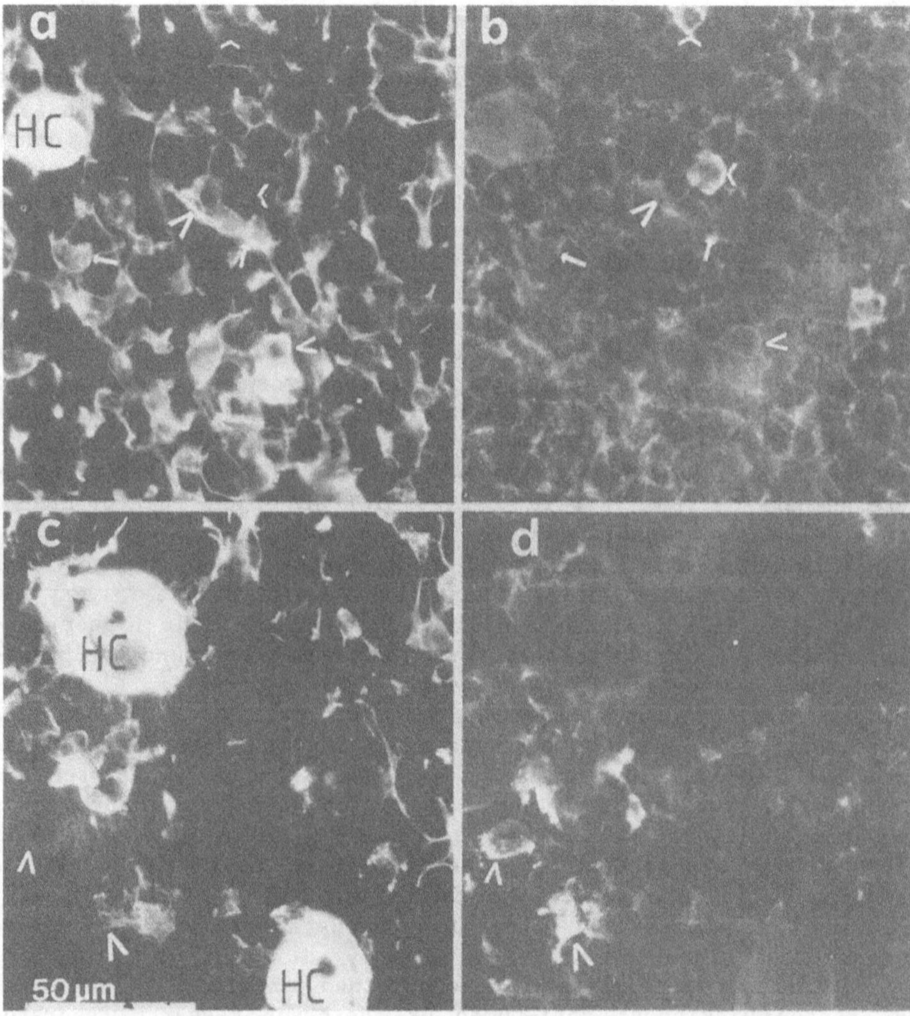

Fig. 3. Two-color immunofluorescence analysis with anti-keratin (left panel; a,c) and anti-HLA-DR (right panel; b,d) on frozen thymus tissue sections. In normal thymus medulla (a,b), part of keratin-positive epithelial cells are HLA-DR positive (Λ), but the majority are negative (↑). Conversely, there are HLA-DR positive cells which are keratin negative (〉).
In thymus from KD (c,d), there are some HLA-DR positive elements, which all are keratin-negative (Λ).
HC, Hassall's corpuscle.

proved to be negative for keratin expression (Figs. 3a,b). In the biopsy from TF, there was no detectable HLA class II expression. In the biopsy from KD anti-HLA-DC/DS was negative, but HLA-DR expression was observed on some nonlymphoid elements in the medulla (Fig. 2d). In two-color immunofluorescence, these HLA-DR positive cells were keratin-negative: the cells presumably are of interdigitating cell origin (Figs. 3c,d).

So, the major abnormality in the thymus biopsies from these patients with "bare lymphocyte" syndrome is the absence of HLA class I and class II on epithelial cells in the cortex. In the medulla there is no HLA class II expression on epithelial cells, and only for KD there is HLA-DR expression on some non-epithelial stromal components in this area.

The thymus cortex is commonly considered as the thymus compartment where cell-cell interactions between the epithelial meshwork and lymphoid elements contribute to T lymphocyte maturation. The presence of "thymic nurse cells" in this area may be the morphological representant of these cell-cell interactions. For both patients, tissue section analysis revealed the presence of these interactions, and also a normal T cell phenotype expression in cortex and medulla. These observations lead us to conclude that the expression of HLA on epithelial cells in the cortex has no physiological significance in intrathymic T cell maturation processes including lymphocyte-epithelial cell interactions and changes in T cell phenotypes. It remains to be established, whether this also applies for processes yielding the antigen recognition repertoire or for restriction or amplification of cells with self-HLA receptors.

Acknowledgements

The authors are grateful to Dr Thea M. Vroom, Slotervaart Hospital, Amsterdam, The Netherlands, for providing the thymus biopsy from patient KD.

References

1. R.K.B. Schuurman, J.J. van Rood, J.M. Vossen, P.Th.A. Schellekens, Th.M. Feltkamp-Vroom, E. Doyer, F. Gmelig-Meyling, and H.K.A. Visser, Failure of lymphocyte-membrane HLA-A and -B expression in two siblings with combined immunodeficiency, Clin. Immunol. Immunopathol. 14: 418 (1979).
2. W. Kuis, J. Roord, B.J.M. Zegers, R.K.B. Schuurman, C.J. Heijnen, W.M. Baldwin, E. Goulmy, F. Claas, R.J. van de Griend, G.T. Rijkers, J.J. van Rood, J.M. Vossen, R.E. Ballieux, and J.W. Stoop, Clinical and immunological studies in a patient with the "bare lymphocyte" syndrome, in: "Bone marrow transplantation in Europe," J.-L. Touraine, E. Gluckman, and C. Griscelli, eds., Excerpta Medica Elsevier North-Holland, Amsterdam, Vol. II: 201 (1981).

3. B.J.M. Zegers, C.J. Heijnen, J.J. Roord, W. Kuis, R.K.B. Schuurman, J.W. Stoop, and R.E. Ballieux, Defective expression of mononuclear cell membrane HLA-antigens associated with combined immunodeficiency: impaired cellular interaction, in: "Primary immunodeficiency diseases," R.J. Wedgwood, F.S. Rosen, and N.W. Paul, eds., Birth defects, original article series vol 19, A.R. Liss Inc., New York, 93 (1983).
4. F.P. van de Wijngaert, L.H.P.M. Rademakers, H.J. Schuurman, R.A. de Weger, and L. Kater, Identification and in situ localization of the "Thymic Nurse Cell" in man, J. Immunol. 130: 2348 (1983).
5. F.P. van de Wijngaert, M.D. Kendall, H.-J. Schuurman, L.H.P.M. Rademakers, and L. Kater, Heterogeneity of epithelial cells in the human thymus. An ultrastructural study, Cell Tissue Res. 237: 227 (1984).
6. H.J. Schuurman, F.P. van de Wijngaert, L. Delvoye, R. Broekhuizen, J.E. McClure, A.L. Goldstein, and L. Kater, Heterogeneity and age dependency of human thymus reticulo-epithelium in production of thymosin components, Thymus in press.
7. R. Oosterom, L. Kater, and J. Oosterom, Effects of human thymic epithelial conditioned medium on mitogen responsiveness of human and mouse lymphocytes, Clin. Immunol. Immunopathol. 12: 460 (1979).

THE THYMUS IN MYASTHENIA GRAVIS: AN IMMUNOHISTOLOGIC STUDY

Michael J. Kornstein, John J. Brooks, Arthur O. Anderson, Arnold I. Levinson, Robert P. Lisak, and Burton Zweiman

Depts. of Pathology & Laboratory Medicine, Internal Medicine (Allergy & Immunology section), and Neurology
University of Pennsylvania
Philadelphia, PA. 19104

INTRODUCTION

Abnormalities of the thymus have long been noted in patients with myasthenia gravis (MG).[1,2,3] Morphologically, the presence of germinal centers in the medulla of the MG thymus was described by Castleman in 1949.[1] Functionally, the MG thymus cells have altered immunologic activity.[2,3] Recently the development of monoclonal antibodies to T cell subsets allows for the furthur investigation of the MG thymus.[4] We have used such monoclonal antibodies to investigate the MG thymus in tissue sections. We have compared these results to findings obtained in studies of suspended cells from the same thymus specimens.

METHODS

Fresh tissue was obtained from twelve MG patients (age range 20-40) undergoing thymectomy and from three controls (ages 4/12, 33, 42 years) undergoing surgery for localized cardiac disease. Tissue sections were stained using the avidin-biotin immunoperoxidase system as described[5] with monoclonal antibodies to the following antigens: T11, T4 (Coulter), T6, T8 (Ortho), IgM, and IgD (Cappel), or with rabbit anti-keratin antibody (Dako). Appropriate controls were included in each run. In microscopic examination (using an ocular grid) of replicate sections, the percentage of cells bearing each phenotypic marker was determined for the cortical and medullary areas. At least 100 cells were counted in each area of the thymus in each section.

For electron microscopy of tissue sections, small pieces (0.5 mm) of thymic tissue from three MG patients were fixed in 2% buffered gluteraldehyde (Karnovsky's fixative) for four hours and then washed in cacodylate buffer (0.1 M pH 7.4) for three days. Tissue for immunoelectron microscopy was then incubated for three days at 4 C with either PBS or monoclonal antibody to Ia antigen (Ortho) diluted 1:50 in PBS. The antigen was then visualized using an avidin-biotin immunoperoxidase technique as described.[5] Sections were observed and photographed with a Philips 200 electron microscope at 60 KV.

For electron microscopy of cell suspensions, thymic lymphocyte suspensions were processed for immunoelectron microscopy using an avidin-biotin immunoperoxidase technique as described.[5] One MG patient and one control were studied.

RESULTS

In tissue sections of MG and control thymus, all cortical lymphocytes appeared positive for T11, T4, T6, and T8 (fig. 1). None of the cortical cells were IgM or IgD positive.

In the medulla of both MG and control thymus, approximately half of the cells were T11 positive (Table I). About one third were T8 positive, half were T4 positive, and 10% were T6 positive. We found no difference in the pattern of T cell subsets when MG was compared to control thymuses.

Germinal centers were frequent and prominent in the medulla of the thymic tissue from 10 of the 12 MG specimens. A single germinal center was seen in one of the control thymus specimens. Mantle zone lymphocytes were both IgM and IgD positive whereas cells in the germinal center were generally only IgM positive. This pattern is similar to that seen in germinal centers of lymph node sections. Apart from the germinal centers, scattered medullary cells positive for IgM and IgD were identified in the thymus of both MG and control thymuses (Table II). T11 positive cells surrounded the germinal centers with occasional T11 and T4 positive cells within the center. T8 positive cells were less frequent and T6 positive cells non-existent in the germinal center.

Localization of Ia antigen was similar for MG and control thymus (fig. 2). The staining of dendritic processes in the cortex and a confluent pattern in the medulla precluded meaningful enumeration and identification of Ia positive cells by light microscopy. In electron microscopic examination of MG thymus tissue sections, Ia antigen was localized to the surface membrane of epithelial cells and interdigitating dendritic cells. In addition, Ia positive portions of the surface membranes of lymphoid cells were noted adjacent to Ia positive non-lymphoid cells (fig.3). Electron microscopy of MG thymus cell suspensions showed occasional

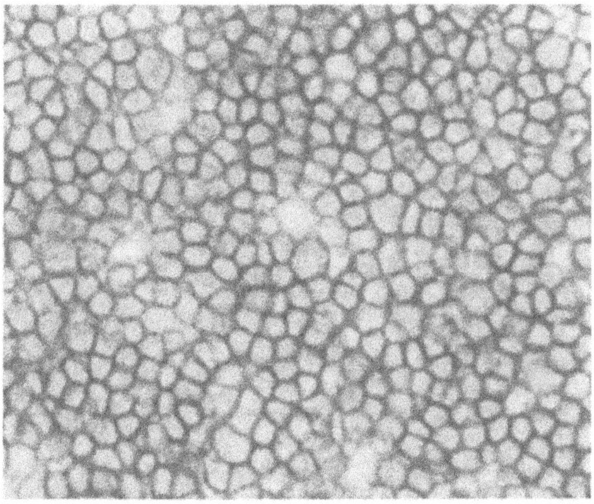

Fig. 1. MG thymic cortex examined for T11. Virtually all cortical lymphocytes are T11 positive. T11 negative cells appear to be non-lymphoic. This same pattern is seen in the cortex for T11, T4, T6, and T8. (400 x)

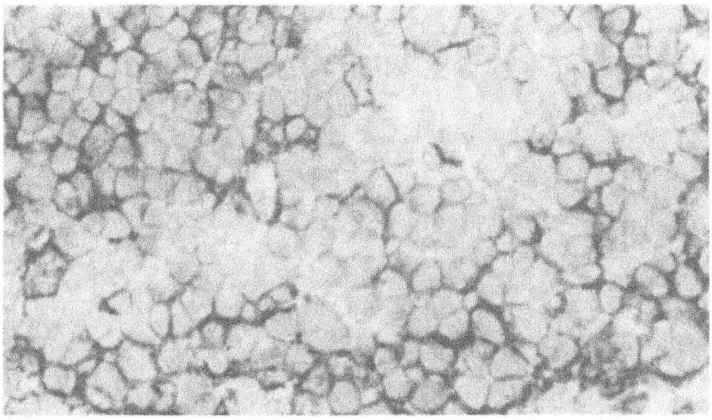

Fig. 2. Frozen section of MG thymic cortex demonstrates localization of Ia antigen. The dendritic pattern precludes enumeration of positive cells. (400 x)

lymphocytes with Ia antigen on their entire surface.

Frozen sections of thymus from 5 MG patients and 1 control were examined by immunoperoxidase technique for keratin. (fig. 4) In the cortex, keratin was localized to long dendritic processes and occasional large cells. In the medulla, numerous large cells and all Hassall's corpuscles were keratin positive. No difference in

TABLE I

T CELL SUBSETS IN THE THYMUS

TISSUE SECTION DATA (% POSITIVE CELLS)

	T11		T4		T8		T6	
MG PATIENTS	CORTEX	MEDULLA	CORTEX	MEDULLA	CORTEX	MEDULLA	CORTEX	MEDULLA
1	99	60	99	50	99	32	99	10
2	99	45	95	50	95	20	ND*	ND*
3	99	45	85	45	85	30	90	8
4	99	45	99	60	99	25	99	3
5	99	62	ND*	ND*	99	44	99	10
6	99	62	99	35	99	37	99	8
7	99	50	99	45	99	28	99	5
8	99	68	99	60	99	20	99	1
9	99	65	99	71	99	36	99	9
10	99	55	99	61	99	41	99	10
11	99	75	99	75	99	22	99	9
12	99	62	99	77	99	27	99	5
MEAN	99 ± 0**	58 ± 9.9	97 ± 13	57 ± 13.5	97 ± 1.3	30 ± 8.0	98 ± 0.8	7.1 ± 3.1
CONTROL								
1	99	90	99	50	99	30	99	11
2	99	70	99	75	99	30	99	10
3	99	52	99	60	99	22	99	4
	99 ± 0	71 ± 19.0	99 ± 0	62 ± 12.6	99 ± 0	27 ± 4.6	99 ± 0	8.3 ± 3.8

*NOT DONE

**Mean ± SD

TABLE II

QUANTITATION OF B CELLS IN THE THYMUS (% POSITIVE CELLS)*

	# CASES	IgD (%)		IgM (%)	
		CORTEX	MEDULLA	CORTEX	MEDULLA
MG	12	0	$9 \pm 7.0^{+}$	0	6 ± 5.4
CONTROLS	3	0	6 ± 3.5	0	7 ± 2.9

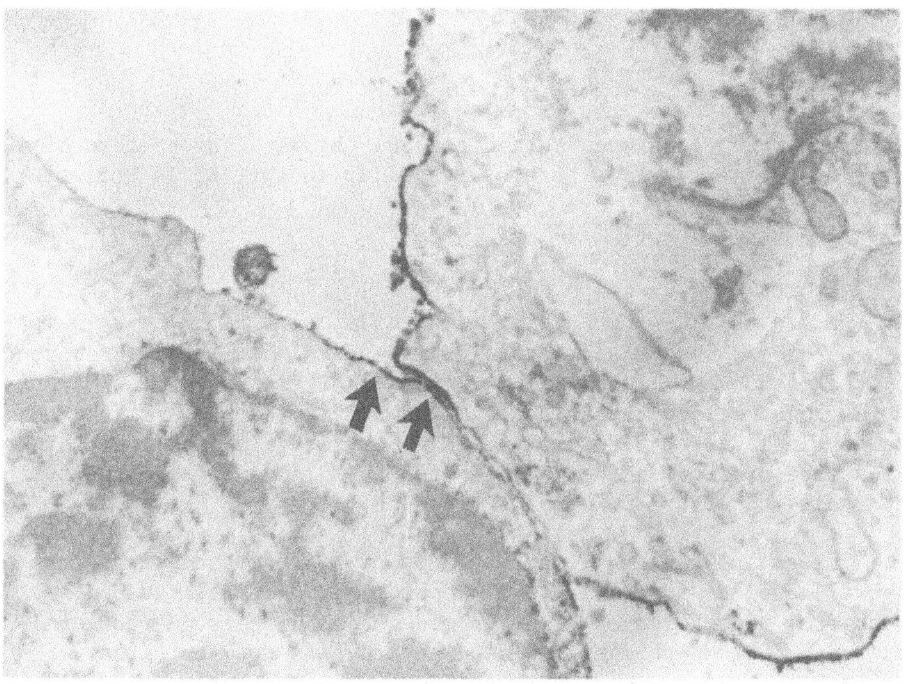

Fig. 3. Electron micrograph of MG thymus tissue sections demonstrating localization of Ia antigen. Ia positive epithelial cell (upper right) and adjacent thymic lymphocyte are shown. Lymphocyte has Ia positivity only on membrane adjacent to epithelial cell (arrows). (11,140 x)

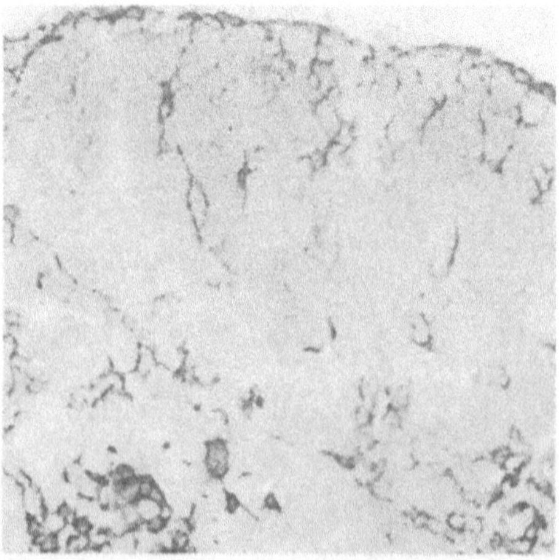

Fig. 4. Frozen section of MG thymus demonstrating localization of keratin. Dendritic pattern is similar to pattern found for Ia antigen (see Figure 2). (100 x)

pattern was noted between MG and control thymus. The pattern of keratin localization was generally similar to that of Ia but not identical.

DISCUSSION

We have found MG and control thymus to be similar with respect to T cells and T cell subsets. The germinal centers of the MG thymus have cells with surface IgM and IgD in the same pattern as the germinal centers of lymph nodes. Scattered IgM and IgD positive cells were present in the medulla of both MG and control thymus apparently unrelated to germinal centers.

These findings in tissue sections of MG thymus differ in several respects from our previous observaions of cell suspensions from these organs.[6] The prominence of B cells in the germinal centers of MG thymus in tissue sections is not reflected in the cells suspensions where the percentage of surface immunoglobulin positive cells was similar in the MG thymus (0.5 %) and control (0.1 %). Also, the predominance of T4 positive cells over T8 and T6 positive cells in tissue sections of medulla of both MG and non-MG thymus was not reflected in the suspension data where T4 positive cells were less numerous than cells positive for either T8 or T6.[6]

The difference between data obtained from tissue sections and cell suspension may be due to several factors. Probably most important is the relative preponderance of cortical cells in the thymus. This preponderance of cortical cells in thymocyte suspensions may obscure differences between MG and control subjects in the prevalence of T4 and T8 positive cells and the germinal center B cells in the medulla of myasthenic patients. Other factors, including loss of more adherent cells in preparing thymus suspensions, heterogeneity among patients, and differences in sensitivity of techniques (immunofluorescence for cell suspension and immunoperoxidase for tissue sections) may also be involved.

In an examination of various phenotypic markers, only the Ia positive cells were significantly greater in MG patients compared to controls.[6] Since the pattern of Ia localization on frozen sections of thymus precluded meaningful identification and enumeration of positive cells by light microscopy, we examined MG thymus sections by immunoelectron microscopy to achieve better resolution. At the ultramicroscopic level, Ia antigen was localized mainly on non-lymphoid cells. Lymphoid cells had Ia positivity only on that portion of the plasma membrane adjacent to Ia positive non-lymphoid cells. This observation is similar to that reported in murine thymus[7,8] and normal human thymus.[9] In thymus cell suspensions examined by electron microscopy for Ia antigen, occasional lymphocytes had diffuse Ia positivity on their entire surface membrane and may represent either B cells or activated T cells. The vast majority of Ia positive cells examined ultrastructurally were non-lymphoid and were either epithelial cells or interdigitating dendritic cells. Using electron microscopy and immunoperoxidase techniques, we were unable to confirm the findings of a recent immunofluorescence study that found increased numbers of interdigitating dendritic cells in the MG thymus compared to control.[10] In agreement with this study, however, we have found no apparent abnormality in the distribution of T cell surface phenotypes in the MG thymus.

The non-quantitative characteristics of electron microscopic examination preclude enumeration of Ia positive cells as compared with light microscopy. Yet, the Ia positive thymic cell population appears to be heterogeneous in both MG and non-MG individuals and includes epithelial cells, interdigitating dendritic cells and some lymphocytes. Keratin appears to be a marker for thymic epithelial cells and keratin positive cells appear to have similar distributions in MG and control thymus. Additional studies using double labeling techniques are in progress and will further investigate the nature of Ia positive cells in the myasthenic thymus. Since Ia antigen is involved in T cell antigen recognition, increased numbers of Ia positive cells may be related to the autoimmune process in myasthenia gravis.

REFERENCES

1. Castleman, B., Norris, E.H., The pathology of the thymus gland in myasthenia gravis: A study of 35 cases, Medicine (Baltimore) 28:27 (1949).
2. Abdou, N.I., Lisak, R.P., Zweiman, B., Abrahamsohn, I., Penn, A.S., The thymus in myasthenia gravis: Evidence of altered cell populations, N Engl J Med 291:1271 (1974).
3. Newson-Davis, J., Wilcox, N., Calder, L., Thymus cells in myasthenia gravis selectively enhance production of anti-acetylcholine-receptor antibody by autologous blood lymphocytes, N Engl J Med 305:1313 (1981).
4. Reinherz, E.L., Schlossman, S.T., The differentiation and function of human T lymphocytes, Cell 19:821 (1980).
5. Kornstein, M., Brooks, J., Anderson, A., Levinson, A., Lisak, R., and Zweiman, B., Immunohistology of the thymus in myasthenia gravis, Amer J Pathol (in press).
6. Lisak, R.P., Zweiman, B., Skolnik, P., Levinson, A.I., Moskovitz, A.R., Guerrero, F., Thymic lymphocyte subpopulations in myasthenia gravis, Neurology (Cleveland) 33:868-872, 1983.
7. Farr, A.G., Nakane, P.K., Cells bearing Ia antigens in the murine thymus: An ultrastructural study, Amer J Pathol 11:88 (1983).
8. Van Ewijk, W., Rouse, R.V., Weissman, I.L., Distribution of H-2 microenvironment in the mouse thymic dendritic cells, J Histochem Cytochem 28:1089 (1980).
9. Bahn, A.K., Reinherz, E.L., Poppema, S., et al., Location of T cell and major histocompatibility complex antigens in the human thymus, J Exp Med 152:771-782 (1980).
10. Thomas, J.A., Wilcox, H.N.A., and Newsom-Davis, J., Immunohistological studies of the thymus in myasthenia gravis. Correlation with clinical state and thymocyte culture responses, J Neuroimmunol 3:319 (1982).

B CELL FUNCTION IN ACUTE MALARIA

D. S. Kumararatne[1], P. Chiodini[2], C. Ellis[2], P. Richardson[1], T. G. Gentle[2], L. Walker[1] and R. P. Stokes[2]

[1]Department of Immunology, University of Birmingham Medical School, Birmingham B15 2TJ [2]Dept of Communicable and Tropical Diseases/Regional Immunology Dept, East Birmingham Hospital, Birmingham B9 5ST

INTRODUCTION

The development of protective immunity to malaria in humans is slow and incomplete[1]. Specific antibody responses to parasite antigens are an important component of protective immunity to parasites[1]. However, secondary humoral immunodeficiency which is known to occur in human and experimental malaria[2] may prejudice the development of effective immunity to this parasite and may also contribute to the susceptibility of children living in endemic areas to infection by other microbial pathogens. The cellular mechanisms contributing to defective antibody production in malaria are incompletely understood[2]. The study of humoral immunosuppression in malaria in endemic areas is complicated by the coexistence of malnutrition and other parasitic diseases, both of which could contribute to the state of anergy. This study investigates possible cellular mechanisms of impaired antibody production in UK residents having acute attacks of malaria.

MATERIALS AND METHODS

Patients: Eleven patients with *P. vivax* malaria, one with *P.ovale* malaria and two with *P.falciparum* infection were studied. Clinical, parasitological and demographic data relating to these patients is summarised in Table 1. Normal control ranges (mean and standard error) have been derived from twenty healthy adult volunteers studied in parallel. Study protocol: lymphocytes were

Table 1. Clinical and demographic data of patients studied

Patient Number	Age in yrs	Sex	Ethnic Group	Country of Residence	Duration of residence in U.K.	Area where malaria was acquired.
1	40	M	Asian	UK	1yr	India
2	51	M	Asian	UK	8yrs	Pakistan
3	59	F	Asian	UK	16yrs	India
4	47	F	Asian	UK	22yrs	India
5	17	F	Asian/Caucasian	UK	17yrs	Pakistan
6	45	M	Asian	UK	5yrs	India
7	51	M	Asian	UK	2yrs	Pakistan
8	26	M	Asian	UK	8yrs	Pakistan
9	20	M	Asian	India	one month	India
10	32	M	Negro	UK	1yr	Nigeria
11	35	M	Asian	UK	18yrs	Pakistan
12	31	M	Caucasian	UK	31yrs	E. & Central Africa
13	21	M	Caucasian	UK	21yrs	E. Africa
14	26	M	Caucasian	UK	26yrs	Central Africa

Table 1 continued.

Patient Number	Species of malaria parasite	% Parasitameia	Severity of Illness	Duration of Symptoms Days	Splenomegaly
1	P.vivax	<1%	Mild	3	+
2	P.vivax	<1%	Moderate	7	-
3	P.vivax	<1%	Moderate	7	-
4	P.vivax	<1%	Mild	5	-
5	P.vivax	<1%	Mild	2	-
6	P.vivax	<1%	Mild	2	-
7	P.vivax	<1%	Mild	21	-
8	P.vivax	<1%	Mild	16	-
9	P.vivax	<1%	Mild	2	+
10	P.vivax	<1%	Mild	1	+
11	P.vivax	<1%	Mild	16	+
12	P.ovale	<1%	Moderate	5	-
13	P.falciparum	2%	Moderate	10	-
14	P.falciparum	50%	Fulminant	7	+

Table 2.

ANTIBODY DESIGNATION[+]	REFERENCE	ANTIGEN DETECTED
UCHT-1	3	T3
B.9.4.1.	3	T8
*	-	T4
Na1/34	16	T6
DA6.164	17	HLA-Dr β-chain
anti-K	3	Kappa light chain
anti-λ	3	Lambda light chain
anti-δ	3	Delta heavy chain
anti-μ	3	Mu heavy chain
anti-γ	3	Gamma heavy chain
anti-α	3	Alpha heavy chain

* T4: bearing cells were computed by subtracting absolute numbers of T8: bearing cells from absolute numbers of T3: bearing cells[3].

\+ All antibodies are mouse monoclonal except the anti-δ and anti-α (polyclonal sheep anti-human reagents prepared by the Immunodiagnostic Research Laboratory, University of Birmingham).

separated from peripheral venous blood of malaria patients and T and B cell subsets enumerated using a sensitive rosetting method[3] employing red cells coated with monoclonal antibodies, capable of detecting T and B cell surface antigens. The monoclonal antibodies used are summarised in table 2. Monocytes were excluded from counts by excluding cells ingesting dyed,opsonised,yeasts[4]. Spontaneous and mitogen induced B-cell activation of peripheral blood lymphocytes was analysed using an optimised immunofluorescence assay for detecting cytoplasmic immunoglobulin containing cells. Epstein-Barr virus (EBV), pokeweed mitogen (PWM) and E. Coli (0111:B4) lipopolysaccharide were used as mitogens[5]. Serial analysis of serum immunoglobulins was also carried out in these patients. Serum autoantibodies were assayed by indirect immunofluorescence to assess in-vivo polyclonal B-cell activation by malarial infection[6]. Rheumatoid factors were detected by standard latex-agglutination and Rose Waaler tests.

RESULTS

Changes in peripheral blood T and B cell subsets in malaria. These results are summarised in Table 3. Patients with benign tertian malaria (data from 11 patients with P.vivax infection and one with P.ovale malaria) had moderate lymphopenia ($P<0.01$) which was due to a reduction in the helper-inducer subset of T-cells ($P<0.05$). T8 bearing cells (suppressor-cytotoxic T cells) and B-cell subsets were within the normal range ($P>0.05$ in each case). Both patients with P.falciparum infection were lymphopenic,mainly due to a reduction in T-helper cells.

In-vitro B cell activation in malaria: See Table 4

13/14 patients had no increase in spontaneous B-cell activation. Only the patient with fulminant P.falciparum infection had evidence of spontaneous polyclonal B-cell activation. None of the patients with malaria (14/14) showed an impairment of B-cell differentiation to antibody forming cells when stimulated with T-independant (EBV, LPS) or T-dependant (PWM) polyclonal activators. ($P>0.05$).

In-vivo polyclonal B-cell activation in malaria

Serum IgG, IgA and IgM levels were within the normal adult range[7] in 12/14 patients. One patient with vivax malaria had a moderately elevated IgG level (20.9g/l; normal range 6-16g/l) and another female patient with vivax malaria had an IgM level of 5.75g/l; normal range 0.5-2.4g/l).

Serum autoantibodies (anti-nuclear antibody, anti gastric-

Table 3a. Changes in Blood Lymphocyte subsets in Malaria

(mean absolute number per litre ± standard error, shown for 'n' individuals)

Donor	Total lymphocytes $\times 10^{-9}$	Total T-cells (T3+ve) $\times 10^{-9}$	Helper-Inducer[2] subset $\times 10^{-9}$	Suppressor Cytotoxic-subset (T8+ve) $\times 10^{-9}$
Group A (n=14)				
Healthy Adults	2.29 ± 0.15	1.64 ± 0.12	1.23 ± 0.10	0.63 ± 0.06
Group B (n=12)				
Benign Tertian Malaria[1]	1.59 ± 0.13	1.23 ± 0.10	0.72 ± 0.08	0.51 ± 0.05
Group C				
P.falciparum malaria				
Case 1:	0.7	0.34	0.19	0.15
Case 2:	0.92	0.64	0.27	0.37
Case 2: (4mth post recovery)	4.8	3.36	1.39	1.92
Results of t-test between GroupsA and B	$P<0.01$	$P<0.02$	$P<0.05$	$P>0.05$ (not significant)

1. consisted of eleven patients with *P.vivax* infection and one with *P.ovale* malaria.
2. obtained by subtracting numbers of T8 positive cells from T3 positive cells.

Table 3b. Changes in Blood Lymphocyte subsets in malaria.

(mean absolute count per litre $\pm$ standard error, shown for 'n' individuals)

Donors	Total B cells* (SmIg +ve cells) $x10^{-9}$	SmIgM+ve cells $x10^{-9}$	SmIgD+ve cells $x10^{-9}$	HLA-DR+ve cells[2] $x\ 10^{-9}$
Group A (n=14)				
Healthy Adults	0.31 $\pm$ 0.04	0.17 $\pm$ 0.02	0.18 $\pm$ 0.05	0.25 $\pm$ 0.04
Group B (n=12)				
Benign Tertian malaria[2]	0.23 $\pm$ 0.04	0.23 $\pm$ 0.03	0.17 $\pm$ 0.04	0.16 $\pm$ 0.03
Group C				
P.falciparum malaria				
Case 1:	0.08	0.11	0.11	0.06
Case 2:	0.04	not done	not done	not done
Case 2: (post recovery)	0.62	0.38	0.29	0.24
Results of t-test between Groups A and B.	P>0.05	P>0.05	P>0.05	P>0.05

* Sum of surface membrane K+ve and λ+ve cells.
1. consisted of eleven patients with *P.vivax* infection and one with *P.ovale* malaria.
2. macrophages excluded from count by omitting yeast ingesting cells.

Table 4a. 'In-Vitro' B cell activation in malaria*

Donors	Polyclonal B-cell activator PBA	N	Number of cells containing following isotype per 10^6 viable cells[†]. IgG	IgA	IgM
Healthy Adults (Group A)	NIL	7	743 ± 248	542 ± 305	1814 ± 634
Benign Tertian Malaria (Group B)	NIL	12	1188 ± 282	1400 ± 528	2708 ± 517
P.falciparum malaria (Group C):	NIL				
Case 1			1000	500	2000
Case 2			18000	15000	15000
Case 2 (post recovery)			5000	1000	5000
Group A	EBV	-7	7429 ± 2232	4929 ± 1076	19429 ± 5702
Group B	"	12	9204 ± 2266	3496 ± 515	15807 ± 3227
Group C:					
Case 1			10000	3000	20000
Case 2			20000	11000	44000
Case 2 (post recovery)			10000	2500	25000

Donors	Polyclonal B cell activator PBA	N	Number of cells containing following isotype per 10^6 viable cells[1].		
			IgG	IgA	IgM
Group A	PWM	7	49943 ± 12400	17457 ± 4575	35786 ± 5414
Group B	"	12	31000 ± 5676	15483 ± 3982	31688 ± 4457
Group C:					
Case 1			19000	1000	5000
Case 2			15000	7500	22500
Case 2 (post recovery)			10000	2500	60000
Group A	LPS	7	17996 ± 2121	16529 ± 3085	29936 ± 5568
Group B	"	12	15483 ± 2042	9400 ± 2052	30408 ± 3433

* 10^6 peripheral blood mononuclear cells were cultured with or without EBV, PWM or LPS for 7 days. Cytoplasmic-Ig containing cells were determined from replicate cultures by immunofluorescence.

1. mean ± standard error, for n individuals.

parietal-cell antibody anti smooth-muscle antibody, anti mitochondrial antibody and rheumatoid factor) were not detected in 12/14 patients. 2/14 patients had autoantibodies to gastric parietal-cells but were 59 years and 45 years old respectively. The 59 year old female patient also had a low titre (1/10) of IgG antinuclear antibody in her serum.

DISCUSSION

Our results confirm the findings reported from Thailand[8,9] and South America[10] that patients with acute malaria (P.vivax or P.falciparum infection) have moderate blood lymphopenia mainly due to a reduction in T-helper cells, while T-suppressor cells and B-cell subsets are within the normal range. However, this T-lymphopenia does not diminish lymphocyte proliferative-capacity, as assessed by lectin induced DNA synthesis, in those adults naturally infected with P.vivax or P.falciparum[9]. The lectins used by these workers were phytohaemagglutinin, Concanavalin A and pokeweed mitogen respectively. Troye-Blomberg et al[10] have reported intact lymphocyte proliferative responses to PWM in adults with P.falciparum infection, although these patients too had reduced T-cell counts. We have extended these findings by showing that T-cell dependant B-cell differentiation (induced by stimulation with PWM) in patients with malaria, was no different from that of normal controls, tested simultaneously. B cell differentiation induced by polyclonal activators EBV[11] or LPS[11] was also within the normal range in the patients we studied.

Evaluating the proliferative response of lymphocytes to lectins and the study of B cell differentiation induced by polyclonal activators have been of value in defining abnormalities of T and B cell function in patients with immunodeficiency[12,13]. However these techniques have failed to provide clear evidence of impaired immunity in this and other studies[8,9]. On the other hand the investigation of specific antibody[14] responses or assessment of specific T-cell responses[15] have consistently revealed significant abnormalities in patients with malaria. Hence, studies of secondary immunosuppression in humans, should concentrate on the evaluation of specific antibody and T cell responses; studies of lymphocyte responses to polyclonal (non-specific) mitogens seems to be of little or no value for this purpose. These studies are of primary importance in view of the possible impact of malaria-induced immunosuppression, on the health of people living in endemic areas[2,15]. UK-residents' acquiring acute malaria on visits to endemic areas appear to be an ideal population for carrying out such studies, since they are unlikely to suffer from malnutrition or other serious parasitic diseases.

REFERENCES

1. S. Cohen. Proc. Roy. Soc. Lond. (B) 203: 323 (1979).
2. W.P. Weidanz. Brit Med. Bull. 38: 167 (1982).
3. A. Abdul Carder, P. Richardson, L. Walsh et. al. Br. J. Cancer 48: 185 (1983).
4. A.Y. Shaala, H.S. Dhaliwal, S. Bishop et. al. J. Imm. Methods. 27: 175 (1979).
5. L. Walker, G.D. Johnson. J. Imm. Methods. 62: 134 (1983).
6. G.D. Johnson, J. Dorling. In Techniques in clinical immunology ed: R.A. Thompson Blackwell Sci. Pub. Oxford p 106 (1981).
7. E.R. Stiehm, H.H. Fudenberg. Paediatrics 37: 717 (1966).
8. R.A. Wells, K. Pavanand Zolyomoi's et. al. Clin. Exp. Immunol. 35: 202 (1979).
9. R.P. MacDermott, R.A. Wells, S. Zolyomi, S. et. al. Infection and Immunity 30: 781 (1980).
10. M. Troy-Blomberg, P.E. Sjöholm, H. Perlmann et.al. Clin. Exp. Immunol. 53: 335 (1983).
11. G. Möller. In B lymphocytes today ed: J.R. Inglis Elsevia Biomedical press, Amsterdam p 59 (1982).
12. Report of IUIS/WHO working group. Clin. Exp. Immunol. 46: 662 (1981).
13. U. Anderson, A.G. Bird, S. Britton et. al. Immunol. Rev. 57: 3 (1981)
14. B.M. Greenwood, A.M. Bradley-Moore, A. Palet, A.D.M. Bryceson. Lancet 1: 169 (1972).
15. D.J. Moss, S.R. Burrows, D.J. Castelino et. al. Int. J. Cancer 31: 727 (1983).
16. G. Brown, L. Walker, N.R. Ling et. al. Scand. J. Immunol. 19: 373 (1984).
17. A.T. McMichael, J.R. Pilch, G. Gaffre et. al. Eur. J. Immunol. 9: 205 (1979).

THE ORIGIN OF B-CHRONIC LYMPHOCYTIC LEUKAEMIA (B-CLL) AND ITS RELATIONSHIP TO HAIRY CELL LEUKAEMIA (HCL)

F.Caligaris-Cappio[1], L.Bergui[1], G.Pizzolo[2], M.Chilosi[2], L.Tesio[1], D.Campana[1], G.Semenzato[3], F.Malavasi[4], and M.Gobbi[5]

Cattedra Clinica Medica A, Università di Torino (1)
Catt. Ematologia e Anatomia Patologica, Università di Verona (2)
Cattedra Clinica Medica, Università di Padova (3)
Cattedra Genetica Medica, Università di Torino (4)
Ist. Ematologia L.& A. Seragnoli, Università di Bologna (5)

B-CLL is the prototype of B-cell chronic lymphoid leukaemias. The malignant cells have well defined features such as weak surface immunoglobulins (sIg), rosetting ability with Mouse red blood cells (Mrbc) and expression of the T-cell associated antigen T1 (reviewed in ref. 1). HCL is an enigmatic disorder classified, in its typical form, among chronic B-cell malignancies. The leukaemic cells express monoclonal sIg [2,3] and exhibit the appropriate Ig gene rearrangement [4]. Unique to HCL is the <u>concomitant</u> presence of four major characters: a) typical membrane irregularities with veils and microprocesses [5,6]; b) strong cytoplasmic tartrate resistant acid phosphatase (TRAP) activity [5,6]; c) positivity with the monoclonal antibody (Mc Ab) SHCL3 which is negative on non-Hodgkin's lymphomas and B-CLL [7]; d) membrane reactivity with the Mc Ab anti-Tac [4,8] which detects the receptor for Interleukin 2 (IL 2; ref. 9).

CLL and HCL have distinct clinical features and patterns of organ involvement (reviewed in ref. 5 and 10). Nevertheless, some links can be envisaged between these two B-cell chronic lymphoid leukaemias. First, cases with intermediate features presenting typical TRAP$^+$ hairy cells in the peripheral blood and the histological features of CLL in the spleen and lymph nodes have been described [11,12,13].

Second, surface markers are shared by the two types of malignant cells. HCL cells form rosettes with Mrbc, though in a lower percentage as compared to CLL [5]. Mc Abs which discriminate between early bone marrow (BM) and peripheral B lymphocytes indicate that CLL and HCL are positive with the same peripheral B-cell reacting Mc Abs [1].

Therefore, the crucial question is whether CLL and HCL are malignant transformations of cells belonging to different lineages or involve different maturation levels along a same lineage. Clearly, any approach to this problem implies the definition of the normal counterparts of these disorders. A $T1^+$, $sIgM^+$ cell population has been identified in fetal lymph nodes where form the bulk of primary nodules clustering around follicular dendritic cells [14]. In adult lymph nodes these cells are diluted out at the edge of germinal centres (GC) and represent a minute population (~ 5%; ref 15). The $T1^+$, $sIgM^+$ B cells are (almost) absent in the BM and the spleen[14,16] and have been purified from adult lymphoid tissues by Mrbc rosetting [15]. Thus, the lymph node $T1^+$, $sIgM^+$ cells are also $Mrbc^+$ and faithfully reproduce the typical features of B-CLL: B-CLL normal equivalent cells. On the contrary, the normal counterpart of HCL has been insofar elusive.

On these bases, the aims of our study have been the investigation of the relationships between CLL and HCL and the search of the normal counterpart of HCL. The experimental approach involved the use of the differentiation promoter 12-O-tetradecanoyl-phorbol-13 acetate (TPA) on purified normal and malignant cell populations. The rationale of our choice lies in the ability of TPA to induce the differentiation in vitro of a variety of cells of lymphohaemopoietic lineages including myeloid [17], T-cell lines [18] and B-CLL [19] thus providing clues as to the cell differentiation programme.

The experimental design was based on the in vitro treatment of separated cell populations with TPA 1.6×10^{-8} M (concentration which previous experiments had shown to induce phenotypic changes without affecting cell viability: ref. 20) for 72-96 hours and the subsequent phenotypic analysis of the induced cells.
Three kinds of experiments were devised:
a) induction of CLL cells (16 cases) to investigate whether they can be transformed into HCL cells;
b) induction of HCL cells (4 cases), to analyse whether they can acquire CLL features;
c) induction of normal B cell populations: tonsil $Mrbc^+$,$T1^+$, $sIgM^+$

(CLL normal counterpart), mantle zone and GC B cells; peripheral blood B lymphocytes.
Tonsil $T1^{+}$, $sIgM^{+}$ B cells were purified by means of Mrbc rosetting techniques. Mantle zone and GC B cells were separated on a fluorescent activated cell sorter (FACS 420, Becton Dickinson) utilizing the RFA-2 Mc Ab which reacts with T cells and B cells in the mantle zone of lymph nodes but is negative on GC blasts (ref 21). The phenotype of induced cells was defined on the bases of ;a) morphology; b) cytoplasmic (cy) TRAP staining according to Ly et al [22] ; c) surface nad cytoplasmic markers. Double immunofluorescence assays were employed utilizing antibodies (both conventional and monoclonal) labelled with fluorescein (FITC:green) and rhodamine (TRITC:red) in various combinations [23]. The expression of Ig isotypes, T1, SHCL3 and Tac defined antigens were evaluated. Mrbc rosetting ability was checked as described by Forbes and Zalewsky [24].

RESULTS AND DISCUSSION

The cell viability in all groups studied was above 85% during the incubation period. The mitotic index was low, not exceeding 1% in malignant samples and 2.5% in normal samples. Control culture cells did not show any appreciable change when compared to the starting population of fresh uncultured cells. The main results can be summarized as follows:

1) B-CLL TPA-induced become undistinguishable from HCL(Table1) Striking membrane irregularities such as veils, protrusions and microprocesses appeared (16/16). These irregularly shaped cells developed strong cytoplasmic TRAP enzyme activity (16/16), reacted with SHCL3 (13/16) and with Tac (9/16). Thus, the four main criteria unique to HCL are fully and concomitantly reproduced by B-CLL cells treated with TPA. Further, these transformed cells have also lost their Mrbc rosetting capacity but retained the T1 antigen.

2) HCL TPA-induced do not develop CLL features (Table 1) These cells became firmly adherent to the culture vessel wall and formed an intricate network of elements with fibroblast-like or stellate features. The transformed cells were still $TRAP^{+}$, $SHCL3^{+}$, Tac^{+}, lost their Mrbc ability and did not acquire " de novo " the T1 antigen. The B cell nature of TPA-induced HCL cells was confirmed by the presence of monoclonal cy Ig.

Table 1. TPA induced changes in B-CLL and HCL[a]

	B-CLL (16 cases)		HCL (4 cases)	
Phenotype	0 hr	72 hr with TPA	0 hr	72 hr with TPA
Morphology	small lymphocytes	hairy cells	hairy cells	fibroblast-like
Mrbc+	61 ± 8[b]	< 5	33 ± 4	< 5
sIg+	80 ± 3 (w.)[c]	< 10(v.w.)[c]	22 ± 5	< 5
CyIg+	< 10	> 50[d]	15 ± 5	> 60
T1+	87 ± 6	79 ± 11	1.5,1,3,65	1,2,3,60
SHCL3+	3.3 ± 2.1	74 ± 11	87 ± 7	44 ± 4
Tac+	4.0 ± 2[e]	64 ± 21	84 ± 4	47 ± 8
TRAP+	< 5	> 70	60 ± 5	> 60

[a]There has been no appreciable change in control culture cells (except a slight decrease in Mrbc rosetting).
See text and ref. 26 for further details.
[b]Percentage (± S.E.M.) of cells reacting with each marker.
[c]w. = weak; v.w. = very weak.
[d]12/16 cases developed cyIg.
[e]Percentages referred to 15 cases. The 16th one had a high percentage of Tac+ cells (80 %: dot-like positivity).

Table 2. TPA induced changes in tonsil Mrbc+, T1+ cells[a]

	Mrbc+ cells (4 cases)	
Phenotype	0 hr	72 hr with TPA
Morphology	small lymphocytes	immunoblast-like
Mrbc+	81 ± 5[b]	< 5
sIg+	85 ± 4	< 10 (v.w.)[c]
CyIg+	< 10	> 80
T1+	75 ± 5	48 ± 5
SHCL3+	< 1	6.8 ± 2
Tac+	< 1	< 1
TRAP+	< 1	18 ± 7

[a]The only appreciable change in control culture cells was the decrease of Mrbc rosetting ability.
See text and ref. 26 for further details.
[b]Percentage (± S.E.M.) of cells reacting with each marker.
[c]w. = weak; v.w. = very weak.

B) Normal Mrbc^{+}, T1^{+} B cells respond differently to TPA as compared to B-CLL cells (Table 2).
TPA-induced Mrbc^{+}, T1^{+} cells acquired immunoblast-like features but did not exhibit membrane perturbations. They lost Mrbc receptor and showed decreased proportions of T1^{+} cells. Few of these B cells became SHCL3^{+} and TRAP^{+} but Tac antigen was not expressed. Similar results were obtained with the other normal (tonsil and peripheral blood) B cell subsets tested.

These data show that TPA induces much more significant membrane disturbances in CLL as compared to their normal equivalent cells. Membrane perturbations are apparent only in CLL and these cells are unable to switch off the T1 antigen production (see also ref 20 and 25). B-CLL cells acquire new markers, whose expression by normal B lymphocytes is significantly less impressive. The expression of Tac antigen is not induced by TPA on normal B cell subsets. However, a small percentage of peripheral blood B lymphocytes (5-10%), stimulated in vitro with pokeweed mitogen for 7 days, express concomitantly cy Ig and membrane Tac reactivity [26]. Therefore, Tac is not an " aberrant " (i.e. leukaemia-specific) antigen, but, together with TRAP enzyme activity, appears to represent a peculiar B-cell activation marker.
TPA is an artificial stimulus and consequently the changes induced on CLL cells could have resulted from a non-specific activation of any malignant B cell. However, prolymphocytic leukaemia and mantle zone lymphoma cells do not acquire HCL-like features after TPA [26]. Therefore, our observations further support the CLL-HCL link.

In conclusion, B-CLL cells TPA-induced reproduce the full set of HCL typical features: morphology, strong cytoplasmic TRAP enzyme activity, membrane reactivity with SHCL3 and Tac. Some of HCL characters (i.e. Tac and TRAP) exhibited also by TPA-treated CLL cells are peculiar markers of activated B cells. Another HCL-specific marker, HC2, is expressed also by normal activated B cells [27]. Therefore, it is suggested that HCL might represent an aberrantly activated variant of B-CLL or of a CLL-related disorder. The HCL normal counterpart might as well represent an activation stage of a B cell subset, thus explaining why the normal equivalent cells of HCL remain elusive when resting B cell populations are investigated. Taking into account the CLL-HCL link, the subset most likely involved in the transforming events which lead either to CLL or to HCL appears to be the lymph node T1^{+}, sIgM^{+}, Mrbc^{+} B cell population.

ACKNOWLEDGEMENTS

We wish to thank Dr. Schwarting and Prof. Stein,Berlin, for the generous gift of SHCL3 Mc Ab. The Mc Ab Tac was kindly provided by Dr. Uchiyama, Tokyo. We are indebted to Prof. G.Janossy, London, for helpful suggestions and encouragement.
This work was supported by PFCCN, CNR, grant N. 830075296.

REFERENCES

1. F.Caligaris-Cappio and G.Janossy, Surface markers in chronic lymphoid leukemias of B cell type,Semin.Hematol in press (1984).
2. D.G.Haegert, J.C.Cawley, R.D.Collins, R.J.Flemans, and J.L.Smith, Leukemic reticuloendotheliosis: a morphological and immunological study in four cases, J.Clin.Pathol.27:967 (1974).
3. S.M.Fu, R.J.Winchester, K.R.Rai, and H.G.Kunkel, Hairy cell leukemia.Proliferation of a cell with phagocytic and B lymphocytic properties, Scand.J.Immunol. 3:847 (1974).
4. S.J.Korsmeyer, W.C.Greene, J.Cossman, S.M.Hsu, J.P.Jensen, L.M. Neckers, S.L.Marshall, A.Bakhshi, J.M.Depper, W.J.Leonard, E.S. Jaffe, and T.A.Waldmann, Rearrangement and expression of immunoglobulin genes and expression of Tac antigen in hairy cell leukemia.,Proc.Natl.Acad.Sci.USA,80:4522 (1983).
5. D.Catowsky, Hairy cell leukemia and prolymphocytic leukemia,Clin. Haematol., 6;245,(1977).
6. L.T.Yam, C.Y.Li, and K.W.Lam, Tartrate resitant acid phosphatase in the reticulum cells of leukemic reticuloendotheliosis,N.Engl. J.Med.284;357 (1961).
7. R.Schwarting, H.Stein, and C.Y.Wang, The monoclonal antibodies SHCL1 and SHCL3 allow the diagnosis of hairy cell leukemia, Blood, in press (1984).
8. T.Uchiyama, S.Broder, and T.A.Waldmann, A monoclonal antibody anti-Tac) reactive with activated and functionally mature human T cells.I.Production of anti-Tac monoclonal antibody and distribution of Tac^+ cells, J.Immunol.126:1393 (1981).
9. W.J.Leonard, J.M.Depper,T.Uchiyama, K.A.Smith, T.A.Waldmann, and W.C.Greene,A monoclonal antibody that appears to recognize the receptor of human T-cell growth factor: partial characterization of the receptor,Nature 300;267 (1982).
10. D.A.G.Galton, and I.C.M.McLennan, Clinical patterns in B lymphoid malignancy,Clin.Haematol.11;561 (1982).
11. R.S.Newman, A.L.Sullivan, and R.Jaffe, Malignant lymphoma simulating malignant reticuloendotheliosis,Cancer,43:329 (1979).

12. M.Palutke, P.Tabaczka, I.Mirchandani, and S.Goldfarb, Lymphocytic lymphoma simulating hairy cell leukemia,Cancer 48:2047 (1981).
13. I.A.Lampert, Splenectomy as a diagnostic technique,Clin.Haematol. 12:535 (1983).
14. M.Bofill, G.Janossy, M.Janossa, G.D.Burdford, G.J.Seymour, and E. Kelemann, Human B cell development.IISubpopulations in the human fetus, in press (1984).
15. F.Caligaris-Cappio,M.Gobbi, M.Bofill, and G.Janossy, Infrequent normal B lymphocytes express features of B chronic lymphocytic leukemia, J.Exp.Med., 155;623,(1982).
16. G.Janossy, F.Caligaris-Cappio, M.Bofill, D.Campana, and M.Janossa Development of B cell subpopulations in man and its relevance to malignancy,in :"Modern trends in human leukemia.VI.,"R.Neth ed., Springer Verlag, in press (1984).
17. J.Lotem, and L.Sachs, Regulation of normal differentiation in mouse and human leukemic cells by phorbol esters and the mecha nism of tumour promotion, Proc.Natl.Acad.Sci.USA 76:5158 (1979)
18. D.Delia, M.F.Greaves,A.Newman,R.Sutherland,J.Minowada, P.Kung and G.Goldstein, Modulation of T leukemic cell phenotype with phorbol ester, Int.J.Canc. 29:23 (1982).
19.T.H.Totterman, K.Nilsson, and C.Sundstrom, Phorbol ester induced differentiation of chronic lymphocytic leukemia cells,Nature 288;176 (1980).
20. F.Caligaris-Cappio, G.Janossy, D.Campana, M.Chilosi, L.Bergui, R. Foa, D.Delia, M.C.Giubellino,P.Preda, and M.Gobbi, Lineage relationship of chronic lymphocytic leukemia and hairy cell leukemia: studies with TPA, Leuk.Res., in press (1984)
21. M. Gobbi, F.Caligaris-Cappio, and G.Janossy, Normal equivalent cells of B cell malignancies: analysis with monoclonal antibodies, Br.J.Haematol. 54:393 (1983)
22. C.Y.Ly, L.T.Yam, and K.W.Lam, Acid phosphatase isoenzymes in human leucocytes in normal and pathologic conditions, J.Histoch. Cytoch. 18:473 (1970).
23. G.Janossy,F.J.Bollum, K.F.Bradstock, and J.Ashley, Cellular phenotypes of normal and leukemic hemopoietic cells determined by analysis with selected antibody combination,Blood 56:430 (1980)
24. I.J.Forbes, and P.D.Zalewsky, A subpopulation of human B lymphocytes that rosette with mouse erythrocytes,Clin.exp.Immunol. 26:99 (1976).
25. D.L.Shawler,M.C.Glassy, S.B.Wormsley, and I.Royston,Alterations in cell surface phenotype of T- and B-cell chronic lymphocytic leukemia cells following in vitro differentiation by phorbol ester J.Natl.Canc.Inst. 72:1059 (1984).

26. F.Caligaris-Cappio,G.Pizzolo, M.Chilosi, L.Bergui, G.P.Semenzato, L.Tesio, D.Campana, L.Morittu, F.Malavasi, M.Gobbi, and G. Janossy, Phorbol ester induces abnormal CLL cells to express features of hairy cell leukaemia, submitted (1984).
27. D.N.Posnett, C.Y.Wang, N.Chiorazzi, M.K.Crow, and H.G.Kunkel, An antigen characteristic of hairy cell leukemia cells is expressed on certain activated B cells, J.Immunol in press (1984).

INCREASED REPRESENTATION OF A SUBPOPULATION OF EARLY B CELLS IN THE PERIPHERAL BLOOD OF PATIENTS WITH RHEUMATOID ARTHRITIS

C. Plater-Zyberk, K. Lam, T. D. Kennedy, R. N. Maini and G. Janossy

Kennedy Institute of Rheumatology
6 Bute Gardens
London W6 7DW

INTRODUCTION

Immune mechanisms play an important role in the autoimmunity seen in patients with rheumatoid arthritis (RA)[1]. Abnormalities in T lymphocytes and monocytes have been reported[2,3]. Recent work has suggested that the autoimmunity arises from excessive stimulation, associated with too little suppression, which leads to a self-perpetuating process of B cell activation[4]. In these studies, it has been assumed that the B lymphocytes in RA are normal and that their abnormal behaviour is due to defects in their regulation. We have already reported one abnormality in the circulating B cell pool in RA by showing an increase in a subpopulation which forms spontaneous rosettes with mouse erythrocytes[5].

In this paper we describe further studies on the phenotype of the mouse rosette forming population in RA and present evidence of an increase in a circulating B cell subpopulation (T1-B cells) which possess the T1 antigen characteristically present only on thymus-derived lymphocytes[6].

RESULTS

Mouse Rosette Forming Cell in RA

The population of cells forming spontaneous rosettes with mouse erythrocytes (MRFC) has been further characterised by indirect immunofluorescence. A preparative enrichment of MRFC and non-MRFC has been obtained from a population of E rosette depleted lymphocytes. MRFC from RA patients are B cells as assessed by surface

Table 1. Surface phenotype of enriched populations of MRFC in RA

MRFC*	$SIgM^+/D^+$ ** (%)	DR^+*** (%)
RA 1	76	ND
RA 2	68	100
RA 3	90	97
RA 4	79	88
	78.3 ± 9.1	95 ± 6.2

* preparative lymphocytes population enriched in MRFC by density centrifugation of cells rosetted with mouse erythrocytes.

** double-fluorochrome immunofluorescence with rhodemine and fluorescein conjugated antibodies

*** single-fluorochrome immunofluorescence

immunoglobulin (SIg) and DR positivity. Most of these cells bear SIgM and SIgD concomitantly and less than 4% are SIgG positive (Table 1). These characteristics are similar to those found on MRFC separated from normal individuals suggesting that the abnormality seen in RA is quantitative rather than qualitative.

T1-B Cells in RA

The T antigen expressed is the p67 T1 antigen identified by a pan-T monoclonal antibody. T1 is normally expressed on the surface of >95% of T cells but not on normal circulating B cells[6]. In normal subjects, this antigen is also expressed on a minor population of B cells from the lymph nodes and tonsils (2-5% of B cells)[7]. There is a clonal expansion of malignant B cells with the T1-B phenotype in chronic lymphocytic leukaemia (CLL)[8]. Thus, B cells expressing T1 appear to be normally a population of resident B cells which thus far have only been reported in the circulation of patients with CLL.

We have studied the expression of T1 on SIg positive cells by double-fluorochrome microscopy where surface Ig was detected with a rhodamine conjugated goat F(ab)2 anti-human Ig (Kallestead, Pershore, Worcestershire) and T1 by indirect immunofluorescence using the RFT-1 monoclonal antibody (produced in Royal Free Hospital) followed by a fluorescein-conjugated goat anti-mouse IgG (Kallestead). Currently, data is available on a B-enriched population from 16 patients with

Table 2. Incidence oi the abnormal T1-B population in the blood of patients with RA, CTD and CLL

Patients	No. tested	Incidence of increased T1-B*
RA	16	13/16 (81%)
CTD	8	2/8 (25%)
CLL	2	2/2 (100%)

* Upper limit of normal representation of T1-B
= 3% of SIg positive cells (Mean + 2 SD)

definite/classical RA, 8 patients with connective tissue disease (6 with SLE, 1 with Sjogren's syndrome and 1 with Raynaud's syndrome) and 8 healthy controls.

Results illustrated in Table 2 show that 13 out of the 16 RA patients studied have an expanded population of B cell expressing the T1 antigen (range from 8% to 48%).

Only 2 out of the 8 patients in the disease control group showed a percentage of T1-B cells higher than the mean + 2 SD obtained with healthy controls (0.6 + 2.4%). Control experiments were done by the same fluorescent technique using another pan-T mouse monoclonal antibody (anti T-3) of the same class than RFT-1. Only 0.6 ± 1.8% of the B cells from 10 RA patients studied stained with this monoclonal. No statistically significant correlation was found between the expansion of the circulating T1-B and disease activity or treatment. However, a trend to a lower representation of the abnormal subpopulation in patients treated with corticosteroid, gold and D-penicillamine was observed.

DISCUSSION

Our studies have shown a significantly increased abnormal sub-population of B cells in RA by utilising two different surface markers. As seen on the schematic representation (Fig 1), in normals, about 25% of the B cells are MRFC and less than 1% are T1-B. In RA, both these populations are expanded to 60-70% MRFC and ~ 20% T1-B cells. In the blood of CLL patients, almost all B cells are MRFC[9] and express the T1 antigen[8]. Whether in RA the T1-B cells are included within MRFC population is not yet known.

In the mouse, the homologue of the T1 is the Ly-1 antigen, Ly-1 shows similar surface pattern expression and biochemical characteristics as the human antigen[10]. A population of B cells expressing Ly 1 antigen forms 1-2% of the spleen cells from normal

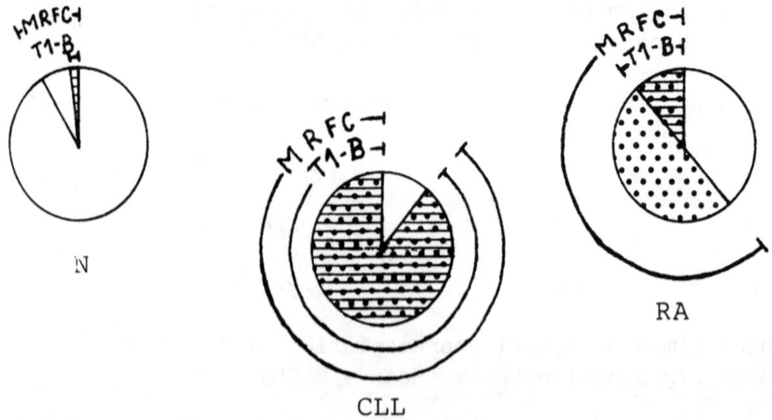

B Cell Sub-sets

mice. This subpopulation is increased to 5-10% of the spleen in some autoimmune strains of mice[11,12]. The concomitant expression of Ly 1 and B cell marker is also found on some murine lymphomas[13]. In functional studies the Ly-1 B population from the spleen of autoimmune mice have been found to spontaneously secrete auto-antibodies in-vitro[14]. The Ly-1 B subpopulation from normal mice have been found to increase an in-vitro antigen specific antibody response via idiotype-antiidiotype interactions and has been designated as a B helper population[15,16].

The functional role of the T1-B population expanded in the blood of RA patients is not yet known. Nevertheless the identification of a subpopulation with similar phenotype in an autoimmune disease and a lymphoproliferative disease may have important repercussions in the understanding of the role of B cells in autoimmunity.

Acknowledgement

We wish to thank the Arthritis and Rheumatism Council of Great Britain for supporting this work.

1. H. G. Kunkel, E. M. Tan, Autoantibodies and disease. Adv. Immunol. 4:351 (1964).
2. V. Corrigall, G. S. Panayi, R. Laurent, Lymphocyte studies in rheumatoid arthritis. Scand. J. Rheumatol. 8:10 (1979).
3. D. T. Y. Yu and J. B. Peter, Cellular immunological aspects of rheumatoid arthritis. Sem. Arthr. Rheumatol. 4:25 (1979).
4. G. Janossy, O. Duke, L. W. Poulter, G. Panayi, M. Bofill, G. Goldstein, Rheumatoid arthritis: a disease of T-lymphocyte/ macrophage immunoregulation. Lancet. ii:839 (1981).
5. G. R. W. Room, C. Plater-Zyberk, M. F. Clarke, R.N. Maini, B-lymphocyte subpopulation which form rosettes with mouse

erythrocytes increased in rheumatoid arthritis. Rheumatol. Int. 2:175 (1982).

6. E. Engleman, R. Warnke, R. Fox, R. Levy, Studies of a human T lymphocyte antigen recognised by a monoclonal antibody. Proc. Natl. Acad. U.S.A. 28:1791 (1981).
7. F. Caligaris-Cappio, M. Gobbi, M. Bofill, G. Janossy, Infrequent normal B lymphocytes express features of B-chronic lymphocytic leukaemia. J. Exp. Med. 155:623 (1982).
8. L. Boumsell, H. Coppin, D. Pham, B. Raynal, J. Lemerle, J. Dausset, A. Bernard, An antigen shared by a human T cell subset and B cell in chronic lymphocytic leukaemic cells. J. Exp. Med. 152:229 (1980).
9. D. Catovsky, M. Cherchi, A. Okos, U. Hedge, D. A. G. Galton, Mouse red-cell rosettes in B-lymphoproliferative disorders. Br. J. Haematol. 33:173 (1976).
10. J. A. Ledbetter, R. L Evans, M. Lipinski, C. Cunningham-Rundles, R. A. Good, L. A. Herzenberg, Evolutionary conservation of surface molecules that distinguish T lymphocyte helper/inducer and cytotoxic/suppressor subpopulations in mouse and man. J. Exp. Med. 153:310 (1981).
11. V. Manohar, E. Brown, W. M. Leiserson, T. M. Chused. Expression of Lyt-1 by a subset of B lymphocytes. J. Immunol. 129:532 (1982).
12. K. Hayakawa, R. R. Hardy, D. R. Parks, L. A. Herzenberg, The "Ly-1B" cell subpopulation in normal immunodefective and autoimmune mice. J. Exp. Med. 157:202 (1983).
13. L. L. Lanier, N. L. Warner, J. A. Ledbetter, L. A. Herzenberg, Expression of Lyt-1 antigen on certain murine B cell lymphomas. J. Exp. Med. 153:998 (1981).
14. K. Hayakawa, R. R. Hardy, M. Honda, L. A. Herzenberg, A. D. Steinberg, L. A. Herzenberg, Ly-1 B cells: Functionally distinct lymphocytes that secrete IgM autoantibodies. Proc. Natl. Acad. Sci. U.S.A. 81:2494 (1984).
15. K. Okumura, K. Hayakawa, T. Tada, Cell-to-cell interaction controlled by immunoglobulin genes. Role of Thy-1$^-$, Lyt-1$^+$, Ig$^+$ (B') cell in allotype-restricted antibody production. J. Exp. Med. 156:443 (1982).
16. D. H. Sherr, M. E. Dorf, An idiotype-specific helper population that bears immunoglobulin, Ia and Lyt-1 determinants. J. Exp. Med. 159:1189 (1984).

NODULAR LYMPHOCYTE PREDOMINANCE TYPE OF HODGKIN'S DISEASE IS A B CELL LYMPHOMA

S. Poppema, W. Timens, and L. Visser

Department of Pathology
University of Groningen, Oostersingel 63
Groningen, 9713 EZ, The Netherlands

INTRODUCTION

In a study on follicular lymphomas, Rappaport et al. (1) distinguished a nodular variant of Hodgkin's paragranuloma from other types of follicular lymphoma. Since then, it has been demonstrated that follicular lymphomas are proliferations of cells normally present in germinal centers of secondary follicles (2,3). With respect to the proliferating cells in Hodgkin's disease no clear evidence has become available. In this study we will demonstrate that the nodular variant of lymphocyte predominance type of Hodgkin's disease is the result of a proliferation of germinal center cells. At least three different cell types can be distinguished in germinal centers. Firstly: transformed B lymphocytes; secondly: regulating T lymphocytes; and thirdly: dendritic reticulum cells. It will be shown that all three constituents are part of the process in nodular lymphocyte predominance type of Hodgkin's disease.

MATERIALS AND METHODS

Frozen tissue sections of lymph nodes in 7 cases, of involved spleen in 2 cases and of uninvolved spleen in 4 cases could be studied. In addition, cytocentrifuge preparations of lymph node suspensions in two cases were available. An indirect immunoperoxidase method was used to demonstrate the reactivity with a series of monoclonal anti-lymphocyte antibodies. The reactivity of these antibodies is summarized in table I.

TABLE 1 MONOCLONAL ANTIBODIES

B cells	B1, B2, BA1, T10
T cells	T1, Leu1, T3, T4, Leu3, T6, T8, T10, T11, Leu7, WT1
DRC	B2, aC3B, aDRC
anti Ig	G, A, M, D, K, λ

Monoclonal antibodies were obtained from Ortho (OKT series), Becton & Dickinson (Leu series), Coulter Clone (B1, B2), Dr. Tucker le Bien (BA1), Dr. Wil Tax (WT1), Dakopatts (aC3B, aDRC) and Dr. Joost Haayman (anti Ig antibodies).

RESULTS

Histology

The histology of nodular lymphocyte predominance type of Hodgkin's disease has been described in detail previously (4,5). The normal lymph node architecture is replaced by large, oval nodules. The predominant cells are small lymphocytes, with intermingled epithelioid histiocytes and so called L&H type Sternberg-Reed cells. These are large cells with frequently lobulated nuclei, sometimes binucleated and a rim of basophilic cytoplasm. Eosinophils and plasma cells are scarce or absent.

Immunohistological staining of B lymphocytes

The majority of the small lymphocytes in the nodules is reactive with monoclonal anti-B cell antibodies B1, B2 and BA1. In addition, they are reactive with anti IgM and anti IgD (fig. 1). The polyclonal nature of these cells is reflected by the presence of kappa as well as lambda positive cells in a ratio of 2:1. Thereby these B cells appear to be equivalent to the mantle zone cells of reactive follicles. The large L&H type Sternberg-Reed cells are also reactive with B1 monoclonal antibody. In cytocentrifuge preparations these cells can be shown to react with anti-immunoglobulin antibodies.
We have shown IgG positive, IgM positive as well as one case of IgD positive L&H type Sternberg-Reed cells. However, kappa as well

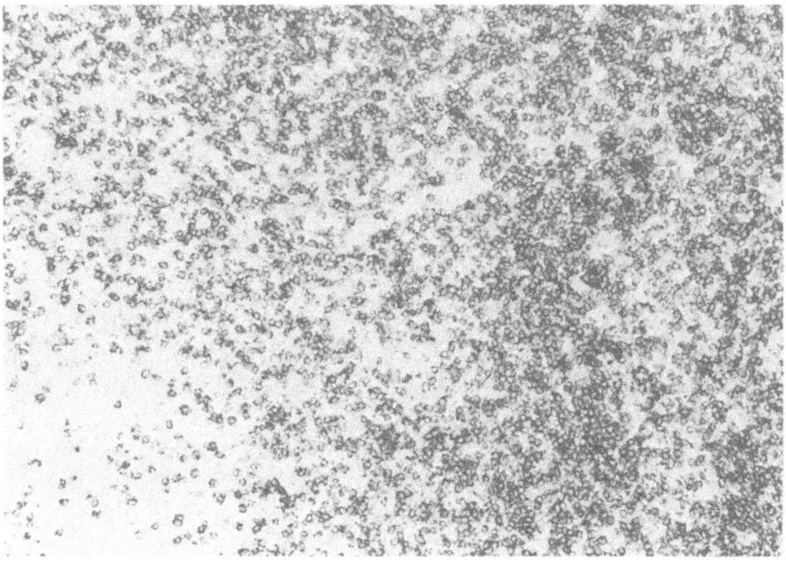

fig. 1. Nodule of L.P. Hodgkin's disease stained for IgM. A majority of small IgM positive lymphocytes can be seen. (x 110).

as lambda positive cells were present, indicating the multiclonal origin of this cell population. Plasma cells are scarce, in anti-Ig as well as in OKT10 stained slides.

Immunohistological staining of T lymphocytes

Inside the nodules a variable amount of T lymphocytes can be found, in some cases up to 50% of the lymphocytes (fig. 2). In between the nodules, the majority of the cells are T lymphocytes. The T lymphocytes in the nodules are reactive with OKT11, OKT1, OKT3, OKT4, Leu3, WT1 antibodies, but also with Leu7 antibody and are thereby similar to the population of T lymphocytes in reactive germinal centers (fig. 3). OKT8 positive cells are very scarce. In between the nodules OKT4, Leu3 as well as OKT8 positive cells are found in a ratio of 2:1, similar to the population in reactive lymph nodes.

Immunohistological staining of dendritic reticulum cells

Staining with antibodies directed against C3b receptor antigen (aC3b) and C3d receptor antigen (B2) shows an extensive dendritic network inside the nodules (fig. 4). In addition, this network is

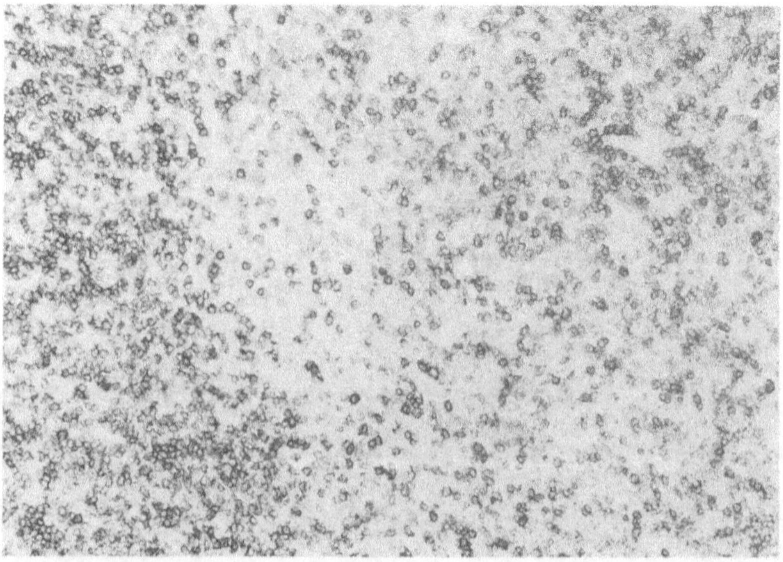

fig. 2. Nodule of L.P. Hodgkin's disease stained for Leu1. Left: Internodular, non involved area with a majority of T cells. Right: Nodule with minority of T cells. (x 110).

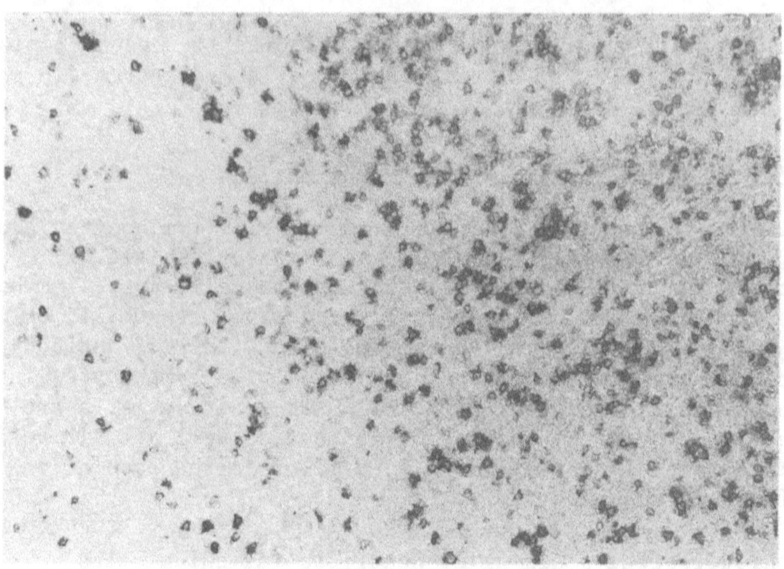

fig. 3. Same area as in fig. 2, stained for Leu7. Left: internodular area, containing only few Leu7 positive cells. Right: Nodule with similar number of positive cells as in fig. 2. (x 110).

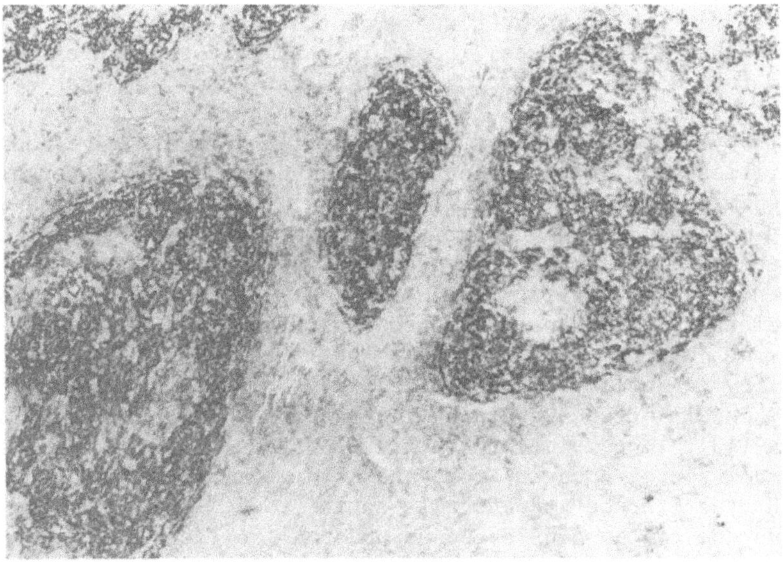

fig. 4. Several nodules of L.P. Hodgkin's disease, stained for B2 (anti C3d receptor), showing strong staining of dendritic cells. (x 45).

immunostained with the anti-dendritic reticulum cell antibody (aDRC).
Compared to reactive germinal centers the dendritic cells in the nodules are missing the reactivity with anti-immunoglobulin antibodies.

DISCUSSION

The nodules of nodular lymphocyte predominance type of Hodgkin's disease appear to exhibit many similarities to reactive follicles. Firstly, they contain polyclonal IgM, IgD positive B lymphocytes, similar to the small B lymphocytes of the mantle zone of follicles. Secondly, the T lymphocytes inside the nodules are OKT4, Leu3, but also Leu7 positive. Such a T cell population is exclusively found in reactive germinal centers and also in the nodules of germinal center cell lymphomas (6,7). The functional activity of this population has not yet been determined. Although it has been shown that T lymphocytes are indispensable for germinal center formation (8,9) and therefore it seems to be probable that the OKT4, Leu7 positive cells may play a role in the development of germinal centers, other arguments have to be taken into account. OKT4, Leu7 positive cells are not prominent in early germinal center reaction, when many blasts are present and also not in the so called dark zone of germinal centers, but in later stages and especially in the light zone where centrocytes (cleaved cells) prevail. Therefore, it seems possible that this subpopulation rather plays a role in the regulation and termination of germinal center reaction than in the initiation of this reaction. Since the nodules of lymphocyte predominance type of Hodgkin's disease have many similarities with germinal centers but lack germinal center blasts, the predominance of OKT4, Leu7 positive T lymphocytes in these nodules may even play a causative role in the development of these large abnormal follicles. The presence of dendritic reticulum cells in the nodules is a further argument for the germinal center nature of nodular lymphocyte predominance type of Hodgkin's disease. The absence of dendritic immunoglobulin staining establishes a similarity to follicular center cell lymphomas which are also usually devoid of such staining in contrast to reactive germinal centers (10). To complete the picture of nodular lymphocyte predominance type of Hodgkin's disease as a germinal center lymphoma, the L&H type Sternberg-Reed cells appear to be immunoglobulin positive and can be considered as abnormal follicular center blasts (4,11).

The findings all support the germinal center origin of nodular lymphocyte predominance type of Hodgkin's disease. However, it is not clear whether it is a malignant or premalignant condition. The polyclonal nature of the proliferation pleads against malignancy. The frequent occurrence of atypical follicular hyperplasia (progressively transformed germinal center) (5) and the presence of hypogamma globulinemia in few patients (unpublished observation) with nodular lymphocyte predominance type of Hodgkin's disease are in agreement with the idea that it may be an abnormal immune reaction. The relatively benign prognosis is also an argument for a nonmalignant disease. On the other hand, the relatively frequent occurrence of transition to non-Hodgkin lymphomas indicates that it has a premalignant condition (12).

In conclusion, nodular lymphocyte predominance type of Hodgkin's disease is a disease entity and is the result of abnormal proliferation of all cell types present in normal follicles and germinal centers. The matter whether it should be considered a malignant, premalignant or benign germinal center lymphomas has not yet been resolved.

REFERENCES

1. H. Rappaport, W. J. Winter and E. B. Hicks, Follicular lymphoma. A reevaluation of its position in the scheme of malignant lymphoma, based on a survey of 253 cases, Cancer 9:792 (1956).
2. R. A. Warnke and R. Levy, Immunopathology of follicular lymphomas: A model of B-lymphocyte homing, N. Engl. J. Med. 298:481 (1978).

3. E. S. Jaffe, E. M. Shevach, M. M. Frank, N. C. Berard and I. Green, Nodular lymphoma - evidence for origin from follicular B lymphocytes, N. Engl. J. Med. 290:813 (1974).
4. S. Poppema, E. Kaiserling and K. Lennert, Nodular paragranuloma and progressively transformed germinal centers, Virchows Arch. B (Cell Pathol.) 31:211 (1979).
5. S. Poppema, E. Kaiserling and K. Lennert, Hodgkin's disease with lymphocyte predominance, nodular type (nodular paragranuloma) and progressively transformed germinal centers: a cytohistological study. Histopathol. 3:295 (1979).
6. S. Poppema, L. Visser and L. de Ley, Reactivity of presumed anti-natural killer cell antibody Leu7 with intrafollicular T lymphocytes. Clin. Exp. Immunol. 54:834 (1983).
7. G. Pizzolo, G. Semenzato, M. Chilosi, L. Morittu, A. Ambrosetti, N. Warner, M. Bofill and G. Janossy, Distribution and heterogeneity of cells detected by HNK-1 monoclonal antibody in blood and tissue in normal, reactive and neoplastic conditions. Clin. Exp. Immunol. 57:195 (1984).
8. M. de Souza and H. Pritchard, The cellular basis of immunological recovery in nude mice after thymus grafting, Immunol. 26:769 (1974).
9. E. B. Jacobson, L. H. Caporale and G. J. Thorbecke, Effect of thymus cell injections on germinal center formation in lymphoid tissues of nude (thymus less) mice, Cell. Immunol. 13:416 (1974).
10. H. Stein, A. Bonk, G. Tolksdorf, K. Lennert, H. Rodt and J. Gerdes, Immunohistologic analysis of the organization of normal lymphoid tissue and non Hodgkin's lymphomas, J. Histochem. Cytochem. 28:746 (1980).
11. S. Poppema, The diversity of the immunohistological staining pattern of Sternberg-Reed cells, J. Histochem. Cytochem. 28:788 (1980).
12. M. Miettinen, K. Franssila and E. Saxen, Hodgkin's disease, Lymphocytic predominance nodular. Increased risk for subsequent non Hodgkin's lymphomas, Cancer 51:2293 (1983).

TWO MONOCLONAL ANTIBODIES RAISED AGAINST A BURKITT LYMPHOMA CELL LINE RECOGNISE SEPARATE COMPONENTS OF LYMPHOID FOLLICLES

Murray L. J. and Habeshaw J. A.

I.C.R.F. Department of Medical Oncology
St Bartholomew's Hospital
West Smithfield, London EC1A 7BE

INTRODUCTION

In peripheral lymphoid tissue, the main sites of B cell localisation are lymphoid follicles[1], from which the majority of non-Hodgkin lymphomas (NHL) arise. Follicles are organised structures, containing a variety of cell types: mantle zone (m.z.) B cells, germinal centre (G.C.) B cells i.e. centrocytes and centroblasts, associated T cells of helper phenotype and dendritic reticular cells (FDRC). Using a panel of monoclonal antibodies (MoAbs) it is possible to demonstrate many differences in the antigens expressed by m.z. and G.C. B cell populations.[2] Distinguishing characteristics of G.C. B cells include expression of the common acute lymphoblastic leukaemia antigen, CALLA,[3] and the p24 recognised by BA-2.[4] Surface IgD, BA-1[5] and RFA-2[6] are not expressed by G.C. B cells.

In an attempt to produce a MoAb which might define a distinct B cell subset in lymphoid follicles, mice were immunised with cells of the Raji Burkitt lymphoma (B.L.) line. MoAb 35.1C5 was selected for further characterisation, as in reactive lymph node (L.N.) it stained a subpopulation of m.z. and interfollicular cells, but was negative on G.C. cells. Screening on tissues biopsied from NHL patients demonstrated reactivity of 35.1C5 with B cells at an intermediate stage of development.

The antigen-presenting cell of the follicle - the FDRC[7] - is of uncertain derivation, but shows a close interaction with G.C. B cells, being present in lymphoid follicles and nodular lymphomas of centroblastic/centrocytic (CB/CC) or centrocytic (CC) type. MoAb FW37.4.D5 showed specificity for FDRC in section, but was also found to react with a minority of cells in some B.L. lines and with blood monocytes. Characterisation of these two MoAbs is described.

METHODS

BALB/C mice were immunised i.p. with 10^7 Raji cells. After two further immunisations, animals were boosted i.v. with 10^7 Raji cells. Fusion was performed 4 days later by a standard protocol. Supernatants from hybrids selectively grown in HAT medium were screened by radioimmunoassay on Raji cells. Cell binding supernatants were screened on frozen sections (F.S.) of reactive lymphoid tissue, using the avidin-biotin-peroxidase technique.[8] Supernatants from hybrids of interest were rescreened after cloning on a more extensive F.S. panel, normal cell suspensions and haematopoietic cell lines. Surface immunofluorescence studies were performed by standard procedures.[9] The molecular weights (m.w.) of the antigens were investigated by lactoperoxidase catalysed I^{125}-iodination of Raji cells, immunoprecipitation with ascites of 35.1C5 and FW37.4.D5, SDS-PAGE and autoradiography.

RESULTS

Normal Lymphoid Tissues (Frozen Section Analysis)

In tonsil and L.N. 35.1C5 stained >60% of m.z. B cells and a smaller proportion (~20%) of interfollicular cells. Positive cells were rarely found within G.C.s (Figure 1). In spleen, the majority of marginal zone B cells were positive, with only rare positive cells in G.C.s, T cell zones and red pulp. Analysis of paediatric thymus revealed staining of a large proportion of medullary thymocytes, while cortical thymocytes were negative.

FW37.4.D5 appeared specific for FDRC in primary and secondary lymphoid follicles of tonsil, L.N. (Figure 2) and spleen. FDRC processes extend into the m.z. No reactivity was found in paediatric thymus. This pattern of reactivity in F.S. was similar to that shown by CR4/23 (kind gift of Dr Stein).

Normal Lymphoid Tissues (Suspension Phenotypes)

A mean of 6% adult bone marrow (B.M.) mononuclear cells were stained with 35.1C5. 4% of peripheral blood mononuclear cells (PBMC) were positive (approx. 60% B1+ P.B. cells).

Less than 1% of B.M. cells expressed the antigen identified by FW37.4.D5. P.B. monocytes were positive.

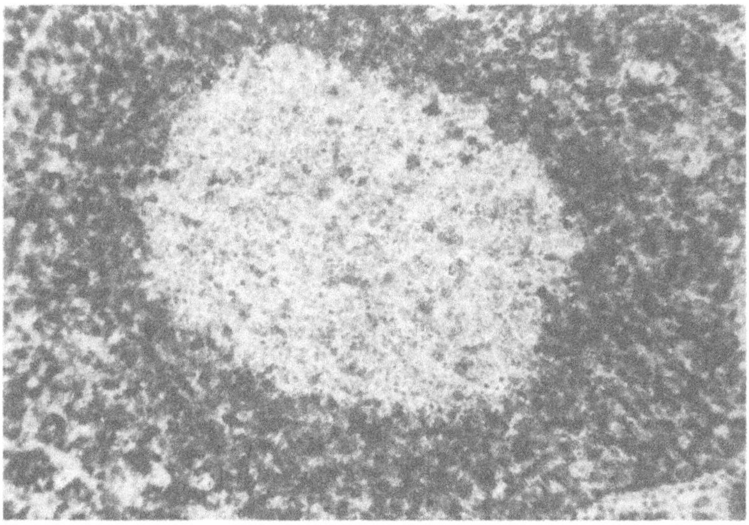

Fig. 1. Reactive L.N. stained with 35.1C5. A subpopulation of m.z. and interfollicular cells are positive, but only rare cells within G.C. are stained.

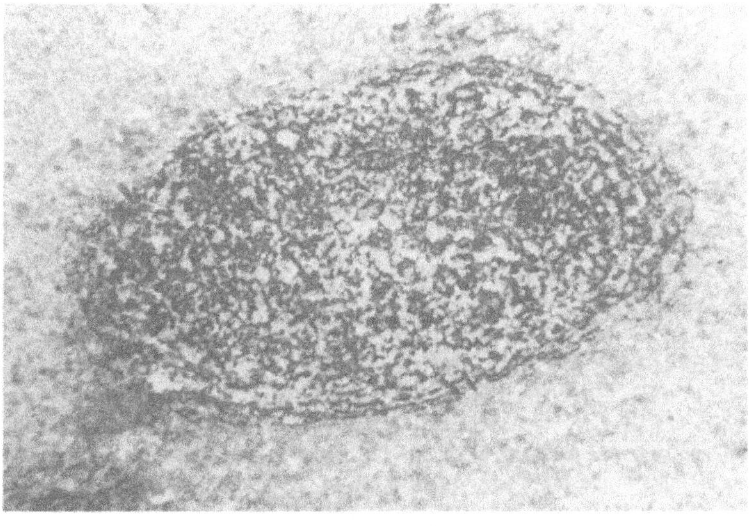

Fig. 2. Reactive L.N. stained with FW37.4.D5. Dendritic reticular cells within the G.C. of a secondary lymphoid follicle are strongly positive.

Non-Hodgkin's Lymphoma Biopsies

Frozen sections from 52 cases of NHL were analysed using the immunoperoxidase technique. Results are shown in Table 1. 35.1C5 reacted strongly with cells of B-CLL, B-HCL, ML,Lpcd and ML,CC. Positivity for FW37.4.D5 was observed in lymphomas of follicular origin (CB/CC), which in 20/26 cases showed FDRC within neoplastic follicles.

Table 1. Frozen section analysis of 35.1C5 and FW37.4.D5 reactivity in 52 cases of Non-Hodgkin lymphoma

Diagnosis	35.1C5	FW37.4.D5
cALL	Scattered + cells	-
B-CLL	7/8 +++ 1/8 small % + (*)	-
ML,Lpcd	2/2 +++	-
B-HCL	2/2 +++	-
ML,CC	6/6 +++	1/3 +
ML,CB/CC/F	nodules	nodules
	6/23 - 14/23 + 3/23 +++	20/26 +
	internodular area	3/26 few nodules only +
	5/23 - 7/23 ~5% + 8/23 20-30%+ 3/23 majority +	3/26 -
ML,CB	1/2 - 1/2 ~50% +	1/2 - 1/2 small number scattered +
T-lymphoblastic	2/3 - 1/3 few cells +	-
T-immunoblastic	2/2 -	-
T zone lymphoma	~20% interfollicular cells +	-
Lennert's lymphoma (T cell)	1/2 - 1/2 ~50% +	-

*CLL undergoing 'immunoblastic' transformation

Abbreviations

cALL = common acute lymphoblastic leukaemia
B-CLL = B-chronic lymphocytic leukaemia
ML,Lpcd = malignant lymphoma, lymphoplasmacytoid
ML,CC = malignant lymphoma, centrocytic
ML,CB/CC/F = malignant lymphoma, centroblastic/centrocytic/follicular
ML,CB = malignant lymphoma, centroblastic

Haematopoietic Cell Lines

The results of screening of the two MoAbs on a variety of cell lines are summarised in Table 2. 35.1C5 recognises an antigen expressed by a subpopulation of cells in B.L. lines. FW37.4.D5, although apparently FDRC-specific in section, reacted weakly with a small percentage of cells of B.L. lines Raji and Daudi. Equivocal weak positivity was observed on ~5% of cells of the HMy2 line.

TABLE 11. Percentage of cells reactive with 35.1C5 and FW37.4.D5 in haematopoietic cell lines

Line	Cell Type	35.1C5	FW37.4.D5
BV173	CML lymphoid blast crisis (CALLA+)	-	-
Nalm-1	pre-B	-	-
Nalm-6	pre-B	-	-
B85	B-lymphoblastoid	-	-
HMy2	B-lymphoblastoid	-	5 weak
Raji	Burkitt lymphoma	65	23 weak
Daudi	Burkitt lymphoma	89	11 weak
Namalwa	Burkitt lymphoma	55	-
BL8A	Burkitt lymphoma	30	-
U-266	plasma cell	-	-
Molt-4	T-ALL	-	-
J6	T-CLL	-	-
K562	erythroleukaemia	-	-
HL-60	promyelocytic	-	-

- means <1% cells positive
CML = chronic myelocytic leukaemia

Biochemical Identification of Antigens

FW37.4.D5 immunoprecipitated a band at 162-168K under both reducing and non-reducing conditions. 35.1C5 did not precipitate an antigen from Raji cells.

DISCUSSION

Monoclonal antibodies raised against the Raji B.L. cell line showed reactivity with distinct cell types in lymphoid follicles. FW37.4.D5 showed specificity for FDRC which present antigen in the form of immune complexes to B lymphocytes.[7] This interaction plays an important role in B cell activation and in the development of immunological memory.[10] The origin and lineage affiliation of the FDRC remain obscure.[11] In contrast to the anti-FDRC MoAb KiM4,[12] FW37.4.D5 did appear to react with P.B. monocytes. A minor subpopulation of B.L. lines Raji and Daudi were also weakly positive. The significance of an antigen shared by FDRC and B.L. B cells is unclear. B.L. are believed to arise from follicular B cells,[13] and it is possible that the antigen is of functional significance in the interaction of FDRC with G.C. B cells. The m.w. (162-168K) and the distribution are not identical to any known complement receptor.

We have previously reported B cell subset distribution in normal lymphoid tissues.[2] The pattern of reactivity of 35.1C5 is distinct, defining a major subpopulation of m.z. and interfollicular cells, as well as splenic marginal zone cells. Staining was different from BA-1,[5] which reacts with almost all m.z. B cells and is also positive on pre-B cells, which appear negative for 35.1C5. Approximately 60% of P.B. B cells (identified by B1) are recognised by 35.1C5. It remains unclear whether such B cell heterogeneity reflects sequential maturation stages within a single B cell developmental pathway, or the existence of two or more separate B cell lineages, diverging at an early stage from a common progenitor.

Among haematopoietic cell lines, 35.1C5 positivity was restricted to B.L. B cell lines. BV173, Nalm-1 and Nalm-6, which represent B cell progenitors, and the plasma cell line U-266 were all completely negative. T cell and two non-lymphoid lines (K562 and HL-60) were also unreactive with 35.1C5. FW37.4.D5 reacted weakly with a minor subpopulation of B.L. lines Raji and Daudi.

NHL biopsies provide a relatively homogeneous population of cells (usually of B lineage), arrested at some point in normal differentiation. Lymphomas which showed strong positivity for 35.1C5 (7/8 B-CLL, 2/2 B-HCL, 6/6 ML,CC) represent intermediate stages of peripheral B lymphocyte development. Common ALL (pre-B stage) and lymphomas of transformed B cells, e.g. ML,CB and 1/8 B-CLL undergoing immunoblastic transformation, comprised mainly 35.1C5 negative cells. This correlates with negativity of G.C. B cells. The accumulated data indicate that 35.1C5 identifies an antigen present on a recirculating B cell subpopulation, which is normally present in the m.z. of follicles and the splenic marginal zone. B lineage specificity of 35.1C5 is in doubt, due to apparent staining of medullary thymocytes, although peripheral T cells, T cell lines and T lymphomas were negative. Clear staining of Reed-Sternberg (R.S.) cells was seen in 1/12 cases of Hodgkin's disease, as previously described with MoAb B1.[14]

The relationship between peripheral B lymphocytes and FDRC is an intimate one. It is, therefore, of interest that two MoAbs raised against a B.L. B cell line, and both reactive with a subpopulation of cells in the immunising line Raji, should recognise two distinct cell types within the lymphoid follicle.

REFERENCES

1. P. Nieuwenhuis, C. E. Van Nouhuijs, J. H. Eggens, F. J. Keuning, Germinal centres and the origin of the B-cell system I. Germinal centres in the rabbit appendix. Immunology 26:497 (1974)
2. L. J. Murray, S. H. Swerdlow, J. A. Habeshaw, Distribution of B lymphocyte subsets in normal lymphoid tissue. Clin. Exp. Immunol. 56:399 (1984)
3. J. Ritz, J. M. Pesando, J. Notis-McConarty, H. Lazarus and S. F. Schlossman, A monoclonal antibody to human acute lymphoblastic leukaemia antigen. Nature 283:583 (1980)
4. J. H. Kersey, T. W. LeBien, C. S. Abramson, R. Newman, R. Sutherland and M. Greaves, p24: a human leukemia-associated and lymphohemopoietic progenitor cell surface structure identified with monoclonal antibody. J. Exp. Med. 153:726 (1981)
5. C. S. Abramson, J. H. Kersey and T. W. LeBien, A monoclonal antibody (BA-1) reactive with cells of human B lymphocyte lineage. J. Immunol. 126:83 (1981)
6. F. Caligaris-Cappio, M. Gobbi, M. Bofill and G. Janossy, Infrequent normal B lymphocytes express features of B-chronic lymphocyte leukaemia. J. Exp. Med. 155:623 (1982)
7. G. J. V. Nossal, G. L. Ada and C. M. Austin, Antigens in immunity IV. Cellular localisation of ^{125}I and ^{131}I labeled flagella in lymph nodes. Aust. J. Biol. Med. Sci. 42:311 (1964)
8. S-M. Hsu, L. Raine and H. Fanger, Use of avidin-biotin-peroxidase complex (ABC) in immunoperoxidase techniques. J. Histochem. Cytochem. 29:577 (1981)
9. J. A. Habeshaw, D. Bailey, A. G. Stansfeld and M. F. Greaves, The cellular content of non-Hodgkin lymphomas: a comprehensive analysis using monoclonal antibodies and other surface marker techniques. Br. J. Cancer 47:327 (1983)
10. G. G. B. Klaus and J. H Humphrey, The generation of B memory cells. Immunology 38:687 (1977)

11. W. G. van Voorhis, M. D. Witmer and R. M. Steinman, The phenotype of dendritic cells and macrophages. Fed. Proc. 42:3114 (1983)
12. M. R. Parwaresch, H. J. Radzun, M. L. Hansmann and K. P. Peters, Monoclonal antibody KiM4 specifically recognises human dendritic reticular cells (follicular dendritic cells) and their possible precursor in blood. Blood 62:585 (1983)
13. K. Lennert and N. Mohri, Histopathology and diagnosis of non-Hodgkin's lymphomas. In "Malignant lymphomas other than Hodgkin's Disease", K. Lennert ed., in collaboration with H. Stein, N. Mohri, E. Kaiserling and H. K. Moller-Hermelink, Springer-Verlag, New York (1978)
14. M. S. Dorreen, J. A. Habeshaw, A. G. Stansfeld, P. F. M. Wrigley and T. A. Lister, Characteristics of Sternberg-Reed and related cells in Hodgkin's disease: an immunohistological study. Br. J. Cancer 49:465 (1984)

ISOLATION OF GERMINAL CENTRE (GC) CELLS IS GREATLY IMPROVED BY USING THE PROTEASE DISPASE TO PREPARE CELL SUSPENSIONS

N Willcox[+], M Schluep[+], M Bofill[*] and J Newsom-Davis[+]

Depts of Neurological Science[+] and Immunology[*]
Royal Free Hospital Medical School, London NW3 2PF
and the National Hospital for Nervous Diseases, WC1

INTRODUCTION

The characteristic muscle weakness in myasthenia gravis (MG) is caused by autoantibodies to the acetylcholine receptor (AChR) of the motor endplate. They can be measured in a very sensitive radio-immunoassay (using detergent extracts of muscle, labelled with ^{125}I-a-bungarotoxin[1]). There appears to be a special relationship with the thymus in MG. About 10% of cases have thymoma, and, in about two thirds of the remainder, there is thymic hyperplasia with medullary germinal centres (GC, reviewed in 1). Cell suspensions prepared by chopping and sieving these thymi (THY_{conv}) spontaneously synthesise anti-AChR antibodies in culture.[1] That is greatly increased if the tissue is dispersed instead by using the proteases collagenase and, especially, dispase[2] (THY_{C+D}). Dispase is a neutral protease that is apparently essential for isolating mouse follicular dendritic (FD) cells.[3] We now report experiments probing the histological and functional basis of this striking bonus.

PATIENTS AND METHODS

Thirty three responder MG patients without thymoma or prior immunosuppressive therapy were studied: all had detectable serum anti-AChR (0.9 - 1,740 nmol/litre). Their ages ranged from 11 to 48 years, and durations of symptoms from 0.2 to 10 years. THY_{conv} suspensions prepared by chopping and sieving were centrifuged at 40 x g for 20-30 secs; the clumps thus pelleted were washed twice (to remove free cells), and incubated for 20-30 mins in 4 mg/ml of dispase (Boehringer Mannheim, grade II) plus 5 units per ml of staphylococcal nuclease (Boehringer), in RPMI/10% FCS, and then washed thoroughly. Immunohistological studies were performed as in refs 4 and 5 using the reagents listed there; all the other methods used are described in detail in ref 1.

RESULTS AND DISCUSSION

Spontaneous Production of anti-AChR and total IgG in Culture

Spontaneous anti-AChR production by THY_{C+D} was, on average, 6.4 times higher than by THY_{conv} in 27 out of the 33 cases studied (eg. Fig 1, Table 1). The difference was even more consistent with spontaneous IgG production, and averaged almost tenfold.

It seemed most likely that, during the preparation of THY_{conv}, activity was lost when the small clumps were discarded. If these were rescued and dispersed with dispase → '$clumps_D$', the resulting suspension synthesised nearly 4 times more anti-AChR (on average) than THY_{conv}, and that was also consistent (eg. Fig 1, Table 1). However, a difference in basal IgG production was found much less regularly.

Table 1 Spontaneous production in culture of anti-AChR and total IgG by THY_{C+D} or $Clumps_D$ as compared with THY_{conv}

		Spontaneous production of:- anti-AChR: no. of cases / total	anti-AChR: mean ratio (s.e.)	total IgG: no. of cases / total	total IgG: mean ratio (s.e.)
THY_{C+D} / THY_{conv}	ratio ≥2	27/33	6.4 (1.2)	23/27	9.3 (1.5)
clear* exceptions		3/33	1	1/27	1
'$Clumps_D$' / THY_{conv}	ratio ≥2	16/21	3.7 (0.6)	3/12	7.8 (4.6)
clear* exceptions		3/21	1	5/12	1

*the remainder were equivocal

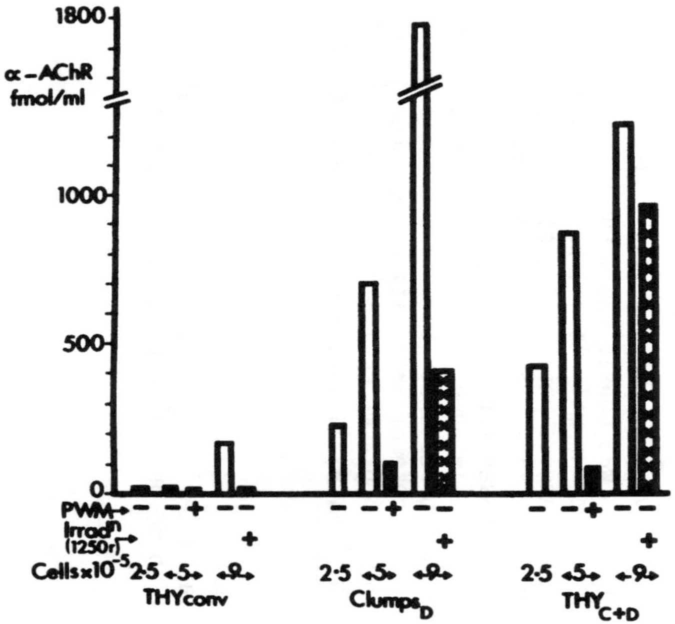

Fig 1. Anti-AChR production by the indicated numbers of MG thymocytes cultured alone (-), with PWM (1:100), or after 1,250 rads 60Co irradiation.

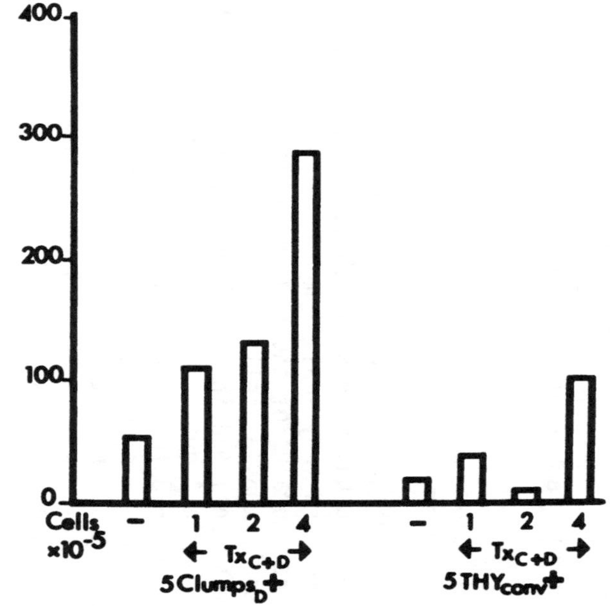

Fig 2. Anti-AChR production by 5.10^5 MG thymocytes cultured alone (-), or with the indicated numbers ($x10^{-5}$) of T_{xC+D}. Amounts of anti-AChR produced by the T_{xC+D} alone have been subtracted from the co-culture values.

Thus these results clearly show that ongoing activity in the MG thymus is largely isolated in structures that tend to remain intact as small clumps on conventional mechanical disruption: their efficient dispersion apparently depends on the use of dispase.

'Feeder' Effects of THY_{C+D}

The appearance in culture of the 'Clumps $_D$' cells was variable, but there were usually less fibroblasts and macrophages than with THY_{C+D}[2], and more very small lymphocytes that die early and thus undermine the 'morale' of the remaining cells. The impression that these 'Clumps'$_D$ cells did not always fulfil their potential *in vitro* was confirmed in co-culture experiments; their responses were enhanced in 4 out of 9 cases - sometimes dramatically, and on average over 3 fold - by addition of irradiated THY_{C+D} (Tx_{C+D}, eg. Fig 2). In addition to 'feeder' effects, this enhancement may also depend in part on specific antigen presenting cells.[2] However, the presence in both cell populations (see below) of radioresistant autonomous plasma cells (which are very susceptible to these 'feeder' effects) makes that difficult to dissect and confirm.

Isolation of plasma cells

In previous work, recovery of plasma cells (detected by staining for cytoplasmic Ig) was considerably enhanced by using dispase, and that applied to 'normal' tonsil as well as MG thymus.[2] Much of the spontaneous anti-AChR production survived 1,250 rads of irradiation* in at least 12 of the 16 cases tested for THY_{C+D}, and in 7 of 8 cases for 'Clumps'$_D$ (eg. fig 1). Such radioresistance is a well-known feature of plasma cells.[6] In sharp contrast, total IgG responses to pokeweed mitogen (PWM) were abolished by the same dose of irradiation (not shown). Another property shared with plasma cells is the depression of anti-AChR production by PWM, as is consistently seen with THY_{C+D} (eg. Fig 1).

---

* Responses of irradiated cells may be underestimates; culture conditions appear not to be optimal because of the inevitable cell death.

Immunohistological Studies

Immunohistological studies in a range of cases showed that the major sites of B cell localisation were the environs of the GC, and these appeared identical to those seen normally in tonsils.[5] They contained FD cells stained by the McAb RFD_3, Ig complex material, RFB_6+ plasmablasts, and some OKT_4^+ T cells. They were surrounded by a corona of RFB_4^+ B cells with some OKT_4^+ T cells, and they were embedded in zones resembling the lymph node paracortex (which we have never seen in non-MG adult control thymus). These contained many T cells, interdigitating cells and high endothelial venules: between these cells there was intense deposition of fibronectin, forming a partial barrier round the GC, similar to that noticed previously with reticulin staining (eg. 1, 7). Finally, these T cell zones alternated with bands of hyperplastic medullary epithelial cells, which partially engulfed them and the GC. In addition to these apparent barriers, the FD cell processes, or other matrix intrinsic to the GC (or perhaps other elements yet to be identified),might restrict the dispersion of GC on mechanical disruption, and thus account for our functional data.

CONCLUSIONS

1. Ongoing activity in the human MG thymus is largely isolated in structures that tend to remain intact on mechanical disruption, but can be dispersed efficiently with dispase. One can take advantage of this property to enrich the cells they contain.
2. At the same time, the yield of plasma cells is much increased, as well as of cells with 'feeder' effects in the cultures.
3. Immunohistological studies show that cells of the B lineage are largely confined to the environs of the GC, though it is not clear exactly what 'barriers' isolate them there.
4. The strong inference that these isolated structures are the GC and include many plasma cells requires confirmation. Once plasma cells can be removed, it will be interesting to study the functions of the accompanying B and T cell subsets.

ACKNOWLEDGEMENTS

This work was supported by grants from the Medical Research Council, the Muscular Dystrophy Group of Great Britain and the Sir Jules Thorn Charitable Trust. We are grateful to Linda Calder and David Pole for skilled technical help.

REFERENCES

1. GK Scadding, A Vincent, J Newsom-Davis and K Henry. Acetylcholine receptor antibody synthesis by thymic lymphocytes: correlation with thymic histology. Neurology (Ny) 31: 935 (1981)
2. HNA Willcox, J Newsom-Davis and LR Calder. Greatly increased autoantibody production in myasthenia gravis by thymocyte suspensions prepared with proteolytic enzymes. Clin exp Immunol, 54: 378 (1983)
3. JH Humphrey and D Grennan. Isolation and properties of spleen follicular dendritic cells. In In vivo Immunology (ed by P Nieuwenhuis, AA van den Broek & MG Hanna) pp 823-837, Plenum Publishing (1982)
4. G Janossy, M Bofill, LK Trejdosiweicz, LW Poulter, M Chilosi & J Newsom-Davis. Cellular differentiation of lymphoid populations and their micro-environment in the human thymus. In The Thymus: Histopathology and Pathology. Current Topics in Pathology. (Ed. by HK Muller-Hermelink) Springer-Verlag, Berlin, Heidelberg, New York (1984).
5. JA Thomas, HNA Willcox and J Newsom-Davis. Immunohistological studies of the thymus in myasthenia gravis. J Neuroimmunol 3: 319 (1982)
6. T Makinodan, P Nettesheim, T Morita, CJ Chadwick. Synthesis of antibody by spleen cells after exposure to kiloroentgen doses of ionising radiation. J Cell Physiol, 69; 355 (1967)
7. MAB de Sousa. Reticulum arrangement related to the behaviour of cell populations in the mouse lymph node. Adv Exp Med Biol, 5: 49 (1969)

FOLLICULAR DENDRITIC CELLS ISOLATED FROM HUMAN TONSILS

Cécile Kinet- Denoël, Ernst Heinen, Daniel Radoux and Léon Simar

Institute of Human Histology - University of Liège
20, rue de Pitteurs,
B-4020 Liège Belgium

INTRODUCTION

Follicular dendritic cells (FDC) are peculiar cells only present in germinal centers and characterized by complex dendritic cytoplasmic processes extending between lymphoid cells (1-4). FDC are mainly studied in relation to their capacity to retain antigens (5-7). This retention apparently occurs in presence of corresponding antibodies via the C_3b or the Fc receptors (8, 9). Many data remain unclear concerning FDC such as their origin or the role they play in germinal centers. Many controversies appear in the literature about their surface antigens namely the presence of Ia or HLA-DR antigens. In order to contribute to the analysis of FDC we isolated them from human tonsils and adenoids (10) and determined their surface Ig, antigens and receptors. The present publication summarizes most of the results obtained on isolated FDC.

MATERIAL AND METHODS

The isolation procedure has been previously described (10). Briefly, lymph follicles were dissected from human tonsils under biomicroscopes and digested with low concentrations of collagenase, dispase and deoxyribonuclease (fig. 1). Cell separation was performed by sedimentation at 1xg on 1,5% bovine serum albumin in a phosphate saline buffer (pH 7.4).

Immunolabellings were realized according either to immunofluorescence or to immunogold labelling techniques (11). Reagents and antibodies also were previously described (11).

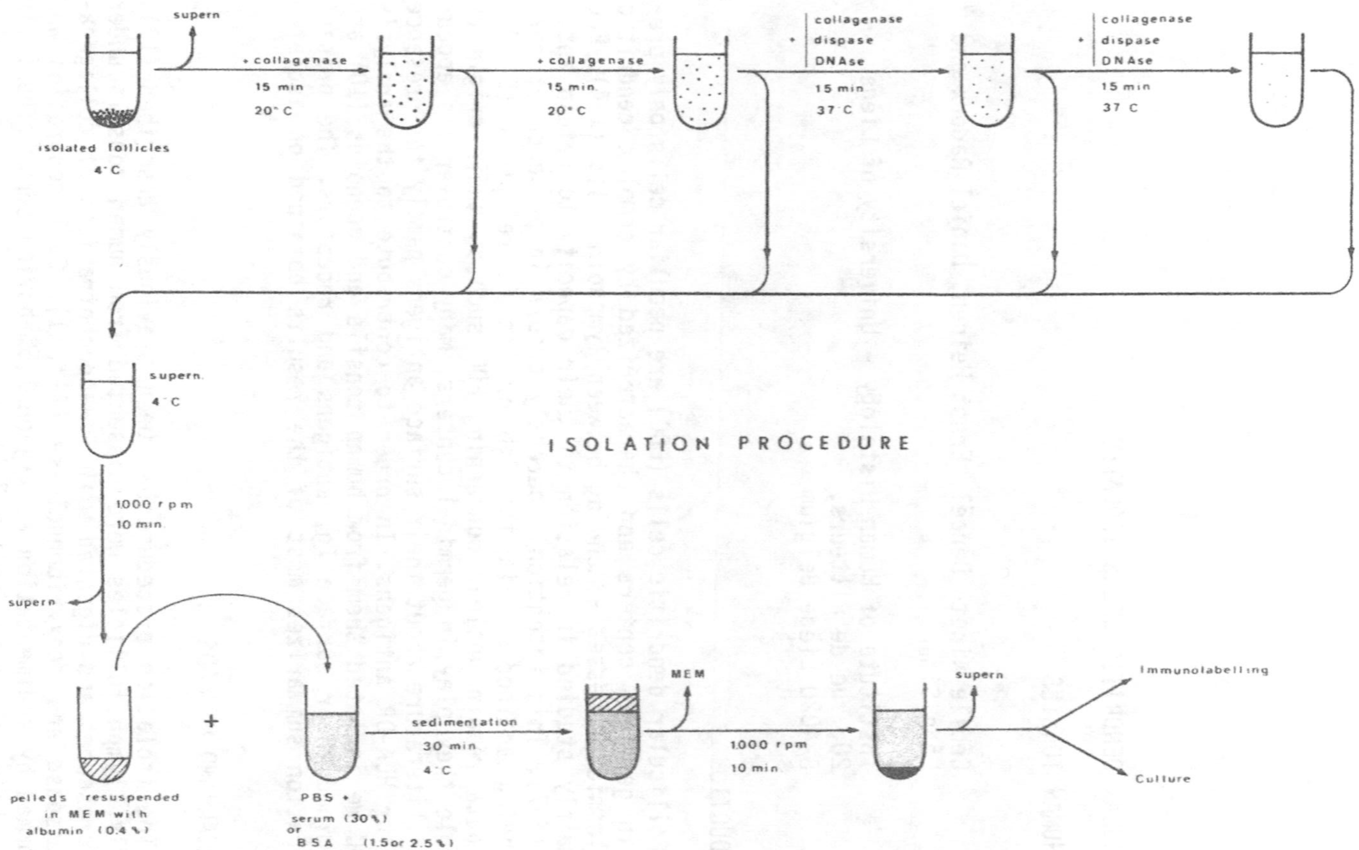

Fig. 1. Schema of the isolation procedure used to obtain FDC enriched cell suspension.

Demonstrations of the presence of Fc receptors were performed with homologous gold labelled anti-tetanus toxoid antibodies on one hand and with protein A on the other (12).

5-nucleotidase reaction was made according to Wachstein and Meisel (1957) using AMP as substrate (13).

RESULTS

FDC appeared as spherical clusters after isolation (fig. 2).

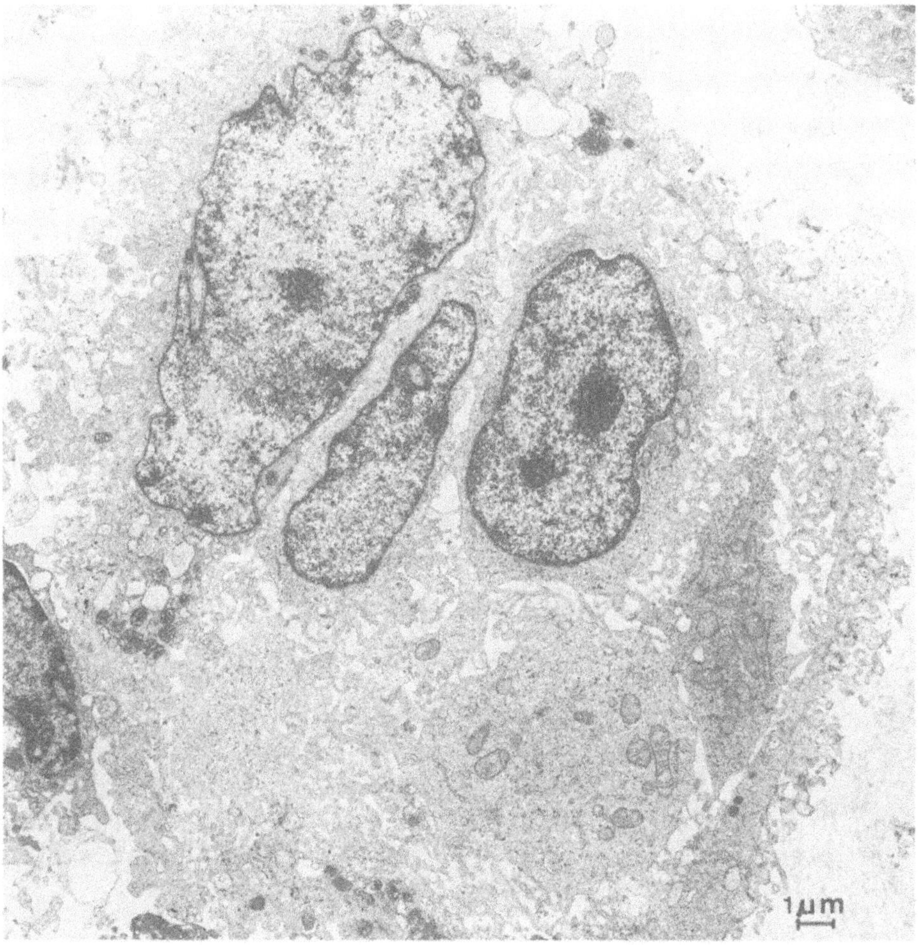

Fig. 2. Ultrastructural aspect of an isolated FDC. It surrounds with its cytoplasmic extensions lymphoid cells.

Their cytoplasmic extensions, covered with dendrites, surrounded lymphoid cells and were sometimes joined together by desmosome-like junctions. Isolated FDC bore one or two nuclei containing dispersed chromatin and a rim of dense chromatin along the nuclear envelope whose large lamina densa was easily observed.

Cells in contact with FDC mainly were B-lymphocytes (fig. 3): centrocytes or centroblasts were enveloped and also found at the periphery of the FDC clusters. These B cells were peanut lectin (PNA) positive and reacted with anti-Pan-B cell antibodies and with anti-$C_{3}b$ receptor antibodies. T-lymphocytes, OKT_3, OKT_4 and rarely OKT_8 positive, were mostly in contact at the periphery. Some macrophages could also be found in close vicinity to FDC.

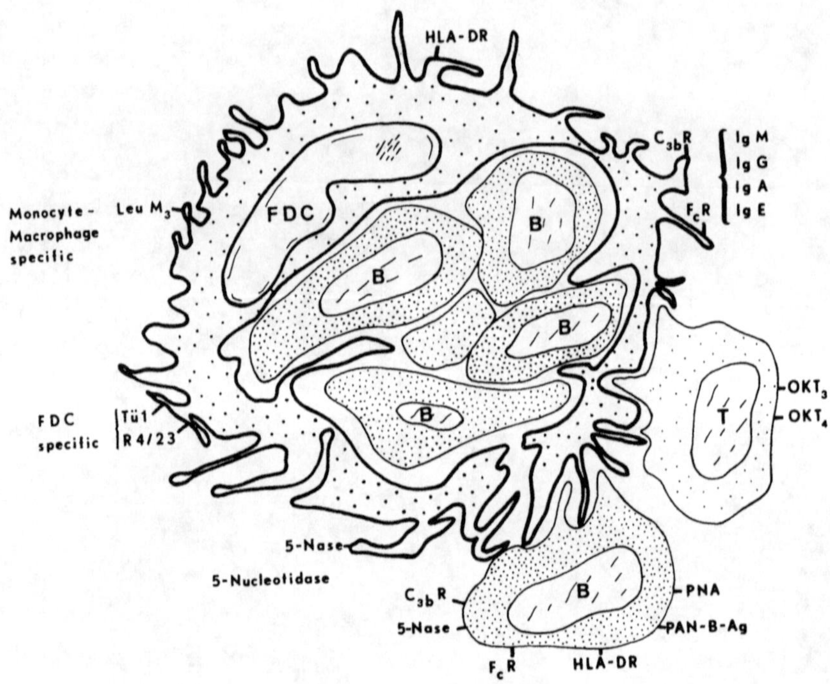

Fig. 3. Schematic representation of a FDC, after isolation. Cells enveloped or in contact with FDC and different surface features are precised.

As represented on figure 3 isolated FDC bore specific FDC surface antigens recognized by R4/23 and TU1 antibodies. They shared some antigens in common with macrophages (Leu-M_3 and HLA-DR antigens). Using anti-isotype specific antibodies we determined that FDC retained on their surface IgM, IgA, IgG and IgE but no IgD.

With the aid of antibodies directed against the C3b receptor we demonstrated the presence of C3b receptors on FDC clusters. These last retained homologous immune complexes (constituted of anti-tetanus toxoid antibodies labelled with colloidal gold and tetanus toxoid) (fig. 4). This retention occured even in absence of serum indicating so the presence of Fc receptors. This was confirmed with inhibition tests performed with protein-A (12).

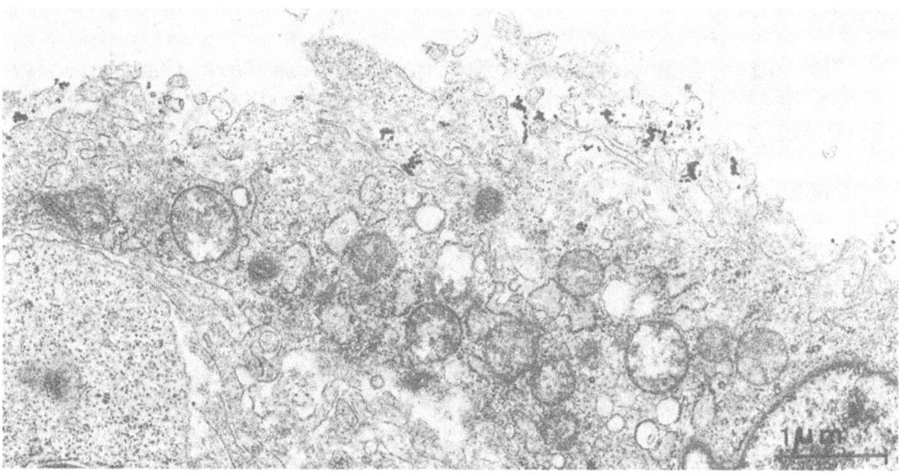

Fig. 4. High magnification of a FDC cytoplasmic portion retaining gold labelled immune complexes (anti-tetanus-toxoid-tetanus toxoid) in absence of serum.

5-nucleotidase reaction was positive at the surface of FDC membranes but also on some lymphoid cells.

DISCUSSION

Isolated FDC envelope lymphoid cells and thus appear as clusters. This aspect seems to correspond to functional entities already present in lymph follicles and described by Curran et al. (14). Their nuclear and cytoplasmic characteristics are the same as those observed in situ (10). FDC are mostly in contact with B lymphocytes but also with T-helper cells, this in accordance with other authors (15). These contacts and the presence of HLA-DR antigens may indicate that FDC act as accessory cells inside the germinal centers. Clear knowledges about their role are however lacking; they may play a part in the induction of B-memory cells

(16) or during antibody secretion (17). But FDC may also regulate the homing and the proliferation of B-cells in secreting immunomodulatory substances (Heinen et al. unpublished results). This secretion activity may be in relation with the presence of C_3b and Fc receptors found in great quantity on FDC membranes.

The presence of 5-nucleotidase at the surface of FDC was already shown by other authors (13); its function remains unclear.

Many features concerning FDC have to be elucidated but the isolation of FDC which conserve their relations to lymphoid cells and their surface antigens and receptors will be a valuable tool to further improve our knowledges concerning these cells.

ACKNOWLEDGMENTS

The authors are grateful to the "Fonds de la Recherche Scientifique Mẹdicale" for their generous financial support (Grand nr 3.4518.81).

REFERENCES

1. M.G. Hanna, and A.K. Szakal. J. Immunol. 101: 949-962 (1968).
2. Z.L. Herd, and G.L. Ada. Aust. J. Exp. Biol. Med. Sci. 47: 63- 72 (1969).
3. P. Groscurth. Path. Res. Pract. 169:212-234 (1980).
4. C. Kinet-Denoël, E. Heinen, D. Radoux and L.J. Simar. Intern. J. Radiat. Biol. 42: 121-130 (1982).
5. R.G. White, in: "The immunological competent cell: its nature and origin", CIBA foundation Study group n°16, p.6, eds., GEW Wolstenhome and J. Knight, Churchill (London) (1963).
6. G.J.V. Nossal, G.L. Ada, C.M. Austin, and J. Pye. Immunol. 9: 349-358 (1965).
7. D. Radoux, E. Heinen, C. Kinet-Denoël, E. Tihange, and L.J. Simar. Cell. Tissue. Res. 235: 267-274 (1984).
8. M. Papamichail, C. Gutierrez, P. Embling, P. Johnson, E.J. Holborow, and M.B. Peys. Scand. J. Immunol. 4: 343-347 (1975).
9. D. Radoux, C. Kinet-Denoël, E. Heinen, J. De Mey, M. Moermans, and L.J. Simar. Submitted for publication (1984).
10. C. Lilet-Leclercq, D. Radoux, E. Heinen, C. Kinet-Denoël, J.O. Defraigne, M.-P. Houben-Defresne, and L.J. Simar. J. Immunol. Meth. 66: 235-244 (1984).
11. E. Heinen, C. Lilet-Leclercq, D.Y. Mason, H. Stein, J. Boniver, D. Radoux, C. Kinet-Denoël, and L.J. Simar. Europ. J. Immunol. 14: 267-273 (1984).
12. E. Heinen, D. Radoux, C. Kinet-Denoël, J. De Mey, and L.J. Simar. Submitted for publication (1984).

13. H.K. Müller-Hermelink. Virchows Arch. B Cell Path. 16: 371-378 (1974).
14. R.C. Curran, J. Gregory, and E.L. Jones. J. Pathol. 136: 307-332 (1982).
15. E.C. Butcher, and I.L. Weissman, in: "Fundamental Immunology" pp. 109-127, W.E. Paul, ed., Raven Press, New York (1984).
16. G.G.B. Klaus, J.H. Humphrey, A. Kunkl, and D.W. Dongworth. Immunol. Rev. 53: 3-28 (1980).
17. J.G. Tew, R.P. Phipps, and T.E. Mandel. Immun. Rev. 53: 175-201 (1980).

FOLLICULAR DENDRITIC CELLS IN HUMAN GERMINAL CENTRES AND LYMPHOMATOUS FOLLICLES; A MORPHOLOGICAL COMPARISON

L.H.P.M. Rademakers, J.P.J. Peters, J.M.M. Roelofs and J.A.M. van Unnik

Institute of Pathology and Department of Medical Physics, University of Utrecht, Utrecht, The Netherlands

INTRODUCTION

Follicular Dendritic Cells (FDC) are non-lymphoid cells which are able to trap immune-complexes. FDC are exclusively present in germinal centres of lymphoid organs and form a meshwork with their extensions within reactive germinal centres. Though their precise function is unknown it is assumed that FDC are involved in the regulation of the germinal centre reaction. FDC are also a constituent of the follicles of follicular lymphomas, the neoplastic counter part of germinal centres (1).
So far no information about the role of these cells in neoplastic follicles is available. To obtain an impression of the functional state of FDC in follicular lymphoma we compared the distribution and morphology of FDC in normal and neoplastic germinal centres at the light and electron microscopic level.

MATERIAL AND METHODS

Human lymphe node biopsies of cervical and axillary lymphe nodes with follicular hyperplasia and follicular centroblastic-centrocytic lymphoma were selected on the basis of histopathological diagnosis. Parts of these biopsies were processed for electron microscopy using routine procedures (2).

The comparison of the ultrastructure of FDC was based on electron micrographs of 100 FDC from both groups.

The following morphometric methods were applied: 1) The relative volume of FDC within 3 follicular structures per biopsy was measured on 5 electron micrographs of each follicular structure by means of a point counting analysis. 2) The number of nucleated FDC per unit area was determined lightmicroscopically using a light microscope connected to an image analysing computer. Five follicular structures of each biopsy were investigated in this way.

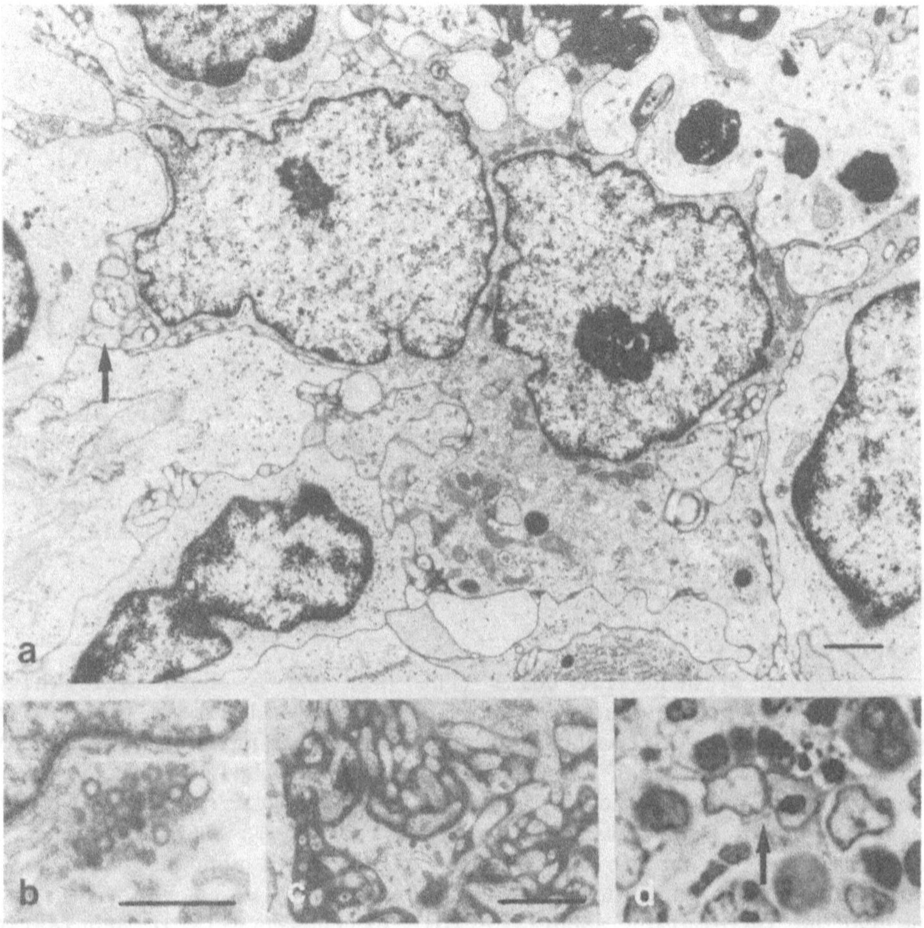

Fig. 1. Normal germinal centre. a Binucleated FDC with abundant cell organelles. Numerous membrane infoldings are located at the periphery of the cell (arrow). b Detail of cluster of ringlinke structures. c Detail of labyrinthine membrane infoldings. d Adjacent 1 um sections; the same FDC of a is recognized on basis of nuclear characteristics (arrow). Bar: 1 um.

3) The frequency of binucleated FDC was determined light microscopically in 1 um sections of at least 300 FDC of each biopsy. A three dimensional model of the typically arranged nuclei of FDC was made by reconstruction of serial 0.5 um sections (3).

RESULTS AND DISCUSSION

Ultrastructurally the elongated cell bodies of FDC and their cytoplasmic extensions are randomly distributed within the normal germinal centres. The indented, euchromatic nuclei have narrow marginal chromatine condensations and contain a distinct central

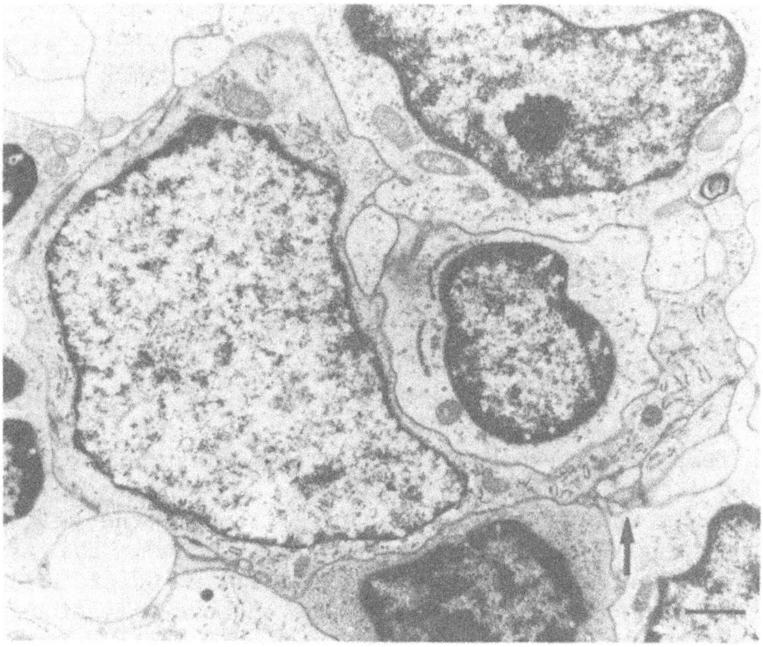

Fig. 2. Neoplastic germinal centre; FDC with scarcity of cell organelles. Labyrinthine membrane infoldings (arrow) are not numerous. Note the absence of electron opaque precipitate in between. Bar: 1 um.

nucleolus (Fig. 1a). Some cells contain two nuclear sections, separated by a small cleft. In the cytoplasm numerous dilated cisterns of rough endoplasmic reticulum (RER) are present containing electron opaque secretory material. Prominent Golgi fields are often located in a nuclear indentation or near the cleft separating the nuclei. In binucleated FDC up till four centrioles are observed in this area. Some FDC contain a cluster of ringlike structures (Fig. 1b). At the periperhy of cell bodies and extensions groups of small labyrinthine processes are present extending between the lymphoid cells. These villous extensions are covered with an electron opaque substance, probably representing fixed immune complexes (Fig. 1c). Cell bodies and extensions of neighbouring FDC are connected by desmosomes. Occasionally, desmosomes are observed within the cytoplasm with no apparent connection with the cell membrane. Lightmicroscopically the same FDC could be identified in the adjacent 1 um section. The pale nuclei of FDC clearly differ by their shape and nuclear contour from other cell types (Fig. 1d). Both mono- and binucleated forms of FDC can be recognized light microscopically in normal as well as in neoplastic germinal centres. Multinucleated forms with more than two nuclear sections are very rare. Observations on isolated FDC confirm that in general FDC are at most binucleated (3,4). Three dimensional

Table 1. Morphometric data on FDC in normal and neoplastic germinal centres (mean value and standard deviation)

	Normal germinal centres	Neoplastic germinal centres
Relative volume (%)	13 ± 4 (n=10)	4 ± 2 (n=9) *
Cell number/0.02 mm^2	49 ± 15 (n=13)	8 ± 3 (n=11)*
Proportion of binucleated cells	26 ± 7 (n=13)	9 ± 6 (n=11)*

* Significant differences (P 0.001)

reconstruction of the typical nuclear doublets in serial 0.5 um sections reveal that these doublets are composed of separate nuclei and are not connected by any bridges.

In neoplastic germinal centres the ultrastructure of FDC is different (Fig. 2). Their cell processes are coarser in shape and the cytoplasm appears more concentrated towards the cell bodies. Nuclear contours are more regular. In general, the cytoplasm contains few cisterns of RER, occasionally Golgi fields and moderate numbers of polyribosomes. The villous extensions of the plasmalemma are larger and less frequent. Their extracellular electron dense coating is consistently absent.

Morphometric analysis of the occurrence of FDC was performed at the light and electron microscopic level (Table 1). The mean relative volume of FDC cell bodies and extensions in reactive germinal centres is 13%. In neoplastic germinal centres this value amounts to 4%. In addition, determination of the number of FDC per unit area of follicular tissue resulted in a substantial lower number of FDC in neoplastic germinal centres. The proportion of binucleated FDC was much lower in neoplastic germinal centres.

Calculation on a stereological basis of the actual frequency of binucleated cells in the tissue, based on the data obtained from tissue section and the dimensions of three dimensional models of the nuclear doublets indicated that in normal germinal centres a large proportion (51-68%) of FDC is binucleated, whereas in neoplastic germincal centres this figure varies between 18 and 23%.

The results show that FDC are present in all follicular lymphomas studied. For several reasons (2) FDC in neoplastic germinal centres cannot be regarded as remnants of preexisting lymphoid tissue. Properties maintained by germinal centre cells may be responsible for the attraction and differentiation of precursor cells in FDC. The degree in which these properties are maintained by the neoplastic cells may differ from normal germinal centre cells and may result in lower numbers of FDC.

The poor cytoplasmic organization of the cytoplasm, the infrequent infoldings of the plasmalemma and the low proportion of binucleated cells suggest that in neoplastic follicles FDC have a low activity. Differences in functional state between FDC in normal and neoplastic germinal centres also indicate that microenvironmental factors, influencing the activity of FDC, are lacking in neoplastic germinal centres. These factors might be mediated by the germinal centre cells or by the accompanying immune reaction following antigenic stimulation. In follicular lymphoma an immune reaction resulting in antibody production is absent. Immune complexes are not detectable in neoplastic follicles by immunocytochemistry (5) or as electron opaque precipitations on the plasmamembrane of FDC. Our observations on FDC in follicular lymphoma are in line with observations made in developing germinal centres which suggest that FDC are stimulated by immune complexes to form numerous membrane infoldings (6).

The majority of FDC in normal germinal centres appears to be binucleated. Binucleated cells may be formed by cell fusion or by nuclear division without cytokinesis. In the material studied we never observed mitotic activity in FDC. In addition, the observation of intracytoplasmic desmosomes strongly suggests that these cells are formed by cell fusion. In normal germinal centres this process might be facilitated by the enlarged cell surface of FDC promoting a higher incidence of cell contacts between mutual FDC. The involvement of C_3b receptors in the fusion process seems to be important, because fixed immune complexes also might promote cell fusion during simultaneous receptor-mediated endocytosis of neighbouring cells as suggested for macrophages (7).

In conclusion, the morphology of FDC in follicular lymphoma reflects a low degree of differentiation of these cells, which may be related to an altered microclimate in the neoplastic germinal centres. A more detailed study of the morphology of FDC may be of value in the histopathology of diseases in which abnormal germinal centre reactions occur.

REFERENCES

1. K. Lennert, H. R. Niedorf, Nachweis von desmosomal verknupften Reticulumzellen follikulären Lymphom (Brill-Symmers), Virchows Arch. (Cell.Pathol.) 4: 148-150 (1969).
2. L. H. P. M. Rademakers, J. P. J. Peters, J. A. M. van Unnik, Histiocytic and dendritic reticulum cells in follicular structures of follicular lymphoma and reactive hyperplasia; a quantitative electron microscopic analysis, Virchows Arch. (Cell. Pathol.) 44: 85-98 (1983).
3. J. P. J. Peters, L. H. P. M. Rademakers, J. M. M. Roelofs, D. de Jong and J. A. M. van Unnik. Distribution of dendritic retiulum cells in follicular lymphoma and reactive hyperplasia; light microscopic identification and general morphology, Virchows Arch. (Cell. Pathol.) in press (1984).

4. L. H. P. M. Rademakers, J. P. J. Peters, D. M. D. S. Go, R. A. de Weger and J. A. M. van Unnik, Morphology and enzyme histochemistry of isolated human follicular dendritic cells, Cell.Biol.Int.Rep. 8: 185 (1984).
5. J. Gerdes, H. Stein, Complement (C_3) receptors and dendritic reticulum cells of normal and malignant lymphoid tissue, Clin.Exp. Immunol. 48: 348-352 (1982).
6. M. G. Hanna, A. K. Szakal, Localization of ^{125}I-labelled antigen in germinal centres of mouse spleen: histological and ultrastructural autoradiographic studies of the secondary immune reaction, J.Immunol. 101: 949-962 (1968).
7. T. J. Chambers, W. G. Spector, Inflammatory giant cells, Immunobiol. 161: 283-289 (1982).

THE NATURE OF THE FOLLICULOID STRUCTURES IN EOSINOPHILIC LYMPHFOLLICULOID GRANULOMA (KIMURA'S DISEASE)

S. Takasaki, H. Tanaka*, K. Takaki,
S. Fukunaga, and E. Ishikawa

Department of Pathology, and * Division of Morphology
Bio-Medical Research Laboratories, The Jikei University
School of Medicine, Tokyo, Japan

INTRODUCTION

Eosinophilic lymphfolliculoid granuloma is a rare disease only seen in the orient. It usually occurs in the soft tissue of the head, neck, and extremities accompanied by hyper IgE-emia and eosinophilia. The main histopathologic features are proliferation of folliculoid structures and granulation tissue with marked infiltration of eosinophils and mast cells. The first report was documented by Kimura et al. in 1948.[1] Since then, many similar cases have been reported, but its etiology is still unknown.[2,3] The folliculoid structures of this disease are morphologically similar to the secondary follicles of the lymph node, but their shape and size are frequently irregular. The margin between the mantle zone (corona) and the germinal center(GC) is sometimes unclear, and small lymphocytes in the mantle zone are frequently found invading into the GC. The light and dark regions, which are regularly arranged in the secondary follicle of the lymph node, lose their polarity.
The helper and suppressor T cells are fairly intermingled in the mantle zone and the GC. In general, as the bulk of folliculoid structure increases, the number of intermingled T lymphocytes decreases.[4] Primary follicle as seen in the lymph node is not found in this lesion. From these findings, it is suggested that the folliculoid structures in this disease might develop in a different way from the secondary follicle in the lymph node. We will mainly report in this paper, how folliculoid structures are formed, and how dendritic reticulum cells(DRC) develop from the granulation tissue in this disease.

MATERIALS AND METHODS

Seven cases typical for Kimura's disease were studied. They were all males, ranging from 13 to 63 years of age.

For histologic study, the specimens were fixed in buffered formalin, embedded in paraffin, and stained with hematoxylin-eosin, periodic acid-Schiff, modified Masson trichrome (Noguchi-Matsumoto), and silver staining. More than 300 serial paraffin sections, each 5 μm thick, were stained by silver impregnation for the stereoangiograms. Immunohistochemical studies were made on two cases with fresh frozen sections. DRC, B cell and B2 were detected by using the monoclonal antibodies for DRC(DAKO-DRC1), B cell(DAKO-pan-B), and B2(COULTER CLONE) for the premature B cell with ABC kit(Vector Inc). IgE was detected by using the anti IgE antibody(TAGO Inc) with the peroxidase-anti-peroxidase(PAP) method. The human $NK\text{-}1^+$ cells were detected by Leu-7(Becton Dickinson Inc) with indirect immunoperoxidase method in the paraffin sections. Electron microscopic observations were made in 4 cases fixed in buffered 1.2% glutaraldehyde at 4°C for 4 hours and post-fixed in 1% osmic acid for 3 hours. After dehydration, specimens were embedded in Epon 812. Ultra-thin sections were doubly stained with uranyl acetate and lead citrate, and observed with the HS-9 electron microscope.

RESULTS

Folliculoid structures of this disease appeared in the granulation tissue, usually accompanying the reticular meshwork with abundant blood vessels, most of which were similar to postcapillary venule(PCV). These structures resembled the paracortical tissue in the lymph node (Fig.1,2). GC were well developed in the folliculoid structures. The ratio of GC to folliculoid structure was $0.55 \pm 0.06 (n=6)$, and larger than that of the lymph node; $0.49 \pm 0.12 (n=5)$. In GC, DRCs were abundant, but the large and medium-sized centroblasts were rather scarce, compared to the lymph node. Figures 3a,b show the distribution of the DRC in GC of this disease(a), and of the lymph node(b). The cytoplasm and the cell process of DRC were much conspicuous in (a). Boundaries between the mantle zone and the GC were frequently unclear (Fig.2). It depends on the irregular arrangement of the reticular cell process, which are lightly seen under the electron microscope, lying and spanning in a lamellar fashion at the mantle zone. Multinucleated giant cells, which resemble Warthin-Finkeldey cells, were frequently found in GC (Fig.4). They are syncytial cells having 3 to 10 round nuclei, sized 5 to 10 μm each, with small clear nucleolus. The cytoplasm was usually scanty, but one side of the cell border sometimes showed a linear contour with a certain connection to the reticular meshwork of the mantle zone. Ultrastructurally, they had euchromatic nuclei, and each of them somewhat interdigitated to each other. The scanty cytoplasm contained many small mitochondria and rough endoplasmic reticula of short sized. Golgi apparatus and microtubular structure

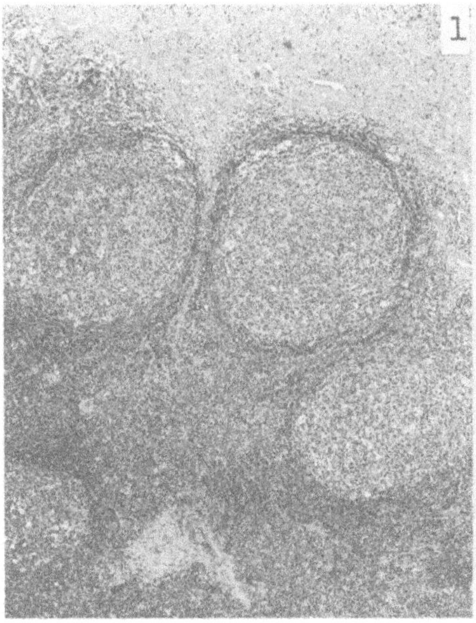

Fig.1 Folliculoid structures of this disease with well eveloped GC.

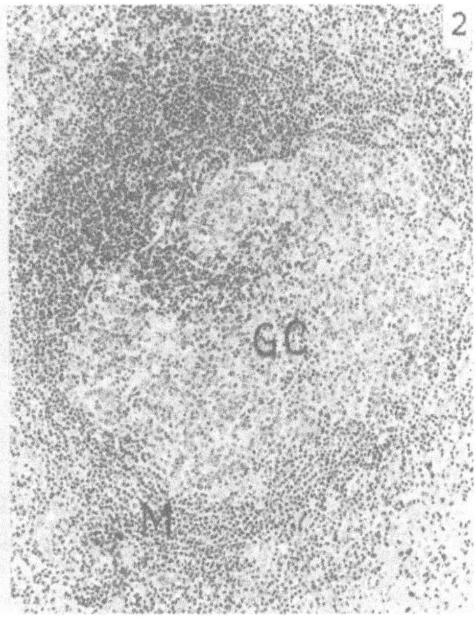

Fig.2 Boundaries between the mantle zone and the GC were frequently irregular.

were also well developed (Fig.5). At the straight-lined cell border, microfilamentous structures were arranged just under the plasma membrane (fig.6). They were similar to the dense patch seen in the reticulum cell. These cells had a connection to the adjoining DRC with a desmosome inbetween. It was considered that multinucleated giant cells were the same kind of DRCs, while they still maintained the nature of fibroblastic reticulum cells in part. A small number of multinucleated giant cells were also found in the granulation tissue, probably originating from fibroblastic reticulum cells.

The distributions of IgE, DRC, and B2 were examined immunohistochemically (Fig.7a,b,c). They were intensely positive in the GC. These reactions made sure that there were a lot of irregular and incomplete folliculoid structures in the lesion. In places, they also revealed the existence of very small folliculoid structures, which were thought to be the beginning of GC, sized less than 200 μm in diameter. They were present in the granulation tissues where several dilated capillaries were gathered together. In these areas, the reactions of DRC and IgE were not always simultaneously positive, while B cells showed positive reaction even to the small cluster of lymphocytes. The human NK-1^+ cells were distributed mainly in GC, though small number of them were also scattered in the granulation tissue (Fig.8). The number of positive cells per fixed area in GC

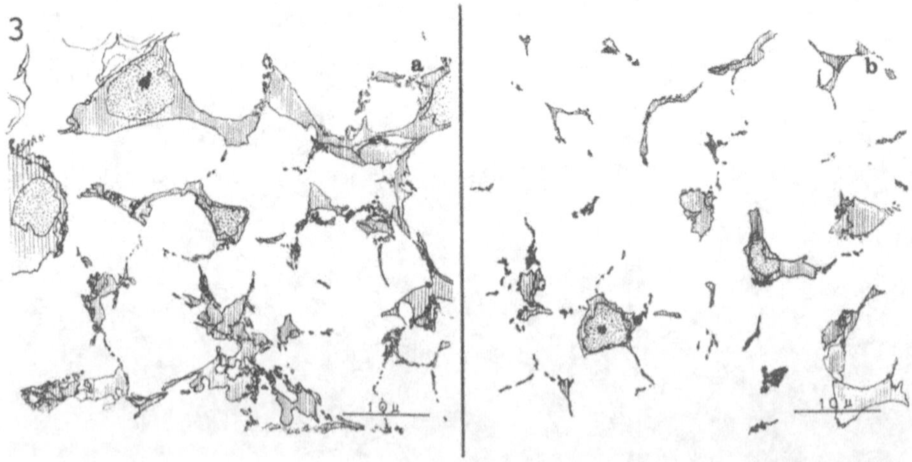

Fig.3a,b Distribution of the DRC in GC of this disease(a), and of the lymph node(b). The cytoplasm and the cell process of DRC were much conspicuous in (a).

were compared to Kimura's disease and the lymph node. The former was 1.91 ± 0.08(n=6), and this was higher than those of the lymph node; 1.23 ± 0.45(n=5), ($p < 0.05$).

The vascular distribution in folliculoid structures were reconstructed by serial sections (Fig.9). Abundant PCV-like vessels, which had high endothelial cells, distributed with many anastomoses to each other around folliculoid structure with smaller vessels branching out into GC. Each folliculoid structure had several blood supplies. At times, the PCV-like vessel itself entered into the mantle zone and even into the GC. Mature lymphocytes easily passed through the PCV-like vessel walls. In general, the vascular distribution was much more conspicuous to the folliculoid structures of irregular formed ones than well shaped ones. Onion skin-like appearances surrounding the capillary were noticed in some place and others among the granulation tissue (Fig.10). They were composed of fibroblastic cell process extending in a lamellar fasion reciprocally with the connective tissue (Fig.11,12). Some of those cells contained a few lymphocytes or lymphoblasts in the cytoplasm (Fig.13). They had a fairly large nuclei with marked nucleolus and abundant rough endoplasmic reticula in the cytoplasm. The outer cell border showed a linear contour and the dense patch was found just under the cell membrane. Microvilli were well developed and the tight junction was also noticed between the adjoining cells. Sometimes, they had trapped more than three lymphocytes encircling with their cytoplasmic process (Fig.14), and gradually they became similar to a DRC (Fig.15) with well developed microvilli.

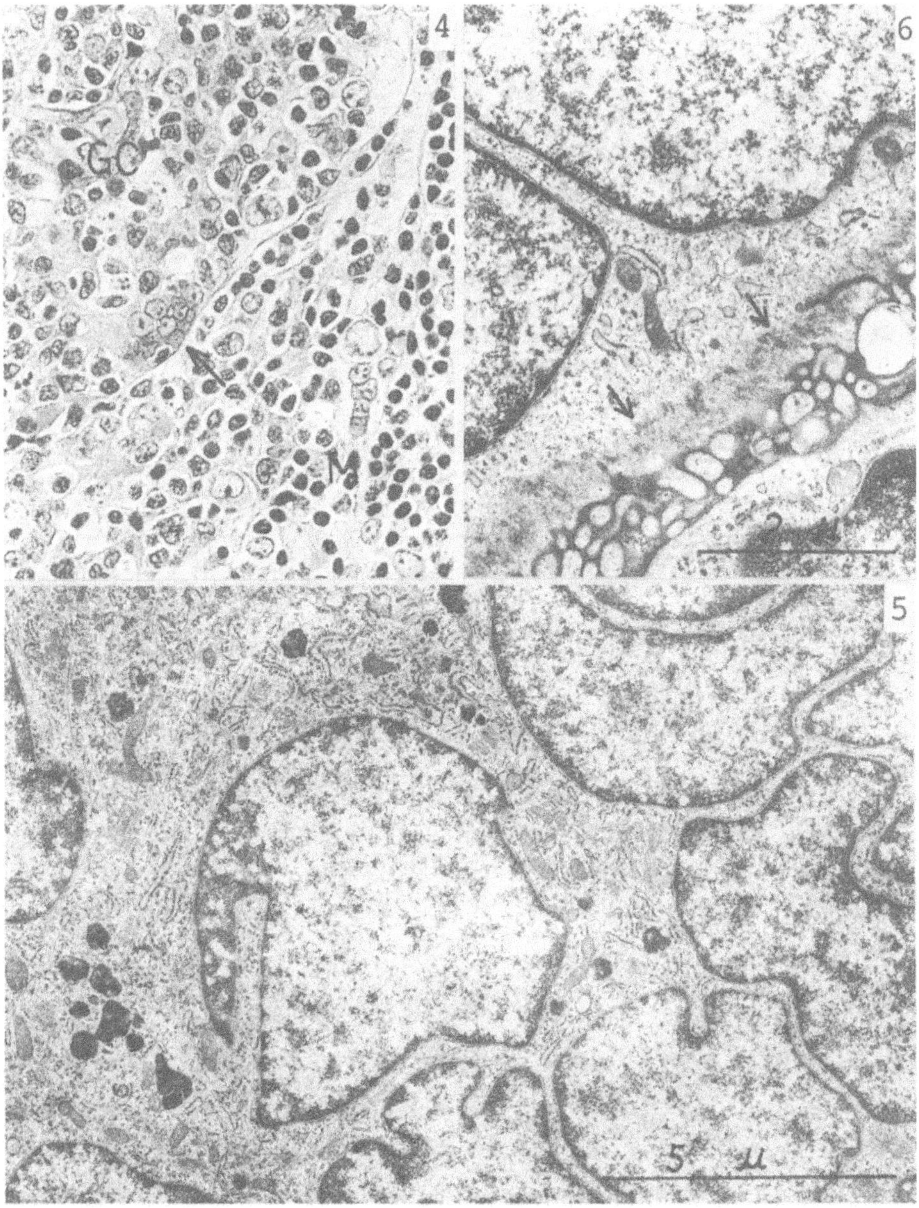

Fig.4 Multinucleated giant cell (arrow) in the germinal center(GC). One side of the cell border shows a linear contour with a certain connection to the reticular meshwork of the mantle zone(M).
Fig.5 Euchromatic nuclei of the giant cell. The scanty cytoplasm contains many small mitochondria and short rER.
Fig.6 At the straight-lined cell border, microfilamentous structure (arrow) arranged just under the plasma membrane.

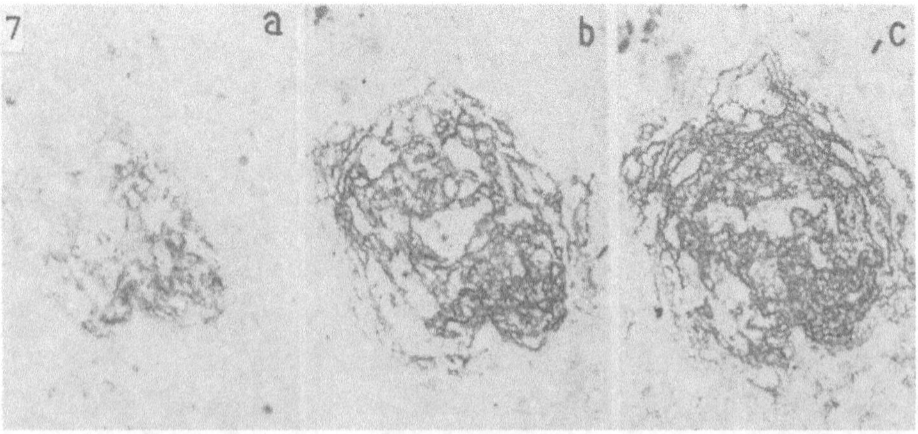

Fig.7 The distribution of IgE (a), DRC (b) and B2 (c) revealed the existence of irregular and incomplete folliculoid structures in the lesion.

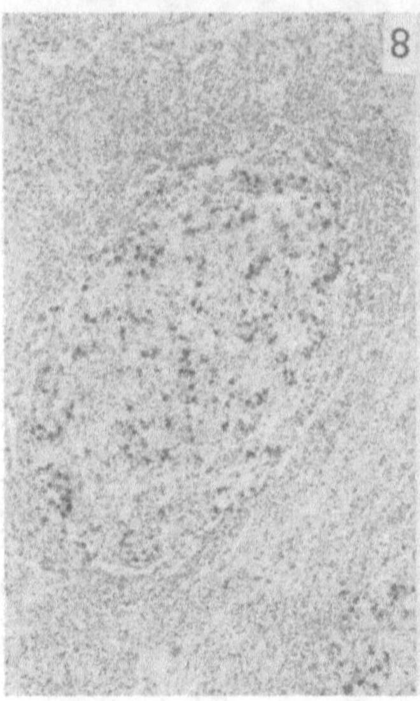

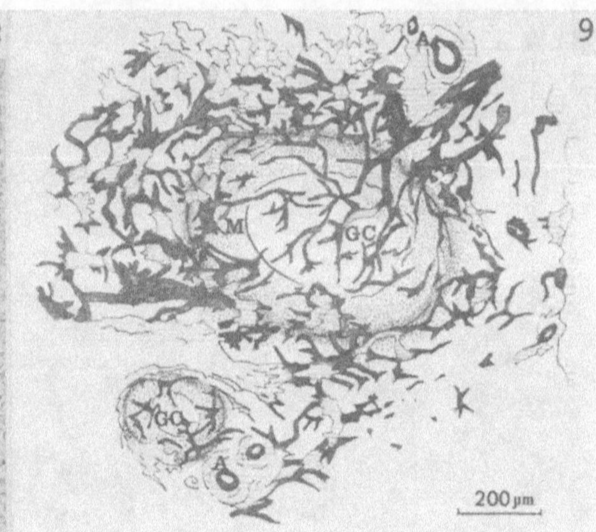

Fig.8 The human NK-1^+ cells were distributed in the germinal center in higher rate than that of the lymph node.

Fig.9 Vascular distribution in folliculoid structure reconstructed by serial sections. Abundant PCV-like vessels are distributed around folliculoid structure with smaller vessels branching out into germinal center.

GC: germinal center. M: mantle zone. A: arteriole.

≡ : connective tissue.

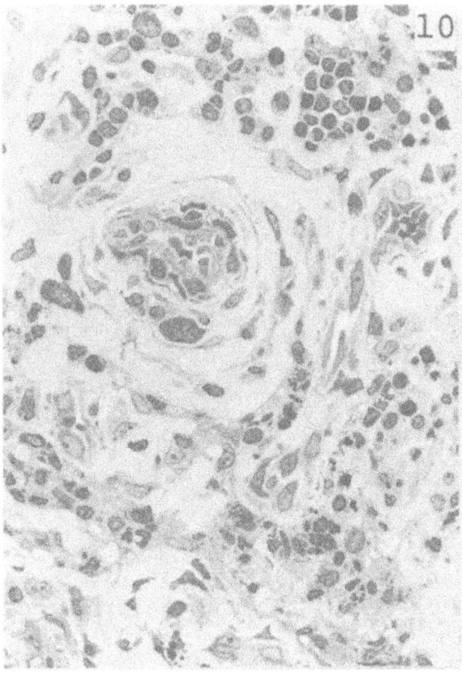

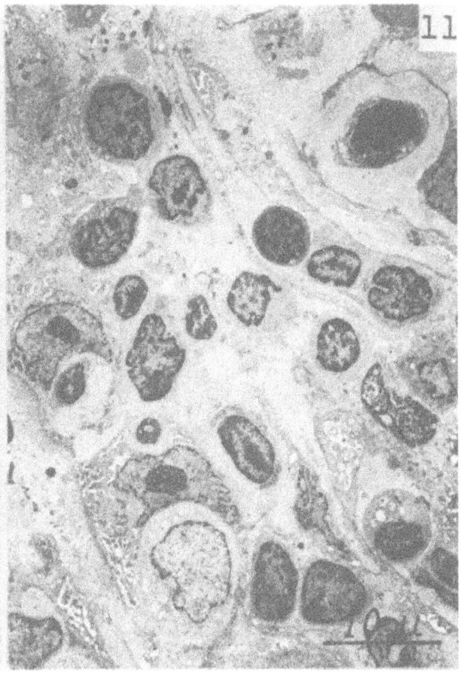

Fig.10 Onion skin-like appearance surrounding the capillary in the granulation tissue. (methylene blue section)
Fig.11 Electron microscopy of the onion skin-like lesion. Fibroblastic cell processes are extending in a lamellar fasion reciprocally with the connective tissue.

DISCUSSION

Folliculoid structures of Kimura's disease were examined, and some of their properties were clarified.
1). There were many folliculoid structures of irregular shape and size in the lesion, and many T lymphocytes were found in the mantle zone and the GC. 2). A small bulk of folliculoid structures had already contained GC-like portion and the GC did not occur from the primary follicle as a secondary follicle of the lymph node. 3). The ratio of GC to folliculoid structure was larger than that of the lymph node. 4). The GC was usually IgE positive showing a network pattern, and contained many HNK-1^+ cells. 5). The DRCs were well developed and abundant in the GC, but the centrocytes or centroblasts were rather scanty compared to the lymph node. 6). Multinucleated giant cells were frequently found in the GC. They had both features of DRC and fibroblastic reticulum cell.

Even though these folliculoid structures were formed under an abnormal condition, they gave us important information on the origin of DRC. Heusermann et al.[5] summerised the possibilities for the

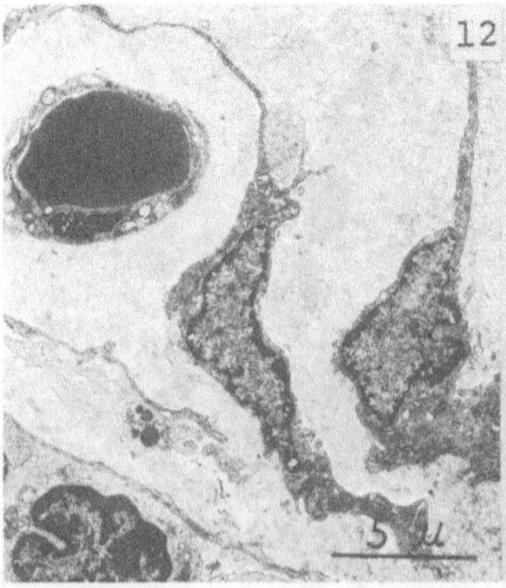

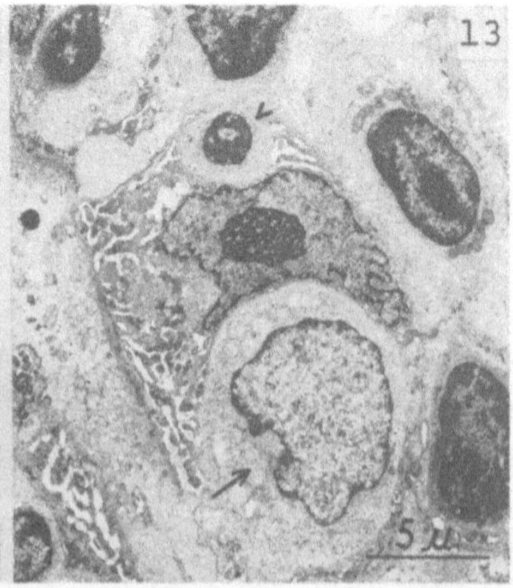

Fig.12 Fibroblastic cells surrounding the capillary.
Fig.13 Some of the fibroblastic cells contain lymphocyte(arrow head) or lymphoblast(arrow) in the cytoplasm.

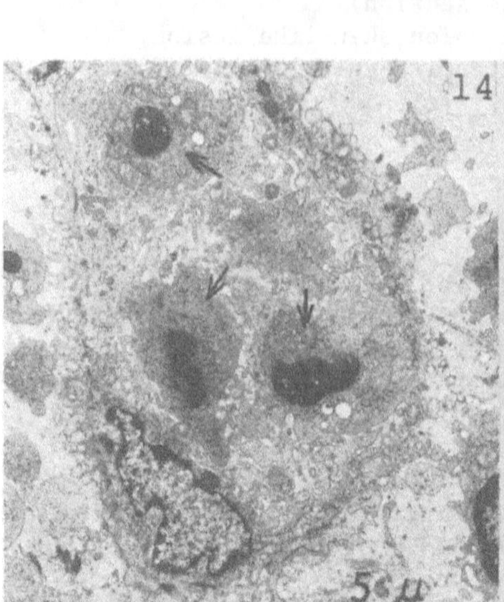

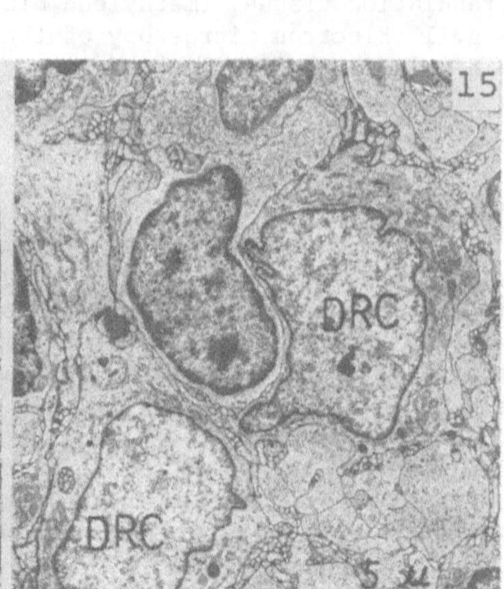

Fig.14 Fibroblastic cell trapping more than three lymphocytes (arrow) encircled with its cytoplasmic process.
Fig.15 DRC in the GC trapping lymphoblasts with its cytoplasmic process with well developed microvilli.

origin of DRC; " 1. Transformation of stroma-forming reticulum cells into DRC. 2. Multiplication of some resting DRC at the sites of prospective germinal centers. 3. Transformation of monocytes into DRC. 4. Immigration of DRC in to germinal centers."

As far as we have observed in these lesions, it was considered that fibroblastic cells transformed into DRC. Dijkstra et al. [6] indicated that the precursor DRC had a possibility of being able to trap B lymphocytes. Fig.13 and 14 might display such a possibility. It is also said that the development of microvilli was stimulated by trapping of antigen-antibody complex.[7] But, such a direct relationship could not clarified in this study. PCV-like vessels around folliculoid structure must be the main entrance of the mature lymphocytes in the lesion. [8] Irregular development of these vessels might be one of the causes in making folliculoid structures to be irregular forms. It is known that multinucleated giant cells appear in many conditions[9,10]; for example, in hyperplastic tonsil, Hodgkin's disease, and so on. Histochemical study has indicated that the nature of these giant cells was close to that of DRC. [11] We showed that these cells had features of both DRC and fibroblastic reticulum cell. A small number of multinucleated giant cells were also found in the granulation tissue. Some unknown factor, which stimulates the stromal cells, might be partaking in this disease.

CONCLUSION

As a result of having examined Kimura's disease, it is concluded that the folliculoid structures of this disease develop not from the primary follicle, but from the beginning as a folliculoid structure with GC-like portion in the granulation tissue. During this process, fibroblastic cells transform into DRC.

REFERENCES

1. T. KIMURA, S. YOCHIMURA, and E. ISHIKAWA, Abnormal Granuloma with Proliferation of Lymphoid Tissue, Tr. Soc. Pathol. Jpn. 37:179 (1948). (in Japanese)
2. H. J. HAMRICK, J. C. JENNETTE and C. F. LAFORCE, Kimura's Disease: Report of a Pediatric Case in the United States, J. Allergy Clin. Immunol. 73:561-566 (1984).
3. E. ISHIKAWA, H. TANAKA, S. KAKIMOTO, S. TAKASAKI, Y. KIRINO, A. SAKATA and M. SUZUKI, A Pathological Study on Eosinophilic Lymphfolliculoid Granuloma (KIMURA'S DISEASE), Acta Pathol.Jpn. 31:767-781 (1981).
4. K. TAKAKI, S. TAKASAKI, S. FUKUNAGA, H. YAMASHITA, H. TANAKA and E. ISHIKAWA, Immunological Study on the Distribution of T and B Lymphocytes in Kimura's Disease, J. Jap. Soc. R. E. S. 24:58 (1984) (in Japanese)
5. U. HEUSERMANN, K. H. ZURBORN, L. SCHROEDER and H. J. STUTTE, The Origin of the Dendritic Reticulum Cell, An Experimental Enzyme-Histochemical and Electron Microscopic Study on the

Rabbit Spleen, Cell Tissue Res. 209:279-294 (1980).
6. C. D. DIJKSTRA and E. A. DÖPP, Ontogenetic Development of T- and B- Lymphocytes and Non-Lymphoid Cells in the White Pulp of the Rat Spleen, Cell Tissue Res. 229:351-363 (1983).
7. E. W. A. KAMPERDIJK, E. M. RAAYMAKERS, J. H. S. de LEEUW and E. Ch. M. HOEFSMIT, Lymph Node Macrophages and Reticulum Cells in the Immune Response, I. The Primary Response to Paratyphoid Vaccine, Cell Tissue Res. 192:1-23 (1978).
8. A. J. FREEMONT and C. J. P. JONES, Light Microscopic, Histochemical and Ultrastructural Studies of Human Lymph Node Paracortical Venules, J. Anat. 136:349-362 (1983).
9. C. R. KJELDSBERG and HUN KIM, Polykaryocytes Resembling Warthin-Finkeldey Giant Cells in Reactive and Neoplastic Lymphoid Disorders, Human Pathology, 12:267-272 (1981).
10. J. GERDES, H. STEIN, D. Y. MASON and A. ZIEGLER, Human Dendritic Reticulum Cells of Lymphoid Follicles: Their Antigen Profile and Their Indentification as Multinucleated Giant Cells, Virchow Arch. [Cell Pathol]. 42:161-172 (1983).
11. H. WECKERLE, U. P. KETELSEN, and M. ERNST, Thymic Nurce Cells: Lymphoepithelial Cell Complexes in Murine Thymuses: Morphological and Serological Characterization, J. Exp. Med. 151:925-944 (1980).

NORMAL AND NEOPLASTIC LANGERHANS CELLS: PHENOTYPIC COMPARISON WITH OTHER TYPES OF MACROPHAGES

Elisabeth Ralfkiaer, H. Stein, N. Ralfkiaer, K. Hou-Jensen and D.Y. Mason

Nuffield Dept. of Pathology, John Radcliffe Hospital, Oxford, UK, and Dept. of Pathology The Finsen Institute, Copenhagen, Denmark

SUMMARY

The antigenic properties of normal epidermal Langerhans cells and their neoplastic counterparts (histiocytosis X cells) were compared with those of other types of macrophages, by immunohistochemical staining of skin and lymph node biopsies with a panel of mono-clonal antibodies. The phenotype of Langerhans cells, histiocytosis X cells, dermal macrophages and interdigitating reticulum cells were similar (HLA-DR^{+}, OKT6$^{+/-}$, Leu3$^{+/-}$, C3-receptor^{-}, Ig-complex^{-}, R4/23^{-}, MO2^{-}), and differed markedly from those of follicular dendritic cells (HLA-DR^{-}, OKT6^{-}, Leu3^{-}, C3 receptor^{+}, Ig^{+}, R4/23^{+}, MO2^{-}), and histiocytic reticulum cells (HLA-DR$^{+/-}$, OKT6^{-}, Leu3$^{+/-}$, C3-receptor^{+}, Ig-complex^{+}, R4/23^{-}, MO2$^{+/-}$). This indicates that Langerhans cells are related to the interdigitating reticulum cells found in the T cell areas of lymphoid organs, and demonstrates that Langerhans cells do not express C3b, C3bi or C3d receptor molecules. Our data also emphasise that the Leu3 antigen is not restricted to cells of the T cell lineage.

INTRODUCTION

Langerhans cells belong to a system of cells which play an antigen-processing, accessory role in the expansion of T cells (1,2), and are considered related to the interdigitating reticulum cells found in the T cell areas of lymphoid organs (3, 4). The relationship of Langerhans cells to other types of macrophages is uncertain (5).

In this study the phenotype of Langerhans cells and their neoplastic counterparts (histiocytosis X cells) have been compared with those of other types of macrophages by immunohistochemical staining of frozen sections with a panel of monoclonal antibodies.

MATERIALS AND METHODS

Tissue Specimens

Frozen biopsies from normal skin, benign inflammatory skin disorders, cutaneous lesions of histiocytosis X and lymph nodes with benign follicular hyperplasia were obtained from the Dept. of Pathology, Finsen Institute.

Processing of Frozen Biopsy Material

Six um cryostat sections were lyophilised at -20°C for 2-4 hr, fixed in acetone for 10 min. and processed for staining (6).

Antibodies

The monoclonal antibodies used in this study are listed in Table 1. A polyclonal rabbit antiserum against C3b, C3bi and C3d receptor molecules was prepared as described elsewhere (7).

Immunoenzymatic Staining Procedures

Sections were stained by 2- or 3-stage immunoperoxidase methods (6,8). Double staining for the detection of antigens on non-overlapping, or partly overlapping populations of cells was performed by a sequential peroxidase/alkaline phosphatase technique (9,10), using the following combinations of monoclonal antibodies: To15 & R4/23; To15 & To5; To15 and anti-Ig; To15 & anti-HLA-DR; To15 & PD7/26; anti-Leu4 & anti-Leu3; anti-Leu3 & R4/23; R4/23 & anti-Leu3 (lymph node biopsies), and anti-HLA-DR & OKT6 (skin biopsies).

Immunofluorescent Staining Procedures

Coexpression of antigens was detected by double fluorescence staining using the following combinations of fluorescein (F), and phycoerythrin (P) conjugated antibodies: F-anti-Leu4 & P-anti-Leu3; F-OKT6 & P-anti-Leu3; F-OKT6 & P-anti-HLA-DR; F-anti-HLA-DR & P-anti-Leu3.

Immunological Control Preparations

Negative controls were performed by omitting the primary antibodies. Positive controls were performed by staining normal

Table 1. Monoclonal Antibodies Used in This Study

Antibody:	Specificity:	Reference and/or Source:
PD7/26	Leucocyte common antigen	DAKO
To15	Pan B cell	DAKO
Anti-Kappa	Kappa light chains	BD
Anti-Lambda	Lambda light chains	DAKO
Lyt3	Pan T cell (E rosette receptor)	NEN
Anti-Leu4	Peripheral T cells	BD
Anti-Leu3a	T helper/inducer cells	BD
OKT6	Cortical thymocytes Langerhans cells	11; ORTHO
CR3/43	HLA-DR	Mason D.Y. (unpublished)
Anti-HLA-DR	HLA-DR	BD
To5	C3b receptor	13; DAKO
R4/23	Follicular dendritic cells	14; DAKO
EB11	Tissue macrophages	27
FMC32	Tissue macrophages	12
KB90	Tissue macrophages	Mason D.Y. (unpublished)
Mo2	Monocytes	CC
Ap-7/6/7	Alkaline phosphatase	9; DAKO

+DAKO = Dakopatts; BD = Becton Dickinson; NEN = New England Nuclear; ORTHO = Ortho Diagnostics; CC = Coulter Counter.

thymus, hyperplastic tonsils, and/or malignant lymphomas of known antigenic phenotypes. The specificity of double labelling was estimated by comparing cellular reactivity patterns seen in double and single stained serial sections.

RESULTS AND DISCUSSION

Normal epidermal Langerhans cells, histiocytosis X cells, dermal macrophages and interdigitating reticulum cells in the T cell areas of lymph nodes demonstrated similar phenotypic properties, whereas the phenotypes of follicular dendritic cells in B cell areas, and histiocytic reticulum cells in the sinuses of lymph nodes were markedly different (Table 2).

Table 2. Expression of Antigens by Dendritic and Histiocytic Cells in Skin and Lymph Nodes

Cell Type:	Antigens: LC*	Mac-Ag	HLA-DR	T6	T4	FDC-Ag	C3R	Ig	Mo-Ag
LaC**	+	+	+	+	+	–	–	–	–
HX	+	+	+	+	+	–	–	–	–
DM	+	+	+	+/–	+	–	–	–	–
IRC	+	+	+	+/–	+	–	–	–	–
FDC	+	+	–	–	–	+	+	+	+
HRC	+	+	+/–	–	–	–	+	+	+

Abbreviations:

*LC = leucocyte common; Mac-Ag = macrophage-associated antigen; FDC-Ag = FDC-associated antigens; C3R = C3 receptor (C3b, C3bi, C3d); Ig = Ig complexes; Mo-Ag = monocyte-associated antigen; **LaC = Langerhans cells; HX = histiocytosis X cells; DM = dermal macrophages; IRC = interdigitating reticulum cells; FDC = follicular dendritic cells; HRC = histiocytic reticulum cells.

These data demonstrate that neoplastic Langerhans cells retain a pattern of antigen expression identical with that of their normal counterparts. The observed cross-reactivity of both these cell types with antibody against T helper/inducer cells (anti-Leu3) is in agreement with data from others (15,16), and emphasises that Leu3 expression is not restricted to cells of the T cell lineage. The implication of this finding in relation to subset characterisation of mixed T/macrophage infiltrates is obvious.

The close phenotypic resemblance between Langerhans cells, dermal macrophages and interdigitating reticulum cells strongly indicates that these cell types are closely related. This is supported by experimental, ultrastructural, enzyme cytochemical and immunological data obtained in other studies: i) Langerhans cells are migratory cells which take up and transport antigens from skin to lymph nodes, and localise in close apposition to interdigitating reticulum cells in paracortical areas (3,4); ii) Langerhans cells possess Birbeck granules, but are otherwise indistinguishable from interdigitating reticulum cells, and subsets of HLA-DR-positive dermal macrophages (so-called "indeterminate cells") on ultrastructural examination (3,17,18); iii) The enzyme cytochemical characteristics of Langerhans cells (weak acid phosphatase and strong ATPase reaction) are similar to those of interdigitating reticulum cells, and differ markedly from those of histiocytic macrophages (19); iv) Langerhans cells and interdigitating reticulum cells both express the S-100 protein, whereas follicular dendritic cells and histiocytic reticulum cells are devoid of this protein (20,21).

In a recent study it has been proposed that HLA-DR positive dermal macrophages are not homogeneous, but comprise "indeterminate cells" ($OKT6^+$, $HLA\text{-}DR^+$) and histiocytic macrophages ($OKT6^-$, $HLA\text{-}DR^+$) (22). This suggestion was made on the basis of immunoelectron microscopic examination with a limited number of antibodies. However, our data do not support this suggestion, since the latter population of OKT6 negative, HLA-DR positive dermal macrophages differed markedly from histiocytic reiculum cells with respect to their antigenic properties (Table 2). Further studies of dermal macrophages by immuno-electron microscopy with more extensive panels of antibodies, and/or by in vitro functional assays are clearly necessary.

It has recently been demonstrated that epidermal Langerhans cells are unreactive with monoclonal antibody against the C3bi receptor (23). The lack of labelling of Langerhans cells with monoclonal anti-C3b receptor and polyclonal anti-C3b, C3bi and C3d receptor antibodies observed in this study (Table 2) extends this finding, and suggests that the rosetting of Langerhans cells with complement-coated erythrocytes (EAC rosettes) described recently (24,25) is not due to the presence of C3 receptors on Langerhans cells. In this context, it should be emphasised that the anti-C3 receptor antibodies used in this study reliably identify C3 receptors on a number of different cell types (e.g. erythrocytes, neutrophils, follicular dendritic cells, B cells, glomerular cells) (8,13,26, and this study). The possibility that Langerhans cells may express C3 receptors different from C3b, C3bi and C3d requires isolation and molecular characterisation of the receptor responsible for EAC binding of Langerhans cells.

The present study emphasises the utility of in situ immunohistochemical staining techniques in the characterisation of cells which are not readily accessible for in vitro isolation and analysis.

Acknowledgments

This work was supported by the Danish Cancer Society and the Leukaemia Research Fund, UK.

REFERENCES

1. W.B. Shelly and L. Juhlin, Nature 261: 46 (1976).
2. G. Stingl, S.I. Katz, E.M. Shevach, A.S. Rosenthal and I. Green, J. Invest. Dermatol. 71: 59 (1978).
3. E. Rausch, E. Kaiserling and M. Gross, Virchows Arch. (Cell. Pathol) 25: 327 (1977).
4. I. Silberberg-Sinakin, G.J. Thorbecke, R.L. Baer, A.S. Rosenthal and V. Berezowsky, Cell. Immunol. 25: 137 (1976).
5. G.J. Thorbecke, I. Silberberg-Sinakin, T. Flotte, J. Invest. Dermatol. 75: 34 (1980).
6. H. Stein, J. Gerdes, H. Schwab, H. Lemke, D.Y. Mason, A. Ziegler, W. Schienle and V. Diehl, Inter. J. Cancer 39: 445 (1982).
7. J. Gerdes, M. Naiem, D.Y. Mason and H. Stein, Immunology 45: 645 (1982).
8. H. Stein, J. Gerdes, G. Tolksdorf and V. Klatt, Scand. J. Immunol. 13: 67 (1981).
9. J.L. Cordell, B. Falini, W.N. Erber, A.K.. Ghosh, Z. Abdulaziz, S. MacDonald, K.A.F. Pulford and D.Y. Mason, J. Histochem. Cytochem. 32: 219 (1984).
10. D.Y. Mason, H. Stein, M. Naiem and Z.Abdulaziz, J. Cancer Res. Clin. Oncol. 101: 13 (1981).
11. E. Fithian, P. Kung, G. Goldstein, M. Rubenfeld, C. Fenoglio and R. Edelson, Medical Sciences 78: 2541 (1981).
12. W.W. Hancock, H. Zola and R.C. Atkins, Blood 62: 1271 (1983).
13. J. Gerdes, M. Naiem, D.Y. Mason and H. Stein, Immunology 45: 645 (1982).
14. M. Naiem, J. Gerdes, Z. Abdulaziz, H. Stein and D.Y. Mason, J. Clin. Path. 36: 167 (1983).
15. J.D. Boss, J.H. Sillevis Smitt, S.R. Kreig, P.M. Bakker, G.D. Vos and D.J. van Zaane, J. Cut. Pathol. 11: 59 (1984).
16. G.S. Wood, N.L. Warner and R.A. Warnke, J. Immunol. 131: 212 (1983).
17. G. Rowden, T.M. Philips and M.G. Lewis, Br. J. Dermatol 100: 531 (1979).
18. G. Rowden, Br. J. Dermatol. 97: 593 (1977).
19. K. Lennert, N. Mohri, H. Stein, E. Kaiserling and H.K. Muller-Hermelink, In: Malignant Lymphomas Other than Hodgkin's Disease, Springer Verlag, Berlin, 1978. p. 51.
20. T. Nakajima, S. Watanabe, Y. Sato, Y. Shimosato, M. Motoi and K. Lennert, Gann 73: 429 (1982).

21. K. Takahashi, H. Yamaguchi, J. Ishezeki, T. Nakajima and Y. Nakazato, Virchows Arch. (Cell. Path.) 37: 125 (1981).
22. G.F. Murphy, A.K. Bhan, T.J. Harrist and M.C. Mihm, Br. J. Dermatol. 108: 423 (1983).
23. K.A. Haines, T.J. Flotte, T.A. Springer I. Gigli and G.J. Thorbecke, Proc. Nat. Acad. Sci. USA 80: 3448 (1983).
24. K. Burke and T. Gigli, J. Invest. Dermatol. 75: 46 (1980).
25. G. Stingl, E.C. Wolf-Schreiner W.J. Pichler, E. Gschnait and W. Knapp, Nature 268: 245 (1977).
26. J. Gerdes and H. Stein, Clin. Exp. Immunol 48: 348 (1982).
27. E. Bliss, M. Naiem, K. Bell and J.O.D. McGee, J. Clin. Pathol. (In Press).

Ia^+T10^+ BLAST CELLS IN ACUTE LYMPHOBLASTIC LEUKAEMIA ARE PRECURSORS OF T^+ AND B^+ CELLS*

Ana K. Miljković, Katarina Isaković, Tatjana Matejič and Branka Stojanović

Immunology Research Center, Vojvode Stepe 458, and Institute of Pathophysiology, Dr Subotica 1 11000 Belgrade, Yugoslavia

Acute lymphoblastic leukaemia (ALL) is a lymphoproliferative disorder characterized by a block of cell maturation at very early stages of development. Immunological studies of ALL provided evidence that there are three types of ALL: T, B and non-T, non-B[1,2]. Furthermore, it was demonstrated that non-T, non-B ALL blasts are a heterogenous population of cells as concerns of their maturity rate. Those cells differentiate *in vitro* into T and B lymphocytes[3]. Therefore *in vitro* differentiation studies of non-T, non-B ALL blasts, as well as of lymphoid blasts in acute crisis of chronic myelocytic leukaemia (AML) provided a useful model for evaluating early progenitors of haemopoietic cells[4].

This report deals with the *in vitro* differentiation of ALL blasts of two patients with non-T, non-B ALL. ALL blasts were of a mixed phenotype since some of the blasts expressed only T10 and Ia antigens, while a large number were negative for all cell markers studied. It was demonstrated that the blasts *in vitro* differentiated into T and B cell lineages, and cells bearing M1 antigen specific for myelomonocytic cells.

MATERIALS AND METHODS

Cells and Culture Procedures

Peripheral blood lymphocytes of two patients with ALL were

*Supported by the Republic of Serbia Research Fund, Belgrade.

used to study the in vitro differentiation of ALL blasts. Using May-Gruenwald and Giemsa staining it was shown that more than 90% of the cells harvested from the interphase after Ficoll-Isopaque gradient centrifugation exibited the morphology of blasts. The blasts expressed only Ia and T10 antigens as it was demonstrated by a panel of cell markers. Washed cells were suspended in RPMI 1640 containing 10% of human AB serum, 100 IU/ml of Penicillin, 50 µg/ml of Streptomycin and L-glutamin in a concentration of 2 mM. $12x10^6$ viable cells/3 ml were placed into 35x10 mm culture dishes and incubated at 37°C in a water-saturated atmosphere of 5% CO_2 in the air. One third of the cultures were incubated in the presence of 50 µl/ml of lipopolisaccharide (LPS), another third with crude thymus protein extract (CTPE) of MW greater than 30 kD (30 µg/ml), while the remaining cultures served as unstimulated controls. Cells from all cultures were harvested on days 3, 5, 8 and L3 (patient No 1) and 3, 6 and 10 (patient No 2), and used for all marker analysis.

Cell Marker Analysis

The following cell markers were examined: receptors for SRBC (E receptors) and MRBC (M receptors), T (pan-T), T3, T4, T6, T8, T9, T10, Ia, M1, CALLA and B (pan-B) antigens, membrane Ig (mIg) and cytoplasmic IgM (cIgM) and IgG (cIgG). E receptors (E^+ cells) and M receptors (M^+ cells) were determined by microrosette tests. Pan-T and pan-B antigen-bearing cells (T^+ and B^+ cells) were detected by indirect immunofluorescence using heterologous antisera. Cells bearing antigens recognized by monoclonal antibodies (T3, T4, T6, T8, T9, T10, Ia, M1 and CALLA) were identified by indirect immunofluorescence by an appropriate monoclonal antibody. For detection of mIg-bearing (mIg^+ cells) and cIgM- and cIgG-containing cells ($cIgM^+$ and $cIgG^+$ cells) $F(ab)_2$ fragments of polyvalent human anti-Ig and anti-µ and anti-γ sera conjugated with FITC were employed.

Double Marker Assays

Indirect immunofluorescence combined with immune rosette formation[5] was used for simmultaneous detection of two antigens expressed on the same cell.

RESULTS

Results of cell marker analysis of peripheral blood lymphocytes from two ALL patients are shown in Table 1. It is evident that 24.4% and 22.8% of the cells from patient No 1 expressed T10 and Ia antigens, respectively. In patient No 2 T10 antigen was expressed on 51.8% and Ia antigen on 74.0% of the cells. Since more than 90% of the cells in the tested samples were blasts it is clear that T10 and Ia antigens were present on the membrane of ALL blasts. Cells bearing

Table 1. Different Cell Markers Expressed on Peripheral Blood Lymphocytes of ALL Patients

Patient	Lymphocyte subsets (%)															
No	E^+	M^+	T^+	$T3^+$	$T4^+$	$T6^+$	$T8^+$	$T9^+$	$T10^+$	Ia^+	B^+	$cIgM^+$	$cIgG^+$	mIg^+	$M1^+$	$CALLA^+$
1	0.1	2.7	2.3	3.3	0	0	0.7	1.5	<u>22.4</u>	<u>22.8</u>	5.0	3.8	0	1.7	2.3	n.d.
2	0.4	7.0	0	0	0	0	0	8.5	<u>51.8</u>	<u>74.0</u>	0	0	1.0	4.2	2.6	0.3

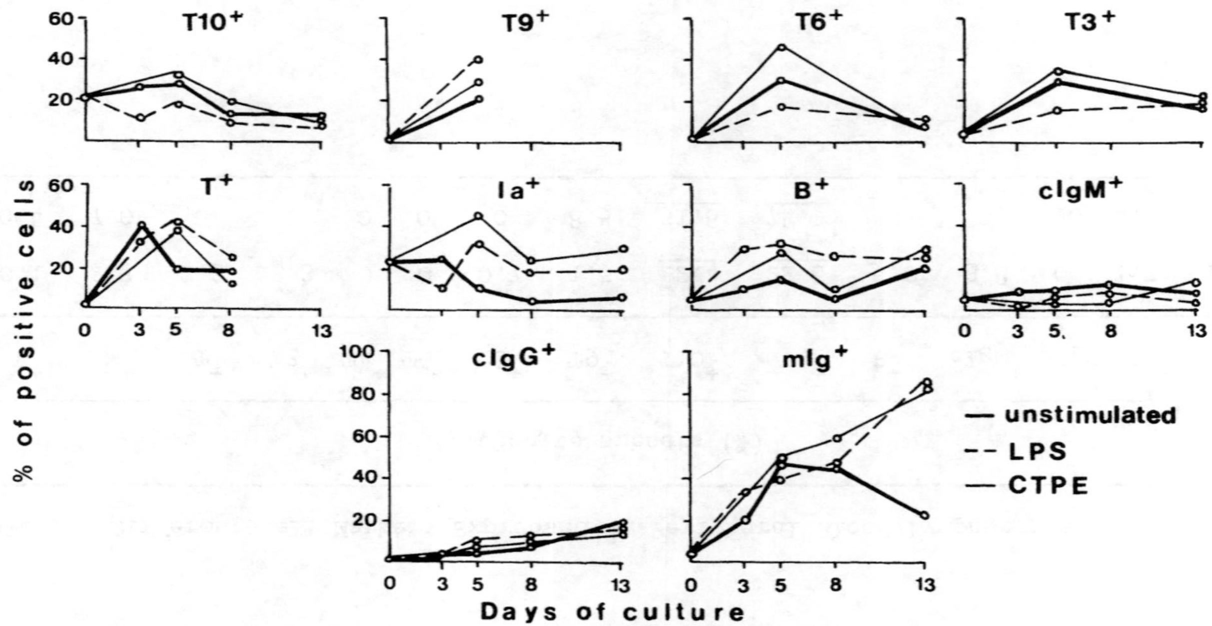

Fig. 1. Occurrence of lymphocytes with different membrane markers in unstimulated cultures and cultures stimulated with LPS and CTPE.

other markers specific for T and B lymphocytes, and myelomonocytic cells were found in a low number. These results also indicate that there is a rather large population of blasts negative for all markers ("null" blasts) in both patients, particularly in patient No 1.

In 3- and 5-day-old cultures of cells from patient No 1 (Fig. 1) T^+, $T3^+$, $T6^+$, $T9^+$, $T10^+$, Ia^+, B^+ and mIg^+ cells showed a significant increase, whereas the presence of only a few $cIgM^+$ and $cIgG^+$ cells were detected. After 8 and 13 days there was an increase in the number of B^+, $cIgM^+$ and $cIgG^+$ cells and a decrease of cells bearing T-specific and early lymphocyte differentiation markers (T10 and T9). LPS and CTPE had no influence on the appearance of cells with T and B characteristics.

In vitro differentiation of ALL blasts from patient No 2 is shown in Fig. 2. It has to be underlined that after an initial decrease of the number of $T10^+$ cells observed on days 3 and 6, the number of these cells increased on the 10th day, and about 40% of the cells were $T10^+$. A slow decrease in the number of Ia^+ cells was noted throughout the incubation period. Cells expressing all other markers increased from the begining of culture and on the 10th day, 30-40% of the cells had the characteristics of T lymphocytes, about 30% exibited B cell characteristics and more than 50% expressed M1 antigen. Again CTPE failed to influence in vitro differentiation of ALL blasts. Using double marker technique, on the 7th day of incubation, it was demonstrated that M1 antigen was not present on T^+ and B^+ lymphocytes, while Ia antigen was expressed on 3.7% of B^+ cells and 1.2% of T^+ cells (Table 2). Only 0.8% of the cells bore both pan-T and pan-B antigens.

Table 2. Lymphocyte Subpopulations Revealed by Double Marker Assay in the 7-day-old Culture of ALL Cells of Patient No 2 Stimulated with CTPE

Cells (%)				
T^+B^+	T^+M1^+	T^+Ia^+	B^+M1^+	B^+Ia^+
0.8	0	1.2	0	3.7

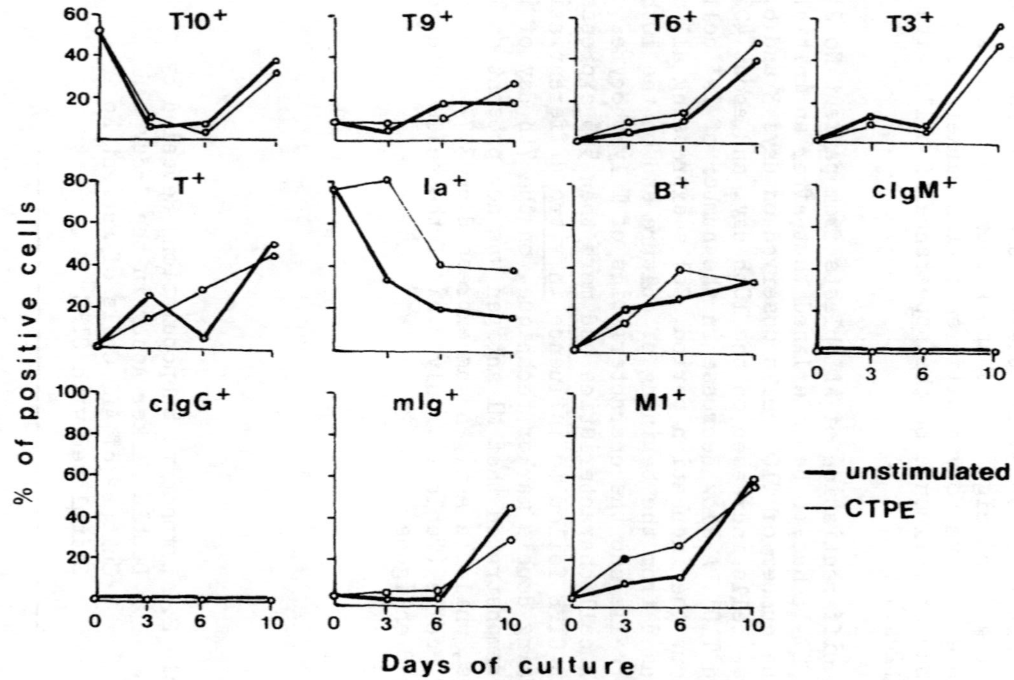

Fig. 2. Occurrence of lymphocytes with different membrane markers in unstimulated cultures and cultures stimulated with CTPE.

DISCUSSION

In this study we demonstrated that leukaemic blasts of two patients with non-T, non-B ALL differentiated *in vitro* spontaneously into cells bearing membrane markers specific for T and B lymphocytes (blasts of patient No 1) and T and B lymphocytes and myelomonocytic cells (blasts of patient No 2). There are data showing *in vitro* differentiation of non-T, non-B ALL blasts after stimulation with phorbol esters[7] and in diffusion chamber[3]. In our patient No 1 22.4% and 22.8% of the blasts expressed T10 and Ia antigens respectively, while the majority of the blasts were negative for all markers studied ("null" blasts). In patient No 2 ALL blasts were of the same membrane phenotype, but the number of Ia^+ cells was significantly greater than the number of $T10^+$ blasts, indicating the presence of a $T10^-Ia^+$ population of blasts. "Null" blasts were also present in patient No 2, but in a smaller number. It is known that Ia antigens are present on pluripotent haemopoietic stem cells and early progenitors of all haemopoietic cell lines[8]. As for T10 antigen, it is expressed on the surface of the progenitors of both T and B lymphocytes[9,10]. Therefore, on the basis of these results we may assume that $T10^+Ia^+$ blasts found in non-T, non-B ALL may be a candidate for a common precursor of T and B lymphocytes. When cells from patient No 1 were cultivated for 13 days, cells of T and B lineages were generated (we did not search for the generation of $M1^+$ cells), while in the culture of cells from patient No 2 cells with T and B specificities were found together with cells bearing M1 antigen. Double marker assays provided strong evidence that T, B and myelomonocytic cell markers were expressed on separate cell lines. Therefore, we feel that the population of blasts from patient No 2 contained precursors of lymphoid and myelomonocytic cells. However, from these results it is not clear whether "null" blasts represent a unique population of cells, or a heterogeneous population of cells containing cells at different stages of differentiation. Also, from these results it could not be concluded whether within "null" blasts there are cells which are common precursors of lymphoid and myeloid cells as it is observed in acute crisis of CML[4]. Moreover, "null" blasts in our ALL patients are very similar to the progenitors of a long term bone marrow culture described in mice[11] which are able to differentiate into all haemopoietic cells including T and B lymphocytes. Nevertheless, on the basis of our results we may speculate that "null" blasts defined in our experiments may represent the "target" for leukaemic alteration as suggested by Janossy et al.[12].

SUMMARY

Leukaemic blasts isolated from peripheral blood of two patients with non-T, non-B ALL were cultivated for 10-13 days and cell markers were assayed on day 0 and during culture period. Using a panel of cell markers it was demonstrated that the populations of ALL blasts of both patients were composed of $T10^+Ia^+$ and "null" blasts. These blasts in the culture differentiated into three cell lines: T and B lymphocytes and myelomonocytic cells. The results thus obtained suggested that within the population of ALL blasts of these two patients there are progenitors of lymphoid and myelomonocytic cells ("null" blasts) and common precursors of T and B lymphocytes ($T10^{+}Ia^{+}$ blasts). *In vitro* differentiation of ALL blasts was not affected by LPS and CTPE.

Acknowledgments

We are especially indebted to Prof.dr Branislav D. Janković for his helpful comments and encouragement throughout this study.

REFERENCES

1. L. Borella, and L. Sen, T cell surface markers on lymphoblasts from acute lymphoblastic leukemia, *J.Immunol.* 111:1257 (1973).
2. M. F. Greaves, G. Brown, N. Rapson, and T. A. Lister, Antisera to acute lymphatic leukemia cells, *Clin.Immunol. Immunopathol.* 4:67 (1975).
3. G. Jäger, B. Lau, K. Pachmann, and P. Dömer, Acute lymphoblastic leukaemia cells in the diffusion chamber: change of surface phenotype during culture periods, *Brit.J. Haematol.* 52:319 (1982).
4. M. F. Greaves, W. E. Verbi, B. R. Reaves, A. V. Hoffbrand, H. C. Draysdale, L. Jones, L. S. Sacker, and I. Samaratunga, "Pre-B" phenotypes in blast crisis of Ph^1 positive CML: evidence for a pluripotential stem cell "target", *Leukemia Res.* 3:181 (1979).
5. J. D. Griffin, R. P. Beveridge, and S. F. Schlossman, Isolation of myeloid progenitor cells from peripheral blood of chronic myelogenous leukemia, *Blood* 60:30 (1982).
6. T. W. LeBien, F. J. Bollum, G. Yasmineh, and J. H. Kersey, Phorbol ester-induced differentiation of a non-T, non-B leukemic cell line: model for human lymphoid progenitor cell development, *J.Immunol.* 128:1316 (1982).
7. K. Agasawa, and T. W. Mak, Phorbol esters induce differentiation in human malignant T lymphoblasts, *Proc.Natl. Acad.Sci.* 77:2964 (1980).
8. M. P. Bodger, C. A. Izaguirre, H. A. Blocklock, and A. V. Hoffbrand, Surface antigenic determinants on human pluripotent and unipotent hemopoietic progenitor cells, *Blood* 61:1008 (1983).
9. E. L. Reinherz, P. C. Kung, G. Goldstein, R. H. Levey, and S. F. Schlossman, Discrete stages of human intrathymic differentiation: analysis of normal thymocytes and leukemic lymphoblasts of T cell lineage, *Proc.Natl.Acad.Sci.* 77:1588 (1980).

10. P. Hokland, P. Rosenthal, J. D. Griffin, L. M. Nadler, J. Doley, M. Hokland, S. F. Schlossman, and J. Ritz, Purification and characterization of fetal hemopoietic cells that expressed the common acute lymphoblastic leukaemia antigen (CALLA), J.Exp.Med. 157:114 (1983).
11. K. Dorshkind, and R. A. Phillips, Characterization of early B lymphocyte precursors present in long-term bone marrow cultures, J.Immunol.131:3340 (1983).
12. G. Janossy, M. Roberts, and M. F. Greaves, Target cell in chronic myeloid leukaemia and its relationship to common ALL, Lancet ii:1058 (1976).

A STUDY OF LYMPH NODE LYMPHOCYTES IN RHEUMATOID ARTHRITIS

Deborah Symmons, P. A. Bacon, M. Salmon and A. D. Blann

Rheumatism Research Wing
The Medical School, Birmingham, U.K.

INTRODUCTION

The role of lymphocytes in the pathogenesis of rheumatoid arthritis (RA) remains controversial. Patients with RA have increased levels of circulating activated lymphocytes, the levels of which correlate with the degree of active synovitis (1,2). It is not known whether these circulating lymphocytes are a spillover from the rheumatoid synovium, which contains lymphocytic infiltrates and follicles (3), or whether the lymphocytes are activated elsewhere and then migrate to the synovium, as the thoracic duct drainage experiments suggest (4).

Lymph nodes (LN) are a potential site of lymphocyte activation outside the synovium. Generalised lymphadenopathy occurs in up to 82% of patients with RA (5). Histology of these lymph nodes shows follicular hyperplasia, lymphocytic infiltration of the capsule and sinus histiocytosis (6). The lymph nodes enlarge during flares of the disease and diminish on succesful treatment (7). Thus LN do appear to be involved in the rheumatoid disease process.

A simple technique of fine needle aspiration has been adopted to study LN lymphocytes in RA. A preliminary study has been conducted to investigate the morphology, surface antigen characteristics and behaviour in culture of LN lymphocytes in RA.

METHODS

12 patients with classical or definite RA and lymphadenopathy were studied.

Lymph Node Aspiration

Enlarged palpable LNs were aspirated up to 5 times using a 21 gauge needle attached to a 10ml syringe. The needle was flushed with RPMI to expel the aspirated cells. Between $0.5x10^6$ and $5x10^6$ lymphocytes were obtained. The following procedures were performed:

1. The percentage of immunoblasts was calculated by examining smears treated with Leishman's stain. Immunoblasts were defined as large lymphoid cells with several nucleoli and deeply basophilic cytoplasm.

2. The surface marker characteristics were determined by labelling lymphocyte smears (8) with the monoclonal antibodies UCHTI (anti T3), OKT9 (anti-transferrin receptor), OKIa and anti-Fab (pooled anti κ and λ chains).

3. Where sufficient cells were obtained cultures were established in microtitre plates (1.5 x 10^5 cells/well) using standard techniques. Spontaneous activity was assessed by adding ^{3}H-thymidine immediately the cultures were started and harvesting 18 hours later. Some samples were also cultured for 3 days with and without interleukin 2 (Biotest diagnostics) (1L2). ^{3}H-thymidine was added for the final 18 hours. The wells were harvested onto glass fibre discs, placed in scintillation fluid and counted for 10 minutes on Searle Inc. Analytic Mark III counter using an external standard.

Peripheral Blood

Peripheral blood lymphocytes (pbl) from the same patients were labelled with monoclonal antibodies. In addition 2 x 10^7 lymphocytes were layered onto a discontinuous Ficoll gradient to determine the proportion of low density lymphocytes (LDL). The LDL fraction contains a high proportion of activated cells (9).

Control Lymph Nodes

5 lymph nodes obtained at routine surgery for non-inflammatory conditions were used as controls. The nodes were treated to simulate aspiration. Each node was pierced many times with a 21 gauge needle and then injected with RPMI to flush out the lymphocytes. These cells were then purified on Ficoll-Paque and the lymphocytes were processed in the same way as the rheumatoid LN lymphocytes.

RESULTS

Immunoblasts and LDLs

The aspirated RA lymphocytes included a mean of 6.67 $\pm$ 4.53 immunoblasts. The mean percentage of LDLs in the peripheral blood was 5.22 $\pm$ 2.58. There was a positive correlation between the proportion of LDLs in the peripheral blood and the proportion of immunoblasts aspirated from the nodes ($r = 0.66$; $p < 0.01$).

Surface Antigens

Table 1 shows the results of lymphocyte labelling with monoclonal antibodies and also the normal range for peripheral blood in our laboratory. There was no difference in the proportion of T or B cells, and no difference in the proportion of 1a or transferrin receptor positive (TRFr, OKT9) cells aspirated from RA lymph nodes and controls. A higher proportion of B cells was aspirated from LNs than was found in peripheral blood. In all patients the percentage of TRF positive cells aspirated from the lymph node was greater than that found in peripheral blood.

Table 1 Surface Marker Characteristics as Determined by Labelling with Monoclonal Antibodies

	T cells	B cells	1a	TRFr (range)
PERIPHERAL BLOOD				
Normal range (%)	61.3-83.7	4.9-18.2	14.3-28.6	<1.3
RA (% ± SD)	59.5±15.3	16.5± 5.4	18.5± 9.1	2.9 (1-9)
LYMPH NODE				
RA (% ± SD)	38.5±13.1	31.2±15.5	35.4±11.5	12.6 (1-36)
Control (% ± SD)	45.4±20.5	20.5±12.4	25.3±10.3	18.8 (3-34)

Table 2 Lymph Node Lymphocyte Behaviour in Culture

	Spontaneous Uptake Day 0	Spontaneous Uptake Day 3	Uptake with Interleukin 2
Control LNs (dpm±SD)	1038.6±606	2105.4±833.1	2077.0±800.4
RA LNs	1121	428	
	453		
	808		
	4235	1997	
		385	1122
Mean RA LNs	1254.2	936.6	1122

Culture Experiments

Sufficient cells for culture were only obtained from a few patients. The results are shown in Table 2.

These preliminary results show a higher spontaneous count from the RA aspirates than from controls at day 0. The addition of interleukin 2 to the control lymphocytes did not augment thymidine uptake, whereas in the one RA patient addition of IL2 caused a three fold rise in thymidine uptake.

DISCUSSION

Lymph nodes are important sites of lymphocyte activation in normal subjects (10). Activated lymphocytes, like all proliferating cells, carry transferrin receptors (11). This study shows that aspirated LN lymphocytes have a higher proportion of TRFr cells than are found in peripheral blood. A similar proportion of TRFr positive cells were aspirated from RA and from control LNs. However the rheumatoid patients had lymphadenopathy so the total number of activated lymphocytes within the lymphoid structures may be greater than in normal subjects.

The percentage of immunoblasts aspirated from RA LNs correlates with the percentage of LDLs in the peripheral blood, which in turn have been shown to correlate with the degree of active synovitis (2). These results extend the results of our previous study (12), in which RA patients with elevated LDLs had 1-6% of immature cells in LN aspirate, while patients with normal LDLs had <1% immature cells (as defined by an independent pathologist).

These results suggest that LNs are involved in the traffic of activated lymphocytes in RA. The findings are in keeping with the hypothesis that lymphocytes in RA are activated in LNs and then migrate into the blood, and could then home to the synovium.

This study is based only on aspirated lymphocytes and it may be that the cells which remain adherent to the LN stroma make an important contribution to lymphocyte trafficking. Further studies are in progress to establish the relationship between aspirated lymphocytes and the total cell population of the node.

ACKNOWLEDGEMENTS

This work was supported financially by the Arthritis and Rheumatism Council for Research.

REFERENCES

1. Bacon PA, Sewell RL and Crowther D. Reactive lymphoid cells (immunoblasts) in autoimmune and haemotological disorders. Clin. Exp. Immunol. (1975) 19: 201-8.
2. Alexander GJM, Bacon PA, Cavalcanti F, Blake DR and Hall ND. Low density lymphocytes : their relationship to disease activity and to anti-rheumatic therapy. Br. J. Rheum. (1984) 23 : 6-14.
3. Gardner DL. Pathology of rheumatoid arthritis in Copeman's Textbook of the Rheumatic Diseases (1978). Ed. J T Scott p. 277-278.
4. Paulus HE, Machleder HI, Levine S, Yu DTY and MacDonald DS. Lymphocyte involvement in rheumatoid arthritis. Arthr. Rheum. (1977) 20 : 1249-62.
5. Robertson MDJ, Hart FD, White WF, Nuki G and Boardman P. Rheumatoid lymphadenopathy. Ann. Rheum. Dis. (1968) 27: 253-260.
6. Cruikshank B. Lesions of lymph nodes in rheumatoid disease and in disseminated lupus erythematosus. Scot. Med. J. (1958) 3: 110-119.
7. Park WM and Ward DJ. An evaluation of rheumatoid lymphadenopathy by follow-up lymphography. Proc. 2nd Congress Eur. Assn. Radiol. (1971) 225-230.
8. Salmon M, Bacon PA, Johnson GD, Walker L and Walton KW. The use of cytocentrifuge preparations for the demonstration of T cell surface antigens. J. Imm. Meths. (1984) 66: 327-330.
9. Carter SD, Bacon PA and Hall ND. Characteristics of activated lymphocytes in the peripheral blood of patients with rheumatoid arthritis. Ann. Rheum. Dis. (1981) 40: 293-298.

10. Hall JG and Morris B. The output of cells in lymph from the popliteal node of sheep. Quart. J. Exp. Physiol. (1962) 47: 360-369.
11. Galbraith G, Galbraith R and Page Faulk W. Transferrin binding by human lymphoblastoid cell lines and other transformed cells. Cell Immunol. (1980) 49: 215-222.
12. Bacon PA, Alexander GJM, Lever J and Elson C. Activated lymphocytes in lymph nodes, blood and synovial fluid in RA. Révue de Rhum. (1981) Numéro Spécial Abstract 0590.

BENIGN AND MALIGNANT LYMPHOCYTIC SKIN INFILTRATES: IMMUNOLOGICAL PATTERNS RECOGNIZED BY IMMUNO-HISTOCHEMICAL STAINING WITH MONOCLONAL ANTIBODIES

Elisabeth Ralfkiaer, K.Hou-Jensen, D.Y.Mason, H.Stein, Gunhild Lange Wantzin, and K.Thomsen

Nuffield Department of Pathology, John Radcliffe Hospital, Oxford, U.K. and Departments of Pathology and Dermatology, the Finsen Institute, Copenhagen, Denmark

ABSTRACT

Skin biopsies from 150 patients with cutaneous lymphoma, premalignant or "pseudolymphomatous" conditions and benign dermatoses were examined by immunoenzymatic staining of frozen sections with a panel of monoclonal antibodies against lymphocytes, macrophages and accessory cells. Two patterns of infiltration were identified. In one of these T cells associated with Langerhans cells and HLA-DR-positive macrophages of dendritic morphology, were seen. In the other B cell follicles were present in addition to the above type of infiltrate. The latter pattern was restricted to patch tests biopsies, cutaneous lymphocytomas, and malignant lymphomas of B-cell origin.

It is concluded that lymphocytic skin infiltrates are divisible into different patterns of immunological reactions similar to those seen in lymphoid organs.

INTRODUCTION

Until recently, the discrimination between B- and T-cell cutaneous infiltrates was based on complex cytological/histological criteria (Braun Falco et al.1981; Burg et al.1978), the unreliability of which have been emphasized in several reports (Winkler & Bunn 1984; Wood et al.1983). The lack of a satisfactory system of classification of lymphocytic skin infiltrates according to their histogenesis has contributed significantly to the controversy concerning the nosological identity of many of these disorders (e.g.lymphomatoid papulosis, cutaneous lymphocytoma (From 1979, Lange Wantzin et al. 1982)).

Data from this and other laboratories have stressed the utility of immunohistochemical staining techniques in the identification of lymphoid cells which are indistinguishable on purely morphological criteria (Ralfkiaer & Lange Wantzin, 1984; Willemze et al. 1983; Wood et al. 1983). In this study the value of these techniques as an aid in the objective classification of lymphocytic skin infiltrates has been examined.

MATERIAL AND METHODS

Tissue specimens

4 mm skin punch biopsies were obtained from 150 patients attending the Department of Dermatology at the Finsen Institute. Their clinical and histological diagnoses are summarized in Table 1. All lymphoma cases were classified according to conventional morphological and immunohistological criteria (Winkler et al.1984; Wood et al.1983).

Table 1 DISTRIBUTION OF PATIENTS ACCORDING TO DIAGNOSIS AND IMMUNOHISTOLOGICAL PATTERN IN SKIN BIOPSIES

Diagnosis	No. of patients	T-cell pattern+	T/B cell pattern++
Cutaneous T-cell lymphoma	45	45	-
Cutaneous B-cell lymphoma	4	-	4
Large plaque parapsoriasis	31	31	-
Lymphomatoid papulosis	10	10	-
Cutaneous lymphocytoma	9	2	7
Patch test	20	18	2
Eczema	10	10	-
Psoriasis	12	12	-
Lichen planus	5	5	-
Atopic dermatitis	4	4	-

+ T-lymphocytes, Langerhans cells and HLA-DR positive dermal macrophages. ++ B-cell follicles (B-cell and follicular dendritic cells) present in addition to the T cell pattern.

Processing of fresh tissue specimens
Biopsies were frozen and stored in liquid nitrogen. 6 µm cryostat sections were lyophilized at -20°C for 2-4 h, fixed in acetone for 10 min, and processed for staining or stored at -20°C until staining. In the latter instance, sections were fixed in acetone for an additional 10 min immediately before staining (Stein et al.1982)

Monoclonal antibodies
The reactivities, sources and/or references of the antibodies used in this study are summarized in Table 2.

Table 2 MONOCLONAL ANTIBODIES

Antibody	Specificity	Reference or source+
Anti-Leu1/Lyt2	pan T-cells	BD/NEN
Anti-Leu5/Lyt3	pan T-cells (E-rosette receptor)	BD/NEN
Anti-Leu4	peripheral T-cells	BD
Anti-Leu3	T-helper/inducer cells	BD
Anti-Leu2	T-suppressor/cytotoxic cells	BD
OKT6	cortical thymocytes, Langerhans cells	Ortho
OKT9	transferrin receptor	Ortho
OKT10	haematopoietic precursor cells, activated B- and T-cells	Ortho
anti-Tac	interleukin-2 receptor	Leonard, 1982
Anti-Leu7	natural killer cells	BD
B4	pan B-cells	CC
B1	pan B-cells	CC
To15	pan B-cells	DAKO
anti-kappa	kappa light chains	BD
anti-lambda	lambda light chains	DAKO
R4/23	dendritic reticulum cells	DAKO
To5	C3b receptor	DAKO
FMC32	tissue macrophages	Brookes, 1983
OKM1	monocytes, tissue macrophhages, granulocytes	Ortho
MO2	monocytes	CC
CR3/43	HLA-DR	own laboratory
Anti-HLA-DR	HLA-DR antigen	BD

+Becton Dickinson (BD); New England Nuclear (NEN); Ortho Diagnostics (Ortho); Coulter Clone (CC); DAKOPATTS (DAKO)

Immunoenzymatic staining techniques
Sections were stained with 2- or 3-stage immunoperoxidase methods (Stein et al.1982) or with an alkaline phosphatase:anti-alkaline phosphatase (APAAP) technique (Cordell et al.1984). Sequential double immunoenzymatic staining for the detection of antigens on non-overlapping or partly overlappping populations of cells was carried out as described previously (Ralfkiaer et al. 1984).

RESULTS

Two patterns of infiltration could be identified by immunohistopathological examination. In one of these many T cells, expressing the antigenic profile of peripheral T cells (Lyt2+, Lyt3+, Leu4+, OKT6-) were present. In all such cases a proportion of the T-cells expressed the activation associated markers HLA-DR, Tac and OKT9 (transferrin receptor) and (less commonly) also OKT10. The relative proportions of helper/inducer to suppressor/cytotoxic T cells varied considerably, in each of the disease categories examined and have been described in detail previously (Ralfkiaer & Lange Wantzin, 1984; Ralfkiaer et al. in press; Ralfkiaer et al submitted). Accessory cells (Langerhans cells (OKT6+, HLA-DR+)), and HLA-DR positive macrophages of dendritic morphology (FMC32+, OKM1+/-) were also present in association with these T cells.

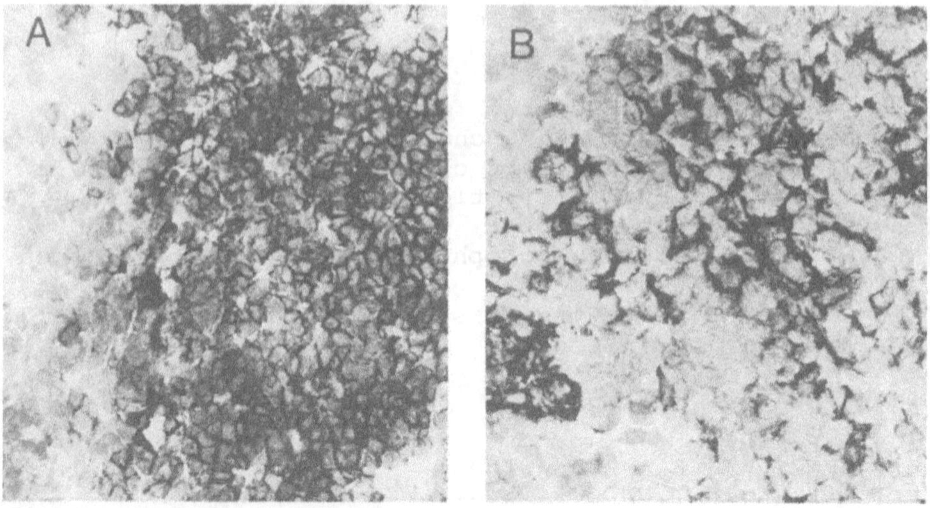

Fig.1: Immunoperoxidase staining of B-lymphocytes (a, To15 x 300), and follicular dendritic cells (b, R4/23 x 400) in patch test biopsies.

The other pattern of infiltration was distinguished by the additional occurrence of clusters of B-cells (B4+,B1+,To15+,SmIg+) in dermal infiltrate. The B-cell clusters contained follicular dendritic cells (R4/23+,To5+) (Fig. 1a and b), and thereby closely resembled the B-cell follicles seen in lymphoid organs.

The distribution of patients according to these two patterns of infiltration is summarized in Table 1. B-cell follilces occurred only infrequently and were restricted to cutaneous lymphomas of B-cell origin, cutaneous lymphocytomas, and some patch tests biopsies. All other disorders (cutaneous T-cell lymphoma, large plaque parapsoriasis, lymphomatoid papulosis, chronic dermatoses) demonstrated T-cell infiltration without detectable B-cells.

DISCUSSION

The immunohistological data from the present study demonstrate that benign lymphocytic skin infilrates and their malignant counterparts are divisible into different patterns of immunological reactions closely resembling those found in the T- and B-cell areas of lymphoid organs (Ralfkiaer et al. 1984). These data extend results obtained by examination of tissue sections or cell suspensions from lymphocytic skin infiltrates with less extensive panels of antibodies and/or with other techniques with a more limited specificity for individual cell types (Burg et al.1978; Knowles et al.1982; Willemze et al. 1983). In particular, our data demonstrate that T- and B-cell skin infiltrates contain T- and B-cell accessory cells (Langerhans cells and follicular dendritic cells) in addition to lymphocytes. In this context it should be emphasized that ultrastructural, enzyme cytochemical, and phenotypic data have provided evidence that Langerhans cells are closely related to the interdigitating reticulum cells found in the T-cell areas of lymphoid organs (Ralfkiaer et al. 1984; Rausch et al.1977; Silberberg-Sinakin et al.1976).

The predominance of T-cell patterns of infiltration observed in the present series of skin disorders supports the concept that T-cells play a central role in immune reactions of the skin (Patterson & Edelson 1982). Our data confirm that the cellular constituents of cutaneous T-cell lymphomas closely resemble those of a range of benign inflammatory disorders (Haynes et al.1982; Ralfkiaer et al. submitted).

The identification of B-cell follicles (i.e.: B-cells associated with follicular dendritic cells) in some lymphocytic skin infiltrates supports the finding that follicular dendritic cells may be attracted to B-lymphoid infiltrates in extranodal areas (Naiem et al.1983; Ralfkiaer & Lange Wantzin, 1984), and provides indirect evidence for an active involvement of B-cells in some immune reactions of the skin. The restricted occurrence of B-cell follicles in certain conditions (patch test biopsies, cutaneous lymphocytomas, malignant lymphomas of B-cell origin) suggests that these disorders share pathogenetic mechanisms which are not operative in other skin diseases and supports the concept that cutaneous lymphocytomas exaggerated B-cell responses of the skin (Knowles et al.1982; Ralfkiaer et al. in press).

The data from the present study demonstrate that lymphocytic skin infiltrates recapitulate patterns of immunological reactions similar to those of lymphoid organs. This emphasizes the value of immunohistochemical staining techniques as an aid in the objective classification of lymphocytic skin infilrates.

Acknowledgements

Anti-Tac was a generous gift from Dr. Thomas Waldmann, National Cancer Institute, NIH, Bethesda, Maryland, U.S.A. This study was supported by a grant (No. 5/83) from the Danish Cancer Society.

REFERENCES

Braun Falco, O., Burg, G and Schomoeckel, C., 1981, Recent advances in the understanding of cutaneous lymphoma. Clin. Exp. Dermat., 6:89.

Brooks, D.A., Zola, H., McNamara, P.J., Bradley, J., Bradstock, K.F., Hancock, W.W. and Atkins, R.C., 1983, Membrane antigens of human cells of the monocyte/macrophage lineage studied with monoclonal antiboides. Pathology, 15:45

Burg, G., Braun Falco, D., Hoffmann-Fezer, G., Rodt, H and Schmoeckel, C., 1978, Patterns of cutaneous lymphomas. Dermatologica, 157:282

Cordell, J.L., Falini, B., Erber, W., Ghosh, A.K., Abdulaziz, A., MacDonald, S., Pulford, K.A.F., Stein, H. and Mason D.Y., 1984, Immunoenzymatic labelling of monoclonal antibodies using immune complexes of alkaline phosphatase and monoclonal anti-alkaline phosphatase (APAAP complexes). J.Histochem. Cytochem., 49:11

From, L., 1979, Lymphocytic infiltrates. in: Dermatology in General Medicine, T.B.Fitspatrick, A.Z.Fisen, K.Wolf, I.M.Freedberg and K.I.Austen, eds., McGraw-Hill, New York, London, Paris.

Haynes, B.F., Hensley, L.L. and Jegatothy, B.V., 1982, Phenotypic characterization of skin-infiltrating T cells in cutaneous T-cell lymphoma: comparison with benign cutaneous T-cell infiltrates. Blood, 60:463.

Knowles, D.M., Halper, J.P. and Jakobiec, F.A., 1982, The immunologic characterisation of 40 extranodal lymphoid infiltrates. Cancer, 49:2321.

Lange Wantzin, G., Hou-Jensen, K., Nielsen, M., Petri, J. and Thomsen, K., 1982, Cutaneous lymphocytomas: clinical and histiological aspects. Acta Dermatovenereologica (Stockholm), 62:119.

Leonard, W.J., Depper, J.M., Uchiyama, T., Smith, K.A., Waldmann, T.A. and Greene, W.C., 1982, A monoclonal antibody that appears to recognise the receptor for human T-cell growth factor; partial characterization of the receptor. Nature, 300:267.

Naiem, M., Gerdes, J., Abdulaziz, Z., Stein, H. and Mason, D.Y., 1983, Production of a monoclonal antibody reactive with human dendritic reticulum cells and its use in the immunohistological analysis of lymphoid tissue. J.Clin.Path., 36:167.

Patterson, J.A.K. and Edelson, R., 1982, Interaction of T-cells with the epidermis. Br.J.Dermatol., 107:117.

Ralfkiaer, E., Plesner, T., Wantzin, G.L., Thomsen, K., Nissen, N.I. and Hou-Jensen, K., 1984, Immunohistochemical identification of lymphocyte subsets and accessory cells in human hyperplastic lymph nodes. Scand.J. Haematol., 32:536.

Ralfkiaer, E. and Lange Wantzin, G., 1984, In situ immunological characterization of the infiltrating cells in positive patch tests. Br.J.Dermatol., 111:13.

Ralfkiaer, E., Lange Wantzin, G., Mason D.Y., Stein, H. and Thomsen, K., 1984, Characterization of benign cutaneous lymphocytic infiltrates by monoclonal antibodies, Br.J.Dermatol., in press.

Ralfkiaer, E., Stein, H., Plesner, T., Hou-Jensen, K. and Mason, D., 1984, In situ immunological characterization of Langerhans cells with monoclonal antibodies: comparison with other dendritic cells in skin and lymph nodes. Virchows Arch. (Pathol.Anal.), 403:407.

Ralfkiaer, E., Hou-Jensen, K., Mason, D.Y., Stein, H., Lange Wantzin, G. and Thomsen, K., 1984, Mycosis fungoides and premycotic lesions (large plaque parapsoriasis): characterization of skin infiltrates by monoclonal antibodies. submitted.

Rausch, E., Kaiserling, E. and Gross, M., 1977, Langerhans cells and interdigitating reticulum cells in the thymus-dependent region in human dermatopathic lymphadentis. Virchows Arch. (Cell Pathol), 25:327.

Silberberg-Sinakin, I., Thornbecke, G.J., Baer, R.L., Rosenthal, S.A. and Berezowsky, V., 1976, Antigen-bearing Langerhans cells in skin, dermal lymphatics and in lymph node. Cell Immunol., 25:137.

Stein, H., Gerdes, J., Schwab, U., Lemke, H., Mason, D.Y., Ziegler, A., Schienle, W. and Diehl, V., 1982, Identification of Hodgkin and Sternberg-Reed cells as a unique cell type derived from a newly-detected small-cell population. Inter.J.Cancer, 30:445.

Willemze, R., de Graaff-Reitsma, C.B., van Volten, W.A. and Meijer, C.J.L.M., 1983, The cell population of cutaneous B-cell lymphomas. Brit.J.Dermatol., 108:395.

Winkler, C.F. and Bunn, P.A., 1984, Cutaneous T-cell lymphoma: A review. Critical Reviews, 1:49.

Wood, G.S., Burke, J.S., Horning, S., Doggett, R.S., Levy, R. and Warnke, R.A., 1983, The immunologic and clinicopathologic heterogeneity of cutaneous lymphomas other than mycosis fungoides. Blood, 62:464.

DIFFERENTIAL EXPRESSION OF CLASS II MHC ANTIGENS ON CELLS DERIVED FROM HUMAN TISSUE

Bruce McDowell, Andrea T. Cifelli, Julia D. Nunn and David R. Katz

Pathology Department, Middlesex and University College London Medical School, London W1P 7PN

INTRODUCTION

The association between expression of Ia (class II MHC) surface products and presentation of antigen has been well documented in the murine system. The strongly Ia positive dendritic cell, when isolated from murine lymphomedullary tissue, is a low density, Fc receptor (FcR) negative cell type, which performs as a highly efficient antigen presenting cell in both allogeneic and syngeneic systems (Sunshine et al, 1980). Cells sharing similar phenotypic and functional characteristics have now been isolated from human peripheral blood (Kuntz Crow and Kunkel, 1982).

However, in particular with respect to cells obtained from human tissue, the expression of class II MHC surface products (HLA-DR, -DC and SB) and their association with antigen presentation has not been fully examined.

This study examines the expression of class II MHC products on cells obtained from human tonsil, thymus, spleen and lymph nodes. Functional assays examining the ability to induce T cell proliferation have been carried out in parallel; these assays indicate that a low density, FcR-ve and HLA-DR+ve cell type, which may be homologous to the murine dendritic cell, can be isolated from human tissue.

METHODS AND MATERIALS

Human Tissue Disaggregation

Fresh tissue was minced and then collagenase digested for 1 hour at 37°C. Tissue was then passed through a nylon mesh, and the resulting cell suspension washed thoroughly. The final pellet was suspended in 35% Bovine Serum Albumin (BSA) (Miles Laboratories Inc.) Using standardised dilutions of BSA in RPMI 1640 at concentrations of 29%, 26%, 23%, 20% and 10% albumin, a five band discontinuous density gradient was generated. Following ultracentrifugation (30 minutes, 4°C at 14,000 rpm with swing-out rotor) the five interface layers were harvested and labelled A (lowest density) to E (highest density). Cells were cultured overnight at 37°C in humidified 5% CO_2 in air, in RPMI 1640 supplemented with 5% heat inactivated Foetal Calf Serum (FCS), 100 I.U. Penicillin-Streptomycin/ml, 2.5 mg/ml Amphotericin B, 0.05mM 2-mercaptoethanol (final concentration), 2mM L-glutamine and 10mM Hepes (Complete Medium).

The overnight non-adherent interface cells were then harvested and subfractionated into different populations by rosetting methods: (i) the test populations were incubated for 5 minutes at 37°C with washed sheep red blood cells (SRBC); they were then pelleted (1000rpm for 5 minutes) and kept at 4°C for 1 hour before gentle resuspension and centrifugation at 1900 rpm for 25 minutes over Ficoll-Hypaque. This separated off cells which bear the SRBC receptor (E+ cells) from those which do not (E- cells).
(ii) FcR rosetting was performed using anti-sheep erythrocyte antibody (7SIgG: Cordis Laboratories Inc) -coated SRBC. These pre-treated SRBC were mixed with the test populations and incubated for one hour at 37°C followed by centrifugation over Fincoll-Hypaque as above. This produced rosetted (FcR+ve) and non-rosetted (FcR-ve) populations.

All populations obtained following Fincoll-Hypaque separation were treated with 0.83% Tris-buffered ammonium chloride for 10 minutes at 37°C to lyse contaminating SRBC.

Binding Assay

Cells were thoroughly washed in PBS and plated out into 96 well flexiplates at 105 cells/well. Following glutaraldehyde fixation (0.25% for 10 minutes, 50 ul/well) and subsequent blocking with PBS + 10% FCS, the first label was added for 1 hour. The second label, 125 I*Sheep anti-Mouse Ig, was added to provide approximately 5×10^4 cpm/well. Background was taken as the level of binding of the second label alone.

Antibodies

DA6.231 and HIG.78 are mouse monoclonal antibodies directed against framework determinants of the human class II MHC surface antigens. DA6.231 recognises determinants common to the DR and SB loci; HIG.78 the DR and DC loci (Guy and Van Heyningen, 1983). In these experiments, culture supernatants obtained from cells kindly supplied by Drs. Guy and Van Heyningen were used.

Proliferation Assay

Cells irradiated with 1.2-1.5 rad were used as stimulators at 5×10^4/well in a 96 well plate. High density E+ cells were used as responders at 10^5/well. All culture was carried out at 37°C in humidified 5% CO_2 in air in Complete medium. Plates were cultured for 96 hours, the final six hours of culture in the presence of $5\text{-}6 \times 10^5$ cpm/well 125 IUdR to assay proliferation.

Concanavalin A (Con A) Pulsing

Stimulator cells at 10^6/ml were pulsed with Con A (22.2ug/ml) for 30 minutes at 37°C. The cells were then thoroughly washed, and irradiated before their use in the proliferation assay. In additional control experiments, -methyl-D-mannoside was added at 20mg/ml immediately prior to the addition of Con A.

RESULTS

In all tissues tested, maximal expression of the products of the HLA-DR and SB loci occurred in the low density cell fractions, with decreasing expression in the higher density layers, as shown by the binding of DA6. 231 (Table 1.). Similarly, the binding of HIG.78, detecting products of the DR and DC loci, was greatest in the low density cell types (data not shown).

In tonsil, FcR-ve and E-ve populations show a marked increase in anti-DR binding when compared to their receptor positive populations (Table 2). Data for spleen, thymus and lymph node indicate a similar pattern.

Pulsing tonsil cells with Con A has two effects: (i) inhibition of binding of DA6.231 (and HIG.78 - data now shown) to the cells. This is true for all lymphoid cells tested (Table 3).

Table 1

PERCENTAGE SPECIFIC[1] BINDING OF DA6.231 SUPERNATANT TO DENSITY SEPARATED HUMAN TONSIL CELLS

Tissue	Density fraction[2] A	B	C	D	E
Tonsil (n=16) (+SEM)	13.7 (0.9)	11.9 (0.8)	10.2 (0.8)	7.3 (0.4)	6.4 (0.8)
Spleen 1	16.1	10.5	2.6	2.8	3.3
Spleen 2	12.5	15.0	14.0	9.6	8.6
Thymus 1	15.2	14.1	4.9	7.7	8.1
Thymus 2	4.6	2.6	2.6	4.6	N.D.[3]
Lymph node 1	11.9	14.4	8.5	7.1	N.D.
Lymph node 2	7.5	3.2	0.9	1.0	N.D.

Notes: 1. Percentage specific binding = $\frac{\text{(Experimental-background)}}{\text{(Input-Background)}} \times 100$

2. Density fractions A (low density) to E (high density) obtained by BSA discontinuous gradient.

3. N.D. : Not determined.

Table 2

PERCENTAGE SPECIFIC BINDING OF DA6.231 SUPERNATANT TO DENSITY SEPARATED HUMAN TONSIL CELLS. Fc AND E RECEPTOR POPULATIONS

Density Fraction	Receptor FcR-ve	FcR+ve	E-ve	E+ve
A	13.7	7.1	17.6	9.9
B	13.5	9.1	17.7	9.4
C	11.2	5.1	14.1	8.7
D	9.4	2.7	13.2	8.2
E	N.D.	N.D.	9.9	6.6

Table 3

EFFECT OF CONCANAVALIN A PULSING ON DA6.231 BINDING TO HUMAN TONSIL CELLS AND ITS MODULATION BY THE ADDITION OF -METHYL-D-MANNOSIDE

Density Fraction	Percentage Specific Binding		
	Unpulsed	Con A Pulsed	Con A pulsed + Mannoside
A	17.8	3.4	15.9
B	19.9	0.6	17.0
C	11.0	4.7	10.8
D	5.4	1.3	10.2
E	1.8	0.0	2.5

Table 4

EFFECT OF PULSING STIMULATOR CELLS WITH CONCANAVALIN A AND ITS MODULATION BY THE ADDITION OF -METHYL-D-MANNOSIDE IN HUMAN TONSIL

Density Fraction*	cpm ^{125}IUdR incorporation + SEM		
	Unpulsed	Con A Pulsed	Con A Pulsed + Mannoside
A	3111+180	10171+489	2721+304
B	1637+112	8911+115	1343+155
C	1367+155	3363+238	1070+114
D	792+ 55	3148+370	1043+304
E	1145+222	1568+ 77	1605+ 87

Table 5

PERCENTAGE INCREASE IN PROLIFERATION DUE TO CONCANAVALIN A PULSING OF STIMULATORS ABOVE THAT OF UNPULSED CONTROLS

Density Fraction*	Whole layer	FcR-ve	FcR+ve
A	60.7	169.0	N.D.
B	140.7	316.1	31.6
C	107.2	176.7	78.3
D	49.1	87.6	47.8
E	7.4	40.0	43.9

* Of stimulator cell type.

(ii) in functional assays, Con A pulsed stimulators induce a marked increase in the proliferation of the responding T cells. The increase is most pronounced when the low density cell layers are used as stimulators (Table 4).

-methyl-D-mannoside inhibits binding of the lectin to the cell surface receptors, by both competing for the receptor (Edelman et al, 1972) and effectively reducing the valency of the Con A molecule (Distasio et al, 1982). The addition of mannoside reverses the effects due to Con A pulsing seen above: (i) binding of DA6.231 returns to normal levels and (ii) no increase in proliferation of the responder cell population is seen when compared to that seen with unpulsed stimulators (Tables 3 and 4).

Finally, Table 5 shows the percentage increase in proliferative uptake of ^{125}IUdR when Con A pulsed stimulators are compared with their unpulsed controls. The most effective populations are the low density FcR-ve populations. With increasing cell density, both overall stimulatory capacity and differences between the FcR-ve and FcR+ve fractions decrease.

DISCUSSION

We are now able to select for a low density, FcR-ve cell type which strongly expresses class II MHC antigens. This cell population proves the most effective in inducing a Con A mediated T cell proliferative response.

With the successful isolation of this human tissue accessory cell, sharing many characteristics with the murine dendritic cell, a system has been developed which can examine the nature of accessory cell/T lymphocyte interaction. Equally important is that the relationship between class II MHC product expression and modulation of expression following surface binding can now be studied in a human system using these effective inducer cells.

In demonstration of this, experiments are at present underway which examine the nature of the inhibition of DA6.231 binding to cells pulsed with Con A. The complete reversal by - methyl-D-mannoside of the effects seen following Con A pulsing indicate that the functional and phenotypic alterations are as a result of interactions at the cell membrane. The primary hypothesis to explain these effects must concern steric hindrance by the Con A molecule binding to, or adjacent to, the class II MHC surface product, preventing subsequent binding of the anti-DR antibody. We are therefore examining the functional consequences of exposing the stimulator cells to anti-DR antibody prior to Con A pulsing.

Correlation of functional and phenotypic characterisation is striking in our system. Cells expressing most class II MHC products perform as the most efficient presenters of Con A. Of additional interest is more detailed examination of the variation between mitogenic responses obtained and the specific class II MHC locus predominantly expressed on the stimulator cell type. Structural homology between the murine I-A and the human HLA-DC (as opposed to -DR) locus (Goyert et al, 1982) raises the question as to the relative functional importance of the various human class II MHC products in antigenic presentation. This system permits examination of this problem.

SUMMARY

Human lymphomedullary tissue cells, separated on the basis of density, Fc and E receptors, show characteristic and consistent differences in their expression of class II MHC antigens. A low density, FcR-ve cell type displays high expression of HLA-DR and acts as the most efficient accessory cell in the induction of a Con A driven proliferative response.

Exposure of cells to Con A inhibits subsequent binding of anti-HLA-DR antibodies. Addition of -methyl-D-mannoside reverses both the proliferative response to Con A and returns binding of anti-class II MHC antibodies to normal levels.

This experimental system can be used to study: (i) the nature of the association between cell bound molecules and class II MHC products, both spatially and functionally; and (ii) the relative functional importance of these products in antigen presentation.

REFERENCES

Distasio, J.A., Cheresh, D.A., Schilder, R.J., Vogel, C.L., Silverman, MA., Lopez, D.M. (1982) Maximising differences in the Concanavalin A-induced blastogenic responses of lymphocytes from breast cancer patients and controls by the use of -methyl-D-mannoside. JNCI: 68:69-74.

Edelman, G.M., Cunningham, B.A., Reeke, G.N., Becker, J.W., Waxdal, M.J., Wang, J.L. (1972) The covalent three-dimensional structure of Concanavalin A. PNAS: 69:2580-2584.

Goyert, S.M., Shively, J.E., Silver, J., (1982) Biochemical characterisation of a second family of human Ia molecules, HLA-DS, equivalent to murine I-A subregion molecules. J. Exp. Med.: 156: 550-556.

Guy, K., Van Heyningen, V., Dewar, E., Steel, C.M. (1983) Enhanced expression of human Ia antigens by chronic lymphocytic leukaemia cells following treatment with 12-0-tetradecanoyl-phorbol-13-acetate. Eur. J. Immunol.: 13: 156-159.

Kuntz Crow, M., and Kunkel, H.G., (1982) Human dendritic cells: major stimulators of the autologous and allogeneic mixed leucocyte reactions. Clin. Exp. Immunol.: 49: 338-346.

Sunshine, G.H., Katz, D.R., Feldman, M. (1980) Dendritic cells induce T cell proliferation to synthetic antigens under Ir gene control. J. Exp. Med.: 180: 1817-1822.

BETA-ADRENERGIC REGULATION OF THE BLOOD LYMPHOCYTE PHENOTYPE DISTRIBUTION IN NORMAL SUBJECTS AND SPLENECTOMIZED PATIENTS

R. Landmann*, M. Dürig*, F. Gudat*,
M. Wesp+ and F. Harder*

Department of Research, Department of Surgery and Institute of Pathology, University Hospital*
Basel, and
Pharmaceutical Division Ciba-Geigy Ltd.
Basel, Switzerland+

ABSTRACT

The beta-adrenergic effect on the release of immunoregulatory cells from the spleen was investigated by physical stress testing (bicycle ergometry up to submaximal work capacity) in 19 normal subjects (15 males, median 21 years) and in 10 male patients splenectomized for trauma (median 29 years). It was repeated in 6 subjects of each group during beta-blockade with 80mg oxprenolol. Blood samples for leucocyte analysis were taken before and at the end of the test. Leucocyte subpopulations were analyzed in a cytofluorograf after staining of buffy coat cells by direct (B cells) or indirect immunofluorescence with monoclonal antibodies directed against the phenotypes of T- (Leu-1), T helper- (Leu-3a), T suppressor/cytotoxic (Leu-2a) cells and natural killer (OKM1+ lymphocytes) cells.

In the controls all leucocyte subsets increased at ergometry, but B-, Leu-2a- and OKM1-cells increased more than Leu-3a cells. During beta-blockade the leucocyte changes reached only 50% of the value without treatment; the B- and Leu-2a cell mobilization was reduced more than the Leu-3a-, OKM1 cell- and monocyte changes. In splenectomized patients the proportional cellular changes were only half of those found in normal subjects, except for the Leu-3a cells which were not released. Beta-blockade during ergometry had no effect on Leu-3a cells, a similar effect on B- and Leu-2a cells as in normal subjects and a stronger effect on granulocytes, monocytes and OKM1 cells than in controls. In conclusion, the B- and Leu-2a cell mobilization from the spleen (50%) was beta-adrenoceptor dependent, while the one from other lymphoid organs was beta-adrenoceptor independent. The strongly spleen dependent Leu-3a cell changes were not beta-adrenoceptor mediated. Granulocyte-, monocyte- and OKM1 cell changes were only partly spleen dependent. The spleen independent changes however were strongly beta-adrenoceptor dependent.

INTRODUCTION

In normal subjects adrenergic activation by injection of adrenergic agents or physical stress leads to a granulocytosis[1] and monocytosis;[2] it produces a lymphocytosis[3,4,5] which is due to an increase of cells with surface immunglobulin (sIg), with Leu-2a- or OKM1 phenotype[2]. Beta-adrenergic stimulation by isoproterenol bolus injections in vivo mimics these changes to a lesser extent[2].

In the present study the beta-adrenergic specificity of the altered blood leucocyte distribution induced by adrenergic stimulation was further investigated by performing physical stress tests in normal subjects without and with beta-blockade. The qualitative and quantitative contribution of the spleen to the cell mobilization induced by sympathetic activation is not known. Therefore the effect of physical stress and beta-blockade upon lymphocyte phenotype distribution was also studied in splenectomized patients.

METHODS

Twenty normal subjects, fifteen males, aged 17 to 27 years (median 21 years) and ten male patients after splenectomy because of trauma were studied, their age ranged between 18 and 44 years (median 29 years). The time interval between splenectomy and the present investigation ranged from 1 to 18 years (median 7 years). At 8 am. an indwelling cannula was placed intravenously. After a rest of 30 minutes in the supine position, bicycle ergometry was carried out up to submaximal work capacity[2]. Blood samples for the determination of the leucocyte distribution were taken before and at the end of ergometry. Heart rate was calculated from the electro-cardiogram which was recorded during ergometry. Six subjects of each group received 80mg of the beta-blocker oxprenolol orally, when their heart rate had reached basal values after the initial ergometry. They were then allowed a rest of 30 minutes in the supine position, after which they underwent a second ergometry with the same work load as before; Blood was again collected before and at the end of the test. To discriminate between changes due to the repetition of the ergometry and those due to beta-blockade, two splenectomized patients underwent the second ergometry without beta-blockade.

Phenotypes of buffy coat cells were determined by direct or indirect immunofluorescence with the following reagents: $10ul/10^6$ cells goat anti-human $F(ab')_2$ immunoglobulin-fluorescein-isothiocyanate (Ig FITC) (Cappel Labs., Cochranville, PA) for B cells, $5ul/10^6$ cells of the biotinylated monoclonal antibodies Leu-1, Leu-2a or Leu-3a for T cells (Becton Dickinson, BD, Sunnyvale, CA) and OKM1 ($10ul/10^6$ cells) (Ortho Diagnostic Systems, Raritan, NY) for natural killer (NK) cells. The second stage reagents were: avidin-FITC (3ul, BD) or rabbit anti-mouse Ig-FITC (5ul, Nordic, Tilburg, The Netherlands). 50'000 cells of each sample were analyzed in a cytofluorograf 50HH with an Ortho computer 2150 (Ortho Instruments, Westwood, MA). Granulocytes, monocytes and lymphocytes were counted in the scatter cytogram. Fluorescence was recorded from the lymphocyte cluster only. The absolute number of cells with a given phenotype was calculated by multiplying the percentage of positive cells with the number of lymphocytes in the buffy coat. The cellular parameters showed a normal distribution. Intra- and interindividual differences were evaluated by the paired and non paired t-test, respectively.

RESULTS

As shown in table 1 and 3, in normal subjects, bicycle ergometry without beta-blockade produced a pronounced granulo-, mono- and lymphocytosis ($p < 0,001$), cell numbers increased by about 100%. Among the lymphocytes sIg- and OKM1 cells showed an increase of 180% and 313% respectively, the Leu-1 cells only increased by 65%. Among the T cells, the Leu-2a cells increased more (100%) than the Leu-3a cells (55%). The T/B- and the Leu-3a/Leu-2a cell ratios both significantly decreased. As shown in table 2, relative proportions of granulocytes, monocytes and lymphocytes were unaltered by ergometry, however among the lymphocytes the B cell fraction strongly increased and the Leu-, particularly the Leu-3a fraction, decreased. The percentage Leu-2a cells remained constant and the relative number of OKM1 cells increased.

In the six subjects undergoing beta-blockade, all the ergometry induced cell numbers (data not specified in table 1) were found significantly higher than the values registered later on during beta-blockade at rest with the exception of the granulocytes (table 1). Resting values without and with treatment did not differ significantly in these subjects (except again for the slightly higher granulocyte number ($p < 0.05$) at rest with beta-blockade than without), possibly as a consequence of the initial ergometry.

Table 1: Leucocyte subsets and lymphocyte phenotypes in normal subjects (C) and splenectomized patients (S) before and after physical stress (bicycle ergometry) without and with beta-blockade. Mean ± S.E.M. " $p < 0.05$, "" $p < 0.01$, """ $p < 0.001$, comparing values before and after ergometry. * $p < 0.05$, ** $p < 0.01$, *** $p < 0.001$, comparing values between C and S.

Cells in buffy coat from 10ml of blood (X10^6)		Beta-Blockade – (C)n=20 (S)n=10 Rest	Beta-Blockade – Ergometry	Beta-Blockade + (C)n=6 (S)n=6 Rest	Beta-Blockade + Ergometry
Total leucocytes	Control=C	15,1 ± 1,1	30,8 ± 3,3 """	19,4 ± 4,1	30,8 ± 6,7 ""
	Splenectomy=S	24,1 ± 4,6 *	35,6 ± 4,5 ""	24,8 ± 2,7	30,6 ± 4,7 "
Granulocytes	C	5,6 ± 0,5	11,8 ± 1,6 """	7,7 ± 1,7	12,6 ± 2,2 """
	S	5,5 ± 1,2	8,1 ± 1,9 "	7,2 ± 2,1	8,1 ± 2,2
Monocytes	C	0,42 ± 0,04	0,89 ± 0,08"""	0,54 ± 0,19	1,05 ± 0,15"
	S	0,70 ± 0,15*	1,23 ± 0,25""	0,69 ± 0,11	0,86 ± 0,16
Lymphocytes	C	7,1 ± 0,7	13,5 ± 1,2 """	9,8 ± 2,9	15,8 ± 4,3 "
	S	14,0 ± 3,1 **	20,3 ± 2,7 ""	12,5 ± 1,9	16,0 ± 2,7 "
- B cells (sIg)	C	1,6 ± 0,2	4,5 ± 0,4 """	2,3 ± 0,6	4,3 ± 1,3 "
	S	5,0 ± 1,3 **	9,7 ± 1,5 """ ***	4,4 ± 1,0	6,2 ± 1,4 "
- T cells (Leu-1)	C	4,9 ± 0,6	8,1 ± 1,0 """	9,1 ± 2,7	13,2 ± 3,1 """
	S	7,3 ± 1,8	8,3 ± 1,1	5,9 ± 0,8	6,9 ± 0,9
- T helper cells (Leu-3a)	C	3,1 ± 0,4	4,8 ± 0,6 ""	6,7 ± 2,2	9,5 ± 2,4 ""
	S	5,2 ± 1,3	5,5 ± 0,6	4,6 ± 0,6	5,2 ± 0,7
- T suppressor cells (Leu-2a)	C	1,8 ± 0,2	3,6 ± 0,4 """	2,6 ± 0,5	4,2 ± 0,6 ""
	S	2,9 ± 0,7	4,4 ± 0,7 "	2,3 ± 0,4	3,1 ± 0,5 "
- Natural killer cells (OKM1)	C	0,32 ± 0,04	1,32 ± 0,21"""	0,25 ± 0,06	0,96 ± 0,42
	S	0,89 ± 0,22***	2,40 ± 0,51"" *	0,91 ± 0,40	1,16 ± 0,23
Ratio T/B cells	C	3,35 ± 0,27	1,81 ± 0,18"""	3,31 ± 0,21	2,93 ± 0,46
	S	1,75 ± 0,29***	0,99 ± 0,18"" **	1,49 ± 0,31**	1,36 ± 0,26*
Ratio T helper/T suppressor cells	C	1,80 ± 0,16	1.39 ± 0,09"""	2,47 ± 0,44	2,22 ± 0,34
	S	1,88 ± 0,21	1,41 ± 0,14"	2,11 ± 0,21	1,73 ± 0,19

Table 2: Fractional distribution of granulocytes, monocytes and lymphocyte phenotypes in normal subjects (C) and splenectomized patients (S) before and after ergometry without and with beta-blockade. Mean ± S.E.M. " p <0.05, "" p <0.01, """ p<0.001, comparing value before and after ergometry. * p <0.05, ** p <0.01, p <0.001, comparing values between C and S.

Relative cell number (%)			(C) n=20 (S) n=10 Rest	Ergometry	Beta-Blockade + (C) n=6 (S) n=6 Rest	Ergometry
Granulocytes		Control=C	36,4 ± 2,5	36,3 ± 3,0	37,0 ± 6,7	37,2 ± 5,6
		Splenectomy=S	23,8 ± 3,7**	21,8 ± 3,7**	28,8 ± 7,1	26,5 ± 4,4
Monocytes		C	2,9 ± 0,3	3,0 ± 0,2	2,6 ± 0,2	3,1 ± 0,5
		S	3,0 ± 0,4	3,4 ± 0,6	2,8 ± 0,3	2,9 ± 0,3
Lymphocytes		C	46,1 ± 2,1	46,0 ± 3,0	43,2 ± 5,8	42,8 ± 4,8
		S	56,1 ± 3,2**	58.1 ± 3,4*	50,7 ± 5,0	52,8 ± 3,3
- B cells	(sIg)	C	23,5 ± 1,4	35,1 ± 1,7"""	24,0 ± 2,9	25,8 ± 3,0
		S	33,8 ± 2,5***	47,4 ± 3,2"""***	33,7 ± 3,1**	37,3 ± 2,9"*
- T cells	(Leu-1)	C	69,0 ± 1,7	55,6 ± 2,2"""	66,3 ± 3,0	64,5 ± 2,6
		S	52,4 ± 3,9***	41,0 ± 3,7"""***	46,0 ± 4,3**	46,2 ± 4,0**
- T helper cells	(Leu-3a)	C	43,7 ± 1,9	33,8 ± 1,3"""	50,8 ± 1,8	48,3 ± 2,0
		S	36,7 ± 2,1*	27,4 ± 1,8"""**	36,6 ± 2,9**	34,0 ± 4,0*
- T suppressor cells	(Leu-2a)	C	25,2 ± 1,1	25,2 ± 1,0	22,3 ± 1,9	22,5 ± 2,5
		S	20,5 ± 1,9*	20,3 ± 2,0*	17,6 ± 2,0	20,4 ± 1,6
- Natural killer cells	(OKM1)	C	4,8 ± 0,5	10,7 ± 1,2"""	2,9 ± 0,3	5,6 ± 1,0
		S	6,5 ± 0,9	11,8 ± 1,8""	4,7 ± 1,3	6,9 ± 1,3

After ergometry during beta-blockade still a significant increase of the number of granulocytes and to a lesser extent of the monocytes and lymphocytes was seen (table 1). The percentage changes were smaller when compared to those without beta-blockade (table 3). The stress induced increase of Leu-1- and particularly Leu-3a cells was weakley influenced, the B- and Leu-2a cell mobilization was more affected by the beta-blockade: The adrenoceptor independent increase of B- and Leu-2a cells reached only about half the value of that without treatment (fig. 1). Beta-blockade completely prevented the ergometry induced reduction of the T/B- and Leu-3a/Leu-2a cell ratios (table 1). As shown in table 2, relative cell numbers remained completely unchanged when beta-blockade was combined with ergometry.

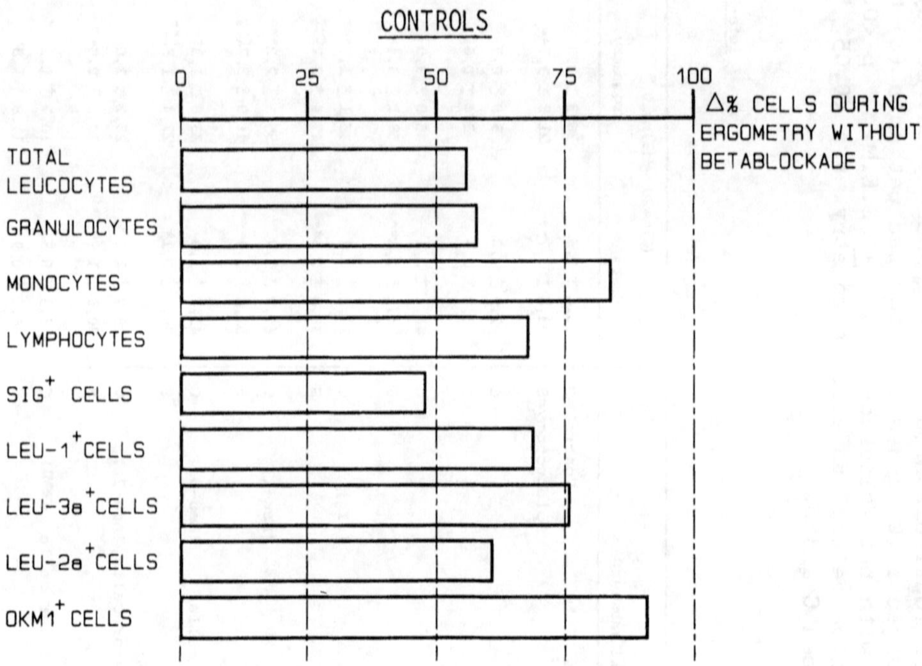

Fig. 1 Extent of cell mobilization after ergometry with beta-blockade as compared to the mobilization without beta-blockade (=100%) in 20 normal subjects.

Table 3: Percentage changes of leucocyte subsets induced by ergometry without and with beta-blockade in normal subjects and patients with splenectomy.

Δ % CELLS		CONTROLS		SPLENECTOMY	
		BETA-BLOCKADE		BETA-BLOCKADE	
		- n=20	+ n=6	- n=10	+ n=6
TOTAL LEUCOCYTES		104	59	48	23
GRANULOCYTES		111	64	47	12
MONOCYTES		112	94	76	24
LYMPHOCYTES		90	61	45	28
- B CELLS	(sIg)	181	87	94	41
- T CELLS	(Leu-1)	65	45	14	17
- T HELPER CELLS	(Leu-3a)	55	42	6	13
- T SUPPRESSOR CELLS	(Leu-2a)	100	61	52	34
- NATURAL KILLER CELLS	(OKM1)	313	284	170	22

As previously described,[6] in splenectomized patients the resting blood leucocyte distribution differed from that observed in normal subjects. Elevated numbers of monocytes ($p < 0,05$), B- ($p < 0,01$) and OKM1 cells ($p < 0,001$) were seen in the patients. The ratio of T/B cells in the blood was reduced and the T cell proportions were unaltered. As shown in table 2, there was a fractional decrease of granulocytes, an increase of the relative number of lymphocytes, a high proportion of B- and a low proportion of T cells.

In the patients the physical stress test led to a qualitatively similar, but quantitatively smaller alteration of the fractional blood cell distribution when compared to that found in normal subjects. As shown in table 1 and 3 and illustrated in fig. 2 in the absence of the spleen less granulocytes, monocytes and lymphocytes were released into the circulation. No significant Leu-3a cell mobilization occurred at all. However B cells increased by nearly 100%, Leu-2a cells by 52% and OKM1 cells by 170%. The ratios of T/B- and of Leu-3a/Leu-2a cells were reduced.

In splenectomized patients the stress induced increase of granulocytes and monocytes was abolished by beta-blockade (table 1 and 3). Leu-1- and Leu-3a cells shown to be insensitive to adrenergic stimulation in the absence of the spleen, as expected, did not show any change when beta-blockade was combined with ergometry (table 1 and 3). B- and Leu-2a cells were still accumulating in the blood after stress with beta-blockade (table 1 and 3). Similar to the observation in normals, the adrenoceptor independent mobilization included only about half of the B- and Leu-2a cells, which had appeared during the initial ergometry (table 1 and 3). T/B cell ratios and Leu-3a/Leu-2a ratio did not change in splenectomized individuals after ergometry with beta-blockade (table 1). Besides a significant proportional increase of the B cells, no changes of the relative cell numbers were noticed after ergometry with beta-blockade (table 2). The two splenectomized patients who repeated ergometry without beta-blockade showed comparable absolute and fractional cellular changes in both the first and second stress test (data not shown).

DISCUSSION

In the present study physiological stimulation of the adrenergic nervous system in normal subjects led to an increase of circulating

SPLENECTOMY

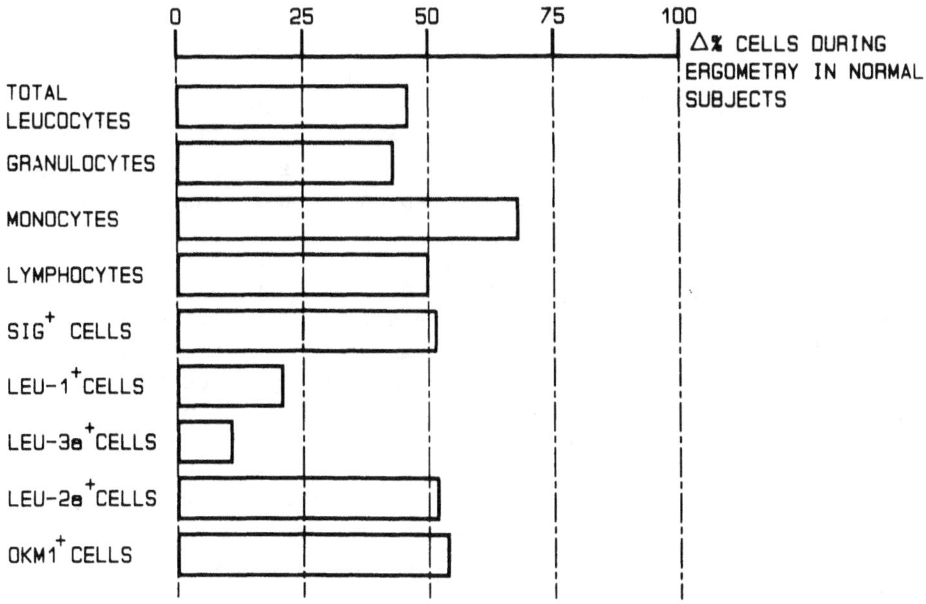

Fig. 2 Extent of cell mobilization after ergometry in 10 splenectomized patients as compared to the mobilization in 20 normal subjects (=100%).

granulocytes, monocytes and lymphocytes. Within the lymphocyte population, Leu-2a -, OKM1- and B cells were preferentially mobilized. The cellular reaction was diminished during beta-blockade, the changes of the B-and Leu-2a cells were more affected than those of the Leu-3a-, OKM1 cells and monocytes. In splenectomized patients the stress induced Leu-3a cell increase was absent, whereas the increase of all other leucocyte subsets reached about half of that found in normal subjects. Beta-blockade reduced the granulocyte-, monocyte- and OKM1 cell increase much more than in normal subjects; the B- and Leu-2a cell changes were affected to the same degree in patients and controls.

We had shown previously, that under the experimental conditions used, the fractional cellular changes in blood were due to the activation of the sympathetic nervous system. Plasma concentrations of both norepinephrine and epinephrine increased during short term physical stress, indicating a stimulation of the alpha- and beta-adrenergic system; plasma cortisol did not change[2]. Our findings at ergometry tally with the results of Crary et al.[7], who observed a decrease of cells with helper- and an increase of cells with natural killer cell phenotype after epinephrine injection in humans. However he did not see any proportional changes of B- and T cells.

The mechanism of cell influx in response to physical stress is unknown, it may be explained by either splanchnic flow changes or by direct binding of catecholamines to lymphocytes in the different lymphoid organs. Hemodynamic phenomena can hardly account for the selective lymphocyte phenotype changes, unless we postulate that certain cells are located in distinct structures of spleen and lymph nodes, which are circumvented by shunts. The present study does not completely exclude this possibility, because the distribution of alpha- and beta-adrenergic receptors in the vessels of spleen and lymph nodes is unknown. Beta-adrenergic receptors have been described in rat spleen homogenates,[8] but their regulation during physical exercise has not been investigated. Alternatively, the fractional cellular changes after stress may be explained by properties of the cells themselves: Mono- and polymorphonuclear leucocytes exhibit adrenergic receptors of alpha- and of beta-2-specificity[9,10,11,12]. Beta-adrenergic receptors of high density or affinity are known to exist on B- and Leu-2a cells rather than on Leu-3a cells[13]. The distribution of alpha-adrenergic receptors in lymphocyte subsets is unknown. At ergometry, norepinephrine released at the sympathetic nerve endings or epinephrine arriving through the bloodstream, may bind preferentially to beta-adrenergic receptors on B- and Leu-2a cells. Neurotransmitter or hormone may thus facilitate the migration of these cells by altering their surface properties. Our earlier results[2] favour the hypothesis of a selective beta-adrenergic effect upon B- and Leu-2a cells with their high density and high affinity receptors, because they were most sensitive to the mobilizing effect of physical stress. The present investigation confirmed these findings since B- and Leu-2a cells were most sensitive to the beta-blocking drug. Similarly, the increase of granulocytes and monocytes could be due to a release of these cells from the intravascular marginal pool[14] by activation of their beta-adrenergic receptors.

In the presence of the spleen, beta-blockade selectively reduced, but did not abolish cell mobilisation, which agrees with the data of French et al.[15], who could not completely inhibit the epinephrine induced leucocyte response with propranolol. Other -perhaps alpha-adrenergic- mechanisms must account for most of the

OKM1 cell and monocyte accumulation in blood and for half of the B-, Leu-2a- and granulocyte changes. By studying the stress induced alterations in splenectomized subjects, it was possible to estimate the contribution from the spleen, which was either dependent or independent of a beta-adrenergic mechanism. Cells circulating in the spleen accounted for approximately 50% of the stress induced leucocytosis. These results are in contrast to those of Steel et al.[16], who found the epinephrine induced lymphocytosis unaffected by splenectomy. Unlike all other lymphocyte subsets, the mobilizable pool of Leu-3a cells originated exclusively from this organ. These cells have a low density of beta-adrenergic receptors;[13] they may therefore need higher catecholamine levels to react, which may not be reached in lymph nodes. Also, the architecture of the lymph nodes may render Leu-3a cells less accessible to sympathetic stimulation.

The beta-blockade had different effects in the presence or absence of the spleen. It reduced the B- and Leu-2a mobilisation to a similar extent in patients and controls. Therefore B- and Leu-2a liberation from the spleen was entirely beta-adrenoceptor dependent; the cells which accumulated by an adrenoceptor independent mechanism probably did not originate from the spleen. The granulocyte-, monocyte- and OKM1 cell changes were weakly influenced by beta-blockade in normals, more affected by splenectomy, but were nearly abolished, when beta-blockade was combined with splenectomy. So we may conclude, that part of these cell changes were spleen dependent. The other, spleen independent but strongly beta-adrenoceptor dependent part may well orginate from the peripheral marginal pool.

The differential beta-adrenergic sensitivity of the lymphocyte subsets and their distribution in various lymphoid compartments may explain the selective cellular changes induced by adrenergic activation.

REFERENCES

1. A. J. Samuels, Primary and secondary leucocyte changes following the intramuscular injection of epinephrine hydrochloride, J. Clin. Invest. 30:941 (1951).
2. R. M. A. Landmannn, F. B. Müller, Ch. Perini, M. Wesp, P. Erne and F. R. Bühler, Changes of immuoregulatory cells induced by psychological and physical stress: relationship to plasma catecholamines, Clin. exp. Immmunol. in press (1984).
3. E. Hedfors, G. Holm and B. Oehnell, Variations of blood lymphocytes during work studied by cell surface markers, DNA synthesis and cytotoxicity, Clin. exp. Immunol. 24:328 (1976).
4. E. Hedfors, G. Holm, M. Ivansen and J. Wahren, Physiological

variation of blood lymphocyte reactivity: T-cell subsets, immunoglobulin production, and mixed-lymphocyte reactivity, Clin. Immunol. Immunopathol. 27:9 (1983).

5. A. M. A. Gader and J. D. Cash, The effect of adrenaline, noradrenaline, isoprenaline and salbutamol on the resting levels of white blood cells in man, Scand. J. Haematol. 14: 5 (1975).
6. M. Dürig, R. M. A. Landmann and F. Harder, Lymphocyte subsets in human peripheral blood after splenectomy and autotransplantation of splenic tissue, J. Lab. Clin. Med. 104:110 (1984).
7. B. Crary, S. L. Hauser, M. Borysenko, I. Kutz, C. Hoban, K. A. A. Ault, H. L. Weiner and H. Benson, Epinephrine induced changes in the distribution of lymphocyte subsets in peripheral blood of humans, J. Immunol. 131: 1178 (1983).
8. S. Yamada, H. I. Yamamura and W. R. Roeske, Characterization of postsynaptic beta-adrenergic receptors and presynaptic muscarinic cholinergic receptors in the rat spleen, Life Sci. 31: 1161 (1982).
9. Y. Sano, K. Krokos, J. B. Cheng, A. Bewtra and R. Townley, Comparison of alpha and beta adrenergic receptors in asthmatics and controls: identification and characterization of alpha adrenergic receptors in human lymphoytes, Clin. Res. 29: 172A (1981).
10. R. Landmann, H. Bittiger and F. R. Bühler, High affinity beta-2-adrenergic receptors in mononuclear leucocytes: Similar density in young and old normal subjects, Life Sci. 29: 1167 (1981).
11. J. O. Panosian and G. V. Marinetti, Alpha-2-adrenergic receptors in human polymorphonuclear leucocyte membranes, Biochem. Pharmacol. 32: 2243 (1983).
12. S. P. Galant and S. Allred, Binding and functional characteristics of beta-adrenergic receptors in the intact neutrophil, J. Lab. Clin. Med. 98: 227 (1981).
13. R. M. A. Landmann, E. Bürgisser, M. Wesp and F. R. Bühler, Beta-adrenergic receptors are different in subpopulations of human circulating lymphocytes, J. Rec. Research in press (1984).
14. J. M. Mishler,IV, and A. A. Sharp, Adrenaline: further discussion of its role in mobilization of neutrophils, Scand. J. Haematol. 17: 78 (1976).
15. E. B. French, C. M. Steel and W. R. C. Aitchison, Studies on adrenaline-induced leucocytosis in normal man. II. The effects of alpha and beta adrenergic blocking agents, Brit. J. Haematol. 21:423 (1971).
16. C. M. Steel, E. B. French and W. R. C. Aitchison, Studies on adrenaline-induced leucocytosis in normal man. I. The role of the spleen and of the thoracic duct, Brit. J. Haematol. 21: 413 (1971).

CHARACTERISATION OF MONONUCLEAR CELL INFILTRATES BY MONOCLONAL ANTIBODIES IN NORMAL AND NEOPLASTIC COLORECTAL TISSUE

Doris Heinemann, H.C. Umpleby, M.O. Symes
and R.C.N. Williamson

Dept. of Surgery, The Medical School,
University of Bristol, Bristol BS8 1TD UK

Monoclonal antibodies to human leukocyte antigens have previously been used to characterise mononuclear cells infiltrating carcinomas of the breast (1-3) and ovary (4). The present report extends this technique to benign and malignant neoplasms of the colorectum. These tumours are of particular interest because adenomas are recognised as pre-cancerous lesions in the large bowel (5). Therefore the study of cell types infiltrating adenomas, in comparison to frank carcinomas, would afford evidence of the nature of the host response to a developing neoplasm.

MATERIALS AND METHODS

Clinical Material

Eight patients with a total of 10 adenomas were studied. A carcinoma co-existed with the adenoma(s) in 5 patients. A specimen of normal bowel was also obtained from 3 of these patients and from a further 8 patients with evidence of colorectal neoplasia. An 0.5g biopsy was obtained from the edge of the freshly resected neoplasm (adenoma or carcinoma) and also from the normal bowel.

Histological Sections

Biopsies were snap frozen in liquid nitrogen, and 5 u sections were cut using a cryostat. After drying in air the sections were fixed in 99% v/v acetone at room temperature for 10 minutes. After washing the sections in TBS (pH 7.6) 50 ul of the appropriate monoclonal antibody was added to the section. Following incubation at room temperature for 45 min, the section was washed twice with TBS, and 50 ul of peroxidase-conjugated rabbit anti-mouse IgG (diluted

1/50 in TBS + 10% normal human serum) was applied for a further 30 min. After a further 2 washes with TBS, 50 ul of TBS/DAB (2.5 ug DAB in 45 ul TBS + 5 ul 3% v/v H_2O_2) was added to the section for 8 min. Finally, the sections were washed with tap water and counter-stained with Mayer's haematoxylin. As a negative control, 3 sections from each biopsy were treated as above, but without the application of a monoclonal antibody.

Monoclonal Antibodies

Sections from each biopsy were stained with one each of the following monoclonal antibodies:

HLe-1: An IgG1 antibody which stains all peripheral blood leukocytes and all thymocytes.

UCHT-1: An IgG1 antibody which recognises the antigen T3 on peripheral T cells and some thymocytes (6).

UCHT-4: An IgG2 antibody which identifies the antigen T8 on suppressor/cytotoxic T cells (7).

OKT-4 (Ortho): An IgG2b antibody which recognises the T4 antigen on helper/inducer T lymphocytes (8). Some sections stained with this antibody were subsequently immersed in 1% osmium tetroxide to enhance their staining (3).

OKIa-1 (Ortho): An IgG2 antibody which recognises HLA-DR framework antigens.

MAS 017 (Sera-Lab): An IgG1 antibody which recognises a common determinant shared by HLA-A-B-C antigens.

Grading of mononuclear cell infiltrates

The whole of each section was examined independently by two observers. The number of mononuclear cells showing membrane staining with peroxidase was estimated following staining with each monoclonal antibody. The following grading scheme was employed to assess the degree of infiltration by a particular mononuclear cell type: 4 (heavy), 3 (moderate), 2 (few), 1 (occasional) and 0 (nil).

RESULTS

The degree of leukocyte infiltration (HLe-1 grade 4) and total T cell infiltration (UCHT-1 grade 3) was similar in all 10 adenomas (Table 1). However, in 9/10 adenomas cytotoxic/suppressor T lymphocytes (UCHT-4 grade 2) predominated over helper T lymphocytes (OKT-4 grade 1). A similar predominance of $T_{c/s}$ over T_H lymph-

Table 1. The degree of membrane staining shown by infiltrating mononuclear cells in 10 adenomas, 5 concomitant carcinomas, and 3 adjacent normal bowel biopsies

Adenoma Number	Adenoma(A) Carcinoma(C) or Normal(N)	HLE-1	Monoclonal Antibody++ UCHT-1	UCHT-4	OKT-4	OKIa-1
1*	A	4**	3	2-3	1	2
2*	A	4	3	2	1	2-3
3	A	4	3	2-3	2-3	2-3
4	A	3	1-2	2	0	1
	C	3	3	1	0	1
5*	A	4	2-3	2	0-1	2
	C	4	4	3	1	1
6*	A	4	3-4	2	1	3
	C	4	4	3	1	1
7	A	4	2-3	2	1	1
	C	3	3	1-2	1	2-3
8	A	4	3	2	1	2-3
	N+	3-4	3	2	2	3-4
9	A	4	3	2	1	1
	N+	2-3	2	0-1	1	2
	C	2	2	1	0-1	1
10	A	4	2-3	2	0-1	3
	N+	4	3	1-2	0-1	1
	C	4	3	2-3	0-1	3

*2 adenomas in the same patient; adenomas 5 and 6 were each compared with the single carcinoma that arose in this patient.

+Biopsies were also obtained from a further 8 specimens of normal large bowel. Staining of these specimens was graded as follows:

HLE-1	4 in two and 3 in six
UCHT-1	3 in seven and 2 in one
UCHT-4	2 in one and 1 in seven
OKT-4	1 in four and 0 in four
OkIa-1	3 in one, 2 in one and 1 in six

++Monoclonal antibodies used for staining.

**Grade of cell infiltration assessed by 2 independent observers over the whole section: 4 heavy, 3 moderate, 2 few, 1 occasional, 0 nil.

Table 2. Histological types of ten adenomas and Dukes stages and differentiation of five concomitant carcinomas

Adenoma Number:	Adenoma(A) or Carcinoma(C)	Adenoma Classif^{n}	Dukes stage/ different^{n}:	Okla-1 staining:
1*	A	tubular		+
2*	A	tubular		+
3	A	tubular		-
4	A	tubulo-villous		-
	C		D+, moderate	
5*	A	villous		-
6*	A	villous		+
	C		A, well	
7	A	tubular		+
	C		A, moderate	
8	A	tubular		+
9	A	tubular		+
	C		C, moderate	
10	A	tubulo-villous		-
	C		C, moderate	

*2 adenomas in the same patient
+with liver metastases

ocytes was seen in all 5 concomitant carcinomas, but not in the specimens of normal bowel. In a subsidiary experiment specimens from a further 7 carcinomas were stained using the antibody OKT-4, with and without osmium tetroxide enhancement, and the degree of T_H infiltration was the same. In both the adenomas and the carcinomas mononuclear cell infiltration was seen in the stromal tissue between the nests or sheets of neoplastic cells. Infiltration among the neoplastic cells did not occur.

Adenomas can be classified according to their tissue architecture as tubular, tubulo-villous or villous (9). The malignancy rate for tubular adenomas is lower (5%), than that for the villous lesions (40%), while that for tubulovillous adenomas is intermediate (22%) (10). The histological classification of the 10 adenomas studied is shown in Table 2. Staining with OKla-1 showed expression of Class II MHC framework antigens on the neoplastic cell membranes in 5/6 tubular adenomas, but in only 1/4 of villous or mixed lesions. Further, the 5 carcinomas all failed to show expression of Class II antigens on the neoplastic cells. Thus, loss of Class II antigen expression would seem to be associated with neoplastic progression.

DISCUSSION

Previously noted for carcinomas of the breast (2), the predominance of $T_{c/s}$ over T_H lymphocytes in both adenomas and carcinomas suggests a response by the cytotoxic cells to tumour associated antigens in association with Class I MHC Ag. The tissues were not examined for Class I antigen in the present experiments, but in further work (unpublished) expression of this antigen in carcinomas has been found to be variable and an excess of cytotoxic over helper cells occurred only in those lesions in which this antigen was demonstrated. However, the failure of the $T_{c/s}$ lymphocytes to infiltrate among the neoplastic cells argues against a response of $T_{c/s}$ to tumour antigens (7), so that they may be acting as suppressor cells. Tumour antigens can also be recognised in association with Class II MHC Ag by helper lymphocytes (11). Such a delayed hypersensitivity type response is associated with lymphocyte infiltration between the areas of neoplastic cells (7), as seen in the present study. Perhaps such a response contributes to the low malignant potential of tubular adenomas, which consistently expressed Class II Ag. As a corollary, the failure of the mixed and villous adenomas, together with the carcinomas to express Class II Ag would lead to a loss of immunologic control in accordance with their greater malignancy.

Abbreviations

DAB = Diaminobenzidine
TBS = Tris-buffered saline
T_H = T helper cell
$T_{c/s}$ = T cytotoxic/suppressor cell

Acknowledgments

We thank Dr. P.C.L. Beverley for his generous gift of monoclonal antibodies HLe-1, UCHT-1 and UCHT-4. H.C.U. is a Wellcome Surgical Fellow.

REFERENCES

1. A.K. Bhan and C.L. Des Maris, J. Natl. Cancer Inst. 71: 1507 (1983).
2. B.J. Rowe and P.C.L. Beverley, Br. J. Cancer 49: 149 (1984).

3. H.L. Whitwell, H.P.A. Hughes, M. Moore and A. Ahmed, Br. J. Cancer 49: 161 (1984).
4. S.E. Kabawat, R.C. Bast, W.R. Welch, R.C. Knapp and A.K. Bhan, Int. J. Cancer 32: 547 (1983).
5. B.C. Morson, Proc. Roy. Soc. Med. 67: 451 (1973).
6. P.C.L. Beverley and R.E. Callard, Eur. J. Immunol. 11: 329 (1981).
7. P.C.L. Beverley, Proc. Roy. Soc. Med. 81B: 221 (1982).
8. J.A. Ledbetter, R.L. Evans, M. Lipinski, C. Cunningham, R.A. Good and L.A. Herzenberg, J. Exp. Med. 153: 310 (1981).
9. T. Muto, H.J.R. Bussey and B.C. Morson, Cancer 36: 2251 (1975).
10. B.C. Morson, Clin. Gastroenterol. 5: 505 (1976).
11. S.C. Meuer, R.F. Hussey, J.C. Hodgson, T. Hergend, S.F. Schlossman and E.L. Reinherz, Science 218: 471 (1982).

HUMAN LYMPHOCYTE SUBPOPULATIONS IN LYMPHADENOPATHIES IN HOMOSEXUAL MEN

P. Rácz*, A. C. Feller°, K. Tenner-Rácz♁,
B. Luther*, K. Lennert°, M. Dietrich*, and P. Kern*

* Institute for Tropical Diseases Hamburg (W.Germany)
° Department of Pathology, University of Kiel "
♁ St. Georg General Hospital, Hamburg "

During the past few years homosexual men have been affected by a syndrome consisting of persistent generalized lymphadenopathy (PGL) (1,2,3). A recent longitudinal study (4) has shown that acquired immunodeficiency syndrome (AIDS) and PGL are integrally related. Most patients with PGL show similar immune abnormalities as patients with AIDS qualitatively, but they are less severe (cutaneous anergy, diminished in vitro lymphocyte transformation by mitogens, lymphopenia and a reduced ratio of circulating T-helper lymphocytes to T-suppressor lymphocytes) (1,5,6).

The architectural distribution of different cell types of the lymphoid organs is of great importance to evaluate the precise nature of this immunodeficiency, because many immune responses are initiated or regulated by these organs.

Patients and Methods

Twenty-one homosexual men with PGL and constitutional symptoms and 5 homosexual men with AIDS from the Hamburg metropolitan area formed the basis of our study. All patients included in the study were between 21 and 50 years old. Fifty cases of nonspecific lymphadenitis served as the control. In all cases, formalin-fixed lymph node tissue was embedded in paraffin wax. Sections were stained with Giemsa, periodic acid-Schiff (PAS), H & E and silver impregnation. Immunoperoxidase evaluation was conducted on 6 μm cryostat sections. The immunoperoxidase technique was used as described by Stein et al. (7,8).

Briefly, sections were first air dried and then fixed twice in acetone. After incubation with monoclonal antibodies (see table 1), the sections were washed and overlayed with peroxidase-conjugated rabbit anti-mouse Ig. The slides were washed again and then incubated with peroxidase-conjugated goat-anti-rabbit IgG (Medac, Hamburg, FRG). Finally, the sections were developed with diaminobenzidine (0.6 mg/ml) and hydrogen peroxidase (0.01 %). The slides were then counterstained with Mayer's modified haematoxylin for 1 min. The number of cells stained with various antibodies was estimated throughout the entire section, and is expressed separately as percentage of stained cells for germinal centres (GC), mantle zones, paracortical areas and sinusoids.

Results and Discussion

Allowing for variations in intensity, the histological picture of the lymph nodes in the cases with PGL alone and in 4 patients with AIDS was essentially similar in all cases. The lesions resembled those of homosexual men described previously by Joachim et al. (3,9,10). One of the dominating histological features was the explosive follicular hyperplasia with irregularly shaped follicles (Fig. 1). Instead of their usual round or oval configuration, the follicles assumed serpentine, hourglass or dumbbell forms with serrated or protruding margins. Their size ranged from 0,2 mm to 3 mm. Some follicles extended almost the whole thickness of the node. The follicles had both cortical and diffuse distributions. The GC-s often contained very small sparsely populated zones. Almost the entire GC showed the histological characteristics of densely populated dark zones, destruction of the lymphocytes, numerous tingible body macrophages and lymphoid cells in mitosis. Centrocytes with plasmacytoid features were not seldom found in sparsely populated areas. Deposition of proteinaceous material within the follicles was occasionaly observed. The large GC-s were surrounded by a partially attenuated mantle zone. Among the small lymphocytes of the mantle zone, immunoblasts, plasma cells and sometimes centrocytes were noted. The submarginal zone, interfollicular area and the paracortex contained increased numbers of immunoblasts, plasma cells, focal aggregates of polymorphonuclear leucocytes, giant cells, multinucleated lymphoid cells, atypical lymphocytes and haemophagocytic macrophages. Increased vascularity was demonstrated. Lymphoid proliferation in the perinodal tissue was often found. One patient with AIDS died 8 month after the lymph node biopsy was performed. The lymph node biopsy showed similar histopathological features as lymph nodes of the other patients. The lymph nodes of the autopsy study had a different histological appearance. The follicles were small, hypocellular and sometimes hyalinized. The nodal parenchyma contained many immunoblasts, plasma cells, and an extensive proliferation of blood vessels.

Table 1. Applied monoclonal antibodies

Antibody	Source of antibody/Reference
Anti-Leu-1 T11	Becton-Dickinson Coulter Electronics, Hialeah
Anti-Leu-2a OKT 8	Becton-Dickinson Ortho Diagnostic Systems
Anti-Leu-3a OKT 4	Becton-Dickinson Ortho Diagnostic Systems
Anti-Leu-6	Becton-Dickinson
Anti-Leu-7	Becton-Dickinson
Anti-Leu-8	Becton-Dickinson
Dako-Pan-B (M 708) TO15 (all B-cells)	Dakopatts Stein et al. (8)
Dako-IgD (M 703) Anti-IgD	Dakopatts Proma
Ki M4 (follicular dendritic cells)	Behringwerke
Ki B1 (follicular mantle lymphocytes)	Behringwerke
Ki M6 (Macrophages)	Behringwerke
T6	Coulter Electronics, Hialeah
Anti-Tac (Interleukin2-Receptor)	Institut of Pathology, University Kiel

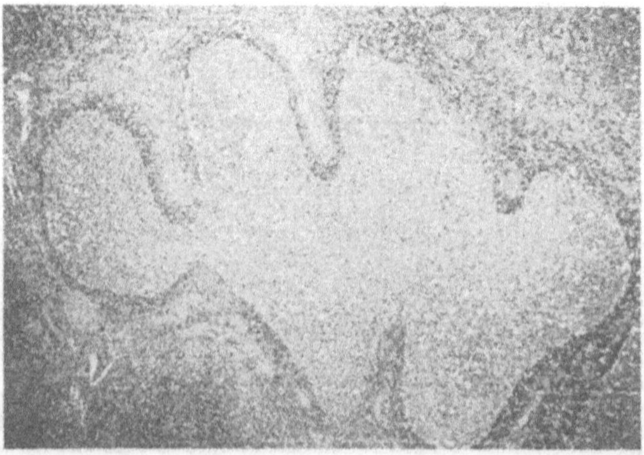

Fig. 1 Hyperplasia of a lymphoid follicle. The germinal center is markedly enlarged, and irregularly shaped. (Giemsa, x 30).

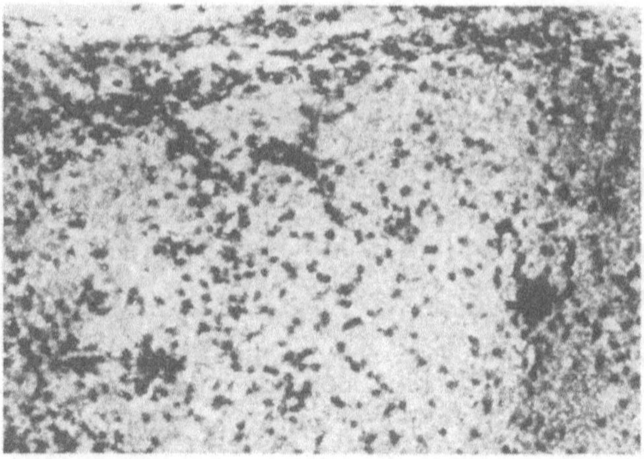

Fig. 2 Increased numbers of suppressor T-cells in the germinal center and in the mantle zone (indirect immunoperoxidase, haematoxylin counterstain, x 48).

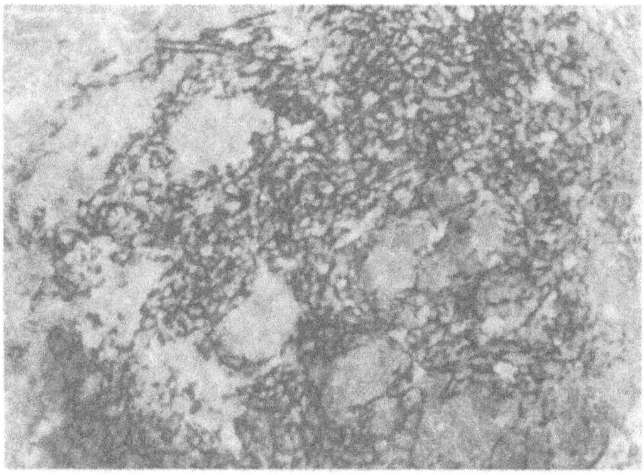

Fig. 3 Immunoperoxidase staining with the monoclonal antibody Ki M4 which reacts nearly selectively with FDC. Note the moth-eaten appearance of the network of FDC (haematoxylin counterstain, x 48).

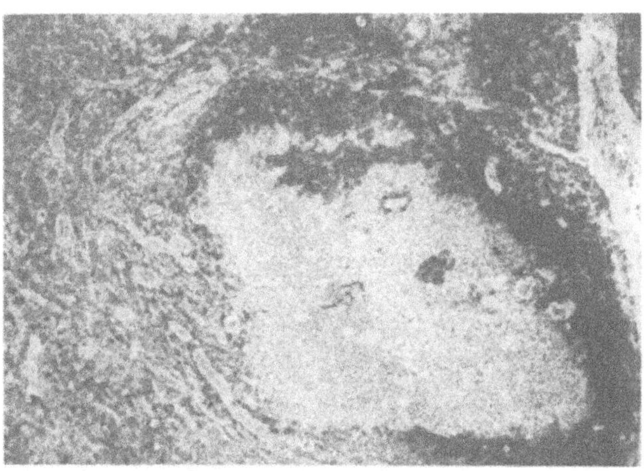

Fig. 4 Immunoperoxidase staining for IgD. Follicular mantle lymphocytes show strong positivity. The mantle zone is irregularly formed (haematoxylin counterstain, x 48).

The homosexual group (PGL and AIDS patients) exhibited several immunohistochemical differences in comparison to the controls. In agreement with other authors (11,12,13,14,15), we found increased numbers of suppressor T-cells in the GC-s and within the mantles. In addition, we noted that the cells stained for Leu2a and OKT8 were often grouped in irregular small clusters scattered thoughout the GC-s (Fig. 2). Often such clusters were located in the neighborhood of small blood vessels entering the densely populated areas of the germinal centers. This finding suggested that the suppressor/cytotoxic T-cells entered the follicles via blood vessels. In the majority of cases, the number of Leu8+ cells in the GC-s slighly (in a few cases strongly) decreased, whereas the number and the distribution of Leu7+, anti Tac+, and OKT6+ cells were within normal ranges. The patient's GC-s contained fewer helper T-cells (Leu3a and OKT4) than the control's GC-s.

Of further interest is our finding that the network of follicular dendritic cells (FDC) in the lymph nodes of our patients has an irregular staining pattern (Fig. 3). This network is present over the entire area of the GC-s in the control nodes. In contrast to the control patients the network showed a moth-eaten appearance according to the localization of groups of Leu2a+ and OKT8 cells in our patients with PGL or AIDS. The FDC are involved in presenting the antigen to the B-cells (16-22). Thus, our findings probably indicate a disturbance of the antigen presenting system of the GC-s in PGL and AIDS.

The IgD+ mantle zones were often irregularly shaped, partially attenuated and eroded (Fig. 4). In the paracortex there was a decreased number of T-helper cells and a concomitant increase in T-suppressor cells. Our findings were in agreement with those of Wood et al. (14). The thymic dependent regions of the lymph nodes are important in cellmediated immunity and thus, changes in these areas may affect the immune response.

Finally, we found a marked immature sinus histiocytosis which is composed of B-lymphocytes as shown by staining for IgM, Tü 66 and Tü 108 (23). In several cases among these lymphocytes we noted the presence of many suppressor/cytotoxic T-cells.

Further investigations are necessary to show whether these findings are specific for PGL and AIDS or whether they may be found also in other conditions involving a decreased T-helper/T-suppressor ratio.

ACKNOWLEDGEMENTS

We are indebted to Ms. A. Schulz-Meyer for her skillfull assistance.

REFERENCES

1. D. Mildvan, U. Mathur, R. E. Enlow, D. Armstrong, J. Gold, C. Sears, B. Wong, A. E. Brown, S. Henry, B. Safai and Z. Arlin, Persistent, generalized lymphadenopathy. Morbid Mortal Weekly Rep 31:249 (1982).
2. A. E. Friedman-Kien, L. J. Laubenstein, P. Rubinstein, E. Buimovici-Klein, M. Marmor, R. Stahl, I. Spigland, K. S. Kim, and S. Zolla-Pazner, Disseminated Kaposi's sarcoma in homosexual men. Am Intern Med 96:693 (1982).
3 H. L. Joachim, C. W. Lerner, and M. L. Tapper, Lymphadenopathies and lymphoma in immunocompromised homosexual men, Lab Invest 48:39A (1983).
4. U. Mathur-Wagh, R. W. Enlow, I. Spigland, R. J. Winchester, H. G. Sacks, E. Rorat, S. R. Yancovitz, M. J. Klein, D. C. William, and D. Mildvan, Longitudinal study of persistent generalized lymphadenopathy in homosexual men: relation to acquired immunodeficiency syndrome. Lancet i:1033 (1984).
5. R. E. Stahl, A. Friedman-Kien, R. Dubin, M. Marmor, and S. Zolla-Pazner, Immunologic abnormalities in homosexual men: relationship to Kaposi's sarcoma. Am J Med 73:171 (1982).
6. J. I. Wallace, F. S. Coral, I. J. Rimm, H. Lane, H. Levine, E. L. Reinherz, S. F. Schlossmann, and J. Sonnabend, T-cell ratios in homosexuals. Lancet i:908 (1982).
7. H. Stein, A. Bonk, G. Tolksdorf, K. Lennert, H. Rodt, and J. Gerdes, Immunohistologic analysis of the organization of normal lymphoid tissue and non Hodgkin's lymphomas. J Histochem Cytochem 28:746 (1980).
8. H. Stein, J. Gerdes, U. Schwab, H. Lemke, D. Y. Mason, A. Ziegler, W. Schienle, and V. Diehl, Identification of Hodgkin and Sternberg-Reed cells as a unique cell type derived from a newly-detected small cell population. Internat J Cancer 30:445 (1982).
9. H. L. Joachim, C. W. Lerner, and M. L. Tapper, Lymphadenopathies in homosexual men. Relationships with the acquired immune deficiency syndrome. JAMA 250:1306 (1983).
10. H. L. Joachim, C. W. Lerner, and M. : Tapper, The lymphoid lesions associated with the acquired immunodeficiency syndrome. Am J Surg Pathol 7:543 (1983).
11. R. L. Modlin, P. R. Meyer, F. M. Hofman, M. Mehlmauer, N. B. Levy, R. J. Lukes, J. W. Parker, A. J. Amman, M. A. Conant, T. H. Rea, and C. R. Taylor, T-lymphocyte subsets in lymph nodes from homosexual men. JAMA 250:1302 (1983).

12. R. L. Modlin, F. M. Hofman, P. R. Meyer, S. A. Vaccaro, A. J. Amman, M. A. Conant, T. H. Rea, and C. R. Taylor, Altered distribution of Band T-lymphocytes in lymph nodes from homosexual men with Kaposi's sarcoma. Lancet ii:768 (1983).
13. W. C. Chan, R. K. Byrnes, P. M. Bank, T. J. Spira, C. Thurmond, E. P. Ewing, and F. W. Chandler, Immunohistochemical studies on lymph nodes from homosexual men with chronic lymphadenopathy. Lab Invest 50:9A (1984).
14. G. S. Wood, B. F. Burns, R. F. Dorfman, and R. A. Warnke, Quantification of helper, suppressor, and cytotoxic T-cell subsets in acquired immune deficiency syndrome. Lab Invest 50:68A (1984).
15. M. Mangkornkanok-Mark, A. S. Mark, and J. Dong, Immunoperoxidase evaluation of lymph nodes from acquired immune deficiency patients. Clin exp Immunol 55:581 (1984)
16. G. J. V. Nossal, A. Abbott, J. Michell, and Z. Lumnus, Antigens in immunity XV. Ultrastructural features of antigen capture in primary and secondary lymphoid follicles. J Exp Med 127:277 (1968).
17. G. G. B. Klaus, H. J. Humphrey, A. Kunkl, and D. W. Dongworth, The follicular dendritic cell: its role in antigen presentation in the generation of immunological memory. Immunol Rev 53:3 (1980).
18. K. Lennert, and H. Stein, The germinal center: Morphology histochemistry and immunobiology. In: Goos, and E. Christophers (eds): Lymphoproliferative Disease of the Skin. Springer Verlag Berlin-Heidelberg-New York, p 3 (1982).
19. M. R. Parwaresch, J. H. Radzun, A. C. Feller, K. P. Peters, and M. L. Hansmann, Peroxidase-positive mononuclear leukocytes as possible precursors of human dendritic reticulum cells. I Immunol 131:2719 (1983).
20. M. R. Parwaresch, J. H. Radzun, M. L. Hansman, and K. P. Peters, Monoclonal antibody Ki-M4 specifically recognizes human dendritic reticulum cells (follicular dendritic cells) and their possible precursors in blood. Blood 62:585 (1983).
21. H. Stein, J. Gerdes, and D. Y. Mason, The normal and malignant germinal centre. The Lymphocytes, in: "Clinics in Haematology," ed., G. Janossy, Saunders Company, London (1982).
22. E. C. Butcher, and I. L. Weissmann, Lymphoid tissues and organs, in: "Fundamental Immunology," ed., W. E. Paul, Raven Press New York (1984)
23. H. Stein, K. Lennert, and D. Y. Mason, Immature sinus histiocytes: Their identification as a novel B-cell population. Am J Pathol in press

INDEX

Accessory cells, 369, 389, 1043
Antibody
  anti-idiotypic, 843
  anti-lymphocyte, 779, 789, 805, 819
  anti-tumor, 779
  dependent cytotoxicity, 331, 797
  feedback, 721
  isotypes, 131, 739
  natural, 131, 487
  repertoire, 131
  response, 153, 369, 463
  secreting cells, 131, 153, 463
  therapy, 779, 789, 805, 813, 819, 833, 843, 855
Antigen
  -antibody complexes, 161, 185
  -binding cells, 145, 477
  Ia, 451, 545, 555, 915, 1043
  I-J, 511
  -laden cells, 401
  -presenting cells, 369, 389, 401
  T-dependent, 369
  Thy-1, 17, 235, 251
  TI-2, 427, 437, 739
  tumor, 779, 833, 863
Arthritis
  rheumatoid, 957, 1027
Autoimunity, 495, 713, 957, 1027

B cell
  development
    chicken, 91
    mouse, 3, 35, 47, 127, 203
    rat, 9, 17, 65, 139, 203
    sheep, 119, 203
  follicular, 65, 81, 139, 203, 437
  germinal center, 3, 145, 203
  marginal zone, 65, 81, 139 203, 437
  markers, 81
  memory, 3, 145
  pre-, 9, 17, 27, 35, 203
    isolation, 9
    markers, 9, 17, 203,
  production, 35, 65, 73
  stimulation, 35, 57
  subpopulations,
    man, 879, 949, 957
    mouse, 3, 203, 451
    rat, 57, 81, 139, 203, 437, 593
  turnover, 35, 73
Bronchus
  -associated lymphoid tissue, 323
Bursa
  of Fabricius, 91, 203, 487

Cadmium, 749
Cell *see* B cell, T cell etc.
Chemotaxis, 683, 827
Complement
  receptors, 603

Cyclophosphamide, 539
Cyclosporine, 539
Cytotoxicity
  antibody-dependent, 331, 797

Dendritic cells, 307, 369, 377, 383, 389, 401, 1043 (see also Follicular dendritic cell)
  development, 377
Dioctyltin, 357
Disease
  Kimura's, 999

Encephalomyelitis
  experimental allergic, 713
Epithelium
  thymic, 245, 261, 273, 283, 289, 299, 915

Follicles
  lymphoid, 3, 65, 101, 111, 139, 161, 203, 879, 993
    T cells in, 443
Follicular
  dendritic cells, 3
    development, 161, 167
    functions, 185
    in disease, 993, 999
    isolation, 171, 193, 985
    markers, 193, 971, 985
    morphology, 171, 985, 993

Germinal centers, 3, 145, 161, 203, 421, 879, 979
Graft
  bone marrow, 805, 813, 869
  heart, 555
  spleen, 565
  -versus-host reaction, 531, 539, 545, 805, 813, 869
Gut
  -associated lymphoid tissue, 91, 101, 111, 119, 127, 203, 487,
Gut
  -associated lymphoid tissue (continued) 579, 647, 655, 661, 701

Hypersensitivity
  delayed type, 463, 471, 629

Ia antigen, 451, 545, 555, 915, 1043
IgD
  effects on immunity, 729
IgE, 127, 757
I-J antigen, 511
Immunity
  to tumors, 779, 819, 827, 833, 843, 1063
Immunodeficiency
  acquired syndrome, 1069
  X-linked, 47
Immunoglobulin
  gene rearrangements, 27
  isotypes, 127
Immunotoxin, 779
Indium 114^{m}, 675
Interdigitating cell, 307, 383, 1009
Interleukin 2, 693, 701, 713
Interleukin 3, 701
Irradiation
  total lymphoid, 503

K cell, 331, 797

Langerhans cell, 1009
Leukemia
  acute lymphoblastic, 1017
  chronic lymphocytic, 949
  hairy cell, 949
  pre-B cell, 27
Liver
  fetal, 215, 683
Lymphadenopathy
  persistent generalised, 1069
Lymphocyte
  bare, 921
  intestinal, 647, 701

Lymphocyte (continued)
  migration, 593-675
  (see also B cell, T cell, etc.)
Lymphoma
  B cell, 833, 843, 963, 971, 993
  cutaneous, 1033

Macrophage
  marginal zone, 167, 421, 427
  subpopulations, 409, 1009
Malaria, 937
Marrow
  transplantation, 805, 813, 869
Maternal
  -fetal relationship, 495
Mice
  transgenic, 341
Mutation
  xid, 47
Myasthenia, 879, 929, 979

Neuropeptide, 637
NK cell, 331

Parabiosis, 521, 593
Peanut agglutinin, 3, 119, 145, 621
Peptide
  neuro-, 637
  vasoactive intestinal, 637
Peyer's patch
  development, 101, 111
  heterogeneity, 101, 661
  ileal, 101, 119, 655, 661
  jejunal, 101, 661
  lymphocytes, 119, 127, 647
Phagocytosis, 749
Pre-B cell, 9, 17, 27, 35, 203
Pre-T cell
  man, 889, 901, 909
  mouse, 215, 223, 261, 683
  rat, 251

Radiosensitivity
  T cells, 629
Receptors
  Complement, 603
  M, 901, 909
Reticulum cell, 161
Ricin, 675, 779

Splenectomy, 427, 1051
Sympathectomy, 349

Targetting
  drugs, 779
  radiation, 675, 779
  ricin, 675, 779
T cell
  -B cell cooperation, 369
  clones, 621
  development
    guinea pig, 693
    man, 889, 901, 909
    mouse, 215, 223, 229, 261, 299, 471, 511, 683
    rat, 235, 251, 313, 357
  growth factor, 693, 701, 713
  helper, 369, 443
  markers, 235, 251, 313, 889
  mitogens, 215
  pre-,
    man, 889, 901, 909
    mouse, 215, 223, 261, 683
    rat, 251
  recirculation, 603, 609, 615, 621
  subpopulations, 323, 531, 721, 879, 889
  suppressor, 495, 503, 511, 521, 701, 721, 739
Thy-1 antigen, 17, 235, 251
Thymoma, 915
Thymus, 223, 235, 879
  cortex, 229
  development, 245

Thymus (continued)
  epithelium, 245, 261, 273, 283, 289, 299, 915
  fetal, 261, 349
  grafts, 261, 349
  medulla, 229
  microenvironmental factor, 683

Thymus (continued)
  nurse cells, 229, 299
  stroma, 245, 273

Veiled cell, 119, 395
Venule
  high endothelial, 609, 615
Virus
  herpes simplex, 855

GPSR Compliance
The European Union's (EU) General Product Safety Regulation (GPSR) is a set of rules that requires consumer products to be safe and our obligations to ensure this.

If you have any concerns about our products, you can contact us on

ProductSafety@springernature.com

In case Publisher is established outside the EU, the EU authorized representative is:

Springer Nature Customer Service Center GmbH
Europaplatz 3
69115 Heidelberg, Germany

www.ingramcontent.com/pod-product-compliance
Lightning Source LLC
LaVergne TN
LVHW081956200926
848481LV00010B/2750
9781461294955